Algebra 2

Common Core

PART 1

Randall I. Charles
Basia Hall
Dan Kennedy
Allan E. Bellman
Sadie Chavis Bragg
William G. Handlin
Stuart J. Murphy
Grant Wiggins

PEARSON

Boston, Massachusetts • Chandler, Arizona • Glenview, Illinois • Upper Saddle River, New Jersey

Cover Art: Back Cover © Gary Bell/zefa/Corbis

Taken from:

Algebra 2, Common Core Edition
Copyright © 2012 by Pearson Education, Inc.
Published by Pearson Prentice Hall
Upper Saddle River, New Jersey 07458

This special edition published in cooperation with Pearson Learning Solutions.

Pearson Learning Solutions, 501 Boylston Street, Suite 900
Boston, MA 02116
A Pearson Education Company
www.pearsoned.com

Printed in the United States of America

2 3 4 5 6 7 8 9 10 0BRV 16 15 14 13

000200010271275829

CT

Acknowledgments appear on page 1165, which constitutes an extension of this copyright page.

PEARSON

ISBN 10: 1-256-33188-0
ISBN 13: 978-1-256-33188-9

Contents *in Brief*

Welcome to ***Pearson Algebra 2 Common Core Edition*** student book. Throughout this textbook, you will find content that has been developed to cover many of the High School Standards for Mathematical Content and all of the Standards for Mathematical Practice. The End-of-Course Assessment provides students with practice with all of the Standards for Mathematical Content listed on pages xvi to xix.

Series *Authors*

Randall I. Charles, Ph.D., is Professor Emeritus in the Department of Mathematics and Computer Science at San Jose State University, San Jose, California. He began his career as a high school mathematics teacher, and he was a mathematics supervisor for five years. Dr. Charles has been a member of several NCTM committees and is the former Vice President of the National Council of Supervisors of Mathematics. Much of his writing and research has been in the area of problem solving. He has authored more than 75 mathematics textbooks for kindergarten through college.

Dan Kennedy, Ph.D., is a classroom teacher and the Lupton Distinguished Professor of Mathematics at the Baylor School in Chattanooga, Tennessee. A frequent speaker at professional meetings on the subject of mathematics education reform, Dr. Kennedy has conducted more than 50 workshops and institutes for high school teachers. He is coauthor of textbooks in calculus and precalculus, and from 1990 to 1994 he chaired the College Board's AP Calculus Development Committee. He is a 1992 Tandy Technology Scholar and a 1995 Presidential Award winner.

Basia Hall currently serves as Manager of Instructional Programs for the Houston Independent School District. With 33 years of teaching experience, Ms. Hall has served as a department chair, instructional supervisor, school improvement facilitator, and professional development trainer. She has developed curricula for Algebra 1, Geometry, and Algebra 2 and co-developed the Texas state mathematics standards. A 1992 Presidential Awardee, Ms. Hall is past president of the Texas Association of Supervisors of Mathematics and is a state representative for the National Council of Supervisors of Mathematics (NCSM).

Consulting *Authors*

Stuart J. Murphy is a visual learning author and consultant. He is a champion of helping students develop learning skills so they become more successful students. He is the author of MathStart, a series of children's books that presents mathematical concepts in the context of stories. A graduate of the Rhode Island School of Design, he has worked extensively in educational publishing and has been on the authorship teams of a number of elementary and high school mathematics programs. He is a frequent presenter at meetings of the National Council of Teachers of Mathematics, the International Reading Association, and other professional organizations.

Grant Wiggins, Ed.D., is the President of Authentic Education in Hopewell, New Jersey. He earned his Ed.D. from Harvard University and his B.A. from St. John's College in Annapolis. Dr. Wiggins consults with schools, districts, and state education departments on a variety of reform matters; organizes conferences and workshops; and develops print materials and Web resources on curricular change. He is perhaps best known for being the coauthor, with Jay McTighe, of *Understanding by Design* and *The Understanding by Design Handbook*[1], the award-winning and highly successful materials on curriculum published by ASCD. His work has been supported by the Pew Charitable Trusts, the Geraldine R. Dodge Foundation, and the National Science Foundation.

[1] ASCD, publisher of the "Understanding by Design Handbook" co-authored by Grant Wiggins and registered owner of the trademark "Understanding by Design", has not authorized or sponsored this work and is in no way affiliated with Pearson or its products.

Program *Authors*

Algebra 1 and Algebra 2

Allan E. Bellman, Ph.D., is a Lecturer/Supervisor in the School of Education at the University of California, Davis. Before coming to Davis, he was a mathematics teacher for 31 years in Montgomery County, Maryland. He has been an instructor for both the Woodrow Wilson National Fellowship Foundation and the T³ program. He has been involved in the development of many products from Texas Instruments. Dr. Bellman has a particular expertise in the use of technology in education and speaks frequently on this topic. He was a 1992 Tandy Technology Scholar and has twice been listed in Who's Who Among America's Teachers.

Sadie Chavis Bragg, Ed.D., is Senior Vice President of Academic Affairs at the Borough of Manhattan Community College of the City University of New York. A former professor of mathematics, she is a past president of the American Mathematical Association of Two-Year Colleges (AMATYC), co-director of the AMATYC project to revise the standards for Introductory college mathematics before calculus, and an active member of the Benjamin Banneker Association. Dr. Bragg has coauthored more than 50 mathematics textbooks for kindergarten through college.

William G. Handlin, Sr., is a classroom teacher and Department Chairman of Technology Applications at Spring Woods High School in Houston, Texas. Awarded Life Membership in the Texas Congress of Parents and Teachers for his contributions to the well-being of children, Mr. Handlin is also a frequent workshop and seminar leader in professional meetings throughout the world.

Geometry

Laurie E. Bass is a classroom teacher at the 9–12 division of the Ethical Culture Fieldston School in Riverdale, New York. A classroom teacher for more than 30 years, Ms. Bass has a wide base of teaching experience, ranging from Grade 6 through Advanced Placement Calculus. She was the recipient of a 2000 Honorable Mention for the Radio Shack National Teacher Awards. She has been a contributing writer for a number of publications, including software-based activities for the Algebra 1 classroom. Among her areas of special interest are cooperative learning for high school students and geometry exploration on the computer. Ms. Bass is a frequent presenter at local, regional, and national conferences.

Art Johnson, Ed.D., is a professor of mathematics education at Boston University. He is a mathematics educator with 32 years of public school teaching experience, a frequent speaker and workshop leader, and the recipient of a number of awards: the Tandy Prize for Teaching Excellence, the Presidential Award for Excellence in Mathematics Teaching, and New Hampshire Teacher of the Year. He was also profiled by the Disney Corporation in the American Teacher of the Year Program. Dr. Johnson has contributed 18 articles to NCTM journals and has authored over 50 books on various aspects of mathematics.

Reviewers *National*

Tammy Baumann
K-12 Mathematics Coordinator
School District of the City of Erie
Erie, Pennsylvania

Sandy Cowgill
Mathematics Department Chair
Muncie Central High School
Muncie, Indiana

Sheryl Ezze
Mathematics Chairperson
DeWitt High School
Lansing, Michigan

Dennis Griebel
Mathematics Coordinator
Cherry Creek School District
Aurora, Colorado

Bill Harrington
Secondary Mathematics Coordinator
State College School District
State College, Pennsylvania

Michael Herzog
Mathematics Teacher
Tucson Small School Project
Tucson, Arizona

Camilla Horton
Secondary Instruction Support
Memphis School District
Memphis, Tennessee

Gary Kubina
Mathematics Consultant
Mobile County School System
Mobile, Alabama

Sharon Liston
Mathematics Department Chair
Moore Public Schools
Oklahoma City, Oklahoma

Ann Marie Palmeri Monahan
Mathematics Supervisor
Bayonne Public Schools
Bayonne, New Jersey

Indika Morris
Mathematics Department Chair
Queen Creek School District
Queen Creek, Arizona

Jennifer Petersen
K-12 Mathematics Curriculum Facilitator
Springfield Public Schools
Springfield, Missouri

Tammy Popp
Mathematics Teacher
Mehlville School District
St. Louis, Missouri

Mickey Porter
Mathematics Teacher
Dayton Public Schools
Dayton, Ohio

Steven Sachs
Mathematics Department Chair
Lawrence North High School
Indianapolis, Indiana

John Staley
Secondary Mathematics Coordinator
Office of Mathematics, PK-12
Baltimore, Maryland

Robert Thomas, Ph.D.
Mathematics Teacher
Yuma Union High School District #70
Yuma, Arizona

Linda Ussery
Mathematics Consultant
Alabama Department of Education
Tuscumbia, Alabama

Denise Vizzini
Mathematics Teacher
Clarksburg High School
Montgomery County, Maryland

Marcia White
Mathematics Specialist
Academic Operations, Technology and Innovations
Memphis City Schools
Memphis, Tennessee

Merrie Wolf
Mathematics Department Chair
Tulsa Public Schools
Tulsa, Oklahoma

From the **Authors**

Welcome

Math is a powerful tool with far-reaching applications throughout your life. We have designed a unique and engaging program that will enable you to tap into the power of mathematics and mathematical reasoning. This award-winning program has been developed to align fully to the Common Core State Standards.

Developing mathematical skills and problem-solving strategies is an ongoing process—a journey both inside and outside the classroom. This course is designed to help make sense of the mathematics you encounter in and out of class each day and to help you develop mathematical proficiency.

You will learn important mathematical principles. You will also learn how the principles are connected to one another and to what you already know. You will learn to solve problems and learn the reasoning that lies behind your solutions. You will also develop the key mathematical practices of the Common Core State Standards.

Each chapter begins with the "big ideas" of the chapter and some essential questions that you will learn to answer. Through this question-and-answer process you will develop your ability to analyze problems independently and solve them in different applications.

Your skills and confidence will increase through practice and review. Work through the problems so you understand the concepts and methods presented and the thinking behind them. Then do the exercises. Ask yourself how new concepts relate to old ones. Make the connections!

Everyone needs help sometimes. You will find that this program has built-in opportunities, both in this text and online, to get help whenever you need it.

This course will also help you succeed on the tests you take in class and on other tests like the SAT, ACT, and state exams. The practice exercises in each lesson will prepare you for the format and content of such tests. No surprises!

The problem-solving and reasoning habits and skills you develop in this program will serve you in all your studies and in your daily life. They will prepare you for future success not only as a student, but also as a member of a changing technological society.

Best wishes,

PowerAlgebra.com

Welcome to Algebra 2. *Pearson Algebra 2 Common Core Edition* is part of a blended digital and print environment for the study of high school mathematics. Take some time to look through the features of our mathematics program, starting with **PowerAlgebra.com,** the site of the digital features of the program.

In each chapter opener, you will be invited to visit the **PowerAlgebra.com** site to access these online features. Look for these buttons throughout the lessons.

The Common Core State Standards have a similar organizing structure. They begin with **Conceptual Categories,** such as Algebra or Functions. Within each category are **domains** and **clusters.**

Common Core State Standards

Big *Ideas*

We start with **Big Ideas.** Each chapter is organized around Big Ideas that convey the key mathematics concepts you will be studying in the program. Take a look at the Big Ideas on pages xx and xxi.

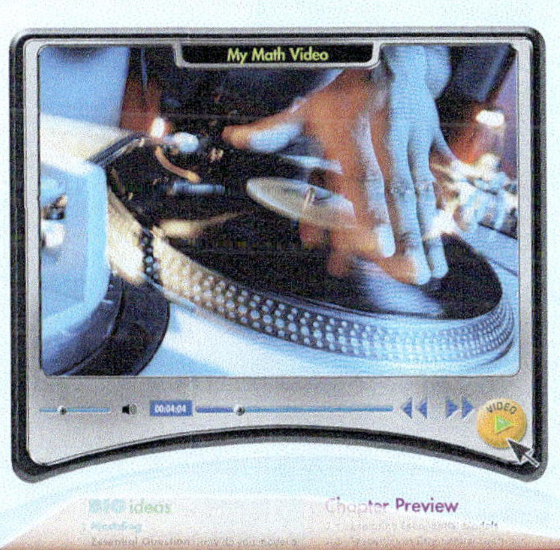

BIG ideas

1 **Modeling**
Essential Question How do you model a quantity that changes regularly over time by the same percentage?

2 **Equivalence**
Essential Question How are exponents and logarithms related?

3 **Function**
Essential Question How are exponential functions and logarithmic functions related?

The **Big Ideas** are organizing ideas for all of the lessons in the program. At the beginning of each chapter, we'll tell you which Big Ideas you'll be studying. We'll also present an **Essential Question** for each Big Idea.

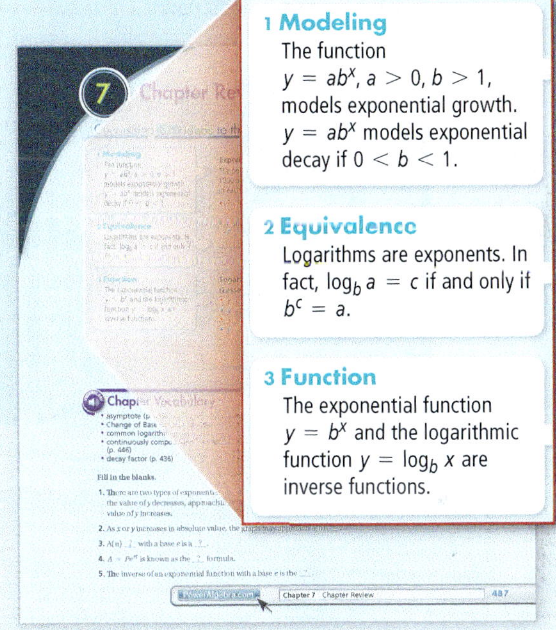

1 **Modeling**
The function $y = ab^x$, $a > 0$, $b > 1$, models exponential growth. $y = ab^x$ models exponential decay if $0 < b < 1$.

2 **Equivalence**
Logarithms are exponents. In fact, $\log_b a = c$ if and only if $b^c = a$.

3 **Function**
The exponential function $y = b^x$ and the logarithmic function $y = \log_b x$ are inverse functions.

In the **Chapter Review** at the end of the chapter, you'll find the answers to the Essential Question for each Big Idea. We'll also remind you of the lesson(s) where you studied the concepts that support the Big Ideas.

Exploring *Concepts*

The lessons offer many opportunities to explore concepts in different contexts and through different media.

Hi, I'm Serena. I never have to power down when I am in math class now.

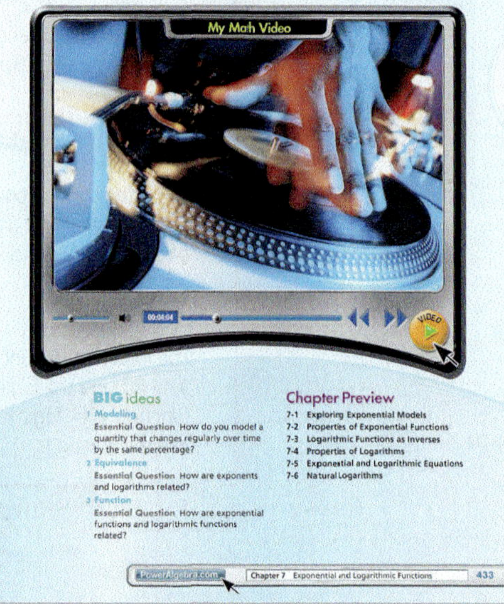

For each chapter, there is a video that you can access at **PowerAlgebra.com.** The video presents concepts in a real-life context. And you can contribute your own math video.

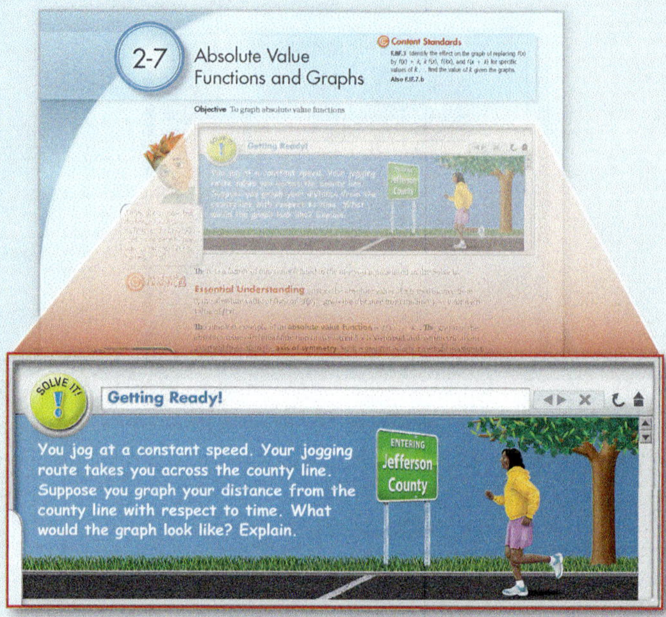

Here's another cool feature. Each lesson opens with a **Solve It,** a problem that helps you connect what you know to an important concept in the lesson. Do you notice how the Solve It frame looks like it comes from a computer? That's because all of the Solve Its can be found at **PowerAlgebra.com.**

Developing Mathematical Proficiency

The **Standards for Mathematical Practice** describe processes, practices, and habits of mind of mathematically proficient students. Many of the features in *Pearson Algebra 2 Common Core Edition* help you develop proficiency in math.

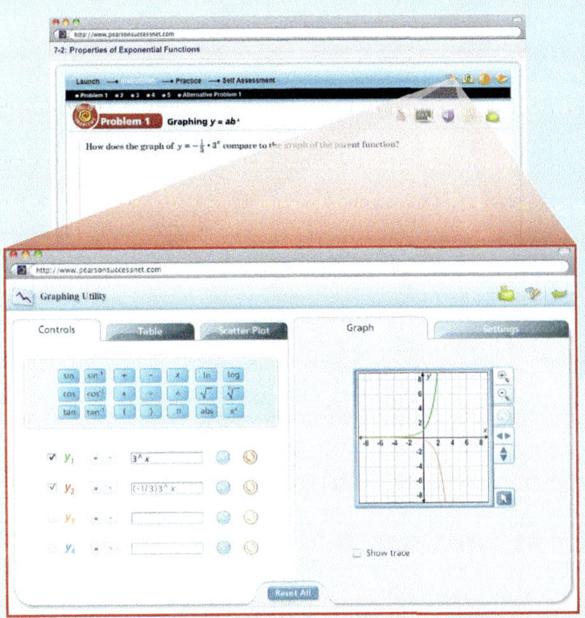

Want to do some more exploring? Try the **Math Tools** at **PowerGeometry.com.** Click on this icon to access these tools, Graphing Utility, Number Line, Algebra Tiles, and 2D and 3D Geometric Constructor. With the Math Tools, you can continue to explore the concepts presented in the lesson.

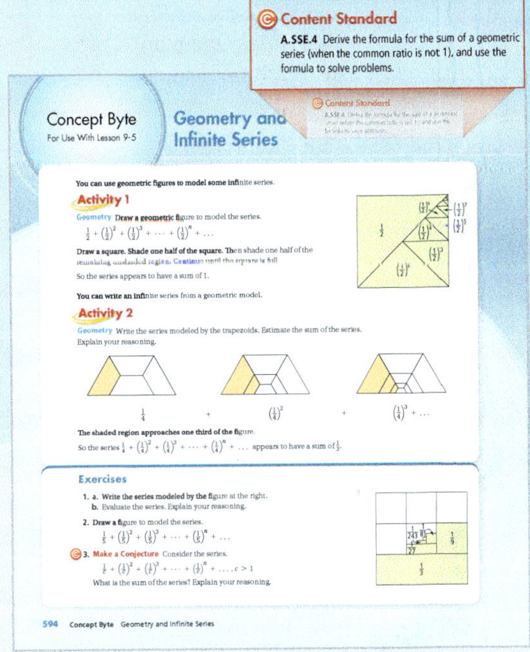

Try a **Concept Byte!** In a Concept Byte, you might explore technology, do a hands-on activity, or try a challenging extension. The text in the top right corner of the first page of a lesson or Concept Byte tells you the **Standard for Mathematical Content** that you will be studying.

Thinking *Mathematically*

Mathematical reasoning is the key to solving problems and making sense of math. Throughout the program you'll learn strategies to develop mathematical reasoning habits.

Hello, I'm Tyler. These Think-Write and Know-Need-Plan boxes help me plan my work.

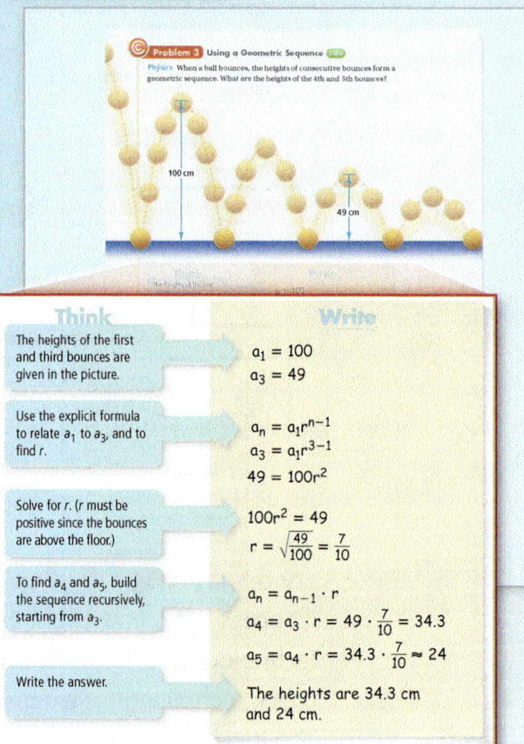

Problem 3 Using a Geometric Sequence

Physics When a ball bounces, the heights of consecutive bounces form a geometric sequence. What are the heights of the 4th and 5th bounces?

Think	Write
The heights of the first and third bounces are given in the picture.	$a_1 = 100$ $a_3 = 49$
Use the explicit formula to relate a_1 to a_3, and to find r.	$a_n = a_1r^{n-1}$ $a_3 = a_1r^{3-1}$ $49 = 100r^2$
Solve for r. (r must be positive since the bounces are above the floor.)	$100r^2 = 49$ $r = \sqrt{\frac{49}{100}} = \frac{7}{10}$
To find a_4 and a_5, build the sequence recursively, starting from a_3.	$a_n = a_{n-1} \cdot r$ $a_4 = a_3 \cdot r = 49 \cdot \frac{7}{10} = 34.3$ $a_5 = a_4 \cdot r = 34.3 \cdot \frac{7}{10} \approx 24$
Write the answer.	The heights are 34.3 cm and 24 cm.

The worked-out problems include call-outs that reveal the strategies and reasoning behind the solution. The **Think-Write** problems model the thinking behind each step of a solution.

Also, look for the boxes labeled **Plan** and **Think.**

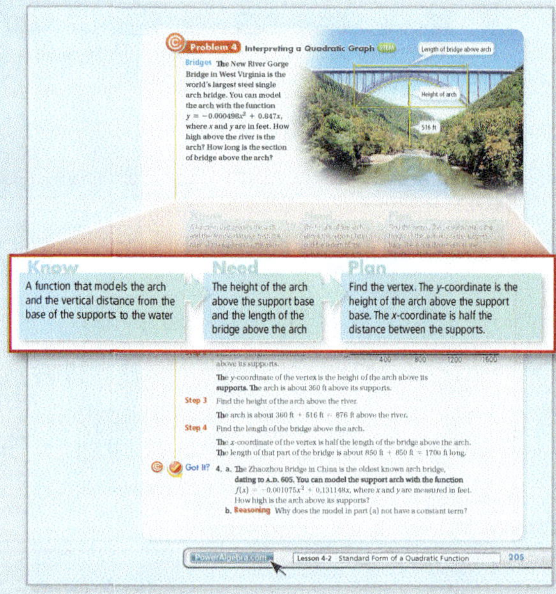

Problem 4 Interpreting a Quadratic Graph

Bridges The New River Gorge Bridge in West Virginia is the world's largest steel single arch bridge. You can model the arch with the function $y = -0.000498x^2 + 0.847x$, where x and y are in feet. How high above the river is the arch? How long is the section of bridge above the arch?

Know	Need	Plan
A function that models the arch and the vertical distance from the base of the supports to the water	The height of the arch above the support base and the length of the bridge above the arch	Find the vertex. The y-coordinate is the height of the arch above the support base. The x-coordinate is half the distance between the supports.

The y-coordinate of the vertex is the height of the arch above its supports. The arch is about 360 ft above its supports.

Step 3 Find the height of the arch above the river.

The arch is about 360 ft + 516 ft = 876 ft above the river.

Step 4 Find the length of the bridge above the arch.

The x-coordinate of the vertex is half the length of the bridge above the arch. The length of that part of the bridge is about 850 ft + 850 ft = 1700 ft long.

Got It? 4. a. The Zhaozhou Bridge in China is the oldest known arch bridge, dating to A.D. 605. You can model the support arch with the function $f(x) = -0.001075x^2 + 0.131148x$, where x and y are measured in feet. How high is the arch above its supports?

b. Reasoning Why does the model in part (a) not have a constant term?

PowerAlgebra.com Lesson 4-2 Standard Form of a Quadratic Function 205

Other worked-out problems model a problem-solving plan that includes the steps of stating what you **Know,** identifying what you **Need,** and developing a **Plan.**

Standards for Mathematical Practice

The **Standards for Mathematical Practice** emphasize sense-making, reasoning, and critical reasoning. Many features in *Pearson Algebra 2 Common Core Edition* provide opportunities for you to develop these skills and dispositions.

Standards for Mathematical Practice

1. Make sense of problems and persevere in solving them.
2. Reason abstractly and quantitatively.
3. Construct viable arguments and critique the reasoning of others.
4. Model with mathematics.
5. Use appropriate tools strategically.
6. Attend to precision.
7. Look for and make use of structure.
8. Look for and express regularity in repeated reasoning.

take note **Key Concepts** Roots, Zeros, and x-intercepts

The following are equivalent statements about a real number b and a polynomial $P(x) = a_nx^n + a_{n-1}x^{n-1} + \cdots + a_1x + a_0$.

- $x - b$ is a linear factor of the polynomial $P(x)$.
- b is a zero of the polynomial function $y = P(x)$.
- b is a root (or solution) of the polynomial equation $P(x) = 0$.
- b is an x-intercept of the graph of $y = P(x)$.

A **Take Note** box highlights key concepts in a lesson. You can use these boxes to review concepts throughout the year.

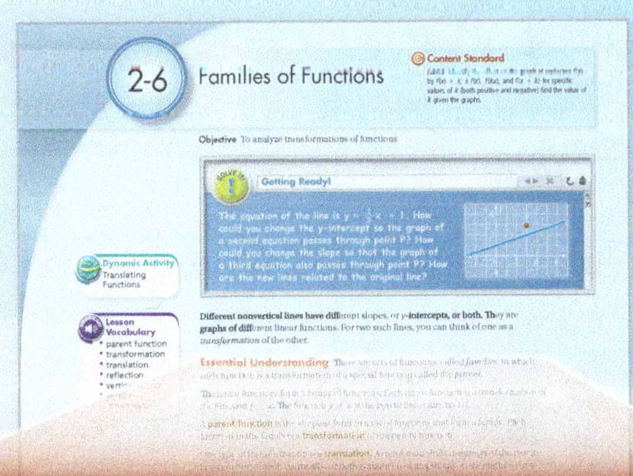

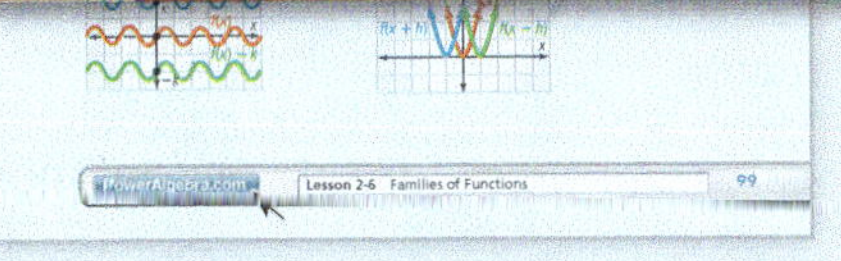

Essential Understanding There are sets of functions, called *families*, in which each function is a transformation of a special function called the *parent*.

Part of thinking mathematically is making sense of the concepts that are being presented. The **Essential Understandings** help you build a framework for the Big Ideas.

COMMON CORE

Practice *Makes Perfect*

Ask any professional and you'll be told that the one requirement for becoming an expert is practice, practice, practice.

Hello, I'm Anya. I can leave my book at school and still get my homework done. All of the lessons are at PowerAlgebra.com

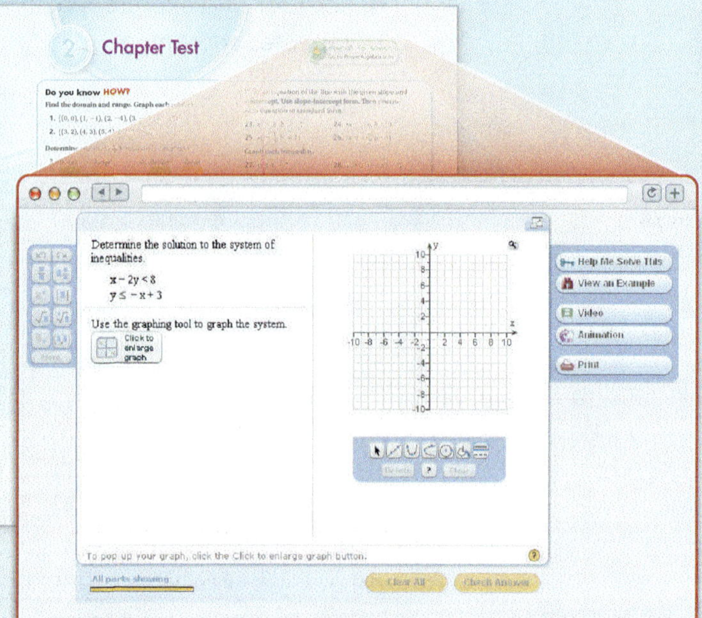

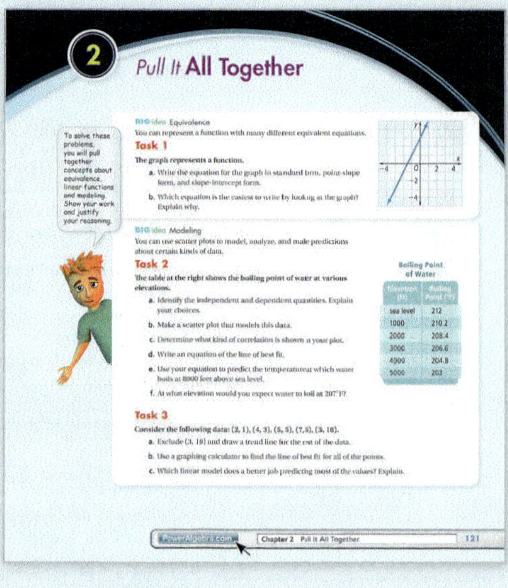

Want more practice? Look for this icon in your book. Check out all of the opportunities in **MathXL® for School.** Your teacher can assign you some practice exercises or you can choose some on your own. And you'll know right away if you got the right answer!

But the best practice occurs when you **Pull It All Together** — understanding of concepts, mathematical thinking, and problem solving — to solve interesting problems. And look: there are those Big Ideas again.

Acing *the Test*

Doing well on tests, whether they are chapter tests or state assessments, depends on a good understanding of math concepts, skill at solving problems, and, just as important, good test-taking skills.

All of these opportunities for practice help you prepare for assessments throughout the year, including the assessments to measure your proficiency with the Common Core State Standards.

Standards for Mathematical Practice

TIPS FOR SUCCESS

Some problems require the selection of an appropriate representation (concrete, pictorial, graphical, verbal, or symbolic) to find a solution.

TIP 1

Make a drawing.

One angle of a right triangle measures 90°. The measure of the second angle is 5 times bigger than the measure of the third. What are the measures of these angles?

Ⓐ 30° and 60°

Ⓑ 30° and 150°

Ⓒ 15° and 75°

Ⓓ 20° and 100°

TIP 2

Write a system:

$x + y + 90 = 180$

$x = 5y$

Think It Through

A triangle can have only one right angle, so the other two angles must each have a measure less than 90°. Use x and y to represent the unknown angles.

$x = 5y$

$5y + y + 90 = 180$

$6y + 90 = 180$

$6y = 90$

$y = 15, x = 75$

The correct answer is C.

Quick Review

Linear programming is used to find a minimum or maximum of an objective function, given constraints as linear inequalities. The maximum or minimum occurs at a vertex of the feasible region, which contains the solutions to the system of constraints.

Example

Graph the system of constraints and name the vertices. $\begin{cases} x \le 8 \\ y \le 5 \\ x \ge 0, y \ge 0 \end{cases}$

Objective function: $P = 2x + y$

Graph the inequalities and shade the area satisfying all inequalities.

The vertices of the feasible region are (0, 0), (0, 5), (8, 5), and (8, 0).

Evaluate the objective function at each vertex:

Exercises

Graph the system of constraints. Name the vertices. Then find the values of x and y that maximize or minimize the objective function.

18. $\begin{cases} x \ge 2 \\ y \ge 0 \\ 3x + 2y \ge 12 \end{cases}$

Minimum for $C = x + 5y$

19. $\begin{cases} 3x + 2y \le 12 \\ x + y \le 5 \\ x \ge 0, y \ge 0 \end{cases}$

Maximum for $P = 3x + 5y$

20. A lunch stand makes $.75 profit on each chef's salad and $1.20 profit on each Caesar salad. On a typical weekday, it sells between 40 and 60 chef's salads and between 35 and 50 Caesar salads. The total number sold has never exceeded 100 salads. How many of each type should be prepared in order to maximize profit?

At the end of the chapter, you'll find a **Quick Review** of the concepts in the chapter and a few examples and exercises so you can check your skill at solving problems related to the concepts.

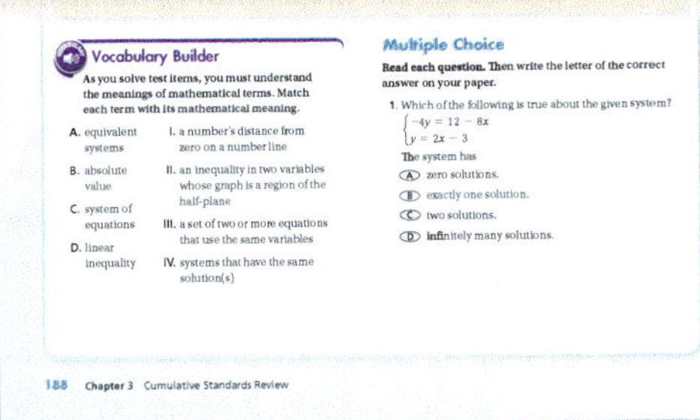
Vocabulary Builder

As you solve test items, you must understand the meanings of mathematical terms. Match each term with its mathematical meaning.

A. equivalent systems

B. absolute value

C. system of equations

D. linear inequality

I. a number's distance from zero on a number line

II. an inequality in two variables whose graph is a region of the half-plane

III. a set of two or more equations that use the same variables

IV. systems that have the same solution(s)

Multiple Choice

Read each question. Then write the letter of the correct answer on your paper.

1. Which of the following is true about the given system?

$\begin{cases} -4y = 12 - 8x \\ y = 2x - 3 \end{cases}$

The system has

Ⓐ zero solutions.

Ⓑ exactly one solution.

Ⓒ two solutions.

Ⓓ infinitely many solutions.

188 Chapter 3 Cumulative Standards Review

In the Cumulative Standards Review at the end of the chapter, you'll also find **Tips for Success** to strengthen your test-taking skills. We include problems of all different formats and types so you can feel comfortable with any test item on your state assessment.

Common Core
State Standards for Mathematics
Algebra 2

Hi, I'm Max. Here is a list of many of the Common Core State Standards that you will study this year. Mastering these topics will help you be ready for the Common Assessment

Number and Quantity

The Complex Number System

Perform arithmetic operations with complex numbers.

N.CN.1 Know there is a complex number i such that $i^2 = -1$, and every complex number has the form $a + bi$ with a and b real.

N.CN.2 Use the relation $i^2 = -1$ and the commutative, associative, and distributive properties to add, subtract, and multiply complex numbers.

Use complex numbers in polynomial identities and equations.

N.CN.7 Solve quadratic equations with real coefficients that have complex solutions.

N.CN.8 (+) Extend polynomial identities to the complex numbers.

N.CN.9 (+) Know the Fundamental Theorem of Algebra; show that it is true for quadratic polynomials.

Vector and Matrix Quantities

Represent and model with vector quantities.

N.VM.1 (+) Recognize vector quantities as having both magnitude and direction. Represent vector quantities by directed line segments, and use appropriate symbols for vectors and their magnitudes (e.g., $\boldsymbol{v}, |\boldsymbol{v}|, \|\boldsymbol{v}\|, v$).

N.VM.2 (+) Find the components of a vector by subtracting the coordinates of an initial point from the coordinates of a terminal point.

N.VM.3 (+) Solve problems involving velocity and other quantities that can be represented by vectors.

Perform operations on vectors.

N.VM.4 (+) Add and subtract vectors.

N.VM.4.a Add vectors end-to-end, component-wise, and by the parallelogram rule. Understand that the magnitude of a sum of two vectors is typically not the sum of the magnitudes.

N.VM.4.b Given two vectors in magnitude and direction form, determine the magnitude and direction of their sum.

N.VM.4.c Understand vector subtraction $\boldsymbol{v} - \boldsymbol{w}$ as $\boldsymbol{v} + (-\boldsymbol{w})$, where $-\boldsymbol{w}$ is the additive inverse of $\boldsymbol{w}$, with the same magnitude as $\boldsymbol{w}$ and pointing in the opposite direction. Represent vector subtraction graphically by connecting the tips in the appropriate order, and perform vector subtraction component-wise.

N.VM.5 (+) Multiply a vector by a scalar.

N.VM.5.a Represent scalar multiplication graphically by scaling vectors and possibly reversing their direction; perform scalar multiplication component-wise, e.g., as $c(v_x, v_y) = (cv_x, cv_y)$.

N.VM.5.b Compute the magnitude of a scalar multiple $c\boldsymbol{v}$ using $\|c\boldsymbol{v}\| = |c|\boldsymbol{v}$. Compute the direction of $c\boldsymbol{v}$ knowing that when $|c|\boldsymbol{v} \neq 0$, the direction of $c\boldsymbol{v}$ is either along $\boldsymbol{v}$ (for c greater than 0) or against $\boldsymbol{v}$ (for c less than 0).

Perform operations on matrices and use matrices in applications.

N.VM.6 (+) Use matrices to represent and manipulate data, e.g., to represent payoffs or incidence relationships in a network.

N.VM.7 (+) Multiply matrices by scalars to produce new matrices, e.g., as when all of the payoffs in a game are doubled.

N.VM.8 (+) Add, subtract, and multiply matrices of appropriate dimensions.

N.VM.9 (+) Understand that, unlike multiplication of numbers, matrix multiplication for square matrices is not a commutative operation, but still satisfies the associative and distributive properties.

N.VM.10 (+) Understand that the zero and identity matrices play a role in matrix addition and multiplication similar to the role of 0 and 1 in the real numbers. The determinant of a square matrix is nonzero if and only if the matrix has a multiplicative inverse.

N.VM.11 (+) Multiply a vector (regarded as a matrix with one column) by a matrix of suitable dimensions to produce another vector. Work with matrices as transformations of vectors.

N.VM.12 (+) Work with 2×2 matrices as a transformations of the plane, and interpret the absolute value of the determinant in terms of area.

Algebra

Seeing Structure in Expressions

Interpret the structure of expressions.

A.SSE.1 Interpret expressions that represent a quantity in terms of its context.

A.SSE.1.a Interpret parts of an expression, such as terms, factors, and coefficients.

A.SSE.1.b Interpret complicated expressions by viewing one or more of their parts as a single entity.

A.SSE.2 Use the structure of an expression to identify ways to rewrite it.

Write expressions in equivalent forms to solve problems.

A.SSE.4 Derive the formula for the sum of a finite geometric series (when the common ratio is not 1), and use the formula to solve problems.

Arithmetic with Polynomials and Rational Expressions

Perform arithmetic operations on polynomials.

A.APR.1 Understand that polynomials form a system analogous to the integers, namely, they are closed under the operations of addition, subtraction, and multiplication; add, subtract, and multiply polynomials.

Understand the relationship between zeros and factors of polynomial.

A.APR.2 Know and apply the Remainder Theorem: For a polynomial $p(x)$ and a number a, the remainder on division by $x - a$ is $p(a)$, so $p(a) = 0$ if and only if $(x - a)$ is a factor of $p(x)$.

A.APR.3 Identify zeros of polynomials when suitable factorizations are available, and use the zeros to construct a rough graph of the function defined by the polynomial.

Use polynomial identities to solve problems.

A.APR.4 Prove polynomial identities and use them to describe numerical relationships.

A.APR.5 (+) Know and apply the Binomial Theorem for the expansion of $(x + y)^n$ in powers of x and y for a positive integer n, where x and y are any numbers, with coefficients determined for example by Pascal's Triangle.

Rewrite rational expressions.

A.APR.6 Rewrite simple rational expressions in different forms; write $a(x)/b(x)$ in the form $q(x) + r(x)/b(x)$, where $a(x)$, $b(x)$, $q(x)$, and $r(x)$ are polynomials with the degree of $r(x)$ less than the degree of $b(x)$, using inspection, long division, or, for the more complicated examples, a computer algebra system.

A.APR.7 (+) Understand that rational expressions form a system analogous to the rational numbers, closed under addition, subtraction, multiplication, and division by a nonzero rational expression; add, subtract, multiply, and divide rational expressions.

Creating Equations

Create equations that describe numbers or relationships.

A.CED.1 Create equations and inequalities in one variable and use them to solve problems. *Include equations arising from linear and quadratic functions, and simple rational and exponential functions.*

A.CED.2 Create equations in two or more variables to represent relationships between quantities; graph equations on coordinate axes with labels and scales.

A.CED.3 Represent constraints by equations or inequalities, and by systems of equations and/or inequalities, and interpret solutions as viable or non-viable options in a modeling context.

A.CED.4 Rearrange formulas to highlight a quantity of interest, using the same reasoning as in solving equations.

Reasoning with Equations and Inequalities

Understand solving equations as a process of reasoning and explain the reasoning.

A.REI.2 Solve simple rational and radical equations in one variable, and give examples showing how extraneous solutions may arise.

Represent and solve equations and inequalities graphically.

A.REI.11 Explain why the x-coordinates of the points where the graphs of the equations $y = f(x)$ and $y = g(x)$ intersect are the solutions of the equation $f(x) = g(x)$; find the solutions approximately, e.g., using technology to graph the functions, make tables of values, or find successive approximations. Include cases where $f(x)$ and/or $g(x)$ are linear, polynomial, rational, absolute value, exponential, and logarithmic functions.

Functions

Interpreting Functions

Understand the concept of a function and use function notation.

F.IF.1 Understand that a function from one set (called the domain) to another set (called the range) assigns to each element of the domain exactly one element of the range. If f is a function and x is and element of its domain, then $f(x)$ denotes the output of f corresponding to the input x. The graph of f is the graph of the equation $y = f(x)$.

F.IF.3 Recognize that sequences are functions, sometimes defined recursively, whose domain is a subset of the integers.

Interpret functions that arise in applications in terms of the context.

F.IF.4 For a function that models a relationship between two quantities, interpret key features of graphs and tables in terms of the quantities, and sketch graphs showing key features given a verbal description of the relationship. *Key features include: intercepts; intervals where the function is increasing, decreasing, positive, or negative; relative maximums and minimums; symmetries; end behavior; and periodicity.*

F.IF.5 Relate the domain of a function to its graph and, where applicable, to the quantitative relationship it describes.

F.IF.6 Calculate and interpret the average rate of change of a function (presented symbolically or as a table) over a specified interval. Estimate the rate of change from a graph.

Analyze functions using different representations.

F.IF.7 Graph functions expressed symbolically, and show key features of the graph, by hand in simple cases and using technology for more complicated cases.

F.IF.7.b Graph square root, cube root, and piecewise-defined functions, including step functions and absolute value functions.

F.IF.7.c Graph polynomial functions, identifying zeros when suitable factorizations are available, and showing end behavior.

F.IF.7.d (+) Graph rational functions, identifying zeros and asymptotes when suitable factorizations are available, and showing end behavior.

F.IF.7.e Graph exponential and logarithmic functions, showing intercepts and end behavior, and trigonometric functions, showing period, midline, and amplitude.

F.IF.8 Write a function defined by an expression in different but equivalent forms to reveal and explain different properties of the function.

F.IF.9 Compare properties of two functions each represented in a different way (algebraically, graphically, numerically in tables, or by verbal descriptions)

Building Functions

Build a function that models a relationship between two quantities.

F.BF.1 Write a function that describes a relationship between two quantities.

F.BF.1.b Combine standard function types using arithmetic operations.

Build new functions from existing functions.

F.BF.3 Identify the effect on the graph of replacing $f(x)$ by $f(x) + k$, $k\,f(x)$, $f(kx)$, and $f(x + k)$ for specific values of k (both positive and negative); find the value of k given the graphs. Experiment with cases and illustrate an explanation of the effects on the graph using technology. *Include recognizing even and odd functions from their graphs and algebraic expressions for them.*

F.BF.4 Find inverse functions.

F.BF.4.a Solve an equation of the form $f(x) = c$ for a simple function f that has an inverse and write an expression for the inverse.

F.BF.4.c (+) Read values of an inverse function from a graph or a table, given that the function has an inverse.

Linear and Exponential Models

Construct and compare linear and exponential models and solve problems.

F.LE.4 For exponential models, express as a logarithm the solution to $ab^{ct} = d$ where *a, c,* and *d* are numbers and the base *b* is 2, 10, or *e*; evaluate the logarithm using technology.

Trigonometric Functions

Extend the domain of trigonometric functions using the unit circle.

F.TF.1 Understand radian measure of an angle as the length of the arc on the unit circle subtended by the angle.

F.TF.2 Explain how the unit circle in the coordinate plane enables the extension of trigonometric functions to all real numbers, interpreted as radian measures of angles traversed counterclockwise around the unit circle.

Model periodic phenomena with trigonometric functions.

F.TF.5 Choose trigonometric functions to model periodic phenomena with specified amplitude, frequency, and midline.

F.TF.6 (+) Understand that restricting a trigonometric function to a domain on which it is always increasing or decreasing allows its inverse to be constructed.

F.TF.7 (+) Use inverse functions to solve trigonometric equations that arise in modeling contexts; evaluate the solutions using technology, and interpret them in terms of the context.

Prove and apply trigonometric identities.

F.TF.8 Prove the Pythagorean identity $\sin^2(\theta) + \cos^2(\theta) = 1$ and use it to calculate trigonometric ratios.

F.TF.9 (+) Prove the addition and subtraction formulas for sine, cosine, and tangent and use them to solve problems.

Statistics and Probability

Interpreting Categorical and Quantitative Data

Summarize, represent, and interpret data on a single count or measurement variable.

S.ID.2 Use statistics appropriate to the shape of the data distribution to compare center (median, mean) and spread (interquartile range, standard deviation) of two or more different data sets.

S.ID.4 Use the mean and standard deviation of a data set to fit it to a normal distribution and to estimate population percentages. Recognize that there are data sets for which such a procedure is not appropriate. Use calculators, spreadsheets, and tables to estimate areas under the normal curve.

Making Inferences and Justifying Conclusions

Understand and evaluate random processes underlying statistical experiments.

S.IC.1 Understand statistics as a process for making inferences to be made about population parameters based on a random sample from that population.

S.IC.2 Decide if a specified model is consistent with results from a given data-generating process, e.g., using simulation.

Make inferences and justify conclusions from sample surveys, experiments, and observational studies.

S.IC.3 Recognize the purposes of and differences among sample surveys, experiments, and observational studies; explain how randomization relates to each.

S.IC.4 Use data from a sample survey to estimate a population mean or proportion; develop a margin of error through the use of simulation models for random sampling.

S.IC.5 Use data from a randomized experiment to compare two treatments; use simulations to decide if differences between parameters are significant.

S.IC.6 Evaluate reports based on data.

Using Probability to Make Decisions

Use probability to evaluate outcomes of decisions.

S.MD.6 (+) Use probabilities to make fair decisions (e.g., drawing by lots, using a random number generator).

S.MD.7 (+) Analyze decisions and strategies using probability concepts (e.g., product testing, medical testing, pulling a hockey goalie at the end of a game).

BIGideas

These Big Ideas are the organizing ideas for the study of important areas of mathematics: algebra, geometry, and statistics.

Algebra

Properties

- In the transition from arithmetic to algebra, attention shifts from arithmetic operations (addition, subtraction, multiplication, and division) to use of the *properties* of these operations.
- All of the facts of arithmetic and algebra follow from certain properties.

Variable

- Quantities are used to form expressions, equations, and inequalities.
- An expression refers to a quantity but does not make a statement about it. An equation (or an inequality) is a statement about the quantities it mentions.
- Using variables in place of numbers in equations (or inequalities) allows the statement of relationships among numbers that are unknown or unspecified.

Equivalence

- A single quantity may be represented by many different expressions.
- The facts about a quantity may be expressed by many different equations (or inequalities).

Solving Equations & Inequalities

- Solving an equation is the process of rewriting the equation to make what it says about its variable(s) as simple as possible.
- Properties of numbers and equality can be used to transform an equation (or inequality) into equivalent, simpler equations (or inequalities) in order to find solutions.
- Useful information about equations and inequalities (including solutions) can be found by analyzing graphs or tables.
- The numbers and types of solutions vary predictably, based on the type of equation.

Proportionality

- Two quantities are *proportional* if they have the same ratio in each instance where they are measured together.
- Two quantities are *inversely proportional* if they have the same product in each instance where they are measured together.

Function

- A function is a relationship between variables in which each value of the input variable is associated with a unique value of the output variable.
- Functions can be represented in a variety of ways, such as graphs, tables, equations, or words. Each representation is particularly useful in certain situations.
- Some important families of functions are developed through transformations of the simplest form of the function.
- New functions can be made from other functions by applying arithmetic operations or by applying one function to the output of another.

Modeling

- Many real-world mathematical problems can be represented algebraically. These representations can lead to algebraic solutions.
- A function that models a real-world situation can be used to make estimates or predictions about future occurrences.

Statistics and Probability

Data Collection and Analysis

- Sampling techniques are used to gather data from real-world situations. If the data are representative of the larger population, inferences can be made about that population.
- Biased sampling techniques yield data unlikely to be representative of the larger population.
- Sets of numerical data are described using measures of central tendency and dispersion.

Data Representation

- The most appropriate data representations depend on the type of data—quantitative or qualitative, and univariate or bivariate.
- Line plots, box plots, and histograms are different ways to show distribution of data over a possible range of values.

Probability

- Probability expresses the likelihood that a particular event will occur.
- Data can be used to calculate an experimental probability, and mathematical properties can be used to determine a theoretical probability.
- Either experimental or theoretical probability can be used to make predictions or decisions about future events.
- Various counting methods can be used to develop theoretical probabilities.

Geometry

Visualization

- Visualization can help you see the relationships between two figures and connect properties of real objects with two-dimensional drawings of these objects.

Transformations

- Transformations are mathematical functions that model relationships with figures.
- Transformations may be described geometrically or by coordinates.
- Symmetries of figures may be defined and classified by transformations.

Measurement

- Some attributes of geometric figures, such as length, area, volume, and angle measure, are measurable. Units are used to describe these attributes.

Reasoning & Proof

- Definitions establish meanings and remove possible misunderstanding.
- Other truths are more complex and difficult to see. It is often possible to verify complex truths by reasoning from simpler ones using deductive reasoning.

Similarity

- Two geometric figures are similar when corresponding lengths are proportional and corresponding angles are congruent.
- Areas of similar figures are proportional to the squares of their corresponding lengths.
- Volumes of similar figures are proportional to the cubes of their corresponding lengths.

Coordinate Geometry

- A coordinate system on a line is a number line on which points are labeled, corresponding to the real numbers.
- A coordinate system in a plane is formed by two perpendicular number lines, called the x- and y-axes, and the quadrants they form. The coordinate plane can be used to graph many functions.
- It is possible to verify some complex truths using deductive reasoning in combination with the distance, midpoint, and slope formulas.

1

Expressions, Equations, and Inequalities

Chapters 1 & 2

Algebra

Seeing Structure in Expressions

Interpret the structure of expressions

Creating Equations

Create equations that describe numbers or relationships

Functions

Interpreting Functions

Interpret functions that arise in applications in terms of the context

Analyze functions using different representations

Building Functions

Build a function that models a relationship between two quantities

2 Functions, Equations, and Graphs

Visual See It!

Reasoning Try It!

Practice Do It!

3

Linear Systems

Chapters 3 & 4

Number and Quantity

The Complex Number System

- Perform arithmetic operations with complex numbers
- Use complex numbers in polynomial identities and equations

Functions

Interpreting Functions

- Interpret functions that arise in applications in terms of the context
- Analyze functions using different representations

Building Functions

- Build new functions from existing functions

Algebra

Seeing Structure in Expressions

- Interpret the structure of expressions

Arithmetic with Polynomials and Rational Expressions

- Understand the relationship between zeros and factors of polynomials

Creating Equations

- Create equations that describe numbers or relationships

Reasoning with Equations and Inequalities

- Solve systems of equations
- Represent and solve equations and inequalities graphically

4 Quadratic Functions and Equations

Visual See It!

Reasoning Try It!

Practice Do It!

5

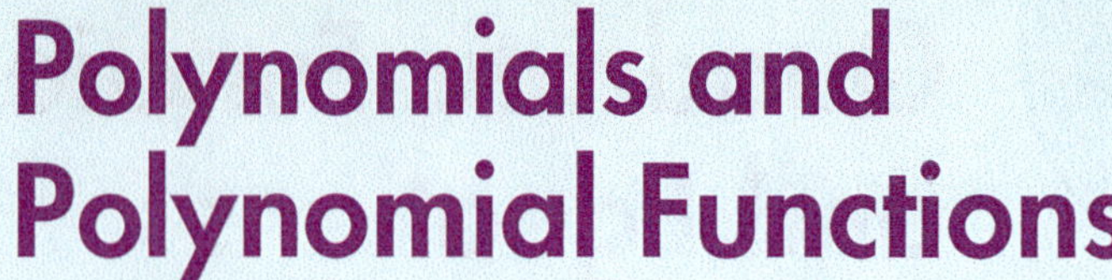

Polynomials and Polynomial Functions

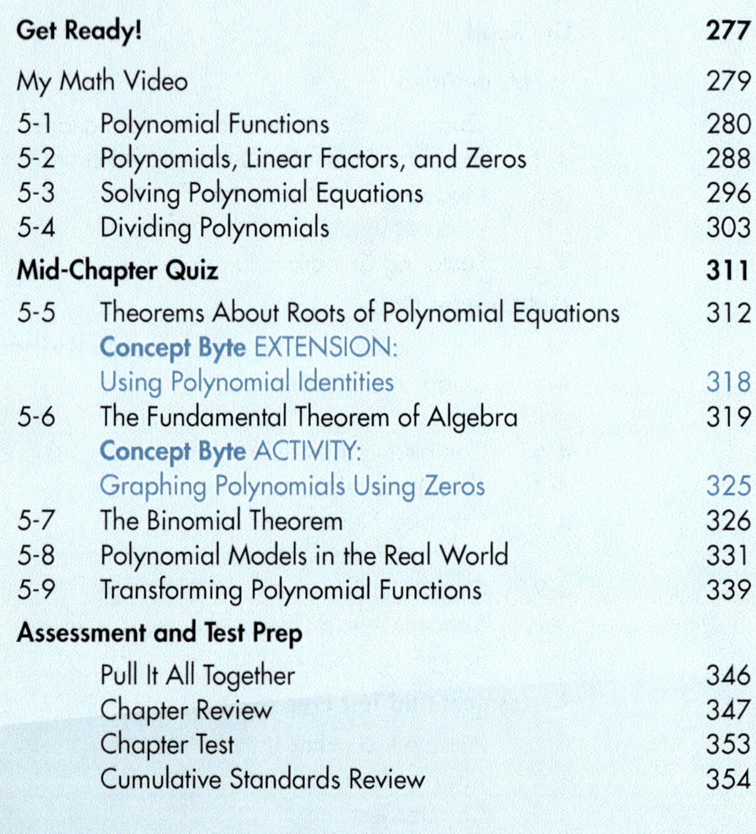

Chapters 5 & 6

Number and Quantity

The Complex Number System

Use complex numbers in polynomial identities and equations

Functions

Interpreting Functions

Interpret functions that arise in applications in terms of the context

Analyze functions using different representations

Building Functions

Build a function that models a relationship between two quantities

Build new functions from existing functions

Algebra

Seeing Structure in Expressions

Interpret the structure of expressions

Creating Equations

Create equations that describe numbers or relationships

Arithmetic with Polynomials and Rational Expressions

Understand the relationship between zeros and factors of polynomials

Use polynomial identities to solve problems

Radical Functions and Rational Exponents

Visual See It!

Reasoning Try It!

Practice Do It!

7

Exponential and Logarithmic Functions

Chapters 7 & 8

Algebra

Seeing Structure in Expressions

Interpret the structure of expressions

Creating Equations

Create equations that describe numbers or relationships

Arithmetic with Polynomials and Rational Expressions

Rewrite rational expressions

Reasoning with Equations and Inequalities

Represent and solve equations and inequalities graphically

Functions

Interpreting Functions

Analyze functions using different representations

Building Functions

Build a function that models a relationship between two quantities

Build new functions from existing functions

Linear and Exponential Models

Construct and compare linear and exponential models and solve problems

Entry-Level Assessment

Multiple Choice

Read each question. Then write the letter of the correct answer on your paper.

1. Let $A = \{1, 2, 3, 4\}$ be a set in the universe $U = \{1, 2, 3, 4, 5, 6, 7, 8\}$. What is the complement of A?

- A $\{2, 3\}$
- B $\{5, 6, 7, 8\}$
- C $\{1, 2, 3, 4\}$
- D $\{2, 3, 7, 8\}$

2. Solve $x^2 + 2x - 3 = 0$ by factoring.

- F $x = -3$ and $x = 1$
- G $x = -1$ and $x = 3$
- H $x = 0$
- I $x = -3$ and $x = 0$

3. Simplify $\frac{3a^2b^3 - 12a^4b^3 + 6a^4b^2}{3a^2b}$.

- A $b^2 - 4a^2b^2 + 2a^2b$
- B $a^2b - 4a^2b^2 + 2a^2b$
- C $3b^2 - 12a^2b + 6b^2$
- D $3ab^2 - 4a^2b + 2ab^2$

4. Which relation is not a function?

- F $\{(1, -5), (2, 4), (1, -4)\}$
- G $\{(1, -5), (2, 4), (3, -3)\}$
- H $\{(1, -5), (2, 4), (3, 2)\}$
- I $\{(1, -5), (2, 4), (3, -4)\}$

5. In the diagram, m and n are parallel.

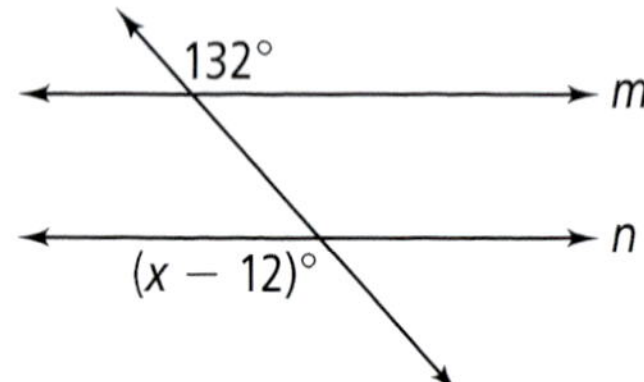

What is the value of x?

- A 36
- B 60
- C 120
- D 144

6. Solve $2(1 - 2w) = 4w + 18$.

- F -4
- G -2
- H 8
- I 16

7. Which of the following lines is perpendicular to the line $3x + y = 2$?

- A $y = 3x + 4$
- B $y = \frac{1}{3}x - 2$
- C $y = -3x + 3$
- D $y = -\frac{1}{3}x + 1$

8. If $y = 1$, then $(x + 5) \cdot y = x + 5$. Which property supports this statement?

- F Inverse Property of Multiplication
- G Identity Property of Multiplication
- H Associative Property of Addition
- I Commutative Property of Addition

9.

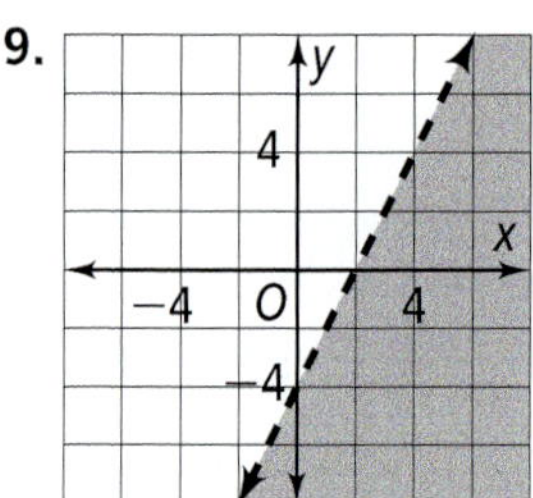

Which inequality does the graph represent?

- A $y < 2x - 4$
- B $y > -4x + 2$
- C $y > 2x - 4$
- D $y < -4x + 2$

10. The area of a trapezoid is $A = \frac{1}{2}h(b_1 + b_2)$. Solve for b_1.

- F $b_1 = \frac{2A - b_2}{h}$
- G $b_1 = \frac{2A - h}{b_2}$
- H $b_1 = \frac{2A}{h} - b_2$
- I $b_1 = 2A - b_2$

11. Let $\overleftrightarrow{AB}$ be parallel to $\overleftrightarrow{CD}$, with $A(-2, 3)$, $B(1, 4)$, and $C(1, 2)$. Which of the following could be the coordinates of point D?

(A) $(4, 1)$ (C) $(-2, 3)$

(B) $(-2, -1)$ (D) $(4, 3)$

12. Solve $3 \geq 4g - 5 \geq -1$.

(F) $-\frac{3}{2} \leq g \leq 2$ (H) $-4 \leq g \leq 8$

(G) $-1 \leq g \leq \frac{3}{4}$ (I) $1 \leq g \leq 2$

13. Which is *not* a solution of $5(2x + 4) \geq 2(x + 34)$?

(A) 48 (C) 6

(B) 8 (D) 3

14. Factor $6x^2 - 216$.

(F) $6(x - 6)(x + 6)$

(G) $(6x - 36)(6x + 36)$

(H) $6(x - 6)$

(I) $6(x - 36)(x + 6)$

15. Mike and Jane leave their home on bikes traveling in opposite directions on a straight road. Mike rides 5 mi/h faster than Jane. After 4 h they are 124 mi apart. At what rate does Mike ride his bike?

(A) 5 mi/h (C) 18 mi/h

(B) 13 mi/h (D) 31 mi/h

16. What is the point-slope form for the equation of the line in the graph?

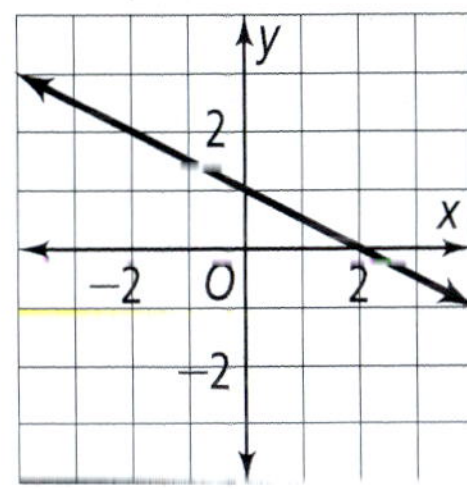

(F) $y - 2 = \frac{3}{2}(x + 2)$

(G) $y - 2 = \frac{1}{2}(x + 2)$

(H) $y - 2 = -\frac{1}{2}(x + 2)$

(I) $y - 2 = -\frac{2}{3}(x + 2)$

17. A rectangular photograph is being enlarged to poster size by making both the length and width six times as large as the original. How many times as large as the area of the original photograph is the area of the poster?

(A) $\frac{1}{6}$ (C) 12

(B) 6 (D) 36

18. A rectangle has a length of $2x + 3$ and a width of $x - 4$. Find the area of the rectangle.

(F) $2x^2 - 12$

(G) $2x^2 - 8x$

(H) $2x^2 - 5x - 12$

(I) $2x^2 - 11x - 12$

19. What is the y-intercept of the line that passes through the points $(-4, 4)$ and $(2, -5)$?

(A) -2 (C) $\frac{3}{2}$

(B) $-\frac{3}{2}$ (D) 2

20. Which of the following is equivalent to $\sqrt{2}(\sqrt{6} - 4)$?

(F) $\sqrt{12} - 4$ (H) $\sqrt{12} - 8$

(G) $2\sqrt{3} - 2\sqrt{2}$ (I) $2\sqrt{3} - 4\sqrt{2}$

21. Which of the following represents the system shown in the graph?

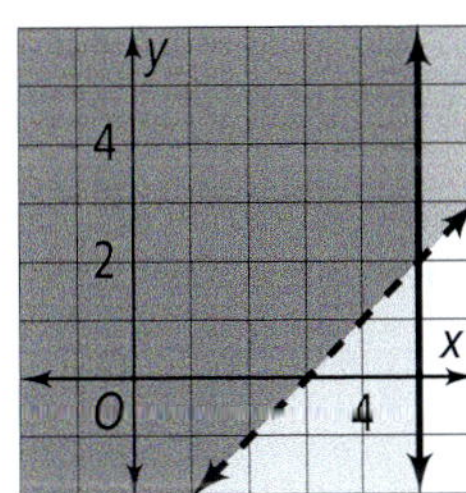

(A) $\begin{cases} y = x - 3 \\ x \geq 5 \end{cases}$ (C) $\begin{cases} y < x - 3 \\ x = 5 \end{cases}$

(B) $\begin{cases} y \leq x - 3 \\ x > 5 \end{cases}$ (D) $\begin{cases} y > x - 3 \\ x \leq 5 \end{cases}$

22. Which of the following equations represents the line that is parallel to the line $y = 5x + 2$ and that passes through the point $(1, -3)$?

(F) $y = -5x + 2$ (H) $y = \frac{1}{5}x - 8$

(G) $y = 5x + 8$ (I) $y = 5x - 8$

23. Which equation represents a line that would be perpendicular to a second line with a slope of $\frac{1}{5}$?

Ⓐ $y = -5x + 2$

Ⓑ $y = -\frac{1}{5}x + 3$

Ⓒ $y = 5x - 2$

Ⓓ $5y + x = 2$

24. $\triangle ABC$ is similar to $\triangle DEF$.

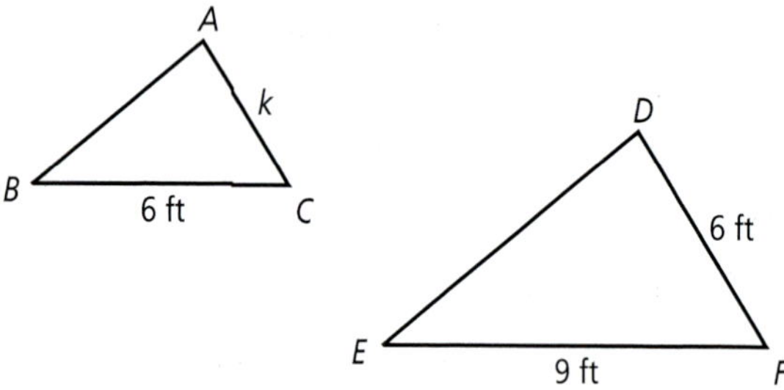

What is the value of k?

Ⓕ 3 ft Ⓗ 6 ft

Ⓖ 4 ft Ⓘ 9 ft

25. Solve the equation using the Quadratic Formula.

$6x^2 - 10x + 3 = 0$

Ⓐ $\frac{5 \pm \sqrt{7}}{6}$

Ⓑ $\frac{3 \pm \sqrt{5}}{6}$

Ⓒ -2 and 5

Ⓓ $\frac{3}{4}$ and $\frac{2}{3}$

26. A rectangle in the coordinate plane has vertices (3, 2), (8, 2), (3, 6), and (8, 6). Which of the following sets of vertices describes a rectangle that is congruent to this one?

Ⓕ (3, −2), (3, −8), (5, −8), (5, −2)

Ⓖ (−2, −4), (−2, −8), (3, −8), (3, −4)

Ⓗ (0, 0), (5, 0), (5, 5), (0, 5)

Ⓘ (−3, 2), (1, 2), (1, 6), (−3, 6)

27. Simplify the expression below.

$(-6y^{-4})^5$

Ⓐ $7776y^{20}$ Ⓒ $-7776y^{20}$

Ⓑ $\frac{7776}{y^{20}}$ Ⓓ $-\frac{7776}{y^{20}}$

28. What is the solution to $\frac{2n + 8}{3} = \frac{n + 7}{2}$?

Ⓕ −9 Ⓗ 5

Ⓖ −1 Ⓘ 13

29. Solve the system of equations below.

$$\begin{cases} 3x + y = -7 \\ 4x - y = -14 \end{cases}$$

Ⓐ (−3, 2) Ⓒ (−3, −2)

Ⓑ (3, 2) Ⓓ no solution

30. Which of the following is equivalent to $\frac{2x - 12}{x^2 - 2x - 24}$?

Ⓕ $\frac{2}{x + 4}$ Ⓗ $\frac{1}{x^2 - 2}$

Ⓖ $\frac{1}{x + 2}$ Ⓘ $\frac{2x - 3}{x - 6}$

31. What is (are) the solution(s) of the graphed function when the value of the function is 0?

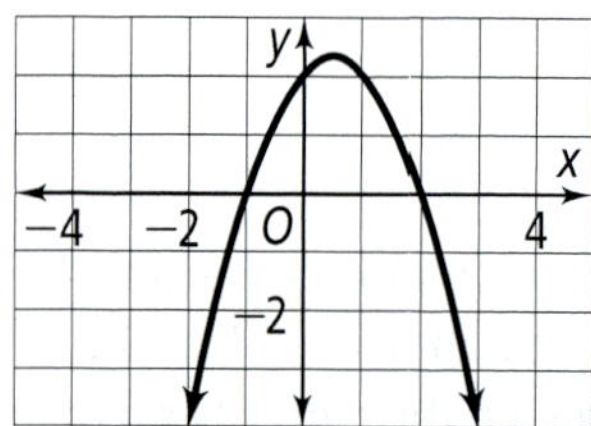

Ⓐ −1 and 2 Ⓒ 2

Ⓑ 1 and −2 Ⓓ 2.2

32. Which of the following is true?

Ⓕ $\sqrt{85} < 9$ Ⓗ $\sqrt{\frac{16}{25}} > \sqrt{\frac{16}{4}}$

Ⓖ $8 < \sqrt{62}$ Ⓘ $\sqrt{121} < \sqrt{144}$

33. A firefighter leans a 30-ft ladder against a building in order to reach a window that is 24 ft high. How far away from the building is the base of the ladder?

Ⓐ 18 ft Ⓒ 24 ft

Ⓑ 20 ft Ⓓ 30 ft

34. What is the number of x-intercepts of the parabola with equation $y = 6x^2 - 4x - 3$?

Ⓕ 0 Ⓗ 2

Ⓖ 1 Ⓘ 3

Get Ready!

Skills Handbook, page 973

Adding Rational Numbers

Find each sum.

1. $6 + (-6)$ **2.** $-8 + 6$ **3.** $5.31 + (-7.40)$ **4.** $-1.95 + 10$

5. $7\frac{3}{4} + \left(-8\frac{1}{2}\right)$ **6.** $-2\frac{1}{3} + 3\frac{1}{4}$ **7.** $6\frac{2}{5} + 4\frac{3}{10}$ **8.** $-1\frac{5}{6} + 5\frac{1}{3}$

Skills Handbook, page 973

Subtracting Rational Numbers

Find each difference.

9. $-28 - 14$ **10.** $61 - (-11)$ **11.** $-16 - (-25)$ **12.** $-6.2 - 3.6$

13. $-5\frac{2}{3} - \left(-2\frac{1}{3}\right)$ **14.** $-2\frac{1}{4} - 3\frac{1}{4}$ **15.** $2\frac{2}{3} - 7\frac{1}{3}$ **16.** $\frac{5}{2} - \frac{13}{4}$

Skills Handbook, page 973

Multiplying and Dividing Rational Numbers

Find each product or quotient.

17. $-3 \cdot 7$ **18.** $-2.1 \cdot (-3.5)$ **19.** $-\frac{2}{3} \div 4$ **20.** $-\frac{3}{8} \div \frac{5}{8}$

Skills Handbook, page 975

Using the Order of Operations

Simplify each expression.

21. $8 \cdot (-3) + 4$ **22.** $3 \cdot 4 - 8 \div 2$ **23.** $1 \div 2^2 - 0.54 + 1.26$

24. $9 \div (-3) - 2$ **25.** $5(3 \cdot 5 - 4)$ **26.** $1 - (1 - 5)^2 \div (-8)$

27. Reasoning Why don't the expressions $3 + 5^2 \cdot 3 \div 15$ and $(3 + 5^2) \cdot 3 \div 15$ yield the same answer?

Looking Ahead Vocabulary

28. The current of a river flows north at a *constant* rate. What is the constant in the mathematical expression $3x + 5y + 3$?

29. Before signing a contract, you must review the *terms* and conditions of the contract. How many terms are there in the surface area formula below?

$$2(\ell w + wh + h\ell)$$

30. Engineers *evaluate* the efficiency of the memory and speed of a computer. What does evaluate mean in mathematics?

31. Smiling is a facial *expression* of happiness or contentment. In math, what is the expression that represents the quotient of 3 and 3 less than a number?

Expressions, Equations, and Inequalities

PowerAlgebra.com

Your place to get all things digital

Download videos connecting math to your world.

Math definitions in English and Spanish

The online Solve It will get you in gear for each lesson.

Interactive! Vary numbers, graphs, and figures to explore math concepts.

Online access to stepped-out problems aligned to Common Core

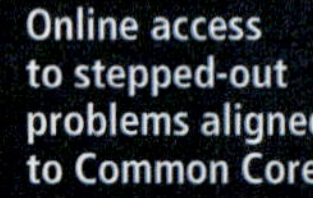

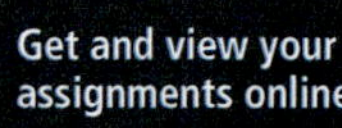

Get and view your assignments online.

Extra practice and review online

DOMAINS

- Seeing Structure in Expressions
- Creating Equations

You can use variables to represent the distance or time of the swimmer in the video.

How can you use algebraic expressions to represent patterns? How can you solve equations and inequalities? How can you solve absolute value equations? You will learn how in this chapter.

Vocabulary

English/Spanish Vocabulary Audio Online:

English	Spanish
absolute value, *p. 41*	valor absoluto
algebraic expression, *p. 5*	expresión algebraica
compound inequality, *p. 36*	desigualdad compuesta
like terms, *p. 21*	términos semejantes
literal equation, *p. 29*	ecuación literal
term, *p. 20*	término
variable, *p. 5*	variable

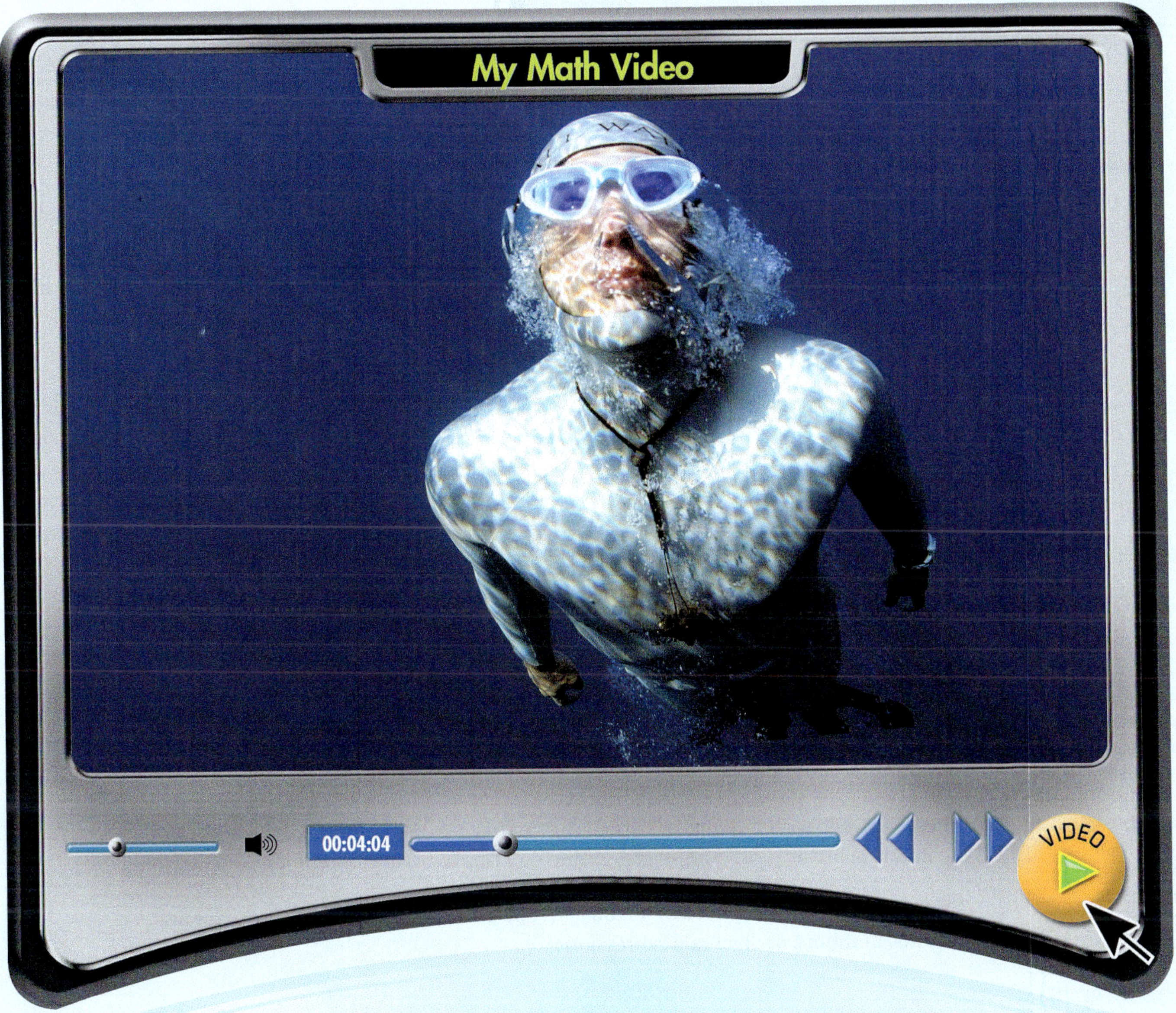

BIG ideas

1 **Variable**

Essential Question How do variables help you model real-world situations?

2 **Properties**

Essential Question How can you use the properties of real numbers to simplify algebraic expressions?

3 **Solving Equations and Inequalities**

Essential Question How do you solve an equation or inequality?

Chapter Preview

- 1-1 Patterns and Expressions
- 1-2 Properties of Real Numbers
- 1-3 Algebraic Expressions
- 1-4 Solving Equations
- 1-5 Solving Inequalities
- 1-6 Absolute Value Equations and Inequalities

Patterns and Expressions

Content Standard

Reviews A.SSE.3 Choose and produce an equivalent form of an expression to reveal and explain properties of the quantity represented by the expression.

Objective To identify and describe patterns

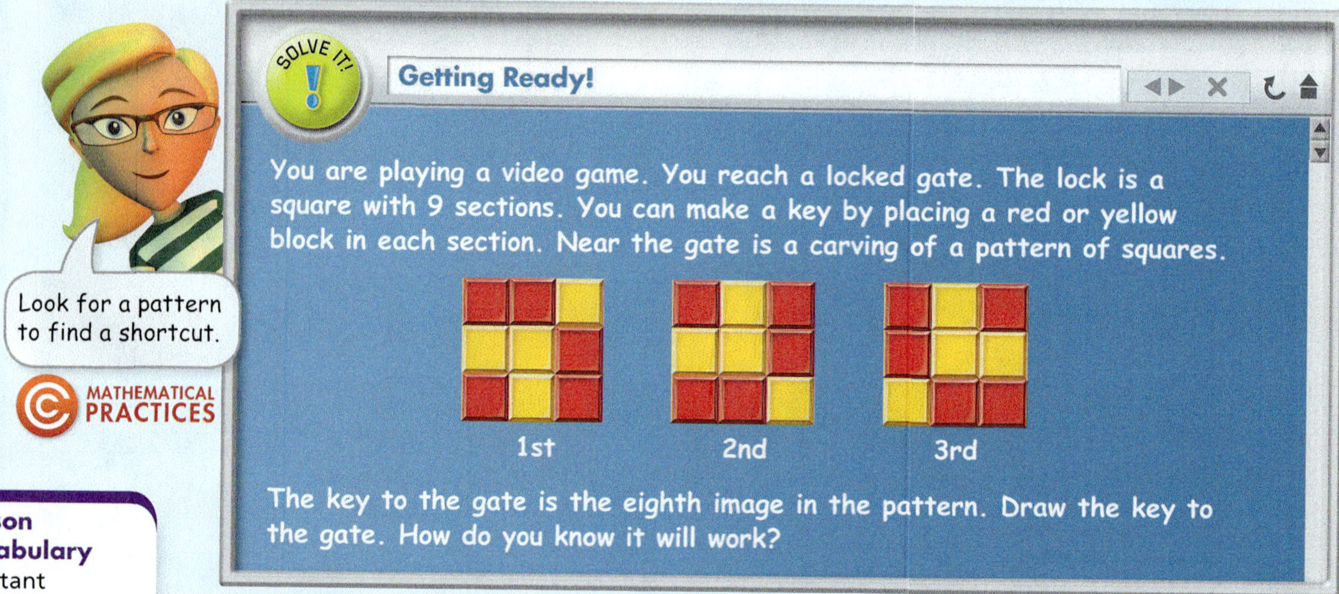

Lesson Vocabulary
- constant
- variable quantity
- variable
- numerical expression
- algebraic expression

In the Solve It, you identified and used a geometric pattern. In this lesson, you will identify patterns in pictures, tables, and graphs and describe them using numbers and variables.

Essential Understanding You can represent some patterns using diagrams, words, numbers, or algebraic expressions.

Problem 1 Identifying a Pattern

Think

How can you identify a pattern?
Look for the same type of change between consecutive figures.

Look at the figures from left to right. What is the pattern? What would the next figure in the pattern look like?

The pattern shows regular polygons with the number of sides increasing by one.

The last figure shown above has six sides, so the next figure would have seven sides. This is a heptagon: ⬡.

Got It? 1. Look at the figures from left to right. What is the pattern? Draw the next figure in the pattern.

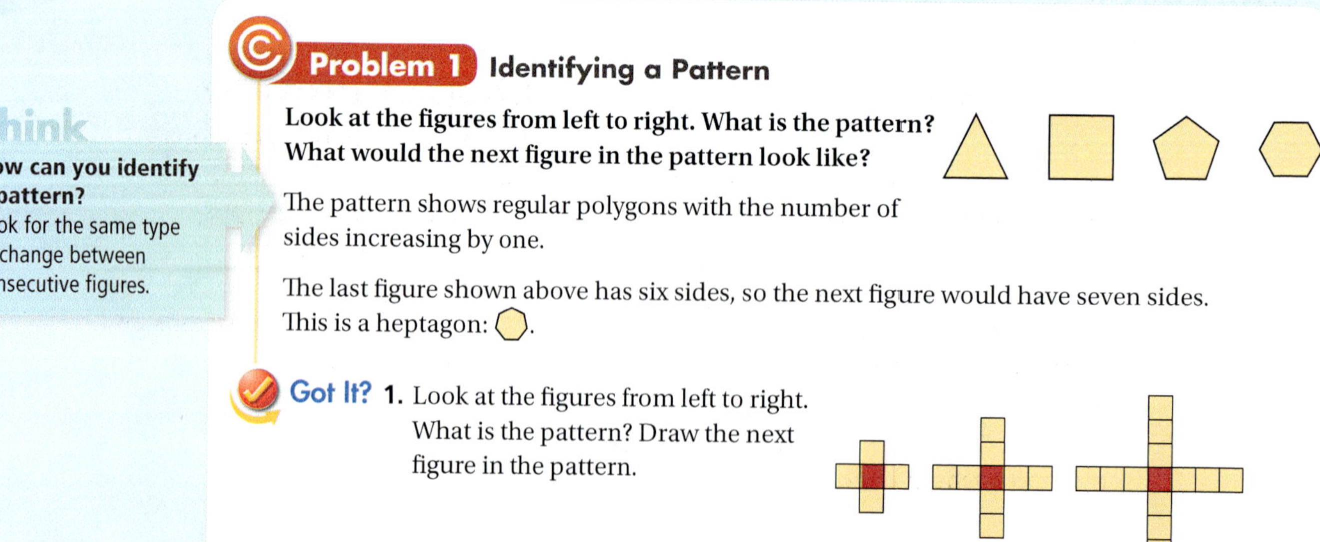

A mathematical *quantity* is anything that can be measured or counted. The *value* of the quantity is its measure or the number of items that are counted. Quantities whose values do not change are called **constants**. In other situations, the value of a quantity can change. Quantities whose values change or vary are called **variable quantities**.

take note

Key Concepts Variables and Expressions

Definition	Examples	
A **variable** is a symbol, usually a letter, that represents one or more numbers.	n	x
A **numerical expression** is a mathematical phrase that contains numbers and operation symbols.	$3 + 5$	$(8 - 2) + 5$
An **algebraic expression** is a mathematical phrase that contains one or more variables.	$3n + 5$	$(8x - 2) + 5n$

Tables are a convenient way to organize data and discover patterns. They work much like an "input/output" machine: a machine that takes one value as an input, processes it, and gives a value as an output. A process column in the table provides a way to understand what happens to the input values.

Problem 2 Expressing a Pattern With Algebra

Use a pattern to answer each question.

A How many toothpicks are in the 20th figure?

Use a table. Look for a pattern that relates the figure number to the number of toothpicks.

Figure Number (Input)	Process Column	Number of Toothpicks (Output)
1	1(4)	4
2	2(4)	8
3	3(4)	12
⋮	⋮	⋮
n	■	■

To get the output, multiply the input by 4.

Pattern: Multiply the figure number by 4 to get the number of toothpicks. So, there are $20(4) = 80$ toothpicks in the 20th figure.

Think

What would the process look like for the *n*th row?

Multiply the figure number, n, by 4.

B What is an expression that describes the number of toothpicks in the *n*th figure?

Use the pattern from part (A). There are $4n$ toothpicks in the nth figure.

Got It? 2. How many tiles are in the 25th figure in this pattern? Show a table of values with a process column.

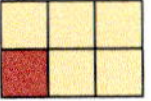
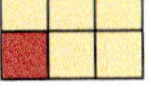
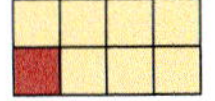
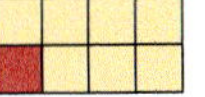

Problem 3 Using a Graph

Aquarium You want to set up an aquarium and need to determine what size tank to buy. The graph shows tank sizes using a rule that relates the capacity of the tank to the combined lengths of the fish it can hold.

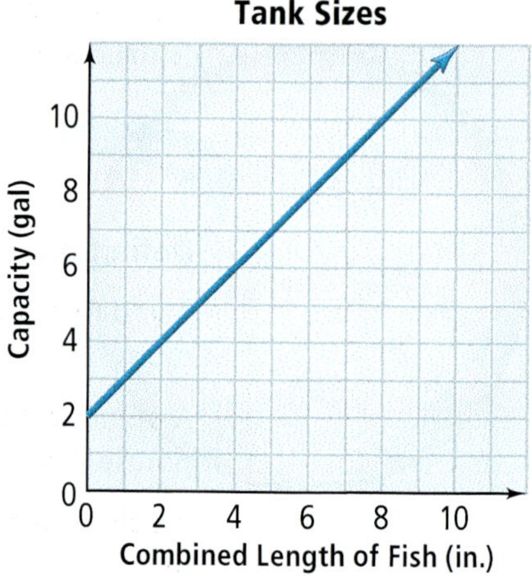

If you want five 2-in. platys, four 1-in. guppies, and a 3-in. loach, which is the smallest capacity tank you can buy: 15-gallon, 20-gallon, or 25-gallon? Use a table to find the answer.

Plan

How can you use the given graph?
You can use the graph to make a table and find a pattern relating capacity and combined fish length.

Think	Write
Choose some points on the graph.	(0, 2), (5, 7), (10, 12)
Make a table using the input and output values shown in the ordered pairs. Find a pattern in the process column. Each output is 2 more than the corresponding input.	See table below
You want 5 platys, 4 guppies, and 1 loach. So, you will have a total of 17 in. of fish. Find the output when the input is 17.	output = input + 2 = 17 + 2 = 19
Write the answer in words.	You need to buy the 20-gal tank.

Input	Process Column	Output
0	0 + 2	2
5	5 + 2	7
10	10 + 2	12

Got It? **3.** The graph at the right shows the total cost of platys at the aquarium shop. Use a table to answer the questions.

a. How much do six platys cost?

b. How much do ten platys cost?

c. **Reasoning** Why is the graph in Problem 3 a line while the graph at the right is a set of points?

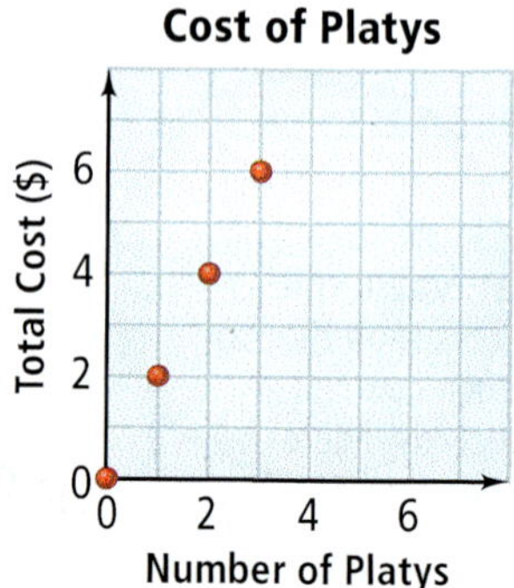

Lesson Check

Do you know HOW?

Describe a rule for each pattern.

1. 35, 70, 105, 140, . . .

2. 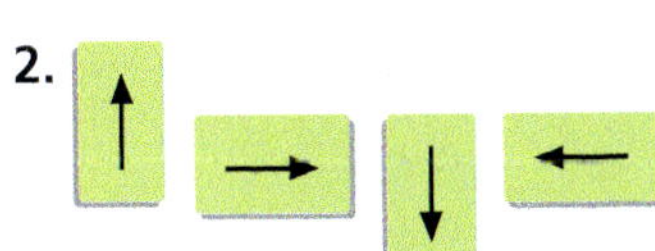

Make a table to represent each pattern. Use a process column.

3. 2, 4, 6, 8, . . .

4. ||| |||||| |||||||| |||||||||||

Do you UNDERSTAND?

5. Explain the strategy you use to identify a pattern.

6. **Compare and Contrast** How are tables of values like pictorial representations? How are they different?

7. **Error Analysis** Your friend looks for a pattern in the table below and claims that the output equals the input divided by 2. Is your friend correct? Explain.

Input	3	6.8	8	10	25
Output	2	3.4	4	5	12.5

Practice and Problem-Solving Exercises

Describe each pattern using words. Draw the next likely figure in each pattern. See Problem 1.

8. 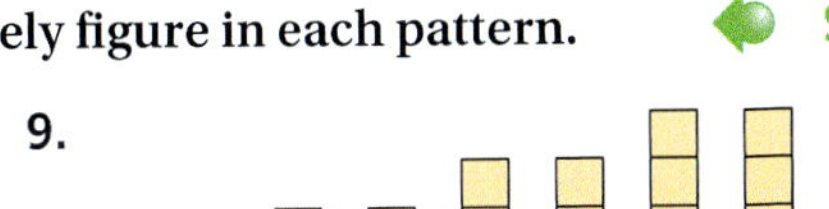

9. 

10. 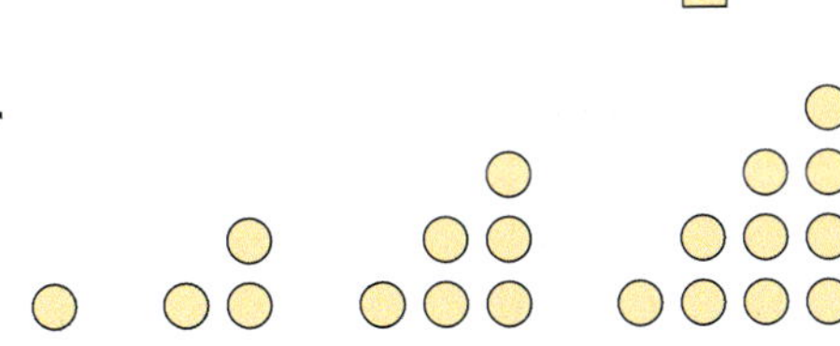

11.

Make a table with a process column to represent each pattern. Write an expression for the number of tiles or circles in the nth figure. See Problem 2.

12. 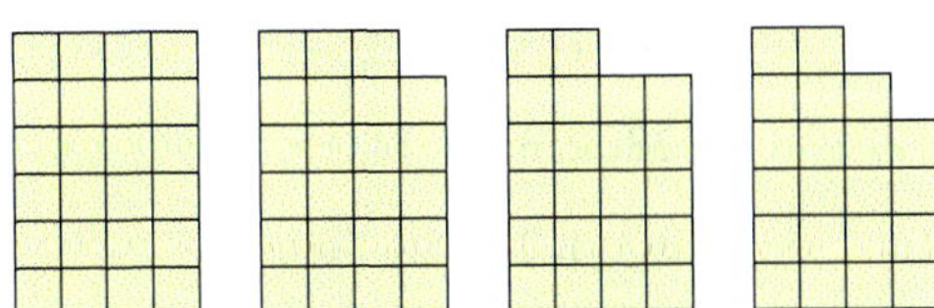

13. 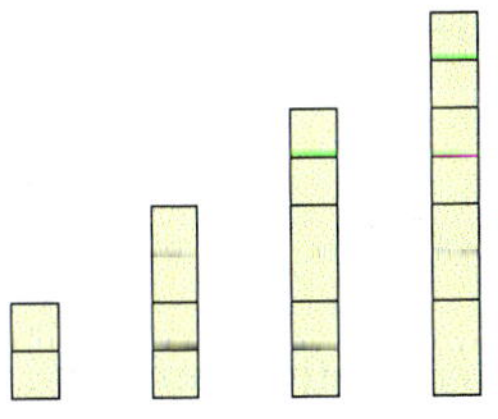

14. 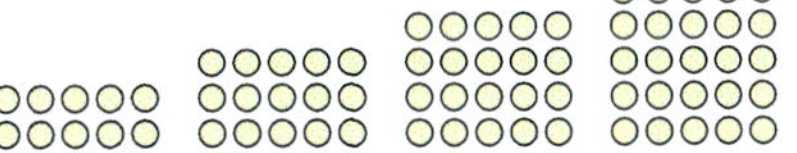

15. 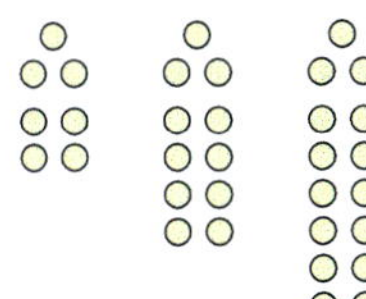

Copy and complete each table. Include a process column.

16.

Input	Output
1	0
2	1
3	2
4	3
5	■
6	■
⋮	⋮
n	■

17.

Input	Output
1	3
2	4
3	5
4	6
5	■
6	■
⋮	⋮
n	■

18.

Input	Output
1	−3
2	−6
3	−9
4	−12
5	■
6	■
⋮	⋮
n	■

Identify a pattern by making a table. Include a process column.

See Problem 3.

19.

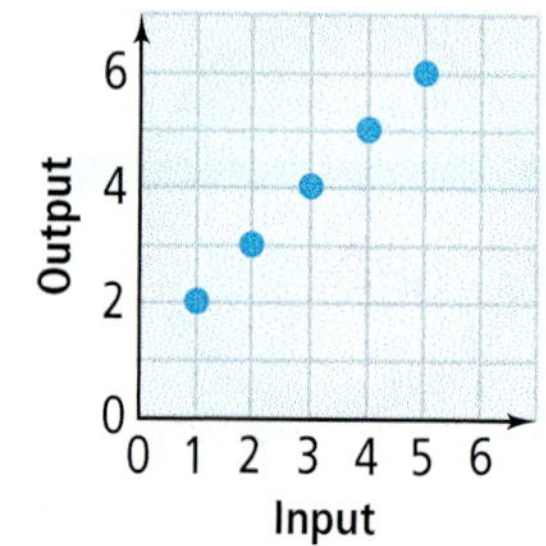

20.

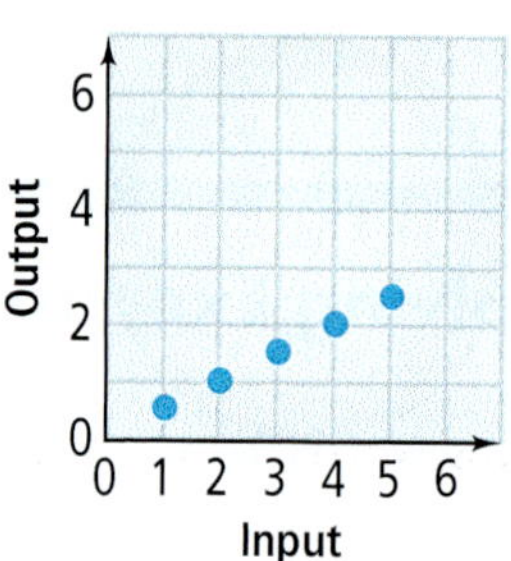

21.

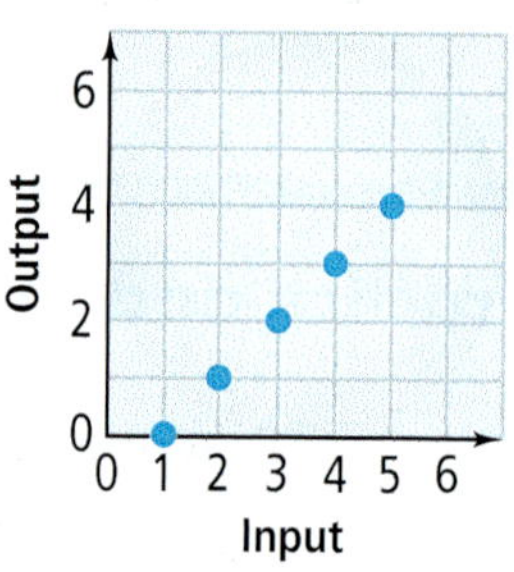

The graph shows the number of bottles of water needed for students going on a field trip.

Class Field Trip

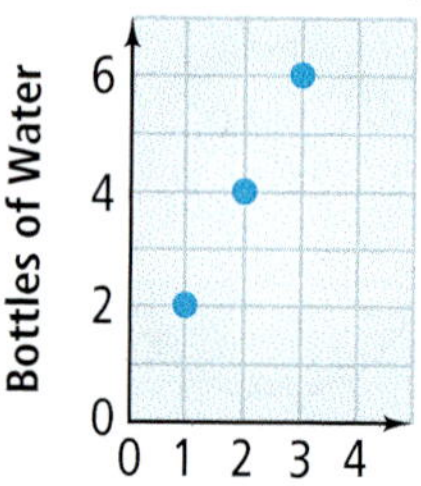

22. How many bottles of water are needed if 5 students attend?

23. How many bottles of water are needed if 20 students attend?

24. How many bottles of water are needed if n students attend?

Identify a pattern and find the next three numbers in the pattern.

25. 6, 12, 18, 24, . . .

26. 1, 4, 3, 6, 5, . . .

27. 3, 6, 10, 15, . . .

28. 2, 6, 10, 14, . . .

29. 1, 3, 9, 27, 81, . . .

30. 4, 20, 100, 500, . . .

31. Identify a pattern and draw the next three figures in the pattern.

32. **Open-Ended** Write a rule so that for every input, the output is an even number.

33. **Think About a Plan** A moving company sells different sizes of boxes as shown. The extra-large box is one size larger than the third box shown. What is its volume?

- Identify a pattern of the dimension changes.
- What is the formula for the volume of a rectangular prism?

12 in. 16 in. 12 in. 16 in. 16 in. 16 in. 20 in. 16 in. 20 in.

34. Use the graph shown.
   a. Identify a pattern of the graph by making a table of the inputs and outputs.
   b. What are the outputs for inputs 6, 7, and 8?

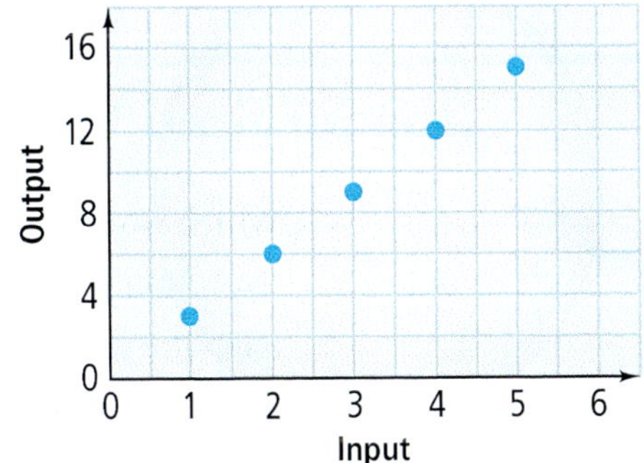

35. **Collecting** Jay has a rare baseball card collection. He currently owns 10 baseball cards. Each month, he purchases a new card for his collection. Write a model to represent the number of cards in Jay's collection after *n* months.

36. Use the figures below.

 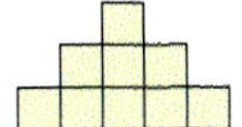

   a. Draw the next two figures.
   b. Copy and complete the table to find the number of squares in each figure.
   c. What is the number of squares in the *n*th figure? Explain your reasoning.

Figure (Input)	Process Column	Number of Squares (Output)
1	■	■
2	■	■
3	■	■
4	■	■
5	■	■

Copy and complete each table.

37.

Input	Output
1	5
2	9
3	13
4	17
5	■
⋮	⋮
n	■

38.

Input	Output
1	2
2	−3
3	−8
4	−13
5	■
⋮	⋮
n	■

39.

Input	Output
1	3
2	−1
3	−5
4	−9
5	■
⋮	⋮
n	■

40. **Open-Ended** Write the first five numbers of two different patterns in which 12 is the third number.

41. **Reasoning** For the past 4 years, Jesse has grown 3 inches each year. He is now 15 years old and is 5 feet 6 inches tall. He predicts that when he is 20 years old, he will be 6 feet 9 inches tall. What would you tell Jesse about his prediction?

42. This pattern shows the first five steps in constructing the Sierpinski Triangle. Use a pattern to describe the figures.

Standardized Test Prep

43. Which of the following is the best statement about the graph?

Ⓐ After 4 weeks, the plant will be 7 inches tall.
Ⓑ The plant was 1 inch tall at the beginning of the experiment.
Ⓒ After 2 weeks, the plant was 3 inches tall.
Ⓓ After 6 weeks, the plant will be 8 inches tall.

Plant Growth

Height of Plant (in.): 0, 2, 4, 6
Week: 0, 1, 2, 3, 4

44. Which is the 7th number in this pattern?

8, 13, 18, 23, . . .

Ⓕ 28 Ⓗ 38
Ⓖ 33 Ⓘ 43

45. Look at the pattern shown.

144, 72, 36, . . .

a. What is a rule for the pattern?
b. What is the first non-integer number in this pattern?

Mixed Review

Simplify each expression. See p. 975.

46. $3.6 + (-1.7)$ 47. $1.2 - 5$

48. $(-3)(-9)$ 49. $0(-8)$

50. $-2.8 \div 7$ 51. $-35 \div (-5)$

Get Ready! **To prepare for Lesson 1-2, do Exercises 52–57.**

Write each number as a percent. See p. 972.

52. 0.5 53. 0.25

54. $\frac{1}{3}$ 55. $1\frac{2}{5}$

56. 1.72 57. 1.23

1-2 Properties of Real Numbers

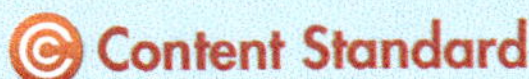

Content Standard

Reviews N.RN.3 Explain why sums and products of rational numbers are rational, that the sum of a rational number and an irrational number is irrational, and that the product of a nonzero rational number and an irrational number is irrational.

Objectives To graph and order real numbers
To identify properties of real numbers

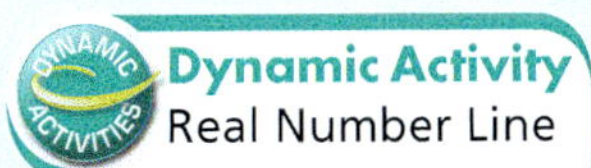

Lesson Vocabulary
- opposite
- additive inverse
- reciprocal
- multiplicative inverse

In the Solve It, you classified sets of emoticons. In this lesson, you will classify real numbers into special subsets.

Essential Understanding The set of real numbers has several special subsets related in particular ways.

Algebra involves operations on and relations among numbers, including real numbers and imaginary numbers. (You will learn about imaginary numbers in Chapter 4.) Rational numbers and irrational numbers form the set of real numbers.

You can graph every real number as a point on the number line.

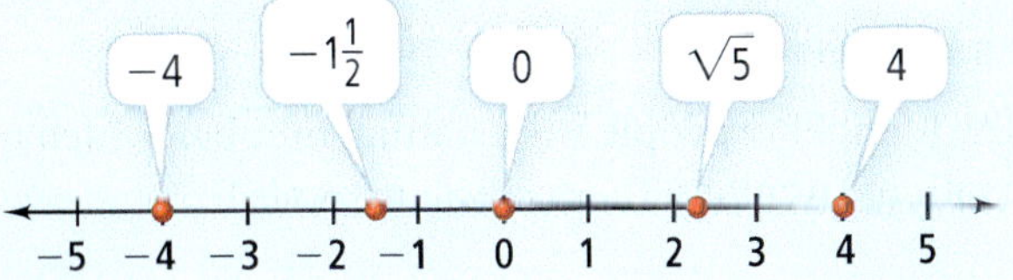

The diagram shows how subsets of the real numbers are related.

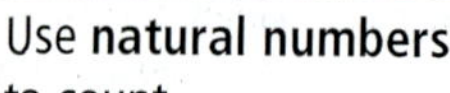

Rational numbers

- are all numbers you can write as a quotient of integers $\frac{a}{b}$, $b \neq 0$.
- include terminating decimals. For example, $\frac{1}{8} = 0.125$.
- include repeating decimals. For example, $\frac{1}{3} = 0.\overline{3}$.

Irrational numbers

- have decimal representations that neither terminate nor repeat. For example, $\sqrt{2} = 1.414213\ldots$.
- cannot be written as quotients of integers.

You classify a variable by naming the subset that gives you the most information about the numbers the variable represents.

Problem 1 Classifying a Variable

Think

What are some examples of possible values of *p*?

The number of people *p* must be represented by a whole number. Determine which other sets of numbers include the whole numbers.

Multiple Choice **Your school is sponsoring a charity race. Which set of numbers does not contain the number of people *p* who participate in the race?**

(A) natural numbers

(B) integers

(C) rational numbers

(D) irrational numbers

The number of people *p* is a natural number, which means that it is also an integer and a rational number. The correct answer is D.

Got It? **1.** In Problem 1, if each participant made a donation *d* of \$15.50 to a local charity, which subset of real numbers best describes the amount of money raised?

Problem 2 Graphing Numbers on the Number Line

What is the graph of the numbers $-\frac{5}{2}$, $\sqrt{2}$, and $2.\overline{6}$?

How do you graph a number on the number line?
If the number is an integer, determine whether it is positive or negative. If it's not an integer, determine which integer it's closest to.

Since $-\frac{5}{2} = -2\frac{1}{2}$, $-\frac{5}{2}$ is between -3 and -2.

Use a calculator. $\sqrt{2} \approx 1.4$.

Think: $2.\overline{6} = 2\frac{2}{3}$.

$-3 \quad -2 \quad -1 \quad 0 \quad 1 \quad 2 \quad 3$

Got It? **2.** What is the graph of the numbers $\sqrt{3}$, $-1.\overline{4}$, and $\frac{1}{3}$?

The number line is helpful for ordering several real numbers. For two numbers, however, it is easier to show order, or compare, using one of the inequality symbols $>$ or $<$.

Problem 3 Ordering Real Numbers

How do $\sqrt{17}$ and 3.8 compare? Use $>$ or $<$.

Think

Why compare $\sqrt{17}$ to the square root of a perfect square?
It makes it easier to determine which two integers $\sqrt{17}$ is between.

Compare both numbers to $\sqrt{16}$.

$\sqrt{16} < \sqrt{17}$ 16 is less than 17.

$3.8 < \sqrt{16}$ $\sqrt{16} = 4$ and $3.8 < 4$.

Therefore, $3.8 < \sqrt{17}$, or $\sqrt{17} > 3.8$.

Check Use a calculator.

$\sqrt{17} \approx 4.123$

$3.8 < 4.123$ ✔

Got It? **3. a.** How do $\sqrt{26}$ and 6.25 compare? Use $>$ or $<$.

b. Reasoning Let a, b, and c be real numbers such that $a < b$ and $b < c$. How do a and c compare? Explain.

Essential Understanding The properties of real numbers are relationships that are true for all real numbers (except, in one case, zero).

One property of real numbers excludes a single number, zero. Zero is the *additive identity* for the real numbers, and zero is the one real number that has no *multiplicative inverse*.

The **opposite** or **additive inverse** of any number a is $-a$.
The sum of a number and its opposite is 0, the additive identity.

Examples $12 + (-12) = 0$ $-7 + 7 = 0$

The **reciprocal** or **multiplicative inverse** of any nonzero number a is $\frac{1}{a}$.
The product of a number and its reciprocal is 1, the multiplicative identity.

Examples $8\left(\frac{1}{8}\right) = 1$ $-5\left(-\frac{1}{5}\right) = 1$

Properties Properties of Real Numbers

Let a, b, and c represent real numbers.

Property	Addition	Multiplication
Closure	$a + b$ is a real number.	ab is a real number.
Commutative	$a + b = b + a$	$ab = ba$
Associative	$(a + b) + c = a + (b + c)$	$(ab)c = a(bc)$
Identity	$a + 0 = a, 0 + a = a$ 0 is the additive identity.	$a \cdot 1 = a, 1 \cdot a = a$ 1 is the multiplicative identity.
Inverse	$a + (-a) = 0$	$a \cdot \frac{1}{a} = 1, a \neq 0$
Distributive	$a(b + c) = ab + ac$	

Problem 4 Identifying Properties of Real Numbers

Plan

How can you analyze an equation?
Determine whether it
- uses addition or multiplication
- reorders or regroups the numbers
- uses an identity

Which property does the equation illustrate?

A $\left(-\frac{2}{3}\right)\left(-\frac{3}{2}\right) = 1$

The product of the numbers is 1.

Inverse Property of Multiplication

B $(3 \cdot 4) \cdot 5 = (4 \cdot 3) \cdot 5$

The equation reorders 3 and 4.

Commutative Property of Multiplication

  **Got It?** **4. a.** Which property does the equation $3(g + h) + 2g = (3g + 3h) + 2g$ illustrate?

b. Reasoning Use properties of real numbers to show that $a + [3 + (-a)] = 3$. Justify each step of your solution.

Lesson Check

Do you know HOW?

Write an example from daily life that uses each type of real number.

1. whole numbers
2. integers
3. rational numbers

Identify the property illustrated by the equation.

4. $5 + (-5) = 0$
5. $2 \cdot (4 \cdot 5) = (2 \cdot 4) \cdot 5$

Do you UNDERSTAND?

6. **Vocabulary** Identify another name for a reciprocal.
7. **Compare and Contrast** How is the Additive Identity Property similar to the Multiplicative Identity Property? How is it different?
8. **Reasoning** There are grouping symbols in the equation $(5 + w) + 8 = (w + 5) + 8$, but it does not illustrate the Associative Property of Addition. Explain.
9. Give an example of a number that is not a rational number. Explain why it is not rational.

Practice and Problem-Solving Exercises

Classify each variable according to the set of numbers that best describes its values. See Problem 1.

10. the number of times n a ball bounces; the height h from which the ball is dropped
11. the year y; the median selling price p for a house that year
12. the circumference C of a circle found by using the formula $C = 2\pi r$

Graph each number on a number line. See Problem 2.

13. 0	14. $-\sqrt{24}$	15. -2	16. $2\frac{1}{2}$	17. $-4\frac{2}{3}$
18. 3.5	19. -1.4	20. $\sqrt{10}$	21. $-2\frac{1}{5}$	22. 4.8

Compare the two numbers. Use > or <. See Problem 3.

23. $16, \sqrt{16}$	24. $-4, -\sqrt{4}$	25. $\sqrt{5}, \sqrt{7}$
26. $-\sqrt{3}, -\sqrt{5}$	27. $5, \sqrt{22}$	28. $-\sqrt{38}, 6$
29. $4, \sqrt{12}$	30. $-8, \sqrt{70}$	31. $\sqrt{63}, 7.5$
32. $4.7, \sqrt{26}$	33. $\sqrt{75}, 9$	34. $12, -\sqrt{150}$

Name the property of real numbers illustrated by each equation. See Problem 4.

35. $\pi(a + b) = \pi a + \pi b$	36. $-10 + 4 = 4 + (-10)$
37. $(2\sqrt{7}) \cdot \sqrt{3} = 2(\sqrt{7} \cdot \sqrt{3})$	38. $29 \cdot \pi = \pi \cdot 29$
39. $-\sqrt{5} + 0 = -\sqrt{5}$	40. $\frac{4}{7} \cdot \frac{7}{4} = 1$

Estimate the numbers graphed at the labeled points.

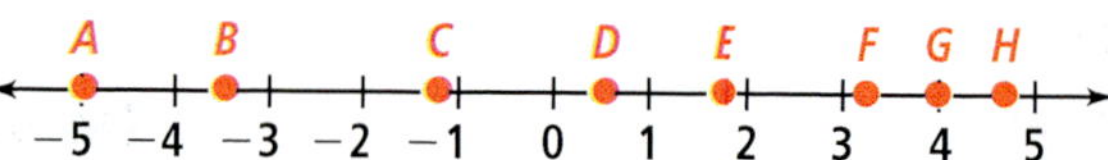

41. point A **42.** point B **43.** point C **44.** point D

45. point E **46.** point F **47.** point G **48.** point H

49. Think About a Plan A cube-shaped jewelry box has a surface area of 300 square inches. What are the dimensions of the jewelry box?

- Write an algebraic expression to find the total surface area of a cube. What is the surface area of one side of a cube?
- How is the side length of a square related to its area?

50. Error Analysis A student labeled the points on the number line as shown. Explain the student's error.

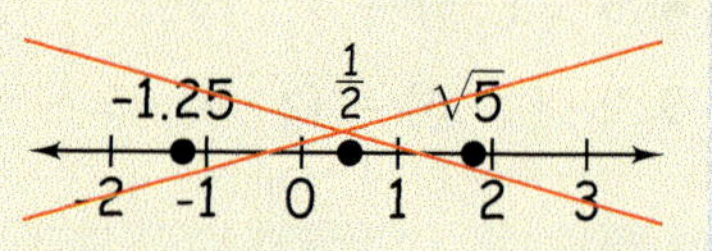

STEM Science **The formula $I = \sqrt{\frac{W}{R}}$ gives the electric current I in amperes that flows through an appliance, where W is the power in watts and R is the resistance in ohms. Which set of numbers best describes the value of I for the given values of W and R?**

51. $W = 100, R = 25$ **52.** $W = 100, R = 5$ **53.** $W = 500, R = 100$

54. $W = 50, R = 200$ **55.** $W = 250, R = 100$ **56.** $W = 240, R = 100$

Write the numbers in decreasing order.

57. $1, -3, -\sqrt{2}, 8, \frac{1}{3}$ **58.** $\sqrt{14}, \frac{5}{2}, -\frac{9}{16}, 1, 11$ **59.** $-17, -0.06, -3\sqrt{3}, 5.73, \frac{1}{4}$

Reasoning **An example is a *counterexample* to a general statement if it makes the statement false. Show that each of the following statements is false by finding a counterexample.**

60. The reciprocal of each natural number is a natural number.

61. The opposite of each whole number is a whole number.

62. There is no integer that has a reciprocal that is an integer.

63. The product of two irrational numbers is an irrational number.

64. All square roots are irrational numbers.

65. Writing Write an example of each of the 11 properties of real numbers shown on page 14.

66. Restaurant Five friends each ordered a sandwich and a drink at a restaurant. Each sandwich costs the same amount and each drink costs the same amount. What are two ways to compute the bill? What property of real numbers is illustrated by the two methods?

67. Open-Ended Write an algebraic problem that requires the use of the real-number properties to solve. Then solve the problem.

68. **Writing** Are there two integers with a product of -12 and a sum of -3? Explain.

69. Your friend used the Distributive Property and got the expression $5x + 10y - 35$. What algebraic expression could your friend have started with?

70. **Geometry** π is an irrational number you can use to calculate the circumference or area of a circle.

a. Find the value of π on your calculator. Can you obtain an exact representation? Explain.

b. The value of π is often represented as $\frac{22}{7}$. How does this representation compare to the decimal representation your calculator gives using the π key?

71. Does zero have a multiplicative inverse? Explain.

Standardized Test Prep

SAT/ACT

72. Which of the following shows the numbers π, $\sqrt{8}$, and 3.5 in the correct order from greatest to least?

Ⓐ $\pi, \sqrt{8}, 3.5$ Ⓑ $3.5, \pi, \sqrt{8}$ Ⓒ $\sqrt{8}, \pi, 3.5$ Ⓓ $\sqrt{8}, 3.5, \pi$

73. Which of the following is the best statement about the graph?

Ⓕ A 400-minute plan costs \$40.

Ⓖ A 100-minute plan costs \$10.

Ⓗ A 1000-minute plan costs \$110.

Ⓘ A 200-minute plan costs \$35.

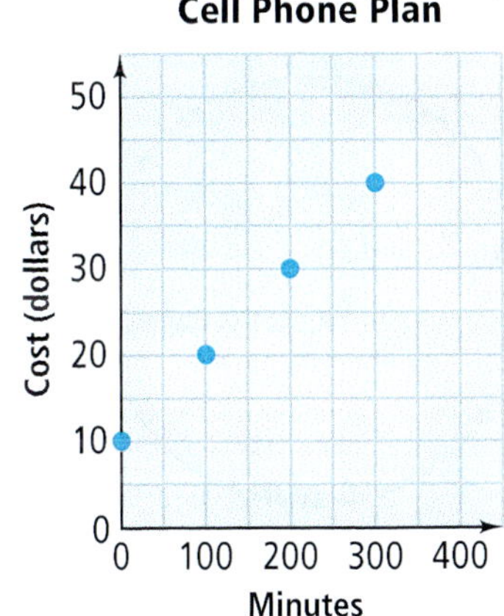

Short Response

74. Why is the opposite of the reciprocal of 5 the same as the reciprocal of the opposite of 5?

Mixed Review

Identify a pattern and find the next three terms in the pattern.

See Lesson 1-1.

75. 4, 8, 12, 16, . . . **76.** 8, 9, 10, 11, . . . **77.** $-4, -3, -2, -1, \ldots$

Get Ready! **To prepare for Lesson 1-3, do Exercises 78–83.**

Use the order of operations to simplify each expression.

See p. 975.

78. $3 \div 4 + 6 \div 4$

79. $5[(2 + 5) \div 3]$

80. $\frac{8 + 5 \times 2}{12}$

81. $(40 + 24) \div 8 - (2^2 - 1)$

82. $40 + 24 \div 8 - 2^2 - 1$

83. $(40 + 24) \div (8 - 2^2) - 1$

1-3 Algebraic Expressions

Content Standard

Reviews A.SSE.1.a Interpret parts of an expression, such as terms, factors, and coefficients.

Objectives To evaluate algebraic expressions
To simplify algebraic expressions

Ten weeks is a long time! Perhaps you can solve a simpler problem first.

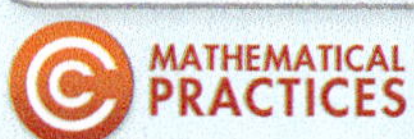

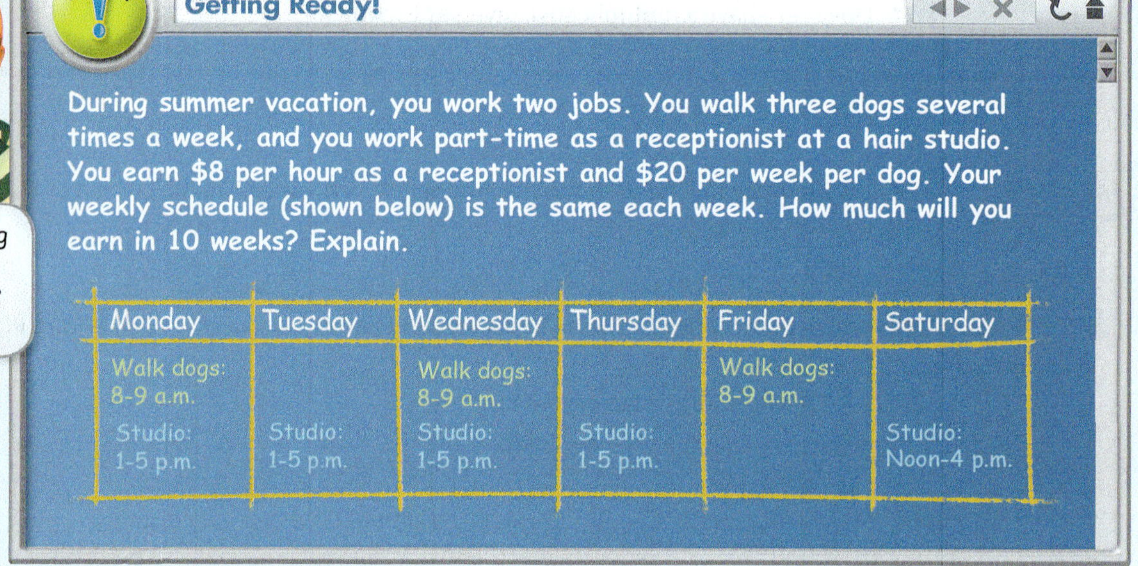

During summer vacation, you work two jobs. You walk three dogs several times a week, and you work part-time as a receptionist at a hair studio. You earn \$8 per hour as a receptionist and \$20 per week per dog. Your weekly schedule (shown below) is the same each week. How much will you earn in 10 weeks? Explain.

Monday	Tuesday	Wednesday	Thursday	Friday	Saturday
Walk dogs: 8-9 a.m.		Walk dogs: 8-9 a.m.		Walk dogs: 8-9 a.m.	
Studio: 1-5 p.m.	Studio: 1-5 p.m.	Studio: 1-5 p.m.	Studio: 1-5 p.m.		Studio: Noon-4 p.m.

Lesson Vocabulary
- evaluate
- term
- coefficient
- constant term
- like terms

Essential Understanding You can represent some mathematical phrases and real-world quantities using algebraic expressions.

Problem 1 Modeling Words With an Algebraic Expression

Think

What does "seven fewer than t" mean?
"Seven fewer than t" means your answer will be less than t.

Multiple Choice Which algebraic expression models the word phrase *seven fewer than a number t*?

Ⓐ $t + 7$ Ⓑ $-7t$ Ⓒ $t - 7$ Ⓓ $7 - t$

"Seven fewer than" suggests subtraction. Begin with the number t and subtract 7. This can be represented by the expression $t - 7$. The correct answer is C.

Got It? 1. Which algebraic expression models the word phrase *two times the sum of a and b*?

Ⓕ $a + b$ Ⓗ $2(a + b)$

Ⓖ $2a + b$ Ⓘ $a + 2b$

To model a situation with an algebraic expression, do the following:

- Identify the actions that suggest operations.
- Define one or more variables to represent the unknown(s).
- Represent the actions using the variables and the operations.

Problem 2 Modeling a Situation

Plan

How can you identify the variable? Determine which quantity is unknown.

Savings **You start with \$20 and save \$6 each week. What algebraic expression models the total amount you save?**

Relate	starting amount	plus	amount saved	times	number of weeks
Define	Let w = the number of weeks.				
Write	20	$+$	6	$\cdot$	w

The expression $20 + 6w$ models the situation.

Got It? 2. You had \$150, but you are spending \$2 each day. What algebraic expression models this situation?

To **evaluate** an algebraic expression, substitute a number for each variable in the expression. Then simplify using the order of operations.

Problem 3 Evaluating Algebraic Expressions

Plan

What operations should you start with? Do operations that occur in grouping symbols first. Parentheses are grouping symbols.

What is the value of the expression for the given values of the variables?

A $7(a + 4) + 3b - 8$ **for** $a = -4$ **and** $b = 5$

$7(-4 + 4) + 3(5) - 8$ Substitute the value for each variable.

$= 7(0) + 3(5) - 8$ Perform operations within grouping symbols.

$= 0 + 15 - 8$ Multiply.

$= 15 - 8$ Add and subtract from left to right.

$= 7$

B $\frac{x}{2} + y^2$ **for** $x = 1$ **and** $y = \frac{1}{2}$

$\frac{1}{2} + \left(\frac{1}{2}\right)^2$ Substitute the value for each variable.

$= \frac{1}{2} + \frac{1}{4}$ Simplify the power.

$= \frac{3}{4}$ Add.

Got It? 3. a. What is the value of the expression $\frac{2(x^2 - y^2)}{3}$ for $x = 6$ and $y = -3$?

b. Reasoning Will the value of the expression change if the parentheses are removed? Explain.

Problem 4 Writing and Evaluating an Expression

Sports In football, a touchdown (TD) is worth six points, an extra-point kick (EPK) one point, and a field goal (FG) three points. What algebraic expression models the total number of points that a football team scores in a game, assuming each scoring play is one of the three given types? Suppose a football team scores 3 touchdowns, 2 extra-point kicks, and 4 field goals. How many points did the team score?

Know	Need	Plan
• Number of points each scoring play is worth • Number of each type of score	• Algebraic expression to model points scored • Total number of points scored	• Determine the variables. • Write an expression. • Evaluate the expression.

Think

How many points come from touchdowns?
The number of points from touchdowns is six times the number of touchdowns.

Relate points per TD · number of TDs + points per EPK · number of EPKs + points per FG · number of FGs

Define Let t = the number of touchdowns.
Let k = the number of extra-point kicks.
Let f = the number of field goals.

Write $6 \cdot t + 1 \cdot k + 3 \cdot f$

The expression $6t + 1k + 3f$ models the team's total score.

The football team scores 3 touchdowns, 2 extra-point kicks, and 4 field goals, so $t = 3$, $k = 2$, and $f = 4$.

$6(3) + 1(2) + 3(4)$ Substitute the value for each variable.
$= 18 + 2 + 12$ Multiply.
$= 32$ Add.

The team scored 32 points.

Got It? 4. In basketball, teams can score by making two-point shots, three-point shots, and one-point free throws. What algebraic expression models the total number of points that a basketball team scores in a game? If a team makes 10 two-point shots, 5 three-point shots, and 7 free throws, how many points does it score in all?

An expression that is a number, a variable, or the product of a number and one or more variables is a **term**. A **coefficient** is the numerical factor of a term. A **constant term** is a term with no variables. You can add terms to form longer expressions. The expression below has three terms.

$$-4ax + 7w - 6$$

coefficients
The numerical coefficient of $-4ax$ is -4.

constant term
Think of $7w - 6$ as $7w + (-6)$.
The constant term is -6.

Like terms have the same variables raised to the same powers.

like terms — like terms

$$3x^2 + 5x^2 + 9y^3z + 2yz - 4y^3z$$

You can simplify an algebraic expression that has like terms. You combine like terms using the properties of real numbers (Lesson 1-2). An expression and its simplified form are equivalent. Their values are equal for all values of their variables.

take note

Concept Summary Properties for Simplifying Algebraic Expressions

Let a, b, and c represent real numbers.

Definition of Subtraction	$a - b = a + (-b)$
Definition of Division	$a \div b = \frac{a}{b} = a \cdot \frac{1}{b}, b \neq 0$
Distributive Property for Subtraction	$a(b - c) = ab - ac$
Multiplication by 0	$0 \cdot a = 0$
Multiplication by −1	$-1 \cdot a = -a$
Opposite of a Sum	$-(a + b) = -a + (-b) = -a - b$
Opposite of a Difference	$-(a - b) = -a + b = b - a$
Opposite of a Product	$-(ab) = -a \cdot b = a \cdot (-b)$
Opposite of an Opposite	$-(-a) = a$

Problem 5 Simplifying Algebraic Expressions

Think

Are $7x^2$ and $3y^2$ like terms?

No; they have different variables.

Combine like terms. What is a simpler form of each expression?

A $7x^2 + 3y^2 + 2y^2 - 4x^2$

$7x^2 + 3y^2 + 2y^2 - 4x^2$	Identify like terms.
$= 7x^2 - 4x^2 + 3y^2 + 2y^2$	Commutative Property of Addition
$= (7 - 4)x^2 + (3 + 2)y^2$	Distributive Property
$= 3x^2 + 5y^2$	Combine like terms.

B $-(3k + m) + 2(k - 4m)$

$-3k - m + 2k - 8m$	Opposite of a Sum and Distributive Property
$= -k - 9m$	Combine like terms.

Got It? **5.** Combine like terms. What is a simpler form of each expression?

a. $-4j^2 - 7k + 5j + j^2$ **b.** $-(8a + 3b) + 10(2a - 5b)$

Lesson Check

Do you know HOW?

Write an algebraic expression that models each word phrase.

1. the quotient of the sum of 2 and a number b, and 3
2. the sum of the product of a number k and 4, and a number m

Evaluate each algebraic expression for $x = 3$ and $y = -2$.

3. $2x - 3y$
4. $5x + y$
5. $y - x$
6. $x + 4y$

Do you UNDERSTAND?

 7. **Error Analysis** A student simplified the expression as shown.

$$3p^2q + 2p - (5q + p - 2p^2q) = q^2p + 3p - 5q$$

Identify the errors and correct them.

 8. **Vocabulary** Explain the difference between a constant and a coefficient.

 9. **Compare and Contrast** How are algebraic expressions and numerical expressions alike? How are they different? Include examples to justify your reasoning.

Practice and Problem-Solving Exercises

Write an algebraic expression that models each word phrase. See Problem 1.

10. four more than a number b
11. the product of 8 and the sum of a number x and 3
12. the quotient of the difference between 5 and a number n, and 2

Write an algebraic expression that models each situation. See Problem 2.

13. Jenny had \$130, but she is spending \$10 per week.
14. The piggy bank contained \$25, and \$1.50 is added each day.
15. You had 250 minutes left on your cell phone, and you talk an hour a week.

Evaluate each expression for the given values of the variables. See Problem 3.

16. $4a + 7b + 3a - 2b + 2a$; $a = -5$ and $b = 3$
17. $-k^2 - (3k - 5n) + 4n$; $k = -1$ and $n = -2$
18. $-5(x + 2y) + 15(x + 2y)$; $x = 7$ and $y = -7$
19. $4(2m - n) - 3(2m - n)$; $m = -15$ and $n = -18$

STEM **Physics** **The expression $16t^2$ models the distance in feet that an object falls during the first t seconds after being dropped. What is the distance the object falls during each time?**

20. 0.25 second
21. 0.5 second
22. 2 seconds
23. 10 seconds

Investing The expression $1000(1.1)^t$ represents the value of a \$1000 investment that earns 10% interest per year, compounded annually for t years. What is the value of a \$1000 investment at the end of each period?

24. 2 years **25.** 3 years **26.** 4 years **27.** 5 years

Write an algebraic expression to model the total score in each situation. Then evaluate the expression to find the total score.  **See Problem 4.**

28. In the first set, the volleyball team made only 8 shots worth one point each.

29. In the last baseball game, there were two 3-run home runs and 4 hits that each scored 2 runs.

Simplify by combining like terms.  **See Problem 5.**

30. $5a - a$ **31.** $5 + 10s - 8s$ **32.** $-5a - 4a + b$

33. $2a + 3b + 4a$ **34.** $6r + 3s + 2s + 4r$ **35.** $0.5x - x$

36. $7b - (3a - 8b)$ **37.** $5 + (4g - 7)$ **38.** $-(3x - 4y) + x$

Evaluate each expression for the given value of the variable.

39. $x + 2x - x - 1; x = 2$ **40.** $2z + 3 + 5 - 3z; z = -3$ **41.** $3(2a + 5) + 2(3 - a); a = 4$

42. $\frac{5(2k - 3) - 3(k + 4)}{3k + 2}; k = -2$ **43.** $y^2 + 3; y = \sqrt{7}$ **44.** $5c^3 - 6c^2 - 2c; c = -5$

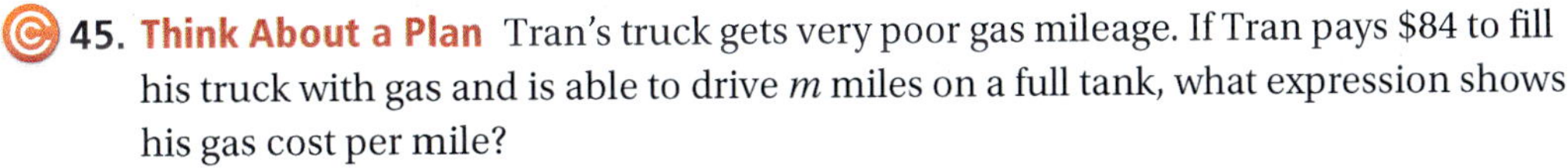

45. Think About a Plan Tran's truck gets very poor gas mileage. If Tran pays \$84 to fill his truck with gas and is able to drive m miles on a full tank, what expression shows his gas cost per mile?

- What operation does "per" indicate?
- Check your expression by substituting 200 miles for m. Does your answer make sense?

Simplify by combining like terms.

46. $a^2 + 2b^2 + \frac{1}{4}a^2$ **47.** $x + \frac{x^2}{2} + 2x^2 - x$ **48.** $\frac{y^2}{4} + \frac{y}{3} + \frac{y^2}{3} - \frac{y}{5}$

49. $-(2x + y) - 2(-x - y)$ **50.** $x(3 - y) + y(x + 6)$ **51.** $\frac{1}{2}(x^2 - y^2) - \frac{5}{2}(x^2 - y^2)$

Write an algebraic expression to model each situation.

52. Class Project The freshman class will be selling carnations as a class project. What is the class's income after it pays the florist a flat fee of \$200 and sells x carnations for \$2 each?

53. Jobs You have a summer job at a car wash. You earn \$8.50 per hour and are expected to pay a one-time fee of \$15 for the uniform. If you work x hours per week, how much will you make during the first week?

54. Reasoning Suppose you need to subtract a from b but mistakenly subtract b from a instead. How is the answer you get related to the correct answer? Explain.

55. **Error Analysis** John simplified the expression as shown. Is his work correct? Explain.

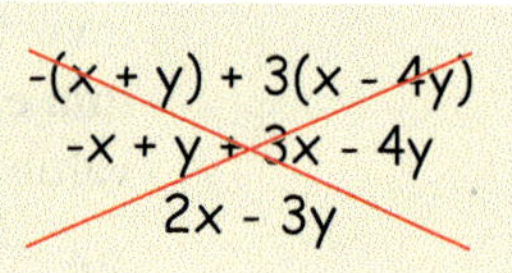

56. **Open-Ended** Write an example of an algebraic expression that has a nonnegative value regardless of the value of the variable.

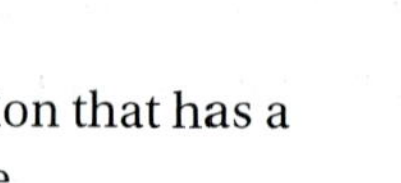

Name the property of real numbers illustrated by the equation.

57. $2(s - t) = 2s - 2t$

58. $-[-(x - 10)] = x - 10$

59. $-(2t - 11) = 11 - 2t$

60. $-(a - b) = (-1)(a - b)$

61. Simplify $2(b - a) + 5(b - a)$ and justify each step in your simplification.

62. **a.** Evaluate the expression $2(2x^2 - x) - 3(x^2 - x) + x^2 - x$ for $x = 3$. Do *not* simplify the expression before evaluating it.
b. Simplify the expression in part (a) and then evaluate your answer for $x = 3$.
c. **Writing** Explain why the values in parts (a) and (b) should be the same.

Standardized Test Prep

SAT/ACT

63. Which expression best represents a simpler form of $4m + 3(m + n)$?

Ⓐ $7m + 3n$ Ⓑ $4m + 3mn$ Ⓒ $3m + 4n$ Ⓓ $7m^2 + 3n$

64. A driver drove 12 miles and made a pit stop. After that, the driver continued driving at a constant speed of 65 miles per hour for t hours. Which of the following represents the total distance driven?

Ⓕ $12 + 65t$ Ⓖ $65t$ Ⓗ $12t + 65$ Ⓘ $12(t + 65)$

65. What is the value of r when $s = -1$, $t = 4$, and $u = \frac{1}{5}$?

$$r = 3s^2 + 5(t - 2u)$$

Ⓐ 7 Ⓑ 15 Ⓒ 21 Ⓓ $22\frac{1}{5}$

Short Response

66. Compare $\sqrt{26}$ and 4.9. Explain your answer.

Mixed Review

Order the numbers from least to greatest. See Lesson 1-2.

67. $-1.5, -0.5, -\sqrt{2}, -1.4$

68. $-\frac{3}{8}, \frac{1}{2}, -\frac{3}{4}, -\frac{5}{6}$

69. $\sqrt{2}, -20, 0.2, \frac{1}{2}$

70. $\frac{3}{4}, -3, -0.5, -\frac{1}{4}$

Get Ready! **To prepare for Lesson 1-4, do Exercises 71–74.**

Simplify each expression. See Lesson 1-3.

71. $4x + 3x - 4$

72. $-\frac{p}{3} + \frac{q}{3} - \frac{2p}{3} - q$

73. $-2(4 + b) + 4(b - 5)$

74. $(k - m) - (m - k)$

1 Mid-Chapter Quiz

MathXL® for School
Go to PowerAlgebra.com

Do you know HOW?

1. Draw the next figure in the pattern.

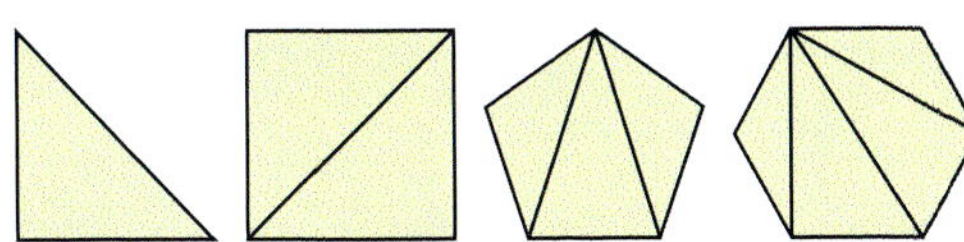

Identify a pattern and find the next number in the pattern.

2. −405, −135, −45, −15, . . .

3. $\frac{2x}{3}, \frac{x}{3}, \frac{x}{6}, \frac{x}{12}, \ldots$

4. 101, 92, 83, 74, . . .

5. 0.4, 1.2, 3.6, 10.8, . . .

Name the property of real numbers illustrated by the equation.

6. $7(x - y) = 7x - 7y$

7. $\sqrt{7} \cdot 1 = \sqrt{7}$

8. $-2 \cdot \left(-\frac{1}{2}\right) = 1$

9. $2.3(3.4 \cdot 12.9) = (2.3 \cdot 3.4)(12.9)$

Write an algebraic expression to model each word phrase.

10. eight times the sum of a and b

11. four more than the product of x and y

12. six less than the quotient of d and g

13. ten less than twice the product of s and t

Simplify each expression.

14. $-x^2 + 2y - 3x^2 + 10$

15. $-2(d + 2e) + 5(3d - 8e)$

16. $-(a + 2b) + 4(a + 2b) - 2(a + 2b)$

17. $-3x + 14x + 7x^2 - 3x + 4x(x + 1)$

Identify a pattern by making a table of the inputs and outputs. Include a process column.

18.

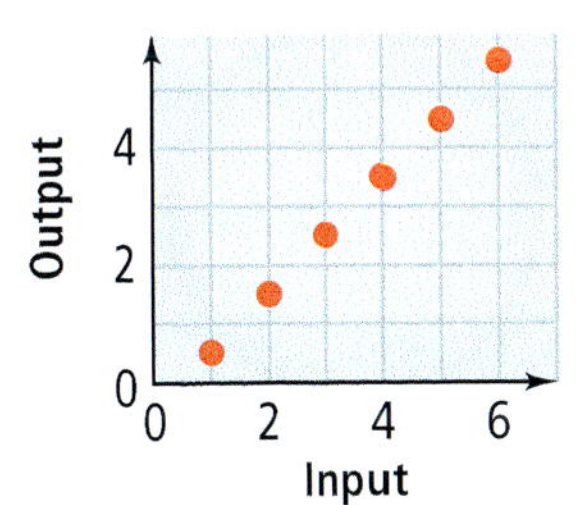

19.

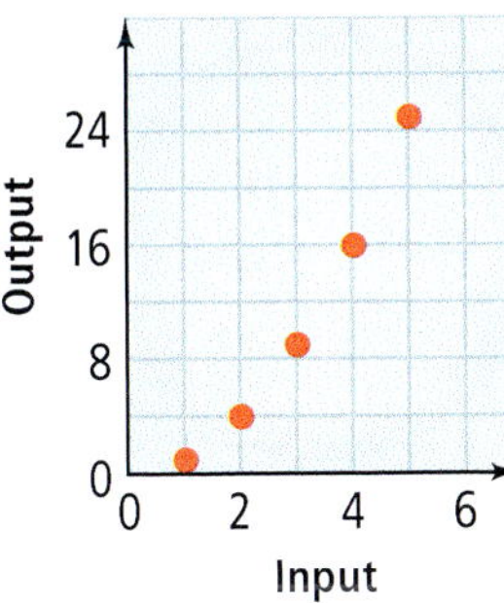

Evaluate each expression for $a = 4$, $b = -3$, and $c = 10$.

20. $7a - 5b$

21. $4a + b - |2c|$

22. $|a - b - c^2|$

Write an algebraic expression to model each situation.

23. You have 16 tomatoes, and your tomato plants produce 5 tomatoes each day.

24. Your car's gas tank holds 25 gallons, and you use 1.5 gallons of gas each day.

Do you UNDERSTAND?

25. **Writing** Explain why every integer is also a rational number.

26. **Reasoning** What expression describes the number of squares in the nth figure?

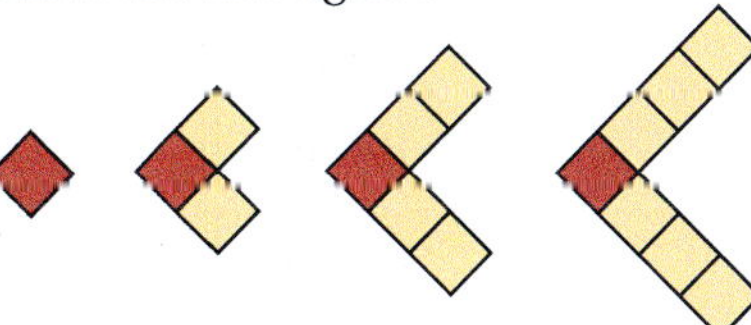

27. **Reasoning** Is there a Closure Property of Subtraction that applies to whole numbers? Explain.

1-4 Solving Equations

Content Standards

A.CED.1 Create equations and inequalities in one variable and use them to solve problems.

A.CED.4 Rearrange formulas to highlight a quantity of interest, using the same reasoning as in solving equations.

Objectives To solve equations

To solve problems by writing equations

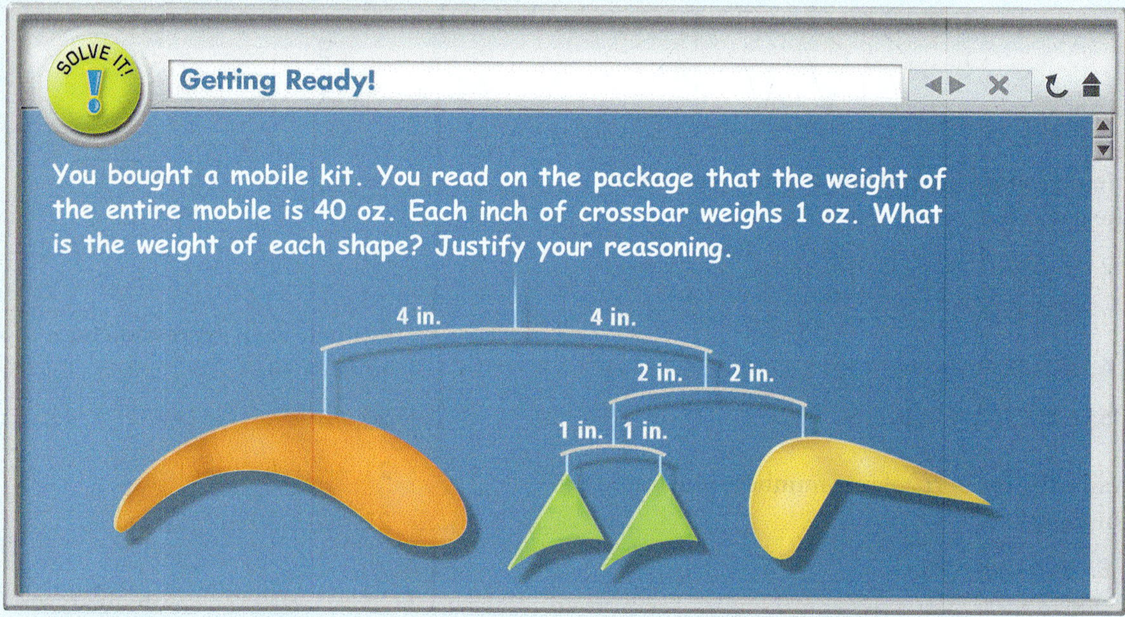

Lesson Vocabulary
- equation
- solution of an equation
- inverse operations
- identity
- literal equation

An **equation** is a statement that two expressions are equal. In this lesson you will use equations to model and solve problems.

Essential Understanding You can use the properties of equality and inverse operations to solve equations.

take note **Properties** Properties of Equality

Assume a, b, and c represent real numbers.

Property	Definition	Example
Reflexive	$a = a$	$5 = 5$
Symmetric	If $a = b$, then $b = a$.	If $\frac{1}{2} = 0.5$, then $0.5 = \frac{1}{2}$.
Transitive	If $a = b$ and $b = c$, then $a = c$.	If $2.5 = 2\frac{1}{2}$ and $2\frac{1}{2} = \frac{5}{2}$, then $2.5 = \frac{5}{2}$.
Substitution	If $a = b$, then you can replace a with b and vice versa.	If $a = b$ and $9 + a = 15$, then $9 + b = 15$.

Properties — Properties of Equality, Continued

Assume a, b, and c represent real numbers.

Property	Definition	Example
Addition	If $a = b$, then $a + c = b + c$.	If $x = 12$, then $x + 3 = 12 + 3$.
Subtraction	If $a = b$, then $a - c = b - c$.	If $x = 12$, then $x - 3 = 12 - 3$.
Multiplication	If $a = b$, then $a \cdot c = b \cdot c$.	If $x = 12$, then $x \cdot 3 = 12 \cdot 3$.
Division	If $a = b$, then $a \div c = b \div c$ (with $c \neq 0$).	If $x = 12$, then $x \div 3 = 12 \div 3$.

Solving an equation that contains a variable means finding all values of the variable that make the equation true. Such a value is a **solution of the equation**. To find a solution, isolate the variable on one side of the equation using *inverse operations*.

Inverse operations are operations that "undo" each other. Addition and subtraction have this inverse relationship, as do multiplication and division.

Problem 1 Solving a One-Step Equation

Plan

How can you isolate the variable?
To isolate the variable, you have to remove the $+4$ from the left side of the equation.

What is the solution of $x + 4 = -12$?

Subtraction is the inverse operation of addition, so subtract 4 from each side.

$x + 4 = -12$

$x + 4 - 4 = -12 - 4$ Subtraction Property of Equality

$x = -16$ Simplify.

Check $-16 + 4 \stackrel{?}{=} -12$

$-12 = -12$ ✔

Got It? 1. What is the solution of $12b = 18$?

Problem 2 Solving a Multi-Step Equation

GRIDDED RESPONSE

Plan

How do you solve an equation with the variable on both sides?
Choose a side for the variable and remove it from the other side.

What is the solution of $-27 + 6y = 3(y - 3)$?

$-27 + 6y = 3(y - 3)$

$-27 + 6y = 3y - 9$ Distributive Property

$6y = 3y + 18$ Add 27 to each side.

$3y = 18$ Subtract $3y$ from each side.

$y = 6$ Divide each side by 3.

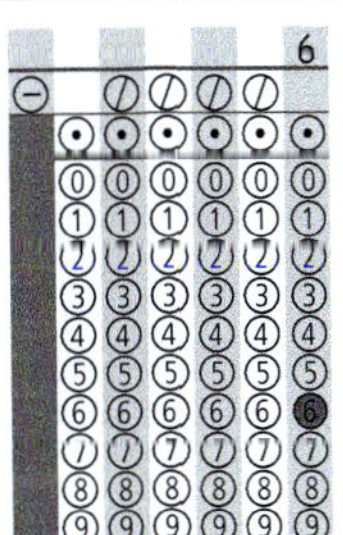

Got It? 2. What is the solution of $3(2x - 1) - 2(3x + 4) = 11x$?

Problem 3 Using an Equation to Solve a Problem

Flowers "Flower carpets" incorporate hundreds of thousands of brightly-colored flowers as well as grass, tree bark, and sometimes fountains to form intricate designs and motifs. The flower carpet shown here, from Grand Place in Brussels, Belgium, has a perimeter of 200 meters. What are the dimensions of the flower carpet?

Plan

How can you relate the dimensions to the perimeter?
Use the formula for the perimeter of a rectangle.

Relate $2 \cdot$ width plus $2 \cdot$ length equals perimeter

Define Let x = the width.

Then $3x$ = the length.

Write $2 \cdot x + 2 \cdot 3x = 200$

$2x + 2 \cdot 3x = 200$

$2x + 6x = 200$ Multiply.

$8x = 200$ Combine like terms.

$\frac{8x}{8} = \frac{200}{8}$ Divide each side by 8.

$x = 25$ Simplify.

Find the length: $3x = 3 \cdot 25 = 75$.

The width is 25 meters. The length is 75 meters.

Got It? **3.** Suppose the flower carpet from Problem 3 had a perimeter of 320 meters. What would the dimensions of the flower carpet be?

An equation does not always have one solution. An equation has no solution if no value of the variable makes the equation true. An equation that is true for every value of the variable is an **identity**.

Essential Understanding Sometimes, no value of the variable makes an equation true. For identities, all values of the variable make the equation true.

Problem 4 Equations With No Solution and Identities

Think

What does it mean for an equation to be sometimes true?
An equation is sometimes true if it is true for some, but not all, values of the variable.

Is the equation *always, sometimes,* or *never* true?

A $11 + 3x - 7 = 6x + 5 - 3x$

$$4 + 3x = 3x + 5$$

$$4 = 5$$

Never true!

The last equation is not true, so no value of x makes the first two equations true. The original equation has no solution. It is never true.

B $6x + 5 - 2x = 4 + 4x + 1$

$$4x + 5 = 4x + 5$$

$$4x = 4x$$

$$0 = 0 \checkmark$$

Always true!

The last equation is true, so any value of x makes the first three equations true. The original equation is always true. It is an identity.

Got It? **4.** Is the equation *always, sometimes,* or *never* true?

a. $7x + 6 - 4x = 12 + 3x - 8$ **b.** $2x + 3(x - 4) = 2(2x - 6) + x$

A **literal equation** is an equation that uses at least two different letters as variables. You can solve a literal equation for any one of its variables by using the properties of equality. You solve for a variable "in terms of" the other variables.

Problem 5 Solving a Literal Equation

How do you solve a literal equation for one of its variables?
Use inverse operations to isolate the indicated variable.

The equation $C = \frac{5}{9}(F - 32)$ relates temperatures in degrees Fahrenheit F and degrees Celsius C. What is F in terms of C?

$C = \frac{5}{9}(F - 32)$	
$\frac{9}{5}C = F - 32$	Multiply each side by $\frac{9}{5}$.
$\frac{9}{5}C + 32 = F$	Add 32 to each side.
$F = \frac{9}{5}C + 32$	Symmetric Property

Got It? **5. a.** The equation $K = C + 273$ relates temperatures kelvins K and degrees Celsius C. What is C in terms of K?

b. Reasoning Is the equation relating temperatures in kelvins and degrees Celsius *always, sometimes,* or *never* true? Explain your answer.

Lesson Check

Do you know HOW?

Solve each equation.

1. $w - 15 = 8.2$
2. $\frac{x}{3} = -30$
3. $2y - 1 = y + 11$

Solve each equation for k.

4. $r - 2k = 15$
5. $6k - 2z = 12$
6. $4k + h = -2k - 14$

Do you UNDERSTAND?

7. **Vocabulary** Explain what it means to find a solution of an equation.
8. **Reasoning** Suppose you solve an equation and find that your school needs 4.3 buses for a class trip. Explain how to interpret this solution.
9. **Error Analysis** Find the error(s) in the steps shown.

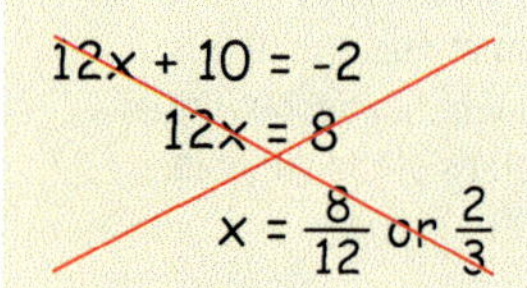

Practice and Problem-Solving Exercises

Solve each equation. See Problem 1.

10. $h - 12 = 6$
11. $-\frac{x}{3} = 27$
12. $4t = 48$
13. $22 + r = 36$

Solve each equation. Check your answer. See Problem 2.

14. $7w + 2 = 3w + 94$
15. $15 - g = 23 - 2g$
16. $43 - 3d = d + 9$
17. $5y + 1.8 = 4y - 3.2$
18. $6a - 5 = 4a + 2$
19. $7y + 4 = 3 - 2y$
20. $5c - 9 = 8 - 2c$
21. $4y - 8 - 2y + 5 = 0$
22. $6(n - 4) = 3n$
23. $2 - 3(x + 4) = 8$
24. $5(2 - g) = 0$
25. $2(x + 4) = 8$

Write an equation to solve each problem.  See Problem 3.

26. **Bus Travel** Two buses leave Houston at the same time and travel in opposite directions. One bus averages 55 mi/h and the other bus averages 45 mi/h. When will they be 400 mi apart?
27. **Aviation** Two planes left an airport at noon. One flew east and the other flew west at twice the speed. After 3 hours the planes were 2700 mi apart. How fast was each plane flying?
28. **Geometry** The length of a rectangle is 3 cm greater than its width. The perimeter is 24 cm. What are the dimensions of the rectangle?

Determine whether the equation is *always*, *sometimes*, or *never* true.  See Problem 4.

29. $5x + 3 - 2x = 7x + 3$
30. $2(5x + 4) = 10x + 6$
31. $\frac{2}{3}x + 4 = 2x$
32. $6x - 12 + 2x = 3 + 8x - 15$

Solve each formula for the indicated variable.  **See Problem 5.**

33. $A = \frac{1}{2}bh$, for h **34.** $s = \frac{1}{2}gt^2$, for g **35.** $V = lwh$, for w **36.** $I = prt$, for r

Solve each equation for x.

37. $ax + bx = c$ **38.** $\frac{x}{a} - 5 = b$ **39.** $\frac{x-2}{2} = m + n$ **40.** $\frac{2}{5}(x + 1) = g$

Solve each equation.

41. $0.2(x + 3) - 4(2x - 3) = 0.9$ **42.** $12 - 3(2w + 1) = 7w - 3(7 + w)$

43. $(m - 2) - 5 = 8 - 2(m - 4)$ **44.** $7(a + 1) - 3a = 5 + 4(2a - 1)$

45. Think About a Plan The measures of an angle and its complement differ by 22°. What are the measures of the angles?

- What is true about the sum of the measures of an angle and its complement?
- When modeling the problem with an equation, how can you algebraically represent that the two angle measures differ by 22°?

Solve each formula for the indicated variable.

46. $R(r_1 + r_2) = r_1r_2$, for R **47.** $A = \frac{1}{2}h(b_1 + b_2)$, for b_2 **48.** $S = 2\pi r^2 + 2\pi rh$, for h

49. $h = vt - 5t^2$, for v **50.** $v = s^2 + \frac{1}{2}sh$, for h **51.** $R(r_1 + r_2) = r_1r_2$, for r_2

52. Writing Suppose you write and solve an equation to determine the amount of money m you have in your bank account after several weeks. You find that $m = -36$. What does this solution mean?

53. Geometry The measure of the supplement of an angle is 20° more than three times the measure of the original angle. Find the measures of the angles.

54. Find 4 consecutive odd integers with a sum of 184.

Solve each equation for x.

55. $c(x + 2) - 5 - b(x - 3)$ **56.** $a(3tx - 2b) - c(dx - 2)$ **57.** $b(5px - 3c) = a(qx - 4)$

58. $\frac{a}{b}(2x - 12) = \frac{c}{d}$ **59.** $\frac{3ax}{5} - 4c = \frac{ax}{5}$ **60.** $\frac{a}{x} - \frac{c}{a} = m$

Write an equation to solve each problem.

61. Swimming A city park is opening a new swimming pool. You can pay a daily entrance fee of \$3 or purchase a membership for the 12-week summer season for \$82 and pay only \$1 per day to swim. How many days would you have to swim to make the membership worthwhile?

STEM **62. Rocket** The first stage of a rocket burns 28 s longer than the second stage. If the total burning time for both stages is 152 s, how long does each stage burn?

63. Error Analysis Your friend says that the equations shown are two ways to write the same formula. Is your friend correct? Explain your answer.

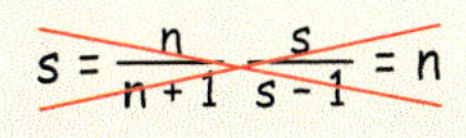

64. Assume that a, b, and c are integers and $a \neq 0$.
   a. **Proof** Prove that the solution of the linear equation $ax - b = c$ must be a rational number.
   b. **Writing** Describe the values of a, b, and c for which the solutions of $ax^2 + b = c$ are rational.

65. A tortoise crawling at a rate of 0.1 mi/h passes a resting hare. The hare wants to rest another 30 min before chasing the tortoise at a rate of 5 mi/h. How many feet must the hare run to catch the tortoise?

Standardized Test Prep

GRIDDED RESPONSE

SAT/ACT

66. A mural made of triangular tiles has an area of 1750 square centimeters. Each triangle has a height that is 3 centimeters longer than its base of 2 centimeters. How many triangular tiles are there?

67. The table shows the population of bacteria in a petri dish at various times. If the pattern continues, what will the bacteria population be at 6:00 P.M.?

Bacteria Population

Time	8:00 A.M.	10:00 A.M.	12:00 P.M.	2:00 P.M.
Population	100	200	400	800

68. To the nearest tenth, what is the value of t in the following equation?

$$4(t - 30) = 5 - 3t$$

69. A 10-foot-tall basketball hoop is 4 feet shorter than twice the height of a flag pole. What is the height in feet of the flag pole?

Mixed Review

Evaluate each expression for $x = -4$ and $y = 3$. See Lesson 1-3.

70. $x - 2y + 3$ 71. $x + x \div y$ 72. $3x - 4y - x$ 73. $x + 2y \div x$

Write an algebraic expression for each phrase. See Lesson 1-3.

74. 5 more than a number x

75. the product of 16 and a number x

76. 3 times the difference of 12 and a number x

Get Ready! **To prepare for Lesson 1-5, do Exercises 77–79.**

State whether each inequality is *true* or *false*. See Lesson 1-2.

77. $5 < 12$ 78. $5 < -12$ 79. $5 \geq 5$

Solving Inequalities

Content Standard

A.CED.1 Create equations and inequalities in one variable and use them to solve problems.

Objectives To solve and graph inequalities
To write and solve compound inequalities

MATHEMATICAL PRACTICES

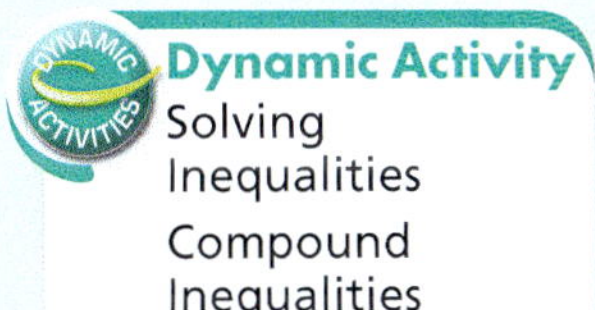

Words like "at most" and "at least" suggest a relationship in which two quantities may not be equal. You can represent such a relationship with a mathematical inequality.

Essential Understanding Just as you use properties of equality to solve equations, you can use properties of inequality to solve inequalities.

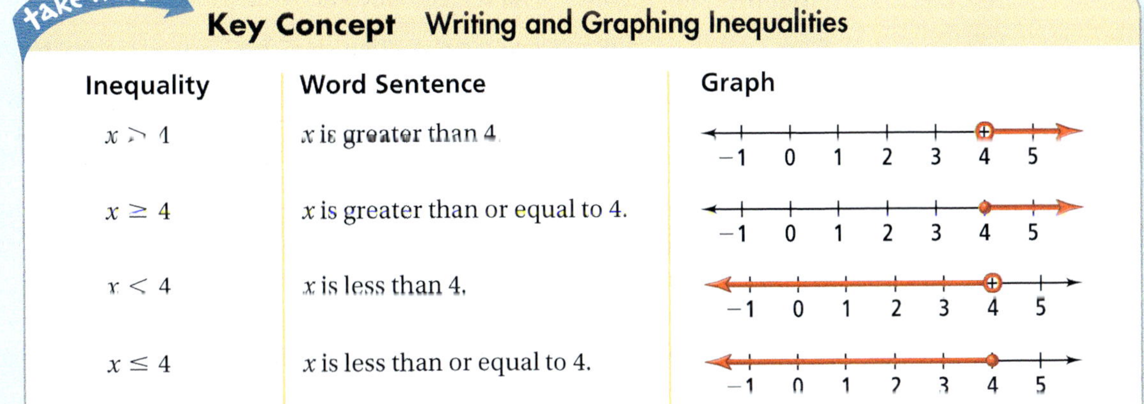

take note **Key Concept** Writing and Graphing Inequalities

Inequality	Word Sentence	Graph
$x > 4$	x is greater than 4.	
$x \geq 4$	x is greater than or equal to 4.	
$x < 4$	x is less than 4.	
$x \leq 4$	x is less than or equal to 4.	

In the graphs above, the point at 4 is a boundary point because it separates the graph of the inequality from the rest of the number line. An open dot at 4 means that 4 *is not* a solution. A closed dot at 4 means that 4 *is* a solution.

Problem 1 Writing an Inequality From a Sentence

Plan

How can you translate a sentence into an inequality? Look for key words, such as "at least" or "greater than."

What inequality represents the sentence, "5 fewer than a number is at least 12."?

5 fewer than a number is at least 12.

"Fewer" indicates subtraction.

"At least" indicates greater than or equal to.

$x - 5 \geq 12$

Got It? **1.** What inequality represents the sentence, "The quotient of a number and 3 is no more than 15."?

The solutions of an inequality are the numbers that make it true. The properties you use for solving inequalities are similar to the properties you use for solving equations. However, when you multiply or divide each side of an inequality by a negative number, you must reverse the inequality symbol.

take note **Properties** Properties of Inequalities

Let a, b, c, and d represent real numbers.

Property	Definition	Example
Transitive	If $a > b$ and $b > c$, then $a > c$.	$5 > 3$ and $3 > 1$, so $5 > 1$
Addition	If $a > b$, then $a + c > b + c$.	$4 > 2$, so $4 + 1 > 2 + 1$
Subtraction	If $a > b$, then $a - c > b - c$.	$7 > 4$, so $7 - 3 > 4 - 3$
Multiplication	If $a > b$ and $c > 0$, then $ac > bc$.	$6 > 5$ and $3 > 0$, so $6(3) > 5(3)$
	If $a > b$ and $c < 0$, then $ac < bc$.	$3 > 2$ and $-4 < 0$, so $3(-4) < 2(-4)$
Division	If $a > b$ and $c > 0$, then $\frac{a}{c} > \frac{b}{c}$.	$9 > 3$ and $3 > 0$, so $\frac{9}{3} > \frac{3}{3}$
	If $a > b$ and $c < 0$, then $\frac{a}{c} < \frac{b}{c}$.	$12 > 6$ and $-6 < 0$, so $\frac{12}{-6} < \frac{6}{-6}$

Here's Why It Works The steps below show that if $a > b$, then $-a < -b$. Therefore, you need to reverse the inequality symbol when multiplying each side of the inequality $a > b$ by -1.

$a > b$

$a - b > 0$ Subtract b from each side.

$-b - (-a) > 0$ $a - b = -b + a = -b - (-a)$.

$-b > -a$ Add $-a$ to each side.

$-a < -b$ Rewrite the inequality with $-a$ on the left side.

Problem 2 Solving and Graphing an Inequality

Plan

How is solving an inequality like solving an equation? You isolate the variable by doing the same things to each side of the inequality.

What is the solution of $-3(2x - 5) + 1 \geq 4$? Graph the solution.

$-3(2x - 5) + 1 \geq 4$	
$-6x + 15 + 1 \geq 4$	Distributive Property
$-6x + 16 \geq 4$	Simplify.
$-6x \geq -12$	Subtraction Property of Inequality
$x \leq 2$	Divide each side by -6. Reverse the inequality symbol.

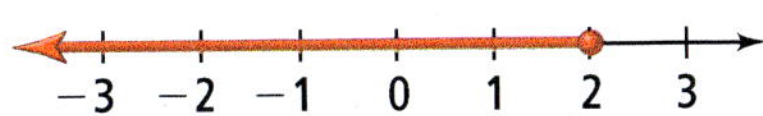

Got It? **2.** What is the solution of $-2(x + 9) + 5 \geq 3$? Graph the solution.

Problem 3 Using an Inequality

Plan

How will an inequality help answer this question? The cost of the first plan must be less than the cost of the second plan, so an inequality can be used to determine the number of movies.

Movie Rentals **A movie rental company offers two subscription plans. You can pay \$36 a month and rent as many movies as desired, or you can pay \$15 a month and \$1.50 to rent each movie. How many movies must you rent in a month for the first plan to cost less than the second plan?**

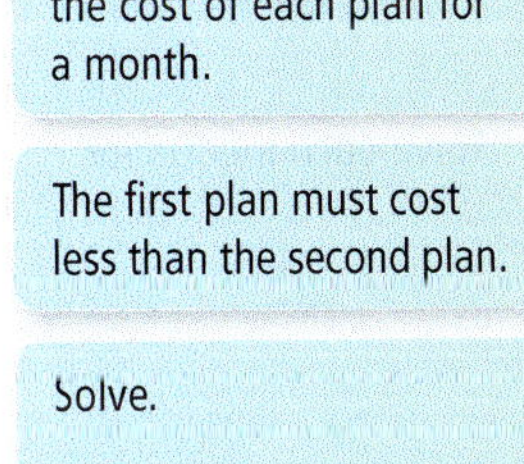

Think	Write
Assign a variable.	Let n = the number of movie rentals in one month.
Write an expression for the cost of each plan for a month.	first plan: 36 second plan: 15 + 1.5n
The first plan must cost less than the second plan.	$36 < 15 + 1.5n$
Solve.	$21 < 1.5n$ $14 < n$
Write the answer in words.	You must rent more than 14 movies in a month for the first plan to cost less.

Got It? **3.** A digital music service offers two subscription plans. The first has a \$9 membership fee and charges \$1 per download. The second has a \$25 membership fee and charges \$.50 per download. How many songs must you download for the second plan to cost less than the first plan?

Problem 4 No Solution or All Real Numbers as Solutions

Is the inequality *always, sometimes,* or *never* true?

A $-2(3x + 1) > -6x + 7$

$-6x - 2 > -6x + 7$ Distributive Property

$-2 > 7$ Add $6x$ to each side.

The last inequality $-2 > 7$ is false, so $-2(3x + 1) > -6x + 7$ is always false. It has no solution, so it is never true.

Think

How did you determine that an *equation* has no solution?
If you solve an equation and obtain a false statement, then the equation has no solution.

B $5(2x - 3) - 7x \leq 3x + 8$

$10x - 15 - 7x \leq 3x + 8$ Distributive Property

$3x - 15 \leq 3x + 8$ Combine like terms.

$-15 \leq 8$ Subtract $3x$ from each side.

The inequality $-15 \leq 8$ is true, so $5(2x - 3) - 7x \leq 3x + 8$ is always true. All real numbers are solutions.

Got It? **4.** Is $4(2x - 3) < 8(x + 1)$ *always, sometimes,* or *never* true?

You can join two inequalities with the word *and* or the word *or* to form a **compound inequality**. To solve a compound inequality containing *and,* find all values of the variable that make both inequalities true.

Problem 5 Solving an *And* Inequality

Think

How do you graph a compound inequality with *and*?
Find the intersection of the solutions of the two inequalities.

What is the solution of $7 < 2x + 1$ and $3x \leq 18$? Graph the solution.

$7 < 2x + 1$ and $3x \leq 18$

$3 < x$ and $x \leq 6$ Solve each inequality.

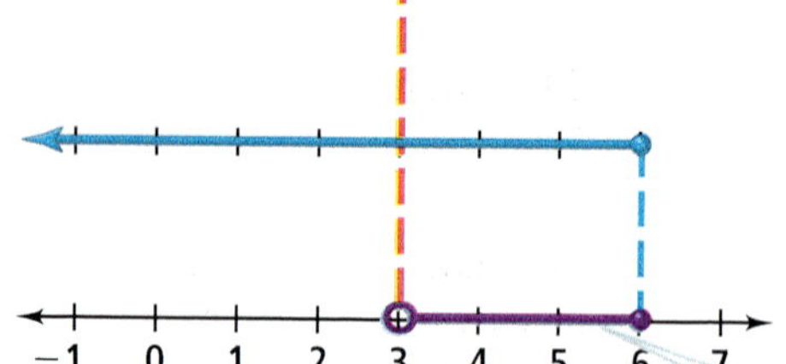

Got It? **5. a.** What is the solution of $5 \leq 3x - 1$ and $2x < 12$? Graph the solution.

b. Reasoning Is the compound inequality in Problem 5 *always, sometimes,* or *never* true? Explain your reasoning.

You can collapse a compound *and* inequality, like $5 < x + 1$ and $x + 1 < 13$, into a simpler form, $5 < x + 1 < 13$. You read $5 < x + 1 < 13$ as "$x + 1$ is greater than 5 and less than 13."

To solve a compound inequality containing *or*, find all values of the variable that make at least one of the inequalities true.

Problem 6 Solving an *Or* Inequality

Think

How does the solution to an *or* inequality differ from the solution to an *and* inequality?
The solution to an *or* inequality includes all of the solutions of each inequality, not just the solutions of both inequalities.

What is the solution of $7 + k \geq 6$ or $8 + k < 3$? Graph the solution.

$7 + k \geq 6$ or $8 + k < 3$

$k \geq -1$ or $k < -5$ Subtraction Property of Inequality

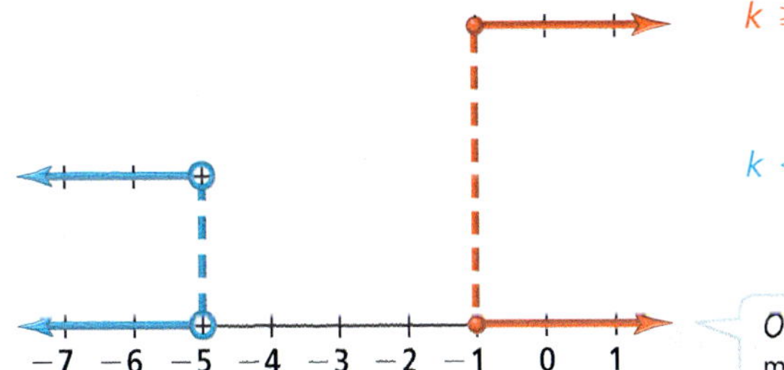

$k \geq -1$

$k < -5$

OR means that a solution makes EITHER inequality true.

Got It? 6. What is the solution of each compound inequality? Graph the solution.

a. $7w + 3 > 11$ or $4w - 1 < -13$ **b.** $16 < 5x + 1$ or $3x + 9 < 6$

Lesson Check

Do you know HOW?

Write an inequality that represents each sentence.

1. Rachel's hair is at least as long as Julia's.

2. The wind speeds of tropical storms are at least 40 mi/h, but less than 74 mi/h.

Solve each inequality. Graph the solution.

3. $-4(3x + 2) \geq 16$

4. $3 < 5x - 2 < 7$

5. $7x \quad 3 > 10$ or $3x \quad 2 \leq \quad 2$

Do you UNDERSTAND?

6. Reasoning Make up an example to help explain why you must reverse the inequality symbol when you multiply or divide by a negative number.

7. Compare and Contrast Describe how the properties of inequality are similar to the properties of equality and how they differ.

8. Write an inequality for which the solution is the set of all real numbers.

9. Error Analysis Your classmate says that you cannot write a compound inequality that has no solution. Do you agree? If so, explain why. If not, give a counterexample.

Practice and Problem-Solving Exercises

Write the inequality that represents the sentence.

See Problem 1.

10. The sum of a number and 5 is less than −7.

11. The product of a number and 8 is at least 25.

12. Six less than a number is greater than 54.

13. The quotient of a number and 12 is no more than 6.

Solve each inequality. Graph the solution.

See Problem 2.

14. $-12 \geq 24x$

15. $-7k < 63$

16. $8a - 15 > 73$

17. $57 - 4t \geq 13$

18. $-18 - 5y \geq 52$

19. $14 - 4y \geq 38$

20. $4(x + 3) \leq 44$

21. $2(m - 3) + 7 < 21$

22. $4(n - 2) - 6 > 18$

23. $-2(w + 4) + 9 < -11$

Solve each problem by writing an inequality.

See Problem 3.

24. The length of a picture frame is 3 in. greater than the width. The perimeter is less than 52 in. Describe the dimensions of the frame.

25. The lengths of the sides of a triangle are in the ratio 5 : 6 : 7. Describe the length of the longest side if the perimeter is less than 54 cm.

26. Find the lesser of two consecutive integers with a sum greater than 16.

27. The cost of a field trip is $220 plus $7 per student. If the school can spend at most $500, how many students can go on the field trip?

Is the inequality *always*, *sometimes*, or *never* true?

See Problem 4.

28. $9(x + 2) > 9(x - 3)$

29. $6x - 13 < 6(x - 2)$

30. $-6(2x - 10) + 12x \leq 180$

31. $-7(3x - 7) + 21x \geq 50$

32. $3 + 5x < 5(x + 1)$

33. $2(x + 6) < 30$

34. $4x - 8 > 1 + 4(x + 3)$

35. $9x + 2(2 + x) < 5 + 9x$

Solve each compound inequality. Graph the solution.

See Problems 5 and 6.

36. $2x > -10$ and $9x < 18$

37. $3x \geq -12$ and $8x \leq 16$

38. $6x \geq -24$ and $9x < 54$

39. $7x > -35$ and $5x \leq 30$

40. $4x < 16$ or $12x > 144$

41. $3x \geq 3$ or $9x < 54$

42. $8x > -32$ or $-6x \geq 48$

43. $9x \leq -27$ or $4x \geq 36$

44. Think About a Plan The diagram shows the scores in seconds of a skater's first three trials in a speed-skating event. What is the maximum time she can score on her last trial so that her average time on all four trials is under 36 seconds?

- What do you need to find an average?
- What inequality can you use to model the situation?

Solve each inequality. Graph the solution.

45. $2 - 3z \geq 7(8 - 2z) + 12$

46. $6(x - 2.5) \geq 8 - 6(3.5 + x)$

47. $\frac{2}{3}(x - 12) \leq x + 8$

48. $\frac{3}{5}(x - 12) > x - 24$

49. $3[4x - (2x - 7)] < 2(3x - 5)$

50. $6[5y - (3y - 1)] \geq 4(3y - 7)$

51. Grades Your math test scores are 68, 78, 90, and 91. What is the lowest score you can earn on the next test and still achieve an average of at least 85?

STEM **52. Chemistry** The pH level of a popular shampoo is between 6.0 and 6.5 inclusive. What compound inequality shows the pH levels of this shampoo? Graph the solution.

53. Geometry The sum of the lengths of any two sides of a triangle is greater than the length of the third side. In $\triangle ABC$, $BC = 4$ and $AC = 8 - AB$. What can you conclude about AB?

54. Writing Write a word problem that can be solved using $25 + 0.5x \leq 60$.

55. Error Analysis A classmate solved the inequality $\frac{1}{2}(y - 16) \geq y + 2$ as shown. Prove that his answer is incorrect by checking a number that is less than -20. (Select a number that makes the computation easy.) What was his error?

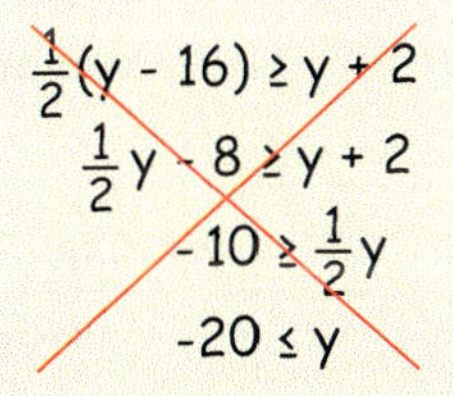

56. Construction A contractor estimated that her expenses for a construction project would be between \$700,000 and \$750,000. She has already spent \$496,000. How much more can she spend and remain within her estimate?

Justifying Steps Justify each step by identifying the property used.

57.
$3x \leq 4(x - 1) - 8$
$3x \leq 4x - 4 - 8$
$3x \leq 4x - 12$
$-x \leq -12$
$x \geq 12$

58.
$\frac{1}{2}(y + 3) > \frac{1}{3}(4 - y)$
$3(y + 3) > 2(4 - y)$
$3y + 9 > 8 - 2y$
$5y + 9 > 8$
$5y > -1$
$y > -0.2$

Solve each compound inequality. Graph the solution.

59. $-6 < 2x - 4 < 12$

60. $4x \leq 12$ and $-7x \leq 21$

61. $15x > 30$ or $18x < -36$

Challenge **Open-Ended** Write an inequality with a solution that matches the graph. At least two steps should be needed to solve your inequality.

62. 

63. −4 −3 −2 −1 0 1 2 3 4

64. −4 −3 −2 −1 0 1 2 3 4

65. −4 −3 −2 −1 0 1 2 3 4

 66. **Reasoning** Consider the compound inequality $x < 8$ and $x > a$.

a. Are there any values of a such that all real numbers are solutions of the compound inequality? If so, what are they?

b. Are there any values of a such that no real numbers are solutions of the compound inequality? If so, what are they?

c. Repeat parts (a) and (b) for the compound inequality $x < 8$ or $x > a$.

Standardized Test Prep

SAT/ACT

67. What is the solution of $1 < 2x + 3 < 9$?

Ⓐ $-1 > x < 2$ Ⓑ $2 < x < 3$ Ⓒ $-1 < x < 2$ Ⓓ $-1 < x < 3$

68. Which expression best represents the value of x in $y = mx + b$?

Ⓕ $\frac{b - y}{m}$ Ⓖ $\frac{y + b}{m}$ Ⓗ $m(y - b)$ Ⓘ $\frac{y - b}{m}$

69. The hourly rate of a waiter is \$4 plus tips. On a particular day, the waiter worked 8 hours and received more than \$150 in pay. Which could be the amount of tips the waiter received?

Ⓐ \$18.75 Ⓑ \$32 Ⓒ \$118 Ⓓ \$120.75

Extended Response

70. Solve $3(x - 2) + 8 = 12$. Identify each property of real numbers or equality you use.

Mixed Review

Simplify each expression. See Lesson 1-3.

71. $(2a - 4) + (5a + 9)$

72. $3(x + 3y) - 5(x - y)$

73. $\frac{1}{3}(b + 12) - \frac{1}{4}(b + 12)$

74. $0.4(k - 0.1) + 0.5(3.3 - k)$

Get Ready! **To prepare for Lesson 1-6, do Exercises 75–78.**

Solve each equation. Check your answers. See Lesson 1-4.

75. $7x - 6(11 - 2x) = 10$

76. $10x - 7 = 2(13 + 5x)$

77. $4y - \frac{1}{10} = 3y + \frac{4}{5}$

78. $0.4x + 1.18 = -3.1(2 - 0.01x)$

1-6 Absolute Value Equations and Inequalities

Content Standards

A.SSE.1.b Interpret complicated expressions by viewing one or more of their parts as a single entity.

A.CED.1 Create equations and inequalities in one variable and use them to solve problems.

Objective To write and solve equations and inequalities involving absolute value

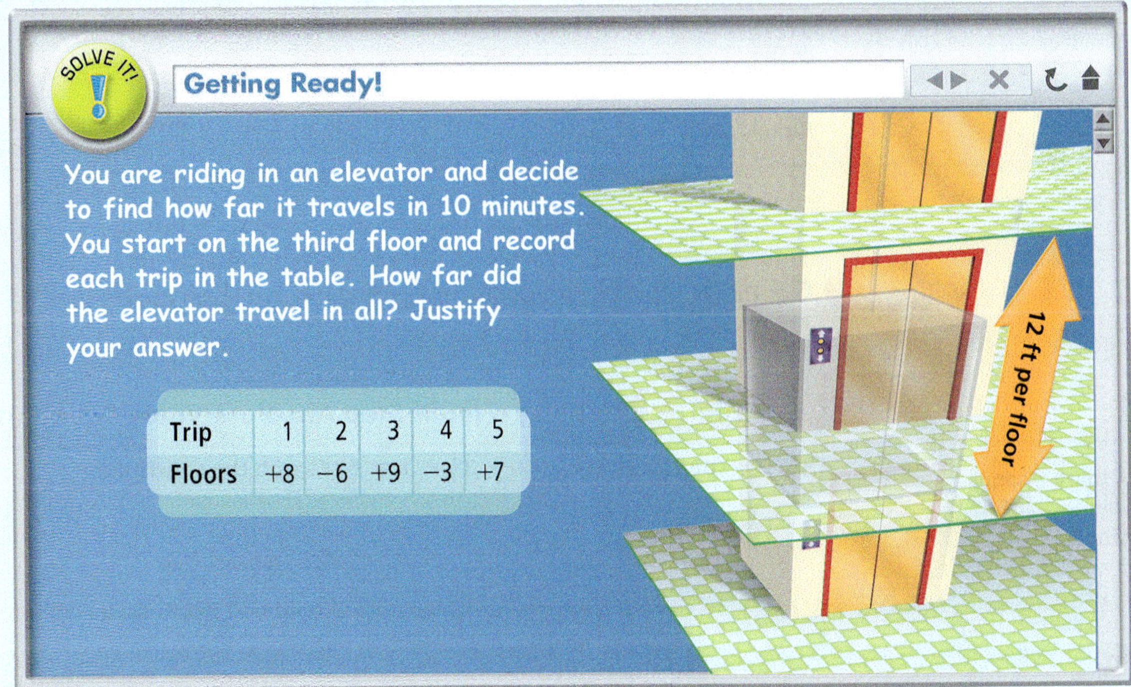

Trip	1	2	3	4	5
Floors	+8	−6	+9	−3	+7

In the Solve It, signed numbers represent distance and direction. Sometimes, only the size of a number (its *absolute value*), not the direction, is important.

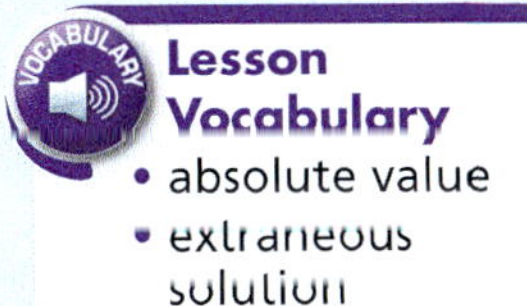

Lesson Vocabulary
- absolute value
- extraneous solution

Essential Understanding An absolute value quantity is nonnegative. Since opposites have the same absolute value, an absolute value equation can have two solutions.

take note

Key Concept Absolute Value

Definition	Numbers	Symbols
The **absolute value** of a real number x, written $\lvert x \rvert$, is its distance from zero on the number line.	$\lvert 4 \rvert = 4$ $\lvert -4 \rvert = 4$	$\lvert x \rvert = x$, if $x \geq 0$ $\lvert x \rvert = -x$, if $x < 0$

An absolute value equation has a variable within the absolute value sign. For example, $|x| = 5$. Here, the value of x can be 5 or -5 since $|5|$ and $|-5|$ both equal 5.

Both 5 and −5 are 5 units from 0.

−6 −5 −4 −3 −2 −1 0 1 2 3 4 5 6

Think

How is solving this equation different from solving a linear equation?
In the absolute value equation, $2x - 1$ can represent two opposite quantities.

Problem 1 Solving an Absolute Value Equation

What is the solution of $|2x - 1| = 5$? Graph the solution.

$|2x - 1| = 5$

$2x - 1 = 5$ or $2x - 1 = -5$ — Rewrite as two equations. $2x - 1$ could be 5 or -5.

$2x = 6$ | $2x = -4$ — Add 1 to each side of both equations.

$x = 3$ or $x = -2$ — Divide each side of both equations by 2.

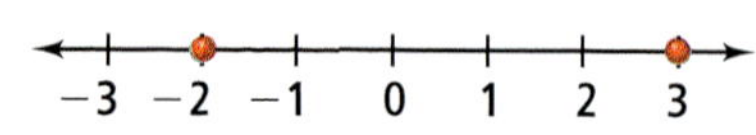

Check $|2(3) - 1| \stackrel{?}{=} 5$ $\quad$ $|2(-2) - 1| \stackrel{?}{=} 5$

$|6 - 1| \stackrel{?}{=} 5$ $\quad$ $|-4 - 1| \stackrel{?}{=} 5$

$|5| = 5$ ✔ $\quad$ $|-5| = 5$ ✔

Got It? **1.** What is the solution of $|3x + 2| = 4$? Graph the solution.

Plan

Is there a simpler way to think of this problem?
Solving $3|x + 2| - 1 = 8$ is similar to solving $3y - 1 = 8$.

Problem 2 Solving a Multi-Step Absolute Value Equation

What is the solution of $3|x + 2| - 1 = 8$? Graph the solution.

$3|x + 2| - 1 = 8$

$3|x + 2| = 9$ — Add 1 to each side.

$|x + 2| = 3$ — Divide each side by 3.

$x + 2 = 3$ or $x + 2 = -3$ — Rewrite as two equations.

$x = 1$ or $x = -5$ — Subtract 2 from each side of both equations.

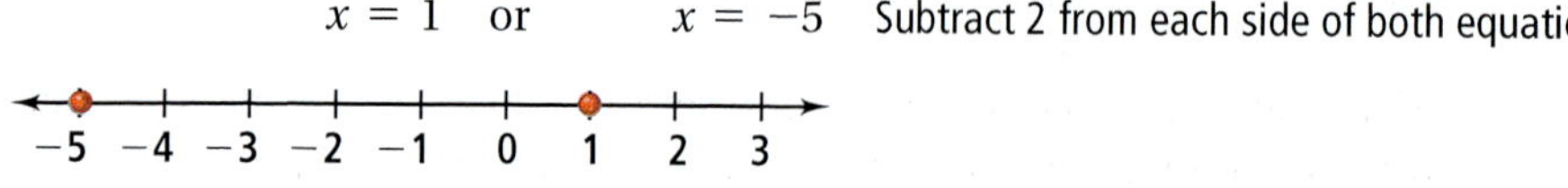

Check $3|(1) + 2| - 1 \stackrel{?}{=} 8$ $\quad$ $3|(-5) + 2| - 1 \stackrel{?}{=} 8$

$3|3| - 1 \stackrel{?}{=} 8$ $\quad$ $3|-3| - 1 \stackrel{?}{=} 8$

$8 = 8$ ✔ $\quad$ $8 = 8$ ✔

Got It? **2.** What is the solution of $2|x + 9| + 3 = 7$? Graph the solution.

Distance from 0 on the number line cannot be negative. Therefore, some absolute value equations, such as $|x| = -5$, have no solution. It is important to check the possible solutions of an absolute value equation. One or more of the possible solutions may be *extraneous*.

An **extraneous solution** is a solution derived from an original equation that is *not* a solution of the original equation.

Think

Can you solve this the same way as you solved Problem 1?
Yes, let $3x + 2$ equal $4x + 5$ and $-(4x + 5)$.

Problem 3 Checking for Extraneous Solutions

What is the solution of $|3x + 2| = 4x + 5$? Check for extraneous solutions.

$|3x + 2| = 4x + 5$

$3x + 2 = 4x + 5$ or $3x + 2 = -(4x + 5)$ — Rewrite as two equations.

$-x = 3$ | $3x + 2 = -4x - 5$ — Solve each equation.

$7x = -7$

$x = -3$ or $x = -1$

Check $|3(-3) + 2| \stackrel{?}{=} 4(-3) + 5$ $\quad$ $|3(-1) + 2| \stackrel{?}{=} 4(-1) + 5$

$|-9 + 2| \stackrel{?}{=} -12 + 5$ $\quad$ $|-3 + 2| \stackrel{?}{=} -4 + 5$

$|-7| \neq -7$ ✗ $\quad$ $|-1| = 1$ ✔

Since $x = -3$ does not satisfy the orginal equation, -3 is an extraneous solution. The only solution to the equation is $x = -1$.

Got It? 3. What is the solution of $|5x - 2| = 7x + 14$? Check for extraneous solutions.

The solutions of the absolute value inequality $|x| < 5$ include values greater than -5 *and* less than 5. This is the compound inequality $x > -5$ *and* $x < 5$, which you can write as $-5 < x < 5$. So, $|x| < 5$ means x is between -5 and 5.

The graph of $|x| < 5$ is all values of x between -5 and 5.

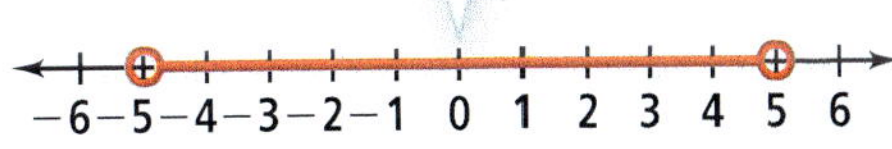

Essential Understanding You can write an absolute value inequality as a compound inequality without absolute value symbols.

Problem 4 Solving the Absolute Value Inequality $|A| < b$

Plan

Is this an *and* problem or an *or* problem?
$2x - 1$ is less than 5 and greater than -5. It is an *and* problem.

What is the solution of $|2x - 1| < 5$? Graph the solution.

$|2x - 1| < 5$

$-5 < 2x - 1 < 5$ — $2x - 1$ is between -5 and 5.

$-4 < 2x < 6$ — Add 1 to each part.

$-2 < x < 3$ — Divide each part by 2.

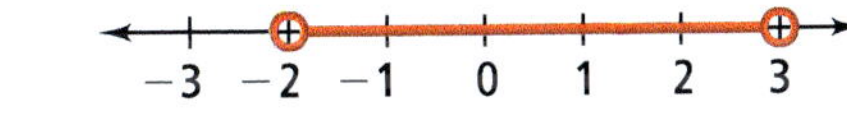

Got It? 4. What is the solution of $|3x - 4| \leq 8$? Graph the solution.

$|x| < 5$ means x is between -5 and 5. So, $|x| > 5$ means x is outside the interval from -5 to 5. You can say $x < -5$ *or* $x > 5$.

Problem 5 Solving the Absolute Value Inequality $|A| \geq b$

What is the solution of $|2x + 4| \geq 6$? Graph the solution.

$|2x + 4| \geq 6$

$2x + 4 \leq -6$ or $2x + 4 \geq 6$ — Rewrite as a compound inequality.

$2x \leq -10$ | $2x \geq 2$ — Subtract 4 from each side of both inequalities.

$x \leq -5$ or $x \geq 1$ — Divide each side of both inequalities by 2.

−6 −5 −4 −3 −2 −1 0 1 2

Think

How do you determine the boundary points?
To find the boundary points, find the solutions of the related equation.

Got It? 5. a. What is the solution of $|5x + 10| > 15$? Graph the solution.

b. Reasoning Without solving $|x - 3| \geq 2$, describe the graph of its solution.

take note

Concept Summary Solutions of Absolute Value Statements

Symbols	Definition	Graph
$\lvert x \rvert = a$	The distance from x to 0 is a units.	−a 0 a $x = -a$ or $x = a$
$\lvert x \rvert < a$ $(\lvert x \rvert \leq a)$	The distance from x to 0 is less than a units.	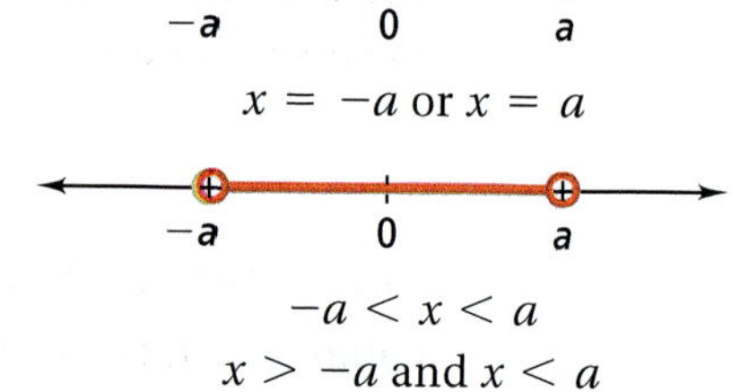 −a 0 a $-a < x < a$ $x > -a$ and $x < a$
$\lvert x \rvert > a$ $(\lvert x \rvert \geq a)$	The distance from x to 0 is greater than a units.	−a 0 a $x < -a$ or $x > a$

A manufactured item's actual measurements and its target measurements can differ by a certain amount, called *tolerance*. Tolerance is one half the difference of the maximum and minimum acceptable values. You can use absolute value inequalities to describe tolerance.

Problem 6 Using an Absolute Value Inequality

Car Racing In car racing, a car must meet specific dimensions to enter a race. Officials use a template to ensure these specifications are met. What absolute value inequality describes heights of the model of race car shown within the indicated tolerance?

Plan

How does *tolerance* relate to an inequality?
Tolerance allows the height to differ from a desired height by no less and no more than a small amount.

$\frac{53 - 51}{2} = \frac{2}{2} = 1$ Find the tolerance.

$-1 \le h - 52 \le 1$ Use h for the height of the race car. Write a compound inequality.

$|h - 52| \le 1$ Rewrite as an absolute value inequality.

Got It? **6.** Suppose the least allowable height of the race car in Problem 6 was 52 in. and the desirable height was 52.5 in. What absolute value inequality describes heights of the model of race car shown within the indicated tolerance?

Lesson Check

Do you know HOW?

Solve each equation. Check your answers.

1. $|-6x| = 24$

2. $|2x + 8| - 4 = 12$

3. $|x - 2| = 4x + 8$

Solve each inequality. Graph the solution.

4. $|2x + 2| - 5 < 15$

5. $|4x - 6| \ge 10$

Do you UNDERSTAND?

6. Vocabulary Explain what it means for a solution of an equation to be extraneous.

7. Reasoning When is the absolute value of a number equal to the number itself?

8. Give an example of a compound inequality that has no solution.

9. Compare and Contrast Describe how absolute value equations and inequalities are like linear equations and inequalities and how they differ.

Practice and Problem-Solving Exercises

Solve each equation. Check your answers. See Problems 1 and 2.

10. $|3x| = 18$ **11.** $|-4x| = 32$ **12.** $|x - 3| = 9$

13. $2|3x - 2| = 14$ **14.** $|3x + 4| = -3$ **15.** $|2x - 3| = -1$

16. $|x + 4| + 3 = 17$ **17.** $|y - 5| - 2 = 10$ **18.** $|4 - z| - 10 = 1$

Solve each equation. Check for extraneous solutions.

19. $|x - 1| = 5x + 10$ **20.** $|2z - 3| = 4z - 1$ **21.** $|3x + 5| = 5x + 2$

22. $|2y - 4| = 12$ **23.** $3|4w - 1| - 5 = 10$ **24.** $|2x + 5| = 3x + 4$

Solve each inequality. Graph the solution.

25. $3|y - 9| < 27$ **26.** $|6y - 2| + 4 < 22$ **27.** $|3x - 6| + 3 < 15$

28. $\frac{1}{4}|x - 3| + 2 < 1$ **29.** $4|2w + 3| - 7 \leq 9$ **30.** $3|5t - 1| + 9 \leq 23$

Solve each inequality. Graph the solution.

31. $|x + 3| > 9$ **32.** $|x - 5| \geq 8$ **33.** $|y - 3| \geq 12$

34. $|2x + 1| \geq -9$ **35.** $3|2x - 1| \geq 21$ **36.** $|3z| - 4 > 8$

Write each compound inequality as an absolute value inequality.

37. $1.3 \leq h \leq 1.5$ **38.** $50 \leq k \leq 51$ **39.** $27.25 \leq C \leq 27.75$

40. $50 \leq b \leq 55$ **41.** $1200 \leq m \leq 1300$ **42.** $0.1187 \leq d \leq 0.1190$

Solve each equation.

43. $-|4 - 8b| = 12$ **44.** $4|3x + 4| = 4x + 8$

45. $|3x - 1| + 10 = 25$ **46.** $\frac{1}{2}|3c + 5| = 6c + 4$

47. $5|6 - 5x| = 15x - 35$ **48.** $7|8 - 3h| = 21h - 49$

49. $2|3x - 7| = 10x - 8$ **50.** $6|2x + 5| = 6x + 24$

51. $\frac{1}{4}|4x + 7| = 8x + 16$ **52.** $\frac{2}{3}|3x - 6| = 4(x - 2)$

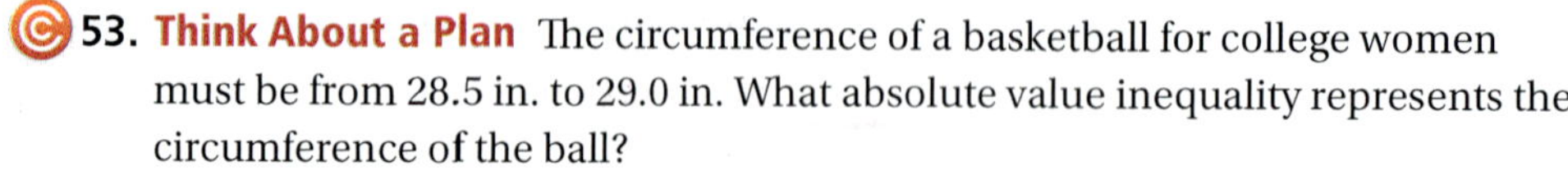

53. Think About a Plan The circumference of a basketball for college women must be from 28.5 in. to 29.0 in. What absolute value inequality represents the circumference of the ball?

- What is the tolerance?
- What is the inequality without using absolute value?

Write an absolute value equation or inequality to describe each graph.

54. −4 −2 0 2 4

55. −4 −2 0 2 4

56.

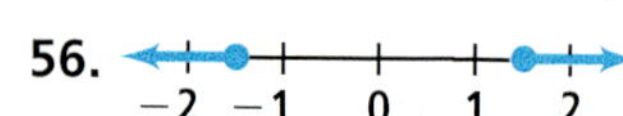

Solve each inequality. Graph the solutions.

57. $|3x - 4| + 5 \leq 27$

58. $|2x + 3| - 6 \geq 7$

59. $-2|x + 4| < 22$

60. $2|4t - 1| + 6 > 20$

61. $|3z + 15| \geq 0$

62. $|-2x + 1| > 2$

63. $\frac{1}{9}|5x - 3| - 3 \geq 2$

64. $\frac{1}{11}|2x - 4| + 10 \leq 11$

65. $\left|\frac{x - 3}{2}\right| + 2 < 6$

66. $\left|\frac{x + 5}{3}\right| - 3 > 6$

67. Writing Describe the differences in the graphs of $|x| < a$ and $|x| > a$, where a is a positive real number.

68. Open-Ended Write an absolute value inequality for which every real number is a solution. Write an absolute value inequality that has no solution.

Write an absolute value inequality to represent each situation.

69. Cooking Suppose you used an oven thermometer while baking and discovered that the oven temperature varied between +5 and −5 degrees from the setting. If your oven is set to 350°, let t be the actual temperature.

70. Time Workers at a hardware store take their morning break no earlier than 10 A.M. and no later than noon. Let c represent the time the workers take their break.

71. Climate A friend is planning a trip to Alaska. He purchased a coat that is recommended for outdoor temperatures from −15°F to 45°F. Let t represent the temperature for which the coat is intended.

Write an absolute value inequality and a compound inequality for each length x with the given tolerance.

72. a length of 36.80 mm with a tolerance of 0.05 mm

73. a length of 9.55 mm with a tolerance of 0.02 mm

74. a length of 100 yd with a tolerance of 4 in.

Is the absolute value inequality or equation *always*, *sometimes*, or *never* true? Explain.

75. $|x| = 6$

76. $0 > |x|$

77. $|x| = x$

78. $|x| + |x| = 2x$

79. $|x + 2| = x + 2$

80. $(|x|)^2 < x^2$

81. Error Analysis A classmate wrote the solution to the inequality $|-4x + 1| > 3$ as shown. Describe and correct the error.

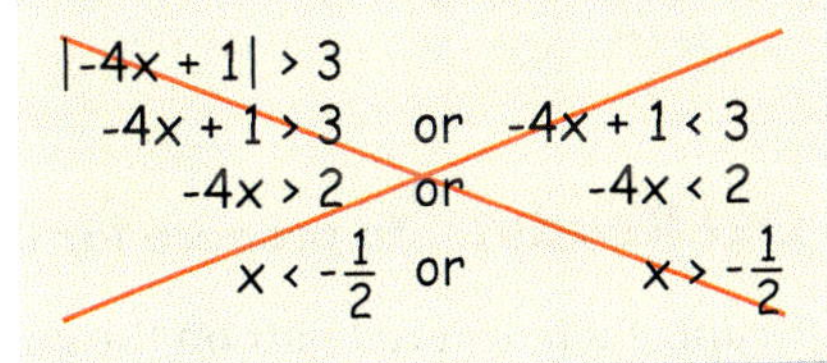

Solve each equation for x.

82. $|ax| - b = c$

83. $|cx - d| = ab$

84. $a|bx - c| = d$

Graph each solution.

85. $|x| \geq 5$ and $|x| \leq 6$

86. $|x| \geq 6$ or $|x| < 5$

87. $|x - 5| \leq x$

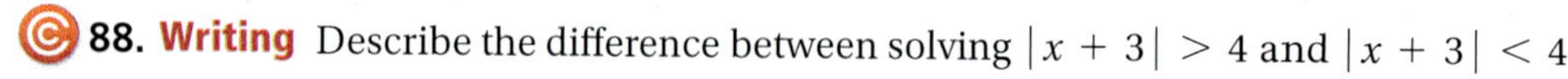

88. **Writing** Describe the difference between solving $|x + 3| > 4$ and $|x + 3| < 4$.

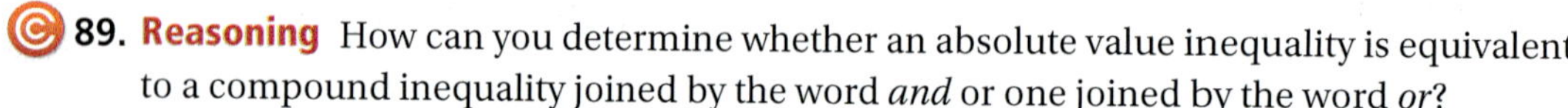

89. **Reasoning** How can you determine whether an absolute value inequality is equivalent to a compound inequality joined by the word *and* or one joined by the word *or*?

Standardized Test Prep

90. What is the positive solution of $|3x + 8| = 19$?

91. If p is an integer, what is the least possible value of p in the following inequality?

$$|3p - 5| \leq 7$$

92. In wood shop, you have to drill a hole that is 2 inches deep into a wood panel. The tolerance for drilling a hole is described by the inequality $|t - 2| \leq 0.125$. What is the shallowest hole allowed?

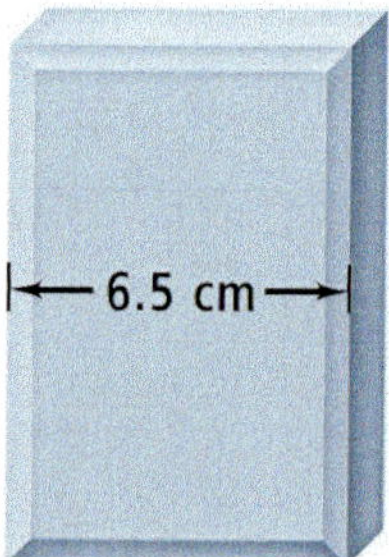

93. The normal thickness of a metal structure is shown. It expands to 6.54 centimeters when heated and shrinks to 6.46 centimeters when cooled down. What is the maximum amount in cm that the thickness of the structure can deviate from its normal thickness?

Mixed Review

Solve each inequality. Graph the solution. See Lesson 1-5.

94. $5y - 10 < 20$

95. $15(4s + 1) < 23$

96. $4a + 6 > 2a + 14$

Describe each pattern using words. Draw the next figure in each pattern.  See Lesson 1-1.

97.

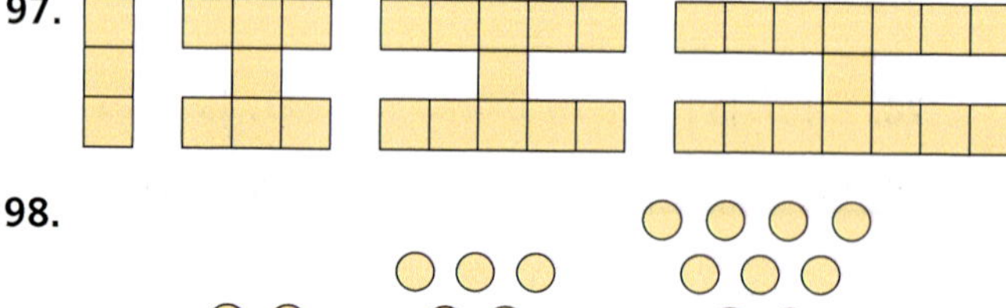

98.

Get Ready! To prepare for Lesson 2-1, do Exercises 99–102. See p. 977.

Graph each ordered pair on the coordinate plane.

99. $(-4, -8)$

100. $(3, 6)$

101. $(0, 0)$

102. $(-1, 3)$

Pull It All Together ASSESSMENT

These problems will challenge you to pull together many concepts and skills of algebra that you have learned.

BIG idea Variable

You can use variables to represent variable quantities in real-world situations and in patterns.

BIG idea Properties

You can use the properties of real numbers to simplify algebraic expressions.

Performance Task 1

Your Basal Metabolic Rate (*BMR*) is the measure of how many Calories you burn if you rest all day. Physicians and other health professionals use this rate as an important tool for determining a person's daily caloric needs. The Harris-Benedict Formula is commonly used to calculate *BMR* for an individual.

The formulas are slightly different for males and females. A female's *BMR* can be expressed as 655 more than the sum of 4.35 times her weight (lb) and 4.7 times her height (in.) and then less 4.7 times her age (years). The formula for a male is 66 more than 6.23 times his weight (lb) plus the product of 12.7 and his height (in.) minus 6.8 times his age (years).

a. Use w = weight, h = height, and a = age to write the formula for the *BMR* of a female and the formula for the *BMR* of a male.

b. Calculate your *BMR*. What does that tell you?

c. Look at each term of the formula. Make a conjecture about what happens to your *BMR* as you get older. Explain.

d. Which term of the expression affects the *BMR* the most? Explain.

e. What is the weight of a 16-year-old male with a *BMR* of 2600 and a height of 5 feet 10 inches?

BIG idea Solving Equations and Inequalities

You can use properties of numbers and equality to solve an equation by finding increasingly simpler equations that have the same solution as the original equation.

Performance Task 2

Find all possible values of a and b in the figures below, given the following conditions. Show all your work.

- The perimeters are equal.
- The rectangle has an area between 80 and 100 square units.
- The values of a and b are integers.

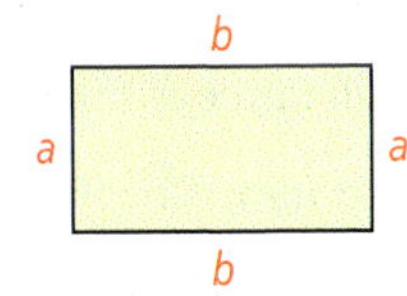

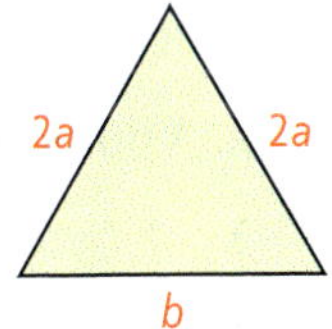

1 Chapter Review

Connecting BIG ideas and Answering Essential Questions

1 Variable
You can use variables to represent variable quantities in real-world situations and in patterns.

Patterns and Expressions (Lesson 1-1)
6, 12, 18, 24, . . . $6n$
8, 9, 10, 11, . . . $(n + 7)$

2 Properties
The properties that apply to real numbers also apply to variables that represent them.

Properties of Real Numbers (Lesson 1-2)
$4(6 - 1) = 4(6) - 4(1)$
$23 + 0 = 23$
$5 + 12 = 12 + 5$

Algebraic Expressions (Lesson 1-3)
$6n$
$n + 7$
$5x - x = (5 - 1)x$
$a + 0 = a$
$h + k = k + h$

3 Solving Equations and Inequalities
You can use properties of numbers and equality (or inequality) to solve an equation (or inequality) by finding increasingly simpler equations (or inequalities) which have the same solution as the original equation (or inequality).

Solving Equations and Inequalities (Lessons 1-4 and 1-5)

$4x - 1 = 5$	$7 > -3h - 2$
$4x = 6$	$9 > -3h$
$x = \frac{3}{2}$	$-3 < h$

Absolute Value Equations and Inequalities (Lesson 1-6)
$|2b + 7| = 15$
$2b + 7 = 15$ or $2b + 7 = -15$
$2b = 8$ | $2b = -22$
$b = 4$ or $b = -11$

Chapter Vocabulary

- absolute value (p. 41)
- additive inverse (p. 14)
- algebraic expression (p. 5)
- coefficient (p. 20)
- compound inequality (p. 36)
- constant (p. 5)
- constant term (p. 20)
- equation (p. 26)
- evaluate (p. 19)
- extraneous solution (p. 42)
- identity (p. 28)
- inverse operations (p. 27)
- like terms (p. 21)
- literal equation (p. 29)
- multiplicative inverse (p. 14)
- numerical expression (p. 5)
- opposite (p. 14)
- reciprocal (p. 14)
- solution of an equation (p. 27)
- term (p. 20)
- variable (p. 5)
- variable quantity (p. 5)

Choose the correct term to complete each sentence.

1. The _?_ makes an equation true.

2. A number's distance from zero on the number line is its _?_.

3. _?_ is another name for the multiplicative inverse of a number.

4. A pair of inequalities joined by *and* or *or* are called a _?_.

1-1 Patterns and Expressions

Quick Review

You can represent patterns using words, diagrams, numbers, and **algebraic expressions**. You can identify a pattern by looking for the same type of change between consecutive figures or numbers. It often helps to make a table.

Example

Identify a pattern by making a table of inputs and outputs. Include a process column. 7, 14, 21, 28, 35, . . .

Input	Process Column	Output
1	$1 \cdot 7$	7
2	$2 \cdot 7$	14
3	$3 \cdot 7$	21
⋮	⋮	⋮
n	$n \cdot 7$	$7n$

The nth output is $7n$.

Exercises

Identify a pattern and find the next three numbers in the pattern.

5. 5, 10, 15, 20, . . .

6. 3, 4, 5, 6, . . .

Copy and complete the table. Then find the output when the input is n.

7.

Input	Output
1	9
2	10
3	11
4	■
⋮	⋮
n	■

8.

Input	Output
1	19
2	38
3	57
4	■
⋮	⋮
n	■

9. Finance If you put $20 in your savings account each week, how much have you saved after n weeks?

1-2 Properties of Real Numbers

Quick Review

The natural numbers, whole numbers, integers, rational numbers, and irrational numbers are all subsets of the real numbers.

You can use properties such as the ones listed below to simplify and evaluate expressions.

Commutative Properties	$-3 + 5 = 5 + (-3)$ $2 \times 9 = 9 \times 2$
Associative Properties	$3 + (5 + 7) = (3 + 5) + 7$ $4 \times (8 \times 11) = (4 \times 8) \times 11$
Distributive Property	$5(7 + 9) = 5(7) + 5(9)$

Example

Identify the property illustrated by the equation.

$4 \cdot x = x \cdot 4$ Commutative Property of Multiplication

Exercises

Name the subset(s) of real numbers to which each number belongs.

10. 0.1π

11. -79

12. $\sqrt{121}$

13. $12\frac{7}{8}$

Compare the two numbers. Use < or >.

14. $-\sqrt{60}, -8$

15. $5, \sqrt{32}$

Name the property of real numbers illustrated by each equation.

16. $\frac{9}{4} \cdot \frac{4}{9} = 1$

17. $\left(8 \cdot \frac{1}{3}\right) \cdot 12 = 8 \cdot \left(\frac{1}{3} \cdot 12\right)$

1-3, 1-4, and 1-5 Expressions, Equations, and Inequalities

Quick Review

You **evaluate** an algebraic expression by substituting numbers for the variables. You simplify an algebraic expression by combining **like terms**. To find the **solution of an equation** or inequality, use the properties of equality or inequality. Some **equations** and inequalities are true for all real numbers, and some have no solution.

Example

Evaluate $3(x - 4) + 2x - x^2$ for $x = 6$.

$3(6 - 4) + 2(6) - 6^2$	Substitute.
$= 3(2) + 2(6) - 6^2$	Simplify inside parentheses.
$= 6 + 12 - 36$	Multiply.
$= 18 - 36$	Add.
$= -18$	Subtract.

Exercises

18. Evaluate $3t(t + 2) - 3t^2$ for $t = 19$.

19. Simplify $-(3a - 2b) - 3(-a - b)$.

Solve each equation. Check your answer.

20. $2x - 5 = 17$

21. $3(x + 1) = 9 + 2x$

Solve each inequality. Graph the solution.

22. $4 - 5z \geq 2$

23. $2(5 - 3x) < x - 4(3 - x)$

Solve each compound inequality. Graph the solution.

24. $10 \geq 7 + 3x$ and $9 - 4x \leq 1$

25. $3 \geq 2x$ or $x - 4 > 2$

Write an equation to solve the problem.

26. **Geometry** The length and width of a rectangle are in the ratio 5 : 3. The perimeter of the rectangle is 32 cm. Find the length and width.

1-6 Absolute Value Equations and Inequalities

Quick Review

To rewrite an equation or inequality that involves the **absolute value** of an algebraic expression, you must consider both cases of the definition of absolute value.

Example

Solve $|3x - 5| = 4 + 2x$. Check for extraneous solutions.

$3x - 5 = 4 + 2x$ or $3x - 5 = -(4 + 2x)$

$3x - 5 = -4 - 2x$

$5x = 1$

$x = 9$ or $x = \frac{1}{5}$

Check $|3(9) - 5| \stackrel{?}{=} 4 + 2(9)$ $\quad\quad$ $\left|3\left(\frac{1}{5}\right) - 5\right| \stackrel{?}{=} 4 + 2\left(\frac{1}{5}\right)$

$|27 - 5| \stackrel{?}{=} 22$ $\quad\quad$ $\left|\frac{3}{5} - 5\right| \stackrel{?}{=} 4 + \frac{2}{5}$

$|22| = 22$ ✔ $\quad\quad$ $\left|-\frac{22}{5}\right| = \frac{22}{5}$ ✔

Exercises

Solve each equation. Check for extraneous solutions.

27. $|2x + 8| = 3x + 7$

28. $|x - 4| + 3 = 1$

29. $3|x + 10| = 6$

30. $2|x - 7| = x - 8$

Solve each inequality. Graph the solution.

31. $|3x - 2| + 4 \leq 7$

32. $4|y - 9| > 36$

33. $|7x| + 3 \leq 21$

34. $\frac{1}{2}|x + 2| > 6$

35. The specification for a length x is 43.6 cm with a tolerance of 0.1 cm. Write the specification as an absolute value inequality.

1 Chapter Test

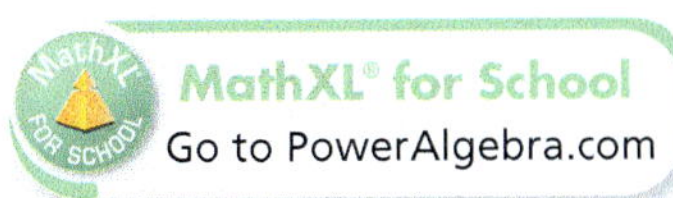

Do you know HOW?

Evaluate each expression for $x = 5$.

1. $\frac{5}{3}(3x - 6) - (6 - 4x)$

2. $3(x^2 - 4) + 7(x - 2)$

3. $x - 2x + 3x - 4x + 5x$

Simplify each expression.

4. $a^2 + a + a^2$

5. $2x + 3y - 5x + 2y$

6. $5(a - 2b) - 3(a - 2b)$

7. $3[2(x - 3) + 2] + 5(x - 3)$

Solve each equation.

8. $4y - 6 = 2y + 8$

9. $3(2z + 1) = 35$

10. $5(3w - 2) - 7 = 23$

11. $t - 2(3 - 2t) = 2t + 9$

12. $5(s - 12) - 24 = 3(s + 2)$

13. The lateral surface area of a cylinder is given by the formula $S = 2\pi rh$. Solve the equation for r.

14. **Savings** Briana and her sister Molly both want to buy the same model bicycle. Briana needs \$73 more before she can afford the bike. Molly needs \$65 more. If they combine their money, they will have just enough to buy one bicycle that they could share. What is the cost of the bicycle?

15. **Musical** There is only one freshman in the cast of a high school musical. There are 6 sophomores and 11 juniors. One third of the cast are seniors. How many seniors are in the musical?

Determine whether each equation is *always*, *sometimes*, or *never* true.

16. $2x + 7 - x = 3 + x + 4$

17. $5a - 1 - 3a = 2a + 1$

Solve each equation or inequality. Graph the solution.

18. $3x + 17 \geq 5$

19. $25 - 2x < 11$

20. $\frac{3}{8}x < -6$ or $5x > 2$

21. $2 < 10 - 4d < 6$

22. $4 - x = |2 - 3x|$

23. $5|3w + 2| - 3 > 7$

Do you UNDERSTAND?

24. **Writing** Describe the relationships among these sets of numbers: natural numbers, whole numbers, integers, rational numbers, irrational numbers, and real numbers.

25. **Reasoning** Justify each step by identifying the property used.

$$\begin{aligned} t + 5(t + 1) &= t + (5t + 5) \\ &= (t + 5t) + 5 \\ &= (1t + 5t) + 5 \\ &= (1 + 5)t + 5 \\ &= 6t + 5 \end{aligned}$$

26. **Reasoning** The first four figures of a pattern are shown.

Describe the tenth figure in the pattern.

1 Cumulative Standards Review ASSESSMENT

TIPS FOR SUCCESS

Some questions on tests ask you to write a short response. To get full credit for an answer, you must give the correct answer (including appropriate units, if applicable) and justify your reasoning or show your work. Read the sample question at the right. Then follow the tips to answer it.

Short Response Your school is having a bake sale to raise money for a field trip. The ingredients for each batch of bran muffins cost \$3.25, and the cost for the energy to bake them is \$.50. You plan to sell each batch for \$5.

Write an expression to represent the total amount of money your school will have after selling n batches of muffins. Evaluate your expression for 25 batches. Show your work.

TIP 1

Think about the total cost to make a batch of muffins.

TIP 2

You need to make a profit in the bake sale in order to raise money. Consider the total sales and the cost to make the muffins. What operation do you use to find the profit?

Think It Through

It costs \$3.25 + \$.50 = \$3.75 to bake each batch. The school makes \$5 − (\$3.75) = \$1.25 after selling one batch. So, it makes $1.25n$ after selling n batches. It earns \$1.25(25) = \$31.25 after selling 25 batches. This answer is complete and earns full credit.

Vocabulary Builder

As you solve test items, you must understand the meanings of mathematical terms. Match each term with its mathematical meaning.

A. inequality

B. compound inequality

C. extraneous solution

D. expression

E. equation

I. a mathematical sentence that contains $>$, $<$, $\geq$, $\leq$, or $\neq$

II. a solution of an equation derived from an original equation but not a solution of the original equation

III. a pair of inequalities joined by *and* or *or*

IV. a mathematical sentence that contains an equals sign

V. a mathematical phrase that uses numbers, variables, and operational symbols

Multiple Choice

Read each question. Then write the letter of the correct answer on your paper.

1. A person receives a salary of \$600 a month and a 10% commission of all sales made. Which expression can be used to find the person's income when the sales amount is x?

(A) $x + 600$

(B) $600 - 0.10x$

(C) $10x + 600$

(D) $0.10x + 600$

2. Which expression can be used to find the next term of the sequence 1, 4, 9, 16, . . . ?

(F) $2n$

(G) $n + 3$

(H) $n^2 + 1$

(I) n^2

3. Which inequality has a solution that matches the graph below?

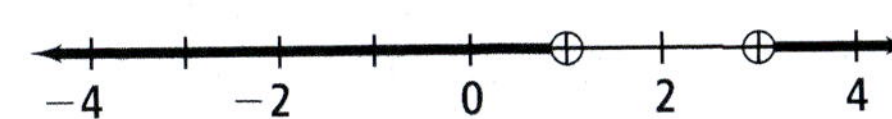

Ⓐ $|x - 2| - 3 > -2$

Ⓑ $|x - 2| < -5$

Ⓒ $|x + 2| + 3 < 2$

Ⓓ $|x + 4| + 2 > 1$

4. A worker is taking boxes of nails on an elevator. Each box weighs 54 lb, and the worker weighs 170 lb. The elevator has a weight limit of 2500 lb. Which inequality describes the number of boxes b that he can safely take on each trip?

Ⓕ $54b - 170 \leq 2500$

Ⓖ $54b + 170 \leq 2500$

Ⓗ $54(b - 170) \leq 2500$

Ⓘ $54(b + 170) \leq 2500$

5. An electrical circuit is connected in series as shown. The total voltage V can be calculated by using the equation shown, where I is the total current and R is the resistance across the circuit. $\left(\textit{Hint: } 1\text{A} = 1\frac{\text{volt}}{\text{ohm}}\right)$

$V = I(R_1 + R_2 + R_3)$

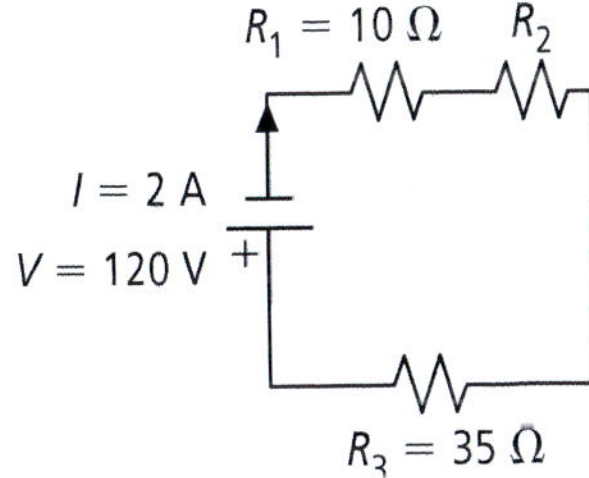

A = amperes
V = volts
Ω = ohms

What is the value of R_2?

Ⓐ 15 Ω

Ⓑ 45 Ω

Ⓒ 60 Ω

Ⓓ 75 Ω

6. Solve $3(x - 2) + 4 \geq -3x + 1$.

Ⓕ $x \geq -3$

Ⓖ $x \geq \frac{1}{2}$

Ⓗ $x \geq \frac{3}{2}$

Ⓘ $x \leq 3$

7. You used an oven thermometer while baking and found out that the oven temperature varied between +7 degrees and −7 degrees from the setting. If your oven is set to 325°F, let t be the actual temperature. What is the absolute value inequality that represents this situation?

Ⓐ $|t - 325| \geq 7$

Ⓑ $|t - 325| < 7$

Ⓒ $|t - 7| \leq 325$

Ⓓ $|t - 325| \leq 7$

8. A designer is designing a handbag. The height of the handbag must be between 16 in. and 18 in. The desirable height is 17 in. Which absolute value inequality represents the height of the handbag?

Ⓕ $|h - 16| \leq 1$

Ⓖ $|h - 17| \leq 1$

Ⓗ $|h - 17| \geq 2$

Ⓘ $|h - 18| \geq 2$

9. A company makes gift boxes in different sizes following the pattern shown below. What is the volume of the fourth gift box to the nearest cubic inch?

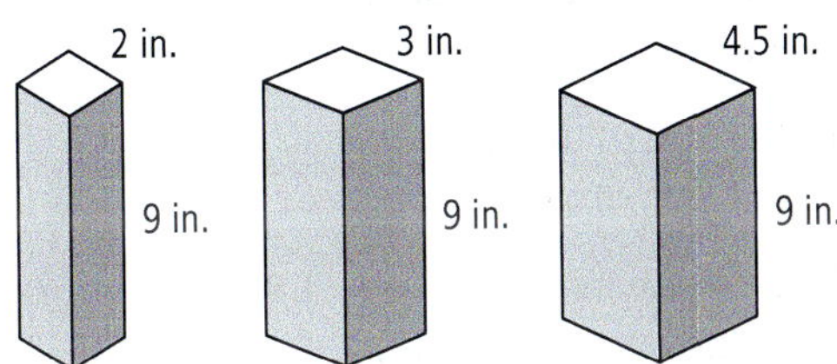

Ⓐ 324 in.^{3}

Ⓑ 352 in.^{3}

Ⓒ 376 in.^{3}

Ⓓ 410 in.^{3}

10. How many negative solutions does $2|3x - 6| < 6$ have?

Ⓕ 0

Ⓖ 1

Ⓗ 2

Ⓘ infinitely many

11. For which value of a does $4 = a + |x - 4|$ have no solution?

Ⓐ −6

Ⓑ 0

Ⓒ 4

Ⓓ 6

GRIDDED RESPONSE

12. The graph shows the amount of paint needed (in gallons) to paint the walls (in square feet) of an office building.

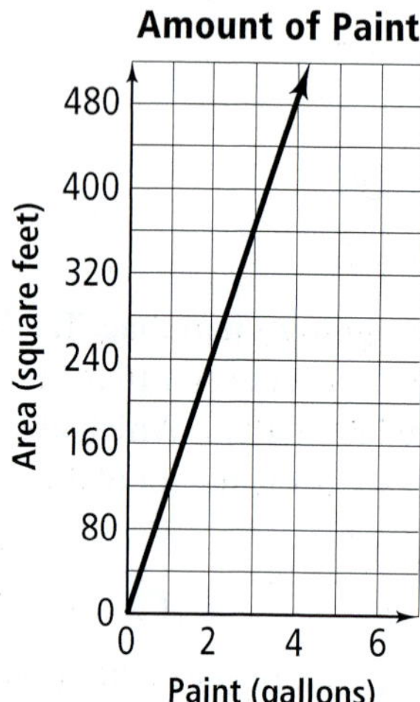

The pattern can be represented by $y = 120x$. How many gallons of paint will be needed to paint 900 square feet?

13. What is the sum of the solutions of $|2x + 4| - 6 = 8$?

14. What is the value of $4x^2 + 2x - 1$ when $x = \frac{3}{4}$? Express the answer as a decimal.

15. The cost for taking a taxi is \$1.80 plus \$.10 per eighth of a mile. What is the cost of a ride that is 4.5 miles long?

16. What is the coefficient of b in the simplified form of the expression $-8(a - 3b) + 2(-a + 4b + 1)$?

17. The expression $21 + 0.5n$ describes the length in inches of a baby n months old. How long is the baby at 6 months?

Short Response

18. A new 10-lb dumbbell will pass inspection if it is between 9.95 lb and 10.05 lb. What is the tolerance of the weight of the dumbbell? What absolute value inequality describes acceptable weights of the dumbbell within an indicated tolerance? Show all work.

19. Is $7\sqrt{56}$ rational or irrational? Explain.

Extended Response

20. A trapezoidal deck has dimensions as shown.

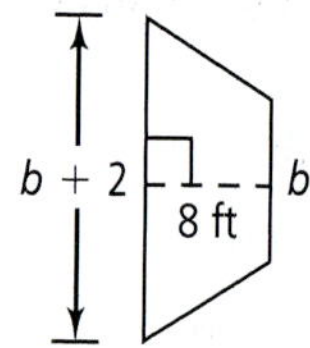

a. What is a formula for the area of this trapezoid?

b. Rearrange the formula so that it is solved for b. Show all work.

c. What is the longer base length if the area is 88 square feet (ft^2)?

21. The two cylinders below have identical bases.

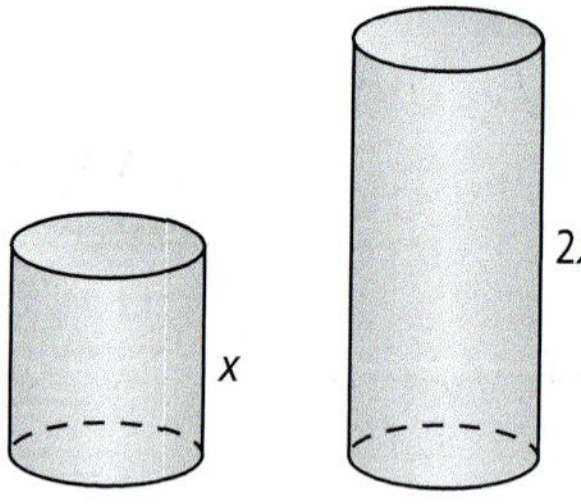

a. Write an equation for the volume of each cylinder in terms of x.

b. Solve each equation you wrote in (a) for x. Show all work.

c. Compare the volume of the shorter cylinder with the volume of the taller cylinder.

CHAPTER 2

Get Ready!

Lesson 1-3

Simplifying Expressions

Simplify by combining like terms.

1. $7s - s$ **2.** $3a + b + a$ **3.** $xy - y + x$

4. $0.5g + g$ **5.** $4t - (t + 3t)$ **6.** $b - 2(1 + c - b)$

7. $5f - (5d - f)$ **8.** $2(h + 2g) - (g - h)$ **9.** $-(3z - 5) + z$

10. $(2 - d)g - 3d(4 + g)$ **11.** $5v - 3(2 - v)$ **12.** $7t - 3s(2 + t) + s$

Lesson 1-4

Solving Equations

Solve each equation.

13. $4 + x = -52$ **14.** $-y + 13 = -67$ **15.** $12 = 2 - k$

16. $3x = -72$ **17.** $\frac{h}{5} = 215$ **18.** $64 = 4 + 12g$

19. $5 - 4t = 12$ **20.** $7x - 9 = x$ **21.** $3(w - 8) = 36$

22. $-10p = 2(p - 12)$ **23.** $3(2 - c) = -(c + 4)$ **24.** $7 + b = 11(b - 3)$

Lesson 1-6

Solving Absolute Value Inequalities

Solve each absolute value inequality. Graph the solution.

25. $|x - 3| < 5$ **26.** $|2a - 1| \geq 2a + 1$ **27.** $|3x + 4| > -4x - 3$

28. $|3x + 1| + 1 > 12$ **29.** $3|d - 4| \leq 13 - d$ **30.** $-\frac{1}{3}|f + 3| + 2 \geq -5$

VOCABULARY

Looking Ahead Vocabulary

31. A person's field of study is often called that person's *domain*. If your domain is American history, what topics might you be interested in?

32. When is a person's height likely to show a greater *rate of change*, from 1 to 2 years of age or from 30 to 31 years of age? Explain.

33. When you look in the mirror, you see your *reflection*. How does the image in the mirror differ from the way other people see you? How is it the same?

34. The boundaries of a country determine the limit of the country's land. How does an inequality form a *boundary* on a number line?

Functions, Equations, and Graphs

PowerAlgebra.com

Your place to get all things digital

Download videos connecting math to your world.

Math definitions in English and Spanish

The online Solve It will get you in gear for each lesson.

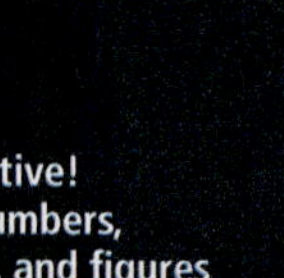

Interactive! Vary numbers, graphs, and figures to explore math concepts.

Online access to stepped-out problems aligned to Common Core

Get and view your assignments online.

Extra practice and review online

DOMAINS

- Interpreting Functions
- Building Functions
- Creating Equations

You can use functions to model all kinds of real-world situations. A function can model something as simple as a line between two points or as complex as the curves of a roller coaster. You will learn how to work with functions in this chapter.

Vocabulary

English/Spanish Vocabulary Audio Online:

English	Spanish
correlation, *p. 92*	correlación
direct variation, *p. 68*	variación directa
domain, *p. 61*	dominio
function, *p. 62*	función
linear equation, *p. 75*	ecuación lineal
range, *p. 61*	rango
relation, *p. 60*	relación
slope, *p. 74*	pendiente

BIG ideas

1 **Equivalence**

Essential Question Does it matter which form of a linear equation you use?

2 **Function**

Essential Question How do you use transformations to help graph absolute value functions?

3. **Modeling**

Essential Question How can you model data with a linear function?

Chapter Preview

Relations and Functions

Content Standards

Reviews F.IF.1 Understand that a function . . . assigns to each element of the domain exactly one element of the range.

Reviews F.IF.2 . . . Evaluate functions for inputs in their domains, and interpret statements . . .

Objectives To graph relations
To identify functions

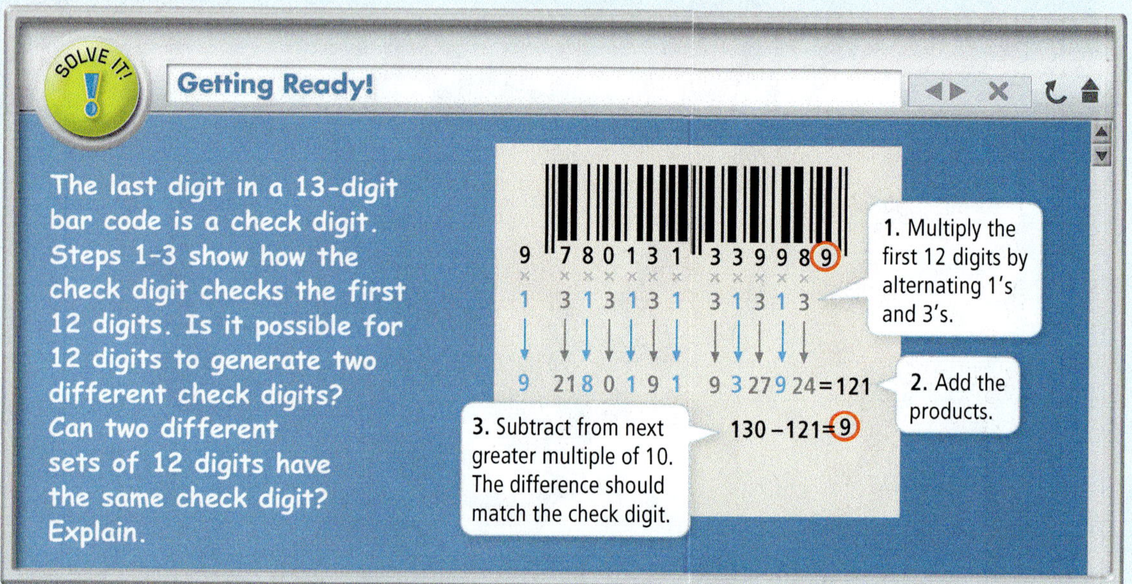

- relation
- domain
- range
- function
- vertical-line test
- function rule
- function notation
- independent variable
- dependent variable

You can use mappings to describe relationships between sets of numbers.

Essential Understanding A pairing of items from two sets is special if each item from one set pairs with exactly one item from the second set.

A **relation** is a set of pairs of input and output values. You can represent a relation in four different ways as shown below.

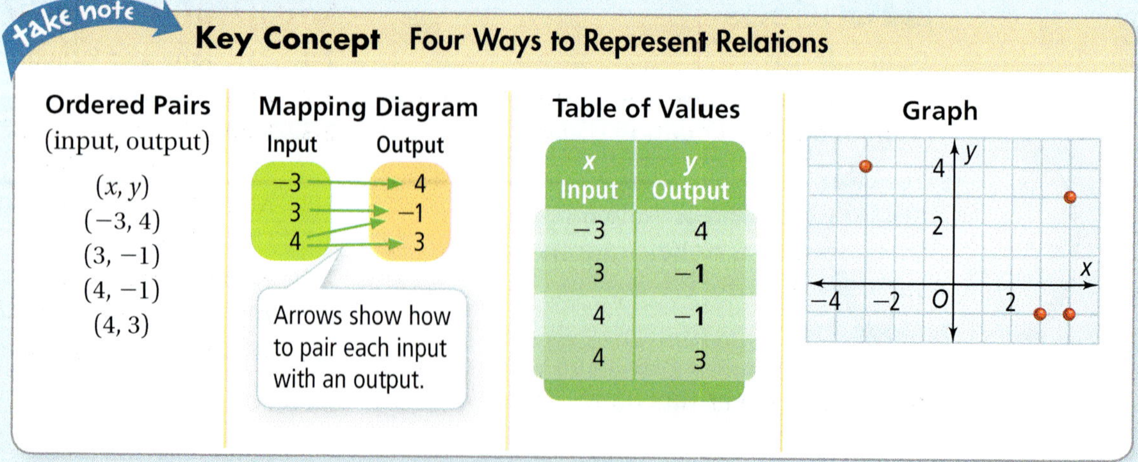

x Input	y Output
−3	4
3	−1
4	−1
4	3

Problem 1 Representing a Relation

Skydiving When skydivers jump out of an airplane, they experience free fall. The photos show various heights of a skydiver at different times during free fall, ignoring air resistance. How can you represent this relation in four different ways?

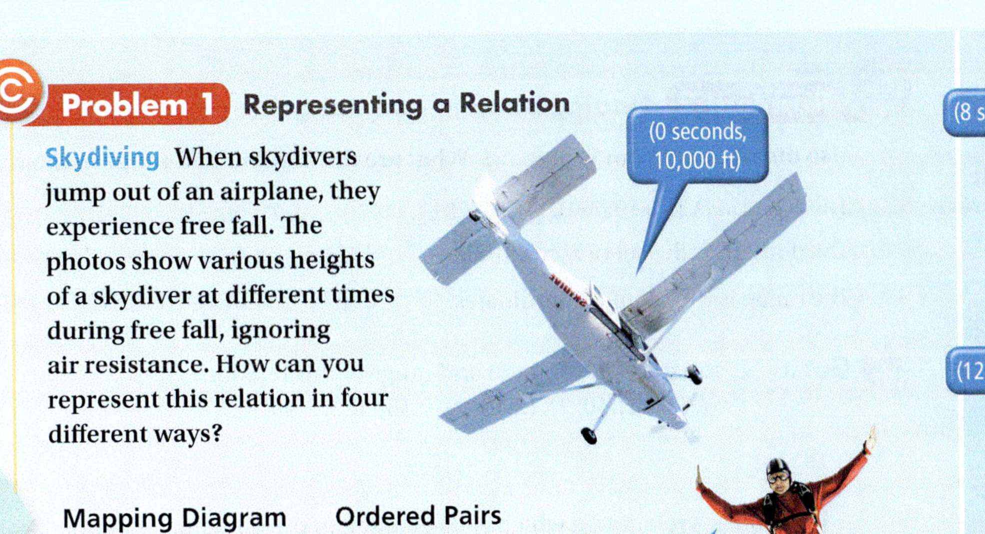

Think

What is the input? The output?
The input is the time. The output is the height above the ground.

Mapping Diagram

Input		Output
0	→	10,000
4	→	9744
8	→	8976
12	→	7696
16	→	5904

Ordered Pairs

{(0, 10,000), (4, 9744), (8, 8976), (12, 7696), (16, 5904)}

Table of Values

Time (s)	Height (ft)
0	10,000
4	9744
8	8976
12	7696
16	5904

Each time value represents an input, which is paired with its corresponding output value (height).

Graph

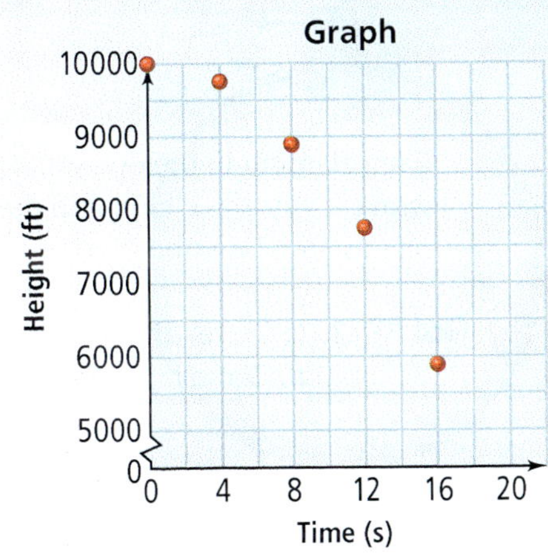

Got It? **1.** The monthly average water temperature of the Gulf of Mexico in Key West, Florida varies during the year. In January, the average water temperature is 69°F, in February, 70°F, in March, 75°F, and in April, 78°F. How can you represent this relation in four different ways?

The **domain** of a relation is the set of inputs, also called x-coordinates, of the ordered pairs. The **range** is the set of outputs, also called y-coordinates, of the ordered pairs.

Problem 2 Finding Domain and Range

Think

How could you use the mapping diagram in Problem 1 to find the domain and range?
The *input* corresponds to the domain of the relation. The *output* corresponds to the range.

Use the relation from Problem 1. What are the domain and range of the relation?

The relation is {(0, 10,000), (4, 9744), (8, 8976), (12, 7696), (16, 5904)}.

The domain is the set of x-coordinates. {0, 4, 8, 12, 16}

The range is the set of y-coordinates. {10,000, 9744, 8976, 7696, 5904}

Got It? **2.** What are the domain and range of this relation?

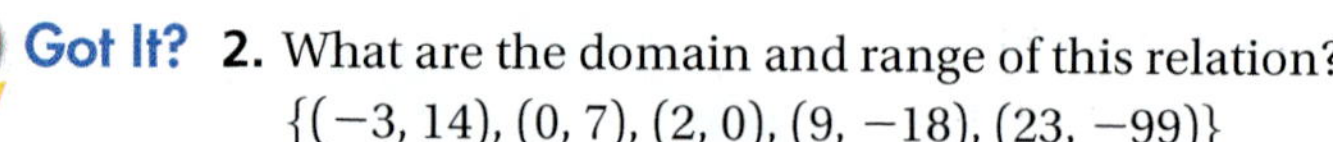

{(−3, 14), (0, 7), (2, 0), (9, −18), (23, −99)}

A **function** is a relation in which each element of the domain corresponds with exactly one element of the range.

Problem 3 Identifying Functions

Plan

How can you use a mapping diagram to determine whether a relation is a function?
A function has only one arrow from each element of the domain.

Is the relation a function?

A

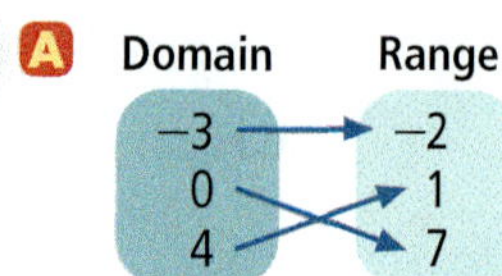

Each element in the domain corresponds with exactly one element in the range. This relation is a function.

B {(4, −1), (8, 6), (1, −1), (6, 6), (4, 1)}

Each x-coordinate must correspond to only one y-coordinate. The x-coordinate 4 corresponds to −1 and 1. The relation is *not* a function.

Got It? **3.** Is the relation a function?

a. Domain: 2, 3, 4, 6 Range: −3, −1, 3, 6

b. {(−7, 14), (9, −7), (14, 7), (7, 14)}

c. Reasoning How does a mapping diagram of a relation that is not a function differ from a mapping diagram of a function?

You can use the *vertical-line test* to determine whether a relation is a function. The **vertical-line test** states that if a vertical line passes through more than one point on the graph of a relation, then the relation is *not* a function.

Here's Why It Works If a vertical line passes through a graph at more than one point, there is more than one value in the range that corresponds to one value in the domain.

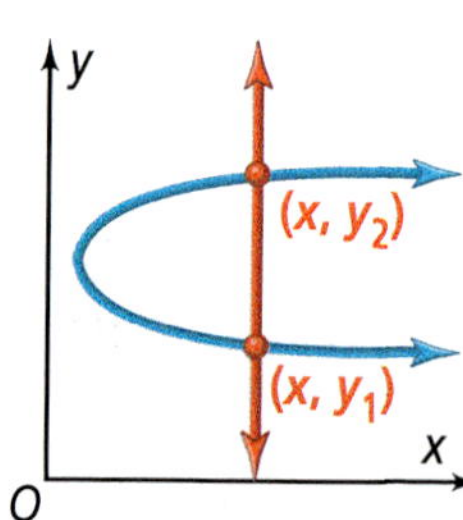

Problem 4 Using the Vertical-Line Test

Use the vertical-line test. Which graph(s) represent functions?

Ⓐ

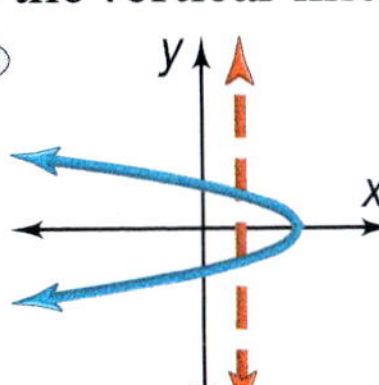

Ⓑ

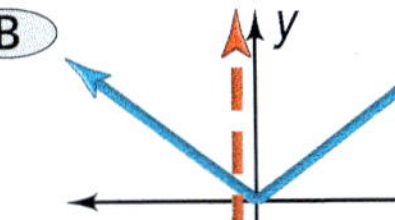

Ⓒ

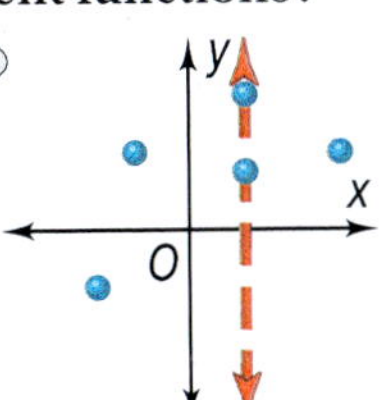

Think

Is a relation a function if it passes through the y-axis twice?
No; the y-axis is a vertical line so the relation fails the vertical-line test.

Graphs A and C fail the vertical-line test because for each graph, a vertical line passes through more than one point. They do not represent functions. Graph B does not fail the vertical-line test so it represents a function.

Got It? **4.** Use the vertical-line test. Which graph(s) represent functions?

a.

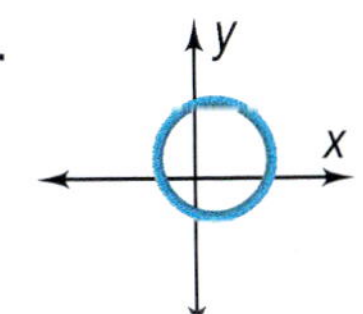

b.

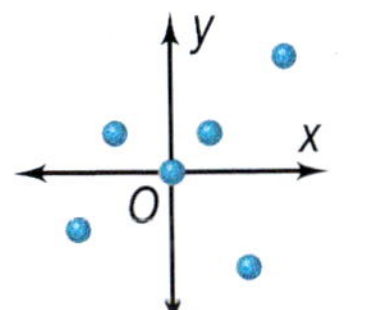

c.

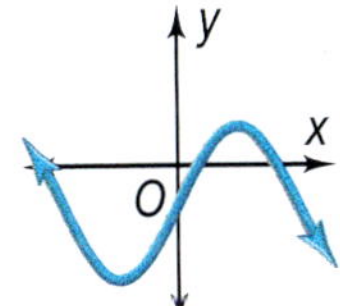

A **function rule** is an equation that represents an output value in terms of an input value. You can write a function rule in **function notation**. Shown below are examples of function rules.

$y = 3x + 2$ — Output (y), Input (x)

$f(x) = 3x + 2$ — Read as "f of x" or "function f of x."

$f(1) = 3(1) + 2$ — "f of 1" is the output when 1 is the input.

The **independent variable**, x, represents the input of the function. The **dependent variable**, $f(x)$, represents the output of the function. It is called the dependent variable because its value depends on the input value.

Problem 5 Using Function Notation

For $f(x) = -2x + 5$, what is the output for the inputs, -3, 0, and $\frac{1}{4}$?

x Input	Function Rule $f(x) = -2x + 5$	$f(x)$ Output
-3	$f(-3) = -2(-3) + 5$	11
0	$f(0) = -2(0) + 5$	5
$\frac{1}{4}$	$f\left(\frac{1}{4}\right) = -2\left(\frac{1}{4}\right) + 5$	$4\frac{1}{2}$

How do you find the output?
Substitute the input into the function rule and simplify.

Got It? **5.** For $f(x) = -4x + 1$, what is the output for the given input?

a. -2 **b.** 0 **c.** 5

To model a real-world situation using a function rule, you need to identify the dependent and independent quantities. One way to describe the dependence of a variable quantity is to use a phrase such as, "distance is a function of time." This means that distance *depends* on time.

Problem 6 Writing and Evaluating a Function

Think

Why is cost the dependent quantity?
Cost is dependent because the cost depends on the number of tickets bought.

Ticket Price **Tickets to a concert are available online for \$35 each plus a handling fee of \$2.50. The total cost is a function of the number of tickets bought. What function rule models the cost of the concert tickets? Evaluate the function for 4 tickets.**

Cost is the dependent quantity and the number of tickets is the independent quantity.

Relate	Total cost	is	cost per ticket	times	number of tickets bought	plus	handling fee
Define	Let t = number of tickets bought.						
	Let $C(t)$ = the total cost.						
Write	$C(t)$	=	35	$\cdot$	t	+	2.50

$C(t) = 35t + 2.50$

$C(4) = 35 \cdot 4 + 2.50$ Substitute 4 for t.

$= 142.50$ Simplify.

The cost of 4 tickets is \$142.50.

Got It? **6.** You are buying bottles of a sports drink for a softball team. Each bottle costs \$1.19. What function rule models the total cost of a purchase? Evaluate the function for 15 bottles.

Lesson Check

Do you know HOW?

List the domain and range of each relation.

1. $\{(3, -2), (4, 4), (0, -2), (4, 1), (3, 2)\}$

2. $\{(0, 4), (4, 0), (-3, -4), (-4, -3)\}$

Determine whether each relation is a function.

3. $\{(3, -8), (-9, 1), (3, 2), (-4, 1), (-11, -2)\}$

4. $\{(1, 1), (2, 0), (3, 1), (4, 3), (0, 2)\}$

Do you UNDERSTAND?

5. Vocabulary Can you have a relation that is not a function? Can you have a function that is not a relation? Explain.

6. Error Analysis Your friend writes, "In a function, every vertical line must intersect the graph in exactly one point." Explain your friend's error and rewrite the statement so that it is correct.

7. Reasoning Why is there no horizontal-line test for functions?

Practice and Problem-Solving Exercises

Every year, the Rock and Roll Hall of Fame and Museum inducts legendary musicians and musical acts to the Hall. The table shows the number of inductees for each year.

See Problems 1 and 2.

Rock and Roll Hall of Fame Inductees

Year	Number of Inductees	Year	Number of Inductees
2001	11	2004	8
2002	8	2005	7
2003	9	2006	6

SOURCE: Rock and Roll Hall of Fame

8. Represent the data using each of the following:
   a. a mapping diagram
   b. ordered pairs
   c. a graph on the coordinate plane

9. What are the domain and range of this relation?

Determine whether each relation is a function.

See Problem 3.

10. Domain Range

−9, 3, 5, 11 → −6, −2, 8, 21

11. Domain Range

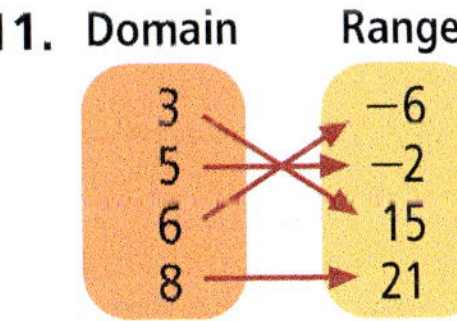

12. Domain Range

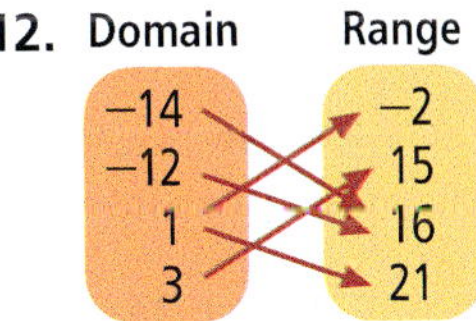

13. $\{(3, -9), (11, 21), (121, 34), (34, 1), (23, 45)\}$

Use the vertical-line test to determine whether each graph represents a function.

See Problem 4.

14.

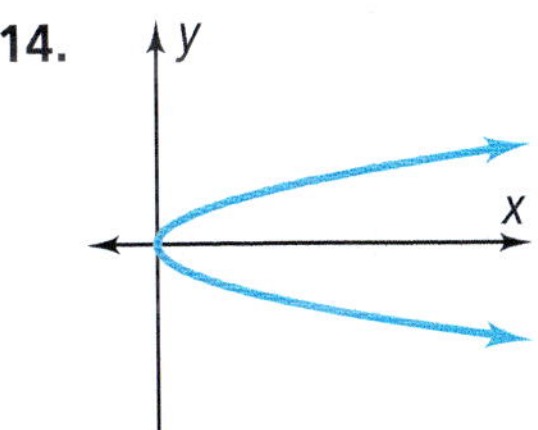

15.

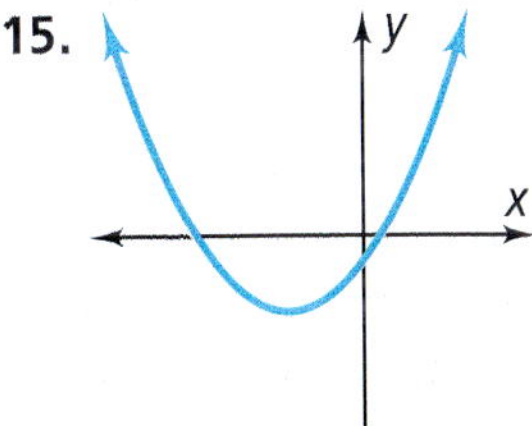

16.

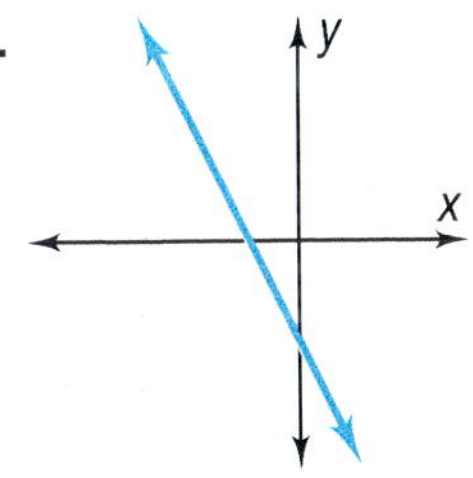

Evaluate each function for the given value of x, and write the input x and output $f(x)$ as an ordered pair.

See Problem 5.

17. $f(x) = 17x + 3$ for $x = 4$

18. $f(x) = -\frac{2x + 1}{3}$ for $x = -5$

19. $f(x) = 2x - 33$ for $x = 9$

20. $f(x) = -9x - 2$ for $x = 7$

21. $f(x) = \frac{7}{3}x - 9$ for $x = 3$

22. $f(x) = -\frac{12x}{5}$ for $x = -1$

23. $f(x) = 11x - 11$ for $x = -11$

24. $f(x) = \frac{2}{9}x - \frac{9}{2}$ for $x = 9$

Write a function rule to model the cost per month of a long-distance cell phone calling plan. Then evaluate the function for the given number of minutes.

See Problem 6.

25. Monthly service fee: $4.52
Rate: $.12 per minute
Minutes used: 250

26. Monthly service fee: $3.12
Rate: $.18 per minute
Minutes used: 175

27. Think About a Plan A cube is a solid figure with six square faces. If the edges of a cube have length 1.5 cm, what is the surface area of the cube?

- What is the relationship between the length of the edges and the area of each face?
- What is the relationship between the area of one face and the surface area of the whole cube?

28. Geometry Suppose you have a box with a 4×4-in. square base and variable height h. The surface area of this box is a function of its height. Write a function to represent the surface area. Evaluate the function for $h = 6.5$ in.

Find the domain and range of each relation, and determine whether it is a function.

29.

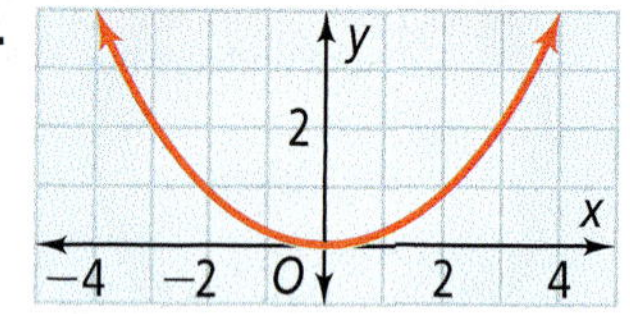

30.

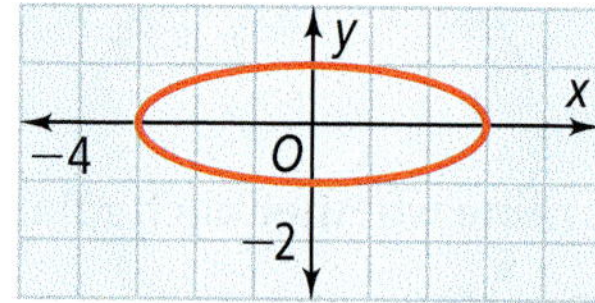

31. Geometry The volume of a sphere is a function of its radius, $V = \frac{4}{3}\pi r^3$. Evaluate the function for the volume of a volleyball with radius 10.5 cm.

32. Car Rental You are considering renting a car from two different rental companies. Proxy car rental company charges \$.32 per mile plus an \$18 surcharge. YourPal rental company charges \$.36 per mile plus a \$12 surcharge.

a. Write a function that shows the cost of renting a car from Proxy.
b. Write a function that shows the cost of renting a car from YourPal.
c. Which company offers the better deal for an 820-mile trip?

33. Temperature The relation between degrees Fahrenheit F and degrees Celsius C is described by the function $F = \frac{9}{5}C + 32$. In the following ordered pairs, the first element is degrees Celsius and the second element is its equivalent in degrees Fahrenheit. Find the unknown measure in each ordered pair.

a. $(43, m)$ **b.** $(-12, n)$ **c.** $(p, 12)$ **d.** $(q, 19)$

34. Reasoning Suppose a function pairs items from set A with items from set B. You can say that the function maps *into* set B. If the function uses every item from set B, the function maps *onto* set B. Does each function below map the set of whole numbers *into* or *onto* the set of whole numbers?

a. Function f doubles every number.
b. Function g maps every number to 1 more than that number.
c. Function h maps every number to itself.
d. Function j maps every number to its square.

35. Reasoning Given the functions $f(x) = 3x - 21$ and $g(x) = 3x + 21$, show that the function $f(x) - g(x)$ is a constant for all the values of x.

Determine whether y is a function of x. Explain.

36. $y = \frac{3}{x} - 11$ **37.** $y^2 = 3x - 7$ **38.** $x^2 = 3y + y$

STEM **39. Chemistry** The time required for a certain chemical reaction is related to the amount of catalyst present during the reaction. The domain of the relation is the number of grains of catalyst, and the range is the number of seconds required for a fixed amount of the chemical to react. The table shows the data from several reactions.

a. Is the relation a function?

b. If the domain and range were interchanged, would the relation be a function? Explain.

Catalyst and Reaction Time

Number of Grains	Number of Seconds
2.0	180
2.5	6
2.7	0.05
2.9	0.001
3.0	6
3.1	15
3.2	37
3.3	176

Standardized Test Prep

SAT/ACT

40. If $f(x) = -3x + 7$ and $g(x) = -7x + 3$, what is the value of $f(-3) - g(3)$?

(A) 40 (B) 34 (C) 8 (D) −8

41. What is the formula for the volume of a cylinder, $V = \pi r^2 h$, solved for h?

(F) $h = \frac{r^2}{\pi V}$ (G) $h = \frac{\pi V}{r^2}$ (H) $h = \frac{V}{\pi r^2}$ (I) $h = \frac{\pi r^2}{V}$

42. Which of the following statements are true?

I. $-(-6) = 6$ and $-(-4) > -4$

II. $-(-4) < 4$ or $-10 > 10 - 10$

III. $5 + 6 = 11$ or $9 - 2 = 11$

IV. $17 > 2$ or $6 < 9$

(A) I and II only (B) I, II, and III only (C) I, III, and IV only (D) III and IV only

Short Response

43. What are the numbers 1.9, $\frac{5}{4}$, −1.2, and $\sqrt{3}$ in order from greatest to least?

Mixed Review

Solve each equation or inequality.

See Lessons 1-5 and 1-6.

44. $|3x + 9| = 11$

45. $19 + |x - 1| = 33$

46. $2 - 3x < 11$

47. $5x - 3 \leq 12 - 5x$

48. $|2x| + 4 < 7$

49. $4x + 6 \geq -6$

Get Ready! **To prepare for Lesson 2-2, do Exercises 50–52.**

See Lesson 1-4.

Solve each equation for *y*.

50. $12y = 3x$

51. $-10y = 5x$

52. $\frac{3}{4}y = 15x$

2-2 Direct Variation

Content Standard

A.CED.2 Create equations in two or more variables to represent relationships between quantities; graph equations on coordinate axes with labels and scales.

Objective To write and interpret direct variation equations

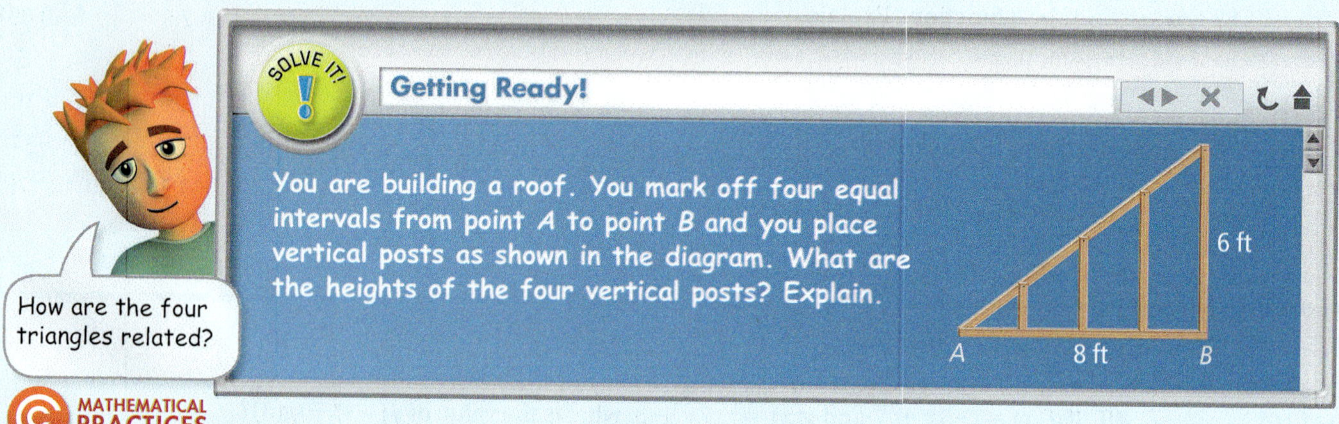

The post heights in the Solve It satisfy a relationship called *direct variation.*

Lesson Vocabulary
- direct variation
- constant of variation

Essential Understanding Some quantities are in a relationship where the ratio of corresponding values is constant.

You can write a formula for a **direct variation** function as $y = kx$, or $\frac{y}{x} = k$, where $k \neq 0$. x represents input values, and y represents output values. The formula $\frac{y}{x} = k$ says that, except for (0, 0), the ratio of all output-input pairs equals the constant k, the **constant of variation**.

Problem 1 Identifying Direct Variation From Tables

For each function, determine whether y varies directly with x. If so, what is the constant of variation and the function rule?

A

x	y
1	2
3	6
4	8

$\frac{y}{x} = \frac{2}{1} = \frac{6}{3} = \frac{8}{4} = 2,$

so y varies directly with x.

The constant of variation is 2.
The function rule is $y = 2x$.

B

x	y
1	4
2	8
3	11

$\frac{y}{x} = \frac{4}{1} = \frac{8}{2} \neq \frac{11}{3}$

so $\frac{y}{x}$ is *not* constant.

y does *not* vary directly with x.

Think

How do you find the constant of variation?
The constant of variation is the ratio of any y-value to the corresponding x-value.

Got It? **1.** For each function, determine whether y varies directly with x. If so, what are the constant of variation and the function rule?

a.

x	3	2	1
y	−21	−14	−7

b.

x	2	3	6
y	5	7	13

Problem 2 Identifying Direct Variation From Equations

For each function, determine whether y varies directly with x. If so, what is the constant of variation?

A $3y = 7x$

Divide each side of the equation $3y = 7x$ by 3 to get $y = \frac{7}{3}x$. Since you can write the equation in the form $y = kx$, y varies directly with x. The constant of variation is $\frac{7}{3}$.

Think

How is the form of this function different from the function in part (A)?
This function includes a nonzero constant term.

B $7y = 14x + 7$

Divide each side of the equation $7y = 14x + 7$ by 7 to get $y = 2x + 1$. Since you cannot write the equation in the form $y = kx$, y does not vary directly with x.

Got It? **2.** For each function, determine whether y varies directly with x. If so, what is the constant of variation?

a. $5x + 3y = 0$ **b.** $y = \frac{x}{9}$

In a direct variation, $\frac{y}{x}$ is the same for all pairs of data where $x \neq 0$. So, $\frac{y_1}{x_1} = \frac{y_2}{x_2}$ is true for the ordered pairs (x_1, y_1) and (x_2, y_2), where neither x_1 nor x_2 is zero.

Problem 3 Using a Proportion to Solve a Direct Variation

Suppose y varies directly with x, and $y = 9$ when $x = -15$. What is y when $x = 21$?

Know
y varies directly with x.
$\frac{y}{x}$ is constant.

Need
The value of y when x is 21.

Plan
Use two forms of $\frac{y}{x}$ in a proportion.

$\frac{9}{-15} = \frac{y}{21}$ In a direct variation, $\frac{y}{x}$ is constant.

$9(21) = -15(y)$ Write the cross products.

$\frac{9(21)}{-15} = \frac{-15y}{-15}$ Divide each side by −15.

$-12.6 = y$ Simplify.

So y is −12.6 when x is 21.

Got It? **3.** Suppose y varies directly with x, and $y = 15$ when $x = 3$. What is y when $x = 12$?

Problem 4 Using Direct Variation to Solve a Problem

Think

Could you use the method in Problem 3 to solve this problem?
Yes; you could solve this problem by using the proportion $\frac{c_1}{s_1} = \frac{c_2}{s_2}$.

A salesperson's commission varies directly with sales. For \$1000 in sales, the commission is \$85. What is the commission for \$2300 in sales?

Step 1 Use $y = kx$ to find k.

Let $c =$ commission.
Let $s =$ sales.

$c = k(s)$ — Commission varies directly with sales, so it is the dependent variable.

$85 = k(1000)$

$0.085 = k$

Step 2 Write the direct variation for the situation and find the commission when sales = \$2300.

$c = k(s)$

$c = 0.085(s)$ — Write the direct variation using k.

$c = 0.085(2300)$ — Substitute 2300 for s.

$c = 195.5$ — Simplify.

The commission for \$2300 in sales is \$195.50.

  **Got It?** **4. a.** The number of Calories varies directly with the mass of cheese. If 50 grams of cheese contain 200 Calories, how many Calories are in 70 grams of cheese?

b. Reasoning If y^2 varies directly with x^2, does that mean y must vary directly with x? Explain.

The graph of a direct variation function is always a line through the origin.

Problem 5 Graphing Direct Variation Equations

Think

What x-values should you use to make a table of values?
The constant of variation is a fraction. Use multiples of the denominator for x. This ensures integer values for y.

What is the graph of each direct variation equation?

A $y = \frac{3}{4}x$

x	y
4	3
8	6
12	9

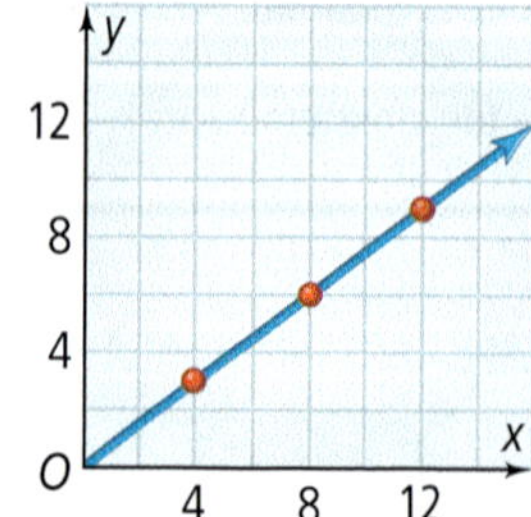

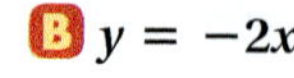
B $y = -2x$

x	y
−2	4
−1	2
1	−2

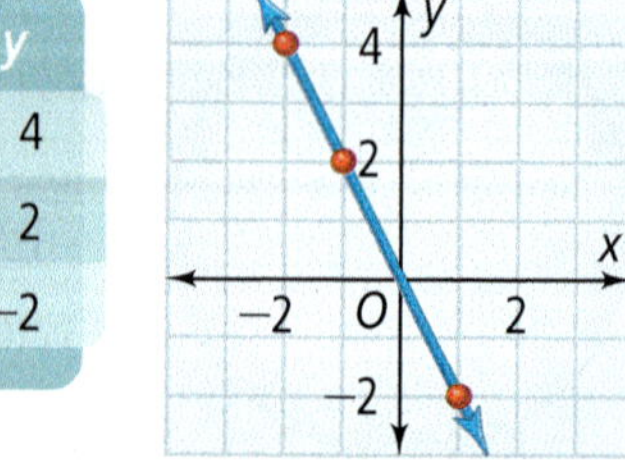

 **Got It?** **5.** What is the graph of each direct variation equation?

a. $y = -\frac{2}{3}x$ **b.** $y = 3x$

Lesson Check

Do you know HOW?

1. Write a function rule for the direct variation in the table.

x	y
2	−1
4	−2
6	−3

Identify the constant of variation.

2. $y = \frac{3}{2}x$
3. $4y - 5x = 0$

Do you UNDERSTAND?

4. **Vocabulary** Explain what it means for two variables to be directly related.
5. **Reasoning** Explain why the graph of a direct variation function always passes through the origin.
6. Give an example of a function that represents a direct variation.

Practice and Problem-Solving Exercises

For each function, determine whether y varies directly with x. If so, find the constant of variation and write the function rule. See Problem 1.

7.

x	y
2	14
3	21
5	35

8.

x	y
27	9
30	10
60	20

9.

x	y
11	22
16	32
7	42

10.

x	y
3	9
4	10
5	11

Determine whether y varies directly with x. If so, find the constant of variation. See Problem 2.

11. $y = 12x$
12. $y = 6x$
13. $y = -2x$
14. $y = 4x + 1$
15. $y = 4x - 3$
16. $y = -5x$
17. $y - 6x = 0$
18. $y + 3 = -3x$

For Exercises 19–24, y varies directly with x. See Problem 3.

19. If $y = 4$ when $x = -2$, find x when $y = 6$.
20. If $y = 6$ when $x = 2$, find x when $y = 12$.
21. If $y = 7$ when $x = 2$, find x when $y = 3$.
22. If $y = 5$ when $x = -3$, find x when $y = -1$.
23. If $y = -7$ when $x = -3$, find y when $x = 9$.
24. If $y = 25$ when $x = 15$, find y when $x = 6$.

25. **Distance** For a given speed, the distance traveled varies directly with the time. Kate's school is 5 miles away from her home and it takes her 10 minutes to reach the school. If Josh lives 2 miles from school and travels at the same speed as Kate, how long will it take him to reach the school? See Problem 4.

26. **Conservation** A dripping faucet wastes a cup of water if it drips for three minutes. The amount of water wasted varies directly with the amount of time the faucet drips. How long will it take for the faucet to waste $4\frac{1}{2}$ cups of water?

Make a table of x- and y-values and use it to graph the direct variation equation.

See Problem 5.

27. $x = \left(-\frac{1}{3}\right)y$ **28.** $y = -9x$ **29.** $x = y$

Determine whether y varies directly with x. If so, find the constant of variation and write the function rule.

30.

x	y
1	−2
3	−8
5	14

31.

x	y
9	6
12	8
15	10

32.

x	y
4	1
6	2
8	3

33.

x	y
23	24
55	56
66	67

34. Think About a Plan Suppose you make a 4-minute local call using a calling card and are charged 7.6 cents. The cost of a local call varies directly with the length of the call. How much more will it cost to make a 30-minute local call?

- Which quantity is the dependent quantity?
- How does the word "more" affect the method needed to solve the problem?

Write and graph a direct variation equation that passes through each point.

35. (1, 2) **36.** (−3, −7) **37.** (2, −9) **38.** (−0.1, 50)

39. (−5, −3) **40.** (9, −1) **41.** (7, 2) **42.** (−3, 14)

For Exercises 43–46, y varies directly with x.

43. If $y = \frac{1}{2}$ when $x = 4$, find y when $x = 5$.

44. If $y = \frac{3}{4}$ when $x = \frac{1}{2}$, find y when $x = 3$.

45. If $y = \frac{5}{3}$ when $x = \frac{3}{4}$, find x when $y = \frac{1}{2}$.

46. If $y = -\frac{5}{8}$ when $x = \frac{3}{2}$, find x when $y = \frac{2}{5}$.

47. Reasoning Explain why you cannot answer the following question.

If $y = 0$ when $x = 0$, what is x when $y = 13$?

Open-Ended Choose a value of k within the given range. Then write and graph a direct variation function using your value for k.

48. $0 < k < 1$ **49.** $3 < k < 4.5$ **50.** $-1 < k < -\frac{1}{2}$

51. Error Analysis Identify the error in the statement shown at the right.

~~If y varies directly with x^2, and y = 2 when x = 4, then y = 3 when x = 9.~~

52. Sports The number of rotations of a bicycle wheel varies directly with the number of pedal strokes. Suppose that in the bicycle's lowest gear, 6 pedal strokes move the cyclist about 357 in. In the same gear, how many pedal strokes are needed to move 100 ft?

53. Writing Suppose you use the origin to test whether a linear equation is a direct variation function. Does this method work? Support your answer with an example.

In Exercises 54–55, y varies directly with x. Explain your answer.

54. If x is doubled, what happens to y?

55. If x is divided by 7, what happens to y?

56. If z varies directly with the product of x and y ($z = kxy$), then z is said to vary jointly with x and y.
   a. **Geometry** The area of a triangle varies jointly with its base and height. What is the constant of variation?
   b. Suppose q varies jointly with v and s, and $q = 24$ when $v = 2$ and $s = 3$. Find q when $v = 4$ and $s = 2$.
   c. **Reasoning** Suppose z varies jointly with x and y, and x varies directly with w. Show that z varies jointly with w and y.

Standardized Test Prep

GRIDDED RESPONSE

SAT/ACT

57. A speed of 75 mi/h is equal to a speed of 110 ft/s. To the nearest mile per hour, what is the speed of an aircraft traveling at a speed of 1600 ft/s?

58. What number is a solution to both $|x - 3| = 2$ and $|9 - x| = 8$?

59. If $f(x) = 7 - 3x$ and $g(x) = 3x - 7$, what is the value of $f(1) + g(1)$?

60. Look at the pattern. How many circles are in the 6th figure of this pattern?

61. What is the solution of $4(x - 5) + x = 8x - 10 - x$?

Mixed Review

Graph each relation. Find the domain and range.

See Lesson 2-1.

62. $\{(0, 1), (1, -3), (-2, -3), (3, -3)\}$

63. $\{(4, 0), (7, 0), (4, -1), (7, -1)\}$

64. $\{(1, -2), (2, -1), (4, 1), (5, 2)\}$

65. $\{(1, 7), (2, 8), (3, 9), (4, 10)\}$

Identify a pattern and find the next three numbers in the pattern.

See Lesson 1-1.

66. 8, 16, 24, 32, . . .

67. 5, 3, 1, −1, . . .

68. 144, 132, 120, 108, . . .

69. 30, 45, 60, 75, . . .

Get Ready! **To prepare for Lesson 2-3, do Exercises 70–73.**

Evaluate each expression for $x = -2, 0, 1,$ and 4.

See Lesson 1-3.

70. $\frac{2}{3}x + 7$

71. $\frac{3}{5}x - 2$

72. $3x + 1$

73. $\frac{1}{2}x - 8$

2-3 Linear Functions and Slope-Intercept Form

Content Standards

A.CED.2 Create equations in two or more variables to represent relationships between quantities; graph equations on coordinate axes with labels and scales.

Also F.IF.4

Objectives To graph linear equations
To write equations of lines

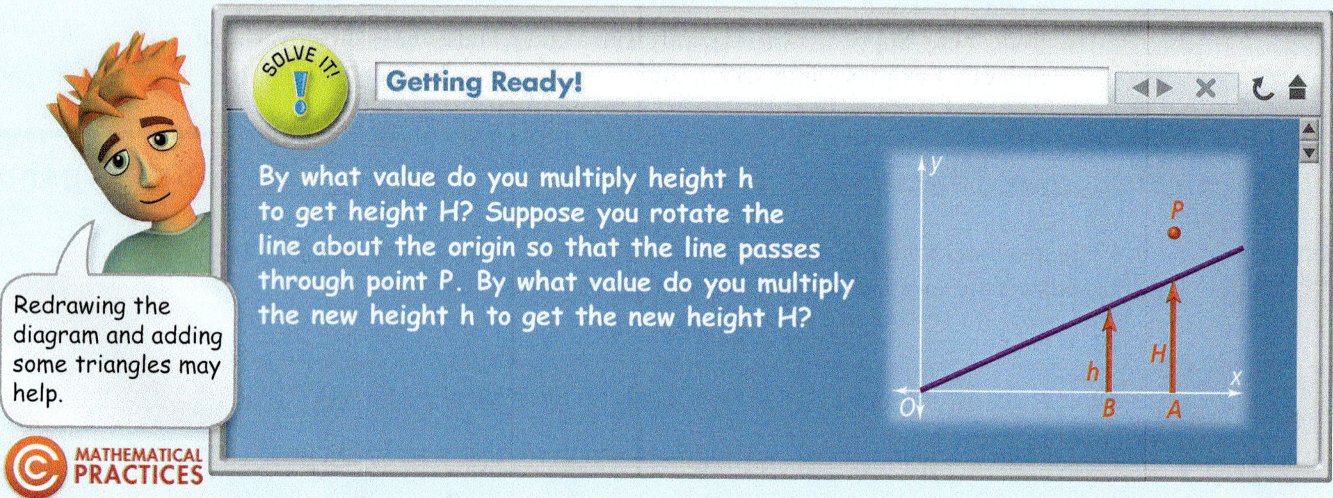

Lesson Vocabulary
- slope
- linear function
- linear equation
- *y*-intercept
- *x*-intercept
- slope-intercept form

You can describe movement in a coordinate plane by describing how far you need to move vertically and how far you need to move horizontally to get from one point to another point.

Essential Understanding Consider a nonvertical line in the coordinate plane. If you move from any point on the line to any other point on the line, the ratio of the vertical change to the horizontal change is constant. That constant ratio is the slope of the line.

The **slope** of a nonvertical line is the ratio of the vertical change to the horizontal change between two points. You can calculate slope by finding the ratio of the difference in the *y*-coordinates to the difference in the *x*-coordinates for any two points on the line.

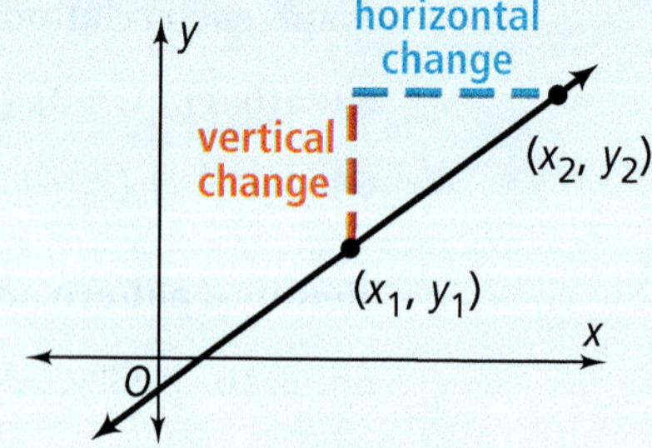

take note

Key Concept Slope

The slope of a nonvertical line through points (x_1, y_1) and (x_2, y_2) is the ratio of the vertical change to the corresponding horizontal change.

$$\text{slope} = \frac{\text{vertical change (rise)}}{\text{horizontal change (run)}} = \frac{y_2 - y_1}{x_2 - x_1}, \text{ where } x_2 - x_1 \neq 0$$

Problem 1 Finding Slope

Think

Does it matter which point you choose for (x_1, y_1)?

No; you can choose either point for (x_1, y_1).

What is the slope of the line that passes through the given points?

A $(-3, 7)$ and $(-2, 4)$

$$m = \frac{y_2 - y_1}{x_2 - x_1} = \frac{4 - 7}{-2 - (-3)} = \frac{-3}{1}$$

The slope of the line that passes through $(-3, 7)$ and $(-2, 4)$ is $\frac{-3}{1}$ or -3.

B $(3, 1)$ and $(-4, 1)$

$$m = \frac{y_2 - y_1}{x_2 - x_1} = \frac{1 - 1}{-4 - 3} = \frac{0}{-7} = 0$$

The slope of the line that passes through $(3, 1)$ and $(-4, 1)$ is $\frac{0}{-7}$ or 0.

C $(7, -3)$ and $(7, 1)$

$$m = \frac{y_2 - y_1}{x_2 - x_1} = \frac{1 - (-3)}{7 - 7} = \frac{4}{0}$$

Division by zero is undefined, so slope is undefined for the line that passes through $(7, -3)$ and $(7, 1)$.

Got It? **1.** What is the slope of the line that passes through the given points?

a. $(5, 4)$ and $(8, 1)$ **b.** $(2, 2)$ and $(-2, -2)$ **c.** $(9, 3)$ and $(9, -4)$

d. Reasoning Use the slope formula to show in part (a) that it does not matter which point you choose for (x_1, y_1).

take note

Concept Summary Slope of a Line

Positive Slope

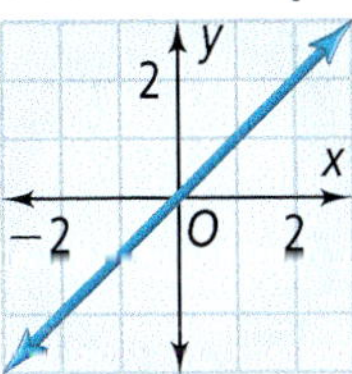

Line rises from left to right

Negative Slope

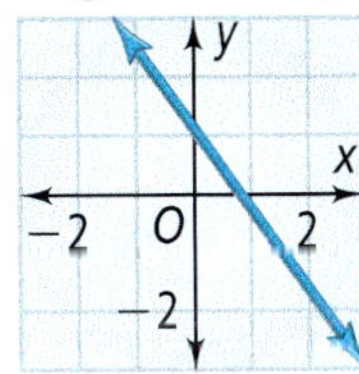

Line falls from left to right

Zero Slope

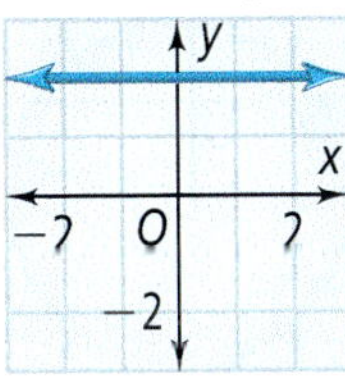

Horizontal line

Undefined Slope

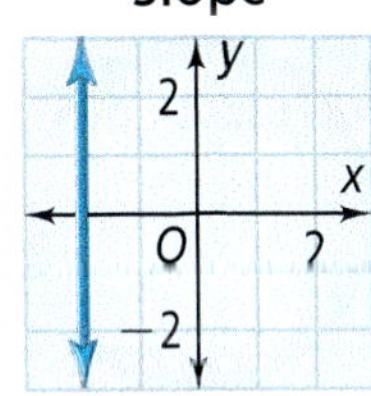

Vertical line

A function whose graph is a line is a **linear function**. You can represent a linear function with a **linear equation**, such as $y = 6x - 4$. A solution of a linear equation is any ordered pair (x, y) that makes the equation true.

A special form of a linear equation is called *slope-intercept form.*

An *intercept* of a line is a point where a line crosses an axis. The **y-intercept** of a nonvertical line is the point at which the line crosses the y-axis. The **x-intercept** of a nonhorizontal line is the point at which the line crosses the x-axis.

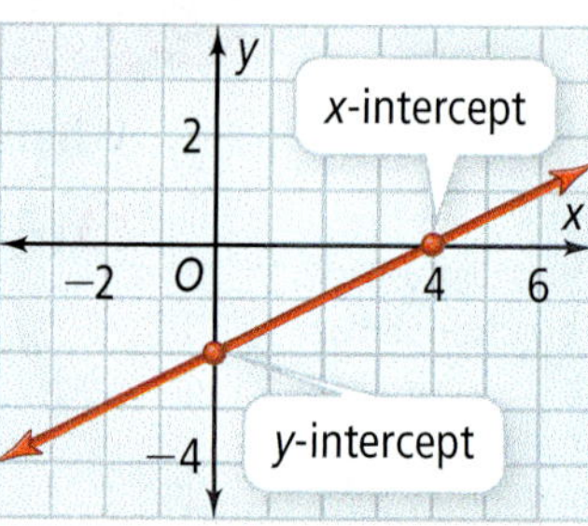

take note

Key Concept Slope-Intercept Form

The **slope-intercept form** of an equation of a line is $y = mx + b$, where m is the slope of the line and $(0, b)$ is the y-intercept.

Problem 2 Writing Linear Equations

What is an equation of each line?

A $m = \frac{1}{5}$ **and the** y**-intercept is** $(0, -3)$

$y = mx + b$ Use the slope-intercept form.

$y = \frac{1}{5}x + (-3)$ Substitute $m = \frac{1}{5}$ and $b = -3$.

$y = \frac{1}{5}x - 3$ Simplify.

Plan

What information do you need from the graph?
You need the y-intercept and another point to find the slope.

B

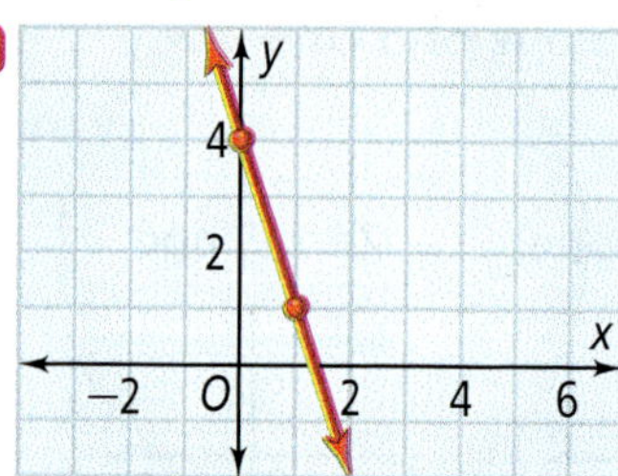

Look at the line shown in the graph. The y-intercept is the point where the line crosses the y-axis, $(0, 4)$, so $b = 4$.

Use the second point $(1, 1)$ to find the slope.

$$m = \frac{y_2 - y_1}{x_2 - x_1} = \frac{1 - 4}{1 - 0} = -3.$$

So, $y = -3x + 4$.

Got It? **2.** What is an equation of each line?

a. $m = 6$, y-intercept is $(0, 5)$

b.

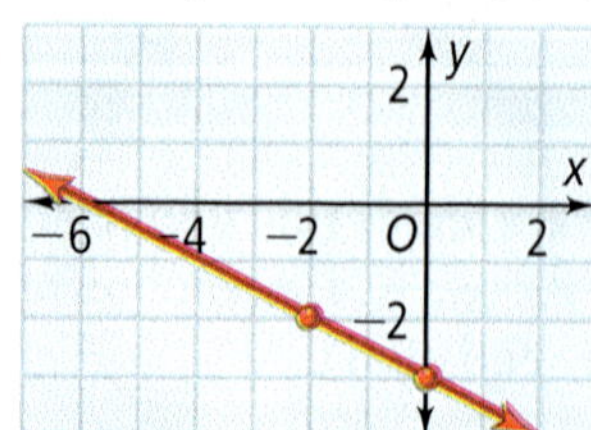

c. Reasoning Using the graph from part (b), do you get a different equation if you use $(-6, 0)$ and the y-intercept to find the slope of the line? Explain.

You can rewrite a linear equation in slope-intercept form by solving for y.

Problem 3 Writing Equations in Slope-Intercept Form

Write the equation in slope-intercept form. What are the slope and y-intercept?

A $5x - 4y = 16$

$$-4y = -5x + 16 \quad \text{Subtract } 5x \text{ from each side.}$$

$$\frac{-4y}{-4} = \frac{-5x}{-4} + \frac{16}{-4} \quad \text{Divide each side by } -4.$$

$$y = \frac{5}{4}x - 4 \quad \text{Compare the equation with } y = mx + b$$

The slope $m = \frac{5}{4}$. The y-intercept is $(0, -4)$.

Think

Is there another way to solve this problem?

Yes; clear the fractions by multiplying all terms by 4, the LCM of the denominators.

B $-\frac{3}{4}x + \frac{1}{2}y = -1$

$$\frac{1}{2}y = \frac{3}{4}x - 1 \quad \text{Add } \tfrac{3}{4}x \text{ to each side.}$$

$$y = \frac{3}{2}x - 2 \quad \text{Multiply each side by 2.}$$

The slope $m = \frac{3}{2}$. The y-intercept is $(0, -2)$.

Got It? **3.** Write the equation in slope-intercept form. What are the slope and y-intercept?

a. $3x + 2y = 18$

b. $-7x - 5y = 35$

Problem 4 Graphing a Linear Equation

What is the graph of $-2x + y = 1$?

Know	Need	Plan
• The equation of a line	Two points to draw the line	• Write the equation in slope-intercept form. • Plot the y-intercept. • Use the slope to find a second point. • Draw a line through the two points.

Write the equation in slope-intercept form.

$$-2x + y = 1$$

$$y = 2x + 1$$

The slope is 2 and the y-intercept is $(0, 1)$.

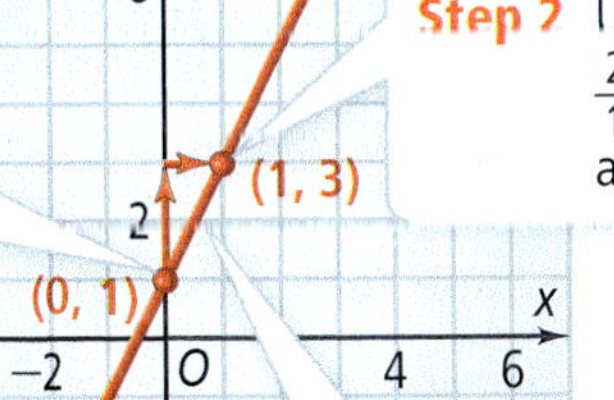

Got It? **4.** What is the graph of $4x - 7y = 14$?

Lesson Check

Do you know HOW?

Write each equation in slope-intercept form.

1. $x - 2y + 3 = 1$
2. $-4x + 3y = 1$

What is the slope of the line passing through the following points?

3. $(2, 4)$ and $(4, 2)$
4. $(-1, -3)$ and $(3, 1)$

Do you UNDERSTAND?

5. **Vocabulary** What is a y-intercept? How is a y-intercept different from an x-intercept?
6. Explain why the slope of a vertical line is called "undefined."
7. **Error Analysis** A classmate found the slope between two points. What error did she make?

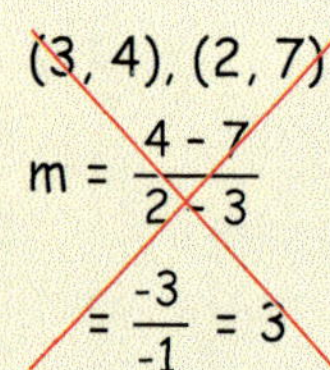

(3, 4), (2, 7)

$m = \frac{4-7}{2-3}$

$= \frac{-3}{-1} = 3$

Practice and Problem-Solving Exercises

Find the slope of the line through each pair of points.

See Problem 1.

8. $(1, 6)$ and $(8, -1)$
9. $(-3, 9)$ and $(0, 3)$
10. $(0, 0)$ and $(2, 6)$
11. $(-4, -3)$ and $(7, 1)$
12. $(-2, -1)$ and $(8, -3)$
13. $(1, 2)$ and $(2, 3)$
14. $(2, 7)$ and $(-3, 11)$
15. $(-3, 5)$ and $(4, 5)$
16. $(-5, -7)$ and $(0, 10)$

Write an equation for each line.

See Problem 2.

17. $m = 3$ and the y-intercept is $(0, 2)$.
18.

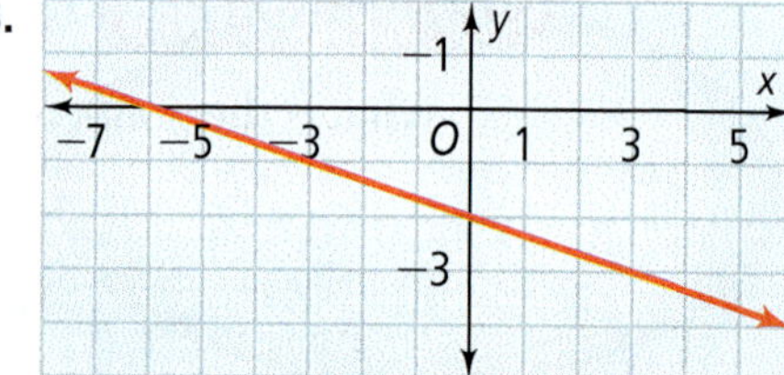

19. $m = \frac{5}{6}$ and the y-intercept is $(0, 12)$.
20. $m = 0$ and the y-intercept is $(0, -2)$.
21. $m = -5$ and the y-intercept is $(0, -7)$.

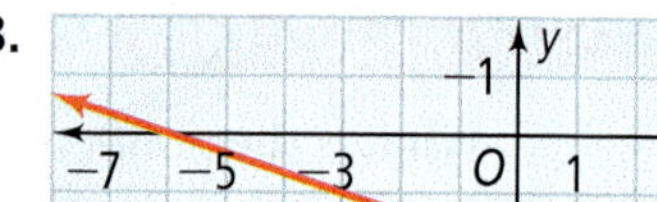

Write each equation in slope-intercept form. Then find the slope and y-intercept of each line.

See Problem 3.

22. $5x + y = 4$
23. $-3x + 2y = 7$
24. $-\frac{1}{2}x - y = \frac{3}{4}$
25. $8x + 6y = 5$
26. $9x - 2y = 10$
27. $y = 7$

Graph each equation.

See Problem 4.

28. $y = 2x$
29. $y = -3x - 1$
30. $y = 3x - 2$
31. $y = -4x + 5$
32. $5x - 2y = -4$
33. $-2x + 5y = -10$
34. $y - 3 = -2x$
35. $y + 4 = -3x$
36. $-y + 5 = -2x$

Graph each equation.

37. $y = -\frac{1}{2}x - \frac{3}{2}$

38. $y = -2x + 3$

39. $y = -x + 7$

40. $3y - 2x = -12$

41. $4x + 5y = 20$

42. $4x - 3y = -6$

43. $\frac{2}{3}x + \frac{y}{3} = -\frac{1}{3}$

44. $x = 5$

45. $2.4 = -3.6x - 0.4y$

46. **Think About a Plan** Suppose the equation $y = 12 + 10x$ models the amount of money in your wallet, where y is the total in dollars and x is the number of weeks from today. If you graphed this equation, what would the slope represent in the situation? Explain.
- Is the equation in slope-intercept form?
- What units make sense for the slope?

Find the slope and y-intercept of each line.

47. 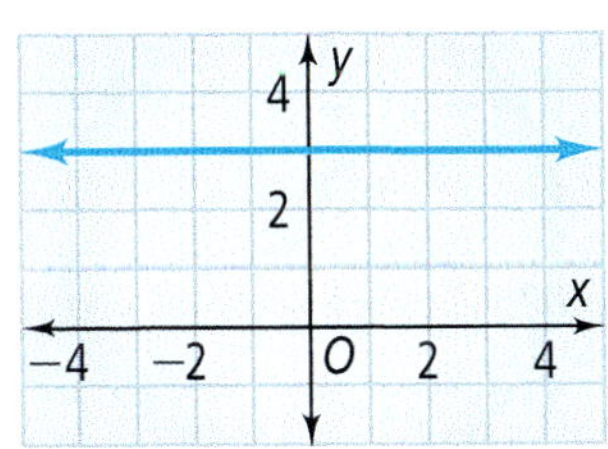

48. 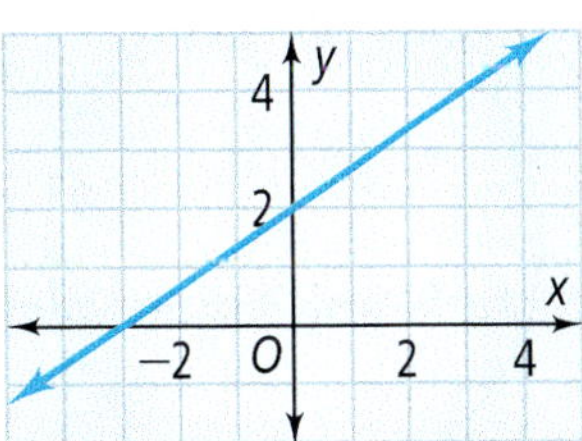

49. 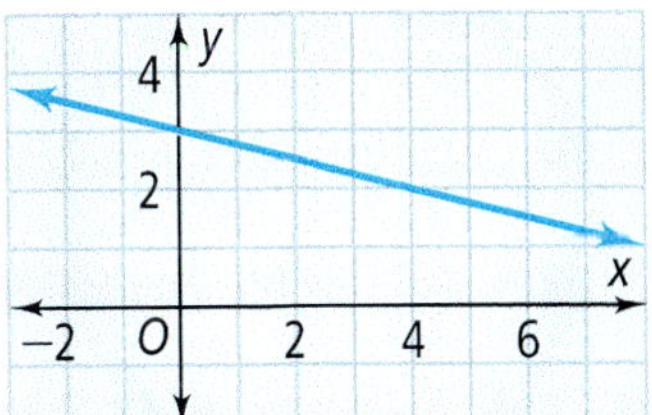

Find the slope and y-intercept of each line.

50. $y = 0.4 - 0.8x$

51. $x = -3$

52. $y = 0$

53. $-\frac{1}{3}x - \frac{2}{3}y = \frac{5}{3}$

54. $-Ax + By = -C$

55. $\frac{A}{D}x + \frac{B}{D}y = \frac{C}{D}$

56. The equation $d = 4 - \frac{1}{15}t$ represents your distance from home d for each minute you walk t.
   a. If you graphed this equation, what would the slope represent? Explain.
   b. Are you walking towards or away from your home? Explain.

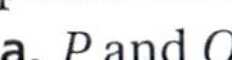

57. **Reasoning** Use the graph to find the slope between the following points on the line.
   a. P and Q
   b. Q and S
   c. S and P
   d. R and Q
   e. Make a conjecture based on your answers to parts (a)–(d).

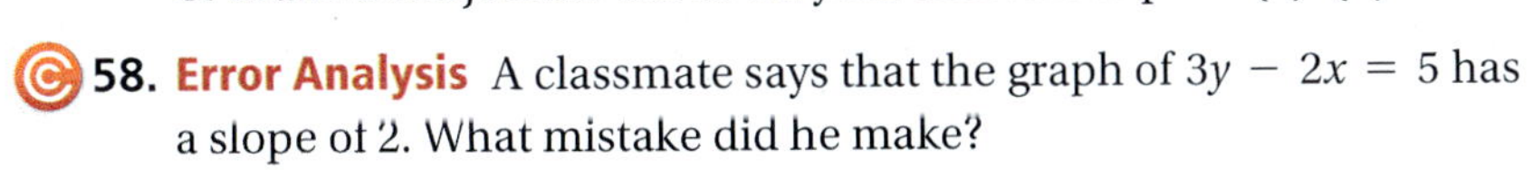

58. **Error Analysis** A classmate says that the graph of $3y - 2x = 5$ has a slope of 2. What mistake did he make?

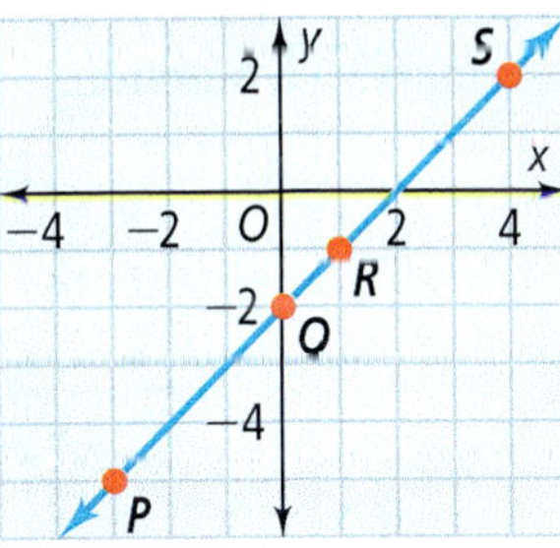

Find the slope of the line through each pair of points.

59. $\left(\frac{3}{2}, -\frac{1}{2}\right)$ and $\left(-\frac{2}{3}, \frac{1}{3}\right)$

60. $\left(-\frac{1}{2}, -\frac{1}{2}\right)$ and $(-3, -4)$

61. $\left(0, -\frac{1}{2}\right)$ and $\left(\frac{7}{5}, 10\right)$

Challenge

62. You can find the equation of a line through two points even if one point is not the y-intercept.

- Find the slope m of the line passing through the two points.
- Using either point, substitute for x, y, and m into $y = mx + b$.
- Solve for b and rewrite $y = mx + b$ for the values of m and b.

Write an equation in slope-intercept form for the line passing through each pair of points.

a. $(2, 5)$ and $(6, 7)$ **b.** $(-4, 16)$ and $(3, -5)$ **c.** $(-2, 17)$ and $(2, 1)$

Standardized Test Prep

SAT/ACT

63. Which equation does NOT represent a direct variation?

(A) $y - 3x = 0$ (B) $y + 2 = \frac{1}{2}x$ (C) $\frac{y}{x} = \frac{2}{3}$ (D) $y = \frac{x}{17}$

64. Which equation models the data in the table?

(F) $y = x^2 - 1$ (H) $y = -x^2 + 3$

(G) $y = x^2 + 3$ (I) $y = x^2 + 1$

x	y
1	4
2	7
3	12
4	19

65. In the formula $V = \frac{1}{3}\pi r^2 h$, which expression is equal to h?

(A) $\frac{V}{3\pi r^2}$ (C) $\frac{3V}{\pi r^2}$

(B) $V - \frac{1}{3}\pi r^2$ (D) $\frac{3V\pi}{r^2}$

Short Response

66. Graph the relation $\{(-2, 1), (0, 2), (-1, -1), (-2, -2)\}$. What are the domain and range?

Mixed Review

Find the domain and range of each relation, and determine whether it is a function.

See Lesson 2-1.

67.

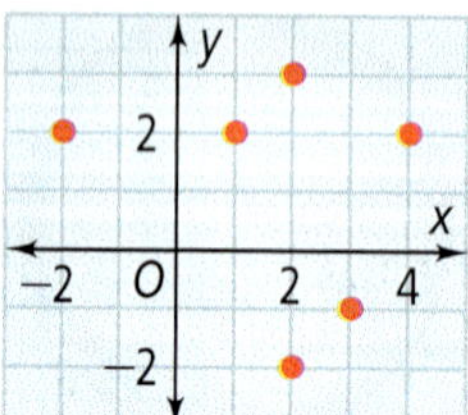

68.

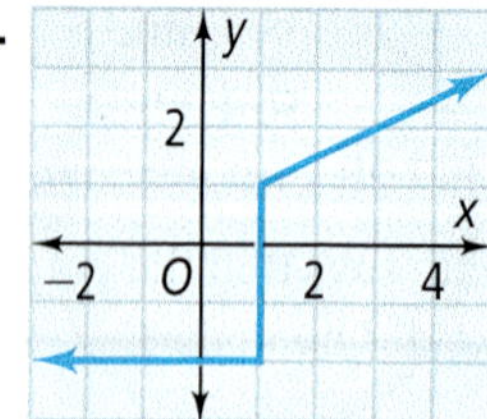

69.

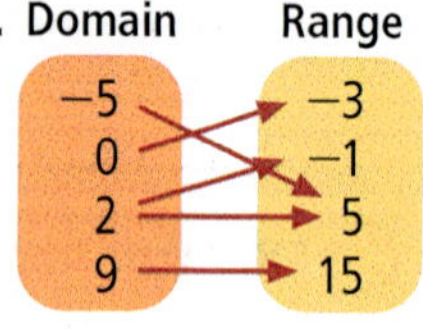

Get Ready! **To prepare for Lesson 2-4, do Exercises 70–72.**

Evaluate each expression for $x = 0$.

See Lesson 1-3.

70. $5x + 2$ **71.** $(x - 4) + 12$ **72.** $13 - 6.5x$

2-4 More About Linear Equations

Content Standards

F.IF.8 Write a function defined by an expression in different but equivalent forms . . .

F.IF.9 Compare properties of two functions each represented in a different way . . .

Also F.IF.2, A.CED.2

Objective To write an equation of a line given its slope and a point on the line

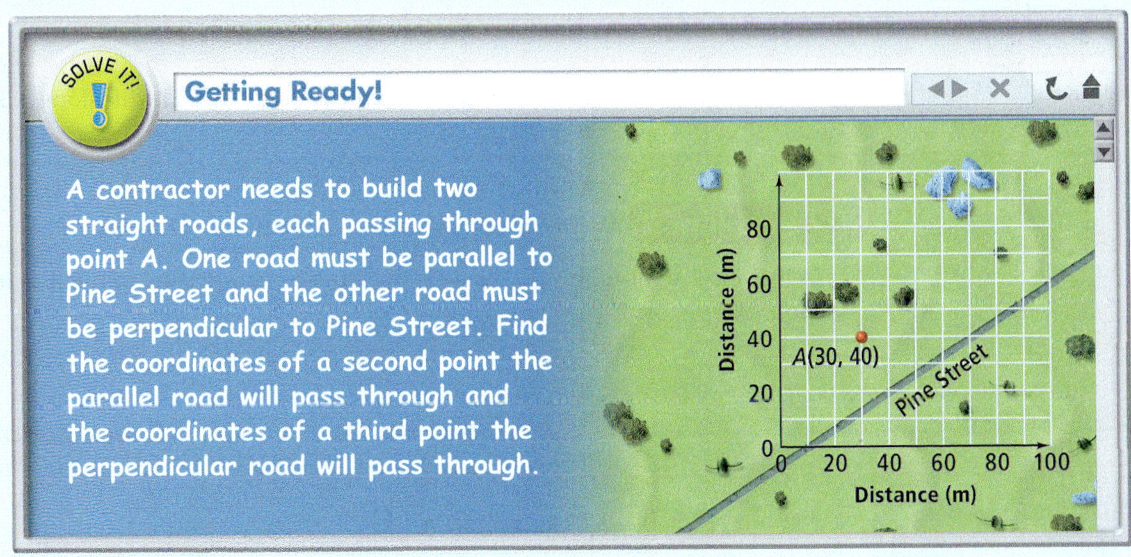

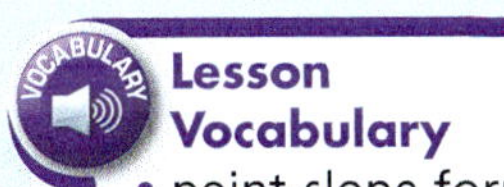

Lesson Vocabulary
- point-slope form
- standard form of a linear equation
- parallel lines
- perpendicular lines

If you travel along a line that is parallel to a given line, you will stay the same distance from the given line. If you travel along a line that is perpendicular to a given line, you will travel either toward or away from the given line along the most direct path.

Essential Understanding The slopes of two lines in the same plane indicate how the lines are related.

Given the slope and y-intercept, you can write the equation of a line in slope-intercept form. You can also write the equation of a line in *point-slope form*.

take note

Key Concept Point-Slope Form

The equation of a line in **point-slope form** through point (x_1, y_1) with slope m:

$$y - y_1 = m(x - x_1)$$

Here's Why It Works By substituting the general point (x, y), for (x_2, y_2) in the slope formula, you can rewrite the slope formula in point-slope form.

$$m = \frac{y_2 - y_1}{x_2 - x_1} = \frac{y - y_1}{x - x_1}$$

$$m(x - x_1) = \frac{y - y_1}{x - x_1}(x - x_1)$$

$$m(x - x_1) = y - y_1$$

$$y - y_1 = m(x - x_1)$$

Problem 1 Writing an Equation Given a Point and the Slope

Is slope-intercept or point-slope form more helpful for writing this equation?
Since the slope and a point (not the y-intercept) are given, point-slope form is more helpful.

A line passes through $(-5, 2)$ with slope $\frac{3}{5}$. What is an equation of the line?

$y - y_1 = m(x - x_1)$ Use point-slope form.

$y - 2 = \frac{3}{5}[x - (-5)]$ Substitute $m = \frac{3}{5}$ and $(x_1, y_1) = (-5, 2)$.

$y - 2 = \frac{3}{5}(x + 5)$ Simplify.

An equation for the line is $y - 2 = \frac{3}{5}(x + 5)$.

Got It? **1.** What is an equation of the line through $(7, -1)$ with slope -3?

Problem 2 Writing an Equation Given Two Points

A line passes through $(3, 2)$ and $(5, 8)$. What is an equation of the line in point-slope form?

Know	Need	Plan
Two points	An equation written in point-slope form	Substitute the slope and either point in the point-slope form.

Think

Does it matter which point you substitute into point-slope form?
No; you can choose either point as (x_1, y_1).

Let $(x_1, y_1) = (3, 2)$ and $(x_2, y_2) = (5, 8)$.

$m = \frac{8 - 2}{5 - 3} = \frac{6}{2} = 3$ Substitute into the slope formula and simplify.

$y - 2 = 3(x - 3)$ Substitute into point-slope form.

Got It? **2. a.** A line passes through $(-5, 0)$ and $(0, 7)$. What is an equation of the line in point-slope form?

b. Reasoning What is another equation in point-slope form of the line through the points $(-5, 0)$ and $(0, 7)$? Explain.

Another form of the equation of a line is *standard form*, in which the sum of the x and y terms are set equal to a constant. When possible, you write the coefficients of x and y and the constant term as integers.

Key Concept Standard Form of a Linear Equation

A **standard form of a linear equation** is $Ax + By = C$, where A, B, and C are real numbers and A and B are not *both* zero.

Think

How can you rewrite the equation using only integer values? Multiply each side of the equation by the least common denominator of all fraction coefficients.

Problem 3 Writing an Equation in Standard Form

What is an equation of the line $y = \frac{3}{4}x - 5$ in standard form? Use integer coefficients.

$$y = \frac{3}{4}x - 5$$

$-\frac{3}{4}x + y = -5$ Subtract $\frac{3}{4}x$ from each side.

$-3x + 4y = -20$ Multiply each side by 4.

Got It? **3.** What is an equation of the line $y = 9.1x + 3.6$ in standard form?

take note

Concept Summary Writing Equations of Lines

Slope-Intercept Form	Point-Slope Form	Standard Form
$y = mx + b$	$y - y_1 = m(x - x_1)$	$Ax + By = C$
Use this form when you know the slope and the y-intercept.	Use this form when you know the slope and a point, or when you know two points.	A, B, and C are real numbers. A and B cannot both be zero.

You can graph an equation in standard form quickly by determining the x- and y-intercepts and then drawing the line through them.

Think

Why set $x = 0$ to find the y-intercept? The x-coordinate of any point on the y-axis is zero.

Problem 4 Graphing an Equation Using Intercepts

What are the intercepts of $3x + 5y = 15$? Graph the equation.

Set $x = 0$ to find the y-intercept.

$3(0) + 5y = 15$

$5y = 15$

$y = 3$

Set $y = 0$ to find the x-intercept.

$3x + 5(0) = 15$

$3x = 15$

$x = 5$

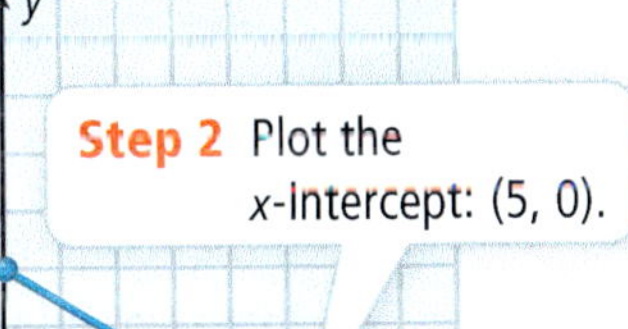

Got It? **4.** What are the intercepts of $2x - 4y = 8$? Graph the equation.

Problem 5 Drawing and Interpreting a Linear Graph STEM

Biology The number of times a cricket chirps per minute depends on the temperature. The number of chirps in 2 seconds for two temperatures are shown at the bottom right.

Think

How can you find the number of chirps in a minute given the number of chirps in 2 seconds?

There are 60 seconds in 1 minute. Multiply the number of chirps in 2 seconds by $\frac{60}{2}$ or 30.

A What graph models the situation?

First, find the number of chirps per minute.

40°F: $30(0) = 0$
93°F: $30(8) = 240$

Let x = temperature in degrees Fahrenheit.
Let y = number of times a cricket chirps.

Plot (40,0) and (93, 240).
Draw a line through the points.

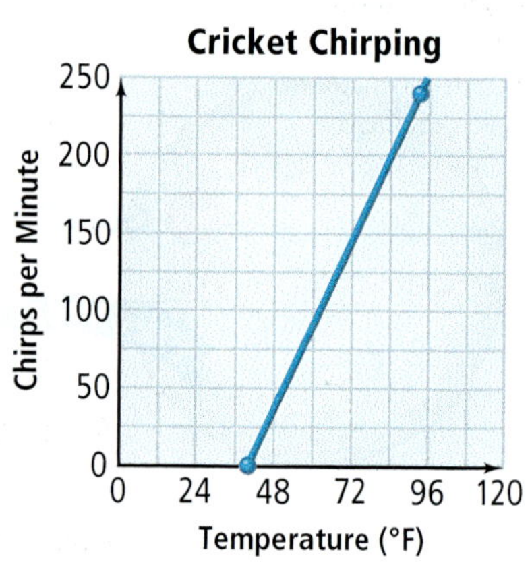

B What is an equation of the line in standard form?

$m = \frac{240 - 0}{93 - 40}$	Use the slope formula $m = \frac{y_2 - y_1}{x_2 - x_1}$.
$= \frac{240}{53} \approx 4.5$	Subtract and simplify.
$y - y_1 = m(x - x_1)$	Use point-slope form.
$y - 0 = 4.5(x - 40)$	Substitute one of the points: (40, 0).
$y = 4.5x - 180$	Simplify.
$4.5x - y = 180$	Write in standard form.

C If the temperature is 70°F, how many times would a cricket be expected to chirp in one minute?

Let $x = 70$.

$y = 4.5x - 180$	Use an equation from part (b).
$y = 4.5(70) - 180$	Substitute.
$y = 135$	Simplify.

If the temperature is 70°F, the cricket would be expected to chirp 135 times in one minute.

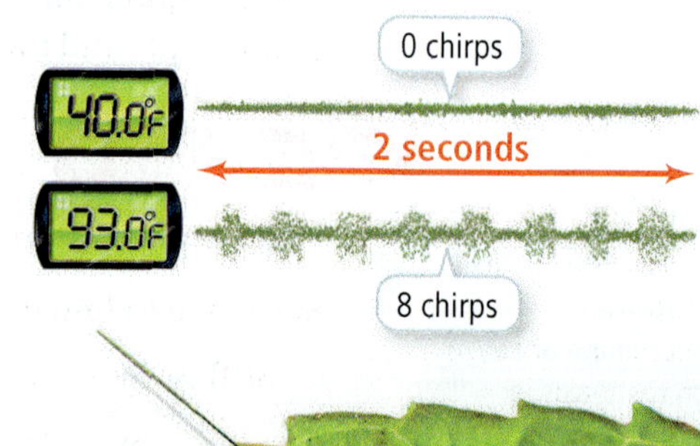

Got It? **5.** The office manager of a small office ordered 140 packs of printer paper. Based on average daily use, she knows that the paper will last about 80 days.

a. What graph represents this situation?

b. What is the equation of the line in standard form?

c. How many packs of printer paper should the manager expect to have after 30 days?

take note

Key Concepts Parallel and Perpendicular Lines

The slopes of **parallel lines** are equal.

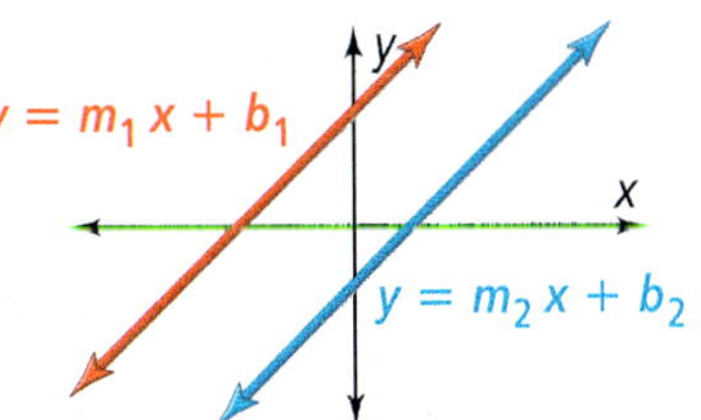

$m_1 = m_2$

$b_1 \neq b_2$

No line can be vertical.

The slopes of **perpendicular lines** are negative reciprocals of each other.

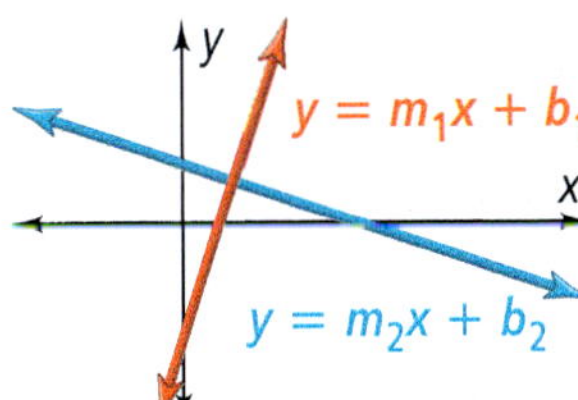

$m_1 \cdot m_2 = -1$

$m_1 = -\frac{1}{m_2}$

$m_2 = -\frac{1}{m_1}$

m_1 and m_2 are negative reciprocals of each other.

Problem 6 Writing Equations of Parallel and Perpendicular Lines

What is the equation of each line in slope-intercept form?

A the line parallel to $y = 6x - 2$ through $(1, -3)$

Identify the slope, use point-slope form, and rewrite in slope-intercept form.

$m = 6$	Parallel lines have the same slope. The slope of the line with equation $y = 6x - 2$ is 6.
$y - y_1 = m(x - x_1)$	Point-slope form
$y - (-3) = 6(x - 1)$	Substitute 6 for m and $(1, -3)$ for (x_1, y_1).
$y + 3 = 6x - 6$	Distributive Property
$y = 6x - 9$	Write in slope-intercept form.

B the line perpendicular to $y = -4x + \frac{2}{5}$ through $(8, 5)$

Identify the slope, use point-slope form, and rewrite in slope-intercept form.

$m = -\frac{1}{-4} = \frac{1}{4}$	The slopes of perpendicular lines are negative reciprocals.
$y - y_1 = m(x - x_1)$	Point-slope form
$y - 5 = \frac{1}{4}(x - 8)$	Substitute $\frac{1}{4}$ for m and $(8, 5)$ for (x_1, y_1).
$y - 5 = \frac{1}{4}x - 2$	Distributive Property
$y = \frac{1}{4}x + 3$	Write in slope-intercept form.

Plan

How can you find the slope of a perpendicular line? Slopes of perpendicular lines are negative reciprocals, so use the equation $m_1 = -\frac{1}{m_2}$.

Got It? **6.** What is the equation of each line in slope-intercept form?

a. the line parallel to $4x + 2y = 7$ through $(4, -2)$

b. the line perpendicular to $y = \frac{2}{3}x - 1$ through $(0, 6)$

Lesson Check

Do you know HOW?

Write an equation of each line in slope-intercept form.

1. slope -3; through $(1, -4)$

2. slope $\frac{1}{2}$; through $(2, 3)$

3. What are the intercepts of $3x + y = 6$? Graph the equation.

Write an equation of each line in standard form.

4. the line parallel to $y = -3x + 4$ through $(0, -1)$

5. the line perpendicular to $-2x + 3y = 9$ through $(-1, -3)$

Do you UNDERSTAND?

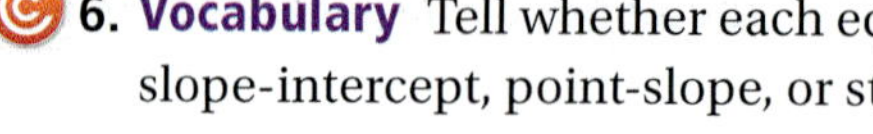

6. Vocabulary Tell whether each equation is in slope-intercept, point-slope, or standard form.

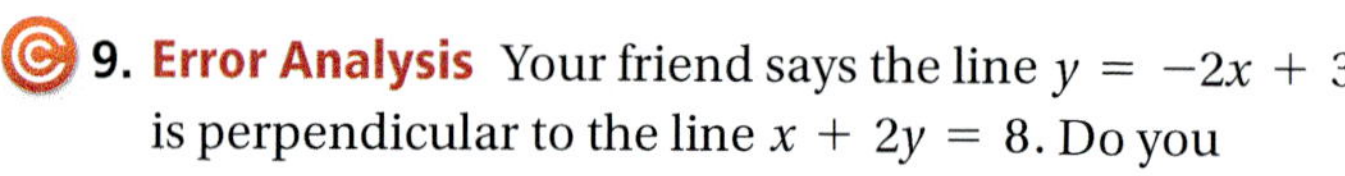

a. $y + 2 = -2(x - 1)$ **b.** $y = -\frac{1}{4}x + 9$

c. $-x - 2y = 1$ **d.** $y - 3 = 4x$

7. Which form would you use to write the equation of a line if you knew its slope and x-intercept? Explain.

8. If the intercepts of a line are $(a, 0)$ and $(0, b)$, what is the slope of the line?

9. Error Analysis Your friend says the line $y = -2x + 3$ is perpendicular to the line $x + 2y = 8$. Do you agree? Explain.

Practice and Problem-Solving Exercises

Write an equation of each line. See Problem 1.

10. slope $= 3$; through $(1, 5)$ **11.** slope $= \frac{5}{6}$; through $(22, 12)$ **12.** slope $= -\frac{3}{5}$; through $(-4, 0)$

13. slope $= 0$; through $(4, -2)$ **14.** slope $= -1$; through $(-3, 5)$ **15.** slope $= 5$; through $(0, 2)$

Write in point-slope form an equation of the line through each pair of points. See Problem 2.

16. $(-10, 3)$ and $(-2, -5)$ **17.** $(1, 0)$ and $(5, 5)$ **18.** $(-4, 10)$ and $(-6, 15)$

19. $(0, -1)$ and $(3, -5)$ **20.** $(7, 11)$ and $(13, 17)$ **21.** $(1, 9)$ and $(6, 2)$

Write an equation of each line in standard form with integer coefficients. See Problem 3.

22. $y = \frac{1}{2}x - 2$ **23.** $y = -7x - 9$ **24.** $y = -\frac{3}{5}x + 3$ **25.** $y = 4.2x + 7.9$

Find the intercepts and graph each line. See Problem 4.

26. $x - 4y = -4$ **27.** $2x + 5y = -10$ **28.** $-3x + 2y = 6$ **29.** $5x + 7y = 14$

Write and graph an equation to represent each situation. See Problem 5.

30. You put 15 gallons of gasoline in your car. You know that this amount of gasoline will allow you to drive about 450 miles.

31. A meal plan lets students buy \$20 meal cards. Each meal card lasts about 8 days.

Write the equation of the line through each point. Use slope-intercept form. See Problem 6.

32. $(1, -1)$; parallel to $y = \frac{2}{5}x - 3$ **33.** $(-3, 1)$; perpendicular to $y = -\frac{2}{5}x - 4$

34. $(-7, 10)$; parallel to $2x - 3y = -3$ **35.** $(-2, 1)$; perpendicular to $3x + y = 1$

B Apply

Graph each equation.

36. $3x + 5y = 12$
37. $y = \frac{2}{3}x + 4$
38. $x + 3 = 0$
39. $3y - x = -6$
40. $y + 3 = 3$
41. $2x - \frac{3}{2}y = -3$

Write an equation of the line through each pair of points. Use point-slope form.

42. $\left(\frac{3}{2}, -\frac{1}{2}\right)$ and $\left(-\frac{2}{3}, \frac{1}{3}\right)$
43. $\left(-\frac{1}{2}, -\frac{1}{2}\right)$ and $(-3, -4)$
44. $\left(0, \frac{1}{2}\right)$ and $\left(\frac{5}{7}, 0\right)$

45. **Think About a Plan** Write an equation for the line shown here. Each interval is 1 unit.
- What do you know from the graph?
- Which form of the equation of a line could you use with the information you have?

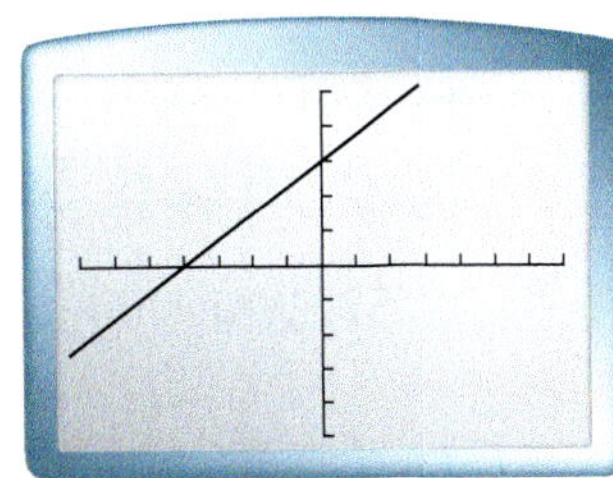

Write an equation for each line. Each interval is 1 unit.

46.

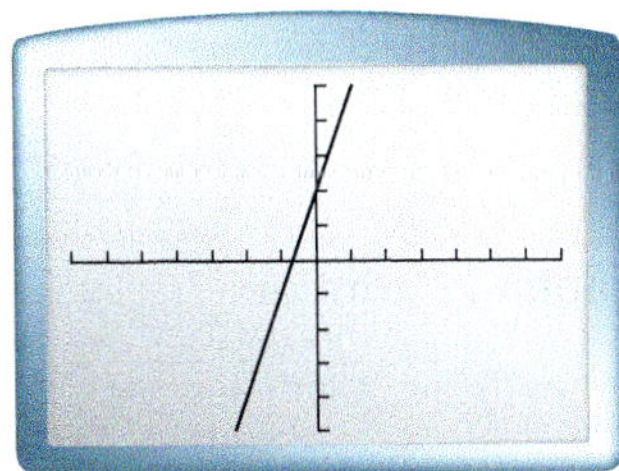

47.

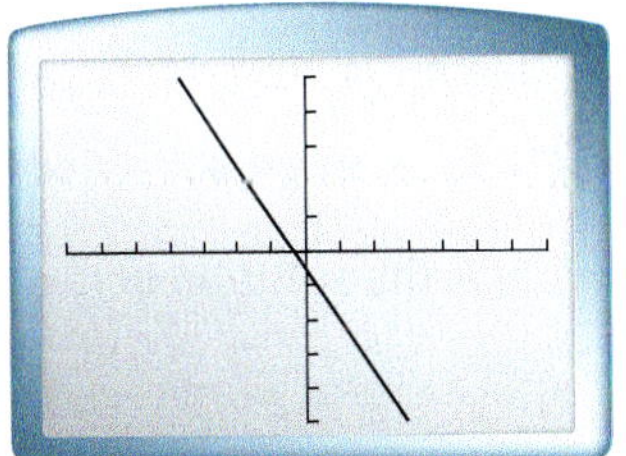

48. The equation $5x - 2y = -6$ and the table each represent linear functions. Which has the greater slope? Explain.

x	−2	−1	0	1	2
y	1	3	5	7	9

49. a. Write the point-slope form of the line that passes through $A(-3, 12)$ and $B(9, -4)$. Use point A in the equation.
b. Write the point-slope form of the same line using point B in the equation.
c. Rewrite each equation in standard form. What do you notice?

Write an equation for each line. Then graph the line.

50. $m = 0$, through $(5, -1)$
51. $m = -\frac{3}{2}$, through $(0, -1)$

52. **Reasoning** Suppose lines ℓ_1 and ℓ_2 intersect at the origin. Also, ℓ_1 has slope $\frac{y}{x}$ $(x > 0, y > 0)$ and ℓ_2 has slope $-\frac{x}{y}$. Then ℓ_1 contains (x, y) and ℓ_2 contains $(-y, x)$.
a. Explain why the two right triangles are congruent.
b. Complete each equation about the angle measures a, b, c, and d.

$a = \blacksquare$ $\qquad c = \blacksquare$

$a + c = \blacksquare$ $\qquad b + d = \blacksquare$

c. What must be true about $a + b$? Why?
d. What must be true about ℓ_1 and ℓ_2? Why?

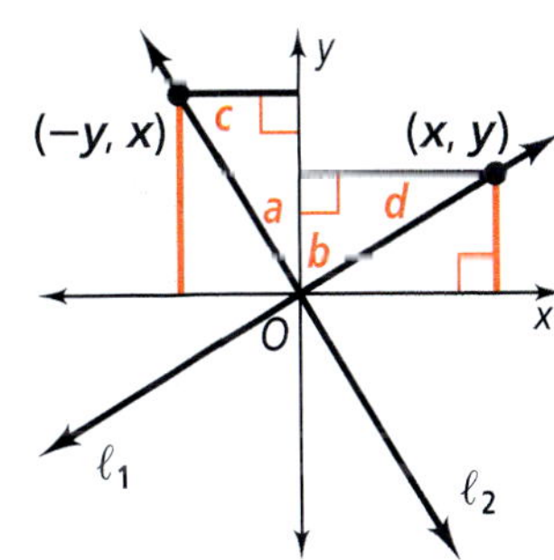

Challenge **Points that are on the same line are *collinear*. Use the definition of slope to determine whether the given points are collinear.**

53. $(-2, 6), (0, 2), (1, 0)$

54. $(3, -5), (-3, 3), (0, 2)$

55. **Geometry** Prove that the triangle with vertices $(3, 5)$, $(-2, 6)$, and $(1, 3)$ is a right triangle.

56. **Geometry** Prove that the quadrilateral with vertices $(2, 5)$, $(4, 8)$, $(7, 6)$, and $(5, 3)$ is a rectangle.

Standardized Test Prep

SAT/ACT

57. Which line is perpendicular to the line shown in the graph?

Ⓐ $-3x + y = 5$

Ⓑ $3x + y = -1$

Ⓒ $x - 3y = 0$

Ⓓ $x + 3y = -3$

58. Which line is parallel to the line $5x + 6y = 30$?

Ⓕ $5x + 9y = 30$

Ⓖ $9x + 6y = 30$

Ⓗ $5x + 6y = 9$

Ⓘ $6x + 5y = 30$

Short Response

59. What is the solution of the inequality $|x - 3| \geq 5$? Graph the solution.

Mixed Review

Find the domain and range of each relation and determine whether it is a function.

 **See Lesson 2-1.**

60. $\{(-3, 4), (-1, 2), (0, -2), (1, 0), (2, 2)\}$

61.

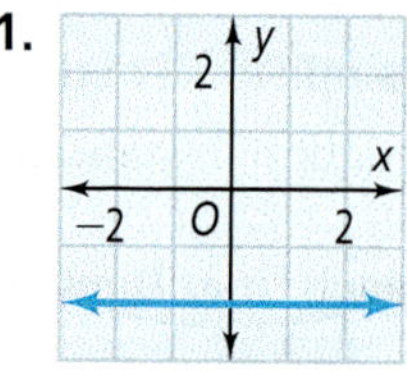

62.

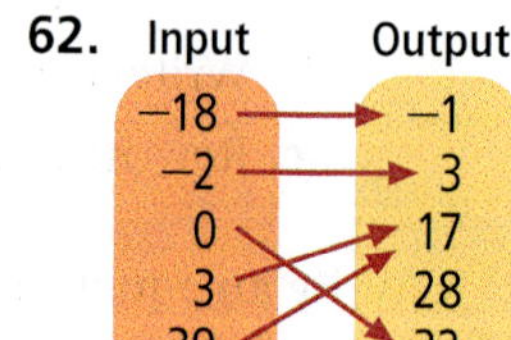

Name the property of real numbers illustrated by each equation.

See Lesson 1-2.

63. $\frac{2}{5} + \frac{27}{5} \cdot \frac{5}{27} = \frac{2}{5} + 1$

64. $97(7) = 100(7) - 3(7)$

65. $21 + 19.7 - 19.7 = 21$

Get Ready! **To prepare for Lesson 2-5, do Exercises 66–69.**

Write the equation for each line. Use slope-intercept form.

See Lesson 2-3.

66. $m = 3$ and the y-intercept is $(0, -5)$.

67. $m = \frac{1}{2}$ and the y-intercept is $(0, 0)$.

68. $m = \frac{4}{5}$ and the y-intercept is $(0, 7)$.

69. $m = -\frac{3}{8}$ and the y-intercept is $(0, 12)$.

2 Mid-Chapter Quiz

MathXL® for School
Go to PowerAlgebra.com

Do you know HOW?

Determine whether each relation is a function.

1.

x	y
3	7
4	2
3	2
5	1

2.

x	y
1	1
2	2
3	3
4	4

Find the x- and y-intercepts of each line.

3. $x - 3y = 9$

4. $y = 7x + 5$

5. $y = 6x$

6. $-4x + y = 10$

Write the equation of each line in slope-intercept form and identify the slope.

7. $2x - y = 9$

8. $4x = 2 + y$

9. $5y = -3x - 10$

10. $4x + 6y = 12$

Write an equation of each line in standard form with integer coefficients.

11. the line through $(2, 3)$ and $(4, 5)$

12. the line through $(-4, 6)$ and $(2, -2)$

13. the line through $(-4, 2)$ with slope 3

14. the line through $(1, 2)$ with slope $\frac{4}{5}$

15. a line through $(3, 1)$ with slope 0

16. a line with slope of $\frac{2}{3}$ and y-intercept $(0, 5)$

17. $2y = -4x - 12$

18. $\frac{2}{3}x + 3 = 6y - 15$

Write an equation of each line in point-slope form.

19. $(-4, 2)$ and $(-3, 5)$

20. $(0, 0)$ and $(-4, -5)$

21. $(-4, -3)$ and $(2, 7)$

Graph each equation.

22. $2y = 4x + 8$

23. $2x - 3y = 6$

24. $4y - x = 16$

For each function, determine whether y varies directly with x. If so, identify the constant of variation.

25. $2y = 3x$

26. $4y - 7x = 0$

27. $y + \frac{3}{4}x = 12$

Do you UNDERSTAND?

28. **a.** A group of friends is going to the movies. Each ticket costs \$8.00. Write an equation to model the total cost of the group's tickets.

b. Graph the equation. Explain what the x- and y-intercepts represent.

c. What would be the cost for 12 tickets?

d. **Writing** Could the domain include fractions? Explain.

29. Which line is perpendicular to $3x + 2y = 6$?

(A) $4x - 6y = 3$

(B) $y = -\frac{3}{2}x + 4$

(C) $2x + 3y = 12$

(D) $y = \frac{3}{2}x + 1$

30. **Reasoning** Why is the slope of a vertical line undefined?

31. Suppose $m = 25 - 0.15n$ describes the amount of money remaining on a \$25 phone card m, as a function of the number of minutes of calls you make n. What are a reasonable domain and range?

Concept Byte

For Use With Lesson 2-4

Piecewise Functions

Content Standard

F.IF.7.b Graph . . . piecewise-defined functions, including step functions and absolute value functions.

Recall from Lesson 1-6 that $|x|$, the absolute value of x, is the distance of x from zero. When $x \geq 0$, $|x| = x$. When $x < 0$, $|x| = -x$. The absolute value function is an example of a *piecewise function*. A **piecewise function** has different rules for different parts of its domain.

Example 1

Graph the absolute value function $f(x) = \begin{cases} x, & \text{for } x \geq 0 \\ -x, & \text{for } x < 0 \end{cases}$.

Graph each piece separately.

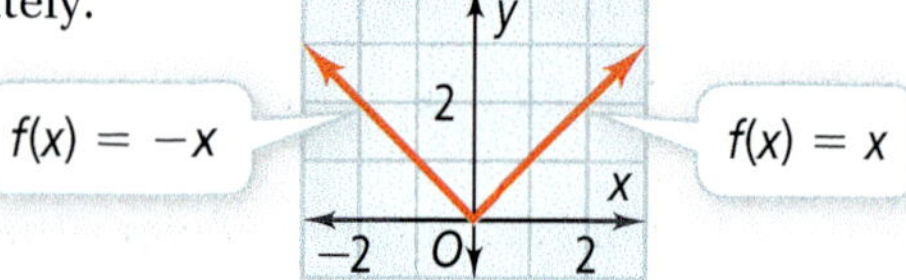

Example 2

Use a graphing calculator to graph the function $f(x) = \begin{cases} -2x + 3, \text{ if } x < 2 \\ x - 1, \text{ if } x \geq 2 \end{cases}$.

Define the function $f(x)$. Use the brackets from the math expression templates to enter the piecewise function.

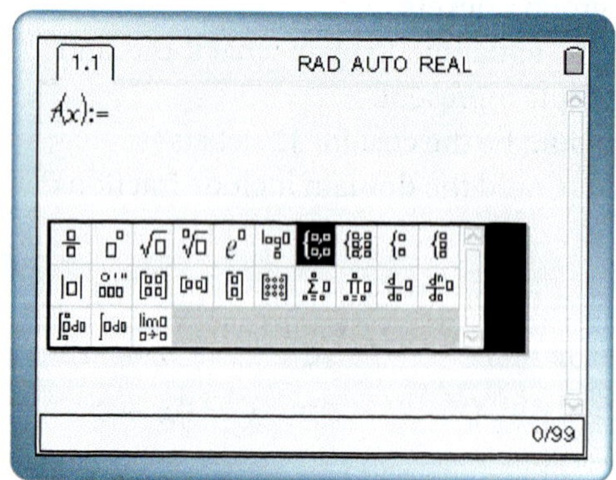

Enter the rules for the two branches. Set $f_1(x) = f(x)$ and graph.

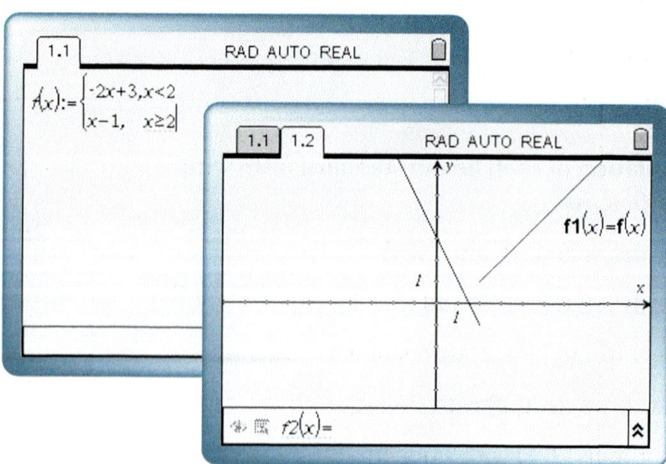

Step functions are piecewise functions. A **step function** pairs every number in an interval with a single value. The graph of a step function can look like the steps of a staircase. One step function is the **greatest integer function** $y = [x]$, where $[x]$ represents the greatest integer less than or equal to x.

Example 3

What is the graph of the function $f(x) = [x]$?

Each piece of the graph is a horizontal segment that is missing its right endpoint. The open circle indicates that the right endpoint is not part of the graph.

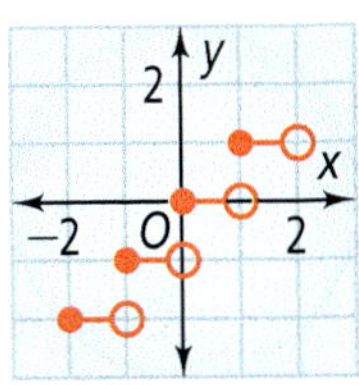

Given a piecewise function in one form, you can represent it in each of the other forms, including a table, graph, algebraic function, or verbal statement.

Example 4

Media Postage **You want to mail a book that weighs 2.5 lb. The table lists postage for a book weighing up to 3 lb. Define and graph the media-postage function. How much will you pay in postage?**

Media Postage

Weight (lb)	Price ($)
$x \leq 1$	2.23
$1 < x \leq 2$	2.58
$2 < x \leq 3$	2.93

$$f(x) = \begin{cases} 2.23, & \text{for } 0 < x \leq 1 \\ 2.58, & \text{for } 1 < x \leq 2 \\ 2.93, & \text{for } 2 < x \leq 3 \end{cases}$$

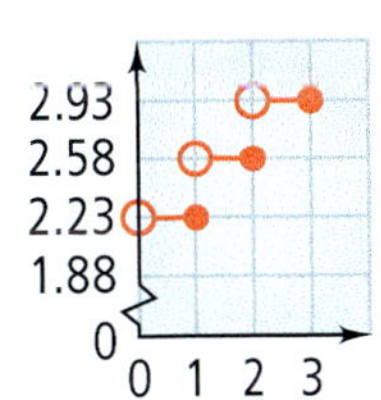

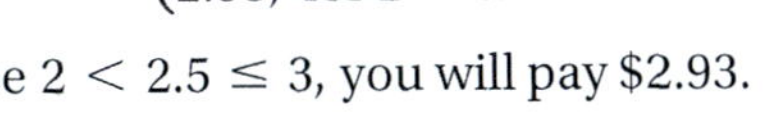

Since $2 < 2.5 \leq 3$, you will pay $2.93.

Example 5

What piecewise function represents the graph?

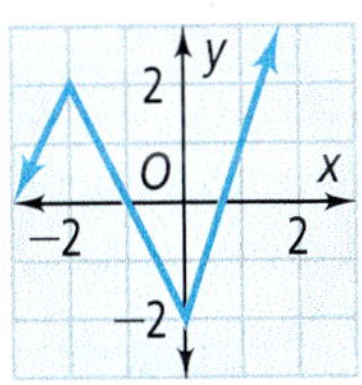

Piece 1 When $x \leq -2$, the rule is $f(x) = 2x + 6$.

Piece 2 When $-2 < x \leq 0$, the rule is $f(x) = -2x - 2$.

Piece 3 When $x > 0$, the rule is $f(x) = 3x - 2$.

$$f(x) = \begin{cases} 2x + 6, & \text{for } x \leq -2 \\ -2x - 2, & \text{for } -2 < x \leq 0 \\ 3x - 2, & \text{for } x > 0 \end{cases}$$

Exercises

1. Graph the sign function. $f(x) = \begin{cases} -1, & \text{for } x < 0 \\ 0, & \text{for } x = 0. \\ 1, & \text{for } x > 0 \end{cases}$
2. Graph the function $f(x) =$ the least integer greater than x.
3. What piecewise function represents the graph at the right?
4. **Postage** In 2008, first-class letter postage was $.42 for up to one ounce and $.17 for each additional ounce up to 3.5 oz. Graph this postage function.

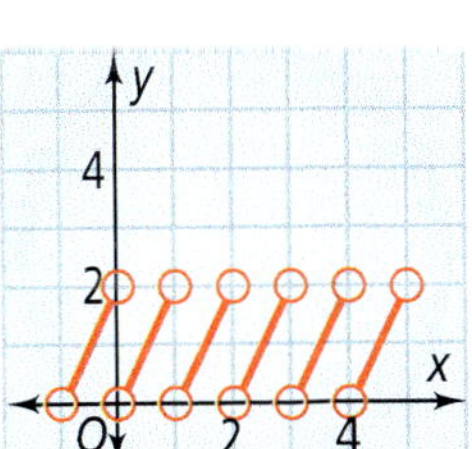

Using Linear Models

Content Standards

F.IF.4 For a function that models a relationship between two quantities, interpret key features of graphs . . . and sketch graphs showing key features . . .

Also A.CED.2, F.IF.6

Objectives To write linear equations that model real-world data
To make predictions from linear models

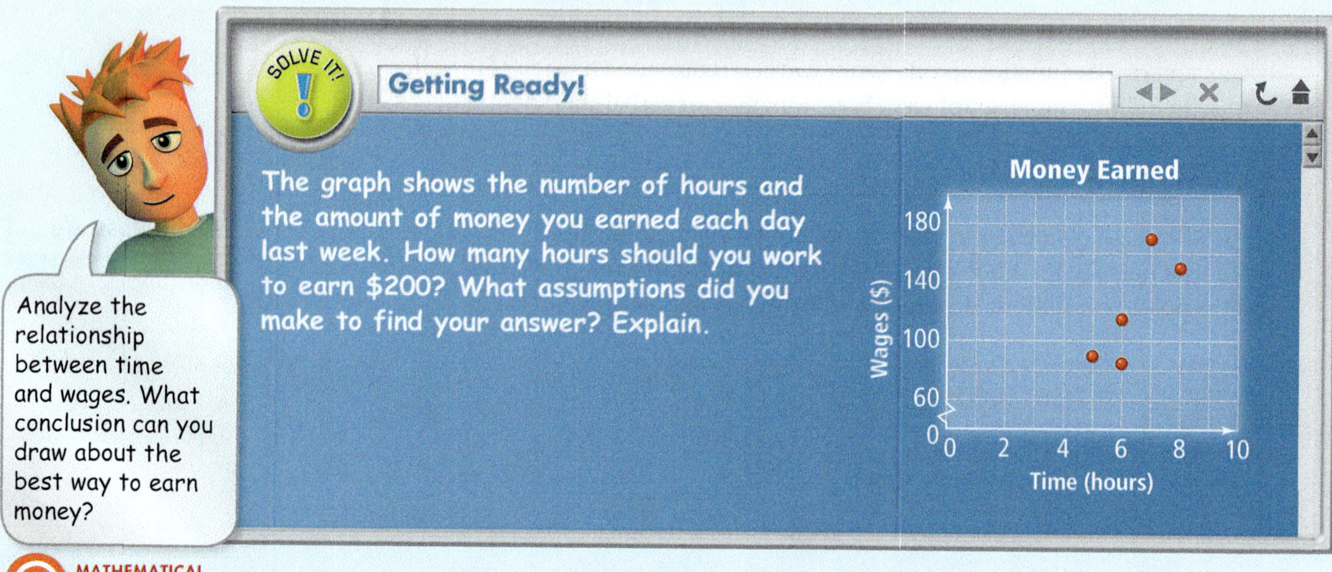

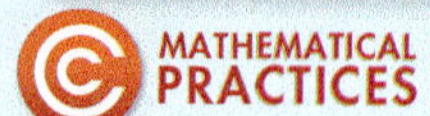

Graphs of data pairs for a real-world situation rarely fall in a line. Their arrangement, however, can suggest a relationship that you can model with a linear function.

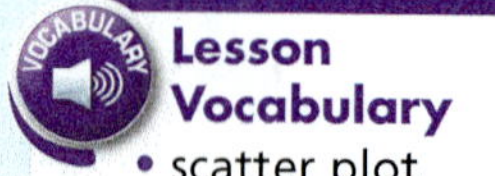

Lesson Vocabulary
- scatter plot
- correlation
- line of best fit
- correlation coefficient

Essential Understanding Sometimes it is possible to model data from a real-world situation with a linear equation. You can then use the equation to draw conclusions about the situation.

A **scatter plot** is a graph that relates two sets of data by plotting the data as ordered pairs. You can use a scatter plot to determine the strength of the relationship, or **correlation**, between data sets. The closer the data points fall along a line,

- the stronger the relationship and
- the stronger the positive or negative correlation

between the two variables.

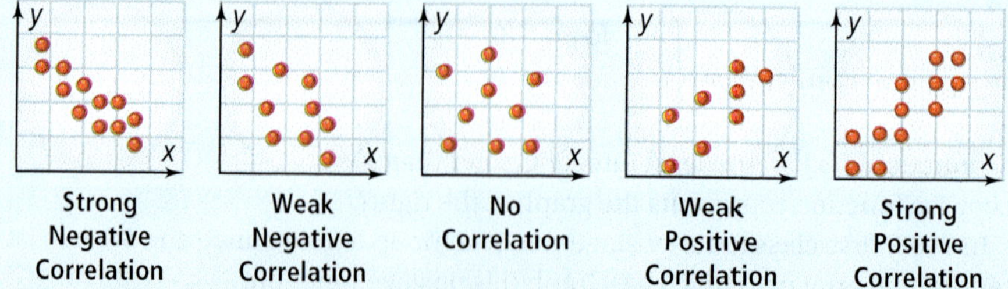

Problem 1 Using a Scatter Plot

Utilities The table lists average monthly temperatures and electricity costs for a Texas home in 2008. The table displays the values rounded to the nearest whole number. Make a scatter plot. How would you describe the correlation?

Average Temperatures and Electricity Costs

Month	Average Temp. (°F)	Electricity Bill ($)	Month	Average Temp. (°F)	Electricity Bill ($)
January	61	150	July	84	255
February	58	139	August	85	245
March	67	172	September	81	210
April	75	205	October	76	183
May	79	170	November	65	132
June	83	234	December	58	110

Plan

Which variable is the independent variable?
Temperature does not depend on the electric bill, so temperature is the independent variable.

Step 1 Make a scatter plot.

Step 2 Describe the correlation.

As the temperature increases, the electricity cost also increases. The points are relatively tightly clustered around a line. There is a strong positive correlation between temperature and electricity cost.

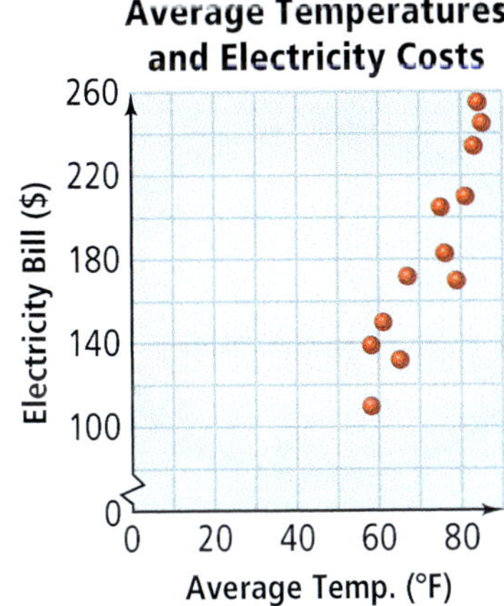

Got It? **1. a.** The table shows the numbers of hours students spent online the day before a test and the scores on the test. Make a scatter plot. How would you describe the correlation?

b. Reasoning Using the graph from Problem 1, how much would you expect to pay for electricity if the average temperature was 70°F? Explain.

Computer Use and Test Scores

Number of Hours Online	0	0	1	1	1.5	1.75	2	2	3	4	4.5	5
Test score	100	94	98	88	92	89	75	70	78	72	57	60

A *trend line* is a line that approximates the relationship between the variables, or data sets, of a scatter plot. You can use a trend line to make predictions from the data. In a previous lesson you learned how to use two points to write the equation of a line to model a real-world problem. You can use this method to write the equation of a trend line.

Problem 2 Writing the Equation of a Trend Line

Finance **The table shows the median home prices in Florida. What is the equation of a trend line that models a relationship between time and home prices? Use the equation to predict the median home price in 2020.**

Florida Median Home Prices

Year	1940	1950	1960	1970	1980	1990	2000
Median Price ($)	23,100	40,100	58,100	57,600	89,300	98,500	105,500

Know

Seven data points

Need

The equation of a trend line

Plan

- Plot the given points.
- Find a trend line.
- Write the equation.

Step 1 Make a scatter plot. Let $x = 0$ correspond to 1940.

Step 2 Sketch a trend line.

Step 3 Choose two points on the trend line, (10, 40,000) and (55, 110,000). Use slope-intercept form to write an equation for the line.

$y = 1556x + 24{,}440$

Step 4 Use the equation to predict the median home price in 2020.

$$y = 1556(80) + 24{,}440 = 148{,}920$$

Based on the trend line, the median home price in 2020 will be around $149,000.

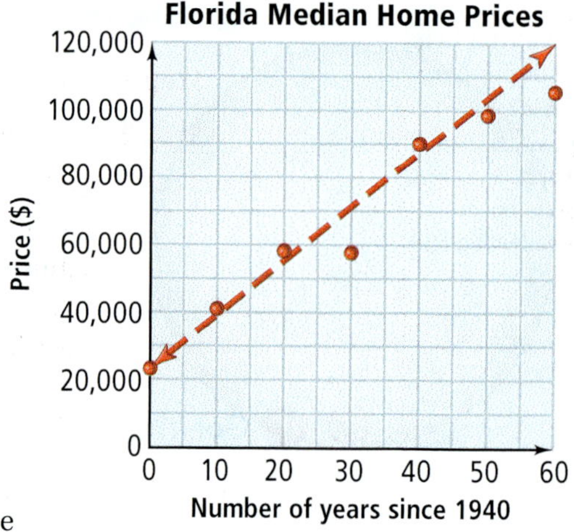

Got It? 2. The table shows median home prices in California. What is an equation for a trend line that models the relationship between time and home prices?

California Median Home Prices

Year	1940	1950	1960	1970	1980	1990	2000
Median Price ($)	36,700	57,900	74,400	88,700	167,300	249,800	211,500

The trend line that gives the most accurate model of related data is the **line of best fit**. One method for finding a line of best fit is *linear regression*. You can use the **LinReg** function on your graphing calculator to find the line of best fit. The **correlation coefficient**, r, indicates the strength of the correlation. The closer r is to 1 or -1, the more closely the data resembles a line and the more accurate your model is likely to be.

Problem 3 Finding the Line of Best Fit

Food You research the average cost of whole milk for several recent years to look for trends. The table shows your data.

Cost of Whole Milk

Year	1998	2000	2002	2004	2006	2008
Average cost for one gallon ($)	2.65	2.89	3.00	3.01	3.20	3.77

SOURCE: U.S. Department of Agriculture

A **What is the equation for the line of best fit? How accurate is your line of best fit?**

Step 1
Use the **STAT** feature to enter the data in your graphing calculator. Enter the x-values (year) in **L1** and the y-values (price) in **L2**. Let 1997 = year 0.

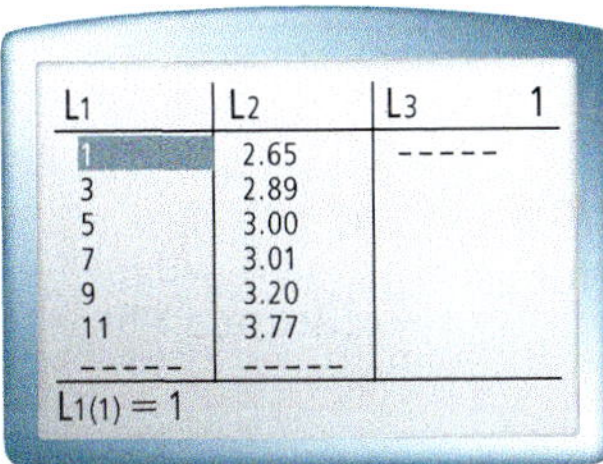

Step 2
Use **LinReg** to find the linear regression line of best fit for the data.

$y = 0.09x + 2.53$

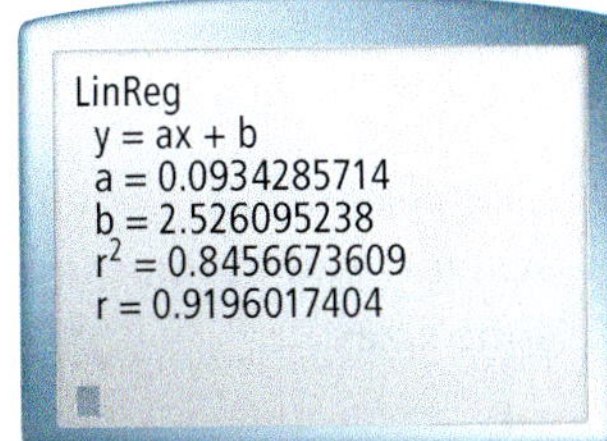

The correlation coefficient, r, is approximately 0.92. Since r is close to 1, the line of best fit is quite accurate.

Think

What factors could affect the accuracy of your prediction?
Predictions based on strongly correlated data are likely to be more reliable than predictions based on weakly correlated data.

B **Based on your linear model, how much would you expect to pay for a gallon of whole milk in 2020?**

$y = 0.09x + 2.53$ Use the line of best fit.

$y = 0.09(23) + 2.53$ Substitute 23 for x.

$y = 4.60$

In 2020, you would expect to pay about $4.60 for a gallon of whole milk.

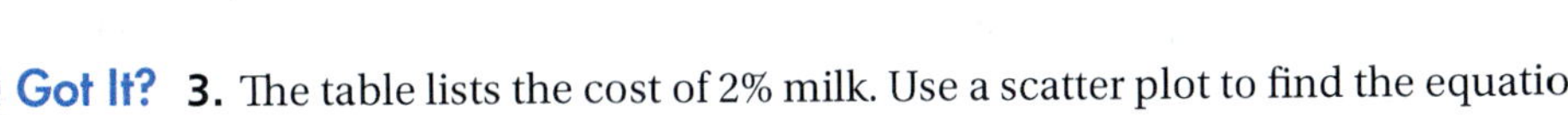

Got It? **3.** The table lists the cost of 2% milk. Use a scatter plot to find the equation of the line of best fit. Based on your linear model, how much would you expect to pay for a gallon of 2% milk in 2025?

Cost of 2% Milk

Year	1998	2000	2002	2004	2006	2008
Average cost for one gallon ($)	2.57	2.83	2.93	2.93	3.10	3.71

SOURCE: U.S. Department of Agriculture

Lesson Check

Do you know HOW?

Make a scatter plot of each set of points and describe the correlation.

1. $\{(1.2, 1), (2.5, 6), (2.5, 7.5), (4.1, 11), (7.9, 19)\}$
2. $\{(1, 55), (2, 38), (3, 54), (4, 37), (5, 53), (6, 40), (7, 53), (8, 36)\}$
3. Make a scatter plot for the following set of points. Describe the correlation and sketch a trend line. $\{(2, 58), (6, 105), (8, 88), (8, 118), (12, 117), (16, 137), (20, 157), (20, 169)\}$

Do you UNDERSTAND?

4. **Writing** How can you determine whether two variables x and y for a real-life situation are correlated?
5. Do you think a trend line on a graph is always the same as the line of best fit? Why or why not?
6. **Compare and Contrast** What is the difference between a positive correlation and a negative correlation? How might you relate positive correlation with direct variation?

Practice and Problem-Solving Exercises

Make a scatter plot and describe the correlation.

See Problem 1.

7. $\{(0, 11), (2, 8), (3, 7), (7, 2), (8, 0)\}$
8. **Manufacturing** The table shows the numbering system used in Europe and the United States for shoe sizes.

Shoe Sizes						
U.S. Size	1	3	5	7	9	11
European Size	31	34	36	39	41	44

Write the equation of a trend line.

See Problem 2.

9. $\{(-10, 3), (-5, 1), (-1, -4), (3, -7), (12, -12)\}$
10. $\{(-15, 8), (-8, 7), (-3, 0), (0, 0), (7, -3)\}$
11. The table shows the number of hours you studied before your eight math tests and your percent score on each test.

Studying Hours and Test Score								
Number of Hours	8	5	12	10	2	9	11	14
Score (%)	75	62	80	85	35	70	82	95

See Problem 3.

12. a. **Food Production** The table below shows pork production in China from 2000 to 2007. Use a calculator to find the line of best fit.
    b. Use your linear model to predict how many metric tons of pork will be produced in 2025.
    c. Use your linear model to predict when production is likely to reach 100,000 metric tons.

Pork Production in China								
Year	2000	2001	2002	2003	2004	2005	2006	2007
Production (metric tons)	40,475	42,010	43,413	45,331	47,177	50,254	52,407	54,491

SOURCE: USDA Foreign Agricultural Service GAIN Report

13. **Think About a Plan** The table shows the relationship between the production and the export of rice in Vietnam from 1985 to 2005.

Rice Production and Export

Production (1000 tonnes)	15,875	19,225	24,964	32,554	35,600
Export (1000 tonnes)	59	1624	1988	3400	5100

SOURCE: International Rice Research Institute

How much rice would you expect Vietnam to export in 2015 if the production that year is 42,250,000 tonnes?

- How can you use a scatter plot to find a linear model?
- How can you use your model to make a prediction?

14. **Nutrition** The table shows the relationship between Calories and fat in various fast-food hamburgers.

Fast Food Calories

Restaurant	A	B	C	D	E	F	G	H	I
Number of Calories	720	530	510	500	305	410	440	320	598
Grams of fat	46	30	27	26	13	20	25	13	26

a. Find the line of best fit for the relationship between Calories and fat.
b. How much fat would you expect a 330-Calorie hamburger to have?
c. **Error Analysis** Which estimate is *not* reasonable: 10 g of fat for a 200-Calorie hamburger or 36 g of fat for a 660-Calorie hamburger? Explain.

Reasoning For any correlation, people often assume that change in one quantity *causes* change in the second quantity. This is not always true. For each situation, do you think that change in the first quantity causes change in the second quantity? What else may have affected the change in the second quantity?

15. number of miles driven and fuel expenses

16. the size of a car's engine and the number of passengers it is designed for

17. a person's age and the number of cassette tapes he or she owns

18. **Data Analysis** The table shows population and licensed driver statistics from a recent year.
a. Make a scatter plot.
b. Draw a trend line.
c. The population of Michigan was approximately 10 million that year. About how many licensed drivers lived in Michigan that year?
d. **Writing** Is the correlation between population and number of licensed drivers strong or weak? Explain.

Licensed Drivers

State	Population (millions)	Number of Drivers (millions)
Arkansas	2.8	2.0
Illinois	12.8	8.1
Kansas	2.8	2.0
Massachusetts	6.4	4.7
Pennsylvania	12.4	8.5
Texas	23.5	14.9

19. **Social Studies** The table shows per capita revenues and expenditures for selected states for a recent year.
   a. Show the data on a scatter plot. Draw a trend line.
   b. If a state collected revenue of $3000 per capita, how much would you expect it to spend per capita?
   c. Ohio spent $5142 per capita during that year. According to your model, how much did it collect in taxes per capita?
   d. In that same year, New Jersey collected $5825 per capita in taxes and spent $5348 per capita. Does this information follow the trend? Explain.

Per Capita Revenue and Expenditure

State	Per Capita Revenue ($)	Per Capita Expenditure ($)
Arizona	4144	3789
Georgia	3904	3834
Maryland	5109	4557
Mississippi	5292	4871
New Mexico	6205	5793
Nevada	4345	3723
New York	7081	6891
Texas	4030	3442
Utah	5439	4459

Standardized Test Prep

SAT/ACT

20. What is the equation of the line shown in the graph?

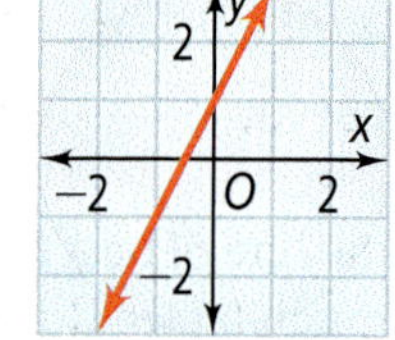

   (A) $y = -2x + 2$ (C) $y = 2x + 1$
   (B) $y = 2x$ (D) $y = 2x + 2$

21. Shauna drove 75 miles in 3 hours at a constant speed. How many miles did she drive in 2 hours?

   (F) 25 miles (G) 50 miles (H) 75 miles (I) 100 miles

22. Which equation does NOT represent a direct variation?

   (A) $y = x$ (C) $2x - y = 0$
   (B) $2x - y = 5$ (D) $2x - 5y = 0$

Short Response

23. The line $(y - 1) = \frac{2}{3}(x + 1)$ contains point $(a, -3)$. What is the value of a? Show your work.

Mixed Review

Graph the following linear equations. See Lesson 2-3.

24. $y = -7.5x + 11$ 25. $-\frac{2}{9}x - \frac{5}{9}y = 10$ 26. $5x - 4y = 3$

Write the equation of each line in standard form. See Lesson 2-4.

27. slope $= 2$; $(2, 6)$ 28. slope $= -1$; $(-3, 3)$ 29. slope $= 0$; $(0, 2)$

Get Ready! **To prepare for Lesson 2-6, do Exercises 30–32.**

Graph each pair of functions on the same coordinate plane. See Lesson 2-3.

30. $y = -x; y = x$ 31. $y = x + 1; y = 2x - 1$ 32. $y = -\frac{1}{4}x; y = -\frac{1}{4}x + 2$

2-6 Families of Functions

Content Standard

F.BF.3 Identify the effect on the graph of replacing $f(x)$ by $f(x) + k$, $k\,f(x)$, $f(kx)$, and $f(x + k)$ for specific values of k (both positive and negative) find the value of k given the graphs.

Objective To analyze transformations of functions

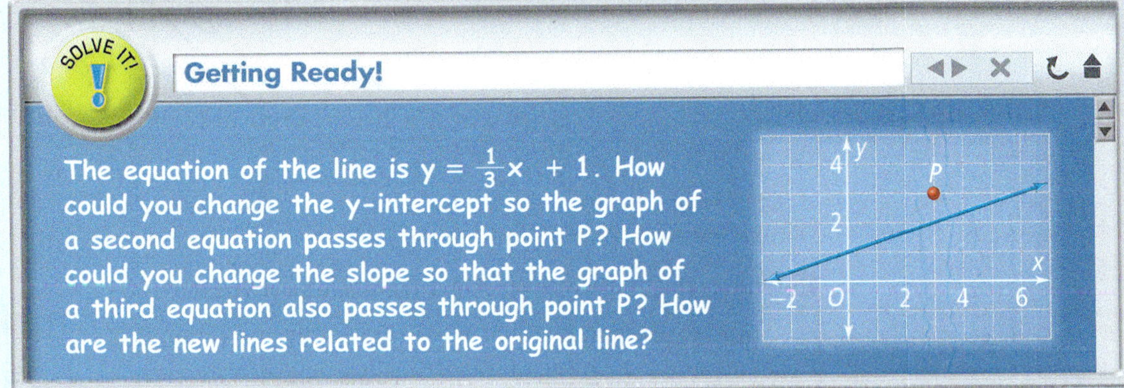

Dynamic Activity
Translating Functions

Lesson Vocabulary
- parent function
- transformation
- translation
- reflection
- vertical stretch
- vertical compression

Different nonvertical lines have different slopes, or y-intercepts, or both. They are graphs of different linear functions. For two such lines, you can think of one as a *transformation* of the other.

Essential Understanding There are sets of functions, called *families*, in which each function is a transformation of a special function called the *parent*.

The linear functions form a family of functions. Each linear function is a transformation of the function $y = x$. The function $y = x$ is the *parent* linear function.

A **parent function** is the simplest form in a set of functions that form a family. Each function in the family is a **transformation** of the parent function.

One type of transformation is a **translation**. A translation shifts the graph of the parent function horizontally, vertically, or both without changing shape or orientation. For a positive constant k and a parent function $f(x)$, $f(x) \pm k$ is a vertical translation. For a positive constant h, $f(x \pm h)$ is a horizontal translation.

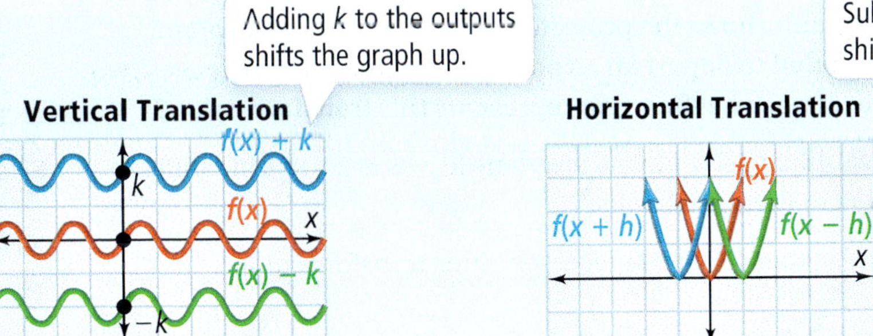

Plan

What is one way to compare two functions?
Use a table to compare their values.

Problem 1 Vertical Translation

A How are the functions $y = x$ and $y = x - 2$ related? How are their graphs related?

Make a table of values.

x	$y = x$	$y = x - 2$
−2	−2	−4
−1	−1	−3
0	0	−2
1	1	−1
2	2	0

Each output for $y = x - 2$ is two less than the corresponding output for $y = x$.

Draw their graphs.

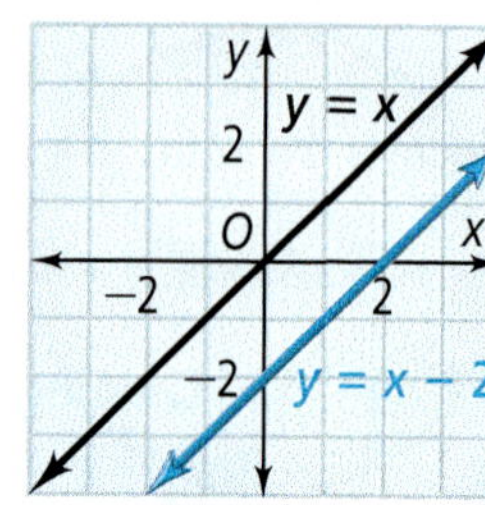

The graph of $y = x - 2$ is the graph of $y = x$ translated down two units.

B What is the graph of $y = x^2 - 1$ translated up 5 units?

Translate the graph of $y = x^2 - 1$ up 5 units to get the blue parabola. The equation of the blue parabola is $y = x^2 + 4$.

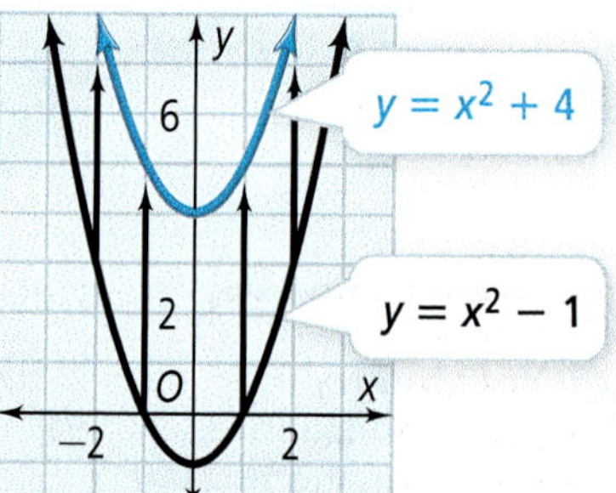

Check Every value in the $y = x^2 + 4$ column is 5 greater than the corresponding value in the $y = x^2 - 1$ column.

x	$y = x^2 - 1$	$y = x^2 + 4$
−2	3	8
−1	0	5
0	−1	4
1	0	5
2	3	8

Got It? **1. a.** How are the functions $y = 2x$ and $y = 2x - 3$ related? How are their graphs related?

b. What is the graph of $y = 3x$ translated up 2 units?

Problem 2 Horizontal Translation

Think

If the time of the flight is later, why do you subtract rather than add in $f(x - 2)$?
The y-values of the delayed flight correspond to x-values 2 hours *earlier* than the delayed flight.

The graph shows the projected altitude $f(x)$ of an airplane scheduled to depart an airport at noon. If the plane leaves two hours late, what function represents this transformation?

A two-hour delay means the plane leaves at 2 P.M. This shifts the graph to the right 2 units.

The function $f(x - 2)$ represents this transformation.

Airplane Altitude

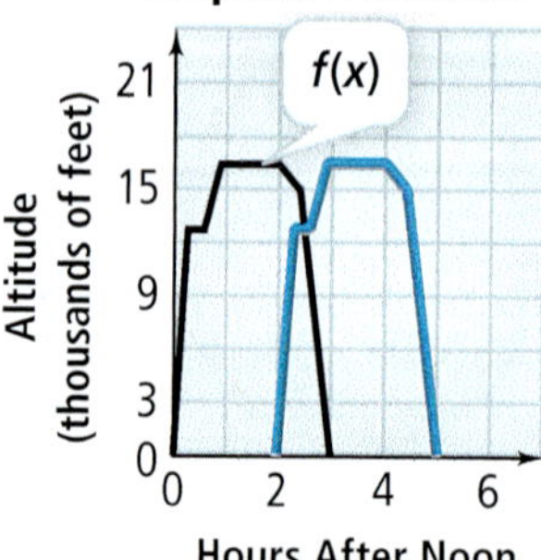

Got It? **2.** Suppose the flight leaves 30 minutes early. What function represents this transformation?

A **reflection** flips the graph of a function across a line, such as the x- or y-axis. Each point on the graph of the reflected function is the same distance from the line of reflection as its corresponding point on the graph of the original function.

When you reflect a graph in the y-axis, the x-values change signs and the y-values stay the same.

When you reflect a graph in the x-axis, the x-values stay the same and the y-values change signs.

For a function $f(x)$, the reflection in the y-axis is $f(-x)$ and the reflection in the x-axis is $-f(x)$.

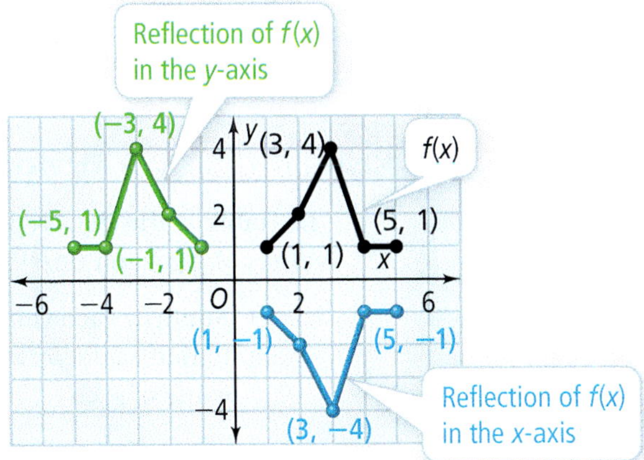

Problem 3 Reflecting a Function Algebraically

Let $g(x)$ be the reflection of $f(x) = 3x + 3$ in the y-axis. What is a function rule for $g(x)$?

Think	Write
For a reflection in the y-axis, change the sign of x.	$g(x) = f(-x)$
Evaluate $f(-x)$ and simplify.	$g(x) = f(-x)$ $= 3(-x) + 3$ $g(x) = -3x + 3$
You can check by graphing $f(x)$ and $g(x)$.	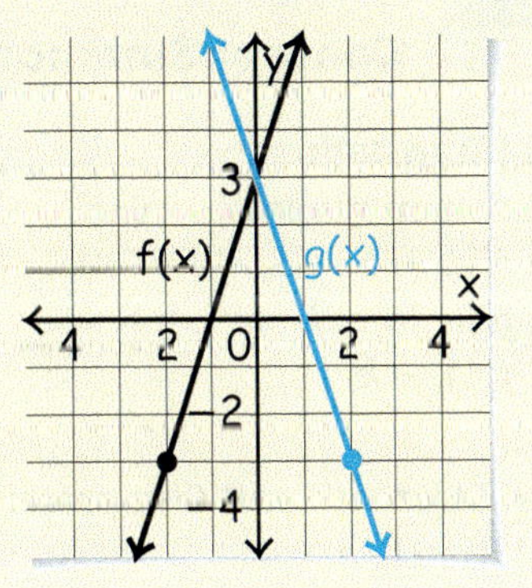

Got It? **3.** Let $h(x)$ be the reflection of $f(x) = 3x + 3$ in the x-axis. What is a function rule for $h(x)$?

A **vertical stretch** multiplies all y-values of a function by the same factor greater than 1. A **vertical compression** reduces all y-values of a function by the same factor between 0 and 1. For a function $f(x)$ and a constant a, $y = af(x)$ is a vertical stretch when $a > 1$ and a vertical compression when $0 < a < 1$.

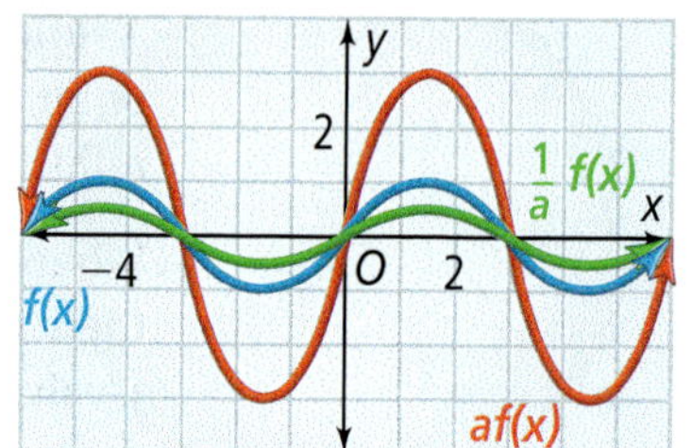

Problem 4 Stretching and Compressing a Function

Think

Is this a vertical stretch or compression?
3 is greater than 1, so this is a vertical stretch.

The table at the right represents the function $f(x)$. What are corresponding values of $g(x)$ and possible graphs for the transformation $g(x) = 3f(x)$?

x	$f(x)$
−5	2
−2	2
0	−3
3	1
5	−2

Step 1 Multiply each value of $f(x)$ by 3 to find each corresponding value of $g(x)$.

x	$f(x)$	$3f(x)$	$g(x)$
−5	2	3(2)	6
−2	2	3(2)	6
0	−3	3(−3)	−9
3	1	3(1)	3
5	−2	3(−2)	−6

Step 2 Use the values from the table in Step 1. Draw simple graphs for $f(x)$ and $g(x)$.

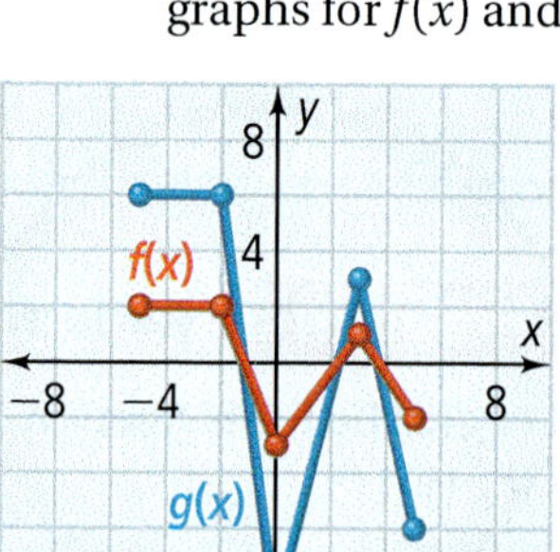

Got It? 4. a. For the function $f(x)$ shown in Problem 4, what are the corresponding table and graph for the transformation $h(x) = \frac{1}{3}f(x)$?

b. Reasoning If several transformations are applied to a graph, will changing the order of transformations change the resulting graph? Explain.

take note

Concept Summary Transformations of $f(x)$

Vertical Translations

Translation up k units, $k > 0$

$y = f(x) + k$

Translation down k units, $k > 0$

$y = f(x) - k$

Horizontal Translations

Translation right h units, $h > 0$

$y = f(x - h)$

Translation left h units, $h > 0$

$y = f(x + h)$

Vertical Stretches and Compressions

Vertical stretch, $a > 1$

$y = af(x)$

Vertical compression, $0 < a < 1$

$y = af(x)$

Reflections

In the x-axis

$y = -f(x)$

In the y-axis

$y = f(-x)$

Problem 5 Combining Transformations

A The graph of $g(x)$ is the graph of $f(x) = 4x$ compressed vertically by the factor $\frac{1}{2}$ and then reflected in the y-axis. What is a function rule for $g(x)$?

$\frac{1}{2}(4x) = 2x$ Compress $f(x)$.

$2(-x) = -2x$ Reflect the new function in the y-axis.

The function rule is $g(x) = -2x$.

B What transformations change the graph of $f(x)$ to the graph of $g(x)$?

Think

How can you write $g(x)$ in terms of $f(x)$? Factor out 3 from the $6x^2$ term.

$f(x) = 2x^2 \quad g(x) = 6x^2 - 1$

$$g(x) = 6x^2 - 1$$
$$= 3(2x^2) - 1$$
$$= 3(f(x)) - 1$$

The graph of $g(x)$ is the graph of $f(x)$ stretched vertically by a factor of 3 and then translated down 1 unit.

Got It? 5. a. The graph of $g(x)$ is the graph of $f(x) = x$ stretched vertically by a factor of 2 and then translated down 3 units. What is the function rule for $g(x)$?

b. What transformations change the graph of $f(x) = x^2$ to the graph of $g(x) = (x + 4)^2 - 2$?

Lesson Check

Do you know HOW?

Describe the transformation of the parent function $f(x)$.

1. $g(x) = f(x) + 6$

2. $h(x) = 0.25f(x)$

3. $j(x) = f(x - 4)$

4. $k(x) = f(-x)$

The graph of $f(x) = -2x$ is shown. Describe and graph each transformation.

5. $g(x) = f(x + 1) - 2$

6. $h(x) = 2f(x) + 1$

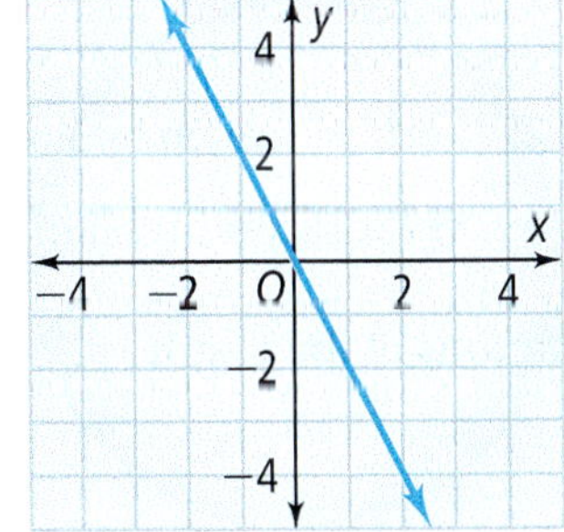

Do you UNDERSTAND?

7. Compare and Contrast The graph shows $f(x) = 0.5x - 1$.

Graph $g(x)$ by translating $f(x)$ up 2 units and then stretching it vertically by the factor 2. Graph $h(x)$ by stretching $f(x)$ vertically by the factor 2 and then translating it up 2 units. Compare the graphs of $g(x)$ and $h(x)$.

8. Reasoning Can you give an example of a function for which a horizontal translation gives the same resulting graph as a vertical translation? Explain.

9. Find a new function $g(x)$ transformed from $f(x) = -x - 2$ such that $g(x)$ is perpendicular to $f(x)$.

Practice and Problem-Solving Exercises

How is each function related to $y = x$? Graph the function by translating the parent function.

See Problem 1.

10. $y = x - 3$

11. $y = x + 4.5$

12. $y = x + 1.5$

Make a table of values for $f(x)$ after the given translation.

13. 3 units up

x	f(x)
−2	3
0	1
1	−2
3	−1

14. 1 unit down

x	f(x)
−1	1
0	0
2	−4
3	2

15. 4 units up

x	f(x)
−3	1
−1	−2
1	0
4	3

Write an equation for each vertical translation of $y = f(x)$.

16. $\frac{2}{3}$ unit down

17. 4 units up

18. 2 units up

For each function, identify the horizontal translation of the parent function, $f(x) = x^2$. Then graph the function.

See Problems 2 and 3.

19. $y = (x - 4)^2$

20. $y = (x + 1)^2$

21. $y = (x + 3)^2$

22. The graph of the function $f(x)$ is shown at the right.
- **a.** Make a table of values for $f(x)$ and $f(x + 3)$.
- **b.** Graph $f(x)$ and $f(x + 3)$ on the same coordinate grid.

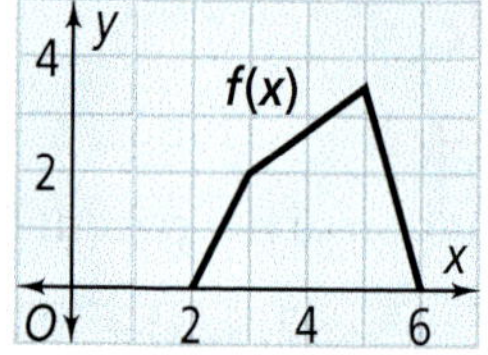

Write the function rule for each function reflected in the given axis.

23. $f(x) = x + 1$; x-axis

24. $f(x) = 3x$; y-axis

25. $f(x) = 2x - 4$; x-axis

Write an equation for each transformation of $y = x$.

See Problem 4.

26. vertical stretch by a factor of 4

27. vertical stretch by a factor of 2

28. vertical compression by a factor of $\frac{1}{2}$

29. vertical compression by a factor of $\frac{1}{4}$

Write the function rule $g(x)$ after the given transformations of the graph of $f(x) = 4x$.

See Problem 5.

30. translation up 5 units; reflection in the x-axis

31. reflection in the y-axis; vertical compression by a factor of $\frac{1}{8}$

Describe the transformations of $f(x)$ that produce $g(x)$.

32. $f(x) = \frac{x}{2}$; $g(x) = -2x + 4$

33. $f(x) = 3x$; $g(x) = \left(\frac{3x}{4} - 2\right)$

34. **Think About a Plan** Suppose you are playing with a yo-yo during a school talent show. The string is 3 ft long and you hold your hand 4 ft above the stage. The stage is 3.5 ft above the floor of the auditorium. Make a graph of the yo-yo's distance from the auditorium floor with respect to time during the show.
- How could you graph the position of the yo-yo with respect to the stage, if you let time $t = 0$ when you start your routine?
- How could you transform this graph to show the position with respect to the auditorium floor?

35. If someone started to take a video of your yo-yo routine when you were introduced, 10 seconds before you actually started, what transformation would you have to make to your graph to match their video?

Write the equations for $f(x)$ and $g(x)$. Then identify the reflection that transforms the graph of $f(x)$ to the graph of $g(x)$.

36. 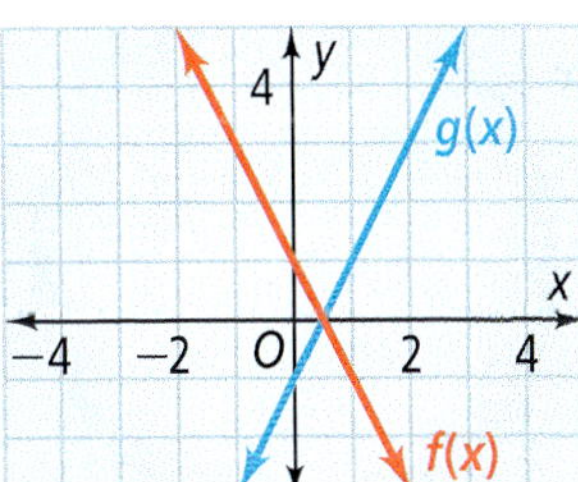

37. 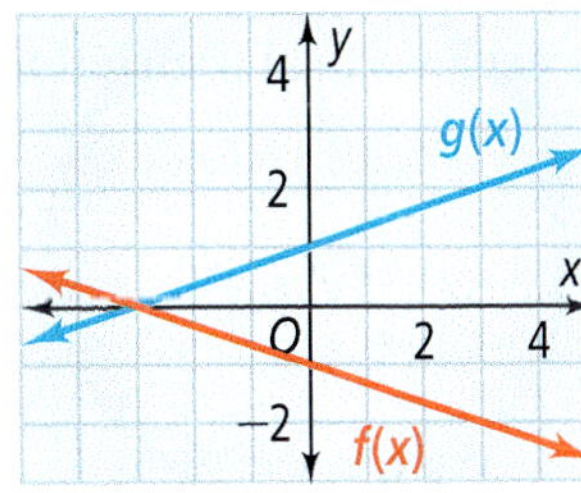

38. 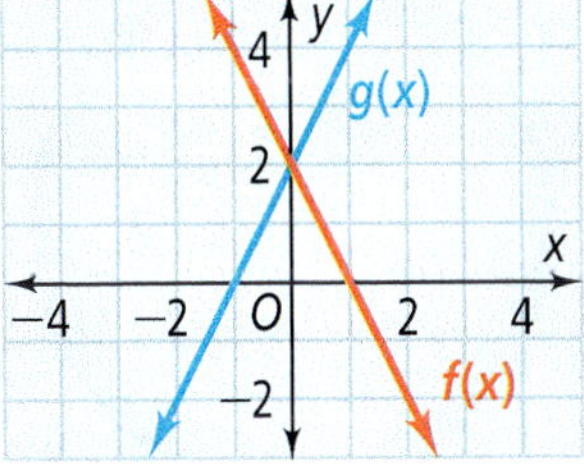

39. **Open-Ended** Draw a figure in Quadrant I. Use a translation to move your figure into Quadrant III. Describe your translation.

40. **Writing** The graph of $f(x)$ is shown at the right. Suppose each transformation of $f(x)$ results in the given functions.

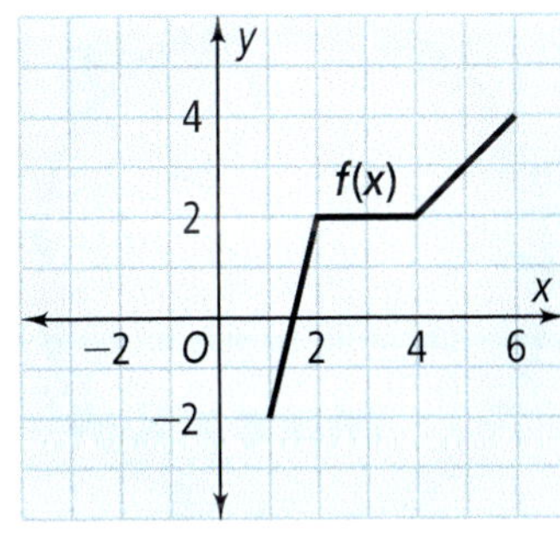

i. vertical translation; $g(x)$
ii. reflection in the x-axis; $h(x)$
iii. vertical stretch; $k(x)$,
iv. horizontal translation; $m(x)$

a. Describe how the domain and range of the four new functions compare with the domain and range of $f(x)$.

b. **Reasoning** Do you think these effects on the domain and range of the original function hold true for all functions? Explain.

Graph each pair of functions on the same coordinate plane. Describe a transformation that changes $f(x)$ to $g(x)$.

41. $f(x) = x + 1; g(x) = x - 5$

42. $f(x) = -x + 3; g(x) = x - 4$

43. $f(x) = x - 3; g(x) = x + 1$

44. $f(x) = -x - 1; g(x) = -x + 2$

45. **Error Analysis** Your friend wrote the transformations shown to describe how to change the graph of $f(x) = x^2$ to the graph of $g(x) = 2(x + 1)^2 - 3$. Explain the error and give the correct transformations.

· shift vertically 1 unit up
· shift horizontally 2 units right
· shift vertically 3 units down

Challenge Using the graph of the function $f(x)$ shown, sketch the graph of each transformed function.

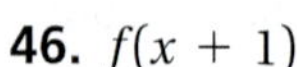

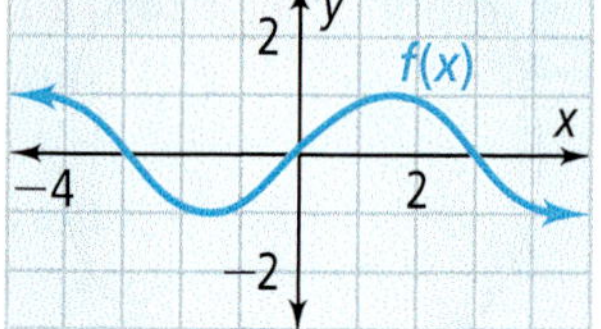

46. $f(x + 1)$

47. $f(x) - 2$

48. $f(x + 2) + 1$

49. $-2f(x)$

50. Graph all of the following functions in the same viewing window. After you enter each new function, view its graph.

i. $y = x^2$ ii. $y = x^2 + 2$ iii. $y = (x + 2)^2$ iv. $y = (x - 1)^2 - 4$ v. $y = -x^2 - 2$

Based on your results, make a sketch of the graph of $f(x) = -(x + 2)^2 + 1$ and check your prediction on your calculator.

Standardized Test Prep

SAT/ACT

What is an equation for each vertical translation of $y = 2x - 1$?

51. 3 units down

Ⓐ $y = 2x - 7$ Ⓑ $y = 2x + 2$ Ⓒ $y = 2x + 5$ Ⓓ $y = 2x - 4$

52. $\frac{3}{5}$ units up

Ⓕ $y = 2x - \frac{2}{5}$ Ⓖ $y = 2x - \frac{11}{5}$ Ⓗ $y = 2x - \frac{8}{5}$ Ⓘ $y = 2x + \frac{1}{5}$

53. What is the slope of the line in the graph at the right?

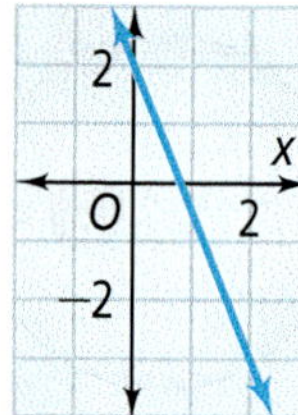

Ⓐ $-\frac{5}{2}$ Ⓒ $\frac{2}{5}$

Ⓑ $-\frac{2}{5}$ Ⓓ $\frac{5}{2}$

Short Response

54. The weight of a gold bar varies directly with its volume. If a 40 cm^3 bar weighs 772 grams, how much will a 100 cm^3 bar weigh?

Mixed Review

55. A musician's manager keeps track of the ticket prices and the attendance at recent performances. Use a graphing calculator to determine the equation of the line of best fit for the given data. See Lesson 2-5.

Ticket Prices($)	41.00	41.50	42.00	43.00	43.50	44.00	44.50	45.00	45.00	47.00
Number Sold	256	276	250	241	210	235	195	194	205	180

Get Ready! **To prepare for Lesson 2-7, do Exercises 56–58.**

Solve each absolute value equation.

See Lesson 1-6.

56. $|x - 3| + 2 = 7$

57. $|2x + 1| - 14 = 9$

58. $\frac{1}{3}|5x - 3| = 6$

2-7 Absolute Value Functions and Graphs

Content Standards

F.BF.3 Identify the effect on the graph of replacing $f(x)$ by $f(x) + k$, $k\,f(x)$, $f(kx)$, and $f(x + k)$ for specific values of k . . . find the value of k given the graphs.

Also F.IF.7.b

Objective To graph absolute value functions

Think about how the distance changes before and after you cross the county line. Can a distance ever be negative?

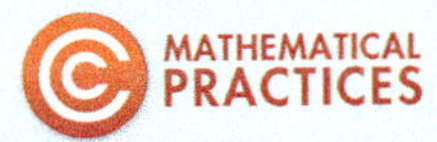

There is a family of functions related to the one you represented in the Solve It.

Essential Understanding Just as the absolute value of x is its distance from 0, the absolute value of $f(x)$, or $|f(x)|$, gives the distance from the line $y = 0$ for each value of $f(x)$.

The simplest example of an **absolute value function** is $f(x) = |x|$. The graph of the absolute value of a linear function in two variables is V-shaped and symmetric about a vertical line called the **axis of symmetry**. Such a graph has either a single maximum point or a single minimum point, called the **vertex**.

Lesson Vocabulary
- absolute value function
- axis of symmetry
- vertex

take note

Key Concept Absolute Value Parent Function $f(x) = |x|$

Table

x	$y = \|x\|$
−3	3
−2	2
−1	1
0	0
1	1
2	2
3	3

Function

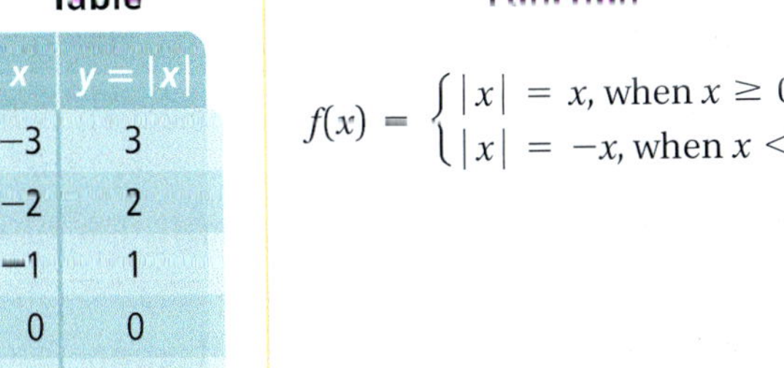

$$f(x) = \begin{cases} |x| = x, \text{ when } x \geq 0 \\ |x| = -x, \text{ when } x < 0 \end{cases}$$

Graph

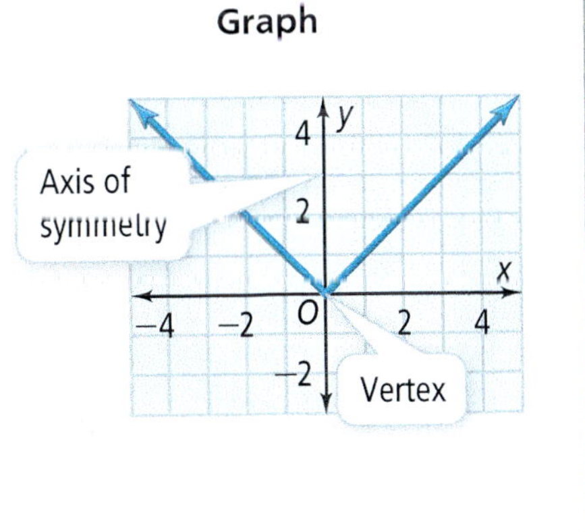

Problem 1 Graphing an Absolute Value Function

What is the graph of the absolute value function $y = |x| - 4$? How is this graph different from the graph of the parent function $f(x) = |x|$?

Think

Make a table of values and graph the function.

Write

x	y
-3	-1
-1	-3
0	-4
1	-3
3	-1

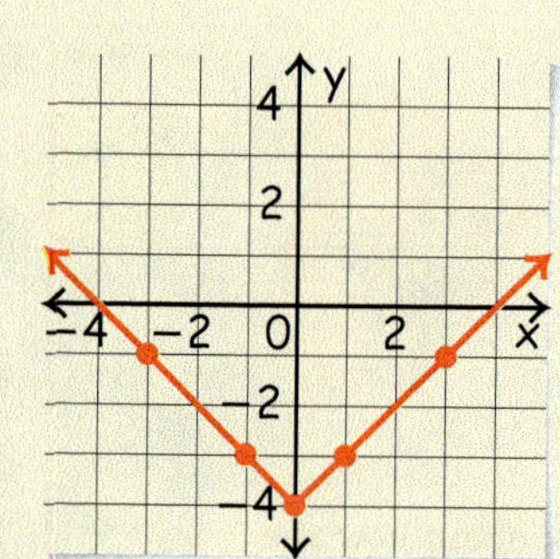

Use the location of the vertex to see how the graph has been translated. The parent function was not multiplied by a number, so the graph wasn't stretched, compressed, or reflected.

Since the vertex is at (0, −4), you translated the graph of $y = |x|$ down 4 units.

  **Got It?** 1. a. What is the graph of the function $y = |x| + 2$? How is this graph different from the parent function?

b. **Reasoning** Do transformations of the form $y = |x| + k$ affect the axis of symmetry? Explain.

The transformations you studied in Lesson 2-6 also apply to absolute value functions.

take note

Key Concept The Family of Absolute Value Functions

Parent Function $y = |x|$

Vertical Translation	**Horizontal Translation**
Translation up k units, $k > 0$ $y = \|x\| + k$	Translation right h units, $h > 0$ $y = \|x - h\|$
Translation down k units, $k > 0$ $y = \|x\| - k$	Translation left h units, $h > 0$ $y = \|x + h\|$
Vertical Stretch and Compression	**Reflection**
Vertical stretch, $a > 1$ $y = a\|x\|$	In the x-axis $y = -\|x\|$
Vertical compression, $0 < a < 1$ $y = a\|x\|$	In the y-axis $y = \|-x\|$

Problem 2 Combining Translations

Multiple Choice Which of the following is the graph of $y = |x + 2| + 3$?

Ⓐ
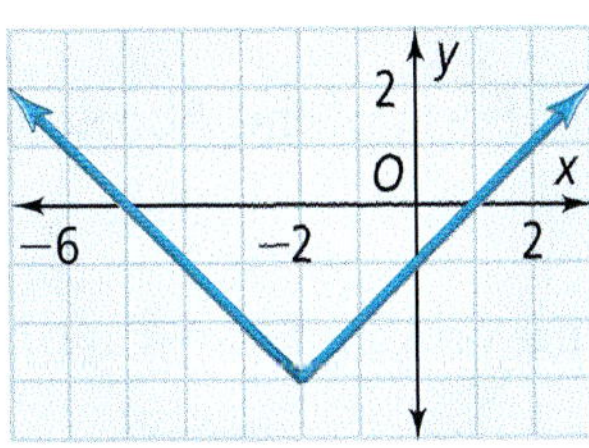

Ⓒ
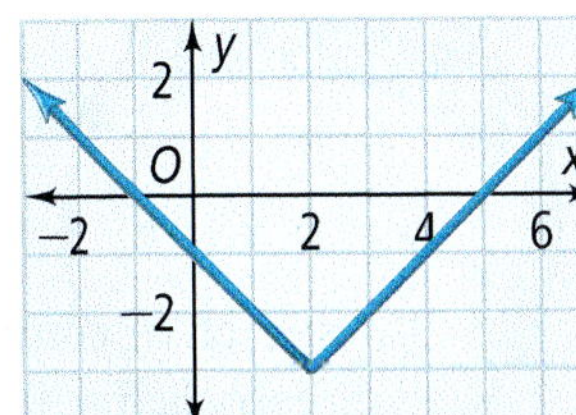

Ⓑ
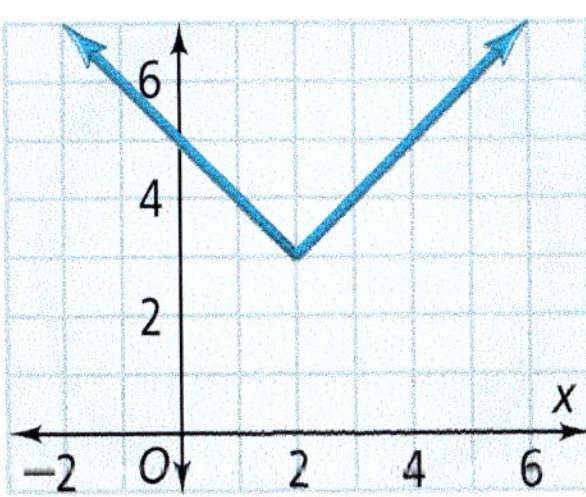

Ⓓ
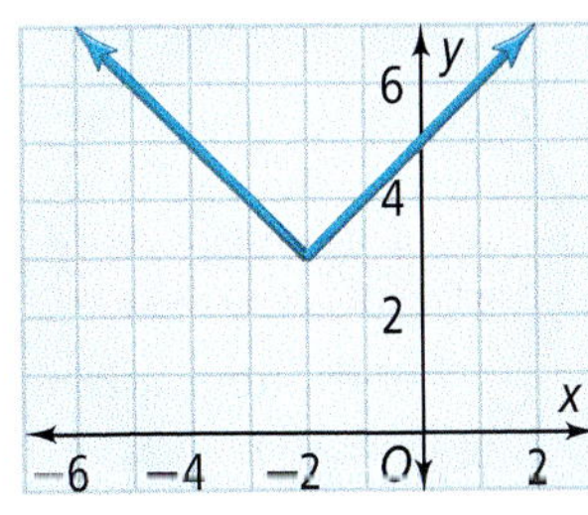

Think

Can you eliminate any answers after this comparison?
Only choices A and D show translations of $y = |x|$ to the left.

Compare $y = |x + 2| + 3$ to each form, $y = |x + h|$ and $y = |x| + k$.

$y = |x + h|$ The parent function, $y = |x|$, is translated left 2 units.

$y = |x| + k$ The parent function, $y = |x|$, is translated up 3 units.

The parent function $y = |x|$ is translated left 2 units and up 3 units. The vertex will be at $(-2, 3)$. The correct choice is D.

Got It? **2.** What is the graph of the function $y = |x - 2| + 1$?

The right branch of the graph of $y = |x|$ has slope 1. The graph of $y = a|x|, a > 0$, is a stretch or compression of the graph of $y = |x|$. Its right branch has slope a. The graph of $y = -a|x|$ is a reflection of $y = a|x|$ in the x-axis and its right branch has slope $-a$.

Problem 3 Vertical Stretch and Compression

What is the graph of $y = \frac{1}{2}|x|$?

The graph is a vertical compression of the graph of $f(x) = |x|$ by the factor $\frac{1}{2}$. Graph the right branch and use symmetry to graph the left branch.

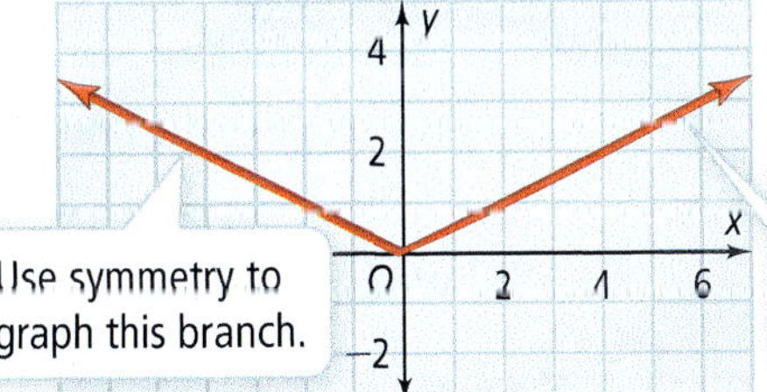

Got It? **3.** What is the graph of each function?

a. $y = 2|x|$ **b.** $y = -\frac{2}{3}|x|$

You can combine the equations for stretches and compressions with the equations for translations to write a general form for absolute value functions.

take note

Key Concept General Form of the Absolute Value Function

$y = a|x - h| + k$

The stretch or compression factor is $|a|$, the vertex is located at (h, k), and the axis of symmetry is the line $x = h$.

Problem 4 Identifying Transformations

Plan

To what should you compare $y = 3|x - 2| + 4$? Compare it to the general form, $y = a|x - h| + k$.

Without graphing, what are the vertex and axis of symmetry of the graph of $y = 3|x - 2| + 4$? How is the parent function $y = |x|$ transformed?

Compare $y = 3|x - 2| + 4$ with the general form $y = a|x - h| + k$.

$a = 3$, $h = 2$, and $k = 4$.

The vertex is $(2, 4)$ and the axis of symmetry is $x = 2$.

The parent function $y = |x|$ is translated 2 units to the right, vertically stretched by the factor 3, and translated 4 units up.

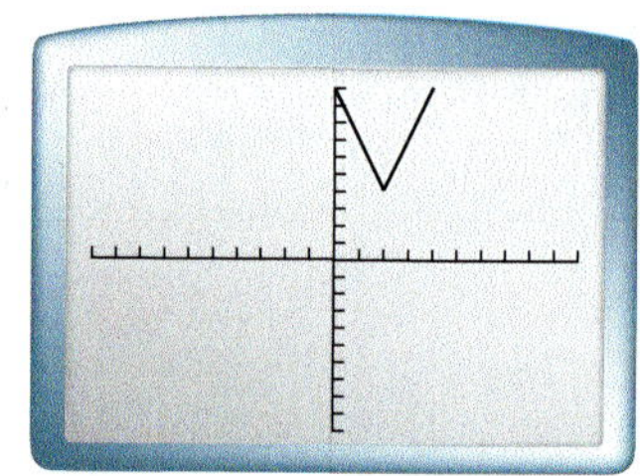

Check Check by graphing the equation on a graphing calculator.

Got It? **4.** What are the vertex and axis of symmetry of $y = -2|x - 1| - 3$? How is $y = |x|$ transformed?

Problem 5 Writing an Absolute Value Function

Think

What does the graph tell you about a? The upside-down V suggests that $a < 0$.

What is the equation of the absolute value function?

Step 1 Identify the vertex.

The vertex is at $(-1, 4)$, so $h = -1$ and $k = 4$.

Step 2 Identify a.

The slope of the branch to the right of the vertex is $-\frac{1}{3}$, so $a = -\frac{1}{3}$.

Step 3 Write the equation.

Substitute the values of a, h, and k into the general form $y = a|x - h| + k$. The equation that describes the graph is $y = -\frac{1}{3}|x + 1| + 4$.

Got It? **5.** What is the equation of the absolute value function?

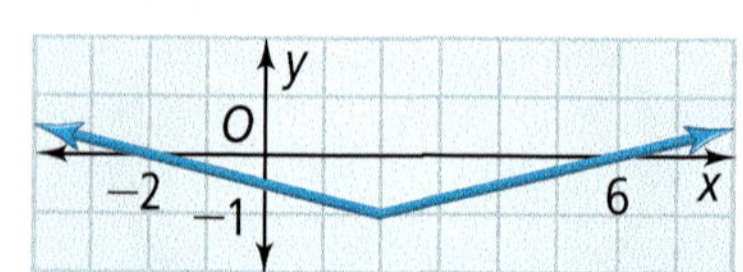

Lesson Check

Do you know HOW?

Find the vertex and the axis of symmetry of the graph of each function.

1. $y = 2|x + 4| - 3$

2. $y = |-x - 3| + 9$

Determine if each function is a vertical stretch or vertical compression of the parent function $y = |x|$.

3. $y = -\frac{7}{2}|x|$

4. $y = \frac{3}{2}|x|$

Do you UNDERSTAND?

5. Is it true that without making a graph of an absolute value function, you can describe its position in the coordinate plane? Explain with an example.

6. Write two absolute value functions such that they have a common vertex in Quadrant III and one is the reflection of the other in a horizontal line.

7. Compare and Contrast How is the graph of $y = x$ different from the graph of $y = |x|$?

Practice and Problem-Solving Exercises

Make a table of values for each equation. Then graph the equation.

See Problems 1 and 2.

8. $y = |x| + 1$

9. $y = |x| - 1$

10. $y = |x| - 3$

11. $y = |x + 2|$

12. $y = |x + 4|$

13. $y = |x + 5|$

14. $y = |x - 1| + 3$

15. $y = |x + 6| - 1$

16. $y = |x - 5| + 4$

Graph each equation. Then describe the transformation from the parent function $f(x) = |x|$.

See Problem 3.

17. $y = 3|x|$

18. $y = -\frac{1}{2}|x|$

19. $y = -2|x|$

20. $y = \frac{1}{3}|x|$

21. $y = \frac{3}{2}|x|$

22. $y = -\frac{3}{4}|x|$

Without graphing, identify the vertex, axis of symmetry, and transformations from the parent function $f(x) = |x|$.

See Problem 4.

23. $y = |x + 2| - 4$

24. $y = \frac{3}{2}|x - 6|$

25. $y = 3|x + 6|$

26. $y = 4 - |x + 2|$

27. $y = -|x - 5|$

28. $y = |x - 2| - 6$

Write an absolute value equation for each graph.

See Problem 5.

29.

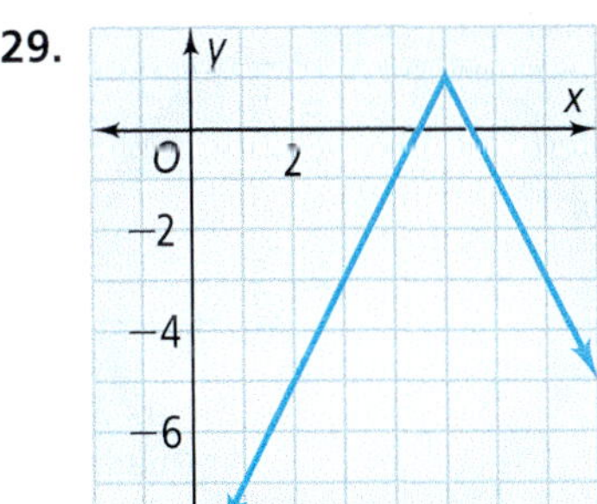

30.

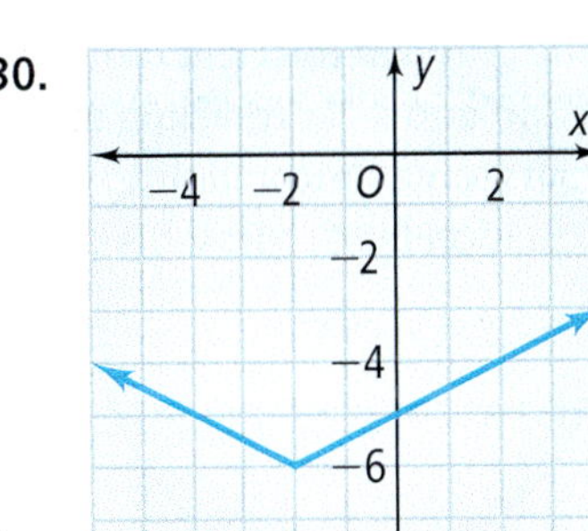

31. Think About a Plan Graph $y = -2|x + 3| + 4$. List the x- and y-intercepts, if any.

- What is the vertex?
- What does y equal at the x-intercept(s)? What does x equal at the y-intercept(s)?

32. Graph $y = 4|x - 3| + 1$. List the vertex and the x- and y-intercepts, if any.

33. Error Analysis A classmate says that the graphs of $y = -3|x|$ and $y = |-3x|$ are identical. Graph each function and explain why your classmate is not correct.

34. Graph each pair of equations on the same coordinate grid.

a. $y = 2|x + 1|$; $y = |2x + 1|$ **b.** $y = 5|x - 2|$; $y = |5x - 2|$

c. Reasoning Explain why each pair of graphs in parts (a) and (b) are different.

35. The graphs of the absolute value functions $f(x)$ and $g(x)$ are given.

a. Describe a series of transformations that you can use to transform $f(x)$ into $g(x)$.

b. Reasoning If you change the order of the transformations you found in part(a), could you still transform $f(x)$ into $g(x)$? Explain.

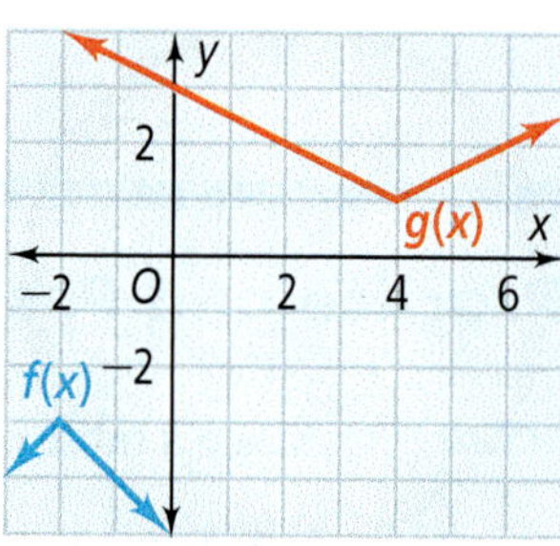

Graph each absolute value equation.

36. $y = \left|-\frac{1}{4}x - 1\right|$ **37.** $y = \left|\frac{5}{2}x - 2\right|$ **38.** $y = \left|\frac{3}{2}x + 2\right|$

39. $y = |3x - 6| + 1$ **40.** $y = -|x - 3|$ **41.** $y = |2x + 6|$

42. $y = 2|x + 2| - 3$ **43.** $y = 6 - |3x|$ **44.** $y = 6 - |3x + 1|$

45. a. Graph the equations $f(x) = -\frac{1}{2}|x - 3|$ and $g(x) = \left|-\frac{1}{2}(x - 3)\right|$ on the same set of axes.

b. Writing Describe the similarities and differences in the graphs.

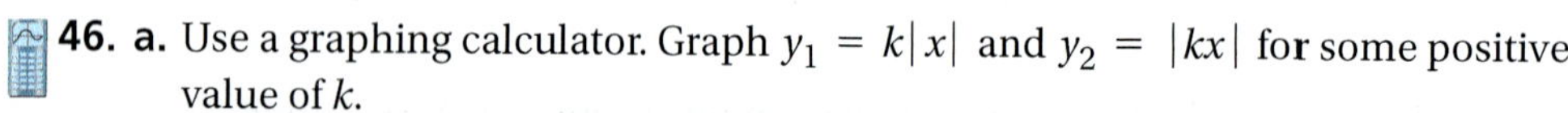

46. a. Use a graphing calculator. Graph $y_1 = k|x|$ and $y_2 = |kx|$ for some positive value of k.

b. Graph $y_1 = k|x|$ and $y_2 = |kx|$ for some negative value of k.

c. What conclusion can you make about the graphs of $y_1 = k|x|$ and $y_2 = |kx|$?

Graph each absolute value equation.

47. $y = |3x| - \frac{x}{3}$ **48.** $y = \frac{1}{2}|x| + 4|x - 1|$ **49.** $y = |2x| + |x - 4|$

50. The graph at the right models the distance between a roadside stand and a car traveling at a constant speed. The x-axis represents time and the y-axis represents distance. Which equation best represents the relation shown in the graph?

(A) $y = |60x|$ (C) $y = |x| + 60$

(B) $y = |40x|$ (D) $y = |x| + 40$

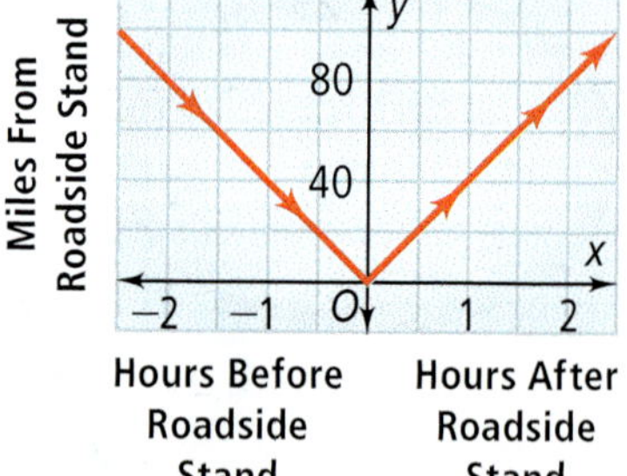

Standardized Test Prep

SAT/ACT

51. The graph shows which equation?

Ⓐ $y = |3x - 1| + 2$

Ⓑ $y = |x - 1| + 2$

Ⓒ $y = |x - 1| - 2$

Ⓓ $y = |3x - 3| - 2$

52. How are the graphs of $y = 2x$ and $y = 2x + 2$ related?

Ⓕ The graph of $y = 2x + 2$ is the graph of $y = 2x$ translated down two units.

Ⓖ The graph of $y = 2x + 2$ is the graph of $y = 2x$ translated up two units.

Ⓗ The graph of $y = 2x + 2$ is the graph of $y = 2x$ translated to the left two units.

Ⓘ The graph of $y = 2x + 2$ is the graph of $y = 2x$ translated to the right two units.

53. What is the equation of a line parallel to $y = x$ that passes through the point $(0, 1)$?

Ⓐ $y = x + 1$

Ⓑ $y = 2x + 2$

Ⓒ $y = x - 1$

Ⓓ $y = -x$

Short Response

54. Is $|y| = x$ a function? Explain.

Mixed Review

Write an equation for each transformation of the graph of $y = x + 2$. See Lesson 2-6.

55. 2 units up, 3 units right

56. vertical compression by a factor of $\frac{1}{2}$, reflection in the y-axis

Write the function rule for each function reflected in the given axis.

57. $f(x) = x - 7$; y-axis

58. $f(x) = 2x - 6$; y-axis

59. $f(x) = 4 + x$; x-axis

Find a trend line for each scatter plot. Write the equation for each trend line. See Lesson 2-5.

60.

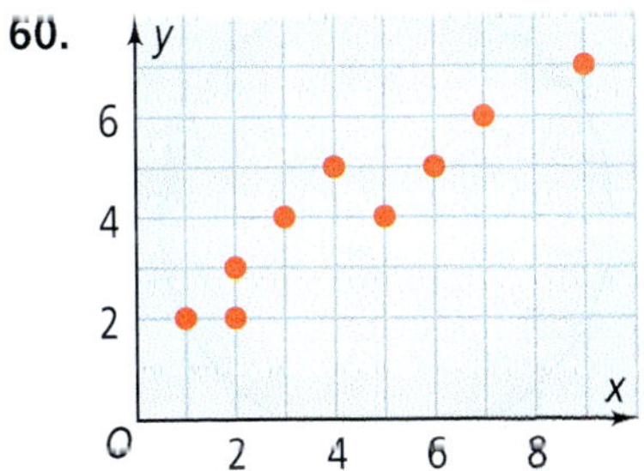

61.

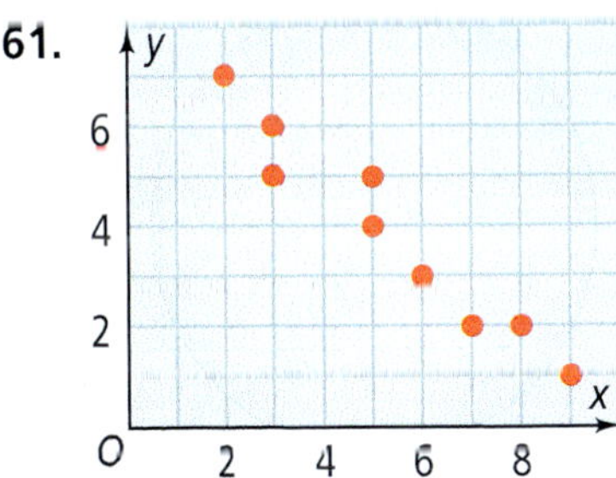

Get Ready! **To prepare for Lesson 2-8, do Exercises 62–64.**

Solve each inequality. Graph the solution on a number line. See Lesson 1-5.

62. $12p \leq 15$

63. $4 + t > 17$

64. $5 - 2t \geq 11$

Two-Variable Inequalities

Content Standards

A.CED.2 Create equations in two or more variables to represent relationships between quantities; graph equations on coordinate axes with labels and scales.

Also F.IF.7.b

Objective To graph two-variable inequalities

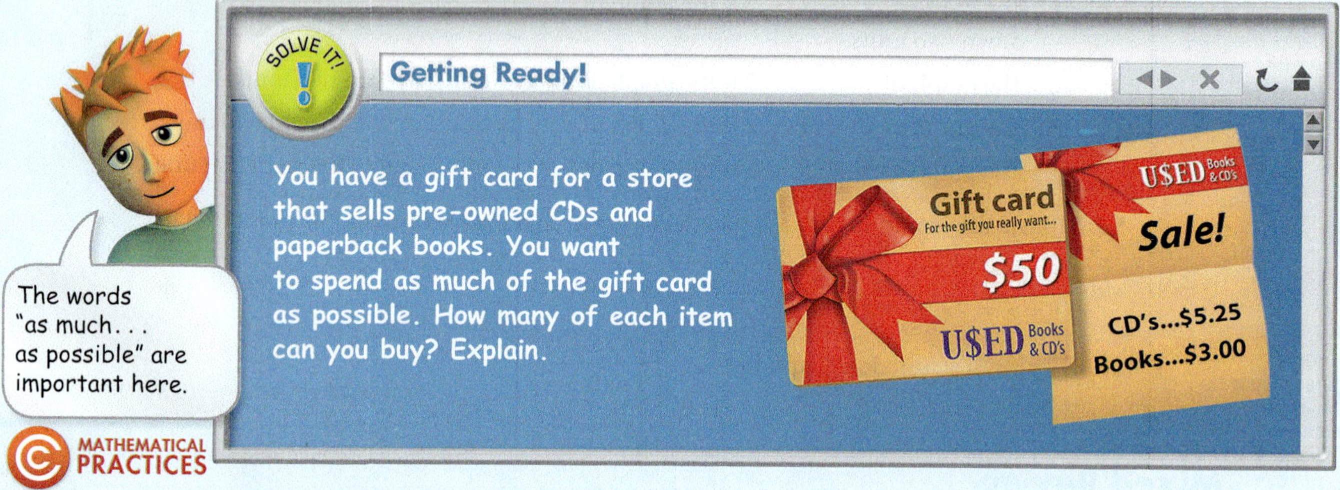

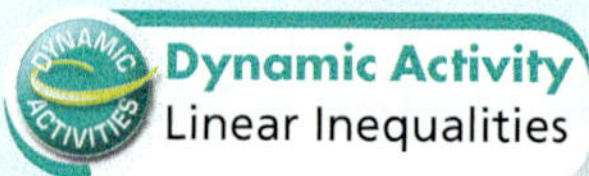

In some situations you need to compare quantities. You can use inequalities for situations that involve these relationships: *less than, less than or equal to, greater than,* and *greater than or equal to.*

Essential Understanding Graphing an inequality in two variables is similar to graphing a line. The graph of a linear inequality contains all points on one side of the line and may or may not include the points on the line.

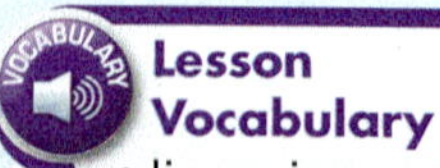

- linear inequality
- boundary
- half-plane
- test point

A **linear inequality** is an inequality in two variables whose graph is a region of the coordinate plane bounded by a line. This line is the **boundary** of the graph. The boundary separates the coordinate plane into two **half-planes**, one of which consists of solutions of the inequality.

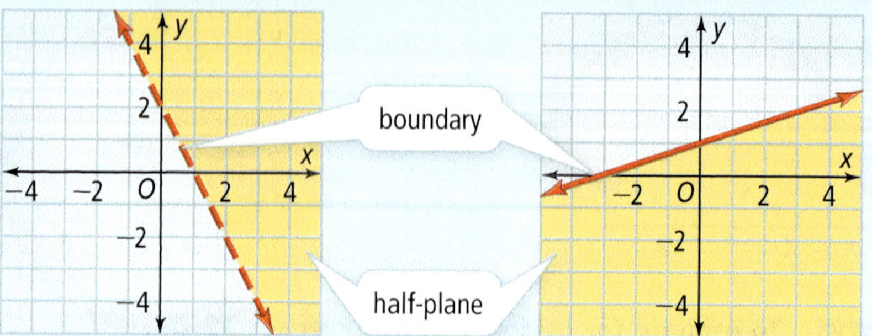

To determine which half-plane to shade, pick a **test point** that is *not* on the boundary. Check whether that point satisfies the inequality. If it does, shade the half-plane that includes the test point. If not, shade the other half-plane. The origin, (0, 0), is usually an easy test point as long as it is not on the boundary.

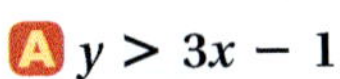

Problem 1 Graphing Linear Inequalities

What is the graph of each inequality?

A $y > 3x - 1$

Step 1

Graph the boundary line $y = 3x - 1$. Use a dashed boundary line because the inequality is *greater than*, and the points on the line do not satisfy the inequality.

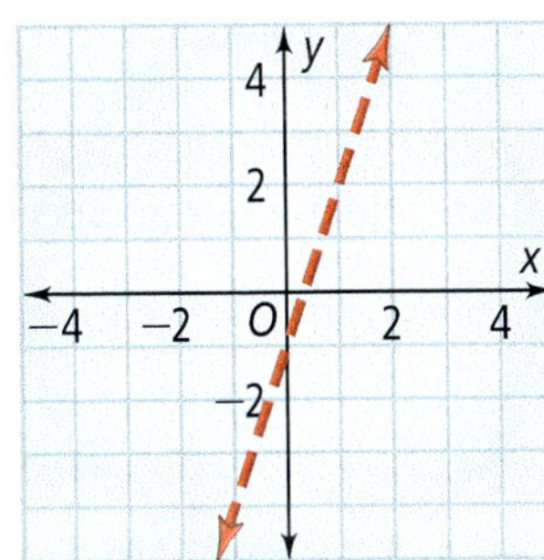

Step 2

Choose a test point, (0, 0). Substitute $x = 0$ and $y = 0$ into $y > 3x - 1$.

$$0 > 3(0) - 1$$
$$0 > -1$$

Since $0 > -1$ is true, shade the half plane that includes (0, 0).

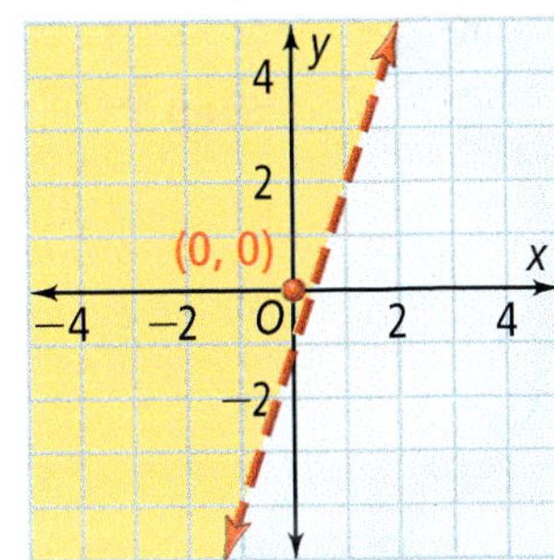

B $y \leq 3x - 1$

The boundary line is again $y = 3x - 1$, but it is solid because the inequality is less than or *equal to*.

Shade the region opposite the region shaded above (for $>$) because the inequality is *less than* or equal to.

You can also check the point (0, 0).

$$0 \leq 3(0) - 1$$
$$0 \leq -1$$

Since $0 \leq -1$ is false, (0, 0) is not part of the solution.

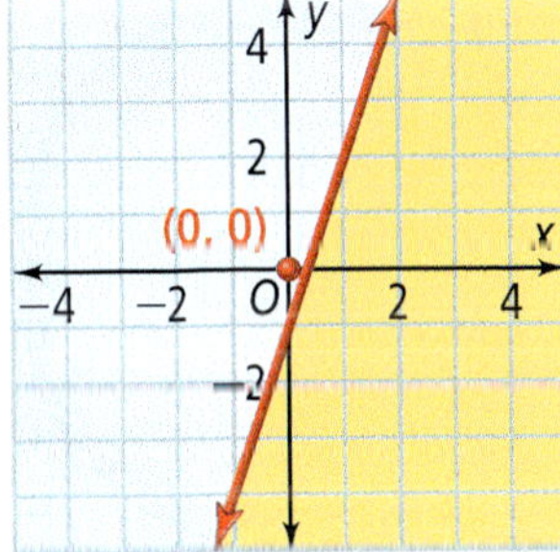

Plan

Can you use the graph of $y > 3x - 1$ to help graph $y \leq 3x - 1$?

If you shaded above the line for $y > 3x - 1$, then shade below the line for $y \leq 3x - 1$.

Got It? **1.** What is the graph of each inequality?

a. $y \geq -2x + 1$ **b.** $y < -2x + 1$

You can also inspect inequalities solved for *y*, such as $y > mx + b$ to determine which half-plane describes the solution. Since *y* describes vertical position, the solution of $y > mx + b$ will be *above* the boundary line. The solution of $y < mx + b$ will be *below* the boundary line.

Problem 2 Using a Linear Inequality

Entertainment The map shows the number of tickets needed for small or large rides at the fair. You do not want to spend more than \$15 on tickets. How many small or large rides can you ride?

Think

What are the unknowns?
The unknowns are the number of small rides and the number of large rides you can get on.

You can buy 60 tickets with \$15.

Relate	the number of tickets for small rides	plus	the number of tickets for large rides	is less than or equal to	60
Define	Let x = the number of small rides.				
	Let y = the number of large rides.				
Write	$3x$	$+$	$5y$	$\leq$	60

Step 1

Find the intercepts of the boundary line. Use the intercepts to graph the boundary line.

When $y = 0$, $3x + 5(0) = 60$.
$$3x = 60$$
$$x = 20$$

When $x = 0$, $3(0) + 5y = 60$.
$$5y = 60$$
$$y = 12$$

Graph the line that connects the intercepts (20, 0) and (0, 12). Since the inequality is $\leq$, use a solid boundary line.

Step 2

The region above the boundary line represents combinations of rides that require more than 60 tickets. You purchased a *finite* number of tickets, 60, so you will not be able to go on an infinite number of rides. Shade the region below the boundary line.

The number of small rides x and the number of large rides y are whole numbers. In math, such a situation is called *discrete*. All points with whole number coordinates in the shaded region represent possible combinations of small and large rides.

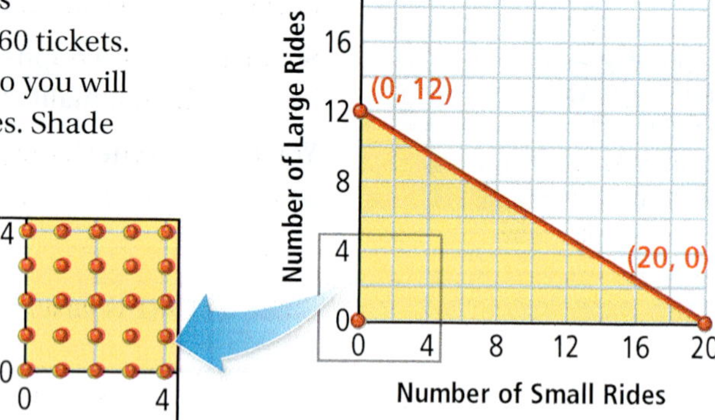

**Got It?** 2. a. Suppose that you decide to spend no more than \$30 for tickets. What are the possible combinations of small and large rides that you can ride now? Use a graph to find your answer.

b. **Reasoning** Why did the graph of the solution in Problem 2 only include Quadrant I?

You can graph two-variable absolute value inequalities in the same way that you graph linear inequalities.

Problem 3 Graphing an Absolute Value Inequality

What is the graph of $1 - y < |x + 2|$?

Know	Need	Plan
Absolute value inequality	Boundary	• Solve the inequality for y. • Graph the related equation. • Shade the solution.

$$1 - y < |x + 2|$$

$-y < |x + 2| - 1$ Subtract 1 from each side.

$y > -|x + 2| + 1$ Multiply both sides by -1.

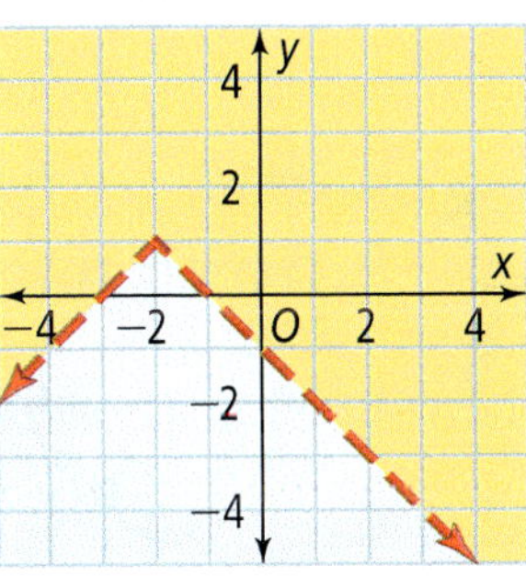

The graph of $y = -|x + 2| + 1$ is the graph of $y = |x|$, reflected in the x-axis and translated left 2 units and up 1 unit.

Since the inequality is solved for y and $y > -|x + 2| + 1$, shade the region above the boundary.

Got It? **3.** What is the graph of $y - 4 \geq 2|x - 1|$?

You can use the transformations discussed in previous lessons to help draw the boundary graphs more quickly. You can also use them to write an inequality based on a graph.

Problem 4 Writing an Inequality Based on a Graph

Plan

How can you tell that the graph is not a stretch or compression of the graph of $y = |x|$?
The slopes of the branches are 1 and -1.

What inequality does this graph represent?

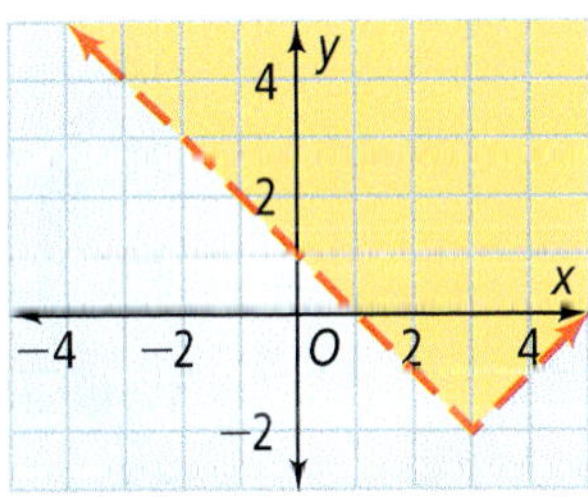

The boundary is the graph of the absolute value function $y = |x|$, translated. The vertex of $y = |x|$ is translated to $(3, -2)$, so the boundary is the graph of $y = |x - 3| - 2$.

The solution is shaded above the boundary, so the inequality is either $>$ or $\geq$. Since the boundary is a dashed line, the correct inequality is $y > |x - 3| - 2$.

Got It? **4. a.** What inequality does this graph represent?

b. Reasoning You can tell from looking at the inequality $y > 5x - 3$ to shade above the boundary line to represent the solution. Can you use the same technique to show the solution of an inequality like $2x - y > 1$? Explain.

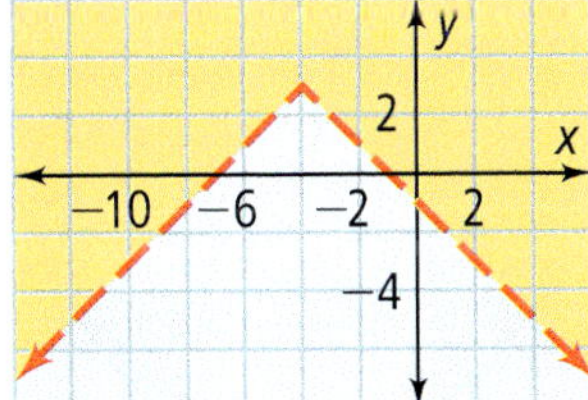

Lesson Check

Do you know HOW?

What is the graph of each inequality?

1. $9y \leq 12x$

2. $7x + y \geq 8$

What is the graph of each absolute value inequality?

3. $y \leq |x + 1|$

4. $y \geq |2x - 3|$

Do you UNDERSTAND?

5. Do the points on the boundary line of the graph of an inequality help determine the shaded area of the graph? Explain.

6. Compare and Contrast How is graphing a linear inequality in two variables different from graphing a linear equation in two variables?

7. Reasoning Is the ordered pair $\left(\frac{3}{4}, 0\right)$ a solution of $3x + y > 3$? Explain.

Practice and Problem-Solving Exercises

Graph each inequality. See Problem 1.

8. $y > 2x + 1$ **9.** $y < 3$ **10.** $x \leq 0$

11. $y \leq x - 5$ **12.** $2x + 3y \geq 12$ **13.** $2y \geq 4x - 6$

14. $3x - 2y \leq 9$ **15.** $-y < 2x + 2$ **16.** $5 - y \geq x$

17. Cooking The time needed to roast a chicken depends on its weight. Allow at least 20 min/lb for a chicken weighing as much as 6 lb. Allow at least 15 min/lb for a chicken weighing more than 6 lb. See Problem 2.

a. Write two inequalities to represent the time needed to roast a chicken.

b. Graph the inequalities.

Graph each absolute value inequality. See Problem 3.

18. $y \geq |2x - 1|$ **19.** $y \leq |3x| + 1$ **20.** $y \leq |4 - x|$

21. $y > |-x + 4| + 1$ **22.** $y - 7 > |x + 2|$ **23.** $y + 2 \leq \left|\frac{1}{2}x\right|$

24. $3 - y \geq -|x - 4|$ **25.** $1 - y < |2x - 3|$ **26.** $y + 3 \leq |3x| - 1$

Write an inequality for each graph. The equation for the boundary line is given. See Problem 4.

27. $y = -x - 2$ **28.** $5x + 3y = 9$ **29.** $2y = |2x + 6|$

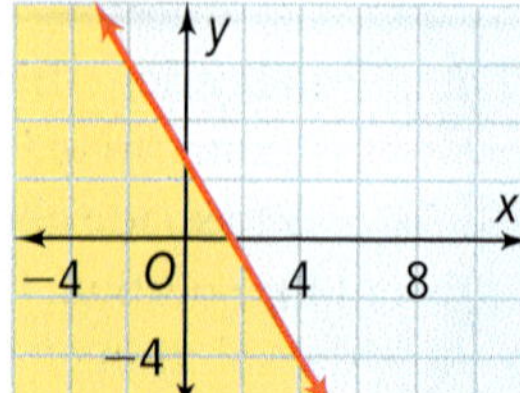

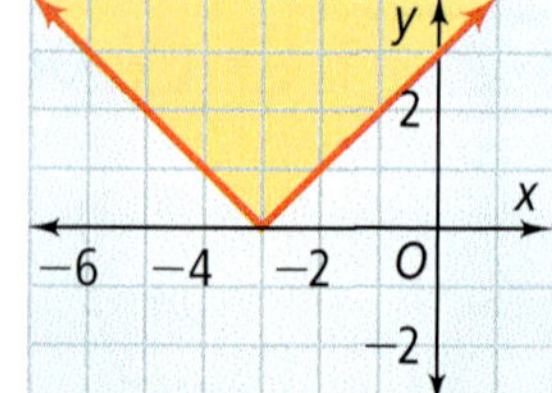

Graph each inequality on a coordinate plane.

30. $5x - 2y \geq -10$ **31.** $2x - 5y < -10$ **32.** $\frac{3}{4}x + \frac{2}{3}y > \frac{5}{2}$ **33.** $3(x - 2) + 2y \leq 6$

34. $|x - 1| > y + 7$ **35.** $y - |2x| \leq 21$ **36.** $\frac{2}{3}x + 2 \leq \frac{2}{9}y$ **37.** $0.25y - 1.5x \geq -4$

38. Think About a Plan The graph at the right relates the number of hours you spend on the phone to the number of hours you spend studying per week. Describe the domain for this situation. Write an inequality for the graph.

- What is the least amount of time you can spend on the phone per week? What is the most?
- What is the least amount of time you can spend studying per week? What is the most?
- What is the greatest amount of time you can spend either on the phone or studying per week?

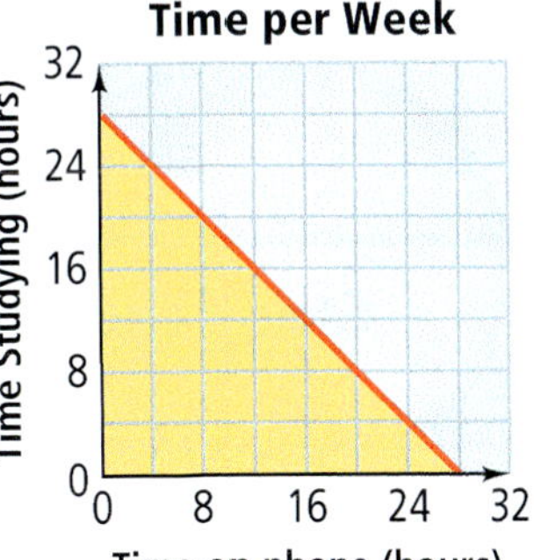

Write an inequality for each graph.

39.
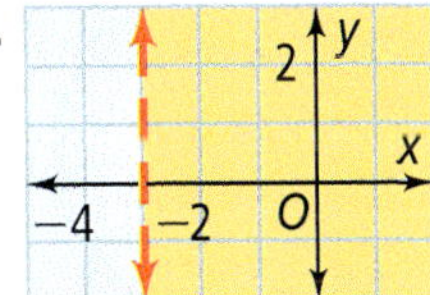

40.
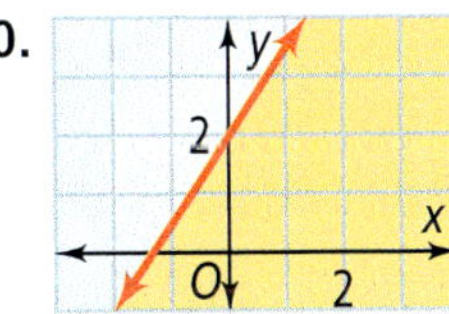

41.
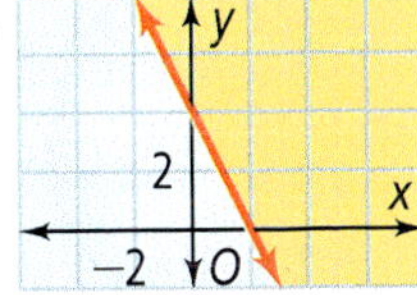

42.
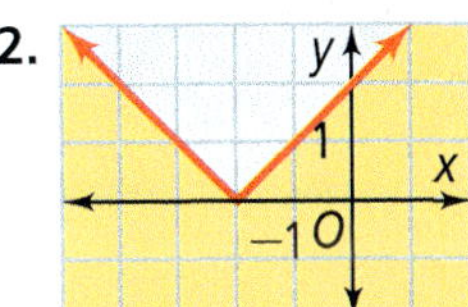

43.
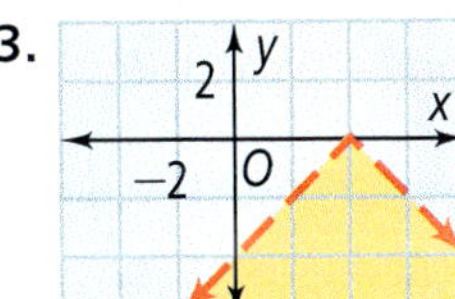

44.
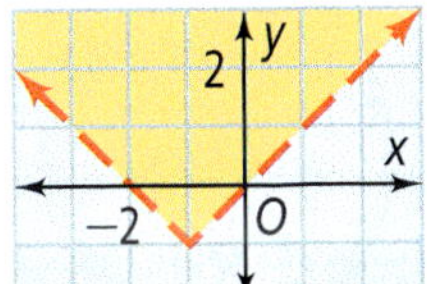

45. Which graph best represents the solution of the inequality $y \geq 2|x - 1| - 2$?

(A)
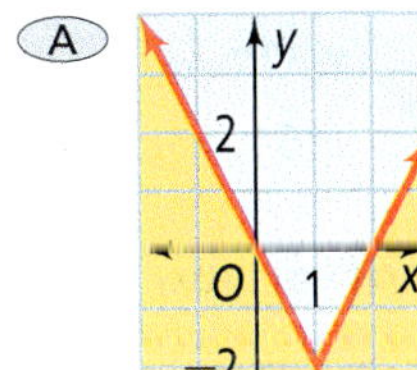

(B)
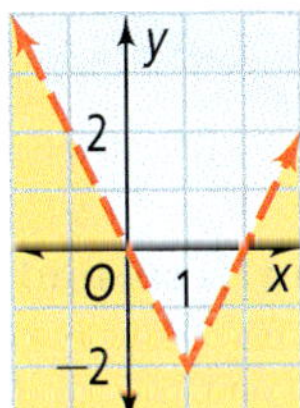

(C)
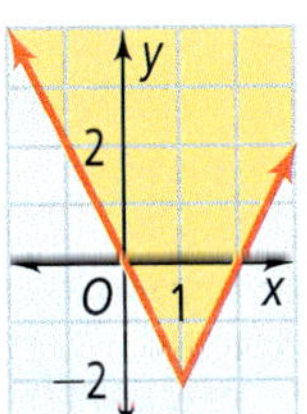

(D)
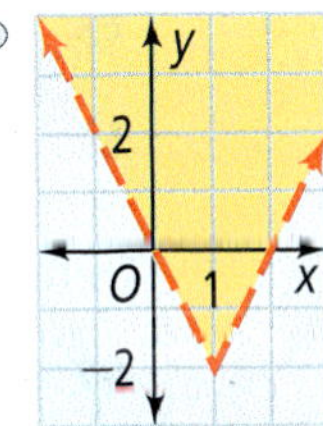

46. The graph at the right relates the amount of gas in the tank of your car to the distance you can drive.

a. Describe the domain for this situation.

b. Why does the graph stop?

c. Why is only the first quadrant shown?

d. Reasoning Would every point in the solution region be a solution?

e. Write an inequality for the graph.

f. What does the coefficient of x represent?

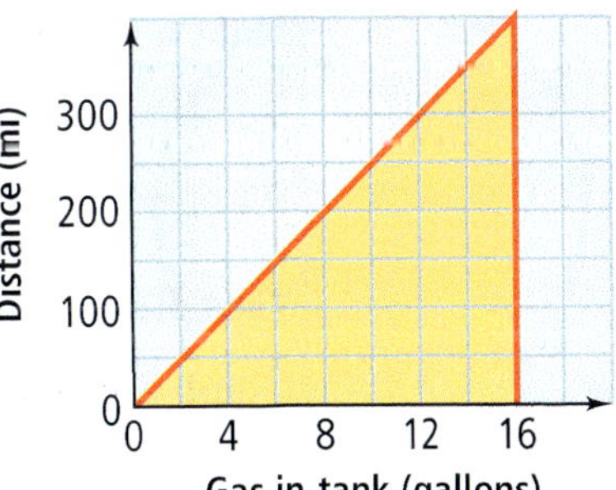

47. Writing When you graph an inequality, you can often use the point (0, 0) to test which side of the boundary line to shade. Describe a situation in which you could not use (0, 0) as a test point.

Graphing Calculator **Graph each inequality on a graphing calculator. Then sketch the graph.**

48. $y \leq |x + 1| - |x - 1|$

49. $y > |x| + |x + 3|$

50. $y < |x - 3| - |x + 3|$

51. $y < 7 - |x - 4| + |x|$

Standardized Test Prep

SAT/ACT

52. Suppose y varies directly with x. If x is 30 when y is 10, what is x when y is 9?

(A) 3 (B) 27 (C) 29 (D) $\frac{300}{9}$

53. Which equation represents a line with slope -2 and y-intercept 3?

(F) $3y = x - 2$ (G) $3y = -2x + 1$ (H) $y = 2x - 3$ (I) $y = -2x + 3$

54. What is the vertex of $y = |x| - 5$?

(A) (5, 0) (B) (−5, 0) (C) (0, 5) (D) (0, −5)

Extended Response

55. The amount of a commission is directly proportional to the amount of a sale. A realtor received a commission of \$48,000 on the sale of an \$800,000 house. How much would the commission be on a \$650,000 house?

Mixed Review

Graph each function by translating its parent function. See Lesson 2-7.

56. $y = |2x + 5|$

57. $y = |x| - 3$

58. $f(x) = |x + 6|$

59. $f(x) = |x| - 2$

60. $y = |x + 2|$

61. $y = |x - 1| + 5$

Determine whether y varies directly with x. If so, find the constant of variation. See Lesson 2-2.

62. $y = x + 1$

63. $y = 100x$

64. $5x + y = 0$

65. $y - 2 = 2x$

66. $x = \frac{y}{3}$

67. $-4 = y - x$

68. $y = -10x$

69. $xy = 1$

Make a scatter plot and describe the correlation. See Lesson 2-5.

70. $\{(0, 6), (1, 4), (2, 4), (4, 1), (5, 0)\}$

71. $\{(-10, 5), (-5, -5), (-2, 0), (0, 3), (5, -2)\}$

Get Ready! **To prepare for Lesson 3-1, do Exercises 72–74.**

Graph each equation. Use one coordinate plane for all three graphs. See Lesson 2-3.

72. $3x - y = 2$

73. $3x - y = -2$

74. $x + 3y = -2$

Pull It All Together ASSESSMENT

To solve these problems, you will pull together concepts about equivalence, linear functions and modeling. Show your work and justify your reasoning.

BIG idea Equivalence

You can represent a function with many different equivalent equations.

Performance Task 1

The graph represents a function.

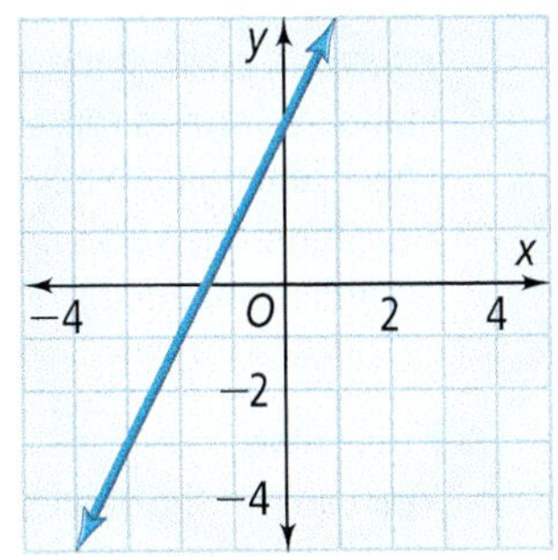

a. Write the equation for the graph in standard form, point-slope form, and slope-intercept form.

b. Which equation is the easiest to write by looking at the graph? Explain why.

BIG idea Modeling

You can use scatter plots to model, analyze, and make predictions about certain kinds of data.

Performance Task 2

The table at the right shows the boiling point of water at various elevations.

a. Identify the independent and dependent quantities. Explain your choices.

b. Make a scatter plot that models this data.

c. Determine what kind of correlation is shown in your plot.

d. Write an equation of the line of best fit.

e. Predict the temperature at which water boils at 8000 feet above sea level. Justify your method for finding the answer.

f. At what elevation would you expect water to boil at 207°F? Explain.

Boiling Point of Water

Elevation (ft)	Boiling Point (°F)
0 (sea level)	212
1000	210.2
2000	208.4
3000	206.6
4000	204.8
5000	203

Performance Task 3

Consider the following data: (2, 1), (4, 3), (5, 5), (7, 6), (3, 18).

a. Graph the data.

b. Which data point(s) do not seem to fit in with the rest of the data? Explain.

c. Find and graph a line of best fit that includes the data points(s) identified in part (b) Then find and graph a line of best fit that does not include the data point(s).

d. Which equation better models the data? Justify your answer.

2 Chapter Review

Connecting BIG ideas and Answering the Essential Questions

1 Equivalence
You can use either slope-intercept, point-slope, or standard form to represent linear functions. (You can transform one version to another as needed.)

Slope-Intercept Form (Lesson 2-3)
$y = mx + b$
$y = 2x - 1$

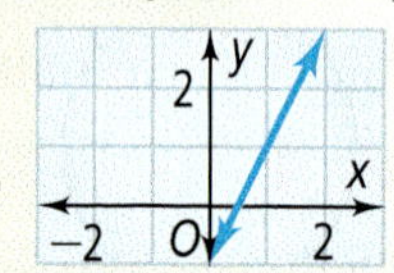

More Linear Equations (Lesson 2-4)
$y - y_1 = m(x - x_1)$ $Ax + By = C$
$y - 5 = 2(x - 3)$ $2x - y = 1$

2 Function
You can use the values of a, h, and k in the form $y = a|x - h| + k$ to determine how the parent function $y = |x|$ has been transformed.

Families of Functions (Lesson 2-6)

$f(x) + k$	vertical translation
$f(x - h)$	horizontal translation
$af(x)$	stretch or compression
$-f(x)$	reflection in the x-axis
$f(-x)$	reflection in the y-axis

Absolute Value Functions and Graphs (Lesson 2-7)
Parent: $y = |x|$
General form:
$y = a|x - h| + k$
vertex: (h, k)

3 Modeling
You can use the equation of a trend line or line of best fit to model data that cluster in a linear pattern.

Using Linear Models (Lesson 2-5)

Positive correlation

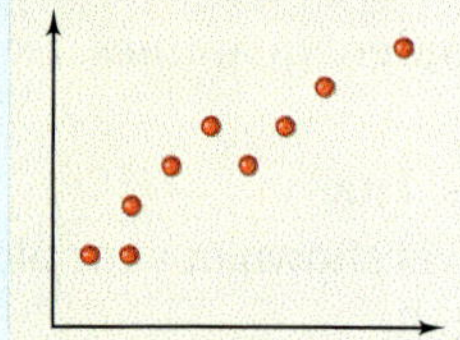

Trend Line

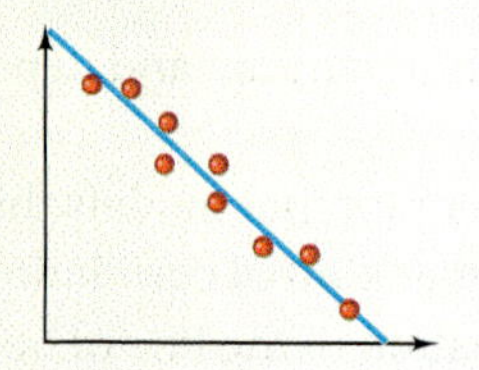

Chapter Vocabulary

- absolute value function (p. 107)
- axis of symmetry (p. 107)
- boundary (p. 114)
- constant of variation (p. 68)
- correlation (p. 92)
- correlation coefficient (p. 94)
- dependent variable (p. 63)
- direct variation (p. 68)
- domain (p. 61)
- function (p. 62)
- function notation (p. 63)
- function rule (p. 63)
- half-plane (p. 114)
- independent variable (p. 63)
- line of best fit (p. 94)
- linear equation (p. 75)
- linear function (p. 75)
- linear inequality (p. 114)
- parallel lines (p. 85)
- parent function (p. 99)
- perpendicular lines (p. 85)
- point-slope form (p. 81)
- range (p. 61)
- reflection (p. 101)
- relation (p. 60)
- scatter plot (p. 92)
- slope (p. 74)
- slope-intercept form (p. 76)
- standard form of a linear equation (p. 82)
- test point (p. 115)
- transformation (p. 99)
- translation (p. 99)
- vertex (p. 107)
- vertical compression (p. 102)
- vertical stretch (p. 102)
- vertical-line test (p. 62)
- x-intercept (p. 76)
- y-intercept (p. 76)

Choose the correct term to complete each sentence.

1. The graph of a function is (*always*/*sometimes*) a line.
2. The equation $y - 5 = 3(x + 2)$ is in (*point-slope*/*slope-intercept*) form.

2-1 Relations and Functions

Quick Review

A **relation** is a set of ordered pairs. The **domain** of a relation is the set of x-coordinates. The **range** is the set of y-coordinates. When each element of the domain is paired with exactly one element of the range, the relation is a **function**.

Example

Determine whether the relation is a function. Find the domain and range.

$\{(5, 0), (8, 1), (1, 3), (5, 2), (3, 8)\}$

In this relation, the x-coordinate 5 is paired with both 0 and 2. This relation is not a function.

The domain is the set of x-coordinates, which is $\{5, 8, 1, 3\}$.

The range is the set of y-coordinates, which is $\{0, 1, 3, 2, 8\}$.

Exercises

Determine whether each relation is a function. Find the domain and range.

3. $\{(10, 2), (-10, 2), (6, 4), (5, 3), (-6, 7)\}$

4. $\{(4, 5), (1, 5), (3, 8), (4, 6), (10, 12)\}$

5.

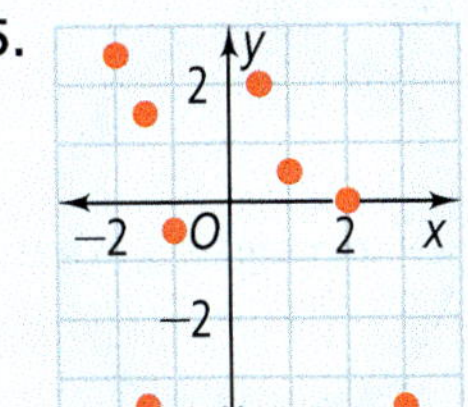

6.

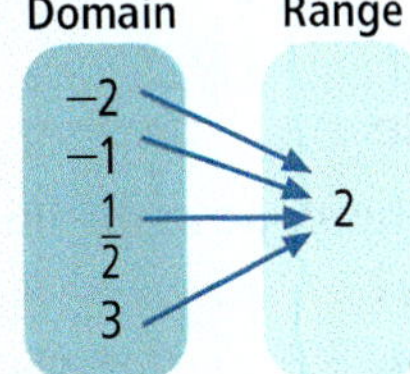

For each function, find $f(-2)$, $f(-0.5)$, and $f(3)$.

7. $f(x) = -x + 4$

8. $f(x) = \frac{3}{8}x - 3$

2-2 Direct Variation

Quick Review

A linear equation of the form $y = kx$, $k \neq 0$, represents **direct variation**. The **constant of variation** is k. You can use proportions to solve direct variation problems.

Example

In the table, determine whether y varies directly with x. If so, what is the constant of variation and the function rule?

x	y
2	6
3	9
8	24

$\frac{6}{2} = \frac{9}{3} = \frac{24}{8} = 3$, so y varies directly with x, and the constant of variation is 3.

The function rule is $y = 3x$.

Exercises

For each function, determine whether y varies directly with x. If so, find the constant of variation and write the function rule.

9.

x	y
−2	3
1	4
2	7

10.

x	y
4	5
6	9
10	17

11.

x	y
1	1
2	2
5	5

For each function, y varies directly with x. Find each constant of variation. Then find the value of y when $x = -0.3$.

12. $y = 2$ when $x = -\frac{1}{2}$

13. $y = \frac{2}{3}$ when $x = 0.2$

14. $y = 7$ when $x = 2$

15. $y = 4$ when $x = -3$

2-3 Linear Functions and Slope-Intercept Form

Quick Review

The graph of a **linear function** is a line. You can represent a linear function with a **linear equation**. Given two points on a line, the **slope** of the line is the ratio of the change in the y-coordinates to the change in the corresponding x-coordinates. The slope is the coefficient of x when you write a linear equation in **slope-intercept form**.

Example

What is the slope of the line that passes through (3, 5) and (−1, −2)?

$m = \frac{y_2 - y_1}{x_2 - x_1}$ Find the difference between the coordinates.

$= \frac{5 - (-2)}{3 - (-1)} = \frac{7}{4}$ Simplify.

Exercises

Identify the slope of the line that passes through the given points.

16. (1, 3) and (6, 1)

17. (4, 4) and (−2, −3)

18. (3, 2) and (−3, −2)

19. (5, 2) and (−4, 6)

Write an equation for each line in slope-intercept form.

20. slope $= -3$ and the y-intercept is (0, 4)

21. slope $= \frac{1}{2}$ and the y-intercept is (0, 6)

Rewrite each equation in slope-intercept form. Graph each line.

22. $4x - 2y = 3$

23. $-4x + 6y = 18$

24. $3y + 3x = 15$

25. $3y + x = 5$

2-4 More About Linear Equations

Quick Review

You write the equation of a line in **point-slope form** when you have a point and the slope or when you have two points. The **standard form** of an equation has both variables and no constants on the left side.

When two lines have the same slope, they are **parallel**. When two lines have slopes that are negative reciprocals of each other, they are **perpendicular**.

Example

Write an equation in standard form for the line with a slope of 2, going through (1, 6).

$y - 6 = 2(x - 1)$ Write the equation in point-slope form, substituting the given point and slope.

$y = 2x - 2 + 6$ Simplify.

$-2x + y = 4$ Write in standard form.

Exercises

Write an equation for each line in point-slope form and then convert it to standard form.

26. slope $= -3$, through (4, 0)

27. slope $= 5$, through (1, −1)

28. through (0, 0) and (3, −7)

29. through (2, 3) and (3, 5)

30. **a.** Write an equation of the line parallel to $x + 2y = 6$ through (8, 3).

b. Write an equation of the line perpendicular to $x + 2y = 6$ through (8, 3).

c. Graph the three lines on the same coordinate plane.

2-5 Using Linear Models

Quick Review

You can use a **scatter plot** to show relationships between data sets. You can make predictions using a trend line, which approximates the relationship between two data sets. The most accurate trend line is a **line of best fit**.

Example

Draw a scatter plot of the data. Is a linear model reasonable? If so, predict the value of y when $x = 9$.

$\{(0, 6), (1, 7), (2, 5), (3, 4), (4, 2), (5, 1)\}$

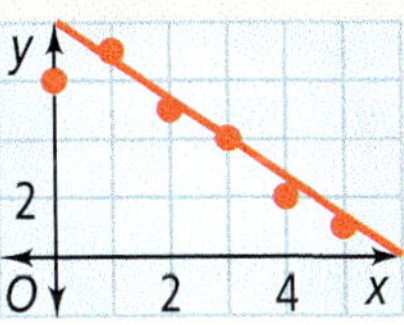

The points are close to the line $y = -\frac{4}{3}x + 8$, so a linear model is reasonable. When $x = 9$,

$$y = -\frac{4}{3}(9) + 8$$
$$= -4$$

Exercises

Draw a scatter plot of each set of data. Decide whether a linear model is reasonable. If so, describe the correlation. Then draw a trend line and write its equation. Predict the value of y when x is 15.

31. $\{(3, 5), (4, 7), (5, 9), (7, 10), (8, 10), (9, 11), (10, 13)\}$

32. $\{(6, 15.5), (7, 14.0), (8, 13.0), (9, 12.5), (10, 12.0), (11, 11.5), (12, 10.0)\}$

33.

x	0	3	6	9	12
y	17.5	35.4	50.5	60.6	66.3

2-6 Families of Functions

Quick Review

A **parent function** is the simplest form of a function in a family of functions. Each member is a **transformation** of the parent function.

Translations shift the graph horizontally, vertically, or both. A **reflection** flips the graph over a line of symmetry. **Vertical stretches** and **compressions** change the shape of the graph by a factor.

Example

Write the equation of the transformation of the graph of $f(x) = x^2$ translated 3 units up, vertically stretched by a factor of 6, and reflected across the y-axis.

$y = x^2 + 3$	Translated 3 units up.
$y = 6(x^2 + 3)$	Vertically stretched.
$y = 6(-x)^2 + 18$	Reflected across the y-axis.
$y = 6x^2 + 18$	

Exercises

Write the equation for the transformation of the graph of $y = f(x)$.

34. translated 2 units left, 7 units down

35. translated 5 units right, reflected across the x-axis

36. translated 3 units up, reflected across the y-axis

Describe the transformation(s) of the parent function $f(x)$.

37. $g(x) = f(x) - 4$

38. $h(x) = 12f(x) + 2$

39. $k(x) = -2f(-x)$

2-7 Absolute Value Functions and Graphs

Quick Review

The **absolute value function** $y = |x|$ is the **parent function** for the family of functions of the form $y = a|x - h| + k$. The maximum or minimum point of the graph is the vertex of the graph.

$y = 2|x + 3| + 1$
$a = 2, h = -3, k = 1$

- Vertex is at $(-3, 1)$
- Translated left 3 units
- Stretched by a factor of 2
- Translated up 1 unit

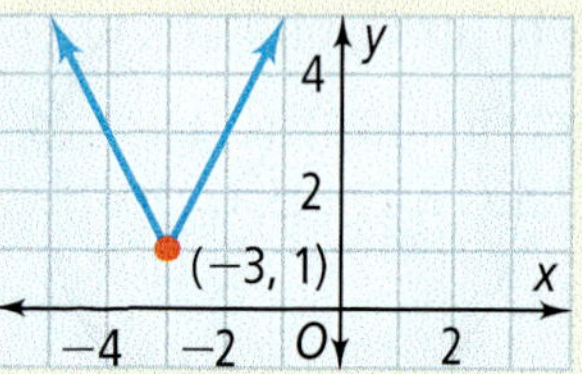

Example

Write an equation for the translation of the graph $y = |x|$ up 5 units.

Because the graph is translated up, k is positive, so the equation of the translated graph is $y = |x| + 5$.

Exercises

Write an equation for each translation of the graph of $y = |x|$.

40. up 4 units, right 2 units
41. vertex $(-3, 0)$
42. vertex $(5, 2)$
43. vertex $(4, 1)$

Graph each function.

44. $f(x) = |x| - 8$
45. $f(x) = 2|x - 5|$
46. $y = -\frac{1}{4}|x - 2| + 3$
47. $y = -2|x + 1| - 1$

Without graphing, identify the vertex and axis of symmetry of each function.

48. $y = 2|x - 4|$
49. $y = -|x| + 2$

2-8 Two Variable Inequalities

Quick Review

An inequality describes a region of the coordinate plane that has a **boundary**. To graph an inequality involving two variables, first graph the boundary. Then determine which side of the boundary contains the solutions. Points on a dashed boundary are not solutions. Points on a solid boundary are solutions.

Example

Graph the inequality $y \geq 2x + 3$.

Graph the solid boundary line $y = 2x + 3$.

Since y is *greater than* $2x + 3$, shade above the boundary.

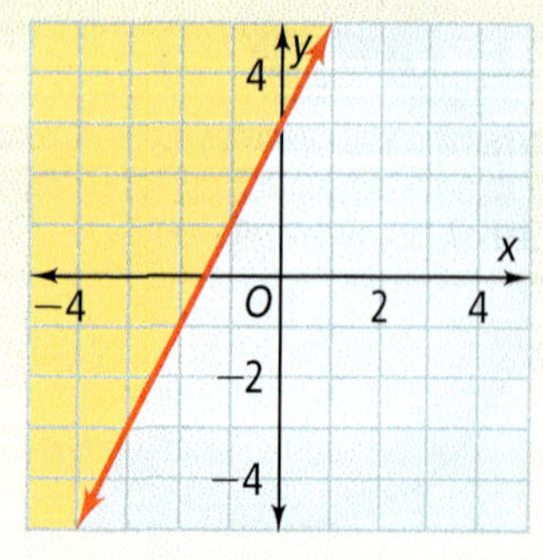

Exercises

Graph each inequality.

50. $y \geq -2$
51. $y < 3x + 1$
52. $y < -|x - 5|$
53. $y > |2x + 1|$

54. **Transportation** An air cargo plane can transport as many as 15 regular shipping containers. One super-size container takes up the space of 3 regular containers.

a. Write an inequality to model the number of regular and super size containers the plane can transport.
b. Describe the domain and range.
c. Graph the inequality you wrote in part (a).

55. **Open-Ended** Write an absolute value inequality with a solid boundary that only has solutions below the x-axis.

2 Chapter Test

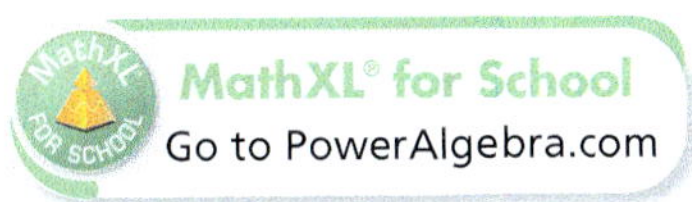

Do you know HOW?

Find the domain and range. Graph each relation.

1. $\{(0, 0), (1, -1), (2, -4), (3, -9), (4, -16)\}$

2. $\{(3, 2), (4, 3), (5, 4), (6, 5), (7, 6)\}$

Determine whether each relation is a function.

3. Domain Range

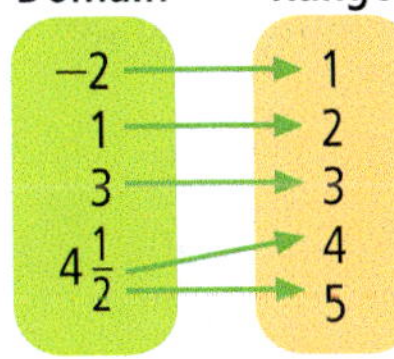

4. Domain Range

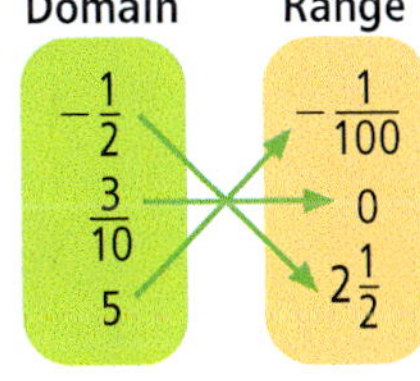

Suppose $f(x) = 2x - 5$ and $g(x) = |-3x - 1|$. Find each value.

5. $f(3)$ **6.** $f(1) + g(2)$ **7.** $g(0)$

8. $g(2) - f(0)$ **9.** $f(-1) - g(3)$ **10.** $2g(-4)$

Find the slope of each line.

11. through $(3, 5)$, parallel to $y = 5x - 1$

12. through $(-0.5, 0.5)$, perpendicular to $y = -2x - 4$

Write an equation of the line in standard form with the given slope through the given point.

13. slope $= -3$, $(0, 0)$ **14.** slope $= \frac{2}{5}$, $(6, 7)$

15. slope $= 4$, $(-2, -5)$ **16.** slope $= -0.5$, $(0, 6)$

Write an equation of the line in point-slope form through each pair of points.

17. $(0, 0)$ and $(-4, 7)$ **18.** $(-1, -6)$ and $(-2, 10)$

19. $(3, 0)$ and $(-1, -2)$ **20.** $(9, 5)$ and $(8, 2)$

For each direct variation, find the constant of variation. Then find the value of y when $x = -0.5$.

21. $y = 4$ when $x = 0.5$ **22.** $y = 2$ when $x = 3$

Write an equation of the line with the given slope and y-intercept. Use slope-intercept form. Then rewrite each equation in standard form.

23. $m = 3, b = -7$ **24.** $m = -6, b = 9$

25. $m = \frac{1}{4}, b = 11$ **26.** $m = -\frac{1}{2}, b = 4$

Graph each inequality.

27. $y \geq x + 7$ **28.** $y > 2|x + 3| - 3$

29. $4x - 3y < 2$ **30.** $y \leq -\frac{1}{2}|x + 2| - 3$

Do you UNDERSTAND?

31. **Open-Ended** Graph a relation that is *not* a function. Find its domain and range.

32. **Writing** Explain how point-slope form is related to the formula for slope.

Describe each transformation of the parent function $y = |x|$. Then, graph each function.

33. $y = |x| - 4$ **34.** $y = |x - 1| - 5$

35. $y = -|x + 4| + 3$ **36.** $y = 2|x + 1|$

37. **Recreation** The table displays the amounts the Jackson family spent on vacations during the years 2000–2009.

Family Vacations

Year	Cost	Year	Cost
2000	\$1750	2005	\$2750
2001	\$1750	2006	\$3200
2002	\$2000	2007	\$2900
2003	\$2200	2008	\$3100
2004	\$2700	2009	\$3300

a. Make a scatter plot of the data.
b. Draw a trend line. Write its equation.
c. Estimate the amount the Jackson family will spend on vacations in 2015.
d. **Writing** Explain how to use a trend line to make a prediction.

2 Cumulative Standards Review

ASSESSMENT

TIPS FOR SUCCESS

Some problems require you to use direct variation to solve for an unknown quantity. Read the question at the right. Then follow the tips to answer the sample question.

A salad dressing recipe calls for $1\frac{1}{3}$ cups of buttermilk, 1 egg, $\frac{1}{2}$ cup of orange juice, and 1 tablespoon of lemon juice. Dan plans to use 2 cups of buttermilk instead. How much orange juice should he use?

(A) $\frac{3}{8}$ cup

(B) $\frac{3}{4}$ cup

(C) $1\frac{1}{6}$ cup

(D) $1\frac{1}{3}$ cup

TIP 1

Some problems give more information than you need. Decide what information you need to answer the question.

TIP 2

Use some of the information to find k, the constant of variation in $y = kx$.

Think It Through

The ratio that shows how the amount of buttermilk changes is $k = \frac{2}{1\frac{1}{3}}$.

You can simplify this ratio.

$$2 \div \frac{4}{3} = \frac{2}{1} \cdot \frac{3}{4} = \frac{3}{2}$$

Use $k = \frac{3}{2}$ in the direct variation equation $y = \frac{3}{2}x$.

Let x represent the original amount of orange juice.

$$y = \frac{3}{2}x = \frac{3}{2} \cdot \frac{1}{2} = \frac{3}{4}$$

The correct answer is B.

Vocabulary Builder

As you solve test items, you must understand the meanings of mathematical terms. Match each term with its mathematical meaning.

A. linear function

B. direct variation

C. range

D. translation

I. the set of all outputs, or y-coordinates, of a relation

II. a transformation that shifts a graph horizontally, vertically, or both

III. a function that can be written in the form $y = mx + b$

IV. a function that can be written in the form $y = kx$, $k \neq 0$

Multiple Choice

Read each question. Then write the letter of the correct answer on your paper.

1. Which of the following absolute value inequalities has no solutions in Quadrant IV?

(A) $y + 2 \geq |x - 3|$

(B) $y > 3 - |5 - x|$

(C) $y - 1 > |2x + 6|$

(D) $y \leq |4x| - 7$

2. For which value of b would the equation $3|x - 2| = bx - 6$ have infinitely many solutions?

(F) -6

(G) 3

(H) -3

(I) 6

3. A meteorologist predicts the daily high and low temperatures as 91°F and 69°F. If t represents the temperature, then this situation can be described with the inequality $69 \le t \le 91$. Which of the following absolute value inequalities is an equivalent way of expressing this?

(A) $69 \le |t| \le 91$

(B) $|t - 80| \le 11$

(C) $|t - 69| \le 91$

(D) $|t - 11| \le 80$

4. Which inequality has a solution that matches the graph below?

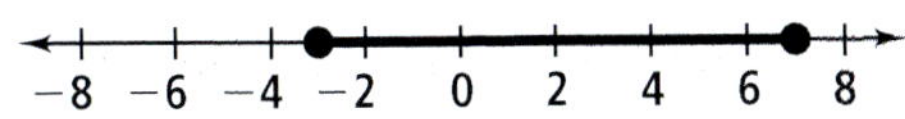

(F) $|x - 2| - 3 \ge 2$

(G) $|x - 2| - 3 \le 2$

(H) $|x - 3| + 2 \ge 7$

(I) $|x - 3| + 2 \le 7$

5. Which inequality best describes the graph?

(A) $y \le \frac{1}{2}|x + 1| - 1$

(B) $y \ge \frac{1}{2}|x - 1| - 1$

(C) $y \le \frac{1}{2}|x - 1| - 1$

(D) $y \ge \frac{1}{2}|x + 1| - 1$

6. Which relation is a function?

(F)

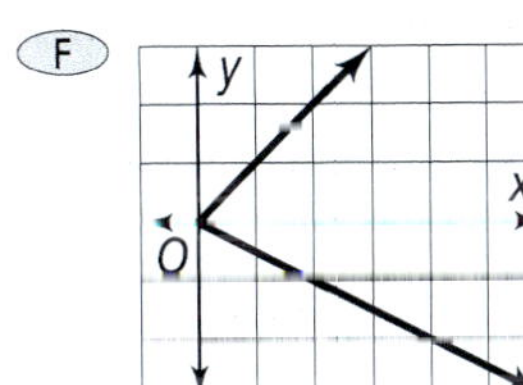

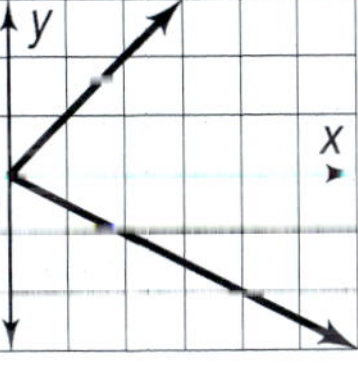

(H)

(G)

(I)

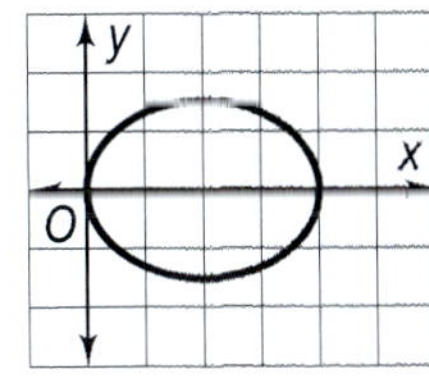

7. For $f(x) = 2x - 3$ find $f\left(-\frac{1}{4}\right)$.

(A) $-\frac{3}{2}$ (B) -2 (C) $2\frac{1}{2}$ (D) $-\frac{7}{2}$

8. Which equation is graphed?

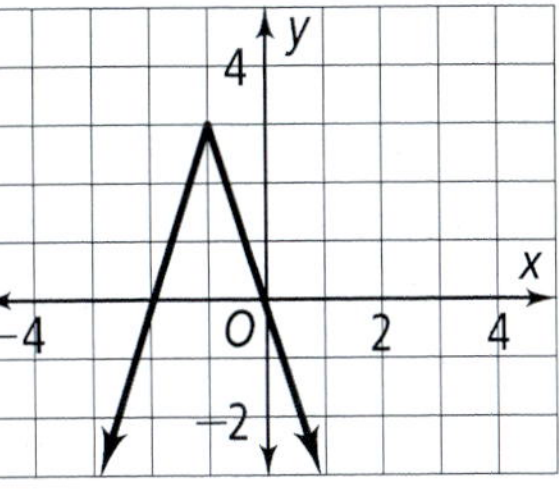

(F) $y = -3|x + 1| + 3$

(G) $y = 3|x + 1| + 3$

(H) $y = -3|x - 1| + 3$

(I) $y = 3|x + 1| - 3$

9. Which describes the translation of $y = |x - 3| + 5$?

(A) $y = |x|$ translated 3 units left and 5 units up

(B) $y = |x|$ translated 3 units right and 5 units up

(C) $y = |x|$ translated 5 units left and 3 units up

(D) $y = |x|$ translated 5 units right and 3 units up

10. If a rate of speed r is constant, then distance, rate, and time are related by the direct variation equation $d = rt$, where d represents distance and t represents time. If $r = 30$ miles per hour, which of the following best describes the graph of $d = rt$?

(F) A straight line through the point (0, 0)

(G) A straight line through the point (0, 30)

(H) A parabola through the point (0, 0)

(I) A parabola through the point (0, 30)

11. Which function has the graph shown?

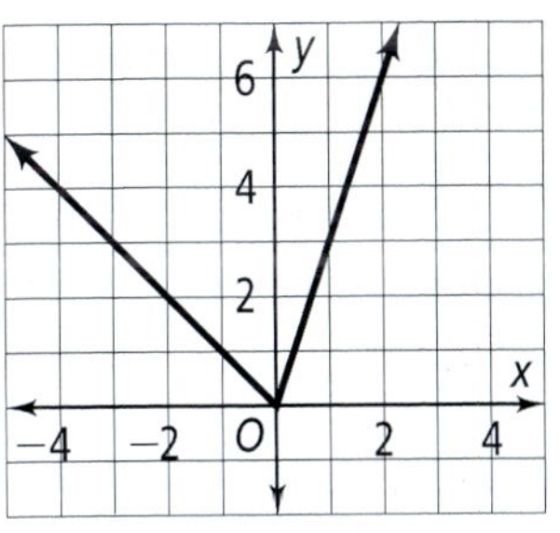

(A) $f(x) = \begin{cases} -x, & x < 0 \\ 3x, & x \ge 0 \end{cases}$

(B) $f(x) = \begin{cases} x, & x < 0 \\ -3x, & x \ge 0 \end{cases}$

(C) $f(x) = \begin{cases} x, & x > 0 \\ -3x, & x \le 0 \end{cases}$

(D) $f(x) = \begin{cases} -x, & x > 0 \\ 3x, & x \le 0 \end{cases}$

12. Which equation has the same graph as

$f(x) = \begin{cases} \frac{1}{3}x, & x > 6 \\ -\frac{1}{3}x + 4, & x \le 6 \end{cases}$?

(F) $f(x) = -\frac{1}{3}|x - 6| + 4$

(G) $f(x) = \frac{1}{3}x$

(H) $f(x) = \frac{1}{3}|x - 6| + 2$

(I) $f(x) = -\frac{1}{3}x + 10$

GRIDDED RESPONSE

13. What is the slope of the line $5y + 3 = \frac{2}{5}x$?

14. A recipe for custard sauce calls for 6 egg yolks, $\frac{2}{3}$ cup of sugar, and $1\frac{1}{2}$ cups of hot milk. Chelsea needs more sauce than the recipe yields. She plans to use 1 cup of sugar. How many cups of hot milk should she use?

15. What is the y-coordinate of the point through which the graph of every direct variation passes?

16. Six members of the math club will participate in a regional competition. A processing fee of $15 is added to the registration cost. If the math coach sends in a check for $87, how much does he pay for each registration?

17. What is the y-coordinate of the y-intercept of the line $4x + 3y = 12$?

18. If $4(x + 2) - 2(x - 10) = 0$, what is the value of x?

19. Mr. Wong traveled 45 miles on a business trip. The cost to rent a car is $30.00 plus $.75 per mile. How much did Mr. Wong pay for the car rental?

20. In the expression $38 + 27y$, which number is a coefficient?

21. What is the greatest integer solution of $|-2x - 5| - 3 \leq 2$?

22. What is the sum of the solutions of $3|y - 4| = 9$ and $|y - 4| = 3$?

23. What is the value of x in the equation $6(x - 4) = 3x$?

24. What is the slope of the line represented by the function $3x - 2y = -12$?

25. What is $f(-3)$ for the function $f(x) = -3x + 6$?

26. What is the y-coordinate of the y-intercept of the line?

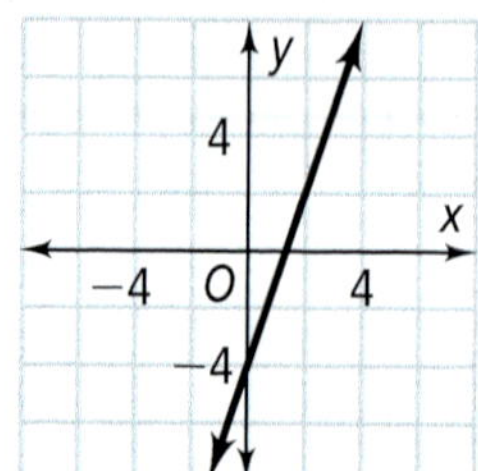

Short Response

27. Matt drove at a steady speed during the first morning of his road trip. The table shows data about his driving.

Time Driving (hours) t	Total Distance (miles) d
1.5	87
2.25	130.5
3	174

Determine whether distance d varies directly with time t. If so, what are the constant of variation and the function rule?

28. Find $f(-3)$, $f(0)$, and $f(1)$ for the function $f(x) = \frac{2}{5}x - 2$.

29. Graph $y < |x + 3|$. Identify the parent function of the boundary and describe the translation.

30. Suppose y varies directly with x, and $y = 2$ when $x = -2$. Find the constant of variation. Then find the value of x when $y = 3$.

31. The points $(-3, 2)$, $(-1, 3)$, $(0, 0)$, $(-2, -1)$ represent a function. What are the domain and range?

Extended Response

32. a. Write an equation of the line through $(-2, 6)$ with slope 2.

b. Write an equation of the line through $(1, 1)$ and perpendicular to the line in part (a).

c. Graph the two lines on the same set of axes.

33. Will the product of an integer and a natural number always be an integer? Why or why not? Justify your answer with two examples.

34. Sketch a graph through the point $(1, 1)$ such that as an x-value increases by 2, the y-value decreases by 3.

CHAPTER 3

Get Ready!

Lesson 1-3

Evaluating Algebraic Expressions

Evaluate each expression for the given values of the variables.

1. $9t + 6(2v - t) - 7v$; $t = 1$ and $v = 5$

2. $11(a + 2b) + 2(a - 2b)$; $a = -3$ and $b = 4$

3. $\frac{3}{5}d + \frac{1}{10}h - \frac{7}{10}d - \frac{4}{5}h$; $d = 5$ and $h = 10$

4. $12\left(\frac{3}{4}x - \frac{1}{2}y\right) - 6\left(\frac{1}{2}x - \frac{3}{4}y\right)$; $x = 2$ and $y = -2$

Lesson 2-3

Writing Linear Equations in Slope-Intercept Form

Write the equation of each line in slope-intercept form.

5. $2x - 4y = 10$ **6.** $3y + 9 = -6x$ **7.** $y - 5x = 16$ **8.** $-7 - y = -3x$

9. $\frac{x}{6} - \frac{5}{12}y = \frac{5}{8}$ **10.** $4x = y - 11$ **11.** $2y = -12x - 16$ **12.** $\frac{y}{9} + \frac{x}{3} = 2$

Lesson 2-3

Graphing Linear Equations

Graph each equation.

13. $3x = y - 1$ **14.** $x - 5y = 10$ **15.** $12 + 2y = 3x$ **16.** $y = 4x$

Lesson 2-8

Graphing Inequalities

Graph each inequality.

17. $4y \leq 24x$ **18.** $y \geq 2|x - 1.5|$ **19.** $x + 5y \geq 20$ **20.** $y > |x + 6| - 2$

Looking Ahead Vocabulary

21. A *system* of mountains is a group of mountains that share similar geographic and geological features. What are some mountain systems in the United States?

22. Two things are *consistent* if they are in agreement with each other. What does it mean for your actions to be consistent with your words?

23. How many books does Jeff own if he has more than 14 books? Describe the number of books that Jeff owns if you add the *constraint* that he owns fewer than his sister, who owns 19 books.

CHAPTER 3

Linear Systems

Your place to get all things digital

Download videos connecting math to your world.

Math definitions in English and Spanish

The online Solve It will get you in gear for each lesson.

Interactive! Vary numbers, graphs, and figures to explore math concepts.

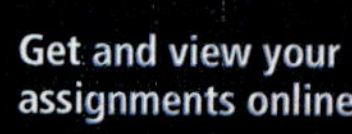

Online access to stepped-out problems aligned to Common Core

Get and view your assignments online.

ONLINE HOMEWORK

Extra practice and review online

MathXL FOR SCHOOL

DOMAINS

- Creating Equations
- Reasoning with Equations and Inequalities

The cables in the photo intersect at certain points, just like the graphs of linear equations might intersect.

How can you solve a system of linear equations? How can you use linear programming to solve real-world problems? How can you use a matrix to represent a system of equations? You will learn how in this chapter.

Vocabulary

English/Spanish Vocabulary Audio Online:

English	Spanish
dependent system, *p. 137*	sistema dependiente
equivalent systems, *p. 144*	sistemas equivalentes
independent system, *p. 137*	sistema independiente
linear system, *p. 134*	sistema lineal
matrix, *p. 174*	matriz
matrix element, *p. 174*	elemento matricial
row operation, *p. 176*	operación de filas
system of equations, *p. 134*	sistema de ecuaciones

BIG ideas

1 **Function**

Essential Question How does representing functions graphically help you solve a system of equations?

2 **Equivalence**

Essential Question How does writing equivalent equations help you solve a system of equations?

3 **Solving Equations and Inequalities**

Essential Question How are the properties of equality used in the matrix solution of a system of equations?

Chapter Preview

Solving Systems Using Tables and Graphs

Content Standards

A.CED.2 Create equations in two or more variables to represent relationships between quantities; graph equations on coordinate axes with labels and scales.

Also A.REI.6, A.REI.11, A.CED.3

Objective To solve a linear system using a graph or a table

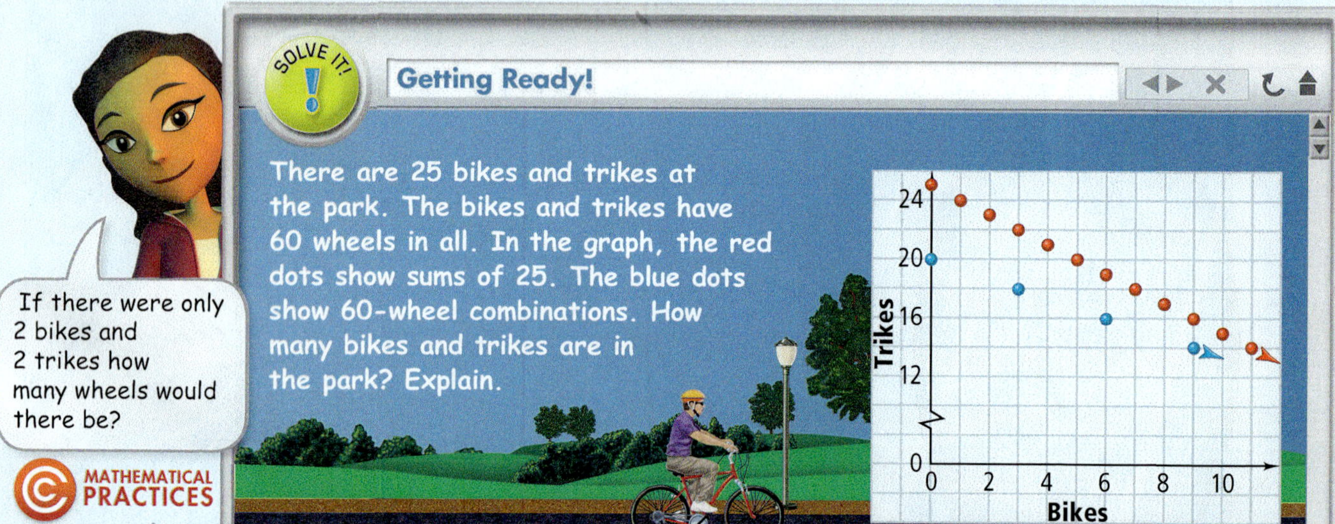

Lesson Vocabulary

- system of equations
- linear system
- solution of a system
- inconsistent system
- consistent system
- independent system
- dependent system

When you have two or more related unknowns, you may be able to represent their relationship with a **system of equations**—a set of two or more equations.

Essential Understanding To solve a system of equations, find a set of values that replace the variables in the equations and make each equation true.

A **linear system** consists of linear equations. A **solution of a system** is a set of values for the variables that makes all the equations true. You can solve a system of equations graphically or by using tables.

Think

How can you use a graph to find the solution of a system? Find the point where the two lines intersect.

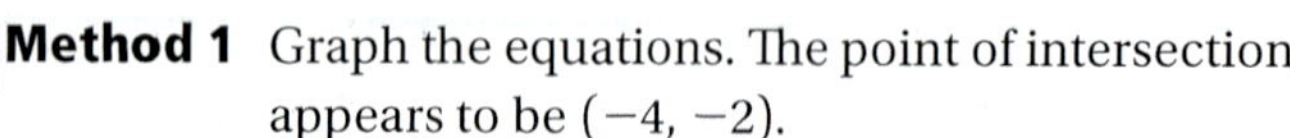

Problem 1 Using a Graph or Table to Solve a System

What is the solution of the system? $\begin{cases} -3x + 2y = 8 \\ x + 2y = -8 \end{cases}$

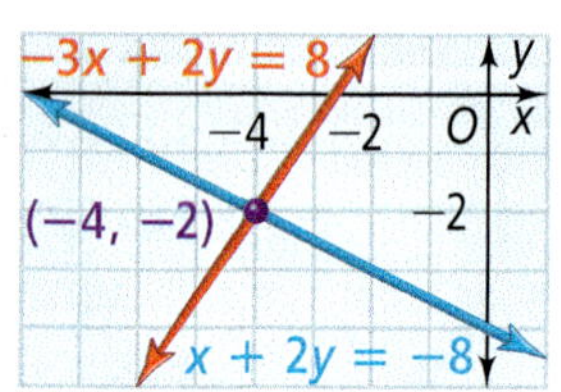

Method 1 Graph the equations. The point of intersection appears to be $(-4, -2)$.

Check by substituting the values into both equations.

$$-3x + 2y = 8 \qquad x + 2y = -8$$

$$-3(-4) + 2(-2) = 8 \checkmark \qquad -4 + 2(-2) = -8 \checkmark$$

Both equations are true so $(-4, -2)$ is the solution of the system.

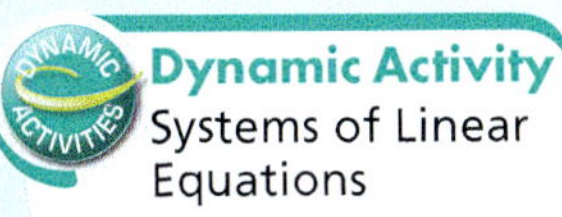

Dynamic Activity Systems of Linear Equations

Method 2 Use a table. Write the equations in slope-intercept form.

$$\begin{aligned} -3x + 2y &= 8 \\ 2y &= 3x + 8 \\ y_1 &= \tfrac{3}{2}x + 4 \end{aligned} \qquad \begin{aligned} x + 2y &= -8 \\ 2y &= -x - 8 \\ y_2 &= -\tfrac{1}{2}x - 4 \end{aligned}$$

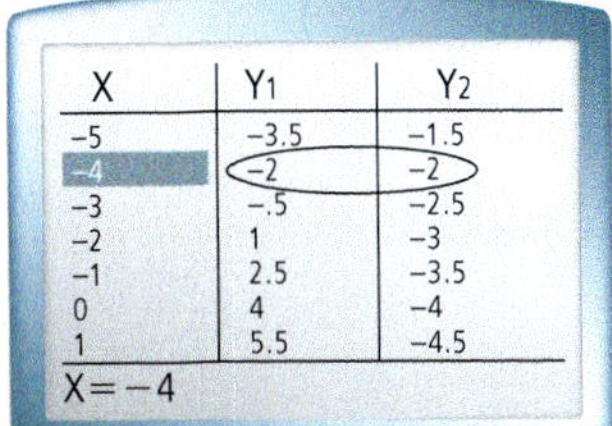

X	Y1	Y2
−5	−3.5	−1.5
−4	−2	−2
−3	−.5	−2.5
−2	1	−3
−1	2.5	−3.5
0	4	−4
1	5.5	−4.5

X=−4

Enter the equations in the **Y=** screen as **Y1** and **Y2**.
View the table. Adjust the x-values until you see $y_1 = y_2$.

When $x = -4$, both y_1 and y_2 equal -2. So, $(-4, -2)$ is the solution of the system.

Got It? **1.** What is the solution of the system? $\begin{cases} x - 2y = 4 \\ 3x + y = 5 \end{cases}$

Problem 2 Using a Table to Solve a Problem STEM

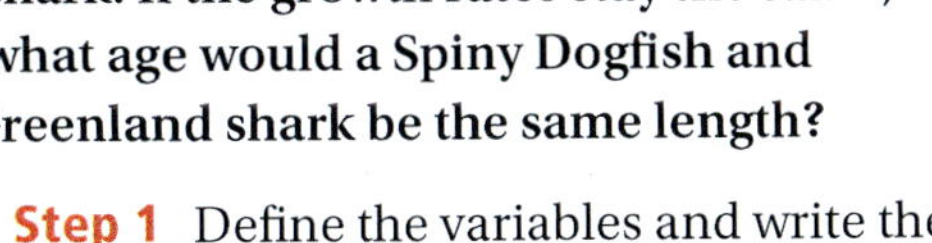

Biology **The diagrams show the birth lengths and growth rates of two species of shark. If the growth rates stay the same, at what age would a Spiny Dogfish and a Greenland shark be the same length?**

Step 1 Define the variables and write the equation for the length of each shark.

Let x = age in years.
Let y = length in centimeters.

Length of Greenland: $y_1 = 0.75x + 37$
Length of Spiny Dogfish: $y_2 = 1.5x + 22$

Step 2 Use the table to solve the problem.

List x-values until the corresponding y-values match.

Think
How can you use slope-intercept form to write each equation?
Use the growth rate for m and the length at birth for b.

Shark Length in cm

Age	Greenland	Spiny Dogfish
x	$y_1 = 0.75x + 37$	$y_2 = 1.5x + 22$
15	48.25	44.5
16	49	46
⋮	⋮	⋮
20	52	52

The sharks will be the same length when they are 20 years old.

Got It? **2. a.** If the growth rates continue, how long will each shark be when it is 25 years old?

b. Reasoning Explain why growth rates for these sharks may not continue indefinitely.

Problem 3 Using Linear Regression

Population The table shows the populations of the New York City and Los Angeles metropolitan regions from the census reports for 1950 through 2000. Assuming these linear trends continue, when will the populations of these regions be equal? What will that population be?

Populations of New York City and Los Angeles Metropolitan Regions (1950–2000)

	1950	1960	1970	1980	1990	2000
New York City	12,911,994	14,759,429	16,178,700	16,121,297	18,087,251	21,199,865
Los Angeles	4,367,911	6,742,696	7,032,075	11,497,568	14,531,529	16,373,645

SOURCE: U.S. Census Bureau

Know
Population data for two regions

Need
The point in time when their populations will be the same

Plan
- Use a calculator to find linear regression models.
- Plot the models.
- Find the point of intersection.

Enter all the numbers as millions, rounded to the nearest hundred thousand. For example, enter 12,911,994 as 12.9.

Step 1 Enter the data into lists on your calculator.
L1: number of years since 1950
L2: New York City populations
L3: Los Angeles populations

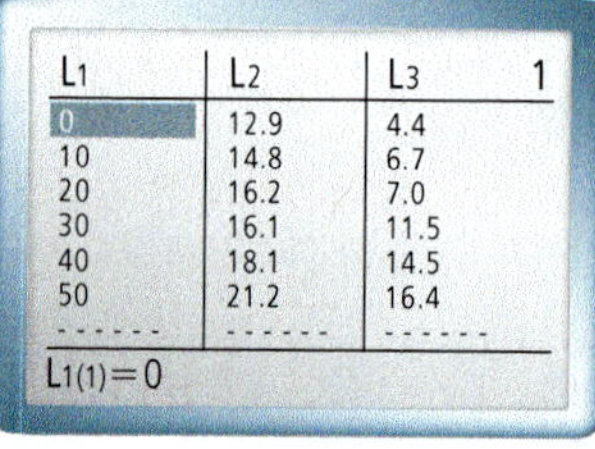

Step 2 Use **LinReg(ax + b)** to find lines of best fit.
Use **L1** and **L2** for New York City.
Use **L1** and **L3** for Los Angeles.

Step 3 Graph the linear regression lines.
Use the **Intersect** feature.

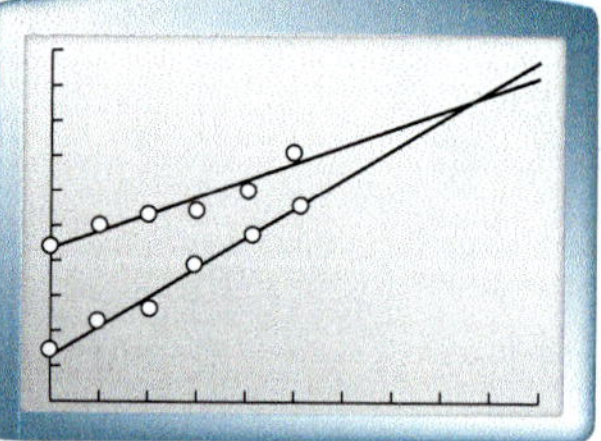

Think
What does x represent?
The x-value is the number of years *since* the zero year.

The x-value of the point of intersection is about 87, which represents the year 2037. The data suggest that the populations of the New York City and Los Angeles metropolitan regions will each be about 25.6 million in 2037.

Got It? 3. The table shows the populations of the San Diego and Detroit metropolitan regions. When were the populations of these regions equal? What was that population?

Populations of San Diego and Detroit Metropolitan Regions (1950–2000)

	1950	1960	1970	1980	1990	2000
San Diego	334,387	573,224	696,769	875,538	1,110,549	1,223,400
Detroit	1,849,568	1,670,144	1,511,482	1,203,339	1,027,974	951,270

SOURCE: U.S. Census Bureau

You can classify a system of two linear equations by the number of solutions.

A **consistent system** has at least one solution. — **Consistent system** → **Independent** / **Dependent**

An **independent system** has one solution.

A **dependent system** has infinitely many solutions.

An **inconsistent system** has no solution. — **Inconsistent system**

The graphs for an inconsistent system are parallel lines. So, there are no solutions. For a dependent system, the two equations represent the same line.

take note

Concept Summary Graphical Solutions of Linear Systems

Intersecting Lines	Coinciding Lines	Parallel Lines
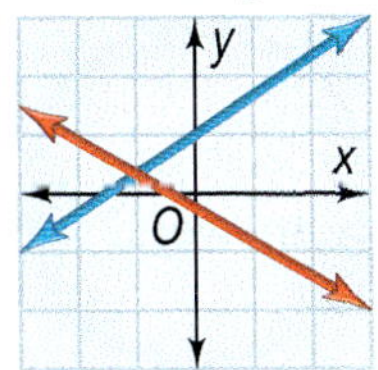	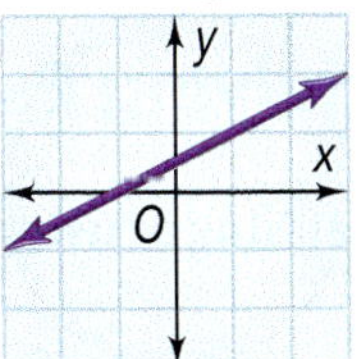	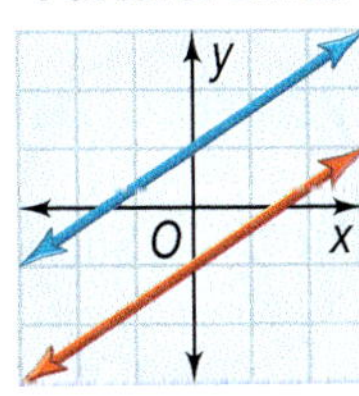
one solution Consistent Independent	infinitely many solutions Consistent Dependent	no solution Inconsistent

Problem 4 Classifying a System Without Graphing

Plan

What should you compare to classify the system?
Compare the slopes and y-intercepts of each line.

Without graphing, is the system *independent*, *dependent*, or *inconsistent*?

$$\begin{cases} 4y - 2x = 6 \\ 8y = 4x - 12 \end{cases}$$

Rewrite each equation in slope-intercept form. Compare slopes and y-intercepts.

$4y - 2x = 6$ $\qquad$ $8y = 4x - 12$

$y = \frac{1}{2}x + \frac{3}{2}$ $\qquad$ $y = \frac{1}{2}x - \frac{3}{2}$

$m = \frac{1}{2}$; y-intercept is $\frac{3}{2}$ $\qquad$ $m = \frac{1}{2}$; y-intercept is $-\frac{3}{2}$

The slopes are equal and the y-intercepts are different. The lines are different but parallel. The system is inconsistent.

Got It? **4.** Without graphing, is each system *independent*, *dependent*, or *inconsistent*?

a. $\begin{cases} -3x + y = 4 \\ x - \frac{1}{3}y = 1 \end{cases}$ **b.** $\begin{cases} 2x + 3y = 1 \\ 4x + y = -3 \end{cases}$ **c.** $\begin{cases} y = 2x - 3 \\ 6x - 3y = 9 \end{cases}$

Lesson Check

Do you know HOW?

Solve each system of equations by graphing. Check your solution.

1. $\begin{cases} y = x - 1 \\ y = -x + 3 \end{cases}$

2. $\begin{cases} 2x + y = 4 \\ x - y = 2 \end{cases}$

3. You bought a total of 6 pens and pencils for \$4. If each pen costs \$1 and each pencil costs \$.50, how many pens and pencils did you buy?

Do you UNDERSTAND?

4. **Vocabulary** Is it possible for a system of equations to be both independent and inconsistent? Explain.

5. **Open-Ended** Write a system of linear equations that has no solution.

6. **Reasoning** In a system of linear equations, the slope of one line is the negative reciprocal of the slope of the other line. Is this system *independent*, *dependent*, or *inconsistent*? Explain.

Practice and Problem-Solving Exercises

Solve each system by graphing or using a table. Check your answers. See Problem 1.

7. $\begin{cases} y = x - 2 \\ y = -2x + 7 \end{cases}$

8. $\begin{cases} y = -x + 3 \\ y = \frac{3}{2}x - 2 \end{cases}$

9. $\begin{cases} 2x + 4y = 12 \\ x + y = 2 \end{cases}$

10. $\begin{cases} x = -3 \\ y = 5 \end{cases}$

11. $\begin{cases} 2x - 2y = 4 \\ y - x = 6 \end{cases}$

12. $\begin{cases} 3x + y = 5 \\ x - y = 7 \end{cases}$

Write and solve a system of equations for each situation. Check your answers. See Problem 2.

13. A store sells small notebooks for \$8 and large notebooks for \$10. If you buy 6 notebooks and spend \$56, how many of each size notebook did you buy?

14. A shop has one-pound bags of peanuts for \$2 and three-pound bags of peanuts for \$5.50. If you buy 5 bags and spend \$17, how many of each size bag did you buy?

Graphing Calculator **Find linear models for each set of data. In what year will the two quantities be equal?** See Problem 3.

15. **U.S. Life Expectancy at Birth (1970–2000)**

Year	1970	1975	1980	1985	1990	1995	2000
Men (years)	67.1	68.8	70.0	71.1	71.8	72.5	74.3
Women (years)	74.7	76.6	77.4	78.2	78.8	78.9	79.7

Source: U.S. Census Bureau

16. **Annual U.S. Consumption of Vegetables**

Year	1980	1985	1990	1995	1998	1999	2000
Broccoli (lb/person)	1.5	2.6	3.4	4.3	5.1	6.5	6.1
Cucumbers (lb/person)	3.9	4.4	4.7	5.6	6.5	6.8	6.4

Source: U.S. Census Bureau

Without graphing, classify each system as *independent*, *dependent*, or *inconsistent*.

See Problem 4.

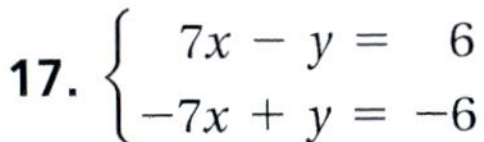

17. $\begin{cases} 7x - y = 6 \\ -7x + y = -6 \end{cases}$

18. $\begin{cases} -3x + y = 4 \\ x - \frac{1}{3}y = 1 \end{cases}$

19. $\begin{cases} 4x + 8y = 12 \\ x + 2y = -3 \end{cases}$

20. $\begin{cases} y = 2x - 1 \\ y = -2x + 5 \end{cases}$

21. $\begin{cases} x = 6 \\ y = -2 \end{cases}$

22. $\begin{cases} 2y = 5x + 6 \\ -10x + 4y = 8 \end{cases}$

23. $\begin{cases} x - 3y = 2 \\ 4x - 12y = 8 \end{cases}$

24. $\begin{cases} y - x = 0 \\ y = -x \end{cases}$

25. $\begin{cases} 2y - x = 4 \\ \frac{1}{2}x + y = 2 \end{cases}$

26. $\begin{cases} x + 4y = 12 \\ 2x - 8y = 4 \end{cases}$

27. $\begin{cases} 4x + 8y = -6 \\ 6x + 12y = -9 \end{cases}$

28. $\begin{cases} 4y - 2x = 6 \\ 8y = 4x - 12 \end{cases}$

B Apply

Graph and solve each system.

29. $\begin{cases} 3 = 4y + x \\ 4y = -x + 3 \end{cases}$

30. $\begin{cases} y = \frac{1}{2}x + \frac{1}{2} \\ y = \frac{1}{4}x + \frac{3}{2} \end{cases}$

31. $\begin{cases} 3x + 6y - 12 = 0 \\ x + 2y = 8 \end{cases}$

32. $\begin{cases} 3x = -5y + 4 \\ 250 + 150x = 300y \end{cases}$

33. $\begin{cases} y = -\frac{1}{2}x + 8 \\ y = 2x - 6 \end{cases}$

34. $\begin{cases} x + 3y = 6 \\ 6y + 2x = 12 \end{cases}$

Without graphing, classify each system as *independent*, *dependent*, or *inconsistent*.

35. $\begin{cases} 3x - 2y = 8 \\ 4y = 6x - 5 \end{cases}$

36. $\begin{cases} 2x + 8y = 6 \\ x = -4y + 3 \end{cases}$

37. $\begin{cases} 3m = -5n + 4 \\ n - \frac{6}{5} = -\frac{3}{5}m \end{cases}$

38. **Reasoning** Find the solution of the system of equations $f(x) = 3x - 1$ and $g(x) = |x - 3|$. Explain why the x-coordinates of the points where the graphs of the equations $y = f(x)$ and $y = g(x)$ intersect are solutions of $3x - 1 = |x - 3|$.

39. **Think About a Plan** You and a friend are both reading a book. You read 2 pages each minute and have already read 55 pages. Your friend reads 3 pages each minute and has already read 35 pages. Graph and solve a system of equations to find when the two of you will have read the same number of pages. Since the number of pages you have read depends on how long you have been reading, let x represent the number of minutes it takes to read y pages.
    - How can you describe the relationship between x and y for you?
    - How can you describe the relationship between x and y for your friend?
    - How can a graph help you solve this problem?

40. **Sports** You can choose between two tennis courts at two university campuses to learn how to play tennis. One campus charges \$25 per hour. The other campus charges \$20 per hour plus a one-time registration fee of \$10.
    a. Write a system of equations to represent the cost c for h hours of court use at each campus.
    b. **Graphing Calculator** Find the number of hours for which the costs are the same.
    c. **Reasoning** If you want to practice for a total of 10 hours, which university campus should you choose? Explain.

41. Error Analysis Your friend used a graphing calculator to solve a system of linear equations, shown below. After using the **TABLE** feature, your friend says that the system has no solution. Explain what your friend did wrong. What is the solution of the system?

$$2x + y = 6 \qquad 3x + 2y = 8$$
$$y = 6 - 2x \qquad y = \frac{8 - 3x}{2}$$

X	Y1	Y2
−4	14	10
−3	12	8.5
−2	10	7
−1	8	5.5
0	6	4
1	4	2.5
2	2	1

X=2

42. Reasoning Is it possible for an inconsistent linear system to contain two lines with the same *y*-intercept? Explain.

43. Writing Summarize the possible relationships for the *y*-intercepts, slopes, and number of solutions in a system of two linear equations in two variables.

Reasoning **Determine whether each statement is *always, sometimes* or *never* true for the following system.**

$$\begin{cases} y = x + 3 \\ y = mx + b \end{cases}$$

44. If $m = 1$, the system has no solution.

45. If $b = 3$, the system has exactly one solution.

46. If $m \neq 1$, the system has no solution.

47. If $m \neq 1$ and $b = 2$, the system has infinitely many solutions.

Open-Ended **Write a second equation for each system so that the system will have the indicated number of solutions.**

48. infinite number of solutions

$$\begin{cases} \frac{x}{4} + \frac{y}{3} = 1 \\ \underline{\quad ? \quad} \end{cases}$$

49. no solutions

$$\begin{cases} 5x + 2y = 10 \\ \underline{\quad ? \quad} \end{cases}$$

50. Write a system of linear equations with the solution set $\{(x, y) \mid y = 5x + 2\}$.

51. Reasoning What relationship exists between the equations in a dependent system?

52. Economics Research shows that in a certain market only 2000 widgets can be sold at \$8 each, but if the price is reduced to \$3, then 10,000 can be sold.

a. Let *p* represent price and *n* represent the number of widgets. Identify the independent and dependent variables.

b. Write a linear equation that relates price and the quantity demanded. This type of equation is called a *demand* equation.

c. A shop can make 2000 widgets for \$5 each and 20,000 widgets for \$2 each. Use this information to write a linear equation that relates price and the quantity supplied. This type of equation is called a *supply* equation.

d. Find the equilibrium point where supply is equal to demand. Explain the meaning of the coordinates of this point within the context of the exercise.

Standardized Test Prep

SAT/ACT

53. Which graph shows the solution of the following system? $\begin{cases} 4x + y = 1 \\ x + 4y = -11 \end{cases}$

Ⓐ 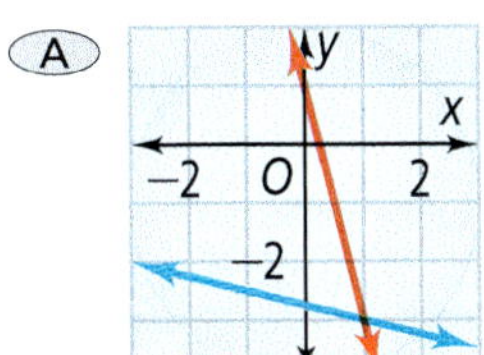

Ⓑ

Ⓒ 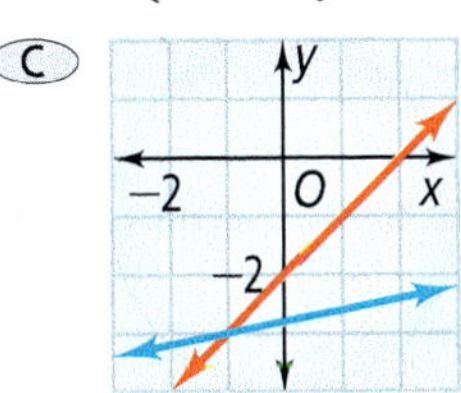

Ⓓ

54. Which is the equation of a line that is perpendicular to the line in the graph?

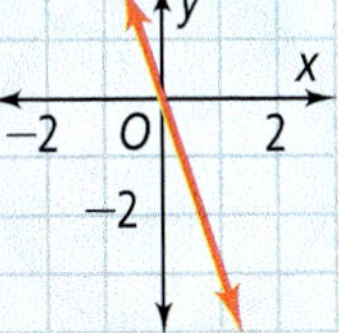

Ⓕ $y = -3x + 2$

Ⓖ $y = \frac{1}{3}x + 5$

Ⓗ $y = -\frac{1}{3}x - 4$

Ⓘ $y = 3x - 1$

55. Which inequality represents the graph at the right?

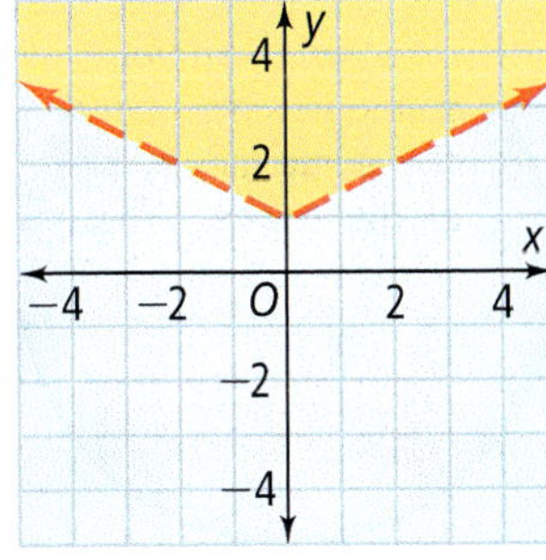

Ⓐ $y \geq \frac{1}{2}|x| + 1$

Ⓑ $y \leq \frac{1}{2}|x| + 1$

Ⓒ $y > \frac{1}{2}|x| + 1$

Ⓓ $y < \frac{1}{2}|x| + 1$

Extended Response

56. Amy ordered prints of a total of 6 photographs in two different sizes, 5×7 and 4×6, from an online site. She paid \$7.50 for her order. The cost of a 5×7 print is \$1.75 and the cost of a 4×6 print is \$.25. Explain how to solve a system of equations using tables to find the number of 4×6 prints Amy ordered.

Mixed Review

Graph each inequality on a coordinate plane. See Lesson 2-8.

57. $3x - 4y \geq 16$ **58.** $-5x > 8y + 4$ **59.** $x < -4$

Solve each inequality. Check your solution. See Lesson 1-5.

60. $3n < -4(2 + n)$ **61.** $\frac{x}{3} + 5 \geq \frac{1}{6}$ **62.** $4x - 2 > \frac{1}{2}$

Find the slope of the line through each pair of points. See Lesson 2-3.

63. $(-2, -4)$ and $(1, 2)$ **64.** $(0, 0)$ and $(5, -3)$ **65.** $(1, 3)$ and $(4, 9)$

Get Ready! To prepare for Lesson 3-2, do Exercises 66 and 67.

66. What is the value of $a + b - 2c$ for $a = 3$, $b = 1$, and $c = -3$? See Lesson 1-3.

67. Substitute -3 for x in each of the following equations. What is the value of y?

a. $y = 2x + 3$
b. $y = -x + 5$
c. $y = 3x - 1$

Solving Systems Algebraically

Content Standards

A.REI.6 Solve systems of linear equations exactly and approximately (e.g., with graphs), focusing on pairs of linear equations in two variables.

Also A.REI.5, A.CED.2

Objective To solve linear systems algebraically

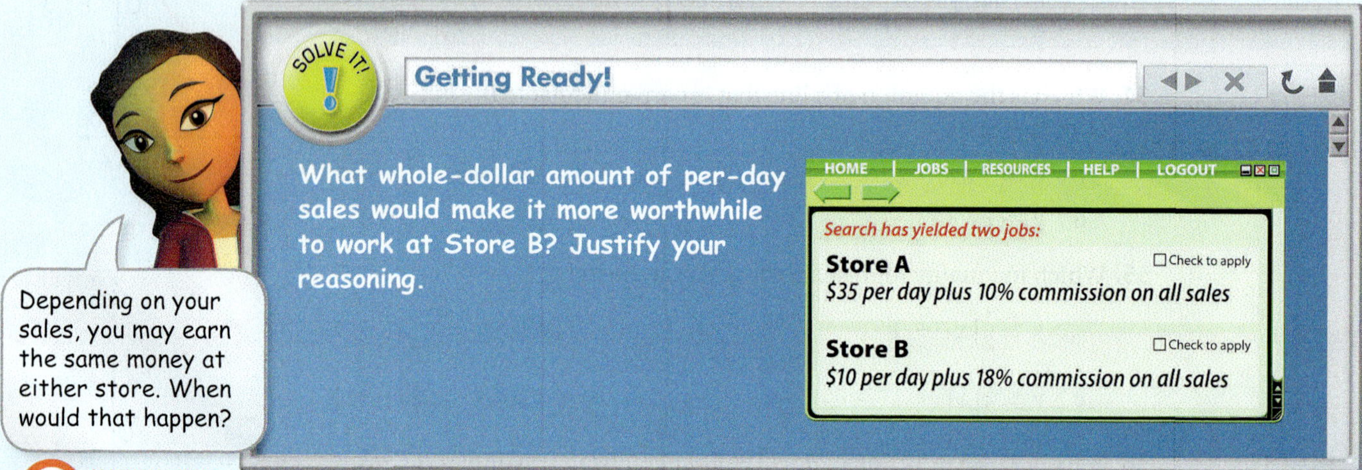

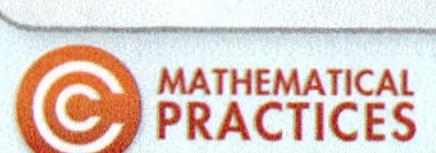

When you try to solve a system of equations by graphing, the coordinates of the point of intersection may not be obvious.

Lesson Vocabulary
- equivalent systems

Essential Understanding You can solve a system of equations by writing equivalent systems until the value of one variable is clear. Then substitute to find the value(s) of the other variable(s).

You can use the substitution method to solve a system of equations when it is easy to isolate one of the variables. After isolating the variable, substitute for that variable in the other equation. Then solve for the other variable.

Problem 1 Solving by Substitution

What is the solution of the system of equations? $\begin{cases} 3x + 4y = 12 \\ 2x + y = 10 \end{cases}$

Think

Which variable should you solve for first?

In the second equation, the coefficient of y is 1. It is the easiest variable to isolate.

Step 1

Solve one equation for one of the variables.

$$\begin{aligned} 2x + y &= 10 \\ y &= -2x + 10 \end{aligned}$$

Step 2

Substitute the expression for y in the other equation. Solve for x.

$$\begin{aligned} 3x + 4y &= 12 \\ 3x + 4(-2x + 10) &= 12 \\ 3x - 8x + 40 &= 12 \\ x &= 5.6 \end{aligned}$$

Step 3

Substitute the value for x into one of the original equations. Solve for y.

$$\begin{aligned} 2x + y &= 10 \\ 2(5.6) + y &= 10 \\ 11.2 + y &= 10 \\ y &= -1.2 \end{aligned}$$

The solution is $(5.6, -1.2)$.

Got It? **1.** What is the solution of the system of equations? $\begin{cases} x + 3y = 5 \\ -2x - 4y = -5 \end{cases}$

Problem 2 Using Substitution to Solve a Problem

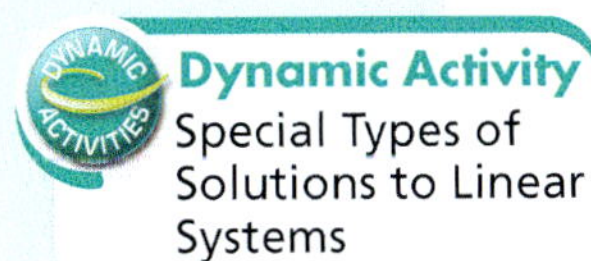

Dynamic Activity
Special Types of Solutions to Linear Systems

Music A music store offers piano lessons at a discount for customers buying new pianos. The costs for lessons and a one-time fee for materials (including music books, CDs, software, etc.) are shown in the advertisement. What is the cost of each lesson and the one-time fee for materials?

Piano Lessons
6 Lessons: $300
12 Lessons: $480
(Prices include one-time fee)

Relate $6 \cdot$ cost of one lesson $+$ one-time fee $= \$300$

$12 \cdot$ cost of one lesson $+$ one-time fee $= \$480$

Define Let c = the cost of one lesson.

Let f = the one-time fee.

Write $\begin{cases} 6 \cdot c + f = 300 \\ 12 \cdot c + f = 480 \end{cases}$

$6c + f = 300$
$f = 300 - 6c$

Choose one equation. Solve for f in terms of c.

$12c + (300 - 6c) = 480$
$c = 30$

Substitute the expression for f into the other equation, $12c + f = 480$. Solve for c.

$6(30) + f = 300$
$f = 120$

Substitute the value of c into one of the equations. Solve for f.

Think
Which equation should you use to find f?
Use the equation with numbers that are easier to work with.

Check Substitute $c = 30$ and $f = 120$ in the original equations.

$6c + f = 300$
$6(30) + 120 \stackrel{?}{=} 300$
$180 + 120 \stackrel{?}{=} 300$
$300 = 300$ ✔

$12c + f = 480$
$12(30) + 120 \stackrel{?}{=} 480$
$360 + 120 \stackrel{?}{=} 480$
$480 = 480$ ✔

The cost of each lesson is $30. The one-time fee for materials is $120.

Got It? **2.** An online music company offers 15 downloads for $19.75 and 40 downloads for $43.50. Each price includes the same one-time registration fee. What is the cost of each download and the registration fee?

You can use the Addition Property of Equality to solve a system of equations. If you add a pair of additive inverses or subtract identical terms, you can eliminate a variable.

Problem 3 Solving by Elimination

What is the solution of the system of equations? $\begin{cases} 4x + 2y = 9 \\ -4x + 3y = 16 \end{cases}$

Think

How can you use the Addition Property of Equality?
Since $-4x + 3y$ is equal to 16, you can add the same value to each side of $4x + 2y = 9$.

$$\begin{aligned} 4x + 2y &= 9 \\ -4x + 3y &= 16 \\ \hline 5y &= 25 \end{aligned}$$

One equation has $4x$ and the other has $-4x$. Add to eliminate the variable x.

$y = 5$ Solve for y.

$4x + 2y = 9$ Choose one of the original equations.

$4x + 2(5) = 9$ Substitute for y.

$4x = -1$ Solve for x.

$x = -\frac{1}{4}$

The solution is $\left(-\frac{1}{4}, 5\right)$.

Got It? **3.** What is the solution of the system of equations? $\begin{cases} -2x + 8y = -8 \\ 5x - 8y = 20 \end{cases}$

When you multiply each side of one or both equations in a system by the same nonzero number, the new system and the original system have the same solutions. The two systems are called **equivalent systems**. You can use this method to make additive inverses.

Problem 4 Solving an Equivalent System

What is the solution of the system of equations? $\begin{cases} ① & 2x + 7y = 4 \\ ② & 3x + 5y = -5 \end{cases}$

Think

By multiplying ① by 3 and ② by -2, the x-terms become opposites, and you can eliminate them. Add ③ and ④. Solve for y.

Write

① $2x + 7y = 4$ ③ $6x + 21y = 12$

② $3x + 5y = -5$ ④ $-6x - 10y = 10$

$11y = 22$

$y = 2$

Now that you know the value of y, use either equation to find x.

① $2x + 7(2) = 4$

$2x + 14 = 4$

$2x = -10$

$x = -5$

The solution is $(-5, 2)$.

  **Got It?** **4. a.** What is the solution of this system of equations? $\begin{cases} 3x + 7y = 15 \\ 5x + 2y = -4 \end{cases}$

b. Reasoning In Problem 4, you found that $y = 2$. Substitute this value into equation ② instead of equation ①. Do you still get the same value for x? Explain why.

Solving a system algebraically does not always provide a unique solution. Sometimes you get infinitely many solutions. Sometimes you get no solutions.

Problem 5 Solving Systems Without Unique Solutions

Think

How are the two equations in this system related?
Multiplying both sides of the first equation by -1 results in the second equation.

What are the solutions of the following systems? Explain.

A $\begin{cases} -3x + y = -5 \\ 3x - y = 5 \end{cases}$

$$0 = 0$$

Elimination gives an equation that is always true. The two equations in the system represent the same line. This is a dependent system with infinitely many solutions.

B $\begin{cases} 4x - 6y = 6 \\ -4x + 6y = 10 \end{cases}$

$$0 = 16$$

Elimination gives an equation that is always false. The two equations in the system represent parallel lines. This is an inconsistent system. It has no solutions.

 **Got It?** **5.** What are the solutions of the following systems? Explain.

a. $\begin{cases} -x + y = -2 \\ 2x - 2y = 0 \end{cases}$

b. $\begin{cases} 4x + y = 6 \\ 12x + 3y = 18 \end{cases}$

Lesson Check

Do you know HOW?

Solve each system by substitution.

1. $\begin{cases} 3x + 5y = 13 \\ 2x + y = 4 \end{cases}$

2. $\begin{cases} 2x - 3y = 6 \\ x + y = -12 \end{cases}$

Solve each system by elimination.

3. $\begin{cases} 2x + 3y = 7 \\ -2x + 5y = 1 \end{cases}$

4. $\begin{cases} x + 2y = -1 \\ x - y = 8 \end{cases}$

5. $\begin{cases} x - y = -4 \\ 3x + 2y = 7 \end{cases}$

6. $\begin{cases} 3x + 4y = 10 \\ 2x + 3y = 7 \end{cases}$

Do you UNDERSTAND?

7. Vocabulary Give an example of two equivalent systems.

8. Compare and Contrast Explain how the substitution method of solving a system of equations differs from the elimination method.

9. Writing A café sells a regular cup of coffee for \$1 and a large cup for \$1.50. Melissa and her friends buy 5 cups of coffee and spend a total of \$6. Explain how to write and solve a system of equations to find the number of large cups of coffee they bought.

Practice and Problem-Solving Exercises

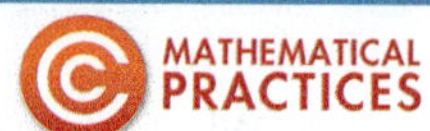

Solve each system by substitution. Check your answers.

 See Problem 1.

10. $\begin{cases} 4x + 2y = 7 \\ y = 5x \end{cases}$

11. $\begin{cases} 3c + 2d = 2 \\ d = 4 \end{cases}$

12. $\begin{cases} x + 12y = 68 \\ x = 8y - 12 \end{cases}$

13. $\begin{cases} 4p + 2q = 8 \\ q = 2p + 1 \end{cases}$

14. $\begin{cases} x + 3y = 7 \\ 2x - 4y = 24 \end{cases}$

15. $\begin{cases} x + 6y = 2 \\ 5x + 4y = 36 \end{cases}$

16. $\begin{cases} t = 2r + 3 \\ 5r - 4t = 6 \end{cases}$

17. $\begin{cases} y = 2x - 1 \\ 3x - y = -1 \end{cases}$

18. $\begin{cases} r + s = -12 \\ 4r - 6s = 12 \end{cases}$

19. Money A student has some \$1 bills and \$5 bills in his wallet. He has a total of 15 bills that are worth \$47. How many of each type of bill does he have?

20. A student took 60 minutes to answer a combination of 20 multiple-choice and extended-response questions. She took 2 minutes to answer each multiple-choice question and 6 minutes to answer each extended-response question.

a. Write a system of equations to model the relationship between the number of multiple choice questions m and the number of extended-response questions r.

b. How many of each type of question was on the test?

21. Transportation A youth group with 26 members is going skiing. Each of the five chaperones will drive a van or sedan. The vans can seat seven people, and the sedans can seat five people. Assuming there are no empty seats, how many of each type of vehicle could transport all 31 people to the ski area in one trip?

Solve each system by elimination.

22. $\begin{cases} x + y = 12 \\ x - y = 2 \end{cases}$

23. $\begin{cases} x + 2y = 10 \\ x + y = 6 \end{cases}$

24. $\begin{cases} 3a + 4b = 9 \\ -3a - 2b = -3 \end{cases}$

25. $\begin{cases} 4x + 2y = 4 \\ 6x + 2y = 8 \end{cases}$

26. $\begin{cases} 2w + 5y = -24 \\ 3w - 5y = 14 \end{cases}$

27. $\begin{cases} 3u + 3v = 15 \\ -2u + 3v = -5 \end{cases}$

28. $\begin{cases} 3x + 2y = 6 \\ 3x + 3 = y \end{cases}$

29. $\begin{cases} 5x - y = 4 \\ 2x - y = 1 \end{cases}$

30. $\begin{cases} 2r + s = 3 \\ 4r - s = 9 \end{cases}$

Solve each system by elimination.

See Problems 4 and 5.

31. $\begin{cases} 4x - 6y = -26 \\ -2x + 3y = 13 \end{cases}$

32. $\begin{cases} 9a - 3d = 3 \\ -3a + d = -1 \end{cases}$

33. $\begin{cases} 2a + 3b = 12 \\ 5a - b = 13 \end{cases}$

34. $\begin{cases} 2x - 3y = 6 \\ 6x - 9y = 9 \end{cases}$

35. $\begin{cases} 20x + 5y = 120 \\ 10x + 7.5y = 80 \end{cases}$

36. $\begin{cases} 6x - 2y = 11 \\ -9x + 3y = 16 \end{cases}$

37. $\begin{cases} 2x - 3y = -1 \\ 3x + 4y = 8 \end{cases}$

38. $\begin{cases} 5x - 2y = -19 \\ 2x + 3y = 0 \end{cases}$

39. $\begin{cases} r + 3s = 7 \\ 2r - s = 7 \end{cases}$

40. $\begin{cases} y = 4 - x \\ 3x + y = 6 \end{cases}$

41. $\begin{cases} 3x + 2y = 10 \\ 6x + 4y = 15 \end{cases}$

42. $\begin{cases} 3m + 4n = -13 \\ 5m + 6n = -19 \end{cases}$

43. **Think About a Plan** Suppose you have a part-time job delivering packages. Your employer pays you a flat rate of \$9.50 per hour. You discover that a competitor pays employees \$2 per hour plus \$3 per delivery. How many deliveries would the competitor's employees have to make in four hours to earn the same pay you earn in a four-hour shift?
- How can you write a system of equations to model this situation?
- Which method should you use to solve the system?
- How can you interpret the solution in the context of the problem?

Solve each system.

44. $\begin{cases} 5x + y = 0 \\ 5x + 2y = 30 \end{cases}$

45. $\begin{cases} 2m = -4n - 4 \\ 3m + 5n = -3 \end{cases}$

46. $\begin{cases} 7x + 2y = -8 \\ 8y = 4x \end{cases}$

47. $\begin{cases} 2m + 4n = 10 \\ 3m + 5n = 11 \end{cases}$

48. $\begin{cases} -6 = 3x - 6y \\ 4x = 4 + 5y \end{cases}$

49. $\begin{cases} \frac{x}{3} + \frac{4y}{3} = 300 \\ 3x - 4y = 300 \end{cases}$

50. $\begin{cases} 0.02a - 1.5b = 4 \\ 0.5b - 0.02a = 1.8 \end{cases}$

51. $\begin{cases} 4y = 2x \\ 2x + y = \frac{x}{2} + 1 \end{cases}$

52. $\begin{cases} \frac{1}{2}x + \frac{2}{3}y = 1 \\ \frac{3}{4}x - \frac{1}{3}y = 2 \end{cases}$

53. **Error Analysis** Identify and correct the error shown in finding the solution of $\begin{cases} 3x - 4y = 14 \\ x + y = -7 \end{cases}$ using substitution.

$x + y = -7$
$y = -7 - x$

$3x - 4y = 14$
$3x - 4(-7 - x) = 14$
$3x - 28 - 4x = 14$
$-x - 28 = 14$
$x = -42$

$y = -7 - (-42)$
$y = 35$

54. **Break-Even Point** Jenny's Bakery sells carrot muffins at \$2 each. The electricity to run the oven is \$120 per day and the cost of making one carrot muffin is \$1.40. How many muffins need to be sold each day to break even?

55. **Open-Ended** Write a system of equations in which both equations must be multiplied by a number other than 1 or −1 before using elimination. Solve the system.

STEM 56. **Chemistry** A scientist wants to make 6 milliliters of a 30% sulfuric acid solution. The solution is to be made from a combination of a 20% sulfuric acid solution and a 50% sulfuric acid solution. How many milliliters of each solution must be combined to make the 30% solution?

57. **Writing** Explain how you decide whether to use substitution or elimination to solve a system.

58. The equation $3x - 4y = 2$ and which equation below form a system with no solutions?

A. $2y = 1.5x - 2$
B. $2y = 1.5x - 1$
C. $3x + 4y = 2$
D. $4y - 3x = -2$

For each system, choose the method of solving that seems easier to use. Explain why you made each choice. Solve each system.

59. $\begin{cases} 3x - y = 5 \\ y = 4x + 2 \end{cases}$

60. $\begin{cases} 2x - 3y = 4 \\ 2x - 5y = -6 \end{cases}$

61. $\begin{cases} 6x - 3y = 3 \\ 5x - 5y = 10 \end{cases}$

62. Entertainment In the final round of a singing competition, the audience voted for one of the two finalists, Luke or Sean. Luke received 25% more votes than Sean received. Altogether, the two finalists received 5175 votes. How many votes did Luke receive?

STEM **63. Weather** The equation $F = \frac{9}{5}C + 32$ relates temperatures on the Celsius and Fahrenheit scales. Does any temperature have the same number reading on both scales? If so, what is the number?

Find the value of a that makes each system a dependent system.

64. $\begin{cases} y = 3x + a \\ 3x - y = 2 \end{cases}$

65. $\begin{cases} 3y = 2x \\ 6y - a - 4x = 0 \end{cases}$

66. $\begin{cases} y = \frac{x}{2} + 4 \\ 2y - x = a \end{cases}$

Standardized Test Prep

GRIDDED RESPONSE

SAT/ACT

67. What is the slope of the line at the right?

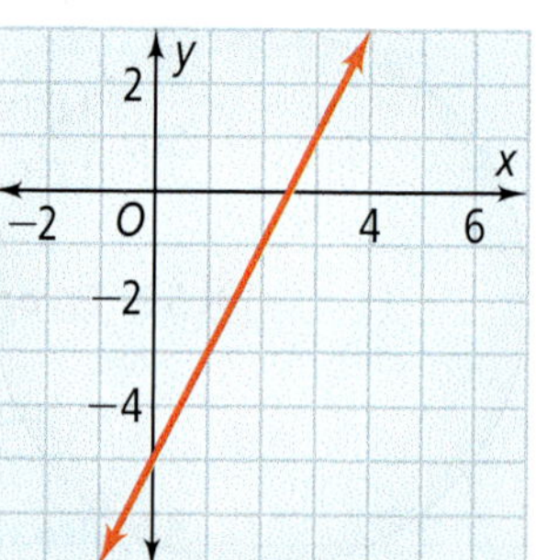

68. What is the x-value of the solution of $\begin{cases} x + y = 7 \\ 3x - 2y = 11 \end{cases}$?

69. Solve $9(x + 7) - 6(x - 3) = 99$. What is the value of x?

70. Georgia has only dimes and quarters in her bag. She has a total of 18 coins that are worth \$3. How many more dimes than quarters does she have?

71. The graph of $g(x)$ is a horizontal translation of $f(x) = 2|x + 1| + 3$, 5 units to the right. What is the x-value of the vertex of $g(x)$?

Mixed Review

Solve each system of equations by graphing.

See Lesson 3-1.

72. $\begin{cases} y = 3x + 4 \\ 2y = 6x - 2 \end{cases}$

73. $\begin{cases} -3y = 9x + 1 \\ 6y = -18x - 2 \end{cases}$

74. $\begin{cases} 4x - y = -5 \\ -8x + 2y = 15 \end{cases}$

Use the vertical line test to determine whether each graph represents a function.

See Lesson 2-1.

75.

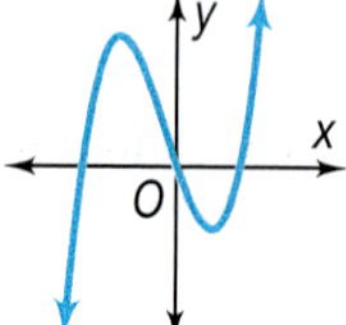

76.

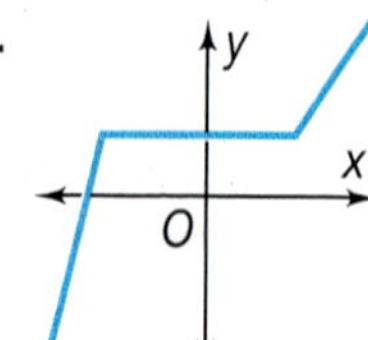

77.

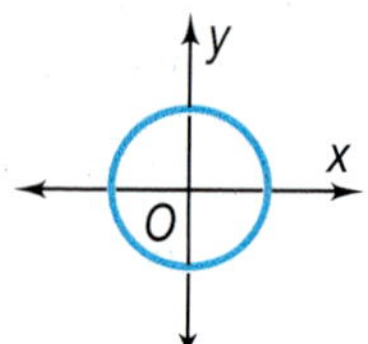

Get Ready! **To prepare for Lesson 3-3, do Exercises 78–80.**

Solve each inequality. Graph the solution.

See Lesson 1-5.

78. $-3(2x + 1) > 3$

79. $4x > -2$

80. $8y + 4 < 12$

3-3 Systems of Inequalities

Content Standards

A.REI.12 Graph the solutions to a linear inequality in two variables as a half-plane . . . and graph the solution set to a system of linear inequalities in two variables as the intersection of the corresponding half-planes.

Also A.REI.6, A.CED.3

Objective To solve systems of linear inequalities

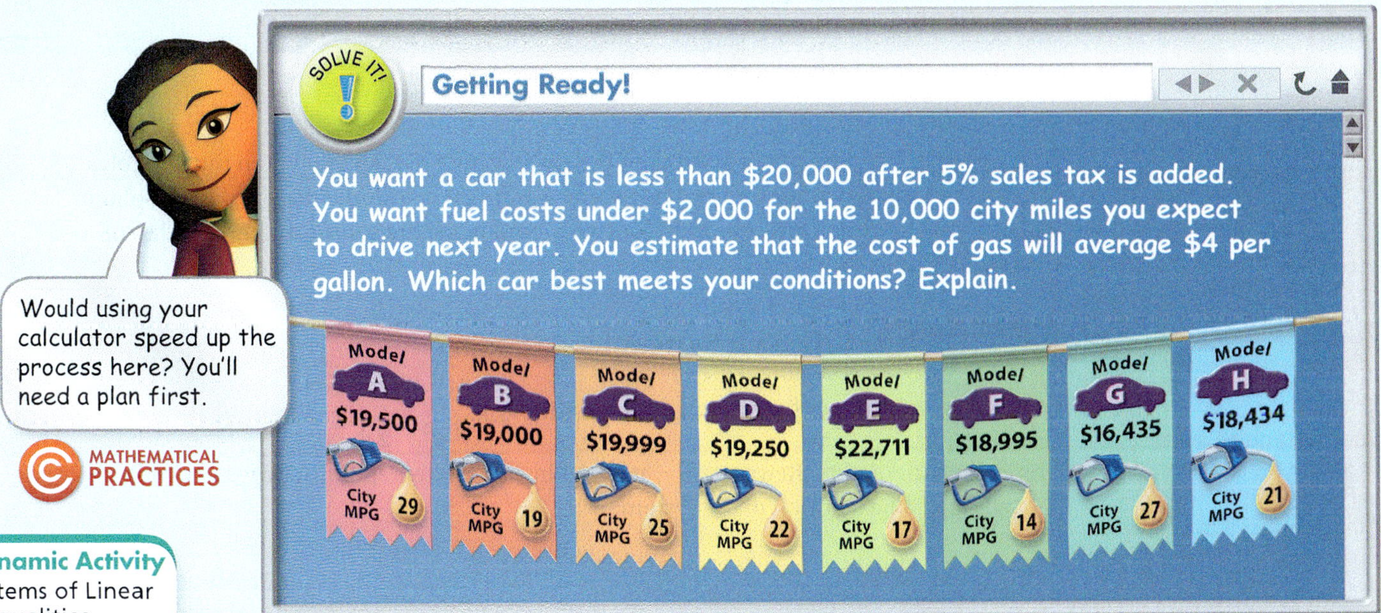

MATHEMATICAL PRACTICES

Dynamic Activity Systems of Linear Inequalities

An inequality and a system of inequalities can each have many solutions. A solution of a system of inequalities is a solution for each inequality in the system.

Essential Understanding You can solve a system of inequalities in more than one way. Graphing the solution is usually the most appropriate method. The solution is the set of all points that are solutions of each inequality in the system.

Problem 1 Solving a System by Using a Table

Assume that g and m are whole numbers. What is the solution of the system of inequalities? $\begin{cases} g + m \geq 6 \\ 5g + 2m \leq 20 \end{cases}$

Plan

Which inequality should you use to build a table?
The first inequality has an infinite number of whole number solutions. The second one has a finite number of solutions. Use the second inequality.

Make a table of values for g and m that satisfy the second inequality. The values for g and m must be whole numbers.

If $g = 0$, then $5(0) + 2m \leq 20$, and $m \leq 10$.
If $m = 0$, then $5g + 2(0) \leq 20$, and $g \leq 4$.

g	m
0	0, 1, 2, 3, 4, 5, 6, 7, 8, 9, 10
1	0, 1, 2, 3, 4, 5, 6, 7
2	0, 1, 2, 3, 4, 5
3	0, 1, 2
4	0

In the table, highlight each pair of values that satisfies the first inequality. The highlighted pairs are the solutions of both inequalities.

g	m
0	0, 1, 2, 3, 4, 5, 6, 7, 8, 9, 10
1	0, 1, 2, 3, 4, 5, 6, 7
2	0, 1, 2, 3, 4, 5
3	0, 1, 2
4	0

Got It? 1. Assume that x and y are whole numbers. What is the solution of the system of inequalities?

$$\begin{cases} x + y > 4 \\ 3x + 7y \le 21 \end{cases}$$

You can solve a system of linear inequalities by graphing. Recall that when the variables of a linear inequality represent real numbers, a graphed solution consists of a half-plane and possibly its boundary line. Thus, for two inequalities, the solution is the overlap of the two half-planes.

Problem 2 Solving a System by Graphing

Plan

How can you be sure to shade the correct half-planes?
If the inequality is in slope-intercept form, shade above the boundary if $y >$ or $y \ge$ and shade below the boundary if $y <$ or $y \le$. If not, use a test point.

What is the solution of the system of inequalities? $\begin{cases} 2x - y \ge -3 \\ y \ge -\frac{1}{2}x + 1 \end{cases}$

Graph each inequality. Rewrite $2x - y \ge -3$ in slope-intercept form as $y \le 2x + 3$. The overlap is the solution of the system.

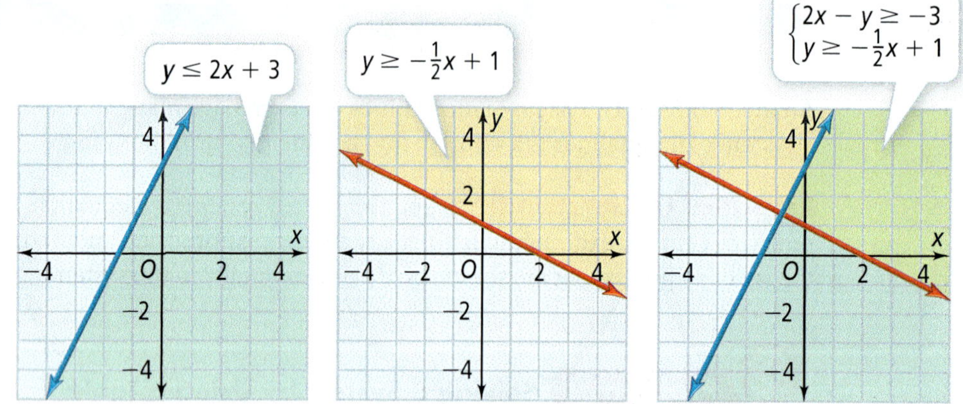

Check Pick a point in the overlap region, such as (0, 2), and check it in both inequalities of the system.

$$2x - y \ge -3 \qquad y \ge \frac{1}{2}x + 1$$

$$2(0) - 2 \ge -3 \qquad 2 \ge \frac{1}{2}(0) + 1$$

$$-2 \ge -3 \checkmark \qquad 2 \ge 1 \checkmark$$

Got It? 2. What is the solution of the system of inequalities? $\begin{cases} x + 2y \le 4 \\ y \ge -x - 1 \end{cases}$

Sometimes, you can model a real situation with a system of linear inequalities. Solutions to real-world problems are often whole numbers, so only certain points in the region of overlap will solve the problem.

Problem 3 Using a System of Inequalities

Fundraising **Your city's cultural center is sponsoring a concert to raise at least \$30,000 for the city's Youth Services. Tickets are \$20 for balcony seats and \$30 for orchestra seats. If the center has 500 orchestra seats, how many of each type of seat must they sell?**

Know

Must raise at least \$30,000. There are at most 500 orchestra seats.

Need

The possible sales of balcony and orchestra seats

Plan

- Model the problem with a system of inequalities.
- Graph the inequalities on your calculator.

Relate 20 · balcony seats + 30 · orchestra seats ≥ 30,000

orchestra seats ≤ 500

Define Let x = the number of balcony seats sold.

Let y = the number of orchestra seats sold.

Write $20 \cdot x + 30 \cdot y \geq 30{,}000$

$y \leq 500$

Rewrite $20x + 30y \geq 30{,}000$ in slope intercept form as $y \geq -\frac{2}{3}x + 1000$.

The system of inequalities is $\begin{cases} y \geq -\frac{2}{3}x + 1000 \\ y \leq 500 \end{cases}$.

Think

What do points in the overlap represent?

The points represent combinations of balcony and orchestra seats that have a total value of at least \$30,000.

Use your graphing calculator to graph the inequalities.

The solution is the overlap.

Test a point. If the cultural center sells 900 balcony and 450 orchestra tickets, will the Youth Services meet its goal?

$20(900) + 30(450) \overset{?}{\geq} 30{,}000$

$18{,}000 + 13{,}250 \overset{?}{\geq} 30{,}000$

$31{,}250 \geq 30{,}000$ ✔

$450 \overset{?}{\leq} 500$

$450 < 500$ ✔

1.1
RAD AUTO REAL
$y \geq -\frac{2}{3} \cdot x + 1000$
$y \leq 500$
(900, 450)
f1(x)=

Because the number of seats must be a whole number, only the points in the overlap that represent whole numbers are solutions.

Got It? **3.** A pizza parlor charges \$1 for each vegetable topping and \$2 for each meat topping. You want at least five toppings on your pizza. You have \$10 to spend on toppings. How many of each type of topping can you get on your pizza?

A system of inequalities can include nonlinear inequalities. You can also solve these systems graphically.

Problem 4 Solving a Linear/Absolute-Value System

What is the solution of the system of inequalities? $\begin{cases} y \le 3 \\ y \ge |x - 1| \end{cases}$

Graph each inequality.

$y \le 3$

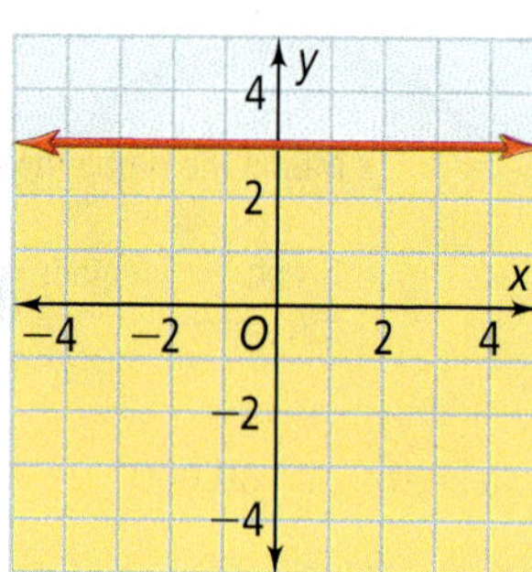

$y \ge |x - 1|$

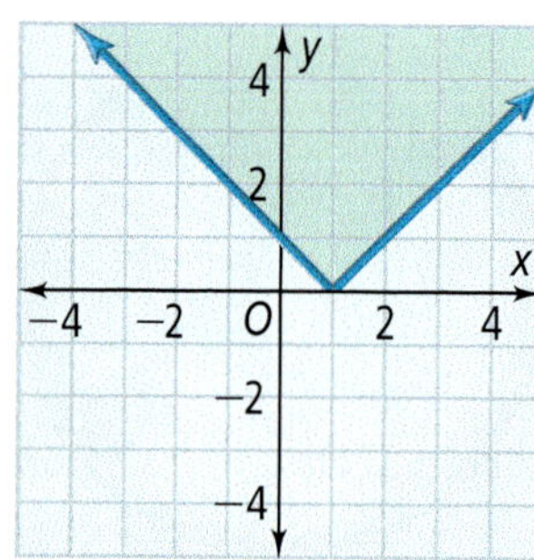

$\begin{cases} y \le 3 \\ y \ge |x - 1| \end{cases}$

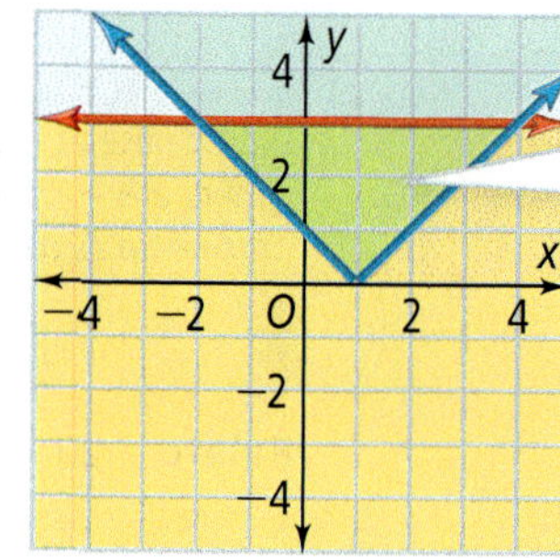

The region of overlap represents the solution.

Think

Why is the shape of the overlap different from that of a system of linear inequalities? The overlap is not formed by 2 half-planes so it is more varied in shape.

Got It? 4. What is the solution of the system of inequalities? $\begin{cases} y < -\frac{1}{3}x + 1 \\ y > 2|x - 1| \end{cases}$

Lesson Check

Do you know HOW?

Solve each system of inequalities by graphing.

1. $\begin{cases} x + y \ge 2 \\ 2x + y \le 5 \end{cases}$

2. $\begin{cases} y > x \\ y < x + 1 \end{cases}$

3. $\begin{cases} y \ge -3x - 1 \\ y < x + 2 \end{cases}$

4. You spend no more than 3 hours each day watching TV and playing football. You play football for at least 1 hour each day. What are the possible numbers of hours you can spend on each activity in one day?

Do you UNDERSTAND?

5. Reasoning Is the solution of a system of linear inequalities the union or intersection of the solutions of the two inequalities? Justify your answer.

6. Compare and Contrast Explain how the graphical solution of a system of inequalities is different from the graphical solution of a system of equations.

7. Error Analysis Describe and correct the error made in solving this system of inequalities.

$\begin{cases} y < \frac{1}{2}x - 1 \\ y > -3x + 3 \end{cases}$

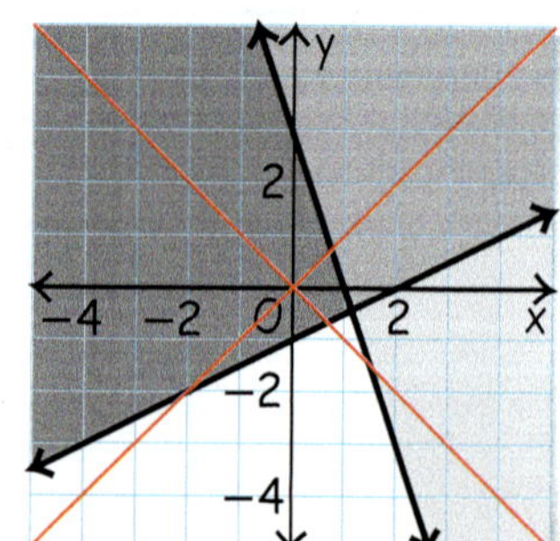

Practice and Problem-Solving Exercises

Practice

Find all whole number solutions of each system using a table. See Problem 1.

8. $\begin{cases} y + 3x \leq 8 \\ y - 3 > 2x \end{cases}$

9. $\begin{cases} x + y < 8 \\ 3x \leq y + 6 \end{cases}$

10. $\begin{cases} y \geq x + 2 \\ 3y < -6x + 6 \end{cases}$

Solve each system of inequalities by graphing. See Problem 2.

11. $\begin{cases} y \leq 2x + 2 \\ y < -x + 1 \end{cases}$

12. $\begin{cases} y > -2 \\ x < 1 \end{cases}$

13. $\begin{cases} y \leq 3 \\ y \leq \frac{1}{2}x + 1 \end{cases}$

14. $\begin{cases} y \leq 3x + 1 \\ -6x + 2y > 5 \end{cases}$

15. $\begin{cases} x + 2y \leq 10 \\ x + y \leq 3 \end{cases}$

16. $\begin{cases} -x - y \leq 2 \\ y - 2x > 1 \end{cases}$

17. $\begin{cases} y > -2x \\ 2x - y \geq 2 \end{cases}$

18. $\begin{cases} c \geq d - 3 \\ c < \frac{1}{2}d + 3 \end{cases}$

19. $\begin{cases} 2x + y < 1 \\ y > -2x + 3 \end{cases}$

20. You want to decorate a party hall with a total of at least 40 red and yellow balloons, with a minimum of 25 yellow balloons. Write and graph a system of inequalities to model the situation. See Problem 3.

21. A gardener wants to plant at least 50 tulips and rose plants in a garden, but no more than 20 rose plants. Write and graph a system of inequalities to model the situation.

Solve each system of inequalities by graphing.  See Problem 4.

22. $\begin{cases} y > 4 \\ y < |x - 1| \end{cases}$

23. $\begin{cases} y < -\frac{1}{3}x + 1 \\ y > |2x - 1| \end{cases}$

24. $\begin{cases} y > x - 2 \\ y \geq |x + 2| \end{cases}$

25. $\begin{cases} y \leq -\frac{4}{3}x \\ y \geq -|x| \end{cases}$

26. $\begin{cases} 3y < -x - 1 \\ y \leq |x + 1| \end{cases}$

27. $\begin{cases} y > -2 \\ y \leq -|x - 3| \end{cases}$

28. $\begin{cases} -2y < 4x + 2 \\ y > |2x + 1| \end{cases}$

29. $\begin{cases} -x \geq 4 - y \\ y \geq |3x - 6| \end{cases}$

30. $\begin{cases} y \leq x - 4 \\ y > |x - 6| \end{cases}$

Apply

31. Think About a Plan The food pyramid suggests that you eat 4–6 servings of fruits and vegetables a day for a healthy diet. It also says that the number of servings of vegetables should be greater than the number of servings of fruits. Find the number of servings of fruits and vegetables that could make a healthy diet. Use whole numbers only.

- How can you write two inequalities that model the information in the problem?
- How can you use a graph to find combinations of fruits and vegetable servings that may help in having a healthy diet?

32. College Admissions An entrance exam has two sections, a verbal section and a mathematics section. You can score a maximum of 1600 points. For admission, the school of your choice requires a math score of at least 600. Write a system of inequalities to model scores that meet the school's requirements. Then solve the system by graphing.

33. **Open-Ended** Write and graph a system of inequalities for which the solution is bounded by a dashed vertical line and a solid horizontal line.

34. **Writing** Explain how you determine where to shade when solving a system of inequalities.

35. Given a system of two linear inequalities, explain how you can pick test points in the plane to determine where to shade the solution set.

In Exercises 36–45, identify the inequalities A, B, and C for which the given ordered pair is a solution.

A. $x + y \leq 2$ **B.** $y \leq \frac{3}{2}x - 1$ **C.** $y > -\frac{1}{3}x - 2$

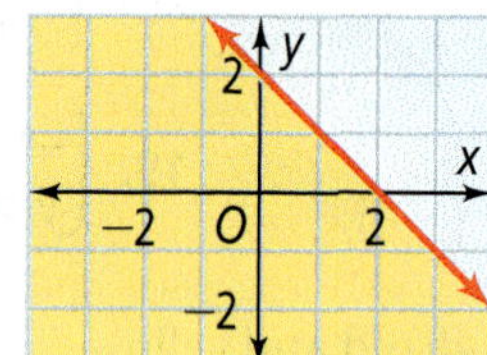
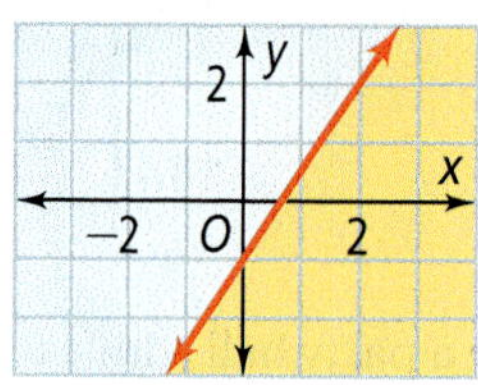
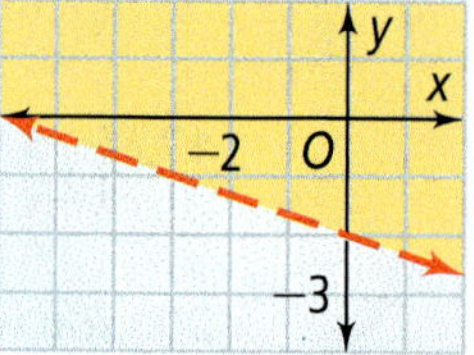

36. $(0, 0)$	**37.** $(-2, -5)$	**38.** $(-2, 0)$	**39.** $(0, -2)$	**40.** $(-15, 15)$
41. $(3, 2)$	**42.** $(2, 0)$	**43.** $(-6, 0)$	**44.** $(4, -1)$	**45.** $(-8, -11)$

Solve each system of inequalities by graphing.

46. $\begin{cases} x + y < 8 \\ x \geq 0 \\ y \geq 0 \end{cases}$ **47.** $\begin{cases} 2y - 4x \leq 0 \\ x \geq 0 \\ y \geq 0 \end{cases}$ **48.** $\begin{cases} y \geq -2x + 4 \\ x > -3 \\ y \geq 1 \end{cases}$

49. $\begin{cases} y \leq \frac{2}{3}x + 2 \\ y \geq |x| + 2 \end{cases}$ **50.** $\begin{cases} y < x - 1 \\ y > -|x - 2| + 1 \end{cases}$ **51.** $\begin{cases} 2x + y \leq 3 \\ y > |x + 3| - 2 \end{cases}$

52. $\begin{cases} y < |x - 1| + 2 \\ x \geq 0 \end{cases}$ **53.** $\begin{cases} y > |x - 1| + 1 \\ y \leq -|x - 3| + 4 \end{cases}$ **54.** $\begin{cases} y \leq |x| - 2 \\ y \leq |x| + 2 \end{cases}$

Geometry **Write a system of inequalities to describe each shaded figure.**

55.
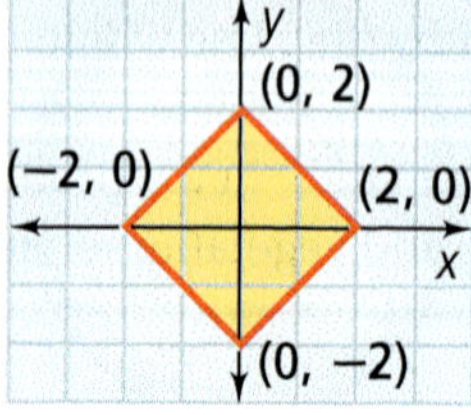

56.
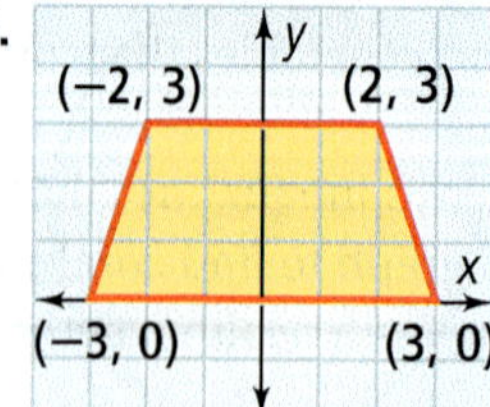

57.
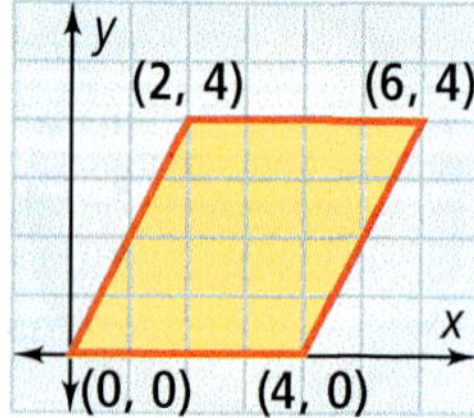

58. **a.** Graph the "bowtie" inequality, $|y| \leq |x|$.
b. Write a system of inequalities to describe the graph shown at the right.

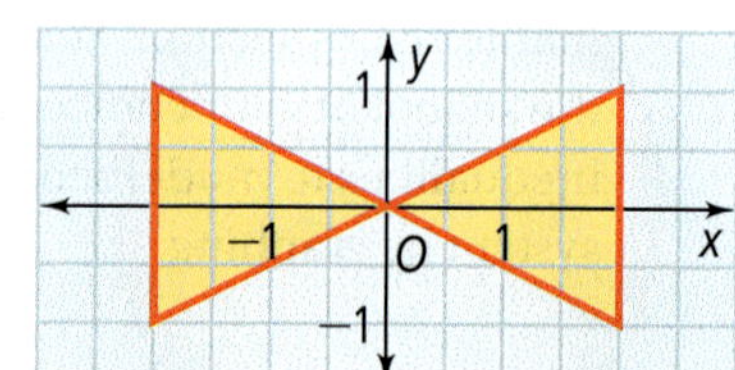

Standardized Test Prep

SAT/ACT

59. Which system of inequalities is shown in the graph?

Ⓐ $\begin{cases} x \geq 4 \\ 3x - 2y > 5 \end{cases}$

Ⓑ $\begin{cases} x > 4 \\ 3x - 2y \leq 5 \end{cases}$

Ⓒ $\begin{cases} x < 4 \\ 3x - 2y \geq 5 \end{cases}$

Ⓓ $\begin{cases} x \leq 4 \\ 3x - 2y < 5 \end{cases}$

60. What is the equation of the line that passes through the point $(4, -3)$ and has slope $\frac{1}{2}$?

Ⓕ $y = \frac{1}{2}x + 3$

Ⓖ $y = \frac{1}{2}x - 1$

Ⓗ $y = \frac{1}{2}x - 5$

Ⓘ $y = x - \frac{1}{2}$

61. Which equation is a vertical translation of $y = -5x$?

Ⓐ $y = -\frac{5}{2}x$

Ⓑ $y = -5x + 2$

Ⓒ $y = -10x$

Ⓓ $y = 5x - 2$

Short Response

62. The cost of renting a pool at an aquatic center is either $30 per hour or $20 per hour with a $40 non-refundable deposit. For how many hours is the cost of renting a pool the same for both plans?

Mixed Review

Solve each system by elimination or substitution.

See Lesson 3-2.

63. $\begin{cases} y = 3x + 1 \\ 2x - y = 8 \end{cases}$

64. $\begin{cases} 3x + y = 4 \\ 2x - 4y = 7 \end{cases}$

65. $\begin{cases} -x + 5y = 3 \\ 2x - 10y = 4 \end{cases}$

66. $\begin{cases} 2x + 4y = -8 \\ -5x + 4y = 6 \end{cases}$

67. $\begin{cases} y - 3 = x \\ 4x + y = -2 \end{cases}$

68. $\begin{cases} 2 = 4y - 3x \\ 5x - 2y = 3 \end{cases}$

Get Ready! **To prepare for Lesson 3-4, do Exercises 69–72.**

Write an ordered pair that is a solution of each system of inequalities.

See Lesson 3-3.

69. $\begin{cases} x + y > 2 \\ 3x + 2y \leq 6 \end{cases}$

70. $\begin{cases} 2y > 4 \\ 3x + 4y \leq 14 \end{cases}$

71. $\begin{cases} x \geq 2 \\ 5x + 2y \leq 9 \end{cases}$

72. $\begin{cases} x + 3y < 6 \\ y < x \end{cases}$

3 Mid-Chapter Quiz

Do you know HOW?

Solve each system by graphing.

1. $\begin{cases} 3x - y = 8 \\ 10 + 2y = 4x \end{cases}$

2. $\begin{cases} y + 5 = 2x \\ 3y + 6x = -3 \end{cases}$

3. $\begin{cases} 14x - 2y = 6 \\ 6y - 9x = 15 \end{cases}$

Without graphing, classify each system as *independent*, *dependent*, or *inconsistent*.

4. $\begin{cases} 3y + 2x = 12 \\ 36 - 9y = -6x \end{cases}$

5. $\begin{cases} -2y = 20 - 2x \\ 3y - 6x = -30 \end{cases}$

6. $\begin{cases} 15x = 10y - 20 \\ 18 + 9x = 6y \end{cases}$

Solve each system by substitution.

7. $\begin{cases} 5m - n = 7 \\ 3 + 3n = 6m \end{cases}$

8. $\begin{cases} 4y - 6 = 2x \\ y - 3x = 9 \end{cases}$

9. $\begin{cases} 3u + 8 = 4v \\ 24v = 6 - 3u \end{cases}$

Solve each system by elimination.

10. $\begin{cases} 5c - 4t = 8 \\ 14 + 4t = 3c \end{cases}$

11. $\begin{cases} 8y + 10 = 6x \\ 8y - 4x = -12 \end{cases}$

12. $\begin{cases} 11 - 2c = 3d \\ 2c - 7d = -9 \end{cases}$

Graph the solutions to each of the following systems.

13. $\begin{cases} y < 2 + 3x \\ y \geq |x - 3| \end{cases}$

14. $\begin{cases} 2x + y > 7 \\ x < 4 \\ y \leq 5 \end{cases}$

Do you UNDERSTAND?

15. Which equation below combines with the equation $-4x + 6y = 3$ to form a system with an infinite number of solutions?

    (A) $0.5 + x = 1.5y$

    (B) $0.75 + 2x = 1.5y$

    (C) $0.5 + 2x = 1.5y$

    (D) $0.75 + x = 1.5y$

16. Write a system of inequalities to describe the shaded region.

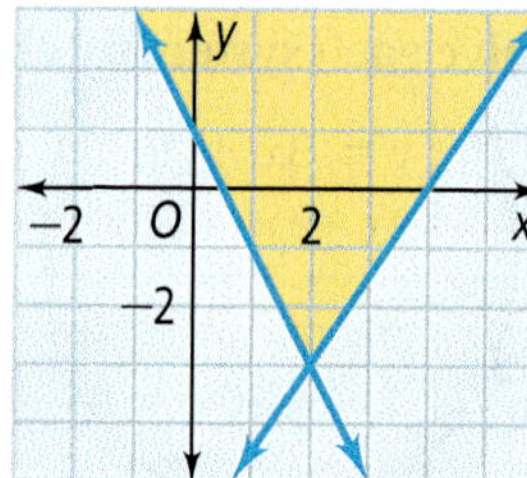

17. An ordinary refrigerator costs \$503 and has an estimated annual operating cost of \$92. An energy-efficient model costs \$615 with an estimated annual operating cost of \$64. Write a system of equations to represent this situation.

    a. What is the solution of the system?

    b. What does the solution mean?

    c. Which model would you choose for your family? Why?

18. **Writing** Explain how to classify a linear system as independent, dependent, or inconsistent without graphing.

3-4 Linear Programming

Content Standard

A.CED.3 Represent constraints by equations or inequalities, and by systems of equations and/or inequalities, and interpret solutions as viable or nonviable options in a modeling context.

Objective To solve problems using linear programming

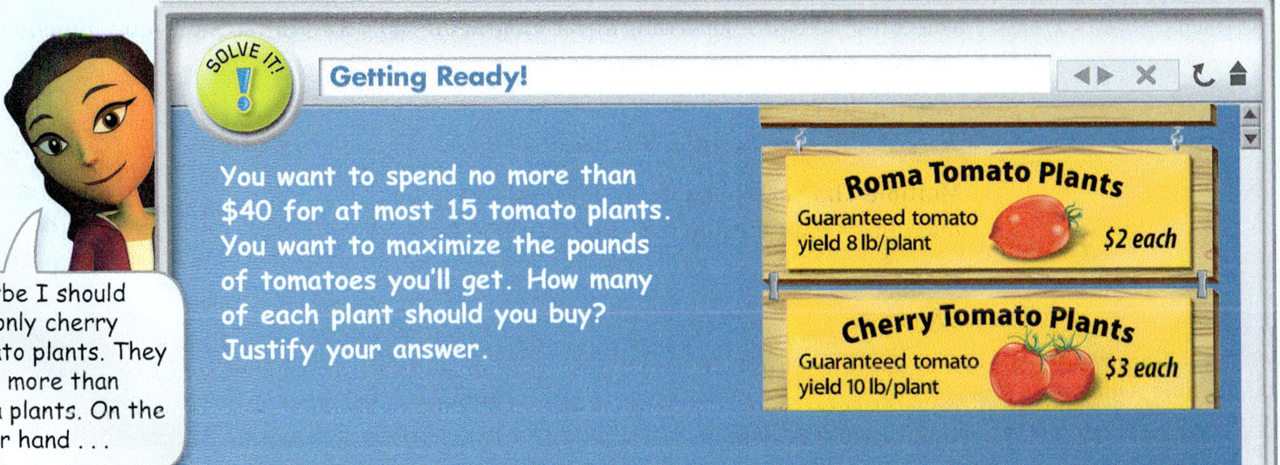

In the Solve It, you maximized your tomato production given some limits, or **constraints. Linear programming** is a method for finding a minimum or maximum value of some quantity, given a set of constraints.

Lesson Vocabulary
- constraint
- linear programming
- feasible region
- objective function

Essential Understanding Some real-world problems involve multiple linear relationships. Linear programming accounts for all of these linear relationships and gives the solution to the problem.

The constraints in a linear programming situation form a system of inequalities, like the one at the right. The graph of the system is the **feasible region**. It contains all the points that satisfy all the constraints.

$$\begin{cases} x \geq 2 \\ y \geq 3 \\ y \leq 6 \\ x + y \leq 10 \end{cases}$$

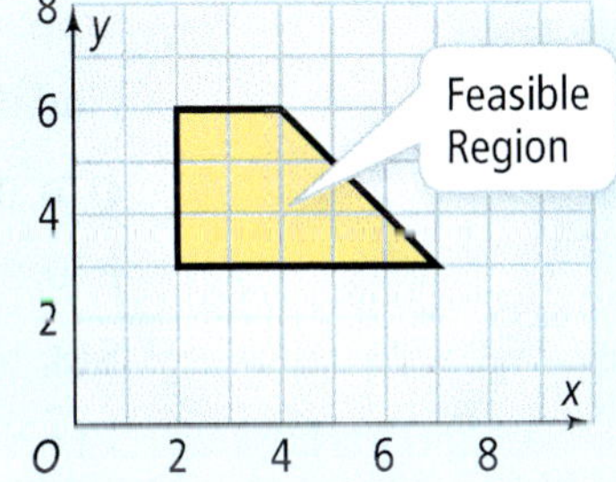

The quantity you are trying to maximize or minimize is modeled with an **objective function**. Often this quantity is cost or profit. Suppose the objective function is $C = 2x + y$.

Graphs of the objective function for various values of C are parallel lines. Lines closer to the origin represent smaller values of C.

The graphs of the equations $7 = 2x + y$ and $17 = 2x + y$ intersect the feasible region at (2, 3) and (7, 3). These vertices of the feasible represent the least and the greatest values for the objective function.

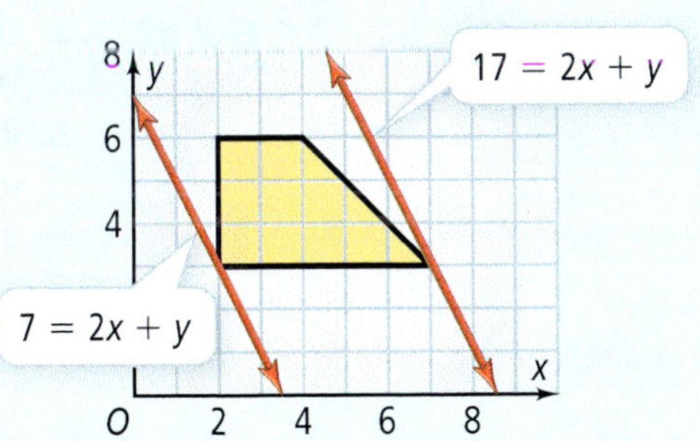

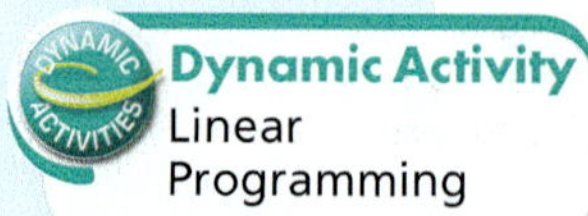

take note

Key Concept Vertex Principle of Linear Programming

If there is a maximum or a minimum value of the linear objective function, it occurs at one or more vertices of the feasible region.

You can solve a problem using linear programming by testing in the objective function all of the vertices of the feasible region.

Problem 1 Testing Vertices

Multiple Choice What point in the feasible region maximizes P for the objective function $P = 2x + y$?

Constraints $\begin{cases} x + 2y \leq 5 \\ x - y \leq 2 \\ x \geq 0 \\ y \geq 0 \end{cases}$

Ⓐ (2, 0) Ⓑ (0, 0) Ⓒ (3, 1) Ⓓ (0, 2.5)

Think

What quadrant will the feasible region be in?
The constraints $x \geq 0$ and $y \geq 0$ indicate the first quadrant.

Step 1

Graph the inequalities.

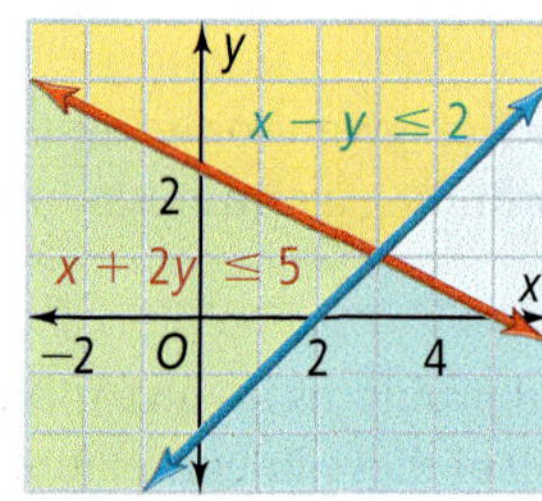

Step 2

Form the feasible region.

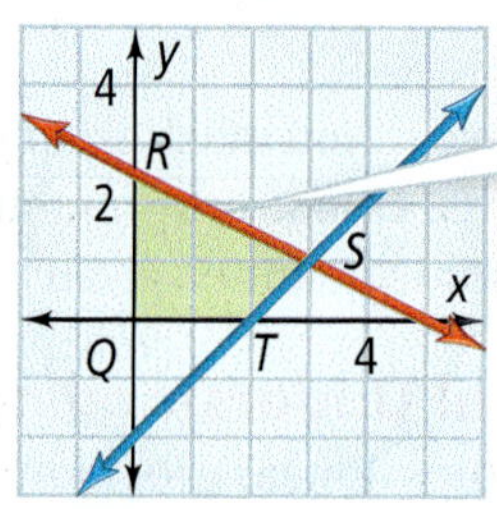

The intersections of the boundaries are the vertices of the feasible region.

Step 3

Find the coordinates of each vertex.

$Q(0, 0)$

$R(0, 2.5)$

$S(3, 1)$

$T(2, 0)$

Step 4

Evaluate P at each vertex.

$P = 2(0) + 0 = 0$

$P = 2(0) + 2.5 = 2.5$

$P = 2(3) + 1 = 7$ Maximum Value

$P = 2(2) + 0 = 4$

P has a maximum value of 7 when $x = 3$ and $y = 1$. The correct choice is C.

  **Got It?** **1. a.** Use the constraints in Problem 1 with the objective function $P = x + 3y$. What values of x and y maximize P?

b. Reasoning Can an objective function $P = ax + by + c$ have (the same) maximum value at all four vertex points Q, R, S, and T? At points R and S only? Explain using examples.

Problem 2 Using Linear Programming to Maximize Profit

Business You are screen-printing T-shirts and sweatshirts to sell at the Polk County Blues Festival and are working with the following constraints.

- You have at most 20 hours to make shirts.
- You want to spend no more than $600 on supplies.
- You want to have at least 50 items to sell.

How many T-shirts and how many sweatshirts should you make to maximize your profit? How much is the maximum profit?

Organize the information in a table.

	T-Shirts, x	Sweatshirts, y	Total
Minutes	$10x$	$30y$	1200
Number	x	y	50
Cost	$4x$	$20y$	600
Profit	$6x$	$20y$	$6x + 20y$

Write the constraints and the objective function.

Constraints: $\begin{cases} 10x + 30y \leq 1200 \\ x + y \geq 50 \\ 4x + 20y \leq 600 \\ x \geq 0 \\ y \geq 0 \end{cases}$

Objective Function: $P = 6x + 20y$

Think

How do you find the coordinates of the vertices if they are hard to read off the graph?

Solve the system of equations related to the lines that intersect to form the vertex.

Step 1

Graph the constraints to form the feasible region.

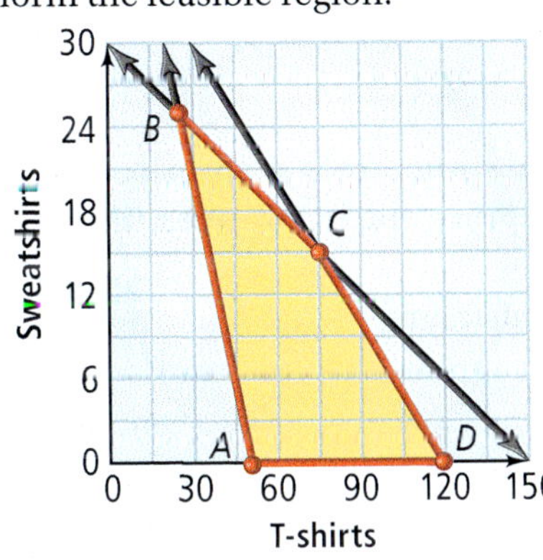

Step 2

Find the coordinates of each vertex.

$A(50, 0)$

$B(25, 25)$

$C(75, 15)$

$D(120, 0)$

Step 3

Evaluate P.

$P = 6(50) + 20(0) = 300$

$P = 6(25) + 20(25) = 650$

$P = 6(75) + 20(15) = 750$

$P = 6(120) + 20(0) = 720$

You can maximize your profit by selling 75 T-shirts and 15 sweatshirts. The maximum profit is $750.

Got It? 2. If it took you 20 minutes to make a sweatshirt, how many of each type of shirt should you make to maximize your profit?

Lesson Check

Do you know HOW?

Graph each system of inequalities.

1. $\begin{cases} x + y \leq 6 \\ x \geq 0 \\ y \geq 0 \end{cases}$

2. $\begin{cases} 2x - y \leq 4 \\ x \geq 0 \\ y \geq 0 \end{cases}$

3. $\begin{cases} x + 2y \leq 10 \\ x \geq 1 \\ y \geq 2 \end{cases}$

4. $\begin{cases} 2x + 3y \leq 18 \\ 0 \leq x \leq 5 \\ 0 \leq y \leq 4 \end{cases}$

Graph each system of constraints. Then name the vertices of the feasible region.

5. $\begin{cases} x \leq 5 \\ y \leq 4 \\ x \geq 0 \\ y \geq 0 \end{cases}$

6. $\begin{cases} x + y \leq 8 \\ y \geq 5 \\ x \geq 0 \end{cases}$

Do you UNDERSTAND?

7. **Vocabulary** Explain why the inequalities of a linear programming problem are called constraints. *Hint*: Use the definition of *constraint* as part of your answer.

8. **Compare and Contrast** What are some similarities between solving a linear programming problem and solving a system of linear inequalities? What are some differences?

9. **Open-Ended** Write a system of constraints whose graphs determine a trapezoid. Write an objective function and evaluate it at each vertex.

Practice and Problem-Solving Exercises

Graph each system of constraints. Name all vertices. Then find the values of x and y that maximize or minimize the objective function.

See Problem 1.

10. $\begin{cases} x + y \leq 8 \\ 2x + y \leq 10 \\ x \geq 0 \\ y \geq 0 \end{cases}$

Maximum for
$N = 100x + 40y$

11. $\begin{cases} x + 2y \geq 8 \\ x \geq 2 \\ y \geq 0 \end{cases}$

Minimum for
$C = x + 3y$

12. $\begin{cases} 2 \leq x \leq 6 \\ 1 \leq y \leq 5 \\ x + y \leq 8 \end{cases}$

Maximum for
$P = 3x + 2y$

13. **Air Quality** A city wants to plant maple and spruce trees to absorb carbon dioxide. It has \$2100 to spend on planting spruce and maple trees. The city has 45,000 ft^2 available for planting.

See Problem 2.

a. Use the data from the table. Write the constraints for the situation.
b. Write the objective function.
c. Graph the feasible region and find the vertices.
d. How many of each tree should the city plant to maximize carbon dioxide absorption?

Spruce and Maple Tree Data

	Spruce	Maple
Planting Cost	\$30	\$40
Area Required	600 ft^2	900 ft^2
Carbon Dioxide Absorption	650 lb/yr	300 lb/yr

SOURCE: Auburn University and Anderson Associates

B Apply

14. Think About a Plan A biologist is developing two new strains of bacteria. Each sample of Type I bacteria produces four new viable bacteria, and each sample of Type II produces three new viable bacteria. Altogether, at least 240 new viable bacteria must be produced. At least 30, but not more than 60, of the original samples must be Type I. Not more than 70 of the original samples can be Type II. A sample of Type I costs \$5 and a sample of Type II costs \$7. How many samples of Type II bacteria should the biologist use to minimize the cost?

- What are the unknowns?
- What constraints do you get from each condition in the problem?
- Are there any implicit constraints?

15. Error Analysis Your friend is trying to find the maximum value of $P = -x + 3y$ subject to the following constraints.

$$\begin{cases} y \leq -2x + 6 \\ y \leq x + 3 \\ x \geq 0, y \geq 0 \end{cases}$$

What error did your friend make? What is the correct solution?

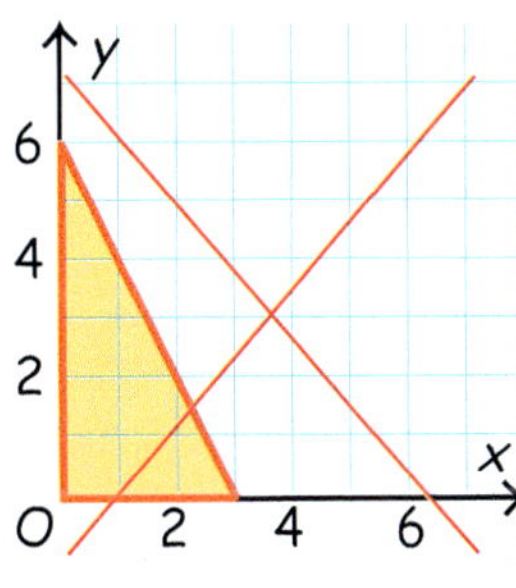

16. Cooking Baking a tray of corn muffins takes 4 cups of milk and 3 cups of wheat flour. Baking a tray of bran muffins takes 2 cups of milk and 3 cups of wheat flour. A baker has 16 cups of milk and 15 cups of wheat flour. He makes \$3 profit per tray of corn muffins and \$2 profit per tray of bran muffins. How many trays of each type of muffin should the baker make to maximize his profit?

Graph each system of constraints. Name all vertices. Then find the values of x and y that maximize or minimize the objective function. Find the maximum or minimum value.

17. $\begin{cases} 3x + y \leq 7 \\ x + 2y \leq 9 \\ x \geq 0, y \geq 0 \end{cases}$

Maximum for
$P = 2x + y$

18. $\begin{cases} 25 \leq x \leq 75 \\ y \leq 110 \\ 8x + 6y \geq 720 \end{cases}$

Minimum for
$C = 8x + 5y$

19. $\begin{cases} x + y \leq 11 \\ 2y \geq x \\ x \geq 0, y \geq 0 \end{cases}$

Maximum for
$P = 3x + 2y$

20. $\begin{cases} 2x + y \leq 300 \\ x + y \leq 200 \\ x \geq 0, y \geq 0 \end{cases}$

Maximum for
$P = x + 2y$

21. $\begin{cases} 5x + y \geq 10 \\ x + y \leq 6 \\ x + 4y \geq 12 \\ x \geq 0, y \geq 0 \end{cases}$

Minimum for
$C = 10{,}000x + 20{,}000y$

22. $\begin{cases} 6 \leq x + y \leq 13 \\ x \geq 3 \\ y \geq 1 \end{cases}$

Maximum for
$P = 4x + 3y$

23. A vertex of a feasible region does not always have whole-number coordinates. Sometimes you may need to round coordinates to find the solution. Using the objective function and the constraints at the right, find the whole-number values of x and y that minimize C. Then find C for those values of x and y.

$$C = 6x + 9y$$
$$\begin{cases} x + 2y \geq 50 \\ 2x + y \geq 60 \\ x \geq 0, y \geq 0 \end{cases}$$

24. **Reasoning** Sometimes two corners of a graph both yield the maximum profit. In this case, many other points may also yield the maximum profit. Evaluate the profit formula $P = x + 2y$ for the graph shown. Find four points that yield the maximum profit.

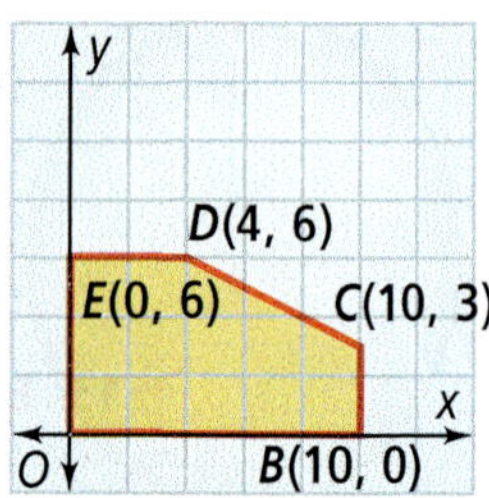

Standardized Test Prep

SAT/ACT

25. Solve the equation $\frac{1}{2}(a + b) = c$ for b.

(A) $b = \frac{1}{2}c - a$ (B) $b = 2a - c$ (C) $b = 2c - a$ (D) $b = 2ca$

26. Which is the graph of $y \leq |x - 3|$?

(F) 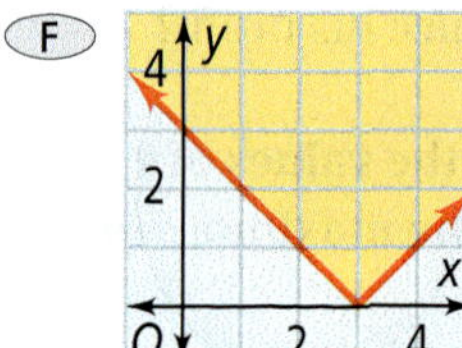
(G) 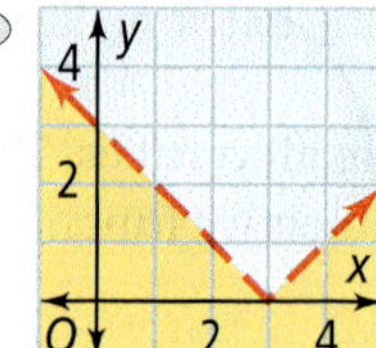
(H) 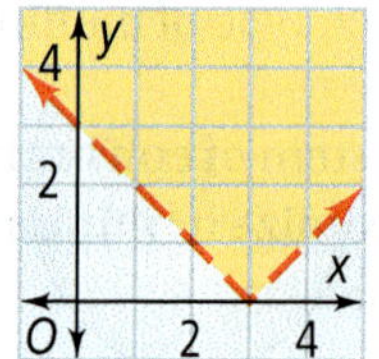
(I) 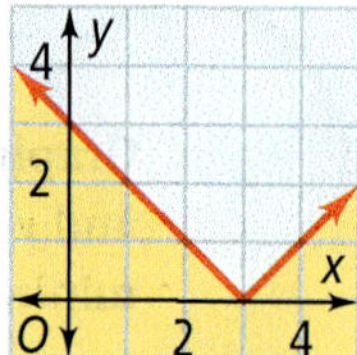

Short Response

27. What are the vertices of the feasible region bounded by the constraints at the right?

$$\begin{cases} x + y \leq 3 \\ 2x + y \leq 4 \\ x \geq 0, y \geq 0 \end{cases}$$

Mixed Review

Solve each system of inequalities by graphing. See Lesson 3-3.

28. $\begin{cases} y < -2x + 8 \\ 3y \geq 4x - 6 \end{cases}$ 29. $\begin{cases} x - 2y \geq 11 \\ 5x + 4y < 27 \end{cases}$ 30. $\begin{cases} 2x + 6y > 12 \\ 3x + 9y \leq 27 \end{cases}$

Evaluate each expression for $a = 3$ and $b = -5$. See Lesson 1-3.

31. $2a + b$ 32. $-4 + 2ab$ 33. $3(a - b)$ 34. $b(2b - a)$

Get Ready! **To prepare for Lesson 3-5, do Exercises 35–37.**

Find the x- and y-intercepts of the graph of each linear equation. See Lesson 2-4.

35. $y = 2x + 6$ 36. $2x + 9y = 36$ 37. $y = x - 1$

Concept Byte

For Use With Lesson 3-4

TECHNOLOGY

Linear Programming

Content Standard

A.CED.3 Represent constraints by equations or inequalities, and by systems of equations and/or inequalities, and interpret solutions as viable or nonviable options in a modeling context.

You can solve linear programming problems using your graphing calculator.

MATHEMATICAL PRACTICES

Activity

Find the values of x and y that will maximize the objective function $P = 13x + 2y$ for the constraints at the right. What is the value of P at this maximum point?

$$\begin{cases} -3x + 2y \leq 8 \\ -8x + y \geq -48 \\ x \geq 0, y \geq 0 \end{cases}$$

Step 1 Rewrite the first two inequalities to isolate y. Enter the inequalities.

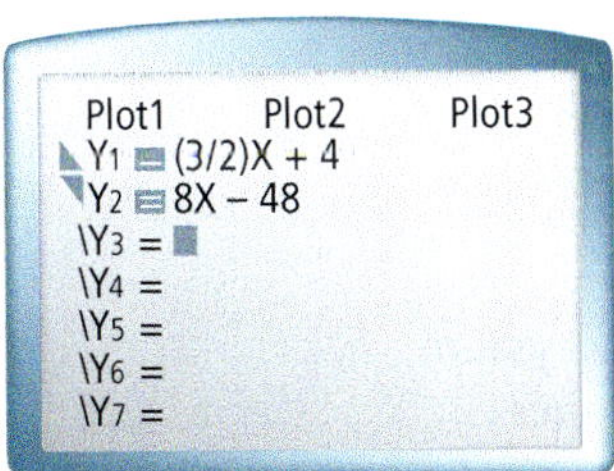

Step 2 Use the **VALUE** option of **CALC** to find the upper left vertex. Press 0 enter.

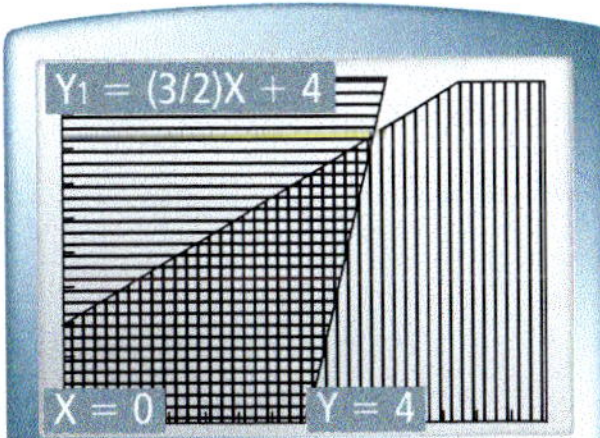

Step 3 Enter the objective function on the home screen. Press enter for the value of P at the vertex.

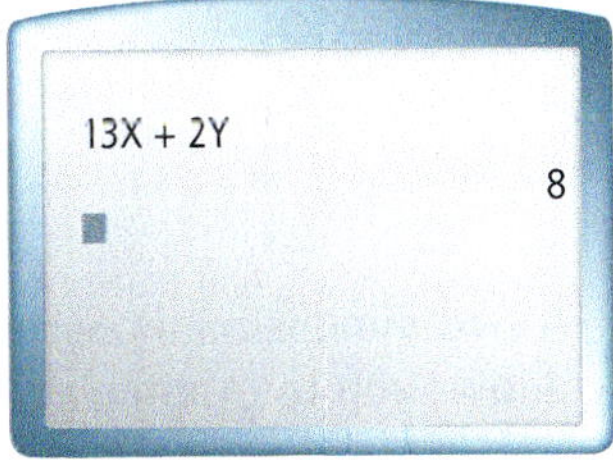

Step 4 Use the **INTERSECT** option of **CALC** to find the upper right vertex. Go to the home screen and press enter for the value of P.

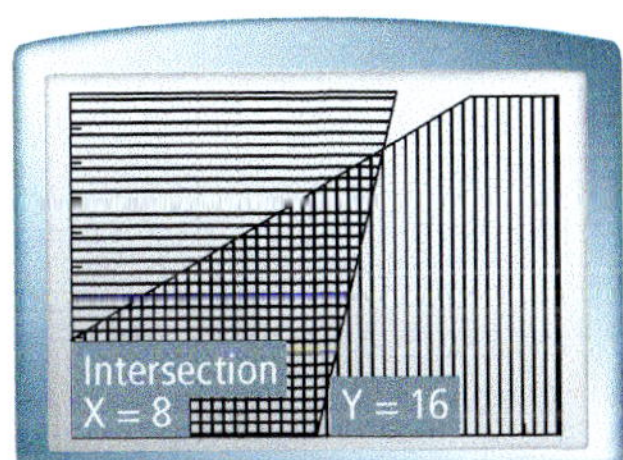

Step 5 Use the **ZERO** option of **CALC** to find the lower right vertex. Go to the home screen and press enter for the value of P. The objective function has a value of 0 when the vertex is at the origin. The maximum value of P is 136.

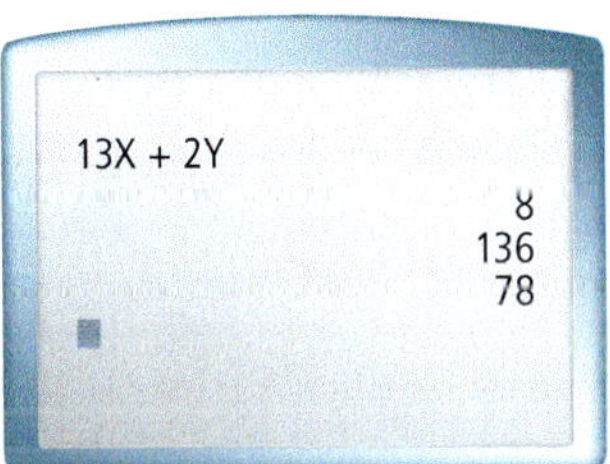

Exercises

Find the values of x and y that maximize or minimize the objective function.

1. $\begin{cases} 4x + 3y \geq 30 \\ x + 3y \geq 21 \\ x \geq 0, y \geq 0 \end{cases}$

   Minimum for $C = 5x + 8y$

2. $\begin{cases} 3x + 5y \geq 35 \\ 2x + y \leq 14 \\ x \geq 0, y \geq 0 \end{cases}$

   Maximum for $P = 3x + 2y$

3. $\begin{cases} x + y \geq 8 \\ x + 5y \geq 20 \\ x \geq 0, y \geq 2 \end{cases}$

   Minimum for $C = 3x + 4y$

4. $\begin{cases} x + 2y \leq 24 \\ 3x + 2y \leq 34 \\ 3x + y \leq 29 \\ x \geq 0 \end{cases}$

   Maximum for $P = 2x + 3y$

Concept Byte

For Use With Lesson 3-5

ACTIVITY

Graphs in Three Dimensions

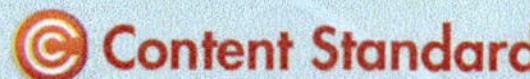

Extends A.REI.6 Solve systems of linear equations exactly and approximately (e.g., with graphs), focusing on pairs of linear equations in two variables.

To describe positions in space, you need a three-dimensional coordinate system. You have learned to graph on an *xy*-coordinate plane using ordered pairs. Adding a third axis, the *z*-axis, to the *xy*-coordinate plane creates **coordinate space**. In coordinate space you graph points using **ordered triples** of the form (x, y, z).

Points in a Plane

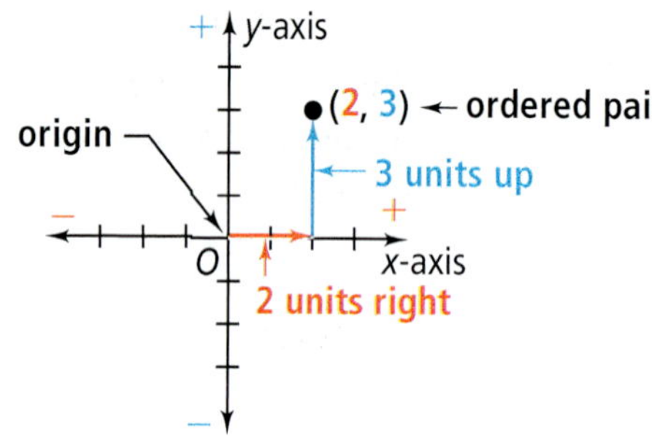

A two-dimensional coordinate system allows you to graph points in a plane.

Points in Space

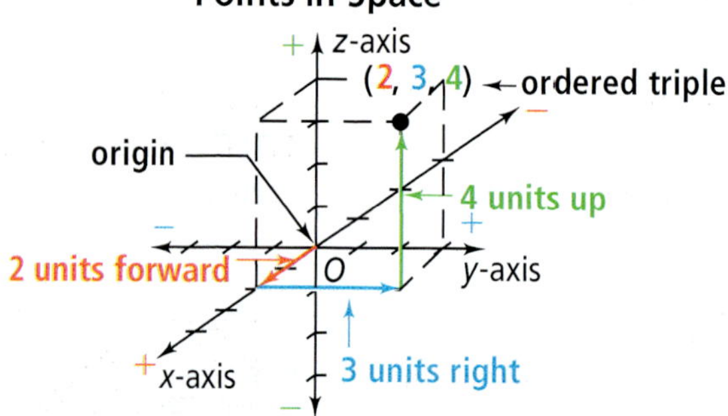

A three-dimensional coordinate system allows you to graph points in space.

In the coordinate plane, point (2, 3) is two units right and three units up from the origin. In coordinate space, point (2, 3, 4) is two units forward, three units right, and four units up.

Activity 1

Define one corner of your classroom as the origin of a three-dimensional coordinate system like the classroom shown. Write the coordinates of each item in your coordinate system.

1. each corner of your classroom
2. each corner of your desk
3. one corner of the blackboard
4. the clock
5. the waste-paper basket
6. Pick 3 items in your classroom and write the coordinates of each.

An equation in two variables represents a line in a plane. An equation in three variables represents a plane in space.

Activity 2

Given the following equation in three variables, draw the plane in a coordinate space. $x + 2y - z = 6$

7. Let $x = 0$. Graph the resulting equation in the *yz* plane.

8. Let $y = 0$. Graph the resulting equation in the *xz* plane.

From geometry you know that two lines determine a plane.

9. Sketch the plane $x + 2y - z = 6$.
(If you need help, find a third line by letting $z = 0$ and then graph the resulting equation in the *xy* plane.)

Activity 3

Two equations in three variables represent two planes in space.

10. Draw the two planes determined by the following equations:
$2x + 3y - z = 12$
$2x - 4y + z = 8$

11. Describe the intersection of the two planes above.

Exercises

Find the coordinates of each point in the diagram.

12. *A* **13.** *B* **14.** *C*

15. *D* **16.** *E* **17.** *F*

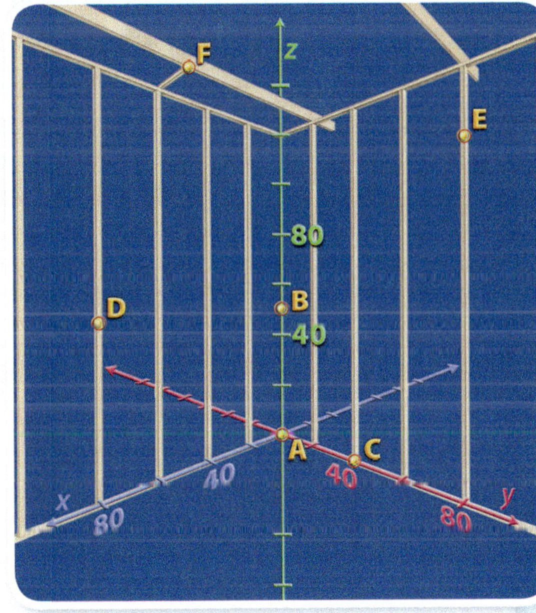

Sketch the graph of each equation.

18. $x - y - 4z = 8$

19. $x + y + z = 2$

20. $-3x + 5y + 10z = 15$

21. $6x + 6y - 12z = 36$

Graph the following pairs of equations in the same coordinate space and describe their intersection, if any.

22. $-x + 3y + z = 6$
$-3x + 5y - 2z = 60$

23. $-2x - 3y + 5z = 7$
$2x - 3y - 4z = -4$

Systems With Three Variables

Content Standard

Extends A.REI.6 Solve systems of linear equations exactly and approximately (e.g., with graphs), focusing on pairs of linear equations in two variables.

Objectives To solve systems in three variables using elimination
To solve systems in three variables using substitution

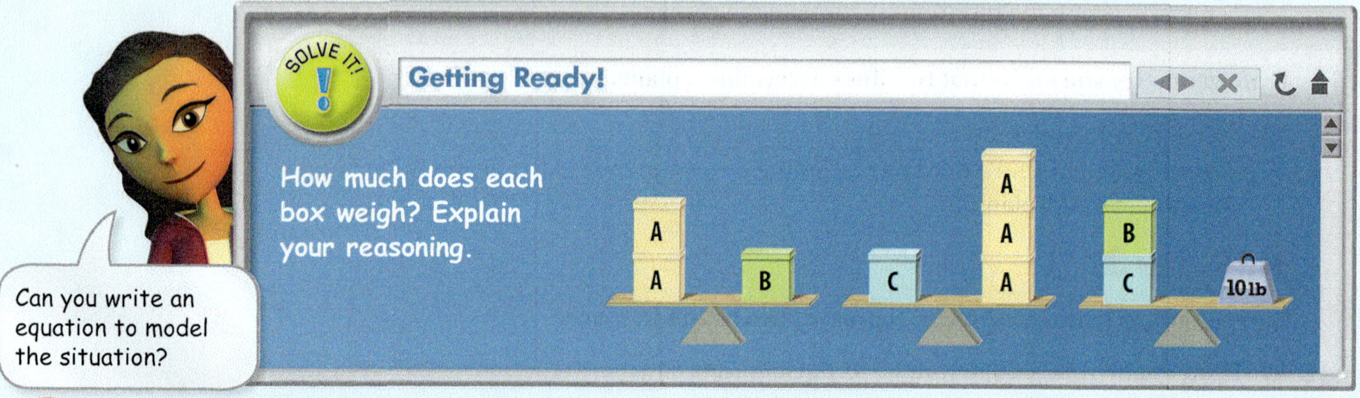

You can represent three relationships involving three unknowns with a system of equations.

Essential Understanding To solve systems of three equations in three variables, you can use some of the same algebraic methods you used to solve systems of two equations in two variables.

You can represent systems of equations in three variables as graphs in three dimensions. The graph of an equation of the form $Ax + By + Cz = D$, where A, B, and C are not all zero, is a plane. You can show the solutions of a three-variable system graphically as the intersection of planes.

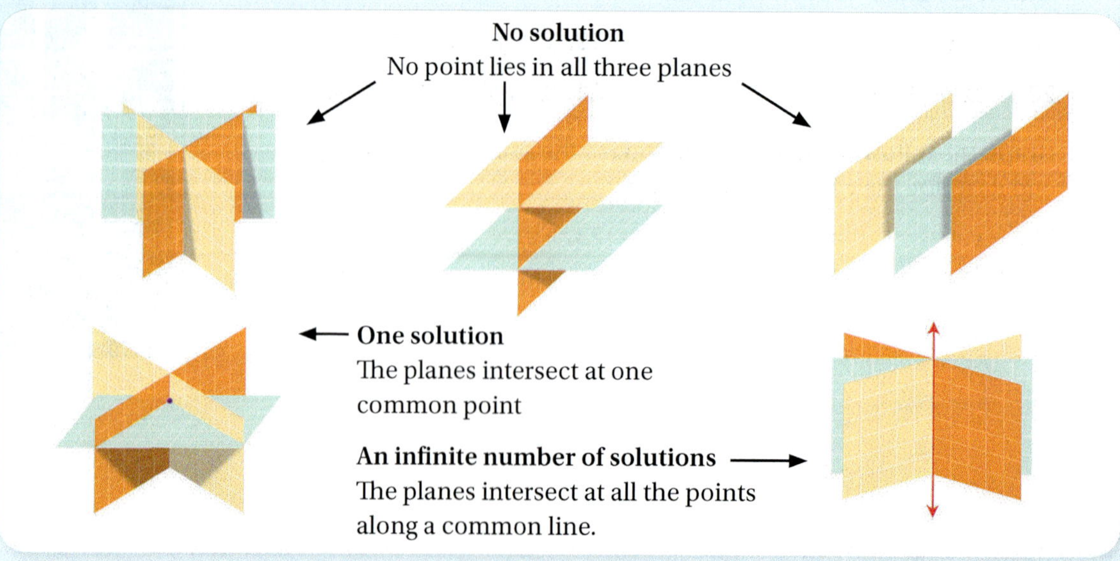

You can use the elimination and substitution methods to solve a system of three equations in three variables by working with the equations in pairs. You will use one of the equations *twice*. When one point represents the solution of a system of equations in three variables, write it as an ordered triple (x, y, z).

Problem 1 Solving a System Using Elimination

Think

Which variable do you eliminate first?
Eliminate the variable for which the process requires the fewest steps.

What is the solution of the system? Use elimination. The equations are numbered to make the procedure easy to follow.

$$\begin{cases} 2x - y + z = 4 & ① \\ x + 3y - z = 11 & ② \\ 4x + y - z = 14 & ③ \end{cases}$$

Step 1 Pair the equations to eliminate z. Then you will have two equations in x and y.

Add.

$$\begin{aligned} ①\quad & 2x - y + z = 4 \\ ②\quad & x + 3y - z = 11 \\ \hline ④\quad & 3x + 2y = 15 \end{aligned}$$

Subtract.

$$\begin{aligned} ②\quad & x + 3y - z = 11 \\ ③\quad & 4x + y - z = 14 \\ \hline ⑤\quad & -3x + 2y = -3 \end{aligned}$$

Step 2 Write the two new equations as a system. Solve for x and y.

Add and solve for y.

$$\begin{aligned} ④\quad & 3x + 2y = 15 \\ ⑤\quad & -3x + 2y = -3 \\ \hline & 4y = 12 \\ & y = 3 \end{aligned}$$

Substitute $y = 3$ and solve for x.

$$\begin{aligned} ④\quad 3x + 2y &= 15 \\ 3x + 2(3) &= 15 \\ 3x &= 9 \\ x &= 3 \end{aligned}$$

Think

Does it matter which equation you substitute into to find z?
No, you can substitute into any of the original three equations.

Step 3 Solve for z. Substitute the values of x and y into one of the original equations.

① $2x - y + z = 4$	Use equation ①.
$2(3) - 3 + z = 4$	Substitute.
$6 - 3 + z = 4$	Simplify.
$z = 1$	Solve for z.

Step 4 Write the solution as an ordered triple. The solution is (3, 3, 1).

Got It? 1. What is the solution of the system? Use elimination. Check your answer in all three original equations.

$$\begin{cases} x - y + z = -1 \\ x + y + 3z = -3 \\ 2x - y + 2z = 0 \end{cases}$$

You can apply the method in Problem 1 to most systems of three equations in three variables. You may need to multiply in one, two, or all three equations by one, two, or three nonzero numbers. Your goal is to obtain a system—equivalent to the original system—with coefficients that allow for the easy elimination of variables.

Problem 2 Solving an Equivalent System

What is the solution of the system? Use elimination.

$$\begin{cases} x + y + 2z = 3 & ① \\ 2x + y + 3z = 7 & ② \\ -x - 2y + z = 10 & ③ \end{cases}$$

Think

You are trying to get two equations in x and z. Multiply ① so you can add it to ② and eliminate y. Do the same with ② and ③.

Write

$$① \begin{cases} x + y + 2z = 3 \\ 2x + y + 3z = 7 \end{cases} ② \longrightarrow \begin{aligned} -x - y - 2z &= -3 \\ 2x + y + 3z &= 7 \\ \hline ④ \quad x + z &= 4 \end{aligned}$$

$$② \begin{cases} 2x + y + 3z = 7 \\ -x - 2y + z = 10 \end{cases} ③ \longrightarrow \begin{aligned} 4x + 2y + 6z &= 14 \\ -x - 2y + z &= 10 \\ \hline ⑤ \quad 3x + 7z &= 24 \end{aligned}$$

Think

Multiply ④ so you can add it to ⑤ and eliminate x.

Write

$$④ \begin{cases} x + z = 4 \\ 3x + 7z = 24 \end{cases} ⑤ \longrightarrow \begin{aligned} -3x - 3z &= -12 \\ 3x + 7z &= 24 \\ \hline 4z &= 12 \\ z &= 3 \end{aligned}$$

Think

Substitute $z = 3$ into ④. Solve for x.

Write

$$x + 3 = 4$$
$$x = 1$$

Think

Substitute the values for x and z into ① to find y. Check the answer in the three original equations.

Write

$$x + y + 2z = 3$$
$$1 + y + 2(3) = 3$$
$$y = -4$$

Check

$1 + (-4) + 2(3) = 3$ ✔

$2(1) + (-4) + 3(3) = 7$ ✔

$-(1) - 2(-4) + 3 = 10$ ✔

The solution is $(1, -4, 3)$.

Got It? 2. a. What is the solution of the system? Use elimination.

$$\begin{cases} x - 2y + 3z = 12 \\ 2x - y - 2z = 5 \\ 2x + 2y - z = 4 \end{cases}$$

b. **Reasoning** Could you have used elimination in another way? Explain.

Here is a graphical representation of the solution of Problem 2. The graphs are enclosed in a 10-by-10-by-10 cube with the origin of the coordinate axes at the center.

The graphs of equations ①, ②, and ③, are planes ①, ②, and ③, respectively.

Each pair of planes intersects in a line. The three lines intersect in $(1, -4, 3)$, the solution of the system.

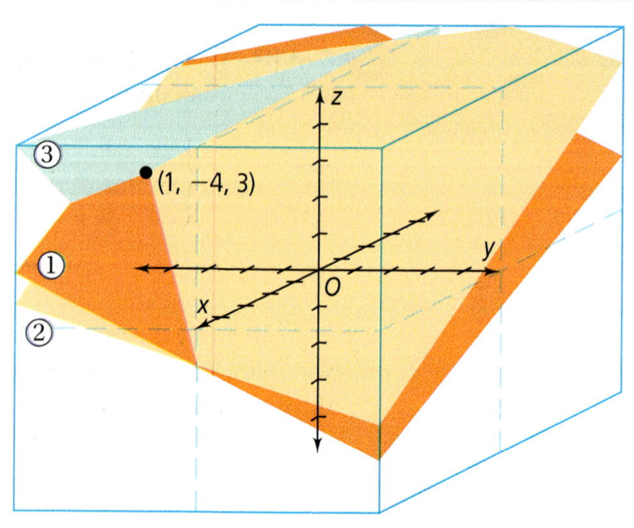

You can also use substitution to solve a system of three equations. Substitution is the best method to use when you can easily solve one of the equations for a single variable.

Problem 3 Solving a System Using Substitution

Multiple Choice What is the x-value in the solution of the system?

$$\begin{cases} 2x + 3y - 2z = -1 & ① \\ x + 5y = 9 & ② \\ 4z - 5x = 4 & ③ \end{cases}$$

(A) 1 (B) 4 (C) 6 (D) 10

Think

Which equation should you solve for one of its variables?
Look for an equation that has a variable with coefficient 1.

Step 1 Choose equation ②. Solve for x.

② $x + 5y = 9$

$x = 9 - 5y$

Step 2 Substitute the expression for x into equations ① and ③ and simplify.

①
$$\begin{aligned} 2x + 3y - 2z &= -1 \\ 2(9 - 5y) + 3y - 2z &= -1 \\ 18 - 10y + 3y - 2z &= -1 \\ 18 - 7y - 2z &= -1 \\ -7y - 2z &= -19 \quad ④ \end{aligned}$$

③
$$\begin{aligned} 4z - 5x &= 4 \\ 4z - 5(9 - 5y) &= 4 \\ 4z - 45 + 25y &= 4 \\ 4z + 25y &= 49 \\ 25y + 4z &= 49 \quad ⑤ \end{aligned}$$

Step 3 Write the two new equations as a system. Solve for y and z.

$$\begin{cases} -7y - 2z = -19 & ④ \\ 25y + 4z = 49 & ⑤ \end{cases}$$

$$\begin{aligned} -14y - 4z &= -38 \quad \text{Multiply by 2.} \\ 25y + 4z &= 49 \quad \text{Then add.} \\ \hline 11y &= 11 \\ y &= 1 \end{aligned}$$

④
$$\begin{aligned} -7y - 2z &= -19 \\ -7(1) - 2z &= -19 \quad \text{Substitute the value of } y \text{ into ④.} \\ -2z &= -12 \\ z &= 6 \end{aligned}$$

Step 4 Use one of the original equations to solve for x.

②
$$\begin{aligned} x + 5y &= 9 \\ x + 5(1) &= 9 \quad \text{Substitute the value of } y \text{ into ②.} \\ x &= 4 \end{aligned}$$

The solution of the system is (4, 1, 6), and $x = 4$.

The correct answer is B.

Got It? 3. a. What is the solution of the system? Use substitution.

$$\begin{cases} x - 2y + z = -4 \\ -4x + y - 2z = 1 \\ 2x + 2y - z = 10 \end{cases}$$

b. **Reasoning** In Problem 3, was it necessary to find the value of z to solve the problem? Explain.

Problem 4 Solving a Real-World Problem

Business You manage a clothing store and budget $6000 to restock 200 shirts. You can buy T-shirts for $12 each, polo shirts for $24 each, and rugby shirts for $36 each. If you want to have twice as many rugby shirts as polo shirts, how many of each type of shirt should you buy?

Relate T-shirts + polo shirts + rugby shirts = 200

rugby shirts = 2 · polo shirts

12 · T-shirts + 24 · polo shirts + 36 · rugby shirts = 6000

Think

How many unknowns are there?
There are three unknowns: the number of each type of shirt.

Define
Let x = the number of T-shirts.
Let y = the number of polo shirts.
Let z = the number of rugby shirts.

Write

$$\begin{cases} x + y + z = 200 & ① \\ z = 2 \cdot y & ② \\ 12 \cdot x + 24 \cdot y + 36 \cdot z = 6000 & ③ \end{cases}$$

Step 1 Since 12 is a common factor of all the terms in equation ③, write a simpler equivalent equation.

$$\begin{cases} 12x + 24y + 36z = 6000 & ③ \\ x + 2y + 3z = 500 & ④ \end{cases}$$ Divide by 12.

Step 2 Substitute $2y$ for z in equations ① and ④. Simplify to find equations ⑤ and ⑥.

① $x + y + z = 200$
$x + y + (2y) = 200$
⑤ $x + 3y = 200$

④ $x + 2y + 3z = 500$
$x + 2y + 3(2y) = 500$
⑥ $x + 8y = 500$

Step 3 Write ⑤ and ⑥ as a system. Solve for x and y.

$$\begin{cases} x + 3y = 200 & ⑤ \\ x + 8y = 500 & ⑥ \end{cases}$$

$-x - 3y = -200$ Multiply by -1.
$x + 8y = 500$ Then add.
$5y = 300$
$y = 60$

⑤ $x + 3y = 200$
$x + 3(60) = 200$ Substitute the value of y into ⑤.
$x = 20$

Step 4 Substitute the value of y in ② and solve for z.

② $z = 2y$
$z = 2(60) = 120$

You should buy 20 T-shirts, 60 polo shirts, and 120 rugby shirts.

Got It? **4.** Suppose you want to have the same number of T-shirts as polo shirts. Buying 200 shirts with a budget of $5400, how many of each shirt should you buy?

Lesson Check

Do you know HOW?

Solve each system.

1. $\begin{cases} 2y - 3z = 0 \\ x + 3y = -4 \\ 3x + 4y = 3 \end{cases}$

2. $\begin{cases} 3x + y - 2z = 22 \\ x + 5y + z = 4 \\ x = -3z \end{cases}$

3. $\begin{cases} 2x + 3y - 2z = 1 \\ -x - y + 2z = 5 \\ 3x + 2y - 3z = -6 \end{cases}$

4. $\begin{cases} 2x - y + z = -2 \\ x + 3y - z = 10 \\ x + 2z = -8 \end{cases}$

Do you UNDERSTAND?

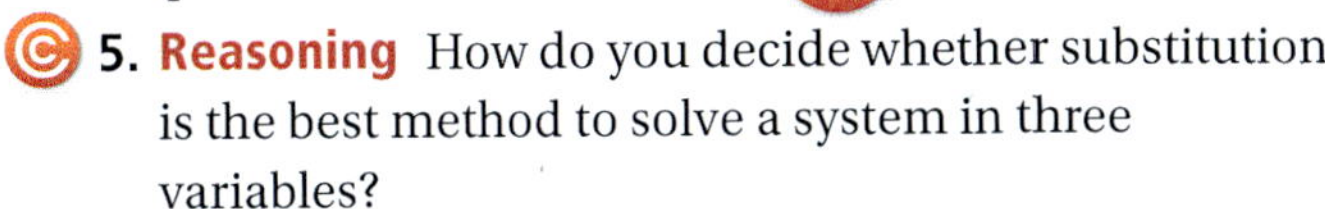

5. **Reasoning** How do you decide whether substitution is the best method to solve a system in three variables?

6. **Error Analysis** A classmate says that the system consisting of $x = 0$, $y = 0$, and $z = 0$ has no solution. Explain the student's error.

7. **Writing** How many solutions does this system have? Explain your answer in terms of intersecting planes. (*Hint:* Is the system dependent? inconsistent?)

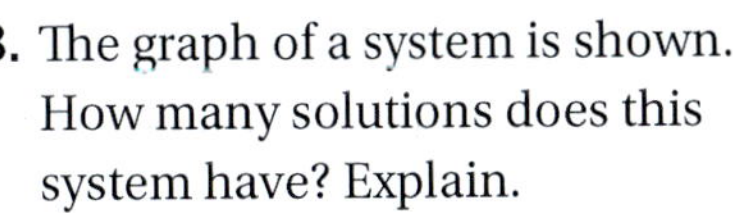

$\begin{cases} ① \ 2x - 3y + z = 5 \\ ② \ 2x - 3y + z = -2 \\ ③ \ -4x + 6y - 2z = 10 \end{cases}$

8. The graph of a system is shown. How many solutions does this system have? Explain.

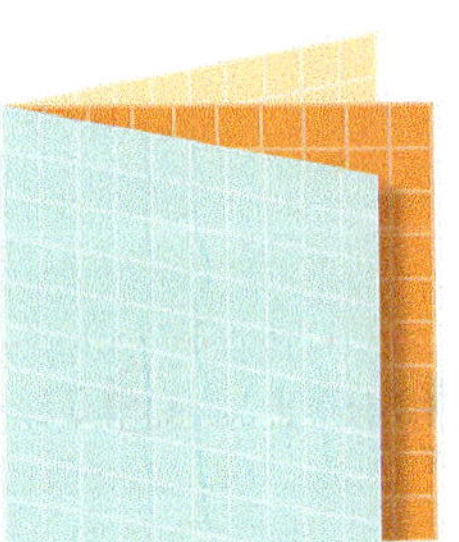

Practice and Problem-Solving Exercises

A Practice

Solve each system by elimination. Check your answers.

See Problems 1 and 2.

9. $\begin{cases} x - y + z = -1 \\ x + y + 3z = -3 \\ 2x - y + 2z = 0 \end{cases}$

10. $\begin{cases} x - y - 2z = 4 \\ -x + 2y + z = 1 \\ -x + y - 3z = 11 \end{cases}$

11. $\begin{cases} -2x + y - z = 2 \\ -x - 3y + z = -10 \\ 3x + 6z = -24 \end{cases}$

12. $\begin{cases} a + b + c = -3 \\ 3b - c = 4 \\ 2a - b - 2c = -5 \end{cases}$

13. $\begin{cases} 8q - r + 2s = 8 \\ 2q + 3r - s = -9 \\ 4q + 2r + 5s = 1 \end{cases}$

14. $\begin{cases} x - y + 2z = -7 \\ y + z = 1 \\ x = 2y + 3z \end{cases}$

15. $\begin{cases} 3x + 3y + 6z = 9 \\ 2x + y + 3z = 7 \\ x + 2y - z = -10 \end{cases}$

16. $\begin{cases} 3x - y + z = 3 \\ x + y + 2z = 4 \\ x + 2y + z = 4 \end{cases}$

17. $\begin{cases} x - 2y + 3z = 12 \\ 2x - y - 2z = 5 \\ 2x + 2y - z = 4 \end{cases}$

18. $\begin{cases} x + 2y = 2 \\ 2x + 3y - z = -9 \\ 4x + 2y + 5z = 1 \end{cases}$

19. $\begin{cases} 3x + 2y + 2z = -2 \\ 2x + y - z = -2 \\ x - 3y + z = 0 \end{cases}$

20. $\begin{cases} x + 4y - 5z = -7 \\ 3x + 2y + 3z = 7 \\ 2x + y + 5z = 8 \end{cases}$

Solve each system by substitution. Check your answers.

21. $\begin{cases} x + 2y + 3z = 6 \\ y + 2z = 0 \\ z = 2 \end{cases}$

22. $\begin{cases} 3a + b + c = 7 \\ a + 3b - c = 13 \\ b = 2a - 1 \end{cases}$

23. $\begin{cases} 5r - 4s - 3t = 3 \\ t = s + r \\ r = 3s + 1 \end{cases}$

24. $\begin{cases} 13 = 3x - y \\ 4y - 3x + 2z = -3 \\ z = 2x - 4y \end{cases}$

25. $\begin{cases} x + 3y - z = -4 \\ 2x - y + 2z = 13 \\ 3x - 2y - z = -9 \end{cases}$

26. $\begin{cases} x - 4y + z = 6 \\ 2x + 5y - z = 7 \\ 2x - y - z = 1 \end{cases}$

27. $\begin{cases} x - y + 2z = 7 \\ 2x + y + z = 8 \\ x - z = 5 \end{cases}$

28. $\begin{cases} x + y + z = 2 \\ x + 2z = 5 \\ 2x + y - z = -1 \end{cases}$

29. $\begin{cases} 5x - y + z = 4 \\ x + 2y - z = 5 \\ 2x + 3y - 3z = 5 \end{cases}$

STEM 30. **Manufacturing** In a factory there are three machines, *A*, *B*, and *C*. When all three machines are working, they produce 287 bolts per hour. When only machines *A* and *C* are working, they produce 197 bolts per hour. When only machines *A* and *B* are working, they produce 202 bolts per hour. How many bolts can each machine produce per hour?

31. **Think About a Plan** In triangle *PQR*, the measure of angle *Q* is three times the measure of angle *P*. The measure of angle *R* is 20° more than the measure of angle *P*. Find the measure of each angle.
   - What are the unknowns in this problem?
   - What system of equations represents this situation?
   - Which method of solving looks easier for this problem?

32. **Sports** A stadium has 49,000 seats. Seats sell for \$25 in Section A, \$20 in Section B, and \$15 in Section C. The number of seats in Section A equals the total number of seats in Sections B and C. Suppose the stadium takes in \$1,052,000 from each sold-out event. How many seats does each section hold?

Solve each system using any method.

33. $\begin{cases} x - 3y + 2z = 11 \\ -x + 4y + 3z = 5 \\ 2x - 2y - 4z = 2 \end{cases}$

34. $\begin{cases} x + 2y + z = 4 \\ 2x - y + 4z = -8 \\ -3x + y - 2z = -1 \end{cases}$

35. $\begin{cases} 4x - y + 2z = -6 \\ -2x + 3y - z = 8 \\ 2y + 3z = -5 \end{cases}$

36. $\begin{cases} 4a + 2b + c = 2 \\ 5a - 3b + 2c = 17 \\ a - 5b = 3 \end{cases}$

37. $\begin{cases} 4x - 2y + 5z = 6 \\ 3x + 3y + 8z = 4 \\ x - 5y - 3z = 5 \end{cases}$

38. $\begin{cases} 2\ell + 2w + h = 72 \\ \ell = 3w \\ h = 2w \end{cases}$

39. $\begin{cases} 6x + y - 4z = -8 \\ \frac{y}{4} - \frac{z}{6} = 0 \\ 2x - z = -2 \end{cases}$

40. $\begin{cases} 4y + 2x = 6 - 3z \\ x + z - 2y = -5 \\ x - 2z = 3y - 7 \end{cases}$

41. $\begin{cases} 4x - y + z = -5 \\ -x + y - z = 5 \\ 2x - z - 1 = y \end{cases}$

42. **Finance** A worker received a \$10,000 bonus and decided to split it among three different accounts. He placed part in a savings account paying 4.5% per year, twice as much in government bonds paying 5%, and the rest in a mutual fund that returned 4%. His income from these investments after one year was \$455. How much did the worker place in each account?

43. **Open-Ended** Write your own system with three variables. Begin by choosing the solution. Then write three equations that are true for your solution. Use elimination to solve the system.

44. **Geometry** Refer to the regular five-pointed star at the right. Write and solve a system of three equations to find the measure of each labeled angle.

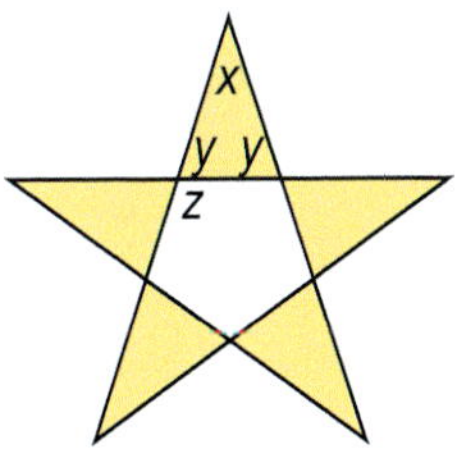

45. **Geometry** In the regular polyhedron described below, all faces are congruent polygons. Use a system of three linear equations to find the numbers of vertices, edges, and faces.

Every face has five edges and every edge is shared by two faces. Every face has five vertices and every vertex is shared by three faces. The sum of the number of vertices and faces is two more than the number of edges.

Standardized Test Prep

GRIDDED RESPONSE

SAT/ACT

46. What is the value of z in the solution of the system? $\begin{cases} y = -2x + 10 \\ -x + y - 2z = -2 \\ 3x - 2y + 4z = 7 \end{cases}$

47. What is the x-intercept of the line at the right after it is translated up 3 units?

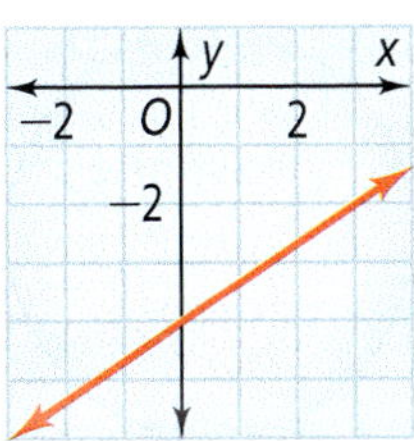

48. Suppose y varies directly with x, and $y = 15$ when $x = 10$. What is y when $x = 22$?

49. A theater has 490 seats. Seats sell for \$25 on the floor, \$20 in the mezzanine, and \$15 in the balcony. The number of seats on the floor equals the total number of seats in the mezzanine and balcony. Suppose the theater takes in \$10,520 from each sold-out event. How many seats does the mezzanine section hold?

Mixed Review

50. Maximize the objective function $P = x + 3y$ under the given constraints. At what vertex does this maximum value occur? $\begin{cases} x + y \leq 5 \\ x + 2y \leq 8 \\ x \geq 0, y \geq 0 \end{cases}$

See Lesson 3-4.

Solve each inequality. Graph the solution on a number line.

See Lesson 1-5.

51. $-4x + 3 \leq 9$

52. $-(x + 4) - 3 \geq 11$

53. $2(3x - 1) < x - 7$

Get Ready! **To prepare for Lesson 3-6, do Exercises 54–56.**

Solve each system using elimination.

See Lesson 3-2.

54. $\begin{cases} x + 4y = 12 \\ 2x - 8y = 4 \end{cases}$

55. $\begin{cases} 4x + 8y = -6 \\ 6x + 12y = -9 \end{cases}$

56. $\begin{cases} 4y - 2x = 6 \\ 8y = 4x - 12 \end{cases}$

Solving Systems Using Matrices

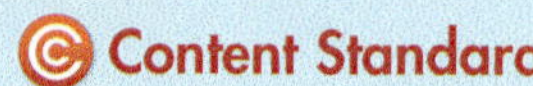

A.REI.8 Represent a system of linear equations as a single matrix equation in a vector variable.

Objectives To represent a system of linear equations with a matrix
To solve a system of linear equations using matrices

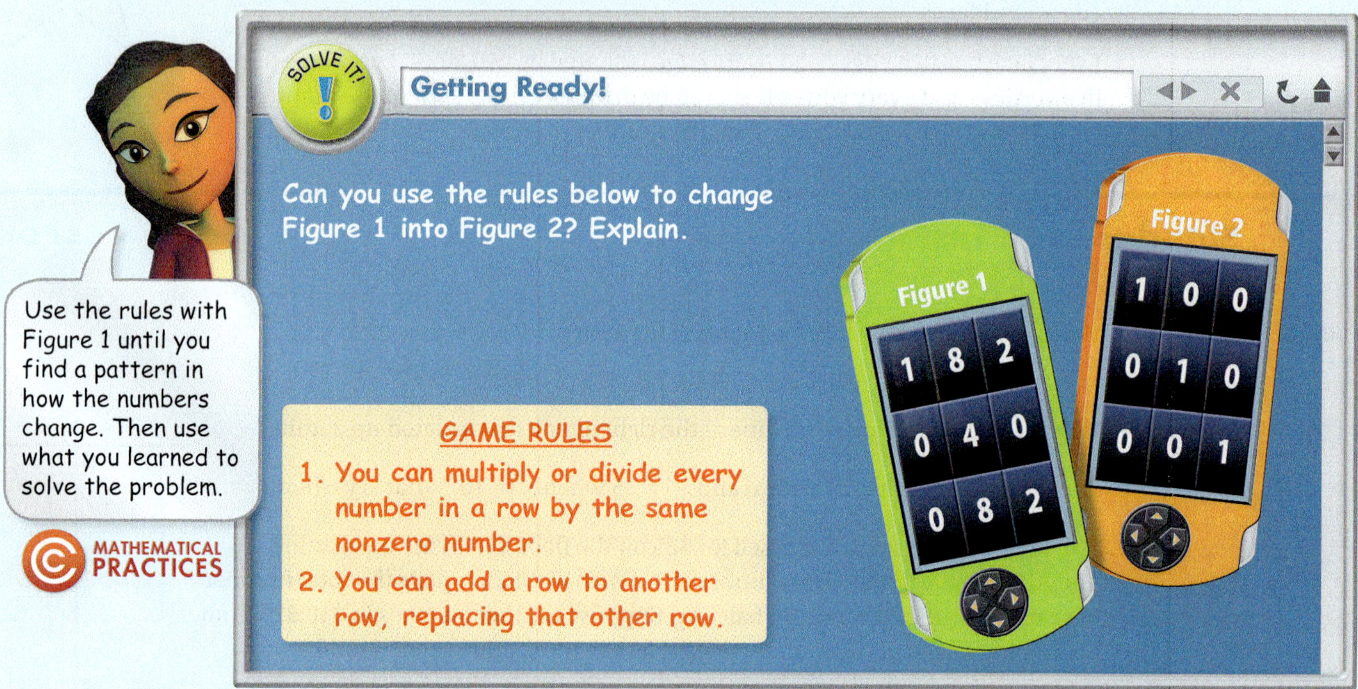

Lesson Vocabulary
- matrix
- matrix element
- row operation

An array of numbers, such as each of those suggested by the tile arrangements in the Solve It, is a matrix.

Essential Understanding You can use a *matrix* to represent and solve a system of equations without writing the variables.

A **matrix** is a rectangular array of numbers. You usually display the array within brackets. The dimensions of a matrix are the numbers of rows and columns in the array.

$$A = \begin{bmatrix} 2 & 4 & 1 \\ 6 & 5 & 3 \end{bmatrix}$$

3 columns; 2 rows

Matrix A has 2 rows and 3 columns and is a 2×3 matrix, read "2 by 3." You can write it as A or $A_{2 \times 3}$.

Each number in a matrix is a **matrix element**. You can identify a matrix element by its row and column numbers. In matrix A, a_{12} is the element in row 1 and column 2. a_{12} is the element 4.

Problem 1 Identifying a Matrix Element

GRIDDED RESPONSE

Does the order of the subscript in a_{23} matter?
Yes. a_{23} and a_{32} are different elements.

What is element a_{23} in matrix A?

$$A = \begin{bmatrix} 4 & -9 & 17 & 1 \\ 0 & 5 & 8 & 6 \\ -3 & -2 & 10 & 0 \end{bmatrix}$$

A_{23} is in Row 2 and Column 3.

a_{23} is 8.

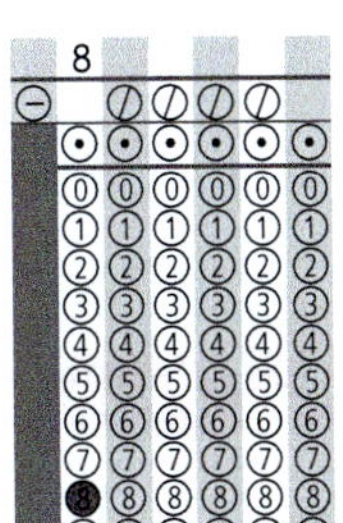

Got It? **1.** What is element a_{13} in matrix A?

You can represent a system of equations efficiently with a matrix. Each matrix row represents an equation. The last matrix column shows the constants to the right of the equal signs. Each of the other columns shows the coefficients of one of the variables.

System of Equations

$$\begin{aligned} x &+ 3y = 7 \\ 3x &+ y = -8 \end{aligned}$$

x-coefficients | y-coefficients | constants

Matrix

$$\left[\begin{array}{cc|c} 1 & 3 & 7 \\ 3 & 1 & -8 \end{array}\right]$$

The 1's are coefficients of x and y.

Draw a vertical bar to replace the equal signs and separate the coefficients from the constants.

Problem 2 Representing Systems With Matrices

Think

Why is the order of elements important in a matrix?
Different orders of elements could correspond to different systems of equations.

How can you represent the system of equations with a matrix?

A $\begin{cases} 2x + y = 9 \\ x - 6y = -1 \end{cases}$

The matrix $\left[\begin{array}{cc|c} 2 & 1 & 9 \\ 1 & -6 & -1 \end{array}\right]$ represents the system above.

B $\begin{cases} x - 3y + z = 6 \\ x + 3z = 12 \\ y = -5x + 1 \end{cases}$

Step 1 Write each equation in the same variable order. Line up the variables. Leave space where a coefficient is 0.

$$\begin{cases} x - 3y + z = 6 \\ x \qquad + 3z = 12 \\ 5x + y \qquad = 1 \end{cases}$$

Step 2 Write the matrix using the coefficients and constants. Notice the 1's and 0's.

$$\left[\begin{array}{ccc|c} 1 & -3 & 1 & 6 \\ 1 & 0 & 3 & 12 \\ 5 & 1 & 0 & 1 \end{array}\right]$$

 **Got It?** **2.** How can you represent the system of equations with a matrix?

a. $\begin{cases} -4x - 2y = 7 \\ 3x + y = -5 \end{cases}$ **b.** $\begin{cases} 4x - y + 2z = 1 \\ y + 5z = 20 \\ 2x = -y + 7 \end{cases}$

 **Problem 3** **Writing a System From a Matrix**

What linear system of equations does this matrix represent? $\left[\begin{array}{cc|c} 5 & 2 & 7 \\ 0 & 1 & 9 \end{array}\right]$

Think	Write
Each row shows coefficient-coefficient-constant of one equation.	$5x + 2y = 7$ $0x + 1y = 9$
Simplify. Write the system.	$\begin{cases} 5x + 2y = 7 \\ y = 9 \end{cases}$

 **Got It?** **3.** What linear system does $\left[\begin{array}{cc|c} 2 & 0 & 6 \\ 5 & -2 & 1 \end{array}\right]$ represent?

You can use a matrix that represents a system of equations to solve the system. In this way, you do not have to write the variables. To solve the system using the matrix, use the steps for solving by elimination. Each step is a **row operation**.

Your goal is to use row operations to get a matrix in the form $\left[\begin{array}{cc|c} 1 & 0 & a \\ 0 & 1 & b \end{array}\right]$ or $\left[\begin{array}{ccc|c} 1 & 0 & 0 & a \\ 0 & 1 & 0 & b \\ 0 & 0 & 1 & c \end{array}\right]$

Notice that the first matrix represents the system $x = a, y = b$, which then will be the solution of a system of two equations in two unknowns. The second matrix represents the system $x = a, y = b$, and $z = c$.

take note

Key Concept **Row Operations**

Switch any two rows. $\begin{bmatrix} 2 & -1 & 3 \\ 3 & 2 & 5 \end{bmatrix}$ becomes $\begin{bmatrix} 3 & 2 & 5 \\ 2 & -1 & 3 \end{bmatrix}$

Multiply a row by a constant. $\begin{bmatrix} 3 & 2 & 5 \\ 2 & -1 & 3 \end{bmatrix}$ becomes $\begin{bmatrix} 3 & 2 & 5 \\ 2 \cdot 2 & -1 \cdot 2 & 3 \cdot 2 \end{bmatrix} = \begin{bmatrix} 3 & 2 & 5 \\ 4 & -2 & 6 \end{bmatrix}$

Add one row to another. $\begin{bmatrix} 3 & 2 & 5 \\ 4 & -2 & 6 \end{bmatrix}$ becomes $\begin{bmatrix} 3+4 & 2-2 & 5+6 \\ 4 & -2 & 6 \end{bmatrix} = \begin{bmatrix} 7 & 0 & 11 \\ 4 & -2 & 6 \end{bmatrix}$

Combine any of these steps.

Problem 4 Solving a System Using a Matrix

Think

How is solving a system using row operations similar to using elimination?
You use the same steps but the variables don't appear in the matrices.

What is the solution of the system? $\begin{cases} x + 4y = -1 \\ 2x + 5y = 4 \end{cases}$

$$\left[\begin{array}{cc|c} 1 & 4 & -1 \\ 2 & 5 & 4 \end{array}\right]$$

Write the matrix for the system.

$$\begin{array}{r} -2\,(1 \quad 4 \quad -1) \\ +\quad 2 \quad 5 \quad 4 \\ \hline 0 \quad -3 \quad 6 \end{array}$$

Multiply Row 1 by -2. Add to Row 2. Replace Row 2 with the sum. Write the new matrix.

$$\left[\begin{array}{cc|c} 1 & 4 & -1 \\ 0 & -3 & 6 \end{array}\right]$$

$$-\frac{1}{3}\,(0 \quad -3 \quad 6) = 0 \quad 1 \quad -2$$

Multiply Row 2 by $-\frac{1}{3}$. Write the new matrix.

$$\left[\begin{array}{cc|c} 1 & 4 & -1 \\ 0 & 1 & -2 \end{array}\right]$$

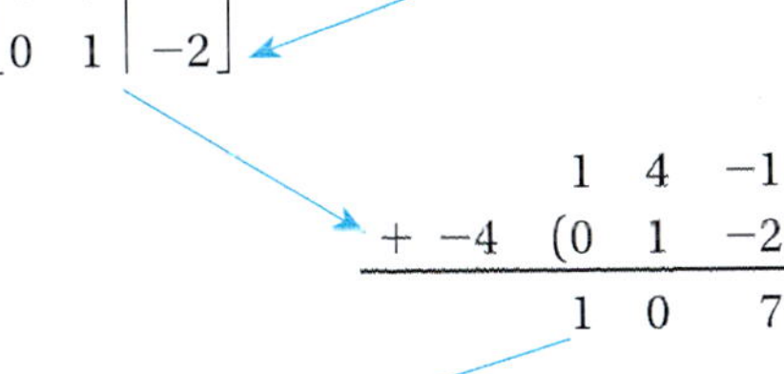

$$\begin{array}{r} 1 \quad 4 \quad -1 \\ +\ -4\,(0 \quad 1 \quad -2) \\ \hline 1 \quad 0 \quad 7 \end{array}$$

Multiply Row 2 by -4. Add to Row 1. Replace Row 1 with the sum. Write the new matrix.

$$\left[\begin{array}{cc|c} 1 & 0 & 7 \\ 0 & 1 & -2 \end{array}\right]$$

The solution to the system is $(7, -2)$.

Check

$x + 4y = -1$	$2x + 5y = 4$	Use the original equations.
$7 + 4(-2) \stackrel{?}{=} -1$	$2(7) + 5(-2) \stackrel{?}{=} 4$	Substitute.
$7 + (-8) \stackrel{?}{=} -1$	$14 + (-10) \stackrel{?}{=} 4$	Multiply.
$-1 = -1$ ✔	$4 = 4$ ✔	Simplify.

Got It? **4. a.** What is the solution of the system? $\begin{cases} 9x - 2y = 5 \\ 3x + 7y = 17 \end{cases}$

b. Reasoning Which method is more similar to solving a system using row operations: *elimination* or *substitution*? Justify your reasoning.

Matrices that represent the solution of a system are in *reduced row echelon form*. Many calculators have a **rref** (reduced row echelon form) function for working with matrices. This function will do all the row operations for you. You can use **rref** to solve a system of equations.

Problem 5 Using a Calculator to Solve a Linear System

What is the solution of the system of equations? $\begin{cases} 2a + 3b - c = 1 \\ -4a + 9b + 2c = 8 \\ -2a + 2c = 3 \end{cases}$

Think

How do you enter missing variables into a matrix?
If a variable is not present in an equation, enter its coefficient as 0 in the matrix.

Step 1 Enter the system into a calculator as a matrix.

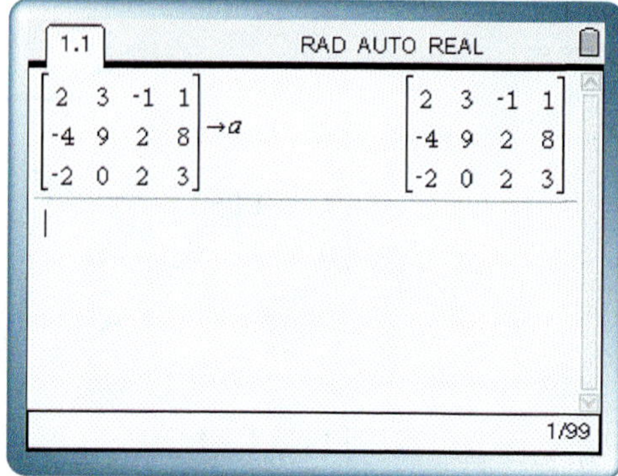

Step 2 Apply the **rref()** function to the matrix. Put the matrix elements in fraction form if some are not integers.

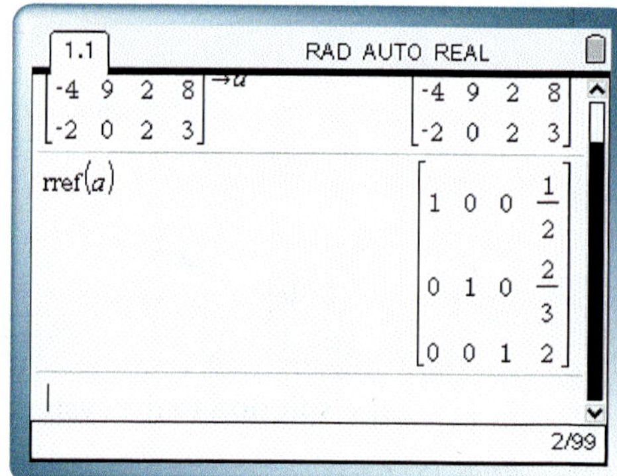

Step 3 List the solution.

The solution of the system is $a = \frac{1}{2}$, $b = \frac{2}{3}$, $c = 2$.

Check

$2a + 3b - c = 1$

$2\left(\frac{1}{2}\right) + 3\left(\frac{2}{3}\right) - 2 \stackrel{?}{=} 1$

$1 + 2 - 2 \stackrel{?}{=} 1$

$1 = 1$ ✔

$-4a + 9b + 2c = 8$

$-4\left(\frac{1}{2}\right) + 9\left(\frac{2}{3}\right) + 2(2) \stackrel{?}{=} 8$

$-2 + 6 + 4 \stackrel{?}{=} 8$

$8 = 8$ ✔

$-2a + 2c = 3$

$-2\left(\frac{1}{2}\right) + 2(2) \stackrel{?}{=} 3$

$-1 + 4 \stackrel{?}{=} 3$

$3 = 3$ ✔

Got It? 5. What is the solution of the system of equations?

$\begin{cases} a + 4b + 6c = 21 \\ 2a - 2b + c = 4 \\ -8b + c = -1 \end{cases}$

Lesson Check

Do you know HOW?

State the dimensions of each matrix.

1. $\begin{bmatrix} 2 \\ 5 \end{bmatrix}$

2. $\left[\begin{array}{ccc|c} 6 & 9 & 0 & 3 \\ 4 & 6 & 2 & 7 \end{array}\right]$

Write a matrix to represent each system.

3. $\begin{cases} 3a + 5b = 0 \\ a + b = 2 \end{cases}$

4. $\begin{cases} x + 3y - z = 2 \\ x + 2z = 8 \\ 2y - z = 1 \end{cases}$

Do you UNDERSTAND?

5. How many elements are in a 4×4 matrix?

6. Writing Using Matrix A in Problem 1, describe the difference in identifying element a_{21} and element a_{12}.

7. Open-Ended Write a situation that can be modeled by the matrix. $\left[\begin{array}{cc|c} 4 & 2 & 8 \\ 0 & 1 & 2 \end{array}\right]$

Practice and Problem-Solving Exercises

Identify the indicated element. $A = \begin{bmatrix} 3 & 12 & 6 \\ 1 & 0 & 9 \\ 8 & 7 & 4 \end{bmatrix}$

See Problem 1.

8. a_{32} **9.** a_{21} **10.** a_{13} **11.** a_{31}

Write a matrix to represent each system.

See Problem 2.

12. $\begin{cases} x + 2y = 11 \\ 2x + 3y = 18 \end{cases}$

13. $\begin{cases} 3x + 2y = 16 \\ y = 5 \end{cases}$

14. $\begin{cases} 2a - 3b = 6 \\ a + b = 2 \end{cases}$

15. $\begin{cases} r - s + t = 150 \\ 2r + t = 425 \\ s + 3t = 0 \end{cases}$

16. $\begin{cases} y = 3x - 7 \\ x = 2 \end{cases}$

17. $\begin{cases} x - y + z = 0 \\ x - 2y - z = 5 \\ 2x - y + 2z = 8 \end{cases}$

Write the system of equations represented by each matrix.

See Problem 3.

18. $\left[\begin{array}{cc|c} 1 & 0 & 4 \\ 0 & 1 & -6 \end{array}\right]$

19. $\left[\begin{array}{cc|c} 5 & 1 & 3 \\ -2 & 2 & 4 \end{array}\right]$

20. $\left[\begin{array}{cc|c} -1 & 2 & -8 \\ 1 & 1 & 7 \end{array}\right]$

21. $\left[\begin{array}{ccc|c} 2 & 1 & 1 & 1 \\ 1 & 1 & 1 & 2 \\ 1 & -1 & 1 & -2 \end{array}\right]$

22. $\left[\begin{array}{ccc|c} 0 & 1 & 2 & 4 \\ -2 & 3 & 6 & 9 \\ 1 & 0 & 1 & 3 \end{array}\right]$

23. $\left[\begin{array}{ccc|c} 5 & 2 & 1 & 5 \\ 4 & 1 & 2 & 8 \\ 1 & 3 & -6 & 2 \end{array}\right]$

Solve the system of equations using a matrix.

See Problems 4 and 5.

24. $\begin{cases} x + 3y = 5 \\ x + 4y = 6 \end{cases}$

25. $\begin{cases} p - 3q = -1 \\ -5p + 16q = 5 \end{cases}$

26. $\begin{cases} 300x - y = 130 \\ 200x + y = 120 \end{cases}$

27. $\begin{cases} x + 3y = 22 \\ 2x - y = 2 \end{cases}$

28. $\begin{cases} x + 3y = 6 \\ 2x + 4y = 12 \end{cases}$

29. $\begin{cases} x + y = 5 \\ -2x + 4y = 8 \end{cases}$

B Apply

30. Business A manufacturer sells pencils and erasers in packages. The price of a package of five erasers and two pencils is \$.23. The price of a package of seven erasers and five pencils is \$.41. Write a system of equations to represent this situation. Then write a matrix to represent the system.

31. Think About a Plan Last year your town invested a total of \$25,000 into two separate funds. The return on one fund was 4% and the return on the other was 6%. If the town earned a total of \$1300 in interest, how much money was invested in each fund?

- What variables will you use? What will they represent?
- What equations can you write to model this situation?
- How can you use a matrix to solve this system?

Graphing Calculator Solve each system.

32. $\begin{cases} x + y + z = 2 \\ 2y - 2z = 2 \\ x - 3z = 1 \end{cases}$

33. $\begin{cases} x - y + z = 3 \\ x + 3z = 6 \\ y - 2z = -1 \end{cases}$

34. $\begin{cases} x + y + z = -1 \\ 3x + 4y - z = 8 \\ 6x + 8y - 2z = 16 \end{cases}$

35. $\begin{cases} x - y + 3z = 9 \\ x + 2z = 3 \\ 2x + 2y + z = 10 \end{cases}$

36. $\begin{cases} 2x + 3y + z = 13 \\ 5x - 2y - 4z = 7 \\ 4x + 5y + 3z = 25 \end{cases}$

37. $\begin{cases} -2w + x + y = 0 \\ -w + 2x - y + z = 1 \\ -2w + 3x + 3y + 2z = 6 \\ w + x + 2y + z = 5 \end{cases}$

38. Snacks Suppose you want to fill nine 1-lb tins with a snack mix. You have \$15 and plan to buy almonds for \$2.45 per lb, hazelnuts for \$1.85 per lb, and raisins for \$.80 per lb. You want the mix to contain an equal amount of almonds and hazelnuts and twice as much of the nuts as the raisins by weight.

a. Writing Explain how each equation to the right relates to the problem. What does each variable represent?

b. Solve the system.

c. How many of each ingredient should you buy?

$\begin{cases} x + y + z = 9 \\ 2.45x + 1.85y + 0.8z = 15 \\ x + y = 2z \end{cases}$

39. Geometry The coordinates (x, y) of a point in a plane are the solution of the system $\begin{cases} 2x + 3y = 13 \\ 5x + 7y = 31 \end{cases}$. Find the coordinates of the point.

40. Error Analysis A classmate writes the matrix at the right to represent a system and says that the solution is $x = 2$, $y = 0$. Explain your classmate's error and describe how to correct it.

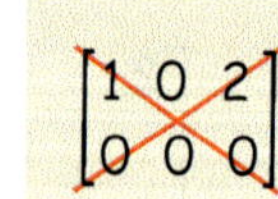

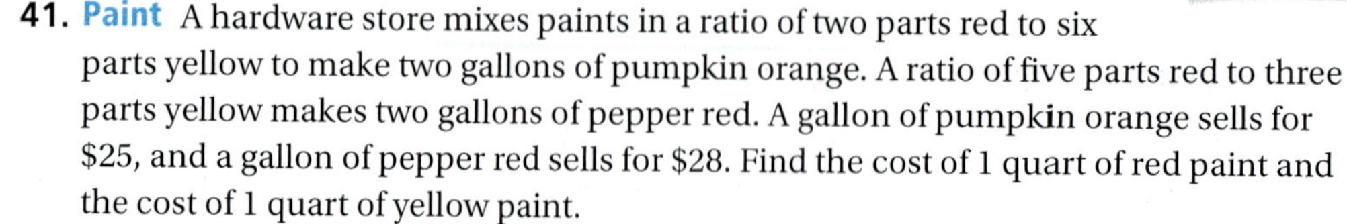

41. Paint A hardware store mixes paints in a ratio of two parts red to six parts yellow to make two gallons of pumpkin orange. A ratio of five parts red to three parts yellow makes two gallons of pepper red. A gallon of pumpkin orange sells for \$25, and a gallon of pepper red sells for \$28. Find the cost of 1 quart of red paint and the cost of 1 quart of yellow paint.

Open-Ended Complete each system for the given number of solutions.

42. infinitely many

$$\begin{cases} x + y = 7 \\ 2x + 2y = \blacksquare \end{cases}$$

43. one solution

$$\begin{cases} x + y + z = 7 \\ y + z = \blacksquare \\ z = \blacksquare \end{cases}$$

44. no solution

$$\begin{cases} x + y + z = 7 \\ y + z = \blacksquare \\ y + z = \blacksquare \end{cases}$$

Solve the system of equations using a matrix. (Hint: Start by substituting $m = \frac{1}{x}$ and $n = \frac{1}{y}$.)

45. $\begin{cases} \frac{4}{x} + \frac{1}{y} = 1 \\ \frac{8}{x} + \frac{4}{y} = 3 \end{cases}$

46. $\begin{cases} \frac{4}{x} - \frac{2}{y} = 1 \\ \frac{10}{x} + \frac{20}{y} = 0 \end{cases}$

47. $\begin{cases} \frac{7}{x} + \frac{3}{y} = 5 \\ \frac{2}{x} + \frac{1}{y} = -1 \end{cases}$

Standardized Test Prep

SAT/ACT

48. Which equation represents a line with a slope of $\frac{1}{2}$ and a y-intercept of $\frac{3}{4}$?

Ⓐ $y = \frac{1}{2}x - \frac{3}{4}$ Ⓑ $y = \frac{3}{4}x - \frac{1}{2}$ Ⓒ $y = \frac{1}{2}x + \frac{3}{4}$ Ⓓ $y = \frac{3}{4}x + \frac{1}{2}$

49. Which graph best represents the solution of the inequality $y \le 2|x - 1| - 4$?

Ⓕ 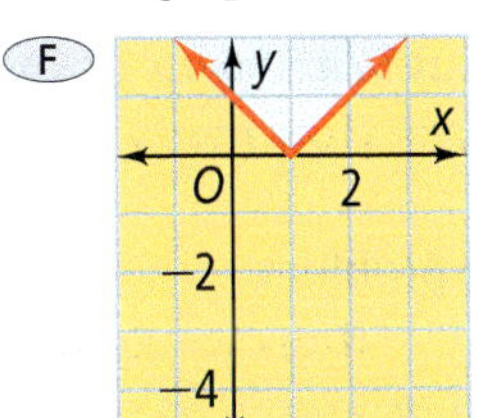

Ⓖ 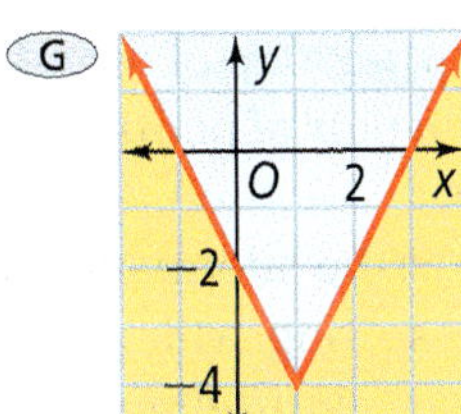

Ⓗ 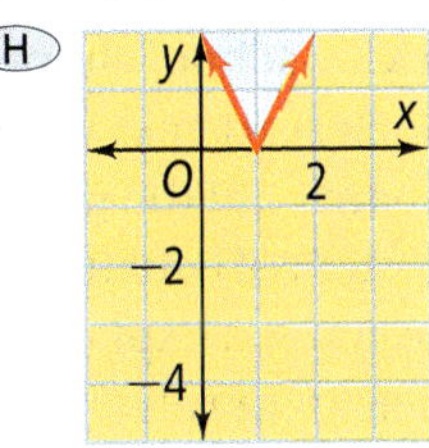

Ⓘ 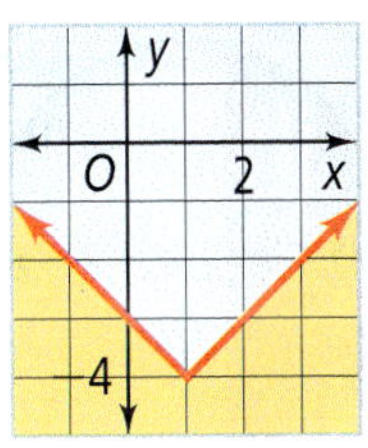

Short Response

50. At what point do the graphs of the equations $y = 7x - 3$ and $-6x + y = 2$ intersect?

Mixed Review

Solve each inequality. Graph the solution.

See Lesson 1-5.

51. $12 \ge 2(4x + 1) + 22$ **52.** $2x - (3x + 5) \le 30$ **53.** $4x + 5 - 3x \le 2x + 1$

Solve each equation. Check your answers.

See Lesson 1-6.

54. $|2y - 3| = 12$ **55.** $|4x| = 40$ **56.** $|2y - 4| = 16$

Get Ready! **To prepare for Lesson 4-1, do Exercises 57 and 58.**

Write an equation for each transformation of $y = x$.

See Lesson 2-6.

57. vertical stretch by a factor of 2. **58.** vertical compression by a factor of $\frac{1}{3}$

Pull It All Together ASSESSMENT

BIG idea Function

The solution of a system of two linear equations corresponds in general to the intersection of the graphs of the corresponding functions.

Performance Task 1

You are given a linear system of two equations in two unknowns. Before solving, describe how you can mentally check whether the system is independent and consistent. In which order would you do your check? Why?

BIG idea Equivalence

You can solve a system of equations by representing the system in some form that is equivalent to the original form but easier to solve. There are different ways to do this.

Performance Task 2

During a back-to-school shopping trip, a group of friends spent \$245.86 on 14 shirts and pants. Each shirt cost \$11.99. Each pair of pants cost \$24.99. How many shirts and pairs of pants did the group buy?

a. Write a system of equations to model the information in the problem.

b. Study the system. Explain, without solving, which method you think would be most efficient for solving the system: *substitution, elimination, graphing*, or *making a table*. Explain why the other methods would be less efficient.

c. How could you simplify the numbers used in this system to simplify the system? Does this new system change your answers to part (b)? Explain.

BIG idea Solving Equations and Inequalities

You can represent a system of equations with a matrix. Transforming the matrix to reduced row echelon form gives you an equivalent system for which the solution is obvious.

Performance Task 3

Solve this system using a matrix. $\begin{cases} 4x + 10y = 3 \\ 7x - 2y = 2 \end{cases}$

Make three columns on your paper. In the first column, show each step, changing one matrix row at a time. In the second column, write the two equations that correspond to each matrix in the first column. In the third column, describe how you could transform each set of equations to the next.

3 Chapter Review

Connecting BIG ideas and Answering the Essential Questions

1 Function
Find a point of intersection (x, y) of the graphs of functions f and g and you have found a solution of the system $y = f(x)$, $y = g(x)$.

Solving Systems Using Tables and Graphs (Lesson 3-1)

$$\begin{cases} y = -2x + 3 \\ y = 2x - 1 \end{cases}$$

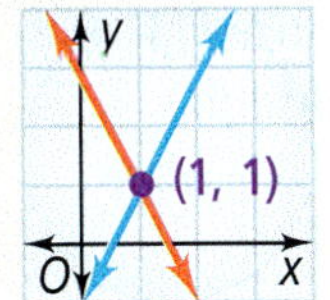

The solution is (1, 1).

Systems of Inequalities and Linear Programming (Lessons 3-3 and 3-4)

$$\begin{cases} y > -2x + 3 \\ y \leq 2x - 1 \end{cases}$$

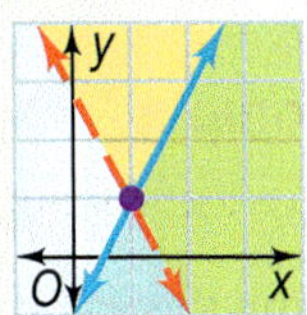

2 Equivalence
If the equations of two systems are equivalent, then a solution of the system that is easier to solve is also a solution of the more difficult system.

Solving Systems Algebraically (Lesson 3-2)

$$\begin{cases} -y = -x + 2 \\ 3y = 2x - 2 \end{cases} \rightarrow \begin{aligned} -2y &= -2x + 4 \\ 3y &= 2x - 2 \\ \hline y &= 2 \end{aligned}$$

$$3(2) = 2x - 2 \quad \rightarrow \quad x = 4$$

The solution is $x = 4, y = 2$

Systems With Three Variables (Lesson 3-5)

$$\begin{cases} -2x + y + z = -3 \\ 2x - y + z = -1 \\ -2x, - y - z = -1 \end{cases}$$

$x = 1, y = 1, z = -2$

3 Solving Equations and Inequalities
The matrix row operations of adding rows and multiplying a row by a constant are equivalent to addition and multiplication properties of equality.

Solving Systems Using Matrices (Lesson 3-6)

$$\left[\begin{array}{cc|c} -2 & 3 & 1 \\ 2 & -1 & 1 \end{array}\right]$$

$$\left[\begin{array}{cc|c} 1 & 0 & 1 \\ 0 & 1 & 1 \end{array}\right] \rightarrow x = 1, y = 1$$

Chapter Vocabulary

- consistent system (p. 137)
- constraint (p. 157)
- dependent system (p. 137)
- equivalent systems (p. 144)
- feasible region (p. 157)
- inconsistent system (p. 137)
- independent system (p. 137)
- linear programming (p. 157)
- linear system (p. 134)
- matrix (p. 174)
- matrix element (p. 174)
- objective function (p. 157)
- row operation (p. 176)
- solution of a system (p. 134)
- system of equations (p. 134)

Fill in the blank.

1. A consistent system with exactly one solution is a(n) ___________.

2. ___________ is a method for finding a minimum or maximum value, given a system of limits called ___________.

3-1 Solving Systems Using Tables and Graphs

Quick Review

A **system of equations** has two or more equations. Points of intersection are solutions. A **linear system** has linear equations. A **consistent system** can be **dependent**, with infinitely many solutions, or **independent**, with one solution. An **inconsistent system** has no solution.

Example

Solve the system $\begin{cases} 3x + 2y = 4 \\ 2x - 4y = 8 \end{cases}$

Graph the equations.

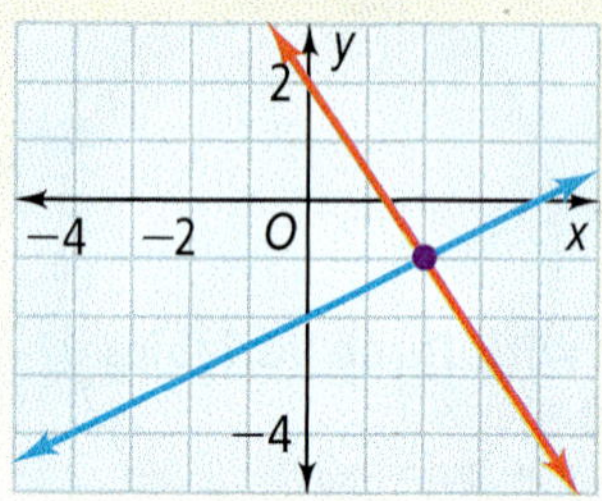

The only solution, where the lines intersect, is $(2, -1)$.

Exercises

Without graphing, classify each system of equations as *independent*, *dependent*, or *inconsistent*. Solve independent systems by graphing.

3. $\begin{cases} 6x - 2y = 2 \\ 2 + 6x = y \end{cases}$

4. $\begin{cases} 5 - y = 2x \\ 6x - 15 = -3y \end{cases}$

5. $\begin{cases} 6y + 2x = 8 \\ 12y + 4x = 4 \end{cases}$

6. $\begin{cases} 1.5 + 3x = 0.5y \\ 6 - 2y = -12x \end{cases}$

7. $\begin{cases} 2 - 0.25x = 0.5y \\ -1.5y = 1.5x - 3 \end{cases}$

8. $\begin{cases} 1 + y = x \\ x + y = 1 \end{cases}$

9. For \$7.52, you purchased 8 pens and highlighters from a local bookstore. Each highlighter cost \$1.09 and each pen cost \$.69. How many pens did you buy?

3-2 Solving Systems Algebraically

Quick Review

To solve an independent system by substitution, solve one equation for a variable. Then substitute that expression into the other equation and solve for the remaining variable. To solve by elimination, add two equations with additive inverses as coefficients to eliminate one variable and solve for the other. In both cases you solve for one of the variables and use substitution to solve for the remaining variable.

Example

Solve $\begin{cases} 10 - y = 4x \\ x = 4 + 0.5y \end{cases}$ by substitution.

Substitute for x: $10 - y = 4(4 + 0.5y) = 16 + 2y$.
Solve for y: $y = -2$.
Substitute into the first equation:
$10 - (-2) = 4x$.
Solve for x: $x = 3$. The solution is $(3, -2)$.

Exercises

Solve each system by substitution.

10. $\begin{cases} x - 2y = 3 \\ 3x + y = -5 \end{cases}$

11. $\begin{cases} 14x - 35 = 7y \\ -25 - 6x = 5y \end{cases}$

Solve each system by elimination.

12. $\begin{cases} 11 - 5y = 2x \\ 5y + 3 = -9x \end{cases}$

13. $\begin{cases} 2x + 3y = 4 \\ 4x + 6y = 9 \end{cases}$

14. Roast beef has 25 g of protein and 11 g of calcium per serving. A serving of mashed potatoes has 2 g of protein and 25 g of calcium. How many servings of each are needed to supply exactly 29 g of protein and 61 g of calcium?

3-3 Systems of Inequalities

Quick Review

To solve a system of inequalities by graphing, first graph the boundaries for each inequality. Then shade the region(s) of the plane containing solutions valid for both inequalities.

Example

Solve the system of inequalities by graphing.

$$\begin{cases} y > -3 \\ y \leq -|x - 1| \end{cases}$$

Graph both inequalities and shade the region valid for both inequalities.

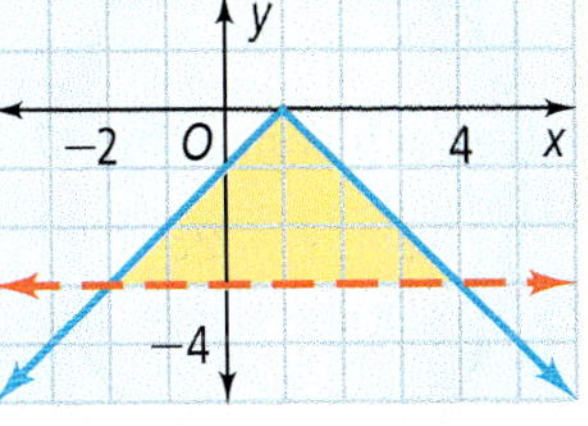

Exercises

Solve each system of inequalities by graphing.

15. $\begin{cases} y < 4x \\ 3x + y \geq 5 \end{cases}$

16. $\begin{cases} y < |2x - 4| \\ x + 5y \geq -1 \end{cases}$

17. $\begin{cases} y \leq |x + 2| - 3 \\ y \geq 1 + \frac{1}{4}x \end{cases}$

18. $\begin{cases} 2x + 3y > 6 \\ x \leq -1 \\ y \geq 4 \end{cases}$

19. For a community breakfast there should be at least three times as much regular coffee as decaffeinated coffee. A total of ten gallons is sufficient for the breakfast. Write and graph a system of inequalities to model the problem.

3-4 Linear Programming

Quick Review

Linear programming is used to find a minimum or maximum of an **objective function**, given **constraints** as linear inequalities. The maximum or minimum occurs at a vertex of the **feasible region**, which contains the solutions to the system of constraints.

Example

Graph the system of constraints and name the vertices.

Objective function: $P = 2x + y$

$$\begin{cases} x \leq 8 \\ y < 5 \\ x \geq 0, y \geq 0 \end{cases}$$

Graph the inequalities and shade the area satisfying all inequalities.

The vertices of the feasible region are (0, 0), (0, 5), (8, 5), and (8, 0).

Evaluate the objective function at each vertex:

$2(0) + 0 = 0$ $\quad$ $2(0) + 5 = 5$

$2(8) + 5 = 21$ $\quad$ $2(8) + 0 = 16$

The maximum value occurs at (8, 5).

Exercises

Graph the system of constraints. Name the vertices. Then find the values of x and y that maximize or minimize the objective function.

20. $\begin{cases} x \geq 2 \\ y \geq 0 \\ 3x + 2y \geq 12 \end{cases}$

Minimum for $C = x + 5y$

21. $\begin{cases} 3x + 2y \leq 12 \\ x + y \leq 5 \\ x \geq 0, y \geq 0 \end{cases}$

Maximum for $P = 3x + 5y$

22. A lunch stand makes \$.75 profit on each chef's salad and \$1.20 profit on each Caesar salad. On a typical weekday, it sells between 40 and 60 chef's salads and between 35 and 50 Caesar salads. The total number sold has never exceeded 100 salads. How many of each type should be prepared in order to maximize profit?

3-5 Systems With Three Variables

Quick Review

To solve a system of three equations, either pair two equations and eliminate the same variable from both equations, using one equation twice, or choose an equation, solve for one variable, and substitute the expression for that variable into the other two equations. Then, solve the remaining system.

Example

Solve by elimination. $\begin{cases} x + y + z = 10 & ① \\ 2x - y + z = 9 & ② \\ -3x + 2y + 2z = 5 & ③ \end{cases}$

Add equations ① and ② to eliminate y. ④ $3x + 2z = 19$

Add 2 times ② to ③ to eliminate y. ⑤ $x + 4z = 23$

Add -3 times ⑤ to ④ to eliminate x. $z = 5$

Substitute $z = 5$ into ⑤. $x = 3$

Substitute $z = 5$ and $x = 3$ into ① or ②. $y = 2$

The solution to the system is (3, 2, 5).

Exercises

Solve each system by elimination.

23. $\begin{cases} x + y - 2z = 8 \\ 5x - 3y + z = -6 \\ -2x - y + 4z = -13 \end{cases}$

24. $\begin{cases} -x + y + 2z = -5 \\ 5x + 4y - 4z = 4 \\ x - 3y - 2z = 3 \end{cases}$

Solve each system by substitution.

25. $\begin{cases} 3x + y - 2z = 22 \\ x + 5y + z = 4 \\ x = -3z \end{cases}$

26. $\begin{cases} x + 2y + z = 14 \\ y = z + 1 \\ x = -3z + 6 \end{cases}$

3-6 Solving Systems Using Matrices

Quick Review

A **matrix** can represent a system of equations where each row stands for a different equation. The columns contain the coefficients of the variables and the constants.

Example

Solve using a matrix. $\begin{cases} 6x + 3y = -15 \\ 2x + 4y = 10 \end{cases}$

Enter coefficients as matrix elements $\left[\begin{array}{cc|c} 6 & 3 & -15 \\ 2 & 4 & 10 \end{array}\right]$.

Divide the first row by 3 to get $\left[\begin{array}{cc|c} 2 & 1 & -5 \\ 2 & 4 & 10 \end{array}\right]$. Subtract the first row from the second row to get $\left[\begin{array}{cc|c} 2 & 1 & -5 \\ 0 & 3 & 15 \end{array}\right]$. Multiply the second row by $\frac{1}{3}$ to get $\left[\begin{array}{cc|c} 2 & 1 & -5 \\ 0 & 1 & 5 \end{array}\right]$. Subtract the second row from the first row to get $\left[\begin{array}{cc|c} 2 & 0 & -10 \\ 0 & 1 & 5 \end{array}\right]$. Divide the first row by 2 to get $\left[\begin{array}{cc|c} 1 & 0 & -5 \\ 0 & 1 & 5 \end{array}\right]$. The solution to the system is $(-5, 5)$.

Exercises

Solve each system using a matrix.

27. $\begin{cases} 4x - 12y = -1 \\ 6x + 4y = 4 \end{cases}$

28. $\begin{cases} 7x + 2y = 5 \\ 13x + 14y = -1 \end{cases}$

29. $\begin{cases} -5x + 3y + 4z = 2 \\ 3x - y - z = 4 \\ x - 6y - 5z = -4 \end{cases}$

30. $\begin{cases} x + y + z = 4 \\ 2x - y + z = 5 \\ x + y - 2z = 13 \end{cases}$

3 Chapter Test

Do you know HOW?

Without graphing, classify each system. Then find the solution to each system using a graph.

1. $\begin{cases} y = 5x - 2 \\ y = x + 4 \end{cases}$

2. $\begin{cases} 3x + 2y = 9 \\ 3x + 2y = 4 \end{cases}$

Solve the system by substitution.

3. $\begin{cases} 0.3x - y = 0 \\ y = 2 + 0.25x \end{cases}$

Solve the system by elimination.

4. $\begin{cases} 4x - 2y = 3 \\ y - 2x = -\frac{3}{2} \end{cases}$

5. $\begin{cases} 3x + 4y = 9 \\ 2x + y = 6 \end{cases}$

Graph the solution of each system.

6. $\begin{cases} 2x + y < 3 \\ x < y + 3 \end{cases}$

7. $\begin{cases} |x + 3| > y \\ y > 2x - 1 \end{cases}$

Graph the system of constraints. Identify all vertices. Then find the values of x and y that maximize or minimize the objective function.

8. $\begin{cases} x \leq 5 \\ y \leq 4 \\ x \geq 0 \\ y \geq 0 \end{cases}$

Maximum for $P = 2x + y$

Solve each system.

9. $\begin{cases} x - y + z = 0 \\ 3x - 2y + 6z = 9 \\ -x + y - 2z = -2 \end{cases}$

10. $\begin{cases} 2x + y + z = 8 \\ x + 2y - z = -5 \\ z = 2x - y \end{cases}$

Do you UNDERSTAND?

Write a matrix that represents the system. Then solve the system. Tell what method you used and why.

11. $\begin{cases} -a + 4b + 2c = -8 \\ 3a + b - 4c = 9 \\ b = -1 \end{cases}$

12. Sales A pizza shop makes \$1.50 on each small pizza and \$2.15 on each large pizza. On a typical Friday, it sells between 70 and 90 small pizzas and between 100 and 140 large pizzas. The shop can make no more than 210 pizzas in a day. How many of each size pizza must be sold in order to maximize profit?

13. Investing Your teacher invested \$5000 in three funds. After a year they had \$5450. The growth fund had a return rate of 12%, the income fund had a return rate of 8%, and the money market fund had a return rate of 5%. Your teacher invested twice as much in the income fund as in the money market fund. How much money was invested in each fund?

14. Writing Describe how to identify situations in which substitution may be the best method for solving a system of equations.

15. Open-Ended Write a system of constraints whose graph is a parallelogram.

3 Cumulative Standards Review

ASSESSMENT

TIPS FOR SUCCESS

Some problems require the selection of an appropriate representation (concrete, pictorial, graphical, verbal, or symbolic) to find a solution.

One angle of a right triangle measures 90°. The measure of the second angle is 5 times the measure of the third. What are the measures of these angles?

(A) 30° and 60°

(B) 30° and 150°

(C) 15° and 75°

(D) 20° and 100°

TIP 1

Make a drawing.

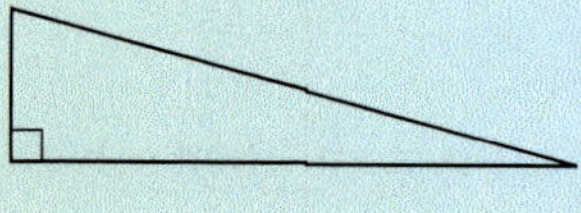

TIP 2

Write a system:

$x + y + 90 = 180$

$x = 5y$

Think It Through

A triangle can have only one right angle, so the other two angles must each have a measure less than 90°. Use x and y to represent the unknown angles.

$$x = 5y$$
$$5y + y + 90 = 180$$
$$6y + 90 = 180$$
$$6y = 90$$
$$y = 15, x = 75$$

The correct answer is C.

Vocabulary Builder

As you solve test items, you must understand the meanings of mathematical terms. Match each term with its mathematical meaning.

A. equivalent systems

B. absolute value

C. system of equations

D. linear inequality

I. a number's distance from zero on a number line

II. an inequality in two variables whose graph is a region of the half-plane

III. a set of two or more equations that use the same variables

IV. systems that have the same solution(s)

Multiple Choice

Read each question. Then write the letter of the correct answer on your paper.

1. Which of the following is true about the given system?

$$\begin{cases} -4y = 12 - 8x \\ y = 2x - 3 \end{cases}$$

The system has

(A) zero solutions.

(B) exactly one solution.

(C) two solutions.

(D) infinitely many solutions.

2. Which is the graph of $y = -|x - 2| + 1$?

F
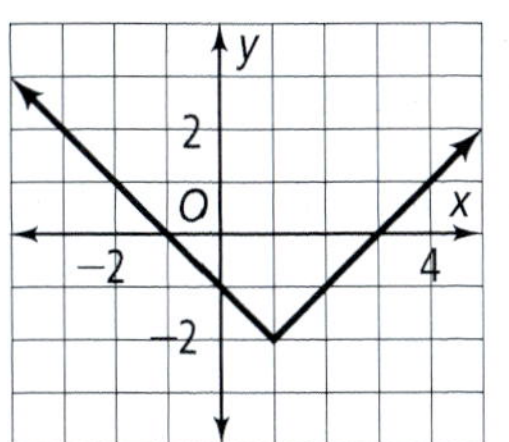

H
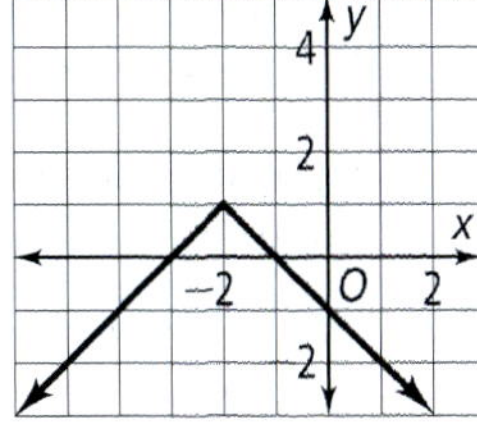

G
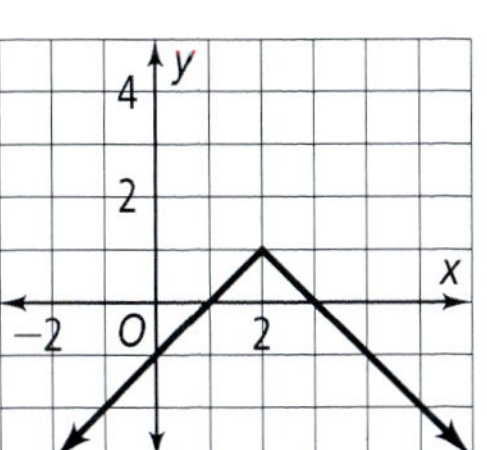

I
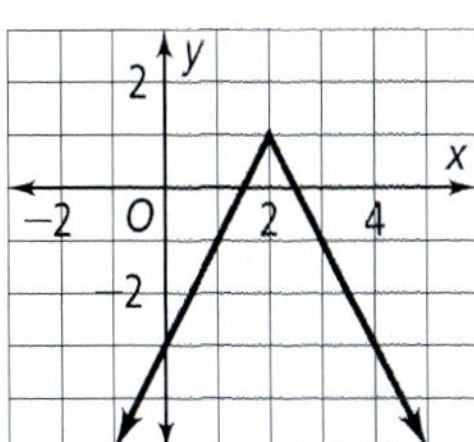

3. Which graph represents the solution of the inequality $|3x + 12| \geq 3$?

A
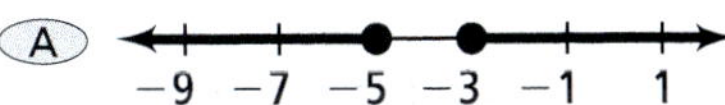

B
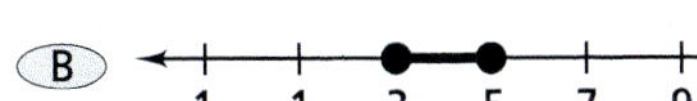

C −5 −3 −1 1 3 5

D −5 −3 −1 1 3 5

4. Josea wants to solve the system using substitution.

$$\begin{cases} x = -2y + 4 \\ 2x - 3y = 5 \end{cases}$$

Which of the following is the best way for Josea to proceed?

F Solve the first equation for y, then substitute into the second equation.

G Solve the second equation for y, then substitute into the first equation.

H Substitute $-2y + 4$ for x in the second equation.

I Substitute $-2y + 4$ for y in the second equation.

5. A board must be cut so that its length is 40.50 cm. The tolerance is 0.25 cm. Which inequality describes the allowable lengths for the board?

A $|x - 0.25| \leq 40.50$

B $|x + 0.25| \leq 40.50$

C $|x - 40.50| \leq 0.25$

D $|x - 0.25| \leq 40.75$

6. Which graph shows the solution to the given system?

$$\begin{cases} \frac{1}{2}x - y = 1 \\ x = 3 \end{cases}$$

F
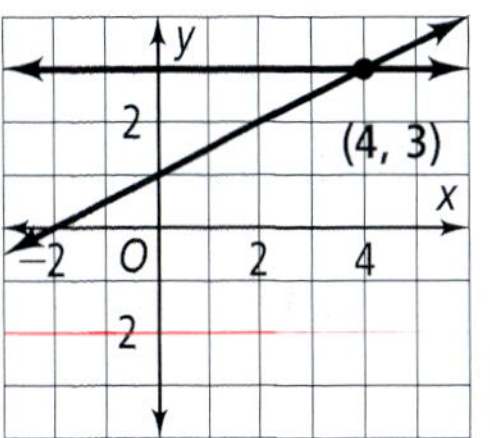

H
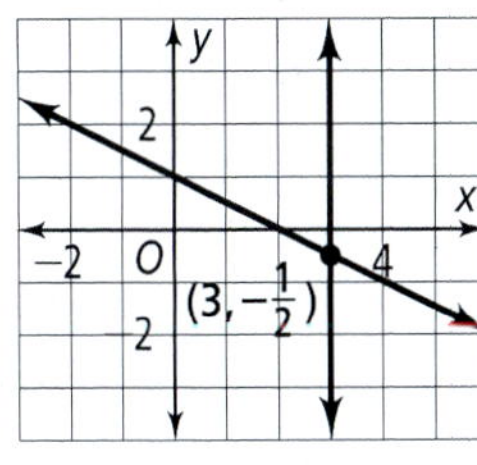

G
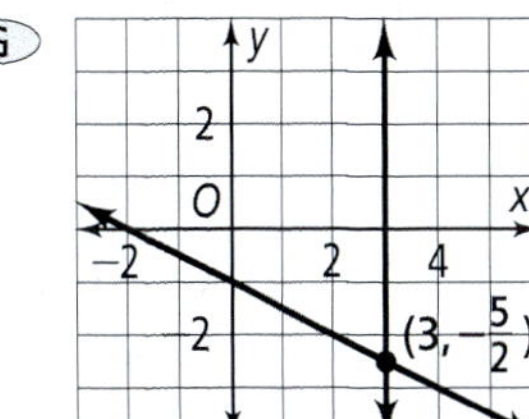

I
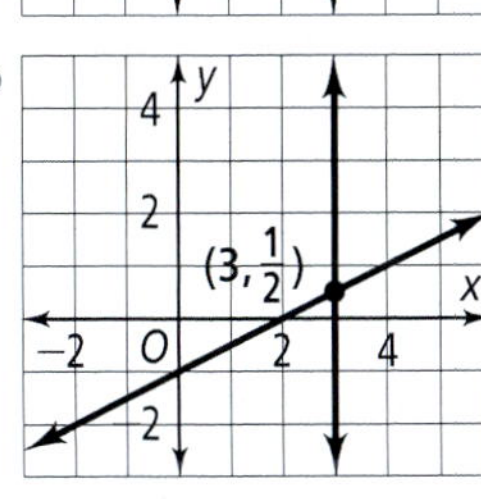

7. The formula for the area of a trapezoid is $A = \frac{h}{2}(b_1 + b_2)$. Solve this equation for b_2.

A $b_2 = \frac{2A}{hb_1}$

B $b_2 = \frac{2A}{h} - b_1$

C $b_2 = \frac{2A}{h - b_1}$

D $b_2 = \frac{2A}{h + b_1}$

8. What is the equation of the line that passes through $(-2, 4)$ and $(2, 7)$?

F $y - 7 = \frac{3}{4}(x + 2)$

G $y + 7 = \frac{3}{4}(x - 2)$

H $y - 7 = \frac{3}{4}(x - 2)$

I $y - 2 = \frac{3}{4}(x - 7)$

9. Which of the following describes the translation of $y = |x|$ to $y = |x + 2| - 1$?

A $y = |x|$ translated 2 units to the left and 1 unit down

B $y = |x|$ translated 2 units to the right and 1 unit down

C $y = |x|$ translated 1 unit to the left and 2 units down

D $y = |x|$ translated 1 unit to the right and 2 units down

GRIDDED RESPONSE

10. The nutrition label on a package of crackers shows there are 80 Calories in 16 grams of crackers. How many grams are in a package labeled 100 Calories?

11. A family with 4 adults and 3 children spends $47 for movie tickets at the theater. Another family with 2 adults and 4 children spends $36. What is the price of a child's ticket in dollars?

12. What is the sum of the solutions of $|5 - 3x| = x + 1$?

13. The graph of $y = x$ is translated up two units. What is the x-intercept of the new graph?

14. What is the value of x in the solution of the system of equations? Round your answer to the nearest tenth.

$$\begin{cases} 5x = -3y - 7 \\ 5y = -4x - 7 \end{cases}$$

15. What is the slope of the line $3y - 4 = \frac{1}{2}x$?

16. The line $(y - 2) = \frac{2}{7}(x - 1)$ contains point $(a, 4)$. What is the value of a?

Short Response

17. An ice cream shop has regular mix-ins for $.50 each and premium mix-ins for $1 each. You have $2.50 to spend on mix-ins, and you want at least 4 mix-ins. How many of each type of mix-in can you get in your ice cream?

18. Solve the following system by graphing.

$$\begin{cases} y < -\frac{1}{3}x + 1 \\ y \leq \frac{2}{3}x + 4 \end{cases}$$

19. Find $f(-4)$, $f(0)$, and $f(3)$ for the function $f(x) = \frac{1}{4}x + 2$.

20. How can you use the graphs of $f(x) = -|3x| + 6$ and $g(x) = 2x + 1$ to find the solutions of $-|3x| + 6 = 2x + 1$?

21. The equation of line m is $y = 3x - 1$. What is the equation of a line that goes through the point $(3, -2)$ and is perpendicular to line m? Show your work.

22. The graph below shows the boundaries for the system of linear inequalities.

$$\begin{cases} y \leq 0.5x + 5 \\ y \leq -5x - 6 \end{cases}$$

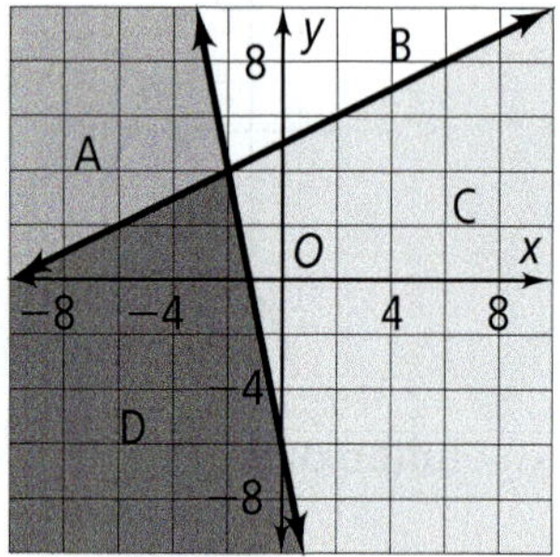

a. Of the shaded areas A, B, C and D, which area represents a solution to the first inequality but not the second?

b. Which represents a solution to the second inequality but not the first?

c. Which represents a solution to both inequalities?

d. Which represents a solution to neither inequality?

Extended Response

23. Jenna is trying to break her school's record for doing the most push-ups in ten minutes. The current record holder did 350 push-ups in ten minutes. The table shows the number of push-ups Jenna completed in the first 6 minutes.

a. Draw a scatter plot and find the line of best fit.

b. Will Jenna beat the current record? Justify your reasoning.

Time (min)	1	2	3	4	5	6
Number of push-ups	37	70	99	132	169	207

CHAPTER 4

Get Ready!

Lesson 1-4

Solving Linear Equations

Solve each equation. Check your answer.

1. $9x - 16 = 8 + 5x$

2. $4(y + 2) + 1 = -5(3 - 2y)$

Lesson 1-6

Solving Absolute Value Inequalities

Solve each inequality. Graph the solution.

3. $|6x - 12| + 6 < 30$

4. $6|4y - 2| \geq 42$

Lesson 2-3

Writing and Graphing Equations in Slope-Intercept Form

Graph the line passing through the given points. Then write its equation in slope-intercept form.

5. $(1, -1)$ and $(3, 17)$

6. $(2, 9)$ and $(6, 11)$

Lesson 2-6

Identifying Translations

Identify each horizontal and vertical translation of the parent function $y = |x|$.

7. $y = |x - 4| + 2$

8. $y = |x + 10| - 3$

Lesson 3-2

Solving Systems of Equations

Solve each system of equations by substitution.

9. $\begin{cases} 2x + 6y = 14 \\ 4x - 8y = 48 \end{cases}$

10. $\begin{cases} x + 2y = -18 \\ 2x - 4y = 12 \end{cases}$

VOCABULARY

Looking Ahead Vocabulary

11. A *form* is a document with blank spaces to fill in. What types of forms might you use?

12. Something is *imaginary* if it has no factual reality. What are some examples of imaginary items?

13. Many items have a specific *function*, or purpose for use. What is the function of a pencil?

Quadratic Functions and Equations

Your place to get all things digital

Download videos connecting math to your world.

Math definitions in English and Spanish

The online Solve It will get you in gear for each lesson.

Interactive! Vary numbers, graphs, and figures to explore math concepts.

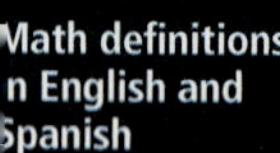

Online access to stepped-out problems aligned to Common Core

Get and view your assignments online.

Extra practice and review online

- Interpreting Functions
- Creating Equations
- The Complex Numbering System

A parabola is the graph of a quadratic function.

Parabolas can be seen all over the place. You can find them in the designs of buildings like the one on the next page.

Vocabulary

English/Spanish Vocabulary Audio Online:

English	Spanish
axis of symmetry, *p. 194*	eje de simetría
complex number, *p. 249*	números complejos
discriminant, *p. 242*	discriminante
greatest common factor, *p. 218*	máximo factor común de una expresión
imaginary number, *p. 249*	número imaginario
parabola, *p. 194*	parábola
Quadratic Formula, *p. 240*	fórmula cuadrática
quadratic function, *p. 194*	función cuadrática
standard form, *p. 202*	forma normal
vertex form, *p. 194*	forma del vértice
zero of a function, *p. 226*	cero de una función

BIG ideas

1 **Equivalence**

Essential Question What are the advantages of a quadratic function in vertex form? In standard form?

2 **Function**

Essential Question How is any quadratic function related to the parent quadratic function $y = x^2$?

3 **Solving Equations and Inequalities**

Essential Question How are the real solutions of a quadratic equation related to the graph of the related quadratic function?

Chapter Preview

Quadratic Functions and Transformations

Content Standards

F.BF.3 Identify the effect on the graph of replacing $f(x)$ by $f(x) + k$, $k\,f(x)$, $f(kx)$, and $f(x + k)$ for specific values of k (both positive and negative) find the value of k given the graphs.

Also A.CED.1, F.IF.4, F.IF.6

Objective To identify and graph quadratic functions

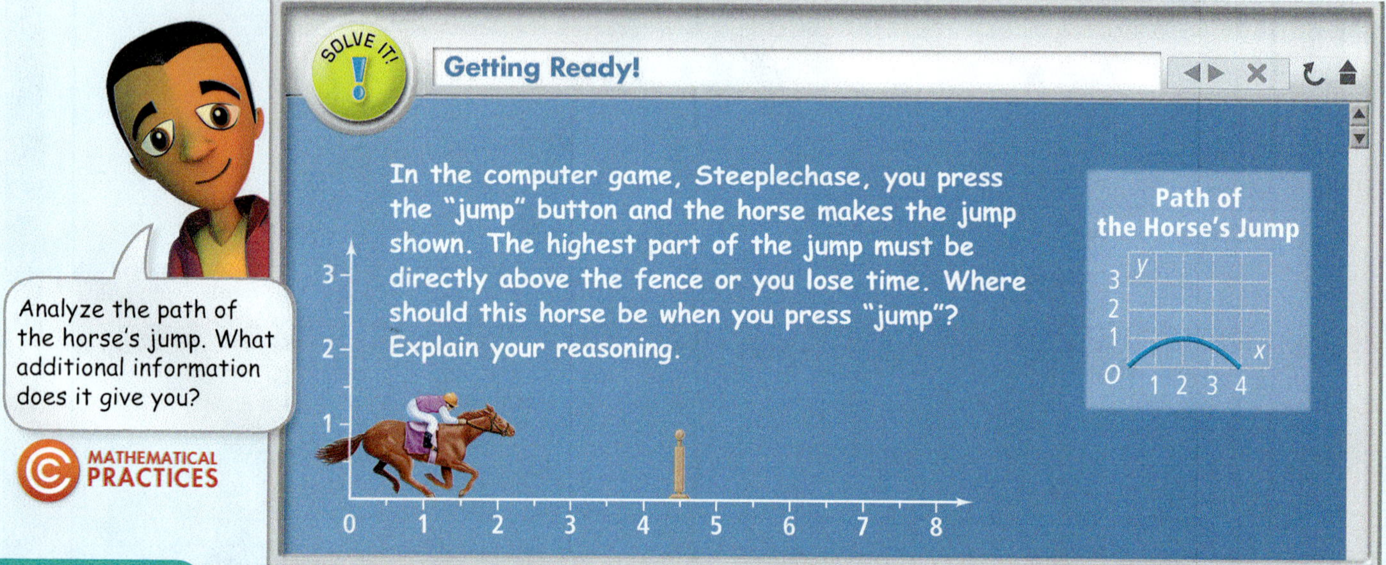

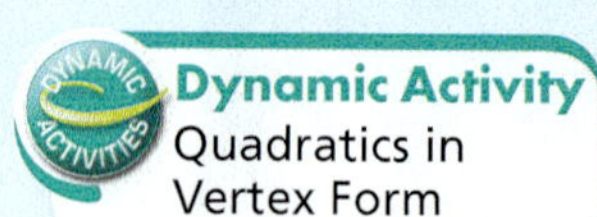

In the Solve It, you used the *parabolic* shape of the horse's jump. A **parabola** is the graph of a **quadratic function**, which you can write in the form $f(x) = ax^2 + bx + c$, where $a \neq 0$.

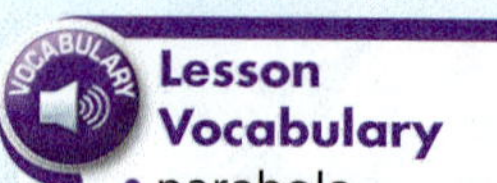

Lesson Vocabulary
- parabola
- quadratic function
- vertex form
- axis of symmetry
- vertex of the parabola
- minimum value
- maximum value

Essential Understanding The graph of any quadratic function is a transformation of the graph of the parent quadratic function, $y = x^2$.

The **vertex form** of a quadratic function is $f(x) = a(x - h)^2 + k$, where $a \neq 0$. The **axis of symmetry** is a line that divides the parabola into two mirror images. The equation of the axis of symmetry is $x = h$. The **vertex of the parabola** is (h, k), the intersection of the parabola and its axis of symmetry.

take note

Key Concept The Parent Quadratic Function

The parent quadratic function is $f(x) = x^2$. Its graph is the parabola shown. The axis of symmetry is $x = 0$. The vertex is $(0, 0)$.

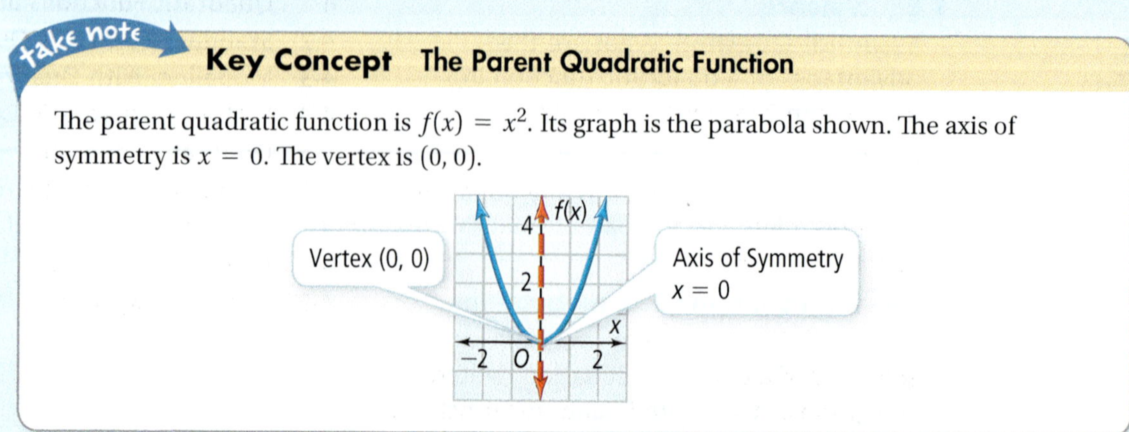

Problem 1 Graphing a Function of the Form $f(x) = ax^2$

Plan

How do you choose points to plot?

Choose the vertex and two points on one side of the axis of symmetry that give integer values of $f(x)$.

What is the graph of $f(x) = \frac{1}{2}x^2$?

Step 1 Plot the vertex (0, 0). Draw the axis of symmetry, $x = 0$.

Step 2 Find and plot two points on one side of the axis of symmetry.

x	$f(x) = \frac{1}{2}x^2$	$(x, f(x))$
0	$\frac{1}{2}(0)^2 = 0$	(0, 0)
2	$\frac{1}{2}(2)^2 = 2$	(2, 2)
4	$\frac{1}{2}(4)^2 = 8$	(4, 8)

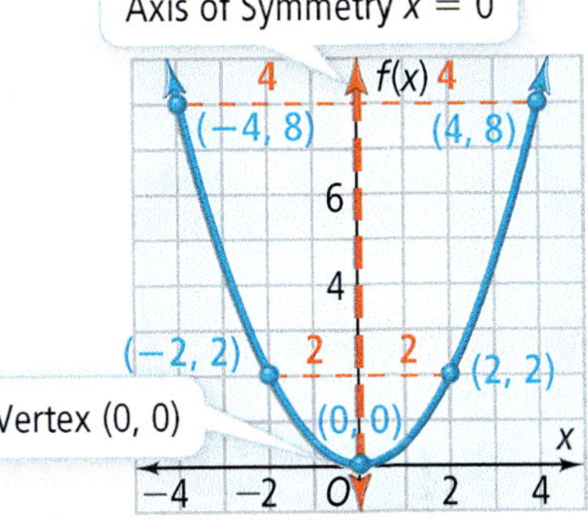

Step 3 Plot the corresponding points on the other side of the axis of symmetry.

Step 4 Sketch the curve.

Got It? 1. a. What is the graph of $f(x) = -\frac{1}{3}x^2$?

b. **Reasoning** What can you say about the graph of the function $f(x) = ax^2$ if a is a negative number? Explain.

Graphs of $y = ax^2$ and $y = -ax^2$ are reflections of each other across the x-axis. Increasing $|a|$ stretches the graph vertically and narrows it horizontally. Decreasing $|a|$ compresses the graph vertically and widens it horizontally.

take note Key Concept Reflection, Stretch, and Compression

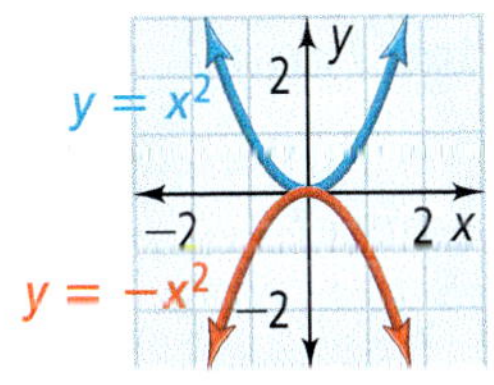

Reflection, a and $-a$

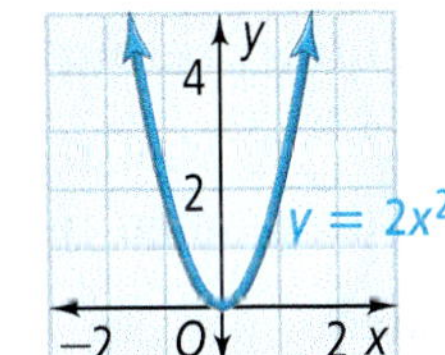

Stretch, $a > 1$

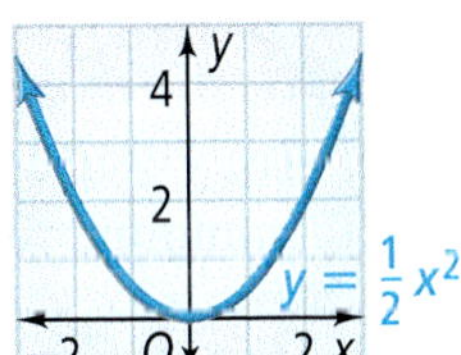

Compression, $0 < a < 1$

If $a > 0$, the parabola opens upward. The y-coordinate of the vertex is the **minimum value** of the function. If $a < 0$, the parabola opens downward. The y-coordinate of the vertex is the **maximum value** of the function.

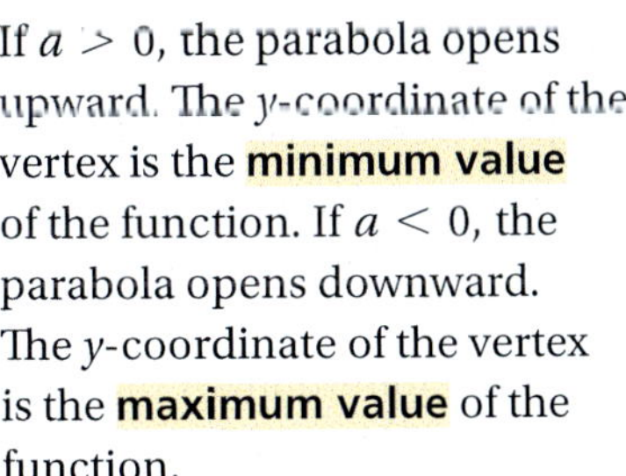

Minimum Value

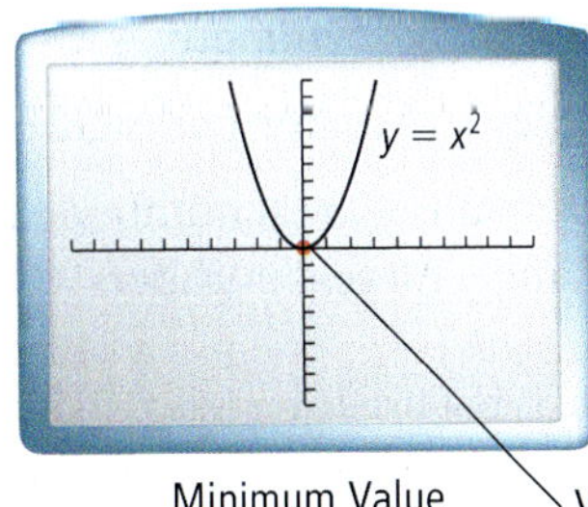

Vertex

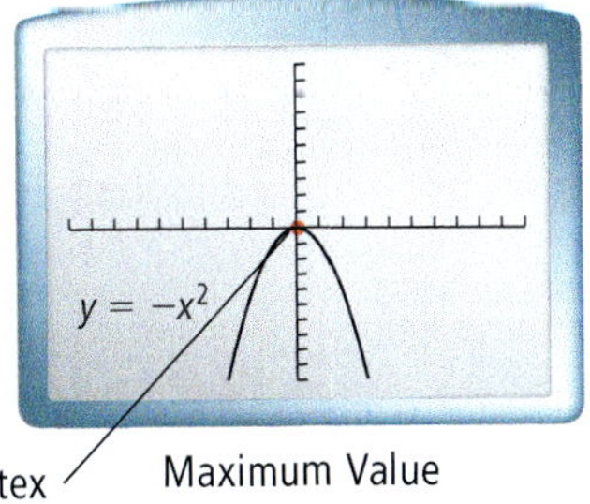

Maximum Value

Problem 2 Graphing Translations of $f(x) = x^2$

Think

How does $g(x)$ differ from $f(x)$?
For each value of x, the value of $g(x)$ is 5 less than the value of $f(x)$.

Graph each function. How is each graph a translation of $f(x) = x^2$?

A $g(x) = x^2 - 5$

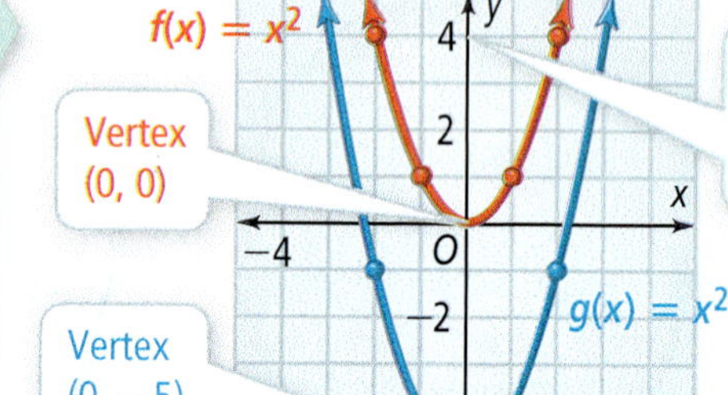

Translate the graph of f down 5 units to get the graph of $g(x) = x^2 - 5$.

B $h(x) = (x - 4)^2$

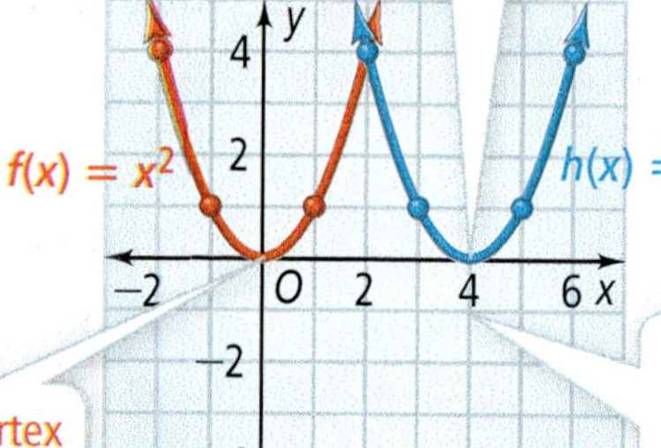

Translate the graph of f to the right 4 units to get the graph of $h(x) = (x - 4)^2$.

Got It? 2. Graph each function. How is it a translation of $f(x) = x^2$?

a. $g(x) = x^2 + 3$

b. $h(x) = (x + 1)^2$

The vertex form, $f(x) = a(x - h)^2 + k$, gives you information about the graph of f without drawing the graph. If $a > 0$, k is the minimum value of the function. If $a < 0$, k is the maximum value.

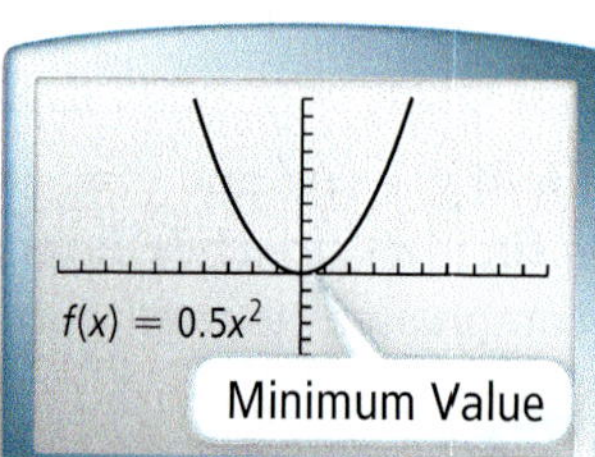

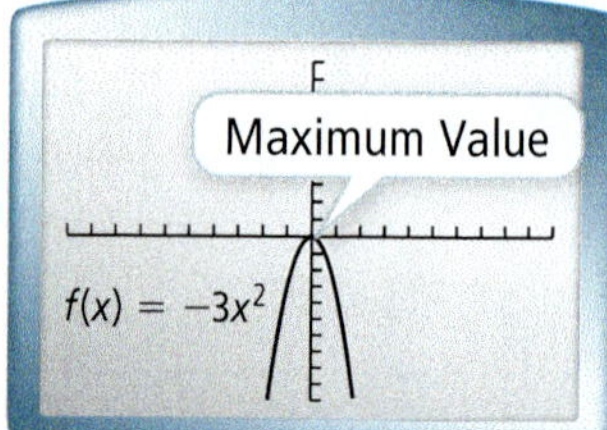

Problem 3 Interpreting Vertex Form

Plan

How do you use vertex form?
Compare $y = 3(x - 4)^2 - 2$ to vertex form $y = a(x - h)^2 + k$ to find values for a, h, and k.

For $y = 3(x - 4)^2 - 2$, what are the vertex, the axis of symmetry, the maximum or minimum value, the domain and the range?

Step 1 Compare: $y = 3(x - 4)^2 - 2$
$y = a(x - h)^2 + k$

Step 2 The vertex is $(h, k) = (4, -2)$.

Step 3 The axis of symmetry is $x = h$, or $x = 4$.

Step 4 Since $a > 0$, the parabola opens upward. $k = -2$ is the minimum value.

Step 5 Domain: All real numbers. There is no restriction on the value of x.
Range: All real numbers ≥ -2, since the minimum value of the function is -2.

Got It? 3. What are the vertex, axis of symmetry, minimum or maximum, and domain and range of the function $y = -2(x + 1)^2 + 4$?

You can use the vertex form of a quadratic function, $f(x) = a(x - h)^2 + k$, to transform the graph of the parent function $f(x) = x^2$.

- Stretch or compress the graph of $f(x) = x^2$ vertically by the factor $|a|$.
- If $a < 0$, reflect the graph across the x-axis.
- Shift the graph $|h|$ units horizontally and $|k|$ units vertically.

take note

Key Concept Translation of the Parabola

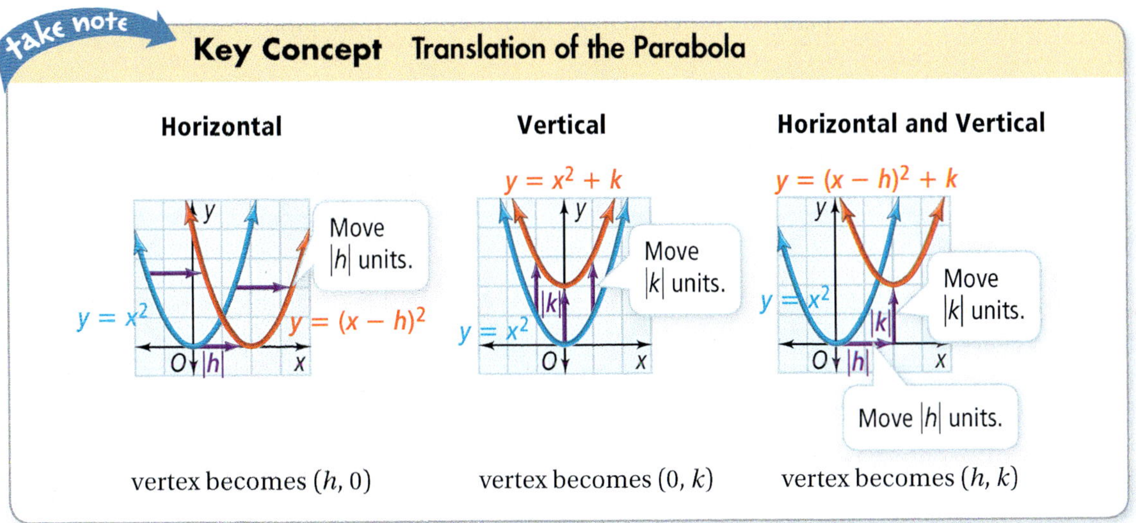

Problem 4 Using Vertex Form

Plan

What do the values of a, h, and k tell you about the graph?
The graph is a stretched reflection of $y = x^2$, shifted 1 unit right and 3 units up.

A **What is the graph of $f(x) = -2(x - 1)^2 + 3$?**

Step 1 Identify the constants $a = -2$, $h = 1$, and $k = 3$. Because $a < 0$, the parabola opens downward.

Step 2 Plot the vertex $(h, k) = (1, 3)$ and draw the axis of symmetry $x = 1$.

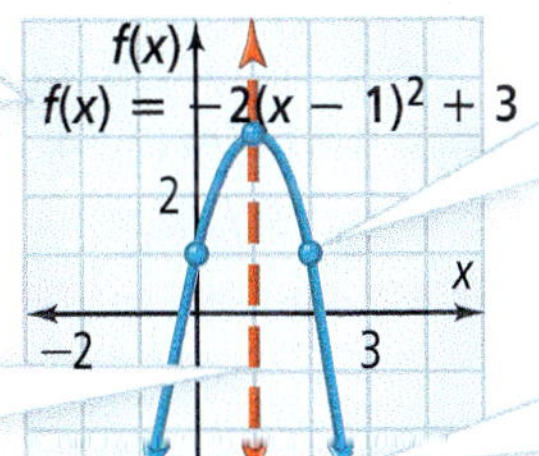

Step 3 Plot two points. $f(2) = -2(2 - 1)^2 + 3 = 1$. Plot (2, 1) and the symmetric point (0, 1).

Step 4 Sketch the curve.

B **Multiple Choice** **What steps transform the graph of $y = x^2$ to $y = -2(x + 1)^2 + 3$?**

(A) Reflect across the x-axis, stretch by the factor 2, translate 1 unit to the right and 3 units up.

(B) Stretch by the factor 2, translate 1 unit to the right and 3 units up.

(C) Reflect across the x-axis, translate 1 unit to the left and 3 units up.

(D) Stretch by the factor 2, reflect across the x-axis, translate 1 unit to the left and 3 units up.

The correct choice is D.

Got It? **4.** What steps transform the graph of $y = x^2$ to $y = 2(x + 2)^2 - 5$?

You can use the vertex form of a quadratic function to model a real-world situation.

Problem 5 Writing a Quadratic Function in Vertex Form

Nature **The picture shows the jump of a dolphin. What quadratic function models the path of the dolphin's jump?**

Think	Write
What is the vertex?	The vertex is $(3, 7)$. $h = 3, k = 7$
Choose another point, $(9, 4)$, from the path. Substitute in the vertex form. Solve for a.	$f(x) = a(x - h)^2 + k$ $4 = a(9 - 3)^2 + 7$ $4 = 36a + 7$ $-3 = 36a$ $a = -\frac{1}{12}$
Substitute in the vertex form.	$f(x) = -\frac{1}{12}(x - 3)^2 + 7$ models the path of the dolphin's jump.

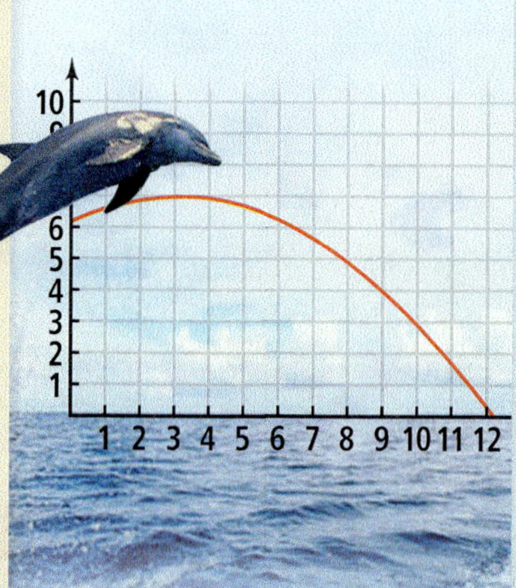

Got It? **5.** Suppose the path of the jump changes so that the axis of symmetry becomes $x = 2$ and the height stays the same. If the path of the jump also passes through the point $(5, 5)$, what quadratic function would model this path?

Lesson Check

Do you know HOW?

1. Graph the function $f(x) = -3x^2$.

2. Determine whether the function $f(x) = 0.25\,(2x - 15)^2 + 150$ has a maximum or a minimum value.

3. Rewrite $y = -2x^2 + 35$ in vertex form.

Do you UNDERSTAND?

4. Vocabulary When does the graph of a quadratic function have a minimum value?

5. Reasoning Is $y = 0(x - 4)^2 + 3$ a quadratic function? Explain.

6. Compare and Contrast Describe the differences between the graphs of $y = (x + 6)^2$ and $y = (x - 6)^2 + 7$.

Practice and Problem-Solving Exercises

Graph each function.

See Problem 1.

7. $y = -x^2$
8. $f(x) = 5x^2$
9. $y = \frac{2}{5}x^2$
10. $y = 2x^2$
11. $f(x) = 2\frac{1}{4}x^2$
12. $y = -\frac{4}{9}x^2$
13. $y = -7x^2$
14. $f(x) = 3\frac{2}{5}x^2$

Graph each function. Describe how it was translated from $f(x) = x^2$.

See Problem 2.

15. $f(x) = x^2 + 3$
16. $f(x) = (x - 2)^2$
17. $f(x) = x^2 - 6$
18. $f(x) = (x + 3)^2$
19. $f(x) = x^2 - 9$
20. $f(x) = (x + 5)^2$
21. $f(x) = x^2 + 1.5$
22. $f(x) = (x - 2.5)^2$

Identify the vertex, the axis of symmetry, the maximum or minimum value, and the domain and the range of each function.

See Problem 3.

23. $y = -1.5(x + 20)^2$
24. $f(x) = 0.1(x - 3.2)^2$
25. $f(x) = 24(x + 5.5)^2$
26. $y = 0.0035(x + 1)^2 - 1$
27. $f(x) = -(x - 4)^2 - 25$
28. $y = (x - 125)^2 + 125$

Graph each function. Identify the axis of symmetry.

See Problem 4.

29. $f(x) = (x - 1)^2 + 2$
30. $y = (x + 3)^2 - 4$
31. $f(x) = 2(x - 2)^2 + 5$
32. $y = -3(x + 7)^2 - 8$
33. $y = -(x - 1)^2 + 4$
34. $f(x) = -(x - 7)^2 + 10$

Write a quadratic function to model each graph.

35.

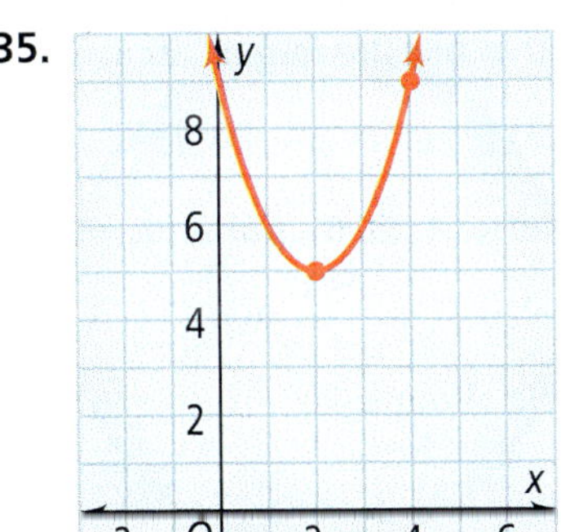

36.

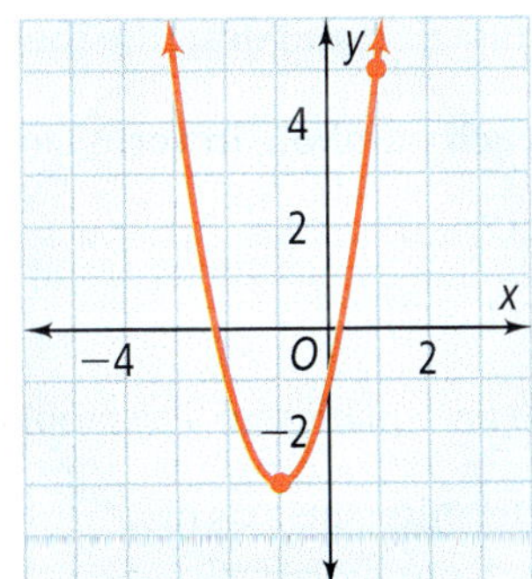

37.

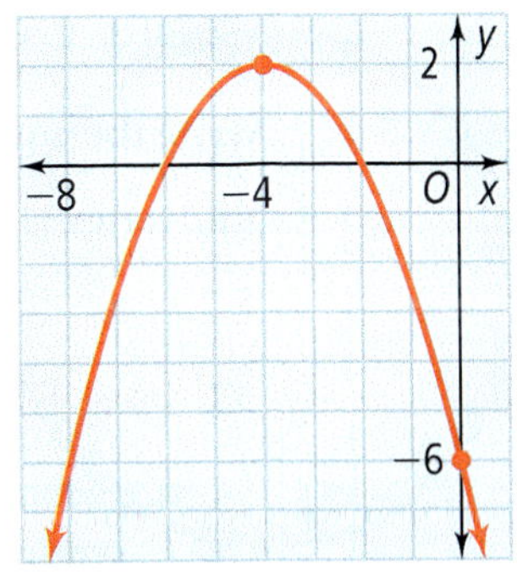

38. **Think About a Plan** A gardener is putting a wire fence along the edge of his garden to keep animals from eating his plants. If he has 20 meters of fence, what is the largest rectangular area he can enclose?
   - To find the area of a rectangle, what two quantities do you need? Choose one to be your variable and write the other in terms of this variable.
   - How can a graph help you solve this problem?
   - What quadratic function represents the area of the garden?

STEM 39. **Manufacturing** The equation for the cost in dollars of producing computer chips is $C = 0.000015x^2 - 0.03x + 35$, where x is the number of chips produced. Find the number of chips that minimizes the cost. What is the cost for that number of chips?

In Chapter 2, you graphed absolute value functions as transformations of their parent function $y = |x|$. Similarly, you can graph a quadratic function as a transformation of the parent function $y = x^2$. Graph the following pairs of functions on the same set of axes. Determine how they are similar and how they are different.

40. $y = -|x - 2| + 1; y = -(x - 2)^2 + 1$

41. $y = 3|x + 1| - 2; y = 3(x + 1)^2 - 2$

42. $y = -2|x| + 4; y = -2x^2 + 4$

43. $y = |x + 3|; y = (x + 3)^2$

Describe how to transform the parent function $y = x^2$ to the graph of each function below. Graph both functions on the same axes.

44. $y = -2(x - 1)^2$

45. $y = -2(x + 1)^2 + 1$

46. $y = -0.25x^2 + 3$

47. You can find the rate of change for an interval between two points of a function by finding the slope between the points. Use the graph to find the y-value for each x-value. Then find the rate of change for each interval.

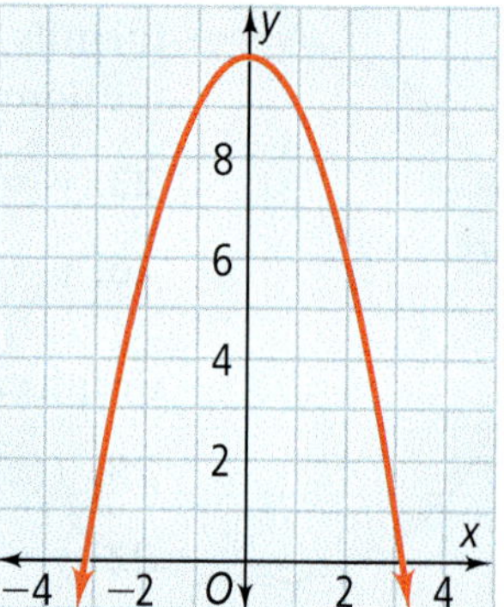

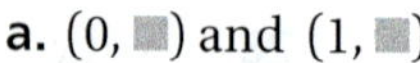

a. (0, ■) and (1, ■)

b. (1, ■) and (2, ■)

c. (2, ■) and (3, ■)

d. Reasoning. What do you notice about the rate of change as the interval gets further away from the vertex?

e. Would your answer to part (d) change if the intervals were on the left side of the graph? Explain.

48. Write a quadratic function to represent the areas of all rectangles with a perimeter of 36 ft. Graph the function and describe the rectangle that has the largest area.

Write the equation of each parabola in vertex form.

49. vertex (1, 2), point (2, −5)

50. vertex (−3, 6), point (1, −2)

51. vertex (0, 5), point (1, −2)

52. vertex $\left(\frac{1}{4}, -\frac{3}{2}\right)$, point (1, 3)

53. Open-Ended Write an equation of a parabola symmetric about $x = -10$.

54. a. Technology Determine the axis of symmetry for each parabola defined by the spreadsheet values at the right.

b. How could you use the spreadsheet columns to verify that the axes of symmetry are correct?

c. What functions in vertex form model the data? Check that the axes of symmetry are correct.

	A	B
1	X1	Y1
2	1	−35
3	2	−15
4	3	−3
5	4	1
6	5	−3

	A	B
1	X2	Y2
2	1	10
3	2	2
4	3	2
5	4	10
6	5	26

Challenge **Determine a and k so the given points are on the graph of the function.**

55. $(0, 1), (2, 1); y = a(x - 1)^2 + k$

56. $(-3, 2), (0, 11); y = a(x + 2)^2 + k$

57. $(1, 11), (2, -19); y = a(x + 1)^2 + k$

58. $(-2, 6), (3, 1); y = a(x - 3)^2 + k$

59. a. In the function $y = ax^2 + bx + c$, c represents the y-intercept. Find the value of the y-intercept in the function $y = a(x - h)^2 + k$.

b. Under what conditions does k represent the y-intercept?

Find the quadratic function $y = a(x - h)^2$ whose graph passes through the given points.

60. $(-2, 1)$ and $(2, 1)$

61. $(-5, 2)$ and $(-1, 2)$

62. $(-1, -4)$ and $(7, -4)$

63. $(2, -1)$ and $(4, 0)$

64. $(-2, 18)$ and $(1, 0)$

65. $(1, -64)$ and $(-3, 0)$

Standardized Test Prep

SAT/ACT

66. One parabola at the right has the equation $y = (x - 4)^2 + 2$. Which equation represents the second parabola?

(A) $y = -(x - 4)^2 + 2$

(B) $y = (-x - 4)^2 + 2$

(C) $y = (x + 4)^2 - 2$

(D) $y = -(x + 4)^2 - 2$

67. Which system has the unique solution $(1, 4)$?

(F) $\begin{cases} y = x - 3 \\ x + y = 5 \end{cases}$

(G) $\begin{cases} y = -x + 5 \\ x - y = -3 \end{cases}$

(H) $\begin{cases} x + y = 5 \\ y = -x + 3 \end{cases}$

(I) $\begin{cases} -x + y = 3 \\ 2x - 2y = -6 \end{cases}$

68. What is the formula for the surface area of a right circular cylinder, $S = 2\pi rh + 2\pi r^2$, solved for h?

(A) $h = \frac{S}{4\pi r}$

(B) $h = \frac{S}{2\pi r^2}$

(C) $h = \frac{S}{2\pi r} - r$

(D) $h = r - \frac{S}{2\pi r}$

Short Response

69. An athletic club has 225 feet of fencing to enclose a tennis court. What quadratic function can be used to find the maximum area of the tennis court? Find the maximum area, and the lengths of the sides of the resulting fence.

Mixed Review

Solve each system of equations using a matrix.

See Lesson 3-6.

70. $\begin{cases} 3x - y = 7 \\ 2x + 2y = 10 \end{cases}$

71. $\begin{cases} 2x + 5y = 10 \\ -3x + y = 36 \end{cases}$

72. $\begin{cases} 3x + y - 2z = -3 \\ x - 3y - z = -2 \\ 2x + 2y + 3z = 11 \end{cases}$

Get Ready! **To prepare for Lesson 4-2, do Exercises 73–75.**

Find the vertex of the graph of each function.

See Lesson 2-7.

73. $y = -2|x|$

74. $y = |-x - 1|$

75. $y = 5|x - 5|$

Standard Form of a Quadratic Function

Content Standards

A.CED.2 Create equations in two or more variables to represent relationships between quantities; graph equations on coordinate axes with labels and scales.

Also F.IF.4, F.IF.6, F.IF.8, F.IF.9

Objective To graph quadratic functions written in standard form

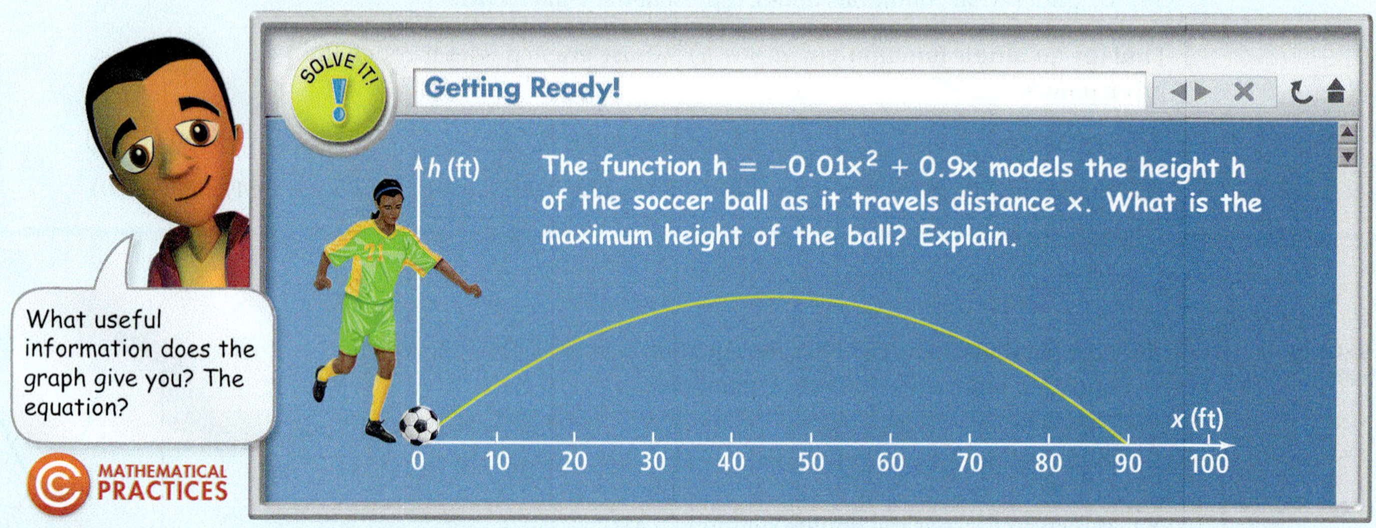

In Lesson 4-1, you worked with quadratic functions written in vertex form. Now you will use quadratic functions in *standard form*. The **standard form** of a quadratic function is $f(x) = ax^2 + bx + c$, where $a \neq 0$.

Essential Understanding For any quadratic function $f(x) = ax^2 + bx + c$, the values of a, b, and c provide key information about its graph.

You can find information about the graph of a quadratic function (such as the vertex) easily from the vertex form. Such information is "hidden" in standard form. However, standard form is easier to enter into a graphing calculator.

Problem 1 Finding the Features of a Quadratic Function

Graphing Calculator What are the vertex, the axis of symmetry, the maximum or minimum value, and the range of $y = 2x^2 + 8x - 2$?

Plan

How can you use a calculator to find the features of a quadratic function in standard form? Graph the function. Then use the **CALC** and **TABLE** features.

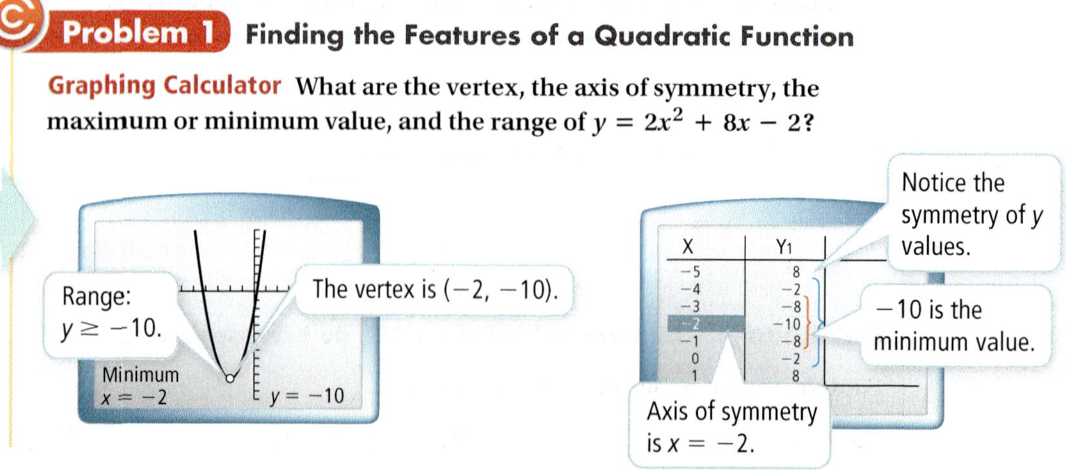

Got It? **1.** What are the vertex, axis of symmetry, maximum or minimum value, and range of $y = -3x^2 - 4x + 6$?

You can find information about the quadratic function $f(x) = ax^2 + bx + c$ from the coefficients a and b, and from the constant term c.

Dynamic Activity
Quadratic Equations in Polynomial Form

take note

Properties Quadratic Function in Standard Form

- The graph of $f(x) = ax^2 + bx + c, a \neq 0$, is a parabola.
- If $a > 0$, the parabola opens upward. If $a < 0$, the parabola opens downward.
- The axis of symmetry is the line $x = -\frac{b}{2a}$.
- The x-coordinate of the vertex is $-\frac{b}{2a}$. The y-coordinate of the vertex is the y-value of the function for $x = -\frac{b}{2a}$, or $y = f\left(-\frac{b}{2a}\right)$.
- The y-intercept is $(0, c)$.

$y = ax^2 + bx + c, a > 0$

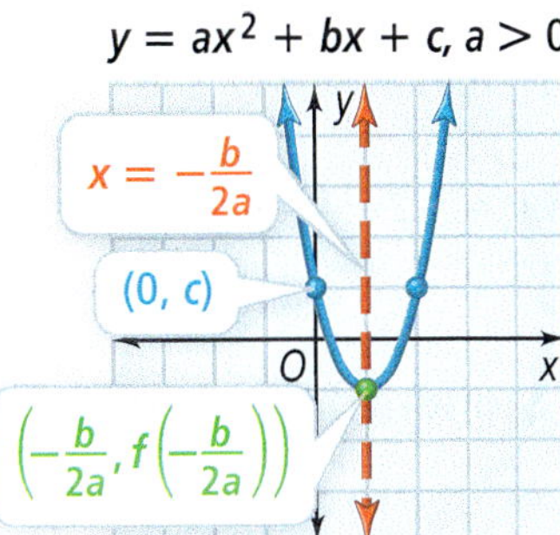

$y = ax^2 + bx + c, a < 0$

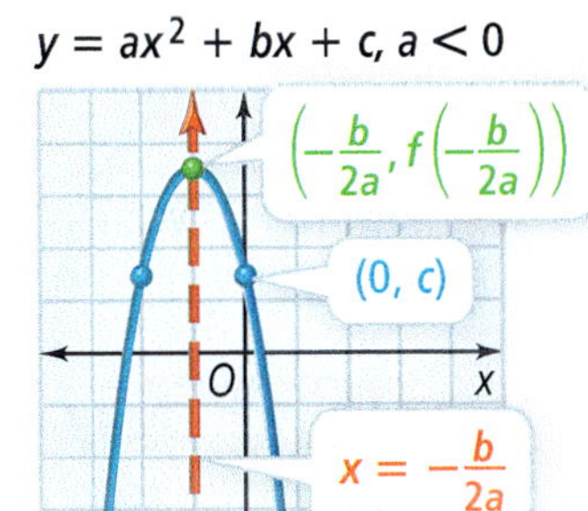

Here's Why It Works You can expand the vertex form of a quadratic function to determine properties of the graph of a quadratic function written in standard form.

$$\begin{aligned} f(x) &= a(x - h)^2 + k \\ &= a(x^2 - 2hx + h^2) + k \\ &= ax^2 - 2ahx + ah^2 + k \\ &= ax^2 + (-2ah)x + (ah^2 + k) \end{aligned}$$

Compare to the standard form, $f(x) = ax^2 + bx + c$.

$a = a$ — a in standard form is the same as a in vertex form.

$b = -2ah$

$-\frac{b}{2a} = h$ — Solve for h.

Since, $h = -\frac{b}{2a}$, the axis of symmetry is $x = -\frac{b}{2a}$ and the vertex is $(h, k) = \left(-\frac{b}{2a}, f\left(-\frac{b}{2a}\right)\right)$.

Problem 2 Graphing a Function of the Form $y = ax^2 + bx + c$

What is the graph of $y = x^2 + 2x + 3$?

Think

How can you use the axis of symmetry? The entire curve on one side of the axis is the mirror image of the curve on the other side.

Step 1 Identify a, b, and c.
$a = 1, b = 2, c = 3$

Step 2 The axis of symmetry is $x = -\frac{b}{2a}$.
$$x = -\frac{2}{2(1)}$$
Lightly sketch the line $x = -1$.

Step 3 The x-coordinate of the vertex is also $-\frac{b}{2a}$, or -1.
The y-coordinate is
$y = (-1)^2 + 2(-1) + 3 = 2$.
Plot the vertex $(-1, 2)$.

Step 4 Since $c = 3$, the y-intercept is $(0, 3)$. The reflection of $(0, 3)$ across $x = -1$ is $(-2, 3)$. Plot both points.

Step 5 $a > 0$ confirms that the graph opens upward. Draw a smooth curve through the points you found in Steps 3 and 4.

Got It? **2.** What is the graph of $y = -2x^2 + 2x - 5$?

Problem 3 Converting Standard Form to Vertex Form

How do you find h, k, and a? Find the vertex. This gives you h and k. The value for a is the same in both forms.

What is the vertex form of $y = 2x^2 + 10x + 7$?

$y = 2x^2 + 10x + 7$	Identify a and b.
$x = -\frac{b}{2a}$	Find the x-coordinate of the vertex.
$= -\frac{10}{2(2)}$	
$= -2.5$	
$y = 2(-2.5)^2 + 10(-2.5) + 7$	Substitute $x = -2.5$ into the equation.
$= -5.5$	

The vertex is $(-2.5, -5.5)$.

$y = a(x - h)^2 + k$	Write the vertex form.
$y = 2[x - (-2.5)]^2 + (-5.5)$	Substitute $a = 2$, $h = -2.5$, $k = -5.5$.
$y = 2(x + 2.5)^2 - 5.5$	Simplify.

The vertex form is $y = 2(x + 2.5)^2 - 5.5$.

Got It? **3.** What is the vertex form of $y = -x^2 + 4x - 5$?

Problem 4 Interpreting a Quadratic Graph STEM

Bridges The New River Gorge Bridge in West Virginia is the world's largest steel single arch bridge. You can model the arch with the function $y = -0.000498x^2 + 0.847x$, where x and y are in feet. How high above the river is the arch? How long is the section of bridge above the arch?

Know
A function that models the arch and the vertical distance from the base of the supports to the water

Need
The height of the arch above the support base and the length of the bridge above the arch

Plan
Find the vertex. The y-coordinate is the height of the arch above the support base. The x-coordinate is half the distance between the supports.

Think
How can you tell that the quadratic function has a maximum value?
Since $a < 0$, the graph of the function opens down. The function has a maximum value.

Step 1 Find the vertex of the arch.

$$x = -\frac{b}{2a} = -\frac{0.847}{2(-0.000498)} \approx 850$$

$$y = -0.000498(850)^2 + 0.847(850) \approx 360$$

The vertex is about (850, 360).

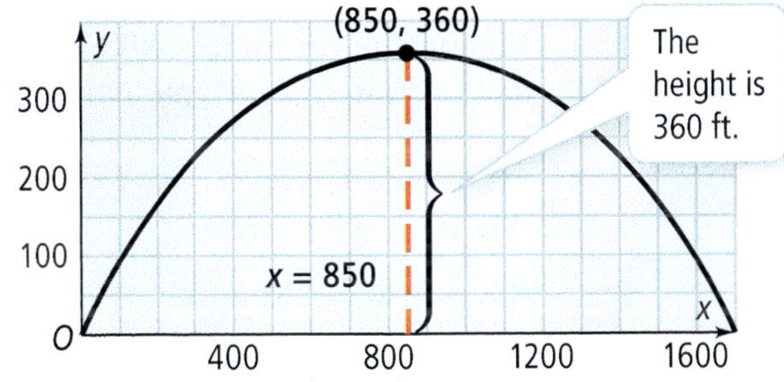

Step 2 Find the height of the arch above its supports.

The y-coordinate of the vertex is the height of the arch above its supports. The arch is about 360 ft above its supports.

Step 3 Find the height of the arch above the river.

The arch is about 360 ft + 516 ft = 876 ft above the river.

Step 4 Find the length of the bridge above the arch.

The x-coordinate of the vertex is half the length of the bridge above the arch. The length of that part of the bridge is about 850 ft + 850 ft = 1700 ft long.

Got It? 4. a. The Zhaozhou Bridge in China is the oldest known arch bridge, dating to A.D. 605. You can model the support arch with the function $f(x) = -0.001075x^2 + 0.131148x$, where x and y are measured in feet. How high is the arch above its supports?

b. Reasoning Why does the model in part (a) not have a constant term?

Lesson Check

Do you know HOW?

1. Identify the vertex, axis of symmetry, and the maximum or minimum value of the parabola at the right.

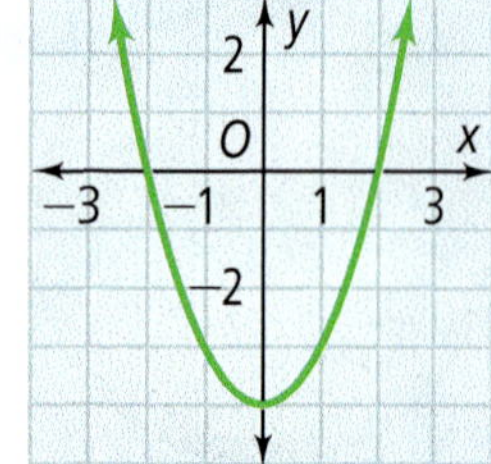

Graph each function.

2. $y = x^2 - 2x + 4$

3. $y = -x^2 - 3x + 6$

Write each function in vertex form.

4. $y = x^2 - 2x + 9$

5. $y = -x^2 + 3x - 1$

Do you UNDERSTAND?

6. **Error Analysis** A student graphed the function $y = 2x^2 - 4x - 3$. Find and correct the error.

$x = \frac{-4}{2(2)} = -1$

$y = 2(-1)^2 - 4(-1) - 3$

$= 2 + 4 - 3$

$= 3$

vertex (-1, 3)

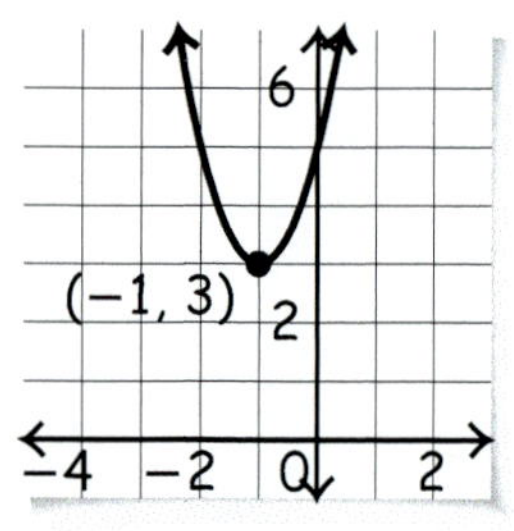

7. **Compare and Contrast** Explain the difference between finding the vertex of a function written in vertex form and finding the vertex of a function written in standard form.

Practice and Problem-Solving Exercises

Identify the vertex, the axis of symmetry, the maximum or minimum value, and the range of each parabola. See Problem 1.

8. $y = x^2 + 2x + 1$
9. $y = -x^2 + 2x + 1$
10. $y = x^2 + 4x + 1$
11. $y = -x^2 + 2x + 5$
12. $y = 3x^2 - 4x - 2$
13. $y = -2x^2 - 3x + 4$
14. $y = 2x^2 - 6x + 3$
15. $y = -x^2 - x$
16. $y = 2x^2 + 5$

Graph each function. See Problem 2.

17. $y = x^2 + 6x + 9$
18. $y = -x^2 - 3x + 6$
19. $y = 2x^2 + 4x$
20. $y = 4x^2 - 12x + 9$
21. $y = -6x^2 - 12x - 1$
22. $y = -\frac{3}{4}x^2 + 6x + 6$
23. $y = 3x^2 - 12x + 10$
24. $y = \frac{1}{2}x^2 + 2x - 8$
25. $y = -4x^2 - 24x - 36$

Write each function in vertex form. See Problem 3.

26. $y = x^2 - 4x + 6$
27. $y = x^2 + 2x + 5$
28. $y = 4x^2 + 7x$
29. $y = 2x^2 - 5x + 12$
30. $y = -2x^2 + 8x + 3$
31. $y = \frac{9}{4}x^2 + 3x - 1$

32. **Economics** A model for a company's revenue from selling a software package is $R = -2.5p^2 + 500p$, where p is the price in dollars of the software. What price will maximize revenue? Find the maximum revenue. See Problem 4.

B Apply

Sketch each parabola using the given information.

33. vertex (3, 6), y-intercept 2

34. vertex $(-1, -4)$, y-intercept 3

35. vertex (0, 5) point $(1, -2)$

36. vertex (2, 3), point (6, 9)

37. Think About a Plan Suppose you work for a packaging company and are designing a box that has a rectangular bottom with a perimeter of 36 cm. The box must be 4 cm high. What dimensions give the maximum volume?

- How can you model the volume of the box with a quadratic function?
- What information can you get from the function to find the maximum volume?

38. Landscaping A town is planning a playground. It wants to fence in a rectangular space using an existing wall. What is the greatest area it can fence in using 100 ft of donated fencing?

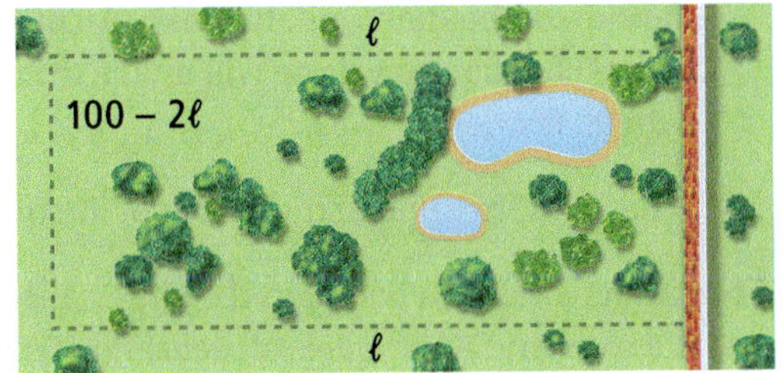

For each function, the vertex of the function's graph is given. Find the unknown coefficients.

39. $y = x^2 + bx + c$; $(3, -4)$

40. $y = -3x^2 + bx + c$; $(1, 0)$

41. $y = ax^2 + 10x + c$; $(-5, -27)$

42. $y = c - ax^2 - 2x$; $(-1, 3)$

STEM **43. Physics** The height of a projectile fired straight up in the air with an initial velocity of 64 ft/s is $h = 64t - 16t^2$, where h is height in feet and t is time in seconds. The table represents the data for another projectile.

a. Which projectile goes higher? How much higher?

b. At what times t will each projectile be at a height of 16 feet?

Time (t)	Height (h)
0.5	20
1	32
1.5	36
2	32

44. A student says that the graph of $y = ax^2 + bx + c$ gets wider as a increases.

a. **Error Analysis** Use examples to show that the student is wrong.

b. **Writing** Summarize the relationship between $|a|$ and the width of the graph of $y = ax^2 + bx + c$.

For each function, find the y-intercept.

45. $y = (x - 1)^2 + 2$

46. $y = -3(x + 2)^2 - 4$

47. $y = -\frac{2}{3}(x - 9)^2$

48. Use the functions $f(x) = 4x + 3$ and $g(x) = \frac{1}{2}x^2 + 2$, to answer parts (a)–(c).

a. Which function has a greater rate of change from $x = 0$ to $x = 1$?

b. Which function has a greater rate of change from $x = 2$ and $x = 3$?

c. Does $g(x)$ ever have a greater rate of change than $f(x)$? Explain.

For each function, the vertex of the function's graph is given. Find a and b.

49. $y = ax^2 + bx - 27; (2, -3)$

50. $y = ax^2 + bx + 5; (-1, 4)$

51. $y = ax^2 + bx + 8; (2, -4)$

52. $y = ax^2 + bx; (-3, 2)$

53. Sketch the parabola with an axis of symmetry $x = 2$, y-intercept 1, and point (3, 2.5).

Standardized Test Prep

GRIDDED RESPONSE

SAT/ACT

54. The time it takes to chalk a baseball diamond varies directly with the length of the side of the diamond. If it takes 10 minutes to chalk a little league diamond with 60 ft sides, how long will it take to chalk a major league baseball diamond with 90 ft sides?

55. What is the x-value of the vertex of the quadratic function $y = -5x^2 + \frac{4}{7}$?

56. Sarah works as a nanny and charges different rates for working during the week and the weekend. One week, she earned \$902.50 working 45 hours, of which 5 hours were during the weekend. The following week she earned \$1045 working 50 hours, of which 10 hours were during the weekend. What does Sarah charge per hour, in dollars, for working during the week?

Mixed Review

Solve each equation.

See Lesson 1-4.

57. $0.6(y + 2) - 0.2(2 - y) = 1$

58. $3(a + 4) + 2(a - 1) = a$

For each system, choose the method of solving that seems easier to use. Explain why you made each choice. Solve each system.

See Lessons 3-2 and 3-6.

59. $\begin{cases} 3x - 5y = 26 \\ -2x - 3y = -11 \end{cases}$

60. $\begin{cases} y = \frac{2}{3}x - 3 \\ -x + 3y = 18 \end{cases}$

61. $\begin{cases} 2m + 3n = 12 \\ -5m + n = -13 \end{cases}$

Get Ready! **To prepare for Lesson 4-3, do Exercises 62–67.**

Identify the vertex, the axis of symmetry, the maximum or minimum value, and the domain and range of each function.

See Lesson 4-2.

62. $y = 2(x + 2)^2 - 1$

63. $y = -(x - 1)^2 + 3$

64. $y = \frac{1}{2}(x - 3)^2 - 2$

65. $y = -\frac{2}{5}(x + 3)^2 + 5$

66. $y = 3(x + 4)^2$

67. $y = -7(x - 4)^2 + 6$

4-3 Modeling With Quadratic Functions

Content Standards

F.IF.5 Relate the domain of a function to its graph and, where applicable, to the quantitative relationship it describes.

Also F.IF.4

Objective To model data with quadratic functions

Try making a sketch of the path of the ball based on what you know about projectile motion.

When you know the vertex and a point on a parabola, you can use vertex form to write an equation of the parabola. If you do not know the vertex, you can use standard form and any three points of the parabola to find an equation.

Essential Understanding Three noncollinear points, no two of which are in line vertically, are on the graph of exactly one quadratic function.

Problem 1 Writing an Equation of a Parabola

Plan

How do you use the 3 given points? Use them to write a system of 3 equations. Solve the system to get *a*, *b*, and *c*.

A parabola contains the points (0, 0), (−1, −2), and (1, 6). What is the equation of this parabola in standard form?

Substitute the (x, y) values into $y = ax^2 + bx + c$ to write a system of equations.

Use (0, 0).	Use (−1, −2).	Use (1, 6).
$y = ax^2 + bx + c$	$y = ax^2 + bx + c$	$y = ax^2 + bx + c$
$0 = a(0)^2 + b(0) + c$	$-2 = a(-1)^2 + b(-1) + c$	$6 = a(1)^2 + b(1) + c$
$0 = c$	$-2 = a - b + c$	$6 = a + b + c$

Since $c = 0$, the resulting system has two variables. $\begin{cases} a - b = -2 \\ a + b = 6 \end{cases}$ Use elimination. $a = 2$ and $b = 4$.

Substitute $a = 2$, $b = 4$, and $c = 0$ into standard form: $y = 2x^2 + 4x + 0$.

$y = 2x^2 + 4x$ is the equation of the parabola that contains the given points.

Got It? **1.** What is the equation of a parabola containing the points (0, 0), (1, −2), and (−1, −4)?

Problem 2 Comparing Quadratic Models STEM

Physics **Campers at an aerospace camp launch rockets on the last day of camp. The path of Rocket 1 is modeled by the equation $h = -16t^2 + 150t + 1$ where t is time in seconds and h is the distance from the ground. The path of Rocket 2 is modeled by the graph at the right. Which rocket flew higher?**

800
700
600
500
400
300
200
100
0
y
x
Height (ft)
2 4 6 8 10 12 14 16
Time (s)

Think

What property of the quadratic tells you how high the rocket flew?
The parabolas model the rockets' paths, so the maximums of each parabola describe how high the rockets flew.

Find the maximum height of each rocket by using the models of their paths.

Rocket 1

The maximum height of Rocket 1 is at the vertex of the parabola.

$-\frac{b}{2a}, f\left(-\frac{b}{2a}\right)$	Use the vertex formula.
$-\frac{150}{2(-16)}, f\left(-\frac{150}{2(-16)}\right)$	$a = -16, b = 150$
$(4.7, 352.6)$	Simplify.

The maximum height of Rocket 1 is 352.6 feet.

Rocket 2

The maximum height of Rocket 2 is at the vertex of the parabola.

You can use the graph to find the approximate maximum height of the rocket.

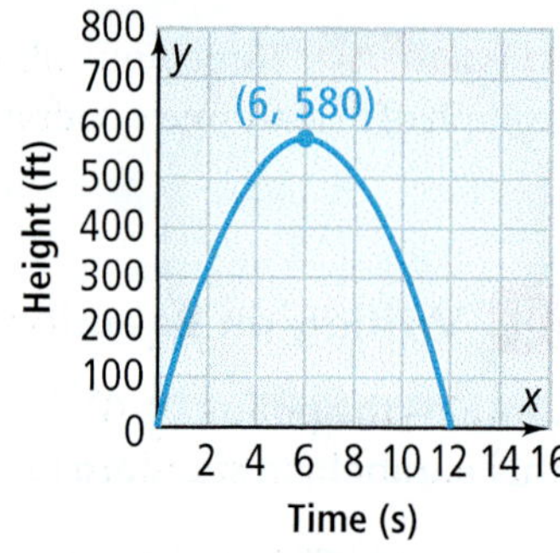

The maximum height of Rocket 2 is at about 580 feet.

Rocket 2 flew higher than Rocket 1.

  **Got It?** **2. a.** Which rocket stayed in the air longer?

b. What is the reasonable domain and range for each quadratic model?

c. Reasoning Describe what the domains tell you about each of the models and why the domains for the models are different.

When more than three data points suggest a quadratic function, you can use the quadratic regression feature of a graphing calculator to find a quadratic model.

Problem 3 Using Quadratic Regression

The table shows a meteorologist's predicted temperatures for an October day in Sacramento, California.

Sacramento, CA

Time	Predicted Temperature (°F)
8 A.M.	52
10 A.M.	64
12 P.M.	72
2 P.M.	78
4 P.M.	81
6 P.M.	76

Think

How do you write times using a 24-hour clock?

Add 12 to the number of hours past noon. So, 2 P.M. is 14:00 in the 24-hour clock.

A What is a quadratic model for this data?

Step 1 Enter the data. Use the 24-hour clock to represent times after noon.

L1	L2	L3 3
8	52	------
10	64	
12	72	
14	78	
16	81	
18	76	
------	------	

L3 =

Step 2 Use **QuadReg**.

QuadReg
y = ax² + bx + c
a = −.46875
b = 14.71607143
c = −36.12142857
R² = .9919573999

Step 3 Graph the data and the function.

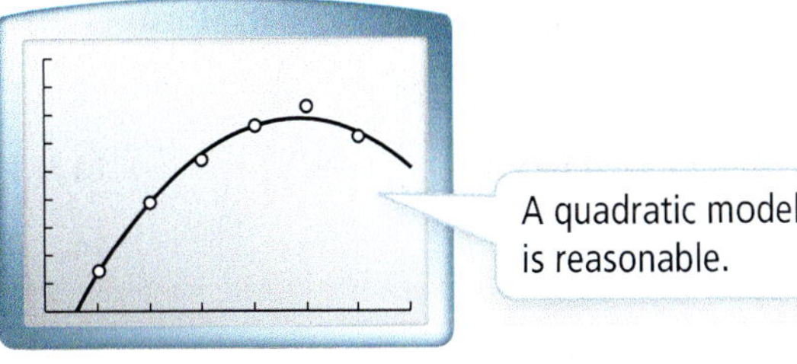

A quadratic model for temperature is $y = -0.469x^2 + 14.716x - 36.121$.

B Use your model to predict the high temperature for the day. At what time does the high temperature occur?

Use the **Maximum** feature or tables.

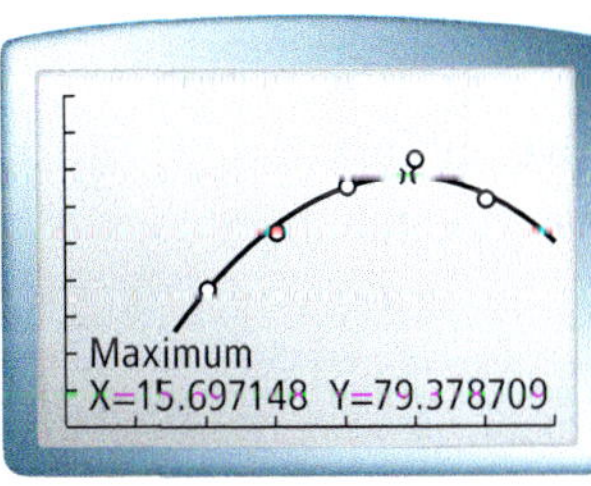

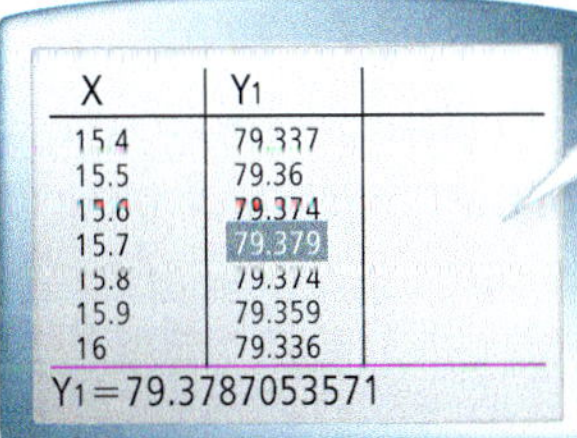

X	Y1
15.4	79.337
15.5	79.36
15.6	79.374
15.7	79.379
15.8	79.374
15.9	79.359
16	79.336

Y1=79.3787053571

16 represents 4 P.M. The maximum occurs at approximately 15.7, or about 3:42 P.M.

Predict the high temperature for the day to be 79.4°F at about 3:42 P.M.

Got It? 3. The table shows a meteorologist's predicted temperatures for a summer day in Denver, Colorado. What is a quadratic model for this data? Predict the high temperature for the day. At what time does the high temperature occur?

Denver, CO

Time	Predicted Temperature (°F)
6 A.M.	63
9 A.M.	76
12 P.M.	86
3 P.M.	89
6 P.M.	85
9 P.M.	76

Lesson Check

Do you know HOW?

Find a quadratic function that includes each set of values.

1. (1, 0), (2, −3), (3, −10)

2.

x	−2	−1	0	1	2
y	3.5	3.5	7.5	15.5	27.5

3.

x	−2	−1	0	1
y	−41.5	−25.5	−13.5	−5.5

Do you UNDERSTAND?

4. **Compare and Contrast** How do you know whether to perform a linear regression or a quadratic regression for a given set of data?

5. **Reasoning** Explain how you can determine which of the quadratic functions in Exercise 1 and Exercise 2 attains the greatest values.

6. **Error Analysis** Your classmate says he can write the equation of a quadratic function that passes through the points (3, 4), (5, −2), and (3, 0). Explain his error.

Practice and Problem-Solving Exercises

A Practice

Find an equation in standard form of the parabola passing through the points.

See Problem 1.

7. (1, −2), (2, −2), (3, −4)

8. (1, −2), (2, −4), (3, −4)

9. (−1, 6), (1, 4), (2, 9)

10. (1, 1), (−1, −3), (−3, 1)

11. (3, −6), (1, −2), (6, 3)

12. (−2, 9), (−4, 5), (1, 0)

13.

x	f(x)
−1	−1
1	3
2	8

14.

x	f(x)
−1	17
1	17
2	8

15.

x	f(x)
−1	−4
1	−2
2	−4

See Problems 2 and 3.

16. A player throws a basketball toward a hoop. The basketball follows a parabolic path that can be modeled by the equation $y = -0.125x^2 + 1.84x + 6$. The table models the parabolic path of another basketball thrown from somewhere else on the court. If the center of the hoop is located at (12, 10), will each ball pass through the hoop?

x	y
2	10
4	12
10	12

STEM 17. **Physics** A man throws a ball off the top of a building and records the height of the ball at different times, as shown in the table.

a. Find a quadratic model for the data.

b. Use the model to estimate the height of the ball at 2.5 seconds.

c. What is the ball's maximum height?

Height of a Ball

Time (s)	Height (ft)
0	46
1	63
2	48
3	1

Apply

Determine whether a quadratic model exists for each set of values. If so, write the model.

18. $f(-2) = 16, f(0) = 0, f(1) = 4$

19. $f(0) = 5, f(2) = 3, f(-1) = 0$

20. $f(-1) = -4, f(1) = -2, f(2) = -1$

21. $f(-2) = 7, f(0) = 1, f(2) = 0$

22. a. Geometry Copy and complete the table. It shows the total number of segments whose endpoints are chosen from x points, no three of which are collinear.

Number of points, x	2	3	■	■
Number of segments, y	1	3	■	■

b. Write a quadratic model for the data.

c. Predict the number of segments that can be drawn using 10 points.

23. Think About a Plan The table shows the height of a column of water as it drains from its container. Use a quadratic model of this data to estimate the water level at 30 seconds.

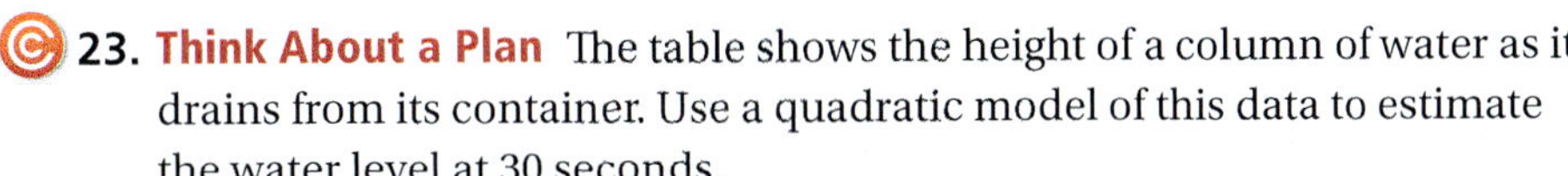

- What system of equations can you use to solve this problem?
- How can you determine if your answer is reasonable?

Water Levels

Elapsed Time (s)	Water Level (mm)
0	120
20	83
40	50

24. A parabola contains the points $(-1, 8)$, $(0, 4)$, and $(1, 2)$. Name another point also on the parabola.

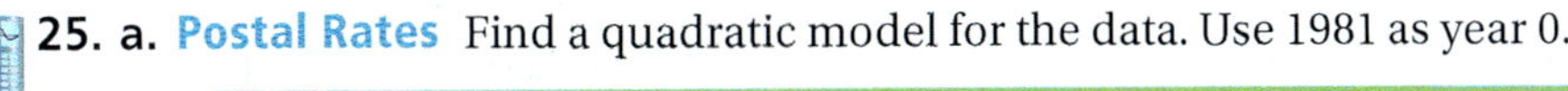

25. a. Postal Rates Find a quadratic model for the data. Use 1981 as year 0.

Price of First-Class Stamp

Year	1981	1991	1995	1999	2001	2006	2007	2008
Price (cents)	18	29	32	33	34	39	41	42

SOURCE: United States Postal Service

b. Describe a reasonable domain and range for your model. (*Hint*: This is a discrete, real situation.)

c. Estimation Estimate when first-class postage was 37 cents.

d. Use your model to predict when first-class postage will be 50 cents. Explain why your prediction may not be valid.

26. Road Safety The table and graph below give the stopping distance for an automobile under certain road conditions.

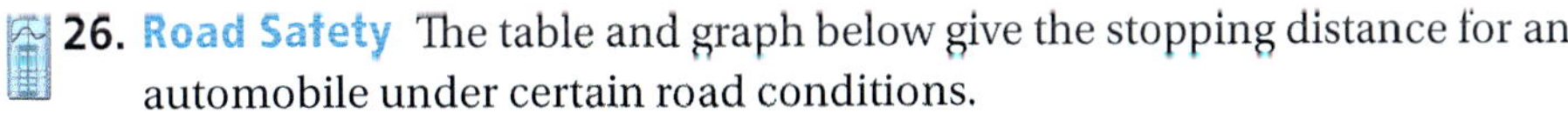

Speed (mi/h)	20	30	40	50	55
Stopping Distance (ft)	17	38	67	105	127

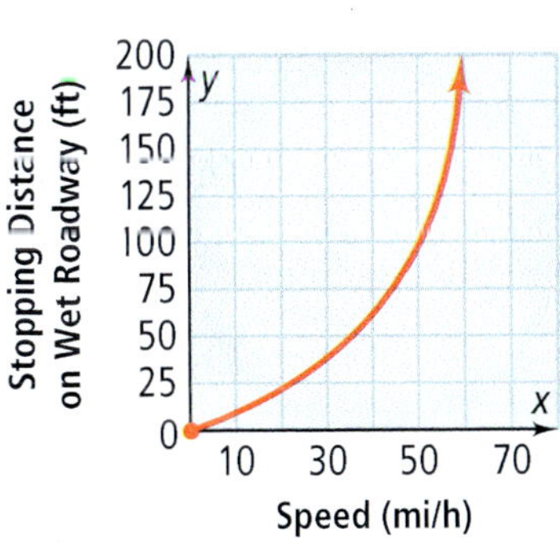

a. Compare the models. What are the reasonable domain and range for each road condition?

b. Writing Explain what that means about stopping distances under certain road conditions.

27. a. A parabola contains the points $(0, -4)$, $(2, 4)$, and $(4, 4)$. Find the vertex.
b. **Reasoning** What is the minimum number of data points you need to find a single quadratic model for a data set? Explain.

28. A model for the height of an arrow shot into the air is $h(t) = -16t^2 + 72t + 5$, where t is time and h is height. Without graphing, answer the following questions.
a. What can you learn by finding the graph's intercept with the h-axis?
b. What can you learn by finding the graph's intercept(s) with the t-axis?

Standardized Test Prep

SAT/ACT

29. The graph of a quadratic function has vertex $(-3, -2)$. What is the axis of symmetry?

(A) $x = -3$ (B) $x = 3$ (C) $y = -2$ (D) $y = 2$

30. Which function is NOT a quadratic function?

(F) $y = (x - 1)(x - 2)$
(G) $y = x^2 + 2x - 3$
(H) $y = 3x - x^2$
(I) $y = -x^2 + x(x - 3)$

31. Which is the composition $f(g(x))$, if $f(x) = -x - 3$ and $g(x) = 7 + 5x$?

(A) $f(g(x)) = 4x + 4$
(B) $f(g(x)) = 4x - 10$
(C) $f(g(x)) = -5x - 8$
(D) $f(g(x)) = -5x - 10$

Extended Response

32. Mark has 42 coins consisting of dimes and quarters. The total value of his coins is \$6. How many of each type of coin does he have? Show all your work and explain what method you used to solve the problem.

Mixed Review

Graph each function. See Lesson 4-2.

33. $y = x^2 - 6x - 3$
34. $y = 2x^2 + 9x - 4$
35. $y = 3x^2 - 4x + 1$

Solve each system by elimination. See Lesson 3-2.

36. $\begin{cases} x + y = 7 \\ 5x - y = 5 \end{cases}$
37. $\begin{cases} 2x - 3y = -14 \\ 3x - y = 7 \end{cases}$
38. $\begin{cases} x - 3y = 2 \\ x - 2y = 1 \end{cases}$

For Exercises 39–40, y varies directly with x. See Lesson 2-2.

39. If $y = 2$ when $x = 5$, find y when $x = 2$.
40. If $y = -2$ when $x = 4$, find y when $x = 7$.

Get Ready! **To prepare for Lesson 4-4, do Exercises 41–43.**

Simplify by combining like terms. See Lesson 1-3.

41. $x^2 + x + 4x - 1$
42. $6x^2 - 4(3)x + 2x - 3$
43. $4x^2 - 2(5 - x) - 3x$

Concept Byte

For Use With Lesson 4-3

Identifying Quadratic Data

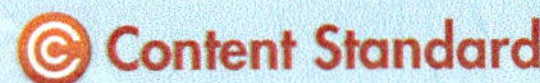

F.IF.6 Calculate and interpret the average rate of change of a function . . . Estimate the rate of change from a graph.

You can identify perfect quadratic data when x-values are evenly spaced using the pattern in the differences between y-values.

Example

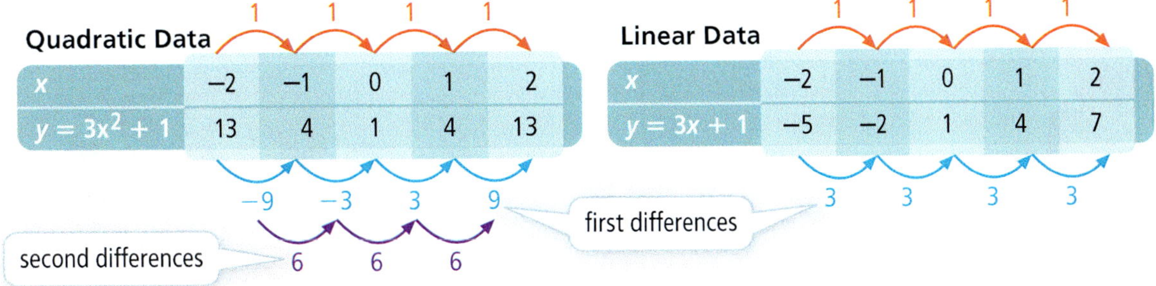

For linear data, the *first* differences of adjacent y-values are constant.
For quadratic data, the *second* differences are constant and not equal to 0.

Exercises

Determine if each data set represents perfect quadratic data.

1.

x	1	2	3	4	5	6	7
y	6	12	22	36	54	76	102

2.

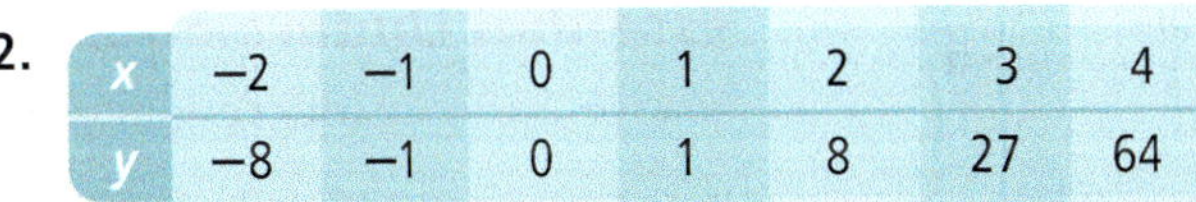

x	−2	−1	0	1	2	3	4
y	−8	−1	0	1	8	27	64

3. Reasoning Can you use the method above to determine if a data set represents perfect linear or quadratic data if the x-values are *not* evenly spaced? Explain.

4. Recall that the average rate of change over any interval of a function is the slope of the line segment joining the endpoints of the interval.

x	−4	−2	0	2	4
y	16	4	0	4	16

a. The table indicates points on a quadratic function. Use the table to find the average rate of change between the vertex and each of the other points in the table.

b. As one of the endpoints gets further away form the vertex, what do you notice about the average rate of change?

4-4 Factoring Quadratic Expressions

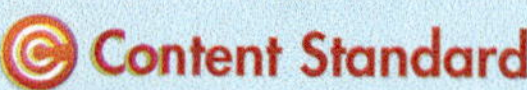

Content Standard

A.SSE.2 Use the structure of an expression to identify ways to rewrite it.

Objectives To find common and binomial factors of quadratic expressions
To factor special quadratic expressions

Getting Ready!

If you have cards numbered 1 to 50, could you play the game with <u>any</u> two cards?

In a game, you see the two cards shown. You get two other cards with numbers. You win if

1. the product of your two numbers equals the number on one card shown, AND
2. the sum of your two numbers equals the number on the other card shown.

What should your two cards be for you to win the game? Is there more than one answer? Explain.

MATHEMATICAL PRACTICES

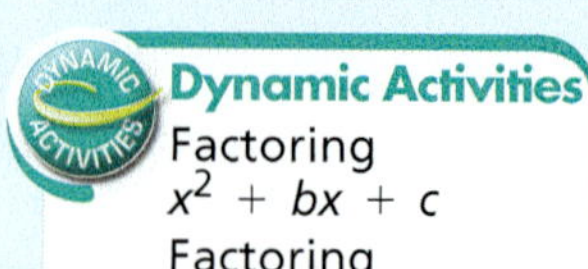

Dynamic Activities
Factoring $x^2 + bx + c$
Factoring $ax^2 + bx + c$

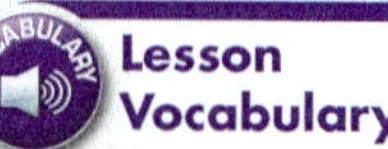

Lesson Vocabulary
- factoring
- greatest common factor (GCF) of an expression
- perfect square trinomial
- difference of two squares

Factors of a given number are numbers that have a product equal to the given number. Factors of a given expression are expressions that have a product equal to the given expression. **Factoring** is rewriting an expression as a product of its factors.

Essential Understanding You can factor many quadratic trinomials $(ax^2 + bx + c)$ into products of two binomials.

You can use the Distributive Property or the FOIL method to multiply two binomials. You can use FOIL in reverse to help you factor.

$$\begin{aligned}(x + 4)(x + 2) &= (x + 4)(x) + (x + 4)(2) \quad \text{Use the Distributive Property.}\\ &= x(x) + 4(x) + x(2) + 4(2)\\ &= x^2 + 6x + 8\end{aligned}$$

$$\begin{aligned}(x + 4)(x + 2) &= x(x) + x(2) + 4(x) + 4(2) \quad \text{F: First; O: Outer; I: Inner; L: Last}\\ &= x^2 + 6x + 8\end{aligned}$$

To factor $x^2 + 6x + 8$, think of FOIL in reverse. Find two binomials for which the first terms have the product x^2, the products of the outer and inner terms have the sum $6x$, and the last terms have the product 8.

$$x^2 + 6x + 8 = (x + 4)(x + 2)$$

When you factor, a table of the different possible factors of the constant term may be helpful.

Problem 1 Factoring $ax^2 + bx + c$ when $a = \pm 1$

Plan

How can you *make a table* to find factors?
Use the first row to list sets of factors of the constant. Use the second row to find the sum of each set of factors.

What is the expression in factored form?

A $x^2 + 9x + 20$

Step 1 Find factors of 20 with sum 9.
Since both 20 and 9 are positive, both factors are positive.

Factors of 20	1, 20	2, 10	4, 5
Sum of factors	21	12	9

Step 2 Use the factors you found. Write the expression as the product of two binomials.

$x^2 + 9x + 20 = (x + 4)(x + 5)$ Use the factors 4 and 5.

B $x^2 + 14x - 72$

Step 1 Find factors of -72 with sum 14.

Since $c < 0$, one factor is positive and the other is negative.
Since $b > 0$, the factor with greater absolute value is positive.

Factors of −72	−1, 72	−2, 36	−3, 24	−4, 18	−6, 12	−8, 9
Sum of factors	71	34	21	14	6	1

Step 2 Use the factors you found, -4 and 18. Write
$x^2 + 14x - 72 = (x - 4)(x + 18)$.

C $-x^2 + 13x - 12$

Think

Will factoring out −1 change the answer?
No; because the final factored expression will include −1 as a factor.

Step 1 Rewrite the expression to show a trinomial with leading coefficient 1.

$-(x^2 - 13x + 12)$ Factor out −1.

Step 2 Find factors of 12 with sum -13.

Since $c > 0$, both factors have the same sign.
Since $b < 0$, both factors must be negative.

Factors of 12	−1, −12	−2, −6	−3, −4
Sum of factors	−13	−8	−7

Step 3 Use the factors you found, -1 and -12. Write
$-x^2 + 13x - 12 = -(x^2 - 13x + 12) = -(x - 1)(x - 12)$.

Got It? **1.** What is the expression in factored form?

a. $x^2 + 14x + 40$ **b.** $x^2 - 11x + 30$ **c.** $-x^2 + 14x + 32$

The **greatest common factor (GCF) of an expression** is a common factor of the terms in the expression. It is the common factor with the greatest coefficient and the greatest exponent. You can factor any expression that has a GCF not equal to 1.

Problem 2 Finding Common Factors

Plan

Should you factor out a number, a variable, or both?
Both; the two terms have numerical and variable common factors.

What is the expression in factored form?

A $6n^2 + 9n$

$6n^2 + 9n = 3n(2n) + 3n(3)$ — Factor out the GCF, $3n$.

$= 3n(2n + 3)$ — Use the Distributive Property.

B $4x^2 + 20x - 56$

$4x^2 + 20x - 56 = 4(x^2) + 4(5x) - 4(14)$ — Factor out the GCF, 4.

$= 4(x^2 + 5x - 14)$ — Use the Distributive Property.

$= 4(x - 2)(x + 7)$ — Factor the trinomial.

Got It? **2.** What is the expression in factored form?

a. $7n^2 - 21$ **b.** $9x^2 + 9x - 18$ **c.** $4x^2 + 8x + 12$

To factor a quadratic trinomial of the form $ax^2 + bx + c$ where $a \neq 1$, and there is no common factor, rewrite the middle term, bx, as two terms. The coefficients of these two terms will be factors of ac that have sum b.

Problem 3 Factoring $ax^2 + bx + c$ when $|a| \neq 1$

What is the expression in factored form?

A $2x^2 + 11x + 12$

Step 1 Since there is no common factor, find ac.

$ac = 2(12) = 24$

Step 2 Since both b and ac are positive, find positive factors of 24 that have sum 11.

Think

How should you make your table in this case?
Use the first row to list sets of factors of ac. Use the second row as before, to find the sum of each set of factors.

Factors of 24	1, 24	2, 12	3, 8	4, 6
Sum of factors	25	14	11	10

Step 3 Factor the trinomial as follows.

$2x^2 + 11x + 12$

$2x^2 + 3x + 8x + 12$ — Rewrite bx. Since $ac = 24$, try 3 and 8 as coefficients of x.

$x(2x + 3) + 4(2x + 3)$ — Find a common factor for the first two terms and another common factor for the last two terms.

$(x + 4)(2x + 3)$ — Rewrite using the Distributive Property.

B $4x^2 - 4x - 3$

Think

Do you need positive or negative factors?
You need one negative factor and one positive factor.

Step 1 Since there is no common factor, rewrite bx as the sum of two terms with coefficients that are factors of ac, and have sum b.

$$ac = 4(-3) = -12$$

Step 2 Since $ac = -12 < 0$, find factors of ac with opposite signs. Since $b < 0$, the factor with greater absolute value is negative.

Factors of −12	1, −12	2, −6	3, −4
Sum of factors	−11	−4	−1

Step 3 Factor the trinomial as follows.

$4x^2 - 4x - 3$	
$4x^2 + 2x - 6x - 3$	Rewrite bx using $b = 2 - 6$.
$2x(2x + 1) - 3(2x + 1)$	Find a common factor for the first two terms and another common factor for the last two terms.
$(2x - 3)(2x + 1)$	Rewrite using the Distributive Property.

Check $(2x - 3)(2x + 1) = 4x^2 + 2x - 6x - 3$
$= 4x^2 - 4x - 3$ ✔

  **Got It? 3.** What is the expression in factored form? Check your answers.

a. $4x^2 + 7x + 3$ **b.** $2x^2 - 7x + 6$

c. Reasoning Can you factor the expression $2x^2 + 2x + 2$ into a product of two binomials? Explain.

A **perfect square trinomial** is a trinomial that is the square of a binomial. For example, $x^2 + 10x + 25 = (x + 5)^2$ is a perfect square trinomial.

If $ax^2 + bx + c$ is a perfect square trinomial, then ax^2 and c are squares of the terms of the binomial and thus are both positive. bx is twice the product of the terms of the binomial. b is negative if the binomial terms have opposite signs.

Here is another way to represent the two forms of a perfect square trinomial.

Key Concept Factoring Perfect Square Trinomials

$$a^2 + 2ab + b^2 = (a + b)^2 \qquad a^2 - 2ab + b^2 = (a - b)^2$$

Problem 4 Factoring a Perfect Square Trinomial

What is $4x^2 - 24x + 36$ in factored form?

Think	Write
Write the expression.	$4x^2 - 24x + 36$
You can do this in one step. Is the first term a perfect square? Yes, $4x^2$ is $(2x)^2$. Is the last term a perfect square? Yes, 36 is 6^2. Is the middle term twice the product of 6 and $2x$? Yes, $2 \cdot 6 \cdot 2x = 24x$ There's a minus sign in the middle. You are done.	$(2x - 6)^2$
You should check.	$(2x - 6)^2 = 4x^2 - 24x + 36$ ✔

Got It? **4.** What is $64x^2 - 16x + 1$ in factored form?

The expression $a^2 - b^2$ is the **difference of two squares**. There is a pattern to its factors.

take note

Key Concept Factoring a Difference of Two Squares

$$a^2 - b^2 = (a + b)(a - b)$$

Think

How can a binomial be the product of two binomials?
If the outer and inner products of the binomials are opposites, their sum is zero.

Problem 5 Factoring a Difference of Two Squares

What is $25x^2 - 49$ in factored form?

$25x^2 - 49 = (5x)^2 - 7^2$ Write as the difference of two squares.

$= (5x + 7)(5x - 7)$ Use the pattern for factoring a difference of two squares.

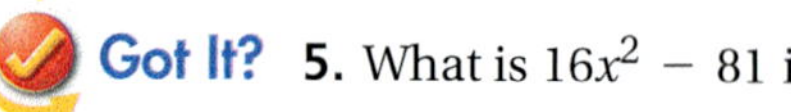

Got It? **5.** What is $16x^2 - 81$ in factored form?

Lesson Check

Do you know HOW?

Factor each expression.

1. $x^2 + 6x + 8$
2. $x^2 - 13x + 12$
3. $x^2 - 81$
4. $25y^2 - 36$
5. $y^2 - 6y + 9$
6. $4x^2 - 4x + 1$

Find the GCF of each expression.

7. $15x^2 - 25x$
8. $4a^3 + 8a^2$
9. $18b^2 - 12b + 24$
10. $21h^3 + 35h^2 - 28h$

Do you UNDERSTAND?

11. **Vocabulary** Is $4b^2 - 26b + 169$ a perfect square trinomial? Explain.
12. **Compare and Contrast** How is factoring a trinomial $ax^2 + bx + c^2$ when $a \neq 1$ different from factoring a trinomial when $a = 1$? How is it similar?
13. **Reasoning** Explain how to rewrite the expression $a^2 - 2ab + b^2 - 25$ as the product of two trinomial factors. (*Hint:* Group the first three terms. What type of expression is this?)

Practice and Problem-Solving Exercises

Factor each expression.

See Problem 1.

14. $x^2 + 3x + 2$
15. $x^2 + 5x + 6$
16. $x^2 + 7x + 10$
17. $x^2 + 10x + 16$
18. $y^2 + 15y + 36$
19. $x^2 + 22x + 40$
20. $x^2 - 3x + 2$
21. $-x^2 + 13x - 12$
22. $-r^2 + 11r - 18$
23. $x^2 - 10x + 24$
24. $d^2 - 12d + 27$
25. $x^2 - 13x + 36$
26. $x^2 - 5x - 14$
27. $-x^2 - x + 20$
28. $-x^2 + 3x + 40$
29. $c^2 + 2c - 63$
30. $x^2 + 10x - 75$
31. $-t^2 + 7t + 44$

Find the GCF of each expression. Then factor the expression.

See Problem 2.

32. $3a^2 + 9$
33. $25b^2 - 20b$
34. $x^2 - 2x$
35. $5t^2 - 5t - 10$
36. $14y^2 + 7y - 21$
37. $27p^2 - 9p + 18$

Factor each expression.

See Problem 3.

38. $3x^2 + 31x + 36$
39. $2x^2 - 19x + 24$
40. $5r^2 + 23r + 26$
41. $2m^2 - 11m + 15$
42. $5t^2 + 28t + 32$
43. $2x^2 - 27x + 36$
44. $3x^2 + 7x - 20$
45. $5y^2 + 12y - 32$
46. $7x^2 - 8x - 12$

Factor each expression that can be factored. For an expression that cannot be factored into a product of two binomials, explain why.

See Problems 4 and 5

47. $x^2 + 2x + 1$
48. $t^2 - 14t + 49$
49. $k^2 - 18k + 81$
50. $4z^2 - 20z + 25$
51. $4x^2 + 16x + 8$
52. $81z^2 + 36z + 4$
53. $x^2 - 4$
54. $25a^2 - 120a + 144$
55. $81y^2 + 49$

56. **Think About a Plan** Suppose you cut a small square from a square sheet of cardboard. Find the sides of one rectangle whose area is equal to the area of the remaining part.

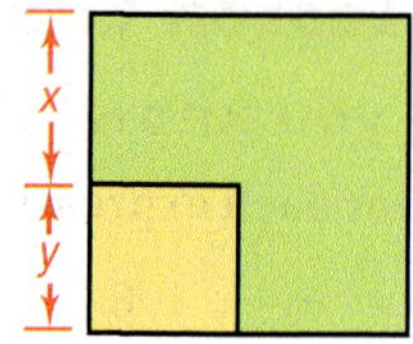

- How can you represent the remaining part as a combination of rectangles with known sides?
- Can you factor the resulting expression?

57. The area in square centimeters of a square area rug is $25x^2 - 10x + 1$. What are the dimensions of the rug in terms of x?

Factor each expression completely.

58. $9x^2 - 36$
59. $18z^2 - 8$
60. $4n^2 - 20n + 24$
61. $64t^2 - 16$
62. $12x^2 + 36x + 27$
63. $3y^2 + 24y + 45$
64. $2a^2 - 16a + 32$
65. $3x^2 - 24x - 27$
66. $-x^2 + 5x - 4$
67. $4x^2 - 22x + 10$
68. $-6z^2 - 600$
69. $-\frac{1}{16}s^2 + 1$

70. a. Multiply $(a + b)(a - b)(a^2 + b^2)$.
    b. Use your result from part (a) to completely factor $81x^4 - 256y^4$.

71. **Error Analysis** Your friend attempted to factor an expression as shown. Find the error in your friend's work. Then factor the expression correctly.

$2x^2 - 7x + 5$
$2x^2 - 5x - 2x + 5$
$x(2x - 5) + (2x - 5)$
$(x + 1)(2x - 5)$

72. **Agriculture** The area in square feet of a rectangular field is $x^2 - 120x + 3500$. The width, in feet, is $x - 50$. What is the length, in feet?

Find the GCF of each expression. Then factor the expression.

73. $y^2 - y$
74. $ab^2 - b$
75. $10x^2 - 90$
76. $3t^2 - 24t$
77. $2x^2 - 74x + 12$
78. $x^2y^2 + xy$

79. What is the factored form of $4x^2 + 15x - 4$?
    (A) $(2x + 2)(2x - 2)$
    (B) $(2x - 4)(2x + 1)$
    (C) $(4x + 1)(x - 4)$
    (D) $(4x - 1)(x + 4)$

80. **Geometry** What is the volume of the shaded pipe with outer radius R, inner radius r, and height h as shown? Express your answer in completely factored form.

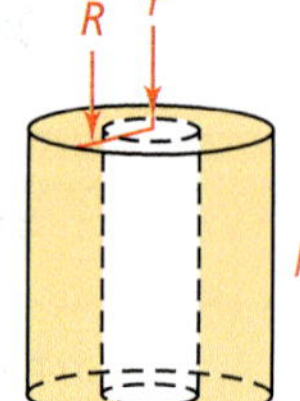

81. **Open-Ended** Write a quadratic trinomial that you can factor, where $a \neq 1$, $ac > 0$, and $b < 0$. Factor the expression.

82. **Writing** Explain how to factor $3x^2 + 6x - 72$ completely.

Factor each expression completely.

83. $0.25t^2 - 0.16$
84. $8100x^2 - 10{,}000$
85. $(x + 3)^2 + 3(x + 3) - 54$
86. $(x - 2)^2 - 15(x - 2) + 56$
87. $6(x + 5)^2 - 5(x + 5) + 1$
88. $3(2a - 3)^2 + 17(2a - 3) + 10$

89. Reasoning Explain how to factor $4x^4 + 24x^3 + 32x^2$.

Factor each expression completely.

90. $\frac{1}{16}x^4 - y^4$ **91.** $16x^4 - 625y^4$ **92.** $243a^5 - 3a$

93. When the expression $x^2 + bx - 24$ is factored completely, the difference of the factors is 11. Find both factors if it is known that b is negative.

94. Prove that $n^3 - n$ is divisible by 3 for all positive integer values of n. (*Hint:* Factor the expression completely.)

Standardized Test Prep

SAT/ACT

95. How can you write $(m - 5)(m + 4) + 8$ as a product of two binomials?

(A) $(m - 1)(m + 8)$
(B) $(m - 4)(m + 3)$
(C) $(m + 8)(m + 8)$
(D) $(m - 5)(8m + 32)$

96. The graph of a quadratic function has vertex (7, 6). What is the axis of symmetry?

(F) $x = 6$ (G) $y = 6$ (H) $x = 7$ (I) $y = 7$

Extended Response

97. Suppose you hit a baseball and its flight takes a parabolic path. The height of the ball at certain times appears in the table below.

Time (s)	0.5	0.75	1	1.25
Height (ft)	10	10.5	9	5.5

a. Find a quadratic model for the ball's height as a function of time.
b. Write the quadratic function in factored form.

Mixed Review

98. Find a quadratic model for the values in the table.

x	0	5	10	15	20
y	17	39	54	61	61

See Lesson 4-3.

99. Coins The combined mass of a penny, a nickel, and a dime is 9.0 g. Ten nickels and three pennies have the same mass as 25 dimes. Fifty dimes have the same mass as 18 nickels and 10 pennies. Write and solve a system of equations to find the mass of each type of coin.

See Lesson 3-6.

Get Ready! **To prepare for Lesson 4-5, do Exercises 100–102.**

Graph each function.

See Lesson 4-2.

100. $y = x^2 - 2x - 5$ **101.** $y = x^2 - 4x + 4$ **102.** $y = -x^2 - 3x + 8$

4 Mid-Chapter Quiz

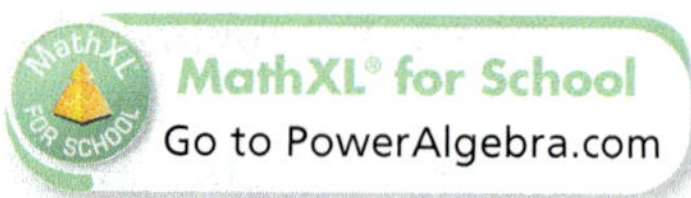

Do you know HOW?

Graph each function.

1. $y = 4x^2 + 16x + 7$
2. $y = (x + 8)^2 - 3$
3. $y = -(x + 2)^2 - 7$
4. $y = -3x^2 - 2x + 1$

Identify the axis of symmetry, maximum or minimum value, and the domain and range of each function.

5. $y = -x^2 + 6x + 5$
6. $y = \frac{1}{2}(x - 6)^2 + 7$
7. $y = -3(x + 2)^2 + 1$
8. $y = 4x^2 - 8x$
9. Rewrite the equation $y = -3x^2 - 6x - 8$ in vertex form. Identify the vertex and the axis of symmetry of the graph.

Write each expression in factored form.

10. $16 - 2m^2$
11. $-x^2 + 3x$
12. $y^2 - 13y + 12$
13. $k^2 - 5k - 24$
14. $4y^2 - 9$
15. $-10n + 25 + n^2$
16. $2x^2 + 7x + 6$

Find a quadratic model in standard form for each set of values.

17. (0, 3), (1, 10), (2, 19)
18. (0, 0), (1, −5), (2, 0)

Write the equation of each parabola in vertex form.

19.

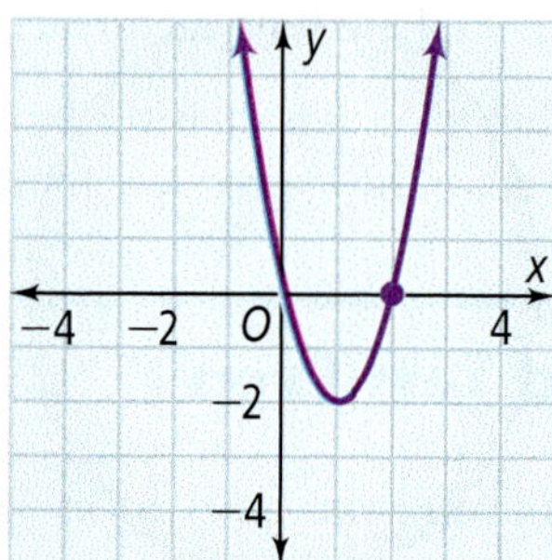

20.

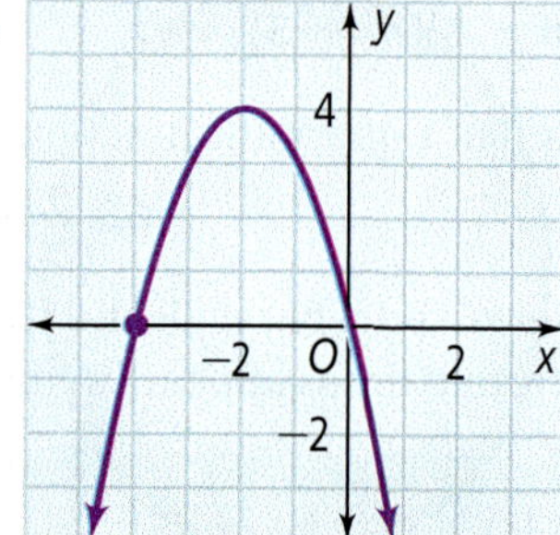

Do you UNDERSTAND?

21. Write the expression $3x^4 - 12x^3 - 36x^2$ in factored form. Explain how you know the expression is completely factored.
22. **Open-Ended** Write the equation of a parabola with a vertex at (3, 2). Name the axis of symmetry and the coordinates of two other points on the graph.
23. **Writing** Explain how to factor $25x^2 - 30x + 9$.
24. **Reasoning** Write the equation of two parabolas such that they have a common vertex and are reflections of each other across the x-axis.
25. Write the equation of a parabola in standard form and explain how to convert it to vertex form. How would you reverse the process?
26. What is the relationship between the x-intercepts of the graph of a quadratic function and the x-coordinate of the vertex of that graph? Explain how you determined your answer.

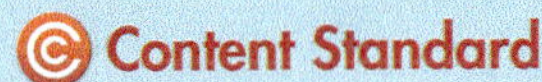

Algebra Review

For Use With Lesson 4-5

Square Roots and Radicals

Content Standard

Reviews N.RN.2 Rewrite expressions involving radicals and rational exponents using the properties of exponents.

A radical symbol $\sqrt{}$ indicates a square root. In general, $\sqrt{x^2} = |x|$ for all real numbers x.

Square Roots

Multiplication Property of Square Roots
For any numbers $a \geq 0$ and $b \geq 0$,
$\sqrt{ab} = \sqrt{a} \cdot \sqrt{b}$.

Division Property of Square Roots
For any numbers $a \geq 0$ and $b > 0$,
$\sqrt{\frac{a}{b}} = \frac{\sqrt{a}}{\sqrt{b}}$.

Example

Simplify each expression.

A $\sqrt{50}$

$\sqrt{50} = \sqrt{25} \cdot \sqrt{2}$ Multiplication Property of Square Roots

$= 5\sqrt{2}$ Simplify.

B $\sqrt{\frac{5}{11}}$

$\sqrt{\frac{5}{11}} = \frac{\sqrt{5}}{\sqrt{11}}$ Division Property of Square Roots

$= \frac{\sqrt{5}}{\sqrt{11}} \cdot \frac{\sqrt{11}}{\sqrt{11}}$ Multiply both the numerator and denominator by $\sqrt{11}$.

$= \frac{\sqrt{55}}{\sqrt{121}}$ Multiplication Property of Square Roots.

$= \frac{\sqrt{55}}{11}$ Simplify.

Exercises

Simplify each radical expression.

1. $\sqrt{18}$
2. $\sqrt{75}$
3. $-\sqrt{32}$
4. $\sqrt{\frac{-5}{7}}$
5. $\sqrt{\frac{7}{13}}$
6. $\sqrt{\frac{3}{15}}$
7. $-\sqrt{200}$
8. $5\sqrt{320}$
9. $(2\sqrt{27})^2$
10. $-\sqrt{10^4}$
11. $\sqrt{x^2y^2}$
12. $\sqrt{\frac{8}{x^2}}$
13. $-\sqrt{\frac{7x^3}{5x}}$
14. $\sqrt{\frac{(-3)^4}{12}}$
15. $\sqrt{\frac{200}{28}}$
16. $\sqrt{120x}$

4-5 Quadratic Equations

Content Standards

A.CED.1 Create equations and inequalities in one variable and use them to solve problems.

A.APR.3 Identify zeros of polynomials when suitable factorizations are available . . .

Also A.SSE.1.a

Objectives To solve quadratic equations by factoring
To solve quadratic equations by graphing

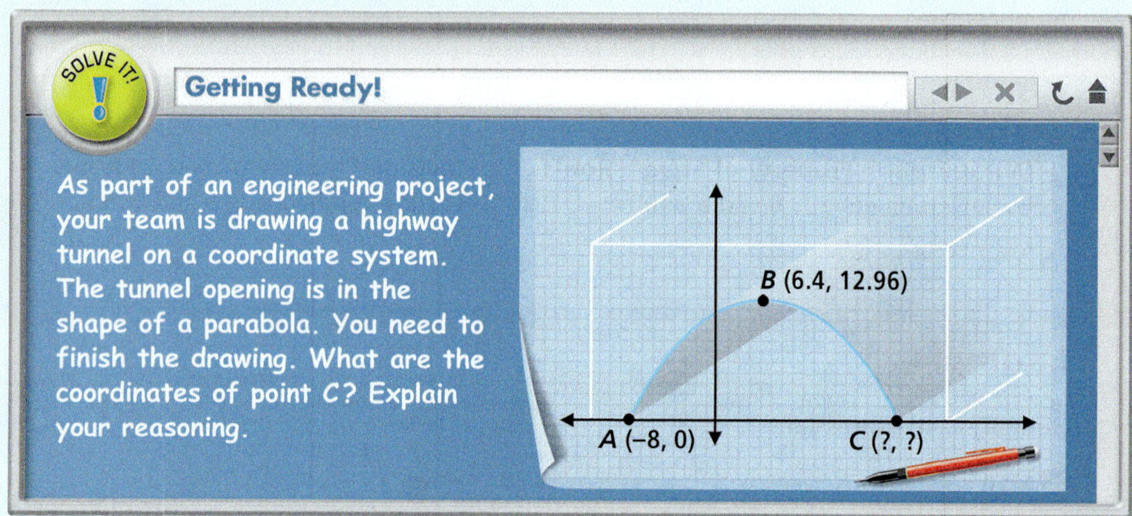

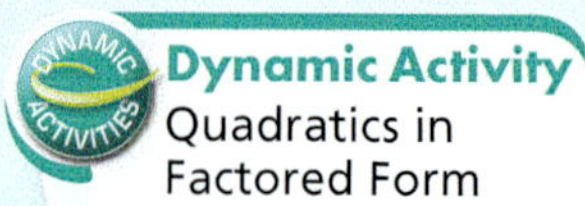

Dynamic Activity
Quadratics in Factored Form

Lesson Vocabulary
- zero of a function
- Zero-Product Property

Wherever the graph of a function $f(x)$ intersects the x-axis, $f(x) = 0$. A value of x for which $f(x) = 0$ is a **zero of the function**.

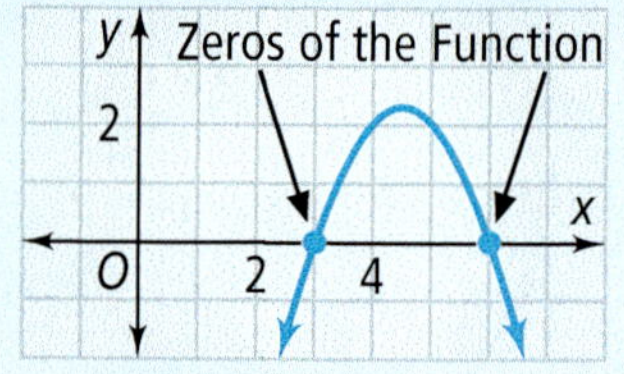

Essential Understanding To find the zeros of a quadratic function $y = ax^2 + bx + c$, solve the related quadratic equation $0 = ax^2 + bx + c$.

You can solve some quadratic equations in standard form by factoring the quadratic expression and using the **Zero-Product Property**.

take note

Property Zero-Product Property

If $ab = 0$, then $a = 0$ or $b = 0$.

Think

What do you know about the factors of $x^2 - bx + c$?
The product of their constant terms is c. The sum is $-b$.

Problem 1 Solving a Quadratic Equation by Factoring

What are the solutions of the quadratic equation $x^2 - 5x + 6 = 0$?

$(x - 2)(x - 3) = 0$	Factor the quadratic expression.
$x - 2 = 0$ or $x - 3 = 0$	Use the Zero-Product Property.
$x = 2$ or $x = 3$	Solve for x.

The solutions are $x = 2$ and $x = 3$.

 **Got It?** **1.** What are the solutions of the quadratic equation $x^2 - 7x = -12$?

Problem 2 Solving a Quadratic Equation With Tables

What are the solutions of the quadratic equation $5x^2 + 30x + 14 = 2 - 2x$?

$5x^2 + 30x + 14 = 2 - 2x$

$5x^2 + 32x + 12 = 0$ Rewrite in standard form.

Use your calculator's **TABLE** feature to find the zeros.

Think

What should you look for in the calculator table? Look for x-values for which $y = 0$.

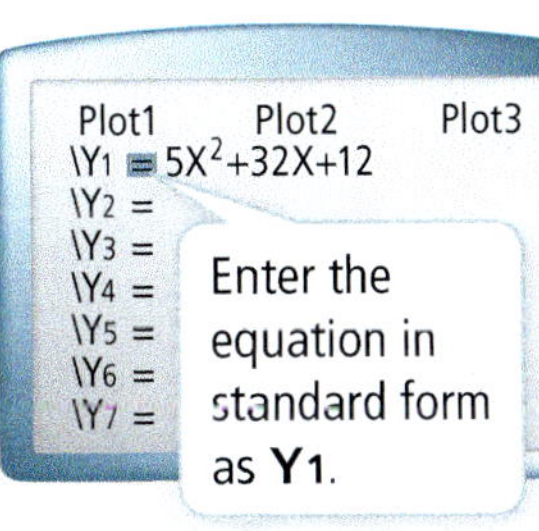

X	Y1
−6	0
−5	−23
−4	−36
−3	−39
−2	−32
−1	−15
0	12

X=−6

Y1 = 0, $x = -6$ is one zero.

Second zero is between $x = -1$ and $x = 0$. Notice change in sign for y-values.

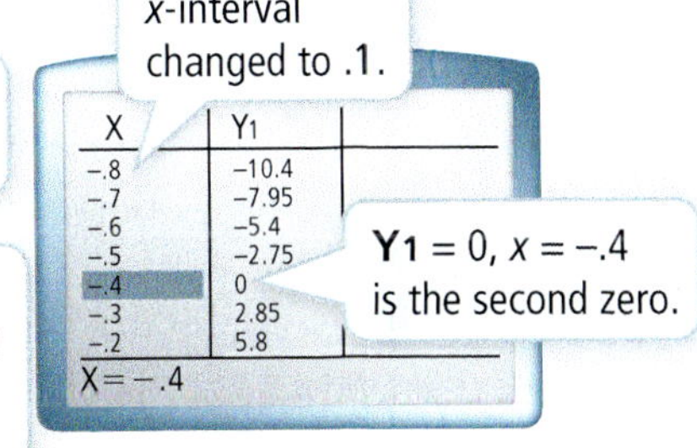

The solutions are $x = -6$ and $x = -0.4$.

 **Got It?** **2.** What are the solutions of the quadratic equation $4x^2 - 14x + 7 = 4 - x$?

Problem 3 Solving a Quadratic Equation by Graphing

What are the solutions of the quadratic equation $2x^2 + 7x = 15$?

$2x^2 + 7x = 15$

$2x^2 + 7x - 15 = 0$ Rewrite in standard form.

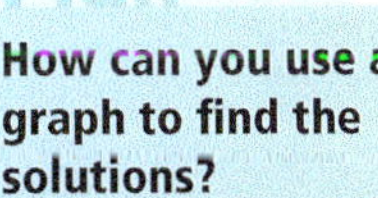

How can you use a graph to find the solutions? Find the zeros of the related quadratic function.

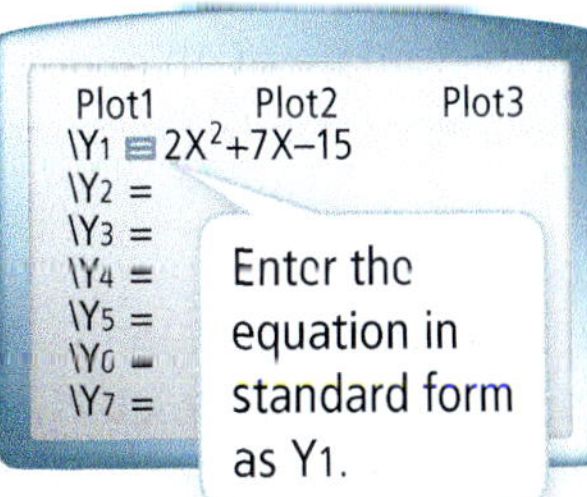

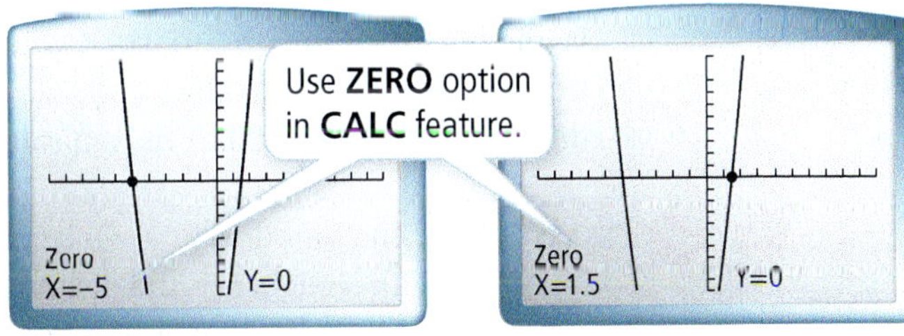

The solutions are $x = -5$ and $x = 1.5$.

 **Got It?** **3.** What are the solutions of the quadratic equation $x^2 + 2x - 24 = 0$?

Problem 4 Using a Quadratic Equation

Competition From the time Mark Twain wrote *The Celebrated Jumping Frog of Calaveras County* in 1865, frog-jumping competitions have been growing in popularity. The graph shows a function modeling the height of one frog's jump, where x is the distance, in feet, from the jump's start.

Think

How can you use a graphing calculator to determine the distance?
Graph the function and locate the point where the graph crosses the x-axis.

A How far did the frog jump?

The height of the jump is 0 at the start and end of the jump. Find the zeros of the function. Use a graphing calculator to find the zeros of the related function $y = -0.029x^2 + 0.59x$.

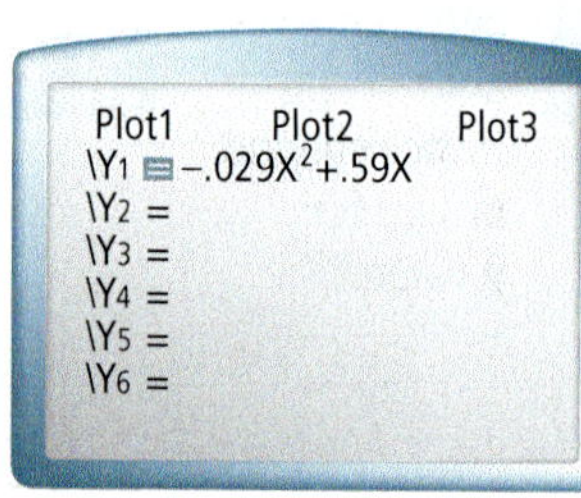

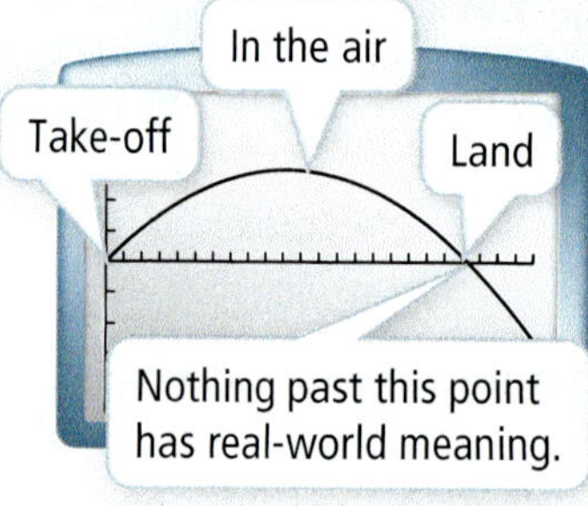

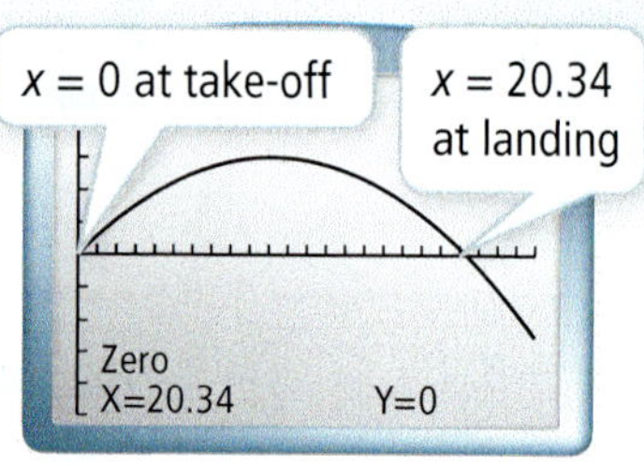

The frog jumped about 20.34 ft.

B How high did the frog jump?

The maximum height of the jump is the maximum value of the function. This occurs midway, at 10.17 ft from the start. Find y for $x = 10.17$.

$$y = -0.029(10.17)^2 + 0.59(10.17) \approx 3.0$$

The frog jumped to a height of about 3.0 ft.

C What is a reasonable domain and range for such a frog-jumping function?

While the function $y = -0.029x^2 + 0.59x$ has a domain of all real numbers, actual frog jumping does not allow negative values. So, a reasonable domain for frog-jumping distances is $0 \leq x \leq 30$. A reasonable range is $0 \leq y \leq 5$.

Got It? **4. a.** The function $y = -0.03x^2 + 1.60x$ models the path of a kicked soccer ball. The height is y, the distance is x, and the units are meters. How far does the soccer ball travel? How high does the soccer ball go? Describe a reasonable domain and range for the function.

b. Reasoning Are all domains and ranges reasonable for real-world situations? Explain.

Lesson Check

Do you know HOW?

Solve each equation by factoring.

1. $x^2 - 9 = 0$
2. $x^2 + 13x = -36$
3. $3x^2 - x - 2 = 0$

Solve by graphing.

4. $x^2 - 3x = 6$
5. $2x^2 - x = 11$

Do you UNDERSTAND?

6. **Vocabulary** If 5 is a zero of the function $y = x^2 + bx - 20$, what is the value of b? Explain.
7. **Compare and Contrast** When is it easier to solve a quadratic equation by factoring than to solve it using a table?
8. **Reasoning** Using tables, how might you recognize that a quadratic equation likely has exactly one solution? no solutions?

Practice and Problem-Solving Exercises

Solve each equation by factoring. Check your answers. See Problem 1.

9. $x^2 + 6x + 8 = 0$
10. $x^2 + 18 - 9x$
11. $2x^2 \quad x = 3$
12. $x^2 - 10x + 25 = 0$
13. $2x^2 + 6x = -4$
14. $3x^2 = 16x + 12$
15. $x^2 - 4x = 0$
16. $6x^2 + 4x = 0$
17. $2x^2 = 8x$

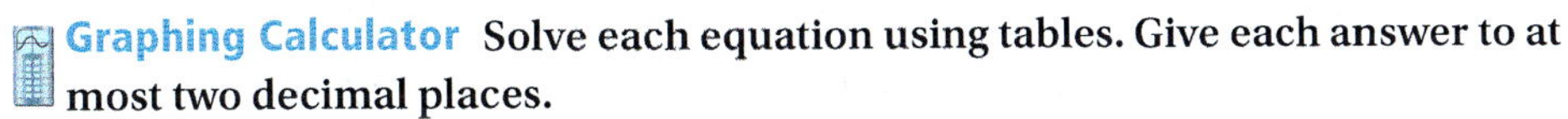

Graphing Calculator Solve each equation using tables. Give each answer to at most two decimal places. See Problem 2.

18. $x^2 + 5x + 3 = 0$
19. $x^2 - 11x + 24 = 0$
20. $x^2 - 7x = 11$
21. $2x^2 - x = 2$
22. $x^2 - 16x = 36$
23. $x^2 + 6x = 40$
24. $4x^2 = x + 3$
25. $5x^2 + x = 4$
26. $10x^2 + 3 = 11x$

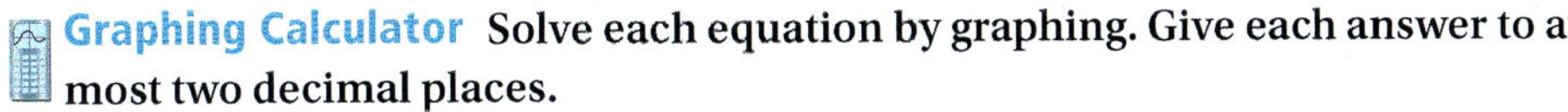

Graphing Calculator Solve each equation by graphing. Give each answer to at most two decimal places.

See Problem 3.

27. $6x^2 = -19x - 15$
28. $3x^2 - 5x - 4 = 0$
29. $5x^2 - 7x - 3 = 8$
30. $6x^2 + 31x = 12$
31. $1 = 4x^2 + 3x$
32. $\frac{1}{2}x^2 - x = 8$
33. $x^2 = 4x + 8$
34. $x^2 + 4x = 6$
35. $2x^2 - 2x - 5 = 0$

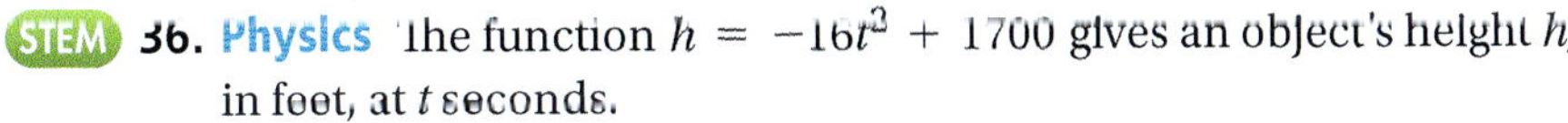

See Problem 4.

STEM 36. **Physics** The function $h = -16t^2 + 1700$ gives an object's height h, in feet, at t seconds.

a. What does the constant 1700 tell you about the height of the object?
b. What does the coefficient of t^2 tell you about the direction the object is moving?
c. When will the object be 1000 ft above the ground?
d. When will the object be 940 ft above the ground?
e. What are a reasonable domain and range for the function h?

Apply

37. Think About a Plan Suppose you want to put a frame around the painting shown at the right. The frame will be the same width around the entire painting. You have 276 in.^{2} of framing material. How wide should the frame be?

- What does 276 in.^{2} represent in this situation?
- How can you write the dimensions of the frame using two binomials?

38. The period of a pendulum is the time the pendulum takes to swing back and forth. The function $L = 0.81t^2$ relates the length L in feet of a pendulum to the time t in seconds that it takes to swing back and forth. A convention center has a pendulum that is 90 feet long. Find the period.

39. Landscaping Suppose you have an outdoor pool measuring 25 ft by 10 ft. You want to add a cement walkway around the pool. If the walkway will be 1 ft thick and you have 304 ft^3 of cement, how wide should the walkway be?

40. Error Analysis A classmate solves the quadratic equation as shown. Find and correct the error. What are the correct solutions?

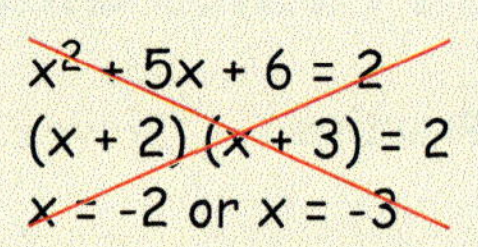

41. Open-Ended Write an equation with the given solutions.

a. 3 and 5 **b.** −3 and 2 **c.** −1 and −6

Solve each equation by factoring, using tables, or by graphing. If necessary, round your answer to the nearest hundredth.

42. $x^2 + 2x = 6 - 6x$

43. $6x^2 + 13x + 6 = 0$

44. $2x^2 + x - 28 = 0$

45. $2x^2 + 8x = 5x + 20$

46. $3x^2 + 7x = 9$

47. $2x^2 - 6x = 8$

48. $(x + 3)^2 = 9$

49. $x^2 + 4x = 0$

50. $x^2 = 8x - 7$

51. $x^2 - 3x = 6$

52. $4x^2 + 5x = 4$

53. $7x - 3x^2 = -10$

Reasoning **The graphs of each pair of functions intersect. Find their points of intersection without using a calculator. (*Hint:* Solve as a system using substitution.)**

54. $y = x^2$
$y = -\frac{1}{2}x^2 + \frac{3}{2}x + 3$

55. $y = x^2 - 2$
$y = 3x^2 - 4x - 2$

56. $y = -x^2 + x + 4$
$y = 2x^2 - 6$

Challenge

57. The equation $x^2 - 10x + 24 = 0$ can be written in factored form as $(x - 4)(x - 6) = 0$. How can you use this fact to find the vertex of the graph of $y = x^2 - 10x + 24$?

58. a. Let $a > 0$. Use algebraic or arithmetic ideas to explain why the lowest point on the graph of $y = a(x - h)^2 + k$ must occur when $x = h$.

b. Suppose that the function in part (a) is $y = a(x - h)^3 + k$. Is your reasoning still valid? Explain.

 **59. Physics** When serving in tennis, a player tosses the tennis ball vertically in the air. The height h of the ball after t seconds is given by the quadratic function $h(t) = -5t^2 + 7t$ (the height is measured in meters from the point of the toss).

a. How high in the air does the ball go?

b. Assume that the player hits the ball on its way down when it's 0.6 m above the point of the toss. For how many seconds is the ball in the air between the toss and the serve?

Standardized Test Prep

SAT/ACT

60. What are the solutions of the equation $6x^2 + 9x - 15 = 0$?

(A) $1, -15$ (B) $1, -\frac{5}{2}$ (C) $-1, -5$ (D) $3, \frac{5}{2}$

61. The vertex of a parabola is (3, 2). A second point on the parabola is (1, 7). Which point is also on the parabola?

(F) $(-1, 7)$ (G) $(3, 7)$ (H) $(5, 7)$ (I) $(3, -2)$

62. For which quadratic function is -3 the constant term?

(A) $y = (3x + 1)(-x - 3)$ (C) $f(x) = (x - 3)(x - 3)$

(B) $y = x^2 - 3x + 3$ (D) $g(x) = -3x^2 + 3x + 9$

Short Response

63. What transformations are needed to go from the parent function $f(x) = x^2$ to the new function $g(x) = -3x^2 + 2$? Graph $g(x)$.

Mixed Review

Factor each expression. See Lesson 4-4.

64. $16x^2 - 1$ **65.** $5x^2 - 26x + 5$ **66.** $2x^2 + 13x - 7$

Solve each system by elimination. Check your answer. See Lesson 3-5.

67. $\begin{cases} 7x - 2y - 5z = 24 \\ x + 3y + 4z = 10 \\ x - y - z = 4 \end{cases}$ **68.** $\begin{cases} -2x + 9y - z = 8 \\ 3x - 4y + z = -5 \\ 5x + 5y - z = -10 \end{cases}$ **69.** $\begin{cases} x - 9y + 8z = -10 \\ x + y - z = 9 \\ -x - 9z = 2 \end{cases}$

Without graphing, identify the vertex, axis of symmetry, and transformations from the parent function $f(x) = |x|$. See Lesson 2-7.

70. $y = |x + 9| + 4$ **71.** $y = |2x - 7|$ **72.** $y = \frac{3}{4}|x| - 1$

Get Ready! **To prepare for Lesson 4-6, do Exercises 73–75.**

Simplify each expression. See Lesson 4-4.

73. $(x + 4)(x + 4) - 3$ **74.** $(2x - 1)(2x - 1)$ **75.** $(x - 3)(x - 3)$

Concept Byte

For Use With Lesson 4-5

Writing Equations From Roots

Content Standard

A.CED.2 Create equations in two or more variables to represent relationships between quantities; graph equations on coordinate axes with labels and scales.

The **root** of an equation is a value that makes the equation true. You can use the Zero-Product Property to write a quadratic function from its zeros or a quadratic equation from its roots.

Activity 1

1. a. Write a nonzero linear function $f(x)$ that has a zero at $x = 3$.

b. Write a nonzero linear function $g(x)$ that has a zero at $x = 4$.

2. a. For f and g from Exercise 1, write the product function $h(x) = f(x) \cdot g(x)$.

b. What kind of function is $h(x)$?

c. Solve the equation $h(x) = 0$.

Mental Math **Write a quadratic equation with each pair of values as roots.**

3. 5 and 3 **4.** 2.5 and 4 **5.** −4 and 4 **6.** 5 and 10 7. $\frac{3}{2}$ and −2

You can also use zeros or roots to write quadratic expressions in standard form.

Activity 2

8. a. Copy and complete the table. Write the product $(x - a)(x - b)$ in standard form for each pair a and b.

b. Is there a pattern in the table? Explain.

9. a. If you know the roots, you can write a quadratic function or equation in standard form. Explain how.

b. Demonstrate your method for each pair of values in Exercises 3–7.

a	b	$a + b$	ab	$(x - a)(x - b)$
4	5	9	20	$x^2 - 9x + 20$
−4	5	1	−20	■
4	−5	■	■	■
−4	−5	■	■	■
−9	−1	■	■	■
−2	7	■	■	■

Exercises

10. Explain how to write a quadratic equation that has −6 as its only root.

11. Describe the family of quadratic functions that have zeros at r and s. Sketch several members of the family in the coordinate plane.

Find the sum and product of the roots for each quadratic equation.

12. $2x^2 + 3x - 2 = 0$ **13.** $x^2 - 2x + 1 = 0$ **14.** $x^2 - 5x + 6 = 0$

Given the sum and product of the roots, write a quadratic equation in standard form.

15. sum = −3, product = −18 **16.** sum = 4, product = 3 **17.** sum = 2, product = $\frac{3}{4}$

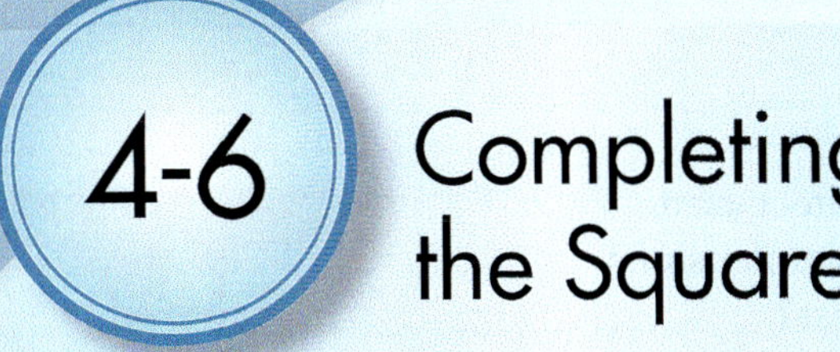

Completing the Square

Content Standard

Reviews A.REI.4.b Solve quadratic equations by . . . completing the square . . .

Objectives To solve equations by completing the square
To rewrite functions by completing the square

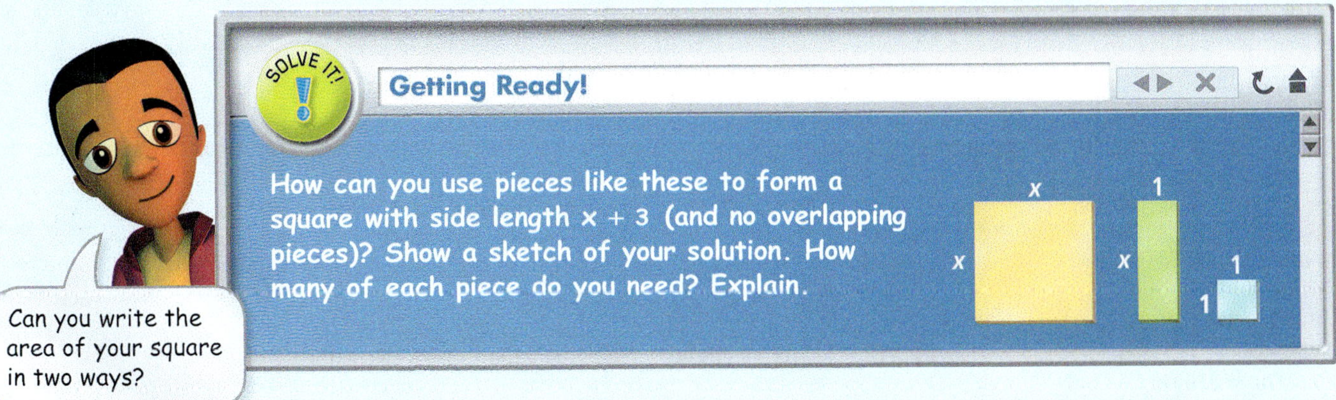

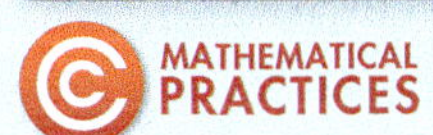

Forming a square with model pieces provides a useful geometric image for completing a square algebraically.

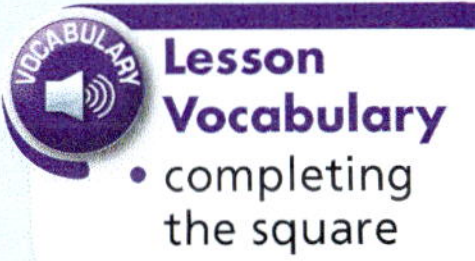

- completing the square

Essential Understanding Completing a perfect square trinomial allows you to factor the completed trinomial as the square of a binomial.

You can solve an equation that contains a perfect square by finding square roots. The simplest of this type of equation has the form $ax^2 = c$.

Problem 1 Solving by Finding Square Roots

What is the solution of each equation?

Plan
How is solving this equation like solving a linear equation? You isolate the variable term.

A $4x^2 + 10 = 46$		**B** $3x^2 - 5 = 25$
$4x^2 = 36$	← Rewrite in $ax^2 = c$ form. →	$3x^2 = 30$
$\frac{4x^2}{4} = \frac{36}{4}$	← Isolate x^2. →	$\frac{3x^2}{3} = \frac{30}{3}$
$x^2 = 9$		$x^2 = 10$
$x = \pm 3$	← Find square roots. →	$x = \pm\sqrt{10}$

Got It? **1.** What is the solution of each equation?
a. $7x^2 - 10 = 25$ **b.** $2x^2 + 9 = 13$

Problem 2 Determining Dimensions STEM

Dynamic Activity
Completing the Square

Architecture **While designing a house, an architect used windows like the one shown here. What are the dimensions of the window if it has 2766 square inches of glass?**

Step 1 Find the area of the window.

The area of the rectangular part is $(2x)(x) = 2x^2$ in.^{2}.

The area of the semicircular part is

$\frac{1}{2}\pi r^2 = \frac{1}{2}\pi\left(\frac{x}{2}\right)^2 = \frac{1}{2}\pi\frac{x^2}{4} = \frac{\pi}{8}x^2$ in.^{2}.

So, the total amount of glass used is

$2x^2 + \frac{\pi}{8}x^2 = 2766$ in.^{2}.

Step 2 Solve for x.

$$\left(2 + \frac{\pi}{8}\right)x^2 = 2766$$ Write the equation in $ax^2 = c$ form.

$$x^2 = \frac{2766}{2 + \frac{\pi}{8}}$$ Isolate x^2.

$$x \approx \pm 34$$ Find square roots. Use a calculator.

Think

Is the answer reasonable?
Yes; the rectangular part is about $30 \times 70 = 2100$ in.^{2}. This leaves enough glass for the semicircle.

Length cannot be negative. So the rectangular portion of the window is 34 in. wide by 68 in. long. The semicircular top has a radius of 17 in.

Got It? **2.** The lengths of the sides of a rectangular window have the ratio 1.6 to 1. The area of the window is 2822.4 in.^{2}. What are the window dimensions?

Sometimes an equation shows a perfect square trinomial equal to a constant. To solve, factor the perfect square trinomial into the square of a binomial. Then find square roots.

Problem 3 Solving a Perfect Square Trinomial Equation

What is the solution of $x^2 + 4x + 4 = 25$?

Think	Write
Factor the perfect square trinomial.	$x^2 + 4x + 4 = 25$ $(x + 2)^2 = 25$
Find square roots.	$x + 2 = \pm 5$
Rewrite as two equations.	$x + 2 = 5$ or $x + 2 = -5$
Solve for x.	$x = 3$ or $x = -7$

Got It? **3.** What is the solution of $x^2 - 14x + 49 = 25$?

If $x^2 + bx$ is not part of a perfect square trinomial, you can use the coefficient b to find a constant c so that $x^2 + bx + c$ is a perfect square. When you do this, you are **completing the square**. The diagram models this process.

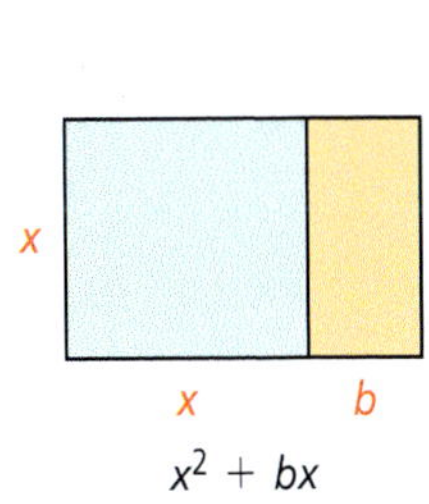

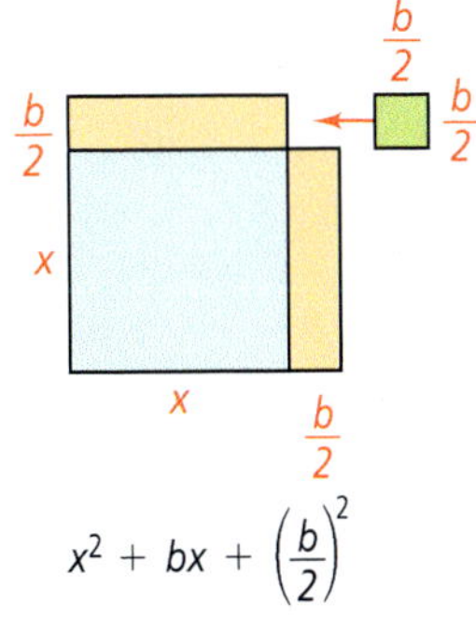

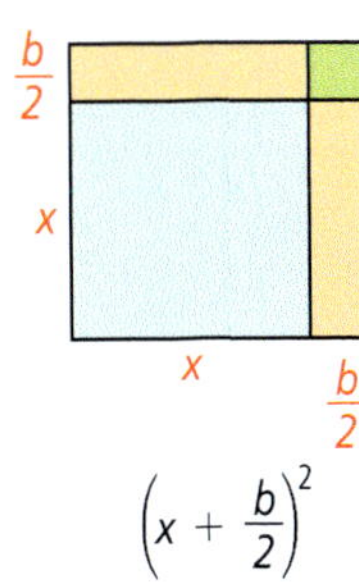

take note

Key Concept Completing the Square

You can form a perfect square trinomial from $x^2 + bx$ by adding $\left(\frac{b}{2}\right)^2$.

$$x^2 + bx + \left(\frac{b}{2}\right)^2 = \left(x + \frac{b}{2}\right)^2$$

Problem 4 Completing the Square

What value completes the square for $x^2 - 10x$? Justify your answer.

$x^2 - 10x$	Identify b; $b = -10$
$\left(\frac{b}{2}\right)^2 = \left(\frac{-10}{2}\right)^2 = (-5)^2 = 25$	Find $\left(\frac{b}{2}\right)^2$.
$x^2 - 10x + 25$	Add the value of $\left(\frac{b}{2}\right)^2$ to complete the square.
$x^2 - 10x + 25 = (x - 5)^2$	Rewrite as the square of a binomial.

Think

Why do you want a perfect square trinomial?
You can factor a perfect square trinomial into the square of a binomial.

  **Got It?** **4. a.** What value completes the square for $x^2 + 6x$?

b. Reasoning Is it possible for more than one value to complete the square for an expression? Explain.

take note

Key Concept Solving an Equation by Completing the Square

1. Rewrite the equation in the form $x^2 + bx = c$. To do this, get all terms with the variable on one side of the equation and the constant on the other side. Divide all the terms of the equation by the coefficient of x^2 if it is not 1.
2. Complete the square by adding $\left(\frac{b}{2}\right)^2$ to each side of the equation.
3. Factor the trinomial.
4. Find square roots.
5. Solve for x.

Problem 5 Solving by Completing the Square

What is the solution of $3x^2 - 12x + 6 = 0$?

$3x^2 - 12x + 6 = 0$	
$3x^2 - 12x = -6$	Rewrite. Get all terms with x on one side of the equation.
$\frac{3x^2}{3} - \frac{12x}{3} = \frac{-6}{3}$	Divide each side by 3 so the coefficient of x^2 will be 1.
$x^2 - 4x = -2$	Simplify.
$\left(\frac{b}{2}\right)^2 = \left(\frac{-4}{2}\right)^2 = (-2)^2 = 4$	Find $\left(\frac{b}{2}\right)^2 = 4$.
$x^2 - 4x + 4 = -2 + 4$	Add 4 to each side.
$(x - 2)^2 = 2$	Factor the trinomial.
$x - 2 = \pm\sqrt{2}$	Find square roots.
$x = 2 \pm \sqrt{2}$	Solve for x.

Think

Is there a way to check without a calculator?
Yes; you can check that your solutions are reasonable by estimating.

Check your results on your calculator. Replace x in the original equation with $2 + \sqrt{2}$ and $2 - \sqrt{2}$.

Got It? 5. What is the solution of $2x^2 - x + 3 = x + 9$?

You can complete a square to change a quadratic function to vertex form.

Problem 6 Writing in Vertex Form

Plan

What should be your first step?
Complete the square.

What is $y = x^2 + 4x - 6$ in vertex form? Name the vertex and y-intercept.

$y = x^2 + 4x - 6$	
$y = x^2 + 4x + 2^2 - 6 - 2^2$	Add $\left(\frac{4}{2}\right)^2 = 2^2$ to complete the square. Also, subtract 2^2 to leave the function unchanged.
$y = (x + 2)^2 - 6 - 2^2$	Factor the perfect square trinomial.
$y = (x + 2)^2 - 10$	Simplify.

The vertex is $(-2, -10)$. The y-intercept is $(0, -6)$.

Check with a graphing calculator.

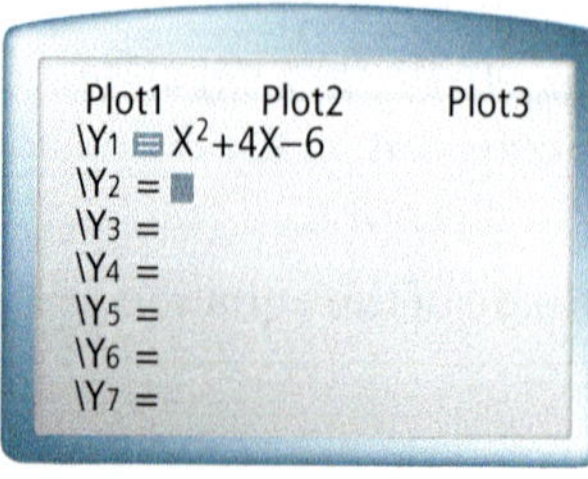

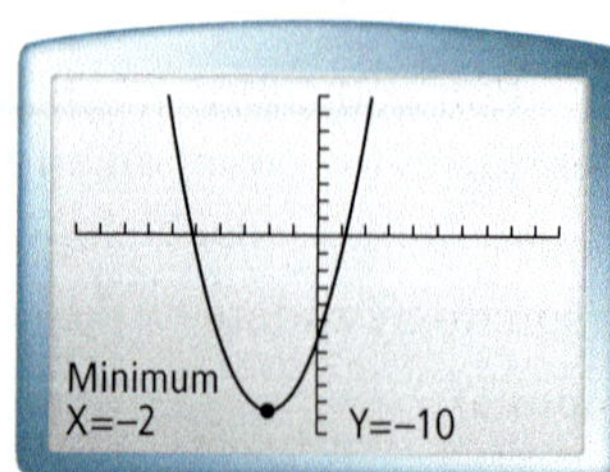

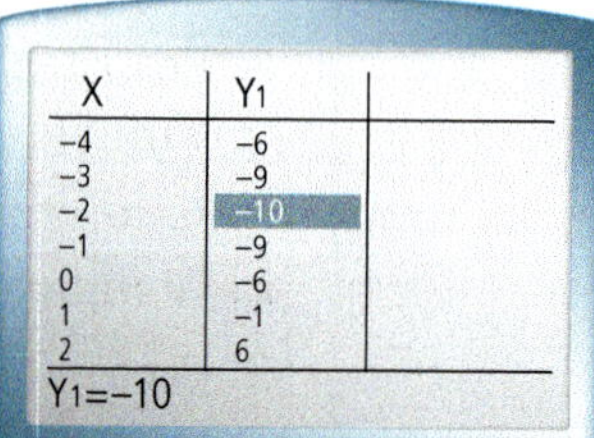

Got It? 6. What is $y = x^2 + 3x - 6$ in vertex form? Name the vertex and y-intercept.

Lesson Check

Do you know HOW?

Solve each equation by finding square roots.

1. $2x^2 = 72$

2. $6x^2 = 54$

Complete the square.

3. $x^2 + 2x + \blacksquare$

4. $x^2 + 10x + \blacksquare$

5. $x^2 - 4x + \blacksquare$

6. $x^2 + 12x + \blacksquare$

7. $x^2 + 100x + \blacksquare$

8. $x^2 - 32x + \blacksquare$

Do you UNDERSTAND?

9. Vocabulary Explain the process of completing the square.

10. How can you rewrite the equation $x^2 + 12x + 5 = 3$ so the left side of the equation is in the form $(x + a)^2$?

11. Error Analysis Your friend completed the square and wrote the expression shown. Explain your friend's error and write the expression correctly.

$x^2 - 14x + 36$
$x^2 - 14x + 49 + 36$
$(x - 7)^2 + 36$

Practice and Problem-Solving Exercises

Solve each equation by finding square roots.

See Problem 1.

12. $5x^2 = 80$

13. $x^2 - 4 = 0$

14. $2x^2 = 32$

15. $9x^2 = 25$

16. $3x^2 - 15 = 0$

17. $5x^2 - 40 = 0$

See Problem 2.

18. Fitness A rectangular swimming pool is 6 ft deep. One side of the pool is 2.5 times longer than the other. The amount of water needed to fill the swimming pool is 2160 cubic feet. Find the dimensions of the pool.

Solve each equation.

See Problem 3.

19. $x^2 + 6x + 9 = 1$

20. $x^2 - 4x + 4 = 100$

21. $x^2 - 2x + 1 = 4$

22. $x^2 + 8x + 16 = \frac{16}{9}$

23. $4x^2 + 4x + 1 = 49$

24. $x^2 - 12x + 36 = 25$

25. $25x^2 + 10x + 1 = 9$

26. $x^2 - 30x + 225 = 400$

27. $9x^2 + 24x + 16 = 36$

Complete the square.

See Problem 4.

28. $x^2 + 18x + \blacksquare$

29. $x^2 - x + \blacksquare$

30. $x^2 - 24x + \blacksquare$

31. $x^2 + 20x + \blacksquare$

32. $m^2 - 3m + \blacksquare$

33. $x^2 + 4x + \blacksquare$

Solve each quadratic equation by completing the square.

See Problem 5.

34. $x^2 + 6x - 3 = 0$

35. $x^2 - 12x + 7 = 0$

36. $x^2 + 4x + 2 = 0$

37. $x^2 - 2x = 5$

38. $x^2 + 8x = 11$

39. $x^2 + 12 = 10x$

40. $x^2 - 3x = x - 1$

41. $x^2 + 2 = 6x + 4$

42. $2x^2 + 2x - 5 = x^2$

43. $4x^2 + 10x - 3 = 0$

44. $9x^2 - 12x - 2 = 0$

45. $25x^2 + 30x = 12$

Rewrite each equation in vertex form.

See Problem 6.

46. $y = x^2 + 4x + 1$ **47.** $y = 2x^2 - 8x + 1$ **48.** $y = -x^2 - 2x + 3$

49. $y = x^2 + 4x - 7$ **50.** $y = 2x^2 - 6x - 1$ **51.** $y = -x^2 + 4x - 1$

52. Think About a Plan The area of the rectangle shown is 80 square inches. What is the value of x?

- How can you write an equation to represent 80 in terms of x?
- How can you find the value of x by completing the square?

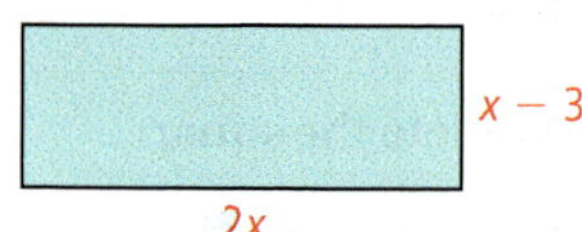

Find the value of k that would make the left side of each equation a perfect square trinomial.

53. $x^2 + kx + 25 = 0$ **54.** $x^2 - kx + 100 = 0$ **55.** $x^2 - kx + 121 = 0$

56. $x^2 + kx + 64 = 0$ **57.** $x^2 - kx + 81 = 0$ **58.** $25x^2 - kx + 1 = 0$

59. $x^2 + kx + \frac{1}{4} = 0$ **60.** $9x^2 - kx + 4 = 0$ **61.** $36x^2 - kx + 49 = 0$

62. Geometry The table shows some possible dimensions of rectangles with a perimeter of 100 units. Copy and complete the table.

a. Plot the points (width, area). Find a model for the data set.
b. What is another point in the data set? Use it to verify your model.
c. What is a reasonable domain for this function? Explain.
d. Find the maximum possible area. What dimensions yield this area?
e. Find a function for area in terms of width without using the table. Do you get the same model as in part (a)? Explain.

Width	Length	Area
1	49	49
2	48	■
3	■	■
4	■	■
5	■	■

Solve each quadratic equation by completing the square.

63. $x^2 + 5x - 3 = 0$ **64.** $x^2 + 3x = 2$

65. $x^2 - x = 5$ **66.** $x^2 + x - 1 = 0$

67. $3x^2 - 4x = 2$ **68.** $5x^2 - x = 4$

69. $x^2 + \frac{3}{4}x = \frac{1}{2}$ **70.** $2x^2 - \frac{1}{2}x = \frac{1}{8}$

71. $3x^2 + x = \frac{2}{3}$ **72.** $-x^2 + 2x + 4 = 0$

73. $-x^2 - 6x = 2$ **74.** $-0.25x^2 - 0.6x + 0.3 = 0$

75. Football The quadratic function $h = -0.01x^2 + 1.18x + 2$ models the height of a punted football. The horizontal distance in feet from the point of impact with the kicker's foot is x, and h is the height of the ball in feet.

a. Write the function in vertex form. What is the maximum height of the punt?
b. The nearest defensive player is 5 ft horizontally from the point of impact. How high must the player reach to block the punt?
c. Suppose the ball was not blocked but continued on its path. How far down the field would the ball go before it hit the ground?

Solve for x in terms of a.

76. $2x^2 - ax = 6a^2$ **77.** $3x^2 + ax = a^2$ **78.** $2a^2x^2 - 8ax = -6$

79. $4a^2x^2 + 8ax + 3 = 0$ **80.** $3x^2 + ax^2 = 9x + 9a$ **81.** $6a^2x^2 - 11ax = 10$

82. Solve $x^2 = (6\sqrt{2})x + 7$ by completing the square.

Rewrite each equation in vertex form. Then find the vertex of the graph.

83. $y = -4x^2 - 5x + 3$ **84.** $y = \frac{1}{2}x^2 - 5x + 12$ **85.** $y = -\frac{1}{5}x^2 + \frac{4}{5}x + \frac{11}{5}$

Standardized Test Prep

SAT/ACT

86. The graph of which inequality has its vertex at $\left(2\frac{1}{2}, -5\right)$?

(A) $y < |2x - 5| + 5$
(B) $y < |2x + 5| - 5$
(C) $y > |2x + 5| - 5$
(D) $y > |2x - 5| - 5$

87. Which number is a solution of $|9 - x| = 9 + x$?

(F) -3 (G) 0 (H) 3 (I) 6

88. Joanne tosses an apple seed on the ground. It travels along a parabola with the equation $y = -x^2 + 4$. Assume the seed was thrown from a height of 4 ft. How many feet away from Joanne will the apple seed land?

(A) 1 ft (B) 2 ft (C) 4 ft (D) 8 ft

Extended Response

89. List the steps for solving the equation $x^2 - 9 = -8x$ by the completing the square method. Explain each step.

Mixed Review

Solve each equation by factoring. Check your answers. See Lesson 4-5.

90. $2x^2 - 3x + 1 = 0$ **91.** $x^2 - 4 = -3x$ **92.** $16 + 22x = 3x^2$

Determine whether a quadratic model exists for each set of values. If so, write the model. See Lesson 4-3.

93. $(-4, 3), (-3, 3), (-2, 4)$ **94.** $\left(-1, \frac{1}{2}\right), (0, 2), (2, 2)$ **95.** $(0, 2), (1, 0), (2, 4)$

Solve each system by elimination. See Lesson 3-2.

96. $\begin{cases} 2x + y = 4 \\ 3x - y = 6 \end{cases}$ **97.** $\begin{cases} 2x + y = 7 \\ -2x + 5y = -1 \end{cases}$ **98.** $\begin{cases} 2x + 4y = 10 \\ 3x + 5y = 14 \end{cases}$

Get Ready! **To prepare for Lesson 4-7, do Exercises 99–100.**

Evaluate each expression for the given values of the variables. See Lesson 1-3.

99. $b^2 - 4ac$; $a = 1, b = 6, c = 3$ **100.** $b^2 - 4ac$; $a = -5, b = 2, c = 4$

The Quadratic Formula

Content Standard

Reviews A.REI.4.b Solve quadratic equations by . . . the quadratic formula . . .

Objectives To solve quadratic equations using the Quadratic Formula
To determine the number of solutions by using the discriminant

To make a smile restrict your domain to $2 \le x \le 6$.

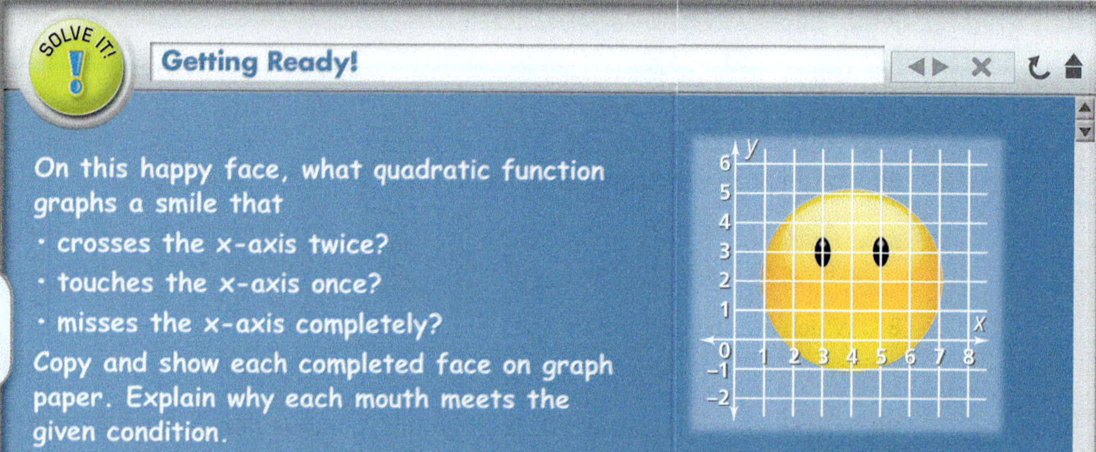

MATHEMATICAL PRACTICES

Dynamic Activity Roots of a Quadratic

Lesson Vocabulary
- Quadratic Formula
- discriminant

Another way to solve a quadratic equation $ax^2 + bx + c = 0$ is by completing a square and then factoring.

Essential Understanding You can solve a quadratic equation $ax^2 + bx + c = 0$ in more than one way. In general, you can find a formula that gives values of x in terms of a, b, and c.

Here's how to solve $ax^2 + bx + c = 0$ to get the *Quadratic Formula*.

$$ax^2 + bx + c = 0$$

$$x^2 + \frac{b}{a}x + \frac{c}{a} = 0 \qquad \text{Divide each side by } a.$$

$$x^2 + \frac{b}{a}x = -\frac{c}{a} \qquad \text{Rewrite so all terms containing } x \text{ are on one side.}$$

$$x^2 + \frac{b}{a}x + \left(\frac{b}{2a}\right)^2 = \left(\frac{b}{2a}\right)^2 - \frac{c}{a} \qquad \text{Complete the square.}$$

$$\left(x + \frac{b}{2a}\right)^2 = \frac{b^2 - 4ac}{4a^2} \qquad \text{Factor the perfect square trinomial. Also, simplify.}$$

$$x + \frac{b}{2a} = \pm\sqrt{\frac{b^2 - 4ac}{4a^2}} \qquad \text{Find square roots.}$$

$$x = -\frac{b}{2a} \pm \frac{\sqrt{b^2 - 4ac}}{2a} \qquad \text{Solve for } x. \text{ Also, simplify the radical.}$$

$$x = \frac{-b \pm \sqrt{b^2 - 4ac}}{2a} \qquad \text{Simplify.}$$

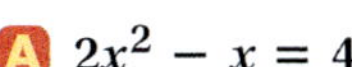

Key Concept The Quadratic Formula

To solve the quadratic equation $ax^2 + bx + c = 0$, use the **Quadratic Formula**.

$$x = \frac{-b \pm \sqrt{b^2 - 4ac}}{2a}$$

Problem 1 Using the Quadratic Formula

What are the solutions? Use the Quadratic Formula.

Plan

Should you write the equation in standard form?
Yes; write the equation in standard form to identify *a*, *b*, and *c*.

A $2x^2 - x = 4$

$2x^2 - x = 4$

$2x^2 - x - 4 = 0$ — Write in standard form.

$a = 2, b = -1, c = -4$ — Find the values of *a*, *b*, and *c*.

$x = \frac{-b \pm \sqrt{b^2 - 4ac}}{2a}$ — Write the Quadratic Formula.

$= \frac{-(-1) \pm \sqrt{(-1)^2 - 4(2)(-4)}}{2(2)}$ — Substitute for *a*, *b*, and *c*.

$= \frac{1 \pm \sqrt{33}}{4}$ — Simplify.

$= \frac{1 + \sqrt{33}}{4}$ or $\frac{1 - \sqrt{33}}{4}$

Check Use a graphing calculator to graph $y = 2x^2 - x - 4$. The x-intercepts are about $-1.186 \approx \frac{1 - \sqrt{33}}{4}$ and $1.686 \approx \frac{1 + \sqrt{33}}{4}$, as expected.

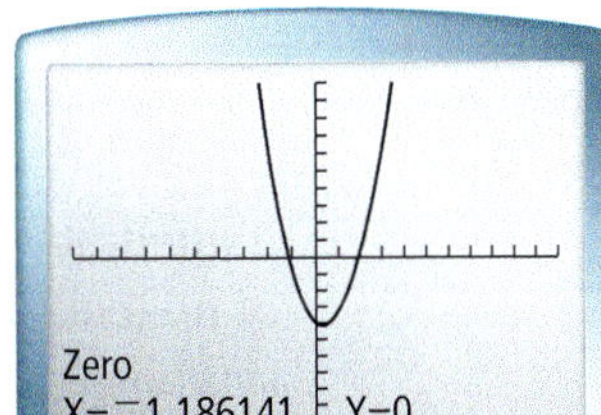

B $x^2 + 6x + 9 = 0$

$x^2 + 6x + 9 = 0$

$a = 1, b = 6, c = 9$ — Find the values of *a*, *b*, and *c*.

$x = \frac{-6 \pm \sqrt{6^2 - 4(1)(9)}}{2(1)}$ — Substitute into $\frac{-b \pm \sqrt{b^2 - 4ac}}{2a}$.

$= \frac{-6 \pm \sqrt{36 - 36}}{2}$ — Simplify.

$= \frac{-6 \pm \sqrt{0}}{2}$

$= -3$

Think

Why is there only one solution?
Because if you add or subtract zero you get the same number.

Got It? **1.** What are the solutions? Use the Quadratic Formula.

a. $x^2 + 4x = -4$ **b.** $x^2 + 4x - 3 = 0$

Problem 2 Applying the Quadratic Formula

GRIDDED RESPONSE

Fundraising **Your school's jazz band is selling CDs as a fundraiser. The total profit p depends on the amount x that your band charges for each CD. The equation $p = -x^2 + 48x - 300$ models the profit of the fundraiser. What is the least amount, in dollars, you can charge for a CD to make a profit of \$200?**

$p = -x^2 + 48x - 300$

$200 = -x^2 + 48x - 300$ Substitute 200 for p.

$0 = -x^2 + 48x - 500$ Write the equation in standard form.

$a = -1, b = 48, c = -500$ Find the values of a, b, and c.

$x = \frac{-48 \pm \sqrt{48^2 - 4(-1)(-500)}}{2(-1)}$ Substitute into $\frac{-b \pm \sqrt{b^2 - 4ac}}{2a}$.

$x = \frac{-48 \pm \sqrt{304}}{-2}$ Simplify.

$x \approx 15.282$ or $x \approx 32.717$ Use a calculator.

The least amount you can charge is \$15.29 for each CD to make a profit of \$200.

The answer is 15.29.

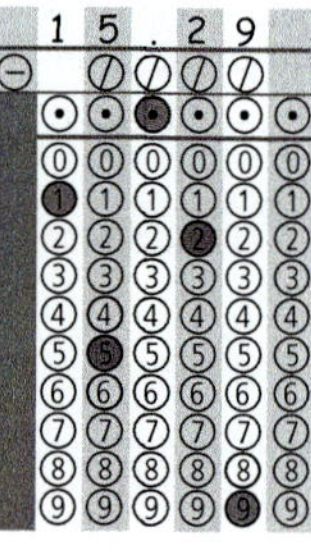

Think

Does it make sense that two different prices can yield the same profit?
Yes. You can generate a given profit either by selling many CDs at a low price, or fewer CDs at a high price.

  **Got It?** **2. a.** In Problem 2, what is the least amount you can charge for each CD to make a \$100 profit?

b. Reasoning Would a negative profit make sense in this problem? Explain.

A quadratic equation can have two real solutions ($x^2 = 4$), one real solution ($x^2 = 0$), or no real solutions ($x^2 = -4$). In the Quadratic Formula, the value under the radical symbol, $b^2 - 4ac$, tells you how many real-number solutions exist.

In Problem 1(a), $b^2 - 4ac > 0$ and there are two real solutions. In Problem 1(b), $b^2 - 4ac = 0$ and there is only one real solution.

take note

Key Concept Discriminant

The **discriminant** of a quadratic equation in the form $ax^2 + bx + c = 0$ is the value of the expression $b^2 - 4ac$.

$$x = \frac{-b \pm \sqrt{b^2 - 4ac}}{2a} \leftarrow \text{discriminant}$$

Discriminants and Solutions of Quadratic Equations		
Value of the Discriminant	**Number of Solutions for $ax^2 + bx + c = 0$**	**x-intercepts of Graph of Related Function $y = ax^2 + bx + c$**
$b^2 - 4ac > 0$	two real solutions	two x-intercepts
$b^2 - 4ac = 0$	one real solution	one x-intercept
$b^2 - 4ac < 0$	no real solutions	no x-intercepts

Problem 3 Using the Discriminant

What is the number of real solutions of $-2x^2 - 3x + 5 = 0$?

Think	Write
Find the values of a, b, and c.	$a = -2, b = -3, c = 5$
Evaluate $b^2 - 4ac$.	$b^2 - 4ac = (-3)^2 - 4(-2)(5)$ $= 49$
Interpret the discriminant.	The discriminant is positive. The equation has two real solutions.

Got It? 3. What is the number of real solutions of each equation?

a. $2x^2 - 3x + 7 = 0$ **b.** $x^2 = 6x + 5$

Problem 4 Using the Discriminant to Solve a Problem STEM

Projectile Motion **You hit a golf ball into the air from a height of 1 in. above the ground with an initial vertical velocity of 85 ft/s. The function $h = -16t^2 + 85t + \frac{1}{12}$ models the height, in feet, of the ball at time t, in seconds. Will the ball reach a height of 115 ft?**

Plan

What value should you substitute for h? You are trying to determine whether the ball will reach 115 ft. Replace h with 115.

$h = -16t^2 + 85t + \frac{1}{12}$

$115 = -16t^2 + 85t + \frac{1}{12}$ Substitute 115 for h.

$0 = -16t^2 + 85t - 114\frac{11}{12}$ Write the equation in standard form.

$a = -16, b = 85, c = -114\frac{11}{12}$ Find the values of a, b, and c.

$b^2 - 4ac = 85^2 - 4(-16)\left(-114\frac{11}{12}\right)$ Evaluate the discriminant.

$= 7225 - 7354\frac{2}{3}$ Simplify.

$= -129\frac{2}{3}$

The discriminant is negative. The equation $115 = -16t^2 + 85t + \frac{1}{12}$ has no real solutions. The golf ball will not reach a height of 115 feet.

  **Got It?** **4. Reasoning** Without solving an equation, will the golf ball in Problem 4 reach a height of 110 ft? Explain.

Lesson Check

Do you know HOW?

Solve each equation using the Quadratic Formula.

1. $x^2 - 5x - 7 = 0$

2. $x^2 + 3x - 13 = 0$

3. $2x^2 - 5x - 3 = 0$

4. $3x^2 - 4x + 3 = 0$

Find the discriminant of each quadratic equation. Determine the number of real solutions.

5. $-x^2 + 2x - 9 = 0$

6. $x^2 + 17x + 4 = 0$

7. $x^2 - 6x + 9 = 0$

Do you UNDERSTAND?

8. Reasoning For what values of k does the equation $x^2 + kx + 9 = 0$ have one real solution? two real solutions?

9. Error Analysis Your friend concluded that because two discriminants are equal, the solutions to the two equations are the same. Explain your friend's error. Give an example of two quadratic equations that disprove this conclusion.

10. Reasoning If one quadratic equation has a positive discriminant, and another quadratic equation has a discriminant equal to 0, can the two quadratic equations share a solution? Explain why or why not. If so, give two quadratic equations that meet this criterion.

Practice and Problem-Solving Exercises

Solve each equation using the Quadratic Formula.

See Problem 1.

11. $x^2 - 4x + 3 = 0$
12. $x^2 + 8x + 12 = 0$
13. $2x^2 + 5x = 7$
14. $3x^2 + 2x - 1 = 0$
15. $x^2 + 10x = -25$
16. $2x^2 - 5 = -3x$
17. $x^2 = 3x - 1$
18. $6x - 5 = -x^2$
19. $3x^2 = 2(2x + 1)$
20. $2x(x - 1) = 3$
21. $x(x - 5) = -4$
22. $12x + 9x^2 = 5$

See Problem 2.

23. **Fundraising** Your class is selling boxes of flower seeds as a fundraiser. The total profit p depends on the amount x that your class charges for each box of seeds. The equation $p = -0.5x^2 + 25x - 150$ models the profit of the fundraiser. What's the smallest amount, in dollars, that you can charge and make a profit of at least \$125?

24. **Baking** Your local bakery sells more bagels when it reduces prices, but then its profit changes. The function $y = -1000x^2 + 1100x - 2.5$ models the bakery's daily profit in dollars, from selling bagels, where x is the price of a bagel in dollars. What's the highest price the bakery can charge, in dollars, and make a profit of at least \$200?

Evaluate the discriminant for each equation. Determine the number of real solutions.

See Problem 3.

25. $x^2 + 4x + 5 = 0$
26. $x^2 - 4x - 5 = 0$
27. $-4x^2 + 20x - 25 = 0$
28. $-2x^2 + x - 28 = 0$
29. $2x^2 + 7x - 15 = 0$
30. $6x^2 - 2x + 5 = 0$
31. $-2x^2 + 7x = 6$
32. $x^2 - 12x + 36 = 0$
33. $x^2 + 8x = -16$
34. $3x^2 + x = -3$
35. $x + 2 = -3x^2$
36. $12x(x + 1) = -3$

See Problem 4.

37. **Business** The weekly revenue for a company is $r = -3p^2 + 60p + 1060$, where p is the price of the company's product. Use the discriminant to find whether there is a price for which the weekly revenue would be \$1500.

STEM 38. **Physics** The equation $h = 80t - 16t^2$ models the height h in feet reached in t seconds by an object propelled straight up from the ground at a speed of 80 ft/s. Use the discriminant to find whether the object will ever reach a height of 90 ft.

39. **Think About a Plan** The area of a rectangle is 36 in.^{2}. The perimeter of the rectangle is 36 in. What are the dimensions of the rectangle to the nearest hundredth of an inch?
    - How can you write an equation using one variable to find the dimensions of the rectangle?
    - How can the discriminant of the equation help you solve the problem?

40. **Writing** Summarize how to use the discriminant to analyze the types of solutions of a quadratic equation.

Solve each equation using any method. When necessary, round real solutions to the nearest hundredth.

41. $6x^2 - 5x - 1 = 0$

42. $7x^2 - x - 12 = 0$

43. $5x^2 + 8x - 11 = 0$

44. $4x^2 + 4x = 22$

45. $2x^2 - 1 = 5x$

46. $2x^2 + x = \frac{1}{2}$

47. $x^2 = 11x - 10$

48. $5x^2 = 210x$

49. $4x^2 + 4x = 3$

50. $2x^2 + 4x = 10$

51. $x^2 - 3x - 8 = 0$

52. $-3x^2 + 147 = 0$

53. $x^2 + 8x = 4$

54. $4x^2 - 4x - 3 = 0$

55. $x^2 = 11 - 6x$

STEM **56. Air Pollution** The function $y = 0.4409x^2 - 5.1724x + 99.0321$ models the emissions of carbon monoxide in the United States since 1987, where y represents the amount of carbon monoxide released in a year in millions of tons, and $x = 0$ represents the year 1987.

a. How can you use a graph to estimate the year in which more than 100 million tons of carbon monoxide were released into the air?

b. How can you use the Quadratic Formula to estimate the year in which more than 100 million tons of carbon monoxide were released into the air?

c. Which method do you prefer? Explain why.

57. Sports A diver dives from a 10 m springboard. The equation $f(t) = -4.9t^2 + 4t + 10$ models her height above the pool at time t in seconds. At what time does she enter the water?

Without graphing, determine how many x-intercepts each function has.

58. $y = -2x^2 + 3x - 1$

59. $y = 0.25x^2 + 2x + 4$

60. $y = x^2 + 3x + 5$

61. $y = -x^2 + 3x + 10$

62. $y = 3x^2 - 10x + 6$

63. $y = 10x^2 + 13x - 3$

64. $y = x^2 + 17x - 2$

65. $y = -5x^2 - 4x + 3$

66. $y = 7x^2 - 2x + 9$

67. Reasoning Determine the value(s) of k for which $3x^2 + kx + 12 = 0$ has each type of solution.

a. no real solutions

b. exactly one real solution

c. two real solutions

68. Use the discriminant to match each function with its graph.

a. $f(x) = x^2 - 4x + 2$

b. $f(x) = x^2 - 4x + 4$

c. $f(x) = x^2 - 4x + 6$

I.

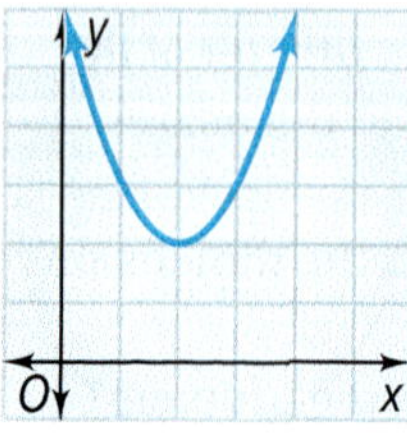

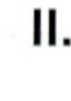

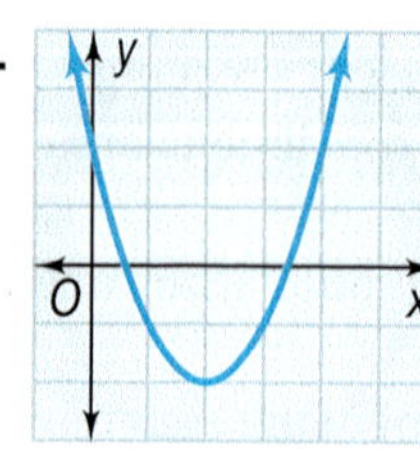

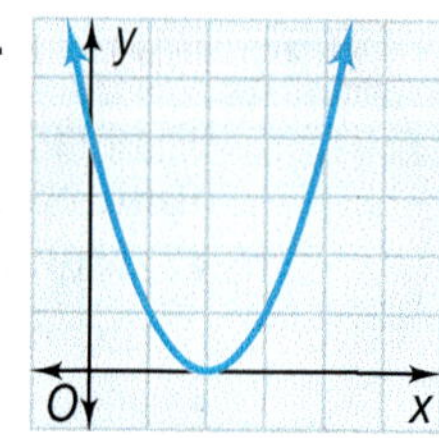

69. a. Geometry Write an equation to find the dimensions of a square that has the same area as a circle with a radius of 10 cm.

b. Find the length of a side of the square, to the nearest hundredth centimeter.

Write a quadratic equation with the given solutions.

70. $\frac{3+\sqrt{5}}{2}, \frac{3-\sqrt{5}}{2}$

71. $\frac{-5+\sqrt{13}}{2}, \frac{-5-\sqrt{13}}{2}$

72. $\frac{-5+\sqrt{17}}{4}, \frac{-5-\sqrt{17}}{4}$

Solve each equation.

73. $|5 - 2x^2| = 5$

74. $|x^2 - 4x| = 3$

75. $|x^2 + 4x + 3| = 8$

76. Use the Quadratic Formula to prove each statement.

a. The sum of the solutions of the quadratic equation $ax^2 + bx + c = 0$ is $-\frac{b}{a}$.

b. The product of the solutions of the quadratic equation $ax^2 + bx + c = 0$ is $\frac{c}{a}$.

77. Explain the meaning of the value $\frac{\sqrt{b^2 - 4ac}}{2a}$ in terms of the graph of the standard quadratic function $y = ax^2 + bx + c$.

Standardized Test Prep

GRIDDED RESPONSE

78. How many different real solutions are there for $2x^2 - 3x + 5 = 0$?

79. What is the y-value of the y-intercept of the quadratic function $y = 2(x + 2)^2 - 5$?

80. What is the x-value in the solution to the system $\begin{cases} 3x + y = -7 \\ 2x - 2y = -10 \end{cases}$?

81. The graph of the system of inequalities $\begin{cases} y \le \frac{1}{2}x + 3 \\ y \ge 6x - 30 \\ x \ge 0 \\ y \ge 0 \end{cases}$ is shown at the right.

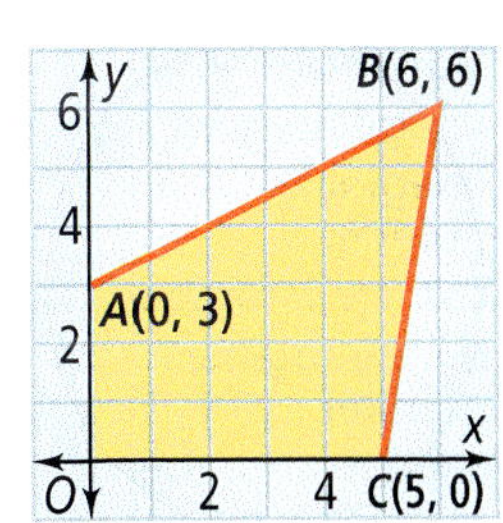

What is the maximum value of the function $P = 3x - 4y$ for the (x, y) pairs in the bounded region shown?

Mixed Review

Solve each equation by completing the square. See Lesson 4-6.

82. $x^2 - 8x - 20 = 0$

83. $2y^2 = 4y - 1$

84. $x^2 - 3x - 8 = 0$

Simplify by combining like terms. See Lesson 1-3.

85. $z^2 + 8z^2 - 2z + 5z$

86. $4k - x - 3k + 5x$

87. $4y - (2y + 3x) - 5x$

Get Ready! **To prepare for Lesson 4-8, do Exercises 88–90.**

See p. 981.

Simplify each expression.

88. $\sqrt{(-2)^2 + 8^2}$

89. $\sqrt{3^2 + 4^2}$

90. $\sqrt{5^2 + (-12)^2}$

4-8 Complex Numbers

Content Standards

N.CN.1 Know there is a complex number i such that $i^2 = -1$, and every complex number has the form $a + bi$ with a and b real.

Also N.CN.2, N.CN.7, N.CN.8

Objectives To identify, graph, and perform operations with complex numbers
To find complex number solutions of quadratic equations

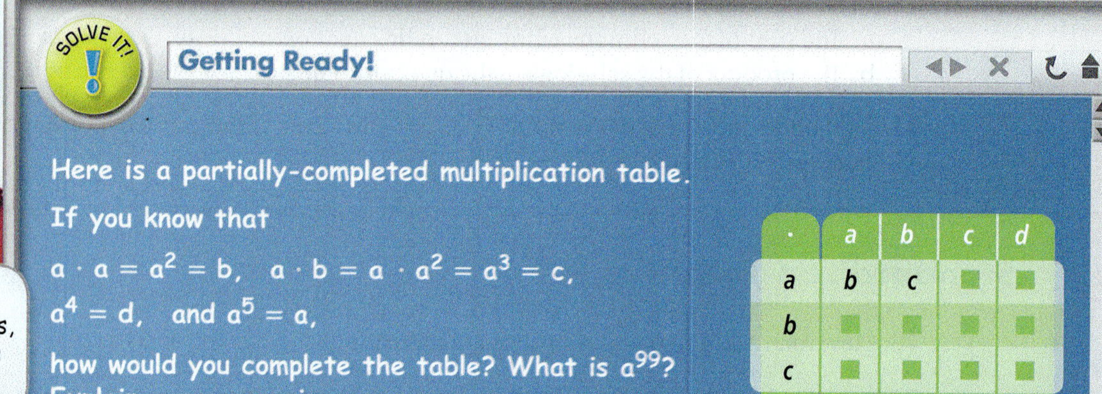

Once you find the pattern in the boxes, finding a^{99} becomes simple.

MATHEMATICAL PRACTICES

Lesson Vocabulary
- imaginary unit
- imaginary number
- complex number
- pure imaginary number
- complex number plane
- absolute value of a complex number
- complex conjugates

In Chapter 1, you learned about different subsets of real numbers. The set of real numbers is itself a subset of a larger set of numbers, the *complex numbers*. Curiously, the complex numbers include a number like a in the Solve It. Its fifth power is itself.

Essential Understanding The complex numbers are based on a number whose square is -1.

The **imaginary unit** i is the complex number whose square is -1. So, $i^2 = -1$, and $i = \sqrt{-1}$.

take note

Key Concept Square Root of a Negative Real Number

Algebra

For any positive number a,
$\sqrt{-a} = \sqrt{-1 \cdot a} = \sqrt{-1} \cdot \sqrt{a} = i\sqrt{a}$.

Example

$\sqrt{-5} = i\sqrt{5}$

Note that $(\sqrt{-5})^2 = (i\sqrt{5})^2 = i^2(\sqrt{5})^2 = -1 \cdot 5 = -5$ (not 5).

Problem 1 Simplifying a Number Using *i*

Think

Is $\sqrt{-18}$ a real number?
No. There is no real number that when multiplied by itself gives −18. You must use the imaginary unit *i* to write $\sqrt{-18}$.

How do you write $\sqrt{-18}$ by using the imaginary unit *i*?

$$\sqrt{-18} = \sqrt{-1 \cdot 18}$$

$= \sqrt{-1} \cdot \sqrt{18}$ Multiplication Property of Square Roots

$= i \cdot \sqrt{18}$ Definition of *i*

$= i \cdot 3\sqrt{2}$ Simplify.

$= 3i\sqrt{2}$

Got It? **1.** How do you write each number in parts (a)–(c) by using the imaginary unit *i*?

a. $\sqrt{-12}$ **b.** $\sqrt{-25}$ **c.** $\sqrt{-7}$

d. Reasoning Explain why $\sqrt{-64} \neq -\sqrt{64}$.

An **imaginary number** is any number of the form $a + bi$, where a and b are real numbers and $b \neq 0$. Imaginary numbers and real numbers together make up the set of *complex numbers.*

take note — Key Concept Complex Numbers

You can write a **complex number** in the form $a + bi$, where a and b are real numbers.

If $b = 0$, the number $a + bi$ is a real number.

If $a = 0$ and $b \neq 0$, the number $a + bi$ is a **pure imaginary number**.

a + bi
↑ Real part ↑ Imaginary part

Complex Numbers ($a + bi$)

- **Real Numbers** ($a + 0i$)
- **Imaginary Numbers** ($a + bi$, $b \neq 0$)
  - **Pure Imaginary Numbers** ($0 + bi$, $b \neq 0$)

In the **complex number plane**, the point (a, b) represents the complex number $a + bi$. To graph a complex number, locate the real part on the horizontal axis and the imaginary part on the vertical axis.

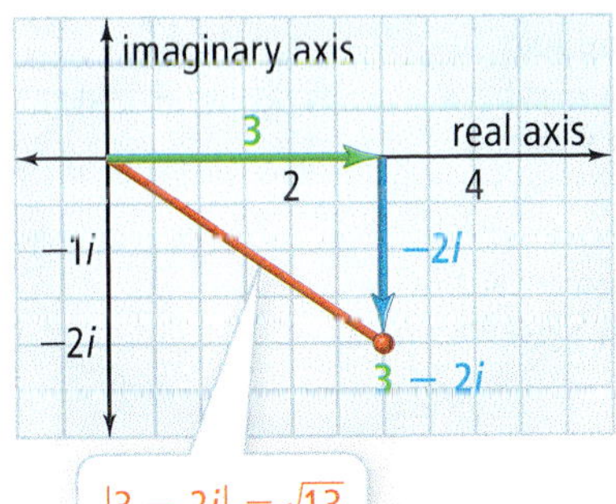

The **absolute value of a complex number** is its distance from the origin in the complex plane.

$$|a + bi| = \sqrt{a^2 + b^2}$$

Problem 2 Graphing in the Complex Number Plane

What are the graph and absolute value of each number?

A $-5 + 3i$

$$|-5 + 3i| = \sqrt{(-5)^2 + 3^2}$$
$$= \sqrt{34}$$

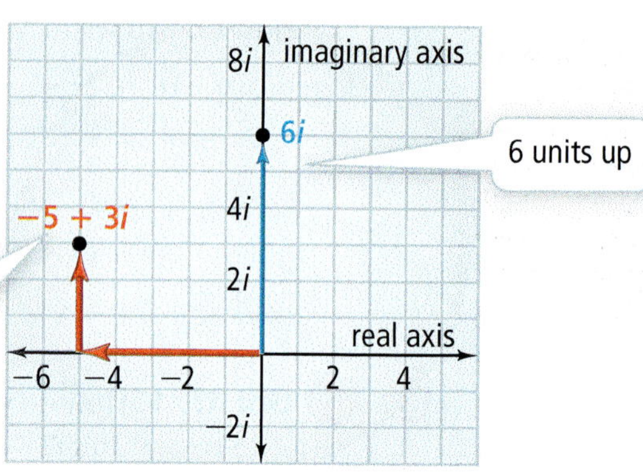

Think

Where is a pure imaginary number in the complex plane? The real part of a pure imaginary number is 0. The number must be on the imaginary axis.

B $6i$

$$|6i| = |0 + 6i|$$
$$= \sqrt{0^2 + 6^2}$$
$$= \sqrt{36}$$
$$= 6$$

Got It? **2.** What are the graph and absolute value of each number?

a. $5 - i$ **b.** $-3 - 2i$ **c.** $1 + 4i$

Essential Understanding You can define operations on the set of complex numbers so that when you restrict the operations to the subset of real numbers, you get the familiar operations on the real numbers.

To add or subtract complex numbers, combine the real parts and the imaginary parts separately. If the sum of two complex numbers is 0, or $0 + 0i$, then each number is the opposite, or additive inverse, of the other. The associative and commutative properties apply to complex numbers as well.

Problem 3 Adding and Subtracting Complex Numbers

What is each sum or difference?

Plan

How is adding complex numbers similar to adding algebraic expressions? Adding the real parts and imaginary parts separately is like adding like terms.

A $(4 - 3i) + (-4 + 3i)$

$4 + (-4) + (-3i) + 3i$ Use the commutative and associative properties.

$0 + 0 = 0$ $4 - 3i$ and $-4 + 3i$ are additive inverses.

B $(5 - 3i) - (-2 + 4i)$

$5 - 3i + 2 - 4i$ To subtract, add the opposite.

$5 + 2 - 3i - 4i$ Use the commutative and associative properties

$7 - 7i$ Simplify.

Got It? **3.** What is each sum or difference?

a. $(7 - 2i) + (-3 + i)$ **b.** $(1 + 5i) - (3 - 2i)$

c. $(8 + 6i) - (8 - 6i)$ **d.** $(-3 + 9i) + (3 + 9i)$

You multiply complex numbers $a + bi$ and $c + di$ as you would multiply binomials. For imaginary parts bi and di, $(bi)(di) = bd(i)^2 = bd(-1) = -bd$.

Problem 4 Multiplying Complex Numbers

What is each product?

A $(3i)(-5 + 2i)$

$-15i + 6i^2$	Distributive Property
$-15i + 6(-1)$	Substitute -1 for i^2.
$-6 - 15i$	Simplify.

Think
How do you multiply two binomials?
Multiply each term of one binomial by each term of the other binomial.

B $(4 + 3i)(-1 - 2i)$

$-4 - 8i - 3i - 6i^2$

$-4 - 8i - 3i - 6(-1)$

$2 - 11i$

Substitute -1 for i^2.

C $(-6 + i)(-6 - i)$

$36 + 6i - 6i - i^2$

$36 + 6i - 6i - (-1)$

37

Got It? **4.** What is each product?

a. $(7i)(3i)$ **b.** $(2 - 3i)(4 + 5i)$ **c.** $(-4 + 5i)(-4 - 5i)$

In Problem 4(c), the product is a real number. Number pairs of the form $a + bi$ and $a - bi$ are **complex conjugates**. The product of complex conjugates is a real number.

$$(a + bi)(a - bi) = a^2 - (bi)^2 = a^2 - b^2i^2 = a^2 - b^2(-1) = a^2 + b^2$$

You can use complex conjugates to simplify quotients of complex numbers.

Problem 5 Dividing Complex Numbers

What is each quotient?

Plan
What is the goal?
Write the quotient in the form $a + bi$.

A $\dfrac{9 + 12i}{3i}$

$\dfrac{9 + 12i}{3i} \cdot \dfrac{-3i}{-3i}$

$\dfrac{-27i - 36i^2}{-9i^2}$

$\dfrac{-27i - 36(-1)}{-9(-1)}$

$\dfrac{36 - 27i}{9}$

$4 - 3i$

Multiply numerator and denominator by the complex conjugate of the denominator.

Substitute -1 for i^2.

B $\dfrac{2 + 3i}{1 - 4i}$

$\dfrac{2 + 3i}{1 - 4i} \cdot \dfrac{1 + 4i}{1 + 4i}$

$\dfrac{2 + 8i + 3i + 12i^2}{1 + 4i - 4i - 16i^2}$

$\dfrac{2 + 8i + 3i + 12(-1)}{1 + 4i - 4i - 16(-1)}$

$\dfrac{-10 + 11i}{17}$

$-\dfrac{10}{17} + \dfrac{11}{17}i$

Got It? **5.** What is each quotient?

a. $\dfrac{5 - 2i}{3 + 4i}$ **b.** $\dfrac{4 - i}{6i}$ **c.** $\dfrac{8 - 7i}{8 + 7i}$

Problem 6 Factoring Using Complex Conjugates

Think

Is the expression factorable using real numbers?
No. Look for factors using complex numbers.

What is the factored form of $2x^2 + 32$?

$2x^2 + 32$	
$2(x^2 + 16)$	Factor out the GCF.
$2(x + 4i)(x - 4i)$	Use $a^2 + b^2 = (a + bi)(a - bi)$ to factor $(x^2 + 16)$.

Check

$2(x^2 + 4xi - 4xi - 16i^2)$	Multiply the binomials.
$2(x^2 - 16(-1))$	$i^2 = -1$
$2(x^2 + 16)$	Simplify within the binomial.
$2x^2 + 32$	Multiply.

Got It? **6.** What are the factored forms of each expression?

a. $5x^2 + 20$ **b.** $x^2 + 81$

Essential Understanding Every quadratic equation has complex number solutions (that sometimes are real numbers).

Problem 7 Finding Imaginary Solutions

What are the solutions of $2x^2 - 3x + 5 = 0$?

Think	Write
Use the Quadratic Formula with $a = 2$, $b = -3$, and $c = 5$.	$x = \dfrac{-b \pm \sqrt{b^2 - 4ac}}{2a}$
	$= \dfrac{-(-3) \pm \sqrt{(-3)^2 - 4(2)(5)}}{2(2)}$
	$= \dfrac{3 \pm \sqrt{9 - 40}}{4}$
Simplify.	$= \dfrac{3 \pm \sqrt{-31}}{4}$
	$= \dfrac{3}{4} \pm \dfrac{\sqrt{31}}{4}i$

Got It? **7.** What are the solutions of each equation?

a. $3x^2 - x + 2 = 0$ **b.** $x^2 - 4x + 5 = 0$

Lesson Check

Do you know HOW?

1. Simplify $\sqrt{-75}$ by using the imaginary number i.
2. Find the absolute value of $4 - 3i$.
3. Find the complex factors of $x^2 + 16$. Check your answers.

Simplify each expression.

4. $(4 - 2i) - (-3 + i)$
5. $(2 + i)(4 - 5i)$

Do you UNDERSTAND?

6. **Vocabulary** Explain the difference between the additive inverse of a complex number and a complex conjugate.
7. **Error Analysis** Describe and correct the error made in simplifying the expression $(4 - 7i)(4 + 7i)$.

$$(4 - 7i)(4 + 7i) = 16 + 28i - 28i + 49i^2$$
$$= 16 - 49$$
$$= -33$$

Practice and Problem-Solving Exercises

Simplify each number by using the imaginary number i.

See Problem 1.

8. $\sqrt{-4}$
9. $\sqrt{-7}$
10. $\sqrt{-15}$
11. $\sqrt{-81}$
12. $\sqrt{-50}$

Plot each complex number and find its absolute value.

See Problem 2.

13. $2i$
14. $5 + 12i$
15. $2 - 2i$
16. $1 - 4i$
17. $3 - 6i$

Simplify each expression.

See Problems 3 and 4.

18. $(2 + 4i) + (4 - i)$
19. $(-3 - 5i) + (4 - 2i)$
20. $(7 + 9i) + (-5i)$
21. $(12 + 5i) - (2 - i)$
22. $(-6 - 7i) - (1 + 3i)$
23. $(8 + i)(2 + 7i)$
24. $(-6 - 5i)(1 + 3i)$
25. $(-6i)^2$
26. $(9 + 4i)^2$

Write each quotient as a complex number.

See Problem 5.

27. $\frac{3 - 2i}{5i}$
28. $\frac{-2i}{1 + i}$
29. $\frac{4 - 3i}{-1 - 4i}$
30. $\frac{i + 2}{i - 2}$
31. $\frac{4}{2 - 3i}$
32. $\frac{3 + 2i}{(1 + i)^2}$

Find the factored forms of each expression. Check your answer.

See Problem 6.

33. $x^2 + 25$
34. $x^2 + 1$
35. $3s^2 + 75$
36. $x^2 + \frac{1}{4}$
37. $4b^2 + 1$
38. $-9x^2 - 100$

Find all solutions to each quadratic equation.

See Problem 7.

39. $x^2 + 2x + 3 = 0$
40. $-3x^2 + x - 3 = 0$
41. $2x^2 - 4x + 7 = 0$
42. $x^2 - 2x + 2 = 0$
43. $x^2 + 5 = 4x$
44. $2x(x - 3) = -5$

45. a. Name the complex number represented by each point on the graph at the right.
b. Find the additive inverse of each number.
c. Find the complex conjugate of each number.
d. Find the absolute value of each number.

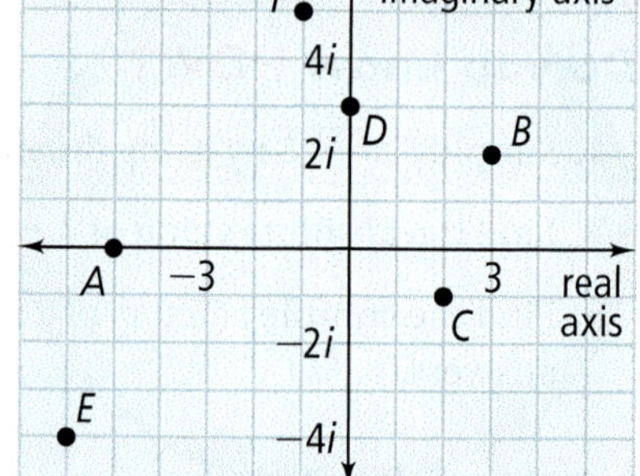

46. **Think About a Plan** In the complex number plane, what geometric figure describes the complex numbers with absolute value 10?
- What does the absolute value of a complex number represent?
- How can you use the complex number plane to solve this problem?

47. Solve $(x + 3i)(x - 3i) = 34$.

Simplify each expression.

48. $(8i)(4i)(-9i)$

49. $(2 + \sqrt{-1}) + (-3 + \sqrt{-16})$

50. $(4 + \sqrt{-9}) + (6 - \sqrt{-49})$

51. $(10 + \sqrt{-9}) - (2 + \sqrt{-25})$

52. $(8 - \sqrt{-1}) - (-3 + \sqrt{-16})$

53. $2i(5 - 3i)$

54. $-5(1 + 2i) + 3i(3 - 4i)$

55. $(3 + \sqrt{-4})(4 + \sqrt{-1})$

56. **Open-Ended** In the equation $x^2 - 6x + c = 0$, find values of c that will give:
a. two real solutions
b. two imaginary solutions
c. one real solution

57. A student wrote the numbers 1, 5, $1 + 3i$, and $4 + 3i$ to represent the vertices of a quadrilateral in the complex number plane. What type of quadrilateral has these vertices?

The multiplicative inverse of a complex number z is $\frac{1}{z}$ where $z \neq 0$. Find the multiplicative inverse, or reciprocal, of each complex number. Then use complex conjugates to simplify. Check each answer by multiplying it by the original number.

58. $2 + 5i$

59. $8 - 12i$

60. $a + bi$

Find the sum and product of the roots of each equation.

61. $x^2 - 2x + 3 = 0$

62. $5x^2 + 2x + 1 = 0$

63. $-2x^2 + 3x - 3 = 0$

For $ax^2 + bx + c = 0$, the sum of the roots is $-\frac{b}{a}$ and the product of the roots is $\frac{c}{a}$. Find a quadratic equation for each pair of roots. Assume $a = 1$.

64. $-6i$ and $6i$

65. $2 + 5i$ and $2 - 5i$

66. $4 - 3i$ and $4 + 3i$

Two complex numbers $a + bi$ and $c + di$ are equal when $a = c$ and $b = d$. Solve each equation for x and y.

67. $2x + 3yi = -14 + 9i$

68. $3x + 19i = 16 - 8yi$

69. $-14 - 3i = 2x + yi$

70. Show that the product of any complex number $a + bi$ and its complex conjugate is a real number.

71. For what real values of x and y is $(x + yi)^2$ an imaginary number?

72. **Reasoning** True or false: The conjugate of the additive inverse of a complex number is equal to the additive inverse of the conjugate of that complex number. Explain your answer.

Standardized Test Prep

SAT/ACT

73. How can you rewrite the expression $(8 - 5i)^2$ in the form $a + bi$?

(A) $39 + 80i$ (B) $39 - 80i$ (C) $69 + 80i$ (D) $69 - 80i$

74. How many solutions does the quadratic equation $4x^2 - 12x + 9 = 0$ have?

(F) two real solutions
(G) one real solution
(H) two imaginary solutions
(I) one imaginary solution

75. What are the solutions of $3x^2 - 2x - 4 = 0$?

(A) $\frac{1 \pm \sqrt{13}}{3}$ (B) $\frac{1 \pm i\sqrt{11}}{3}$ (C) $\frac{-1 \pm \sqrt{13}}{3}$ (D) $\frac{-1 \pm i\sqrt{11}}{3}$

Short Response

76. Using factoring, what are all four solutions to $x^4 - 16 = 0$? Show your work.

Mixed Review

Solve each equation using the Quadratic Formula. See Lesson 4-7.

77. $2x^2 + 3x - 4 = 0$

78. $4x^2 + x = 1$

79. $x^2 = -7x - 8$

Graph each function. Identify the axis of symmetry. See Lesson 4-1.

80. $y = -2(x + 1)^2 - 3$

81. $y = \frac{1}{2}(x - 4)^2 + 1$

82. $y = 3(x - 1)^2 - 5$

Write an equation for each line. See Lesson 2-3.

83. $m = 3$ and the y-intercept is -4

84. $m = -0.5$ and the y-intercept is -2

85. $m = -7$ and the y-intercept is 10

86. $m = 2$ and the y-intercept is 8

Get Ready! **To prepare for Lesson 4-9, do Exercises 87–89.** See Lesson 3-3.

Solve each system of inequalities by graphing.

87. $\begin{cases} y < 2x + 4 \\ y \geq |x - 3| + 2 \end{cases}$

88. $\begin{cases} y > -x \\ y < -|x + 1| \end{cases}$

89. $\begin{cases} y \leq |x| + 2 \\ y \geq -\frac{1}{2}x + 4 \end{cases}$

Concept Byte

For Use With Lesson 4-9

Quadratic Inequalities

Content Standard

A.APR.3 Identify zeros of polynomials when suitable factorizations are available, and use the zeros to construct a rough graph of the function defined by the polynomial.

To solve some quadratic inequalities, relate the quadratic expression to 0 and factor. To determine the sign of each factor, use what you know about multiplying positive and negative numbers.

Example 1

Solve each inequality algebraically.

a. $2x^2 - 14x < 0$

$2x(x - 7) < 0$	Factor.
$2x > 0$ and $(x - 7) < 0$, or $2x < 0$ and $(x - 7) > 0$	The product is negative, so the two factors must have *different* signs.
$x > 0$ and $x < 7$, or $x < 0$ and $x > 7$	Simplify.
$0 < x < 7$	No value can be both greater than 7 *and* less than 0.

b. $2x^2 - 14x > 0$

$2x(x - 7) > 0$	Factor.
$2x > 0$ and $(x - 7) > 0$, or $2x < 0$ and $(x - 7) < 0$	The product is positive, so the two factors must have the *same* signs.
$x > 0$ and $x > 7$, or $x < 0$ and $x < 7$	Simplify.
$x > 7$ or $x < 0$	A value that is greater than both 0 *and* 7 is always greater than 7. A value that is less than both 0 *and* 7 is always less than 0.

You can use a table to solve inequalities by analyzing the values of y around 0.

Activity

Use a table to find the solutions of $x^2 - 6x + 5 < 0$.

1. What happens to the value of y when $0 \le x \le 6$?
2. Does this make sense when you think of the shape of the graph of $y = x^2 - 6x + 5$? Explain.
3. What x-values in the table make the inequality $x^2 - 6x + 5 < 0$ true?
4. What are the solutions of $x^2 - 6x + 5 < 0$?

x	y
0	5
1	0
2	−3
3	−4
4	−3
5	0
6	5

You can determine the solution of a quadratic inequality based on how many times and where the graph of the related function crosses the x-axis. The graph could open upward or downward, and could intersect the x-axis at 0, 1, or 2 points.

You can solve inequalities of the form $ax^2 + bx + c > 0$ or $ax^2 + bx + c < 0$ by graphing the corresponding function and seeing where the graph is above or below the x-axis.

Example 2

Find the solution sets for $\frac{1}{4}(x - 2)^2 - 1 > 0$ and $\frac{1}{4}(x - 2)^2 - 1 < 0$.

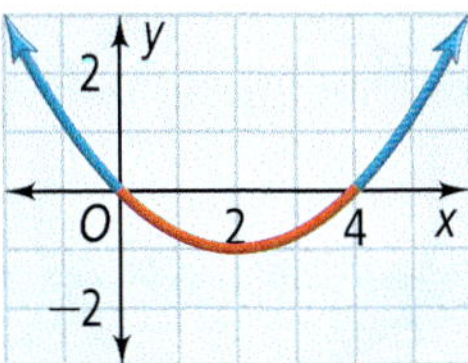

The solution set for $\frac{1}{4}(x - 2)^2 - 1 > 0$ is all x-values of points on the parabola that lie above the x-axis.

$x < 0$ or $x > 4$

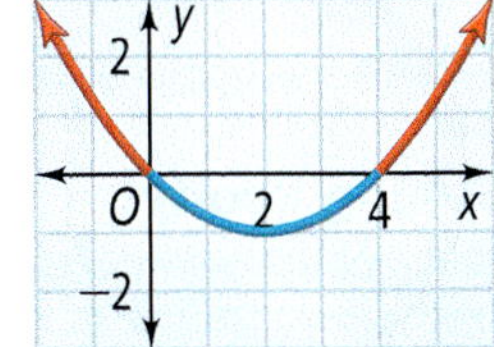

The solution set for $\frac{1}{4}(x - 2)^2 - 1 < 0$ is all x-values of points on the parabola that lie below the x-axis.

$0 < x < 4$

Example 3

Solve $-2x^2 - 8x - 6 < 0$.

Think: Since the coefficient of x^2 is less than zero, the graph of $y = -2x^2 - 8x - 6$ opens downward.

Solve: Find where $-2x^2 - 8x - 6$ equals 0.

$$-2x^2 - 8x - 6 = 0$$
$$-2(x^2 + 4x + 3) = 0$$
$$-2(x + 3)(x + 1) = 0$$
$$x = -3 \text{ or } x = -1$$

The graph of $y = -2x^2 - 8x - 6$ opens down and crosses the x-axis at $x = -3$ and $x = -1$. The solution of $-2x^2 - 8x - 6 < 0$ is $x < -3$ or $x > -1$.

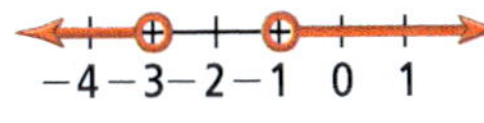

Exercises

5. Solve each inequality. Graph your solution on a number line.

   a. $x^2 < 36$ b. $x^2 - 9 > 0$ c. $x^2 < -4$ d. $x^2 - 3x - 18 > 0$

6. How can you use the graph of $y = 3x - 4$ to solve the linear inequality $3x - 4 < 0$? Graph the solution.

7. How can you solve the absolute value inequality $|-3x + 4| > 0$?

8. Example 2 shows two possible graphs for a quadratic inequality. What other possibilities are there?

9. Describe the graphs of possible solutions of $ax^3 + bx^2 + cx + d > 0$.

4-9 Quadratic Systems

Content Standards

A.CED.3 Represent constraints by equations or inequalities, and by systems of equations and/or inequalities, and interpret solutions as viable or nonviable options in a modeling context.

Also A.REI.7, A.REI.11

Objectives To solve and graph systems of linear and quadratic equations
To solve and graph systems of quadratic inequalities

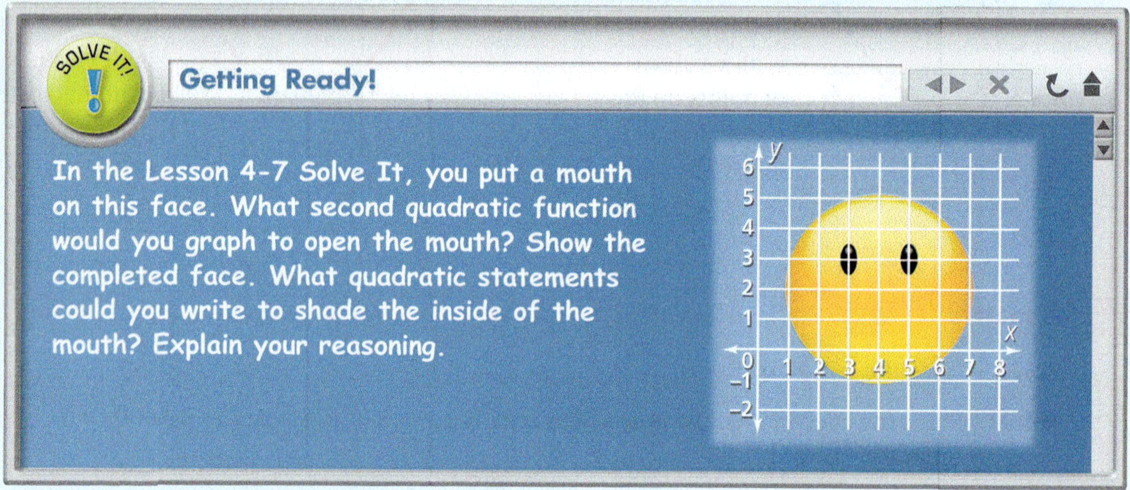

By drawing a second parabola in the Solve It, you created a quadratic system.

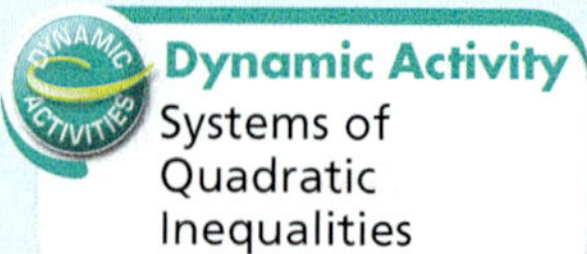

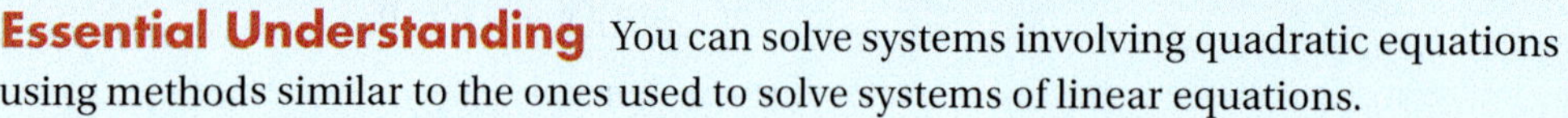

Essential Understanding You can solve systems involving quadratic equations using methods similar to the ones used to solve systems of linear equations.

The points where the graphs of the equations intersect represent the solutions of a system.

take note

Key Concept Solutions of a Linear-Quadratic System

A system of one quadratic equation and one linear equation can have two solutions, one solution, or no solution.

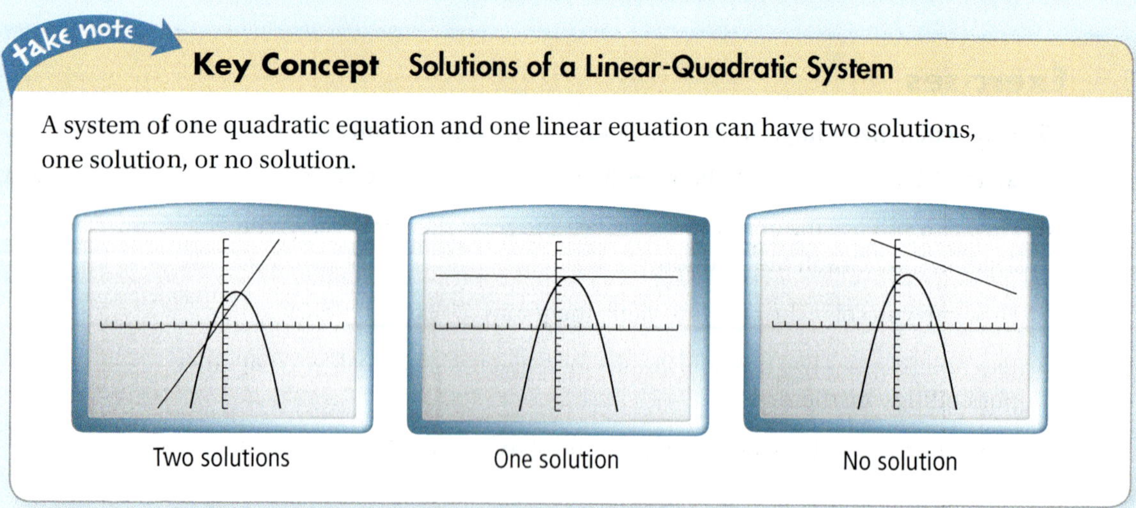

Problem 1 Solving a Linear-Quadratic System by Graphing

Multiple Choice Which numbers are y-values of the solutions of the system of equations? $\begin{cases} y = -x^2 + 5x + 6 \\ y = x + 6 \end{cases}$

Ⓐ 4 only　　Ⓑ 6 only　　Ⓒ 4 and 6　　Ⓓ 6 and 10

Plan

How can you graph these two equations? Use slope-intercept form to graph the linear equation. Make a table of values to graph the quadratic equation.

Graph the equations. Find their intersections.

The solutions appear to be (0, 6) and (4, 10).

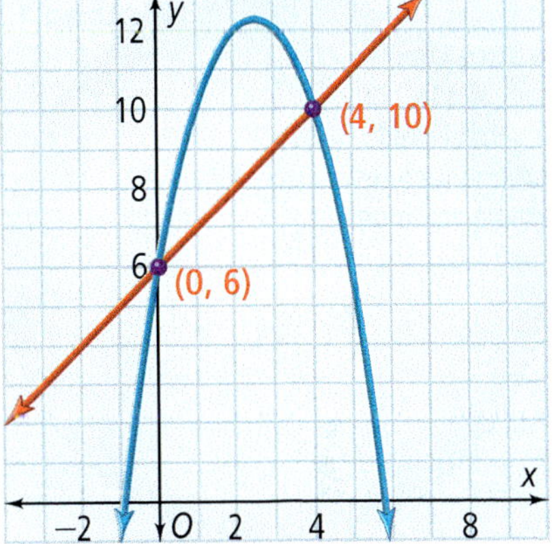

Check

$y = -x^2 + 5x + 6$	$y = x + 6$
$6 \stackrel{?}{=} -(0)^2 + 5(0) + 6$	$6 \stackrel{?}{=} 0 + 6$
$6 = 6$ ✔	$6 = 6$ ✔
$y = -x^2 + 5x + 6$	$y = x + 6$
$10 \stackrel{?}{=} -(4)^2 + 5(4) + 6$	$10 \stackrel{?}{=} 4 + 6$
$10 = 10$ ✔	$10 = 10$ ✔

The y-values of the solutions are 6 and 10, choice D.

Got It? 1. What is the solution of the system? $\begin{cases} y = x^2 + 6x + 9 \\ y = x + 3 \end{cases}$

Problem 2 Solving a Linear-Quadratic System Using Substitution

What is the solution of the system of equations? $\begin{cases} y = -x^2 - x + 6 \\ y = x + 3 \end{cases}$

Think	Write
Substitute $x + 3$ for y in the quadratic equation.	$x + 3 = -x^2 - x + 6$
Write in standard form.	$x^2 + 2x - 3 = 0$
Factor. Solve for x.	$(x - 1)(x + 3) = 0$ $x = 1$ or $x = -3$
Substitute for x in $y = x + 3$.	$x = 1 \rightarrow y = 1 + 3 = 4$ $x = -3 \rightarrow y = -3 + 3 = 0$ The solutions are (1, 4) and (−3, 0).

Got It? 2. What is the solution of the system? $\begin{cases} y = -x^2 - 3x + 10 \\ y = x + 5 \end{cases}$

You can solve quadratic–quadratic systems using the same methods you used for linear–quadratic systems.

Problem 3 Solving a Quadratic System of Equations

What is the solution of the system? $\begin{cases} y = -x^2 - x + 12 \\ y = x^2 + 7x + 12 \end{cases}$

Plan

Which variable should you substitute for?
You can substitute for either variable, but substituting for *y* results in a simple equation.

Method 1 Use substitution.

Substitute $y = -x^2 - x + 12$ for y in the second equation. Solve for x.

$-x^2 - x + 12 = x^2 + 7x + 12$ Substitute for *y*.

$-2x^2 - 8x = 0$ Write in standard form.

$-2x(x + 4) = 0$ Factor.

$x = 0$ or $x = -4$ Solve for *x*.

Substitute each value of x into either equation. Solve for y.

$y = x^2 + 7x + 12$	$y = x^2 + 7x + 12$
$y = (0)^2 + 7(0) + 12$	$y = (-4)^2 + 7(-4) + 12$
$y = 0 + 0 + 12 = 12$	$y = 16 - 28 + 12 = 0$

The solutions are (0, 12) and (−4, 0).

Method 2 Graph the equations.

Use a graphing calculator.
Define functions $\mathbf{Y_1}$ and $\mathbf{Y_2}$.

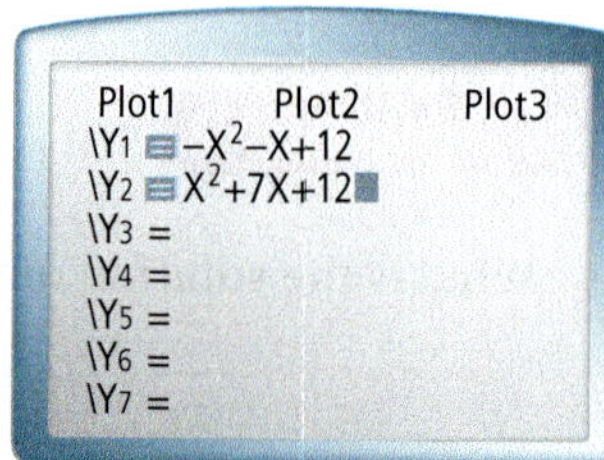

Use the **INTERSECT** feature to find the points of intersection.

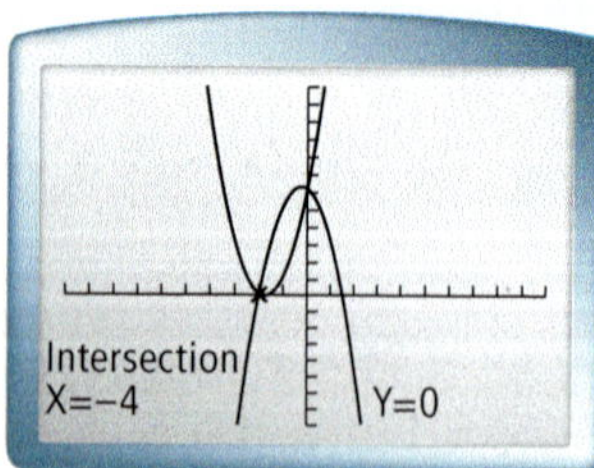

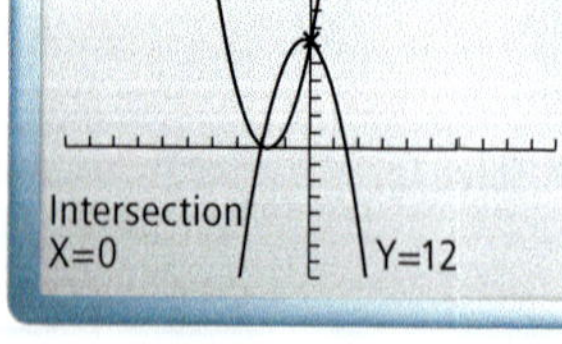

The solutions are (−4, 0) and (0, 12).

Got It? 3. What is the solution of each system of equations?

a. $\begin{cases} y = x^2 - 4x + 5 \\ y = -x^2 + 5 \end{cases}$ **b.** $\begin{cases} y = x^2 - 4x + 5 \\ y = -x^2 - 5 \end{cases}$

You can use the techniques for solving a linear system of inequalities to solve a quadratic system of inequalities.

Problem 4 Solving a Quadratic System of Inequalities

Plan

How can you find the solution?
Graph each inequality and find the region where the graphs overlap.

What is the solution of the system of inequalities? $\begin{cases} y < -x^2 - 9x - 2 \\ y > x^2 - 2 \end{cases}$

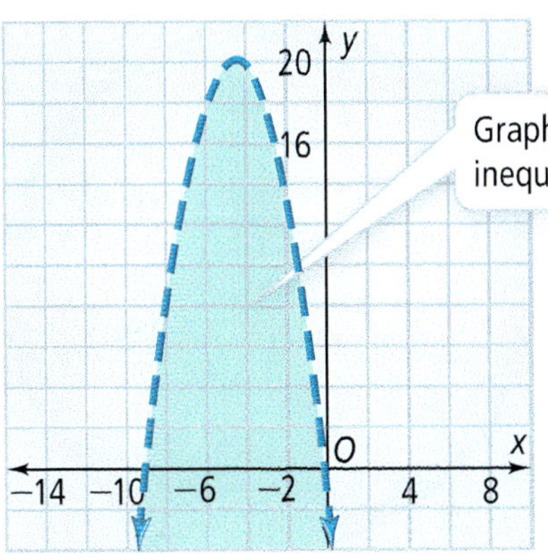

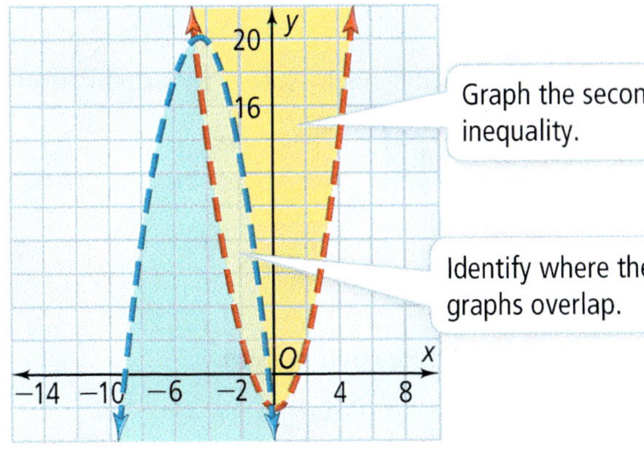

The solution of this system is the region where the graphs overlap. The region contains no boundary points.

Got It? 4. a. What is the solution of this system of inequalities? $\begin{cases} y \leq -x^2 - 4x + 3 \\ y > x^2 + 3 \end{cases}$

b. Reasoning How many solutions can a system of inequalities have?

Lesson Check

Do you know HOW?

Solve each system by substitution.

1. $\begin{cases} y = x^2 - 2x + 3 \\ y = x + 1 \end{cases}$

2. $\begin{cases} y = 2x^2 - 5x + 2 \\ y = x - 2 \end{cases}$

3. $\begin{cases} y = x^2 - 3x - 3 \\ y = -2x^2 - x + 5 \end{cases}$

Solve each system by graphing.

4. $\begin{cases} y > 2x^2 + x + 3 \\ y < -x^2 - 4x + 1 \end{cases}$

5. $\begin{cases} y > -3x^2 - 6x + 1 \\ y < -2x^2 - 3x + 5 \end{cases}$

Do you UNDERSTAND?

6. **Compare and Contrast** How are solving systems of two linear equations or inequalities and solving systems of two quadratic equations or inequalities alike? How are they different?

7. **Reasoning** How many points of intersection can graphs of the following types of functions have? Draw graphs to justify your answers.
   a. a linear function and a quadratic function
   b. two quadratic functions
   c. a quadratic function and an absolute value function (*Hint:* Graph $y = x^2$ and $y = |x|$ together. Can you transform one of the graphs slightly to increase the number of intersections?)

Practice and Problem-Solving Exercises

Solve each system by graphing. Check your answers.

See Problem 1.

8. $\begin{cases} y = -x^2 + 2x + 1 \\ y = 2x + 1 \end{cases}$

9. $\begin{cases} y = x^2 - 2x + 1 \\ y = 2x + 1 \end{cases}$

10. $\begin{cases} y = x^2 - x + 3 \\ y = -2x + 5 \end{cases}$

11. $\begin{cases} y = 2x^2 + 3x + 1 \\ y = -2x + 1 \end{cases}$

12. $\begin{cases} y = -x^2 - 3x + 2 \\ y = x + 6 \end{cases}$

13. $\begin{cases} y = -x^2 - 2x - 2 \\ y = x - 4 \end{cases}$

Solve each system by substitution. Check your answers.

See Problem 2.

14. $\begin{cases} y = x^2 + 4x + 1 \\ y = x + 1 \end{cases}$

15. $\begin{cases} y = -x^2 + 2x + 10 \\ y = x + 4 \end{cases}$

16. $\begin{cases} y = -x^2 + x - 1 \\ y = -x - 1 \end{cases}$

17. $\begin{cases} y = 2x^2 - 3x - 1 \\ y = x - 3 \end{cases}$

18. $\begin{cases} y = x^2 - 3x - 20 \\ y = -x - 5 \end{cases}$

19. $\begin{cases} y = -x^2 - 5x - 1 \\ y = x + 2 \end{cases}$

Solve each system.

See Problem 3.

20. $\begin{cases} y = x^2 + 5x + 1 \\ y = x^2 + 2x + 1 \end{cases}$

21. $\begin{cases} y = x^2 - 2x - 1 \\ y = -x^2 - 2x - 1 \end{cases}$

22. $\begin{cases} y = -x^2 - 3x - 2 \\ y = x^2 + 3x + 2 \end{cases}$

23. $\begin{cases} y = -x^2 - x - 3 \\ y = 2x^2 - 2x - 3 \end{cases}$

24. $\begin{cases} y = -3x^2 - x + 2 \\ y = x^2 + 2x + 1 \end{cases}$

25. $\begin{cases} y = x^2 + 2x + 1 \\ y = x^2 + 2x - 1 \end{cases}$

Solve each system by graphing.

See Problem 4.

26. $\begin{cases} y > x^2 + 2x \\ y > x^2 - 1 \end{cases}$

27. $\begin{cases} y > x^2 - 3x \\ y < 2x^2 - 3x \end{cases}$

28. $\begin{cases} y < -x^2 - 3x \\ y > x^2 - 1 \end{cases}$

29. **Think About a Plan** A manufacturer is making cardboard boxes by cutting out four equal squares from the corners of the rectangular piece of cardboard and then folding the remaining part into a box. The length of the cardboard piece is 1 in. longer than its width. The manufacturer can cut out either 3×3 in. squares, or 4×4 in. squares. Find the dimensions of the cardboard for which the volume of the boxes produced by both methods will be the same.
   - How can you represent the volume of the box using one variable?
   - What system of equations can you write?
   - Which method can you use to solve the system?

30. **Open-Ended** Can you solve the system of equations shown by graphing? Justify your answer. Can you solve this system using another method? If so, solve the system and explain why you chose that method. $\begin{cases} x = y^2 + 2y + 1 \\ y = x - 4 \end{cases}$

Solve each system by substitution.

31. $\begin{cases} x + y = 3 \\ y = x^2 - 8x - 9 \end{cases}$

32. $\begin{cases} y - 2x = x + 5 \\ y + 1 = x^2 + 5x + 3 \end{cases}$

33. $\begin{cases} y - \frac{1}{2}x^2 = 1 + 3x \\ y + \frac{1}{2}x^2 = x \end{cases}$

34. $\begin{cases} x + y - 2 = 0 \\ x^2 + y - 8 = 0 \end{cases}$

35. $\begin{cases} x^2 - y = x + 4 \\ x - 1 = y + 3 \end{cases}$

36. $\begin{cases} 2y = y - x^2 + 1 \\ y = x^2 - 5x - 2 \end{cases}$

Graph the solution to each set of inequalities.

37. $\begin{cases} y < -3x^2 + 1 \\ y > x^2 - x - 5 \end{cases}$

38. $\begin{cases} y < 3x^2 + 2x + 1 \\ y > 2x^2 - x + 1 \end{cases}$

39. $\begin{cases} y > x^2 - 5x + 4 \\ y > x^2 + 3x + 2 \end{cases}$

40. Error Analysis A classmate graphed the system of inequalities and concluded that because the shaded regions do not intersect, there are no solutions to the system. Describe and correct the error.

$\begin{cases} y \le x^2 - 4x + 6 \\ y \ge x^2 - 4x + 2 \end{cases}$

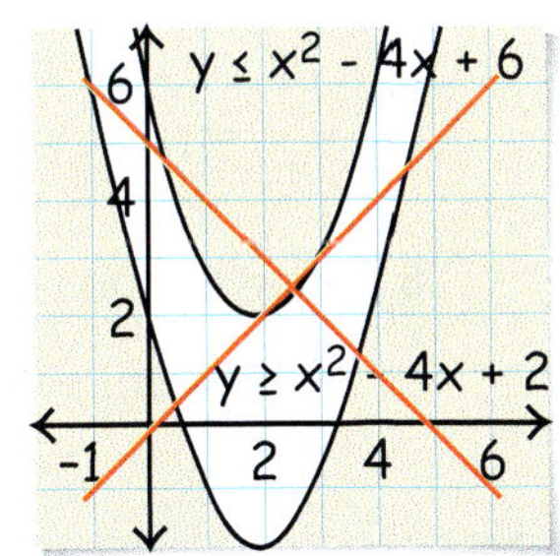

Solve each system.

41. $\begin{cases} y = 3x^2 - 2x - 1 \\ y = x - 1 \end{cases}$

42. $\begin{cases} y = -x^2 + x - 5 \\ y = x - 5 \end{cases}$

43. $\begin{cases} y = x^2 - 3x - 2 \\ y = 4x + 28 \end{cases}$

44. $\begin{cases} y = \frac{1}{2}x^2 + 4x + 4 \\ y = -4x + 12\frac{1}{2} \end{cases}$

45. $\begin{cases} y = -\frac{3}{4}x^2 - 4x \\ y = 3x + 8 \end{cases}$

46. $\begin{cases} y = -\frac{1}{4}x^2 + x + 1 \\ y = x - 4 \end{cases}$

47. Business A company's weekly revenue R is given by the formula $R = -p^2 + 30p$, where p is the price of the company's product. The company is considering hiring a distributor, which will cost the company $4p + 25$ per week.

a. Use a system of equations to find the values of the price p for which the product will still remain profitable if they hire this distributor.

b. Which value of p will maximize the profit after including the distributor cost?

Solve each system.

48. $\begin{cases} y = 5x^2 + 9x + 4 \\ y = -5x + 3 \end{cases}$

49. $\begin{cases} y = -7x^2 - 9x + 6 \\ y = \frac{1}{2}x + 11 \end{cases}$

50. $\begin{cases} y = x^2 + 3x + 6 \\ y = -x + 2 \end{cases}$

51. $\begin{cases} y = -4x^2 + 7x + 1 \\ y = 3x + 2 \end{cases}$

52. Reasoning Sketch the graphs of $y = 2x^2 + 4x - 5$ and $y = x^2 + 2x - 3$. Change these equations into inequalities so the system has solutions that comprise:

a. two non-overlapping regions

b. one bounded region

Solve the systems by graphing. For each system indicate one point in the solution set.

53. $\begin{cases} y < x^2 - 1 \\ y > 3x^2 - 3 \end{cases}$

54. $\begin{cases} y > x^2 \\ y < -x^2 + 1 \end{cases}$

55. $\begin{cases} y > (x - 3)^2 + 4 \\ y < -(x - 3)^2 + 5 \end{cases}$

56. Find a value of a for which the line $y = x + a$ separates the parabolas $y = x^2 - 3x + 2$ and $y = -x^2 + 8x - 15$.

Determine whether the following systems *always*, *sometimes*, or *never* have solutions. (Assume that different letters refer to unequal constants.) Explain.

57. $\begin{cases} y = x^2 + c \\ y = x^2 + d \end{cases}$

58. $\begin{cases} y = ax^2 + c \\ y = bx^2 + c \end{cases}$

59. $\begin{cases} y = (x + a)^2 \\ y = (x + b)^2 \end{cases}$

60. $\begin{cases} y = a(x + m)^2 + c \\ y = b(x + n)^2 + d \end{cases}$

61. Find the side of the square with vertical and horizontal sides inscribed in the region representing the solution of the system $\begin{cases} y \leq -x^2 + 1 \\ y \geq x^2 - 1 \end{cases}$.

Standardized Test Prep

SAT/ACT

62. How many solutions does the system have? $\begin{cases} y = -\frac{1}{4}x^2 - 2x \\ y = x^2 + \frac{3}{4} \end{cases}$

(A) 0 (B) 1 (C) 2 (D) 3

63. Which expression is equivalent to $(-3 + 2i)(2 - 3i)$?

(F) $13i$ (G) 12 (H) $12 + 13i$ (I) -12

64. Which expression is equivalent to $(2 - 7i) \div (2i)^3$?

(A) $\frac{7}{8} - \frac{1}{4}i$ (B) $\frac{1}{4} - \frac{7}{8}i$ (C) $\frac{7}{8} + \frac{1}{4}i$ (D) $\frac{1}{4} + \frac{7}{8}i$

Short Response

65. Solve the equation $-3x^2 + 5x + 4 = 0$. Show your work.

Mixed Review

Find the sum or difference. See Lesson 4-8.

66. $(1 - i) + (-5 + 4i)$ **67.** $(3 + 4i) - (-4 - 3i)$ **68.** $(1 + i) + (2 + 2i)$

Solve each equation using the Quadratic Formula. See Lesson 4-7.

69. $2m^2 + 5m + 3 = 0$ **70.** $p^2 - 4p + 3 = 0$ **71.** $25x^2 - 30x + 9 = 0$

Rewrite each equation in vertex form. See Lesson 4-6.

72. $y = -k^2 + 4k + 6$ **73.** $y = x^2 + 6x + 1$ **74.** $y = 2n^2 - 8n - 3$

Get Ready! **To prepare for Lesson 5-1, do Exercises 75–77.**

Simplify by combining like terms. See Lesson 1-3.

75. $3q + 9q - q$ **76.** $-2ab^2 + 2a^2b + 3ab^2$ **77.** $-4y^2 + 2y + 3y^2$

Concept Byte

For Use With Lesson 4-9

EXTENSION

Powers of Complex Numbers

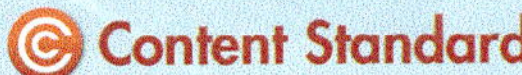

Extends N.CN.2 Use the relation $i^2 = -1$ and the commutative, associative, and distributive properties to add, subtract, and multiply . . .

You can use the rules for multiplying complex numbers to find powers of complex numbers.

Example 1

Compute and graph $(2i)^n$, for $n = 0, 1, 2,$ and 3.

n	$(2i)^n$
0	$(2i)^0 = 1$
1	$(2i)^1 = 2i$
2	$(2i)^2 = 4i^2 = 4(-1) = -4$
3	$(2i)^3 = 8i^3 = 8(i^2 \cdot i) = 8(-1 \cdot i) = -8i$

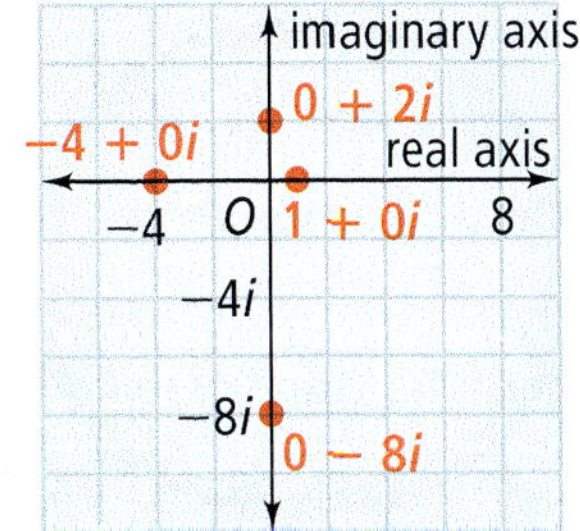

Example 2

Compute and graph $(2 - 3i)^n$, for $n = 0, 1, 2,$ and 3.

n	$(2 - 3i)^n$
0	$(2 - 3i)^0 = 1$
1	$(2 - 3i)^1 = 2 - 3i$
2	$(2 - 3i)^2 = 4 - 6i - 6i + 9i^2 = 4 - 12i + 9(-1) = -5 - 12i$
3	$(2 - 3i)^3 = -10 - 24i + 15i + 36i^2 = -10 - 9i + 36(-1) = -46 - 9i$

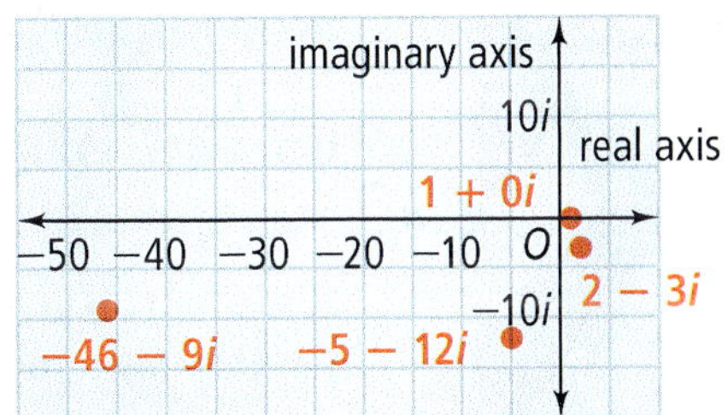

Exercises

1. Based on the graph in Example 1, predict the location of $(2i)^5$.
2. Compute and graph $(-3i)^n$ for $n = 0, 1, 2,$ and 3.
3. a. Connect the points from the graph in Example 1 with a smooth curve. Estimate $(2i)^{\frac{1}{2}}$.
   b. Use a graphing calculator to compute $(2i)^{\frac{1}{2}}$. Does it fall on the curve? Was it close to your estimate?
4. Use a graphing calculator to find values of $(2 - 3i)^n$ for $n = 0.5, 1.5,$ and 2.5. Copy the graph and add these points.
5. Compute and graph $(3 - 4i)^n$ for $n = 0, 1, 2,$ and 3.

Pull It All Together 

BIG idea Equivalence and Function

The parameters a, b, c, h, and k in the standard and vertex forms of a quadratic function give information on how the graph of the function relates to the graph of the parent function $y = x^2$.

Standard form: $y = ax^2 + bx + c$ Vertex form: $y = a(x - h)^2 + k$

Performance Task 1

Refer to the two forms shown above.

a. What information do the parameters, or combinations of parameters, provide about the graph of the quadratic function?

b. Begin with standard form. Transform it to vertex form. What are the values of h and k in terms of a, b, and c?

c. Show how the Quadratic Formula follows from your result in part (b). *Hint:* Set the expression in your vertex form equal to 0. Then solve by factoring.

BIG idea Solving Equations and Inequalities

A problem may require different types of equation solving. You should know when and how to use a graphing calculator to help you with your work.

Performance Task 2

You shoot an arrow at a target. The parabolic path of your arrow passes through the points shown in the table. Answer parts (a) – (d) below. Justify your answers.

x	y
30	6
60	7
100	4

a. Find a quadratic function in standard form that models the path of your arrow. *Hint:* The three points are (x, y)-values that satisfy $y = ax^2 + bx + c$.

b. If the y-value represents height above the ground, for what value of x would your arrow hit the ground if you miss the target?

c. If the target bull's-eye is at $x = 100$, at what height should the bull's-eye be for your arrow to hit it?

d. If the target bull's-eye is at height $y = 2.98$, at what value of x should the bull's-eye be for the arrow to hit it?

4 Chapter Review

Connecting BIG ideas and Answering the Essential Questions

1 Equivalence
Vertex form of a quadratic function shows the vertex of the parabola. Standard form is "calculator ready." Both forms give additional information.

The Different Forms of a Quadratic Function (Lessons 4-1 and 4-2)

$y = 2(x-1)^2 + 3$ has vertex (1, 3) and opens upward ($2 > 0$).

$y = -2x^2 + 4x + 1$ has vertex with x-coordinate $-\frac{4}{2(-2)} = 1$ and opens downward ($-2 < 0$).

Each has axis of symmetry $x = 1$.
Each is a stretch of $y = x^2$ by the factor 2.

Modeling With Quadratics (Lessons 4-3 and 4-9)

$y = -16x^2 + 12x + 4$ can model the height y in feet reached by the coin tossed by the referee before the game. x represents time in seconds.

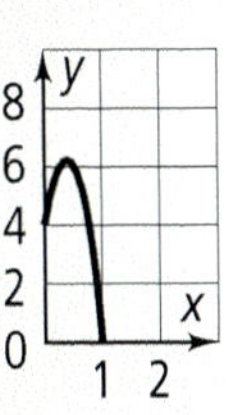

2 Function
Any quadratic function is possibly a stretch or compression, a reflection, and a translation of $y = x^2$.

3 Solving Equations and Inequalities
The real solutions of a quadratic equation show the zeros of the related quadratic function and the x-intercepts of its graph.

Helpful Aids for Solving Quadratic Equations (Lessons 4-4, 4-6, 4-8)

Factor a quadratic: $-16x^2 + 12x + 4$
$= -4(4x + 1)(x - 1)$

Complete the square: $x^2 + 4x + 1$
$= x^2 + 4x + \left(\frac{4}{2}\right)^2 + 1 - \left(\frac{4}{2}\right)^2$
$= (x + 2)^2 - 3$

Complex numbers: $x^2 + 1 = (x + i)(x - i)$ where $i = \sqrt{-1}$.

Solving Quadratic Equations (Lessons 4-5, 4-7)

$-16x^2 + 12x + 4 = 0 \rightarrow$
$-4(4x + 1)(x - 1) = 0 \rightarrow$
$x = -\frac{1}{4}$ or $x = 1$.

$-2x^2 + 4x + 1 = 0 \rightarrow$
$x = \frac{-4 \pm \sqrt{4^2 - 4(-2)(1)}}{2(-2)} \rightarrow$
$x = 1 + \frac{\sqrt{6}}{2}$ or $x = 1 - \frac{\sqrt{6}}{2}$.

Chapter Vocabulary

- absolute value of a complex number (p. 249)
- axis of symmetry (p. 194)
- completing the square (p. 235)
- complex conjugate (p. 251)
- complex number (p. 249)
- complex number plane (p. 249)
- difference of two squares (p. 220)
- discriminant (p. 242)
- factoring (p. 216)
- greatest common factor (p. 218)
- imaginary number (p. 249)
- imaginary unit (p. 248)
- maximum value (p. 195)
- minimum value (p. 195)
- parabola (p. 194)
- perfect square trinomial (p. 219)
- pure imaginary number (p. 249)
- quadratic formula (p. 240)
- quadratic function (p. 194)
- standard form (p. 202)
- vertex form (p. 194)
- vertex of the parabola (p. 194)
- zero of a function (p. 226)
- zero product property (p. 226)

Choose the correct term to complete each sentence.

1. To solve an equation by factoring, the equation should first be written in (standard form/vertex form).

2. The value of $b^2 - 4ac$ for the equation $ax^2 + bx + c = 0$ is called the (discriminant/difference of two squares).

3. The number $a + bi$, where $b = 0$, is an example of a(n)(imaginary/complex) number.

4-1 Quadratic Functions and Transformations

Quick Review

You can write every **quadratic function** in the form $f(x) = ax^2 + bx + c$, where $a \neq 0$. A **parabola** is the graph of a quadratic function. Every parabola has a vertex and an axis of symmetry. Shown below is the graph of the quadratic parent function $f(x) = x^2$.

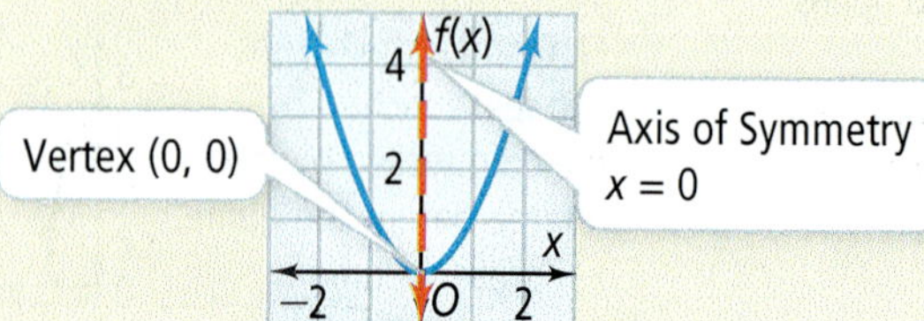

The **vertex form** of a quadratic function is $f(x) = a(x - h)^2 + k$, where $a \neq 0$. The vertex of the parabola formed by a quadratic function is (h, k).

If $a > 0$, k is the **minimum value** of the function.

If $a < 0$, k is the **maximum value** of the function. The axis of symmetry is given by $x = h$.

Example

What is the vertex, axis of symmetry, maximum or minimum, and domain and range of the function $f(x) = 5(x - 7)^2 + 2$?

$a = 5, h = 7, k = 2$	Identify *a*, *h*, and *k*.
vertex: $(7, 2)$	Find the vertex: (h, k).
axis of symmetry: $x = 7$	The axis of symmetry is at $x = h$.
$k = 2$ is a minimum	Since $a > 0$, k is a minimum.
domain: all real numbers	There are no restrictions on x.
range: $y \geq 2$.	Since the minimum is 2, $y \geq 2$.

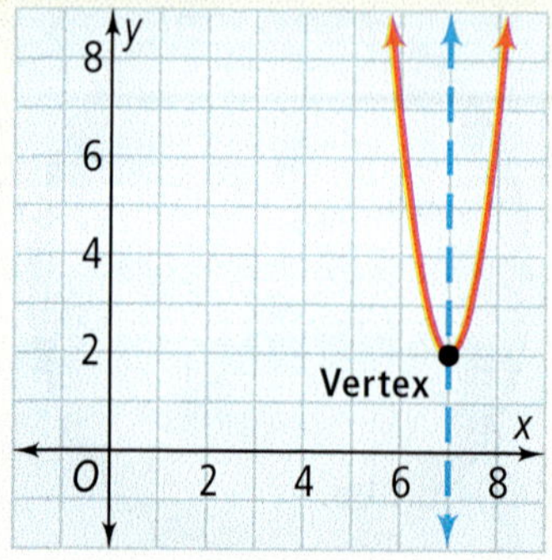

Exercises

Identify the vertex, axis of symmetry, maximum or minimum, and domain and range of each function.

4. $f(x) = 4(x + 2)^2 - 6$

5. $f(x) = -(x - 3)^2 + 2$

6. $f(x) = 10(x - 1)^2 + 5$

7. $f(x) = 2(x + 9)^2 - 4$

Graph each function. Describe each transformation of the parent function $f(x) = x^2$.

8. $f(x) = x^2 + 4$

9. $f(x) = (x - 9)^2 + 2$

10. $f(x) = \frac{1}{2}(x + 1)^2 - 5$

Write the equation of each parabola in vertex form.

11.

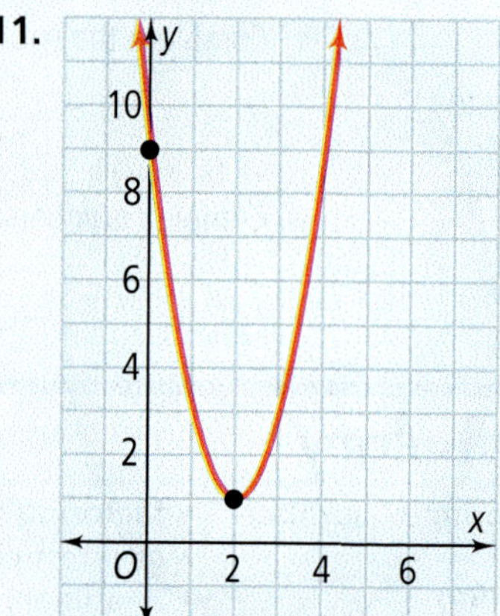

12.

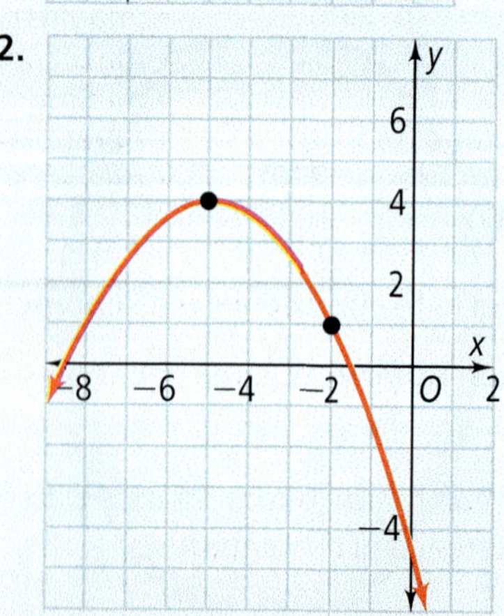

4-2 Standard Form of a Quadratic Function

Quick Review

The **standard form** of a quadratic function is $f(x) = ax^2 + bx + c$, where $a \neq 0$. When $a > 0$, the parabola opens up. When $a < 0$, the parabola opens down.

The axis of symmetry is the line $x = -\frac{b}{2a}$. The vertex is $\left(-\frac{b}{2a}, f\left(-\frac{b}{2a}\right)\right)$, and the y-intercept is $(0, c)$.

Example

What are the vertex, the axis of symmetry and y-intercept of the graph of the function $f(x) = x^2 - 6x + 8$?

axis of symmetry: $x = -\left(\frac{-6}{2(1)}\right) = 3$

vertex: $(3, -1)$

y-intercept: $(0, 8)$

Exercises

Graph each function.

13. $f(x) = x^2 + 6x + 5$

14. $f(x) = x^2 - 7x - 18$

15. $f(x) = x^2 - 7x + 12$

16. $f(x) = x^2 - 9$

Write each function in vertex form.

17. $f(x) = 4x^2 - 8x + 2$

18. $f(x) = x^2 - 8x + 12$

19. $f(x) = 8x^2 + 8x - 12$

20. $f(x) = -2x^2 - 6x + 10$

21. **Physics** The equation $h = -16t^2 + 32t + 9$ gives the height of a ball, h, in feet above the ground, at t seconds after the ball is thrown upward. How many seconds after the ball is thrown will it reach its maximum height? What is its maximum height?

4-3 Modeling With Quadratic Functions

Quick Review

You can use quadratic functions to model real world data. You can find a quadratic function to model data that passes through any three non-collinear points given that no two of the points lie on a vertical line.

Example

Find the equation of the parabola that passes through the points $(-2, 8)$, $(0, -2)$, and $(1, 2)$.

$y = ax^2 + bx + c$ — Use the standard form of a quadratic function.

$$\begin{cases} 8 = a(-2)^2 + b(-2) + c \\ -2 = a(0)^2 + b(0) + c \\ 2 = a(1)^2 + b(1) + c \end{cases}$$

Substitute the (x, y) values to write a system of equations.

$$\begin{cases} 4a - 2b + c = 8 \\ c = -2 \\ a + b + c = 2 \end{cases}$$

$a = 3, b = 1, c = -2$ — Solve the system of equations.

$y = 3x^2 + x - 2$ — Substitute a, b, and c to find the quadratic function.

Exercises

Find the equation of the parabola that passes through each set of points.

22. $(0, 5), (2, -3), (-1, 12)$

23. $(2, 0), (3, -2), (1, -2)$

24. $(4, 10), (0, -18), (-2, -20)$

25. $(0, -7), (7, -14), (-3, -19)$

26. **Track and Field** The table shows the height of a javelin as it is thrown and travels across a horizontal distance. Use your calculator to find a quadratic model to represent the path of the javelin.

Distance (m)	Height (m)
5	2
18	5
33	8
55	6
68	4
74	3

4-4 Factoring Quadratic Expressions

Quick Review

To factor an expression of the form $ax^2 + bx + c$, when $a \neq 1$, you find numbers that have the product ac and sum b. You can also factor an expression using the FOIL method in reverse or by finding the greatest common factor (GCF).

Example

Factor the expression $5x^2 + 13x + 6$.

$ac = (5)(6) = 30$	Find *ac*.
$30 = 1 \cdot 30 = 2 \cdot 15 = 3 \cdot 10 = 5 \cdot 6$	Find the factors of *ac*.
$b = 13 = 3 + 10$	Find two factors that sum to *b*.
$5x^2 + 10x + 3x + 6$	Rewrite *bx*.
$5x(x + 2) + 3(x + 2)$	Find the common factors.
$(5x + 3)(x + 2)$	Rewrite using the Distributive Property

Exercises

Factor each expression.

27. $x^2 - 8x + 12$

28. $3x^2 + 11x - 20$

29. $-4x^2 + 14x - 6$

30. $x^2 + 14x + 40$

Factor each perfect square trinomial.

31. $x^2 - 14x + 49$

32. $9x^2 + 30x + 25$

Factor each difference of two squares.

33. $36x^2 - 16$

34. $25x^2 - 4$

Find the GCF of each expression. Then factor each expression.

35. $6x^2 - 24x$

36. $-14x^2 - 49$

4-5 Solving Quadratic Equations

Quick Review

The **zeros** of a quadratic function are the solutions of the related quadratic equation. You can find the zeros from a table or from the x-intercepts of the parabola that is the graph of the function. You can also find them by **factoring** the **standard form of a quadratic equation**, $ax^2 + bx + c = 0$, and using the **Zero-Product Property**.

Example

Solve $2x^2 + 6x = 8$ by factoring.

$2x^2 + 6x - 8 = 0$	Rewrite the equation in standard form.
$2(x^2 + 3x - 4) = 0$	Factor out the GCF, 2.
$2(x + 4)(x - 1) = 0$	Factor the quadratic expression.
$2(x + 4) = 0$ or $x - 1 = 0$	Use the Zero-Product Property
$x = -4$ or $x = 1$	Solve.

Exercises

Solve each equation by factoring.

37. $x^2 = 4x + 12$

38. $2x^2 - 3x - 14 = 0$

39. $x^2 + 2x = 8$

40. $x^2 + 7x = 18$

Solve each equation by graphing.

41. $5x^2 + 8x - 13 = 0$

42. $9 - 4x = 2x^2$

43. $x^2 - x = 1$

44. $x^2 - 2x - 4 = 0$

Solve each equation by using a table.

45. $x^2 - 6x + 8 = 0$

46. $9x - 14 = 3x^2$

47. $x^2 - 5x + 2 = 0$

48. $2x^2 - 12x = -16$

4-6 Completing the Square

Quick Review

If you cannot solve a quadratic equation by factoring, you can use **completing the square**. You write one side as a perfect square trinomial and then take square roots. You can also convert a quadratic function from standard form to vertex form by completing the square.

Example

Solve $x^2 + 6x - 7 = 0$ by completing the square.

$x^2 + 6x = 7$	Rewrite the equation so the constant is by itself.
$\left(\frac{b}{2}\right)^2 = \left(\frac{6}{2}\right)^2 = 3^2 = 9$	Find $\left(\frac{b}{2}\right)^2$.
$x^2 + 6x + 9 = 7 + 9$	Add $\left(\frac{b}{2}\right)^2$ to each side.
$(x + 3)^2 = 16$	Factor and simplify.
$x + 3 = \pm 4$	Take the square root of each side.
$x = 1$ or $x = -7$	Solve for x.

Exercises

Solve each equation by finding square roots.

49. $4x^2 = 16$

50. $4x^2 - 20 = 0$

51. $5x^2 - 45 = 0$

52. $3x^2 = 36$

What values complete each square?

53. $x^2 - 6x$

54. $x^2 + 3x$

Solve each equation by completing the square.

55. $x^2 + 8x + 6 = 0$

56. $x^2 - 10x = 13$

57. $9x^2 + 6x + 1 = 4$

58. $x^2 - 2x + 4 = 0$

59. $x^2 + 3x = -25$

60. $4x^2 - x - 3 = 0$

4-7 The Quadratic Formula

Quick Review

You can solve a quadratic equation in the form $ax^2 + bx + c = 0$ by using the **Quadratic Formula**, $x = \frac{-b \pm \sqrt{b^2 - 4ac}}{2a}$. The **discriminant** of a quadratic equation in standard form is the value of the expression $b^2 - 4ac$. You can use it to find the quantity and type of solutions of a quadratic equation.

Example

Use the Quadratic Formula to solve $2x^2 - 6x = -3$.

$2x^2 - 6x + 3 = 0$	Write the equation in standard form.
$a = 2, b = -6, c = 3$	Identify a, b, and c.
$x = \frac{-(-6) \pm \sqrt{(-6)^2 - 4(2)(3)}}{2(2)}$	Substitute a, b, and c into the quadratic formula.
$x = \frac{6 \pm \sqrt{12}}{4} = \frac{3 \pm \sqrt{3}}{2}$	Simplify.

Exercises

Solve each equation using the quadratic formula.

61. $3x^2 + 5x = 8$

62. $x^2 = 6x - 9$

63. $x(x - 3) = 4$

64. $5x^2 - 7x - 3 = 0$

Determine the discriminant of each equation. How many real solutions does each equation have?

65. $4x^2 - 2x = 10$

66. $x^2 - 5x + 7 = 0$

67. $3x^2 + 3 - 6x$

68. $7 - 3x = 8x^2$

69. **Gardening** Margaret is planning a rectangular garden. Its length is 4 ft less than twice its width. Its area is 170 ft^2. What are the dimensions of the garden?

4-8 Complex Numbers

Quick Review

A **complex number** is written in the form $a + bi$, where a and b are real numbers, and i is equal to $\sqrt{-1}$. If $b = 0$, $a + bi$ is a real number. If $b \neq 0$, $a + bi$ is an **imaginary number**. You can use the Quadratic Formula or completing the square to find the imaginary solutions of quadratic equations.

Example

Use the Quadratic Formula to solve $3x^2 - 4x + 2 = 0$.

$x = \frac{-(-4) \pm \sqrt{(-4)^2 - 4(3)(2)}}{2(3)}$ Enter a, b, and c into the quadratic formula.

$x = \frac{4 \pm \sqrt{16 - 24}}{6} = \frac{4 \pm \sqrt{-8}}{6}$ Simplify.

$x = \frac{2}{3} \pm \frac{\sqrt{2}}{3}i$ Write the solutions.

Exercises

Simplify each expression using the imaginary unit i.

70. $\sqrt{-24}$

71. $\sqrt{-2} - 3$

72. $(4 + \sqrt{-25})(\sqrt{-100})$

73. $2\sqrt{-24} + 6$

Simplify each expression.

74. $(9 + 7i) - (6 - 2i)$

75. $(3 + 11i) + (10 + 9i)$

76. $(1 - 9i)(3 + 2i)$

77. $(3i)^2 - 3(1 + 5i)$

78. $\frac{4 - 6i}{2i}$

79. $\frac{2 - 3i}{1 + 5i}$

Solve each equation.

80. $x^2 + 9 = 0$

81. $5x^2 - 2x + 1 = 0$

82. $-x^2 + 4x = 10$

83. $7x^2 + 8x = -6$

4-9 Quadratic Systems

Quick Review

A system of quadratic equations can be solved by substitution or by graphing. You can use these methods to solve a linear-quadratic system or a quadratic-quadratic system. Use graphing to solve a quadratic system of inequalities.

Example

Use substitution to solve $\begin{cases} y = 2x^2 + 2x - 10 \\ y = x^2 + 5x - 6 \end{cases}$.

$2x^2 + 2x - 10 = x^2 + 5x - 6$ Substitute for y.

$x^2 - 3x - 4 = 0$ Rewrite in standard form.

$(x + 1)(x - 4) = 0$ Factor.

$x = -1$ or $x = 4$ Solve for x.

$y = (-1)^2 + 5(-1) - 6 = -10$ Substitute for x then solve for y.

$y = (4)^2 + 5(4) - 6 = 30$

$(-1, -10)$ and $(4, 30)$ Write solutions as ordered pairs.

Exercises

Solve each system by substitution.

84. $\begin{cases} y = x^2 - 7x - 6 \\ y = 8 - 2x \end{cases}$

85. $\begin{cases} y = -x^2 - 2x + 8 \\ y = x^2 - 8x - 12 \end{cases}$

Solve each system by graphing.

86. $\begin{cases} y = -x^2 - 10x + 12 \\ y = x^2 - 6x - 18 \end{cases}$

87. $\begin{cases} y = x^2 - x - 18 \\ y = 2x + 3 \end{cases}$

Solve each system of inequalities.

88. $\begin{cases} y < x + 4 \\ y \geq x^2 + 2x + 2 \end{cases}$

89. $\begin{cases} y > 3x^2 - 10x - 8 \\ y < x^2 - 5x + 4 \end{cases}$

Chapter Test

Do you know HOW?

Sketch a graph of the quadratic function with the given vertex and through the given point. Then write the equation of the parabola in vertex form and describe how the function was transformed from the parent function $y = x^2$.

1. vertex $(0, 0)$, point $(-3, 3)$
2. vertex $(1, 5)$, point $(2, 1)$

Graph each quadratic function. Identify the axis of symmetry, the vertex, and the domain and the range of each function.

3. $y = x^2 - 7$
4. $y = x^2 + 2x + 6$
5. $y = -x^2 + 5x - 3$

Simplify each expression.

6. $\sqrt{-16}$
7. $4\sqrt{-9} - 2$
8. $(2 + 3i)(8 - 5i)$
9. $(-3 + 2i) - (6 + i)$
10. $\frac{4 + 2i}{2 - i}$

Factor each expression completely.

11. $2y^2 - 8y$
12. $3x^2 + 8x - 3$
13. $9w^2 - 30w + 25$

Solve each quadratic equation.

14. $x^2 - 25 = 0$
15. $x^2 - 2x + 3 = 0$
16. $x^2 - 8x = -6$

Solve the following systems of equations.

17. $\begin{cases} y = 3x^2 - x + 1 \\ y = 3x^2 + x - 1 \end{cases}$
18. $\begin{cases} y = -x^2 + 2x - 3 \\ y = 4x - 3 \end{cases}$

Solve the following systems of inequalities.

19. $\begin{cases} y > 2x^2 + 5x + 1 \\ y < -2x^2 - 5x - 1 \end{cases}$
20. $\begin{cases} y < x^2 - x + 2 \\ y > x^2 - 1 \end{cases}$

Evaluate the discriminant of each equation. How many real and imaginary solutions does each have?

21. $x^2 + 6x - 7 = 0$
22. $3x^2 - x + 3 = 0$
23. $-4x^2 - 4x + 1 = 0$

Do you UNDERSTAND?

24. **Writing** Compare graphing a number on the complex plane to graphing a point on the coordinate plane. How are they similar? How are they different?
25. **Open-Ended** Sketch the graph of a quadratic function $f(x) = ax^2 + bx + c$ that has no real zeros. How does this relate to the solutions of the related equation $ax^2 + bx + c = 0$?
26. STEM **Physics** A model for the path of a toy rocket is given by $h = 68t - 4.9t^2$, where h is the altitude in meters and t is the time in seconds. Explain how to find both the maximum altitude of the rocket and how long it takes to reach that altitude.
27. How many solutions are possible for:
    a. a system of two quadratic equations?
    b. a system of two quadratic inequalities?
    Explain your answers.

4 Cumulative Standards Review 

TIPS FOR SUCCESS

Some questions on tests require that you model a word problem with a quadratic function.

Roy has a 400 foot roll of wire. He wants to use it to fence in a rectangular area. What is the maximum area of the enclosed space?

(A) 20,000 square feet

(B) 10,000 square feet

(C) 200 square feet

(D) 100 square feet

TIP 1

To identify the function, use what you already know. You know that the perimeter of a rectangle is $2(\ell + w)$ and the area is $\ell \cdot w$.

TIP 2

Use the information from the problem. The perimeter is 400, so $2(\ell + w) = 400$. Solve for w: $w = 200 - \ell$.

Think It Through

Substitute for w in the area formula:

$$A = f(\ell) = \ell \cdot (200 - \ell) = -\ell^2 + 200\ell.$$

The maximum value is the y-coordinate of the vertex, $f\left(-\frac{b}{2a}\right) = f(100) = 10{,}000$. So, the maximum area Roy can enclose is 10,000 square feet.

The correct answer is B.

Vocabulary Builder

As you solve test items, you must understand the meanings of mathematical terms. Match each term with its mathematical meaning.

A. axis of symmetry

B. discriminant

C. imaginary number

D. Zero-Product Property

E. parabola

F. perfect square trinomial

G. completing the square

I. value of $b^2 - 4ac$ for the equation $ax^2 + bx + c = 0$

II. If $ab = 0$, then $a = 0$ or $b = 0$.

III. line that divides a parabola into two parts that are mirror images

IV. $a + bi$, a and b are real numbers and $b \neq 0$

V. square of a binomial

VI. process of finding the last term to make a perfect square trinomial

VII. graph of a quadratic function

Multiple Choice

Read each question. Then write the letter of the correct answer on your paper.

1. Which equation is equivalent to $x^2 + 24x + 100 = -46$?

(A) $(x + 12)^2 = -2$

(B) $(x - 12)^2 = -2$

(C) $(x - 12)^2 = 2$

(D) $(x + 12)^2 = 2$

2. What is the solution of the following system of equations?

$2x - y = 4$
$3x + y = 1$

(F) $(-1, 2)$

(G) $(1, -2)$

(H) $(2, 1)$

(I) $(-2, 1)$

3. What are the factors of the quadratic function graphed below?

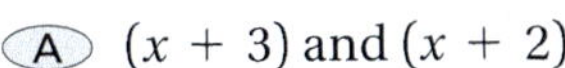

Ⓐ $(x + 3)$ and $(x + 2)$

Ⓑ x and $(x - 6)$

Ⓒ x and $(x + 6)$

Ⓓ $(x - 3)$ and $(x + 2)$

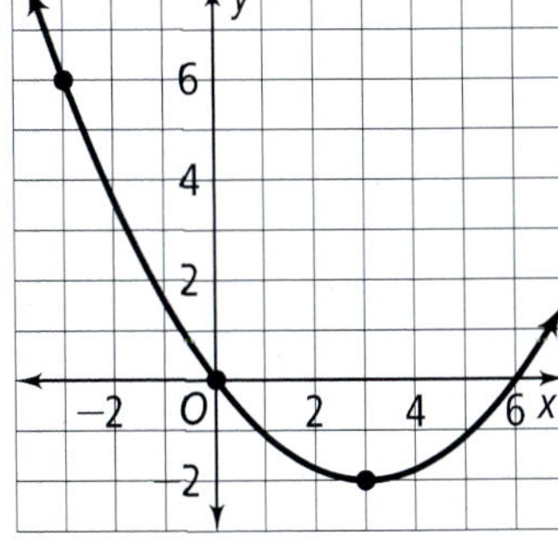

4. Which equation has $-1 \pm i$ as its solution?

Ⓕ $x^2 - 2x - 2 = 0$

Ⓖ $2x^2 - 2x - 1 = 0$

Ⓗ $2x^2 + 2x + 1 = 0$

Ⓘ $x^2 + 2x + 2 = 0$

5. What are the solutions of $|3x - 5| = 2$?

Ⓐ $x = -1$ and $x = \frac{7}{3}$

Ⓑ $x = 1$ and $x = \frac{7}{3}$

Ⓒ $x = 1$ and $x = \frac{1}{5}$

Ⓓ $x = -1$ and $x = \frac{1}{5}$

6. What is the transformation of the graph of $y = (x + 3)^2 - 2$ from its parent function $y = x^2$?

Ⓕ 3 units left and 2 units down

Ⓖ 3 units right and 2 units up

Ⓗ 6 units right and 2 units up

Ⓘ 2 units left and 3 units down

7. What is the axis of symmetry for the graph of the quadratic equation $y = -3x^2 - 12 + 12x$?

Ⓐ $x = -2$

Ⓑ $x = 2$

Ⓒ $x = 12$

Ⓓ $x = -12$

8. What are the domain and range of the function graphed below?

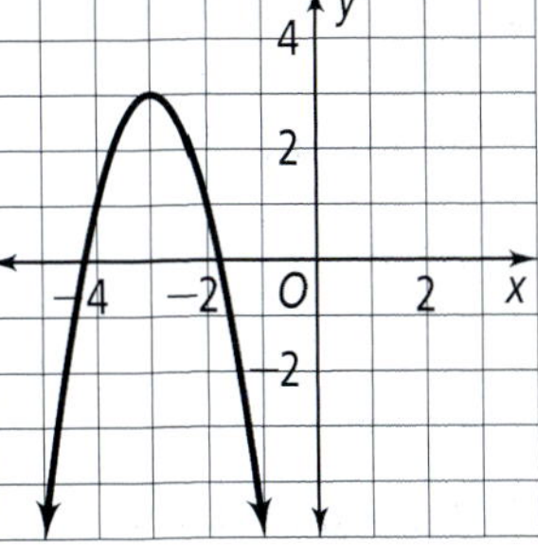

Ⓕ Domain: All real numbers
Range: All real numbers ≤ 3

Ⓖ Domain: All real numbers
Range: All real numbers ≥ 3

Ⓗ Domain: All real numbers between -5 and -1
Range: All real numbers ≤ 3

Ⓘ Domain: All real numbers between -5 and -1
Range: All real numbers ≥ 3

9. The formula for the total surface area of a regular right pentagonal prism is $A = ap + pH$. Solve this equation for p.

Ⓐ $p = \frac{a + H}{A}$

Ⓑ $p = \frac{A}{a + H}$

Ⓒ $p = A - \frac{a}{H}$

Ⓓ $p = \frac{H - a}{A}$

10. What is the solution of $\begin{cases} -y = 3x - 1 \\ 2y = -x - 2 \end{cases}$?

Ⓕ $x = 20, y = -11$

Ⓖ $x = \frac{4}{5}, y = -\frac{7}{5}$

Ⓗ $x = -20, y = 11$

Ⓘ $x = -\frac{4}{5}, y = \frac{7}{5}$

11. Which system of inequalities is graphed below?

Ⓐ $\begin{cases} y \leq -1 \\ y + x \geq 2 \end{cases}$

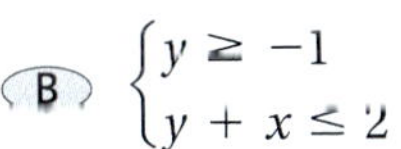

Ⓑ $\begin{cases} y \geq -1 \\ y + x \leq 2 \end{cases}$

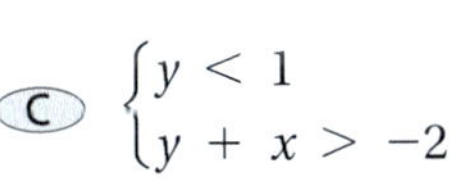

Ⓒ $\begin{cases} y < 1 \\ y + x > -2 \end{cases}$

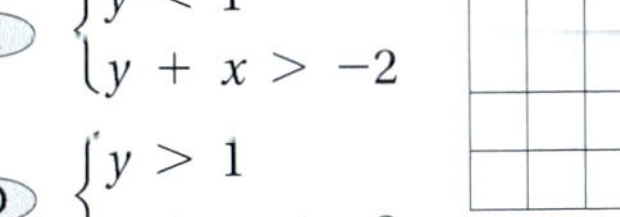

Ⓓ $\begin{cases} y > 1 \\ y + x < -2 \end{cases}$

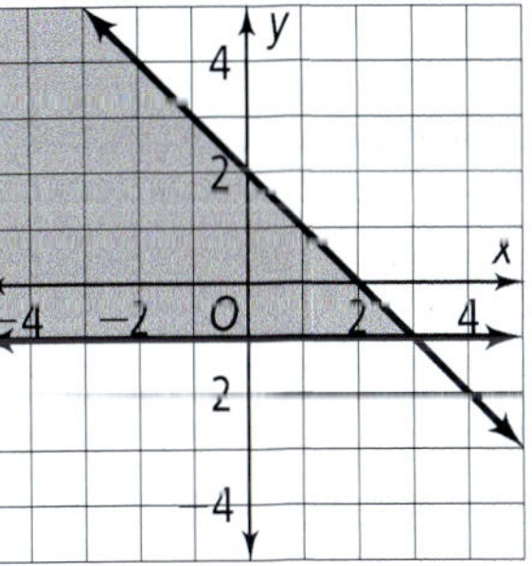

12. What is the vertex of $y = -2|x + 4| - 5$?

Ⓕ $(-2, -5)$

Ⓖ $(-4, -5)$

Ⓗ $(4, -5)$

Ⓘ $(2, -5)$

GRIDDED RESPONSE

13. What is the sum of the solutions of the equation $1.5x^2 - 2.5x - 1.5 = 0$? Round to the nearest hundredth.

14. What is the value of y in the system of equations?

$x + y = 10$
$y = 2x + 1$

15. Zeroy and Darius shop at the mall during the special sale extravaganza. Zeroy spends $120 on 3 pairs of pants and 4 shirts. Darius buys 2 pairs of pants and 3 shirts and spends $85. What is the price of one shirt?

16. What is the sum of the solutions of $|5x - 4| = 8 - x$?

17. Suppose y varies directly with x, and $y = -4$ when $x = 5$. What is the constant of variation?

18. Molly is making a punch for the school picnic. The recipe calls for $\frac{3}{4}$ quart of lemonade, 3 cups of cranberry juice, 4 cans of orange juice concentrate, and 5 cups of water. If Molly uses $1\frac{2}{5}$ quarts of lemonade, how many cups of cranberry juice will she need?

19. What is the value of $3x^2 - 5x + 7$ when $x = \frac{2}{5}$? Express the answer as a decimal.

20. What is the real part of $(11 + 10i)(2 + 3i)$?

21. How many imaginary roots does $2x^2 + 3x - 5 = 0$ have?

22. What is the product of $(4 + 3i)(4 - 3i)$?

23. What is the greatest integer solution of $|2x + 3| - 4 \leq 0$?

24. What is the coefficient of the x-term of the factorization of $25x^2 + 20x + 4$?

25. A piggy bank contains $2.40 in nickels and dimes. If there are 33 coins in all, how many nickels are there?

Short Response

26. What are the solutions of the system? Solve by graphing.

$y = x^2 - x - 2$
$y = -x + 2$

27. A swimmer swam 1000 meters downstream in 15 minutes and swam back in 30 minutes against the current. What was the rate of the swimmer in still water? How fast was the current?

28. Claudia has a rectangular flowerbed. She decided that the original width w, in feet, was too small, so she increased the width by 3 feet. She also changed the length to be 1 foot less than twice the original width. What is an expression that represents the area of the new flower bed?

29. Explain how you would graph $y + 4 < 2|x - 3|$ on a coordinate grid.

Extended Response

30. A hat company is designing a one-size-fits-all hat with a strap in the back that makes the hat smaller or larger. Head sizes normally range from 51 to 64 centimeters. What absolute value inequality models the different sizes of the hat? Graph the solution.

31. Robby decided to earn extra money by making and selling brownies and cookies. He had space in his oven to make at most 80 brownies and cookies. Each brownie cost $.10 to make and each cookie cost $.05 to make. He had $6 to spend on ingredients.
    a. Write a system of inequalities to represent the situation.
    b. Graph the system, choose one point in the feasible region, and explain what the point means in terms of the problem.
    c. If Robby makes a profit of $.25 on each brownie and $.20 on each cookie, how many of each dessert should he make to maximize his profit?

CHAPTER 5

Get Ready!

Lesson 4-2

Graphing Quadratic Functions

Graph each function.

1. $f(x) = x^2 - 8x + 7$ **2.** $f(x) = -\frac{1}{2}x^2 - 4x - 4$ **3.** $f(x) = x^2 + 4x + 4$

Lesson 4-3

Writing Equations of Parabolas

Write in standard form the equation of the parabola passing through the given points.

4. $(-1, -6), (-3, -4), (2, 6)$ **5.** $(3, 4), (-2, 9), (2, 1)$ **6.** $(-5, -8), (4, -8), (-3, 6)$

Lesson 4-5

Solving Quadratic Equations by Graphing

Solve each equation by graphing. Round to the nearest hundredth.

7. $1 = 4x^2 + 3x$ **8.** $\frac{1}{2}x^2 + x - 14 = 0$ **9.** $5x^2 + 30x = 12$

Lesson 4-5

Solving Quadratic Equations by Factoring

Solve each equation by factoring.

10. $x^2 - x - 20 = 0$ **11.** $x^2 + 6x - 27 = 0$ **12.** $3x^2 - 9x + 6 = 0$

Lesson 4-7

Finding the Number and Type of Solutions

Evaluate the discriminant of each equation. Tell how many solutions each equation has and whether the solutions are real or imaginary.

13. $x^2 - 12x + 30 = 0$ **14.** $-4x^2 + 20x - 25 = 0$ **15.** $2x^2 = 8x - 8$

VOCABULARY

Looking Ahead Vocabulary

16. A *turning point* is a place where a graph changes direction. Suppose you start hiking north on a winding trail, and the trail makes a turn and heads south, and then north again. If you make a total of 3 of these 180 degree turns, in which direction will you be hiking after the last turn?

17. A *relative maximum* is the greatest value in a region. The highest point in Maine is Mt. Katahdin at 5267 ft. How might that compare to the highest point in the United States? What might the relative maximum of a graph be?

18. A contraction is a shortened form of a word or phrase. The expanded form of the contraction "don't" is "do not." You can *expand* a math phrase by multiplying it out. For example, $(x - 2)^2 = (x - 2)(x - 2) = x^2 - 4x + 4$. Expand $(2x + 1)^2$.

Polynomials and Polynomial Functions

PowerAlgebra.com

Your place to get all things digital

ownload videos onnecting math o your world.

Math definitions n English and panish

he online olve It will get ou in gear for ach lesson.

nteractive! ary numbers, raphs, and figures o explore math oncepts.

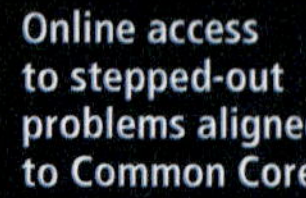

Online access to stepped-out problems aligned to Common Core

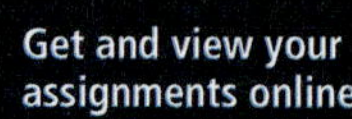

Get and view your assignments online.

Extra practice and review online

DOMAINS

- Interpreting Functions
- Arithmetic with Polynomials and Rational Expressions
- The Complex Number System

Polynomial functions are used to model all kinds of real-world situations, like the energy produced by a turbine.

In this chapter, you will also learn theorems that will help you when working with polynomial functions and equations.

Vocabulary

English/Spanish Vocabulary Audio Online:

English	Spanish
end behavior, *p. 282*	comportamiento extremo
monomial, *p. 280*	monomio
multiplicity, *p. 291*	multiplicidad
Pascal's Triangle, *p. 327*	Triángulo de Pascal
polynomial function, *p. 280*	función polinomial
relative maximum, *p. 291*	máximo relativo
relative minimum, *p. 291*	mínimo relativo
standard form of a polynomial function, *p. 281*	forma normal de una función polinomial
synthetic division, *p. 306*	división sintética
turning point, *p. 282*	punto de giro

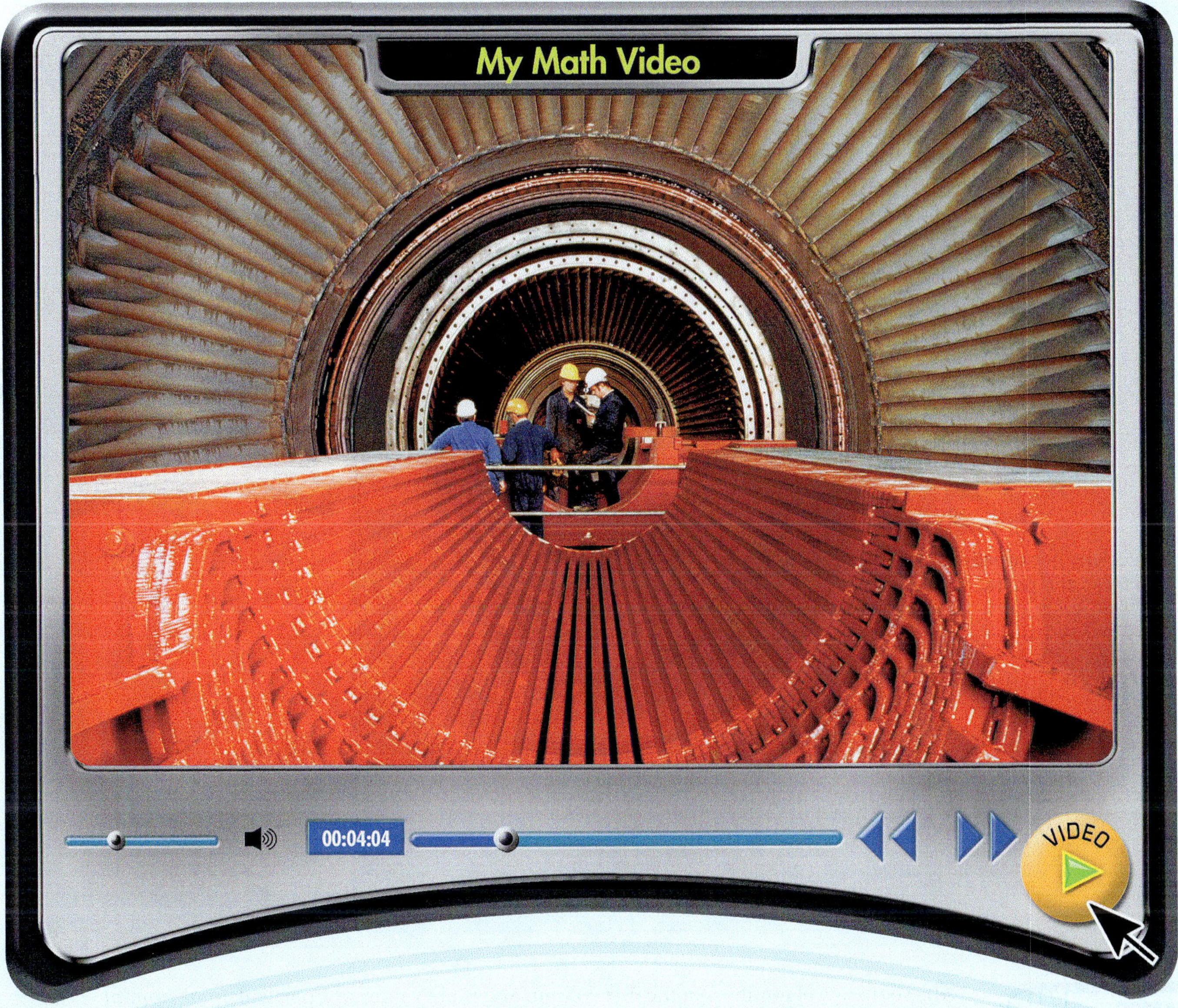

BIG ideas

1 **Function**
Essential Question What does the degree of a polynomial tell you about its related polynomial function?

2 **Equivalence**
Essential Question For a polynomial function, how are factors, zeros, and x-intercepts related?

3 **Solving Equations and Inequalities**
Essential Question For a polynomial equation, how are factors and roots related?

Chapter Preview

5-1 Polynomial Functions

Content Standards

F.IF.7.c Graph polynomial functions, identifying zeros when suitable factorizations are available and showing end behavior.

Also A.SSE.1.a

Objectives To classify polynomials

To graph polynomial functions and describe end behavior

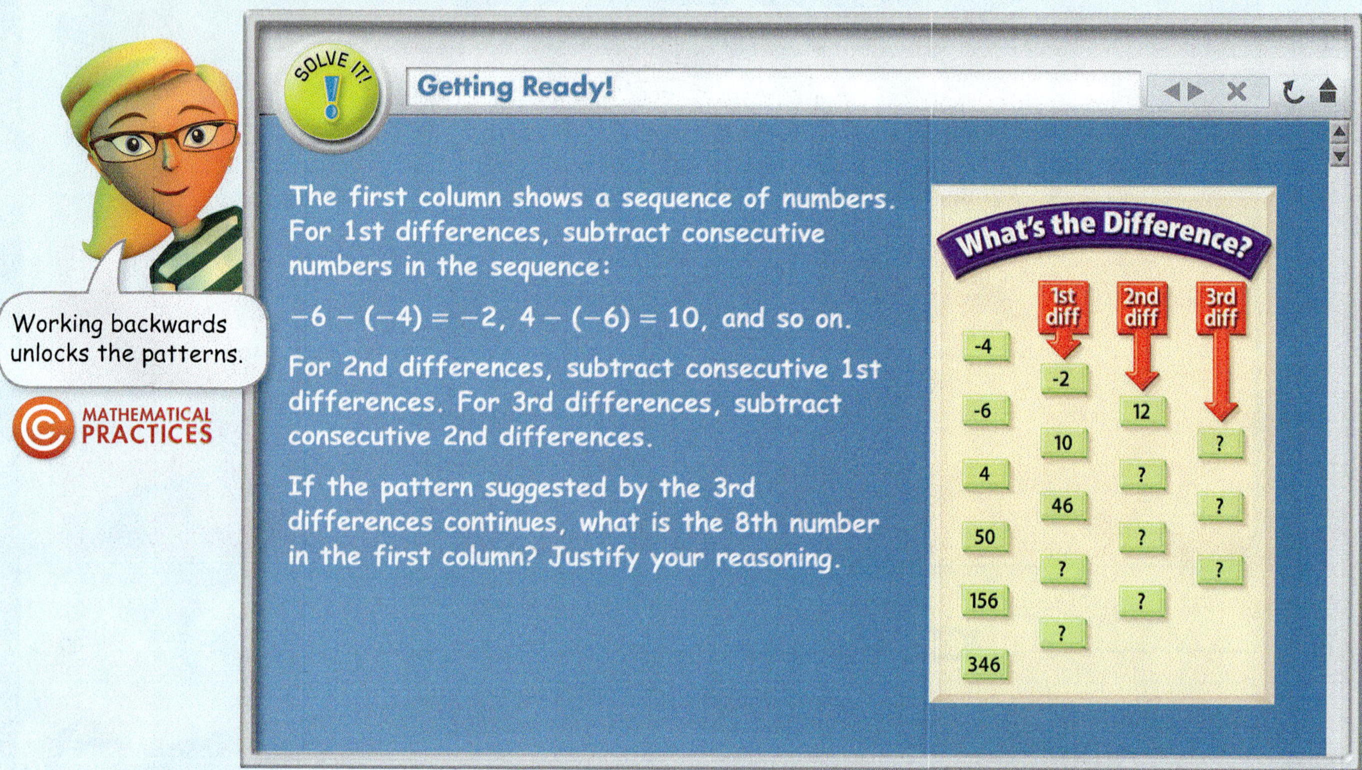

The sequence of numbers in the first column above are values of a particular *polynomial function.* For such a sequence, you can use patterns of 1st differences, 2nd differences, 3rd differences, and so on, to learn more about the polynomial function.

Lesson Vocabulary
- monomial
- degree of a monomial
- polynomial
- degree of a polynomial
- polynomial function
- standard form of a polynomial function
- turning point
- end behavior

Essential Understanding A polynomial function has distinguishing "behaviors." You can look at its algebraic form and know something about its graph. You can look at its graph and know something about its algebraic form.

A **monomial** is a real number, a variable, or a product of a real number and one or more variables with whole-number exponents. The **degree of a monomial** in one variable is the exponent of the variable. A **polynomial** is a monomial or a sum of monomials. The **degree of a polynomial** in one variable is the greatest degree among its monomial terms.

A polynomial with the variable x defines a **polynomial function** of x. The degree of the polynomial function is the same as the degree of the polynomial.

Key Concept Standard Form of a Polynomial Function

The **standard form of a polynomial function** arranges the terms by degree in descending numerical order.

A polynomial function $P(x)$ in standard form is

$$P(x) = a_n x^n + a_{n-1} x^{n-1} + \cdots + a_1 x + a_0,$$

where n is a nonnegative integer and $a_n, \ldots, a_0$ are real numbers.

$$P(x) = 4x^3 + 3x^2 + 5x - 2$$

Cubic term, Quadratic term, Linear term, Constant term

You can classify a polynomial by its degree or by its number of terms. Polynomials of degrees zero through five have specific names, as shown in this table.

Degree	Name Using Degree	Polynomial Example	Number of Terms	Name Using Number of Terms
0	constant	5	1	monomial
1	linear	$x + 4$	2	binomial
2	quadratic	$4x^2$	1	monomial
3	cubic	$4x^3 - 2x^2 + x$	3	trinomial
4	quartic	$2x^4 + 5x^2$	2	binomial
5	quintic	$-x^5 + 4x^2 + 2x + 1$	4	polynomial of 4 terms

Problem 1 Classifying Polynomials

Write each polynomial in standard form. What is the classification of each polynomial by degree? by number of terms?

Think

How do you write a polynomial in standard form?
Combine like terms if possible. Then, write the terms with their degrees in descending order.

A $3x + 9x^2 + 5$

$9x^2 + 3x + 5$

The polynomial has degree 2 and 3 terms. It is a quadratic trinomial.

B $4x - 6x^2 + x^4 + 10x^2 - 12$

$x^4 + 4x^2 + 4x - 12$

The polynomial has degree 4 and 4 terms. It is a quartic polynomial of 4 terms.

Got It? **1.** Write each polynomial in standard form. What is the classification of each by degree? by number of terms?

a. $3x^3 - x + 5x^4$ **b.** $3 - 4x^5 + 2x^2 + 10$

The degree of a polynomial function affects the shape of its graph and determines the maximum number of **turning points**, or places where the graph changes direction. It also affects the **end behavior**, or the directions of the graph to the far left and to the far right.

The table below shows you examples of polynomial functions and the four types of end behavior. The table also shows intervals where the functions are increasing and decreasing. A function is *increasing* when the y-values increase as x-values increase. A function is *decreasing* when the y-values decrease as x-values decrease.

take note

Key Concept Polynomial Functions

$y = 4x^4 + 6x^3 - x$

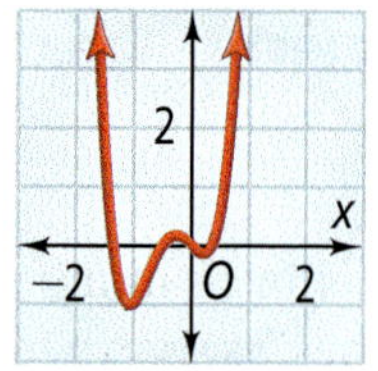

End Behavior: Up and Up

Turning Points: $(-1.07, -1.04)$, $(-0.27, 0.17)$, and $(0.22, -0.15)$

The function is decreasing when $x < -1.07$ and $-0.27 < x < 0.22$. The function increases when $-1.07 < x < -0.27$ and $x > 0.22$.

$y = -x^2 + 2x$

End Behavior: Down and Down

Turning Point: $(1, 1)$

The function is increasing when $x < 1$ and is decreasing when $x > 1$.

$y = x^3$

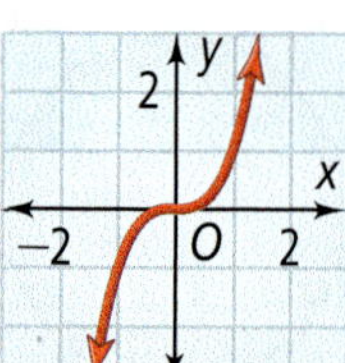

End Behavior: Down and Up

Zero turning points.

The function is increasing for all x.

$y = -x^3 + 2x$

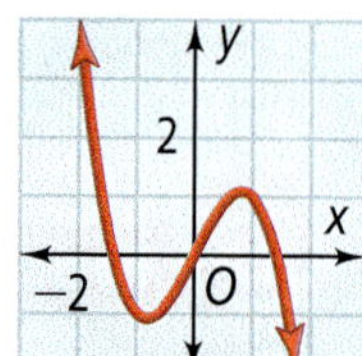

End Behavior: Up and Down

Turning Points: $(-0.82, -1.09)$ and $(0.82, 1.09)$

The function is decreasing when $x < -0.82$ and when $x > 0.82$. The function is increasing when $-0.82 < x < 0.82$.

You can determine the end behavior of a polynomial function of degree n from the leading term ax^n of the standard form.

End Behavior of a Polynomial Function With Leading Term ax^n

	n Even ($n \neq 0$)	n Odd
a Positive	Up and Up	Down and Up
a Negative	Down and Down	Up and Down

In general, the graph of a polynomial function of degree n ($n \geq 1$) has at most $n - 1$ turning points. The graph of a polynomial function of odd degree has an even number of turning points. The graph of a polynomial function of even degree has an odd number of turning points.

Problem 2 Describing End Behavior of Polynomial Functions

Consider the leading term of each polynomial function. What is the end behavior of the graph? Check your answer with a graphing calculator.

Think

What do a and n represent?
a is the coefficient of the leading term. n is the exponent of the leading term.

A $y = 4x^3 - 3x$

The leading term is $4x^3$. Since n is odd and a is positive, the end behavior is down and up.

B $y = -2x^4 + 8x^3 - 8x^2 + 2$

The leading term is $-2x^4$. Since n is even and a is negative, the end behavior is down and down.

Got It? **2.** Consider the leading term of $y = -4x^3 + 2x^2 + 7$. What is the end behavior of the graph?

Problem 3 Graphing Cubic Functions

What is the graph of each cubic function? Describe the graph, including end behavior, turning points, and increasing/decreasing intervals.

Plan

How can you graph a polynomial function?
Make a table of values to help you sketch the middle part of the graph. Use what you know about end behavior to sketch the ends of the graph.

A $y = \frac{1}{2}x^3$

Step 1

x	y
−2	−4
−1	−0.5
0	0
1	0.5
2	4

Step 2

Step 3

The end behavior is down and up. There are no turning points. The function increases from $-\infty$ to ∞.

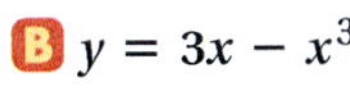

B $y = 3x - x^3$

Step 1

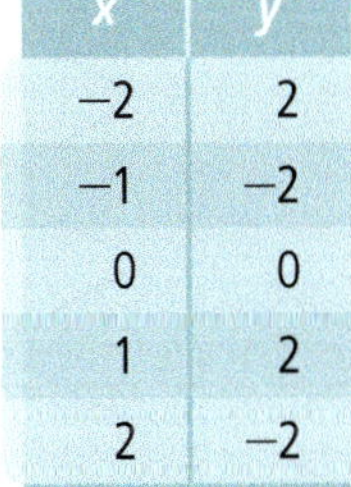

x	y
−2	2
−1	−2
0	0
1	2
2	−2

Step 2

Step 3

The end behavior is up and down. There are turning points at $(-1, -2)$ and $(1, 2)$. The function decreases from $-\infty$ to -1, increases from -1 to 1, and decreases from 1 to ∞.

Got It? **3.** What is the graph of each cubic function? Describe the graph.

a. $y = -x^3 + 2x^2 - x - 2$ **b.** $y = x^3 - 1$

Suppose you are given a set of polynomial function outputs. You know that their inputs are an ordered set of x-values in which consecutive x-values differ by a constant. By analyzing the differences of consecutive y-values, it is possible to determine the least-degree polynomial function that could generate the data.

If the first differences are constant, the function is linear. If the second differences (but not the first) are constant, the function is quadratic. If the third differences (but not the second) are constant, the function is cubic, and so on.

Problem 4 Using Differences to Determine Degree

What is the degree of the polynomial function that generates the data shown at the right?

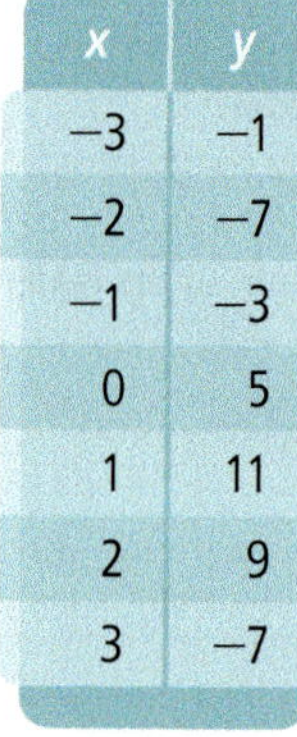

x	y
−3	−1
−2	−7
−1	−3
0	5
1	11
2	9
3	−7

Know
A set of polynomial function values

Need
Degree of the polynomial function

Plan
Check first differences of y-values. Then check second differences, third differences, and so on until they are constant.

Think
How do you find the second differences?
Subtract the consecutive first differences.

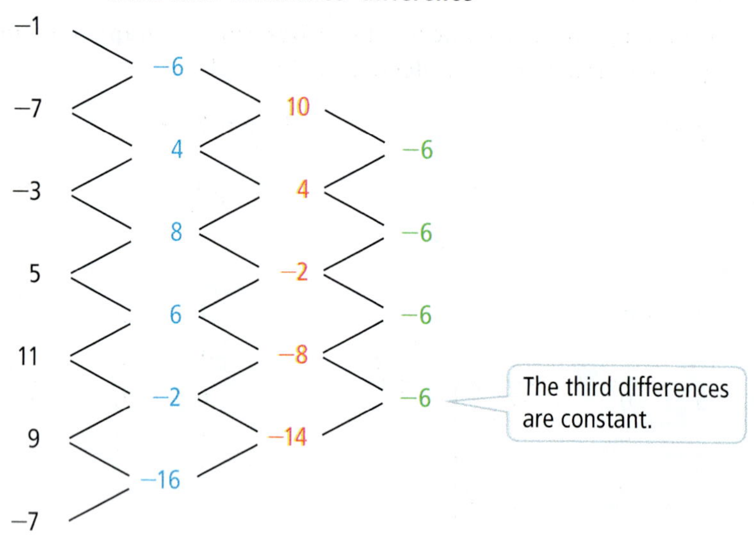

The degree of the polynomial function is 3.

Got It? **4. a.** What is the degree of the polynomial function that generates the data shown at the right?

b. Reasoning What is an example of a polynomial function whose fifth differences are constant but whose fourth differences are not constant?

x	y
−3	23
−2	−16
−1	−15
0	−10
1	−13
2	−12
3	29

Lesson Check

Do you know HOW?

Classify each polynomial by degree and by number of terms.

1. $5x^3$

2. $6x^2 + 4x - 2$

Write each polynomial in standard form.

3. $7x + 3 + 5x^2$

4. $-3 + 9x$

Do you UNDERSTAND?

5. Vocabulary Describe the end behavior of the graph of $y = -2x^7 - 8x$.

6. Reasoning Can the graph of a polynomial function be a straight line? If so, give an example.

7. Error Analysis Your friend claims the graph of the function $y = 4x^3 + 4$ has only one turning point. Describe the error your friend made and give the correct number of turning points.

Practice and Problem-Solving Exercises

Write each polynomial in standard form. Then classify it by degree and by number of terms.

See Problem 1.

8. $7x + 3x + 5$

9. $5 - 3x$

10. $2m^2 - 3 + 7m$

11. $-x^3 + x^4 + x$

12. $-4p + 3p + 2p^2$

13. $5a^2 + 3a^3 + 1$

14. $-x^5$

15. $3 + 12x^4$

16. $6x^3 - x^3$

17. $7x^3 - 10x^3 + x^3$

18. $4x + 5x^2 + 8$

19. $x^2 - x^4 + 2x^2$

Determine the end behavior of the graph of each polynomial function.

See Problem 2.

20. $y = -7x^3 + 8x^2 + x$

21. $y = -3x + 6x^2 - 1$

22. $y = 1 - 4x - 6x^3 - 15x^6$

23. $y = 8x^{11} - 2x^9 + 3x^6 + 4$

24. $y = -x^5 - 15x^7 - 4x^9$

25. $y = -3 - 6x^5 - 9x^8$

26. $y = x^4 - 7x^2 + 3$

27. $y = -8x^7 + 16x^6 + 9$

28. $y = -14x^6 + 11x^5 - 11$

29. $y = -x^3 - x^2 + 3$

30. $y = x^3 - 14x - 4$

31. $y = 5 - 17x^7 + 9x^{10}$

Describe the shape of the graph of each cubic function including end behavior, turning points, and increasing/decreasing intervals.

See Problem 3.

32. $y = 3x^3 - x - 3$

33. $y = -9x^3 - 2x^2 + 5x + 3$

34. $y = 10x^3 + 9$

35. $y = 3x^3$

36. $y = -4x^3 - 5x^2$

37. $y = 8x^3$

Determine the degree of the polynomial function with the given data.

See Problem 4.

38.

x	−2	−1	0	1	2
y	16	7	2	1	4

39.

x	−2	−1	0	1	2
y	−15	−9	−9	−9	−3

40. **Think About a Plan** The data shows the power generated by a wind turbine. The x column gives the wind speed in meters per second. The y column gives the power generated in kilowatts. What is the degree of the polynomial function that models the data?

- What are the first differences of the y-values?
- What are the second differences of the y-values?
- When are the differences constant?

x	y
5	10
6	17.28
7	27.44
8	40.96
9	58.32

Classify each polynomial by degree and by number of terms. Simplify first if necessary.

41. $a^2 + a^3 - 4a^4$

42. 7

43. $2x(3x)$

44. $(2a - 5)(a^2 - 1)$

45. $(-8d^3 - 7) + (-d^3 - 6)$

46. $b(b - 3)^2$

Determine the sign of the leading coefficient and the least possible degree of the polynomial function for each graph.

47. 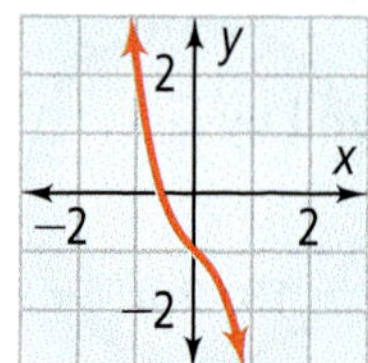

48. 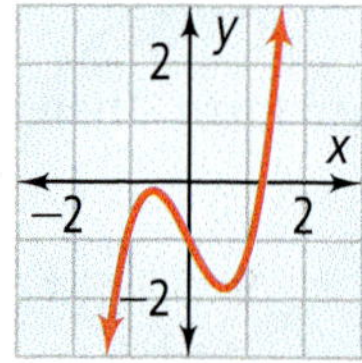

49. 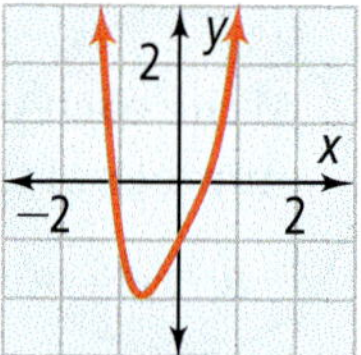

50. **Open-Ended** Write an equation for a polynomial function that has three turning points and end behavior up and up.

51. Show that the third differences of a polynomial function of degree 3 are nonzero and constant. First, use $f(x) = x^3 - 3x^2 - 2x - 6$. Then show third differences are nonzero and constant for $f(x) = ax^3 + bx^2 + cx + d, a \neq 0$.

52. **Reasoning** Suppose that a function pairs elements from set A with elements from set B. A function is called *onto* if it pairs every element in B with at least one element in A. For each type of polynomial function, and for each set B, determine whether the function is *always, sometimes,* or *never* onto.
 a. linear; B = all real numbers
 b. quadratic; B = all real numbers
 c. quadratic; B = all real numbers greater than or equal to 4
 d. cubic; B = all real numbers

53. Make a table of second differences for each polynomial function. Using your tables, make a conjecture about the second differences of quadratic functions.
 a. $y = 2x^2$
 b. $y = 5x^2$
 c. $y = 5x^2 - 2$
 d. $y = 7x^2$
 e. $y = 7x^2 + 1$
 f. $y = 7x^2 + 3x + 1$

54. a. Write the equation for the volume of a box with a length that is 5 in. less than its width and a height that is 3 in. less than its width.
 b. Graph the equation.
 c. For which interval(s) does the graph increase?
 d. For which interval(s) does the graph decrease?

55. Copy and complete the table, which shows the first and second differences in y-values for consecutive x-values for a polynomial function of degree 2.

x	y	1st diff.	2nd diff.
−3	14	−8	2
−2	6	■	2
−1	■	−4	2
0	−4	−2	2
1	■	0	2
2	−6	■	
3	■		

56. The outputs for a certain function are 1, 2, 4, 8, 16, 32, and so on.
   a. Find the first differences of this function.
   b. Find the second differences of this function.
   c. Find the tenth difference of this function.
   d. Can you find a polynomial function that matches the original outputs? Explain your reasoning.

57. **Reasoning** A cubic polynomial function f has leading coefficient 2 and constant term 7. If $f(1) = 7$ and $f(2) = 9$, what is $f(-2)$? Explain how you found your answer.

Standardized Test Prep

58. Which expression is a cubic polynomial?

(A) x^3 (B) $3x + 3$ (C) $2x^2 + 3x - 1$ (D) $3x$

59. Which equation has $-3 \pm 5i$ as its solutions?

(F) $x^2 + 6x = -34$ (G) $x^2 + 6x = -14$ (H) $x^2 + 3x = 4$ (I) $x^2 + 3x = 2$

60. What is the discriminant of $qx^2 + rx + s = 0$?

(A) qrs (B) $q^2 - 4rs$ (C) $r^2 - 4qs$ (D) $s^2 - 4qr$

61. What is a simpler form of $x^2(3x^2 - 2x) - 3x^4$? Classify the polynomial by degree and by number of terms.

Mixed Review

Solve each system of equations.

62. $\begin{cases} y = x^2 - 3x - 7 \\ y = -2x + 3 \end{cases}$ 63. $\begin{cases} y = x^2 + x - 20 \\ y = x^2 + 2x \end{cases}$ 64. $\begin{cases} y = 3x^2 + 8x - 3 \\ y = x^2 - 9 \end{cases}$

Write an equation of each line in standard form with integer coefficients.

65. $y = 7x + 0.4$ 66. $y = -3x - 2.5$ 67. $y = -\frac{2}{7}x + 4$ 68. $y = 1.2x - 0.5$

Get Ready! **To prepare for Lesson 5-2, do Exercises 69–71.**

Factor each quadratic expression.

69. $x^2 + 7x + 12$ 70. $x^2 + 8x - 20$ 71. $x^2 - 14x + 24$

5-2 Polynomials, Linear Factors, and Zeros

Content Standards

F.IF.7.c. Graph polynomial functions, identifying zeros when suitable factorizations are available and showing end behavior.

Also A.APR.3

Objectives To analyze the factored form of a polynomial
To write a polynomial function from its zeros

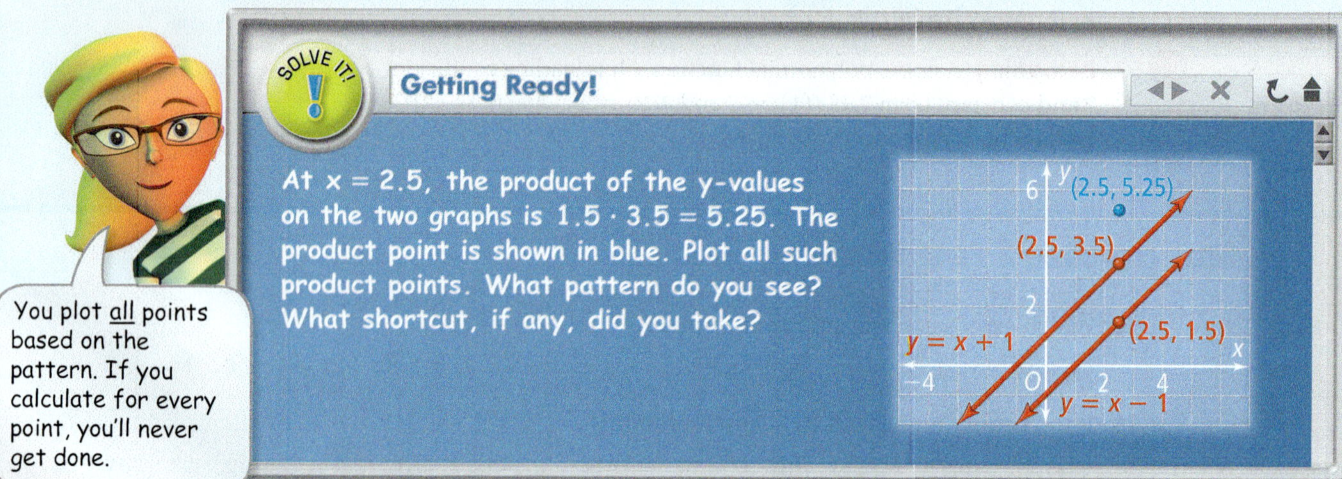

MATHEMATICAL PRACTICES If $P(x)$ is a polynomial function, the solutions of the related polynomial equation $P(x) = 0$ are the zeros of the function.

Lesson Vocabulary
- factor theorem
- multiple zero
- multiplicity
- relative maximum
- relative minimum

Essential Understanding Finding the zeros of a polynomial function will help you factor the polynomial, graph the function, and solve the related polynomial equation.

In Chapter 4, you solved a quadratic equation of the form $x^2 + bx + c = 0$ by factoring. You wrote it using *linear factors* in the form $(x - r_1)(x - r_2) = 0$. Then you applied the Zero-Product Property to find the solutions $x = r_1$ and $x = r_2$. You can solve some polynomial equations $a_n x^n + a_{n-1}x^{n-1} + \cdots + a_0 = 0$ in much the same way.

Plan

How do you write the factored form of a polynomial?
Write the polynomial as a product of factors. Make sure each factor cannot be factored any further.

Problem 1 Writing a Polynomial in Factored Form

What is the factored form of $x^3 - 2x^2 - 15x$?

$x^3 - 2x^2 - 15x = x(x^2 - 2x - 15)$ — Factor out the GCF, x.

$= x(x - 5)(x + 3)$ — Factor $x^2 - 2x - 15$.

Check $x(x - 5)(x + 3) = x(x^2 - 2x - 15)$ — Multiply $(x - 5)(x + 3)$.

$= x^3 - 2x^2 - 15x$ ✔ — Distributive Property

Got It? **1.** What is the factored form of $x^3 - x^2 - 12x$?

take note

Key Concepts Roots, Zeros, and x-intercepts

The following are equivalent statements about a real number b and a polynomial $P(x) = a_n x^n + a_{n-1}x^{n-1} + \cdots + a_1 x + a_0$.

- $x - b$ is a linear factor of the polynomial $P(x)$.
- b is a zero of the polynomial function $y = P(x)$.
- b is a root (or solution) of the polynomial equation $P(x) = 0$.
- b is an x-intercept of the graph of $y = P(x)$.

Dynamic Activity Polynomials and Linear Factors

Problem 2 Finding Zeros of a Polynomial Function

What are the zeros of $y = (x + 2)(x - 1)(x - 3)$? Graph the function.

Know	Need	Plan
Polynomial function	• Zeros • Additional points • End behavior	• Use the Zero-Product Property to find zeros. • Find points between the zeros. • Sketch the graph.

Think

Does knowing the zeros of a function give you enough information to sketch it?
No; several different cubic functions could pass through (−2, 0), (1, 0), and (3,0).

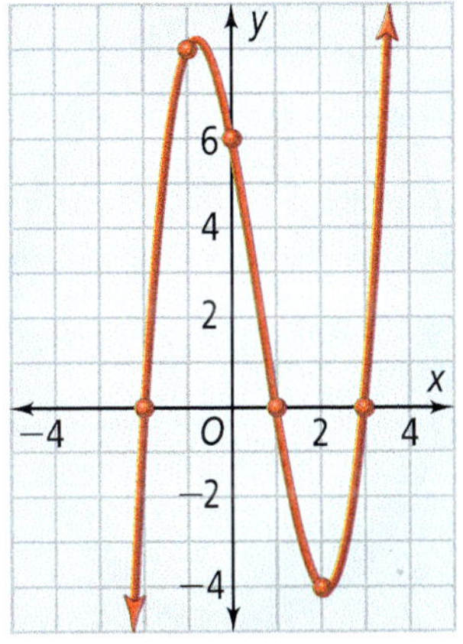

Step 1 Use the Zero-Product Property to find the zeros.

$$(x + 2)(x - 1)(x - 3) = 0$$

so $x + 2 = 0$ or $x - 1 = 0$ or $x - 3 = 0$.
The zeros of the function are −2, 1, and 3.

Step 2 Find points for x-values between the zeros.
Evaluate $y = (x + 2)(x - 1)(x - 3)$ for $x = -1, 0,$ and 2.

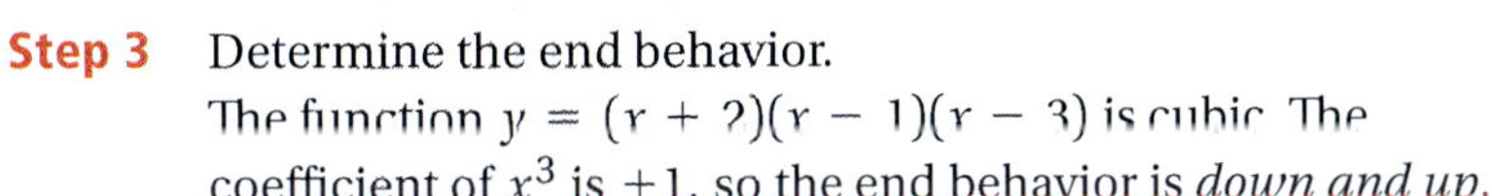

$(-1 + 2)(-1 - 1)(-1 - 3) = 8$ $(-1, 8)$

$(0 + 2)(0 - 1)(0 - 3) = 6$ $(0, 6)$

$(2 + 2)(2 - 1)(2 - 3) = -4$ $(2, -4)$

Step 3 Determine the end behavior.
The function $y = (x + 2)(x - 1)(x - 3)$ is cubic. The coefficient of x^3 is +1, so the end behavior is *down and up*.

Step 4 Use the zeros: (−2, 0), (1, 0), (3, 0); the additional points: (−1, 8), (0, 6), (2, −4); and end behavior to sketch the graph.

Got It? **2.** What are the zeros of $y = x(x - 3)(x + 5)$? Graph the function.

The Factor Theorem describes the relationship between the linear factors of a polynomial and the zeros of a polynomial.

Theorem Factor Theorem

The expression $x - a$ is a factor of a polynomial if and only if the value a is a zero of the related polynomial function.

Problem 3 Writing a Polynomial Function From Its Zeros

Plan

How can you use the zeros to find the function?
By the Factor Theorem, a is a zero means that $x - a$ is a factor of the related polynomial.

A What is a cubic polynomial function in standard form with zeros −2, 2, and 3?

$$\begin{array}{ll} \quad -2 \quad\; 2 \quad\; 3 & -2, 2, \text{ and } 3 \text{ are zeros.} \\ \quad\downarrow \quad\; \downarrow \quad\; \downarrow & \\ f(x) = (x+2)(x-2)(x-3) & \text{Write a linear factor for each zero.} \\ \quad = (x+2)(x^2-5x+6) & \text{Multiply } (x-2) \text{ and } (x-3). \\ \quad = x(x^2-5x+6) + 2(x^2-5x+6) & \text{Distributive Property} \\ \quad = x^3 - 5x^2 + 6x + 2x^2 - 10x + 12 & \text{Distributive Property} \\ \quad = x^3 - 3x^2 - 4x + 12 & \text{Simplify.} \end{array}$$

The cubic polynomial $f(x) = x^3 - 3x^2 - 4x + 12$ has zeros −2, 2, and 3.

B What is a quartic polynomial function in standard form with zeros −2, −2, 2, and 3?

$$\begin{array}{ll} \quad -2 \quad -2 \quad 2 \quad 3 & -2, -2, 2, \text{ and } 3 \text{ are zeros.} \\ \quad\downarrow \quad\;\; \downarrow \quad\; \downarrow \quad \downarrow & \\ g(x) = (x+2)(x+2)(x-2)(x-3) & \text{Write a linear factor for each zero.} \\ \quad = x^4 - x^3 - 10x^2 + 4x + 24 & \text{Simplify.} \end{array}$$

The quartic polynomial $g(x) = x^4 - x^3 - 10x^2 + 4x + 24$ has zeros −2, −2, 2, and 3.

C Graph both functions. How do the graphs differ? How are they similar?

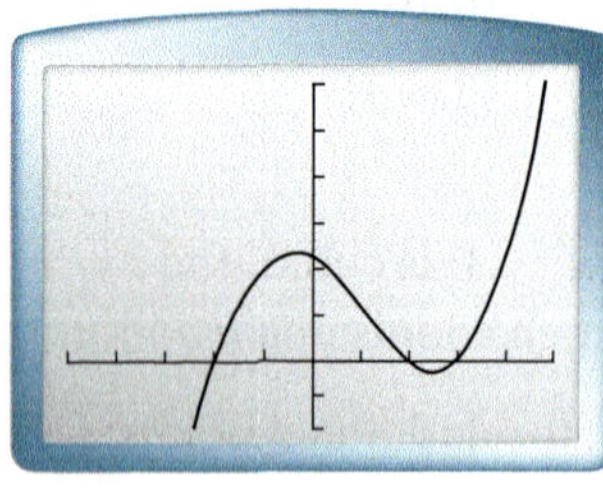

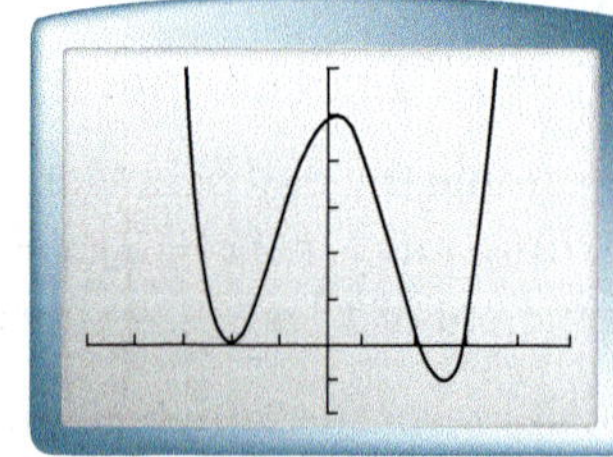

Both screens:
x-scale: 1
y-scale: 5

Both graphs have x-intercepts at −2, 2, and 3. The cubic has down-and-up end behavior. The quartic has up-and-up end behavior.

The cubic function has two turning points, and it crosses the x-axis at −2. The quartic function touches the x-axis at −2 but does not cross it. The quartic function has three turning points.

  **Got It? 3. a.** What is a quadratic polynomial function with zeros 3 and −3?

b. What is a cubic polynomial function with zeros 3, 3, and −3?

c. Reasoning Graph both functions. How do the graphs differ? How are they similar?

You can write the polynomial functions in Problem 3 in factored form as $f(x) = (x + 2)(x - 2)(x - 3)$ and $g(x) = (x + 2)^2(x - 2)(x - 3)$. In $g(x)$ the repeated linear factor $x + 2$ makes -2 a **multiple zero**.

In particular, since the linear factor $x + 2$ appears twice, you can say that -2 is a zero of **multiplicity** 2. In general, *a is a zero of multiplicity n* means that $x - a$ appears n times as a factor.

take note

Key Concept How Multiple Zeros Affect a Graph

If a is a zero of multiplicity n in the polynomial function $y = P(x)$, then the behavior of the graph at the x-intercept a will be close to linear if $n = 1$, close to quadratic if $n = 2$, close to cubic if $n = 3$, and so on.

Problem 4 Finding the Multiplicity of a Zero

What are the zeros of $f(x) = x^4 - 2x^3 - 8x^2$? What are their multiplicities? How does the graph behave at these zeros?

$$f(x) = x^4 - 2x^3 - 8x^2$$

$= x^2(x^2 - 2x - 8)$ Factor out the GCF, x^2.

$= x^2(x + 2)(x - 4)$ Factor $(x^2 - 2x - 8)$.

Think

How can you find the multiplicities?
Factor the polynomial. Find the number of times each linear factor appears.

Since $x^2 = (x - 0)^2$, the number 0 is a zero of multiplicity 2. The numbers -2 and 4 are zeros of multiplicity 1.

The graph looks close to linear at the x-intercepts -2 and 4. It resembles a parabola at the x-intercept 0.

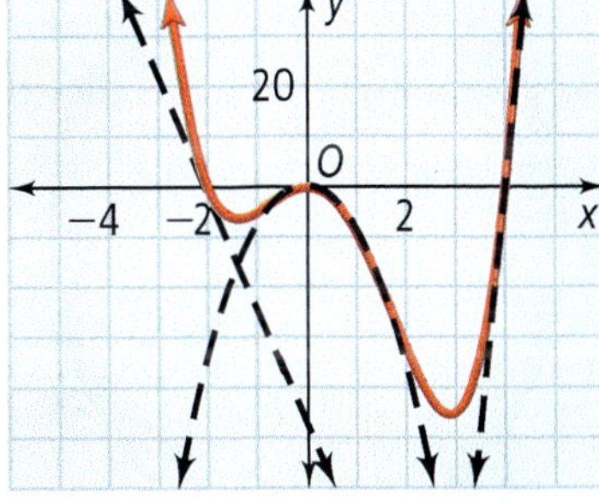

Got It? 4. What are the zeros of $f(x) = x^3 - 4x^2 + 4x$? What are their multiplicities? How does the graph behave at these zeros?

If the graph of a polynomial function has several turning points, the function can have a **relative maximum** and a **relative minimum**. A relative maximum is the value of the function at an up-to-down turning point. A relative minimum is the value of the function at a down-to-up turning point.

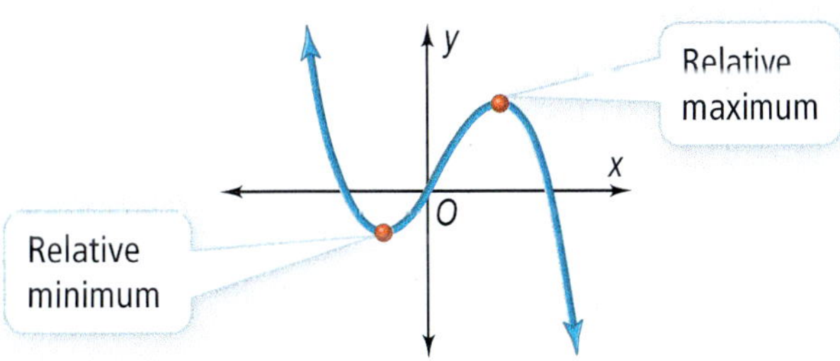

Problem 5 Identifying a Relative Maximum and Minimum

What are the relative maximum and minimum of $f(x) = x^3 + 3x^2 - 24x$?

Think

How is a relative maximum different from a maximum at the vertex of a parabola?
A relative maximum is the greatest *y*-value in the "neighborhood" of its *x*-value. The maximum at the vertex of a parabola is the greatest *y*-value for *all* *x*-values.

Use a graphing calculator to find a relative maximum and a relative minimum.

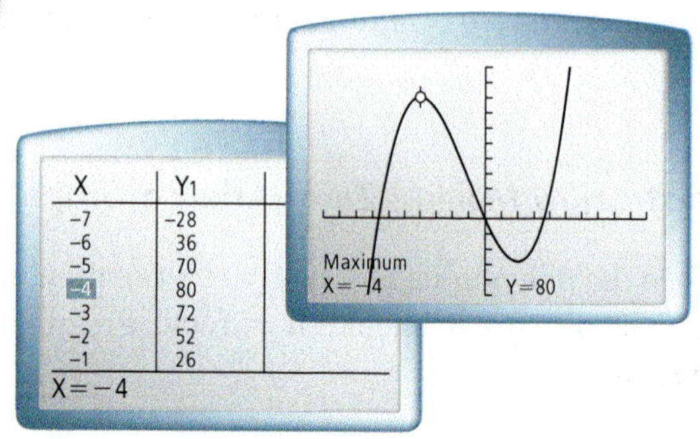

Relative maximum

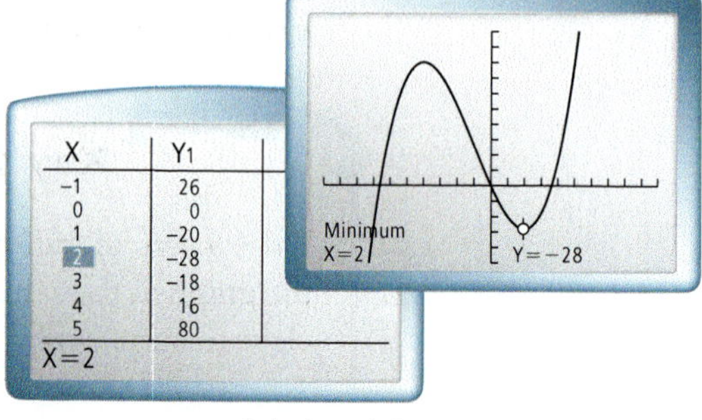

Relative minimum

The relative maximum is 80 at $x = -4$ and the relative minimum is -28 at $x = 2$.

Got It? 5. What are the relative maximum and minimum of $f(x) = 3x^3 + x^2 - 5x$?

Problem 6 Using a Polynomial Function to Maximize Volume

Technology **The design of a digital box camera maximizes the volume while keeping the sum of the dimensions at 6 inches. If the length must be 1.5 times the height, what should each dimension be?**

Think

What is the formula for the volume of a "box"?
$V = \ell wh$

Step 1 Define a variable x.

Let $x =$ the height of the camera.

Step 2 Determine length and width.

length $= 1.5x$; width $= 6 - (x + 1.5x) = 6 - 2.5x$

Step 3 Model the volume.

$$V = \text{(length)(width)(height)} = (1.5x)(6 - 2.5x)(x)$$
$$= -3.75x^3 + 9x^2$$

Step 4 Graph the polynomial function. Use the **MAXIMUM** feature to find that the maximum volume is 7.68 in.^{3} for a height of 1.6 in.

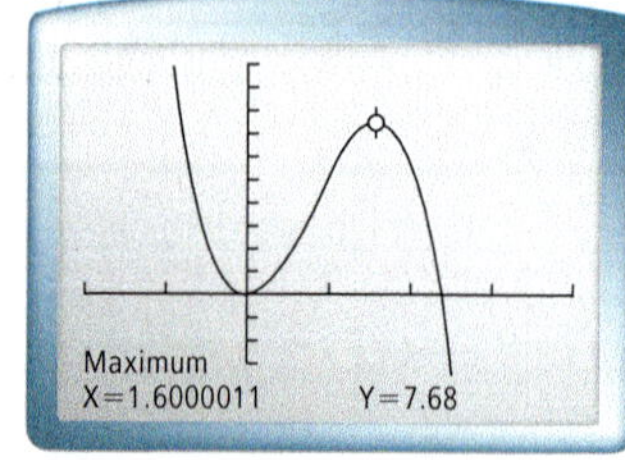

height $= x = 1.6$

length $= 1.5x = 1.5(1.6) = 2.4$

width $= 6 - 2.5x = 6 - 2.5(1.6) = 2$

The dimensions of the camera should be 2.4 in. long by 2 in. wide by 1.6 in. high.

Got It? 6. What is the maximum volume of the camera in Problem 6, if the sum of the dimensions is at most 4 inches?

Lesson Check

Do you know HOW?

Find the zeros of each function.

1. $y = x(x - 6)$
2. $y = (x + 4)(x - 5)$
3. $y = (x + 12)(x - 9)(x - 7)$
4. Write a polynomial function in standard form with zeros -1, 1, and 0.

Do you UNDERSTAND?

5. **Vocabulary** Write a polynomial function h in standard form that has 3 and -5 as zeros of multiplicity 2.
6. **Error Analysis** Your friend says that to write a function that has zeros 3 and -1, you should multiply the two factors $(x + 3)$ and $(x - 1)$ to get $f(x) = x^2 + 2x - 3$. Describe and correct your friend's error.

Practice and Problem-Solving Exercises

Write each polynomial in factored form. Check by multiplication.

See Problem 1.

7. $x^3 + 7x^2 + 10x$
8. $x^3 - 7x^2 - 18x$
9. $x^3 - 4x^2 - 21x$
10. $x^3 - 36x$
11. $x^3 + 8x^2 + 16x$
12. $9x^3 + 6x^2 - 3x$

Find the zeros of each function. Then graph the function.

See Problem 2.

13. $y = (x - 1)(x + 2)$
14. $y = (x - 2)(x + 9)$
15. $y = x(x + 5)(x - 8)$
16. $y = (x + 1)(x - 2)(x - 3)$
17. $y = (x + 1)(x - 1)(x - 2)$
18. $y = x(x + 2)(x + 3)$

Write a polynomial function in standard form with the given zeros.

See Problem 3.

19. $x = 5, 6, 7$
20. $x = -2, 0, 1$
21. $x = -5, -5, 1$
22. $x = 3, 3, 3$
23. $x = 1, -1, -2$
24. $x = 0, 4, -\frac{1}{2}$
25. $x = 0, 0, 2, 3$
26. $x = -1, -2, -3, -4$

Find the zeros of each function. State the multiplicity of multiple zeros.

See Problem 4.

27. $y = (x + 3)^3$
28. $y = x(x - 1)^3$
29. $y = 2x^3 + x^2 - x$
30. $y = 3x^3 - 3x$
31. $y = (x - 4)^2$
32. $y = (x - 2)^2(x - 1)$
33. $y = (2x + 3)(x - 1)^2$
34. $y = (x + 1)^2(x - 1)(x - 2)$

Find the relative maximum and relative minimum of the graph of each function.

See Problem 5.

35. $f(x) = x^3 + 4x^2 - 5x$
36. $f(x) = -x^3 + 16x^2 - 76x + 96$
37. $f(x) = -4x^3 + 12x^2 + 4x - 12$
38. $f(x) = x^3 - 7x^2 + 7x + 15$

 **39. Metalwork** A metalworker wants to make an open box from a sheet of metal, by cutting equal squares from each corner as shown.

See Problem 6.

a. Write expressions for the length, width, and height of the open box.

b. Use your expressions from part (a) to write a function for the volume of the box. (*Hint:* Write the function in factored form.)

c. Graph the function. Then find the maximum volume of the box and the side length of the cut-out squares that generates this volume.

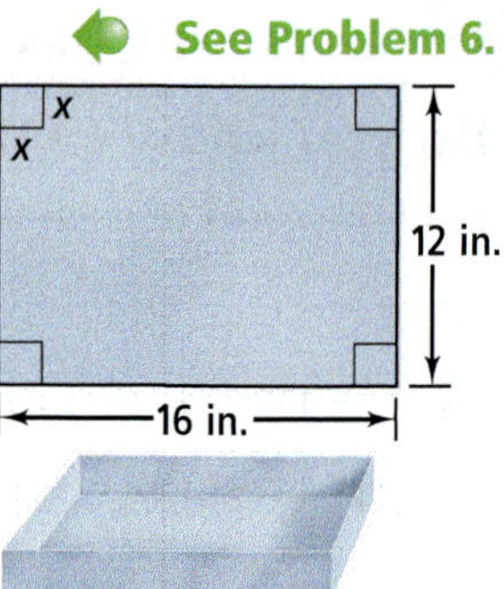

Apply

Write each function in factored form. Check by multiplication.

40. $y = 3x^3 - 27x^2 + 24x$ **41.** $y = -2x^3 - 2x^2 + 40x$ **42.** $y = x^4 + 3x^3 - 4x^2$

43. Think About a Plan A storage company needs to design a new storage box that has twice the volume of its largest box. Its largest box is 5 ft long, 4 ft wide, and 3 ft high. The new box must be formed by increasing each dimension by the same amount. Find the increase in each dimension.

- How can you write the dimensions of the new storage box as polynomial expressions?
- How can you use the volume of the current largest box to find the volume of the new box?

 **44. Carpentry** A carpenter hollowed out the interior of a block of wood as shown at the right.

a. Express the volume of the original block and the volume of the wood removed as polynomials in factored form.

b. What polynomial represents the volume of the wood remaining?

45. Geometry A rectangular box is $2x + 3$ units long, $2x - 3$ units wide, and $3x$ units high. What is its volume, expressed as a polynomial?

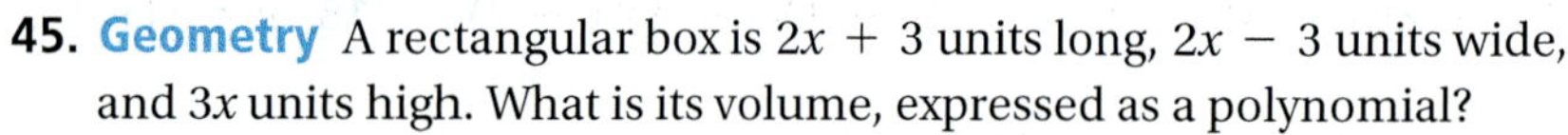

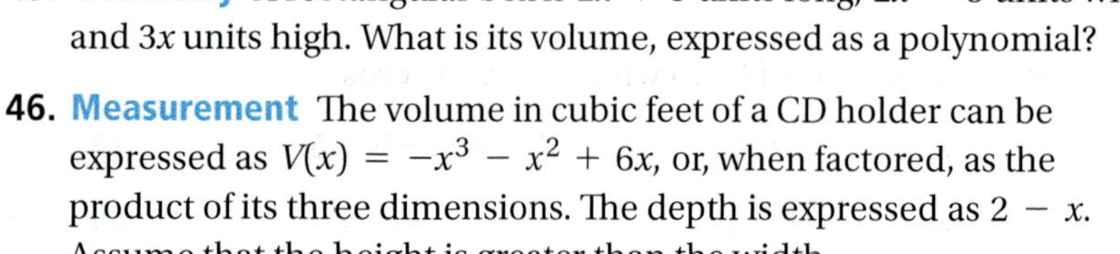

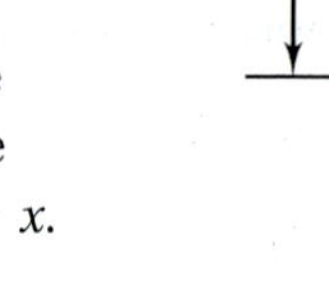

46. Measurement The volume in cubic feet of a CD holder can be expressed as $V(x) = -x^3 - x^2 + 6x$, or, when factored, as the product of its three dimensions. The depth is expressed as $2 - x$. Assume that the height is greater than the width.

a. Factor the polynomial to find linear expressions for the height and the width.

b. Graph the function. Find the x-intercepts. What do they represent?

c. What is a realistic domain for the function?

d. What is the maximum volume of the CD holder?

Find the relative maximum, relative minimum, and zeros of each function.

47. $y = 2x^3 - 23x^2 + 78x - 72$ **48.** $y = 8x^3 - 10x^2 - x - 3$ **49.** $y = (x + 1)^4 - 1$

50. Open-Ended Write a polynomial function with the following features: it has three distinct zeros; one of the zeros is 1; another zero has a multiplicity of 2.

51. Writing Explain how the graph of a polynomial function can help you factor the polynomial.

For each function, determine the zeros. State the multiplicity of any multiple zeros.

52. $f(x) = x^3 - 36x$ **53.** $y = (x + 1)(x - 4)(3 - 2x)$ **54.** $y = (x + 7)(5x + 2)(x - 6)^2$

55. Find a fourth-degree polynomial function with zeros 1, -1, i, and $-i$. Write the function in factored form.

56. a. Compare the graphs of $y = (x + 1)(x + 2)(x + 3)$ and $y = (x - 1)(x - 2)(x - 3)$. What transformation could you use to describe the change from one graph to the other?

b. Compare the graphs of $y = (x + 1)(x + 3)(x + 7)$ and $y = (x - 1)(x - 3)(x - 7)$. Does the transformation that you chose in part (a) still hold true? Explain.

c. Make a Conjecture What transformation could you use to describe the effect of changing the signs of the zeros of a polynomial function?

Standardized Test Prep

SAT/ACT

57. The three most frequent letters in the English language are E, T, and A. They represent, on average, 30% of all letters. The most frequent letter, E, is 4% more frequent than the second most frequent letter, T. The combined frequency of T and A is 4% greater than the frequency of E. Approximately how many E's can you expect to encounter in a 500-letter paragraph?

(A) 49 (B) 65 (C) 72 (D) 88

58. Which expression is the factored form of $x^3 + 2x^2 - 5x - 6$?

(F) $(x + 1)(x + 1)(x - 6)$ (H) $(x + 2)(2x - 5)(x - 6)$

(G) $(x + 3)(x + 1)(x - 2)$ (I) $(x - 3)(x - 1)(x + 2)$

59. A ball with a 3 in. radius has volume V_1. A second ball has a 9 in. radius and volume V_2. Which equation represents the volume of the second ball in terms of the first?

(A) $V_2 = 3V_1$ (B) $V_2 = 27V_1$ (C) $V_2 = {V_1}^2$ (D) $V_2 = 9{V_1}^2$

Extended Response

60. What is the polynomial function, in factored form, whose zeros are -2, 5, and 6, and whose leading coefficient is -2? Graph this function and find any relative minimums or maximums.

Mixed Review

Write each polynomial in standard form. Then classify it by degree and by number of terms.

See Lesson 5-1.

61. $x^2 - 1 - 3x^5 + 2x^2$ **62.** $-2x^3 - 7x^4 + x^3$ **63.** $6x + x^3 - 6x - 2$

Factor each expression. See Lesson 4-4.

64. $x^2 + 5x + 4$ **65.** $x^2 - 2x - 15$ **66.** $x^2 - 12x + 36$

Get Ready! **To prepare for Lesson 5-3, do Exercises 67–69.**

Solve each quadratic equation using any method.

See Lesson 4-7.

67. $x^2 + x - 6 = 0$ **68.** $2x^2 - 7x + 3 = 0$ **69.** $4x^2 - 25 = 0$

5-3 Solving Polynomial Equations

Content Standards

A.REI.11 Explain why the x-coordinates of the points where the graphs of the equations $y = f(x)$ and $y = g(x)$ intersect are the solutions of the equation $f(x) = g(x)$ . . .

Also A.SSE.2

Objectives To solve polynomial equations by factoring
To solve polynomial equations by graphing

I count 2 pieces with area x^2, 11 with area x, and 12 with area 1. The rectangle would have the same total area.

Getting Ready!

Can you arrange all of these pieces to make a rectangle with no pieces overlapping and no gaps? If you can, make a sketch. If you cannot, explain why.

Factoring a polynomial like $ax^2 + bx + c$ can help you solve a polynomial equation like $ax^2 + bx + c = 0$.

Lesson Vocabulary
- sum of cubes
- difference of cubes

Essential Understanding If $(x - a)$ is a factor of a polynomial, then the polynomial has value 0 when $x = a$. If a is a real number, then the graph of the polynomial has $(a, 0)$ as an x-intercept.

To solve a polynomial equation by factoring:

1. Write the equation in the form $P(x) = 0$ for some polynomial function P.
2. Factor $P(x)$. Use the Zero Product Property to find the roots.

Problem 1 Solving Polynomial Equations Using Factors

Plan

What does it mean if x is a common factor of every term in $P(x)$? You can write $P(x)$ as $xQ(x)$, so 0 will be a solution of $P(x) = 0$.

What are the real or imaginary solutions of each polynomial equation?

A $2x^3 - 5x^2 = 3x$

$2x^3 - 5x^2 - 3x = 0$	Rewrite in the form $P(x) = 0$.
$x(2x^2 - 5x - 3) = 0$	Factor out the GCF, x.
$x(2x + 1)(x - 3) = 0$	Factor $2x^2 - 5x - 3$.
$x = 0$ or $2x + 1 = 0$ or $x - 3 = 0$	Zero Product Property
$x = 0$ $\quad x = -\frac{1}{2}$ $\quad x = 3$	Solve each equation for x.

The solutions are $0, -\frac{1}{2}$, and 3.

B $3x^4 + 12x^2 = 6x^3$

$3x^4 - 6x^3 + 12x^2 = 0$	Rewrite in the form $P(x) = 0$.
$x^4 - 2x^3 + 4x^2 = 0$	Multiply by $\frac{1}{3}$ to simplify.
$x^2(x^2 - 2x + 4) = 0$	Factor out the GCF, x^2.
$x^2 = 0$ or $x^2 - 2x + 4 = 0$	Zero Product Property

Think

How will the solution be similar to the solution of the equation in part (a)?
Both equations have 0 as a solution, but here it will have a multiplicity of 2.

$$x = 0 \quad \Big| \quad x = \frac{-(-2) \pm \sqrt{(-2)^2 - 4(1)(4)}}{2(1)}$$

Use the Quadratic Formula to solve $x^2 - 2x + 4 = 0$. Substitute $a = 1$, $b = -2$, and $c = 4$.

$$x = \frac{2 \pm \sqrt{-12}}{2} = \frac{2 \pm 2i\sqrt{3}}{2} = 1 \pm i\sqrt{3}$$

The solutions are 0, $1 + i\sqrt{3}$, and $1 - i\sqrt{3}$.

Got It? **1.** What are the real or imaginary solutions of each equation?

a. $(x^2 - 1)(x^2 + 4) = 0$ **b.** $x^5 + 4x^3 = 5x^4 - 2x^3$

take note

Concept Summary Polynomial Factoring Techniques

Techniques	Examples
Factoring out the GCF Factor out the greatest common factor of all the terms.	$15x^4 - 20x^3 + 35x^2$ $= 5x^2(3x^2 - 4x + 7)$
Quadratic Trinomials For $ax^2 + bx + c$, find factors with product ac and sum b.	$6x^2 + 11x - 10$ $= (3x - 2)(2x + 5)$
Perfect Square Trinomials $a^2 + 2ab + b^2 = (a + b)^2$ $a^2 - 2ab + b^2 = (a - b)^2$	$x^2 + 10x + 25 = (x + 5)^2$ $x^2 - 10x + 25 = (x - 5)^2$
Difference of Squares $a^2 - b^2 = (a + b)(a - b)$	$4x^2 - 15 = (2x + \sqrt{15})(2x - \sqrt{15})$
Factoring by Grouping $ax + ay + bx + by$ $= a(x + y) + b(x + y)$ $= (a + b)(x + y)$	$x^3 + 2x^2 - 3x - 6$ $= x^2(x + 2) + (-3)(x + 2)$ $= (x^2 - 3)(x + 2)$
Sum or Difference of Cubes $a^3 + b^3 = (a + b)(a^2 - ab + b^2)$ $a^3 - b^3 = (a - b)(a^2 + ab + b^2)$	$8x^3 + 1 = (2x + 1)(4x^2 - 2x + 1)$ $8x^3 - 1 = (2x - 1)(4x^2 + 2x + 1)$

The sum and difference of cubes is a new factoring technique.

Here's Why It Works Factoring $a^3 + b^3 = (a + b)(a^2 - ab + b^2)$:

$$a^3 + b^3 = a^3 + a^2b - a^2b - ab^2 + ab^2 + b^3 \quad \text{Add 0.}$$
$$= a^2(a + b) - ab(a + b) + b^2(a + b) \quad \text{Factor out } a^2, -ab, \text{ and } b^2.$$
$$= (a + b)(a^2 - ab + b^2) \quad \text{Factor out } (a + b).$$

For $a^3 - b^3 = (a - b)(a^2 + ab + b^2)$, you can follow steps similar to those above, or you can factor $a^3 - b^3$ as the sum of cubes $a^3 + (-b)^3$.

Problem 2 Solving Polynomial Equations by Factoring

Think

How can you write the polynomial in quadratic form?
Write in terms of x^2: $(x^2)^2 - 3(x^2) - 4 = 0$, which shows the factorable quadratic form $a^2 - 3a - 4 = 0$.

What are the real or imaginary solutions of each polynomial equation?

A $x^4 - 3x^2 = 4$

$$x^4 - 3x^2 - 4 = 0 \quad \text{Rewrite in the form } P(x) = 0.$$
$$a^2 - 3a - 4 = 0 \quad \text{Let } a = x^2.$$
$$(a - 4)(a + 1) = 0 \quad \text{Factor.}$$
$$(x^2 - 4)(x^2 + 1) = 0 \quad \text{Replace } a \text{ with } x^2.$$
$$(x + 2)(x - 2)(x^2 + 1) = 0 \quad \text{Factor } x^2 - 4 \text{ as a difference of squares.}$$

It follows from the Zero Product Property that $x = 2$, $x = -2$, or $x^2 = -1$. Solving $x^2 = -1$ yields two imaginary roots: $x = i$ or $x = -i$.

Check Graph the related function $y = x^4 - 3x^2 - 4$.

The graph shows zeros at $x = 2$ and $x = -2$. It also shows three turning points. This means that there are imaginary roots, which do not appear on the graph.

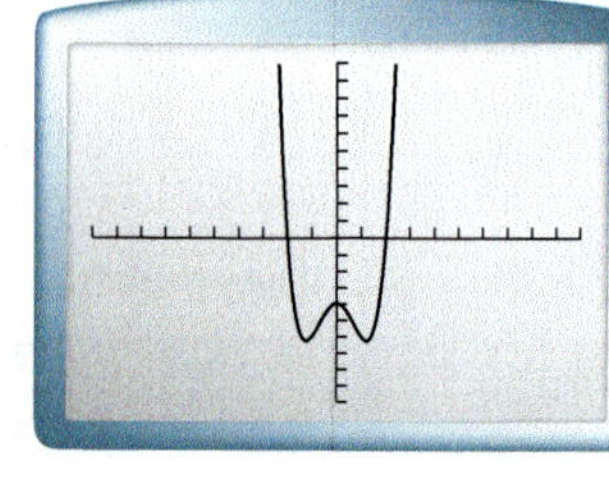

B $x^3 = 1$

$$x^3 - 1 = 0 \quad \text{Rewrite in the form } P(x) = 0.$$
$$(x - 1)(x^2 + x + 1) = 0 \quad \text{Factor the difference of cubes.}$$

It follows from the Zero Product Property that $x = 1$ or $x^2 + x + 1 = 0$. Use the Quadratic Formula to solve $x^2 + x + 1 = 0$.

$$x = \frac{-(1) \pm \sqrt{(1)^2 - 4(1)(1)}}{2(1)} = \frac{-1 \pm \sqrt{-3}}{2} = \frac{-1 \pm i\sqrt{3}}{2}$$

The three solutions of $x^3 = 1$ are 1, $-\frac{1}{2} + i\frac{\sqrt{3}}{2}$, and $-\frac{1}{2} - i\frac{\sqrt{3}}{2}$.

Got It? **2.** What are the real or imaginary solutions of each polynomial equation?

a. $x^4 = 16$ **b.** $x^3 = 8x - 2x^2$ **c.** $x(x^2 + 8) = 8(x + 1)$

While factoring is an effective way to solve a polynomial equation, you can also find the real roots quickly by using a graphing calculator.

Plan

Why is it helpful to graph Y_1 and Y_2? The values of x for which $Y_1 = Y_2$ are the solutions of the original equation.

Problem 3 Finding Real Roots by Graphing

What are the real solutions of the equation $x^3 + 5 = 4x^2 + x$?

Method 1 Graph $Y_1 = x^3 + 5$ and $Y_2 = 4x^2 + x$. Use the **INTERSECT** feature to find the x values of the points of intersection.

These two intersection points are visible in the standard viewing window.

You must adjust the window to obtain this point of intersection.

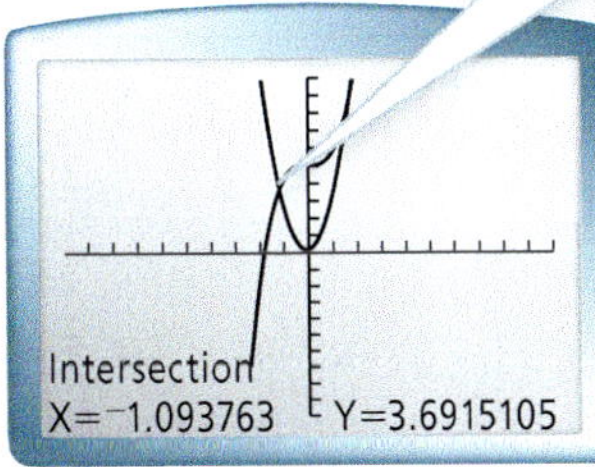

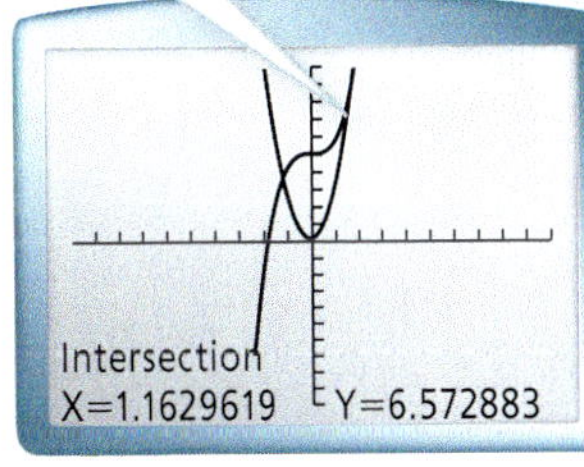

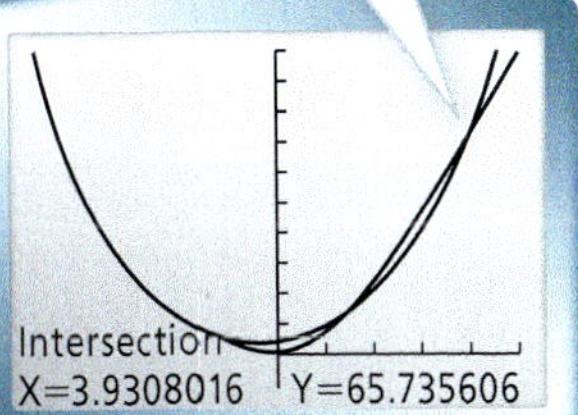

Approximate solutions are $x = -1.09$, $x = 1.16$, and $x = 3.93$.

Method 2 Rewrite the equation as $x^3 - 4x^2 - x + 5 = 0$. Graph the related function $y = x^3 - 4x^2 - x + 5$. Use the **ZERO** feature.

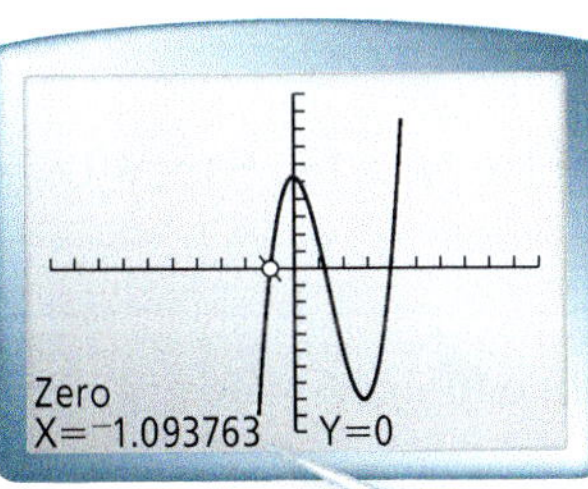

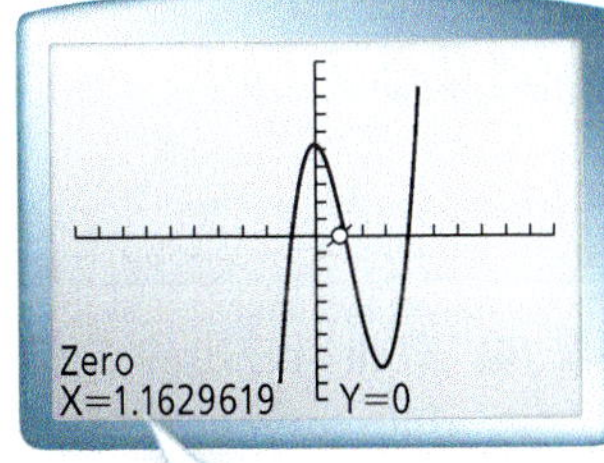

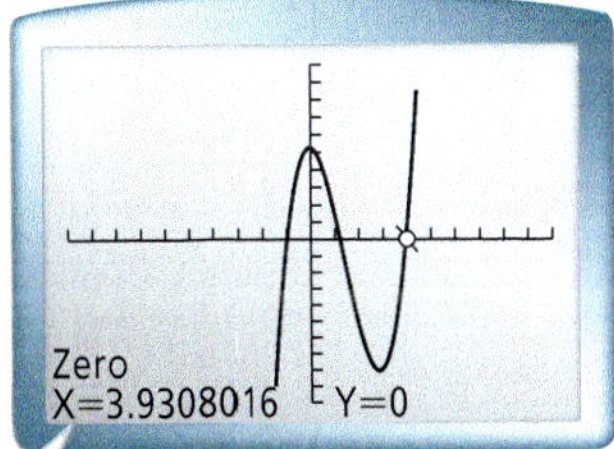

The solutions are the same as identified using Method 1.

Approximate solutions are $x = -1.09$, $x = 1.16$, and $x = 3.93$.

Check Verify the solutions by showing that they satisfy the original equation. Show values of $y_1 = x^3 + 5$ and $y_2 = 4x^2 + x$ in a table.

X	Y1	Y2
⁻1.09	3.69	3.69
1.16	6.57	6.57
3.93	65.7	65.7

X=

The solution checks.

Got It? 3. a. What are the real solutions of the equation $x^3 + x^2 = x - 1$?

b. Reasoning In Problem 3, which method seems to be an easier and more reliable way to find the solutions of an equation? Explain.

Problem 4 Modeling a Problem Situation

Close friends Stacy, Una, and Amir were all born on July 4. Stacy is one year younger than Una. Una is two years younger than Amir. On July 4, 2010, the product of their ages was 2300 more than the sum of their ages. How old was each friend on that day?

Think	Write
Define variables.	Let x = Una's age on July 4, 2010. Stacy's age = $x - 1$. Amir's age = $x + 2$.
Write an equation. Simplify and change it to $P(x) = 0$ form.	$\overbrace{x + (x - 1) + (x + 2)}^{\text{Sum of ages}} + 2300 = \overbrace{x(x - 1)(x + 2)}^{\text{Product of ages}}$ $3x + 2301 = x(x^2 + x - 2)$ $3x + 2301 = x^3 + x^2 - 2x$ $x^3 + x^2 - 5x - 2301 = 0$
Only real solutions make sense, so graphing $y_1 = P(x)$ should show any real solution that exists. Use the zero feature.	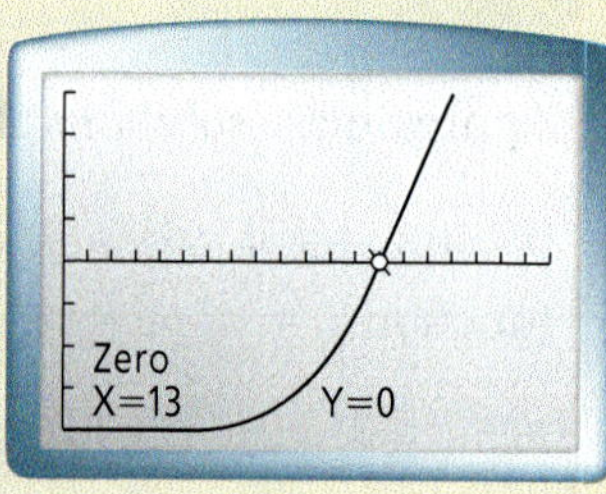 $x = 13$
Write the answer.	Una was 13, Stacy 12, and Amir 15.

Got It? **4.** What are three consecutive integers whose product is 480 more than their sum?

Lesson Check

Do you know HOW?

Factor each polynomial.

1. $x^2 - 3x - 18$

2. $x^3 - 27$

3. $x^3 + 3x^2 + 4x + 12$

4. $x^4 - y^4$

Solve each equation by factoring.

5. $2x^2 + 7x - 4 = 0$

6. $2x^3 + 2x^2 - 4x = 0$

Do you UNDERSTAND?

7. Vocabulary Identify each expression as a sum of cubes, difference of cubes, or difference of squares.

a. $x^2 - 64$

b. $x^3 + 8$

c. $x^3 - 125$

d. $x^2 - 81$

8. Reasoning Which method of solving polynomial equations will not identify the imaginary roots? Explain.

9. Reasoning Show two different ways to find the real roots of the polynomial equation $0 = x^6 - x^2$. Show your steps.

Practice and Problem-Solving Exercises

A Practice

Find the real or imaginary solutions of each equation by factoring. See Problems 1 and 2.

10. $x^3 + 64 = 0$
11. $x^3 - 1000 = 0$
12. $125x^3 - 27 = 0$
13. $64x^3 - 1 = 0$
14. $x^3 + 2x^2 + 5x + 10 = 0$
15. $6x^2 + 13x - 5 = 0$
16. $0 = x^3 - 27$
17. $0 = x^3 - 64$
18. $8x^3 = 1$
19. $64x^3 = -8$
20. $x^4 - 10x^2 = -9$
21. $x^4 - 8x^2 = -16$
22. $x^4 - 12x^2 = 64$
23. $x^4 + 7x^2 = 18$
24. $x^4 + 4x^2 = 12$

Find the real solutions of each equation by graphing. See Problem 3.

25. $x^3 - 4x^2 - 7x = -10$
26. $3x^3 - 6x^2 - 9x = 0$
27. $4x^3 - 8x^2 + 4x = 0$
28. $6x^2 = 48x$
29. $x^3 + 3x^2 + 2x = 0$
30. $2x^3 + 5x^2 = 7x$
31. $4x^3 = 4x^2 + 3x$
32. $2x^4 - 5x^3 - 3x^2 = 0$
33. $x^2 - 8x + 7 = 0$
34. $x^4 - 4x^3 - x^2 + 16x = 12$
35. $x^3 - x^2 - 16x = 20$
36. $3x^3 + 12x^2 - 3x = 12$

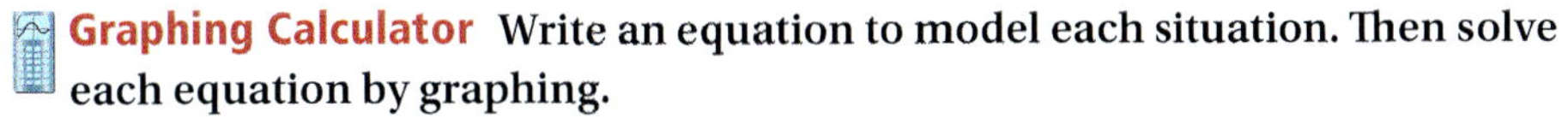

Graphing Calculator **Write an equation to model each situation. Then solve each equation by graphing.** See Problem 4.

37. The Johnson twins were born two years after their older sister. This year, the product of the three siblings ages is exactly 4558 more than the sum of their ages. How old are the twins?

38. The product of three consecutive integers is 210. What are the numbers?

B Apply

Solve each equation.

39. $x^3 + 13x = 10x^2$
40. $x^3 - 6x^2 + 6x = 0$
41. $12x^3 = 60x^2 + 75x$
42. $125x^3 + 216 = 0$
43. $81x^3 - 192 = 0$
44. $x^4 - 64 = 0$
45. $-2x^4 - 100 = 0$
46. $27 = -x^4 - 12x^2$
47. $x^5 - 5x^3 + 4x = 0$
48. $5x^3 = 5x^2 + 12x$
49. $x^3 + x^2 + x + 1 = 0$
50. $x^3 + 1 = x^2 + x$

51. Think About a Plan The width of a plastic storage box is 1 ft longer than the height. The length is 4 ft longer than the height. The volume is 36 ft^3. What are the dimensions of the box?

- What is the formula for the volume of a rectangular prism?
- What variable expressions represent the length, height, and width?
- What equation represents the volume of the plastic storage box?

52. Error Analysis A student claims that 1, 2, 3, and 4 are the zeros of a cubic polynomial function. Explain why the student is mistaken.

53. Geometry The width of a box is 2 m less than the length. The height is 1 m less than the length. The volume is 60 m^3. What is the length of the box?

Graph each function to find the zeros. Rewrite the function with the polynomial in factored form.

54. $y = 2x^2 + 3x - 5$ **55.** $y = x^4 - 10x^2 + 9$ **56.** $y = x^3 - 3x^2 + 4$

57. Open-Ended To solve a polynomial equation, you can use any combination of graphing, factoring, and the Quadratic Formula. Write and solve an equation to illustrate each method.

Challenge

58. The geometric figure at the right has volume $a^3 + b^3$. You can split it into three rectangular blocks (including the long one with side $a + b$). Explain how to use this figure to prove the factoring formula for the sum of cubes, $a^3 + b^3 = (a + b)(a^2 - ab + b^2)$.

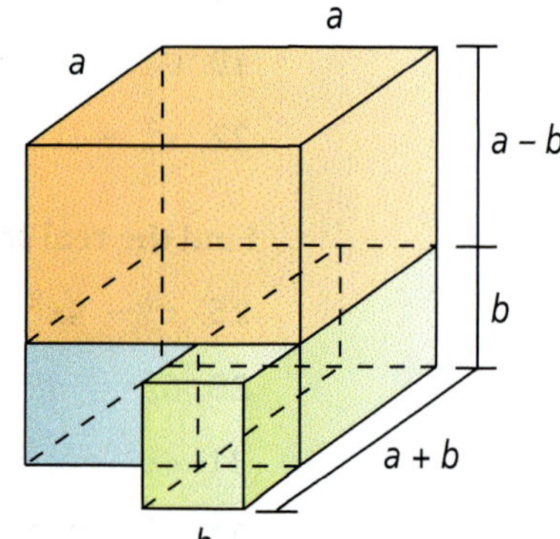

59. Open-Ended Find equations for two different polynomial functions whose zeros include $-12, 0, \frac{1}{4}$, and $\frac{1}{6}$.

60. What are the complex solutions of $x^5 + x^3 + 2x = 2x^4 + x^2 + 1$?

Standardized Test Prep

SAT/ACT

61. Which value is NOT a solution to the equation $x^4 - 3x^2 - 54 = 0$?

(A) -3 (B) 3 (C) $-3i$ (D) $-i\sqrt{6}$

62. Ava drove 3 hours at 45 miles per hour. How many miles did she drive?

(F) 45 miles (G) 48 miles (H) 90 miles (I) 135 miles

63. Which polynomial has the complex roots $1 + i\sqrt{2}$ and $1 - i\sqrt{2}$?

(A) $x^2 + 2x + 3$ (B) $x^2 - 2x + 3$ (C) $x^2 + 2x - 3$ (D) $x^2 - 2x - 3$

Short Response

64. Sam has only quarters and dimes in his pocket. He has a total of 12 coins, totaling \$1.95. How many of each coin does Sam have?

Mixed Review

Write each polynomial in factored form. Check by multiplication.

See Lesson 5-2.

65. $3x^2 - 18x + 24$ **66.** $2x^4 + 6x^3 - 18x^2 - 54x$ **67.** $x^4 - 4x^3 - 5x^2$

Solve each equation by factoring. Check your answers.

See Lesson 4-5.

68. $x^2 - 4x = 12$ **69.** $x^2 + 1 = 37$ **70.** $2x^2 - 5x - 3 = 0$

Get Ready! **To prepare for Lesson 5-4, do Exercises 71 and 72.**

Evaluate each expression for the given values of the variables.

See Lesson 1-3.

71. $\frac{16(x - 4)(y - 2)}{4(x - 3)y}$; $x = 1$ and $y = -2$ **72.** $\frac{2(x + 5)y}{10(x - 4)(y - 2)}$; $x = 1$ and $y = -2$

5-4 Dividing Polynomials

A.APR.2 Know and apply the Remainder Theorem: For a polynomial $p(x)$ and a number a, the remainder on division by $x - a$ is $p(a)$, so $p(a) = 0$ if and only if $(x - a)$ is a factor of $p(x)$.

Also A.APR.1, A.APR.6

Objectives To divide polynomials using long division
To divide polynomials using synthetic division

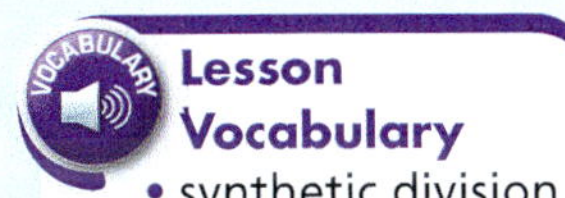

- synthetic division
- Remainder Theorem

Long division is one of many methods you can use to divide whole numbers.

Essential Understanding You can divide polynomials using steps that are similar to the long-division steps that you use to divide whole numbers.

When you try to factor a polynomial, you are trying to find a divisor of the polynomial that gives a quotient (the other factor) and remainder 0. This suggests that being able to divide one polynomial by another could help you factor polynomials.

Numerical long division and polynomial long division are similar.

Numerical Long Division

$\quad 32$	
$21\overline{)672}$	21 divides into
$\underline{63}$	67 3 times
42	21 divides into
$\underline{42}$	42 2 times
0	

Polynomial Long Division

$3x + 2$	
$2x + 1\overline{)6x^2 + 7x + 2}$	$(2x + 1)$ divides into
$\underline{6x^2 + 3x}$	$(6x^2 + 7x)$ $3x$ times
$4x + 2$	$(2x + 1)$ divides into
$\underline{4x + 2}$	$(4x + 2)$ 2 times
0	

The remainder from each division above is 0, so 21 is a factor of 672 and $2x + 1$ is a factor of $6x^2 + 7x + 2$.

Problem 1 Using Polynomial Long Division

Use polynomial long division to divide $4x^2 + 23x - 16$ by $x + 5$. What is the quotient and remainder?

$$\begin{array}{r l}
4x & \\
x+5\,\overline{)\,4x^2 + 23x - 16} & \text{Divide: } \frac{4x^2}{x} = 4x. \\
\underline{4x^2 + 20x} & \text{Multiply: } 4x(x+5) = 4x^2 + 20x. \\
3x - 16 & \text{Subtract to get } 3x. \text{ Bring down } -16.
\end{array}$$

Repeat the process of dividing, multiplying, and subtracting.

$$\begin{array}{r l}
4x + 3 & \\
x+5\,\overline{)\,4x^2 + 23x - 16} & \\
\underline{4x^2 + 20x} & \\
3x - 16 & \text{Divide: } \frac{3x}{x} = 3 \\
\underline{3x + 15} & \text{Multiply: } 3(x+5) = 3x + 15. \\
-31 & \text{Subtract to get } -31.
\end{array}$$

The quotient is $4x + 3$ with remainder -31. You can say: $4x + 3$, R -31.

Think

How can you check your result?
Show that (divisor)(quotient) + remainder = dividend.

Check

$(x + 5)(4x + 3) - 31 = (4x^2 + 3x + 20x + 15) - 31$ Multiply $(x + 5)(4x + 3)$.

$= 4x^2 + 23x - 16$ ✔ Simplify.

Got It? **1.** Use polynomial long division to divide $3x^2 - 29x + 56$ by $x - 7$. What is the quotient and remainder?

take note

Key Concept The Division Algorithm for Polynomials

You can divide polynomial $P(x)$ by polynomial $D(x)$ to get polynomial quotient $Q(x)$ and polynomial remainder $R(x)$. The result is $P(x) = D(x)Q(x) + R(x)$.

$$\begin{array}{r}
Q(x) \\
D(x)\,\overline{)\,P(x)} \\
\vdots \\
\overline{R(x)}
\end{array}$$

If $R(x) = 0$, then $P(x) = D(x)Q(x)$ and $D(x)$ and $Q(x)$ are factors of $P(x)$.

To use long division, $P(x)$ and $D(x)$ should be in standard form with zero coefficients where appropriate. The process stops when the degree of the remainder, $R(x)$, is less than the degree of the divisor, $D(x)$.

Problem 2 Checking Factors

A **Is $x^2 + 1$ a factor of $3x^4 - 4x^3 + 12x^2 + 5$?**

$$\begin{array}{r}
3x^2 - 4x + 9 \\
x^2 + 0x + 1\overline{)3x^4 - 4x^3 + 12x^2 + 0x + 5} \\
\underline{3x^4 + 0x^3 + 3x^2} \qquad\qquad \\
-4x^3 + 9x^2 + 0x \qquad \\
\underline{-4x^3 + 0x^2 - 4x} \qquad \\
9x^2 + 4x + 5 \\
\underline{9x^2 + 0x + 9} \\
4x - 4
\end{array}$$

Include $0x$ terms.

The degree of the remainder is less than the degree of the divisor. Stop!

The remainder is not zero. $x^2 + 1$ is not a factor of $3x^4 - 4x^3 + 12x^2 + 5$.

Plan

Can you use the Factor Theorem to help answer this question?
Yes; recall that if $P(a) = 0$, then $x - a$ is a factor of $P(x)$.

B **Is $x - 2$ a factor of $P(x) = x^5 - 32$? If it is, write $P(x)$ as a product of two factors.**

Step 1 Use the Factor Theorem to determine if $x - 2$ is a factor of $x^5 - 32$.

$$\begin{aligned}
P(2) &= 2^5 - 32 \\
&= 32 - 32 \\
&= 0
\end{aligned}$$

Since $P(2) = 0$, $x - 2$ is a factor of $P(x)$.

Step 2 Use polynomial long division to find the other factor.

$$\begin{array}{r}
x^4 + 2x^3 + 4x^2 + 8x + 16 \\
x - 2\overline{)x^5 + 0x^4 + 0x^3 + 0x^2 + 0x - 32} \\
\underline{x^5 - 2x^4} \qquad\qquad\qquad\qquad\qquad\quad \\
2x^4 + 0x^3 \qquad\qquad\qquad\qquad \\
\underline{2x^4 - 4x^3} \qquad\qquad\qquad\qquad \\
4x^3 + 0x^2 \qquad\qquad\qquad \\
\underline{4x^3 - 8x^2} \qquad\qquad\qquad \\
8x^2 + 0x \qquad\quad \\
\underline{8x^2 - 16x} \qquad\quad \\
16x - 32 \\
\underline{16x - 32} \\
0
\end{array}$$

$$P(x) = (x - 2)(x^4 + 2x^3 + 4x^2 + 8x + 16)$$

Got It? **2. a.** Is $x^4 - 1$ a factor of $P(x) = x^5 + 5x^4 - x - 5$? If it is, write $P(x)$ as a product of two factors.

b. Reasoning Use the fact that $12 \cdot 31 = 372$ to write $3x^2 + 7x + 2$ as the product of two factors.

Synthetic division simplifies the long-division process for dividing by a linear expression $x - a$. To use synthetic division, write the coefficients (including zeros) of the polynomial in standard form. Omit all variables and exponents. For the divisor, reverse the sign (use a). This allows you to add instead of subtract throughout the process.

Problem 3 Using Synthetic Division

Use synthetic division to divide $x^3 - 14x^2 + 51x - 54$ by $x + 2$. What is the quotient and remainder?

Think

To divide by $x + 2$ what number do you use for the synthetic divisor?

$x + 2 = x - (-2)$ so use -2.

Step 1 Reverse the sign of $+2$. Write the coefficients of the polynomial.

$$\begin{array}{r|rrrr} -2 & 1 & -14 & 51 & -54 \end{array}$$

Step 2 Bring down the first coefficient.

$$\begin{array}{r|rrrr} -2 & 1 & -14 & 51 & -54 \\ & & & & \\ \hline & 1 & & & \end{array}$$

Step 3 Multiply the coefficient by the divisor. Add to the next coefficient.

$$\begin{array}{r|rrrr} -2 & 1 & -14 & 51 & -54 \\ & & -2 & & \\ \hline & 1 & -16 & & \end{array}$$

Step 4 Continue multiplying and adding through the last coefficient.

$$\begin{array}{r|rrrr} -2 & 1 & -14 & 51 & -54 \\ & & -2 & 32 & -166 \\ \hline & 1 & -16 & 83 & -220 \end{array}$$

The quotient is $x^2 - 16x + 83$, R -220.

Got It? **3.** Use synthetic division to divide $x^3 - 57x + 56$ by $x - 7$. What is the quotient and remainder?

Problem 4 Using Synthetic Division to Solve a Problem

Crafts **The polynomial $x^3 + 7x^2 - 38x - 240$ expresses the volume, in cubic inches, of the shadow box shown.**

Plan

How can you use the picture to help solve the problem?

The picture gives the width of box. Remember for a rectangular prism, $V = \ell \times w \times h$.

A **What are the dimensions of the box? (*Hint:* The length is greater than the height (or depth).)**

$$\begin{array}{r|rrrr} -5 & 1 & 7 & -38 & -240 \\ & & -5 & -10 & 240 \\ \hline & 1 & 2 & -48 & 0 \end{array}$$

$x^2 + 2x - 48 = (x - 6)(x + 8)$

So, $x^3 + 7x^2 - 38x - 240 = (x + 5)(x^2 + 2x - 48)$
$= (x + 5)(x - 6)(x + 8)$

The length, width, and height (or depth) of the box are $(x + 8)$ in., $(x + 5)$ in., and $(x - 6)$ in., respectively.

B **If the width of the box is 15 in., what are the other two dimensions?**

The width of the box is $x + 5$. So if $x + 5 = 15$, then $x = 10$.

Substitute for x to find the length and height (or depth).

Length: $x + 8 = 10 + 8 = 18$ in.
Height: $x - 6 = 10 - 6 = 4$ in.

Got It? 4. If the polynomial $x^3 + 6x^2 + 11x + 6$ expresses the volume, in cubic inches, of the box, and the width is $(x + 1)$ in., what are the dimensions of the box?

The **Remainder Theorem** provides a quick way to find the remainder of a polynomial long-division problem.

take note

Theorem The Remainder Theorem

If you divide a polynomial $P(x)$ of degree $n \geq 1$ by $x - a$, then the remainder is $P(a)$.

Here's Why It Works When you divide polynomial $P(x)$ by $D(x)$, you find $P(x) = D(x)Q(x) + R(x)$.

$P(x) = (x - a)Q(x) + R(x)$	Substitute $(x - a)$ for $D(x)$.
$P(a) = (a - a)Q(a) + R(a)$	Evaluate $P(a)$. Substitute a for x.
$= R(a)$	Simplify.

Problem 5 Evaluating a Polynomial

GRIDDED RESPONSE

Given that $P(x) = x^5 - 2x^3 - x^2 + 2$, what is $P(3)$?

Think
Is there a way to find $P(3)$ without substituting?
Use synthetic division. $P(3)$ is the remainder.

By the Remainder Theorem, $P(3)$ is the remainder when you divide $P(x)$ by $x - 3$.

3	1	0	−2	−1	0	2
		3	9	21	60	180
	1	3	7	20	60	182

$P(3) = 182$.

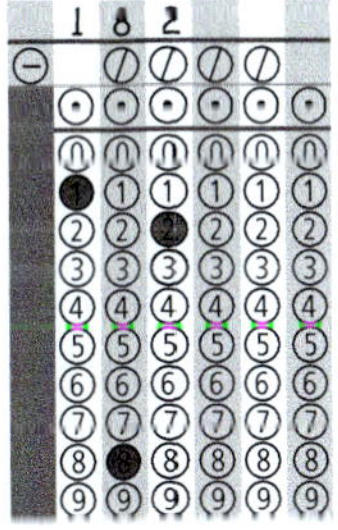

Got It? 5. Given that $P(x) = x^5 - 3x^4 - 28x^3 + 5x + 20$, what is $P(-4)$?

Lesson Check

Do you know HOW?

Divide using any method.

1. $(2x^2 + 7x + 11) \div (x + 2)$
2. $(x^3 + 5x^2 + 11x + 15) \div (x + 3)$
3. $(x^3 - x^2 - 4x + 4) \div (x - 2)$
4. $(4x^3 + 21x^2 - x - 24) \div (x + 5)$
5. $(9x^3 - 15x^2 + 4x) \div (x - 3)$

Do you UNDERSTAND?

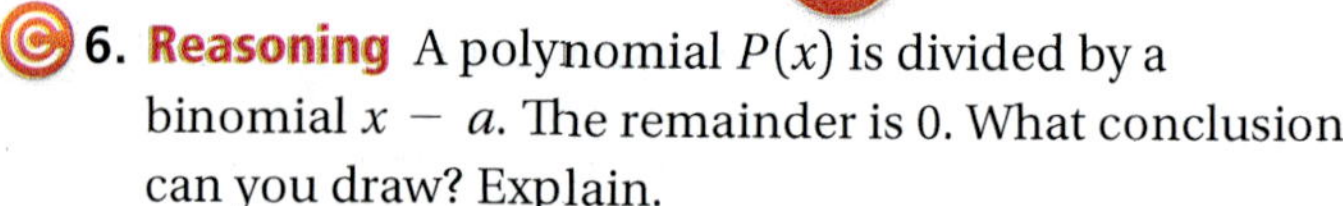

6. **Reasoning** A polynomial $P(x)$ is divided by a binomial $x - a$. The remainder is 0. What conclusion can you draw? Explain.

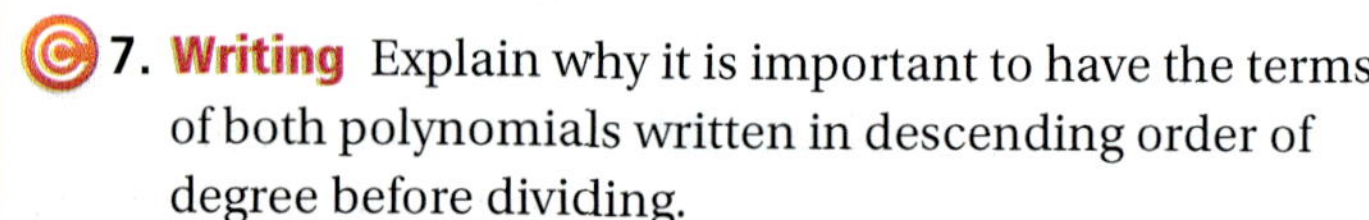

7. **Writing** Explain why it is important to have the terms of both polynomials written in descending order of degree before dividing.

8. **Open-Ended** Write a polynomial division that has a quotient of $x + 3$ and a remainder of 2.

Practice and Problem-Solving Exercises

A Practice

Divide using long division. Check your answers. See Problem 1.

9. $(x^2 - 3x - 40) \div (x + 5)$
10. $(3x^2 + 7x - 20) \div (x + 4)$
11. $(x^3 + 3x^2 - x + 2) \div (x - 1)$
12. $(2x^3 - 3x^2 - 18x - 8) \div (x - 4)$
13. $(3x^3 + 9x^2 + 8x + 4) \div (x + 2)$
14. $(9x^2 - 21x - 20) \div (x - 1)$
15. $(x^2 - 7x + 10) \div (x + 3)$
16. $(x^3 - 13x - 12) \div (x - 4)$

Determine whether each binomial is a factor of $x^3 + 4x^2 + x - 6$. See Problem 2.

17. $x + 1$
18. $x + 2$
19. $x + 3$
20. $x - 3$

Divide using synthetic division. See Problem 3.

21. $(x^3 + 3x^2 - x - 3) \div (x - 1)$
22. $(x^3 - 4x^2 + 6x - 4) \div (x - 2)$
23. $(x^3 - 7x^2 - 7x + 20) \div (x + 4)$
24. $(x^3 - 3x^2 - 5x - 25) \div (x - 5)$
25. $(x^2 + 3) \div (x - 1)$
26. $(3x^3 + 17x^2 + 21x - 9) \div (x + 3)$
27. $(x^3 + 27) \div (x + 3)$
28. $(6x^2 - 8x - 2) \div (x - 1)$

Use synthetic division and the given factor to completely factor each polynomial function. See Problem 4.

29. $y = x^3 + 2x^2 - 5x - 6; (x + 1)$
30. $y = x^3 - 4x^2 - 9x + 36; (x + 3)$

31. **Geometry** The volume, in cubic inches, of the decorative box shown can be expressed as the product of the lengths of its sides as $V(x) = x^3 + x^2 - 6x$. What linear expressions with integer coefficients represent the length and height of the box?

Use synthetic division and the Remainder Theorem to find $P(a)$.

See Problem 5.

32. $P(x) = x^3 + 4x^2 - 8x - 6; a = -2$
33. $P(x) = x^3 + 4x^2 + 4x; a = -2$
34. $P(x) = x^3 - 7x^2 + 15x - 9; a = 3$
35. $P(x) = x^3 + 7x^2 + 4x; a = -2$
36. $P(x) = 6x^3 - x^2 + 4x + 3; a = 3$
37. $P(x) = 2x^3 - x^2 + 10x + 5; a = \frac{1}{2}$
38. $P(x) = 2x^3 + 4x^2 - 10x - 9; a = 3$
39. $P(x) = 2x^4 + 6x^3 + 5x^2 - 45; a = -3$

40. **Think About a Plan** Your friend multiplies $x + 4$ by a quadratic polynomial and gets the result $x^3 - 3x^2 - 24x + 30$. The teacher says that everything is correct except for the constant term. Find the quadratic polynomial that your friend used. What is the correct result of multiplication?
    - What does the fact that all the terms except for the constant are correct tell you?
    - How can polynomial division help you solve this problem?
    - What is the connection between the remainder of the division and your friend's error?

41. **Error Analysis** A student used synthetic division to divide $x^3 - x^2 - 2x$ by $x + 1$. Describe and correct the error shown.

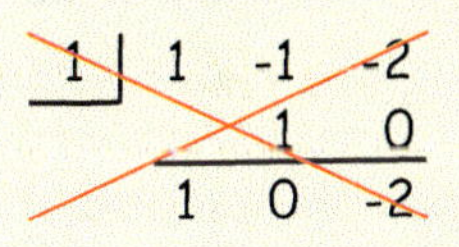

42. **Reasoning** When a polynomial is divided by $(x - 5)$, the quotient is $5x^2 + 3x + 12$ with remainder 7. Find the polynomial.

43. **Geometry** The expression $\frac{1}{3}(x^3 + 5x^2 + 8x + 4)$ represents the volume of a square pyramid. The expression $x + 1$ represents the height of the pyramid. What expression represents the side length of the base? (*Hint:* The formula for the volume of a pyramid is $V = \frac{1}{3}Bh$.)

Divide.

44. $(2x^3 + 9x^2 + 14x + 5) \div (2x + 1)$
45. $(x^4 + 3x^2 + x + 4) \div (x + 3)$
46. $(x^5 + 1) \div (x + 1)$
47. $(x^4 + 4x^3 - x - 4) \div (x^3 - 1)$
48. $(3x^4 - 5x^3 + 2x^2 + 3x - 2) \div (3x - 2)$

Determine whether each binomial is a factor of $x^3 + x^2 - 16x - 16$.

49. $x + 2$
50. $x - 4$
51. $x + 1$
52. $x - 1$

Use synthetic division to determine whether each binomial is a factor of $3x^3 + 10x^2 - x - 12$.

53. $x + 3$
54. $x - 1$
55. $x + 2$
56. $x - 4$

Divide using synthetic division.

57. $(x^4 - 2x^3 + x^2 + x - 1) \div (x - 1)$
58. $(x^4 + 3x^3 + 3x^2 + 4x + 3) \div (x + 1)$
59. $(x^4 + 3x^3 + 7x^2 + 26x + 15) \div (x + 3)$
60. $(x^4 - 6x^2 - 27) \div (x + 2)$
61. $(x^4 - 5x^2 + 4x + 12) \div (x + 2)$
62. $\left(x^4 - \frac{9}{2}x^3 + 3x^2 - \frac{1}{2}x\right) \div \left(x - \frac{1}{2}\right)$

63. **Reasoning** Divide. Look for patterns in your answers.
 a. $(x^2 - 1) \div (x - 1)$ b. $(x^3 - 1) \div (x - 1)$ c. $(x^4 - 1) \div (x - 1)$
 d. Using the patterns, factor $x^5 - 1$.

64. **Reasoning** The remainder from the division of the polynomial $x^3 + ax^2 + 2ax + 5$ by $x + 1$ is 3. Find a.

65. Use synthetic division to find $(x^2 + 4) \div (x - 2i)$.

66. **Writing** Suppose 3, −1, and 5 are zeros of a cubic polynomial function $f(x)$. What is the sign of $f(1) \cdot f(4)$? (*Hint:* Sketch the graph; consider all possibilities.)

Standardized Test Prep

SAT/ACT

67. What is the remainder when $x^2 - 5x + 7$ is divided by $x + 1$?
 (A) 1 (B) 3 (C) 11 (D) 13

68. What is the least degree of a polynomial that has a zero of multiplicity 3 at 1, a zero of multiplicity 1 at 0, and a zero of multiplicity 2 at 2?
 (F) 3 (G) 4 (H) 5 (I) 6

69. The equation $y = 0.17x$ represents your weight, in pounds, on the Moon y in relation to your weight on Earth x. If Al weighs 130 lb on Earth, what would he weigh on the Moon?
 (A) 22.1 lb (B) 92.3 lb (C) 130 lb (D) 764.7 lb

Extended Response

70. The formula for the area of a circle is $A = \pi r^2$. Solve the equation for r. If the area of a circle is 78.5 cm^2, what is the radius? Use 3.14 for π.

Mixed Review

Find the real solutions of each equation by factoring. See Lesson 5-3.

71. $x^3 + 2x^2 + x = 0$ 72. $2x^4 - 2x^3 + 2x^2 = 2x$ 73. $5x^5 = 125x^3$

Solve each equation using the Quadratic Formula. See Lesson 4-7.

74. $x^2 + 3x - 2 = 0$ 75. $2x^2 + 4x - 4 = 0$ 76. $7x^2 - 2x - 5 = 0$

77. $x^2 - 5x = -5$ 78. $x^2 - 6x = -7$ 79. $x^2 + 7x + 11 = 0$

Find the solution of each system by graphing. See Lesson 3-3.

80. $\begin{cases} y < 2x + 3 \\ y > -x \end{cases}$ 81. $\begin{cases} y > x - 4 \\ y > 4 - \frac{1}{3}x \end{cases}$ 82. $\begin{cases} y < -|x| + 3 \\ y > x + 1 \end{cases}$

Get Ready! **To prepare for Lesson 5-5, do Exercises 83–85.**

Simplify each expression. See Lesson 4-8.

83. $(-4i)(6i)$ 84. $(2 + i)(2 - i)$ 85. $(4 - 3i)(5 + i)$

5 Mid-Chapter Quiz

Do you know HOW?

For each polynomial function, describe the end behavior of its graph.

1. $f(x) = x^8 - 8x^4 + 6x^2$

2. $f(x) = -x^4 - x^3 + 1$

3. $f(x) = x^7 - 3x^5 - 5x^3$

4. What is the degree of the function that generates the data shown?

x	y
−3	159
−2	29
−1	−1
0	−3
1	−1
2	29
3	159

Find all the solutions of each equation by factoring.

5. $x^3 - 5x^2 = 36x$

6. $27x^3 = 8$

7. $x^4 - 20x^2 + 64 = 0$

8. $x^3 + 125 = 0$

9. Use the Remainder Theorem and synthetic division to find $P(4)$ for $P(x) = 2x^4 - 3x^2 + 4x - 1$.

You have several boxes with the same dimensions. They have a combined volume of $2x^4 + 4x^3 - 18x^2 - 4x + 16$. Determine whether each binomial below could represent the number of boxes you have.

10. $x - 1$

11. $x + 2$

12. $2x + 8$

Write each polynomial in standard form. Then classify it by degree and number of terms.

13. $-2x^3 + 6 - x^3 + 5x$

14. $3(x - 1)(x + 4)$

Describe the shape of the graph of each cubic function by determining the end behavior and number of turning points.

15. $y = -5x^3$

16. $y = 3x^3 + 4x^2 + 2x - 1$

Do you UNDERSTAND?

17. You buy one container each of strawberries, blueberries, and cherries. Cherries are \$1 more per container than blueberries, which are \$1 more per container than strawberries. The product of the 3 individual prices is 5 times the total cost of one container of each fruit.

a. Write a polynomial function to model the cost of your purchase.

b. Graph to find the price of each container.

c. **Writing** Explain how you used the graph to find the prices.

18. A cylinder has a radius of $3x - 2$ and a height of $3 - 2x$.

a. Use 3.14 as π and graph the equation for the volume.

b. Find the relative maximum.

c. **Reasoning** What kind of limitation on the radius would make your answer in part (b) the maximum possible volume?

19. **Open-Ended** Write a polynomial function in factored form with at least three zeros that are negative, one of which has multiplicity 2.

5-5 Theorems About Roots of Polynomial Equations

Content Standards

N.CN.7 Solve quadratic equations with real coefficients that have complex solutions.

Also N.CN.8

Objectives To solve equations using the Rational Root Theorem
To use the Conjugate Root Theorem

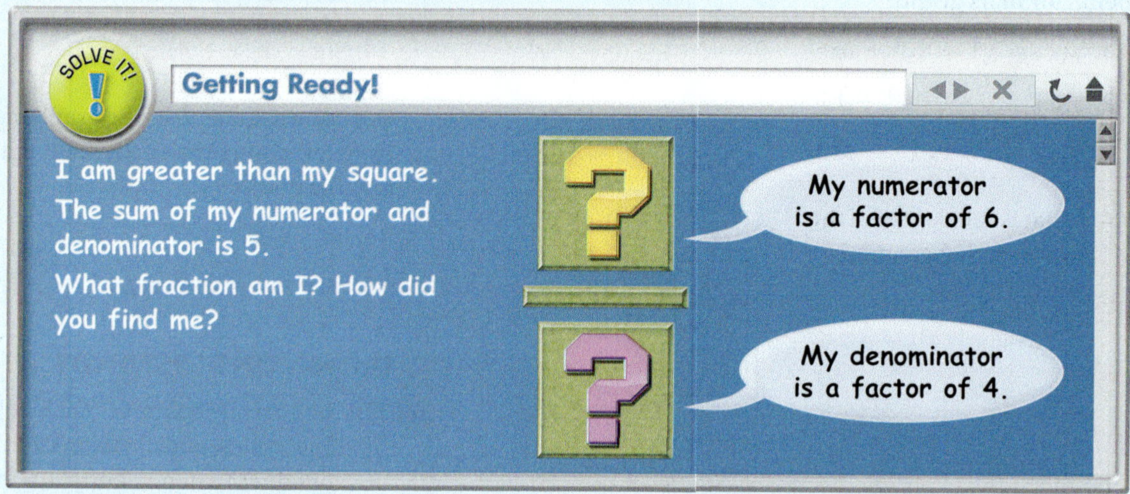

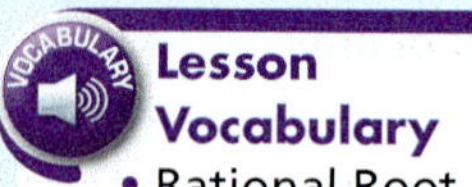

Lesson Vocabulary
- Rational Root Theorem
- Conjugate Root Theorem
- Descartes' Rule of Signs

Factoring the polynomial $P(x) = a_nx^n + a_{n-1}x^{n-1} + \cdots + a_1x + a_0$ can be challenging, especially when both a_n and a_0 have many factors.

Essential Understanding The factors of the numbers a_n and a_0 in $P(x) = a_nx^n + a_{n-1}x^{n-1} + \cdots + a_1x + a_0$ can help you factor $P(x)$ and solve the equation $P(x) = 0$.

One way to find a root of the polynomial equation $P(x) = 0$ is to guess and check. This is inefficient unless there is a way to minimize the number of guesses, or possible roots. The **Rational Root Theorem** does just that.

take note

Theorem Rational Root Theorem

Let $P(x) = a_nx^n + a_{n-1}x^{n-1} + \cdots + a_1x + a_0$ be a polynomial with integer coefficients. There are a limited number of possible roots of $P(x) = 0$:

- Integer roots must be factors of a_0.
- Rational roots must have reduced form $\frac{p}{q}$ where p is an integer factor of a_0 and q is an integer factor of a_n.

$$21x^2 + 29x + 10 = 0$$

$$x^2 + \frac{29}{21}x + \frac{10}{21} = 0$$

$$\left(x + \frac{2}{3}\right)\left(x + \frac{5}{7}\right) = 0$$

Factors of the leading coefficient: ±1, ±3, ±7, and ±21.

Factors of the constant term: ±1, ±2, ±5, and ±10.

The roots are $-\frac{2}{3}$ and $-\frac{5}{7}$.

Problem 1 Finding a Rational Root

What information can you get from the equation?
The equation gives you the leading coefficient and the constant term.

What are the rational roots of $2x^3 - x^2 + 2x + 5 = 0$?

The only possible rational roots have the form $\frac{\text{factor of constant term}}{\text{factor of leading coefficient}}$.

The constant factors are $\pm 1, \pm 5$. The leading coefficient factors are $\pm 1, \pm 2$.
The only possible rational roots are $\pm 1, \pm 5, \pm \frac{1}{2}, \pm \frac{5}{2}$.

The table shows the values of the function $y = P(x)$ for the possible roots.

x	1	−1	5	−5	$\frac{1}{2}$	$-\frac{1}{2}$	$\frac{5}{2}$	$-\frac{5}{2}$
$P(x)$	8	0	240	−280	6	$\frac{7}{2}$	35	$-\frac{75}{2}$

The only rational root of $2x^3 - x^2 + 2x + 5 = 0$ is -1.

Got It? 1. What are the rational roots of $3x^3 + 7x^2 + 6x - 8 = 0$?

Once you find one root, use synthetic division to factor the polynomial. Continue finding roots and dividing until you have a second-degree polynomial. Use the Quadratic Formula to find the remaining roots.

Problem 2 Using the Rational Root Theorem

What are the rational roots of $15x^3 - 32x^2 + 3x + 2 = 0$?

Know	Need	Plan
Coefficients and the constant term of the polynomial	The roots of the polynomial equation	• Find one root. • Factor until you get a quadratic. • Use the Quadratic Formula to find the other roots.

Step 1 The constant term factors are ± 1 and ± 2. The leading coefficient factors are $\pm 1, \pm 3, \pm 5$, and ± 15.

Step 2 The possible rational roots are: $\pm 1, \pm 2, \pm \frac{1}{3}, \pm \frac{2}{3}, \pm \frac{1}{5}, \pm \frac{2}{5}, \pm \frac{1}{15}$, and $\pm \frac{2}{15}$.

Step 3 Test each possible rational root in $15x^3 - 32x^2 + 3x + 2$ until you find a root.

Test 1: $15(1)^3 - 32(1)^2 + 3(1) + 2 = -12 \neq 0$

Test 2: $15(2)^3 - 32(2)^2 + 3(2) + 2 = 0$ So 2 is a root.

Step 4 Factor the polynomial by using synthetic division:
$P(x) = (x - 2)(15x^2 - 2x - 1)$.

$$\begin{array}{r|rrrr} 2 & 15 & -32 & 3 & 2 \\ & & 30 & -4 & -2 \\ \hline & 15 & -2 & -1 & 0 \end{array}$$

Step 5 Since $15x^2 - 2x - 1 = (5x + 1)(3x - 1)$, the other roots are $-\frac{1}{5}$ and $\frac{1}{3}$.

The rational roots of $15x^3 - 32x^2 + 3x + 2 = 0$ are $2, -\frac{1}{5}$, and $\frac{1}{3}$.

Got It? 2. What are the rational roots of $2x^3 + x^2 - 7x - 6 = 0$?

Recall from Lesson 4-8 that the complex numbers $a + bi$ and $a - bi$ are conjugates. Similarly, the irrational numbers $a + \sqrt{b}$ and $a - \sqrt{b}$ are conjugates. If a complex number or an irrational number is a root of a polynomial equation with rational coefficients, so is its conjugate.

take note

Theorem Conjugate Root Theorem

If $P(x)$ is a polynomial with *rational* coefficients, then irrational roots of $P(x) = 0$ that have the form $a + \sqrt{b}$ occur in conjugate pairs. That is, if $a + \sqrt{b}$ is an irrational root with a and b rational, then $a - \sqrt{b}$ is also a root.

If $P(x)$ is a polynomial with *real* coefficients, then the complex roots of $P(x) = 0$ occur in conjugate pairs. That is, if $a + bi$ is a complex root with a and b real, then $a - bi$ is also a root.

Problem 3 Using the Conjugate Root Theorem to Identify Roots

Think

Do you have real coefficients?
All rational numbers are real numbers. Therefore the rational coefficients are real coefficients.

A quartic polynomial $P(x)$ has rational coefficients. If $\sqrt{2}$ and $1 + i$ are roots of $P(x) = 0$, what are the two other roots?

Since $P(x)$ has rational coefficients and $0 + \sqrt{2}$ is a root of $P(x) = 0$, it follows from the Conjugate Root Theorem that $0 - \sqrt{2}$ is also a root.

Since $P(x)$ has real coefficients and $1 + i$ is a root of $P(x) = 0$, it follows that $1 - i$ is also a root.

The two other roots are $-\sqrt{2}$ and $1 - i$.

Got It? **3.** A cubic polynomial $P(x)$ has real coefficients. If $3 - 2i$ and $\frac{5}{2}$ are two roots of $P(x) = 0$, what is one additional root?

Problem 4 Using Conjugates to Construct a Polynomial

Multiple Choice **What is a third-degree polynomial function $y = P(x)$ with rational coefficients so that $P(x) = 0$ has roots -4 and $2i$?**

Ⓐ $P(x) = x^3 - 2x^2 - 16x + 32$

Ⓑ $P(x) = x^3 - 4x^2 + 4x - 16$

Ⓒ $P(x) = x^3 + 4x^2 + 4x + 16$

Ⓓ $P(x) = x^3 + 4x^2 - 4x - 16$

Think

Does the Conjugate Root Theorem apply to −4?
No; the theorem does not apply because −4 is neither irrational nor imaginary.

Since $2i$ is a root, then $-2i$ is also a root.

$P(x) = (x + 2i)(x - 2i)(x + 4)$	Write the polynomial function.
$= (x^2 + 4)(x + 4)$	Multiply the complex conjugates.
$= x^3 + 4x^2 + 4x + 16$	Write the polynomial function in standard form.

The equation $x^3 + 4x^2 + 4x + 16 = 0$ has rational coefficients and has roots -4 and $2i$. The correct answer is C.

Got It? **4.** What quartic polynomial equation has roots $2 - 3i, 8, 2$?

The French mathematician René Descartes (1596–1650) recognized a connection between the roots of a polynomial equation and the $+$ and $-$ signs of the standard form.

take note

Theorem Descartes' Rule of Signs

Let $P(x)$ be a polynomial with real coefficients written in standard form.

- The number of positive real roots of $P(x) = 0$ is either equal to the number of sign changes between consecutive coefficients of $P(x)$ or is less than that by an even number.
- The number of negative real roots of $P(x) = 0$ is either equal to the number of sign changes between consecutive coefficients of $P(-x)$ or is less than that by an even number.

In both cases, count multiple roots according to their multiplicity.

Problem 5 Using Descartes' Rule of Signs

What does Descartes' Rule of Signs tell you about the real roots of $x^3 - x^2 + 1 = 0$?

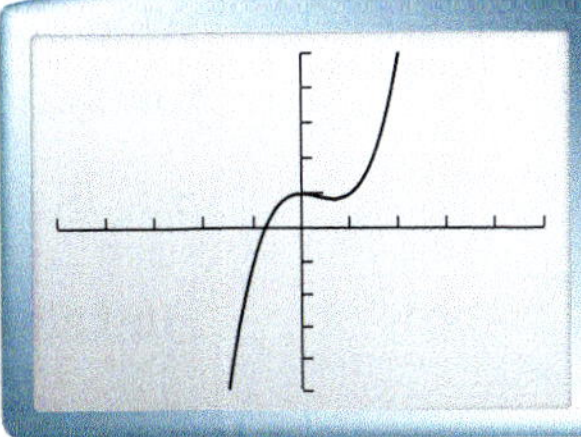

There are two sign changes, $+$ to and to $+$.
Therefore, there are either 0 or 2 positive real roots.

$P(-x) = (-x)^3 - (-x)^2 + 1 = -x^3 - x^2 + 1 = 0$ has only one sign change $-$ to $+$. There is one negative real root.

Recall that graphs of cubic functions have zero or two turning points. Because the graph already shows two turning points, it will not change direction again. So there are no positive real roots.

Think

Why can't there be zero negative real roots?
The number of negative roots is equal to 1 or is less than 1 by an even number. Zero is less than 1 by an odd number.

Got It? 5. a. What does Descartes' Rule of Signs tell you about the real roots of $2x^4 - x^3 + 3x^2 - 1 = 0$?

b. Reasoning Can you confirm real and complex roots graphically? Explain.

Lesson Check

Do you know HOW?

Use the Rational Root Theorem to list all possible rational roots for each equation.

1. $x^2 + x - 2 = 0$
2. $2x^3 - x^2 - 6 = 0$
3. $3x^4 + 2x^2 - 12 = 0$

Write a polynomial function with rational coefficients so that $P(x) = 0$ has the given roots.

4. 5 and 9
5. -4 and $2i$

Do you UNDERSTAND?

6. **Vocabulary** Give an example of a conjugate pair.
7. **Reasoning** In the statements below, r and s represent integers. Is each statement *always, sometimes,* or *never* true? Explain.
   a. A root of the equation $3x^3 + rx^2 + sx + 8 = 0$ could be 5.
   b. A root of the equation $3x^3 + rx^2 + sx + 8 = 0$ could be -2.
8. **Error Analysis** A student claims that $-4i$ is the only imaginary root of a polynomial equation that has real coefficients. What is the student's mistake?

Practice and Problem-Solving Exercises

Use the Rational Root Theorem to list all possible rational roots for each equation. Then find any actual rational roots.

See Problems 1 and 2.

9. $x^3 - 4x + 1 = 0$
10. $x^3 + 2x - 9 = 0$
11. $2x^3 - 5x + 4 = 0$
12. $3x^3 + 9x - 6 = 0$
13. $4x^3 + 2x - 12 = 0$
14. $6x^3 + 2x - 18 = 0$
15. $7x^3 - x^2 + 4x + 10 = 0$
16. $8x^3 + 2x^2 - 5x + 1 = 0$
17. $10x^3 - 7x^2 + x - 10 = 0$

A polynomial function $P(x)$ with rational coefficients has the given roots. Find two additional roots of $P(x) = 0$.

See Problem 3.

18. $-2i$ and $\sqrt{10}$
19. $14 - \sqrt{2}$ and $-6i$
20. i and $7 + 8i$
21. $-\sqrt{3}$ and $5 - \sqrt{11}$

Write a polynomial function with rational coefficients so that $P(x) = 0$ has the given roots.

See Problem 4.

22. 7 and 12
23. -9 and -15
24. $-10i$
25. $3i + 9$
26. 4, 16, and $1 + 19i$
27. $13i$ and $5 + 10i$
28. $11 - 2i$ and $8 + 13i$
29. $17 - 4i$ and $12 + 5i$

What does Descartes' Rule of Signs say about the number of positive real roots and negative real roots for each polynomial function?

See Problem 5.

30. $P(x) = x^2 + 5x + 6$
31. $P(x) = 9x^3 - 4x^2 + 10$
32. $P(x) = 8x^3 + 2x^2 - 14x + 5$

Find all rational roots for $P(x) = 0$.

33. $P(x) = 2x^3 - 5x^2 + x - 1$
34. $P(x) = 6x^4 - 13x^3 + 13x^2 - 39x - 15$
35. $P(x) = 7x^3 - x^2 - 5x + 14$
36. $P(x) = 3x^4 - 7x^3 + 10x^2 - x + 12$
37. $P(x) = 6x^4 - 7x^2 - 3$
38. $P(x) = 2x^3 - 3x^2 - 8x + 12$

Write a polynomial function $P(x)$ with rational coefficients so that $P(x) = 0$ has the given roots.

39. -6, 3, and $-15i$
40. $4 + \sqrt{5}$ and $8i$
41. $-5 - 7i$ and $2 - \sqrt{11}$

42. **Think About a Plan** You are building a square pyramid out of clay and want the height to be 0.5 cm shorter than twice the length of each side of the base. If you have 18 cm^3 of clay, what is the greatest height you could use for your pyramid?
    - How can drawing a diagram help you solve this problem?
    - What is the formula for the volume of a pyramid?
    - What equation can you solve to find the height of the pyramid?

43. **Error Analysis** Your friend is using Descartes' Rule of Signs to find the number of negative real roots of $x^3 + x^2 + x + 1 = 0$. Describe and correct the error.

$P(-x) = (-x)^3 + (-x)^2 + (-x) + 1$
$= -x^3 - x^2 - x + 1$

Because there is only one sign change in P(-x), there must be one negative real root.

44. **Reasoning** A quartic equation with integer coefficients has two real roots and one imaginary root. Explain why the fourth root must be imaginary.

45. **Gardening** A gardener is designing a new garden in the shape of a trapezoid. She wants the shorter base to be twice the height and the longer base to be 4 feet longer than the shorter base. If she has enough topsoil to create a 60 ft^2 garden, what dimensions should she use for the garden?

46. **Open-Ended** Write a fourth-degree polynomial equation with integer coefficients that has two irrational roots and two imaginary roots.

47. a. Find a polynomial equation in which $1 + \sqrt{2}$ is the only root.
    b. Find a polynomial equation with root $1 + \sqrt{2}$ of multiplicity 2.
    c. Find c such that $1 + \sqrt{2}$ is a root of $x^2 - 2x + c = 0$.

48. a. Using *real* and *imaginary* as types of roots, list all possible combinations of root type for a fourth-degree polynomial equation.
    b. Repeat the process for a fifth-degree polynomial equation.
    c. **Make a Conjecture** Make a conjecture about the number of real roots of an odd-degree polynomial equation.

49. **Writing** A student states that $2 + \sqrt{3}$ is a root of $x^2 - 2x - (3 + 2\sqrt{3}) = 0$. The student claims that $2 - \sqrt{3}$ is another root of the equation by the Conjugate Root Theorem. Explain how you would respond to the student.

Standardized Test Prep

GRIDDED RESPONSE

50. What is a positive root of $-5x^3 - 2x^2 + 9x + 30 = 0$?

51. What is the remainder when you divide $x^3 + 2x^2 - x - 6$ by $x - 1$?

52. A polynomial with rational coefficients has roots $-3i$ and $8 + \sqrt{7}$. What is the minimum degree of the polynomial?

53. What is the value of y in the solution of the system of equations? $\begin{cases} 10x + 24y = 9 \\ 8x + 60y = 14 \end{cases}$

54. What is the value of the greater solution of the equation $6x^2 - 17x + 5 = 0$?

Mixed Review

Divide. See Lesson 5-4.

55. $(x^3 + 5x^2 - 3) \div (x - 1)$
56. $(8x^3 + 12x^2 + 7) \div (x + 6)$
57. $(7x^2 + 11x - 4) \div (x + 2)$

Solve. See Lesson 4-8.

58. $7x^2 + 63 = 0$
59. $x^2 + 81 = 0$
60. $2x^2 + 288 = 0$

Get Ready! **To prepare for Lesson 5-6, do Exercises 61 and 62.**

Write each polynomial in standard form. Then classify it by degree and by number of terms. See Lesson 5-1.

61. $6x^2 + 11 - 5x^4 + 9x$
62. $13x - 4x^5 + 7x^3 + 2$

Concept Byte

For Use With Lesson 5-5

EXTENSION

Using Polynomial Identities

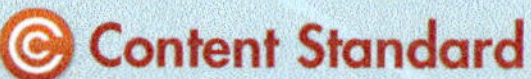

A.APR.4 Prove polynomial identities and use them to describe numerical relationships.

You can use what you know about polynomial identities to discover relationships among numbers.

Example

Use polynomial identities to prove that the sum of the cubes of any two consecutive positive integers is odd.

You can represent any two consecutive positive integers as n and $n + 1$, where n is a positive integer. Use the formula for factoring the sum of two cubes.

$$\begin{aligned}
a^3 + b^3 &= (a + b)(a^2 - ab + b^2) \\
&= (n + (n + 1))(n^2 - n(n + 1) + (n + 1)^2) && \text{Substitute } n \text{ for } a \text{ and } n + 1 \text{ for } b. \\
&= (2n + 1)(n^2 - n^2 - n + n^2 + 2n + 1) && \text{Simplify.} \\
&= (2n + 1)(n^2 + n + 1) \\
&= 2n^3 + 2n^2 + 2n + n^2 + n + 1 \\
&= 2n^3 + 3n^2 + 3n + 1
\end{aligned}$$

You know that $2n^3$ is always even because it has a factor of 2.

If n is even, then $3n^2$ is even and $3n$ is even. So $2n^3 + 3n^2 + 3n$ is even, and $2n^3 + 3n^2 + 3n + 1$ is odd, because it is 1 more than an even number.

If n is odd, then $3n^2$ is odd and $3n$ is odd. The sum of two odd integers is always even. So $2n^3 + 3n^2 + 3n$ is even, and $2n^3 + 3n^2 + 3n + 1$ is odd, because it is 1 more than an even number.

Therefore, $n^3 + (n + 1)^3$ is always odd for consecutive positive integers.

Exercises

1. Use polynomial identities to prove that the difference of the squares of any two consecutive integers is odd.
2. a. Prove that for any two positive consecutive integers a and b, where $a > b, a^3 - b^3 = a^2 + ab + b^2$.
   b. Prove that the difference of the cubes of two consecutive positive integers is always odd.
3. Prove that the square of the sum of two consecutive positive integers is odd.
4. Prove that the reciprocals of any two consecutive integers have a product that is equal to the reciprocal of the smaller integer minus the reciprocal of the larger integer.
5. Use the identity $n^3 - n = n(n - 1)(n + 1)$ to prove that 6 is a factor of $n^3 - n$ for all integers n. (*Hint: n, n* $-$ 1, and $n + 1$ are consecutive integers)

The Fundamental Theorem of Algebra

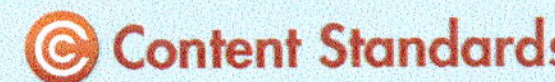

N.CN.7 Solve quadratic equations with real coefficients that have complex solutions.

N.CN.8 Extend polynomial identities to the complex numbers.

Also N.CN.9, A.APR.3

Objective To use the Fundamental Theorem of Algebra to solve polynomial equations with complex solutions

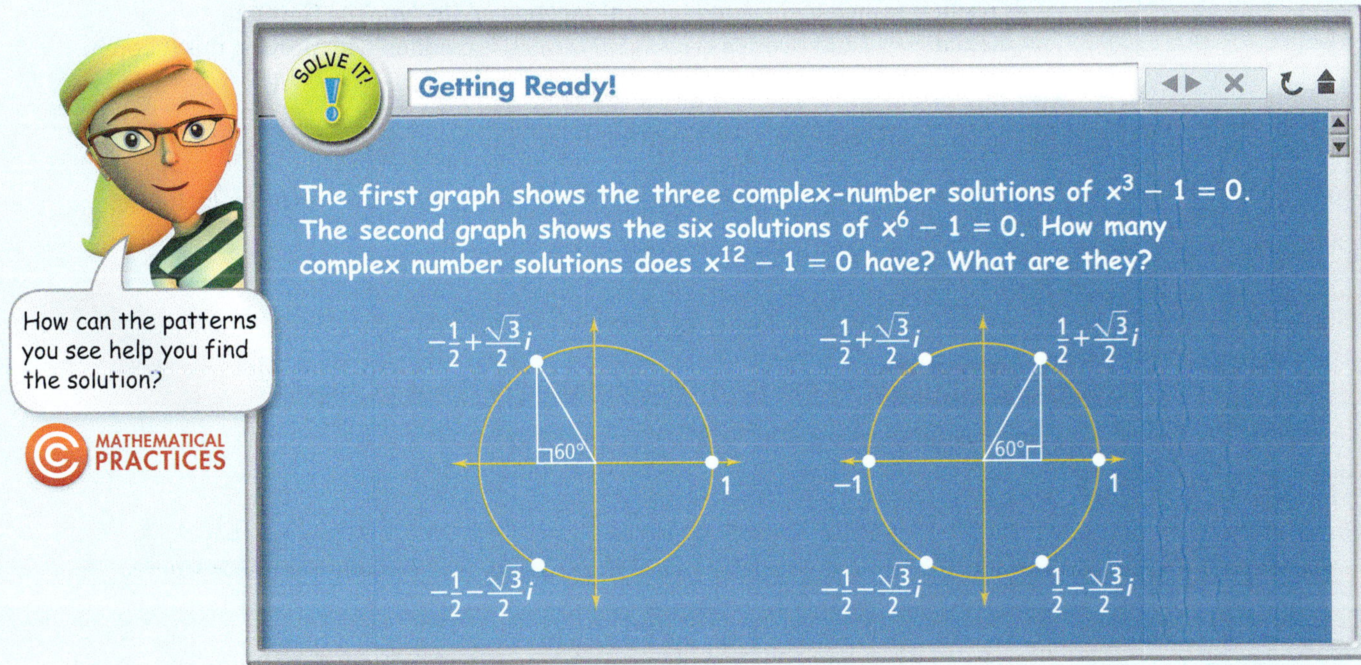

Lesson Vocabulary
- Fundamental Theorem of Algebra

You can factor any polynomial of degree n into n linear factors, but sometimes the factors will involve imaginary numbers.

Essential Understanding The degree of a polynomial equation tells you how many roots the equation has.

It is easy to see graphically that every polynomial function of degree 1 has a single zero, the x-intercept. However, there appear to be three possibilities for polynomials of degree 2. They correspond to these three graphs:

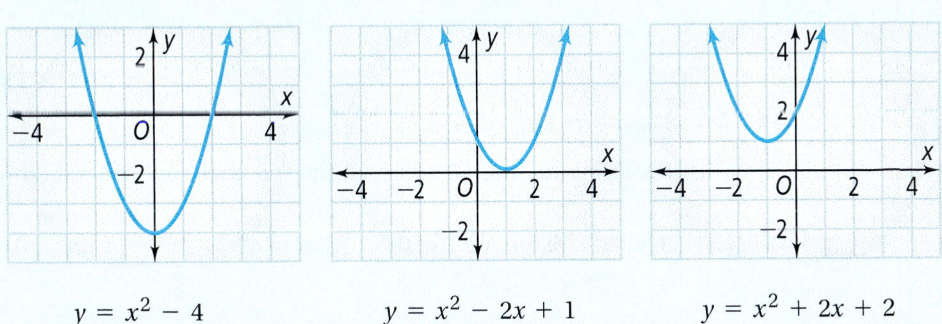

$y = x^2 - 4$ Two real zeros | $y = x^2 - 2x + 1$ One real zero | $y = x^2 + 2x + 2$ No real zeros

However, by factoring, you can see that each related equation has two roots.

$x^2 - 4 = (x - 2)(x + 2) = 0$	*two* real roots, 2 and −2
$x^2 - 2x + 1 = (x - 1)(x - 1) = 0$	a root of multiplicity *two* at 1
$x^2 + 2x + 2 = (x - (-1 + i))(x - (-1 - i)) = 0$	*two* complex roots, $-1 + i$ and $-1 - i$

Every quadratic polynomial equation has two roots, every cubic polynomial equation has three roots, and so on.

This result is related to the *Fundamental Theorem of Algebra*. The German mathematician Carl Friedrich Gauss (1777-1855) is credited with proving this theorem.

take note

Theorem The Fundamental Theorem of Algebra

If $P(x)$ is a polynomial of degree $n \geq 1$, then $P(x) = 0$ has exactly n roots, including multiple and complex roots.

Problem 1 Using the Fundamental Theorem of Algebra

What are all the roots of $x^5 - x^4 - 3x^3 + 3x^2 - 4x + 4 = 0$?

Know
The polynomial equation has degree 5. There are 5 roots.

Need
The zeros of the function

Plan
Use the Rational Root and Factor Theorems, synthetic division, and factoring.

Step 1 The polynomial is in standard form. The possible rational roots are ±1, ±2, ±4.

Step 2 Evaluate the related polynomial function for $x = 1$. Since $P(1) = 0$, 1 is a root and $x - 1$ is a factor. Use synthetic division to factor out $x - 1$:

$$\begin{array}{r|rrrrrr} 1 & 1 & -1 & -3 & 3 & -4 & 4 \\ & & 1 & 0 & -3 & 0 & -4 \\ \hline & 1 & 0 & -3 & 0 & -4 & 0 \end{array}$$

Think
How many linear factors will there be?
If there are five roots, there must be five linear factors.

Step 3 Continue factoring until you have five linear factors.

$$\begin{aligned} x^5 - x^4 - 3x^3 + 3x^2 - 4x + 4 &= (x - 1)(x^4 - 3x^2 - 4) \\ &= (x - 1)(x^2 - 4)(x^2 + 1) \\ &= (x - 1)(x - 2)(x + 2)(x - i)(x + i) \end{aligned}$$

Step 4 The roots are 1, 2, −2, i, and $-i$.

By the Fundamental Theorem of Algebra, these are the only roots.

Got It? **1.** What are all the roots of the equation $x^4 + 2x^3 = 13x^2 - 10x$?

Problem 2 Finding All the Zeros of a Polynomial Function

What are the zeros of $f(x) = x^4 + x^3 - 7x^2 - 9x - 18$?

Think

Does the graph show all of the real roots?
Yes; the graphs of quartic functions have one or three turning points. Since the graph shows three turning points, it will not turn again to cross the x-axis a third time.

Step 1 Use a graphing calculator to find any real roots. The graph of $y = x^4 + x^3 - 7x^2 - 9x - 18$ shows real zeros at $x = -3$ and $x = 3$.

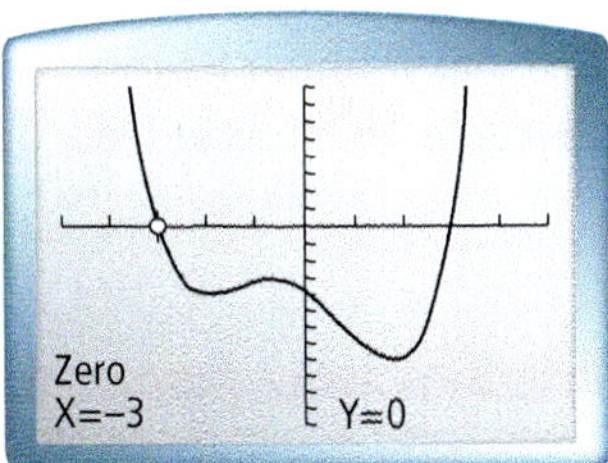

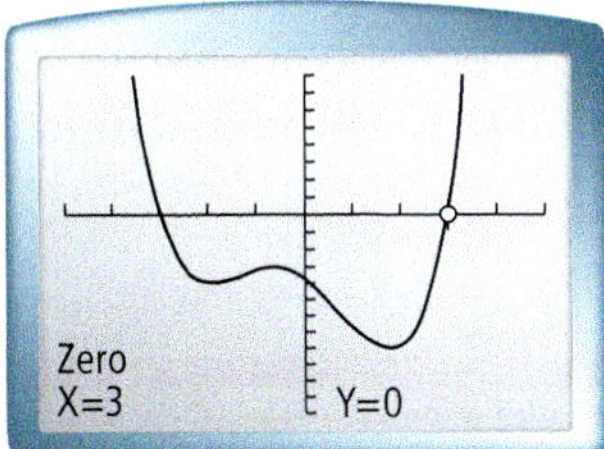

Step 2 Factor out the linear factors $x + 3$ and $x - 3$. Use synthetic division twice.

$$\begin{array}{r|rrrrr} -3 & 1 & 1 & -7 & -9 & -18 \\ & & -3 & 6 & 3 & 18 \\ \hline & 1 & -2 & -1 & -6 & 0 \end{array} \qquad \begin{array}{r|rrrr} 3 & 1 & -2 & -1 & -6 \\ & & 3 & 3 & 6 \\ \hline & 1 & 1 & 2 & 0 \end{array}$$

$$\begin{aligned} x^4 + x^3 - 7x^2 - 9x - 18 &= (x + 3)(x^3 - 2x^2 - x - 6) \\ &= (x + 3)(x - 3)(x^2 + x + 2) \end{aligned}$$

Step 3 Use the Quadratic Formula. Find the complex roots of $x^2 + x + 2 = 0$.

$a = 1, b = 1, c = 2$ — Identify the values of a, b, and c.

$\dfrac{-1 \pm \sqrt{1^2 - 4(1)(2)}}{2(1)}$ — Substitute.

$\dfrac{-1 \pm \sqrt{-7}}{2}$ — Simplify.

The complex roots are $\frac{-1 + i\sqrt{7}}{2}$ and $\frac{-1 - i\sqrt{7}}{2}$.

Step 4 The four zeros of the function are -3, 3, $\frac{-1 + i\sqrt{7}}{2}$, and $\frac{-1 - i\sqrt{7}}{2}$.

By the Fundamental Theorem of Algebra, there can be no other zeros.

Got It? **2. a.** What are all the zeros of the function $g(x) = 2x^4 \quad 3x^3 \quad x \quad 6$?

b. Reasoning The graph of $f(x) = x^5 + 4x^4 - 3x^3 - 12x^2 - 4x - 4$ is shown at the right.

i. Use the turning points to explain why the graph does NOT show all of the real zeros of the function.

ii. The graph of $g(x) = f(x) + 4$ is a translation of the graph of f up 4 units. How many real zeros of g will the graph of g show? Explain.

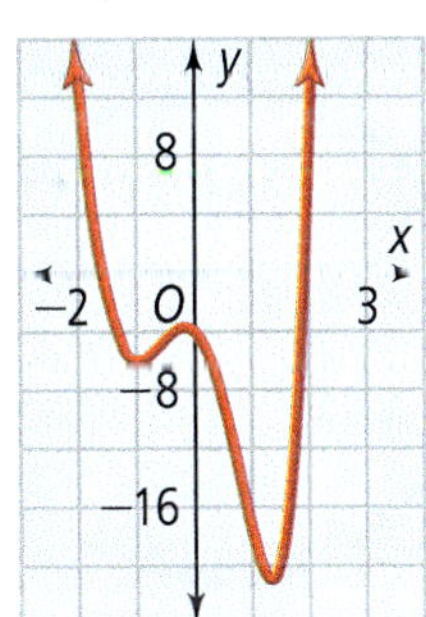

take note

Concept Summary The Fundamental Theorem of Algebra

Here are equivalent ways to state the Fundamental Theorem of Algebra. You can use any one of these statements to prove the others.

- Every polynomial equation of degree $n \geq 1$ has exactly n roots, including multiple and complex roots.
- Every polynomial of degree $n \geq 1$ has n linear factors.
- Every polynomial function of degree $n \geq 1$ has at least one complex zero.

Lesson Check

Do you know HOW?

Find the number of roots for each equation.

1. $5x^4 + 12x^3 - x^2 + 3x + 5 = 0$

2. $-x^{14} - x^8 - x + 7 = 0$

Find all the zeros for each function.

3. $y = x^3 - 5x^2 + 16x - 80$

4. $y = x^4 - 2x^3 + x^2 - 2x$

Do you UNDERSTAND?

5. Vocabulary Given a polynomial equation of degree n, explain how you determine the number of roots of the equation.

6. Open-Ended Write a polynomial function of degree 4 with rational coefficients and two complex zeros of multiplicity 2.

7. Writing Describe when to use synthetic division and when to use the Quadratic Formula to determine the linear factors of a polynomial.

Practice and Problem-Solving Exercises

Without using a calculator, find all the roots of each equation.

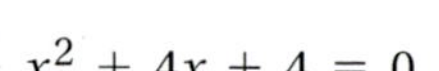

See Problem 1.

8. $x^3 - 3x^2 + x - 3 = 0$

9. $x^3 + x^2 + 4x + 4 = 0$

10. $x^3 + 4x^2 + x - 6 = 0$

11. $x^3 - 5x^2 + 2x + 8 = 0$

12. $x^4 + 4x^3 + 7x^2 + 16x + 12 = 0$

13. $x^4 - 4x^3 + x^2 + 12x - 12 = 0$

14. $x^5 + 3x^3 - 4x = 0$

15. $x^5 - 8x^3 - 9x = 0$

Find all the zeros of each function.

See Problem 2.

16. $y = 2x^3 + x^2 + 1$

17. $f(x) = x^3 - 3x^2 + x - 3$

18. $g(x) = x^3 - 5x^2 + 5x - 4$

19. $y = x^3 - 2x^2 - 3x + 6$

20. $y = x^4 - 6x^2 + 8$

21. $f(x) = x^4 - 3x^2 - 4$

22. $y = x^3 - 3x^2 - 9x$

23. $y = x^3 + 6x^2 + x + 6$

24. $y = x^4 + 3x^3 + x^2 - 12x - 20$

25. $y = x^4 + x^3 - 15x^2 - 16x - 16$

Apply

For each equation, state the number of complex roots, the possible number of real roots, and the possible rational roots.

26. $2x^4 - x^3 + 2x^2 + 5x - 26 = 0$

27. $x^5 - x^3 - 11x^2 + 9x + 18 = 0$

28. $-12 + x + 10x^2 + 3x^3 = 0$

29. $4x^6 - x^5 - 24 = 0$

Find all the zeros of each function.

30. $y = x^3 - 4x^2 + 9x - 36$

31. $f(x) = x^3 + 2x^2 - 5x - 10$

32. $y = 2x^3 - 3x^2 - 18x - 8$

33. $y = 3x^3 - 7x^2 - 14x + 24$

34. $g(x) = x^3 - 4x^2 - x + 22$

35. $y = x^3 - x^2 - 3x - 9$

36. $y = x^4 - x^3 - 5x^2 - x - 6$

37. $y = 2x^4 + 3x^3 - 17x^2 - 27x - 9$

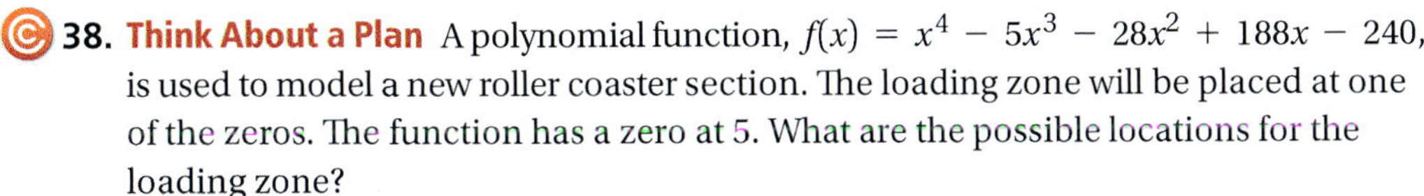

38. Think About a Plan A polynomial function, $f(x) = x^4 - 5x^3 - 28x^2 + 188x - 240$, is used to model a new roller coaster section. The loading zone will be placed at one of the zeros. The function has a zero at 5. What are the possible locations for the loading zone?

- Can you determine how many zeros you need to find?
- How can you use polynomial division?
- What other methods can be helpful?

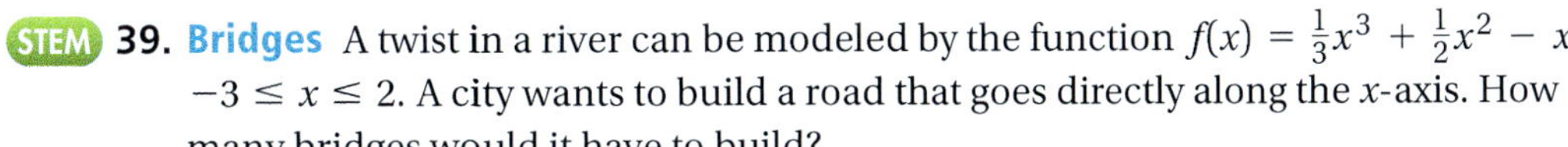

STEM **39. Bridges** A twist in a river can be modeled by the function $f(x) = \frac{1}{3}x^3 + \frac{1}{2}x^2 - x$, $-3 \leq x \leq 2$. A city wants to build a road that goes directly along the x-axis. How many bridges would it have to build?

40. Error Analysis Maurice says: "Every linear function has exactly one zero. It follows from the Fundamental Theorem of Algebra." Cheryl disagrees. "What about the linear function $y = 2$?" she asks. "Its graph is a line, but it has no x-intercept." Whose reasoning is incorrect? Where is the flaw?

Determine whether each of the following statement is *always*, *sometimes*, or *never* true.

41. A polynomial function with real coefficient has real zeros.

42. Polynomial functions with complex coefficients have one complex zero.

43. A polynomial function that does not intercept the x-axis has complex roots only.

44. Reasoning A 4th degree polynomial function has zeros at 3 and $5 - i$. Can $4 + i$ also be a zero of the function? Explain your reasoning.

45. Open-Ended Write a polynomial function that has four possible rational zeros but no actual rational zeros.

46. Reasoning Show that the Fundamental Theorem of Algebra must be true for all quadratic polynomial functions.

47. Use the Fundamental Theorem of Algebra and the Conjugate Root Theorem to show that any odd degree polynomial equation with real coefficients has at least one real root.

48. **Reasoning** What is the maximum number of points of intersection between the graphs of a quartic and a quintic polynomial function?

49. **Reasoning** What is the least possible degree of a polynomial with rational coefficients, leading coefficient 1, constant term 5, and zeros at $\sqrt{2}$ and $\sqrt{3}$? Show that such a polynomial has a rational zero and indicate this zero.

Standardized Test Prep

SAT/ACT

50. How many roots does $f(x) = x^4 + 5x^3 + 3x^2 + 2x + 6$ have?

(A) 5 (B) 4 (C) 3 (D) 2

51. Which translation takes $y = |x + 2| - 1$ to $y = |x| + 2$?

(F) 2 units right, 3 units down
(G) 2 units right, 3 units up
(H) 2 units left, 3 units up
(I) 2 units left, 3 units down

52. What is the factored form of the expression $x^4 - 3x^3 + 2x^2$?

(A) $x^2(x - 1)(x + 2)$
(B) $x^2(x + 1)(x + 2)$
(C) $x^2(x + 1)(x - 2)$
(D) $x^2(x - 1)(x - 2)$

Short Response

53. How would you test whether (2, −2) is a solution of the system? $\begin{cases} y < -2x + 3 \\ y \geq x - 4 \end{cases}$

Mixed Review

54. Find a fourth-degree polynomial equation with real coefficients that has $2i$ and $-3 + i$ as roots. See Lesson 5-5.

Solve each equation using the Quadratic Formula. See Lesson 4-7.

55. $x^2 - 6x + 1 = 0$
56. $2x^2 + 5x = -9$
57. $2(x^2 + 2) = 3x$

Determine whether a quadratic model exists for each set of values. If so, write the model. See Lesson 4-3.

58. $f(-1) = 0, f(2) = 3, f(1) = 4$
59. $f(-4) = 11, f(-5) = 5, f(-6) = 3$

Get Ready! **To prepare for Lesson 5-7, do Exercises 60–65.** See Lesson 4-2.

Write each polynomial in standard form.

60. $(x + 1)^3$
61. $(x - 3)^3$
62. $(x - 2)^4$
63. $(x - 1)^2$
64. $(x + 5)^3$
65. $(4 - x)^3$

Concept Byte

For Use With Lesson 5-6

ACTIVITY

Graphing Polynomials Using Zeros

Content Standard

A.APR.3 Identify zeros of polynomials when suitable factorizations are available, and use the zeros to construct a rough graph of the function defined by the polynomial.

In this activity you will learn how to sketch the graph of a polynomial function by using the zeros, turning points, and end behavior.

Example

Sketch the graph of the function $f(x) = (x - 3)(x + 1)(x - 2)$.

Step 1 Identify the zeros and plot them on a coordinate grid.

$f(x) = (x - 3)(x + 1)(x - 2)$

$0 = (x - 3)(x + 1)(x - 2)$

$0 = x - 3$ or $0 = x + 1$ or $0 = x - 2$

$3 = x$ $\quad -1 = x$ $\quad 2 - x$

The function has zeros as $(3, 0)$, $(-1, 0)$, and $(2, 0)$.

Step 2 Determine whether the polynomial is positive or negative over each interval.

To determine whether $f(x)$ is positive or negative over the interval $x < -1$, choose an x-value within the interval. Let $x = -2$. Then evaluate $f(-2)$.

$f(-2) = (-2 - 3)(-2 + 1)(-2 - 2) = (-5)(-1)(-4) = -20$

$f(x)$ is negative over the interval $x < -1$.

Repeat the process for the intervals $-1 < x < 2$, $2 < x < 3$, and $x > 3$.

Interval	x	$f(x)$
$x < -1$	−2	−20
$-1 < x < 2$	0	6
$2 < x < 3$	1.5	1.875
$x > 3$	4	10

Step 3 Sketch the graph.

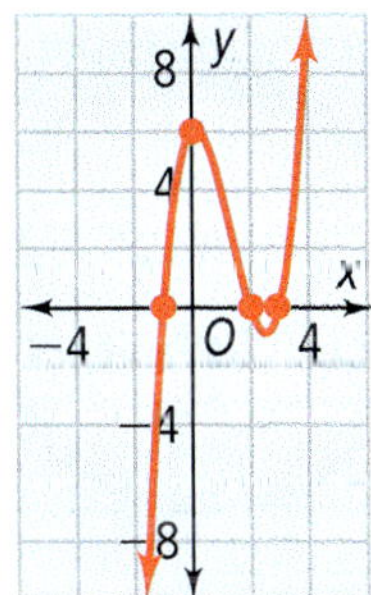

Exercises

Sketch a graph of each function. Check your answer using a graphing calculator.

1. $h(x) = (x + 6)(x - 7)$

2. $g(x) = (x + 1)(x - 3)(x - 5)$

3. $p(x) = x(x + 4)(x - 4)$

4. $h(x) = (x + 2)(x - 3)(x + 1)(x - 1)$

5. $f(x) = x^4 - 8x^2 + 16$

6. $k(x) = x^4 - 10x^2 + 9$

5-7 The Binomial Theorem

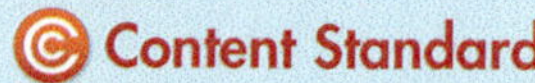

A.APR.5 Know and apply the Binomial Theorem for the expansion of $(x + y)^n$ in powers of x and y for a positive integer n, where x and y are any numbers, with coefficients determined for example by Pascal's Triangle.

Objectives To expand a binomial using Pascal's Triangle
To use the Binomial Theorem

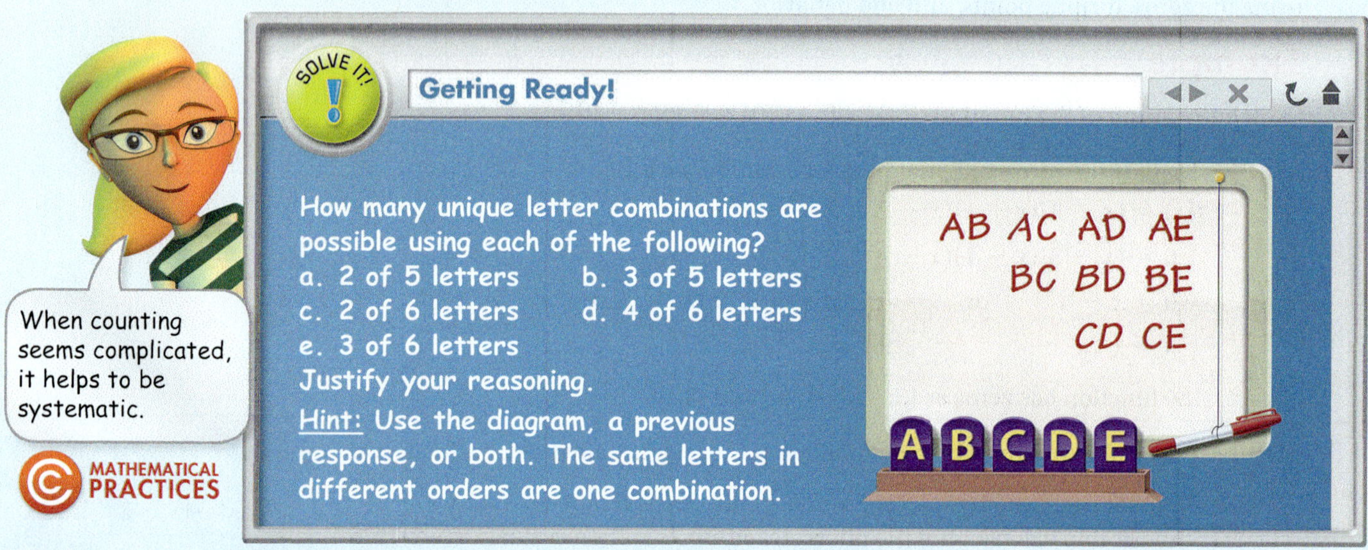

Lesson Vocabulary
- expand
- Pascal's Triangle
- Binomial Theorem

There is a connection between the triangular pattern of numbers in the Solve It and the expansion of $(a + b)^n$.

Essential Understanding You can use a pattern of coefficients and the pattern $a^n, a^{n-1}b, a^{n-2}b^2, \dots, a^2b^{n-2}, ab^{n-1}, b^n$ to write the expansion of $(a + b)^n$.

You can *expand* $(a + b)^3$ using the Distributive Property.

$$(a + b)^3 = (a + b)(a + b)(a + b) = a^3 + 3a^2b + 3ab^2 + b^3$$

To **expand** the power of a binomial in general, first multiply as needed. Then write the polynomial in standard form.

Consider the expansions of $(a + b)^n$ for the first few values of n:

Row	Power	Expanded Form	Coefficients Only
0	$(a + b)^0$	1	1
1	$(a + b)^1$	$1a^1 + 1b^1$	1 1
2	$(a + b)^2$	$1a^2 + 2a^1b^1 + 1b^2$	1 2 1
3	$(a + b)^3$	$1a^3 + 3a^2b^1 + 3a^1b^2 + 1b^3$	1 3 3 1
4	$(a + b)^4$	$1a^4 + 4a^3b^1 + 6a^2b^2 + 4a^1b^3 + 1b^4$	1 4 6 4 1

The "coefficients only" column matches the numbers in *Pascal's Triangle*. **Pascal's Triangle**, named for the French mathematician Blaise Pascal (1623–1662), is a triangular array of numbers in which the first and last number of each row is 1. Each of the other numbers in the row is the sum of the two numbers above it.

For example, to generate row 5, use the sums of the adjacent elements in the row above it.

Row	Pascal's Triangle
0	1
1	1 1
2	1 2 1
3	1 3 3 1
4	1 4 6 4 1
5	1 5 10 10 5 1
6	1 6 15 20 15 6 1
7	1 7 21 35 35 21 7 1
8	1 8 28 56 70 56 28 8 1

1 + 4 = 5, 4 + 6 = 10, 6 + 4 = 10, 4 + 1 = 5

Problem 1 Using Pascal's Triangle

Plan

What row of Pascal's Triangle should you use for this expansion?
The expression is raised to the 6th power so use the 6th row.

What is the expansion of $(a + b)^6$? Use Pascal's Triangle.

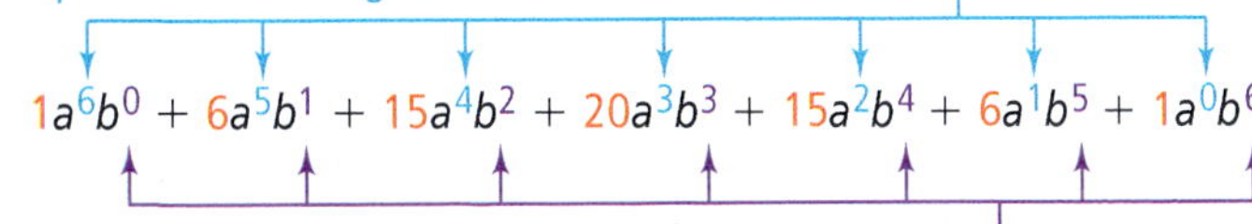

$$(a + b)^6 = a^6 + 6a^5b + 15a^4b^2 + 20a^3b^3 + 15a^2b^4 + 6ab^5 + b^6.$$

Got It? **1.** What is the expansion of $(a + b)^8$? Use Pascal's Triangle.

The **Binomial Theorem** gives a general formula for expanding a binomial.

take note

Theorem Binomial Theorem

For every positive integer n,

$$(a + b)^n = P_0a^n + P_1a^{n-1}b + P_2a^{n-2}b^2 + \cdots + P_{n-1}ab^{n-1} + P_nb^n$$

where $P_0, P_1, \ldots, P_n$ are the numbers in the nth row of Pascal's Triangle.

When you use the Binomial Theorem to expand $(x - 2)^4$, $a = x$ and $b = -2$. To expand a binomial such as $(3x - 2)^5$, $a = 3x$ so remember that $a^4 = (3x)^4$ not $3x^4$.

Problem 2 Expanding a Binomial

What is the expansion of $(3x - 2)^5$? Use the Binomial Theorem.

Think	Write
For $(3x - 2)^5$, use the 5th row of Pascal's Triangle.	**Pascal's Triangle** 1 1 1 1 2 1 1 3 3 1 1 4 6 4 1 1 5 10 10 5 1
The Binomial Theorem uses a binomial sum.	$(3x - 2)^5 = (3x + (-2))^5$
Apply the Binomial Theorem.	$= (3x)^5 + 5(3x)^4(-2)^1 + 10(3x)^3(-2)^2 + 10(3x)^2(-2)^3 + 5(3x)^1(-2)^4 + 1(-2)^5$
Simplify.	$= 243x^5 - 810x^4 + 1080x^3 - 720x^2 + 240x - 32$

Got It? **2. a.** What is the expansion of $(2x - 3)^4$? Use the Binomial Theorem.

b. Reasoning Consider the following:

$11^0 = 1 \quad 11^1 = 11 \quad 11^2 = 121 \quad 11^3 = 1331 \quad 11^4 = 14641$

Why do these powers of 11 have digits that mirror Pascal's Triangle?

Lesson Check

Do you know HOW?

Use Pascal's Triangle to expand each binomial.

1. $(x + a)^3$

2. $(x - 2)^5$

3. $(2x + 4)^2$

4. $(3a - 2)^3$

Do you UNDERSTAND?

5. Vocabulary Tell whether each expression can be expanded using the Binomial Theorem.

a. $(2a - 6)^4$ **b.** $(5x^2 + 1)^5$ **c.** $(x^2 - 3x - 4)^3$

6. Writing Describe the relationship between Pascal's Triangle and the Binomial Theorem.

7. Reasoning Using Pascal's Triangle, determine the number of terms in the expansion of $(x + a)^{12}$. How many terms are there in the expansion of $(x + a)^n$?

Practice and Problem-Solving Exercises

Expand each binomial.

See Problems 1 and 2.

8. $(x - y)^3$ **9.** $(a + 2)^4$ **10.** $(6 + a)^6$ **11.** $(x - 5)^3$

12. $(y + 1)^8$ **13.** $(x + 2)^{10}$ **14.** $(b - 4)^7$ **15.** $(b + 3)^9$

16. $(2x - y)^7$ **17.** $(a + 3b)^4$ **18.** $(4x + 2)^6$ **19.** $(4 - x)^8$

20. $(4x + 5)^2$ **21.** $(3a - 7)^3$ **22.** $(2a + 16)^6$ **23.** $(3y - 11)^4$

24. Think About a Plan The side length of a cube is $\left(x^2 - \frac{1}{2}\right)$. Determine the volume of the cube.

- Rewrite the binomial as a sum.
- Consider $(a + b)^n$. Identify a and b in the given binomial.
- Which row of Pascal's Triangle can be used to expand the binomial?

25. In the expansion of $(2m - 3n)^9$, one of the terms contains m^3.

a. What is the exponent of n in this term?

b. What is the coefficient of this term?

Find the specified term of each binomial expansion.

26. Fourth term of $(x + 2)^5$ **27.** Third term of $(x - 3)^6$

28. Third term of $(3x - 1)^5$ **29.** Fifth term of $(a + 5b^2)^4$

30. Reasoning Explain why the coefficients in the expansion of $(x + 2y)^3$ do not match the numbers in the 3rd row of Pascal's Triangle.

31. Compare and Contrast What are the benefits and challenges of using the Binomial Theorem when expanding $(2x + 3)^2$? Using FOIL? Which method would you choose when expanding $(2x + 3)^6$? Why?

Expand each binomial.

32. $(2x - 2y)^6$ **33.** $(x^2 + 4)^{10}$ **34.** $(x^2 - y^2)^3$ **35.** $(a - b^2)^5$

36. $(3x + 8y)^3$ **37.** $(4x - 7y)^4$ **38.** $(7a + 2y)^{10}$ **39.** $(4x^3 + 2y^2)^6$

40. $(3b - 36)^7$ **41.** $(5a + 2b)^3$ **42.** $(b^2 - 2)^8$ **43.** $(-2y^2 + x)^5$

44. Geometry The side length of a cube is given by the expression $(2x + 8)$. Write a binomial power for the area of a face of the cube and for the volume of the cube. Then use the Binomial Theorem to expand and rewrite the powers in standard form.

45. Writing Explain why the terms of $(x - y)^n$ have alternating positive and negative signs.

46. Error Analysis A student expands $(3x - 8)^4$ as shown below. Describe and correct the student's error.

$$(3x - 8)^4 = (3x)^4 + 4(3x)^3(-8) + 6(3x)^2(-8)^2 + 4(3x)(-8)^3 + (-8)^4$$
$$= 3x^4 - 96x^3 + 1152x^2 - 6144x + 4096$$

Challenge Use the Binomial Theorem to expand each complex expression.

47. $(7 + \sqrt{-16})^5$ **48.** $(\sqrt{-81} - 3)^3$ **49.** $(x^2 - i)^7$

50. The first term in the expansion of a binomial $(ax + by)^n$ is $1024x^{10}$. Find a and n.

51. Determine the coefficient of x^7y in the expansion of $\left(\frac{1}{2}x + \frac{1}{4}y\right)^8$.

52. **a.** Expand $(1 + i)^4$.
b. Verify that $1 - i$ is a fourth root of -4 by repeating the process in part (a) for $(1 - i)^4$.

53. Verify that $-1 + \sqrt{3}i$ is a cube root of 8 by expanding $(-1 + \sqrt{3}i)^3$.

Standardized Test Prep

SAT/ACT

54. What is the fourth term in the expansion of $(2a + 4b)^5$?

Ⓐ $256a^4b$ Ⓑ $768a^3b^2$ Ⓒ $2560a^2b^3$ Ⓓ $2048ab^4$

55. Suppose y varies directly with x. If x is 30 when y is 10, what is x when y is 9?

Ⓕ 3 Ⓖ 27 Ⓗ 29 Ⓘ $\frac{300}{9}$

56. Which of following is a root of $9x^2 - 30x + 25 = 0$?

Ⓐ $x = \frac{3}{5}$ Ⓑ $x = \frac{5}{3}$ Ⓒ $x = -\frac{5}{3}$ Ⓓ $x = -\frac{3}{5}$

Extended Response

57. One company charges a monthly fee of \$7.95 and \$2.25 per hour for Internet access. Another company does not charge a monthly fee, but charges \$2.75 per hour for Internet access. Write a system of equations to represent the cost c for t hours of access in one month for each company. Then find how many hours of use it will take for the costs to be equal.

Mixed Review

Find all the roots of each equation. See Lesson 5-6.

58. $x^4 + 7x^3 + 20x^2 + 29x + 15 = 0$ **59.** $x^5 - x^4 + 10x^3 - 10x^2 + 9x - 9 = 0$

60. $2x^3 + 11x^2 + 14x + 8 = 0$ **61.** $x^4 - x^3 + 6x^2 - 13x + 7 = 0$

Simplify each expression. See Lesson 4-8.

62. $(5i - 4)(-2i + 7)$ **63.** $(-3i)(20i)(10i)$

64. $\frac{-6 - 2i}{3 + i}$ **65.** $\frac{11i + 9}{2 - i}$

Get Ready! **To prepare for Lesson 5-8, do Exercises 66–68.**

Write each polynomial in standard form. Then classify it by degree and by number of terms. 
See Lesson 5-1.

66. $5x^2 - x + 2x^3 + 9$ **67.** $1 + 4x - 7x^2$ **68.** $-9x^2 + x - 3x^3 - 8 + 12x^4$

5-8 Polynomial Models in the Real World

Content Standards

F.IF.5 Relate the domain of a function to its graph and, where applicable, to the quantitative relationship it describes.

Also F.IF.4, F.IF.6

Objective To fit data to linear, quadratic, cubic, or quartic models

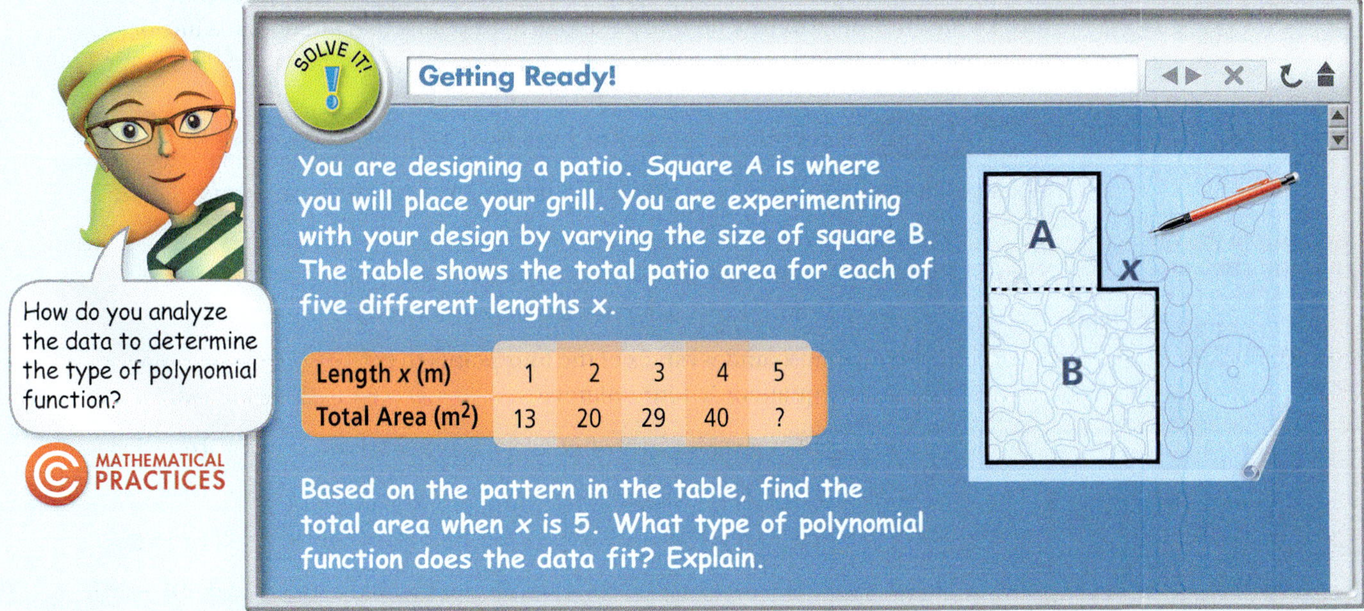

Polynomial functions can be degree 0 (constant), degree 1 (linear), degree 2 (quadratic), degree 3 (cubic), and so on.

Essential Understanding You can use polynomial functions to model many real-world situations. The behavior of the graphs of polynomial functions of different degrees can suggest what type of polynomial will best fit a particular data set.

You can use a graphing calculator to help you find functions that model the data in the table shown here.

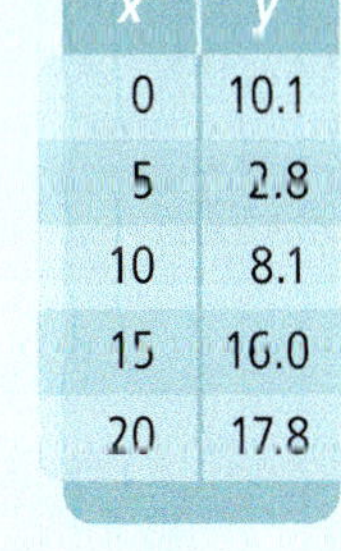

x	y
0	10.1
5	2.8
10	8.1
15	16.0
20	17.8

Enter the data into calculator lists **L1** and **L2**. Use three different regressions: **LINREG**, **QUADREG**, and **CUBICREG**.

Graph each regression function and the scatter plot of the data in the same window. For this data, the cubic function appears to be a perfect fit.

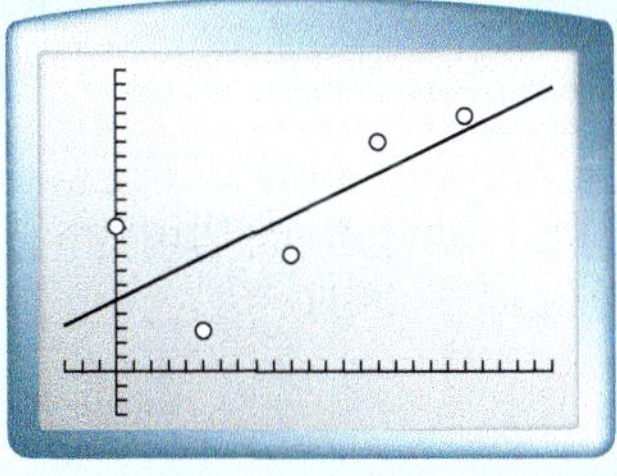

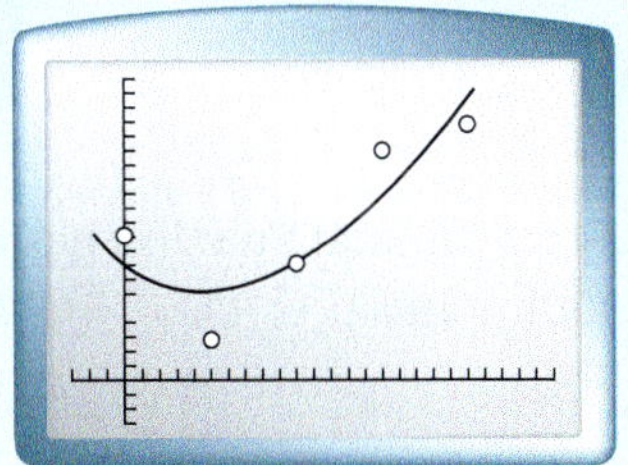

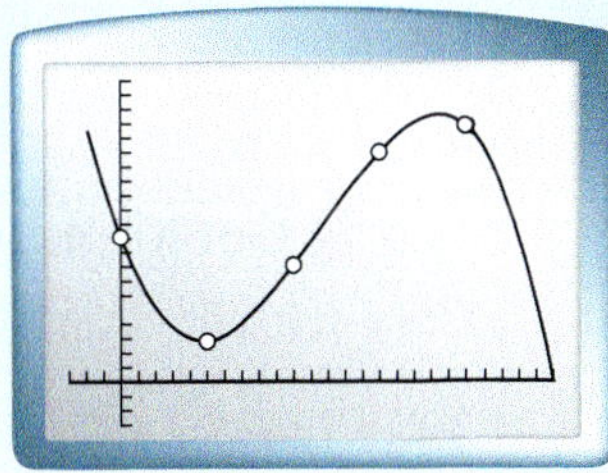

take note

Key Concept The $(n + 1)$ Point Principle

For any set of $n + 1$ points in the coordinate plane that pass the vertical line test, there is a unique polynomial of degree at most n that fits the points perfectly.

This principle confirms that any two points determine a unique line. Three points that are not on a line determine a unique parabola. Four points that are not on a line or a parabola determine a unique cubic, and so forth.

Plan

How can you use the four points to find a system of equations?
Substitute each point into a polynomial function.

Problem 1 Using A Polynomial Function to Model Data

What polynomial function has a graph that passes through the four points $(0, -3)$, $(1, -1)$, $(2, 5)$, and $(-1, -7)$?

Step 1 By the $(n + 1)$ Point Principle, there is a cubic polynomial $y = ax^3 + bx^2 + cx + d$ that fits the points perfectly.

Substitute the x- and y-values of the four points to get four linear equations in four unknowns.

$$-3 = a(0)^3 + b(0)^2 + c(0) + d \qquad 0a + 0b + 0c + 1d = -3$$
$$-1 = a(1)^3 + b(1)^2 + c(1) + d \qquad 1a + 1b + 1c + 1d = -1$$
$$5 = a(2)^3 + b(2)^2 + c(2) + d \qquad 8a + 4b + 2c + 1d = 5$$
$$-7 = a(-1)^3 + b(-1)^2 + c(-1) + d \qquad -1a + 1b - 1c + 1d = -7$$

Step 2 Write the system in augmented matrix form. Use the **RREF()** function to find the coefficient values a, b, c, and d.

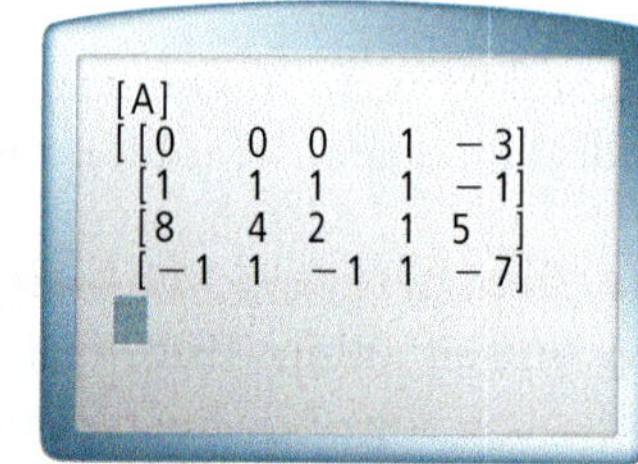

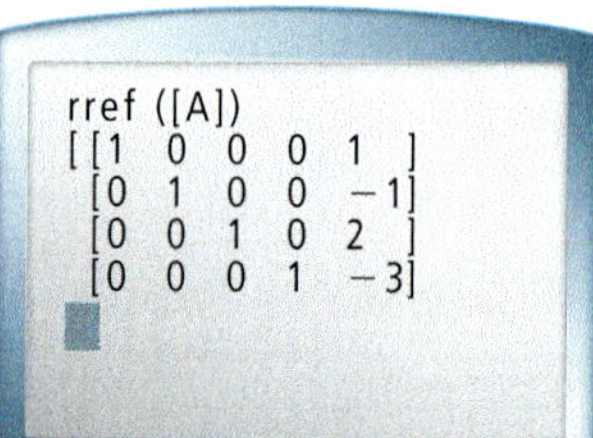

Step 3 $a = 1$, $b = -1$, $c = 2$, and $d = -3$. The polynomial function is $y = x^3 - x^2 + 2x - 3$.

Check Use **CUBICREG** with the four given points.

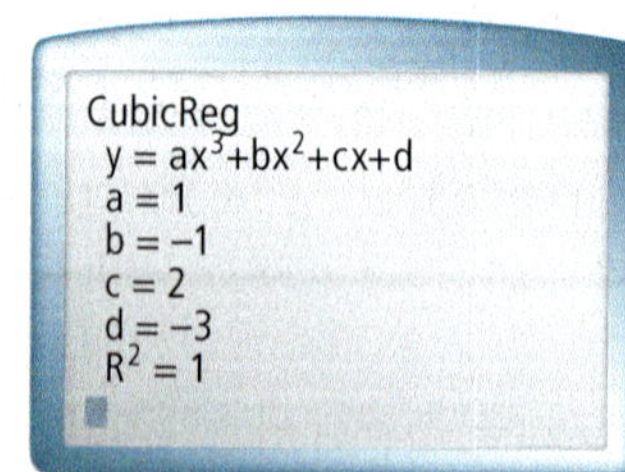

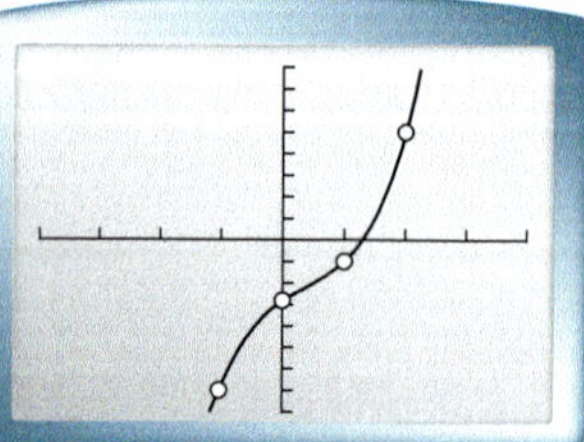

The solution checks.

Got It? **1.** What polynomial function has a graph that passes through the four points $(-2, 1)$, $(0, 5)$, $(2, 9)$, and $(3, 36)$?

Problem 2 Modeling Data STEM

Food Production The chart shows how much milk Wisconsin dairy farms produced in 1955, 1980, and 2005. What linear model best fits this data? Use the model to estimate milk production in 2000.

Milk Production (in billions of lbs)

1955	1980	2005
16.5	22.4	22.9

Think

What data should you enter?
Enter the years, after 1900, and billions of pounds of milk produced.

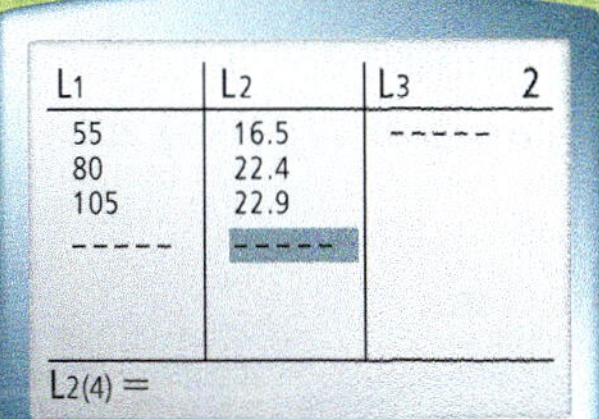

Enter the data. Let x represent years after 1900. Let y represent pounds of milk.

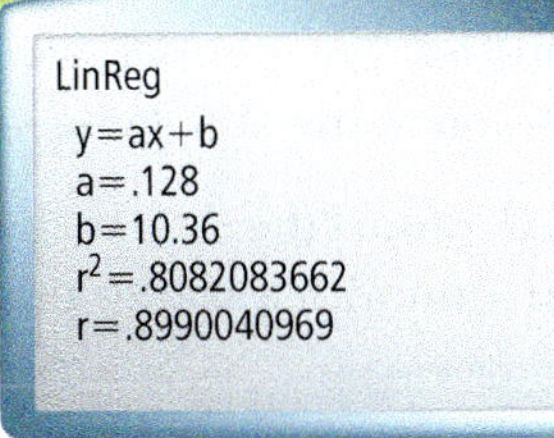

Use **LINREG** to find a linear model: $f(x) = 0.128x + 10.36$. Notice the value of r^2.

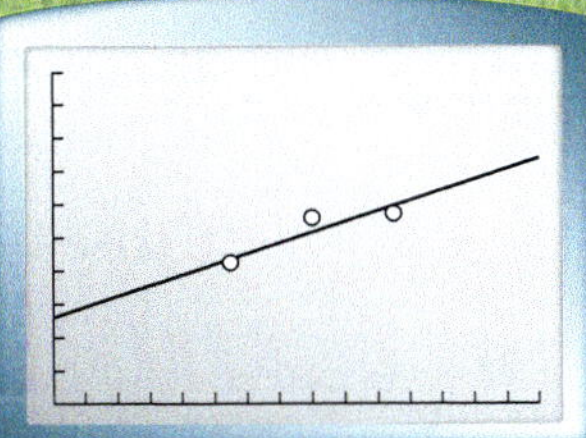

Show the scatter plot of the data and a graph of $f(x)$.

The closer r^2 is to 1, the better the fit.

Use the model to estimate milk production in 2000.

$$f(100) = 0.128(100) + 10.36 \approx 23.2$$

In 2000, Wisconsin produced about 23.2 billion pounds of milk.

Got It? 2. Use the linear model. Estimate Wisconsin milk production in 1995.

Problem 3 Comparing Models STEM

Food Production The graph shows the quadratic model for the milk production data in Problem 2. The quadratic model fits the data points exactly because of the $(n + 1)$ Point Principle. Given that both models are good fits, which seems more likely to represent milk production over time?

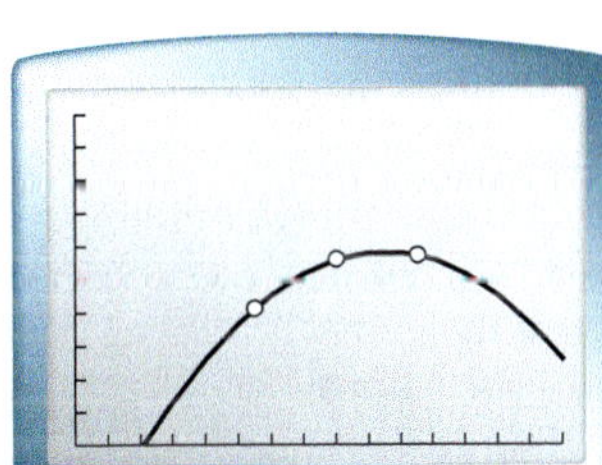

Think

Is the quadratic model reasonable?
No; the quadratic model will show milk production eventually decreasing to below zero.

Linear Model: This model continually rises.

Quadratic Model: This model has down-and-down end behavior. It shows slowing growth, a turning point, and then a decline, eventually to 0 and negative values.

Despite the R^2 value of 1 for the quadratic model, the linear model is more likely to represent milk production over time since it shows a continuing increase.

Got It? 3. If four data points were given, would a cubic function be the best model for the data? Explain your answer.

When deciding whether a model is reliable, it is important to consider the source of the data. Data generated by a law of physics or a geometric formula will have a mathematical model that fits the data and yields accurate predictions. With other data, such as sales records, you can approximate the data only within, or close to, the domain over which it was generated.

Using a model to predict a y-value "outside" the domain of a data set is *extrapolation*. Estimating within the domain is *interpolation*. Interpolation usually yields reliable estimates. Extrapolation becomes less reliable as you move farther away from the data.

Problem 4 Using Interpolation and Extrapolation

Cheese Consumption **The table shows average annual consumption of cheese per person in the U. S. for selected years from 1910 to 2001.**

Cheese Consumption

Year	Pounds Consumed
1910	4
1940	5
1970	8
1975	10
1995	25
2001	30

SOURCE: U.S. Department of Agriculture

A **Use CUBICREG. Model the data with a cubic function. Graph the function with a scatter plot of the data.**

CubicReg
$y = ax^3+bx^2+cx+d$
$a = 9.242151\text{E}^{-}5$
$b = {}^{-}.0095843361$
$c = .3106497055$
$d = 1.802125648$
$R^2 = .9976342$

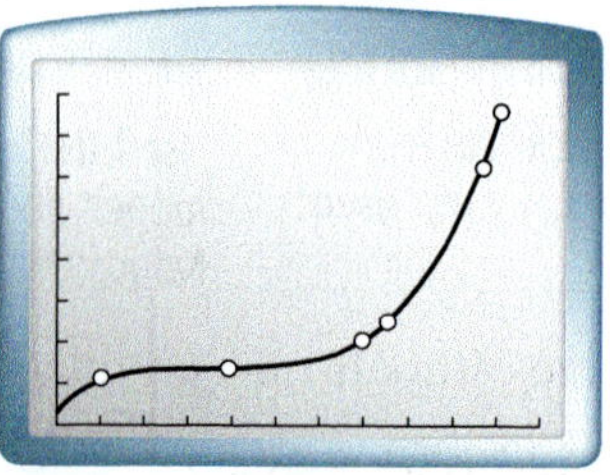

Since R^2 is close to 1, the fit is good. The cubic model is approximately $y = 0.0000924x^3 - 0.00958x^2 + 0.311x + 1.802$.

B **Use the model to estimate cheese consumption for 1980, 2000, and 2012. In which estimate do you have the most confidence? The least confidence? Explain.**

Use the cubic model from part A to estimate the cheese consumption for each year:

1980: 12.6 lb of cheese per person
2000: 29.4 lb of cheese per person
2012: 46.2 lb of cheese per person

You can be confident in interpolating the estimates for 1980 (**Y1(80)**) and 2000 (**Y1(100)**) because the cheese consumption fits the pattern of increase shown in the table. You should have the least confidence in the extrapolated 2012 (**Y1(112)**) estimate because the cubic model increases so quickly beyond 2001.

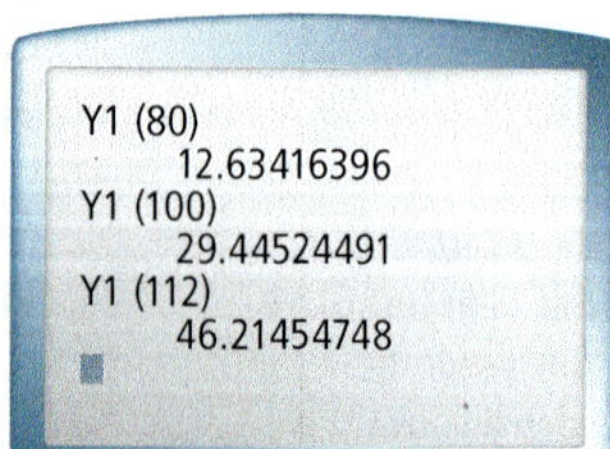

Think

What affects your confidence in drawing conclusions from the model?
Your confidence can waver where model behavior is extreme or where there are large gaps in the data.

Got It? **4. a.** Use **LINREG** to find a linear model for cheese consumption. Graph it with a scatter plot.

b. **Reasoning** Use the model to estimate consumption for 1980, 2000, and 2012. In which of these estimates do you have the most confidence? The least confidence? Explain.

Lesson Check

Do you know HOW?

Determine which type of model best fits each set of points.

1. (−2, −1), (0, 3), and (2, 7)
2. (0, 3), (3, 4), and (5, 6)
3. (2, 3), (4, 2), (6, 4), and (8, 5)
4. (−5, 6), (−4, 3), (0, 2), (2, 4), and (5, 10)

Do you UNDERSTAND?

5. **Vocabulary** Explain which form of estimation, interpolation or extrapolation, is more reliable.
6. **Reasoning** Is it possible to create a cubic function that passes through (0, 0), (−1, 1), (−2, 2), and (−3, 9)? Explain.
7. **Writing** The R^2 value for a quartic model is 0.94561. The R^2 value for a cubic model of the same data is 0.99817. Which model seems to show a better fit? Explain.

Practice and Problem-Solving Exercises

A Practice

Find a polynomial function whose graph passes through each set of points. See Problem 1.

8. (0, 5) and (2, −13)
9. (−2, −4) and (8, 1)
10. (−5, 14) and (1, −16)
11. (7, 13), (10, −11), and (0, 4)
12. (−2, −16), (3, 11), and (0, 2)
13. (−1, 8), (5, −4), and (7, 8)
14. (−1, −15), (1, −7), and (6, −22)
15. (−1, 9), (0, 6), (1, 5), and (2, 18)

For each set of data, compare two models and determine which one best fits the data. Which model seems more likely to represent each set of data over time? See Problems 2 and 3.

16. **U.S. Federal Spending**

Year	Total (billions $)
1965	630
1980	1,300
1995	1,950
2005	2,650

17. **World Population**

Year	Average Growth Rate (%)
1972	1.96
1982	1.73
1992	1.5
2002	1.22

18. **U.S. Homes**

Year	Average Sale Price (thousands $)
1990	149
1995	158
2000	207

19. **U.S. Crude Oil and Petroleum**

Month (2008)	Products Supplied (millions of barrels/day)
2	19.782
4	19.768
6	19.553

Use your models from Exercises 16-19 to make predictions.

See Problem 4.

20. Estimate total U.S. federal spending for 1990 and 2010.

21. Estimate the average annual growth rate of the world population for 1950, 1988, and 2010.

22. Estimate the average sale price of homes sold in the United States for 1985, 1999, and 2020.

23. Estimate the number of barrels of crude oil and petroleum supplied per day for January, March, and October of 2008.

B Apply

Find a cubic and a quartic model for each set of values. Explain why one models the data better.

24.

x	−2	−1	0	2	3
y	−25	−4	3	23	40

25.

x	−2	−1	0	1	2
y	−65	−14	−4	2	90

Find a polynomial function whose graph passes through the points.

26. $(-14, 14)$, $(-10, 0)$, $(0, -1)$, $(8, 0)$, and $(12, 4)$

27. $(-3, -50)$, $(-2, -4)$, $(-1, 10)$, $(0, 7)$, and $(2, -23)$

28. Think About a Plan The table at the right shows the amount of carbon dioxide in the Earth's atmosphere for selected years. Predict the amount of carbon dioxide in the Earth's atmosphere in 2022. How confident are you in your prediction?

- How can you plot the data? (*Hint:* Let x equal the years after 1900.)
- What polynomial model should you use?

Year	CO_2 in atmosphere (ppm)
1968	324.14
1983	343.91
1998	367.68
2003	376.68
2008	385.60

Source: The Weather Channel

Find a cubic model for each set of values. Then use the regression coefficient of each model to determine whether the model is a good fit.

29. $(-5, -60)$, $(-1, -5)$, $(0, 0.5)$, $(1, 8)$, $(5, 17)$, $(10, 32)$

30. $(8, -101)$, $(-1, 10)$, $(-8, 47)$, $(-10, 59)$

31. Air Travel The table shows the percent of on-time flights for selected years. Find a polynomial function to model the data. Use 1998 as Year 0.

Year	1998	2000	2002	2004	2006
On-time Flights (%)	77.20	72.59	82.14	78.08	75.45

Source: U.S. Bureau of Transportation Statistics

32. Writing Explain two ways to find a polynomial function to model a given set of data.

33. **Error Analysis** The table at the right shows the number of students enrolled in a high school personal finance course. A student says that a cubic model would best fit the data based on the $(n + 1)$ Point Principle. Explain why a quadratic model might be more appropriate.

Year	Number of Students Enrolled
2000	50
2004	65
2008	94
2010	110

34. **Compare and Contrast** The table shows the United States gross domestic product for selected years. Construct curves using cubic regression and quartic regression to model the data. Which curve seems most likely to model gross domestic product over the years?

Year	1960	1970	1980	1990	2000
GDP (billions $)	526.4	1038.5	2789.5	5803.1	9817.0

35. The table below shows the percentage of the U.S. labor force in unions for selected years between 1955 and 2005.

Year	1955	1960	1965	1970	1975	1980	1985	1990	1995	2000	2005
%	33.2	31.4	28.4	27.3	25.5	21.9	18.0	16.1	14.9	13.5	12.5

a. What is the average rate of change between 1955 and 1965? Between 1975 and 1985?

b. Make a scatter plot of the data. Which kind of polynomial model seems to be most appropriate?

c. Use a graphing calculator to find the type of model from part (b).

d. Use the model you found in part (c) to predict the percent of the labor force in unions in the year 2020.

e. **Reasoning** Do you have much confidence in this prediction? Explain.

36. Your friend's teacher showed the class a graph of a cubic polynomial in the ZDecimal window, which is $[-4.7, 4.7]$ by $[-3.1, 3.1]$. She then challenged the class to find the polynomial *without using cubic regression on their calculators,* and your friend succeeded. Follow your friend's steps and see if you can find the polynomial.

a. The graph resembles a parabola with vertex $(2, 0)$ near $x = 2$. Find the equation in standard form for that parabola.

b. Find the equation of a line in slope-intercept form through $(-2, 0)$ with slope 1. Multiply the linear expression by the quadratic expression from part (a) to get a cubic. (Leave it in factored form.) Graph the cubic function. What do you notice about the zeros and the y-intercept of the cubic function?

c. Multiply the cubic by a constant to change the y-intercept to 2. Graph the function to see if you've found the right polynomial. What is the function?

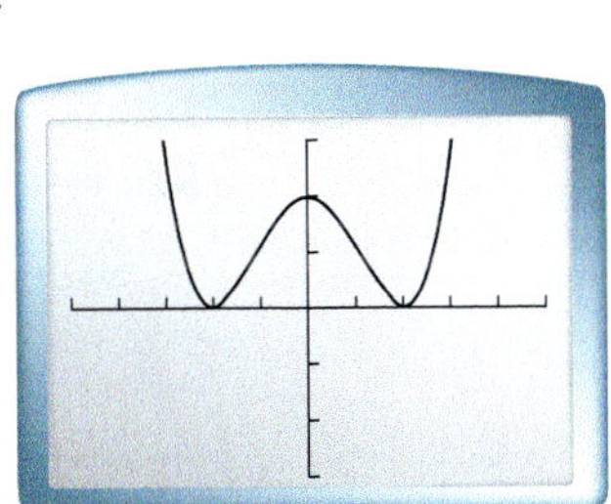

37. The graph at the right is that of a certain quartic polynomial in the ZDecimal window, which is $[-4.7, 4.7]$ by $[-3.1, 3.1]$. Find the equation of the quartic *without using quartic regression on your calculator*. You may leave it in factored form.

Standardized Test Prep

GRIDDED RESPONSE

SAT/ACT

38. The table shows the time it takes a computer program to run, given the number of files used as input. Using a cubic model, what do you predict the run time will be if the input consists of 1000 files?

Files	Time(s)
100	0.5
200	0.9
300	3.5
400	8.2
500	14.8

39. Suppose you hit a ball and its flight follows the graph of $f(x) = -16x^2 + 20x + 3$. How many seconds will it take for the ball to hit the ground? Round your answer to the nearest second.

40. What is the multiplicity of the zeros of $y = 16x^2 - 8x + 1$?

41. What is the slope of the line shown?

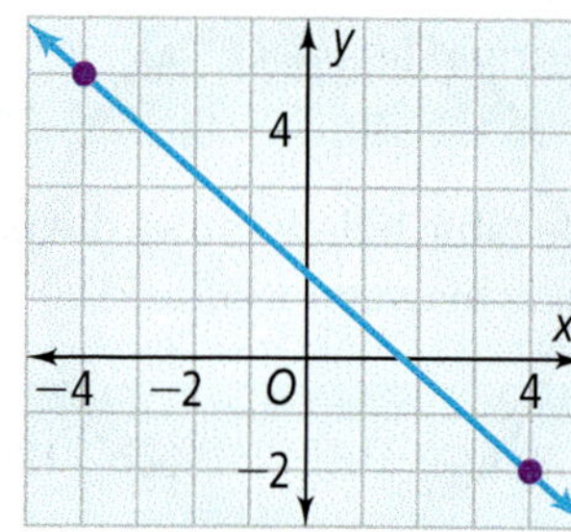

42. What is the degree of the polynomial function $y = -9x^3 - 5x^2 - 2x^5 + 4x + x^4 + 1$?

Mixed Review

Expand each binomial. See Lesson 5-7.

43. $(2x + 3)^5$

44. $(11x - 1)^3$

45. $(8 - 3x)^4$

46. $(4 + 9x)^3$

Write each compound inequality as an absolute value inequality. See Lesson 1-6.

47. $7 < x < 9$

48. $\frac{1}{4} \leq x \leq \frac{1}{2}$

49. $1.7 < y < 3.9$

50. $500 < t < 1000$

Solve each formula for the indicated variable. See Lesson 1-4.

51. $A = s^2$, for s

52. $P = 2(l + w)$, for l

53. $C = 2\pi r$, for r

54. $A = bh$, for b

Get Ready! **To prepare for Lesson 5-9, do Exercises 55–58.**

Graph each function. See Lesson 4-1.

55. $y = x^2$

56. $y = -4x^2$

57. $y = x^2 + 3$

58. $y = -7x^2 - 1$

5-9 Transforming Polynomial Functions

Content Standards

F.BF.3 Identify the effect on the graph of replacing $f(x)$ by $f(x) + k$, $k\,f(x)$, $f(kx)$, and $f(x + k)$ for specific values of k (both positive and negative) find the value of k given the graphs.

Also F.IF.7.c, F.IF.8, F.IF.9

Objective To apply transformations to graphs of polynomials

Remember you transformed the graphs of quadratic and absolute value functions.

SOLVE IT! Getting Ready!

The graph of the parent cubic function $f(x) = x^3$ is one of the graphs at the right. The other graph is a transformation g of the parent function. What is an equation for g? How do you know?

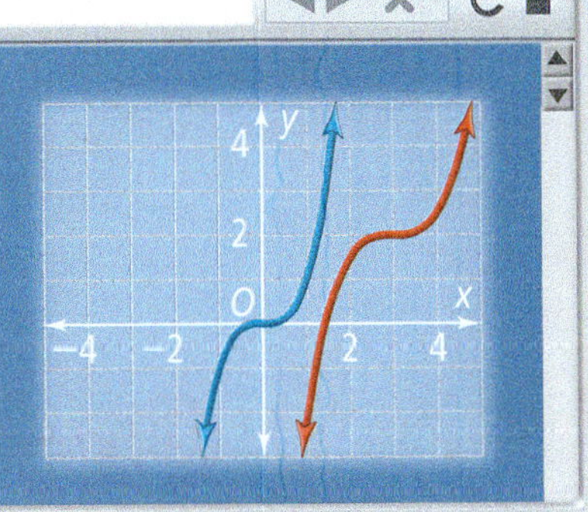

MATHEMATICAL PRACTICES

Recall that you can obtain the graph of any quadratic function from the graph of the parent quadratic function, $y = x^2$, using one or more basic transformations. You will find that this is not true of cubic functions.

Lesson Vocabulary
- power function
- constant of proportionality

Essential Understanding The graph of the function $y = af(x - h) + k$ is a vertical stretch or compression by the factor $|a|$, a horizontal shift of h units, and a vertical shift of k units of the graph of $y = f(x)$.

Problem 1 Transforming $y = x^3$

What is an equation of the graph of $y = x^3$ under a vertical compression by the factor $\frac{1}{2}$ followed by a reflection across the x-axis, a horizontal translation 3 units to the right, and then a vertical translation 2 units up?

Step 1 Multiply by $\frac{1}{2}$ to compress.

$y = x^3 \longrightarrow y = \frac{1}{2}x^3$

Step 2 Multiply by -1 to reflect.

$y = \frac{1}{2}x^3 \longrightarrow y = -\frac{1}{2}x^3$

Step 3 Replace x with $x - 3$ to translate horizontally.

$y = -\frac{1}{2}x^3 \longrightarrow y = -\frac{1}{2}(x - 3)^3$

Step 4 Add 2 to translate vertically.

$y = -\frac{1}{2}(x - 3)^3 \longrightarrow y = -\frac{1}{2}(x - 3)^3 + 2$

Think

How is translating this cubic function like translating a quadratic function?
In each case you replace x with $x - h$ to translate h units to the right.

Got It? **1.** What is an equation of the graph of $y = x^3$ under a vertical stretch by the factor 2 followed by a horizontal translation 3 units to the left and then a vertical translation 4 units down?

The graph shows $y = x^3$ and the graphs that result from the transformations in Problem 1.

In general, $y = a(x - h)^3 + k$ represents all of the cubic functions you can obtain by stretching, compressing, reflecting, or translating the cubic parent function $y = x^3$.

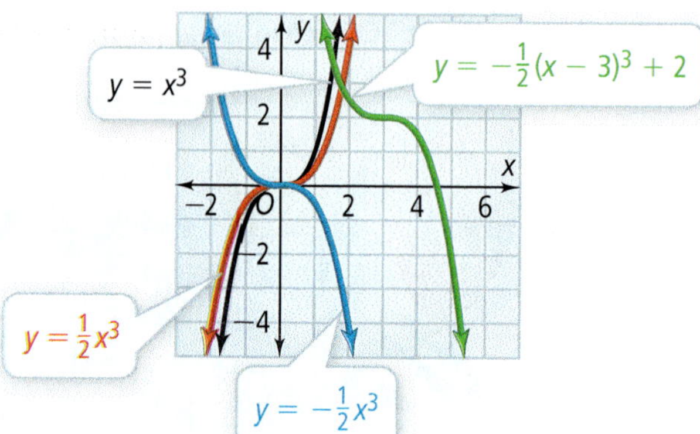

Problem 2 Finding Zeros of a Transformed Cubic Function

Multiple Choice **If a, h, and k are real numbers and $a \neq 0$, how many distinct real zeros does $y = -a(x - h)^3 + k$ have?**

Ⓐ 0 Ⓑ 1 Ⓒ 2 Ⓓ 3

Plan

How can you find the zeros of a function? Set the function equal to 0 and solve for x.

$-a(x - h)^3 + k = 0$	x is a zero means it is an x-intercept, so $y = 0$.
$-a(x - h)^3 = -k$	Subtract k from each side.
$(x - h)^3 = \frac{k}{a}$	Divide each side by $-a$.
$x - h = \sqrt[3]{\frac{k}{a}}$	Take the cube root of each side.
$x = \sqrt[3]{\frac{k}{a}} + h$	Solve for x.

Disregarding multiplicities, the function has a single real zero. The correct answer is B.

Got It? **2.** What are all the real zeros of the function $y = 3(x - 1)^3 + 6$?

Problems 1 and 2 together illustrate that the graph of an "offspring" function of the parent cubic function $y = x^3$ has only one x-intercept.

The graph of the cubic function $y = x^3 - 2x^2 - 5x + 6$ has three x-intercepts. You cannot obtain this function or others like it by transforming the parent cubic function $y = x^3$ using stretches, reflections, and translations.

Similarly, some quartic functions are simple transformations of $y = x^4$ and some are not.

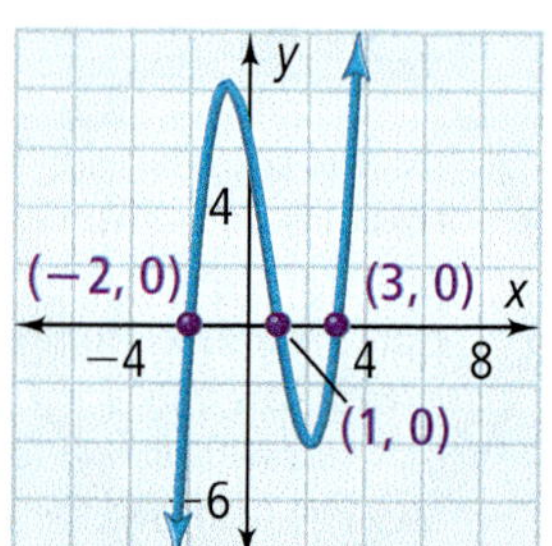

Problem 3 Constructing a Quartic Function with Two Real Zeros

What is a quartic function with only two real zeros, $x = 5$ and $x = 9$?

Method 1 Use transformations.

First, find a quartic with zeros at ± 2.
Translate the basic quartic 16 units down:
$y = x^4 \rightarrow y = x^4 - 16$

9 is 7 units to the right of 2.
Translate 7 units to the right.
$y = x^4 - 16 \rightarrow y = (x - 7)^4 - 16$

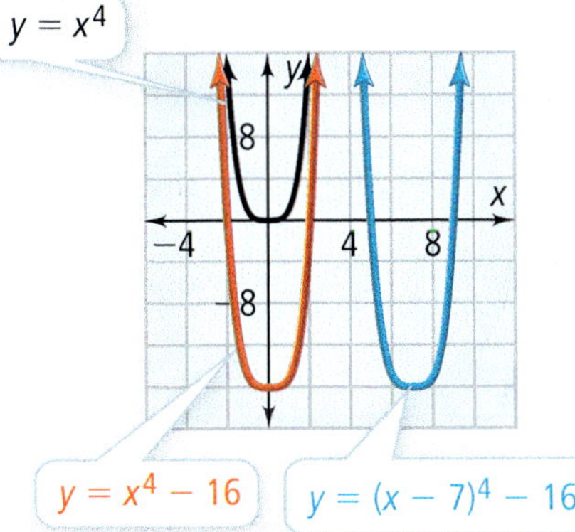

A quartic function with its only real zeros at 5 and 9 is $y = (x - 7)^4 - 16$.

Method 2 Use algebraic methods.

$$\begin{aligned} y &= (x - 5)(x - 9) \cdot Q(x) \\ &= (x - 5)(x - 9)(x^2 + 1) \\ &= (x^2 - 14x + 45)(x^2 + 1) \\ &= x^4 - 14x^3 + 46x^2 - 14x + 45 \end{aligned}$$

Make $Q(x)$ a quadratic with no real zeros.

Think

What should you use for $Q(x)$?
Choose $Q(x)$ to be a quadratic with no real zeros. Keep it simple, such as $x^2 + 1$.

Another quartic function with its only real zeros at 5 and 9 is $y = x^4 - 14x^3 + 46x^2 - 14x + 45$.

Got It? **3. a.** What is a quartic function $f(x)$ with only two real zeros, $x = 0$ and $x = 6$?

b. Reasoning Does the quartic function $-f(x)$ have the same zeros? Explain.

The "offspring" of the parent function $y = x^4$ is a subfamily of all quartic polynomials. This subfamily consists of quartics of the form $y = a(x - h)^4 + k$. These functions also belong to another category of polynomials, and in this category you can generate families as usual.

take note

Key Concept Power Functions

Definition	Examples	
A **power function** is a function of the form $y = a \cdot x^b$, where a and b are nonzero real numbers.	$y = 0.5x^6$	$y = \frac{1}{2}x^2$
	$y = -4x^{\frac{2}{3}}$	$y = x^{0.25}$

If the exponent b in $y = ax^b$ is a positive integer, the function is also a *monomial function*.

If $y = ax^b$ describes y as a power function of x, then *y varies directly with*, or *is proportional to*, the bth power of x. The constant a is the **constant of proportionality**. Power functions arise in many real-world contexts related to the concept of direct variation, which you studied in Chapter 2.

Problem 4 Modeling With a Power Function STEM

Wind-Generated Power Wind farms are a source of renewable energy found around the world. The power P (in kilowatts) generated by a wind turbine varies directly as the cube of the wind speed v (in meters per second). The picture shows the power output of one turbine at one wind speed. To the nearest kilowatt, how much power does this turbine generate in a 10 m/s wind?

Wind 8 m/s

Electric Power 600 kW

Think

How is this problem like the direct variation you studied in Chapter 2?
With the direct variation, $y = kx$, you use a first power. Here you use a third power.

The formula for P as a power function of v is $P = a \cdot v^3$. From the picture, $P = 600$ when $v = 8$, or $600 = a \cdot 8^3$. Solve for a.

$600 = a \cdot 8^3$	Use values of P and v to find a.
$600 = 512a$	
$a \approx 1.1719$	
$P \approx 1.1719v^3.$	Use the value of a in the original formula.
$P \approx 1.1719 \cdot 10^3 = 1171.9.$	Substitute 10 for v and simplify.

This turbine generates about 1172 kW of power in a 10 m/s wind.

Got It? 4. Another turbine generates 210 kW of power in a 12 mi/h wind. How much power does this turbine generate in a 20 mi/h wind?

Lesson Check

Do you know HOW?

Find all the real zeros of each function.

1. $y = -(x + 3)^3 + 1$

2. $y = -8(x - 5)^3 - 64$

3. $y = \frac{9}{2}(x - 1)^3 + \frac{4}{3}$

Do you UNDERSTAND?

4. Vocabulary Is the function $y = 4x^3 + 5$ an example of a power function? Explain.

5. Error Analysis Your friend says that he has found a way to transform the graph of $y = x^3$ to obtain three real roots. Using the graph of the function, explain why this is impossible.

6. Compare and Contrast How are the graphs of $y = x^3$ and $y = 4x^3$ alike? How are they different? What transformation was used to get the second equation?

Practice and Problem-Solving Exercises

Determine the cubic function that is obtained from the parent function $y = x^3$ after each sequence of transformations.

See Problem 1.

7. a vertical stretch by a factor of 3; a reflection across the x-axis; a vertical translation 2 units up; and a horizontal translation 1 unit right

8. a vertical stretch by a factor of 2; a vertical translation 4 units up; and a horizontal translation 3 units left

9. a reflection across the y-axis; a vertical translation 1 unit down; and a horizontal translation 5 units left

10. a vertical translation 3 units down; and a horizontal translation 2 units right

11. a vertical stretch by a factor of 3; a reflection across the y-axis; a vertical translation $\frac{3}{4}$ unit up; and a horizontal translation $\frac{1}{2}$ unit left

12. a vertical stretch by a factor of $\frac{5}{3}$; a reflection across the x-axis; a vertical translation 4 units down; and a horizontal translation 3 units right

Find all the real zeros of each function.

See Problem 2.

13. $y = -27(x - 2)^3 + 8$

14. $y = -\frac{1}{8}(x - 7)^3 - 8$

15. $y = -3\left(x + \frac{4}{5}\right)^3 + \frac{8}{9}$

16. $y = -16(x + 3)^3 + 9$

17. $y = 4(x - 1)^3 + 10$

18. $y = 2(x + 5)^3 + 10$

Find a quartic function with the given x-values as its only real zeros.

See Problem 3.

19. $x = 2$ and $x = -1$

20. $x = -3$ and $x = -4$

21. $x = -1$ and $x = 3$

22. $x = 4$ and $x = 2$

23. $x = -4$ and $x = -1$

24. $x = -3$ and $x = 2$

See Problem 4.

25. **Cooking** The number of pepperoni slices that Kim puts on a pizza varies directly as the square of the diameter of the pizza. If she puts 15 slices on a 10" diameter pizza, how many slices should she put on a 16" diameter pizza?

26. **Volume** The amount of water that a spherical tank can hold varies directly as the cube of its radius. If a tank with radius 7.5 ft holds 1767 ft^3 of water, how much water can a tank with radius 16 ft hold?

27. **Think About a Plan** The kinetic energy generated by a 5 lb ball is represented by the formula $K = \frac{1}{2}(5)v^2$. If the ball is thrown with a velocity of 6 ft/sec, how much kinetic energy is generated?
    - What does 5 represent in the function?
    - What number should you substitute for v?

Determine whether each function can be obtained from the parent function, $y = x^n$, using basic transformations. If so, describe the sequence of transformations.

28. $y = -4x^3$

29. $y = 2(x - 3)^2 + 5$

30. $y = x^3 - x$

31. $y = x^2 - 8x + 7$

Determine the transformations that were used to change the graph of the parent function $y = x^3$ to each of the following graphs.

32.

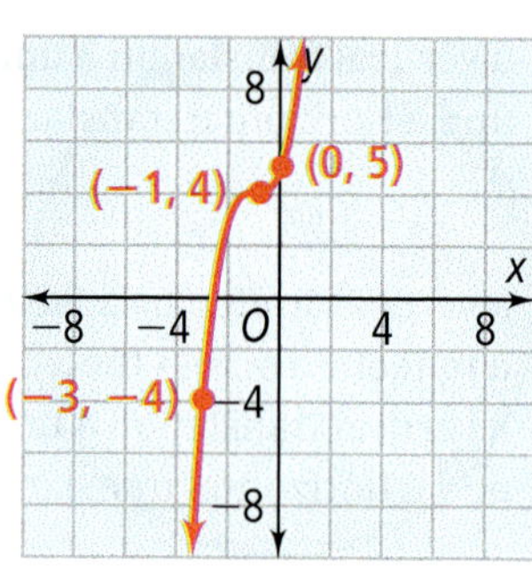

33.

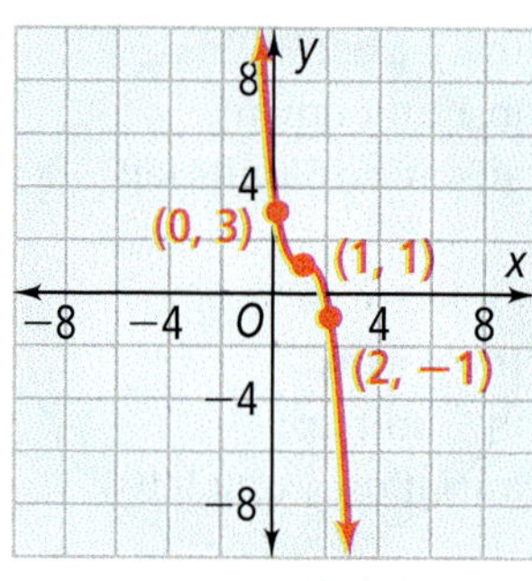

34.

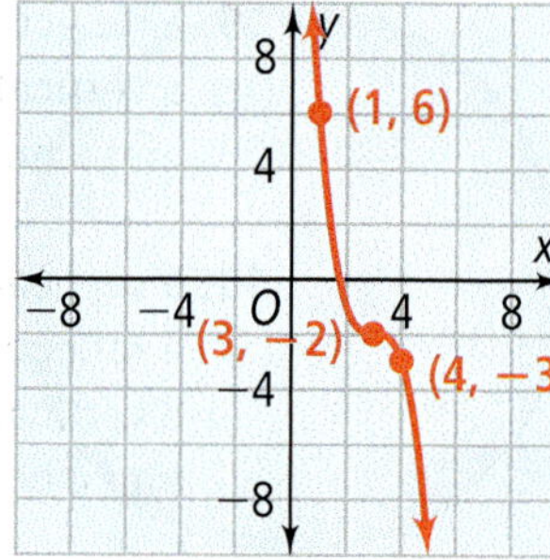

35.

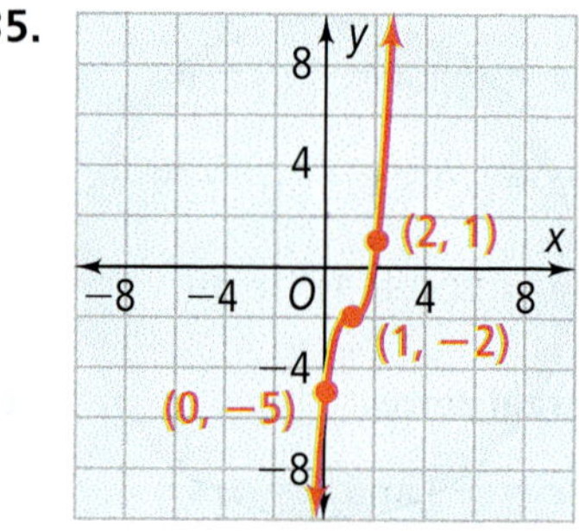

36.

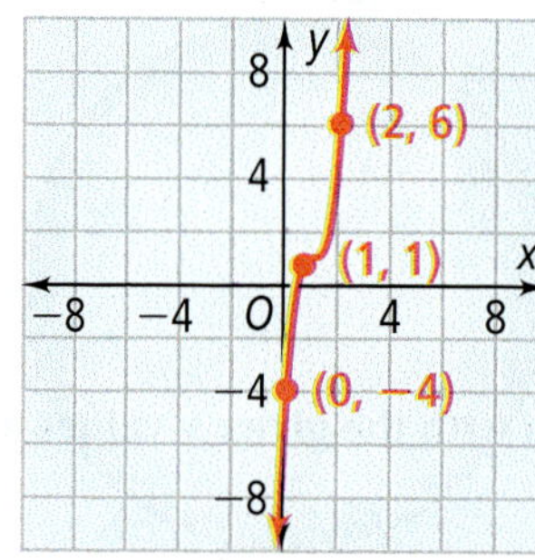

37.

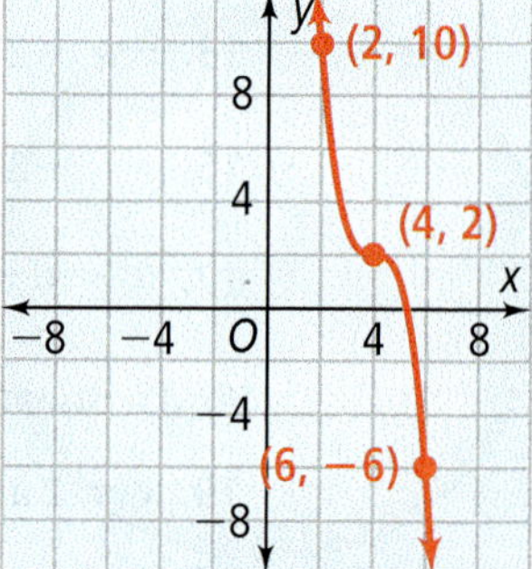

38. Compare the function $y = 3x^3$ to the function shown in the graph at the right. Which function has a greater vertical stretch factor? Explain.

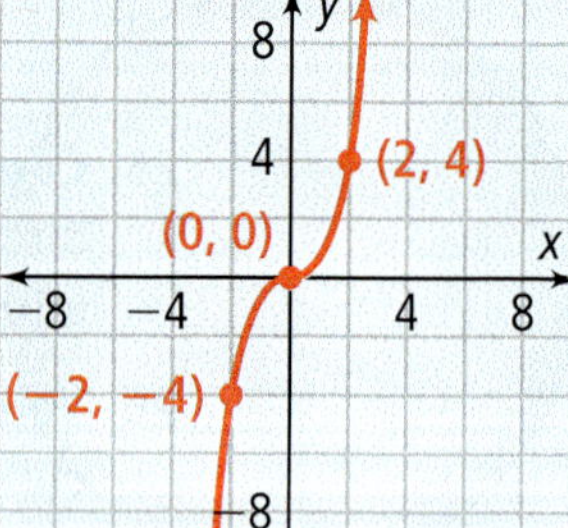

STEM **39. Physics** The formula $K = \frac{1}{2}mv^2$ represents the kinetic energy of an object. If the kinetic energy of a ball is 10 lb-ft^2/s^2 when it is thrown with a velocity of 4 ft/s, how much kinetic energy is generated if the ball is thrown with a velocity of 8 ft/s?

40. Reasoning Explain why the basic transformations of the parent function $y = x^5$ will only generate functions that can be written in the form $y = a(x - h)^5 + k$.

41. Reasoning Explain why some quartic polynomials cannot be written in the form $y = a(x - h)^4 + k$. Give two examples.

42. **Reasoning** Find a sequence of basic transformations by which the polynomial function $y = 2x^3 - 6x^2 + 6x + 5$ can be derived from the cubic function $y = x^3$.

STEM **43.** **Physics** For a constant resistance R (in ohms), the power P (in watts) dissipated across two terminals of a battery varies directly as the square of the current I (in amps). If a battery connected in a circuit dissipates 24 watts of power for 2 amps of current flow, how much power would be dissipated when the current flow is 5 amps?

44. **Writing** Give an argument that shows that *every* polynomial family of degree $n > 2$ contains polynomials that cannot be generated from the basic function $y = x^n$ by using stretches, compressions, reflections, and translations.

Standardized Test Prep

SAT/ACT

Use the graph to answer questions 45–47.

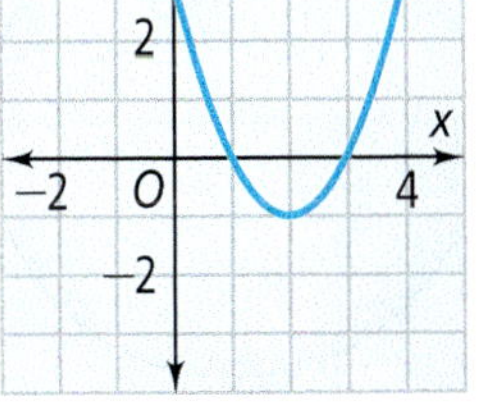

45. Which equation does the graph represent?

Ⓐ $y = (x + 2)^2 - 1$

Ⓑ $y = (x - 2)^2 - 1$

Ⓒ $y = (x - 2)^2 + 1$

Ⓓ $y = (x - 2)^4 - 1$

46. If $y = f(x)$ is an equation for the graph, what are factors of $f(x)$?

Ⓕ $(x - 1)$ and $(x + 3)$

Ⓖ $(x - 1)$ and $(x - 3)$

Ⓗ $(x + 1)$ and $(x - 3)$

Ⓘ $(x + 1)$ and $(x + 3)$

Short Response

47. If $y = ax^2 + bx + c$ is an equation for the graph, what type of number is its discriminant?

Mixed Review

Find a polynomial function whose graph passes through the given points.

See Lesson 5-8.

48. $(-1, 4), (0, -2), (1, -2), (2, -8)$

49. $(-2, -17), (0, -3), (1, -5), (3, 63)$

Write an equation of each line. See Lesson 2-4.

50. slope $= -\frac{4}{5}$; through $(-1, 4)$

51. slope $= -3$; through $(2, -1)$

Determine whether each relation is a function. See Lesson 2-1.

52. $\{(0, -1), (-1, 3), (2, 3), (-3, 3)\}$

53. $\{(-4, 0), (-7, 0), (-4, 1), (-7, 1)\}$

Get Ready! **To prepare for Lesson 6-1, do Exercises 54–56.**

Factor each expression. See Lesson 5-2.

54. $x^{10} + x^2$

55. $x^4 - y^4$

56. $169x^6y^{12} - 13x^3y^6$

Pull It All Together  ASSESSMENT

To solve these problems, you will pull together concepts and skills related to working with polynomials and their related functions and equations.

BIG idea Function

You can represent quantities using variables and algebraic expressions. You can represent some relationships between quantities using equations.

Performance Task 1

The polynomial $2x^3 + 9x^2 + 4x - 15$ represents the volume in cubic feet of a rectangular holding tank at a fish hatchery. The depth of the tank is $(x - 1)$ feet. The length is 13 feet.

a. Use synthetic division to help you factor the volume polynomial. How many linear factors should you look for? What are they?

b. Assume the length is the greatest dimension. Which linear factor represents the 13-ft length? What are the dimensions of the tank?

c. What is its volume? Do you get the same volume if you substitute the value of x into $2x^3 + 9x^2 + 4x - 15$?

BIG idea Equivalence

You can use the Binomial Theorem and properties of algebra to rewrite some powers.

Show that the following equation is true for all values of a and b.

$[(a - b) + 1]^5 = a^5 - 5a^4(b - 1) + 10a^3(b - 1)^2 - 10a^2(b - 1)^3 + 5a(b - 1)^4 - (b - 1)^5$

BIG idea Solving Equations and Inequalities

A polynomial $P(x)$ of degree n, $n \geq 1$, and its related polynomial function $y = P(x)$ have n complex zeros. The zeros are identical to the n complex roots of the related polynomial equation $P(x) = 0$.

Performance Task 3

All of the following polynomials have two zeros in common. Find these zeros. Use that information to completely factor each polynomial. Then find the real and complex roots of each polynomial. Show all your work.

$P_1(x) = x^3 - 6x^2 + 11x - 6$
$P_2(x) = x^4 - 3x^3 - 7x^2 - 3x + 6$
$P_3(x) = 2x^4 - 9x^3 + 6x^2 + 11x - 6$
$P_4(x) = x^4 - 6x^3 + 10x^2 - x - 6$
$P_5(x) = x^4 - 7x^3 + 17x^2 - 17x + 6$

5 Chapter Review

Connecting BIG ideas and Answering the Essential Questions

1 Function
A polynomial of degree n has n linear factors. The graph of the related function crosses the x-axis an even or odd number of times depending on whether n is even or odd.

2 Equivalence
$(x - a)$ is a linear factor if and only if a is a zero, and if and only if $(a, 0)$ is an x-intercept when a is a real number.

3 Solving Equations and Inequalities
$(x - a)$ is a linear factor if and only if a is a root of the related polynomial equation.

Polynomial Functions, Zeros, and Linear Factors (Lessons 5-1 and 5-2)
$y = 2x^3 + 7x^2 - 9$ has 3 linear factors $(x + 3)$, $(2x + 3)$, $(x - 1)$, it crosses the x-axis 3 times—at $(-3, 0)$, $(-\frac{3}{2}, 0)$, and $(1, 0)$. Its end behavior is down and up.

Theorems About Roots of Polynomial Equations (Lesson 5-5)
$P(x) = 2x^3 + 7x^2 - 9$ and $Q(x) = 2x^3 + 5x^2 + 9$ each have degree 3 so $P(x) = 0$ and $Q(x) = 0$ each have 3 complex roots.
Each has $\pm 1, \pm 3, \pm 9, \pm\frac{1}{2}, \pm\frac{3}{2}, \pm\frac{9}{2}$ as its possible rational roots.
$P(-3) = 0$, so $x + 3$ is a factor of $2x^3 + 7x^2 - 9$.
$Q(-3) = 0$, so $x + 3$ is a factor of $2x^3 + 5x^2 + 9$.

Solving Polynomial Equations (Lesson 5-3)
$2x^3 + 7x^2 - 9 = 0$ has factored form $(x + 3)(2x + 3)(x - 1) = 0$. It has 3 roots or solutions, $x = -3$, $x = -\frac{3}{2}$, and $x = 1$.

The Fundamental Theorem of Algebra (Lesson 5-6)
$2x^3 + 5x^2 + 9 = 0$ has factored form $(x + 3)(2x^2 - x + 3) = 0$. It has 3 roots or solutions, $x = -3$, $x = \frac{1}{4} - \frac{\sqrt{23}}{4}i$, and $x = \frac{1}{4} + \frac{\sqrt{23}}{4}i$.

Chapter Vocabulary

- Binomial Theorem (p. 327)
- Conjugate Root Theorem (p. 314)
- constant of proportionality (p. 341)
- degree of a monomial (p. 280)
- degree of a polynomial (p. 280)
- Descartes' Rule of Signs (p. 315)
- difference of cubes (p. 297)
- end behavior (p. 282)
- expand a binomial (p. 326)
- Factor Theorem (p. 290)
- Fundamental Theorem of Algebra (p. 320)
- monomial (p. 280)
- multiple zero (p. 291)
- multiplicity (p. 291)
- Pascal's Triangle (p. 327)
- polynomial (p. 280)
- polynomial function (p. 280)
- power function (p. 341)
- Rational Root Theorem (p. 312)
- relative maximum (p. 291)
- relative minimum (p. 291)
- Remainder Theorem (p. 307)
- standard form of a polynomial function (p. 281)
- sum of cubes (p. 297)
- synthetic division (p. 306)
- turning point (p. 282)

Match each vocabulary term with the description that best fits it.

1. Conjugate Root Theorem
2. Fundamental Theorem of Algebra
3. Rational Root Theorem
4. Remainder Theorem

A. determines $P(a)$ by dividing the polynomial by $x - a$
B. the degree equals the number of roots
C. minimizes guessing fraction and integer solutions
D. complex numbers as roots come in pairs

5-1 Polynomial Functions

Quick Review

The **standard form of a polynomial function** is $P(x) = a_nx^n + a_{n-1}x^{n-1} + \cdots + a_1x + a_0$, where n is a nonnegative integer and the coefficients are real numbers. A **polynomial function** is classified by degree. Its degree is the highest degree among its monomial term(s). The degree determines the possible number of **turning points** in the graph and the **end behavior** of the graph.

Example

Write the polynomial function in standard form and classify it by degree. How many terms does it have? What are the possible numbers of turning points of the graph of $P(x)$ given the degree of the polynomial?

$P(x) = -4x^2 + x^4$

Standard form arranges the terms by decreasing exponents, or $P(x) = x^4 - 4x^2$. Its degree is 4, so $x^4 - 4x^2$ is a quartic binomial. It has two terms. The graph of a quartic polynomial function can have either one or three turning points.

Exercises

Write each polynomial function in standard form, classify it by degree, and determine the end behavior of its graph.

5. $y = 12 - x^4$
6. $y = x^2 + 7 - x$
7. $y = 2x^3 - 6x + 3x^2 - x^4 + 12$
8. $y = 2x^2 + 8 - 4x + x^3$
9. $y = 10 - 3x^3 + 3x^2 + x^4$

10. If the volume of a cube can be represented by a polynomial of degree 9, what is the degree of the polynomial that represents each side length?
11. A polynomial function $P(x)$ has degree n. If n is even, is the number of turning points of the graph of $P(x)$ even or odd? What can you say about the number of turning points if n is odd?

5-2 Polynomials, Linear Factors, and Zeros

Quick Review

For any real number a and polynomial $P(x)$, if $x - a$ is a **factor** of $P(x)$, then a is:

- a **zero** of $y = P(x)$
- a **root** (or **solution**) of $P(x) = 0$, and
- an ***x*-intercept** of the graph of $y = P(x)$.

If a is a **multiple zero**, its **multiplicity** is the same as the number of times $x - a$ appears as a factor.

A turning point is a **relative maximum** or **relative minimum** of a polynomial function.

Example

Find the zeros for $y = 3x^3 - 6x^2 + 3x$ and state the multiplicity of any multiple zeros.

$y = 3x(x^2 - 2x + 1)$ Factor out the GCF, $3x$.

$y = 3x(x - 1)(x - 1)$ Factor the quadratic.

The zeros are 0, and 1 with multiplicity 2.

Exercises

Write a polynomial function with the given zeros.

12. $x = -1, -1, 6$
13. $x = -1, 0, 2$
14. $x = 1, 2, 3$
15. $x = -2, 1, 4$

Find the zeros of each function. State the multiplicity of any multiple zeros.

16. $y = 3x(x + 2)^3$
17. $y = x^4 - 8x^2 + 16$
18. $y = 4x^3 - 2x^2 - 2x$
19. $y = (x - 5)(x + 2)^2$

Use a graphing calculator to find the relative maximum, relative minimum, and zeros of each function.

20. $y = x^4 - 5x^3 + 5x^2 - 3$
21. $y = 5x^3 + x^2 - 9x + 4$
22. $y = x^4 - 4x - 1$
23. $y = x^3 - 3x^2 - 3x - 4$

5-3 Solving Polynomial Equations

Quick Review

One way to solve a polynomial equation is by factoring. First write the equation in the form $P(x) = 0$, where $P(x)$ is the polynomial. Then factor the polynomial. Last, use the Zero-Product Property to find the solutions, or roots. The solutions may be real or imaginary. Real solutions and approximations of irrational solutions can also be found by using a graphing calculator.

Example

Solve $x^3 + 4x^2 = 12x$ by factoring.

$x^3 + 4x^2 - 12x = 0$	Set equal to 0.
$x(x - 2)(x + 6) = 0$	Factor the left side.
$x = 0, x - 2 = 0, x + 6 = 0$	Zero Product Property.
$x = 0, x = 2, x = -6$	Solve each equation.

The solutions are 0, 2, and −6.

Exercises

Find the real or imaginary solutions of each equation by factoring.

24. $x^2 - 11x = -24$

25. $4x^2 = -4x - 1$

26. $3x^3 + 3x^2 = 27x$

27. $2x^2 + 3 = 4x$

Find the real roots of each equation by graphing.

28. $x^4 + 3x^2 - 2x + 5 = 0$

29. $x^2 + 3 = x^3 - 5$

30. The height and width of a rectangular prism are each 2 inches shorter than the length of the prism. The volume of the prism is 40 cubic inches. Approximate the dimensions of the prism to the nearest hundredth.

5-4 Dividing Polynomials

Quick Review

You can divide a polynomial by one of its factors to find another factor. When you divide by a linear factor, you can simplify this division by writing only the coefficients of each term. This is called **synthetic division**. The **Remainder Theorem** says that $P(a)$ is the remainder when you divide $P(x)$ by $x - a$.

Example

Let $P(x) = 3x^2 - 13x + 15$. What is $P(3)$?

According to the Remainder Theorem, $P(3)$ is the remainder when you divide $P(x)$ by $x - 3$.

3	3	−13	15	Put the opposite of the constant in the divisor at the top left.
		9	−12	
	3	−4	3	

The quotient is $3x - 4$ with remainder 3, so $P(3) = 3$.

Exercises

Divide using long division. Check your answers.

31. $(x^3 + 7x^2 + 15x + 9) \div (x + 1)$

32. $(2x^3 - 7x^2 - 7x + 13) \div (x - 4)$

Determine whether each binomial is a factor of $x^3 + x^2 - 10x + 8$.

33. $x - 2$

34. $x - 4$

Divide using synthetic division.

35. $(x^3 + 5x^2 - x - 5) \div (x + 5)$

36. $(2x^3 + 14x^2 - 58x) \div (x + 10)$

37. $(5x^3 + 8x^2 - 60) \div (x - 2)$

Use the Remainder Theorem to determine the value of $P(a)$.

38. $P(x) = 2x^3 + 5x^2 + 7x - 4, a = -2$

39. $P(x) = x^3 - 4x^2 + 2x + 3, a = 1$

5-5 Theorems About Roots of Polynomial Equations

Quick Review

The **Rational Root Theorem** gives a way to determine the possible roots of a polynomial equation $P(x) = 0$. If the coefficients of $P(x)$ are all integers, then every root of the equation can be written in the form $\frac{p}{q}$, where p is a factor of the constant term and q is a factor of the leading coefficient.

The **Conjugate Root Theorem** states that if $P(x)$ is a polynomial with rational coefficients, then irrational roots that have the form $a + \sqrt{b}$ and imaginary roots of $P(x) = 0$ come in conjugate pairs. Therefore, if $a + \sqrt{b}$ is an irrational root, where a and b are rational, then $a - \sqrt{b}$ is also a root. Likewise, if $a + bi$ is a root, where a and b are real and i is the imaginary unit, then $a - bi$ is also a root.

Descartes' Rule of Signs gives a way to determine the possible number of positive and negative real roots by analyzing the signs of the coefficients. The number of positive real roots is equal to the number of sign changes in consecutive coefficients of $P(x)$, or is less than that by an even number. The number of negative real roots is equal to the number of sign changes in consecutive coefficients of $P(-x)$, or is less than that by an even number.

Example

Find the rational roots of $P(x) = 0$ if $P(x) = 2x^3 - 4x^2 - 10x + 12$.

List the possible roots: $\pm\frac{1}{2}, \pm1, \pm\frac{3}{2}, \pm2, \pm3, \pm4, \pm6, \pm12$.
Use synthetic division to test roots.

$$\begin{array}{r|rrrr} 3 & 2 & -4 & -10 & 12 \\ & & 6 & 6 & -12 \\ \hline & 2 & 2 & -4 & 0 \end{array}$$

So $x - 3$ and $(2x^2 + 2x - 4)$ are factors of $P(x)$.

$P(x) = (x - 3)(2x^2 + 2x - 4)$

Factor the quadratic.

$P(x) = 2(x - 3)(x + 2)(x - 1)$

Solve $2(x - 3)(x + 2)(x - 1) = 0$.

$x = 3, x = -2, \text{ or } x = 1$

The rational roots are 3, −2, and 1.

Exercises

List the possible rational roots of $P(x)$ given by the Rational Root Theorem.

40. $P(x) = x^3 + 4x^2 - 10x + 6$

41. $P(x) = 3x^3 - x^2 - 7x + 2$

42. $P(x) = 4x^4 - 2x^3 + x^2 - 12$

43. $P(x) = 3x^4 - 4x^3 - x^2 - 7$

Find any rational roots of $P(x)$.

44. $P(x) = x^3 + 2x^2 + 4x + 21$

45. $P(x) = x^3 + 5x^2 + x + 5$

46. $P(x) = 2x^3 + 7x^2 - 5x - 4$

47. $P(x) = 3x^4 + 2x^3 - 9x^2 + 4$

A polynomial $P(x)$ has rational coefficients. Name additional roots of $P(x)$ given the following roots.

48. $1 - i$ and 5

49. $5 + \sqrt{3}$ and $-\sqrt{2}$

50. $-3i$ and $7i$

51. $-2 + \sqrt{11}$ and $-4 - 6i$

Write a polynomial function with the given roots.

52. 7 and 10

53. −3 and $5i$

54. $6 - i$

55. $3 + i$, 2, and −4

Determine the possible number of positive real zeros and negative real zeros for each polynomial function given by Descartes' Rule of Signs.

56. $P(x) = 5x^3 + 7x^2 - 2x - 1$

57. $P(x) = -3x^3 + 11x^2 + 12x - 8$

58. $P(x) = 6x^4 - x^3 + 5x^2 - x + 9$

59. $P(x) = -x^4 - 3x^3 + 8x^2 + 2x - 14$

5-6 The Fundamental Theorem of Algebra

Quick Review

The **Fundamental Theorem of Algebra** states that if $P(x)$ is a polynomial of degree n, where $n \geq 1$, then $P(x) = 0$ has exactly n roots. This includes multiple and complex roots.

Example

Use the Fundamental Theorem of Algebra to determine the number of roots for $x^4 + 2x^2 - 3 = 0$.

Because the polynomial is of degree 4, it has 4 roots.

Exercises

Find the number of roots for each equation.

60. $x^3 - 2x + 5 = 0$

61. $2 - x^4 + x^2 = 0$

62. $-x^5 - 6 = 0$

63. $5x^4 - 7x^6 + 2x^3 + 8x^2 + 4x - 11 = 0$

Find all the zeros for each function.

64. $P(x) = x^3 + 5x^2 - 4x - 2$

65. $P(x) = x^4 - 4x^3 - x^2 + 20x - 20$

66. $P(x) = 2x^3 - 3x^2 + 3x - 2$

67. $P(x) = x^4 - 4x^3 - 16x^2 + 21x + 18$

5-7 The Binomial Theorem

Quick Review

Rows 0–5 of Pascal's Triangle are shown below.

1
1 1
1 2 1
1 3 3 1
1 4 6 4 1
1 5 10 10 5 1

The **Binomial Theorem** uses Pascal's Triangle to expand binomials. For a positive integer n,
$(a + b)^n = P_0a^n + P_1a^{n-1}b + P_2a^{n-2}b^2 + \cdots + P_{n-1}ab^{n-1} + P_nb^n$, where $P_0, P_1, \ldots P_n$ are the coefficients of the nth row of Pascal's Triangle.

Example

Use the binomial theorem to expand $(2x + 3)^3$.

$(2x + 3)^3$

$= 1(2x)^3 + 3(2x)^2(3) + 3(2x)(3)^2 + 1(3)^3$

$= 8x^3 + 36x^2 + 54x + 27$

Exercises

68. How many numbers are in the eighth row of Pascal's Triangle?

69. List the numbers in the eighth row of Pascal's Triangle.

70. How many numbers are in the fifteenth row of Pascal's Triangle?

71. What is the third number in the fifteenth row of Pascal's Triangle?

Use the Binomial Theorem to expand each binomial.

72. $(x + 9)^3$

73. $(b + 2)^4$

74. $(3a + 1)^3$

75. $(x - 5)^3$

76. $(x - 2y)^3$

77. $(3a + 4b)^5$

78. $(x + 1)^6$

79. $(2x - 1)^6$

Find the coefficient of the x^2 term in each binomial expansion.

80. $(3x + 4)^3$

81. $(ax - c)^4$

5-8 Polynomial Models in the Real World

Quick Review

A data set can be modeled by a polynomial function. Methods of finding a model that fits the data include the $(n + 1)$ Point Principle and **regression**. Linear, quadratic, cubic, and quartic regressions can be performed on a graphing calculator. A higher R^2 value means a better fit. Once the equation that models the data is known, it can be used to make predictions.

Example

For the data set (8, 30), (10, 45), and (11, 65), predict y when $x = 15$.

Enter 8, 10, and 11 in **L1** and 30, 45, and 65 in **L2**. Choose **LINREG** to find the regression model $y \approx 11.071x - 60.357$. The r^2 value is about 0.928.

Now try **QUADREG**. The model is $y \approx 4.17x^2 - 67.5x + 303.3$ with an R^2 value of 1. Assuming the model makes sense in context, it fits the data better.

Using the quadratic model, when $x = 15$, $y \approx 228.3$.

Exercises

82. Write a polynomial function whose graph passes through (0, 5), (2, 10), and (1, 4). Use a regression to check your answer.

83. Find a linear, a quadratic, and a cubic model for the data. Which model best fits the data?

x	3	8	15	21
y	7	11	26	44

84. Use **CUBICREG** to model the data below. Then use the model to estimate the population in 2008. Let x be the number of years after 2000.

Year	2004	2007	2009	2010
Population	457	910	1244	1315

5-9 Transforming Polynomial Functions

Quick Review

A polynomial function can be transformed into other polynomial functions using stretches, reflections, and translations. The monomial function $y = ax^b$ is called a **power function.**

Example

This is the graph of a cubic function. Determine which sequence of transformations you can apply to the graph of the parent function $y = x^3$ to get this graph. Write an equation for the graph.

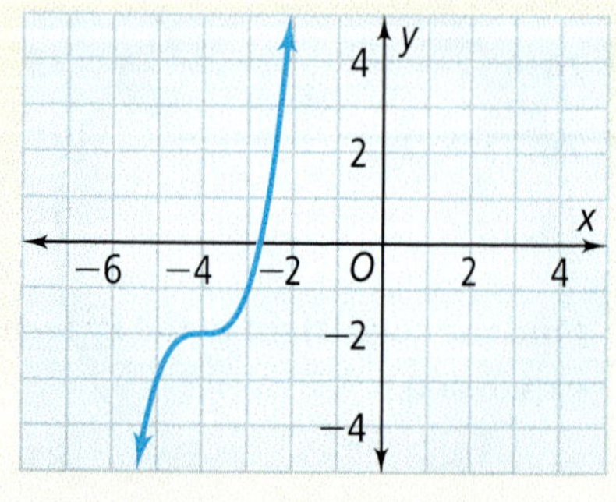

Translate the parent function 4 units left and 2 units down: $y = (x + 4)^3 - 2$

Exercises

Determine the cubic function obtained from the parent function $y = x^3$ after each sequence of transformations.

85. a reflection across the x-axis;
a translation 1 unit up;
and a translation 2 units right

86. a vertical stretch by a factor of 6;
and a translation 3 units left

87. Find a quartic function whose only real zeros are 4 and 6.

88. The parent power function $y = x^5$ is translated 3 units up and is compressed by the factor 0.3. Write the function.

5 Chapter Test

Do you know HOW?

Write each polynomial function in standard form. Then classify it by degree and by number of terms and describe its end behavior.

1. $y = 3x^2 - 7x^4 + 9 - x^4$
2. $y = 2x(x^2 - 3)(x^2 + 2)$
3. $y = (t - 2)(t + 1)(t + 1)$

Write a polynomial function for each set of zeros.

4. $x = 1, 2, \frac{3}{5}$
5. $x = \sqrt{2}, -i$
6. $x = 3 + i, -1 - \sqrt{5}$

Find the quotient and remainder.

7. $(x^2 + 3x - 4) \div (x - 1)$
8. $(x^3 + 7x^2 - 5x - 6) \div (x + 2)$
9. $(2x^3 + 9x^2 + 11x + 3) \div (2x + 3)$

For each equation, state the number of complex roots, the possible number of real roots, and the possible rational roots.

10. $3x^4 + 5x^3 - 2x^2 + x - 9 = 0$
11. $x^7 - 2x^5 - 4x^3 - 2x - 1 = 0$

For each equation, find all the roots.

12. $3x^4 - 11x^3 + 15x^2 - 9x + 2 = 0$
13. $x^3 - x^2 - x - 2 = 0$
14. One x-intercept of the graph of the cubic function $f(x) = x^3 - 2x^2 - 111x - 108$ is -9. What are the other zeros?

Use synthetic division and the Remainder Theorem to find $P(a)$.

15. $P(x) = 6x^4 + 19x^3 - 2x^2 - 44x - 24; a = -\frac{2}{3}$
16. $P(x) = x^4 + 3x^3 - 7x^2 - 9x + 12; a = 3$
17. $P(x) = x^3 + 3x^2 - 5x - 4; a = -1$

Expand each binomial.

18. $(x + z)^5$
19. $(1 - 2t)^2$
20. Graph and write the equation of the cubic function that is obtained from the parent function $y = x^3$ after this sequence of transformations: vertical stretch by a factor of 2, reflection across the x-axis, translation 3 units down and 4 units right.

Do you UNDERSTAND?

21. **Physics** You take measurements of the distance traveled by an object that is increasing its speed at a constant rate. The distance traveled as a function of time can be modeled by a quadratic function.
    a. Write a quadratic function that models distances of 10 ft at 1 sec, 30 ft at 2 sec, and 100 ft at 4 sec.
    b. Find the zeros of the function.
    c. **Reasoning** Describe what each zero represents for this real-world situation.

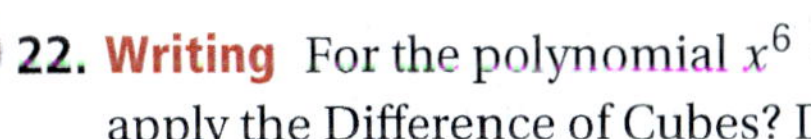

22. **Writing** For the polynomial $x^6 - 64$, could you apply the Difference of Cubes? Difference of Squares? Explain your answers.
23. The number of pairs of shoes Emily buys varies directly as the square of the area of the floor of her closet. If she can fit 12 pairs of shoes when her closet was 10 square feet, how many pairs of shoes will she fit when the area of her closet floor is 18 square feet?

5 Cumulative Standards Review

ASSESSMENT

TIPS FOR SUCCESS

Some problems require you to simplify expressions that contain imaginary numbers.

TIP 1

When multiplying binomials, you should use the Distributive Property.

Which expression is equivalent to $(3 - 4i)(2 + i)$?

(A) $2 - 5i$

(B) $2 + 5i$

(C) $10 - 5i$

(D) $10 + 5i$

TIP 2

Recall that $i^2 = -1$.

Think It Through

$$(3 - 4i)(2 + i)$$
$$= 6 + 3i - 8i - 4i^2$$
$$= 6 - 5i - 4(-1)$$
$$= 10 - 5i$$

The correct answer is C.

Vocabulary Review

As you solve test items, you must understand the meanings of mathematical terms. Match each term with its mathematical meaning.

A. multiplicity

B. synthetic division

C. Conjugate Root Theorem

D. relative maximum

E. polynomial

I. the process of dividing a polynomial by a linear factor, omitting all variables and exponents

II. if $a + bi$ is an irrational root where a and b are real numbers, then $a - bi$ is also a root

III. a monomial or the sum of monomials

IV. the greatest y-value in a region of a graph

V. the number of times a zero is repeated in a polynomial function

Multiple Choice

Read each question. Then write the letter of the correct answer on your paper.

1. Which statement is true about this system of linear equations?

$$\begin{cases} 3x - 4y = 12 \\ 6x - 8y = 12 \end{cases}$$

(A) The solution is $(0, -1)$.

(B) The solution is $(8, 4)$.

(C) There is no solution because the lines are parallel.

(D) There are infinitely many solutions because the lines are coinciding.

2. Which of the following statements is *never* true about a quartic function?

(F) The end behavior of the function is up and up.

(G) The function has 4 zeros.

(H) The function has 4 turning points.

(I) The function has complex roots.

3. Tia deposited x dollars in a bank account that paid 4% interest. She also deposited y dollars in a bank account that paid 8% interest. The system below represents one year's interest on Tia's deposits.

$$\begin{cases} 0.04x + 0.08y = 240 \\ 0.04x = 0.08y \end{cases}$$

Based on the solution of the system of equations, which of the following can you conclude?

(A) Tia deposited \$3000 in each account, and the amounts of interest earned were \$240 and \$120.

(B) Tia deposited \$3000 in each account, and the amount of interest earned in each account was \$120.

(C) The deposit amounts were \$3000 and \$1500, and the amounts of interest earned in each account were \$240 and \$120.

(D) The deposit amounts were \$3000 and \$1500, and the amount of interest earned in each account was \$120.

4. The total area of the parallelogram below is $4x^4 + 3x^3 - 14x^2 + 33x - 35$. Which of the following expressions best represents the length of the base of the parallelogram? (*Hint:* $A = bh$)

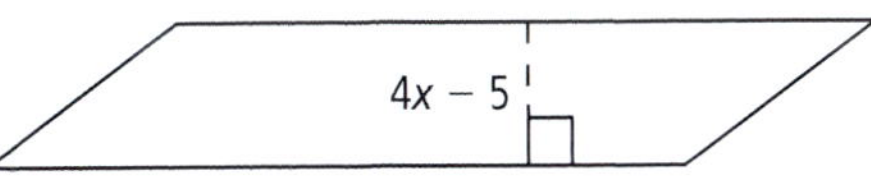

(F) $x^3 + 2x^2 - x + 7$

(G) $x^3 - 2x^2 + x - 7$

(H) $4x^4 + 3x^3 - 14x^2 + 33x - 7$

(I) $x^3 + 5x^2 - x + 5$

5. Which point corresponds to a zero of the function $f(x) = x^2 + 2x - 15$?

(A) $(0, -15)$ (C) $(-5, 0)$

(B) $(5, 0)$ (D) $(-3, 0)$

6. Solve the equation $x^2 + 3w = P$ for x.

(F) $x = P - 3w$ (H) $x = \pm\sqrt{P - 3w}$

(G) $x = \pm\sqrt{\frac{P}{3w}}$ (I) $x = \pm\sqrt{P + 3w}$

7. A bridge supported by a parabolic arch spans a stream of water 180 feet wide. There must be a clearance of at least 40 feet over a 100-foot channel in the middle of the stream. The origin is placed at water level directly below the center of the arch. Which equation best represents the situation?

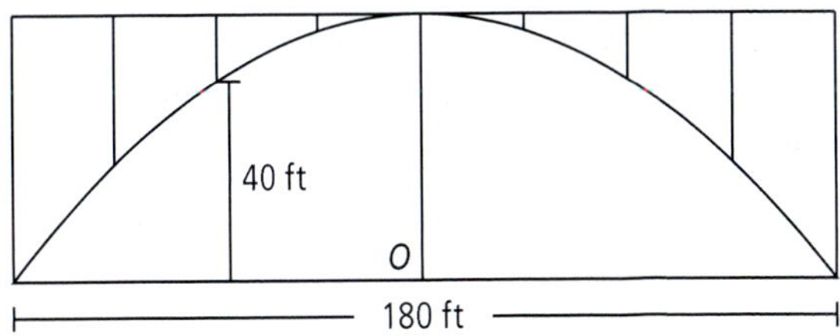

(A) $y = 140(x + 180)(x - 180)$

(B) $y = -\frac{1}{140}(x + 90)(x - 90)$

(C) $y = -\frac{1}{140}(x + 40)(x - 40)$

(D) $y = 140(x + 40)(x - 40)$

8. Sofia has \$25 in her savings account. She plans to deposit between \$5 and \$10 each week into her account. On the graph, line m represents a deposit of exactly \$5 per week and line n represents a deposit of exactly \$10 per week.

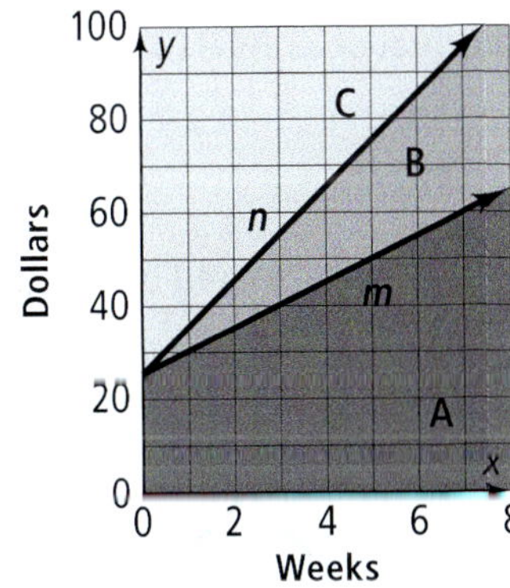

If Sofia deposits between \$5 and \$10 per week, which region on the graph represents all possible balances in her account?

(F) A (H) C

(G) B  (I) A and C combined

GRIDDED RESPONSE

9. The power created by a wind turbine varies directly as the cube of the wind speed in miles per hour. A turbine with 30% efficiency spinning in a 50 mile per hour wind can be expected to produce approximately 10,000 watts of electricity. How many watts would the same turbine produce in a 25-mile-per-hour wind? If necessary, round your answer to the nearest whole number.

10. One root of a cubic equation is $2i$. How many real roots does the equation have?

11. What is the value of the real part of the sum of $(6 + 4i)$ and $(5 - i)$?

12. If the solutions of an equation are -1, 2, and 5, what is the sum of the zeros of the related function?

13. Assume y varies directly with x. If $y = -3$ when $x = -\frac{2}{5}$, what is x when y is 45?

14. What is the x-coordinate of the point where a relative maximum of $g(x) = -2x^3 + 6x^2 - 10$ occurs?

15. Using a graph, find the real zero of the function $y = 2x^3 - 2x^2 + x - 1$.

16. How many imaginary roots does $x^2 - 5x + 10 = 0$ have?

17. What is the product of $(2 + i)(2 - i)$?

Short Response

18. Write the equation represented by the graph. Show your work.

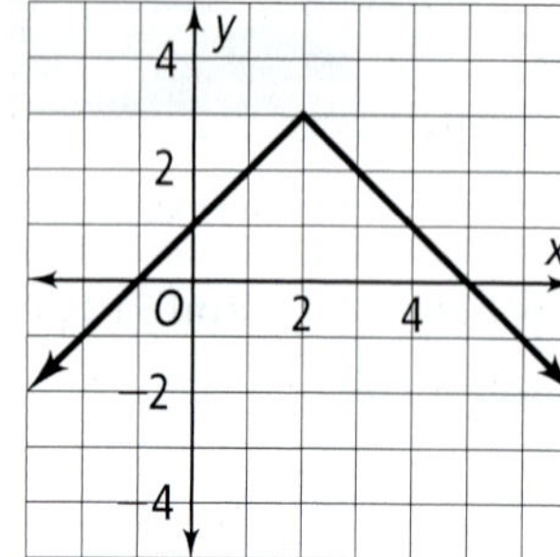

19. Solve the absolute value inequality $-2|x - 3| \leq -16$. Show your work.

20. The graph shows line m and point P. Write the equation of a line n that goes through point P and is perpendicular to line m. Show your work.

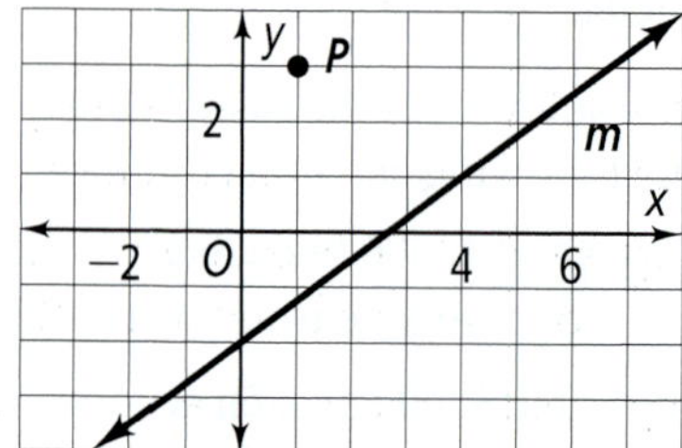

21. A cat ran around part of a telephone pole on a path modeled by the equation $y = -x^2 + 4x - 1$. Graph the cat's path in the first quadrant. If the pole is at (2, 0), at what point is the cat furthest from the pole?

22. Solve the equation $x^2 - 7x = 8$.

Extended Response

23. Use the graph to answer the questions below.

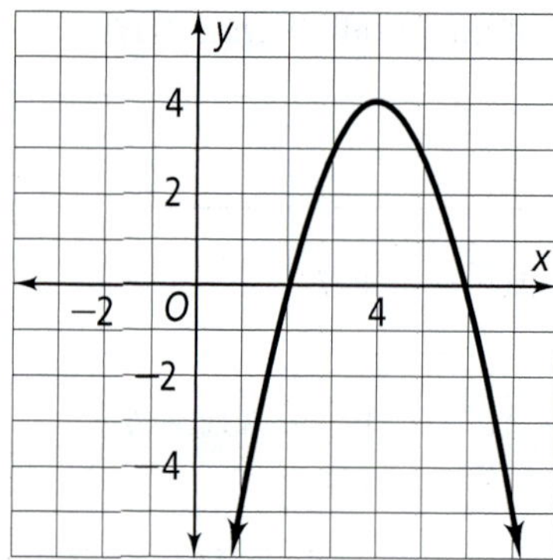

a. Describe the sequence of transformations that would take the graph of the parent function $y = x^2$ to the graph shown.

b. Identify the zeros of the function represented in the graph and explain your reasoning.

c. Write the equation of the function in the graph.

d. Explain how you could check to see that the equation is correct.

Get Ready!

Lessons 2-1 and 4-1

Finding the Domain and Range of Functions

Find the domain and range of each function.

1. $\{(1, 2), (2, 3), (3, 4), (4, 5)\}$
2. $\{(1, 2), (2, 2), (3, 2), (4, 2)\}$
3. $f(x) = (x - 4)^2 - 8$
4. $f(x) = 2x^2 + 3$

Lesson 4-1

Graphing Quadratic Functions

Graph each function.

5. $y = 2x^2 - 4$
6. $y = -3(x^2 + 1)$
7. $y = \frac{1}{2}(x - 3)^2 + 1$

Lesson 4-4

Multiplying Binomials

Multiply.

8. $(3y - 2)(y - 4)$
9. $(7a + 10)(7a - 10)$
10. $(x - 3)(x + 6)(x + 1)$

Lessons 4-5 and 5-3

Solving by Factoring

Solve each equation by factoring.

11. $x^2 - 5x - 14 = 0$
12. $2x^2 - 11x + 15 = 0$
13. $3x^2 + 10x - 8 = 0$
14. $12x^2 - 12x + 3 = 0$
15. $8x^2 - 98 = 0$
16. $x^4 - 14x^2 + 49 = 0$

Looking Ahead Vocabulary

17. Combining two or more elements forms composite chemical mixtures. In some cases, if you change the order in which you mix two chemicals, it can produce very different results. A *composite function* is made by combining two functions. If you are buying a \$60 shirt and there is a 50% off sale and you have a \$10 coupon, does it make a difference which discount is applied first?

18. One-to-one relationships describe situations where people are matched with unique identifiers, such as their social security numbers. A function is a relation that matches *x* values to *y* values. What do you suppose a *one-to-one function* is?

19. In an orchestra, the principal player is chosen among all the other musicians that play a certain instrument to sit in the first chair and lead his section. In math, what do you suppose a *principal root* is?

Radical Functions and Rational Exponents

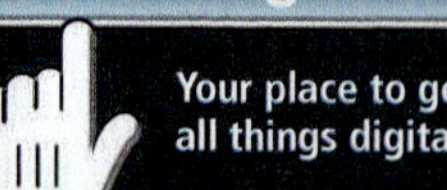

Your place to get all things digital

Download videos connecting math to your world.

Math definitions in English and Spanish

The online Solve It will get you in gear for each lesson.

Interactive! Vary numbers, graphs, and figures to explore math concepts.

Online access to stepped-out problems aligned to Common Core

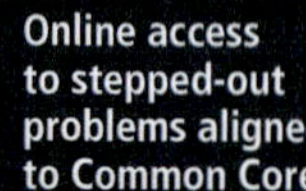

Get and view your assignments online.

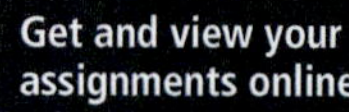

Extra practice and review online

DOMAINS

- Seeing Structure in Expressions
- Interpreting Functions
- Reasoning with Equations and Inequalities

In this chapter, you will learn how to work with radicals, whether they occur by themselves or as parts of functions or equations; whether they appear with the symbol $\sqrt{}$ or as fractional exponents. In a sense, radicals are the inverses of powers. Mountain climbers need to be aware of air pressure, which can be calculated using fractional exponents.

Vocabulary

English/Spanish Vocabulary Audio Online:

English	Spanish
composite function, *p. 399*	función compuesta
inverse function, *p. 405*	función inversa
*n*th root, *p. 361*	raíz *n*-ésima
principal root, *p. 361*	raíz principal
radical equation, *p. 390*	ecuación radical
radicand, *p. 362*	radicando
rational exponent, *p. 382*	exponente racional
rationalize the denominator, *p. 369*	racionalizar el denominador
square root equation, *p. 390*	ecuación de raíz cuadrada
square root function, *p. 415*	función de raíz cuadrada

BIG ideas

1 **Equivalence**

Essential Question To simplify the *n*th root of an expression, what must be true about the expression?

2 **Solving Equations and Inequalities**

Essential Question When you square each side of an equation, is the resulting equation equivalent to the original?

3 **Function**

Essential Question How are a function and its inverse function related?

Chapter Preview

Concept Byte

For Use With Lesson 6-1

REVIEW

Properties of Exponents

Content Standard

Prepares for N.RN.1 Explain how the definition of the meaning of rational exponents follows from extending the properties of integer exponents to those values, allowing for a notation for radicals in terms of rational exponents.

Exponents indicate powers. The table below lists the properties of exponents. Assume that no denominator is equal to zero and that m and n are integers.

take note

Properties Properties of Exponents

- $a^0 = 1, a \neq 0$
- $\frac{a^m}{a^n} = a^{m-n}$
- $a^{-n} = \frac{1}{a^n}$
- $(ab)^n = a^n b^n$
- $(a^m)^n = a^{mn}$
- $a^m \cdot a^n = a^{m+n}$
- $\left(\frac{a}{b}\right)^n = \frac{a^n}{b^n}$

Example

Simplify and rewrite each expression using only positive exponents.

a. $(5a^3)(-3a^{-4})$

$$(5a^3)(-3a^{-4}) = 5(-3)a^{(3+(-4))}$$
$$= -15a^{-1}$$
$$= \frac{-15}{a}, \text{ or } -\frac{15}{a}$$

b. $(-4x^{-3}y^5)^2$

$$(-4x^{-3}y^5)^2 = (-4)^2(x^{-3})^2(y^5)^2$$
$$= 16x^{-6}y^{10}$$
$$= \frac{16y^{10}}{x^6}$$

c. $\frac{4ab^6c^3}{a^5bc^3}$

$$\frac{4ab^6c^3}{a^5bc^3} = 4a^{(1-5)}b^{(6-1)}c^{(3-3)}$$
$$= 4a^{-4}b^5c^0$$
$$= \frac{4b^5}{a^4}$$

Exercises

Simplify each expression. Use only positive exponents.

1. $(2a^3)(5a^4)$
2. $(-3x^2)(-4x^{-2})$
3. $(3x^2y^3)^2$
4. $(3x^{-4}y^3)^2$
5. $\frac{4a^8}{2a^4}$
6. $\frac{12x^5y^3}{4x^{-1}}$
7. $\frac{(6x^3)^0}{3xy^2}$
8. $\left(\frac{2x^4}{3}\right)^3$
9. $(-4m^2n^3)(2mn)$
10. $(2x^3y^7)^{-2}$
11. $\frac{(3r^{-2}s^3t^0)^{-3}}{3rs}$
12. $(h^7k^3)^0$
13. $\frac{r^2s^4t^6}{r^3s^4t^{-6}}$
14. $\frac{x^2y}{4} \cdot \frac{16x}{y}$
15. $(s^4t)^2(st)$
16. $\left(\frac{1}{h^{-2}}\right)^{-1} \cdot h^3$
17. $\frac{1}{a^2b^{-3}}(a^2b^{-3})^{-1}$
18. $\left(\frac{r^{-1}s^2t^{-3}}{r^{-2}s^0t^1}\right)^{-1}$
19. **Reasoning** Your friend tells you that $(k^2)^{-5} = -k^{10}$. Did she apply the properties of exponents correctly? Explain why or why not.

6-1 Roots and Radical Expressions

Content Standard

A.SSE.2 Use the structure of an expression to identify ways to rewrite it.

Objective To find *n*th roots

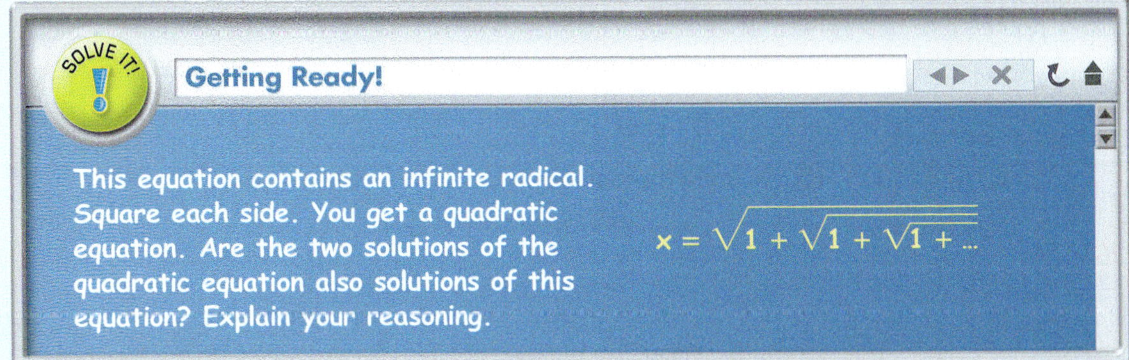

Dynamic Activity
Simplifying Radical Expressions

Lesson Vocabulary
- *n*th root
- principal root
- radicand
- index

In Chapter 5, you used *root* to represent a solution of an equation. For example, 2 is a root of the equation $x^3 = 8$. For such a simple power equation, you can simply refer to 2 as a cube root of 8.

Essential Understanding Corresponding to every power, there is a root. For example, just as there are squares (second powers), there are square roots. Just as there are cubes (third powers), there are cube roots, and so on.

$5^2 = 25$ 5 is a square root of 25.

$5^3 = 125$ 5 is a cube root of 125.

$5^4 = 625$ 5 is a fourth root of 625.

$5^5 = 3125$ 5 is a fifth root of 3125.

This pattern suggests a definition of an *n*th root.

take note

Key Concept The *n*th Root

If $a^n = b$, with a and b real numbers and n a positive integer, then a is an ***n*th root** of b.

If *n* is odd...
there is one real *n*th root of b, denoted in radical form as $\sqrt[n]{b}$.

If *n* is even...
- and b is positive, there are two real *n*th roots of b. The positive root is the **principal root** (or principal *n*th root) and its symbol is $\sqrt[n]{b}$. The negative root is its opposite, or $-\sqrt[n]{b}$.
- and b is negative, there are no real *n*th roots of b.

The only *n*th root of 0 is 0.

You use a radical sign to indicate a root. The number under the radical sign is the **radicand**. The **index** gives the degree of the root.

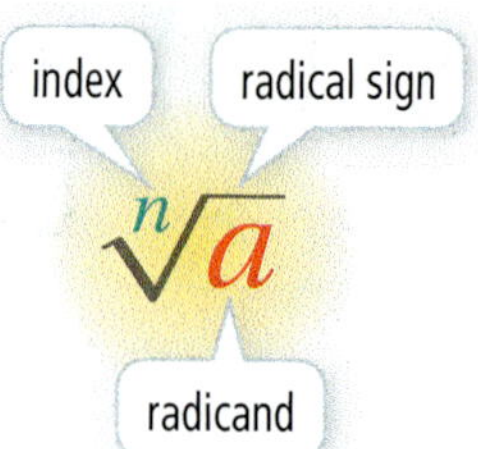

Problem 1 Finding All Real Roots

Plan

How many real cube roots are there? A cube root is the same as a third root, and 3 is odd. So there is only one real cube root of a number.

A What are the real cube roots of 0.008, -1000, and $\frac{1}{27}$?

$0.008 = (0.2)^3$ — 0.2 is the only real cube root of 0.008.

$-1000 = (-10)^3$ — -10 is the only real cube root of -1000.

$\frac{1}{27} = \left(\frac{1}{3}\right)^3$ — $\frac{1}{3}$ is the only real cube root of $\frac{1}{27}$.

B What are the real fourth roots of 1, -0.0001, and $\frac{16}{81}$?

Since 1 is positive, there are two real fourth roots.

$1 = 1^4$ — 1 is a real fourth root of 1.

$1 = (-1)^4$ — -1 is the other real fourth root of 1.

Since -0.0001 is negative, there are no real fourth roots of -0.0001.

Since $\frac{16}{81}$ is positive, there are two real fourth roots.

$\frac{16}{81} = \left(\frac{2}{3}\right)^4$ — $\frac{2}{3}$ is a real fourth root of $\frac{16}{81}$.

$\frac{16}{81} = \left(-\frac{2}{3}\right)^4$ — $-\frac{2}{3}$ is the other real fourth root of $\frac{16}{81}$.

Got It? **1. a.** What are the real fifth roots of 0, -1, and 32?

b. What are the real square roots of 0.01, -1, and $\frac{36}{121}$?

c. Reasoning Explain why a negative real number b has no real nth roots if n is even.

According to the Fundamental Theorem of Algebra, $x^4 - 1 = 0$ has four roots, only two of which are real. In this chapter, the focus is on real roots only.

Problem 2 Finding Roots

Plan

How can you find a cube root? Work backwards. Find a number whose cube is the radicand.

What is each real-number root?

A $\sqrt[3]{-8}$

$(-2)^3 = -8$

So, $\sqrt[3]{-8} = -2$.

B $\sqrt{0.04}$

$(0.2)^2 = 0.04$

So, $\sqrt{0.04} = 0.2$.

$(-0.2)^2 = 0.04$ also, but $\sqrt{0.04}$ represents the positive square root.

C $\sqrt[4]{-1}$

There is no real root because there is no real number whose fourth power is -1.

D $\sqrt{(-2)^2}$

$\sqrt{(-2)^2} = \sqrt{4} = 2$.

Got It? **2.** What is each real-number root?

a. $\sqrt[3]{-27}$ **b.** $\sqrt[4]{-81}$ **c.** $\sqrt{(-7)^2}$ **d.** $\sqrt{-49}$

It is tempting to conclude that $\sqrt[n]{a^n} = a$ for all real numbers a, but part (d) of Problem 2 shows that this is not the case. If n is even, then $\sqrt[n]{a^n}$ is positive even if a itself is negative.

take note

Property ***n*th Roots of *n*th Powers**

For any real number a, $\sqrt[n]{a^n} = \begin{cases} a \text{ if } n \text{ is odd} \\ |a| \text{ if } n \text{ is even} \end{cases}$

It is easy to overlook this rule for simplifying radicals. It is particularly important that you remember it when the radicand contains a variable expression. You must *include* the absolute value when n is even, and you must *omit* it when n is odd.

Problem 3 Simplifying Radical Expressions

Plan

How can you get started?

You're simplifying a square root, so use properties of exponents to write the *entire* radicand as a perfect square.

What is a simpler form of each radical expression?

A $\sqrt{16x^8}$

$\sqrt{16x^8} = \sqrt{4^2(x^4)^2} = \sqrt{(4x^4)^2} = |4x^4| = 4x^4$

You need to include absolute value symbols because the index of a square root is 2, which is even. However, $|4x^4| = 4x^4$ because x^4 is always nonnegative.

B $\sqrt[3]{a^6b^9}$

$\sqrt[3]{a^6b^9} = \sqrt[3]{(a^2)^3(b^3)^3} = \sqrt[3]{(a^2b^3)^3} = a^2b^3$

The index is odd, so you cannot include absolute value symbols here.

C $\sqrt[4]{x^8y^{12}}$

$\sqrt[4]{x^8y^{12}} = \sqrt[4]{(x^2)^4(y^3)^4} = \sqrt[4]{(x^2y^3)^4} = x^2|y^3|$

The index is even. The absolute value symbols ensure that the root is positive when y^3 is negative. Absolute value symbols are not needed for x^2 since x^2 is always nonnegative.

Got It? **3.** What is the simplified form of each radical expression?

a. $\sqrt{81x^4}$ **b.** $\sqrt[3]{a^{12}b^{15}}$ **c.** $\sqrt[4]{x^{12}y^{16}}$

Problem 4 Using a Radical Expression

Academics Some teachers adjust test scores when a test is difficult. One teacher's formula for adjusting scores is $A = 10\sqrt{R}$, where A is the adjusted score and R is the raw score. If the raw scores on one test range from 36 to 90, what is the range of the adjusted scores?

Think	Write
You have to adjust the lowest raw score and the highest raw score.	$10\sqrt{36} = 10(6) = 60$ $10\sqrt{90} \approx 10(9.487) = 94.87 \approx 95$
The other adjusted scores must be between the lowest and highest adjusted scores.	The adjusted scores range from 60 to 95.

Got It? **4.** In Problem 4, what are the adjusted scores for raw scores of 0 and 100?

Lesson Check

Do you know HOW?

Find all the real square roots of each number.

1. 25 **2.** 0.16 **3.** -64

Simplify each radical expression.

4. $\sqrt{9b^2}$ **5.** $\sqrt{a^8b^{18}}$ **6.** $\sqrt[3]{-125a^3}$

Do you UNDERSTAND?

MATHEMATICAL PRACTICES

7. Error Analysis A student said the only fourth root of 16 is 2. Describe and correct his error.

8. Vocabulary Explain the difference between a real root and the principal root.

9. Reasoning A number has only one real nth root. What can you conclude about the index n?

Practice and Problem-Solving Exercises

MATHEMATICAL PRACTICES

A Practice

Find all the real square roots of each number.

See Problem 1.

10. 225 **11.** 0.0049 **12.** $-\frac{1}{121}$ **13.** $\frac{64}{169}$

Find all the real cube roots of each number.

14. -64 **15.** 0.125 **16.** $-\frac{27}{216}$ **17.** 0.000343

Find all the real fourth roots of each number.

18. 16 **19.** -16 **20.** 0.0081 **21.** $\frac{10{,}000}{81}$

Find each real root. See Problem 2.

22. $\sqrt{36}$ 23. $\sqrt{0.25}$ 24. $-\sqrt[3]{64}$ 25. $\sqrt[3]{-27}$

Simplify each radical expression. Use absolute value symbols when needed. See Problem 3.

26. $\sqrt{16x^2}$ 27. $\sqrt[3]{27y^6}$ 28. $\sqrt[4]{x^{20}y^{28}}$ 29. $\sqrt[5]{32y^{10}}$

30. **Grades** In many classes, a passing test grade is 70. Using the formula $A = 10\sqrt{R}$, what raw score would a student need to get a passing grade after her score is adjusted? See Problem 4.

Find the two real solutions of each equation.

31. $x^2 = 100$ 32. $x^4 = 1$ 33. $x^2 = 0.25$ 34. $x^4 = \frac{16}{81}$

35. **Think About a Plan** The radius of a spherical balloon can be expressed as $r = \sqrt[3]{\frac{3V}{4\pi}}$ inches, where r is the radius and V is the volume of the balloon in cubic inches. If air is pumped to inflate the balloon from 500 cubic inches to 800 cubic inches, by how many inches has the radius of the balloon increased?
- What was the radius of the balloon originally?
- What was the radius after inflating the balloon to 800 cubic inches?
- How can you use the two radii to find the amount of increase?

STEM 36. **Electricity** The voltage V of an audio system's speaker can be represented by $V = 4\sqrt{P}$, where P is the power of the speaker. An engineer wants to design a speaker with 400 watts of power. What will the voltage be?

STEM 37. **Boat Building** Boat builders share an old rule of thumb for sailboats. The maximum speed K in knots is 1.35 times the square root of the length L in feet of the boat's waterline.

a. A customer is planning to order a sailboat with a maximum speed of 12 knots. How long should the waterline be?

b. How much longer would the waterline have to be to achieve a maximum speed of 15 knots?

Simplify each radical expression. Use absolute value symbols when needed.

38. $\sqrt[3]{0.125}$ 39. $\sqrt[3]{\frac{8}{216}}$ 40. $\sqrt[4]{0.0016}$ 41. $\sqrt[4]{\frac{1}{256}}$ 42. $\sqrt[4]{16c^4}$

43. **Open-Ended** Write three radical expressions that simplify to $-2x^2$.

44. **Reasoning** For what positive integers n is each of the statements true?

a. If $x^n = b$, then x is an nth root of b.

b. If $x^n = b$, then $x = \sqrt[n]{b}$.

Is each equation *always, sometimes,* or *never* true? Explain your answer.

45. $\sqrt{x^4} = x^2$ 46. $\sqrt{x^6} = x^3$ 47. $\sqrt[3]{x^8} = x^2$ 48. $\sqrt[3]{x^3} = |x|$

Simplify each radical expression if n is even, and then if n is odd.

49. $\sqrt[n]{m^n}$
50. $\sqrt[n]{m^{2n}}$
51. $\sqrt[n]{m^{3n}}$
52. $\sqrt[n]{m^{4n}}$

53. **Reasoning** How many square roots of integers are in the interval between 24 and 25?

54. **Reasoning** The square root of a positive integer is either a positive integer or an irrational number. Is this a true statement or not? Explain your reasoning.

55. **Geometry** Without using a calculator, determine which is greater: the altitude of an equilateral triangle with side 8 or the diagonal of a square with side 5. Show your work.

Standardized Test Prep

56. Which equation has more than one real-number solution?

(A) $x^2 = 0$ (B) $x^2 = 1$ (C) $x^2 = -1$ (D) $x^3 = -1$

57. According to the Rational Root Theorem, which of the following is NOT a possible root of the polynomial equation $7x^5 + 3x^2 - 4x + 21 = 0$?

(F) $\frac{1}{7}$ (G) $\frac{1}{3}$ (H) 3 (I) 7

58. The fuse of a three-break firework rocket is programmed to ignite three times with 2-second intervals between the ignitions. When the rocket is shot vertically in the air, its height h in feet after t seconds is given by the formula $h(t) = -5t^2 + 70t$. At how many seconds after the shot should the firework technician set the timer of the first ignition to make the second ignition occur when the rocket is at its highest point?

(A) 3 (B) 9 (C) 5 (D) 7

Extended Response

59. Write a system of equations to find a cubic polynomial that goes through $(-3, -35)$, $(0, 1)$, $(2, 3)$, and $(4, 7)$.

Mixed Review

Determine the cubic function that is obtained from the parent function $y = x^3$ after each sequence of transformations.

See Lesson 5-9.

60. translation up 3 units and to the left 2 units

61. vertical compression by a factor of $\frac{1}{2}$, translation down 2 units

Solve each equation by using the Quadratic Formula. See Lesson 4-7.

62. $-4x^2 + 7x - 3 = 0$
63. $3x^2 - 5x + 3 = 0$
64. $36x^2 - 132x + 121 = 0$

Get Ready! **To prepare for Lesson 6-2, do Exercises 65–67.**

Simplify each algebraic expression. See Lesson 1-3.

65. $\frac{14x^7y^9}{7x^4y^6}$
66. $\frac{3abc}{9b}$
67. $\frac{20x}{5x^3}$

6-2 Multiplying and Dividing Radical Expressions

Content Standard

A.SSE.2 Use the structure of an expression to identify ways to rewrite it.

Objective To multiply and divide radical expressions

SOLVE IT! Getting Ready!

You can cut the 36-square into four 9-squares or nine 4-squares. What other n-square can you cut into sets of smaller squares in two ways? Is there a square you can cut into smaller squares in three ways? Explain your reasoning.

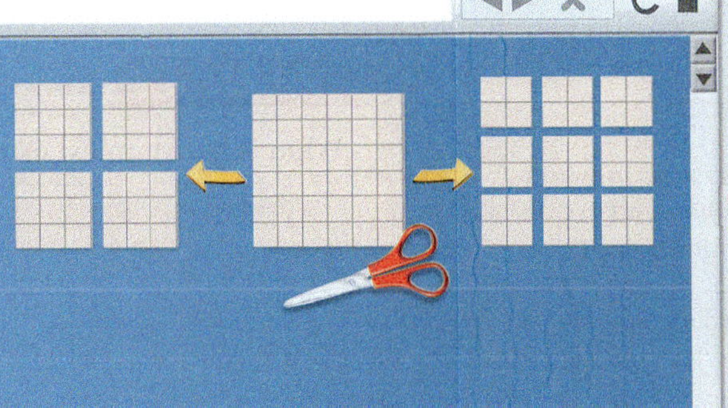

Cutting *n*-squares into 1-squares doesn't count.

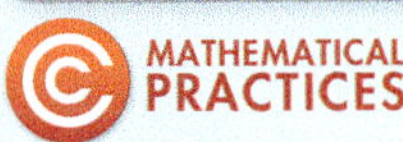

MATHEMATICAL PRACTICES

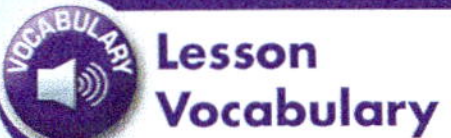

Lesson Vocabulary
- simplest form of a radical
- rationalize the denominator

Knowing the perfect squares greater than 1 (namely, 4, 9, 16, and so on) will help you simplify some radical expressions.

Essential Understanding You can simplify a radical expression when the exponent of one factor of the radicand is a multiple of the radical's index.

You can simplify the product of powers that have the same exponent. Similarly, you can simplify the product of radicals that have the same index.

Same Exponent	Same Index
$2^2 \cdot 3^2 = (2 \cdot 3)^2$	$\sqrt{2} \cdot \sqrt{3} = \sqrt{2 \cdot 3}$
$4^3 \cdot 5^3 = (4 \cdot 5)^3$	$\sqrt[3]{4} \cdot \sqrt[3]{5} = \sqrt[3]{4 \cdot 5}$

take note

Property Combining Radical Expressions: Products

If $\sqrt[n]{a}$ and $\sqrt[n]{b}$ are real numbers, then $\sqrt[n]{a} \cdot \sqrt[n]{b} = \sqrt[n]{ab}$.

Problem 1 Multiplying Radical Expressions

Plan

What allows you to use the property for multiplying radicals?
The radicals must be real numbers. The indexes must be the same.

Can you simplify the product of the radical expressions? Explain.

A $\sqrt[3]{6} \cdot \sqrt{2}$

No. The indexes are different. The property above does not apply.

B $\sqrt[3]{-4} \cdot \sqrt[3]{2}$

Yes. $\sqrt[3]{-4} \cdot \sqrt[3]{2} = \sqrt[3]{-4(2)} = \sqrt[3]{-8} = -2.$

 **Got It?** 1. Can you simplify the product of the radical expressions? Explain.

a. $\sqrt[4]{7} \cdot \sqrt[5]{7}$ b. $\sqrt[5]{-5} \cdot \sqrt[5]{-2}$

If the radicand of $\sqrt[n]{a}$ has a perfect nth power among its factors, you can *reduce* the radical. If you reduce a radical as much as possible, the radical is in **simplest form**. For example, consider $\sqrt{24}$ and $\sqrt[3]{24}$.

$\sqrt{24} = \sqrt{4 \cdot 6} = \sqrt{2^2 \cdot 6} = \sqrt{2^2} \cdot \sqrt{6} = 2\sqrt{6}$ $2\sqrt{6}$ is in simplest form.

$\sqrt[3]{24} = \sqrt[3]{8 \cdot 3} = \sqrt[3]{2^3 \cdot 3} = \sqrt[3]{2^3} \cdot \sqrt[3]{3} = 2\sqrt[3]{3}$ $2\sqrt[3]{3}$ is in simplest form.

Problem 2 Simplifying a Radical Expression

Think

How do you know when you are done simplifying?
You are done when the radicand contains no perfect cube factors.

What is the simplest form of $\sqrt[3]{54x^5}$?

$\sqrt[3]{54x^5} = \sqrt[3]{3^3 \cdot 2 \cdot x^2 \cdot x^3}$	Find all perfect cube factors.
$= \sqrt[3]{3^3x^3} \cdot \sqrt[3]{2x^2}$	$\sqrt[n]{ab} = \sqrt[n]{a} \cdot \sqrt[n]{b}$
$= 3x\sqrt[3]{2x^2}$	Simplify.

 **Got It?** 2. What is the simplest form of $\sqrt[3]{128x^7}$?

Problem 2 involves simplifying a cube root, so absolute value symbols are not needed. Remember that to combine $\sqrt[n]{a}$ and $\sqrt[n]{b}$ by multiplication, both radical expressions must be real numbers.

Problem 3 Simplifying a Product

What is the simplest form of $\sqrt{72x^3y^2} \cdot \sqrt{10xy^3}$?

Think

You need to multiply the radicands and find the perfect square factors.

Write

$\sqrt{72x^3y^2} \cdot \sqrt{10xy^3} = \sqrt{(72x^3y^2)(10xy^3)}$

$= \sqrt{720x^4y^5}$

$= \sqrt{12^2(5)(x^2)^2(y^2)^2y}$

$= \sqrt{12^2(x^2)^2(y^2)^2} \cdot \sqrt{5y}$

$= 12|x^2y^2| \cdot \sqrt{5y}$

$= 12x^2y^2\sqrt{5y}$

Now find square roots. Since $\sqrt{72x^3y^2}$ and $\sqrt{10xy^3}$ must be real numbers, x and y are nonnegative, so no absolute value symbols are needed.

The simplest form is $12x^2y^2\sqrt{5y}$.

 **Got It?** 3. What is the simplest form of $\sqrt{45x^5y^3} \cdot \sqrt{35xy^4}$?

Since you define division in terms of multiplication, you can extend the property for multiplying radical expressions. If the indexes are the same, you can write a quotient of roots as a root of a quotient.

Multiplying	Dividing
$\sqrt{2} \cdot \sqrt{3} = \sqrt{2 \cdot 3}$	$\frac{\sqrt{2}}{\sqrt{3}} = \sqrt{\frac{2}{3}}$
$\sqrt[3]{4} \cdot \sqrt[3]{5} = \sqrt[3]{4 \cdot 5}$	$\frac{\sqrt[3]{4}}{\sqrt[3]{5}} = \sqrt[3]{\frac{4}{5}}$

take note

Property **Combining Radical Expressions: Quotients**

If $\sqrt[n]{a}$ and $\sqrt[n]{b}$ are real numbers and $b \neq 0$, then $\frac{\sqrt[n]{a}}{\sqrt[n]{b}} = \sqrt[n]{\frac{a}{b}}$.

Problem 4 Dividing Radical Expressions

What is the simplest form of the quotient?

Think

Do you need to include absolute value symbols?
No. Both the divisor and dividend already require that x be nonnegative.

A $\frac{\sqrt{18x^5}}{\sqrt{2x^3}}$

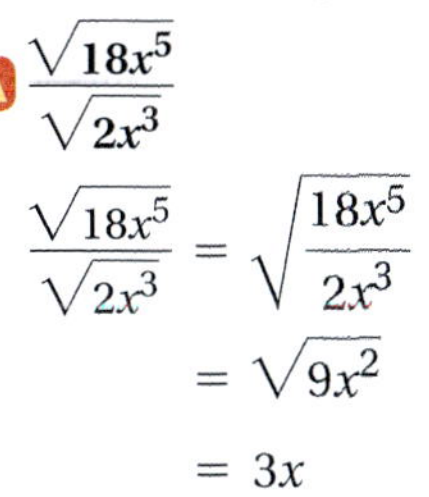

$$\frac{\sqrt{18x^5}}{\sqrt{2x^3}} = \sqrt{\frac{18x^5}{2x^3}}$$
$$= \sqrt{9x^2}$$
$$= 3x$$

B $\frac{\sqrt[3]{162y^5}}{\sqrt[3]{3y^2}}$

$$\frac{\sqrt[3]{162y^5}}{\sqrt[3]{3y^2}} = \sqrt[3]{\frac{162y^5}{3y^2}}$$
$$= \sqrt[3]{54y^3}$$
$$= \sqrt[3]{27y^3} \cdot \sqrt[3]{2}$$
$$= \sqrt[3]{3^3y^3} \cdot \sqrt[3]{2}$$
$$= 3y\sqrt[3]{2}$$

Got It? 4. a. What is the simplest form of $\frac{\sqrt{50x^6}}{\sqrt{2x^4}}$?

b. Reasoning Can you simplify the expression in Problem 4(a) by first simplifying $\sqrt{18x^5}$ and $\sqrt{2x^3}$? Explain.

Another way to simplify a radical expression is to **rationalize the denominator**. You rewrite the expression so that there are no radicals in any denominator and no denominator in any radical.

Multiply by 1

$$\frac{1}{\sqrt{2}} = \frac{1}{\sqrt{2}}\left(\frac{\sqrt{2}}{\sqrt{2}}\right) = \frac{\sqrt{2}}{2}$$

The product of $\sqrt{2}$ and itself is a rational number, 2.

Problem 5 Rationalizing the Denominator

Multiple Choice What is the simplest form of $\sqrt[3]{\frac{5x^2}{12y^2z}}$?

Ⓐ $\frac{\sqrt[3]{90x^2yz^2}}{6yz}$ Ⓑ $\frac{\sqrt[3]{5x^2}}{\sqrt[3]{12y^2z}}$ Ⓒ $\frac{5\sqrt[3]{x^2yz^2}}{yz}$ Ⓓ $5\sqrt[3]{x^2z}$

Think

How do you choose what to multiply by? Choose a cube root with a radicand that will make each factor of the radicand in the denominator a perfect cube.

$$\sqrt[3]{\frac{5x^2}{12y^2z}} = \frac{\sqrt[3]{5x^2}}{\sqrt[3]{2^2 \cdot 3y^2z}}$$

The radicand in the denominator needs 2, 3^2, y, and z^2 to make the factors perfect cubes.

$$= \frac{\sqrt[3]{5x^2}}{\sqrt[3]{2^2 \cdot 3y^2z}} \cdot \frac{\sqrt[3]{2 \cdot 3^2yz^2}}{\sqrt[3]{2 \cdot 3^2yz^2}}$$

Multiply the numerator and denominator by $\sqrt[3]{2 \cdot 3^2yz^2}$.

$$= \frac{\sqrt[3]{90x^2yz^2}}{\sqrt[3]{2^3 \cdot 3^3y^3z^3}}$$

$$= \frac{\sqrt[3]{90x^2yz^2}}{2 \cdot 3yz}$$

Simplify.

$$= \frac{\sqrt[3]{90x^2yz^2}}{6yz}$$

The correct answer is A.

  **Got It? 5. a.** What is the simplest form of $\frac{\sqrt[3]{7x}}{\sqrt[3]{5y^2}}$?

b. Reasoning Which choice in Problem 5 could be eliminated immediately? Explain your reasoning.

Lesson Check

Do you know HOW?

Multiply, if possible. Then simplify.

1. $\sqrt{2} \cdot \sqrt{5}$
2. $\sqrt[3]{-27} \cdot \sqrt[3]{4}$
3. $\sqrt[3]{2} \cdot \sqrt[2]{7}$
4. $\sqrt{3} \cdot \sqrt{-4}$

Divide and simplify.

5. $\frac{\sqrt[3]{15x^2}}{\sqrt[3]{5x}}$
6. $\frac{\sqrt{21x^{10}}}{\sqrt{7x^5}}$

Do you UNDERSTAND?

7. **Vocabulary** Write the simplest form of $\sqrt[3]{32x^4}$.
8. **Reasoning** For what values of x is $\sqrt{-4x^3}$ real? Justify your reasoning.
9. **Error Analysis** Explain the error in this simplification of radical expressions.

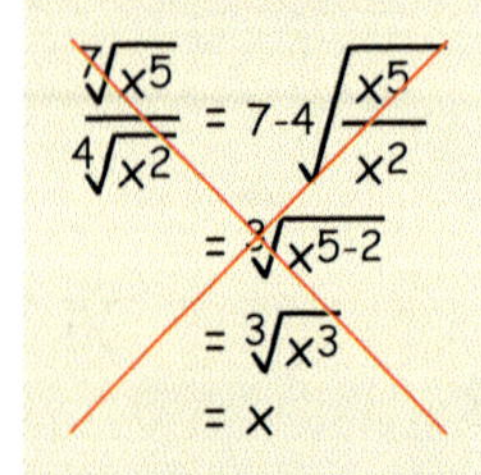

Practice and Problem-Solving Exercises

Multiply, if possible. Then simplify.

See Problem 1.

10. $\sqrt{8} \cdot \sqrt{32}$
11. $\sqrt[3]{4} \cdot \sqrt[3]{16}$
12. $\sqrt[3]{9} \cdot \sqrt[3]{-81}$
13. $\sqrt[4]{8} \cdot \sqrt[3]{32}$
14. $\sqrt{-5} \cdot \sqrt{5}$
15. $\sqrt[3]{-5} \cdot \sqrt[3]{-25}$
16. $\sqrt[3]{9} \cdot \sqrt[3]{-24}$
17. $\sqrt[3]{-12} \cdot \sqrt[3]{-18}$
18. $\sqrt{50} \cdot \sqrt{75}$

Simplify.

See Problem 2.

19. $\sqrt{20x^3}$
20. $\sqrt[3]{81x^3}$
21. $\sqrt{50x^5}$
22. $\sqrt[3]{32a^5}$
23. $\sqrt[3]{54y^{10}}$
24. $\sqrt{200a^6b^7}$
25. $\sqrt[3]{-250x^6y^5}$
26. $\sqrt[4]{64x^3y^6}$
27. $\sqrt[5]{-32x^6y^7}$

Multiply and simplify.

See Problem 3.

28. $\sqrt[3]{6} \cdot \sqrt[3]{16}$
29. $\sqrt{8y^5} \cdot \sqrt{40y^2}$
30. $\sqrt{8x^5} \cdot \sqrt{3x}$
31. $4\sqrt{2x} \cdot 5\sqrt{6xy^2}$
32. $3\sqrt[3]{5y^3} \cdot 2\sqrt[3]{50y^4}$
33. $-\sqrt[3]{2x^2y^2} \cdot 2\sqrt[3]{15x^5y}$
34. $\sqrt[4]{81x^5y^4} \cdot \sqrt[4]{32x^3y}$
35. $2\sqrt[3]{2xy^2} \cdot \sqrt[3]{4x^2y^5}$
36. $3\sqrt[4]{18a^9} \cdot \sqrt[4]{6ab^2}$

Divide and simplify.

See Problem 4.

37. $\dfrac{\sqrt{500}}{\sqrt{5}}$
38. $\dfrac{\sqrt{48x^3}}{\sqrt{3xy^2}}$
39. $\dfrac{\sqrt{56x^5y^5}}{\sqrt{7xy}}$
40. $\dfrac{\sqrt[3]{250x^7y^3}}{\sqrt[3]{2x^2y}}$
41. $\dfrac{\sqrt[3]{48x^3y^2}}{\sqrt[3]{6x^4y}}$
42. $\dfrac{\sqrt{20ab}}{\sqrt{45a^2b^3}}$

Rationalize the denominator of each expression.

See Problem 5.

43. $\dfrac{\sqrt{x}}{\sqrt{2}}$
44. $\dfrac{\sqrt{5}}{\sqrt{8x}}$
45. $\dfrac{\sqrt[3]{x}}{\sqrt[3]{2}}$
46. $\sqrt[3]{\dfrac{5}{3x}}$
47. $\dfrac{\sqrt[4]{2}}{\sqrt[4]{5}}$
48. $\dfrac{15\sqrt{60x^5}}{3\sqrt{12x}}$
49. $\dfrac{\sqrt{3xy^2}}{\sqrt{5xy^3}}$
50. $\dfrac{\sqrt{5x^4y}}{\sqrt{2x^2y^3}}$
51. $\dfrac{\sqrt[3]{12ab^3c^2}}{\sqrt[3]{10a^3bc}}$

52. **Think About a Plan** The formula $t = \sqrt{\frac{2s}{a}}$ shows the time t that any vehicle takes to travel a distance s at a constant acceleration a, starting from rest. What is the difference in time between a car accelerating at 16 m/s^2 and one accelerating at 25 m/s^2 for a distance of 200 m?
    - What is the time that a car accelerating at 16 m/s^2 takes to travel 200 m?
    - What is the time that a car accelerating at 25 m/s^2 takes to travel 200 m?
53. **Geometry** The base of a triangle is $\sqrt{18}$ cm and its height is $\sqrt{8}$ cm. Find its area.

STEM **54. Physics** The formula $F = \frac{mv^2}{r}$ gives the centripetal force F of an object of mass m moving along a circle of radius r, where v is the tangential velocity of the object. Solve the formula for v. Rationalize the denominator.

STEM **55. Satellites** The circular velocity v in miles per hour of a satellite orbiting Earth is given by the formula $v = \sqrt{\frac{1.24 \times 10^{12}}{r}}$, where r is the distance in miles from the satellite to the center of the Earth. How much greater is the velocity of a satellite orbiting at an altitude of 100 mi than the velocity of a satellite orbiting at an altitude of 200 mi? (The radius of the Earth is 3950 mi.)

56. a. Simplify $\frac{\sqrt{2} + \sqrt{3}}{\sqrt{75}}$ by multiplying the numerator and denominator by $\sqrt{75}$.

b. Simplify the expression in (a) by multiplying by $\sqrt{3}$ instead of $\sqrt{75}$.

c. Explain how you would simplify $\frac{\sqrt{2} + \sqrt{3}}{\sqrt{98}}$.

Simplify each expression. Rationalize all denominators.

57. $\sqrt{5} \cdot \sqrt{50}$

58. $\sqrt[3]{4} \cdot \sqrt[3]{80}$

59. $\sqrt{x^5y^5} \cdot 3\sqrt{2x^7y^6}$

60. $5\sqrt{2xy^6} \cdot 2\sqrt{2x^3y}$

61. $\sqrt{2}(\sqrt{50} + 7)$

62. $\sqrt{5}(\sqrt{5} + \sqrt{15})$

63. $\frac{\sqrt{5x^4}}{\sqrt{2x^2y^3}}$

64. $\frac{5\sqrt{2}}{3\sqrt{7x}}$

65. $\frac{1}{\sqrt[3]{9x}}$

66. $\frac{10}{\sqrt[3]{5x^2}}$

67. $\frac{\sqrt[3]{14}}{\sqrt[3]{7x^2y}}$

68. $\frac{3\sqrt{11x^3y}}{-2\sqrt{12x^4y}}$

STEM **69. Physics** The mass m of an object is $\sqrt{80}$ g and its volume V is $\sqrt{5}$ cm^3. Use the formula $D = \frac{m}{V}$ to find the density D of the object.

70. Writing Does $\sqrt{x^3} = \sqrt[3]{x^2}$ for *all*, *some*, or *no* values of x? Explain.

71. Open-Ended Of the equivalent expressions $\sqrt{\frac{2}{3}}$, $\frac{\sqrt{2}}{\sqrt{3}}$, and $\frac{\sqrt{6}}{3}$, which do you prefer to use for finding a decimal approximation with a calculator? Justify your reasoning.

72. Error Analysis Explain the error in this simplification of radical expressions.

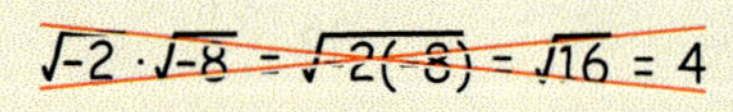
$\sqrt{-2} \cdot \sqrt{-8} = \sqrt{-2(-8)} = \sqrt{16} = 4$

Determine whether each expression is *always*, *sometimes*, or *never* a real number. Assume that x can be any real number.

73. $\sqrt[3]{-x^2}$

74. $\sqrt{-x^2}$

75. $\sqrt{-x}$

Simplify each expression. Rationalize all denominators.

76. $\sqrt{\sqrt{16x^4y^4}}$

77. $\sqrt{\sqrt[3]{8000}}$

78. $\sqrt[6]{\frac{y^{-3}}{x^{-4}}}$

79. Reasoning When $\sqrt{x^ay^b}$ is simplified, the result is $\frac{1}{x^cy^{3d}}$, where c and d are positive integers. Express a in terms of c, and b in terms of d.

Standardized Test Prep

SAT/ACT

80. What is the simplified form of the expression $\frac{3}{\sqrt{18xy^2}}$ if x and y are positive?

(A) $\frac{\sqrt{2x}}{2xy}$ (B) $\frac{\sqrt{2y}}{2xy}$ (C) $\frac{\sqrt{54xy^2}}{2xy}$ (D) $\frac{\sqrt{27xy^2}}{2xy}$

81. What are the solutions, in simplest form, of the quadratic equation $3x^2 + 6x - 5 = 0$?

(F) $\frac{-6 \pm \sqrt{96}}{6}$ (G) $\frac{-6 \pm i\sqrt{24}}{6}$ (H) $\frac{-3 \pm 2\sqrt{6}}{3}$ (I) $\frac{-3 \pm i\sqrt{6}}{3}$

82. Which inequality is shown by the graph at the right?

(A) $y \geq \frac{2}{3}|x - 1| - 2$ (C) $y \geq \frac{3}{2}|x - 1| - 2$

(B) $y \geq \frac{2}{3}|x - 2| - 1$ (D) $y \geq \left|\frac{2}{3}x - 1\right| - 2$

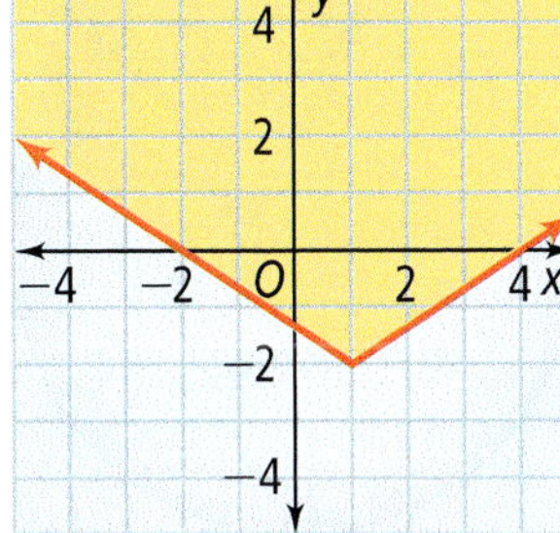

83. A triangle has the dimensions shown below.

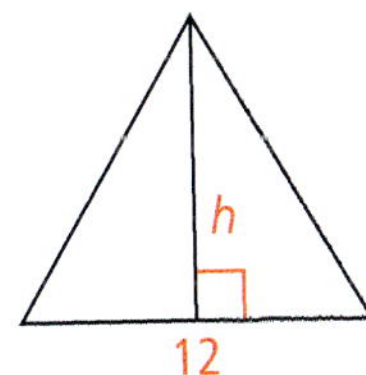

What is the height of a triangle with equal area but a base of 36?

(F) $\frac{h}{3}$ (G) $\frac{2h}{3}$ (H) $2h$ (I) $3h$

Short Response

84. Find the axis of symmetry of the graph of the function $y = -2x^2 - 5x + 4$. Show your work.

Mixed Review

Simplify each radical expression. Use absolute value symbols when needed.

See Lesson 6-1.

85. $\sqrt{121a^{90}}$ **86.** $\sqrt{81c^{48}d^{64}}$ **87.** $\sqrt[3]{64a^{81}}$ **88.** $\sqrt[5]{32y^{25}}$

Divide using synthetic division. See Lesson 5-4.

89. $(y^3 - 64) \div (y + 4)$ **90.** $(6a^3 + a^2 - a + 4) \div (a + 1)$

Complete each square. See Lesson 4-6.

91. $x^2 + 10x + ■$ **92.** $x^2 - 10x + ■$ **93.** $x^2 + 11x + ■$ **94.** $x^2 - 11x + ■$

Get Ready! **To prepare for Lesson 6-3, do Exercises 95–98.**

Write each quotient as a complex number in the form $a \pm bi$. See Lesson 4-8.

95. $\frac{2}{3 - i}$ **96.** $\frac{5}{2 + 3i}$ **97.** $\frac{4}{4 + i}$ **98.** $\frac{-1}{7 - 5i}$

6-3 Binomial Radical Expressions

Content Standard

A.SSE.2 Use the structure of an expression to identify ways to rewrite it.

Objective To add and subtract radical expressions

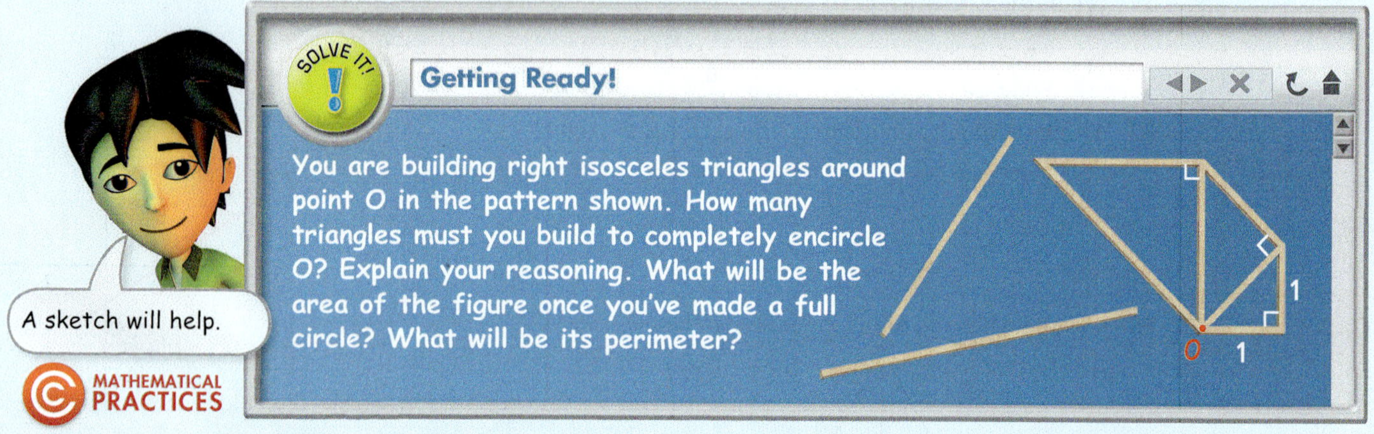

Lesson Vocabulary
- like radicals

Like radicals are radical expressions that have the same index and radicand.

Essential Understanding You can combine like radicals using properties of real numbers.

Here is how you can combine like radicals using the Distributive Property.

Like Radicals With Numbers	Like Radicals With Variables
$\sqrt{2} + 3\sqrt{2} = 4\sqrt{2}$	$\sqrt{5xy} + 8\sqrt{5xy} = 9\sqrt{5xy}$
$\sqrt[3]{7} - 5\sqrt[3]{7} = -4\sqrt[3]{7}$	$\sqrt[3]{9x^2y} - 8\sqrt[3]{9x^2y} = -7\sqrt[3]{9x^2y}$

take note

Property Combining Radical Expressions: Sums and Differences

Use the Distributive Property to add or subtract like radicals.

$$a\sqrt[n]{x} + b\sqrt[n]{x} = (a + b)\sqrt[n]{x} \qquad a\sqrt[n]{x} - b\sqrt[n]{x} = (a - b)\sqrt[n]{x}$$

Combining radical expressions is different from *adding* them. The sum of any two real numbers is a real number, so you can add $\sqrt{2}$ and $\sqrt{3}$ to get the real number $\sqrt{2} + \sqrt{3}$. However, you cannot *combine* the result into a single radical, so $\sqrt{2} + \sqrt{3} \neq \sqrt{5}$.

$$\begin{array}{rl} \sqrt{2} \approx 1.414 & \quad \sqrt{5} \approx 2.236 \\ + \sqrt{3} \approx 1.732 & \\ \hline \sqrt{2} + \sqrt{3} \approx 3.146 & \quad \neq 2.236 \end{array}$$

Problem 1 Adding and Subtracting Radical Expressions

What is the simplified form of each expression?

A $3\sqrt{5x} - 2\sqrt{5x}$

$3\sqrt{5x} - 2\sqrt{5x} = (3 - 2)\sqrt{5x}$ Distributive Property

$= \sqrt{5x}$ Simplify.

Think

Can you always simplify a radical sum?
No. The radicands and the indexes must be the same.

B $6x^2\sqrt{7} + 4x\sqrt{5}$

The radicands are different. You cannot combine the expressions.

C $12\sqrt[3]{7xy} - 8\sqrt[5]{7xy}$

The indexes are different. You cannot combine the expressions.

Got It? **1.** What is the simplified form of each expression?

a. $7\sqrt[3]{5} - 4\sqrt{5}$ **b.** $3x\sqrt{xy} + 4x\sqrt{xy}$ **c.** $17\sqrt[5]{3x^2} - 15\sqrt[5]{3x^2}$

Problem 2 Using Radical Expressions STEM

Architecture **In the stained-glass window design, the side of each small square is 5 in. Find the perimeter of the window to the nearest tenth of an inch.**

Length of the diagonal of a square with side s: $s\sqrt{2}$.
Length of the diagonal of each 5-inch square: $5\sqrt{2}$.

Length of the window: $l = 3(5\sqrt{2}) = 15\sqrt{2}$
Width of the window: $w = 2(5\sqrt{2}) = 10\sqrt{2}$

Think

Does it make sense that you have a radical expression as the answer?
Yes, because perimeter is a linear measure, and there is no squaring in the calculations.

Perimeter $= 2l + 2w$

$= 2(15\sqrt{2}) + 2(10\sqrt{2})$ Substitute for length and width.

$= 30\sqrt{2} + 20\sqrt{2}$ Simplify.

$= 50\sqrt{2}$ Distributive Property

≈ 70.7 Use a calculator to approximate.

The perimeter of the window is about 70.7 inches.

  **Got It?** **2. a.** Find the perimeter of the window if the side of each small square is 6 in.

b. Reasoning Describe a different sequence of steps which you could use to compute the perimeter of the window.

When you have a sum or difference of radical expressions, you should simplify each expression so that you can find all the like radicals.

Problem 3 Simplifying Before Adding or Subtracting

What is the simplest form of the expression? $\sqrt{12} + \sqrt{75} - \sqrt{3}$

Think

To simplify each radical expression, factor each radicand.

These are like radicals. Combine them. Remember $\sqrt{3} = 1\sqrt{3}$.

Write

$$\sqrt{12} + \sqrt{75} - \sqrt{3}$$
$$= \sqrt{4 \cdot 3} + \sqrt{25 \cdot 3} - \sqrt{3}$$
$$= \sqrt{2^2 \cdot 3} + \sqrt{5^2 \cdot 3} - \sqrt{3}$$
$$= \sqrt{2^2}\sqrt{3} + \sqrt{5^2}\sqrt{3} - \sqrt{3}$$
$$= 2\sqrt{3} + 5\sqrt{3} - \sqrt{3}$$
$$= 6\sqrt{3}$$

Got It? **3.** What is the simplest form of the expression? $\sqrt[3]{250} + \sqrt[3]{54} - \sqrt[3]{16}$

You can use the FOIL method to multiply binomials that have radical expressions. Remember that the FOIL method ensures that you multiply each term of one binomial by each term of the other.

Problem 4 Multiplying Binomial Radical Expressions

Plan

How do you multiply two binomials?
Use the FOIL method: **F**irst, **O**uter, **I**nner, **L**ast. Then simplify.

What is the product of each radical expression?

A $(4 + 2\sqrt{2})(5 + 4\sqrt{2})$

$(4 + 2\sqrt{2})(5 + 4\sqrt{2})$

$= 4 \cdot 5 + 4 \cdot 4\sqrt{2} + 2\sqrt{2} \cdot 5 + 2\sqrt{2} \cdot 4\sqrt{2}$ Distribute.

$= 20 + 16\sqrt{2} + 10\sqrt{2} + 16$ Multiply.

$= 36 + 26\sqrt{2}$ Combine like radicals.

B $(3 - \sqrt{7})(5 + \sqrt{7})$

$(3 - \sqrt{7})(5 + \sqrt{7})$

$= 3 \cdot 5 + 3\sqrt{7} - \sqrt{7} \cdot 5 - \sqrt{7} \cdot \sqrt{7}$ Distribute.

$= 15 - 2\sqrt{7} - 7$ Multiply and combine like radicals.

$= 8 - 2\sqrt{7}$ Simplify.

Got It? **4.** What is the product $(3 + 2\sqrt{5})(2 + 4\sqrt{5})$?

Conjugates are expressions, like $\sqrt{a} + \sqrt{b}$ and $\sqrt{a} - \sqrt{b}$, that differ only in the signs of the second terms. When a and b are rational numbers, the product of two radical conjugates is a rational number.

Problem 5 Multiplying Conjugates

Think

Where have you seen conjugates before?
The complex number $a + bi$ has a conjugate, $a - bi$. Multiplying them results in a number with no imaginary part.

What is the product $(5 - \sqrt{7})(5 + \sqrt{7})$?

$$(5 - \sqrt{7})(5 + \sqrt{7}) = 5 \cdot 5 + 5\sqrt{7} - 5\sqrt{7} - (\sqrt{7})^2 \quad \text{Distribute.}$$
$$= 25 - 7 \quad \text{Simplify.}$$
$$= 18$$

Got It? 5. What is each product?

a. $(6 - \sqrt{12})(6 + \sqrt{12})$ **b.** $(3 + \sqrt{8})(3 - \sqrt{8})$

Sometimes a denominator is a sum or difference involving radicals. If the radical expressions are square roots, you can rationalize the denominator by multiplying the numerator and the denominator by the conjugate of the denominator.

Problem 6 Rationalizing the Denominator

Think

What is a rationalized denominator?
A rationalized denominator contains no radicals.

How can you write the expression with a rationalized denominator?

$$\frac{3\sqrt{2}}{\sqrt{5} - \sqrt{2}}$$

$$\frac{3\sqrt{2}}{\sqrt{5} - \sqrt{2}} = \frac{3\sqrt{2}}{\sqrt{5} - \sqrt{2}} \cdot \frac{\sqrt{5} + \sqrt{2}}{\sqrt{5} + \sqrt{2}} \quad \text{Multiply. Use the conjugate of the denominator.}$$

$$= \frac{3\sqrt{2}(\sqrt{5} + \sqrt{2})}{(\sqrt{5})^2 - (\sqrt{2})^2} \quad \text{The radicals in the denominator cancel out.}$$

$$= \frac{3(\sqrt{2} \cdot \sqrt{5} + \sqrt{2} \cdot \sqrt{2})}{5 - 2} \quad \text{Distribute } \sqrt{2} \text{ in the numerator.}$$

$$= \frac{3(\sqrt{10} + 2)}{3} \quad \text{Simplify}$$

$$= \sqrt{10} + 2$$

Got It? 6. How can you write the expression with a rationalized denominator?

a. $\frac{2\sqrt{7}}{\sqrt{3} - \sqrt{5}}$ **b.** $\frac{4x}{3 - \sqrt{6}}$

c. Reasoning Suppose you were going to rationalize the denominator of $\frac{1 - \sqrt{8}}{2 - \sqrt{8}}$. Would you simplify $\sqrt{8}$ before or after rationalizing? Explain your answer.

Lesson Check

Do you know HOW?

Simplify if possible.

1. $10\sqrt{6} + 2\sqrt{6}$
2. $3\sqrt{2} + 4\sqrt[3]{2}$
3. $8\sqrt{3x} - 5\sqrt{3x}$
4. $5\sqrt{3} + \sqrt{12}$

Multiply.

5. $(4 + \sqrt{3})(4 - \sqrt{3})$
6. $(5 + 2\sqrt{5})(7 + 4\sqrt{5})$
7. $(2 + 3\sqrt{2})(1 - 3\sqrt{2})$

Do you UNDERSTAND?

8. **Vocabulary** Determine whether each of the following is a pair of like radicals. If so, add them.
   a. $3x\sqrt{11}$ and $3x\sqrt{10}$
   b. $2\sqrt{3xy}$ and $7\sqrt{3xy}$
   c. $12\sqrt{13y}$ and $12\sqrt{6y}$

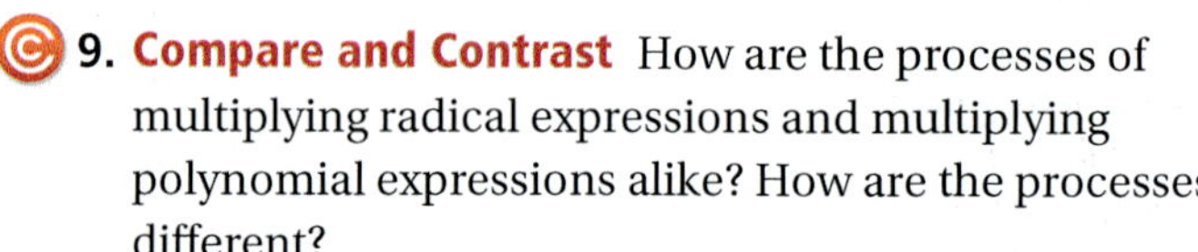

9. **Compare and Contrast** How are the processes of multiplying radical expressions and multiplying polynomial expressions alike? How are the processes different?

Practice and Problem-Solving Exercises

A Practice

Simplify if possible. See Problem 1.

10. $5\sqrt{6} + \sqrt{6}$
11. $6\sqrt[3]{3} - 2\sqrt[3]{3}$
12. $4\sqrt{3} + 4\sqrt[3]{3}$
13. $3\sqrt{x} - 5\sqrt{x}$
14. $14\sqrt{x} + 3\sqrt{y}$
15. $7\sqrt[3]{x^2} - 2\sqrt[3]{x^2}$

See Problem 2.

16. The design of a garden path uses stone pieces shaped as squares with a side length of 15 in. Find the length of the path.

Simplify. See Problem 3.

17. $6\sqrt{18} + 3\sqrt{50}$
18. $14\sqrt{20} - 3\sqrt{125}$
19. $\sqrt{18} + \sqrt{32}$
20. $\sqrt[3]{54} + \sqrt[3]{16}$
21. $3\sqrt[3]{81} - 2\sqrt[3]{54}$
22. $\sqrt[4]{32} + \sqrt[4]{48}$

Multiply. See Problem 4.

23. $(3 + \sqrt{5})(1 + \sqrt{5})$
24. $(2 + \sqrt{7})(1 + 3\sqrt{7})$
25. $(3 - 4\sqrt{2})(5 - 6\sqrt{2})$.
26. $(\sqrt{3} + \sqrt{5})^2$
27. $(\sqrt{13} + 6)^2$
28. $(2\sqrt{5} + 3\sqrt{2})^2$

Multiply each pair of conjugates. See Problem 5.

29. $(5 - \sqrt{11})(5 + \sqrt{11})$
30. $(4 - 2\sqrt{3})(4 + 2\sqrt{3})$
31. $(2\sqrt{6} + 8)(2\sqrt{6} - 8)$
32. $(\sqrt{3} + \sqrt{5})(\sqrt{3} - \sqrt{5})$

Rationalize each denominator. Simplify your answer.

See Problem 6.

33. $\frac{4}{1 + \sqrt{3}}$
34. $\frac{4}{3\sqrt{3} - 2}$
35. $\frac{5 + \sqrt{3}}{2 - \sqrt{3}}$
36. $\frac{3 + \sqrt{8}}{2 - 2\sqrt{8}}$

37. Think About a Plan The design on a parquet floor, shown at the right, is made of equilateral triangles. The side of a large triangle is 6 in., and the side of a small triangle is 3 in. Find the total area of the design to the nearest tenth of a square inch.

- How many large and how many small triangles form the design?
- Can you express the area of an equilateral triangle through its side?

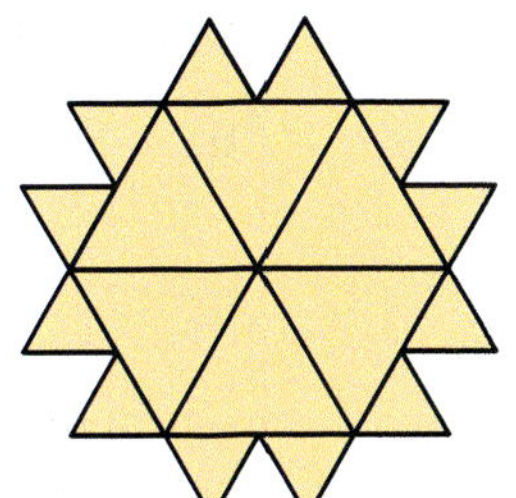

Simplify.

38. $\sqrt{72} + \sqrt{32} + \sqrt{18}$

39. $\sqrt{75} + 2\sqrt{48} - 5\sqrt{3}$

40. $5\sqrt{32x} + 4\sqrt{98x}$

41. $\sqrt{75} - 4\sqrt{18} + 2\sqrt{32}$

42. $4\sqrt{216y^2} + 3\sqrt{54y^2}$

43. $3\sqrt[3]{16} - 4\sqrt[3]{54} + \sqrt[3]{128}$

44. $(1 + \sqrt{72})(5 + \sqrt{2})$

45. $(\sqrt{3} - \sqrt{7})(\sqrt{3} + 2\sqrt{7})$

46. $(\sqrt{y} + \sqrt{2})(\sqrt{y} - 7\sqrt{2})$

47. $(\sqrt{12} + \sqrt{72})^2$

48. $(\sqrt{1.25} - \sqrt{1.8})(\sqrt{5} + \sqrt{0.2})$

49. $(\sqrt{a+1} + \sqrt{a-1})(\sqrt{a+1} - \sqrt{a-1})$

50. Error Analysis Describe and correct the error made while simplifying the expression $\frac{3 + \sqrt{2}}{3 - \sqrt{2}}$.

$$\frac{3+\sqrt{2}}{3-\sqrt{2}} = \frac{3+\sqrt{2}}{3-\sqrt{2}} \cdot \frac{3+\sqrt{2}}{3+\sqrt{2}} = \frac{3^2+(\sqrt{2})^2}{3^2-(\sqrt{2})^2} = \frac{9+2}{9-2} = \frac{11}{7}$$

STEM **51. Chemistry** A scientist found that x grams of Metal A is completely oxidized in $2x\sqrt{3}$ seconds and x grams of Metal B is completely oxidized in $6x\sqrt{3}$ seconds. How much faster is Metal A oxidized than Metal B?

52. Reasoning Describe the possible values of a such that $\sqrt{72} + \sqrt{a}$ simplifies to a single term.

53. Writing Discuss the advantages and disadvantages of first simplifying $\sqrt{72} + \sqrt{32} + \sqrt{18}$ in order to estimate its decimal value.

54. Geometry Show that a right triangle with legs of lengths $\sqrt{2} - 1$ and $\sqrt{2} + 1$ is similar to a right triangle with legs of lengths $6 - \sqrt{32}$ and 2.

55. Open-Ended Find two pairs of conjugates with a product of 3.

Rationalize the denominators and simplify.

56. $\frac{4 + \sqrt{27}}{2 - 3\sqrt{27}}$

57. $\frac{4 + \sqrt{6}}{\sqrt{2} + \sqrt{3}}$

58. $\frac{5 - \sqrt{21}}{\sqrt{3} - \sqrt{7}}$

59. $\frac{\sqrt{44x^2}}{\sqrt{11} + 3}$

60. $\frac{\sqrt{2} + \sqrt{6}}{\sqrt{1.5} + \sqrt{0.5}}$

61. $\frac{\sqrt{27} - \sqrt{5}}{\sqrt{15} - 3}$

62. $\frac{4 + \sqrt[3]{2}}{\sqrt[3]{2}}$

63. $\frac{5 + \sqrt[4]{x}}{\sqrt[4]{x}}$

64. $\frac{4 - 2\sqrt[3]{6}}{\sqrt[3]{4}}$

Challenge Add or subtract.

65. $\frac{1}{1 - \sqrt{5}} + \frac{1}{1 + \sqrt{5}}$

66. $\frac{4}{\sqrt{5} - \sqrt{3}} - \frac{4}{\sqrt{5} + \sqrt{3}}$

67. For what values of a and b does $\sqrt{a} + \sqrt{b} = \sqrt{a + b}$?

68. In the expression $\sqrt[n]{x^m}$, m and n are positive integers and x is a real number. The expression can be simplified.
   a. If $x > 0$, what are the possible values for m and n?
   b. If $x < 0$, what are the possible values for m and n?
   c. If $x < 0$, and an absolute value symbol is needed in the simplified expression, what are the possible values of m and n?

Standardized Test Prep

GRIDDED RESPONSE

SAT/ACT

69. What is the value of the expression $(5 - 2\sqrt{3})(5 + 2\sqrt{3})$?

70. What is the value of z in the solution of the system of equations below?

$$\begin{cases} 2x - 3y + z = 6 \\ -x + y - 2z = -5 \\ 3x - y - 3z = -7 \end{cases}$$

71. What is the y-value of the y-intercept of the line $5x - 7y = -15$?

72. What is the slope of a line perpendicular to the line $2x + 5y = 10$?

73. What is the value of p for which the equation $x^2 - 12x + 4p = 0$ has exactly one real root?

Mixed Review

Simplify each expression. Rationalize all denominators.

See Lesson 6-2.

74. $\sqrt[3]{3} \cdot \sqrt[3]{18}$

75. $\sqrt[3]{\frac{4}{0.5x}}$

76. $\frac{\sqrt{32}}{\sqrt{2}}$

77. $\frac{\sqrt{216}}{\sqrt{6}}$

78. $\sqrt[3]{2x^2} \cdot \sqrt[3]{4x}$

79. $\sqrt{7x} \cdot \sqrt{14x^3}$

80. $\sqrt{3x} \cdot \sqrt{5x}$

81. $\sqrt{9x^2} \cdot \sqrt{25x^2}$

Find the real and imaginary solutions of each equation.

See Lesson 5-3.

82. $2x^3 - 16 = 0$

83. $x^3 + 1000 = 0$

84. $125x^3 - 1 = 0$

85. $x^4 - 14x^2 + 49 = 0$

86. $25x^4 - 40x^2 + 16 = 0$

87. $81x^4 - 1 = 0$

Get Ready! To prepare for Lesson 6-4, do Exercises 88–91.

Simplify.

See p. 978.

88. $(x^2)^3$

89. $(pq)^5$

90. $(2^4)(2^5)$

91. $(3^{-2})(3^5)$

Rational Exponents

Content Standards

Reviews N.RN.2 Rewrite expressions involving radicals and rational exponents using the properties of exponents.

Also reviews N.RN.1

Objective To simplify expressions with rational exponents

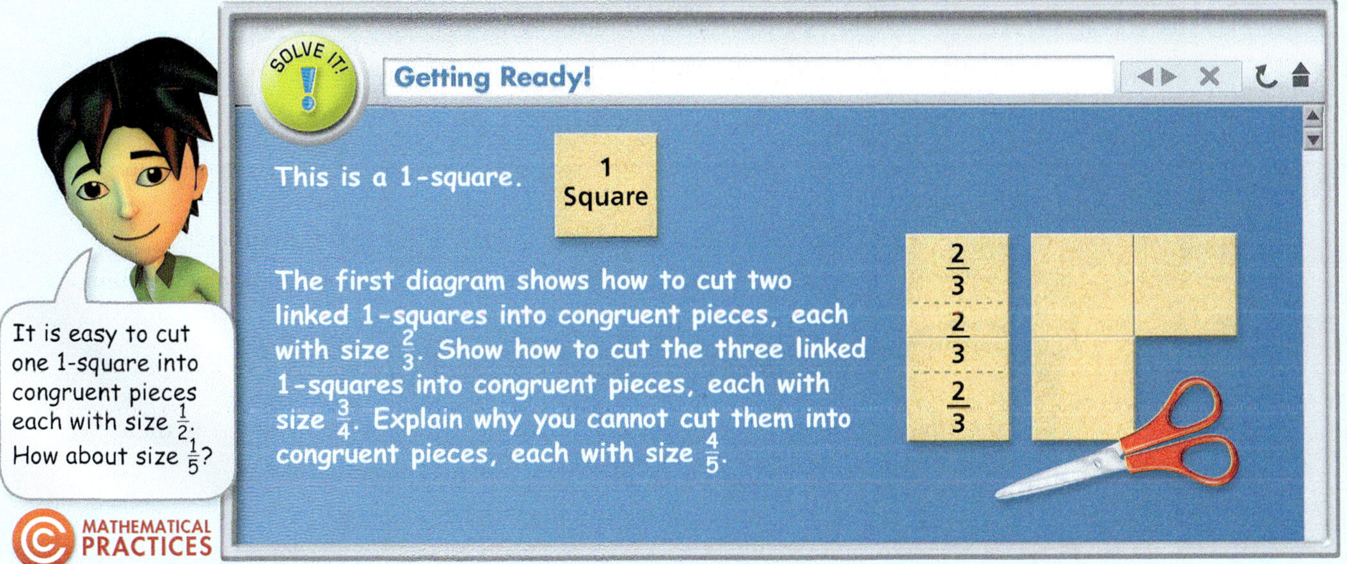

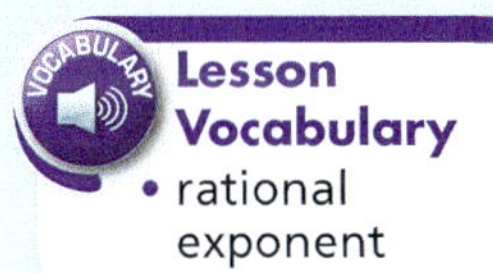

Lesson Vocabulary
- rational exponent

If $a^x = \sqrt[4]{a^3}$, then by definition, $a^x \cdot a^x \cdot a^x \cdot a^x = a^3$. By adding exponents, $a^{4x} = a^3$, then $4x = 3$. So x is $\frac{3}{4}$. This suggests an alternative notation for radical expressions in which, for example, $\sqrt[4]{a^3} = a^{\frac{3}{4}}$.

Essential Understanding You can write a radical expression in an equivalent form using a fractional (rational) exponent instead of a radical sign.

In general, $\sqrt[n]{x} = x^{\frac{1}{n}}$ for any positive integer n. Like the radical form, the exponent form indicates the principal root.

$$\sqrt{36} = 36^{\frac{1}{2}} \qquad \sqrt[3]{64} = 64^{\frac{1}{3}} \qquad \sqrt[4]{16} = 16^{\frac{1}{4}}$$

Problem 1 Simplifying Expressions with Rational Exponents

Think

What does the denominator of the fractional exponent represent?
The denominator of the fraction is the index of the radical.

What is the simplified form of each expression?

A $216^{\frac{1}{3}}$

$$216^{\frac{1}{3}} = \sqrt[3]{216}$$
$$= \sqrt[3]{6^3}$$
$$= 6$$

Rewrite as radicals.

B $7^{\frac{1}{2}} \cdot 7^{\frac{1}{2}}$

$$7^{\frac{1}{2}} \cdot 7^{\frac{1}{2}} = \sqrt{7} \cdot \sqrt{7}$$
$$= \sqrt{7 \cdot 7}$$
$$= \sqrt{7^2}$$
$$= 7$$

You can also solve this problem by adding the exponents.
$7^{\frac{1}{2}} \cdot 7^{\frac{1}{2}} = 7^{\frac{1}{2}+\frac{1}{2}} = 7^1 = 7$

C $5^{\frac{1}{4}} \cdot 125^{\frac{1}{4}}$

$$5^{\frac{1}{4}} \cdot 125^{\frac{1}{4}} = \sqrt[4]{5} \cdot \sqrt[4]{125}$$ Rewrite as radicals.

$$= \sqrt[4]{5 \cdot 125}$$ Property for multiplying radical expressions

$$= \sqrt[4]{625}$$ Multiply.

$$= \sqrt[4]{5^4}$$ Rewrite the radicand.

$$= 5$$ Simplify.

Got It? 1. What is the simplified form of each expression?

a. $64^{\frac{1}{2}}$ **b.** $11^{\frac{1}{2}} \cdot 11^{\frac{1}{2}}$ **c.** $3^{\frac{1}{2}} \cdot 12^{\frac{1}{2}}$

If $\sqrt[n]{x} = x^{\frac{1}{n}}$, it follows from the Laws of Exponents that for all real numbers $\sqrt[n]{x^m} = (x^m)^{\frac{1}{n}} = (x^{\frac{1}{n}})^m = \left(\sqrt[n]{x}\right)^m$. This leads to the definition of a rational exponent.

take note

Key Concept Rational Exponent

If the nth root of a is a real number, m is an integer, and $\frac{m}{n}$ is in lowest terms, then

$a^{\frac{1}{n}} = \sqrt[n]{a}$ and $a^{\frac{m}{n}} = \sqrt[n]{a^m} = (\sqrt[n]{a})^m$. If m is negative, $a \neq 0$.

Problem 2 Converting Between Exponential and Radical Forms

Think

Does the fraction $\frac{3}{7}$ first need to be simplified?

No. The fraction is already in lowest terms.

A **What are $x^{\frac{3}{7}}$ and $y^{-3.5}$ in radical form?**

$$x^{\frac{3}{7}} = \sqrt[7]{x^3} \text{ or } (\sqrt[7]{x})^3$$

$$y^{-3.5} = y^{-\frac{7}{2}}$$

$$= \frac{1}{y^{\frac{7}{2}}}$$

$$= \frac{1}{\sqrt{y^7}} = \frac{1}{\sqrt{y^6 y}} = \frac{1}{y^3\sqrt{y}} \text{ or } \frac{\sqrt{y}}{y^4}$$

B **What are $\sqrt{a^5}$ and $(\sqrt[5]{b})^3$ in exponential form?**

$$\sqrt{a^5} = (a^5)^{\frac{1}{2}} = a^{\frac{5}{2}}$$

$$(\sqrt[5]{b})^3 = (b^{\frac{1}{5}})^3 = b^{\frac{3}{5}}$$

Got It? 2. a. What are the expressions $w^{-\frac{5}{8}}$ and $w^{0.2}$ in radical form?

b. What are the expressions $\sqrt[4]{x^3}$ and $(\sqrt[5]{y})^4$ in exponential form?

c. **Reasoning** Refer to the definition of rational exponent. Explain the need for the restriction that $a \neq 0$ if m is negative.

Problem 3 Using Rational Exponents STEM

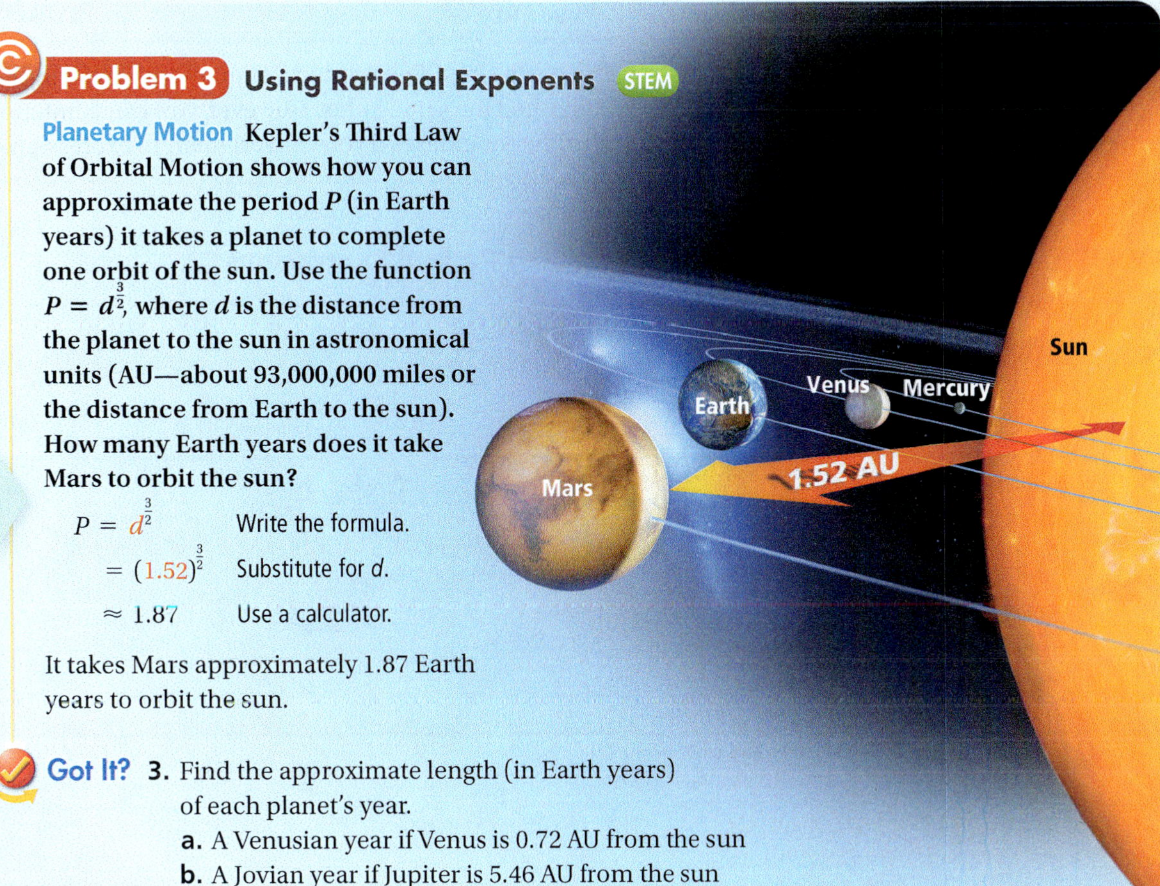

Planetary Motion Kepler's Third Law of Orbital Motion shows how you can approximate the period P (in Earth years) it takes a planet to complete one orbit of the sun. Use the function $P = d^{\frac{3}{2}}$, where d is the distance from the planet to the sun in astronomical units (AU—about 93,000,000 miles or the distance from Earth to the sun). How many Earth years does it take Mars to orbit the sun?

Plan

How can you find a $\frac{3}{2}$ power on a calculator?

You can use ^ (3 ÷ 2). You can also cube the number and then take the square root, or take the square root then cube.

$P = d^{\frac{3}{2}}$	Write the formula.
$= (1.52)^{\frac{3}{2}}$	Substitute for d.
≈ 1.87	Use a calculator.

It takes Mars approximately 1.87 Earth years to orbit the sun.

Got It? **3.** Find the approximate length (in Earth years) of each planet's year.

a. A Venusian year if Venus is 0.72 AU from the sun

b. A Jovian year if Jupiter is 5.46 AU from the sun

All the properties of integer exponents apply to rational exponents.

take note — Properties Properties of Rational Exponents

Let m and n represent rational numbers. Assume that no denominator equals 0.

Property	Example	Property	Example
$a^m \cdot a^n = a^{m+n}$	$8^{\frac{1}{3}} \cdot 8^{\frac{2}{3}} = 8^{\frac{1}{3}+\frac{2}{3}} = 8^1 = 8$	$a^{-m} = \frac{1}{a^m}$	$9^{-\frac{1}{2}} = \frac{1}{9^{\frac{1}{2}}} = \frac{1}{3}$
$(a^m)^n = a^{mn}$	$(5^{\frac{1}{2}})^4 = 5^{\frac{1}{2}\cdot 4} = 5^2 = 25$	$\frac{a^m}{a^n} = a^{m-n}$	$\frac{7^{\frac{3}{2}}}{7^{\frac{1}{2}}} = 7^{\frac{3}{2}-\frac{1}{2}} = 7^1 = 7$
$(ab)^m = a^m b^m$	$(4 \cdot 5)^{\frac{1}{2}} = 4^{\frac{1}{2}} \cdot 5^{\frac{1}{2}} = 2 \cdot 5^{\frac{1}{2}}$	$\left(\frac{a}{b}\right)^m = \frac{a^m}{b^m}$	$\left(\frac{5}{27}\right)^{\frac{1}{3}} = \frac{5^{\frac{1}{3}}}{27^{\frac{1}{3}}} = \frac{5^{\frac{1}{3}}}{3}$

Recall from Lesson 6-2 that you simplified products or quotients involving radical expressions only when they had the same index. However, you can combine radical expressions with different indexes if you convert them to expressions with rational exponents.

Problem 4 Combining Radical Expressions

What is $\frac{\sqrt[4]{x^3}}{\sqrt[8]{x^2}}$ in simplest form?

Think

The radicands are different, but both are powers of the same variable. Write the expressions using exponents.

Use the division property for exponents. Subtract the exponents.

Simplify, and write in either exponential or radical form.

Write

$$\frac{\sqrt[4]{x^3}}{\sqrt[8]{x^2}} = \frac{x^{\frac{3}{4}}}{x^{\frac{2}{8}}}$$

$$= x^{\frac{3}{4}-\frac{2}{8}}$$

$$= x^{\frac{3}{4}-\frac{1}{4}}$$

$$= x^{\frac{1}{2}} = \sqrt{x}$$

Got It? 4. What is each product or quotient in simplest form?

a. $\sqrt{3}(\sqrt[4]{3})$ **b.** $\frac{\sqrt{x^3}}{\sqrt[3]{x^2}}$ **c.** $\sqrt{7}(\sqrt[3]{7})$

You can simplify a number with a rational exponent by using the properties of exponents or by converting the expression to a radical expression.

Problem 5 Simplifying Numbers With Rational Exponents

What is each number in simplest form?

Plan

What is the first step? Rewrite the decimal exponent as a fraction in lowest terms.

A $16^{-2.5}$

Method 1

$$16^{-2.5} = 16^{-\frac{5}{2}}$$

$$= (2^4)^{-\frac{5}{2}}$$

$$= 2^{4 \cdot -\frac{5}{2}}$$

$$= 2^{-10}$$

$$= \frac{1}{2^{10}} = \frac{1}{1024}$$

Method 2

$$16^{-2.5} = 16^{-\frac{5}{2}}$$

$$= \frac{1}{16^{\frac{5}{2}}}$$

$$= \frac{1}{(\sqrt{16})^5}$$

$$= \frac{1}{4^5}$$

$$= \frac{1}{1024}$$

Think

Does it matter that the base is negative?
No; because the denominator of the exponent is odd, there will be a real root.

B $(-32)^{\frac{4}{5}}$

Method 1

$$\begin{aligned}(-32)^{\frac{4}{5}} &= ((-2)^5)^{\frac{4}{5}}\\ &= (-2)^{5\cdot\frac{4}{5}}\\ &= (-2)^4\\ &= 16\end{aligned}$$

Method 2

$$\begin{aligned}(-32)^{\frac{4}{5}} &= \left(\sqrt[5]{-32}\right)^4\\ &= \left(\sqrt[5]{(-2)^5}\right)^4\\ &= (-2)^4\\ &= 16\end{aligned}$$

Got It? **5.** What is each number in simplest form?

a. $32^{-\frac{3}{5}}$ **b.** $16^{\frac{3}{4}}$ **c.** $9^{-3.5}$

To write an expression with rational exponents in simplest form, write every exponent as a positive number.

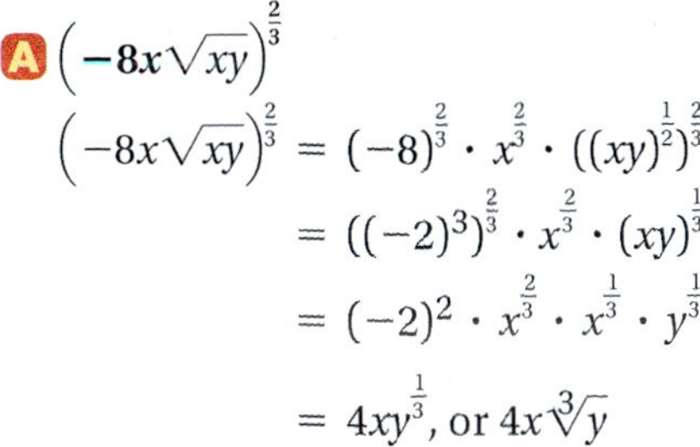

Problem 6 Writing Expressions in Simplest Form

What is each expression in simplest form?

Plan

What is the first step in simplifying a radical expression using the properties of exponents?
Rewrite the radicals using rational exponents.

A $\left(-8x\sqrt{xy}\right)^{\frac{2}{3}}$

$$\begin{aligned}\left(-8x\sqrt{xy}\right)^{\frac{2}{3}} &= (-8)^{\frac{2}{3}}\cdot x^{\frac{2}{3}}\cdot\left((xy)^{\frac{1}{2}}\right)^{\frac{2}{3}}\\ &= ((-2)^3)^{\frac{2}{3}}\cdot x^{\frac{2}{3}}\cdot(xy)^{\frac{1}{3}}\\ &= (-2)^2\cdot x^{\frac{2}{3}}\cdot x^{\frac{1}{3}}\cdot y^{\frac{1}{3}}\\ &= 4xy^{\frac{1}{3}},\text{ or } 4x\sqrt[3]{y}\end{aligned}$$

B $\left(16y^{-8}\right)^{-\frac{3}{4}}$

$$\begin{aligned}\left(16y^{-8}\right)^{-\frac{3}{4}} &= 16^{-\frac{3}{4}}\cdot y^{-8\cdot -\frac{3}{4}}\\ &= (2^4)^{-\frac{3}{4}}\cdot y^6\\ &= 2^{-3}y^6\\ &= \frac{y^6}{8}\end{aligned}$$

Got It? **6.** What is each expression in simplest form?

a. $(8x^{15})^{-\frac{1}{3}}$ **b.** $\left(9x\sqrt[4]{y}\right)^{\frac{3}{2}}$

Lesson Check

Do you know HOW?

Simplify each expression.

1. $125^{\frac{1}{3}}$
2. $5^{\frac{1}{2}}\cdot 5^{\frac{1}{2}}$
3. $25^{-\frac{3}{2}}$
4. $4^{-3.5}$
5. $\sqrt{11}(\sqrt[4]{11})$
6. $\dfrac{\sqrt[3]{x}}{\sqrt[6]{x^5}}$

Do you UNDERSTAND?

7. **Open-Ended** Find a nonzero number q such that $q(1-2^{\frac{1}{2}})$ is a rational number. Explain.
8. **Error Analysis** Explain why this simplification is incorrect.

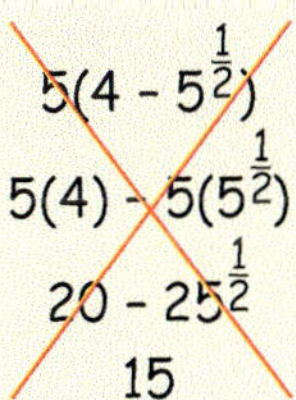

9. **Reasoning** Explain why $(-64)^{\frac{1}{3}} = -64^{\frac{1}{3}}$ but $(-64)^{\frac{1}{2}} \neq -64^{\frac{1}{2}}$.

Practice and Problem-Solving Exercises

Simplify each expression.

See Problem 1.

10. $36^{\frac{1}{2}}$ 11. $27^{\frac{1}{3}}$ 12. $49^{\frac{1}{2}}$

13. $10^{\frac{1}{2}} \cdot 10^{\frac{1}{2}}$ 14. $(-3)^{\frac{1}{3}} \cdot (-3)^{\frac{1}{3}} \cdot (-3)^{\frac{1}{3}}$ 15. $7^{\frac{1}{2}} \cdot 21^{\frac{1}{2}}$

16. $2^{\frac{1}{2}} \cdot 32^{\frac{1}{2}}$ 17. $3^{\frac{1}{3}} \cdot 9^{\frac{1}{3}}$ 18. $3^{\frac{1}{4}} \cdot 27^{\frac{1}{4}}$

Write each expression in radical form.

See Problem 2.

19. $x^{\frac{1}{6}}$ 20. $x^{\frac{1}{5}}$ 21. $x^{\frac{2}{7}}$ 22. $y^{\frac{2}{5}}$

23. $y^{-\frac{9}{8}}$ 24. $t^{-\frac{3}{4}}$ 25. $x^{1.5}$ 26. $y^{1.2}$

Write each expression in exponential form.

27. $\sqrt{-10}$ 28. $\sqrt{7x^3}$ 29. $\sqrt{(7x)^3}$ 30. $(\sqrt{7x})^3$

31. $\sqrt[3]{a^2}$ 32. $(\sqrt[3]{a})^2$ 33. $\sqrt[4]{c^2}$ 34. $\sqrt[3]{(5xy)^6}$

Optimal Height The optimal height h of the letters of a message printed on pavement is given by the formula $h = \frac{0.00252d^{2.27}}{e}$. Here d is the distance of the driver from the letters and e is the height of the driver's eye above the pavement. All of the distances are in meters. Find h for the given values of d and e.

See Problem 3.

35. $d = 100$ m, $e = 1.2$ m 36. $d = 50$ m, $e = 1.2$ m

37. $d = 50$ m, $e = 2.3$ m 38. $d = 25$ m, $e = 2.3$ m

Find each product or quotient.

See Problem 4.

39. $(\sqrt[4]{6})(\sqrt[3]{6})$ 40. $\frac{\sqrt[9]{y^3}}{\sqrt[3]{y^9}}$ 41. $\sqrt{5} \cdot \sqrt[5]{5}$ 42. $\sqrt[7]{7} \cdot \sqrt[3]{7}$

43. $\frac{\sqrt[6]{4}}{\sqrt[3]{4}}$ 44. $\sqrt[4]{18} \cdot \sqrt{12}$ 45. $\frac{\sqrt{6}}{\sqrt[3]{36}}$ 46. $\frac{\sqrt{x^4y}}{\sqrt[4]{x^2y^8}}$

Simplify each number.

See Problem 5.

47. $8^{\frac{2}{3}}$ 48. $64^{\frac{2}{3}} 64^{\frac{2}{3}}$ 49. $(-8)^{\frac{2}{3}}$ 50. $(-32)^{\frac{6}{5}}$

51. $(32)^{-\frac{4}{5}}$ 52. $4^{1.5}$ 53. $16^{1.5}$ 54. $10{,}000^{0.75}$

Write each expression in simplest form.

See Problem 6.

55. $\left(x^{\frac{2}{3}}\right)^{-3}$ 56. $\left(x^{-\frac{4}{7}}\right)^{7}$ 57. $\left(3x^{\frac{2}{3}}\right)^{-1}$ 58. $5\left(x^{\frac{2}{3}}\right)^{-1}$

59. $(-27x^{-9})^{\frac{1}{3}}$ 60. $(-32y^{15})^{\frac{1}{5}}$ 61. $\left(x^{\frac{1}{2}}y^{-\frac{2}{3}}\right)^{-6}$ 62. $\left(x^{\frac{2}{3}}y^{-\frac{1}{6}}\right)^{-12}$

63. $\left(\frac{x^3}{x^{-1}}\right)^{-\frac{1}{4}}$ 64. $\left(\frac{x^2}{x^{-11}}\right)^{\frac{1}{3}}$ 65. $\left(\frac{x^{\frac{1}{4}}}{y^{-\frac{3}{4}}}\right)^{12}$ 66. $\left(\frac{x^{-\frac{2}{3}}}{y^{-\frac{1}{3}}}\right)^{15}$

B Apply

67. Think About a Plan The ratio R of radioactive carbon to nonradioactive carbon left in a sample of an organism that died T years ago can be approximated by the formula $R = A(2.7)^{-\frac{T}{8033}}$. Here A is the ratio of radioactive carbon to nonradioactive carbon in the living organism. What percent of A is left after 2000 years? After 4000 years? After 8000 years?

- What are the known and unknown values?
- How can you use the properties of exponents to solve this problem?

68. The expression $0.036m^{\frac{3}{4}}$ is used in the study of fluids. Which best represents the value of the expression for $m = 46 \times 10^4$?

(A) 636 (B) 1460 (C) 1660 (D) 16,600

Simplify each number.

69. $(-343)^{\frac{1}{3}}$

70. $(-243)^{\frac{1}{5}}$

71. $32^{1.2}$

72. $243^{1.2}$

73. $64^{3.5}$

74. $100^{4.5}$

75. $-(-27)^{-\frac{4}{3}}$

76. $\dfrac{1000^{\frac{4}{3}}}{100^{\frac{3}{2}}}$

77. $25^{\frac{3}{2}}$

STEM 78. Science A desktop world globe has a volume of about 1386 cubic inches. The radius of Earth is approximately equal to the radius of the globe raised to the 10th power. Find the radius of Earth. (*Hint:* Use the formula $V = \frac{4}{3}\pi r^3$ for the volume of a sphere.)

Simplify each expression.

79. $x^{\frac{2}{7}} \cdot x^{\frac{3}{14}}$

80. $y^{\frac{1}{2}} \cdot y^{\frac{3}{10}}$

81. $x^{\frac{3}{5}} \div x^{\frac{3}{10}}$

82. $y^{\frac{5}{7}} \div y^{\frac{3}{14}}$

83. $\dfrac{x^{\frac{2}{3}} y^{-\frac{1}{4}}}{x^{\frac{1}{2}} y^{-\frac{1}{2}}}$

84. $\dfrac{x^{\frac{1}{2}} y^{-\frac{1}{3}}}{x^{\frac{3}{4}} y^{\frac{1}{2}}}$

85. $\left(\dfrac{16x^{14}}{81y^{18}}\right)^{\frac{1}{2}}$

86. $\left(\dfrac{81y^{10}}{16x^{12}}\right)^{\frac{1}{4}}$

87. $\left(\dfrac{8x^{6}}{27y^{9}}\right)^{\frac{1}{3}}$

88. Open-Ended Find three nonzero numbers a such that $a\left(4 + 5^{\frac{1}{2}}\right)$ is a rational number. Can a itself be a rational number? Explain.

89. a. Reasoning Show that $\sqrt[4]{x^2} = \sqrt{x}$ by using the definition of fourth root.

b. Reasoning Show that $\sqrt[4]{x^2} = \sqrt{x}$ by rewriting $\sqrt[4]{x^2}$ in exponential form.

90. Simplify $4^{\frac{1}{2}} \cdot 4^{\frac{1}{2}}$ using the following methods. Show all your work.

a. Use the properties of exponents.

b. Simplify each term in the product, then multiply.

c. Convert to radical form, then simplify.

You can define the rules for irrational exponents so that they have the same properties as rational exponents. Use those properties to simplify each expression.

91. $\left(7^{\sqrt{2}}\right)^{\sqrt{2}}$ **92.** $\frac{3^{3+\sqrt{5}}}{3^{1+\sqrt{5}}}$ **93.** $\frac{x^{4\pi}}{x^{2\pi}}$

94. $5^{2\sqrt{3}} \cdot 25^{-\sqrt{3}}$ **95.** $9^{\frac{1}{\sqrt{2}}}$ **96.** $\left(3^{2+\sqrt{2}}\right)^{2-\sqrt{2}}$

97. Weather Using data for the effect of temperature and wind on an exposed face, the National Weather Service uses the following formula to determine wind chill.

$$\text{Wind Chill Index} = 35.74 + 0.6215T - 35.75V^{0.16} + 0.4275TV^{0.16}$$

T is the temperature in degrees Fahrenheit and V is the velocity of the wind in miles per hour. Frostbite occurs in about 15 minutes when the wind chill index is -20. Find the wind velocity that produces a wind chill index of -20 when the temperature is 5°F.

Standardized Test Prep

GRIDDED RESPONSE

98. What is the simplified value of $\left(\frac{1}{64}\right)^{-\frac{1}{6}}$?

99. What positive value of b makes $9x^2 - bx + 4$ a perfect square trinomial?

100. How many real roots does the cubic polynomial equation $x^3 - 7x^2 + 13x - 4 = 0$ have?

101. What is the y-value of the y-intercept of the graph of $f(x) = 4|x - 2| - 5$?

Mixed Review

Simplify. See Lesson 6-3.

102. $6\sqrt[3]{3} - 2\sqrt[3]{3}$ **103.** $3\sqrt{18} + 2\sqrt{72}$ **104.** $(\sqrt{5} - 1)(\sqrt{5} + 4)$

105. $(1 - 2\sqrt{2})(1 + 2\sqrt{2})$ **106.** $2\sqrt{12} - 4\sqrt{27}$ **107.** $\sqrt[4]{162} + 3\sqrt[4]{32}$

Factor each expression completely. See Lesson 4-4.

108. $4x^3 - 8x^2 + 16x$ **109.** $x^2 + 4x + 4$ **110.** $x^2 - 18x + 81$

111. $16a^2 - 9b^2$ **112.** $25x^2 - 40xy + 16y^2$ **113.** $9x^2 + 48x + 64$

Get Ready! **To prepare for Lesson 6-5, do Exercises 114–119.**

Solve by factoring. Check your answers.

114. $x^2 = -x + 6$ **115.** $x^2 = 5x + 14$ **116.** $2x^2 + x = 3$

117. $3x^2 - 2 = 5x$ **118.** $4x^2 = -8x + 5$ **119.** $6x^2 = 5x + 6$

6

Mid-Chapter Quiz

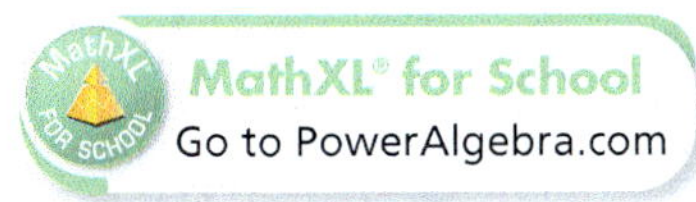

Do you know HOW?

Find all the real square roots of each number.

1. 100
2. 0.49

Simplify each radical expression. Use absolute value symbols when needed.

3. $\sqrt{36x^2}$
4. $\sqrt[3]{0.008y^3x^6}$

Simplify.

5. $\sqrt{50x^4y^8}$
6. $\sqrt[4]{32m^7n^9}$

Multiply and simplify.

7. $6\sqrt{4x^2} \cdot 2\sqrt{9x^2y^2}$
8. $\sqrt[3]{9} \cdot \sqrt[3]{9}$
9. $\sqrt[4]{16x^8} \cdot \sqrt[4]{x^{14}}$

Divide and simplify.

10. $\dfrac{\sqrt{36x^4}}{\sqrt{9x^6}}$
11. $\dfrac{\sqrt[3]{64x^9y^3}}{\sqrt[3]{8x^3}}$

Simplify. Rationalize all denominators.

12. $10\sqrt[3]{81} - 8\sqrt[3]{24}$
13. $\dfrac{4 + \sqrt{12}}{4 - \sqrt{12}}$
14. $\sqrt{40} - 3\sqrt{27} + 2\sqrt{75}$
15. $(3 + \sqrt{63})(1 + \sqrt{7})$
16. $\dfrac{\sqrt{x}}{\sqrt{6y^3}}$

Write each expression in exponential form.

17. $-\sqrt{17}$
18. $\sqrt[3]{y^8}$

Write each expression in radical form.

19. $m^{\frac{3}{7}}$
20. $y^{-\frac{4}{3}}$

Simplify each expression.

21. $(-27)^{\frac{2}{3}}$
22. $(16)^{\frac{3}{4}}$

Write each expression in simplest form.

23. $7\sqrt[3]{2x} - 3\sqrt[3]{2x}$
24. $2\sqrt{32x^2} + 3\sqrt{72x^2}$
25. $\sqrt[3]{125x^6} - \sqrt[3]{27x^6}$
26. $\sqrt[4]{7} - \sqrt[3]{7}$
27. $(\sqrt{y} - \sqrt{3})(\sqrt{y} + 2\sqrt{3})$
28. $\left(16x^{\frac{1}{4}}y^{\frac{3}{4}}\right)^{-4}$
29. $\left(\dfrac{x^{\frac{1}{3}}}{y^{\frac{2}{3}}}\right)^9$
30. $\left(\dfrac{x^{-10}}{x^5}\right)^{\frac{2}{5}}$
31. The radius of a circle can be expressed as $r = \sqrt{\frac{A}{\pi}}$ inches where r is the radius and A is the area of the circle. If the area of a circle is 169π in.^{2}, what is its radius?

Do you UNDERSTAND?

32. What are the real roots of $\sqrt{-16}$? Explain.
33. **Error Analysis** Identify the error in this statement.
$$\frac{\sqrt[3]{x}}{\sqrt[3]{y}} \cdot \frac{\sqrt[3]{y}}{\sqrt[3]{y}} = \frac{\sqrt[3]{xy}}{y}$$
34. **Reasoning** If $0^{\frac{2}{3}} = 0$, why is $0^{-\frac{2}{3}}$ undefined?
35. Given that x and y are integers, explain why the product of $x + \sqrt{y}$ and its conjugate will always be an integer.
36. **Reasoning** Explain why $(-8)^{\frac{1}{2}} \neq -(8)^{\frac{1}{2}}$, but $(-27)^{\frac{1}{3}} = -(27)^{\frac{1}{3}}$.

6-5 Solving Square Root and Other Radical Equations

Content Standards

A.REI.2 Solve simple rational and radical equations in one variable, and . . . show how extraneous solutions may arise.

A.CED.4 Rearrange formulas to highlight a quantity of interest, using the same reasoning as in solving equations.

Objective To solve square root and other radical equations

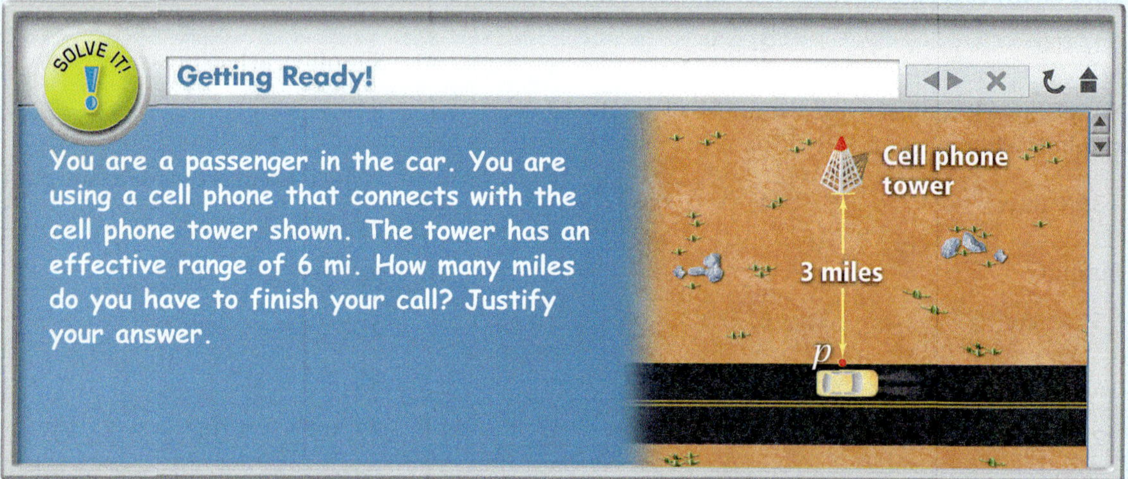

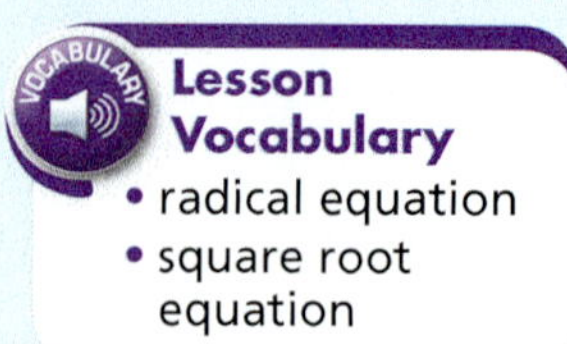

Lesson Vocabulary
- radical equation
- square root equation

A **radical equation** is an equation that has a variable in a radicand or a variable with a rational exponent. If the radical has index 2, the equation is a **square root equation**. In this lesson, assume that all radicals and expressions with rational exponents represent real numbers.

Essential Understanding Solving a square root equation may require that you square each side of the equation. This can introduce extraneous solutions.

To solve a radical equation, isolate the radical on one side of the equation. Then raise each side to the power suggested by the index.

Problem 1 Solving a Square Root Equation

Think

Do you need to introduce a ± sign here?
No, when you take the square root of each side of an equation you do, but here you are squaring both sides of the equation.

What is the solution of $3 + \sqrt{2x - 3} = 8$?

$3 + \sqrt{2x - 3} = 8$

$\sqrt{2x - 3} = 5$ Isolate the radical expression.

$(\sqrt{2x - 3})^2 = 5^2$ Square each side.

$2x - 3 = 25$

$2x = 28$ Add 3 to each side.

$x = 14$ Divide each side by 2.

Check

$3 + \sqrt{2x - 3} = 8$ Write the original equation.

$3 + \sqrt{2(14) - 3} \stackrel{?}{=} 8$ Substitute 14 for *x*.

$3 + \sqrt{25} \stackrel{?}{=} 8$ Simplify.

$3 + 5 \stackrel{?}{=} 8$

$8 = 8$ ✔

Got It? **1.** What is the solution of $\sqrt{4x + 1} - 5 = 0$?

To solve equations of the form $x^{\frac{m}{n}} = k$, raise each side of the equation to the power $\frac{n}{m}$, the reciprocal of $\frac{m}{n}$. If either m or n is even, then $\left(x^{\frac{m}{n}}\right)^{\frac{n}{m}} = |x|$.

Problem 2 Solving Other Radical Equations

A **What is the solution of $3(x + 1)^{\frac{2}{3}} = 12$?**

Know	Need	Plan
• The equation • The power of the exponential expression	Solution of the equation	• Isolate the exponential expression. • Use the inverse of the power to simplify and solve the equation.

Think

How can you get rid of the rational exponent?

Raise each side to the reciprocal power.

$3(x + 1)^{\frac{2}{3}} = 12$

$(x + 1)^{\frac{2}{3}} = 4$ Divide each side by 3.

$\left((x + 1)^{\frac{2}{3}}\right)^{\frac{3}{2}} = 4^{\frac{3}{2}}$ Raise each side to the $\frac{3}{2}$ power.

$(x + 1)^{\frac{6}{6}} = 4^{\frac{3}{2}}$

$|x + 1| = 8$ Since the numerator of $\frac{2}{3}$ is even, $(x^{\frac{2}{3}})^{\frac{3}{2}} = |x|$

$x + 1 = \pm 8$

$x = 7$ or $x = -9$

The solutions are 7 and −9.

Check

$3(x + 1)^{\frac{2}{3}} = 12$

$3(7 + 1)^{\frac{2}{3}} \stackrel{?}{=} 12$

$3(2^3)^{\frac{2}{3}} \stackrel{?}{=} 12$

$3(2)^2 \stackrel{?}{=} 12$

$12 = 12$ ✔

$3(x + 1)^{\frac{2}{3}} = 12$

$3(-9 + 1)^{\frac{2}{3}} \stackrel{?}{=} 12$

$3((-2)^3)^{\frac{2}{3}} \stackrel{?}{=} 12$

$3(-2)^2 \stackrel{?}{=} 12$

$12 = 12$ ✔

B **What is the solution of $3\sqrt[5]{(x+1)^3} + 1 = 25$?**

$$3\sqrt[5]{(x+1)^3} + 1 = 25$$

$$3(x+1)^{\frac{3}{5}} + 1 = 25 \quad \text{Rewrite the radical using a rational exponent.}$$

$$3(x+1)^{\frac{3}{5}} = 24 \quad \text{Subtract 1 from each side.}$$

$$(x+1)^{\frac{3}{5}} = 8 \quad \text{Divide each side by 3.}$$

$$\left((x+1)^{\frac{3}{5}}\right)^{\frac{5}{3}} = 8^{\frac{5}{3}} \quad \text{Raise each side to the } \tfrac{5}{3} \text{ power.}$$

$$x + 1 = 32 \quad \text{Simplify.}$$

$$x = 31 \quad \text{Subtract 1 from each side.}$$

The solution is 31.

Think

Why is isolating the variable important?
If you raise each side of $3\sqrt[5]{(x+1)^3} + 1 = 25$ to the $\frac{5}{3}$ power you will end up with a more complicated equation, not a simpler one.

Got It? **2.** What are the solution(s) of $2(x+3)^{\frac{2}{3}} = 8$?

Problem 3 Using Radical Equations STEM

Earth Science **For Meteor Crater in Arizona, the formula $d = 2\sqrt[3]{\frac{V}{0.3}}$ relates the diameter d of the rim (in meters) to the volume V (in cubic meters). What is the volume of Meteor Crater? (All values are approximate.)**

1.2 km

$$d = 2\sqrt[3]{\frac{V}{0.3}}$$

$$\frac{d}{2} = \sqrt[3]{\frac{V}{0.3}} \quad \text{Solve for } V\text{. First divide each side by 2.}$$

$$\left(\frac{d}{2}\right)^3 = \frac{V}{0.3} \quad \text{Cube each side.}$$

$$0.3\left(\frac{d}{2}\right)^3 = V \quad \text{Multiply each side by 0.3.}$$

$$0.3\left(\frac{1200}{2}\right)^3 = V \quad \text{Substitute 1200 for } d\text{.}$$

$$64{,}800{,}000 = V \quad \text{Simplify.}$$

The volume of Meteor Crater is about 64,800,000 m^3.

Think

What is the diameter in meters?
1.2 km = 1.2 × 1000 m.

Got It? **3.** Suppose the diameter of a similarly shaped crater is 1 km. What is the volume of the crater? Use the formula given in Problem 3.

When you raise each side of an equation to a power, it is possible to introduce extraneous solutions. Therefore, it becomes very important that you check all solutions in the original equation. A correct solution will give a true statement. An extraneous solution will give a false statement.

Problem 4 Checking for Extraneous Solutions

What is the solution of $\sqrt{x+7} - 5 = x$? Check your results.

Think

How do you square a binomial?
Use the formula, $(a+b)^2 = a^2 + 2ab + b^2$.

$\sqrt{x+7} - 5 = x$

$\sqrt{x+7} = x + 5$ Isolate the radical.

$(\sqrt{x+7})^2 = (x+5)^2$ Square each side.

$x + 7 = x^2 + 10x + 25$ Simplify.

$0 = x^2 + 9x + 18$ Combine like terms.

$0 = (x+3)(x+6)$ Factor.

$x = -3$ or $x = -6$ Zero-Product Property

Check

$\sqrt{x+7} - 5 = x$

$\sqrt{-3+7} - 5 \stackrel{?}{=} -3$

$\sqrt{4} - 5 \stackrel{?}{=} -3$

$2 - 5 \stackrel{?}{=} -3$

$-3 = -3$ ✔

$\sqrt{x+7} - 5 = x$

$\sqrt{-6+7} - 5 \stackrel{?}{=} -6$

$\sqrt{1} - 5 \stackrel{?}{=} -6$

$1 - 5 \stackrel{?}{=} -6$

$-4 \neq -6$

false

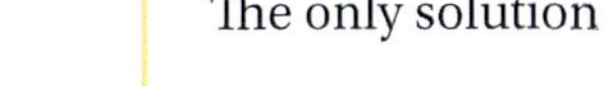

The only solution is -3.

Got It? 4. a. What is the solution of $\sqrt{5x-1} + 3 = x$? Check your results.

b. Reasoning When should you check for extraneous solutions? Explain.

In this lesson you studied algebraic methods of solving square root and radical equations. In Lesson 6-8 you will study the graphs of square root functions. These graphs can help you find solutions and identify extraneous solutions.

The calculator screen shows the graphs **Y1 = √(x + 7) − 5** and **Y2 = x.** From the graph, it is clear that -3 is a solution of $\sqrt{x+7} - 5 = x$, and -6 is not a solution.

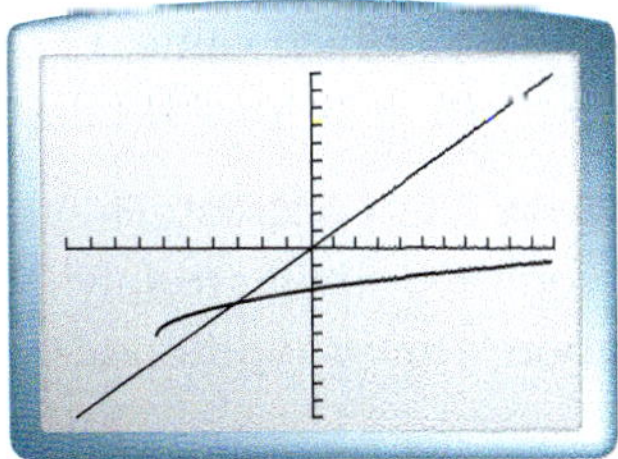

If an equation contains two radical expressions (or two terms with rational exponents), isolate one of the radicals (or one of the terms), then eliminate it (or its rational exponent). Isolate the more complicated radical expression first. In the resulting equation, simplify the expressions before you eliminate the second radical.

Problem 5 Solving an Equation With Two Radicals

Plan

Which radical expression should you isolate first? Isolate the more complicated radical first, $\sqrt{2x+1}$.

What is the solution of $\sqrt{2x+1} - \sqrt{x} = 1$?

$$\begin{aligned}
\sqrt{2x+1} - \sqrt{x} &= 1 \\
\sqrt{2x+1} &= \sqrt{x} + 1 && \text{Isolate the more complicated radical.} \\
(\sqrt{2x+1})^2 &= (\sqrt{x}+1)^2 && \text{Square each side.} \\
2x + 1 &= x + 2\sqrt{x} + 1 \\
x &= 2\sqrt{x} && \text{Isolate } 2\sqrt{x}. \\
x^2 &= (2\sqrt{x})^2 && \text{Square each side.} \\
x^2 &= 4x \\
x^2 - 4x &= 0 && \text{Subtract } 4x \text{ from each side.} \\
x(x-4) &= 0 && \text{Factor.} \\
x = 0 &\text{ or } x = 4 && \text{Zero-Product Property}
\end{aligned}$$

Check

$$\begin{aligned}
\sqrt{2x+1} - \sqrt{x} &= 1 \\
\sqrt{2(0)+1} - \sqrt{0} &\stackrel{?}{=} 1 \\
\sqrt{1} - 0 &\stackrel{?}{=} 1 \\
1 - 0 &\stackrel{?}{=} 1 \\
1 &= 1 \checkmark
\end{aligned}$$

$$\begin{aligned}
\sqrt{2x+1} - \sqrt{x} &= 1 \\
\sqrt{2(4)+1} - \sqrt{4} &\stackrel{?}{=} 1 \\
\sqrt{9} - \sqrt{4} &\stackrel{?}{=} 1 \\
3 - 2 &\stackrel{?}{=} 1 \\
1 &= 1 \checkmark
\end{aligned}$$

The solutions are 0 and 4.

Got It? 5. What is the solution of $\sqrt{5x+4} - \sqrt{x} = 4$?

Lesson Check

Do you know HOW?

Solve. Check for extraneous solutions.

1. $\sqrt{4x - 23} - 3 = 2$

2. $-\sqrt[3]{x} + 3 = 0$

3. $5\sqrt{x} + 7 = 8$

4. $3\sqrt{x} = 6$

5. $5 - 2\sqrt{x} = 3$

6. $\sqrt[3]{x} = 8$

Do you UNDERSTAND?

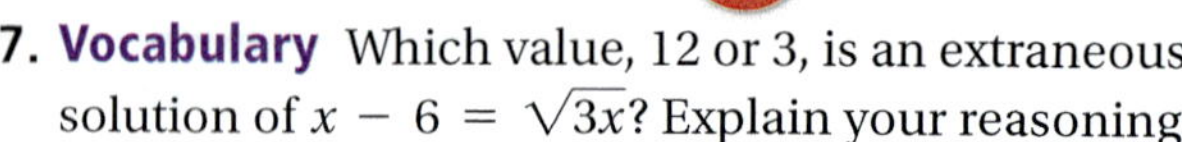

7. Vocabulary Which value, 12 or 3, is an extraneous solution of $x - 6 = \sqrt{3x}$? Explain your reasoning.

8. Compare and Contrast How is solving a square root equation similar to solving an absolute value equation? How is it different?

Practice and Problem-Solving Exercises

Solve. See Problem 1.

9. $3\sqrt{x} + 3 = 15$
10. $4\sqrt{x} - 1 = 3$
11. $\sqrt{x + 3} = 5$
12. $\sqrt{x + 1} = 4$
13. $\sqrt{2x - 1} = 3$
14. $\sqrt{x + 2} - 2 = 0$
15. $\sqrt{3x + 4} = 4$
16. $\sqrt{2x + 3} - 7 = 0$
17. $\sqrt{6 - 3x} - 2 = 0$

Solve. See Problem 2.

18. $(x + 5)^{\frac{2}{3}} = 4$
19. $(x + 2)^{\frac{2}{3}} = 9$
20. $3(x - 2)^{\frac{3}{4}} = 24$
21. $3(x + 3)^{\frac{3}{4}} = 81$
22. $(x + 1)^{\frac{3}{2}} - 2 = 25$
23. $3 + (4 - x)^{\frac{3}{2}} = 11$

See Problem 3.

24. **Volume** A spherical water tank holds 9000 ft^3 of water. What is the diameter of the tank? $\left(Hint: \frac{1}{6}d^3\pi = V\right)$

STEM 25. **Hydraulics** The formula $\frac{\pi d^2 v}{4} = Q$ models the diameter of a pipe where Q is the maximum flow of water in a pipe, and v is the velocity of the water. What is the diameter of a pipe that allows a maximum flow of 30 ft^3/min of water flowing at a velocity of 400 ft/min? Round your answer to the nearest inch.

Solve. Check for extraneous solutions. See Problem 4.

26. $\sqrt{3x + 7} = x - 1$
27. $(5 - x)^{\frac{1}{2}} = x + 1$
28. $\sqrt{-3x - 5} = x + 3$
29. $\sqrt{11x + 3} - 2x = 0$
30. $(5x - 4)^{\frac{1}{2}} - x = 0$
31. $\sqrt{3x + 13} - 5 = x$
32. $\sqrt{x + 7} + 5 = x$
33. $(x + 3)^{\frac{1}{2}} - 1 = x$
34. $\sqrt{x + 7} - x = 1$

Solve. Check for extraneous solutions. See Problem 5.

35. $\sqrt{3x} = \sqrt{x + 6}$
36. $(2x)^{\frac{1}{2}} = (x + 5)^{\frac{1}{2}}$
37. $(7x + 6)^{\frac{1}{2}} - (9 + 4x)^{\frac{1}{2}} = 0$
38. $\sqrt{3x + 2} - \sqrt{2x + 7} = 0$
39. $(x + 5)^{\frac{1}{2}} - (5 - 2x)^{\frac{1}{4}} = 0$
40. $(x - 2)^{\frac{1}{2}} - (28 - 2x)^{\frac{1}{4}} = 0$
41. $\sqrt{5 - x} - \sqrt{x} = 1$
42. $\sqrt{3x + 1} - \sqrt{x + 1} = 2$
43. $\sqrt{2x + 6} - \sqrt{x - 1} = 2$
44. $\sqrt{3 - x} + \sqrt{x + 2} = 3$

45. **Think About a Plan** A hexagonal tray of vegetables has an area of 450 cm^2. What is the length of each side of the hexagon?
    - What is the area of the triangle at the bottom in terms of the side length?
    - How can you use the diagram at the right to find the formula for the area of the hexagon? (*Hint:* Six triangles make one hexagon.)

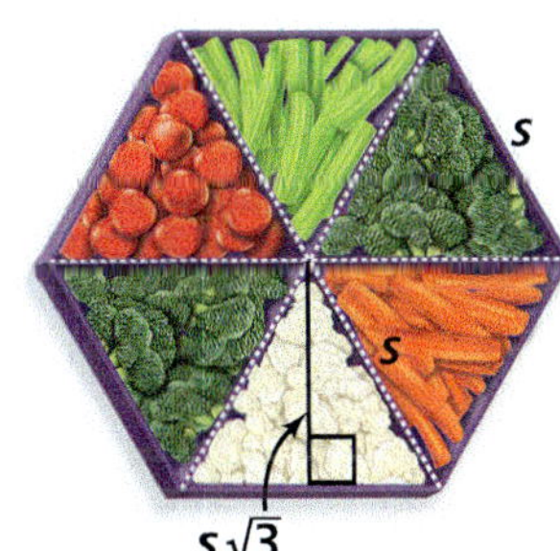

46. **Traffic Signs** A stop sign is a regular octagon, formed by cutting triangles off the corners of a square. If a stop sign measures 36 in. from top to bottom, what is the length of each side?

47. Mental Math What is the solution? $\sqrt{x + 11} = 4$

48. You can find the area A of a square whose side is s units with the formula $A = s^2$. What is the best estimate for the side of a square with an area of 32 m^2?

Ⓐ 4.2 m
Ⓑ 5.7 m
Ⓒ 8.0 m
Ⓓ 16 m

Solve. Check for extraneous solutions.

49. $3\sqrt{2x} - 3 = 9$

50. $2(2x)^{\frac{1}{3}} + 1 = 5$

51. $\sqrt{2x - 1} - 3 = 0$

52. $(2x + 3)^{\frac{1}{2}} - 7 = 0$

53. $\sqrt{x^2 + 3} = x + 1$

54. $(2x + 3)^{\frac{3}{4}} - 3 = 5$

55. $2(x - 1)^{\frac{4}{3}} + 4 = 36$

56. $x^{\frac{1}{2}} - (x - 5)^{\frac{1}{2}} = 2$

57. $\sqrt{x} = \sqrt{x - 8} + 2$

58. $(x - 3)^{\frac{2}{3}} = x - 7$

59. Error Analysis A student said that 4 and 1 are the solutions of the problem shown. Describe and correct the student's error.

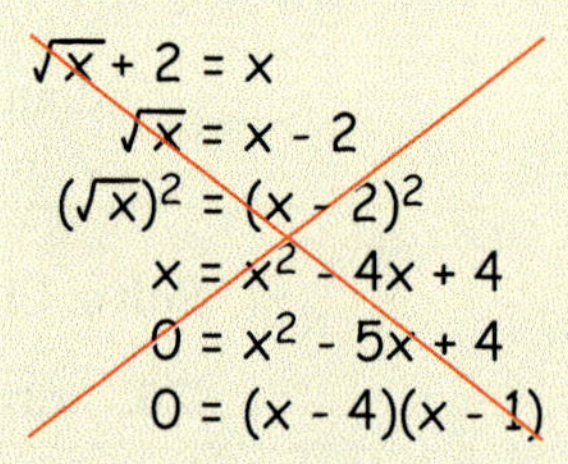

STEM **60. Physics** The velocity v of an object dropped from a tall building is given by the formula $v = \sqrt{64d}$, where d is the distance the object has dropped. Solve the formula for d.

61. Open-Ended Write an equation that has two radical expressions and no real roots.

62. Reasoning You have solved equations containing square roots by squaring each side. You were using the property that if $a = b$ then $a^2 = b^2$. Show that the following statements are *not* true for all real numbers.

a. If $a^2 = b^2$ then $a = b$.

b. If $a \leq b$ then $a^2 \leq b^2$.

63. A teacher asked students why it is necessary to check for extraneous roots when squaring both sides of the equation. Which of the following answers is the best? Is this answer complete? Explain.

Ⓐ Because the squared equation can have negative roots.
Ⓑ Because squaring is multiplication, and any multiplication is a potential source of extraneous roots.
Ⓒ Because when you square both sides of the equation $a = b$, you add to the solution set the roots of the equation $a = -b$.
Ⓓ Because any operation with an equation may result in extraneous roots.

Solve. Check for extraneous solutions.

64. $\sqrt{x + 1} + \sqrt{2x} = \sqrt{5x + 3}$

65. $\sqrt{x + \sqrt{2x}} = \sqrt{2x}$

66. $\sqrt{x + \sqrt{2x}} = 2$

67. $\sqrt{\sqrt{x + 25}} = \sqrt{x + 5}$

68. **Reasoning** Devise a plan to find the value of x.

$$x = \sqrt{2 + \sqrt{2 + \sqrt{2 + \cdots}}}$$

For each set of values, determine which is greater without using a calculator.

69. $\sqrt{6}$ or $\sqrt{2} + 1$

70. $\sqrt{3} + \sqrt{11}$ or 5

71. $\sqrt{10}$ or $\sqrt{2} + \sqrt{3}$

72. $\sqrt{19} + \sqrt{3}$ or $\sqrt{5} + \sqrt{13}$

Standardized Test Prep

SAT/ACT

73. What is the solution of $(x + 2)^{\frac{3}{4}} = 27$?

Ⓐ $x = 27$ Ⓑ $x = 79$ Ⓒ $x = 81$ Ⓓ $x = 83$

74. A problem on a test asked students to solve a fifth-degree polynomial equation with rational coefficients. Adam found the following roots: -11.5, $\sqrt{2}$, $\frac{2i + 6}{2}$, $-\sqrt{2}$ and $3 - i$. His teacher wrote that four of these roots are correct, and one is incorrect. Which root is incorrect?

Ⓕ -11.5 Ⓖ $\sqrt{2}$ Ⓗ $\frac{2i + 6}{2}$ Ⓘ $3 - i$

75. Which expression represents the solution of the equation $\frac{x}{y} = \frac{c}{a + b}$ solved for a?

Ⓐ $\frac{c}{b} - \frac{x}{y}$ Ⓑ $\frac{yc}{a + b}$ Ⓒ $\frac{yc}{x} + b$ Ⓓ $\frac{yc - xb}{x}$

Short Response

76. To rationalize the denominator of $\sqrt[4]{\frac{4}{25}}$, by what number would you multiply the numerator and denominator of the fraction?

Mixed Review

Simplify each number.  See Lesson 6-4.

77. $81^{\frac{1}{4}}$

78. $4^{\frac{1}{2}}$

79. $125 \cdot 125^{\frac{1}{3}}$

80. $32 \cdot 256^{\frac{1}{2}}$

81. $100^{-\frac{3}{2}}$

82. $64^{\frac{4}{3}}$

83. $25^{1.5}$

84. $6^{\frac{1}{2}} \cdot 12^{\frac{1}{2}}$

Solve each equation by factoring. Check your answers.  See Lesson 4-5.

85. $x^2 - 7x + 12 = 0$

86. $x^2 - 8x + 15 = 0$

87. $x^2 + 9x + 20 = 0$

88. $3x^2 + 8x + 4 = 0$

89. $9x^2 + 15x + 4 = 0$

90. $4x^2 + 11x + 6 = 0$

Get Ready! **To prepare for Lesson 6-6, do Exercises 91–96.**

Find the domain and range of each relation, and determine whether it is a function.  See Lesson 2-1.

91. $\{(0, -5), (2, -3), (4, -1)\}$

92. $\{(-1, 2), (0, 0), (1, 1)\}$

93. $\{(-2, -2), (0, 0), (1, 1)\}$

94. $\{(3, -1), (4, -1), (5, -1)\}$

95. $\{(0, 0), (1, 0), (2, 1), (2, 2)\}$

96. $\{(0, -2), (0, 0), (0, 2)\}$

Function Operations

Content Standards

F.BF.1.b Combine standard function types using arithmetic operations.

F.BF.1.c Compose functions.

Objectives To add, subtract, multiply, and divide functions
To find the composite of two functions

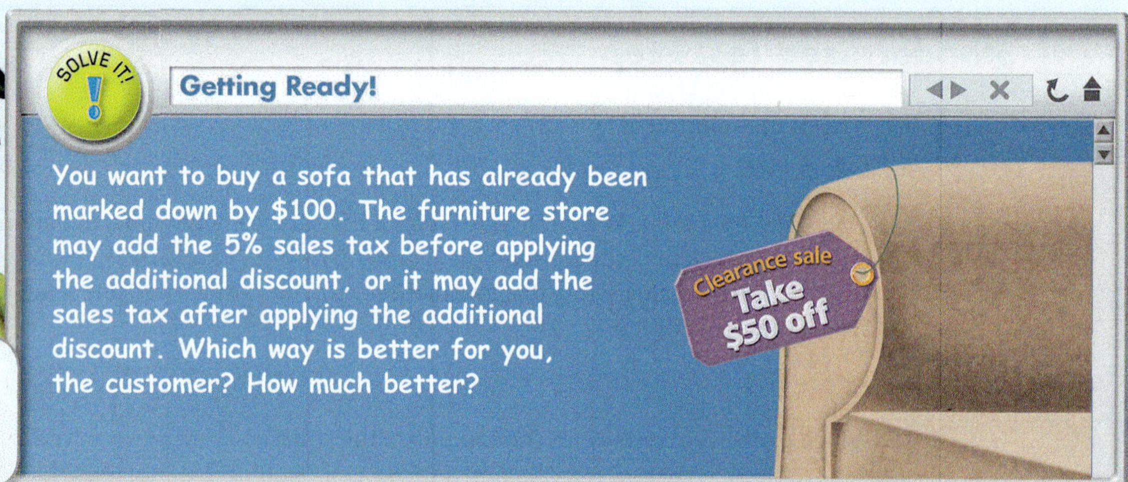

The final cost of the sofa in the Solve It involves two functions: one that gives an additional discount and one that multiplies to find the sales tax.

- composite function

Essential Understanding You can add, subtract, multiply, and divide functions based on how you perform these operations for real numbers. One difference, however, is that you must consider the domain of each function.

take note

Key Concepts Function Operations

Addition	$(f+g)(x) = f(x) + g(x)$
Subtraction	$(f-g)(x) = f(x) - g(x)$
Multiplication	$(f \cdot g)(x) = f(x) \cdot g(x)$
Division	$\left(\frac{f}{g}\right)(x) = \frac{f(x)}{g(x)}, g(x) \neq 0$

The domains of the sum, difference, product, and quotient functions consist of the x-values that are in the domains of *both* f and g. Also, the domain of the quotient function does not contain any x-value for which $g(x) = 0$.

Problem 1 Adding and Subtracting Functions

Let $f(x) = 4x + 7$ and $g(x) = \sqrt{x} + x$. What are $f + g$ and $f - g$? What are their domains?

$$(f + g)(x) = f(x) + g(x) = (4x + 7) + (\sqrt{x} + x) = 5x + \sqrt{x} + 7$$

$$(f - g)(x) = f(x) - g(x) = (4x + 7) - (\sqrt{x} + x) = 3x - \sqrt{x} + 7$$

Think

What determines the domain of g?

Because there is a square root of x, x must be ≥ 0.

The domain of f is the set of all real numbers. The domain of g is all $x \geq 0$. The domain of both $f + g$ and $f - g$ is the set of numbers common to the domains of both f and g, which is all $x \geq 0$.

Got It? **1.** Let $f(x) = 2x^2 + 8$ and $g(x) = x - 3$. What are $f + g$ and $f - g$? What are their domains?

Problem 2 Multiplying and Dividing Functions

Let $f(x) = x^2 - 9$ and $g(x) = x + 3$. What are $f \cdot g$ and $\frac{f}{g}$ and their domains?

$$(f \cdot g)(x) = f(x) \cdot g(x) = (x^2 - 9)(x + 3)$$

$$= x^3 + 3x^2 - 9x - 27$$

$$\left(\frac{f}{g}\right)(x) = \frac{f(x)}{g(x)} = \frac{x^2 - 9}{x + 3} = \frac{(x - 3)(x + 3)}{x + 3} = x - 3, x \neq -3$$

Think

Is the domain of $\frac{f}{g}$ the domain of $x - 3$?

No; The fraction can only be simplified and the function is only defined when $g(x) \neq -3$.

The domain of both f and g is the set of real numbers, so the domain of $f \cdot g$ is also the set of real numbers.

The domain of $\frac{f}{g}$ is the set of all real numbers except $x \neq -3$, because $g(-3) = 0$. The definition of $\frac{f}{g}$ requires that you consider the zero denominator in the *original* expression for $\frac{f(x)}{g(x)}$ despite the fact that the simplified form has the domain all real numbers.

Got It? **2.** Let $f(x) = 3x^2 - 11x - 4$ and $g(x) = 3x + 1$. What are $f \cdot g$ and $\frac{f}{g}$ and their domains?

The diagram shows what happens when you apply one function $g(x)$ after another function $f(x)$.

The output from the first function becomes the input for the second function. When you combine two functions as in the diagram, you form a **composite function**.

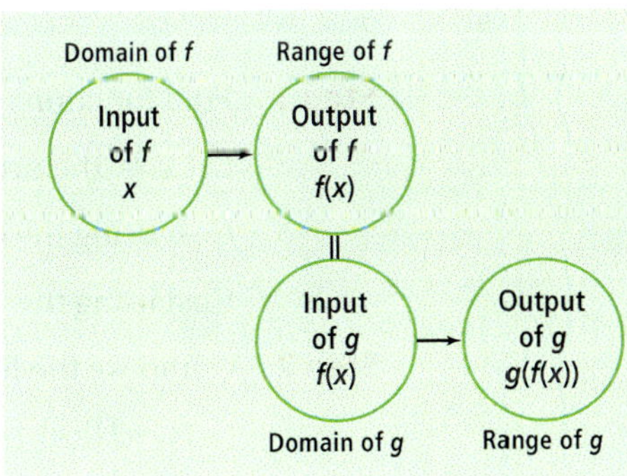

Key Concept Composition of Functions

The composition of function g with function f is written as $g \circ f$ and is defined as $(g \circ f)(x) = g(f(x))$. The domain of $g \circ f$ consists of the x-values in the domain of f for which $f(x)$ is in the domain of g.

$(g \circ f)(x) = g(f(x))$ **1.** Evaluate $f(x)$ first.

2. Then use $f(x)$ as the input for g.

Function composition is not commutative since $f(g(x))$ does not always equal $g(f(x))$.

Problem 3 Composing Functions

GRIDDED RESPONSE

Let $f(x) = x - 5$ and $g(x) = x^2$. What is $(g \circ f)(-3)$?

Method 1

$$(g \circ f)(x) = g(f(x))$$
$$= g(x - 5) = (x - 5)^2$$
$$(g \circ f)(-3) = (-3 - 5)^2$$
$$= (-8)^2$$
$$= 64$$

Method 2

$$(g \circ f)(-3) = g(f(-3))$$
$$= g(-3 - 5)$$
$$= g(-8)$$
$$= (-8)^2$$
$$= 64$$

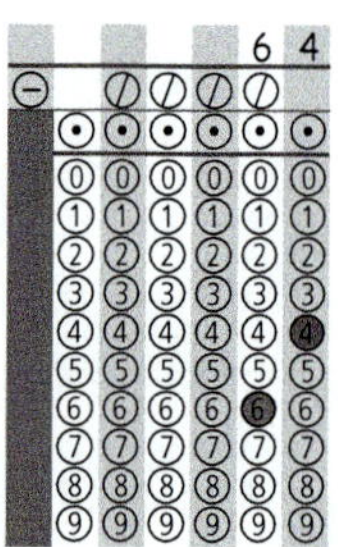

Think

Which function is substituted into the other?
Use $f(x)$ as the input for g.

Got It? **3.** What is $(f \circ g)(-3)$ for the functions f and g defined in Problem 3?

Problem 4 Using Composite Functions

You have a coupon good for \$5 off the price of any large pizza. You also get a 10% discount on any pizza if you show your student ID. How much more would you pay for a large pizza if the cashier applies the coupon first?

Know
The coupon value and the discount rate

Need
The difference between the results of applying the discount or coupon first

Plan
- Compose two functions in two ways.
- Then find the difference in their results.

Step 1 Find functions C and D that model the cost of a large pizza.

Let x = the price of a large pizza.

Cost using the coupon: $C(x) = x - 5$

Cost using the 10% discount: $D(x) = x - 0.1x = 0.9x$

Step 2 Compose the functions to apply the discount and then the coupon.

$(C \circ D)(x) = C(D(x))$ Apply the discount, $D(x)$, first.

$= C(0.9x) = 0.9x - 5$

Step 3 Compose the functions to apply the coupon and then the discount.

$(D \circ C)(x) = D(C(x))$ Apply the coupon, $C(x)$, first.

$= D(x - 5) = 0.9(x - 5) = 0.9x - 4.5$

Step 4 Subtract the functions to find how much more you would pay if the cashier applies the coupon first.

$$(D \circ C)(x) - (C \circ D)(x) = (0.9x - 4.5) - (0.9x - 5)$$
$$= -4.5 + 5$$
$$= 0.5$$

You pay $.50 more if the cashier applies the coupon first.

Got It? **4.** A store is offering a 15% discount on all items. Also, employees get a 20% employee discount. Write composite functions

a. to model taking the 15% discount and then the 20% discount.

b. to model taking the 20% discount and then the 15% discount.

c. Reasoning If you were an employee, which discount would you take first? Why?

Lesson Check

Do you know HOW?

Let $f(x) = 3x - 2$ and $g(x) = x^2 + 1$. Perform each function operation.

1. $(f \cdot g)(x)$ **2.** $(f - g)(x)$

3. $(f \circ g)(x)$ **4.** $f(x) + g(x)$

5. $g(x) - f(x)$ **6.** $f(x) - g(x)$

Do you UNDERSTAND?

7. Error Analysis Your friend used some simple functions and found that $(f \circ g)(x) = (g \circ f)(x)$, and concluded that function composition is commutative. Give an example to show that your friend is mistaken.

8. Open-Ended Find two functions f and g such that $f(g(x)) = x$ for all real numbers x.

Practice and Problem-Solving Exercises

See Problems 1 and 2.

Let $f(x) = 7x + 5$ and $g(x) = x^2$. Perform each function operation and then find the domain of the result.

9. $(f + g)(x)$ **10.** $(f - g)(x)$ **11.** $(g - f)(x)$

12. $(f \cdot g)(x)$ **13.** $\frac{f}{g}(x)$ **14.** $\frac{g}{f}(x)$

Let $f(x) = 2 - x$ and $g(x) = \frac{1}{x}$. Perform each function operation and then find the domain of the result.

15. $(f + g)(x)$ **16.** $(f - g)(x)$ **17.** $(g - f)(x)$

18. $(f \cdot g)(x)$ **19.** $\frac{f}{g}(x)$ **20.** $\frac{g}{f}(x)$

Let $f(x) = 2x^2 + x - 3$ and $g(x) = x - 1$. Perform each function operation and then find the domain.

21. $(f + g)(x)$

22. $(f - g)(x)$

23. $(g - f)(x)$

24. $(f \cdot g)(x)$

25. $\frac{f}{g}(x)$

26. $\frac{g}{f}(x)$

Let $g(x) = 2x$ and $h(x) = x^2 + 4$. Find each value or expression.

See Problem 3.

27. $(h \circ g)(1)$

28. $(h \circ g)(-5)$

29. $(h \circ g)(-2)$

30. $(g \circ h)(-2)$

31. $(g \circ h)(0)$

32. $(g \circ h)(a)$

33. $(g \circ g)(a)$

34. $(h \circ h)(a)$

35. $(h \circ g)(a)$

Let $f(x) = x^2$ and $g(x) = x - 3$. Find each value or expression.

36. $(g \circ f)(-2)$

37. $(f \circ g)(-2)$

38. $(g \circ f)(0)$

39. $(f \circ g)(0)$

40. $(g \circ f)(3.5)$

41. $(f \circ g)(3.5)$

42. $(f \circ g)(a)$

43. $(g \circ f)(-a)$

44. $(f \circ g)(-a)$

See Problem 4.

45. Sales A computer store offers a 5% discount off the list price x for any computer bought with cash, rather than put on credit. At the same time, the manufacturer offers a \$200 rebate for each purchase of a computer.

a. Write a function $f(x)$ to represent the price after the cash discount.

b. Write a function $g(x)$ to represent the price after the \$200 rebate.

c. Suppose the list price of a computer is \$1500. Use a composite function to find the price of the computer if the discount is applied before the rebate.

d. Suppose the list price of a computer is \$1500. Use a composite function to find the price of the computer if the rebate is applied before the discount.

46. Economics Suppose the function $f(x) = 0.15x$ represents the number of U.S. dollars equivalent to x Chinese yuan and the function $g(y) = 14.07y$ represents the number of Mexican pesos equivalent to y U.S. dollars.

a. Write a composite function that represents the number of Mexican pesos equivalent to x Chinese yuan.

b. Find the value in Mexican pesos of an item that costs 15 Chinese yuan.

Let $f(x) = 2x + 5$ and $g(x) = x^2 - 3x + 2$. Perform each function operation and then find the domain.

47. $f(x) + g(x)$

48. $3f(x) - 2$

49. $g(x) - f(x)$

50. $-2g(x) + f(x)$

51. $f(x) - g(x) + 10$

52. $4f(x) + 2g(x)$

53. $-f(x) + 4g(x)$

54. $f(x) - 2g(x)$

55. $f(x) \cdot g(x)$

56. $-3f(x) \cdot g(x)$

57. $\frac{f(x)}{g(x)}$

58. $\frac{5f(x)}{g(x)}$

B Apply

59. **Think About a Plan** A craftsman makes and sells violins. The function $I(x) = 5995x$ represents the income in dollars from selling x violins. The function $P(y) = y - 100{,}000$ represents his profit in dollars if he makes an income of y dollars. What is the profit from selling 30 violins?
- How can you write a composite function to represent the craftsman's profit?
- How can you use the composite function to find the profit earned when he sells 30 violins?

60. Suppose your teacher offers to give the whole class a bonus if everyone passes the next math test. The teacher says she will give everyone a 10-point bonus and increase everyone's grade by 9% of their score.
   a. You earned a 75 on the test. Would you rather have the 10-point bonus first and then the 9% increase, or the 9% increase first and then the 10-point bonus?
   b. **Reasoning** Is this the best plan for all students? Explain.

61. **Sales** A salesperson earns a 3% bonus on weekly sales over $5000. Consider the following functions.

   $g(x) = 0.03x$ $\qquad\qquad$ $h(x) = x - 5000$

   a. Explain what each function above represents.
   b. Which composition, $(h \circ g)(x)$ or $(g \circ h)(x)$, represents the weekly bonus? Explain.

62. If $(f \circ g)(x) = x^2 - 6x + 8$ and $g(x) = x - 3$, what is $f(x)$?

Let $g(x) = 3x + 2$ and $f(x) = \frac{x-2}{3}$. Find each value.

63. $f(g(1))$	64. $g(f(-4))$	65. $f(g(0))$	66. $g(f(2))$
67. $g(g(0))$	68. $(g \circ g)(1)$	69. $(f \circ g)(-2)$	70. $(f \circ f)(0)$

71. **Geometry** You toss a pebble into a pool of water and watch the circular ripples radiate outward. You find that the function $r(x) = 12.5x$ describes the radius r, in inches, of a circle x seconds after it was formed. The function $A(x) = \pi x^2$ describes the area A of a circle with radius x.
   a. Find $(A \circ r)(x)$ when $x = 2$. Interpret your answer.
   b. Find the area of a circle 4 seconds after it was formed.

For each pair of functions, find $f(g(x))$ and $g(f(x))$.

72. $f(x) = 3x,\ g(x) = x^2$	73. $f(x) = x + 3,\ g(x) = x - 5$
74. $f(x) = 3x^2 + 2,\ g(x) = 2x$	75. $f(x) = \frac{x-3}{2},\ g(x) = 2x - 3$
76. $f(x) = -x - 7,\ g(x) = 4x$	77. $f(x) = \frac{x+5}{2},\ g(x) = x^2$

78. **Open-Ended** Write a function rule that approximates each value.
   a. The amount you save is a percent of what you earn. (You choose the percent.)
   b. The amount you earn depends on how many hours you work. (You choose the hourly wage.)
   c. Write and simplify a composite function that expresses your savings as a function of the number of hours you work. Interpret your results.

Challenge Let $f(x) = x^4 + 2x^3 - 5x^2 - 10x$ and $g(x) = x^3 - 3x^2 - 5x + 15$. Perform each function operation and simplify, and then find the domain.

79. $f(x) \cdot g(x)$

80. $\frac{f(x)}{g(x)}$

81. $\frac{g(x)}{f(x)}$

Find each composition of functions. Simplify your answer.

82. Let $f(x) = \frac{1}{x}$. Find $f(f(f(x)))$.

83. Let $f(x) = 2x - 3$. Find $\frac{f(1 + h) - f(1)}{h}$, $h \neq 0$.

84. Let $f(x) = 4x - 1$. Find $\frac{f(a + h) - f(a)}{h}$, $h \neq 0$.

85. Let $f(x) = 4x^2 - 1$. Find $\frac{f(a + h) - f(a)}{h}$, $h \neq 0$.

Standardized Test Prep

SAT/ACT

86. Let $f(x) = x + 5$ and $g(x) = x^2 - 25$. What is the domain of $\frac{f}{g}(x)$?

Ⓐ All real numbers

Ⓑ All real numbers except 5

Ⓒ All real numbers except -5

Ⓓ All real numbers except -5 and 5

87. Let $g(x) = x - 3$ and $h(x) = x^2 + 6$. What is $(h \circ g)(1)$?

Ⓕ -14 Ⓖ 4 Ⓗ 10 Ⓘ 15

88. Which number is a solution of $|3 - 2x| < 5$?

Ⓐ -6 Ⓑ -1 Ⓒ 2 Ⓓ 4

Short Response

89. What is the coefficient of the x^3y^4 term in the expansion of $(3x - y)^7$? Show your work.

Mixed Review

Solve. Check for extraneous solutions. See Lesson 6-5.

90. $\sqrt{x^2 + 3} = x + 1$

91. $x + 8 = (x^2 + 16)^{\frac{1}{2}}$

92. $\sqrt{x^2 + 9} = x + 1$

93. $(x^2 - 9)^{\frac{1}{2}} - x = -3$

94. $\sqrt{x^2 + 12} - 2 = x$

95. $(3x)^{\frac{1}{2}} = (x + 6)^{\frac{1}{2}}$

Expand each binomial. See Lesson 5-7.

96. $(x + 4)^8$

97. $(x + y)^6$

98. $(2x - y)^4$

99. $(2x - 3y)^7$

100. $(9 - 2x)^5$

101. $(4x - y)^5$

102. $(x^2 + x)^4$

103. $(x^2 + 2y^3)^6$

Get Ready! **To prepare for Lesson 6-7, do Exercises 104–106.**

Graph and solve each system. See Lesson 3-1.

104. $\begin{cases} y = x - 6 \\ y = x + 6 \end{cases}$

105. $\begin{cases} y = 0.5x + 1 \\ y = 2x - 2 \end{cases}$

106. $\begin{cases} y = \frac{x + 4}{5} \\ y = 5x - 4 \end{cases}$

6-7 Inverse Relations and Functions

Content Standards

F.BF.4.a. Solve an equation of the form $f(x) = c$ for a simple function f that has an inverse and write an expression for the inverse.

Also F.BF.4.c

Objective To find the inverse of a relation or function

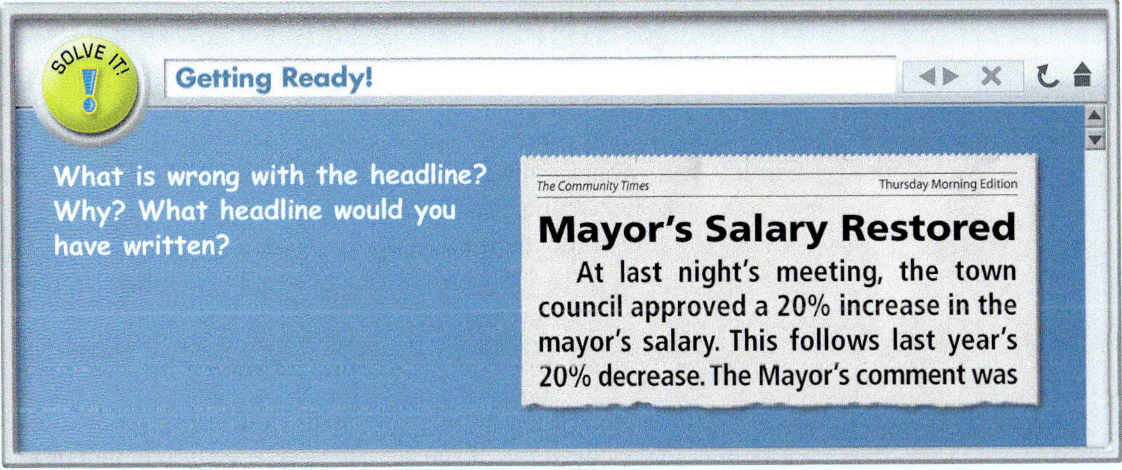

The Community Times — Thursday Morning Edition

Mayor's Salary Restored

At last night's meeting, the town council approved a 20% increase in the mayor's salary. This follows last year's 20% decrease. The Mayor's comment was

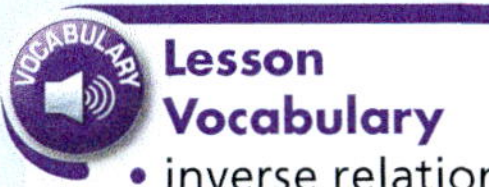

Lesson Vocabulary
- inverse relation
- inverse function
- one-to-one function

If a relation pairs element a of its domain to element b of its range, the **inverse relation** pairs b with a. So, if (a, b) is an ordered pair of a relation, then (b, a) is an ordered pair of its inverse. If both a relation and its inverse happen to be functions, they are **inverse functions**.

Essential Understanding The inverse of a function may or may not be a function.

This diagram shows a relation r (a function) and its inverse (not a function). The range of the relation is the domain of the inverse. The domain of the relation is the range of the inverse.

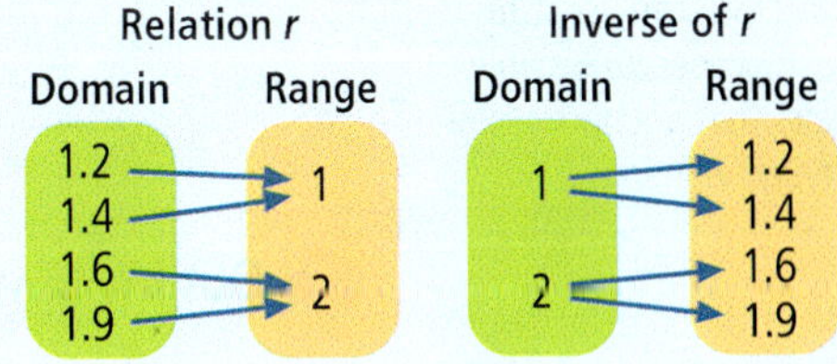

Problem 1 Finding the Inverse of a Relation

Think

(0, −1) is in s. How do you find the corresponding pair in the inverse of s?

Switch the coordinates. (−1, 0) is in the inverse of s.

A What is the inverse of relation s?

Relation s

x	y
0	−1
2	0
3	2
4	3

Switch the x and y values to get the inverse. →

Inverse of Relation s

x	y
−1	0
0	2
2	3
3	4

B What are the graphs of s and its inverse?

Relation s

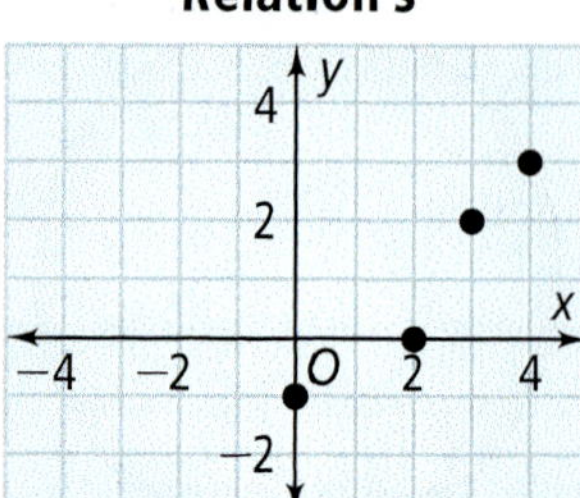

Reversing the Ordered Pairs

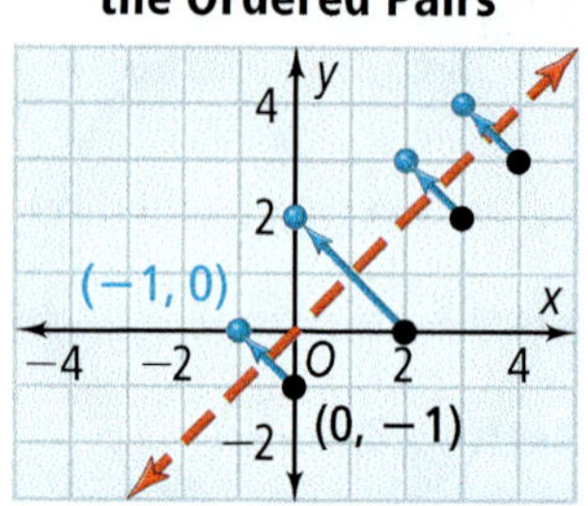

Inverse of s

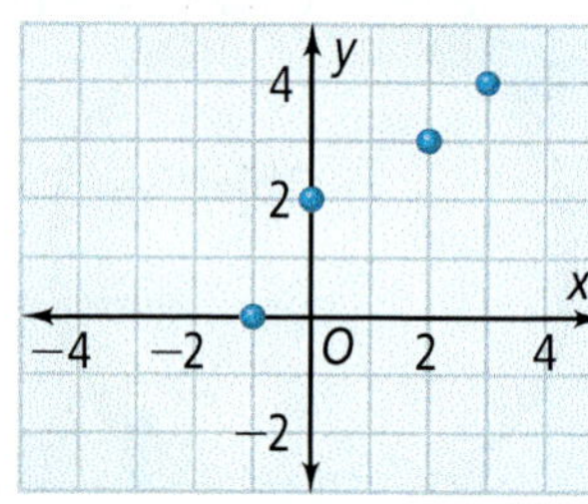

  **Got It?** **1. a.** What are the graphs of t and its inverse?

b. Reasoning Is t a function? Is the inverse of t a function? Explain.

Relation t

x	0	1	2	3
y	−5	−4	−3	−3

As shown in Problem 1, the graphs of a relation and its inverse are the reflections of each other in the line $y = x$. If you describe a relation or function by an equation in x and y, you can switch x and y to get an equation for the inverse.

Problem 2 Finding an Equation for the Inverse

Think

Why do you solve for y?
If you solve the equation for y, you can use it to easily generate ordered pairs that are part of the inverse relation.

What is the inverse of the relation described by $y = x^2 - 1$?

$y = x^2 - 1$

$x = y^2 - 1$ Switch x and y.

$x + 1 = y^2$ Add 1 to each side.

$\pm\sqrt{x + 1} = y$ Find the square root of each side to solve for y.

Got It? **2.** What is the inverse of $y = 2x + 8$?

Problem 3 Graphing a Relation and Its Inverse

Think

What does the graph of $y = x^2 - 1$ look like?
The graph of $y = x^2 - 1$ is a translation of $y = x^2$ down one unit.

What are the graphs of $y = x^2 - 1$ and its inverse, $y = \pm\sqrt{x + 1}$?

The graph of $y = x^2 - 1$ is a parabola that opens upward with vertex $(0, -1)$. The graph of the inverse is the reflection of the parabola in the line $y = x$.

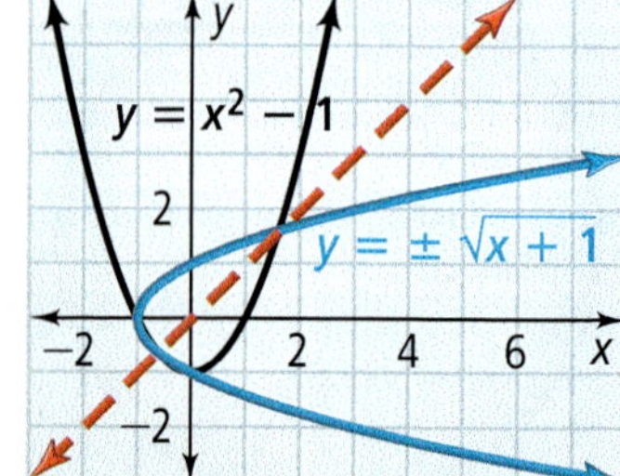

 **Got It?** **3.** What are the graphs of $y = 2x + 8$ and its inverse?

The inverse of a function f is denoted by f^{-1}. You read f^{-1} as "the inverse of f" or as "f inverse." The notation $f(x)$ is used for functions, but the relation f^{-1} may not even be a function.

Problem 4 Finding an Inverse Function

Consider the function $f(x) = \sqrt{x - 2}$.

A What are the domain and range of f?

The radicand cannot be negative, so the numbers $x \geq 2$ make up the domain. The principal square root is nonnegative, so the numbers $y \geq 0$ make up the range.

B What is f^{-1}, the inverse of f?

$f(x) = \sqrt{x - 2}$

$y = \sqrt{x - 2}$ Rewrite the equation using y.

$x = \sqrt{y - 2}$ Switch x and y. Since x equals a principal square root, $x \geq 0$.

$x^2 = y - 2$ Square both sides.

$y = x^2 + 2$ Solve for y.

So, $f^{-1}(x) = x^2 + 2$, for $x \geq 0$.

C What are the domain and range of f^{-1}?

Part (b) shows that the domain of f^{-1} is the range of f—the numbers $x \geq 0$. Since $x^2 \geq 0$, $x^2 + 2 \geq 2$. Therefore, the numbers $y \geq 2$ make up the range of f^{-1}. Note that the range of f^{-1} is the same as the domain of f.

D Is f^{-1} a function? Explain.

For each x in the domain ($x \geq 0$) of f^{-1}, there is only one value of y in the range. So $f^{-1}(x) = x^2 + 2, x \geq 0$, is a function.

Think

How could a graph help you check your answer?
You could graph f^{-1} and see whether the graph passes the vertical line test. If it does, f^{-1} is a function.

Got It? **4.** Let $g(x) = 6 - 4x$.

a. What are the domain and range of g?

b. What is the inverse of g?

c. What are the domain and range of g^{-1}?

d. Is g^{-1} a function? Explain.

Functions that model real-world behavior are often expressed as formulas with meaningful variables, like $A = \pi r^2$ for the area of a circle. Strictly speaking, the inverse formula would be $r = \pi A^2$, but this expresses a false relationship between A and r. It is better to leave the variables in place and solve for r as a function of A.

$A = \pi r^2$ Original formula.

$r = \sqrt{\frac{A}{\pi}}$ Same formula, but inversely expressed.

Problem 5 Finding the Inverse of a Formula

The function $d = 4.9t^2$ represents the distance d, in meters, that an object falls in t seconds due to Earth's gravity. Find the inverse of this function. How long, in seconds, does it take for the cliff diver shown to reach the water below?

Think

Why shouldn't you interchange the variables?
Interchanging the variables leads to a false relationship between distance and time.

$d = 4.9t^2$

$t^2 = \frac{d}{4.9}$ Solve for t. Do not switch the variables.

$t = \sqrt{\frac{d}{4.9}}$ Time must be nonnegative.

$= \sqrt{\frac{24}{4.9}}$ Substitute 24 for d.

≈ 2.2 Use a calculator.

It will take about 2.2 seconds for the diver to reach the water.

Got It? 5. The function $d = \frac{v^2}{19.6}$ relates the distance d, in meters, that an object has fallen to its velocity v, in meters per second. Find the inverse of this function. What is the velocity of the cliff diver in meters per second as he enters the water?

You know that for any function f, each x-value in the domain corresponds to exactly one y-value in the range. For a **one-to-one function**, it is also true that each y-value in the range corresponds to exactly one x-value in the domain. A one-to-one function f has an inverse f^{-1} that is also a function. If f maps a to b, then f^{-1} must map b to a.

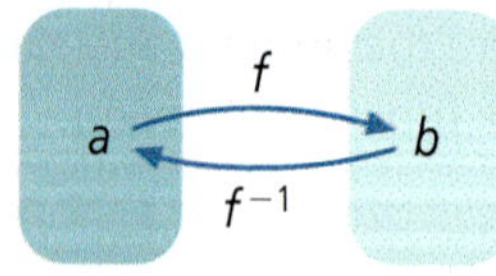

take note

Key Concept Composition of Inverse Functions

If f and f^{-1} are inverse functions, then

$(f^{-1} \circ f)(x) = x$ and $(f \circ f^{-1})(x) = x$ for x in the domains of f and f^{-1}, respectively.

This says that the composition of a function and its inverse is essentially the identity function, $id(x) = x$, or $y = x$.

Problem 6 Composing Inverse Functions

For $f(x) = \frac{1}{x-1}$, what is each of the following?

A $f^{-1}(x)$

$$f(x) = \frac{1}{x-1}$$

$$y = \frac{1}{x-1} \quad \text{Rewrite the equation using } y.$$

$$x = \frac{1}{y-1} \quad \text{Switch } x \text{ and } y.$$

$$x(y-1) = 1 \quad \text{Solve for } y.$$

$$y - 1 = \frac{1}{x}$$

$$y = \frac{1}{x} + 1$$

So $f^{-1}(x) = \frac{1}{x} + 1$.

Think

Is this a function?
Yes. For each value of x, there is only one value for y.

B $(f \circ f^{-1})(1)$

$$(f \circ f^{-1})(1) = f(f^{-1}(1))$$

$$= f\left(\frac{1}{1} + 1\right)$$

$$= f(2)$$

$$= \frac{1}{2-1} = 1$$

C $(f^{-1} \circ f)(1)$

$$(f^{-1} \circ f)(1) = f^{-1}(f(1))$$

$$= f^{-1}\left(\frac{1}{1-1}\right)$$

$$= f^{-1}\left(\frac{1}{0}\right) \quad \text{undefined}$$

1 is not in the domain of f. Therefore $(f^{-1} \circ f)(1)$ does not exist.

Got It? **6.** Let $g(x) = \frac{4}{x+2}$. What is each of the following?

a. $g^{-1}(x)$ **b.** $(g \circ g^{-1})(0)$ **c.** $(g^{-1} \circ g)(0)$

Lesson Check

Do you know HOW?

Find the inverse of each function. Is the inverse a function?

1. $f(x) = 4x + 3$

2. $f(x) = x^2 - 1$

3. $f(x) = (x + 1)^2$

4. For $h(x) = -\frac{1}{x+2}$, find:

a. $h^{-1}(x)$

b. $h^{-1}(4)$

c. Value of x for which the equality $(h \circ h^{-1})(x) = x$ does not hold.

Do you UNDERSTAND?

5. Vocabulary Does every function have an inverse which is a function? Does every relation have an inverse which is a relation?

6. Reasoning A function consists of the pairs (2, 3), $(x, 4)$, and (5, 6). What values, if any, may x not assume?

7. Error Analysis A classmate says that $(f \circ g)^{-1}(x) = (f^{-1} \circ g^{-1})(x)$. Show that this is incorrect by finding examples of $f(x)$ and $g(x)$ for which the equation does not hold.

Practice and Problem-Solving Exercises

Find the inverse of each relation. Graph the given relation and its inverse.

See Problem 1.

8.

x	y
1	0
2	1
3	0
4	2

9.

x	y
1	0
2	1
3	2
4	3

10.

x	y
0	0
1	1
2	4
3	9

11.

x	y
−3	2
−2	2
−1	2
0	2

Find the inverse of each function. Is the inverse a function?

See Problem 2.

12. $y = 3x + 1$ **13.** $y = 2x - 1$ **14.** $y = 4 - 3x$

15. $y = 5 - 2x^2$ **16.** $y = x^2 + 4$ **17.** $y = 3x^2 - 5$

18. $y = (x - 8)^2$ **19.** $y = (3x - 4)^2$ **20.** $y = (1 - 2x)^2 + 5$

Graph each relation and its inverse.

See Problem 3.

21. $y = 2x - 3$ **22.** $y = 3 - 7x$ **23.** $y = -x$

24. $y = 3x^2$ **25.** $y = -x^2$ **26.** $y = 4x^2 - 2$

27. $y = (x - 1)^2$ **28.** $y = (2 - x)^2$ **29.** $y = (3 - 2x)^2 - 1$

For each function, find the inverse and the domain and range of the function and its inverse. Determine whether the inverse is a function.

See Problem 4.

30. $f(x) = 3x + 4$ **31.** $f(x) = \sqrt{x - 5}$

32. $f(x) = \sqrt{x + 7}$ **33.** $f(x) = \sqrt{-2x + 3}$

34. $f(x) = 2x^2 + 2$ **35.** $f(x) = -x^2 + 1$

36. Temperature The formula for converting from Celsius to Fahrenheit temperatures is $F = \frac{9}{5}C + 32$.

See Problem 5.

a. Find the inverse of the formula. Is the inverse a function?

b. Use the inverse to find the Celsius temperature that corresponds to 25°*F*.

37. Geometry The formula for the volume of a sphere is $V = \frac{4}{3}\pi r^3$.

a. Find the inverse of the formula. Is the inverse a function?

b. Use the inverse to find the radius of a sphere that has a volume of 35,000 ft^3.

For Exercises 38–41, $f(x) = 10x - 10$. Find each value.

See Problem 6.

38. $(f^{-1} \circ f)(10)$ **39.** $(f \circ f^{-1})(-10)$

40. $(f^{-1} \circ f)(0.2)$ **41.** $(f \circ f^{-1})(d)$

Find the inverse of each function. Is the inverse a function?

42. $f(x) = x^3$

43. $f(x) = x^4$

44. $f(x) = \frac{2x^2}{5} + 1$

45. $f(x) = 1.5x^2 - 4$

46. $f(x) = \frac{3x^2}{4}$

47. $f(x) = \sqrt{2x - 1} + 3$

48. **Think About a Plan** The velocity of the water that flows from an opening at the base of a tank depends on the height of water above the opening. The function $v(x) = \sqrt{2gx}$ models the velocity v in feet per second where g, the acceleration due to gravity, is about 32 ft/s^2 and x is the height in feet of the water. What is the depth of water when the flow is 40 ft/s, and when the flow is 20 ft/s?

- How can you use inverse functions to help you find the answer?
- What restrictions are on the domain of $v(x)$? of $v^{-1}(x)$?

49. Let $f(x) = 3x^2 - 4$ and $g(x) = x - 2$. Calculate $(f \circ g^{-1})(x)$ for $x = -3$.

50. **Writing** Explain how you can find the range of the inverse of $f(x) = \sqrt{x - 1}$ without finding the inverse itself.

For each function, find the inverse and the domain and range of the function and its inverse. Determine whether the inverse is a function.

51. $f(x) = -\sqrt{x}$

52. $f(x) = \sqrt{x} + 3$

53. $f(x) = \sqrt{-x + 3}$

54. $f(x) = \sqrt{x + 2}$

55. $f(x) = \frac{x^2}{2}$

56. $f(x) = \frac{1}{x^2}$

57. $f(x) = (x - 4)^2$

58. $f(x) = (7 - x)^2$

59. $f(x) = \frac{1}{(x + 1)^2}$

60. $f(x) = 4 - 2\sqrt{x}$

61. $f(x) = \frac{3}{\sqrt{x}}$

62. $f(x) = \frac{1}{\sqrt{-2x}}$

63. a. **Open-Ended** Copy the mapping diagram at the right. Complete it by writing members of the domain and range and connecting them with arrows so that r is a function and r^{-1} is not a function.
   b. Repeat part (a) so that r is not a function and r^{-1} is a function.

Relation r

64. **Reasoning** Relation r has one element in its domain and two elements in its range. Is r a function? Is the inverse of r a function? Explain.

65. **Geometry** Write a function that gives the length of the hypotenuse of an isosceles right triangle with side length s. Evaluate the inverse of the function to find the side length of an isosceles right triangle with a hypotenuse of 6 in.

66. For the function $f(x) = \sqrt[3]{2x}$, find $f^{-1}(x)$. Then determine the value of x when $f(x) = 16$.

67. **Reasoning** To determine if the inverse of function f is also a function, you can use a *horizontal line test.* It says that if no horizontal line intersects the graph of the function f in more than one point, then the inverse of f is a function.
   a. Explain why the horizontal line test works.
   b. The graph of a polynomial function passes through the points $(-1, 1)$, $(0, 4)$ and $(2, 3)$. Can its inverse be a function?

Challenge Find the inverse of each function. Is the inverse a function?

68. $f(x) = \frac{1}{5}x^3$
69. $f(x) = \sqrt[3]{x - 5}$
70. $f(x) = \frac{\sqrt[3]{x}}{3}$
71. $f(x) = (x - 2)^3$
72. $f(x) = \sqrt[4]{x}$
73. $f(x) = 1.2x^4$

74. Function $f(x)$ is defined the following way:
- if x is an integer, then $f(x) = x + 1$;
- for all other x, $f(x) = x + 2$.

Is the inverse of $f(x)$ a function? Explain.

Standardized Test Prep

SAT/ACT

75. Which pair of words make this sentence FALSE?
The product of two ____(I)____ numbers is always a (n) ____(II)____ number.

Ⓐ (I) complex; (II) complex
Ⓑ (I) real; (II) complex
Ⓒ (I) rational; (II) real
Ⓓ (I) imaginary; (II) imaginary

76. If $f(x) = x + 1$ and $g(x) = x^2 - 3x - 4$, what is $(f \circ g)(x)$?

Ⓕ $x^2 - 3x - 3$
Ⓖ $x^2 - x - 6$
Ⓗ $x^2 - x$
Ⓘ $x^2 - x - 3$

77. What is the simplified form of $\left(a^{\frac{2}{3}}b^{\frac{3}{4}}\right)^2$?

Ⓐ $a^{\frac{4}{9}}b^{\frac{9}{16}}$
Ⓑ $a^{\frac{4}{3}}b^{\frac{3}{2}}$
Ⓒ ab
Ⓓ $(ab)^{\frac{17}{6}}$

Extended Response

78. Let $f(x) = (x + 1)^2 - 2$. Find the x- and y-intercepts of $f(x)$ and the inverse of $f(x)$. Is the inverse a function?

Mixed Review

Let $f(x) = 4x$, $g(x) = \frac{1}{2}x + 7$, and $h(x) = -2x + 4$. Perform each function operation.

79. $(g \circ f)(x)$
80. $(h \circ g)(x)$
81. $h(x) + g(x)$
82. $f(x) \cdot g(x)$
83. $(f \circ g)(x) + h(x)$
84. $(f \circ g)(x)$

Find each real root.

See Lesson 6-1.

85. $-\sqrt[4]{16}$
86. $\sqrt[4]{-16}$
87. $\sqrt[5]{243}$
88. $-\sqrt[5]{243}$
89. $\sqrt[5]{-243}$
90. $\sqrt[3]{0.064}$
91. $\sqrt[4]{810{,}000}$
92. $\sqrt[4]{\frac{1}{160{,}000}}$

Get Ready! **To prepare for Lesson 6-8, do Exercises 93–95.**

Graph each function.

See Lesson 4-1.

93. $y = -x^2 - 1$
94. $y = -(x + 1)^2 + 1$
95. $y = 3x^2 + 3$

Concept Byte

For Use With Lesson 6-7

TECHNOLOGY

Graphing Inverses

Content Standard

Extends F.BF.4.a. Solve an equation of the form $f(x) = c$ for a simple function f that has an inverse and write an expression for the inverse.

You can graph inverses of functions on a graphing calculator by using the **DrawInv** feature or by using parametric equations. It takes more keystrokes to set up parametric equations, but once you do you can easily change from one function to another and quickly see the graphs of the new function and its inverse.

Activity

Graph $y = 0.3x^2 + 1$ and its inverse.

Method 1 Use the **DrawInv** feature.

Step 1 Press y= and enter the equation. Press zoom 5 to see a graph of the function with equal x- and y-intervals.

Step 2 Press 2nd draw **8.** You will see **DrawInv** followed by a flashing cursor. Select equation Y_1 by pressing vars ▷ **1 1.** Press enter to see the graph of the function and its inverse.

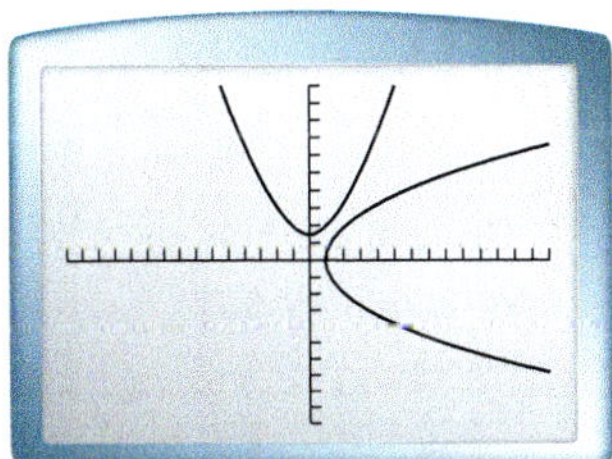

Method 2 Use parametric equations.

Step 1 Set to parametric mode. Press mode, select **Par**, and press 2nd quit.

Step 2 Enter the given equation in parametric form. Press y= and enter the equations $X_{1T} = T$ and $Y_{1T} = .3T^2 + 1$.

Step 3 Now use $X_{2T} = Y_{1T}$ and $Y_{2T} = X_{1T}$ to interchange the x- and y-values of the first parametric equation. Press y= and move the cursor to follow X_{2T} =. Select Y_{1T} by pressing vars ▷ **2 2.** Enter the equation $Y_{2T} = X_{1T}$ in a similar fashion.

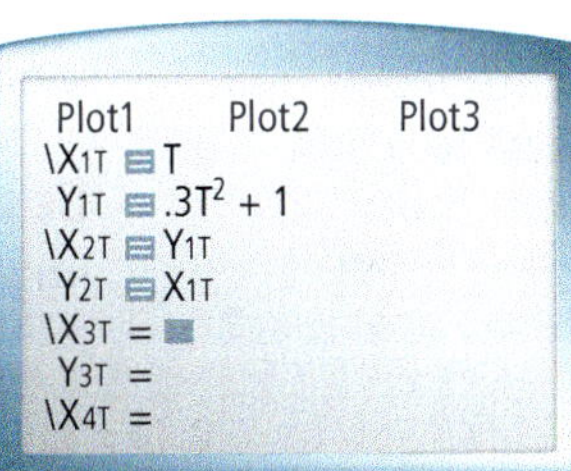

Step 4 Press zoom **5.** Adjust the **Window** so that **Tmin** and **Tmax** approximately agree with **Xmin** and **Xmax**. Press graph to see the graph of the function and its inverse.

Exercises

Graph each function and its inverse with a graphing calculator. Then sketch the graphs.

1. $y = x^2 - 5$ **2.** $y = (x - 3)^2$ **3.** $y = 0.01x^4$ **4.** $y = 0.5x^3 - 3$

5. Writing Change the parametric equation $X_{2T} = Y_{1T}$ in Method 2, Step 3 to $X_{2T} = -Y_{1T}$. Describe the graph that results.

6. Explain how once you set up parametric equations, you can change from one function to another and quickly see the graphs of the new function and its inverse.

Graphing Radical Functions

Content Standards

F.IF.7.b. Graph square root *and* cube root functions . . .

F.IF.8 Write a function defined by an expression in different but equivalent forms . . .

Objective To graph square root and other radical functions

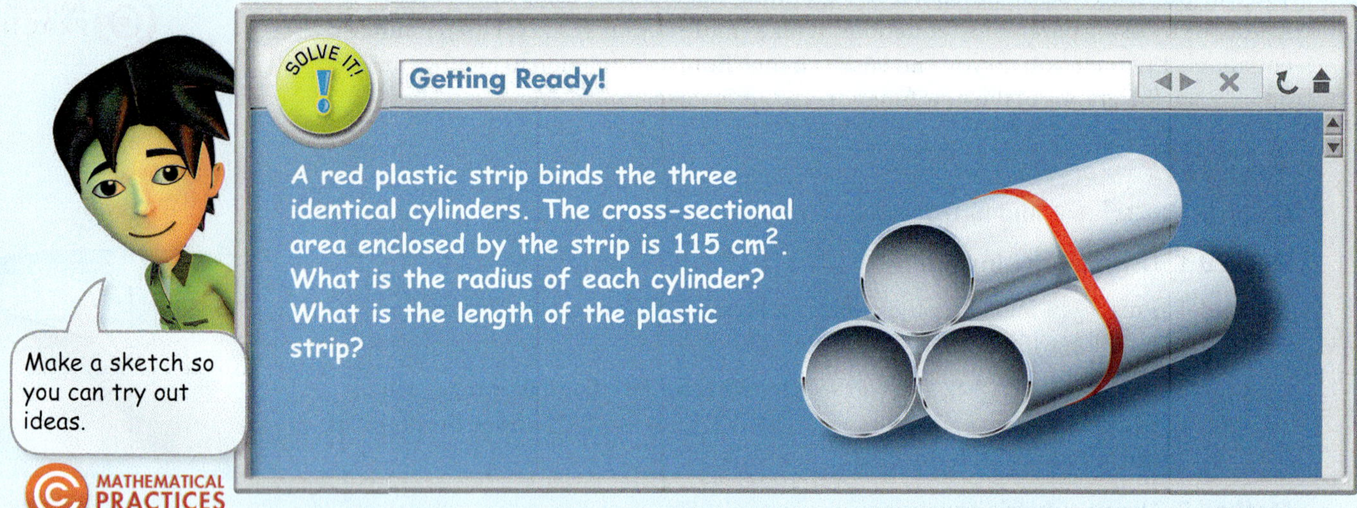

The formula $A = \pi r^2$ shows that area is a quadratic function of the radius of a circle. The formula $r = \frac{1}{\sqrt{\pi}}\sqrt{A}$ shows that the radius of a circle is a square root function of the area.

Lesson Vocabulary
- radical function
- square root function

Essential Understanding A square root function is the inverse of a quadratic function that has a restricted domain.

A horizontal line can intersect the graph of $f(x) = x^2$ in two points—where $f(-2) = f(2)$, for example. Thus, a vertical line can intersect the graph of f^{-1} in two points. f^{-1} is *not* a function because it fails the vertical line test.

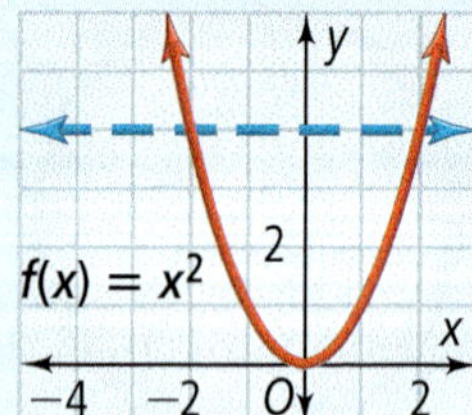

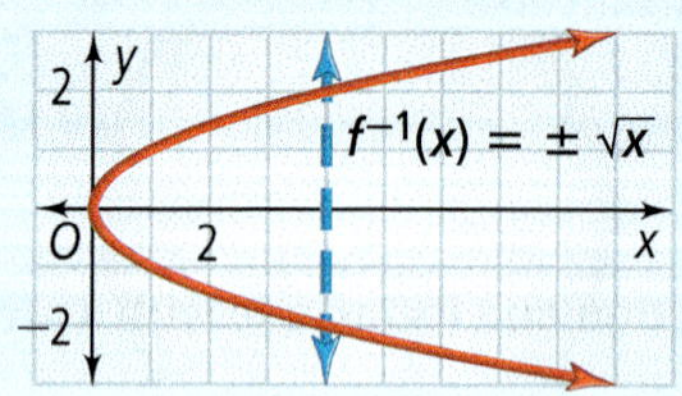

However, you can restrict the domain of f so that the inverse of the restricted function is a function.

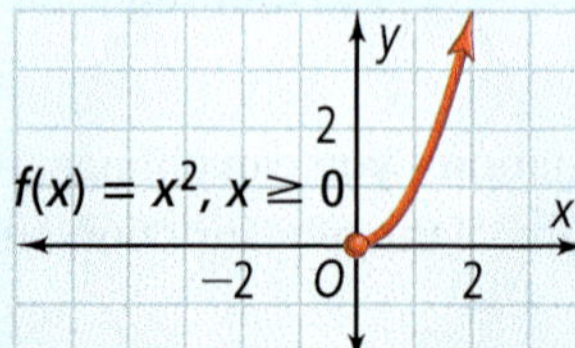

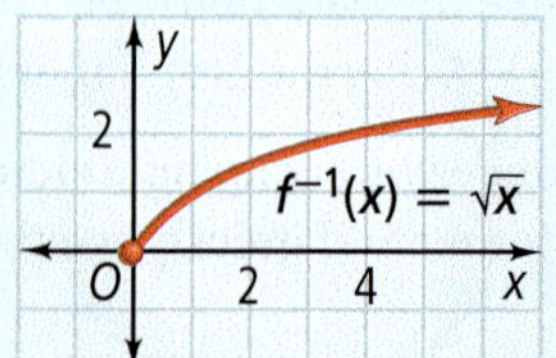

Inverses of the power functions $y = x^n$ (with domains restricted as needed) form parent functions $y = \sqrt[n]{x}$ for families of **radical functions**. In particular, $f(x) = \sqrt{x}$ is the parent for the family of **square root functions**. Members of this family have the general form $f(x) = a\sqrt{x - h} + k$.

take note

Key Concepts Families of Radical Functions

	Square Root	Radical
Parent function:	$y = \sqrt{x}$	$y = \sqrt[n]{x}$
Reflection in x-axis:	$y = -\sqrt{x}$	$y = -\sqrt[n]{x}$
Stretch ($a > 1$), shrink ($0 < a < 1$) by the factor a:	$y = a\sqrt{x}$	$y = a\sqrt[n]{x}$
Translation: Horizontal by h Vertical by k	$y = \sqrt{x - h} + k$	$y = \sqrt[n]{x - h} + k$

Problem 1 Translating a Square Root Function Vertically

Think

How is $y = \sqrt{x} + k$ related to the parent function $y = \sqrt{x}$?
It is related to the parent function in the same way that $y = f(x) + k$ is related to $y = f(x)$. It is a vertical translation of k units.

What are the graphs of $y = \sqrt{x} - 2$ and $y = \sqrt{x} + 1$?

The graph of $y = \sqrt{x} - 2$ is the graph of $y = \sqrt{x}$ shifted down 2 units.

The graph of $y = \sqrt{x} + 1$ is the graph of $y = \sqrt{x}$ shifted up 1 unit.

The domains of both functions are the set of nonnegative numbers, but their ranges differ.

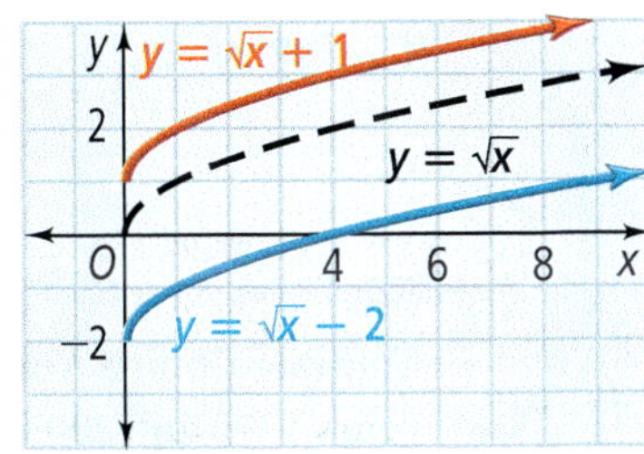

Got It? 1. What are the graphs of $y = \sqrt{x} + 2$ and $y = \sqrt{x} - 3$?

Problem 2 Translating a Square Root Function Horizontally

Think

How is $y = \sqrt{x - h}$ related to the parent function $y = \sqrt{x}$?
It is a horizontal translation of h units.

What are the graphs of $y = \sqrt{x + 4}$ and $y = \sqrt{x - 1}$?

The graph of $y = \sqrt{x + 4}$ is the graph of $y = \sqrt{x}$ shifted left 4 units.

The graph of $y = \sqrt{x - 1}$ is the graph of $y = \sqrt{x}$ shifted right 1 unit.

The ranges of both functions are the set of nonnegative numbers, but their domains differ.

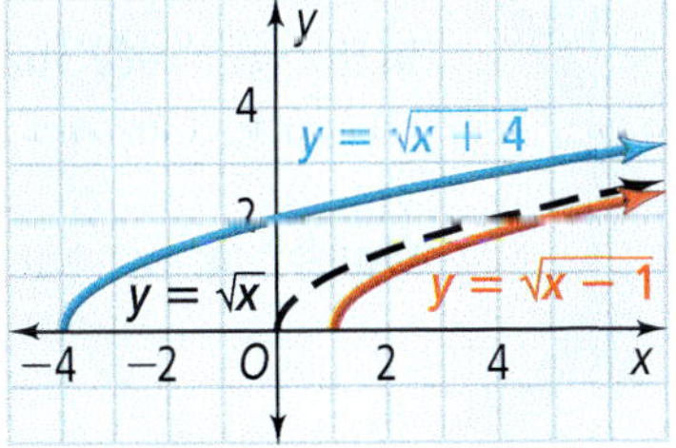

Got It? 2. What are the graphs of $y = \sqrt{x - 3}$ and $y = \sqrt{x + 2}$?

Recall from Lesson 2-7 that for any transformation, $y = af(x - h) + k$ of the parent function $f(x)$, a indicates a vertical stretch or shrink.

Similarly, for the combined transformation $y = a\sqrt{x - h} + k$, a indicates a vertical stretch ($|a| > 1$) or shrink ($|a| < 1$). A negative value of a indicates a reflection in the x-axis.

Problem 3 Graphing a Square Root Function

Think

What would be good points to choose?
Points that have integer x- and y-coordinates.

What is the graph of $y = -\frac{1}{2}\sqrt{x - 3} + 1$?

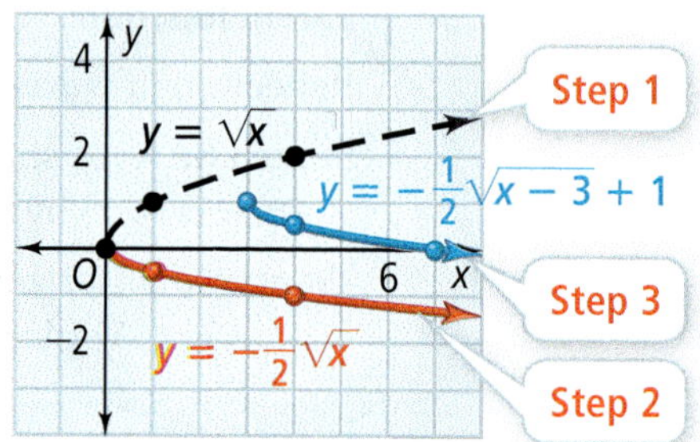

Step 1 Choose several points from the parent function $y = \sqrt{x}$.

Step 2 Multiply the y-coordinates by $a = -\frac{1}{2}$. This shrinks the parent graph vertically by the factor $\frac{1}{2}$ and reflects the result in the x-axis.

Step 3 The values of h and k give the horizontal and vertical translations. Translate the graph from Step 2 to the right 3 units and up 1 unit.

Got It? 3. What is the graph of $y = 3\sqrt{x + 2} - 4$?

Problem 4 Solving a Radical Equation by Graphing

Multiple Choice **You can model the population P of Corpus Christi, Texas, between the years 1970 and 2005 by the radical function $P(x) = 75{,}000\sqrt[3]{x - 1950}$, where x is the year. Using this model, in what year was the population of Corpus Christi 250,000?**

Ⓐ 1980 Ⓑ 1983 Ⓒ 1987 Ⓓ 1990

Think

How can you rewrite a radical function using an exponent?
You can write a radical function $y = \sqrt[n]{x}$ as $y = x^{\frac{1}{n}}$.

For $P = 250{,}000$, solve the equation $250{,}000 = 75{,}000\sqrt[3]{x - 1950}$.

Graph **Y1 = 75000(X − 1950)^(1/3)** and **Y2 = 250000**. Adjust the window to find where the graphs intersect.

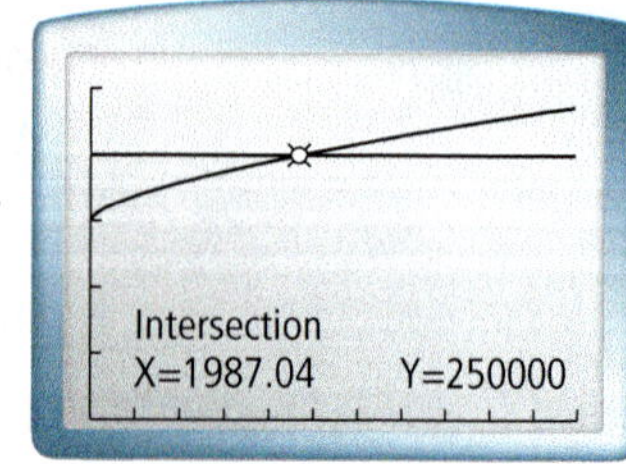

Use the **INTERSECT** feature to find the x-coordinate of the intersection.

In the year 1987, the population of Corpus Christi was 250,000. The correct answer is C.

Got It? 4. In what year was the population of Corpus Christi 275,000?

Problem 4 uses a transformation of $y = \sqrt[3]{x}$. The function $f(x) = \sqrt[3]{x}$ is the inverse of $g(x) = x^3$. Unlike $y = \sqrt{x}$, the domain and range of $f(x) = \sqrt[3]{x}$ are all real numbers.

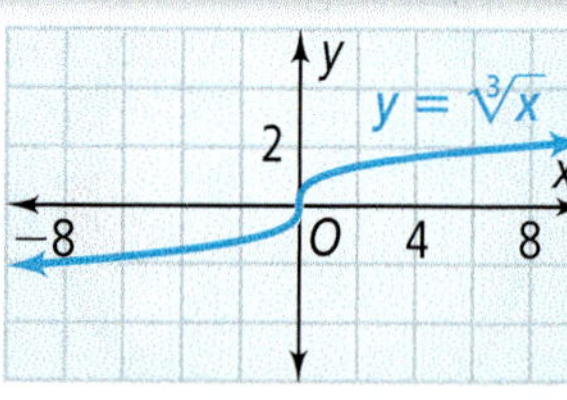

The patterns for graphing square root functions apply to other radical functions.

Problem 5 Graphing a Cube Root Function

Plan

How is $y = a\sqrt[n]{x-h} + k$ related to its parent function?

a stretches or shrinks the parent function and h and k translate it horizontally and vertically.

What is the graph of $y = 2\sqrt[3]{x+1} - 4$?

Step 1 Graph the parent function, $y = \sqrt[3]{x}$.

Step 2 Multiply the y-coordinates by 2. This stretches the graph vertically.

Step 3 Translate the graph from step 2, 1 unit to the left and 4 units down.

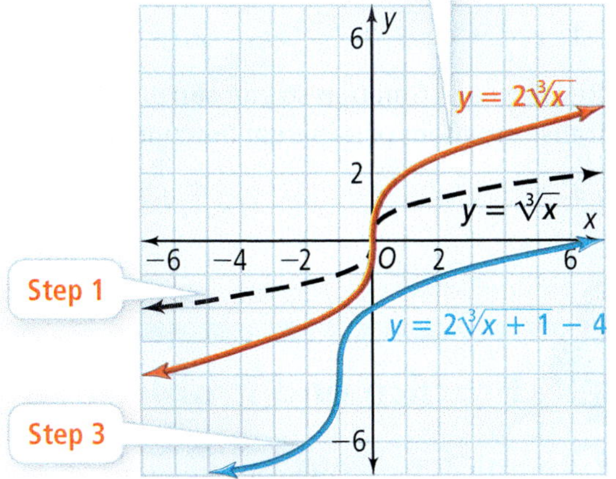

Got It? 5. What is the graph of $y = 3 - \frac{1}{2}\sqrt[3]{x-2}$?

You can graph functions of the form $y = \sqrt[n]{bx + c}$ using transformations, if you can simplify the radicand so that x has a coefficient of 1. This is also true for functions in the form $y = a\sqrt[n]{bx+c} + k$.

Problem 6 Rewriting a Radical Function

How can you rewrite $y = \sqrt{9x + 18}$ so you can graph it using transformations? Describe the graph.

Think	Write
The form $y = a\sqrt{x}$ shows the stretch or shrink. Factor to get $x - h$ in the radicand.	$y = \sqrt{9x + 18}$ $y = \sqrt{9(x + 2)}$ $y = \sqrt{9(x - (-2))}$
Find the square root of 9. Now, you have the form $y = a\sqrt{x - h}$ that you can graph using transformations.	$y = 3\sqrt{x - (-2)}$ The graph of $y = \sqrt{9x + 18}$ is the graph of $y = 3\sqrt{x}$ translated 2 units to the left.

Got It? 6. a. How can you rewrite $y = \sqrt[3]{8x + 32} - 2$ so you can graph it using transformations? Describe the graph.

b. **Reasoning** Describe the graph of $y = |9x + 18|$ by rewriting it in the form $y = a|x - h|$. How is this similar to rewriting $y = \sqrt{9x + 18}$ in Problem 6?

Lesson Check

Do you know HOW?

Graph each function.

1. $y = -\sqrt{x+3}$
2. $y = -\sqrt[3]{x} + 5$

Rewrite each function so you can graph it using transformations of its parent function. Describe the graph.

3. $y = \sqrt{4x-4}$
4. $y = \sqrt[3]{8x+16}$

Do you UNDERSTAND?

5. **Writing** Explain the effect that a has on the graph of $y = a\sqrt{x}$. How does this compare to its effect on other functions you have studied?
6. **Error Analysis** Your friend states that the graph of the function $g(x) = \sqrt{-x-1}$ is a reflection of the graph of the function $f(x) = -\sqrt{x+1}$ across the x-axis. Describe your friend's error.

Practice and Problem-Solving Exercises

Graph each function. See Problems 1 and 2.

7. $y = \sqrt{x} + 1$
8. $y = \sqrt{x} - 2$
9. $y = \sqrt{x} - 4$
10. $y = \sqrt{x} + 5$
11. $y = \sqrt{x-3}$
12. $y = \sqrt{x+1}$
13. $y = \sqrt{x+6}$
14. $y = \sqrt{x-4}$

Graph each function. See Problem 3.

15. $y = 3\sqrt{x}$
16. $y = -\sqrt{x-1}$
17. $y = -5\sqrt{x+2}$
18. $y = -0.5\sqrt{x} + 3$
19. $y = \frac{1}{2}\sqrt{x+2} - 1$
20. $y = 3\sqrt{x+1} + 4$

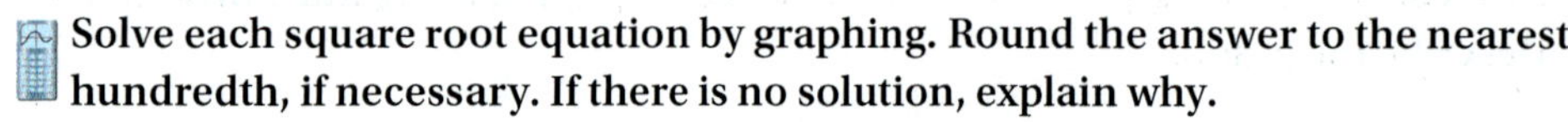

Solve each square root equation by graphing. Round the answer to the nearest hundredth, if necessary. If there is no solution, explain why.  See Problem 4.

21. $\sqrt{x-3} = 12$
22. $\sqrt{2x-3} = 4$
23. $\sqrt{2x+5} = \sqrt{2-x}$

24. **Landscaping** A sprinkler can water between 1 and 130 square yards of a lawn. The length L in inches of rotating pipe needed to water A square yards is given by the function $L = 117.75\sqrt{A}$.
    a. Graph the equation on your calculator. Make a sketch of the graph.
    b. How much area can be watered if the length of the pipe is 500, 800, or 1,300 inches long?

Graph each function.  See Problem 5.

25. $y = \sqrt[3]{x+5}$
26. $y = \sqrt[3]{x} - 4$
27. $y = \sqrt[3]{x+2} - 7$
28. $y = -\sqrt[3]{x+3} - 1$
29. $y = 2\sqrt[3]{x-6} - 9$
30. $y = \frac{1}{2}\sqrt[3]{x-1} + 3$

Rewrite each function to make it easy to graph using transformations of its parent function. Describe the graph.  See Problem 6.

31. $y = \sqrt{9x-9}$
32. $y = -\sqrt{16x+32}$
33. $y = -2\sqrt{4x+16}$
34. $y = \sqrt[3]{64x+128}$
35. $y = \sqrt{25x+125} - 3$
36. $y = \sqrt[3]{8x-24} + 1$

37. **Think About a Plan** The time t in seconds for a pendulum to complete one full cycle is given by the function $t = 1.11\sqrt{l}$, where l is the length of the pendulum in feet. How long is a pendulum that takes 4.5 seconds to complete one full cycle? 6 seconds to complete one full cycle? Round your answers to the nearest hundredth.
   - How can you use a graph to approximate the length of a pendulum?
   - How can you check your answers algebraically?

Graph each function. Find the domain and range.

38. $y = 4\sqrt[3]{x - 2} + 1$

39. $y = \frac{1}{2}\sqrt{x - 1} + 3$

40. $y = 3\sqrt[3]{x - 6} + 2$

41. Suppose that a function pairs elements from set A with elements from set B. Recall that a function is called *onto* if every element in B is paired with at least one element in A.

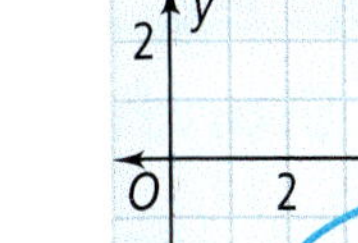

   a. The graph shows a transformation of $y = \sqrt{x}$. Write the function.
   b. What are the domain and range of the function?
   c. For the domain, is the function onto the set of nonnegative real numbers? Explain.

42. **Open-Ended** Write a radical function such that for its domain, the function is onto the set of real numbers such that $y \leq 3$.

Rewrite each function to make it easy to graph using transformations of its parent function. Describe the graph.

43. $y = \sqrt{25x - 100} - 1$

44. $y = \sqrt{36x + 108} + 4$

45. $y = -\sqrt[3]{8x - 2}$

46. $y = \sqrt{\frac{x - 1}{4}} - 2$

47. $y = 10 - \sqrt[3]{\frac{x + 3}{27}}$

48. $y = \sqrt{\frac{x}{9} + 1} + 5$

Graphing Calculator **Solve the following radical equations.**

49. $2\sqrt{x} = \sqrt{(x + 1)}$

50. $\sqrt{(x + 3)} = 4\sqrt{(x)} - 2$

51. $\sqrt[3]{x - 1} = \sqrt{x} - 1$

52. a. Solve $3 - \sqrt{(x - 3)} = x$ algebraically.
   b. Solve the equation from part (a) graphically.
   c. What do you notice about your answer to part (a) compared to your answer to part (b)?

STEM 53. **Electronics** The size of a computer monitor is given as the length of the screen's diagonal d in inches. The equation $d = \frac{5}{6}\sqrt{3A}$ models the length of a diagonal of a monitor screen with area A in square inches.
   a. Graph the equation on your calculator.
   b. Suppose you want to buy a new monitor with a screen that is twice the area of your old screen. Your old screen has a diagonal of 15 inches. What will be the diagonal of your new screen?

STEM 54. **Physics** You can model time t, in seconds, an object takes to reach the ground falling from height H, in meters, by $t(H) = \sqrt{\frac{2H}{g}}$. The value of g is 9.81 m/s^2. If an object takes 7 seconds to fall to the ground, what was its initial height?

Challenge Rewrite each function to make it easy to graph using transformations of its parent function. Describe the graph. Find the domain and range of each function.

55. $y = -\sqrt{2(4x - 3)}$ **56.** $y = \sqrt{3x - 5} + 6$ **57.** $y = -3 - \sqrt{12x + 18}$

58. a. Graph $y = \sqrt{-x}$, $y = \sqrt{1 - x}$, and $y = \sqrt{2 - x}$.

b. Make a Conjecture How does the graph of $y = \sqrt{h - x}$ differ from the graph of $y = \sqrt{x - h}$?

59. For what positive integers n are the domain and range of $y = \sqrt[n]{x}$ the set of real numbers? Assume that x is a real number.

Standardized Test Prep

SAT/ACT

60. How is the graph of $y = \sqrt{x} - 5$ translated from the graph of $y = \sqrt{x}$?

(A) shifted 5 units left
(B) shifted 5 units right
(C) shifted 5 units up
(D) shifted 5 units down

61. Which absolute value inequality has the graph shown here?

(F) $|x - 1| \leq 3$
(G) $|x - 1| \geq 3$
(H) $|x + 1| \leq 3$
(I) $|x + 1| \geq 3$

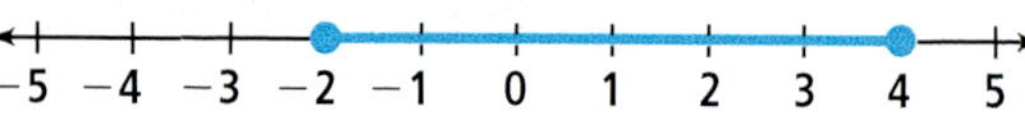

62. Which polynomial cannot be factored in the real number system?

(A) $x^2 - 3x + 2$ (B) $x^2 + 4$ (C) $4x^2 - 1$ (D) $2x^2y - 2xy^2$

Short Response

63. How do the domains and ranges of the functions $f(x) = \sqrt{x - 1}$ and $g(x) = \sqrt{x} - 1$ compare?

Mixed Review

Find the inverse of each function. Is the inverse a function? See Lesson 6-7.

64. $f(x) = \frac{2}{3}x - 3$ **65.** $f(x) = \sqrt{x + 3} - 4$ **66.** $f(x) = (2x + 1)^2$

Rationalize the denominator of each expression. Assume that all variables are positive. See Lesson 6-2.

67. $\frac{\sqrt{36x^3}}{\sqrt{12y}}$ **68.** $\sqrt{\frac{3x}{2y}}$ **69.** $\frac{\sqrt[3]{x}}{\sqrt[3]{3y}}$ **70.** $\sqrt[5]{\frac{3x^3}{2y}}$

Solve using the Quadratic Formula. See Lesson 4-7.

71. $x^2 - 9x + 15 = 0$ **72.** $3x^2 + 9x = 27$ **73.** $5x^2 + x = 3$

Get Ready! **To prepare for Lesson 7-1, do Exercises 74–76.**

Evaluate each expression for the given value of x. See Lesson 1-3.

74. 2^x for $x = 3$ **75.** 4^{x+1} for $x = 1$ **76.** 2^{3x+4} for $x = -1$

Pull It All Together

ASSESSMENT

BIG idea Solving Equations and Inequalities

Solving an equation is the process of rewriting the equation to make what it says about its variables as simple as possible.

Performance Task 1

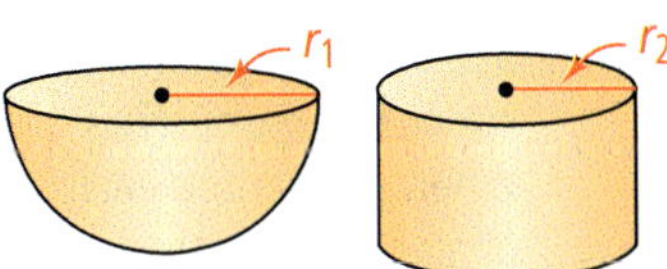

An environmental equipment supplier sells hemispherical holding ponds for treatment of chemical waste. The volume of a pond is $V_1 = \frac{1}{2}\left(\frac{4}{3}\pi r_1^3\right)$, where r_1 is the radius in feet. The supplier also sells cylindrical collecting tanks. A collecting tank fills completely and then drains completely to fill the empty pond. The volume of the tank is $V_2 = 12\pi r_2^2$, where r_2 is the radius of the tank.

a. Since $V_1 = V_2$, write an equation that shows r_1 as a function of r_2. Write an equation that shows r_2 as a function of r_1.

b. You want to double the radius of the pond. How will the radius of the tank change?

Performance Task 2

George jumps off a tire swing into a circular pond causing circles of waves. One of the waves travels at a rate of 16 ft/sec.

a. The radius of the circle increases with time. Write an equation to represent this relationship.

b. The area of the circle depends on the radius. Write an equation to represent this relationship.

c. Write a composite function that represents the relationship between area and time.

d. If the area of the pond is 22,000 square feet, how long will it take the circle to cover the area of the pond? Use this information to determine the diameter of the pond.

6 Chapter Review

Connecting BIG ideas and Answering the Essential Questions

1 Equivalence

You can simplify the nth root of an expression that contains an nth power as a factor.

$$\sqrt[n]{x^n} = x^{\frac{n}{n}} = \begin{cases} x, & n \text{ odd} \\ |x|, & n \text{ even} \end{cases}$$

Radical Expressions and Rational Exponents (Lessons 6-1, 6-2 and 6-4)

$$\sqrt[3]{-8x^5}\,\sqrt[3]{x^2} = \sqrt[3]{-8x^7}$$
$$= \sqrt[3]{(-2)^3x^6 \cdot x}$$
$$= -2x^2\sqrt[3]{x}$$
$$(-8x^5)^{\frac{1}{3}}(x^2)^{\frac{1}{3}} = (-8x^7)^{\frac{1}{3}}$$
$$= ((-2)^3 \cdot x^6 \cdot x)^{\frac{1}{3}}$$
$$= -2x^2\,x^{\frac{1}{3}}$$

Solving Square Root Equations (Lesson 6-5)

$$x - 2 = \sqrt{x}$$
$$x^2 - 4x + 4 = x$$
$$x^2 - 5x + 4 = 0$$
$$(x - 4)(x - 1) = 0$$
$$x = 4 \text{ or } x = 1$$
$$4 - 2 = \sqrt{4}\ ✓$$
$$1 - 2 \neq \sqrt{1}\ ✗$$

2 Solving Equations and Inequalities

When you square each side of an equation, the resulting equation may have more solutions than the original equation.

Inverse Relations and Functions (Lesson 6-7)

The inverse of $y = \sqrt{x} + 2, x \geq 0, y \geq 2$ is $x = \sqrt{y} + 2$, or $\sqrt{y} = x - 2$, or $y = (x - 2)^2, y \geq 0, x \geq 2$.

Graphing Radical Functions (Lesson 6-8)

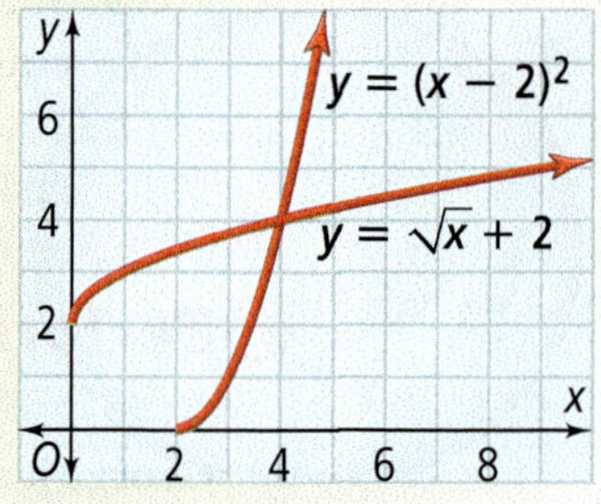

3 Function

If f and f^{-1} are inverse functions and if one maps a to b, then the other maps b to a, i.e.,

$$(f \circ f^{-1})(a) = (f^{-1} \circ f)(a) = a.$$

Chapter Vocabulary

- composite function (p. 399)
- index (p. 362)
- inverse function (p. 405)
- inverse relation (p. 405)
- like radicals (p. 374)
- nth root (p. 361)
- one-to-one function (p. 408)
- principal root (p. 361)
- radical equation (p. 390)
- radical function (p. 415)
- radicand (p. 362)
- rational exponent (p. 382)
- rationalize the denominator (p. 369)
- simplest form of a radical (p. 368)
- square root equation (p. 390)
- square root function (p. 415)

Choose the correct term to complete each sentence.

1. The number under a radical sign is called the (index/radicand).
2. (Radical functions/Inverse functions) are of the form $f(x) = \sqrt[n]{x}$.
3. A radical expression can always be rewritten using a(n) (rational exponent/inverse relation).
4. When two functions are combined so the range of one becomes the domain of the other, the resulting function is called a (square root function/composite function).

6-1 Roots and Radical Expressions

Quick Review

You can simplify a radical expression by finding the roots. The **principal root** of a number with two real roots is the positive root. The principal ***n*th root** of b is written as $\sqrt[n]{b}$, where b is the **radicand** and n is the **index** of the radical expression.

For any real number a, $\sqrt[n]{a^n} = \begin{cases} a \text{ if } n \text{ is odd} \\ |a| \text{ if } n \text{ is even} \end{cases}$.

Example

What is the simplified form of $\sqrt{36x^6}$?

$\sqrt{6^2x^6}$	Find the root of the integer.
$= \sqrt{6^2(x^3)^2}$	Find the root of the variable.
$= 6\|x^3\|$	Take the square root of each term. Since the index is even, include the absolute value symbol to ensure that the root is positive even when x^3 is negative.

Exercises

Find each real root.

5. $\sqrt{25}$
6. $\sqrt{0.49}$
7. $\sqrt[3]{-8}$
8. $-\sqrt[3]{8}$

Simplify each radical expression. Use absolute value symbols when needed.

9. $\sqrt{81x^2}$
10. $\sqrt[3]{64x^6}$
11. $\sqrt[4]{16x^{12}}$
12. $\sqrt[5]{0.00032x^5}$
13. $\sqrt{\frac{9x^4}{36}}$
14. $\sqrt[3]{125x^6y^9}$

6-2 Multiplying and Dividing Radical Expressions

Quick Review

If $\sqrt[n]{a}$ and $\sqrt[n]{b}$ are real numbers, then

$(\sqrt[n]{a})(\sqrt[n]{b}) = \sqrt[n]{ab}$, and, if $b \neq 0$, then $\frac{\sqrt[n]{a}}{\sqrt[n]{b}} = \sqrt[n]{\frac{a}{b}}$.

To **rationalize the denominator** of an expression, rewrite it so that the denominator contains no radical expressions.

Example

What is the simplest form of $\sqrt{32x^2y} \cdot \sqrt{18xy^3}$?

$\sqrt{(32x^2y)(18xy^3)}$	Combine terms.
$= \sqrt{(4^2 \cdot 2x^2y)(3^2 \cdot 2xy^3)}$	Factor.
$= \sqrt{4^2 \cdot 3^2 \cdot 2^2x^3y^4}$	Consolidate like terms.
$= \sqrt{4^2 \cdot 3^2 \cdot 2^2(x^2x)(y^2)^2}$	Identify perfect squares.
$= 4 \cdot 3 \cdot 2xy^2\sqrt{x} = 24xy^2\sqrt{x}$	Extract perfect squares.

Exercises

Multiply if possible. Then simplify.

15. $\sqrt[3]{9} \cdot \sqrt[3]{3}$
16. $\sqrt[3]{-7} \cdot \sqrt[3]{49}$
17. $\sqrt{2} \cdot \sqrt{8}$

Multiply and simplify.

18. $\sqrt{8x^2} \cdot \sqrt{2x^2}$
19. $5\sqrt[3]{9y^2} \cdot \sqrt[3]{24y}$

Divide and simplify.

20. $\sqrt{\frac{128}{8}}$
21. $\frac{\sqrt[3]{81x^5y^3}}{\sqrt[3]{3x^2}}$
22. $\frac{\sqrt[4]{162x^4}}{\sqrt[4]{2y^8}}$

Divide. Rationalize all denominators.

23. $\frac{\sqrt{8}}{\sqrt{6}}$
24. $\frac{\sqrt{3x^5}}{8x^2}$
25. $\frac{\sqrt[3]{6x^2y^4}}{2\sqrt[3]{5x^7y}}$

6-3 Binomial Radical Expressions

Quick Review

Like radicals have the same index and the same radicand. Use the distributive property to add and subtract them. Use the FOIL method to multiply binomial radical expressions. To rationalize a denominator that is a square root binomial, multiply the numerator and denominator by the conjugate of the denominator.

Example

What is the simplified form of $\sqrt{18} + \sqrt{50} - \sqrt{8}$?

$\sqrt{18} + \sqrt{50} - \sqrt{8}$

$= \sqrt{3^2 \cdot 2} + \sqrt{5^2 \cdot 2} - \sqrt{2^2 \cdot 2}$ Factor.

$= 3\sqrt{2} + 5\sqrt{2} - 2\sqrt{2}$ Simplify each radical.

$= (3 + 5 - 2)\sqrt{2}$ Combine like terms.

$= 6\sqrt{2}$ Simplify.

Exercises

Add or subtract if possible.

26. $10\sqrt{27} - 4\sqrt{12}$

27. $3\sqrt{20x} + 8\sqrt{45x} - 4\sqrt{5x}$

28. $\sqrt[3]{54x^3} - \sqrt[3]{16x^3}$

Multiply.

29. $(3 + \sqrt{2})(4 + \sqrt{2})$

30. $(\sqrt{5} + \sqrt{11})(\sqrt{5} - \sqrt{11})$

31. $(10 + \sqrt{6})(10 - \sqrt{3})$

Divide. Rationalize all denominators.

32. $\dfrac{2 + \sqrt{5}}{\sqrt{5}}$

33. $\dfrac{3 + \sqrt{18}}{1 + \sqrt{8}}$

6-4 Rational Exponents

Quick Review

You can rewrite a radical expression with a rational exponent. By definition, if the nth root of a is a real number and m is an integer, then $a^{\frac{m}{n}} = \sqrt[n]{a^m} = (\sqrt[n]{a})^m$; if m is negative then $a \neq 0$. Rational exponents can be used to simplify radical expressions.

Example

Multiply and simplify $\sqrt{x}(\sqrt[4]{x^3})$.

$\sqrt{x}(\sqrt[4]{x^3}) = x^{\frac{1}{2}} \cdot x^{\frac{3}{4}}$ Rewrite with rational exponents.

$= x^{\frac{5}{4}}$ Combine exponents.

$= \sqrt[4]{x^5}$ Rewrite as a radical expression.

Exercises

Simplify each expression.

34. $25^{\frac{1}{2}}$

35. $81^{\frac{1}{4}}$

36. $16^{\frac{1}{3}} \cdot 4^{\frac{1}{3}}$

37. $5^{\frac{3}{2}} \cdot 5^{\frac{1}{2}}$

Write each expression in simplest form.

38. $\left(x^{\frac{1}{4}}\right)^4$

39. $\left(-8y^9\right)^{\frac{1}{3}}$

40. $\left(\sqrt{9xy^2}\right)^4$

41. $\left(x^{\frac{1}{6}} y^{\frac{1}{3}}\right)^{-18}$

42. $\left(\dfrac{x^4}{x^{-1}}\right)^{-\frac{1}{5}}$

43. $\left(\dfrac{x^{\frac{1}{3}}}{y^{-\frac{2}{3}}}\right)^9$

6-5 Solving Square Root and Other Radical Equations

Quick Review

To solve a **radical equation**, you must isolate a radical expression on one side of the equation. You can then rewrite the radical expression using a rational exponent and use the reciprocal of the exponent to solve the equation.

For example, to solve a square root equation, you square each side of the equation. Check all possible solutions in the original equation to eliminate extraneous solutions.

Example

What is the solution of $4(x-2)^{\frac{2}{3}} = 16$?

$(x-2)^{\frac{2}{3}} = 4$	Isolate the radical.
$((x-2)^{\frac{2}{3}})^{\frac{3}{2}} = 4^{\frac{3}{2}}$	Raise both sides to the $\frac{3}{2}$ power.
$(x-2)^{\frac{6}{6}} = 4^{\frac{3}{2}}$	Law of exponents.
$\lvert x-2 \rvert = 8$	Simplify.
$x = 10$ or $x = -6$	Solve for x.

Exercises

Solve each equation. Check for extraneous solutions.

44. $2 + \sqrt{x+5} = 4$

45. $3\sqrt{2x+6} = 18$

46. $5(3x+1)^{\frac{1}{4}} = 10$

47. $4(3x-3)^{\frac{2}{3}} = 36$

48. $\sqrt{3x+3} - 1 = x$

49. $\sqrt{x+6} + 2 = x + 6$

50. $\sqrt{5x+1} - 2\sqrt{x} = 1$

51. $\sqrt{2x+9} - \sqrt{x} = 3$

52. Electricity The power P, in watts, that a circular solar cell produces and the radius of the cell r in centimeters are related by the square root equation $r = \sqrt{\frac{P}{0.02\pi}}$. About how much power is produced by a cell with a radius of 12 cm?

6-6 Function Operations

Quick Review

When performing function operations, you can use the same rules you used for real numbers, but you must take into consideration the domain and range of each function. The composition of function g with function f is defined as $(g \circ f)(x) = g(f(x))$.

Example

Let $f(x) = x + 3$ and $g(x) = x^2 - 2$. What is $(g \circ f)(-2)$?

$g(f(-2)) = g((-2) + 3)$	Evaluate $f(-2)$.
$= g(1)$	Simplify.
$= (1)^2 - 2$	Evaluate $g(f(-2))$.
$= -1$	Simplify.

Therefore, $(g \circ f)(-2) = -1$

Exercises

Let $f(x) = x - 4$ and $g(x) = x^2 - 16$. Perform each function operation and then find the domain.

53. $f(x) + g(x)$

54. $g(x) - f(x)$

55. $f(x) \cdot g(x)$

56. $\frac{g(x)}{f(x)}$

Let $g(x) = 5x - 2$ and $h(x) = x^2 + 1$. Find the value of each expression.

57. $(h \circ g)(-1)$

58. $(h \circ g)(0)$

59. $(g \circ h)(2)$

60. $(g \circ h)(a)$

61. Discounts A grocery store is offering a 50% discount off a \$4.00 box of cereal. You also have a \$1.00 off coupon for the same cereal. Use a composite function to show whether it is better to use the coupon before or after the store discount.

6-7 Inverse Relations and Functions

Quick Review

If a relation or a function is described by an equation in x and y, you can interchange x and y to get the inverse. The domain of a function becomes the range of its inverse, and the range of a function becomes the domain of its inverse.

Example

What is the inverse of $f(x) = \sqrt{x - 10}$?

$y = \sqrt{x - 10}$	Rewrite using y.
$x = \sqrt{y - 10}$	Interchange the x and y values.
$x^2 = y - 10$	Square each side.
$y = x^2 + 10$	Solve for y.
$f^{-1}(x) = x^2 + 10$	Write the inverse function.

The domain of $f(x)$ is $x \geq 10$, which means the range of $f^{-1}(x)$ is $y \geq 10$. Also, since the range of $f(x)$ is $y \geq 0$, the domain of $f^{-1}(x)$ is $x \geq 0$.

Exercises

Find the inverse of each function. Determine whether each inverse is a function.

62. $f(x) = 2x^2 - 8$

63. $f(x) = 15 - 3x$

64. $f(x) = \sqrt{x + 6}$

65. $f(x) = (2x - 3)^2$

Graph each function and its inverse. Describe the domain and range of each.

66. $f(x) = 4x - 1$

67. $f(x) = (x + 3)^2$

68. $f(x) = \sqrt{x - 3}$

69. $f(x) = 6 - 5x^2$

70. Geometry The volume of cube is determined by the formula $V = s^3$, where s is the length of one side. Find the inverse formula. Use it to find the side length of a cube with a volume of 64 ft^3.

6-8 Graphing Radical Functions

Quick Review

The function $f(x) = \sqrt{x}$ is the parent function of the **square root function** $f(x) = a\sqrt{x - h} + k$. The graph of $f(x) = a\sqrt{x}$ is a stretch $(a > 1)$ or a shrink $(0 < a < 1)$ of the parent function. The graph of $f(x) = a\sqrt{x - h} + k$ is a translation h units horizontally and k units vertically of $y = a\sqrt{x}$. The graph of $f(x) = \sqrt[n]{x}$ is transformed by a, h, and k in the same way as the graph of $f(x) = \sqrt{x}$.

Example

Describe the graph of $y = \sqrt{4x + 12}$.

$y = \sqrt{4x + 12}$	
$y = \sqrt{4(x + 3)}$	Factor the polynomial.
$y = 2\sqrt{x + 3}$	Simplify the radical.

The graph of $y = \sqrt{4x + 12}$ is the graph of $y = 2\sqrt{x}$ translated 3 units to the left.

Exercises

Graph each function. Find the domain and range.

71. $y = \sqrt{x} - 5$

72. $y = \sqrt{x + 8}$

73. $y = 5\sqrt{x} + 9$

74. $y = -\sqrt{x - 4}$

75. $y = \sqrt[3]{x + 10}$

76. $y = -\sqrt[3]{x - 2} + 5$

Rewrite each function to make it easy to graph using transformations. Describe each graph.

77. $y = \sqrt{9x - 27} + 4$

78. $y = -3\sqrt{4x - 16}$

79. $y = \sqrt[3]{8x + 24}$

80. $y = \sqrt{\frac{x - 4}{4}} + 6$

Solve each equation by graphing.

81. $5 = -\sqrt{x - 3}$

82. $\sqrt{8x - 16} = 2\sqrt{x + 2}$

6 Chapter Test

Do you know HOW?

Simplify each radical expression. Use absolute value symbols when needed.

1. $\sqrt{54x^3y^5}$
2. $\sqrt[3]{-0.027}$
3. $\sqrt[5]{-64x^{14}y^{20}}$

Simplify each expression. Rationalize all denominators.

4. $\sqrt{7x^3} \cdot \sqrt{14x}$
5. $\frac{1 - \sqrt{3x}}{\sqrt{6x}}$
6. $\sqrt{48} + 2\sqrt{27} + 5\sqrt{12}$
7. $(3 + 2\sqrt{5})(1 - \sqrt{20})$
8. $4\sqrt{7xz} + 2\sqrt{7xz}$
9. $\frac{5\sqrt{2}}{\sqrt{7} - \sqrt{2}}$

Simplify each expression.

10. $(125)^{-\frac{2}{3}}$
11. $x^{\frac{1}{6}} \cdot x^{\frac{1}{3}}$
12. $\left(\frac{8x^9y^3}{27x^2y^{12}}\right)^{\frac{2}{3}}$
13. $\sqrt{8x^5} - \sqrt{18x^5}$

Solve each equation. Check for extraneous solutions.

14. $\sqrt{x} - 3 = x - 5$
15. $\sqrt{x} + 4 = \sqrt{3x}$
16. $2(x - 1)^{\frac{3}{4}} = 16$
17. $\sqrt{x + 3} - 1 = x$

Let $f(x) = x - 2$ and $g(x) = x^2 - 3x + 2$. Perform each function operation and then find the domain.

18. $-2g(x) + f(x)$
19. $-f(x) \cdot g(x)$
20. $\frac{g(x)}{f(x)}$

Find each product or quotient.

21. $\sqrt{5}(\sqrt[4]{5})$
22. $\frac{\sqrt{x^3}}{\sqrt[5]{x^2}}$

For each pair of functions, find $(g \circ f)(x)$ and $(f \circ g)(x)$.

23. $f(x) = x^2 - 2$, $g(x) = 4x + 1$
24. $f(x) = 2x^2 + x - 7$, $g(x) = -3x - 1$

Find the inverse of each function. Is the inverse a function?

25. $f(x) = (x + 3)^2 + 1$
26. $f(x) = \sqrt{2x + 1}$
27. $g(x) = 3x^3 - 4$
28. $f(x) = \frac{1}{4}x$

Rewrite each function to make it easy to graph using transformations. Describe the graph.

29. $y = \sqrt{16x + 80} - 1$
30. $y = \sqrt{9x + 3}$

Graph. Find the domain and range of each function.

31. $y = 2\sqrt{x} + 3$
32. $y = -\sqrt{2x + 3}$
33. $y = \sqrt{x + 3} - 4$

Do you UNDERSTAND?

34. **Writing** Explain why -108 has no real 6th roots.

35. **Open-Ended** Write a relation that is not a function, but whose inverse is a function.

36. **Measurement** The time t in seconds for a swinging pendulum to complete one full cycle is given by the function $t = 0.2\sqrt{l}$, where l is the length of the pendulum in centimeters. To the nearest tenth, how long is a full cycle if the pendulum is 10 cm long? 20 cm long? How long, in centimeters, is a pendulum that takes 2 seconds for one full cycle?

6 Cumulative Standards Review

ASSESSMENT

TIPS FOR SUCCESS

Some problems require you to find the inverse of a function.

What is the inverse of the function $y = x^2 + 3$?

(A) $y = x - 3$

(B) $y = \pm\sqrt{x - 3}$

(C) $y = \pm\sqrt{x^2 + 3}$

(D) $y = (x - 3)^2$

TIP 1

To find the inverse of a function, interchange x and y.

TIP 2

After you interchange x and y, solve for y.

Think It Through

$y = x^2 + 3$

$x = y^2 + 3$

$x - 3 = y^2$

$\pm\sqrt{x - 3} = y$

$y = \pm\sqrt{x - 3}$

The correct answer is B.

Vocabulary Review

As you solve test items, you must understand the meanings of mathematical terms. Match each term with its mathematical meaning.

A. radicand

B. index

C. composite function

D. inverse functions

E. radical function

I. the combination of two functions such that the output from the first becomes the input for the second

II. the degree of a root in a radical expression

III. the number under the radical sign in a radical expression

IV. a function that can be written in the form $f(x) = a\sqrt[n]{x - h} + k$

V. the range of one function is the domain of the other and vice versa

Multiple Choice

Read each question. Then write the letter of the correct answer on your paper.

1. Find all the roots of $2x^4 + x^3 - 8x^2 - 4x = 0$.

(A) $x = -2, x = -0.5, x = 0, x = 2$

(B) $x = -2, x = -0.5, x = 2$

(C) $x = -2, x = 0.5, x = 0, x = 2$

(D) $x = -2, x = 0.5, x = 2$

2. Solve the equation $ax^2 + bx + c = 0$ for b.

(F) $b = -cx - ax^2$

(G) $b = \frac{-c - ax^2}{x}$

(H) $b = -(cx - ax^2)$

(I) $b = \frac{-(c - ax^2)}{x}$

3. Use the sum of cubes formula to factor $x^3 + 64$.

(A) $(x + 4)(x^2 - 4x + 4)$

(B) $(x + 4)(x^2 + 4x + 4)$

(C) $(x + 4)(x^2 - 4x + 16)$

(D) $(x + 4)(x^2 + 4x + 16)$

4. The time it takes to copy pages varies directly with the number of pages being copied. The copier at your office can copy 21 color pages per minute and 40 black and white pages per minute. Approximately how long will it take to copy 60 color pages and 35 black and white pages?

Ⓕ 0.9 minute
Ⓗ 2.9 minutes
Ⓖ 2.5 minutes
Ⓘ 3.7 minutes

5. Which equation is modeled by the graph?

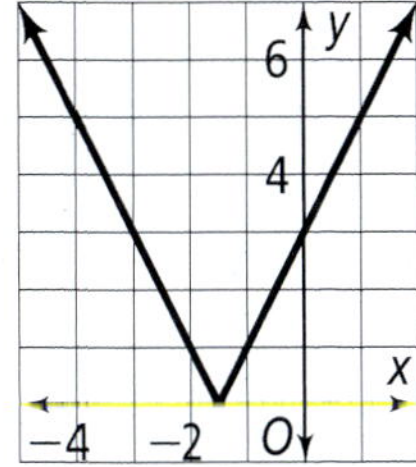

Ⓐ $y = |2x - 3|$
Ⓒ $y = |2x + 3|$
Ⓑ $y = 2|x - 3|$
Ⓓ $y = 2|x + 3|$

6. What are the vertex and axis of symmetry for the given parabola?

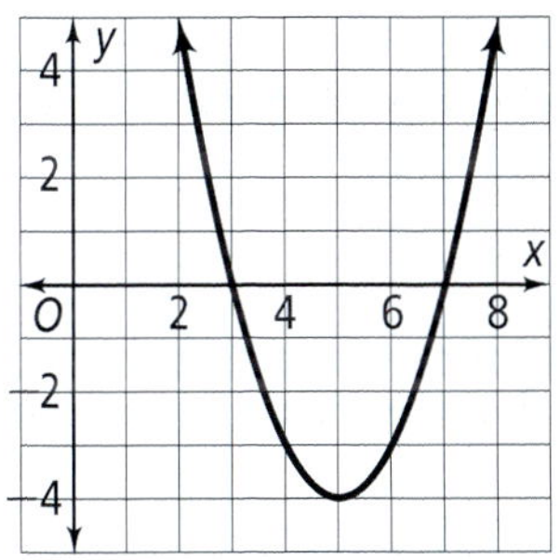

Ⓕ $(-4, 5), y = 5$
Ⓖ $(-4, 5), x = 5$
Ⓗ $(5, -4), y = -4$
Ⓘ $(5, -4), x = 5$

7. What is the product of $\sqrt[3]{3}$ and $\sqrt[5]{3}$?

Ⓐ $\sqrt[8]{3}$
Ⓑ $\sqrt[8]{9}$
Ⓒ $\sqrt[15]{3^8}$
Ⓓ $\sqrt[8]{3^{15}}$

8. Solve $7x^2 + 196 = 0$ for x.

Ⓕ $\pm 4i\sqrt{7}$
Ⓖ $\pm 4\sqrt{7}$
Ⓗ $\pm 2i\sqrt{7}$
Ⓘ $\pm 2\sqrt{7}$

9. Which inequality is modeled by the graph?

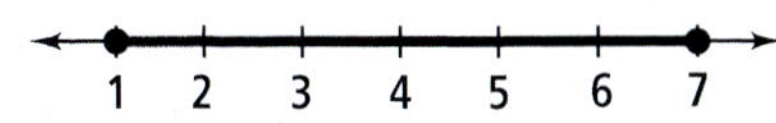

Ⓐ $k + 1 \leq 7$
Ⓑ $|k + 1| \leq 7$
Ⓒ $k - 4 \leq 3$
Ⓓ $|k - 4| \leq 3$

10. What is the solution of the system? $\begin{cases} 4x + 2y = 4 \\ 6x + 2y = 8 \end{cases}$

Ⓕ $(-2, 2)$
Ⓖ $(2, -2)$
Ⓗ $(1, 2)$
Ⓘ $(-1, 2)$

11. A photographer is promoting three photo specials. How much does it cost for each type of print?

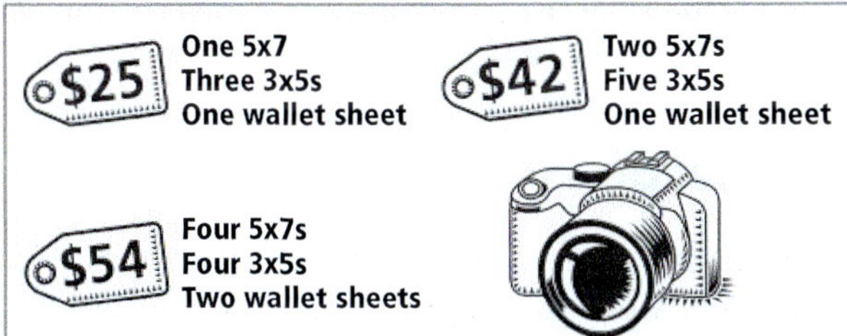

Ⓐ 5×7 costs \$7, 3×5 costs \$5, Wallet costs \$3
Ⓑ 5×7 costs \$11, 3×5 costs \$7, Wallet costs \$3
Ⓒ 5×7 costs \$12, 3×5 costs \$11, Wallet costs \$7
Ⓓ 5×7 costs \$7, 3×5 costs \$5, Wallet costs \$5

12. What is an equivalent form of $\frac{5}{2 + 2i}$?

Ⓕ $\frac{5}{4i}$
Ⓗ $\frac{5 + 5i}{4}$
Ⓖ $\frac{10 - 10i}{4 - 4i}$
Ⓘ $\frac{5 - 5i}{4}$

GRIDDED RESPONSE

13. Let $g(x) = x - 3$ and $h(x) = x^2 + 6$. What is $h(1) \times g(1)$?

14. A laptop comes without any programs installed on it. Each program costs \$20 and the laptop you want costs \$319. What is the greatest number of programs you can buy if you want to spend at most \$500 for the laptop?

15. You are building an entertainment center with shelves that are x in. deep by x in. long. The height of the unit will be twice the depth. If the volume of the unit will be 8,192 in.^{3}, what is the height, in inches, of the entertainment center?

16. What is the solution of the equation $x^2 - 24x + 144 = 0$?

17. What is the number of real roots of the equation $2x^2 + 3x = -4$?

18. What is the quotient $\dfrac{\sqrt[3]{8x^6y^{12}}}{\sqrt{4x^4y^8}}$?

19. What is the solution of $4 + \sqrt{3x + 5} = 7$?

20. All 385 tickets for a high-school play sold in 10 days. The ticket receipts totaled \$1960. If the cost of a child's ticket was \$4 and the cost of an adult's ticket was \$6, how many adult tickets were sold?

21. What is the x-value of the x-intercept of the graph of $f(x) = x^2 + 4x + 4$?

Short Response

22. The graph shows a transformation of $f(x) = x^2$. What is an equation of the graph? Explain your answer by using translations of the parent quadratic function.

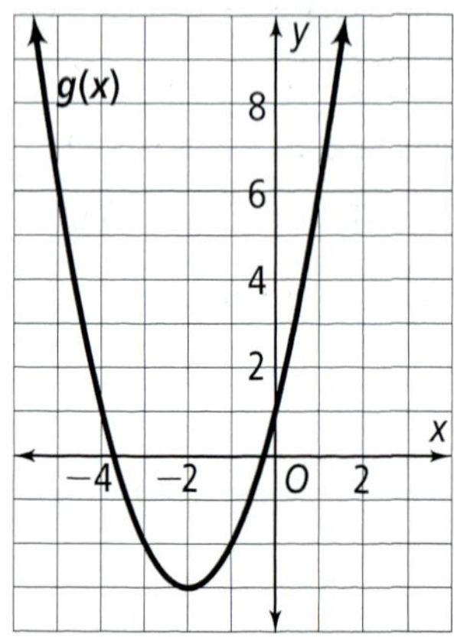

23. The total cost (in cents) of $x + 2$ markers is $x^3 + 5x^2 + 2x - 8$.

a. Write an expression that models the cost of each marker.

b. If you buy 7 markers, how much would each marker cost?

24. You are given that $f(x) = x^2 + 4$ and $g(x) = 3x - 1$.

a. What are the domain and range of $f(x)$ and $g(x)$?

b. Find $f(x) + g(x)$.

c. What is the domain of your answer to part (b)?

25. Two friends went shopping together. One friend bought 2 hats and 1 shirt and spent \$70, while the other friend bought 1 hat and 3 shirts and spent \$85. Use a graph to determine the costs of each shirt and hat.

26. A student found that a cubic function has zeros 16 and $1 - 2i$ with a leading coefficient of 3. What is the constant term of this polynomial function with real coefficients?

27. Describe the graph of the polynomial function $f(x) = -x^6 + 3x^5 + 4x - 10$. What is its end behavior?

Decide whether the following statements are *always*, *sometimes*, or *never* true.

28. If n is a real number, then $0^n = 0$.

29. If a and b are rational numbers, then the product of $(a + \sqrt{b})$ and its conjugate is a rational number.

Extended Response

30. An online music store is having a promotion. Customers receive a \$5 rebate if they buy any regular priced CD at \$13 each. They can also receive 15% off if they register as a store member.

a. What functions model the two discounts?

b. In which order should the discounts be applied for the customer to receive the greatest discount?

c. Use your answer from part (b) to determine the amount a customer will save if she buys 5 CDs.

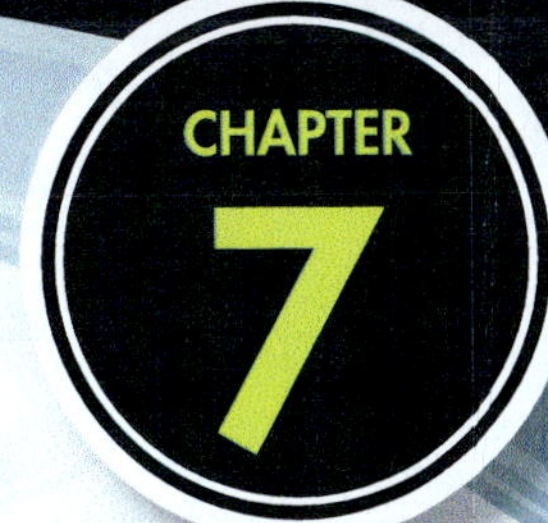

Get Ready!

Lesson 1-3

Evaluating Expressions

Evaluate each expression for $x = -2, 0,$ and 2.

1. 10^{x+1} **2.** $\left(\frac{3}{2}\right)^x$ **3.** -5^{x-2} **4.** $-(3)^{0.5x}$

Lesson 2-5

Using Linear Models

Draw a scatter plot and find the line of best fit for each set of data.

5. (0, 2), (1, 4), (2, 6.5), (3, 8.5), (4, 10), (5, 12), (6, 14)

6. (3, 100), (5, 150), (7, 195), (9, 244), (11, 296), (13, 346), (15, 396)

Lessons 4-1 and 5-9

Graphing Transformations

Identify the parent function of each equation. Graph each equation as a transformation of its parent function.

7. $y = (x + 5)^2 - 3$ **8.** $y = -2(x - 6)^3$

Lesson 6-4

Simplifying Rational Exponents

Simplify each expression.

9. $\left(x^{\frac{1}{5}}\right)^{10}$ **10.** $\left(-8x^3\right)^{\frac{4}{3}}$

Lesson 6-7

Finding Inverses

Find the inverse of each function. Is the inverse a function?

11. $y = 10 - 2x^2$ **12.** $y = (x + 4)^3 - 1$

Looking Ahead Vocabulary

13. In advertising, the *decay factor* describes how an advertisement loses its effectiveness over time. In math, would you expect a decay factor to increase or decrease the value of y as x increases?

14. There are many different kinds of growth patterns. Patterns that increase by a constant rate are linear. Patterns that grow *exponentially* increase by an ever-increasing rate. If your allowance doubles each week, does that represent linear growth or exponential growth?

15. The word *asymptote* comes from a Greek word meaning "not falling together." When looking at the end behavior of a function, do you expect the graph to intersect its asymptote?

Exponential and Logarithmic Functions

PowerAlgebra.com

Your place to get all things digital

Download videos connecting math to your world.

Math definitions in English and Spanish

The online Solve It will get you in gear for each lesson.

Interactive! Vary numbers, graphs, and figures to explore math concepts.

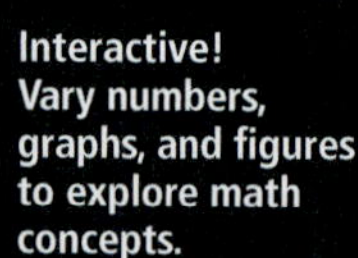

Online access to stepped-out problems aligned to Common Core

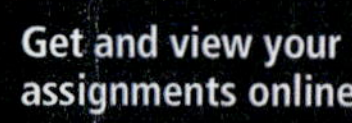

Get and view your assignments online.

Extra practice and review online

- Linear and Exponential Models
- Creating Equations that Describe Numbers
- Interpreting Functions

Logarithms provide a way to work with the inverses of exponential functions. Exponential functions model what some might call "explosive" growth, but logarithmic values grow very slowly. Decibels are logarithms that measure sound, and when sound energy increases dramatically, the decibel values creep upward. A few extra decibels can bust your eardrums!

Vocabulary

English/Spanish Vocabulary Audio Online:

English	Spanish
asymptote, *p. 435*	asíntota
Change of Base Formula, *p. 464*	fórmula de cambio de base
common logarithm, *p. 453*	logaritmo común
exponential equation, *p. 469*	ecuación exponencial
exponential function, *p. 434*	función exponencial
exponential growth, *p. 435*	incremento exponencial
logarithm, *p. 451*	logaritmo
logarithmic equation, *p. 471*	ecuación logarítmica
logarithmic function, *p. 454*	función logarítmica
natural logarithmic function, *p. 478*	función logarítmica natural

BIG ideas

1 **Modeling**

Essential Question How do you model a quantity that changes regularly over time by the same percentage?

2 **Equivalence**

Essential Question How are exponents and logarithms related?

3 **Function**

Essential Question How are exponential functions and logarithmic functions related?

Chapter Preview

Exploring Exponential Models

Content Standards

F.IF.7.e Graph exponential . . . functions, showing intercepts and end behavior . . .

A.CED.2 Create equations in two or more variables to represent relationships between quantities . . .

Also F.IF.8, A.SSE.1.b

Objective To model exponential growth and decay

Hmmm, I wonder if there is a shortcut?

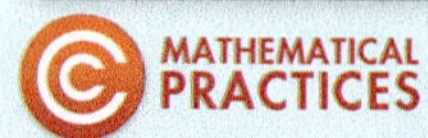

SOLVE IT! Getting Ready!

You are to move the stack of 5 rings to another post. Here are the rules.

- A move consists of taking the top ring from one post and placing it onto another post.
- You can move only one ring at a time.
- Do not place a ring on top of a smaller ring.

What is the fewest number of moves needed? How many moves are needed for 10 rings? 20 rings? Explain.

Lesson Vocabulary
- exponential function
- exponential growth
- exponential decay
- asymptote
- growth factor
- decay factor

The number of moves needed for additional rings in the Solve It suggests a pattern that approximates repeated multiplication.

Essential Understanding You can represent repeated multiplication with a function of the form $y = ab^x$ where b is a positive number other than 1.

An **exponential function** is a function with the general form $y = ab^x$, $a \neq 0$, with $b > 0$, and $b \neq 1$. In an exponential function, the base b is a constant. The exponent x is the independent variable with domain the set of real numbers.

Problem 1 Graphing an Exponential Function

Plan

How does making a table help you sketch the graph?
The table shows coordinates of several points on the graph.

What is the graph of $y = 2^x$?

Step 1 Make a table of values.

x	2^x	y
−4	2^{-4}	$\frac{1}{16} = 0.0625$
−3	2^{-3}	$\frac{1}{8} = 0.125$
−2	2^{-2}	$\frac{1}{4} = 0.25$
−1	2^{-1}	$\frac{1}{2} = 0.5$

x	2^x	y
0	2^0	1
1	2^1	2
2	2^2	4
3	2^3	8

Step 2 Plot and connect the points.

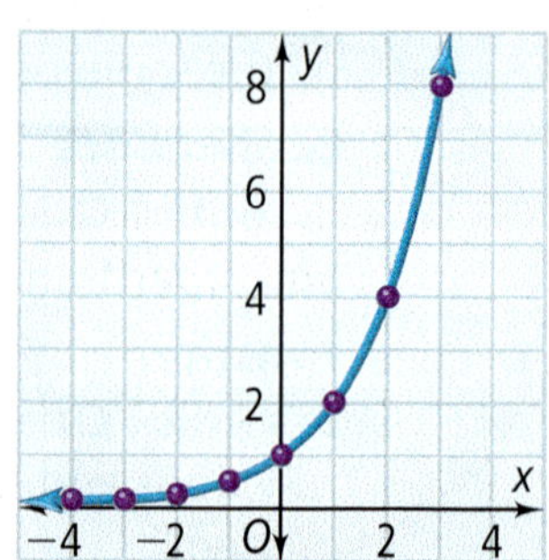

  **Got It?** 1. What is the graph of each function?

a. $y = 4^x$ b. $y = \left(\frac{1}{3}\right)^x$ c. $y = 2(3)^x$

d. **Reasoning** What generalizations can you make about the domain, range, and y-intercepts of these functions?

Two types of exponential behavior are *exponential growth* and *exponential decay*.

For **exponential growth**, as the value of x increases, the value of y increases. For **exponential decay**, as the value of x increases, the value of y decreases, approaching zero.

The exponential functions shown here are *asymptotic* to the x-axis. An **asymptote** is a line that a graph approaches as x or y increases in absolute value.

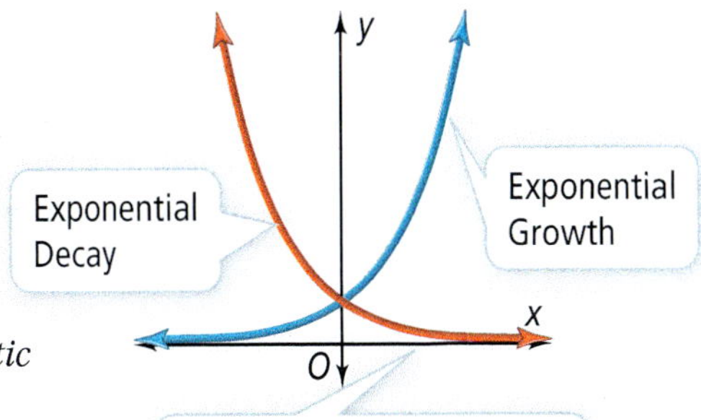

take note

Concept Summary Exponential Functions

For the function $y = ab^x$,

- if $a > 0$ and $b > 1$, the function represents exponential growth.
- if $a > 0$ and $0 < b < 1$, the function represents exponential decay.

In either case, the y-intercept is $(0, a)$, the domain is all real numbers, the asymptote is $y = 0$, and the range is $y > 0$.

Problem 2 Identifying Exponential Growth and Decay

Identify each function or situation as an example of exponential growth or decay. What is the y-intercept?

A $y = 12(0.95)^x$

Since $0 < b < 1$, the function represents exponential decay. The y-intercept is $(0, a) = (0, 12)$.

B $y = 0.25(2)^x$

Since $b > 1$, the function represents exponential growth. The y-intercept is $(0, a) = (0, 0.25)$.

C **You put \$1000 into a college savings account for four years. The account pays 5% interest annually.**

The amount of money in the bank grows by 5% annually. It represents exponential growth. The y-intercept is 1000, which is the dollar value of the initial investment.

Think

What quantity does the y-intercept represent?
The y-intercept is the amount of money at $t = 0$, which is the initial investment.

 **Got It?** 2. Identify each function or situation as an example of exponential growth or decay. What is the y-intercept?

a. $y = 3(4^x)$ b. $y = 11(0.75^x)$

c. You put \$2000 into a college savings account for four years. The account pays 6% interest annually.

For exponential growth $y = ab^x$, with $b > 1$, the value b is the **growth factor**. A quantity that exhibits exponential growth increases by a constant percentage each time period. The percentage increase r, written as a decimal, is the *rate of increase* or *growth rate*. For exponential growth, $b = 1 + r$.

For exponential decay, $0 < b < 1$ and b is the **decay factor**. The quantity decreases by a constant percentage each time period. The percentage decrease, r, is the *rate of decay*. Usually a rate of decay is expressed as a negative quantity, so $b = 1 + r$.

take note

Key Concept Exponential Growth and Decay

You can model exponential growth or decay with this function.

Amount after t time periods

Rate of growth ($r > 0$) or decay ($r < 0$)

$$A(t) = a(1 + r)^t$$

Initial amount

Number of time periods

For growth or decay to be exponential, a quantity changes by a fixed percentage each time period.

Problem 3 Modeling Exponential Growth

You invested $1000 in a savings account at the end of 6th grade. The account pays 5% annual interest. How much money will be in the account after six years?

Step 1 Determine if an exponential function is a reasonable model.

The money grows at a fixed rate of 5% per year. An exponential model is appropriate.

Step 2 Define the variables and determine the model.

Let t = the number of years since the money was invested.
Let $A(t)$ = the amount in the account after each year.

A reasonable model is $A(t) = a(1 + r)^t$.

Step 3 Use the model to solve the problem.

$A(6) = 1000(1 + 0.05)^6$ Substitute $a = 1000$, $r = 0.05$, and $t = 6$.

$= 1000(1.05)^6$ Simplify.

$\approx \$1340.10$

The account contains $1340.10 after six years.

Think

What is the growth rate r?
It is the annual interest rate, written as a decimal: $5\% = 0.05$.

Got It? 3. Suppose you invest $500 in a savings account that pays 3.5% annual interest. How much will be in the account after five years?

Problem 4 Using Exponential Growth

Suppose you invest $1000 in a savings account that pays 5% annual interest. If you make no additional deposits or withdrawals, how many years will it take for the account to grow to at least $1500?

Plan

How can you make a table to solve this problem? Define the variables, write an equation and enter it into a graphing calculator. Then you can inspect a table to find the solution.

Think	Write
Define the variables.	Let t = the number of years. Let A(t) = the amount in the account after t years.
Determine the model.	$A(t) = 1000(1 + 0.05)^t$ $= 1000(1.05)^t$
Make a table using the table feature on a graphing calculator. Find the input when the output is 1500.	(calculator table below)
The account pays interest only once a year. The balance after the 8th year is not yet $1500.	The account will not contain $1500 until the ninth year. After nine years, the balance will be $1551.33.

X	Y1
4	1215.5
5	1276.3
6	1340.1
7	1407.1
8	1477.5
9	1551.3
10	1628.9

Y1=1551.32821598

  **Got It?** **4. a.** Suppose you invest $500 in a savings account that pays 3.5% annual interest. When will the account contain at least $650?

b. Reasoning Use the table in Problem 4 to determine when that account will contain at least $1650. Explain.

Exponential functions are often discrete. In Problem 4, interest is paid only once a year. So the graph consists of individual points corresponding to $t = 1, 2, 3$, and so on. It is not continuous. Both the table and the graph show that there is never *exactly* $1500 in the account and that the account will not contain more than $1500 until the ninth year.

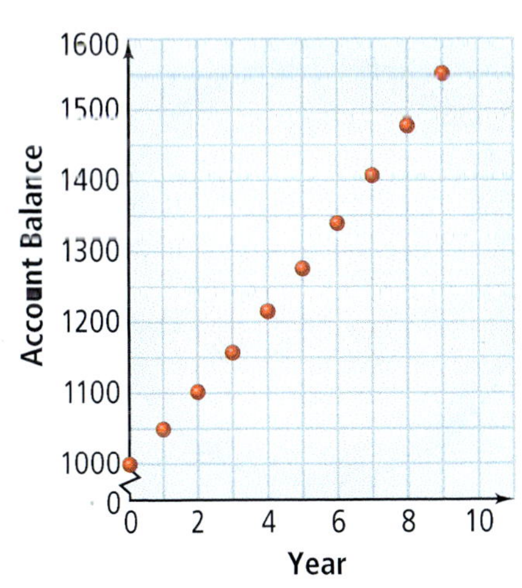

To model a discrete situation using an exponential function of the form $y = ab^x$, you need to find the growth or decay factor b. If you know y-values for two consecutive x-values, you can find the rate of change r, and then find b using $r = \frac{(y_2 - y_1)}{y_1}$ and $b = 1 + r$.

Problem 5 Writing an Exponential Function STEM

Endangered Species **The table shows the world population of the Iberian lynx in 2003 and 2004. If this trend continues and the population is decreasing exponentially, how many Iberian lynx will there be in 2014?**

World Population of Iberian Lynx		
Year	2003	2004
Population	150	120

Use the general form of the exponential equation, $y = ab^x = a(1 + r)^x$.

Step 1 Define the variables.

Let x = the number of years since 2003.
Let y = the population of the Iberian lynx.

Think

How can you find the value of r?
You can use the populations for two consecutive years to find r.

Step 2 Determine r.

Use the populations for 2003 and 2004.

$$r = \frac{y_2 - y_1}{y_1} = \frac{120 - 150}{150} = -0.2$$

Step 3 Use r to determine b.

$$b = 1 + r = 1 + (-0.2) = 0.8$$

Step 4 Write the model.

$$y = ab^x$$
$$150 = a(0.8)^0$$
$$150 = a$$

Solve for a using the initial values $x = 0$ and $y = 150$.

The model is $y = 150(0.8)^x$.

Think

How do you find the x-value corresponding to 2014?
The initial x-value corresponds to 2003, so find the difference.

Step 5 Use the model to find the population in 2014.
For the year 2014, $x = 2014 - 2003 = 11$.

$$y = 150(0.8)^x = 150(0.8)^{11} \approx 13$$

If the 2003–2004 trend continues, there will be approximately 13 Iberian lynx in the wild in 2014.

Got It? **5. a.** For the model in Problem 5, what will be the world population of Iberian lynx in 2020?

b. Reasoning If you graphed the model in Problem 5, would it ever cross the x-axis? Explain.

Lesson Check

Do you know HOW?

Without graphing, determine whether the function represents exponential growth or exponential decay. Then find the y-intercept.

1. $y = 10(0.45)^x$ **2.** $y = 0.75(4)^x$

3. $y = 3^x$ **4.** $y = 0.95^x$

Graph each function.

5. $A(t) = 3(1.04)^t$ **6.** $A(t) = 7(0.6)^t$

Do you UNDERSTAND?

7. Vocabulary Explain how you can tell if $y = ab^x$ represents exponential growth or exponential decay.

8. Reasoning Identify each function as *linear*, *quadratic*, or *exponential*. Explain your reasoning.

a. $y = 3(x + 1)^2$ **b.** $y = 4(3)^x$
c. $y = 2x + 5$ **d.** $y = 4(0.2)^x + 1$

9. Error Analysis A classmate says that the growth factor of the exponential function $y = 15(0.3)^x$ is 0.3. What is the student's mistake?

Practice and Problem-Solving Exercises

Graph each function. See Problem 1.

10. $y = 6^x$ **11.** $y = 3(10)^x$ **12.** $y = 1000(2)^x$ **13.** $y = 9(3)^x$

14. $f(x) = 2(3)^x$ **15.** $s(t) = 1.5^t$ **16.** $y = 8(5)^x$ **17.** $y = 2^{2x}$

Without graphing, determine whether the function represents exponential growth or exponential decay. Then find the y-intercept. See Problem 2.

18. $y = 129(1.63)^x$ **19.** $f(x) = 2(0.65)^x$ **20.** $y = 12\left(\frac{17}{10}\right)^x$ **21.** $y = 0.8\left(\frac{1}{8}\right)^x$

22. $f(x) = 4\left(\frac{5}{6}\right)^x$ **23.** $y = 0.45(3)^x$ **24.** $y = \frac{1}{100}\left(\frac{4}{3}\right)^x$ **25.** $f(x) = 2^{-x}$

26. Interest Suppose you deposit \$2000 in a savings account that pays interest at an annual rate of 4%. If no money is added or withdrawn from the account, answer the following questions. See Problems 3 and 4.

a. How much will be in the account after 3 years?
b. How much will be in the account after 18 years?
c. How many years will it take for the account to contain \$2500?
d. How many years will it take for the account to contain \$3000?

Write an exponential function to model each situation. Find each amount after the specified time. See Problem 5.

27. A population of 120,000 grows 1.2% per year for 15 years.

28. A population of 1,860,000 decreases 1.5% each year for 12 years.

29. a. Sports Before a basketball game, a referee noticed that the ball seemed under-inflated. She dropped it from 6 feet and measured the first bounce as 36 inches and the second bounce as 18 inches. Write an exponential function to model the height of the ball.
b. How high was the ball on its fifth bounce?

30. **Think About a Plan** Your friend invested \$1000 in an account that pays 6% annual interest. How much interest will your friend have after her college graduation in 4 years?

- Is an exponential model reasonable for this situation?
- What equation should you use to model this situation?
- Is the solution of the equation the final answer to the problem?

STEM 31. **Oceanography** The function $y = 20(0.975)^x$ models the intensity of sunlight beneath the surface of the ocean. The output y represents the percent of surface sunlight intensity that reaches a depth of x feet. The model is accurate from about 20 feet to about 600 feet beneath the surface.

a. Find the percent of sunlight 50 feet beneath the surface of the ocean.
b. Find the percent of sunlight at a depth of 370 feet.

32. **Population** The population of a certain animal species decreases at a rate of 3.5% per year. You have counted 80 of the animals in the habitat you are studying.

a. Write a function that models the change in the animal population.
b. **Graphing Calculator** Graph the function. Estimate the number of years until the population first drops below 15 animals.

33. **Sports** While you are waiting for your tennis partner to show up, you drop your tennis ball from 5 feet. Its rebound was approximately 35 inches on the first bounce and 21.5 inches on the second. What exponential function would be a good model for the bouncing ball?

For each annual rate of change, find the corresponding growth or decay factor.

34. +70%	35. +500%	36. −75%	37. −55%
38. +12.5%	39. −0.1%	40. +0.1%	41. +100%

42. **Manufacturing** The value of an industrial machine has a decay factor of 0.75 per year. After six years, the machine is worth \$7500. What was the original value of the machine?

43. **Zoology** Determine which situation best matches the graph.

(A) A population of 120 cougars decreases 98.75% yearly.
(B) A population of 120 cougars increases 1.25% yearly.
(C) A population of 115 cougars decreases 1.25% yearly.
(D) A population of 115 cougars decreases 50% yearly.

44. **Open-Ended** Write a problem that could be modeled with $y = 20(1.1)^x$.

45. **Reasoning** Which function does the graph represent? Explain. (Each interval represents one unit.)

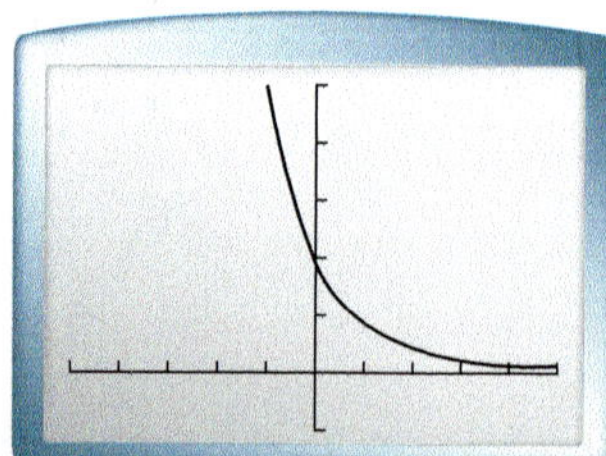

(A) $y = \left(\frac{1}{3}\right)2^x$

(B) $y = 2\left(\frac{1}{3}\right)^x$

(C) $y = -2\left(\frac{1}{3}\right)^x$

Standardized Test Prep

SAT/ACT

46. Which function represents the value after x years of a new delivery van that costs \$25,000 and depreciates 15% each year?

(A) $y = -15(25{,}000)^x$ (C) $y = 25{,}000(0.85)^x$

(B) $y = 25{,}000(0.15)^x$ (D) $y = 25{,}000(1.15)^x$

47. What is $f(x) = 3x^{\frac{1}{3}}$ for $x = \frac{1}{125}$?

(F) 15 (G) $\frac{3}{5}$ (H) $\frac{\sqrt[3]{3}}{5}$ (I) $5\sqrt[3]{3}$

48. What is the simplified form of $\frac{2 + i}{2 - i}$?

(A) -1 (B) $\frac{3 + 4i}{3}$ (C) $\frac{5 + 4i}{5}$ (D) $\frac{3 + 4i}{5}$

49. Which graph represents the equation $y = x^2 - x - 2$?

(F) 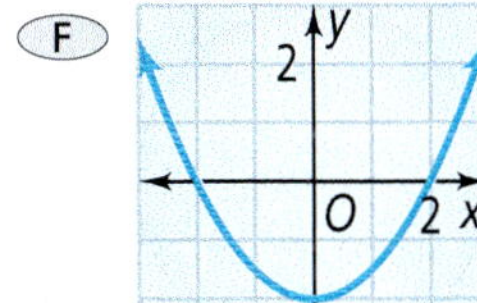

(G) 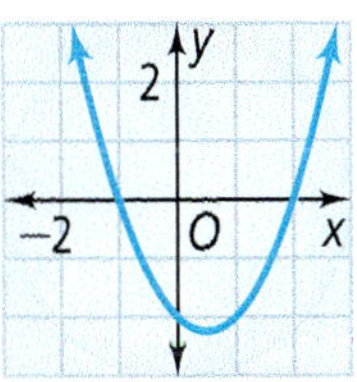

(H) (I) 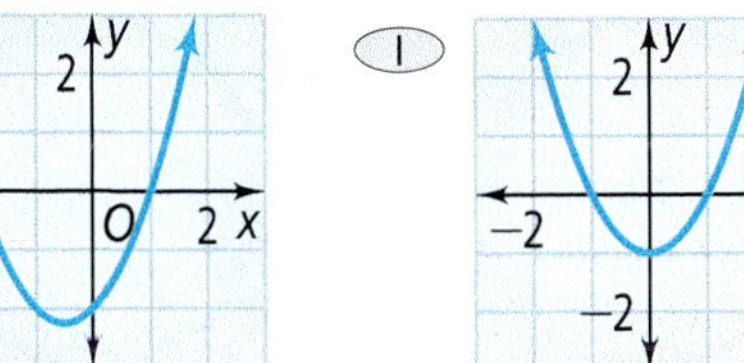

Extended Response

50. You are driving a car when a deer suddenly darts across the road in front of you. Your brain registers the emergency and sends a signal to your foot to hit the brake. The car travels a reaction distance D, in feet, during this time, where D is a function of the speed r, in miles per hour, that the car is traveling when you see the deer, given by $D(r) = \frac{11r + 5}{10}$. Find the inverse and explain what it represents. Is the inverse a function?

Mixed Review

Graph each function. See Lesson 6-8.

51. $y = 3 - 2\sqrt{x + 2}$ **52.** $y = 3\sqrt[3]{2x - 1}$ **53.** $y = -2 + \sqrt{x}$

Factor the expression. See Lesson 4-4.

54. $8 + 27x^3$ **55.** $3x^2 + 11x - 4$ **56.** $25 - 40x + 16x^2$

Solve the system of equations using a matrix. See Lesson 3-6.

57. $\begin{cases} x + 5y = -4 \\ x + 6y = -5 \end{cases}$ **58.** $\begin{cases} 3a + 5b = 0 \\ a + b = 0 \end{cases}$ **59.** $\begin{cases} -x + 2y + z = 0 \\ y = -2x + 3 \\ z = 3x \end{cases}$

Get Ready! **To prepare for Lesson 7-2, do Exercises 60–63.**

Graph each function. See Lesson 7-1.

60. $y = 3^x$ **61.** $y = 4(2)^x$ **62.** $y = 0.75^x$ **63.** $y = 0.5(4)^x$

Properties of Exponential Functions

Content Standards

F.IF.8 Write a function defined by an expression in different but equivalent forms . . .

F.IF.7.e Graph exponential . . . functions, showing intercepts and end behavior . . .

Also F.BF.1.b, A.CED.2, A.SSE.1.b

Objectives To explore the properties of functions of the form $y = ab^x$
To graph exponential functions that have base e

Solve a simpler problem. Use your calculator to experiment with transformations of $y = 2^x$.

Solve It! Getting Ready!

f and g are exponential functions with the same base. Is the graph of g

- a compression,
- a reflection, or
- a translation

of the graph of f? Or is it none of the above? Justify your reasoning.

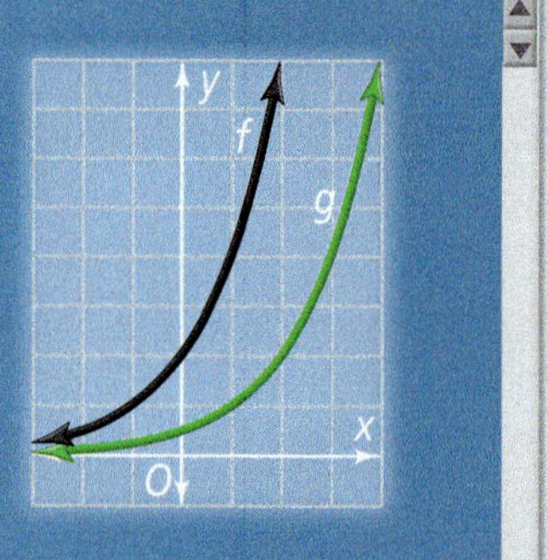

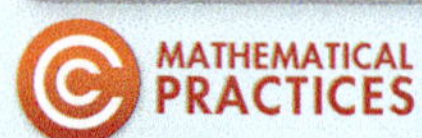

You can apply the four types of transformations—stretches, compressions, reflections, and translations—to exponential functions.

Essential Understanding The factor a in $y = ab^x$ can stretch or compress, and possibly reflect the graph of the parent function $y = b^x$.

- natural base exponential function
- continuously compounded interest

The graphs of $y = 2^x$ (in red) and $y = 3 \cdot 2^x$ (in blue) are shown. Each y-value of $y = 3 \cdot 2^x$ is 3 times the corresponding y-value of the parent function $y = 2^x$.

x	$y = 2^x$	$y = 3 \cdot 2^x$
−2	$\frac{1}{4}$	$\frac{3}{4}$
−1	$\frac{1}{2}$	$\frac{3}{2}$
0	1	3
1	2	6
2	4	12

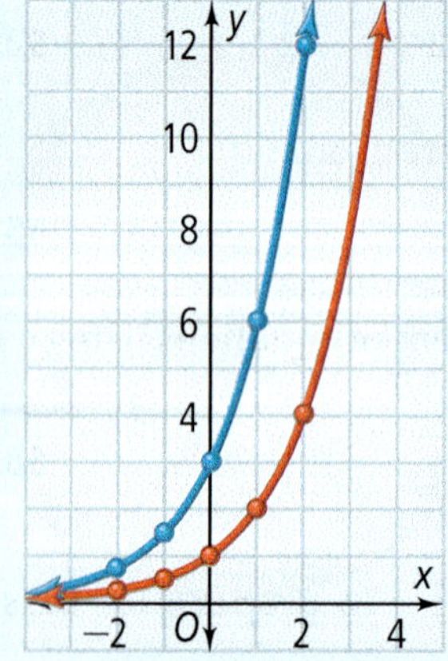

$y = 3 \cdot 2^x$ stretches the graph of the parent function $y = 2^x$ by the factor 3.

Problem 1 Graphing $y = ab^x$

Think
Which x-values should you use to make a table?
Use $x = 0$ and then choose both positive and negative values.

How does the graph of $y = -\frac{1}{3} \cdot 3^x$ compare to the graph of the parent function?

Step 1 Make a table of values.

x	$y = 3^x$	$y = -\frac{1}{3} \cdot 3^x$
−2	$\frac{1}{9}$	$-\frac{1}{27}$
−1	$\frac{1}{3}$	$-\frac{1}{9}$
0	1	$-\frac{1}{3}$
1	3	−1
2	9	−3

Each value is $-\frac{1}{3}$ times the corresponding value of the parent function.

Step 2 Graph the function.

The $-\frac{1}{3}$ in $y = -\frac{1}{3} \cdot 3^x$ reflects the graph of the parent function $y = 3^x$ across the x-axis and compresses it by the factor $\frac{1}{3}$. The domain and asymptote remain unchanged. The y-intercept becomes $-\frac{1}{3}$ and the range becomes $y < 0$.

Got It? 1. How does the graph of $y = -0.5 \cdot 5^x$ compare to the graph of the parent function?

A horizontal shift $y = ab^{(x-h)}$ is the same as the vertical stretch or compression $y = (ab^{-h})b^x$. A vertical shift $y = ab^x + k$ also shifts the horizontal asymptote from $y = 0$ to $y = k$.

Problem 2 Translating the Parent Function $y = b^x$

Think
How is the graph of $y = 2^{(x-4)}$ different from the graph of $y = 2^x$?
The graph of $y = 2^{(x-4)}$ is a horizontal translation of $y = 2^x$ to the right 4 units.

How does the graph of each function compare to the graph of the parent function?

A $y = 2^{(x-4)}$

Step 1 Make a table of values of the parent function $y = 2^x$.

x	$y = 2^x$
2	$\frac{1}{4}$
−1	$\frac{1}{2}$
0	1

x	$y = 2^x$
1	2
2	4
3	8

Step 2 Graph $y = 2^x$ then translate 4 units to the right.

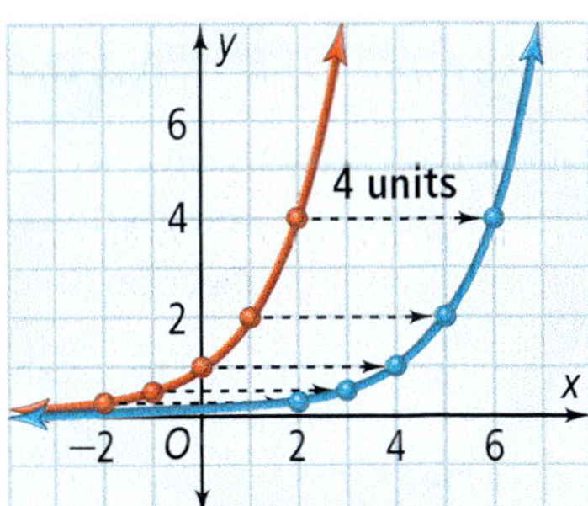

The $(x - 4)$ in $y = 2^{(x-4)}$ translates the graph of $y = 2^x$ to the right 4 units. The asymptote remains $y = 0$. The y-intercept becomes $\frac{1}{16}$.

B $y = 20\left(\frac{1}{2}\right)^x + 10$

Step 1 Make a table of values for $y = 20\left(\frac{1}{2}\right)^x$.

x	$y = 20 \cdot \left(\frac{1}{2}\right)^x$
−1	40
0	20
1	10
2	5
3	2.5

Step 2 Graph $y = 20\left(\frac{1}{2}\right)^x$, then translate 10 units up.

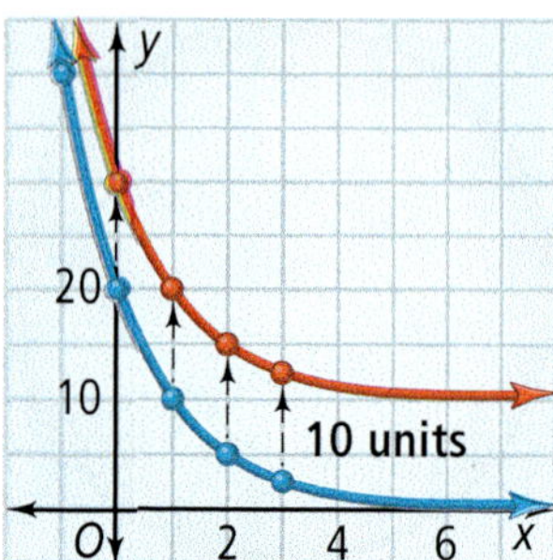

Think

Where have you seen this situation before? The graph of a function like $y = 20\left(\frac{1}{2}\right)^x + 10$ is both a stretch and a vertical translation of its parent function.

The "+ 10" in $y = 20\left(\frac{1}{2}\right)^x + 10$ translates the graph of $y = 20\left(\frac{1}{2}\right)^x$ up 10 units. It also translates the asymptote, the y-intercept, and the range 10 units up. The asymptote becomes $y = 10$, the y-intercept becomes 30, and the range becomes $y > 10$. The domain is unchanged.

Check Use a graphing calculator to graph $y = 20\left(\frac{1}{2}\right)^x + 10$.

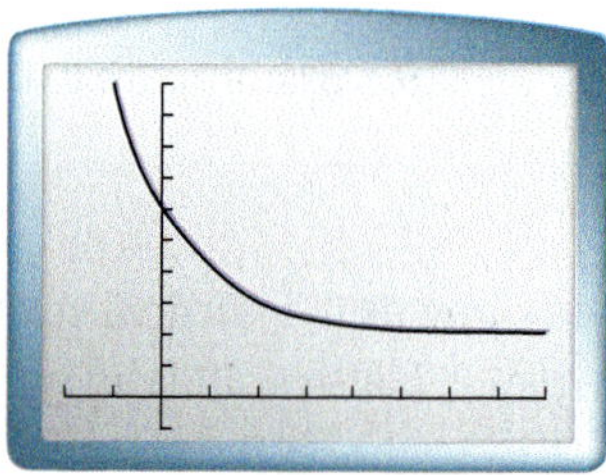

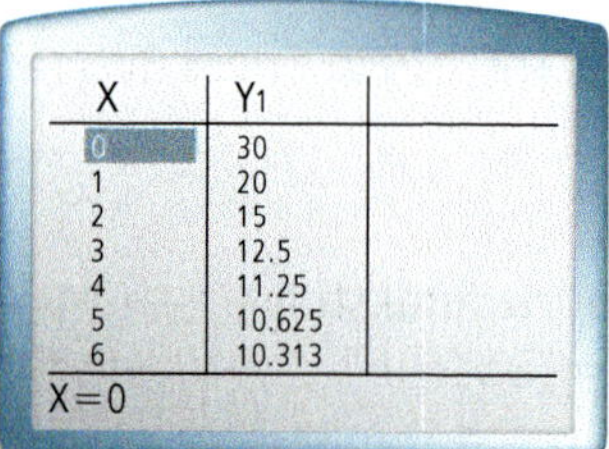

X	Y1
0	30
1	20
2	15
3	12.5
4	11.25
5	10.625
6	10.313

X=0

Got It? **2.** How does the graph of each function compare to the graph of the parent function?

a. $y = 4^{(x+2)}$

b. $y = 5 \cdot 0.25^x + 5$

take note

Concept Summary Families of Exponential Functions

Parent function	$y = b^x$
Stretch ($\lvert a \rvert > 1$) Compression (Shrink) ($0 < \lvert a \rvert < 1$) Reflection ($a < 0$) in x-axis	$y = ab^x$
Translations (horizontal by h; vertical by k)	$y = b^{(x-h)} + k$
All transformations combined	$y = ab^{(x-h)} + k$

Problem 3 Using an Exponential Model STEM

Physics The best temperature to brew coffee is between 195°F and 205°F. Coffee is cool enough to drink at 185°F. The table shows temperature readings from a sample cup of coffee. How long does it take for a cup of coffee to be cool enough to drink? Use an exponential model.

Time (min)	Temp (°F)
0	203
5	177
10	153
15	137
20	121
25	111
30	104

Know
- Set of values
- Best serving temperature

Need
Time it takes for a cup of coffee to become cool enough to drink

Plan
Use an exponential model to find the time it takes for coffee to reach 185°F.

Think

Why does it make sense that a graph of this data would have an asymptote?
The temperature of the hot coffee will get closer and closer to room temperature as it cools, but it cannot cool below room temperature.

Step 1
Plot the data to determine if an exponential model is realistic.

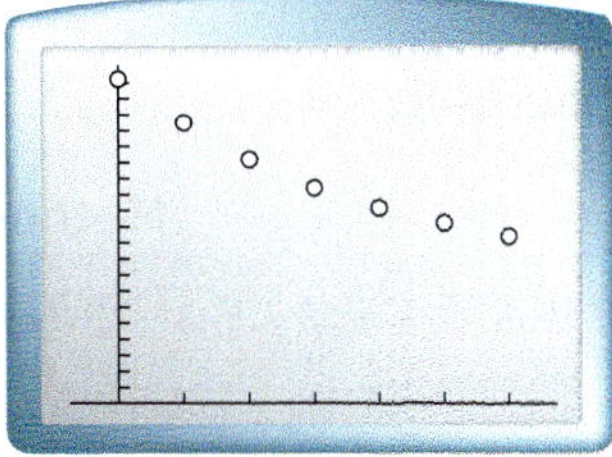

Step 2
The graphing calculator exponential model assumes the asymptote is $y = 0$. Since room temperature is about 68°F, subtract 68 from each temperature value. Calculate the third list by letting **L3 = L2 – 68**.

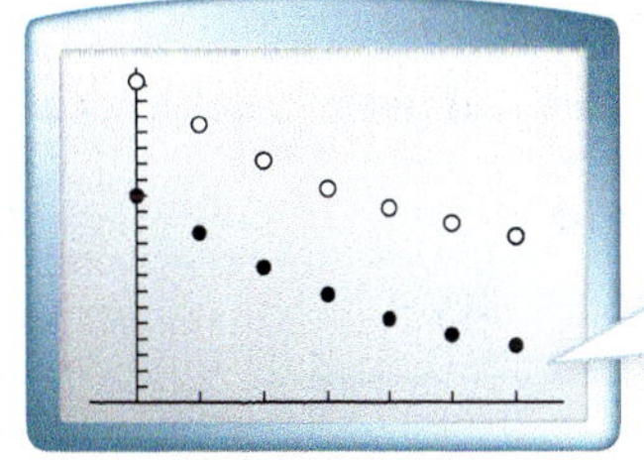

The graphing calculator exponential model assumes the asymptote is $y = 0$.

Step 3
Use the **ExpReg L1**, **L3** function on the transformed data to find an exponential model.

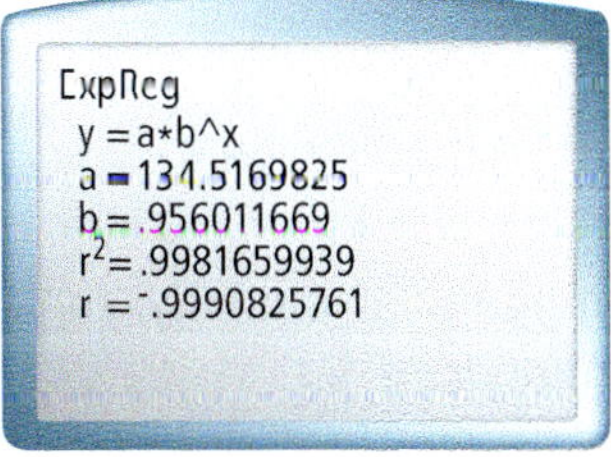

Step 4
Translate $y = 134.5(0.956)^x$ vertically by 68 units to model the original data. Use the model $y = 134.5 \cdot 0.956^x + 68$ to find how long it takes the coffee to cool to 185°F.

X	Y1
2.6	187.65
2.7	187.11
2.8	186.58
2.9	186.05
3	185.52
3.1	184.99
3.2	184.46

X=3.1

The coffee takes about 3.1 min to cool to 185°F.

Got It? 3. a. Use the exponential model. How long does it take for the coffee to reach a temperature of 100 degrees?

b. Reasoning In Problem 3, would the model of the exponential data be useful if you did not translate the data by 68 units? Explain.

Up to this point you have worked with rational bases. However, exponential functions can have irrational bases as well. One important irrational base is the number e. The graph of $y = \left(1 + \frac{1}{x}\right)^x$ has an asymptote at $y = e$ or $y \approx 2.71828$.

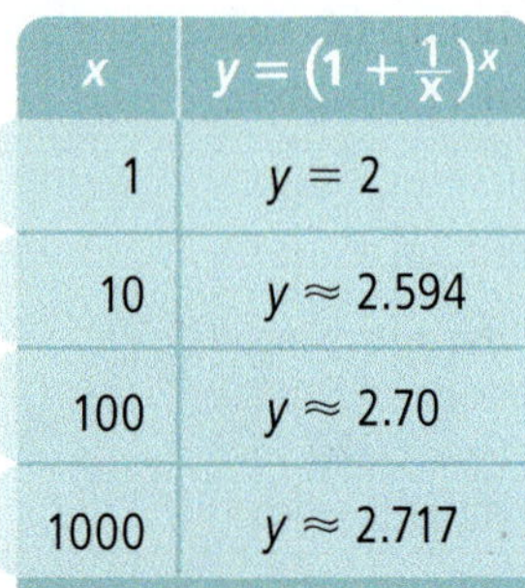

x	$y = \left(1 + \frac{1}{x}\right)^x$
1	$y = 2$
10	$y \approx 2.594$
100	$y \approx 2.70$
1000	$y \approx 2.717$

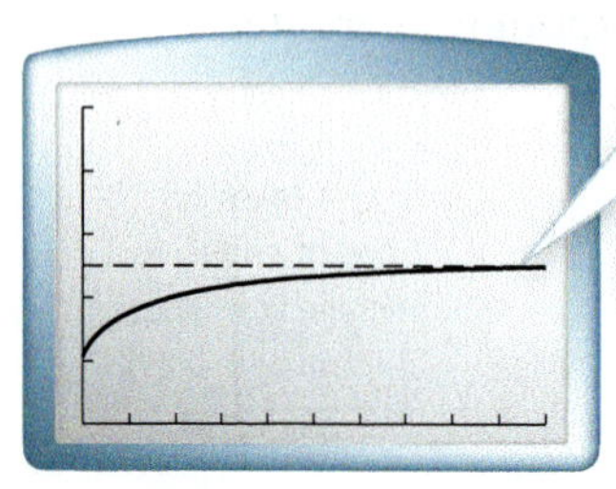

As x approaches infinity the graph approaches the value of e.

Natural base exponential functions are exponential functions with base e. These functions are useful for describing continuous growth or decay. Exponential functions with base e have the same properties as other exponential functions.

Problem 4 Evaluating e^x

Think

After you press the e^x key, what keys should you press?

Press 3,), and enter.

How can you use a graphing calculator to evaluate e^3?

Method 1

Use the e^x key.

```
e^(3)
     20.08553692
```

Method 2

Use the graph of $y = e^x$.

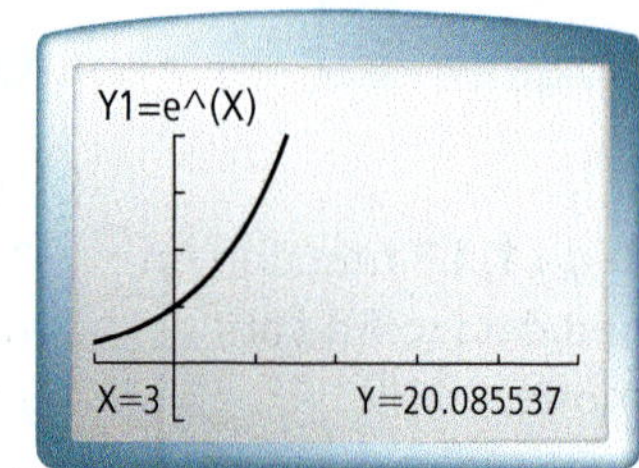

Method 3

Use a table of values for $y = e^x$.

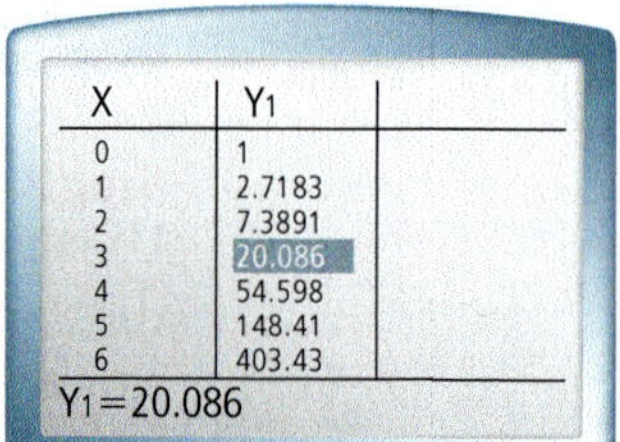

X	Y1
0	1
1	2.7183
2	7.3891
3	20.086
4	54.598
5	148.41
6	403.43

Y1=20.086

$e^3 \approx 20.086$

Got It? 4. How can you use a graphing calculator to calculate e^8?

In Lesson 7-1 you studied interest that was compounded annually. The formula for continuously compounded interest uses the number e.

take note Key Concept Continuously Compounded Interest

$$A(t) = P \cdot e^{rt}$$

- $A(t)$: amount in account at time t
- P: Principal
- r: interest rate (annual)
- t: time in years

Problem 5 Continuously Compounded Interest

GRIDDED RESPONSE

Plan

What is the unknown?
The amount A in the account after 4 years.

Scholarships Suppose you won a contest at the start of 5th grade that deposited $3000 in an account that pays 5% annual interest compounded continuously. How much will you have in the account when you enter high school 4 years later? Express the answer to the nearest dollar.

$A = P \cdot e^{rt}$

$= 3000e^{(0.05)(4)}$ Substitute values for P, r, and t.

$= 3000e^{0.2}$ Simplify.

≈ 3664 Use a calculator. Round to the nearest dollar.

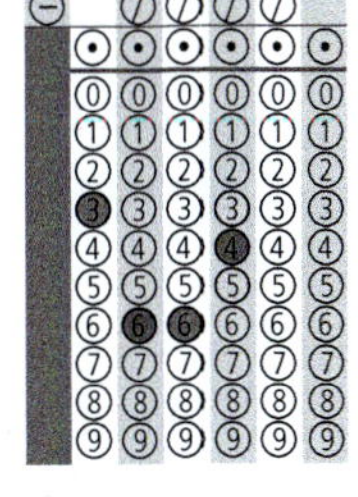

The amount in the account, to the nearest dollar, is $3664. Write your answer, 3664 in the grid.

Got It? 5. About how much will be in the account after 4 years of high school?

Lesson Check

Do you know HOW?

For each function, identify the transformation from the parent function $y = b^x$.

1. $y = -2 \cdot 3^x$

2. $y = \frac{1}{2}(9)^x$

3. $y = 7^{(x-5)}$

4. $y = 5^x + 3$

Do you UNDERSTAND?

5. Vocabulary Is $y = e^{(x+7)}$ a natural base exponential function?

6. Reasoning Is investing $2000 in an account that pays 5% annual interest compounded continuously the same as investing $1000 at 4% and $1000 at 6%, each compounded continuously? Explain.

Practice and Problem-Solving Exercises

Graph each function.

See Problem 1.

7. $y = 5^x$

8. $y = \left(\frac{1}{2}\right)^x$

9. $y = 2(4)^x$

10. $y = -9(3)^x$

11. $y = 3(2)^x$

12. $y = 24\left(\frac{1}{2}\right)^x$

13. $y = -4^x$

14. $y = -\left(\frac{1}{3}\right)^x$

15. $y = 2\left(\frac{3}{2}\right)^x$

Graph each function as a transformation of its parent function.

See Problem 2.

16. $y = 2^x + 5$

17. $y = 5\left(\frac{1}{3}\right)^x - 8$

18. $y = -(0.3)^{x-2}$

19. $y = -2(5)^{x+3}$

20. $y = 3(2)^{x-1} + 4$

21. $y = -2(3)^{x+1} - 5$

See Problem 3.

22. **Baking** A cake recipe says to bake the cake until the center is 180°F, then let the cake cool to 120°F. The table shows temperature readings for the cake.

a. Given a room temperature of 70°F, what is an exponential model for this data set?

b. How long does it take the cake to cool to the desired temperature?

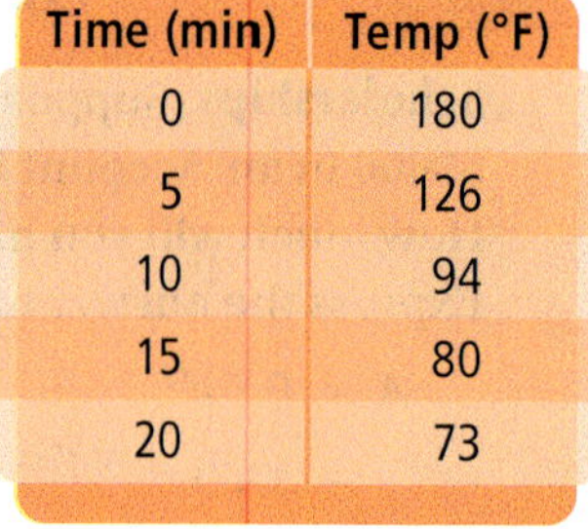

Time (min)	Temp (°F)
0	180
5	126
10	94
15	80
20	73

See Problem 4.

Graphing Calculator **Use the graph of $y = e^x$ to evaluate each expression to four decimal places.**

23. e^6 24. e^{-2} 25. e^0 26. $e^{\frac{5}{2}}$ 27. e^e

See Problem 5.

Find the amount in a continuously compounded account for the given conditions.

28. principal: \$2000
annual interest rate: 5.1%
time: 3 years

29. principal: \$400
annual interest rate: 7.6%
time: 1.5 years

30. principal: \$950
annual interest rate: 6.5%
time: 10 years

31. **Think About a Plan** A student wants to save \$8000 for college in five years. How much should be put into an account that pays 5.2% annual interest compounded continuously?
   - What formula should you use?
   - What information do you know?
   - What do you need to find?

32. **Investment** How long would it take to double your principal in an account that pays 6.5% annual interest compounded continuously?

33. **Error Analysis** A student says that the graph of $f(x) = \left(\frac{1}{3}\right)^{x+2} + 1$ is a shift of the parent function 2 units up and 1 unit to the left. Describe and correct the student's error.

34. Assume that a is positive and $b \geq 1$. Describe the effects of $c > 0$, $c = 0$, and $c < 0$ on the graph of the function $y = ab^{cx}$.

35. **Graphing Calculator** Using a graphing calculator, graph each of the functions below on the same coordinate grid. What do you notice? Explain why the definition of exponential functions has the constraint that $b \neq 1$.

$y = \left(\frac{1}{2}\right)^x$ $y = \left(\frac{8}{10}\right)^x$ $y = \left(\frac{9}{10}\right)^x$ $y = \left(\frac{99}{100}\right)^x$

STEM 36. **Botany** The half-life of a radioactive substance is the time it takes for half of the material to decay. Phosphorus-32 is used to study a plant's use of fertilizer. It has a half-life of 14.3 days. Write the exponential decay function for a 50-mg sample. Find the amount of phosphorus-32 remaining after 84 days.

STEM 37. **Archaeology** Archaeologists use carbon-14, which has a half-life of 5730 years, to determine the age of artifacts in carbon dating. Write the exponential decay function for a 24-mg sample. How much carbon-14 remains after 30 millennia? (*Hint:* 1 millennium = 1000 years)

The parent function for each graph below is of the form $y = ab^x$. Write the parent function. Then write a function for the translation indicated.

38.

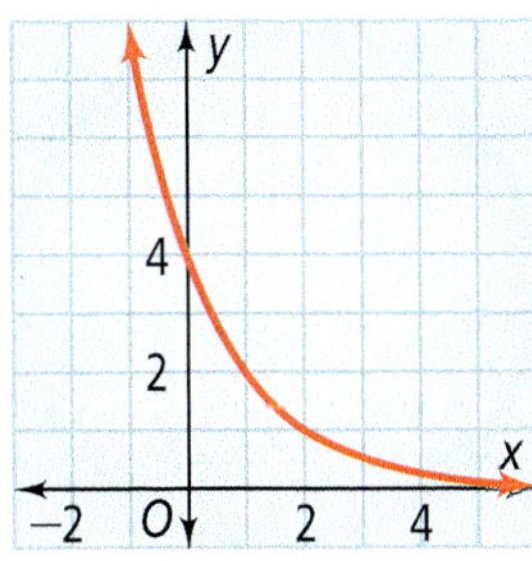

translation: left 4 units, up 3 units

39.

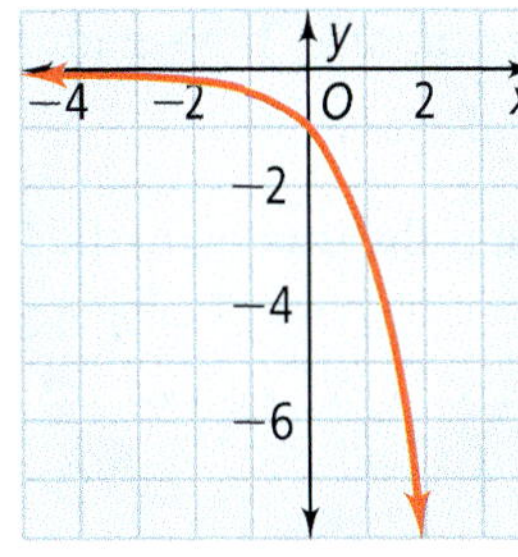

translation: right 8 units, up 2 units

40. Two financial institutions offer different deals to new customers. The first bank offers an interest rate of 3% for the first year and 2% for the next two years. The second bank offers an interest rate of 2.49% for three years. You decide to invest the same amount of principal in each bank. To answer the following, assume you make no withdrawal or deposits during the three-year period.

a. Write a function that represents the total amount of money in the account in the first bank after three years.

b. Write a function that represents the total amount of money in the account in the second bank after three years.

c. Write a function that represents the total amount of money in both accounts at the end of three years.

STEM **41.** **Physics** At a constant temperature, the atmospheric pressure p in pascals is given by the formula $p = 101.3e^{-0.001h}$, where h is the altitude in meters. What is p at an altitude of 500 m?

42. **Landscaping** A homeowner is planting hedges and begins to dig a 3-ft-deep trench around the perimeter of his property. After the first weekend, the homeowner recruits a friend to help. After every succeeding weekend, each digger recruits another friend. One person can dig 405 ft^3 of dirt per weekend. The figure at the right shows the dimensions of the property and the width of the trench.

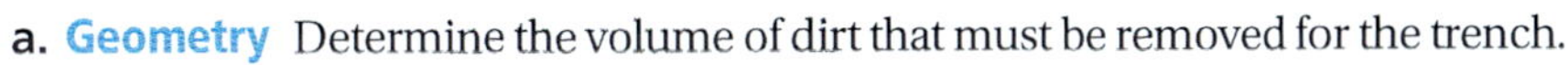

a. **Geometry** Determine the volume of dirt that must be removed for the trench.

b. Write an exponential function to model the volume of dirt remaining to be shoveled after x weekends. Then, use the model to determine how many weekends it will take to complete the trench.

STEM **43.** **Psychology** Psychologists use an exponential model of the learning process, $f(t) = c(1 - e^{-kt})$, where c is the total number of tasks to be learned, k is the rate of learning, t is time, and $f(t)$ is the number of tasks learned.

a. Suppose you move to a new school, and you want to learn the names of 25 classmates in your homeroom. If your learning rate for new tasks is 20% per day, how many complete names will you know after 2 days? After 8 days?

b. **Graphing Calculator** Graph the function on your graphing calculator. How many days will it take to learn everyone's name? Explain.

c. **Open-Ended** Does this function seem to describe your own learning rate? If not, how could you adapt it to reflect your learning rate?

Standardized Test Prep

44. A savings account earns 4.62% annual interest, compounded continuously. After approximately how many years will a principal of \$500 double?

(A) 2 years (B) 10 years (C) 15 years (D) 44 years

45. What is the inverse of the function $f(x) = \sqrt{x-4}$?

(F) $f^{-1}(x) = x^2 - 4, x \geq 0$

(G) $f^{-1}(x) = x^2 + 4, x \geq 0$

(H) $f^{-1}(x) = \sqrt{x+4}$

(I) $f^{-1}(x) = \frac{\sqrt{x-4}}{x-4}$

In Exercises 47 and 48, let $f(x) = x^2 - 4$ and $g(x) = \frac{1}{x+4}$.

46. What is $(g \circ f)(x)$?

(A) $\frac{1}{x^2}$ (B) $\frac{1}{x^2 - 8x + 16} - 4$ (C) $\frac{x^2-4}{x+4}$ (D) $x - 4$

47. What is $(f \circ f)(3)$?

(F) 1 (G) 5 (H) 21 (I) 77

48. What is the equation of the line shown at the right?

(A) $y = -\frac{4}{5}x + 2$

(B) $y = \frac{5}{4}x - 2$

(C) $-4x + 5y = 7$

(D) $4x - 5y = 15$

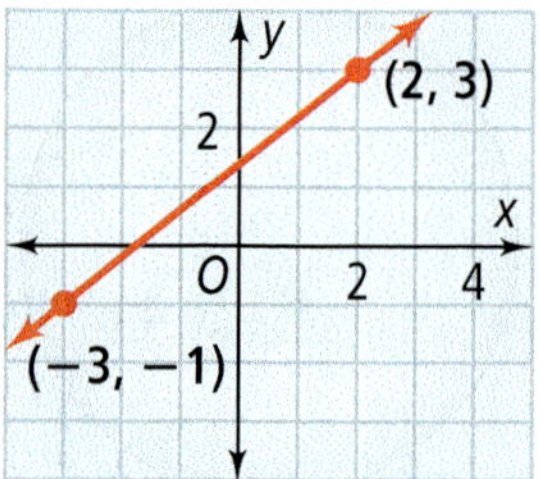

Short Response

49. How much should you invest in an account that pays 6% annual interest compounded continuously if you want exactly \$8000 after four years? Show your work.

Mixed Review

Without graphing, determine whether the function represents exponential growth or exponential decay. Then find the *y*-intercept.

See Lesson 7-1.

50. $y = 23(3.03)^x$ **51.** $f(x) = 3(5)^x$

52. $y = 2\left(\frac{3}{4}\right)^x$ **53.** $y = 5\left(\frac{8}{3}\right)^x$

Simplify.

See Lesson 6-3.

54. $5\sqrt{5} + \sqrt{5}$ **55.** $\sqrt[3]{4} - 2\sqrt[3]{4}$ **56.** $\sqrt{75} + \sqrt{125}$

57. $\sqrt[4]{32} + \sqrt[4]{128}$ **58.** $5\sqrt{3} - 2\sqrt{12}$ **59.** $3\sqrt{63} + \sqrt{28}$

Get Ready! **To prepare for Lesson 7-3, do Exercises 60–62.**

Find the inverse of each function. Is the inverse a function?

See Lesson 6-7.

60. $f(x) = 4x - 1$ **61.** $f(x) = x^7$ **62.** $f(x) = 5x^3 + 1$

Logarithmic Functions as Inverses

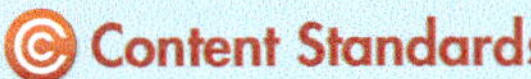

F.BF.4.a Solve an equation of the form $f(x) = c$ for a simple function f that has an inverse and write . . . the inverse.

F.IF.7.e Graph exponential . . . functions, showing intercepts and end behavior . . .

Also A.SSE.1.b, F.IF.8, F.IF.9

Objectives To write and evaluate logarithmic expressions
To graph logarithmic functions

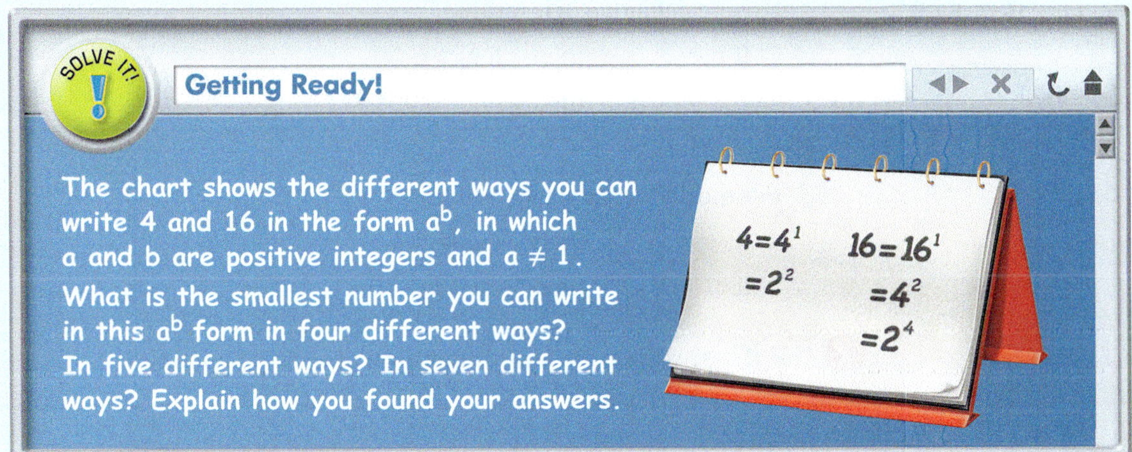

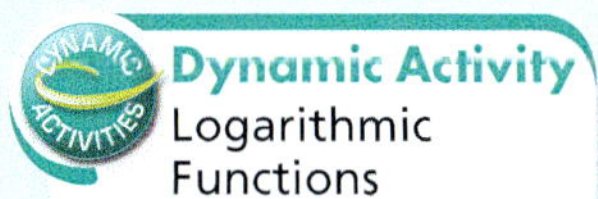

Lesson Vocabulary
- logarithm
- logarithmic function
- common logarithm
- logarithmic scale

Many even numbers can be written as power functions with base 2. In this lesson you will find ways to express all numbers as powers of a common base.

Essential Understanding The exponential function $y = b^x$ is one-to-one, so its inverse $x = b^y$ is a function. To express "y as a function of x" for the inverse, write $y = \log_b x$.

take note

Key Concept Logarithm

A **logarithm** base b of a positive number x satisfies the following definition.

$$\text{For } b > 0,\ b \neq 1,\ \log_b x = y \text{ if and only if } b^y = x.$$

You can read $\log_b x$ as "log base b of x." In other words, the logarithm y is the exponent to which b must be raised to get x.

The exponent y in the expression b^y is the logarithm in the equation $\log_b x = y$. The base b in b^y and the base b in $\log_b x$ are the same. In both, $b \neq 1$ and $b > 0$.

Since $b \neq 1$ and $b > 0$, it follows that $b^y > 0$. Since $b^y = x$ then $x > 0$, so $\log_b x$ is defined only for $x > 0$.

Because $y = b^x$ and $y = \log_b x$ are inverse functions, their compositions map a number a to itself. In other words, $b^{\log_b a} = a$ for $a > 0$ and $\log_b b^a = a$ for all a.

You can use the definition of a logarithm to write exponential equations in logarithmic form.

Problem 1 Writing Exponential Equations in Logarithmic Form

Think

To what power do you raise 10 to get 100?
10 raised to the 2nd power equals 100.

What is the logarithmic form of each equation?

A $100 = 10^2$

Use the definition of logarithm.

If $x = b^y$ then $\log_b x = y$

If $100 = 10^2$ then $\log_{10} 100 = 2$

B $81 = 3^4$

Use the definition of logarithm.

If $x = b^y$ then $\log_b x = y$

If $81 = 3^4$ then $\log_3 81 = 4$

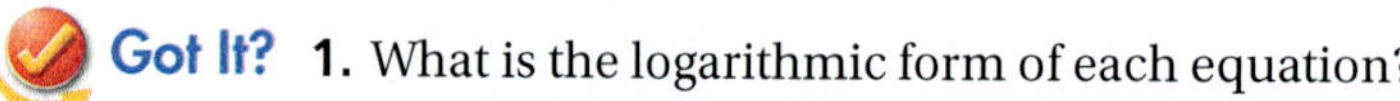

Got It? **1.** What is the logarithmic form of each equation?

a. $36 = 6^2$ **b.** $\frac{8}{27} = \left(\frac{2}{3}\right)^3$ **c.** $1 = 3^0$

You can use the exponential form to help you evaluate logarithms.

Problem 2 Evaluating a Logarithm

Plan

How can you use the definition of logarithm to help you find the value of $\log_8 32$?
If $\log_b x = y$ then $x = b^y$, so to what power must you raise 8 to get 32?

Multiple Choice **What is the value of $\log_8 32$?**

Ⓐ $\frac{3}{5}$ Ⓑ $\frac{5}{3}$  Ⓒ 3 Ⓓ 5

$\log_8 32 = x$	Write a logarithmic equation.
$32 = 8^x$	Use the definition of a logarithm to write an exponential equation.
$2^5 = (2^3)^x$	Write each side using base 2.
$2^5 = 2^{3x}$	Power Property of Exponents
$5 = 3x$	Since the bases are the same, the exponents must be equal.
$\frac{5}{3} = x$	Solve for x.

Since $8^{\frac{5}{3}} = 32$, then $\log_8 32 = \frac{5}{3}$.
The correct answer is B.

Got It? **2.** What is the value of each logarithm?

a. $\log_5 125$ **b.** $\log_4 32$ **c.** $\log_{64} \frac{1}{32}$

A **common logarithm** is a logarithm with base 10. You can write a common logarithm $\log_{10} x$ simply as $\log x$, without showing the 10.

Many measurements of physical phenomena have such a wide range of values that the reported measurements are logarithms (exponents) of the values, not the values themselves. When you use the logarithm of a quantity instead of the quantity, you are using a **logarithmic scale**. The Richter scale is a logarithmic scale. It gives logarithmic measurements of earthquake magnitude.

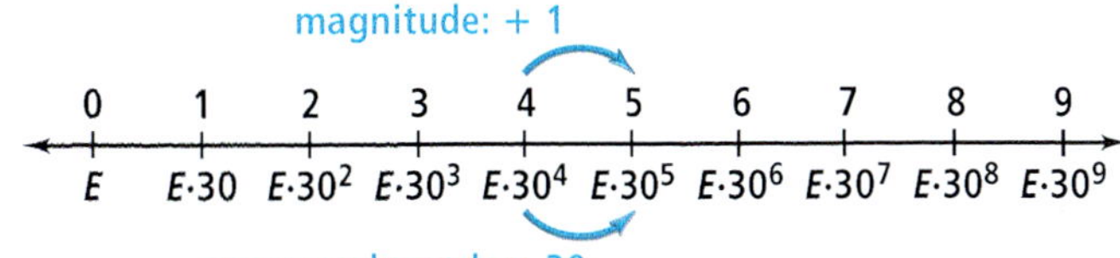

Problem 3 Using a Logarithmic Scale

In December 2004, an earthquake with magnitude 9.3 on the Richter scale hit off the northwest coast of Sumatra. The diagram shows the magnitude of an earthquake that hit Sumatra in March 2005. The formula $\log \frac{I_1}{I_2} = M_1 - M_2$ compares the intensity levels of earthquakes where I is the intensity level determined by a seismograph, and M is the magnitude on a Richter scale. How many times more intense was the December earthquake than the March earthquake?

$\log \frac{I_1}{I_2} = M_1 - M_2$ Use the formula.

$\log \frac{I_1}{I_2} = 9.3 - 8.7$ Substitute $M_1 = 9.3$ and $M_2 = 8.7$.

$\log \frac{I_1}{I_2} = 0.6$ Simplify.

$\frac{I_1}{I_2} = 10^{0.6}$ Apply the definition of common logarithm.

≈ 4 Use a calculator.

The December earthquake was about 4 times as strong as the one in March.

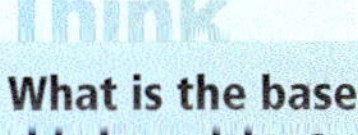

What is the base of this logarithm?
This is the common logarithm. It has base 10.

Got It? **3.** In 1995, an earthquake in Mexico registered 8.0 on the Richter scale. In 2001, an earthquake of magnitude 6.8 shook Washington state. How many times more intense was the 1995 earthquake than the 2001 earthquake?

A **logarithmic function** is the inverse of an exponential function. The graph shows $y = 10^x$ and its inverse $y = \log x$. Note that (0, 1) and (1, 10) are on the graph of $y = 10^x$, and that (1, 0) and (10, 1) are on the graph of $y = \log x$.

Recall that the graphs of inverse functions are reflections of each other across the line $y = x$. You can graph $y = \log_b x$ as the inverse of $y = b^x$.

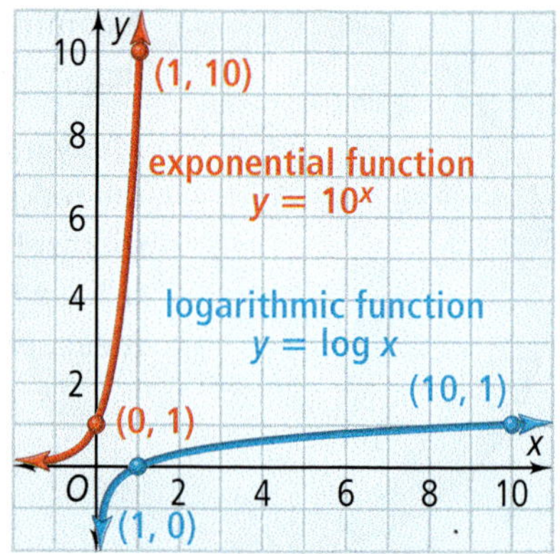

Problem 4 Graphing a Logarithmic Function

What is the graph of $y = \log_3 x$? Describe the domain and range and identify the y-intercept and the asymptote.

$y = \log_3 x$ is the inverse of $y = 3^x$.

Step 1 Graph $y = 3^x$.

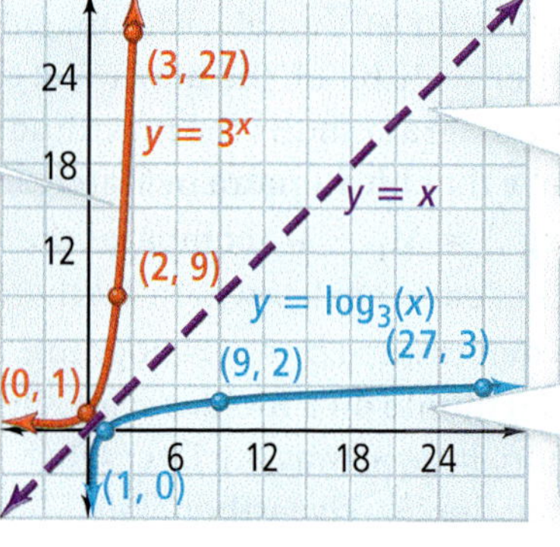

Step 2 Reflecting across the line $y = x$ produces the inverse of $y = 3^x$.

Step 3 Choose a few points on $y = 3^x$ and reverse their coordinates. Plot these new points and graph $y = \log_3 x$.

Think

How are the domain and range of $y = 3^x$ and $y = \log_3 x$ related?

Since they are inverse functions, the domain and range of $y = \log_3 x$ are the same as the *range* and *domain* of $y = 3^x$.

The domain is $x > 0$. The range is all real numbers. There is no y-intercept. The vertical asymptote is $x = 0$.

Got It? 4. a. What is the graph of $y = \log_4 x$? Describe the domain, range, y-intercept and asymptotes.

b. Reasoning Suppose you use the following table to help you graph $y = \log_2 x$. (Recall that if $y = \log_2 x$, then $2^y = x$.) Copy and complete the table. Explain your answers.

x	$2^y = x$	y
−1	$2^y = -1$	■
0	$2^y = 0$	■
1	$2^y = 1$	■
2	$2^y = 2$	■

The function $y = \log_b x$ is the parent for a function family. You can graph $y = \log_b (x - h) + k$ by translating the graph of the parent function, $y = \log_b x$, horizontally by h units and vertically by k units. The a in $y = a \log_b x$ indicates a stretch, a compression, and possibly a reflection.

take note

Concept Summary Families of Logarithmic Functions

Parent functions:	$y = \log_b x, b > 0, b \neq 1$
Stretch ($\lvert a \rvert > 1$) Compression (Shrink) ($0 < \lvert a \rvert < 1$) Reflection ($a < 0$) in x-axis	$y = a \log_b x$
Translations (horizontal by h; vertical by k)	$y = \log_b (x - h) + k$
All transformations together	$y = a \log_b (x - h) + k$

Problem 5 Translating $y = \log_b x$

Think

How is the function $y = \log_4(x - 3) + 4$ similar to other functions you have seen?

Recall that the graph of $y = f(x - h) + k$ is a vertical and horizontal translation of the parent function, $y = f(x)$.

How does the graph of $y = \log_4 (x - 3) + 4$ compare to the graph of the parent function?

Step 1

Make a table of values for the parent function. Use the definition of logarithm.

x	$\log_4 x = y \rightarrow 4^y = x$	y
$\frac{1}{16}$	$4^{-2} = \frac{1}{16}$	-2
$\frac{1}{4}$	$4^{-1} = \frac{1}{4}$	-1
1	$4^0 = 1$	0
4	$4^1 = 4$	1
16	$4^2 = 16$	2

Step 2

Graph the parent function. Shift the graph to the right 3 units and up 4 units to graph $y = \log_4 (x - 3) + 4$.

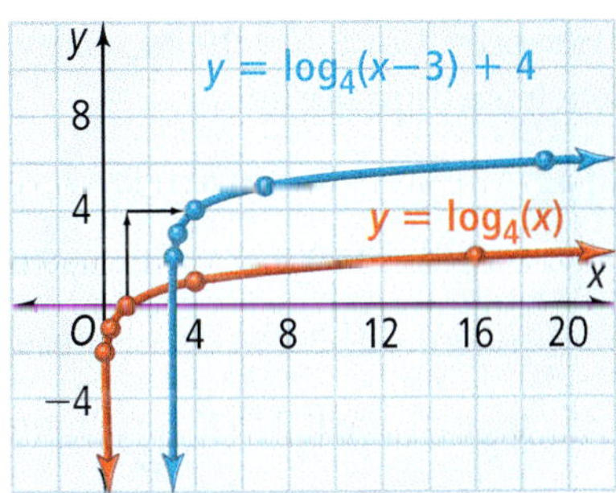

Because $y = \log_4 (x - 3) + 4$ translates the graph of the parent function 3 units to the right, the asymptote changes from $x = 0$ to $x = 3$. The domain changes from $x > 0$ to $x > 3$. The range remains all real numbers.

Got It? 5. How does the graph of each function compare to the graph of the parent function?

a. $y = \log_2(x - 3) + 4$ **b.** $y = 5 \log_2 x$

Lesson Check

Do you know HOW?

Write each equation in logarithmic form.

1. $25 = 5^2$
2. $64 = 4^3$
3. $243 = 3^5$
4. $25 = 5^2$

Evaluate each logarithm.

5. $\log_2 8$
6. $\log_9 9$
7. $\log_7 49$
8. $\log_2 \frac{1}{4}$

Do you UNDERSTAND?

9. **Vocabulary** Determine whether each logarithm is a common logarithm.
   a. $\log_2 4$ b. $\log 64$ c. $\log_{10} 100$ d. $\log_5 5$
10. **Reasoning** Explain how you could use an inverse function to graph the logarithmic function $y = \log_6 x$.
11. **Compare and Contrast** Compare the graph of $y = \log_2 (x + 4)$ to the graph of $y = \log_2 x$. How are the graphs alike? How are they different?

Practice and Problem-Solving Exercises

Write each equation in logarithmic form. See Problem 1.

12. $49 = 7^2$
13. $10^3 = 1000$
14. $625 = 5^4$
15. $\frac{1}{10} = 10^{-1}$
16. $8^2 = 64$
17. $4 = \left(\frac{1}{2}\right)^{-2}$
18. $\left(\frac{1}{3}\right)^3 = \frac{1}{27}$
19. $10^{-2} = 0.01$

Evaluate each logarithm. See Problem 2.

20. $\log_2 16$
21. $\log_4 2$
22. $\log_8 8$
23. $\log_4 8$
24. $\log_2 8$
25. $\log_{49} 7$
26. $\log_5 (-25)$
27. $\log_3 9$
28. $\log_2 2^5$
29. $\log_{\frac{1}{2}} \frac{1}{2}$
30. $\log 10{,}000$
31. $\log_5 125$

STEM Seismology In 1812, an earthquake of magnitude 7.9 shook New Madrid, Missouri. Compare the intensity level of that earthquake to the intensity level of each earthquake below. See Problem 3.

32. magnitude 7.7 in San Francisco, California, in 1906
33. magnitude 9.5 in Valdivia, Chile, in 1960
34. magnitude 3.2 in Charlottesville, Virginia, in 2001
35. magnitude 6.9 in Kobe, Japan, in 1995

Graph each function on the same set of axes. See Problem 4.

36. $y = \log_2 x$
37. $y = 2^x$
38. $y = \log_{\frac{1}{2}} x$
39. $y = \left(\frac{1}{2}\right)^x$

Describe how the graph of each function compares with the graph of the parent function, $y = \log_b x$. See Problem 5.

40. $y = \log_5 x + 1$
41. $y = \log_7 (x - 2)$
42. $y = \log_3 (x - 5) + 3$
43. $y = \log_4 (x + 2) - 1$

Apply

 **44. Think About a Plan** The pH of a substance equals $-\log[H^+]$, where $[H^+]$ is the concentration of hydrogen ions, and it ranges from 0 to 14. A pH level of 7 is neutral. A level greater than 7 is basic, and a level less than 7 is acidic. The table shows the hydrogen ion concentration $[H^+]$ for selected foods. Is each food basic or acidic?

- How can you find the pH value of each food?
- What rule can you use to determine if the food is basic or acidic?

Approximate $[H^+]$ of Foods

Food	$[H^+]$
Apple juice	3.2×10^{-4}
Buttermilk	2.5×10^{-5}
Cream	2.5×10^{-7}
Ketchup	1.3×10^{-4}
Shrimp sauce	7.9×10^{-8}
Strained peas	1.0×10^{-6}

 **45. Chemistry** Find the concentration of hydrogen ions in seawater, if the pH level of seawater is 8.5.

Write each equation in exponential form.

46. $\log_2 128 = 7$ **47.** $\log 0.0001 = -4$ **48.** $\log_6 6 = 1$ **49.** $\log_4 1 = 0$

50. $\log_7 16{,}807 = 5$ **51.** $\log_2 \frac{1}{2} = -1$ **52.** $\log_3 \frac{1}{9} = -2$ **53.** $\log 10 = 1$

Find the greatest integer that is less than the value of the logarithm. Use your calculator to check your answers.

54. $\log 5$ **55.** $\log 0.08$ **56.** $\log 17.52$ **57.** $\log(1.3 \times 10^7)$

58. Compare the graph at the right to the function $y = \log_5 x$. Describe the domain and range and identify the y-intercept of $y = \log_5 x$.

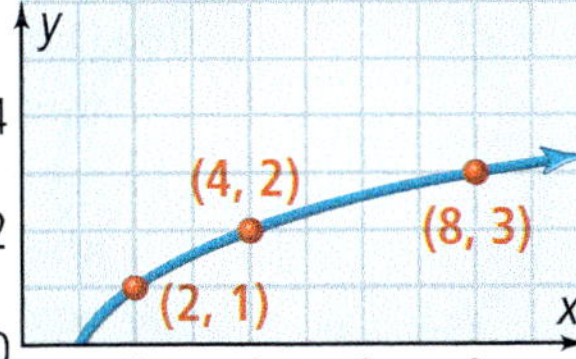

59. Write $5 = \log_{2x+1}(a + b)$ in exponential form.

 **60. Open-Ended** Write a logarithmic function of the form $y = \log_b x$. Find its inverse function. Graph both functions on one set of axes.

Find the inverse of each function.

61. $y = \log_4 x$ **62.** $y = \log_{0.5} x$ **63.** $y = \log_{10} x$ **64.** $y = \log_2 2x$

65. $y = \log(x + 1)$ **66.** $y = \log 10x$ **67.** $y = \log_2 4x$ **68.** $y = \log(x - 6)$

Graph each logarithmic function.

69. $y = \log 2x$ **70.** $y = 2\log_2 x$ **71.** $y = \log_4(2x + 3)$ **72.** $y = \log_3(x + 5)$

Find the domain and the range of each function.

73. $y = \log_5 x$ **74.** $y = 3\log x$ **75.** $y = \log_2(x - 3)$ **76.** $y = 2\log(x - 2)$

You can write $5^3 = 125$ in logarithmic form using the fact that $\log_b b^x = x$.

$\log_5(5^3) = \log_5(125)$ Apply the log base 5 to each side.

$3 = \log_5 125$ Use $\log_b b^x = x$ to simplify.

Use this method to write each equation in logarithmic form. Show your work.

77. $3^4 = 81$ **78.** $x^4 = y$ **79.** $6^8 = a + 1$

Challenge Find the least integer greater than each number. Do not use a calculator.

80. $\log_3 38$ **81.** $\log_{1.5} 2.5$ **82.** $\log_{\sqrt{7}} \sqrt{50}$ **83.** $\log_5 \frac{1}{47}$

84. Match each function with the graph of its inverse.

a. $y = \log_3 x$ b. $y = \log_2 4x$ c. $y = \log_{\frac{1}{2}} x$

I. II. III.

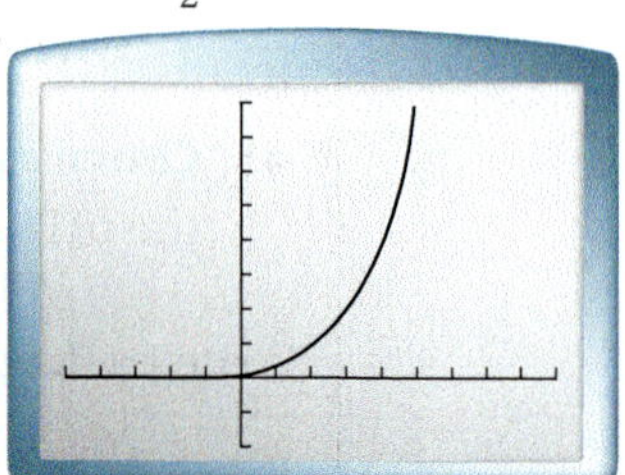

Standardized Test Prep

SAT/ACT

85. Which is the logarithmic form of the exponential equation $2^3 = 8$?

(A) $\log_8 2 = 3$ (B) $\log_8 3 = 2$ (C) $\log_3 8 = 2$ (D) $\log_2 8 = 3$

86. Dan will begin advertising his video production business online using a pay-per-click method, which charges \$30 as an initial fee, plus a fixed amount each time the ad is clicked. Dan estimates that with the cost of 8 cents per click, his ad will be clicked about 150 times per day. Which expression represents Dan's total estimated cost of advertising, in dollars, after x days?

(F) $(30 + 0.08x)150$ (G) $360x$ (H) $30 + 1200x$ (I) $30 + 12x$

87. Which translation takes $y = |x|$ to $y = |x + 3| - 1$?

(A) 3 units right, 1 unit down
(B) 3 units right, 1 unit up
(C) 3 units left, 1 unit down
(D) 3 units left, 1 unit up

Short Response

88. What is the expression $\sqrt[3]{(\sqrt{a})^7}$ written as a variable raised to a single rational exponent?

Mixed Review

Graph each function. See Lesson 7-2.

89. $y = 5^x - 100$ **90.** $y = -10(4)^{x+2}$ **91.** $y = -27(3)^{x-1} + 9$

Factor each expression. See Lesson 4-4.

92. $4x^2 - 8x + 3$ **93.** $4b^2 - 100$ **94.** $5x^2 + 13x - 6$

Get Ready! **To prepare for Lesson 7-4, do Exercises 95–98.**

See Lesson 1-3.

Evaluate each expression for the given value of the variable.

95. $x^2 - x; x = 2$ **96.** $x^3 \cdot x^5; x = 2$ **97.** $\frac{x^8}{x^{10}}; x = 2$ **98.** $x^3 + x^2; x = 2$

Concept Byte

Use With Lesson 7-3

TECHNOLOGY

Fitting Curves to Data

Content Standard

F.IF.4 For a function that models a relationship between two quantities, interpret key features of graphs . . . and sketch graphs showing key features . . .

Example 1

Which type of function models the data best—linear, logarithmic, or exponential?

Connect the points with a smooth curve. Since the points do not fall along a line, the function is not linear. The graph appears to approach a horizontal asymptote, so an exponential function models the data best.

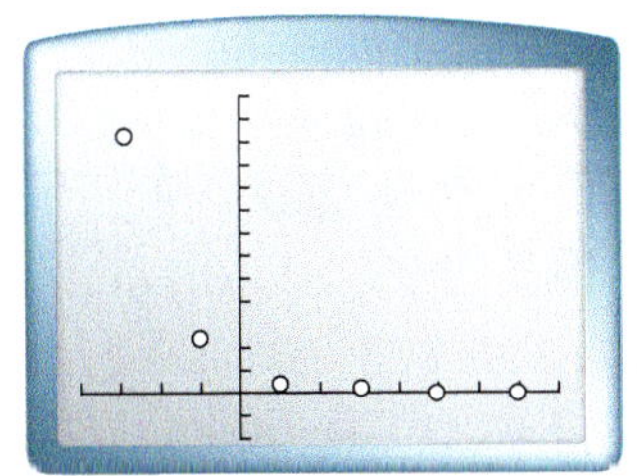

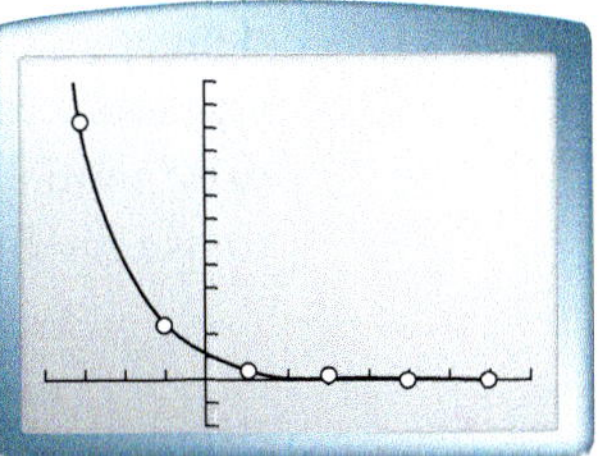

Example 2

Which type of function models the data best—quadratic, logarithmic, or cubic?

Step 1 Press stat enter to enter the data in lists.

Step 2 Use the stat plot feature to draw a scatter plot.

Step 3 If you connect the points with a smooth curve, the end behavior of the graph is up and up. The graph is not cubic or logarithmic, so the quadratic function best models the data.

x	y
0	14
1	7.5
2	4
3	1.8
4	1.8
5	3.9

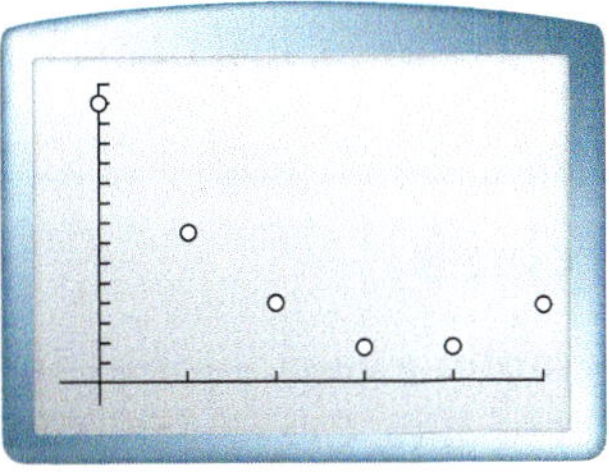

Exercises

1. Which type of function models the data shown in the graphing calculator screen best—*linear*, *quadratic*, *logarithmic*, *cubic*, or *exponential*?

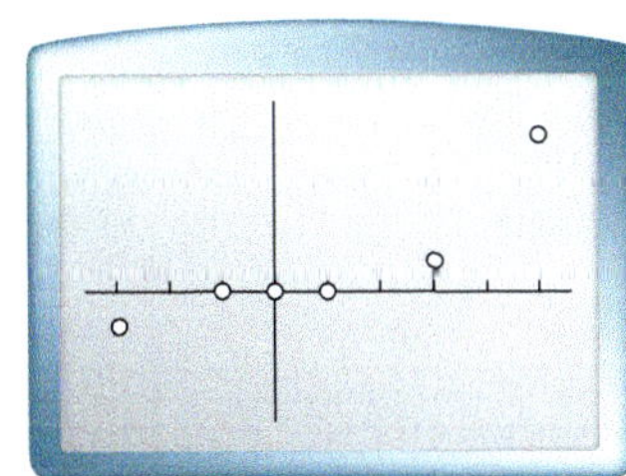

2. Which type of function models the data in the table best—*linear*, *quadratic*, *logarithmic*, *cubic*, or *exponential*?

x	y
−1	0
1	1.4
3	2.09
5	2.53
7	2.81
9	3.12

3. **Reasoning** Could you use a different model for the data in Exercises 1 and 2? Explain.

Example 3

The table shows the number of bacteria in a culture after the given number of hours. Find a good model for the data. Based on the model, how many bacteria will be in the culture after an additional ten hours?

Hour	Bacteria
1	2205
2	2270
3	2350
4	2653
5	3052
6	3417
7	3890
8	4522
9	5107
10	5724

Step 1 Press stat enter to enter the data in lists. Use the stat plot feature to draw a scatter plot.

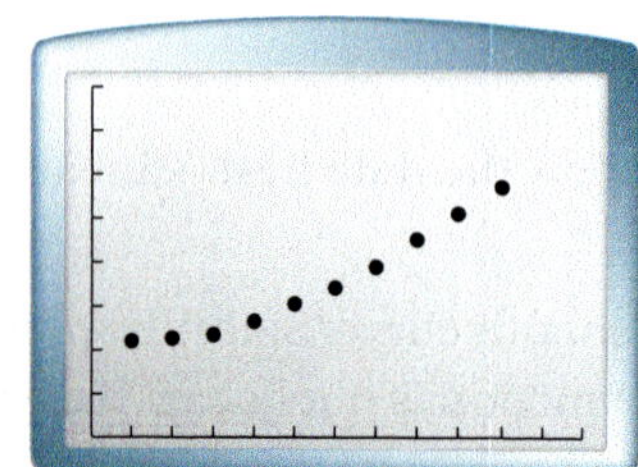

Step 2 Notice from the scatter plot that the data appears exponential. Find the equations for the best-fitting exponential function. Press stat ▷ **0** to use the **ExpReg** feature.

$$y = 1779.404(1.121)^x$$

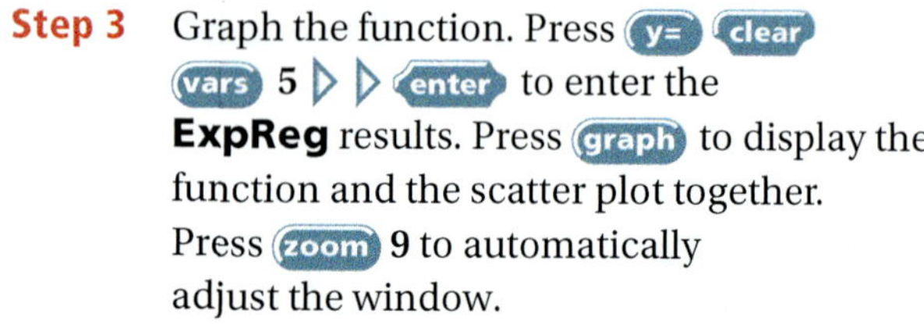

Step 3 Graph the function. Press y= clear vars 5 ▷ ▷ enter to enter the **ExpReg** results. Press graph to display the function and the scatter plot together. Press zoom 9 to automatically adjust the window.

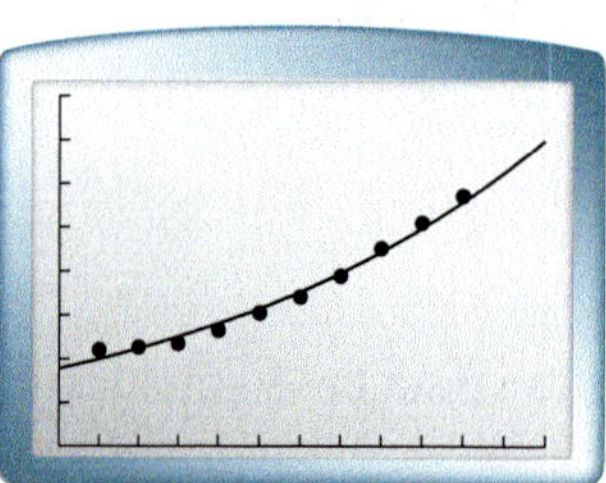

Step 4 In 10 more hours, there will be approximately $y = 1779.404(1.121)^{20} \approx 17{,}474$ bacteria in the culture.

Exercises

Use a graphing calculator to find the exponential or quadratic function that best fits each set of data. Graph each function.

4.

x	y
−1	4.9
0	3.8
1	5.0
2	8.1
3	13.3
4	70.2

5.

x	y
−3	0.1
−1	0.4
1	1.6
3	6.4
5	25.6
7	102.4

6.

x	y
1	3.5
2	2.11
3	1.30
4	0.73
5	0.28
6	0.08

7.

x	y
−1	0.04
0	0.1
1	0.5
2	2.5
3	12.5
4	62.5

8. Writing In Exercise 6 the function appears to level off. Explain why.

9. A savings account begins with \$14.00. After 1 year, the account has a balance of \$16.24. After 2 years, the account has a balance of \$18.84. Assuming no additional deposits or withdraws are made, find the equation for the best-fitting exponential function to represent the balance of the account after x years. How much money will be in the account after 20 years?

7

Mid-Chapter Quiz

MathXL® for School
Go to PowerAlgebra.com

Do you know HOW?

Determine whether each function is an example of exponential growth or decay. Then find the y-intercept.

1. $y = 100(0.25)^x$ **2.** $y = 0.6\left(\frac{1}{10}\right)^x$ **3.** $y = \frac{7}{8}(18)^x$

Graph each function. Then find the domain, range, and y-intercept.

4. $y = -4(2)^x$ **5.** $y = \frac{1}{4}(10)^x$ **6.** $y = 8(0.25)^x$

7. Investment Suppose you deposit $600 into a savings account that pays 3.9% annual interest. How much will you have in the account after 3 years if no money is added or withdrawn?

8. Depreciation The initial value of a car is $25,000. After one year, the value of the car is $21,250. Write an exponential function to model the expected value of the car. Estimate the value of the car after 5 years.

Graph each function as a transformation of its parent function. Write the parent function.

9. $y = 3^x - 2$ **10.** $y = \frac{1}{2}(5)^{x-1} + 4$

11. $y = -(0.5)^{x+3}$ **12.** $y = -6\left(\frac{3}{4}\right)^x - 10$

Evaluate each expression to four decimal places.

13. e^5 **14.** $e^{\frac{3}{2}}$ **15.** e^{-4}

Find the amount in a continuously compounded account for the given conditions.

16. principal: $500; annual interest rate: 4.9%; time: 2.5 years

17. principal: $6000; annual interest rate: 6.8%; time: 10 years

Write each equation in logarithmic form.

18. $10^4 = 10{,}000$ **19.** $\frac{1}{4} = 4^{-1}$ **20.** $8 = \left(\frac{1}{2}\right)^{-3}$

Evaluate each logarithm.

21. $\log_8 64$ **22.** $\log_4(256)$ **23.** $\log_{\frac{1}{5}} 625$

Graph each logarithmic function. Find the domain and range.

24. $y = \log_5(x - 1)$ **25.** $y = 4\log x + 5$

26. Crafts For glass to be shaped, its temperature must stay above 1200°F. The temperature of a piece of glass is 2200°F when it comes out of the furnace. The table shows temperature readings for the glass. Write an exponential model for this data set and then find how long it takes for the piece of glass to cool to 1200°F.

Time (min)	Temp (°F)
0	2200
5	1700
10	1275
15	1000
20	850
25	650

Do you UNDERSTAND?

27. Error Analysis A student claims the y-intercept of the graph of the function $y = ab^x$ is the point $(0, b)$. What is the student's mistake? What is the actual y-intercept?

28. Writing Without graphing, how can you tell whether an exponential function represents exponential growth or exponential decay?

29. Compare and Contrast Compare the graph of $y = \log_3(x + 1)$ to the graph of its inverse $y = 3^x - 1$. How are the graphs alike? How are they different?

30. Vocabulary Explain how the continuously compounded interest formula differs from the annually compounded interest formula.

Properties of Logarithms

Content Standard

Prepares for F.LE.4 For exponential models, express as a logarithm the solution to $ab^{ct} = d$ where *a*, *c*, and *d* are numbers and the base *b* is 2, 10, or *e*; evaluate the logarithm using technology.

Objective To use the properties of logarithms

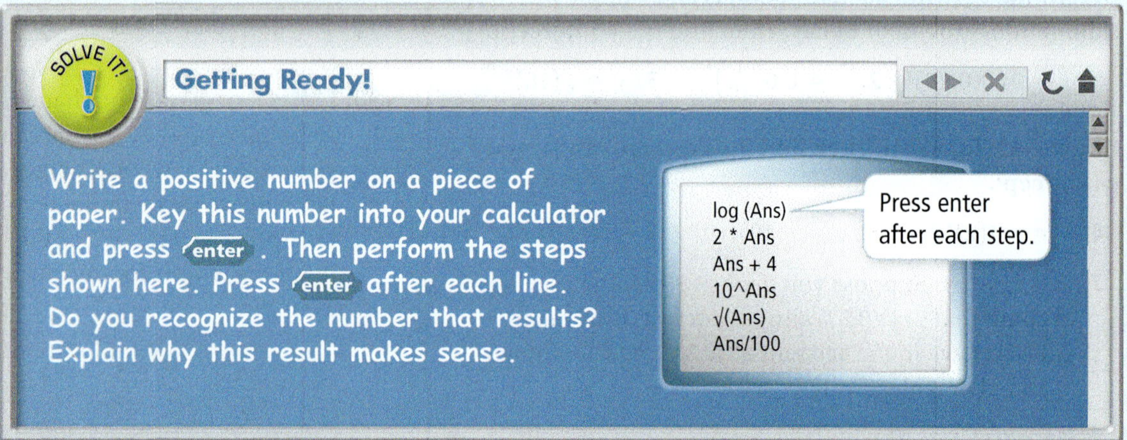

You can derive the properties of logarithms from the properties of exponents.

Essential Understanding Logarithms and exponents have corresponding properties.

Here's Why It Works You can use a product property of exponents to derive a product property of logarithms.

Let $x = \log_b m$ and $y = \log_b n$.

$m = b^x$ and $n = b^y$	Definition of logarithm
$mn = b^x \cdot b^y$	Write *mn* as a product of powers.
$mn = b^{x+y}$	Product Property of Exponents
$\log_b mn = x + y$	Definition of logarithm
$\log_b mn = \log_b m + \log_b n$	Substitute for *x* and *y*.

take note **Properties** **Properties of Logarithms**

For any positive numbers m, n, and b where $b \neq 1$, the following properties apply.

Product Property	$\log_b mn = \log_b m + \log_b n$
Quotient Property	$\log_b \frac{m}{n} = \log_b m - \log_b n$
Power Property	$\log_b m^n = n \log_b m$

Problem 1 Simplifying Logarithms

What is each expression written as a single logarithm?

A $\log_4 32 - \log_4 2$

$$\log_4 32 - \log_4 2 = \log_4 \frac{32}{2}$$ Quotient Property of Logarithms

$$= \log_4 16$$ Divide.

$$= \log_4 4^2$$ Write 16 as a power of 4.

$$= 2$$ Simplify.

Think

What must you do with the numbers that multiply the logarithms?
Apply the Power Property of Logarithms.

B $6 \log_2 x + 5 \log_2 y$

$$6 \log_2 x + 5 \log_2 y = \log_2 x^6 + \log_2 y^5$$ Power Property of Logarithms

$$= \log_2 x^6 y^5$$ Product Property of Logarithms

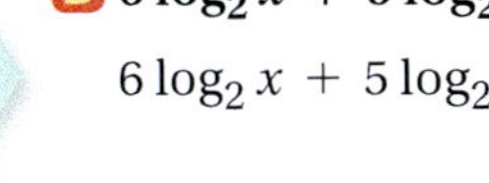

Got It? **1.** What is each expression written as a single logarithm?

a. $\log_4 5x + \log_4 3x$ **b.** $2 \log_4 6 - \log_4 9$

You can expand a single logarithm to involve the sum or difference of two or more logarithms.

Problem 2 Expanding Logarithms

What is each logarithm expanded?

A $\log \frac{4x}{y}$

$$\log \frac{4x}{y} = \log 4x - \log y$$ Quotient Property of Logarithms

$$= \log 4 + \log x - \log y$$ Product Property of Logarithms

Think

Can you apply the Power Property of Logarithms first?
No; the fourth power applies only to x.

B $\log_9 \frac{x^4}{729}$

$$\log_9 \frac{x^4}{729} = \log_9 x^4 - \log_9 729$$ Quotient Property of Logarithms

$$= 4 \log_9 x - \log_9 729$$ Power Property of Logarithms

$$= 4 \log_9 x - \log_9 9^3$$ Write 729 as a power of 9.

$$= 4 \log_9 x - 3$$ Simplify.

Got It? **2.** What is each logarithm expanded?

a. $\log_3 \frac{250}{37}$ **b.** $\log_3 9x^5$

You have seen logarithms with many bases. The **log** key on a calculator finds $\log_{10}$ of a number. To evaluate a logarithm with any base, use the **Change of Base Formula**.

take note

Property Change of Base Formula

For any positive numbers m, b, and c, with $b \neq 1$ and $c \neq 1$,

$$\log_b m = \frac{\log_c m}{\log_c b}.$$

Here's Why It Works

$$\log_b m = \frac{(\log_b m)(\log_c b)}{\log_c b} \qquad \text{Multiply } \log_b m \text{ by } \frac{\log_c b}{\log_c b} = 1.$$

$$= \frac{\log_c b^{\log_b m}}{\log_c b} \qquad \text{Power Property of Logarithms}$$

$$= \frac{\log_c m}{\log_c b} \qquad b^{\log_b m} = m$$

Problem 3 Using the Change of Base Formula

Think

What common base has powers that equal 27 and 81?

3; $3^3 = 27$ and $3^4 = 81$.

What is the value of each expression?

A $\log_{81} 27$

Method 1 Use a common base.

$$\log_{81} 27 = \frac{\log_3 27}{\log_3 81} \qquad \text{Change of Base Formula}$$

$$= \frac{3}{4} \qquad \text{Simplify.}$$

Method 2 Use a calculator.

$$\log_{81} 27 = \frac{\log 27}{\log 81} \qquad \text{Change of Base Formula}$$

$$= 0.75 \qquad \text{Use a calculator.}$$

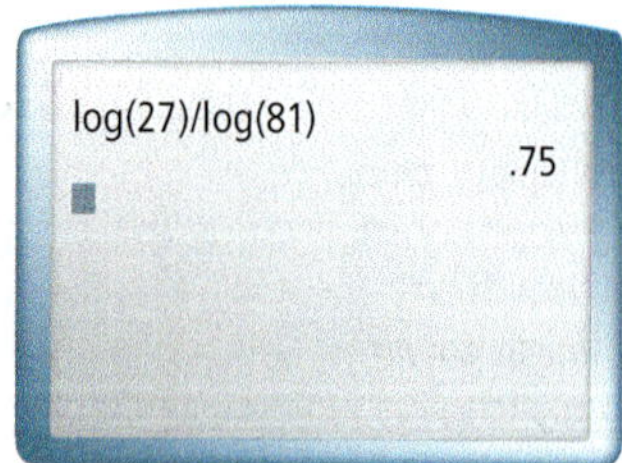

Think

What would be a reasonable result?

$5^2 = 25$ and $5^3 = 125$, so $\log_5 36$ should be between 2 and 3.

B $\log_5 36$

$$\log_5 36 = \frac{\log 36}{\log 5} \qquad \text{Change of Base Formula}$$

$$\approx 2.23 \qquad \text{Use a calculator to evaluate.}$$

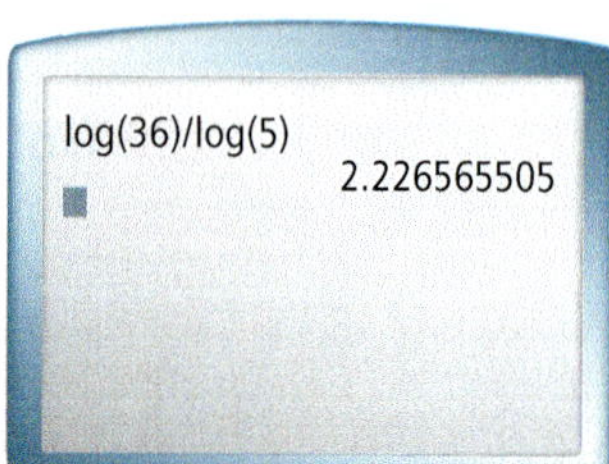

Got It? **3.** Use the Change of Base Formula. What is the value of each expression?

a. $\log_8 32$ **b.** $\log_4 18$

Problem 4 Using a Logarithmic Scale 

Chemistry The pH of a substance equals $-\log[H^+]$, where $[H^+]$ is the concentration of hydrogen ions. $[H^+_a]$ for household ammonia is 10^{-11}. $[H^+_v]$ for vinegar is 6.3×10^{-3}. What is the difference of the pH levels of ammonia and vinegar?

Think	Write
Write the equation for pH.	$\text{pH} = -\log[H^+]$
Write the difference of the pH levels.	$-\log[H^+_a] - (-\log[H^+_v])$ $= -\log[H^+_a] + \log[H^+_v]$ $= \log[H^+_v] - \log[H^+_a]$
Substitute values for $[H^+_v]$ and $[H^+_a]$.	$= \log(6.3 \times 10^{-3}) - \log 10^{-11}$
Use the Product Property of Logarithms, and simplify.	$= \log 6.3 + \log 10^{-3} - \log 10^{-11}$ $= \log 6.3 - 3 + 11$
Use a calculator.	≈ 8.8
Write the answer.	The pH level of ammonia is about 8.8 greater than the pH level of vinegar.

  **Got It?** 4. **Reasoning** Suppose the hydrogen ion concentration for Substance A is twice that for Substance B. Which substance has the greater pH level? What is the greater pH level minus the lesser pH level? Explain.

Lesson Check

Do you know HOW?

Write each expression as a single logarithm.

1. $\log_4 2 + \log_4 8$
2. $\log_6 24 - \log_6 4$

Expand each logarithm.

3. $\log_3 \frac{x}{y}$
4. $\log m^2n^5$
5. $\log_2 \sqrt{x}$

Do you UNDERSTAND? 

6. **Vocabulary** State which property or properties need to be used to write each expression as a single logarithm.
   a. $\log_4 5 + \log_4 5$
   b. $\log_5 4 - \log_5 6$
7. **Reasoning** If $\log x = 5$, what is the value of $\frac{1}{x}$?
8. **Open-Ended** Write log 150 as a sum or difference of two logarithms. Simplify if possible.

Practice and Problem-Solving Exercises

Write each expression as a single logarithm. See Problem 1.

9. $\log 7 + \log 2$ **10.** $\log_2 9 - \log_2 3$ **11.** $5\log 3 + \log 4$

12. $\log 8 - 2\log 6 + \log 3$ **13.** $4\log m - \log n$ **14.** $\log 5 - k\log 2$

15. $\log_6 5 + \log_6 x$ **16.** $\log_7 x + \log_7 y - \log_7 z$ **17.** $\log_3 4 + \log_3 y + \log_3 8x$

Expand each logarithm. See Problem 2.

18. $\log x^3y^5$ **19.** $\log_7 49xyz$ **20.** $\log_b \frac{b}{x}$ **21.** $\log a^2$

22. $\log_5 \frac{r}{s}$ **23.** $\log_3 (2x)^2$ **24.** $\log_3 7(2x - 3)^2$ **25.** $\log \frac{a^2b^3}{c^4}$

26. $\log_4 5\sqrt{x}$ **27.** $\log_8 8\sqrt{3a^5}$ **28.** $\log_5 \frac{25}{x}$ **29.** $\log 10m^4n^{-2}$

Use the Change of Base Formula to evaluate each expression. See Problem 3.

30. $\log_2 9$ **31.** $\log_{12} 20$ **32.** $\log_7 30$ **33.** $\log_5 10$

34. $\log_4 7$ **35.** $\log_3 54$ **36.** $\log_5 62$ **37.** $\log_3 33$

38. Science The concentration of hydrogen ions in household dish detergent is 10^{-12}. What is the pH level of household dish detergent?

Use the properties of logarithms to evaluate each expression.

39. $\log_2 4 - \log_2 16$ **40.** $\log_2 96 - \log_2 3$ **41.** $\log_3 27 - 2\log_3 3$

42. $\log_6 12 + \log_6 3$ **43.** $\log_4 48 - \frac{1}{2}\log_4 9$ **44.** $\frac{1}{2}\log_5 15 - \log_5 \sqrt{75}$

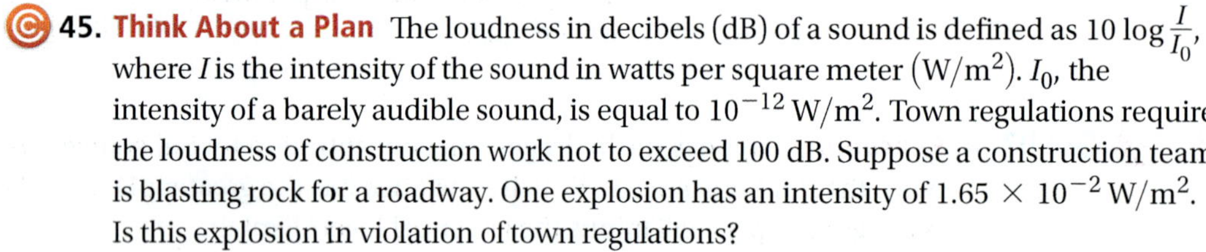

45. Think About a Plan The loudness in decibels (dB) of a sound is defined as $10 \log \frac{I}{I_0}$, where I is the intensity of the sound in watts per square meter (W/m^2). I_0, the intensity of a barely audible sound, is equal to 10^{-12} W/m^2. Town regulations require the loudness of construction work not to exceed 100 dB. Suppose a construction team is blasting rock for a roadway. One explosion has an intensity of 1.65×10^{-2} W/m^2. Is this explosion in violation of town regulations?

- Which physical value do you need to calculate to answer the question?
- What values should you use for I and I_0?

46. Construction The foreman of a construction team puts up a sound barrier that reduces the intensity of the noise by 50%. By how many decibels is the noise reduced? Use the formula $L = 10 \log \frac{I}{I_0}$ to measure loudness. (*Hint:* Find the difference between the expression for loudness for intensity I and the expression for loudness for intensity $0.5I$.)

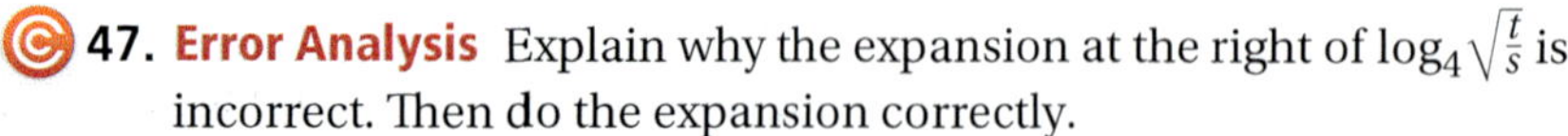

47. Error Analysis Explain why the expansion at the right of $\log_4 \sqrt{\frac{t}{s}}$ is incorrect. Then do the expansion correctly.

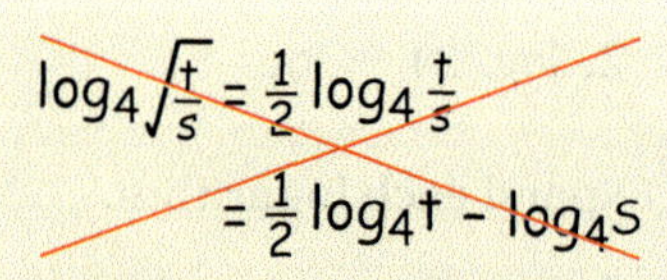

48. Reasoning Can you expand $\log_3 (2x + 1)$? Explain.

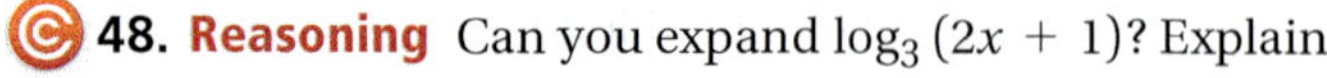

49. Writing Explain why $\log (5 \cdot 2) \neq \log 5 \cdot \log 2$.

Determine if each statment is *true* or *false*. Justify your answer.

50. $\log_2 4 + \log_2 8 = 5$

51. $\log_3 \frac{3}{2} = \frac{1}{2}\log_3 3$

52. $\log(x - 2) = \frac{\log x}{\log 2}$

53. $\frac{\log_b x}{\log_b y} = \log_b \frac{x}{y}$

54. $(\log x)^2 = \log x^2$

55. $\log_4 7 - \log_4 3 = \log_4 4$

Write each logarithmic expression as a single logarithm.

56. $\frac{1}{4}\log_3 2 + \frac{1}{4}\log_3 x$

57. $\frac{1}{2}(\log_x 4 + \log_x y) - 3\log_x z$

58. $x\log_4 m + \frac{1}{y}\log_4 n - \log_4 p$

59. $\left(\frac{2\log_b x}{3} + \frac{3\log_b y}{4}\right) - 5\log_b z$

Expand each logarithm.

60. $\log\sqrt{\frac{2x}{y}}$

61. $\log\frac{s\sqrt{7}}{t^2}$

62. $\log\left(\frac{2\sqrt{x}}{5}\right)^3$

63. $\log\frac{m^3}{n^4p^{-2}}$

64. $\log 4\sqrt{\frac{4r}{s^2}}$

65. $\log_b\frac{\sqrt{x}\sqrt[3]{y^2}}{\sqrt[5]{z^2}}$

66. $\log_4\frac{\sqrt{x^5y^7}}{zw^4}$

67. $\log\frac{\sqrt{x^2 - 4}}{(x + 3)^2}$

Write each logarithm as the quotient of two common logarithms. Do not simplify the quotient.

68. $\log_7 2$

69. $\log_3 8$

70. $\log_5 140$

71. $\log_9 3.3$

72. $\log_4 3x$

STEM **Astronomy** The apparent brightness of stars is measured on a logarithmic scale called magnitude, in which lower numbers mean brighter stars. The relationship between the ratio of apparent brightness of two objects and the difference in their magnitudes is given by the formula $m_2 - m_1 = -2.5\log\frac{b_2}{b_1}$, where m is the magnitude and b is the apparent brightness.

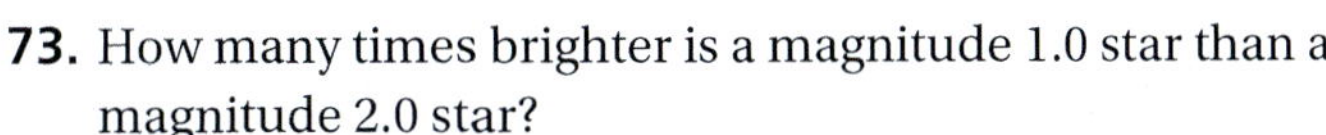

73. How many times brighter is a magnitude 1.0 star than a magnitude 2.0 star?

74. The star Rigel has a magnitude of 0.12. How many times brighter is Capella than Rigel?

Challenge

Expand each logarithm.

75. $\log\sqrt{\frac{x\sqrt{2}}{y^2}}$

76. $\log_3[(xy^{\frac{1}{3}}) + z^2]^3$

77. $\log_7\frac{\sqrt{r + 9}}{s^2t^{\frac{1}{3}}}$

Simplify each expression.

78. $\log_3(x + 1) - \log_3(3x^2 - 3x - 6) + \log_3(x - 2)$

79. $\log(a^2 - 10a + 25) + \frac{1}{2}\log\frac{1}{(a - 5)^3} - \log(\sqrt{a} - 5)$

Standardized Test Prep

80. Which expression is NOT equivalent to $\sqrt[6]{16r^2}$?

Ⓐ $(16r^2)^{\frac{1}{6}}$ Ⓑ $4r^{\frac{1}{3}}$ Ⓒ $(4r)^{\frac{1}{3}}$ Ⓓ $\sqrt[3]{4r}$

81. Assume that there are no more turning points beyond those shown. Which graph CANNOT be the graph of a fourth degree polynomial?

Ⓕ
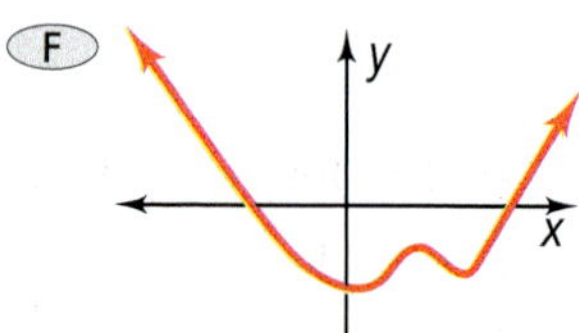

Ⓗ

Ⓖ
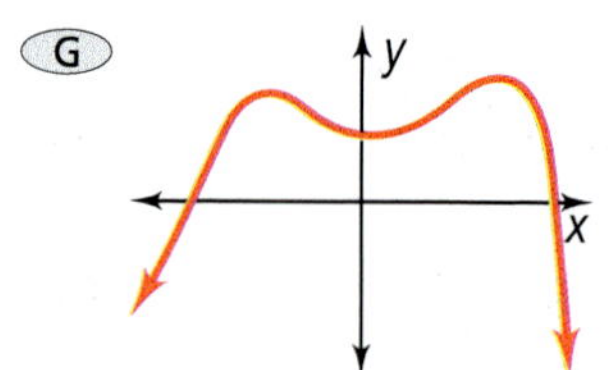

Ⓘ
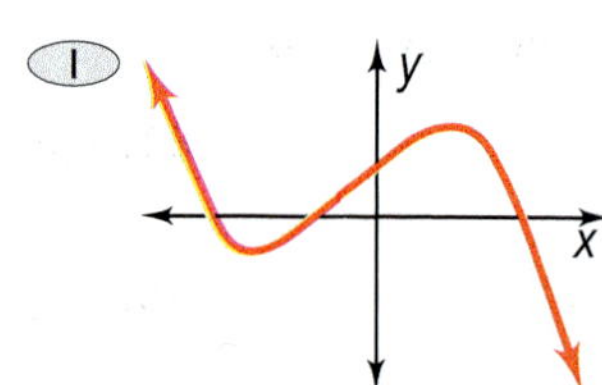

82. A florist is arranging a bouquet of daisies and tulips. He wants twice as many daisies as tulips in the bouquet. If the bouquet contains 24 flowers, how many daisies are in the bouquet?

Ⓐ 8 daisies Ⓑ 12 daisies Ⓒ 16 daisies Ⓓ 24 daisies

Short Response

83. Use the properties of logarithms to write log 18 in four different ways. Name each property you use.

Mixed Review

Write each equation in logarithmic form. See Lesson 7-3.

84. $49 = 7^2$ **85.** $\frac{1}{4} = 8^{-\frac{2}{3}}$ **86.** $5^{-3} = \frac{1}{125}$

Solve. Check for extraneous solutions. See Lesson 6-5.

87. $\sqrt[3]{y^4} = 16$ **88.** $\sqrt[3]{7x} - 4 = 0$ **89.** $2\sqrt{w-1} = \sqrt{w+2}$

Write a polynomial function with rational coefficients and the given roots. See Lesson 5-5.

90. $\sqrt{3}, -5$ **91.** $-i, 4i$ **92.** $-\sqrt{7}, 1 + 2i$

Get Ready! **To prepare for Lesson 7-5, do Exercises 93–95.**

Evaluate each logarithm. See Lesson 7-3.

93. $\log_{12} 144$ **94.** $\log_4 64$ **95.** $\log_{64} 4$

7-5 Exponential and Logarithmic Equations

Content Standards

F.LE.4 For exponential models, express as a logarithm the solution to $ab^{ct} = d$ where a, c, and d are numbers and the base b is 2, 10, or e; evaluate the logarithm using technology.

Also A.REI.11

Objective To solve exponential and logarithmic equations

MATHEMATICAL PRACTICES

Lesson Vocabulary
- exponential equation
- logarithmic equation

Any equation that contains the form b^{cx}, such as $a = b^{cx}$ where the exponent includes a variable, is an **exponential equation**.

Essential Understanding You can use logarithms to solve exponential equations. You can use exponents to solve logarithmic equations.

Problem 1 Solving an Exponential Equation—Common Base

Multiple Choice What is the solution of $16^{3x} = 8$?

Ⓐ $x = \frac{1}{4}$ Ⓑ $x = \frac{3}{7}$ Ⓒ $x = 1$ Ⓓ $x = 4$

Plan

What common base is appropriate?

2 because 16 and 8 are both powers of 2.

$16^{3x} = 8$

$(2^4)^{3x} = 2^3$ Rewrite the terms with a common base.

$2^{12x} = 2^3$ Power Property of Exponents

$12x = 3$ If two numbers with the same base are equal, their exponents are equal.

$x = \frac{1}{4}$ Solve and simplify.

The correct answer is A.

Got It? **1.** What is the solution of $27^{3x} = 81$?

When bases are not the same, you can solve an exponential equation by taking the logarithm of each side of the equation. If m and n are positive and $m = n$, then $\log m = \log n$.

Problem 2 Solving an Exponential Equation—Different Bases

What is the solution of $15^{3x} = 285$?

Think

Which property of logarithms will help isolate x?
The rule $\log a^x = x \log a$ moves x out of the exponent position.

$15^{3x} = 285$

$\log 15^{3x} = \log 285$ Take the logarithm of each side.

$3x \log 15 = \log 285$ Power Property of Logarithms.

$x = \dfrac{\log 285}{3 \log 15}$ Divide each side by 3 log 15 to isolate x.

$x \approx 0.6958$ Use a calculator.

Check $15^{3x} = 285$

$15^{3(0.6958)} \approx 285.0840331 \approx 285$ ✓

Got It? **2. a.** What is the solution of $5^{2x} = 130$?

b. Reasoning Why can't you use the same method you used in Problem 1 to solve Problem 2?

Problem 3 Solving an Exponential Equation With a Graph or Table

What is the solution of $4^{3x} = 6000$?

Method 1 Solve using a graph.

Use a graphing calculator. Graph the equations.

$Y_1 = 4^{3x}$
$Y_2 = 6000$

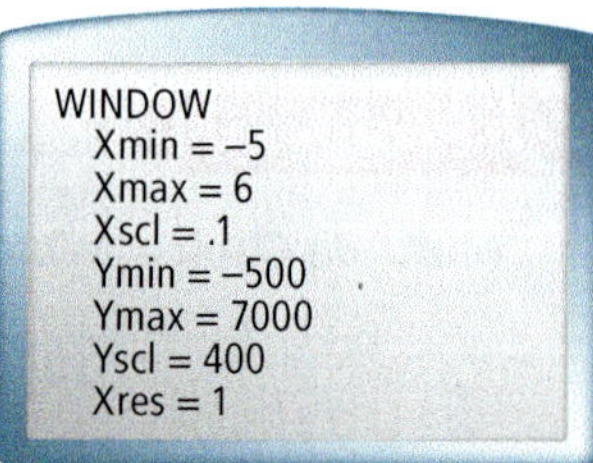

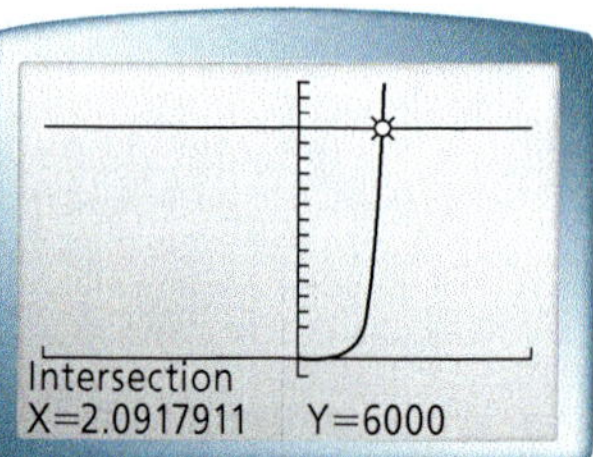

Adjust the window to find the point of intersection. The solution is $x \approx 2.09$.

Method 2 Solve using a table.

Think

How do you choose TblStart and ΔTbl values?
Start with 0 and 1, respectively. Adjust both values as you close in on the solution.

Use the table feature of a graphing calculator. Enter $Y_1 = 4^{3x}$.

Use the **TABLE SETUP** and **ΔTbl** features to locate the x-value that gives the y-value closest to 6000.

The solution is $x \approx 2.09$.

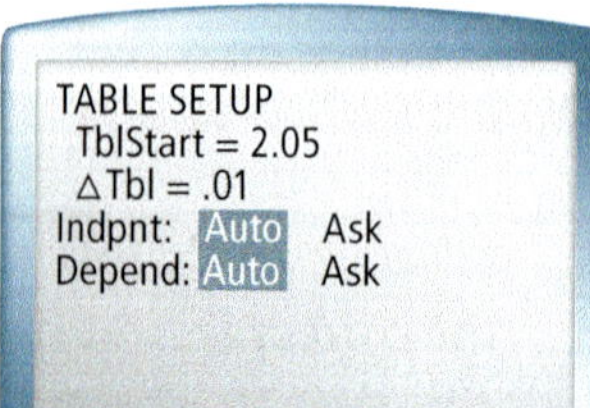

X	Y1
2.05	5042.8
2.06	5256.9
2.07	5480.2
2.08	5712.9
2.09	5955.5
2.1	6208.4
2.11	6472

Y1=5955.47143094

Got It? **3.** What is the solution of each exponential equation? Check your answer.

a. $7^{4x} = 800$ **b.** $5.2^{3x} = 400$

Problem 4 Modeling With an Exponential Equation STEM

Resource Management Wood is a sustainable, renewable, natural resource when you manage forests properly. Your lumber company has 1,200,000 trees. You plan to harvest 7% of the trees each year. How many years will it take to harvest half of the trees?

Know	Need	Plan
• Number of trees • Rate of decay	Number of years it takes to harvest 600,000 trees	• Write an exponential equation. • Use logarithms to solve the equation.

Think

What equation should you use to model this situation? Since you are planning to harvest 7% of the trees each year, you should use $y = ab^x$, where b is the decay factor.

Step 1 Is an exponential model reasonable for this situation?

Yes, you are harvesting a fixed percentage each year.

Step 2 Define the variables and determine the model.

Let n = the number of years it takes to harvest half of the trees.

Let $T(n)$ = the number of trees remaining after n years.

A reasonable model is $T(n) = a(b)^n$.

Step 3 Use the model to write an exponential equation.

$T(n) = 600{,}000$

$a = 1{,}200{,}000$

$r = -7\% = -0.07$

$b = 1 + r = 1 + (-0.07) = 0.93$

So, $1{,}200{,}000(0.93)^n = 600{,}000$.

Step 4 Solve the equation. Use logarithms.

$1{,}200{,}000(0.93)^n = 600{,}000$

$0.93^n = \frac{600{,}000}{1{,}200{,}000}$ Isolate the term with n.

$\log 0.93^n = \log 0.5$ Take the logarithm of each side.

$n \log 0.93 = \log 0.5$ Power Property of Logarithms.

$n = \frac{\log 0.5}{\log 0.93}$ Solve for n.

$n \approx 9.55$ Use a calculator.

It will take about 9.55 years to harvest half of the original trees.

Got It? **4.** After how many years will you have harvested half of the trees if you harvest 5% instead of 7% yearly?

A **logarithmic equation** is an equation that includes one or more logarithms involving a variable.

Problem 5 Solving a Logarithmic Equation

Plan

How do you convert between log form and exponential form?
Use the rule: $\log a = b$ if and only if $a = 10^b$.

What is the solution of $\log(4x - 3) = 2$?

Method 1 Solve using exponents.

$\log(4x - 3) = 2$

$4x - 3 = 10^2$ Write in exponential form.

$4x = 103$ Simplify.

$x = \frac{103}{4} = 25.75$ Solve for x.

Method 2 Solve using a graph.
Graph the equations
$Y_1 = \text{LOG}(4x - 3)$ and $Y_2 = 2$.
Find the point of intersection.
The solution is $x = 25.75$.

Method 3 Solve using a table.
Enter $Y_1 = \text{LOG}(4x - 3)$.
Use the **TABLE SETUP** feature to find the x-value that corresponds to a y-value of 2 in the table.
The solution is $x = 25.75$.

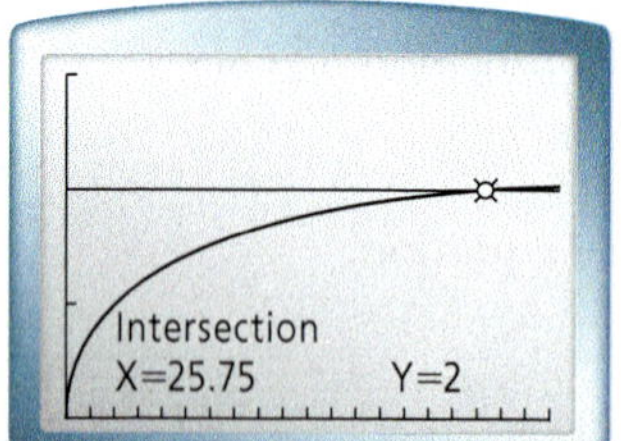

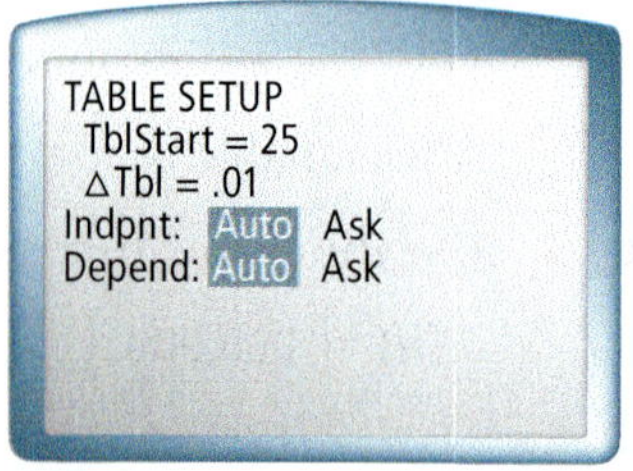

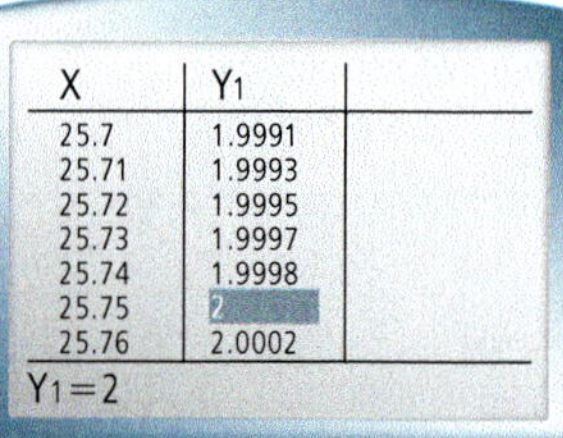

Got It? 5. What is the solution of $\log(3 - 2x) = -1$?

Problem 6 Using Logarithmic Properties to Solve an Equation

What is the solution of $\log(x - 3) + \log x = 1$?

$\log(x - 3) + \log x = 1$

$\log((x - 3)x) = 1$ Product Property of Logarithms

$(x - 3)x = 10^1$ Write in exponential form.

$x^2 - 3x - 10 = 0$ Simplify to a quadratic equation in standard form.

$(x - 5)(x + 2) = 0$ Factor the trinomial.

$x = 5$ or $x = -2$ Solve for x.

Check

$\log(x - 3) + \log(x) = 1$

$\log(-2 - 3) + \log(-2) \stackrel{?}{=} 1$ ✗

$\log(x - 3) + \log(x) = 1$

$\log(5 - 3) + \log(5) \stackrel{?}{=} 1$

$\log 2 + \log 5 \stackrel{?}{=} 1$

$0.3010 + 0.6990 = 1$ ✓

If $\log(x - 3) + \log(x) = 1$, $x = 5$.

Think

What is the domain of the logarithm function?
Logs are defined only for positive numbers. The log of a negative number is undefined.

Got It? 6. What is the solution of $\log 6 - \log 3x = -2$?

Lesson Check

Do you know HOW?

Solve each equation.

1. $3^x = 9$
2. $2^{y+1} = 25$
3. $\log 4x = 2$
4. $\log x - \log 2 = 3$

Do you UNDERSTAND?

MATHEMATICAL PRACTICES

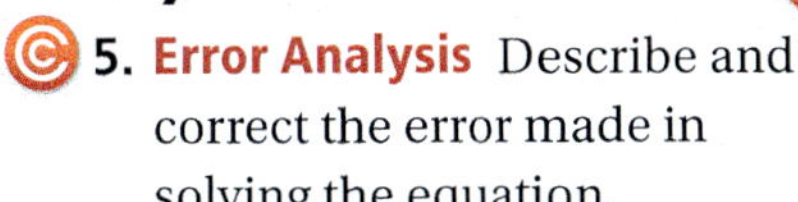

5. **Error Analysis** Describe and correct the error made in solving the equation.

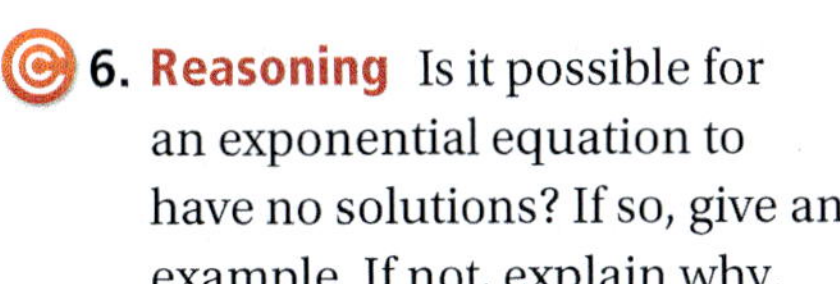

6. **Reasoning** Is it possible for an exponential equation to have no solutions? If so, give an example. If not, explain why.

Practice and Problem-Solving Exercises

MATHEMATICAL PRACTICES

A Practice

Solve each equation. See Problem 1.

7. $2^x = 8$
8. $3^{2x} = 27$
9. $4^{3x} = 64$
10. $5^{3x} = \frac{1}{125}$
11. $2^{5x+1} = 32$
12. $3^{2x+2} = 81$
13. $2^{3x} = 4^{x+1}$
14. $3^{x+2} = 27^{2x}$

Solve each equation. Round to the nearest ten-thousandth. Check your answers. See Problem 2.

15. $2^x = 3$
16. $4^x = 19$
17. $8 + 10^x = 1008$
18. $5 - 3^x = -40$
19. $9^{2y} = 66$
20. $12^{y-2} = 20$
21. $25^{2x+1} = 144$
22. $2^{3x-4} = 5$

Graphing Calculator Solve by graphing. Round to the nearest ten-thousandth. See Problem 3.

23. $4^{7x} = 250$
24. $5^{3x} = 500$
25. $6^x = 4565$
26. $1.5^x = 356$

Use a table to solve each equation. Round to the nearest hundredth.

27. $2^{x+3} = 512$
28. $3^{x-1} = 72$
29. $6^{2x} = 10$
30. $5^{2x} = 56$

31. The equation $y = 6.72(1.014)^x$ models the world population y, in billions of people, x years after the year 2000. Find the year in which the world population is about 8 billion. See Problem 4.

Solve each equation. Check your answers. See Problem 5.

32. $\log 2x = -1$
33. $2 \log x = -1$
34. $\log (3x + 1) = 2$
35. $\log x + 4 = 8$
36. $\log 6x - 3 = -4$
37. $3 \log x = 1.5$
38. $2 \log (x + 1) = 5$
39. $\log (5 - 2x) = 0$

Solve each equation. See Problem 6.

40. $\log x - \log 3 = 8$
41. $\log 2x + \log x = 11$
42. $2 \log x + \log 4 = 2$
43. $\log 5 - \log 2x = 1$
44. $3 \log x - \log 6 + \log 2.4 = 9$
45. $\log (7x + 1) = \log (x - 2) + 1$

46. Think About a Plan An earthquake of magnitude 9.1 occurred in 2004 in the Indian Ocean near Indonesia. It was about 74,900 times as strong as the greatest earthquake ever to hit Texas. Find the magnitude of the Texas earthquake. (Remember that an increase of 1.0 on the Richter scale means an earthquake is 30 times stronger.)

- Can you write an exponential or logarithmic equation?
- How does the solution of your equation help you find the magnitude?

47. Consider the equation $2^{\frac{x}{3}} = 80$.

a. Solve the equation by taking the logarithm base 10 of each side.

b. Solve the equation by taking the logarithm base 2 of each side.

c. Writing Compare your result in parts (a) and (b). What are the advantages of each method? Explain.

STEM **48. Seismology** An earthquake of magnitude 7.7 occurred in 2001 in Gujarat, India. It was about 4900 times as strong as the greatest earthquake ever to hit Pennsylvania. What is the magnitude of the Pennsylvania earthquake? (*Hint*: Refer to the Richter scale on page 453.)

49. As a town gets smaller, the population of its high school decreases by 6% each year. The senior class has 160 students now. In how many years will it have about 100 students? Write an equation. Then solve the equation without graphing.

Mental Math **Solve each equation.**

50. $2^x = \frac{1}{2}$ **51.** $3^x = 27$ **52.** $\log_9 3 = x$ **53.** $\log_4 64 = x$

54. $\log_8 2 = x$ **55.** $10^x = \frac{1}{100}$ **56.** $\log_7 343 = x$ **57.** $25^x = \frac{1}{5}$

58. Demography The table below lists the states with the highest and with the lowest population growth rates. Determine in how many years each event can occur. Use the model $P = P_0(1 + r)^x$, where P_0 is population from the table, as of July, 2007; x is the number of years after July, 2007, P is the projected population and r is the growth rate.

a. Population of Idaho exceeds 2 million.

b. Population of Michigan decreases by 1 million.

c. Population of Nevada doubles.

State	Growth rate (%)	Population (in thousands)	State	Growth rate (%)	Population (in thousands)
1. Nevada	2.93	2,565	46. New York	0.08	19,298
2. Arizona	2.81	6,339	47. Vermont	0.08	621
3. Utah	2.55	2,645	48. Ohio	0.03	11,467
4. Idaho	2.43	1,499	49. Michigan	−0.30	10,072
5. Georgia	2.17	9,545	50. Rhode Island	−0.36	1,058

Source: U.S. Census Bureau

59. **Open-Ended** Write and solve a logarithmic equation.

60. **Reasoning** The graphs of $y = 2^{3x}$ and $y = 3^{x+1}$ intersect at approximately (1.1201, 10.2692). What is the solution of $2^{3x} = 3^{x+1}$?

61. **Reasoning** If $\log 12^{0.5x} = \log 143.6$, then $12^{0.5x} = \underline{\ ?\ }$.

STEM **Acoustics** **In Exercises 62–63, the loudness measured in decibels (dB) is defined by loudness $= 10 \log \frac{I}{I_0}$, where I is the intensity and $I_0 = 10^{-12}$ W/m^2.**

62. The human threshold for pain is 120 dB. Instant perforation of the eardrum occurs at 160 dB.
    a. Find the intensity of each sound.
    b. How many times as intense is the noise that will perforate an eardrum as the noise that causes pain?

63. The noise level inside a convertible driving along the freeway with its top up is 70 dB. With the top down, the noise level is 95 dB.
    a. Find the intensity of the sound with the top up and with the top down.
    b. By what percent does leaving the top up reduce the intensity of the sound?

Solve each equation. If necessary, round to the nearest ten-thousandth.

64. $8^x = 444$

65. $\frac{1}{2}\log x + \log 4 = 2$

66. $4 \log_3 2 - 2 \log_3 x = 1$

67. $\log x^2 = 2$

68. $9^{2x} = 42$

69. $\log_8 (2x - 1) = \frac{1}{3}$

70. $\log(5x - 4) = 3$

71. $12^{4-x} = 20$

72. $5^{3x} = 125$

73. $\log 4 + 2 \log x = 6$

74. $4^{3x} = 77.2$

75. $\log_7 3x = 3$

Use the properties of exponential and logarithmic functions to solve each system. Check your answers.

76. $\begin{cases} y = 2^{x+4} \\ y - 4^{x-1} = 0 \end{cases}$

77. $\begin{cases} 2^{x+y} = 16 \\ 4^{x-y} = 1 \end{cases}$

78. $\begin{cases} \log(2x - y) = 1 \\ \log(x + y) = 3 \log 2 \end{cases}$

Challenge **Solve each equation.**

79. $\log_7 (2x - 3)^2 = 2$

80. $\log_2 (x^2 + 2x) = 3$

81. $\frac{3}{2}\log_2 4 - \frac{1}{2}\log_2 x = 3$

STEM 82. **Meteorology** In the formula $P = P_0\left(\frac{1}{2}\right)^{\frac{h}{4795}}$, P is the atmospheric pressure in millimeters of mercury at elevation h meters above sea level. P_0 is the atmospheric pressure at sea level. If P_0 equals 760 mm, at what elevation is the pressure 42 mm?

STEM **83. Music** The pitch, or frequency, of a piano note is related to its position on the keyboard by the function $F(n) = 440 \cdot 2^{\frac{n}{12}}$, where F is the frequency of the sound waves in cycles per second and n is the number of piano keys above or below Concert A, as shown. If $n = 0$ at Concert A, which of the instruments shown in the diagram can sound notes at the given frequency?

a. 590 **b.** 120 **c.** 1440 **d.** 2093

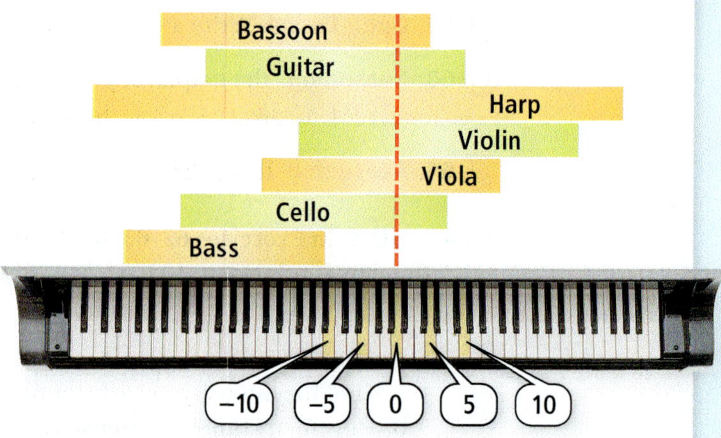

Standardized Test Prep

GRIDDED RESPONSE

SAT/ACT

84. The graph at the right shows the translation of the graph of the parent function $y = |x|$ down 2 units and 3 units to the right. What is the area of the shaded triangle in square units?

85. What does x equal if $\log(1 + 3x) = 3$?

86. Using the change of base formula, what is the value of x for which $\log_9 x = \log_3 5$?

87. The polynomial $x^4 + 3x^3 + 16x^2 - 19x + 8$ is divided by the binomial $x - 1$. What is the coefficient of x^2 in the quotient?

88. What positive value of b makes $x^2 + bx + 81$ a perfect square trinomial?

Mixed Review

Expand each logarithm. See Lesson 7-4.

89. $\log 2x^3y^{-2}$ **90.** $\log_3 \frac{x}{y}$ **91.** $\log_3 \sqrt{9x}$

Let $f(x) = 3x$ and $g(x) = x^2 - 1$. Perform each function operation.  See Lesson 6-6.

92. $(g - f)(x)$ **93.** $(f \circ g)(x)$ **94.** $(g \circ f)(x)$

Find all the zeros of each function. See Lesson 5-6.

95. $y = x^3 - x^2 + x - 1$ **96.** $f(x) = x^4 - 16$ **97.** $f(x) = x^4 - 5x^2 + 6$

Get Ready! **To prepare for Lesson 7-6, do Exercises 98–100.**

Write each logarithmic expression as a single logarithm. See Lesson 7-4.

98. $\log_2 15 - \log_2 5$ **99.** $\log 3 + 4 \log x$ **100.** $5 \log_7 2 - 2 \log_7 y$

Concept Byte

For Use With Lesson 7-5

TECHNOLOGY

Using Logarithms for Exponential Models

Content Standards

F.IF.8 Write a function defined by an expression in different but equivalent forms to reveal and explain different properties of the function.

Also F.IF.7.e

You can transform an exponential function into a linear function by taking the logarithm of each side. Since linear models are easy to recognize, you can then determine whether an exponential function is a good model for a set of values.

$y = ab^x$	Write the general form of an exponential function.
$\log y = \log ab^x$	Take the logarithm of each side.
$\log y = \log a + x(\log b)$	Use the Product Property and the Power Property.

If $\log b$ and $\log a$ are constants, then $\log y = (\log b)x + \log a$ is a linear equation in slope-intercept form when you plot the points as $(x, \log y)$.

Activity

Determine whether an exponential function is a good model for the values in the table.

x	0	2	4	6	8	10
y	0.5	2	7.8	32	127.9	511.7

Step 1 Enter the values into stat lists L_1 and L_2. To enter the values of $\log y$, place the cursor in the heading of L_3 and press log L_2 enter.

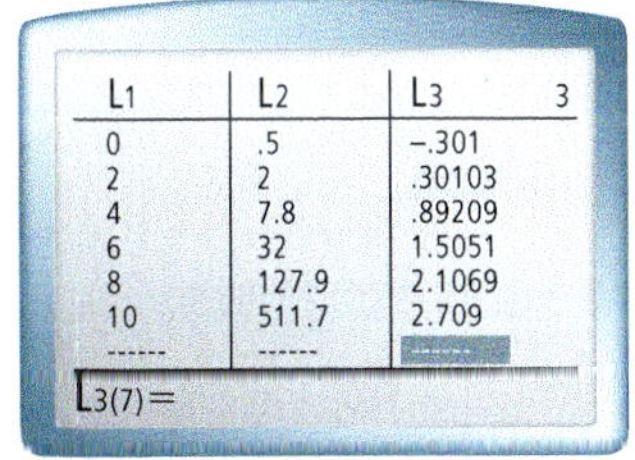

L1	L2	L3
0	.5	−.301
2	2	.30103
4	7.8	.89209
6	32	1.5051
8	127.9	2.1069
10	511.7	2.709
------	------	------

L3(7)=

Step 2 To graph $\log y$, access the stat plot feature and press **1**. Then enter L_3 next to **YLIST**:. Then press zoom **9.**

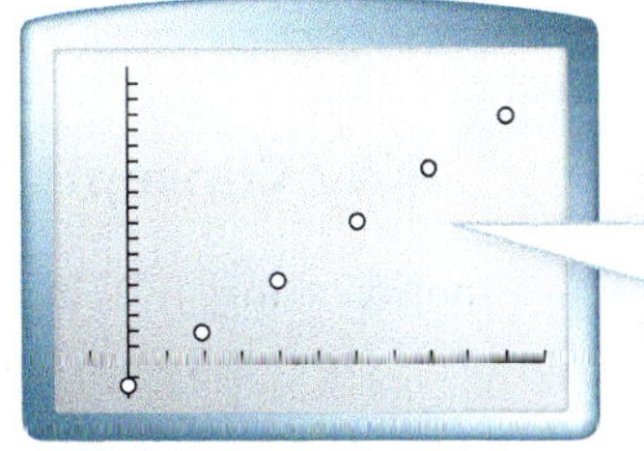

The points $(x, \log y)$ lie on a line, so an exponential model is appropriate.

Step 3 Press stat ▷ **0** enter to find the exponential function $y = 0.5(2)^x$.

Exercises

For each set of values, determine whether an exponential function is a good model. If so, find the exponential function.

1.

x	1	3	5	7	9
y	6	22	54	102	145

2.

x	−1	0	1	2	3
y	40.2	19.8	9.9	5.1	2.5

3. Writing Explain how you could determine whether a logarithmic function is a good model for a set of data.

Natural Logarithms

Content Standard

F.LE.4 For exponential models, express as a logarithm the solution to $ab^{ct} = d$ where *a*, *c*, and *d* are numbers and the base *b* is 2, 10, or *e* . . .

Objectives To evaluate and simplify natural logarithmic expressions
To solve equations using natural logarithms

Getting Ready!

A function f is bounded above if there is some number B that f(x) can never exceed. The exponential function base e shown here is not bounded above. Is the logarithmic function base e bounded above? If so, find a bounding number. If not, explain why.

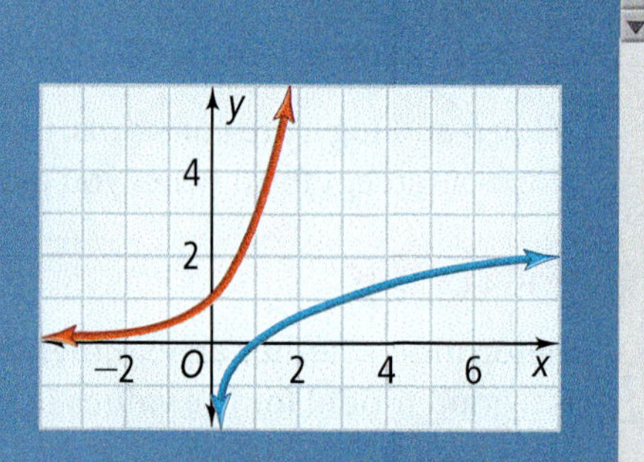

What do you know about inverse functions that could help here?

Lesson Vocabulary
- natural logarithmic function

The function $y = e^x$ has an inverse, the **natural logarithmic function**, $y = \log_e x$, or $y = \ln x$.

Essential Understanding The functions $y = e^x$ and $y = \ln x$ are inverse functions. Just as before, this means that if $a = e^b$, then $b = \ln a$, and vice versa.

take note

Key Concept Natural Logarithmic Function

If $y = e^x$, then $x = \log_e y = \ln y$. The natural logarithmic function is the inverse of $x = \ln y$, so you can write it as $y = \ln x$.

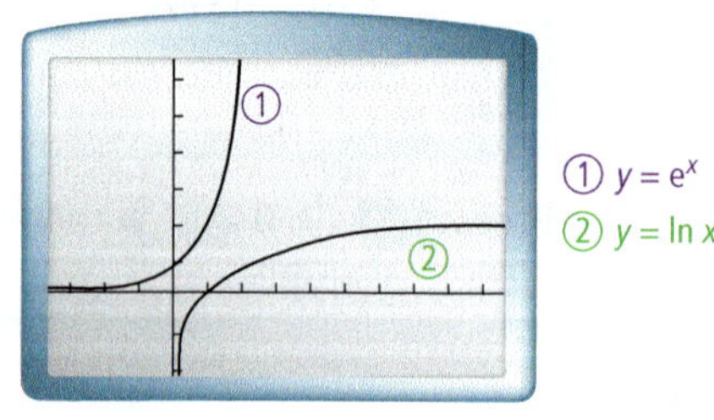

① $y = e^x$
② $y = \ln x$

Problem 1 Simplifying a Natural Logarithmic Expression

Plan

Can you use properties of logarithms?
Yes; the properties you studied in Lesson 7-4 apply to logarithms with any base.

What is $2 \ln 15 - \ln 75$ written as a single natural logarithm?

$2 \ln 15 - \ln 75 = \ln 15^2 - \ln 75$ Power Property of Logarithms

$= \ln \frac{15^2}{75}$ Quotient Property of Logarithms

$= \ln 3$ Simplify.

Got It? 1. What is each expression written as a single natural logarithm?

a. $\ln 7 + 2\ln 5$ **b.** $3\ln x - 2\ln 2x$ **c.** $3\ln x + 2\ln y + \ln 5$

You can use the inverse relationship between the functions $y = \ln x$ and $y = e^x$ to solve certain logarithmic and exponential equations.

Problem 2 Solving a Natural Logarithmic Equation

Think

How do you go from logarithmic form to exponential form?
Use the definition of logarithm: $\ln x = y$ if and only if $x = e^y$.

What are the solutions of $\ln(x-3)^2 = 4$?

$\ln(x-3)^2 = 4$

$(x-3)^2 = e^4$ Rewrite in exponential form.

$x - 3 = \pm e^2$ Find the square root of each side.

$x = 3 \pm e^2$ Solve for x.

$x \approx 10.39$ or -4.39 Use a calculator.

Check

$\ln(10.39 - 3)^2 \stackrel{?}{=} 4$ $\ln(-4.39 - 3)^2 \stackrel{?}{=} 4$

$4.0003 \approx 4$ ✔ $4.0003 \approx 4$ ✔

Got It? 2. What are the solutions of each equation? Check your answers.

a. $\ln x = 2$ **b.** $\ln(3x+5)^2 = 4$ **c.** $\ln 2x + \ln 3 = 2$

Problem 3 Solving an Exponential Equation

Plan

How can you solve this equation?
First get e^x by itself on one side of the equation. Then rewrite the equation in logarithmic form and solve for x.

What is the solution of $4e^{2x} + 2 = 16$?

$4e^{2x} + 2 = 16$

$4e^{2x} = 14$ Subtract 2 from each side.

$e^{2x} = 3.5$ Divide each side by 4.

$2x = \ln 3.5$ Rewrite in logarithmic form.

$x = \frac{\ln 3.5}{2}$ Divide each side by 2.

$x \approx 0.626$ Use a calculator.

Check

$4e^{2x} + 2 = 16$

$4e^{2(0.626)} + 2 \stackrel{?}{=} 16$

$15.99 \approx 16$ ✔

Got It? 3. What is the solution of each equation? Check your answers.

a. $e^{x-2} = 12$ **b.** $2e^{-x} = 20$ **c.** $e^{3x} + 5 = 15$

Natural logarithms are useful because they help express many relationships in the physical world.

Problem 4 Using Natural Logarithms 

Space **A spacecraft can attain a stable orbit 300 km above Earth if it reaches a velocity of 7.7 km/s. The formula for a rocket's maximum velocity v in kilometers per second is $v = -0.0098t + c \ln R$. The booster rocket fires for t seconds and the velocity of the exhaust is c km/s. The ratio of the mass of the rocket filled with fuel to its mass without fuel is R. Suppose the rocket shown in the photo has a mass ratio of 25, a firing time of 100 s and an exhaust velocity as shown. Can the spacecraft attain a stable orbit 300 km above Earth?**

Plan

How can you solve the problem? Use the given formula to find the maximum velocity v of the spacecraft. If $v \geq 7.7$, the spacecraft can attain a stable orbit.

Let $R = 25$, $c = 2.8$, and $t = 100$. Find v.

$v = -0.0098t + c \ln R$	Use the formula.
$= -0.0098(100) + 2.8 \ln 25$	Substitute.
$\approx -0.98 + 2.8(3.219)$	Use a calculator.
≈ 8.0	Simplify.

The maximum velocity of 8.0 km/s is greater than the 7.7 km/s needed for a stable orbit. Therefore, the spacecraft can attain a stable orbit 300 km above Earth.

  **Got It?** **4. a.** A booster rocket for a spacecraft has a mass ratio of about 15, an exhaust velocity of 2.1 km/s, and a firing time of 30 s. Can the spacecraft achieve a stable orbit 300 km above Earth?

b. Reasoning Suppose a rocket, as designed, cannot provide enough velocity to achieve a stable orbit. Could alterations to the rocket make a stable orbit achievable? Explain.

Lesson Check

Do you know HOW?

Write each expression as a single natural logarithm.

1. $4 \ln 3$

2. $\ln 18 - \ln 10$

3. $\ln 3 + \ln 4$

4. $-2 \ln 2$

Solve each equation.

5. $\ln 5x = 4$

6. $\ln (x - 7) = 2$

7. $2 \ln x = 4$

8. $\ln (2 - x) = 1$

Do you UNDERSTAND? 

9. Error Analysis Describe the error made in solving the equation. Then find the correct solution.

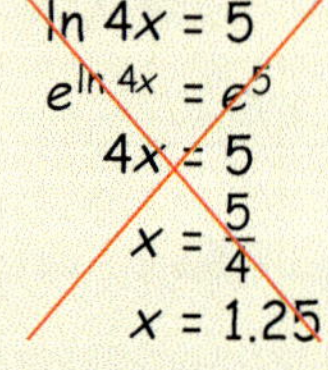

10. Reasoning Can $\ln 5 + \log_2 10$ be written as a single logarithm? Explain your reasoning.

Practice and Problem-Solving Exercises

Write each expression as a single natural logarithm.

See Problem 1.

11. $3 \ln 5$

12. $\ln 9 + \ln 2$

13. $\ln 24 - \ln 6$

14. $5 \ln m - 3 \ln n$

15. $\frac{1}{3}(\ln x + \ln y) - 4 \ln z$

16. $\ln a - 2 \ln b + \frac{1}{3} \ln c$

17. $4 \ln 8 + \ln 10$

18. $\ln 3 - 5 \ln 3$

19. $2 \ln 8 - 3 \ln 4$

Solve each equation. Check your answers.

See Problem 2.

20. $\ln 3x = 6$

21. $\ln x = -2$

22. $\ln (4x - 1) = 36$

23. $1.1 + \ln x^2 = 6$

24. $\ln \frac{x - 1}{2} = 4$

25. $\ln 4r^2 = 3$

26. $2 \ln 2x^2 = 1$

27. $\ln (2m + 3) = 8$

28. $\ln (t - 1)^2 = 3$

Use natural logarithms to solve each equation.

See Problem 3.

29. $e^x = 18$

30. $e^{\frac{x}{5}} + 4 = 7$

31. $e^{2x} = 12$

32. $e^{\frac{x}{2}} = 5$

33. $e^{x+1} = 30$

34. $e^{2x} = 10$

35. $e^{3x} + 5 = 6$

36. $e^{\frac{x}{9}} - 8 = 6$

37. $7 - 2e^{\frac{x}{2}} = 1$

STEM **Space** **For Exercises 38 and 39, use $v = -0.0098t + c \ln R$, where v is the velocity of the rocket, t is the firing time, c is the velocity of the exhaust, and R is the ratio of the mass of the rocket filled with fuel to the mass of the rocket without fuel.**

See Problem 4.

38. Find the velocity of a spacecraft whose booster rocket has a mass ratio of 20, an exhaust velocity of 2.7 km/s, and a firing time of 30 s. Can the spacecraft achieve a stable orbit 300 km above Earth?

39. A rocket has a mass ratio of 24 and an exhaust velocity of 2.5 km/s. Determine the minimum firing time for a stable orbit 300 km above Earth.

40. Think About a Plan By measuring the amount of carbon-14 in an object, a paleontologist can determine its approximate age. The amount of carbon-14 in an object is given by $y = ae^{0.00012t}$, where a is the amount of carbon-14 originally in the object, and t is the age of the object in years. In 2003, a bone believed to be from a dire wolf was found at the La Brea Tar Pits. The bone contains 14% of its original carbon-14. How old is the bone?

- What numbers should you substitute for y and t?
- What properties of logarithms and exponents can you use to solve the equation?

STEM **41. Archaeology** A fossil bone contains 25% of its original carbon-14. What is the approximate age of the bone?

Simplify each expression.

42. $\ln 1$

43. $\frac{\ln e}{4}$

44. $\frac{\ln e^2}{2}$

45. $\ln e^{83}$

46. $\ln e$

47. $\ln e^2$

48. $\ln e^{10}$

49. $10 \ln e$

50. $\ln e^3$

51. $\frac{\ln e^4}{8}$

52. **Error Analysis** A student has broken the natural logarithm key on his calculator, so he decides to use the Change of Base Formula to find ln 100. Explain his error and find the correct answer.

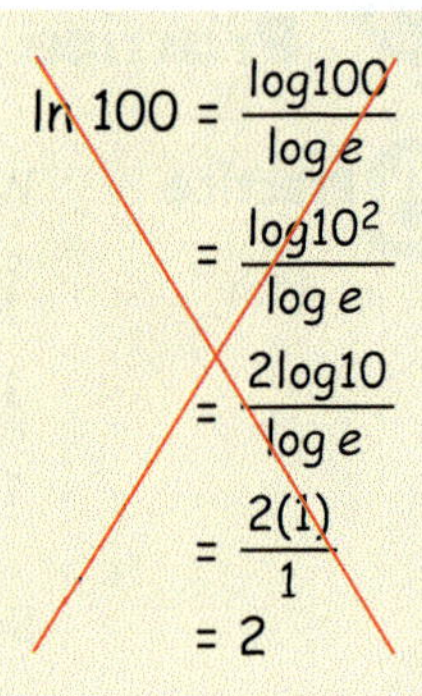

53. **Satellite** The battery power available to run a satellite is given by the formula $P = 50\,e^{-\frac{t}{250}}$, where P is power in watts and t is time in days. For how many days can the satellite run if it requires 15 watts of power?

Determine whether each statement is *always*, *sometimes*, or *never* true.

54. $\ln e^x \geq 1$

55. $\ln e^x = \ln e^x + 1$

56. $\ln t = \log_e t$

STEM 57. **Space** Use the formula for maximum velocity $v = -0.0098t + c \ln R$. Find the mass ratio of a rocket with an exhaust velocity of 3.1 km/s, a firing time of 50 s, and a maximum shuttle velocity of 6.9 km/s.

STEM **Biology** **The formula $H = \frac{1}{r}(\ln P - \ln A)$ models the number of hours it takes a bacteria culture to decline, where H is the number of hours, r is the rate of decline, P is the initial bacteria population, and A is the reduced bacteria population.**

58. A scientist determines that an antibiotic reduces a population of 20,000 bacteria to 5000 in 24 hours. Find the rate of decline caused by the antibiotic.

59. A laboratory assistant tests an antibiotic that causes a rate of decline of 0.14. How long should it take for a population of 8000 bacteria to shrink to 500?

Solve each equation.

60. $\frac{1}{3}\ln x + \ln 2 - \ln 3 = 3$

61. $\ln(x + 2) - \ln 4 = 3$

62. $2e^{x-2} = e^x + 7$

63. **Error Analysis** Consider the solution to the equation $\ln(x - 3)^2 = 4$ at the right. In Problem 2 you saw that there are two solutions to this equation, $3 + e^2$ and $3 - e^2$. Why do you get only one solution using this method?

$\ln(x - 3)^2 = 4$
$2\ln(x - 3) = 4$
$\ln(x - 3) = 2$
$e^{\ln(x - 3)} = e^2$
$x - 3 = e^2$
$x = e^2 + 3$

STEM 64. **Technology** In 2008, there were about 1.5 billion Internet users. That number is projected to grow to 3.5 billion in 2015.

a. Let t represent the time, in years, since 2008. Write a function of the form $y = ae^{ct}$ that models the expected growth in the population of Internet users.

b. In what year are there 2 billion Internet users?

c. In what year are there 5 billion Internet users?

d. Solve your equation for t.

e. **Writing** Explain how you can use your equation from part (d) to verify your answers to parts (b) and (c).

STEM **65. Physics** The function $T(t) = T_r + (T_i - T_r)e^{kt}$ models Newton's Law of Cooling. $T(t)$ is the temperature of a heated substance t minutes after it has been removed from a heat (or cooling) source. T_i is the substance's initial temperature, k is a constant for that substance, and T_r is room temperature.

a. The initial surface temperature of a beef roast is 236°F and room temperature is 72°F. If $k = -0.041$, how long will it take for this roast to cool to 100°F?

b. Graphing Calculator Write and graph an equation that you can use to check your answer to part (a). Use your graph to complete the table below.

Temperature (°F)	225	200	175	150	125	100	75
Minutes Later	■	■	■	■	■	■	■

Standardized Test Prep

GRIDDED RESPONSE

SAT/ACT

66. An investment of \$750 will be worth \$1500 after 12 years of continuous compounding at a fixed interest rate. What percent is the interest rate?

67. What is $\log 33{,}000 - \log 99 + \log 30$?

68. If $f(x) = 5 - x^2$ and $g(x) = x^2 - 3$, what is $(g \circ f)(6)$?

69. What is the positive root of $y = 2x^2 - 35x - 57$?

70. What is the real part of $3 + 2i$?

71. What is $\frac{\sqrt{36}}{\sqrt{4}}$?

Mixed Review

Solve each equation. See Lesson 7-5.

72. $3^{2x} = 6561$ **73.** $7^x - 2 = 252$ **74.** $25^{2x+1} = 144$

75. $\log 3x = 4$ **76.** $\log 5x + 3 = 3.7$ **77.** $\log 9 - \log x + 1 = 6$

Find the inverse of each function. Is the inverse a function? See Lesson 6-7.

78. $y = 5x + 7$ **79.** $y = 2x^3 + 10$ **80.** $y = -x^2 + 5$ **81.** $y = 3x + 2$

Get Ready! To prepare for Lesson 8-1, do Exercises 82–84.

For Exercises 82–84, y varies directly with x. See Lesson 2-2.

82. If $x = 2$ when $y = 4$, find y when $x = 5$.

83. If $x = 1$ when $y = 5$, find y when $x = 3$.

84. If $x = 10$ when $y = 3$, find y when $x = 4$.

Concept Byte

For Use With Lesson 7-6

EXTENSION

Exponential and Logarithmic Inequalities

Content Standard

A.REI.11 Explain why the x-coordinates of the points where the graphs of the equations $y = f(x)$ and $y = g(x)$ intersect are the solutions of the equation $f(x) = g(x)$ . . .

You can use the graphing and table capabilities of your calculator to solve problems involving exponential and logarithmic inequalities.

Example 1

Solve $2(3)^{x+4} > 10$ using a graph.

Step 1 Define **Y1** and **Y2**.

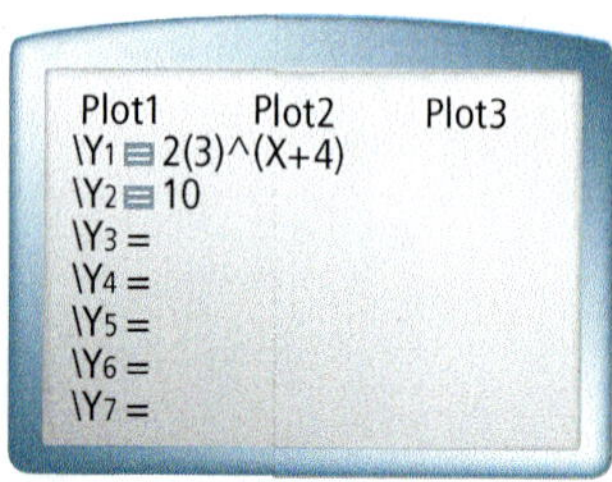

Step 2 Make a graph and find the point of intersection.

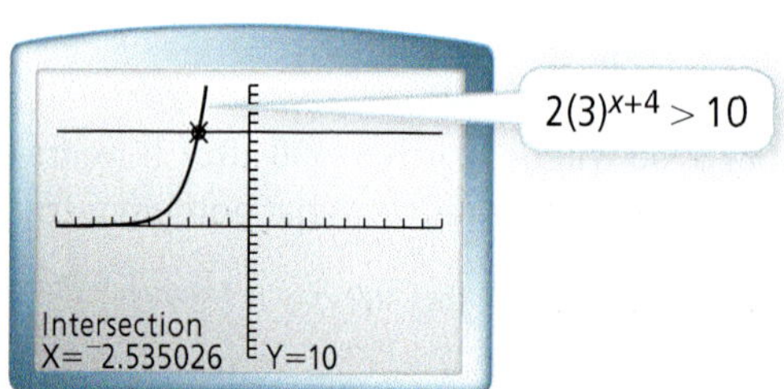

Step 3 Identify the x-values that make the inequality true.

The solution is $x > -2.535$.

Exercises

Solve each inequality using a graph.

1. $4(3)^{x+1} > 6$
2. $\log x + 3\log(x - 1) < 4$
3. $3(2)^{x+2} \geq 5$
4. $x + 1 < 12\log x$
5. $2(3)^{x-4} > 7$
6. $\log x + 2\log(x - 1) < 1$
7. $4(2)^{x-1} \leq 5$
8. $2\log x + 4\log(x + 3) > 3$
9. $5(4)^{x-1} < 2$

STEM 10. **Bacteria Growth** Scientists are growing bacteria in a laboratory. They start with a known population of bacteria and measure how long it takes the population to double.

Bacteria Population

Sample	Initial Population	Doubling Time (in hours)
Sample A	200,000	1
Sample B	50,000	0.5

a. Write an exponential function that models the population in Sample A as a function of time in hours.
b. Write an exponential function that models the population in Sample B as a function of time in hours.
c. Write an inequality that models the population in Sample B overtaking the population in Sample A.
d. Use a graphing calculator to solve the inequality in part (c).

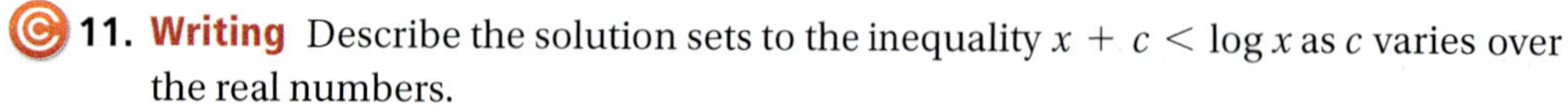

11. **Writing** Describe the solution sets to the inequality $x + c < \log x$ as c varies over the real numbers.

Example 2

Solve $\log x + 2\log(x + 1) < 2$ using a table.

Step 1 Define **Y1** and **Y2**.

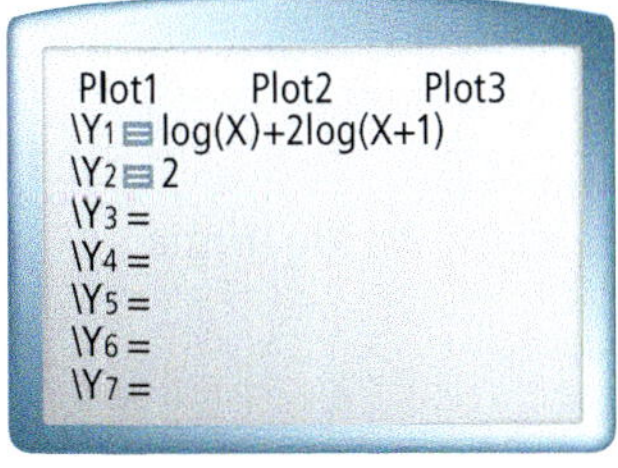

Step 2 Make a table and examine the values.

X	Y1	Y2
0	ERR.	2
1	.60206	2
2	1.2553	2
3	1.6812	2
4	2	2
5	2.2553	2
6	2.4683	2

X=4

$Y_1 < Y_2$ for these x-values.

Step 3 Identify the x-values that make the inequality true.

The solution is $0 < x < 4$.

Exercises

Solve each inequality using a table.
(Hint: For more accurate results, set ΔTbl = 0.001.)

12. $\log x + \log(x + 1) < 3$

13. $3(2)^{x+1} > 5$

14. $\log x + 5\log(x - 1) \geq 3$

15. $5(3)^x \leq 2$

16. $3\log x + \log(x + 2) > 1$

17. $2(4)^{x+3} \leq 8$

Barometric Pressure **Average barometric pressure varies with the altitude of a location. The greater the altitude is, the lower the pressure. The altitude A is measured in feet above sea level. The barometric pressure P is measured in inches of mercury (in. Hg). The altitude can be modeled by the function $A(P) = 90{,}000 - 26{,}500 \ln P$.**

18. What is a reasonable domain of the function? What is the range of the function?

19. **Graphing Calculator** Use a graphing calculator to make a table of function values. Use **TblStart** = 30 and Δ**Tbl** = −1.

20. Write an equation to find what average pressure the model predicts at sea level, or $A = 0$. Use your table to solve the equation.

21. Kilimanjaro is a mountain in Tanzania that formed from three extinct volcanoes. The base of the mountain is at 3000 ft above sea level. The peak is at 19,340 ft above sea level. On Kilimanjaro, $3000 \leq A(P) \leq 19{,}340$ is true for the altitude. Write an inequality from which you can find minimum and maximum values of normal barometric pressure on Kilimanjaro. Use a table and solve the inequality for P.

22. Denver, Colorado, is nicknamed the "Mile High City" because its elevation is about 1 mile, or 5280 ft, above sea level. The lowest point in Phoenix, Arizona, is 1117 ft above sea level. Write an inequality that describes the range of $A(P)$ as you drive from Phoenix to Denver. Then solve the inequality for P. (Assume that you never go lower than 1117 ft and you never go higher than 5280 ft.)

Pull It All Together ASSESSMENT

BIG idea Modeling

You can represent many real-world mathematical problems algebraically. An algebraic model can lead to an algebraic solution.

Performance Task 1

Suppose you invest a dollars to earn an annual interest rate of r percent (as a decimal). After t years, the value of the investment with interest compounded yearly is $A(t) = a(1 + r)^t$. The value with interest compounded continuously is $A(t) = a \cdot e^{rt}$.

a. Explain why you can call $e^r - 1$ the effective annual interest rate for the continuous compounding.

b. Suppose you can earn interest at some rate between 0% and 5%. Use your knowledge of the exponential function to explain why continuous compounding does not give you much of an investment advantage.

c. For each situation find the unknown quantity, such that continuous compounding gives you a \$1 advantage over annually compounded interest. Show your work.

- How much must you invest for 1 year at 2%?
- At what interest rate must you invest \$1000 for 1 year?
- For how long must you invest \$1000 at 2%?

BIG idea Function

You can use transformations such as translations, reflections, and dilations to understand relationships within a family of functions.

Performance Task 2

$f(x) = b^x$ and $g(x) = \log_b x$ are inverse functions. Explain why each of the following is true.

a. The translation $f_1(x) = b^{x-h}$ of f is equivalent to a vertical stretch or compression of f.

b. The inverse of $f_1(x) = b^{x-h}$ is equivalent to a translation of g.

c. The inverse of $f_1(x) = b^{x-h}$ is not equivalent to a vertical stretch or compression of g.

d. The function $k(x) = \log_c x$ is a vertical stretch or compression of g or of its reflection $-g$.

7 Chapter Review

Connecting BIG ideas to the Math You've Learned

1 Modeling
The function $y = ab^x$, $a > 0$, $b > 1$, models exponential growth. $y = ab^x$ models exponential decay if $0 < b < 1$.

Exponential Models (Lesson 7-1)
The population P is 1000 at the start. In each time period,
- P grows by 5%. $P = 1000(1.05)^t$
- P shrinks by 5%. $P = 1000(0.95)^t$

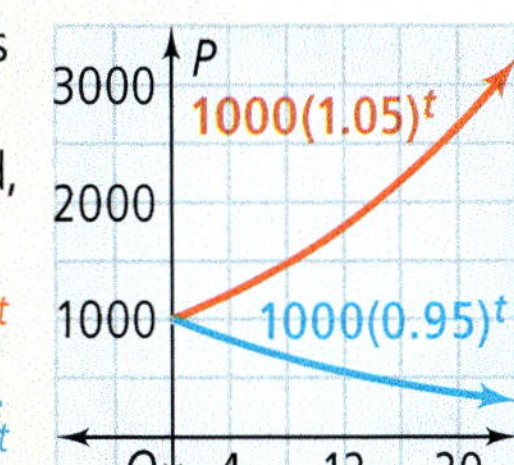

Properties of Logarithms (Lesson 7-4)

$b^a b^c = b^{a+c}$

$\log_b mn = \log_b m + \log_b n$

$\frac{b^a}{b^c} = b^{a-c}$

$\log_b \frac{m}{n} = \log_b m - \log_b n$

$\log_b m^n = n \log_b m$

$\log_n m = \frac{\log_b m}{\log_b n}$

2 Equivalence
Logarithms are exponents. In fact, $\log_b a = c$ if and only if $b^c = a$.

3 Function
The exponential function $y = b^x$ and the logarithmic function $y = \log_b x$ are inverse functions.

Logarithmic Functions as Inverses (Lesson 7-3)
- $y = 2^x$
- $y = \log_2 x$
- $y = 2^{x-1}$
- $y = (\log_2 x) + 1$

$y = 2x$, $y = 2x - 1$, $y = \log_2 x + 1$, $y = \log_2 x$

Exponential and Natural Logarithm Equations (Lessons 7-5 and 7-6)

$e^{x+3} = 4$

$e^x \cdot e^3 = 4$ or $x + 3 = \ln 4$

$e^x = \frac{4}{e^3}$

$x = \ln \frac{4}{e^3}$

$x = \ln 4 - \ln e^3$

$x = (\ln 4) - 3$

Chapter Vocabulary

- asymptote (p. 435)
- Change of Base Formula (p. 464)
- common logarithm (p. 453)
- continuously compounded interest (p. 446)
- decay factor (p. 436)
- exponential decay (p. 435)
- exponential equation (p. 469)
- exponential function (p. 434)
- exponential growth (p. 435)
- growth factor (p. 436)
- logarithm (p. 451)
- logarithmic equation (p. 471)
- logarithmic function (p. 454)
- logarithmic scale (p. 453)
- natural base exponential functions (p. 446)
- natural logarithmic function (p. 478)

Fill in the blanks.

1. There are two types of exponential functions. For ___?___, as the value of x increases, the value of y decreases, approaching zero. For ___?___, as the value of x increases, the value of y increases.

2. As x or y increases in absolute value, the graph may approach a(n) ___?___.

3. A(n) ___?___ with a base e is a ___?___.

4. $A = Pe^{rt}$ is known as the ___?___ formula.

5. The inverse of an exponential function with a base e is the ___?___.

7-1 Exploring Exponential Models

Quick Review

The general form of an **exponential function** is $y = ab^x$, where x is a real number, $a \neq 0$, $b > 0$, and $b \neq 1$. When $b > 1$, the function models **exponential growth**, and b is the **growth factor**. When $0 < b < 1$, the function models **exponential decay**, and b is the **decay factor**. The y-intercept is $(0, a)$.

Example

Determine whether $y = 2(1.4)^x$ is an example of exponential growth or decay. Then, find the y-intercept.

Since $b = 1.4 > 1$, the function represents exponential growth.

Since $a = 2$, the y-intercept is $(0, 2)$.

Exercises

Determine whether each function is an example of exponential growth or decay. Then, find the y-intercept.

6. $y = 5^x$

7. $y = 2(4)^x$

8. $y = 0.2(3.8)^x$

9. $y = 3(0.25)^x$

10. $y = \frac{25}{7}\left(\frac{7}{5}\right)^x$

11. $y = 0.0015(10)^x$

12. $y = 2.25\left(\frac{1}{3}\right)^x$

13. $y = 0.5\left(\frac{1}{4}\right)^x$

Write a function for each situation. Then find the value of each function after five years. Round to the nearest dollar.

14. A \$12,500 car depreciates 9% each year.

15. A baseball card bought for \$50 increases 3% in value each year.

7-2 Properties of Exponential Functions

Quick Review

Exponential functions can be translated, stretched, compressed, and reflected.

The graph of $y = ab^{x-h} + k$ is the graph of the parent function $y = b^x$ stretched or compressed by a factor $|a|$, reflected across the x-axis if $a < 0$, and translated h units horizontally and k units vertically.

The **continuously compounded interest** formula is $A = Pe^{rt}$, where P is the principal, r is the annual interest rate, and t is time in years.

Example

How does the graph of $y = -3^x + 1$ compare to the graph of the parent function?

The parent function is $y = 3^x$.

Since $a = -1$, the graph is reflected across the x-axis.

Since $k = 1$, it is translated up 1 unit.

Exercises

How does the graph of each function compare to the graph of the parent function?

16. $y = 5(2)^{x+1} + 3$

17. $y = -2\left(\frac{1}{3}\right)^{x-2}$

Find the amount in a continuously compounded account for the given conditions.

18. principal: \$1000
annual interest rate: 4.8%
time: 2 years

19. principal: \$250
annual interest rate: 6.2%
time: 2.5 years

Evaluate each expression to four decimal places.

20. e^{-3}

21. e^{-1}

22. e^{5}

23. $e^{-\frac{1}{2}}$

7-3 Logarithmic Functions as Inverses

Quick Review

If $x = b^y$, then $\log_b x = y$. The **logarithmic function** is the inverse of the exponential function, so the graphs of the functions are reflections of one another across the line $y = x$. Logarithmic functions can be translated, stretched, compressed, and reflected, as represented by $y = a\log_b(x - h) + k$, similarly to exponential functions.

When $b = 10$, the logarithm is called a **common logarithm**, which you can write as $\log x$.

Example

Write $5^{-2} = 0.04$ in logarithmic form.

If $y = b^x$, then $\log_b y = x$.

$y = 0.04$, $b = 5$ and $x = -2$.

So, $\log_5 0.04 = -2$.

Exercises

Write each equation in logarithmic form.

24. $6^2 = 36$ **25.** $2^{-3} = 0.125$

26. $3^3 = 27$ **27.** $10^{-3} = 0.001$

Evaluate each logarithm.

28. $\log_2 64$ **29.** $\log_3 \frac{1}{9}$

30. $\log 0.00001$ **31.** $\log_2 1$

Graph each logarithmic function.

32. $y = \log_3 x$ **33.** $y = \log x + 2$

34. $y = 3\log_2 (x)$ **35.** $y = \log_5 (x + 1)$

How does the graph of each function compare to the graph of the parent function?

36. $y = 3\log_4 (x + 1)$ **37.** $y = -\ln x + 2$

7-4 Properties of Logarithms

Quick Review

For any positive numbers, m, n, and b where $b \neq 1$, each of the following statements is true. Each can be used to rewrite a logarithmic expression.

- $\log_b mn = \log_b m + \log_b n$, by the Product Property
- $\log_b \frac{m}{n} = \log_b m - \log_b n$, by the Quotient Property
- $\log_b m^n = n\log_b m$, by the Power Property

Example

Write $2\log_2 y + \log_2 x$ as a single logarithm. Identify any properties used.

$2\log_2 y + \log_2 x$

$= \log_2 y^2 + \log_2 x$ Power Property

$= \log_2 xy^2$ Product Property

Exercises

Write each expression as a single logarithm. Identify any properties used.

38. $\log 8 + \log 3$ **39.** $\log_2 5 - \log_2 3$

40. $4\log_3 x + \log_3 7$ **41.** $\log x - \log y$

42. $\log 5 - 2\log x$ **43.** $3\log_4 x + 2\log_4 x$

Expand each logarithm. State the properties of logarithms used.

44. $\log_4 x^2y^3$ **45.** $\log 4s^4t$

46. $\log_3 \frac{2}{x}$ **47.** $\log (x + 3)^2$

48. $\log_2 (2y - 4)^3$ **49.** $\log \frac{z^3}{5}$

Use the Change of Base Formula to evaluate each expression.

50. $\log_2 7$ **51.** $\log_3 10$

7-5 Exponential and Logarithmic Equations

Quick Review

An equation in the form $b^{cx} = a$, where the exponent includes a variable, is called an **exponential equation**. You can solve exponential equations by taking the logarithm of each side of the equation. An equation that includes one or more logarithms involving a variable is called a **logarithmic equation**.

Example

Solve and round to the nearest ten-thousandth.

$6^{2x} = 75$

$\log 6^{2x} = \log 75$ Take the logarithm of both sides.

$2x \log 6 = \log 75$ Power Property of Logarithms

$x = \frac{\log 75}{2 \log 6}$ Divide both sides by 2 log 6.

$x \approx 1.2048$ Evaluate using a calculator.

Exercises

Solve each equation. Round to the nearest ten-thousandth.

52. $25^{2x} = 125$

53. $3^x = 36$

54. $7^{x-3} = 25$

55. $5^x + 3 = 12$

56. $\log 3x = 1$

57. $\log_2 4x = 5$

58. $\log x = \log 2x^2 - 2$

59. $2 \log_3 x = 54$

Solve by graphing. Round to the nearest ten-thousandth.

60. $5^{2x} = 20$

61. $3^{7x} = 160$

62. $6^{3x+1} = 215$

63. $0.5^x = 0.12$

64. A culture of 10 bacteria is started, and the number of bacteria will double every hour. In about how many hours will there be 3,000,000 bacteria?

7-6 Natural Logarithms

Quick Review

The inverse of $y = e^x$ is the **natural logarithmic function** $y = \log_e x = \ln x$. You solve natural logarithmic equations in the same way as common logarithmic equations.

Example

Use natural logarithms to solve $\ln x - \ln 2 = 3$.

$\ln x - \ln 2 = 3$

$\ln \frac{x}{2} = 3$ Quotient Property

$\frac{x}{2} = e^3$ Rewrite in exponential form.

$\frac{x}{2} \approx 20.0855$ Use a calculator to find e^3.

$x \approx 40.171$ Simplify.

Exercises

Solve each equation. Check your answers.

65. $e^{3x} = 12$

66. $\ln x + \ln (x + 1) = 2$

67. $2 \ln x + 3 \ln 2 = 5$

68. $\ln 4 - \ln x = 2$

69. $4e^{(x-1)} = 64$

70. $3 \ln x + \ln 5 = 7$

71. An initial investment of $350 is worth $429.20 after six years of continuous compounding. Find the annual interest rate.

7

Chapter Test

Do you know HOW?

Determine whether each function is an example of exponential growth or decay. Then find the y-intercept.

1. $y = 3(0.25)^x$

2. $y = 2(6)^{-x}$

3. $y = 0.1(10)^x$

4. $y = 3e^x$

Describe how the graph of each function is related to the graph of its parent function. Then find the domain, range, and asymptotes.

5. $y = 3^x + 2$

6. $y = \left(\frac{1}{2}\right)^{x+1}$

7. $y = -(2)^{x+2}$

Write each equation in logarithmic form.

8. $5^4 = 625$

9. $e^0 = 1$

Evaluate each logarithm.

10. $\log_2 8$

11. $\log_7 7$

12. $\log_5 \frac{1}{125}$

13. $\log_{11} 1$

Graph each logarithmic function. Compare each graph to the graph of its parent function. List each function's domain, range, y-intercept, and asymptotes.

14. $y = \log_3 (x - 1)$

15. $y = \frac{1}{2}\log_5 (x + 2)$

16. $y = 1 - \log_2 x$

Write each logarithmic expression as a single logarithm.

17. $\log_2 4 + 3 \log_2 9$

18. $3 \log a - 2 \log b$

Expand each logarithm.

19. $\log_7 \frac{a}{b}$

20. $\log 3x^3y^2$

Use the properties of logarithms to evaluate each expression.

21. $\log_9 27 - \log_9 9$

22. $2 \log 5 + \log 40$

Solve each equation.

23. $(27)^{3x} = 81$

24. $3^{x-1} = 24$

25. $2e^{3x} = 16$

26. $2 \log x = -4$

Use the Change of Base Formula to rewrite each expression using common logarithms.

27. $\log_3 16$

28. $\log_2 10$

29. $\log_7 8$

30. $\log_4 9$

Use the properties of logarithms to simplify and solve each equation. Round to the nearest thousandth.

31. $\ln 2 + \ln x = 1$

32. $\ln (x + 1) + \ln (x - 1) = 4$

33. $\ln (2x - 1)^2 = 7$

34. $3 \ln x - \ln 2 = 4$

Do you UNDERSTAND?

35. Writing Show that solving the equation $3^{2x} = 4$ by taking the common logarithm of each side is equivalent to solving it by taking the logarithm with base 3 of each side.

36. Open-Ended Give an example of an exponential function that models exponential growth and an example of an exponential function that models exponential decay.

37. Investment You put \$1500 into an account that pays 7% annual interest compounded continuously. How long will it be before you have \$2000 in your account?

7 Cumulative Standards Review ASSESSMENT

TIPS FOR SUCCESS

Some problems ask you to find lengths of arcs or areas of sectors. Read the question at the right. Then follow the tips to answer the sample question.

TIP 1

Draw a diagram

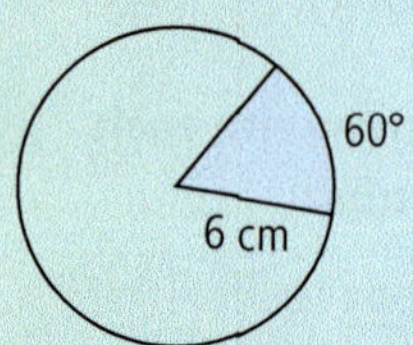

Let $f(x) = \pi x^2$ represent the area of a circle with radius x. Let $g(x) = \frac{60x}{360}$ represent the area of a 60° sector of a circle with area x. A circle with radius 6 centimeters has a sector measuring 60°. What is the area of this sector $(g \circ f)(x)$?

(A) 6π cm^2

(B) 12π cm^2

(C) 24π cm^2

(D) 36π cm^2

TIP 2

Find $f(x)$ first.

$f(x) = \pi x^2$
$= \pi \cdot 6^2$
$= 36\pi$

Think It Through

$$g(x) = \frac{60x}{360}$$

$$(g \circ f)(x) = \frac{60(36\pi)}{360} = \frac{36\pi}{6} = 6\pi$$

The correct answer is A.

Vocabulary Builder

As you solve problems, you must understand the meanings of mathematical terms. Match each term with its mathematical meaning.

A. growth factor

B. asymptote

C. logarithmic function

D. exponential equation

I. the inverse of an exponential function

II. a line that a graph approaches as x or y increases in absolute value

III. the value of b in $y = ab^x$, when $b > 1$

IV. an equation of the form $b^{cx} = a$, where the exponent includes a variable

Multiple Choice

Read each question. Then write the letter of the correct answer on your page.

1. The population of a town is modeled by the equation $P = 16{,}581e^{0.02t}$ where P represents the population t years after 2000. According to the model, what will the population of the town be in 2020?

(A) 16,916 (C) 20,252

(B) 17,258 (D) 24,736

2. If $i = \sqrt{-1}$, then which expression is equal to $9i(13i)$?

(F) -117 (H) 117

(G) $117i$ (I) $-117i$

3. Which expression is equivalent to $\log_5 32$?

(A) $\log 5 + \log 32$

(B) $\log 5 - \log 32$

(C) $(\log 5)(\log 32)$

(D) $\frac{\log 32}{\log 5}$

4. The table shows the height of a ball that was tossed into the air. Which equation best models the relationship between time t and the height of the ball h?

Time (seconds)	0	0.25	0.5	0.75
Height (feet)	4	10.5	15	17.5

Ⓕ $h = 26t + 4$

Ⓖ $h = -16t^2 + 30t + 4$

Ⓗ $h = 4t^2$

Ⓘ $h = -16t^2 + 4$

5. Which is the graph of $y = 3^x$?

Ⓐ
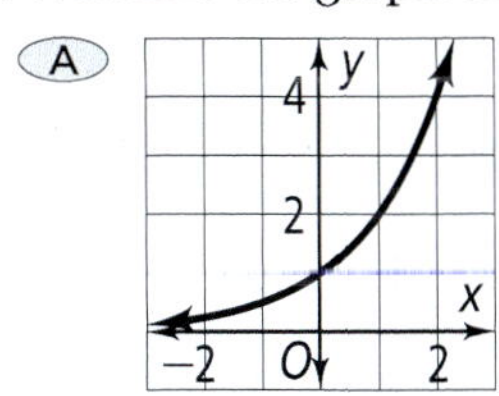

Ⓒ
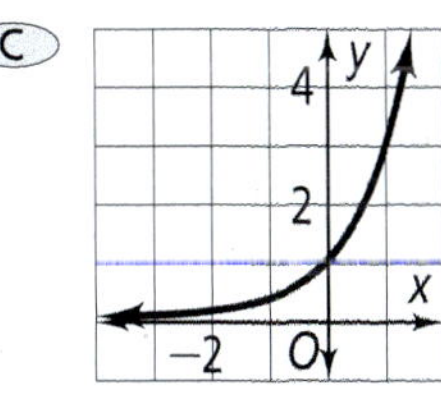

Ⓑ
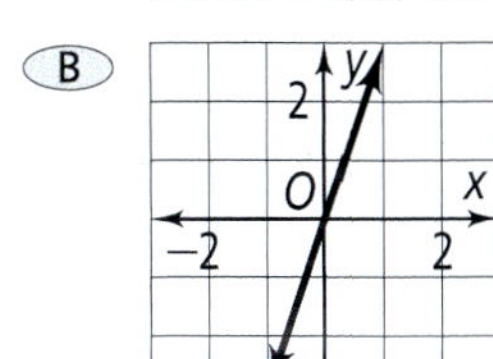

Ⓓ
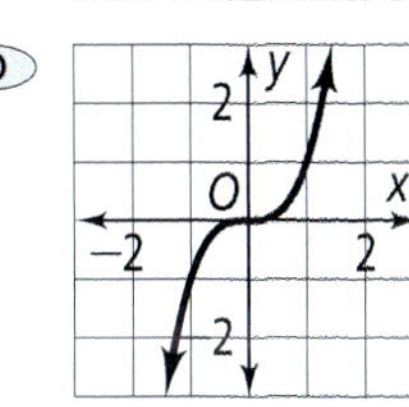

6. Which exponential function is equivalent to $y = \log_3 x$?

Ⓕ $y = 3^x$ Ⓗ $y = x^3$

Ⓖ $y = \frac{x}{3}$ Ⓘ $x = 3^y$

7. What is the simplest form of the quotient $\frac{\sqrt{72x^5y^3z^8}}{\sqrt{3xy^2z^2}}$?

Ⓐ $\sqrt{24x^4yz^6}$ Ⓒ $24x^4yz^6$

Ⓑ $2x^2z^3\sqrt{6y}$ Ⓓ $12x^2yz^6$

8. Solve the mass energy equivalence formula $e = mc^2$ for c.

Ⓕ $c = e^2m$ Ⓗ $c = \sqrt{\frac{e}{m}}$

Ⓖ $c = \sqrt{\frac{m}{e}}$ Ⓘ $c = \sqrt{(e - m)}$

9. What is the quotient of $(x^3 + 2x^2 - x + 6) \div (x + 3)$?

Ⓐ $x^2 + 5x + 14$, R 42 Ⓒ $x^3 + 5x^2 + 14x + 42$

Ⓑ $x^2 - x + 2$ Ⓓ $x^2 + x - 2$

10. On a certain night, a restaurant employs x servers at \$25 per hour and y bus persons at \$8 per hour. The total hourly cost for the restaurant's 12 employees that night is \$249. The following system of equations can be used to find the number of servers and the number of bus persons at work.

$$\begin{cases} 25x + 8y = 249 \\ x + y = 12 \end{cases}$$

Based on the solution of the system of equations, which of the following can you conclude?

Ⓕ Fewer than 2 bus persons were working.

Ⓖ More than ten servers were working.

Ⓗ 50% of the people working were bus persons.

Ⓘ 75% of the people working were servers.

11. Which polynomial equation has the real roots of $-3, 1, 1$, and $\frac{3}{2}$?

Ⓐ $x^4 - \frac{1}{2}x^3 - \frac{13}{2}x^2 + \frac{21}{2}x - \frac{9}{2} - 0$

Ⓑ $x^4 - \frac{1}{2}x^3 - \frac{17}{2}x^2 - 10x - \frac{9}{2} = 0$

Ⓒ $x^4 + x^3 - 5x^2 + 3x - \frac{3}{2} = 0$

Ⓓ $(x - 3)(x + 1)(x + 1)\left(x + \frac{3}{2}\right) = 0$

12. What is the factored form of $2x^3 + 5x^2 - 12x$?

Ⓕ $x(2x - 3)(x + 4)$ Ⓗ $x(2x + 4)(x - 3)$

Ⓖ $(2x^2 - 3)(x + 4)$ Ⓘ $(2x - 4)(x + 3)$

13. If $f(x) = 5\sqrt[3]{x^2}$ and $g(x) = 3\sqrt[3]{x^2}$, what is $f(x) + g(x)$?

Ⓐ $8\sqrt[3]{x^2}$ Ⓑ $8\sqrt[6]{x^2}$ Ⓒ $8\sqrt[3]{x^4}$ Ⓓ $8\sqrt[6]{x^4}$

14. What is the equation of the function graphed below?

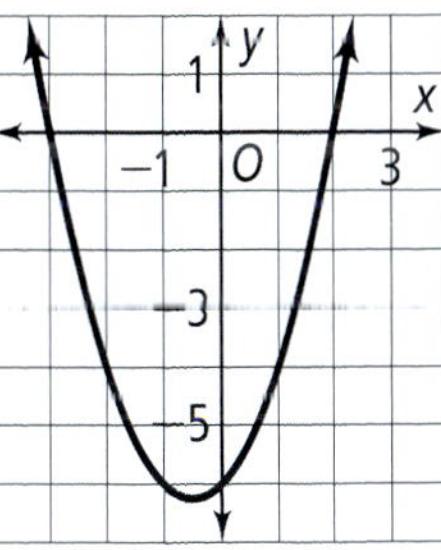

Ⓕ $y = (x + 2)(x - 3)$ Ⓗ $y = (x + 3)(x - 2)$

Ⓖ $y = (x - 6)^2$ Ⓘ $y = (x - 1)(x + 5)$

GRIDDED RESPONSE

15. What is the solution of the equation $\log_9 x = \log_6 x$?

16. A savings account pays 4.62% annual interest, compounded continuously. After approximately how many years will a principal of $500 double?

17. The graph of a polynomial has x-intercepts at $(-3, 0)$, $(-1, 0)$, and $(1, 0)$. What is the least possible degree of the polynomial?

18. Evaluate $\log_4 8$.

19. Solve $4^{2x} = 32$.

20. How many different real solutions are there for the equation $4x^2 = -4x - 4$?

21. Use the Fundamental Theorem of Algebra to determine the total number of complex zeros of $f(x) = x^2 - 3x^5 + 4x - x^7 - 44$.

22. Solve the following system of equations. What is the x-coordinate of the solution?

$$\begin{cases} y - 3 = x \\ \dfrac{y}{x^2 - 9} = 1 \end{cases}$$

23. What is the solution of $\sqrt{x + 2} = x$?

24. What is the x-intercept of $f(x) = x^3 - x^2 + 1$? Round the answer to the nearest hundredth.

25. What is the maximum y-value of the parabola?

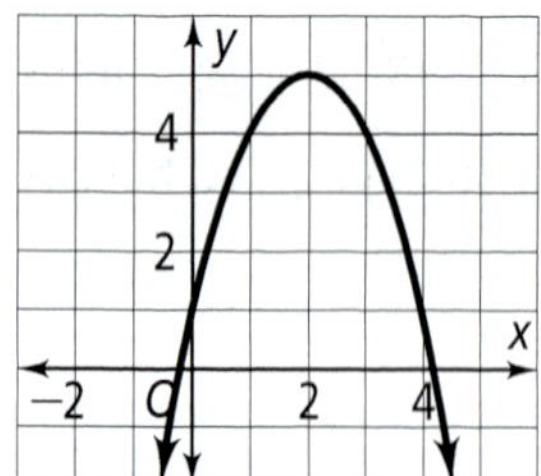

26. What is solution of the equation $-16 = x^2 + 8x$?

Short Response

27. The graph below is transformation of the radical parent function $y = \sqrt{x}$.

a. Describe the transformations used to create the graph shown.

b. What is an equation of the graph shown?

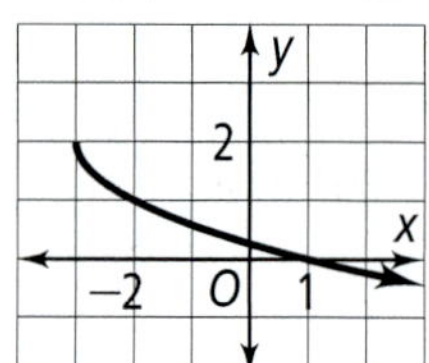

28. Compare the end behaviors of the functions below.

$f(x) = x^3 + 1$ $\qquad$ $g(x) = 5x^2 - 7$

29. To use an outdoor toddler swing, a child must weigh at least 15 pounds, and can weigh no more than 35 pounds. Draw a graph to model this situation.

30. Can a quadratic equation with real coefficients have exactly one imaginary root? Explain your answer.

31. Graph the function $y = -3(2)^{x+1}$ as a translation of its parent function.

Extended Response

32. A transformation of the parent absolute value function is shown in the graph.

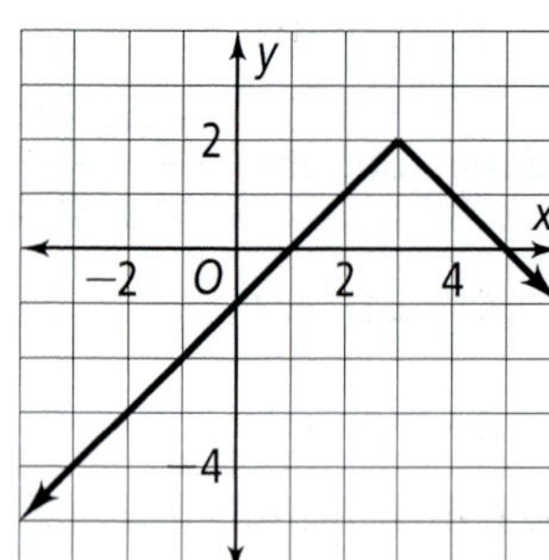

a. What is the transformation from the parent function $y = |x|$?

b. Draw the graph of the transformed function after reflecting it across the y-axis and translating it 4 units down.

End-of-Course Assessment

ASSESSMENT

Complete the following items. For multiple choice items, write the letter of the correct response on your paper. For all other items, show or explain your work.

1. The graph of a quadratic function $f(x)$ is shown below.

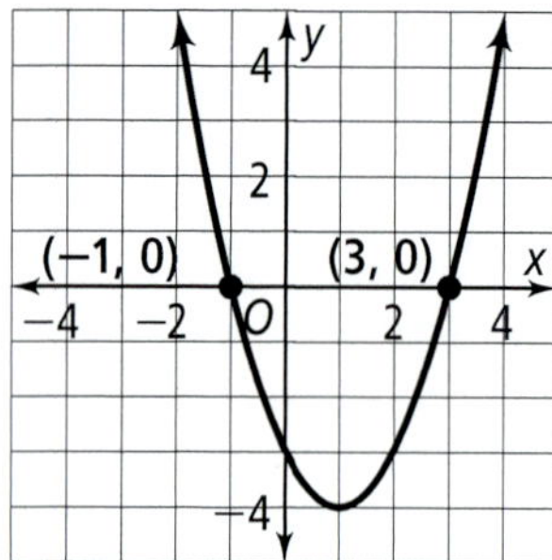

Use the graph to solve $f(x) < 0$.

(A) $-1 < x < 3$

(B) $-1 \le x \le 3$

(C) $x < -1$ or $x > 3$

(D) $x \le -1$ or $x > 3$

2. Newton's Law of Universal Gravitation is $F = \frac{Gm_1m_2}{r^2}$. Solve this equation for r.

(F) $r = \sqrt{\frac{F}{Gm_1m_2}}$

(G) $r = \sqrt{\frac{Gm_1m_2}{F}}$

(H) $r = \frac{F}{2Gm_1m_2}$

(I) $r = \frac{Gm_1m_2}{2F}$

3. Let $f(x) = x^3 - 4x^2 + 9x$ and let $g(x) = 6x^3 + x^2 - 5x - 12$. What is $f(x) - g(x)$?

(A) $-5x^3 - 5x^2 + 14x + 12$

(B) $-5x^3 - 3x^2 + 4x - 12$

(C) $7x^3 - 3x^2 + 4x - 12$

(D) $-5x^4 - 5x^3 + 14x^2 + 12x$

4. Let $f^{-1}(x) = 2x + 3$. What is the solution of $f(x) = f^{-1}(x)$?

(F) $x = -1$ or $x = -2$

(G) $x = -3$

(H) $(x, y) = (-1, -2)$

(I) $(x, y) = (-3, -3)$

5. Which is a simpler form of $\frac{\sqrt{5}}{3 - \sqrt{2}}$?

(A) $\frac{\sqrt{10}}{3\sqrt{2} - 2}$

(B) $\frac{5}{3\sqrt{5} - 10}$

(C) $\frac{3\sqrt{5} - 10}{7}$

(D) $\frac{3\sqrt{5} + \sqrt{10}}{7}$

6. What is the product of $(2 + 5i)$ and $(4 + 3i)$?

(F) $8 + 15i$

(G) $8 + 26i$

(H) $-7 + 15i$

(I) $-7 + 26i$

7. Multiply $\frac{x^3}{x^2 - 4} \cdot \frac{5x + 10}{10x}$.

(A) $\frac{5x}{4}$

(B) $\frac{x^2}{2(x - 2)}$

(C) $\frac{x^2(x + 2)}{2(x^2 - 4)}$

(D) $\frac{5x^3}{10x(x - 2)}$

8. Which is a simpler form of the complex fraction $\frac{\frac{1}{b} + c}{b + \frac{1}{c}}$?

(F) 1

(G) $\frac{c}{b}$

(H) $\left(\frac{1}{b} + c\right)^2$

(I) $(1 + c)(b + 1)$

9. What is the sum of the x-intercepts of the graph of the quadratic function $y = x^2 - 4x - 12$?

(A) 6

(B) 4

(C) −1

(D) −4

10. What is the equation of a parabola with the following characteristics?

Axis of symmetry: $x = -3$

Range: all real numbers less than or equal to 4

Ⓕ $y = -(x - 4)^2 - 3$ Ⓗ $y = -(x + 3)^2 + 4$

Ⓖ $y = (x - 4)^2 - 3$ Ⓘ $y = (x + 3)^2 + 4$

11. The graph of a degree 4 polynomial function with integer zeros is shown below.

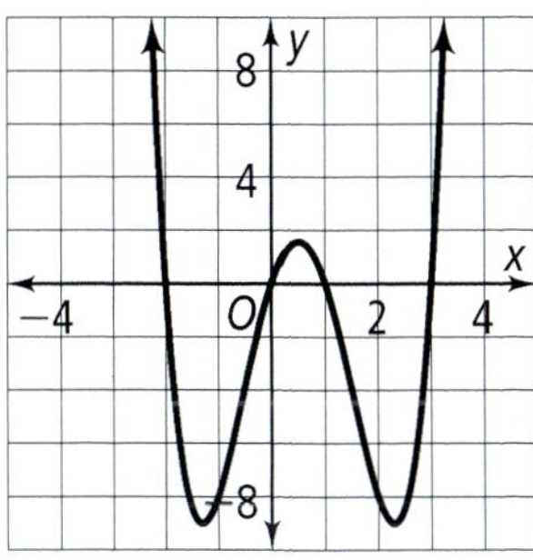

What is the equation of the polynomial function?

Ⓐ $y = x^4 - 6x^3 + 11x^2 - 6x$

Ⓑ $y = x^4 - 2x^3 - 5x^2 + 6x$

Ⓒ $y = x^4 - 2x^3 + x^2 + 3x$

Ⓓ $y = x^4 + 2x^3 - 5x^2 - 6x$

12. Which function is best represented by the graph below?

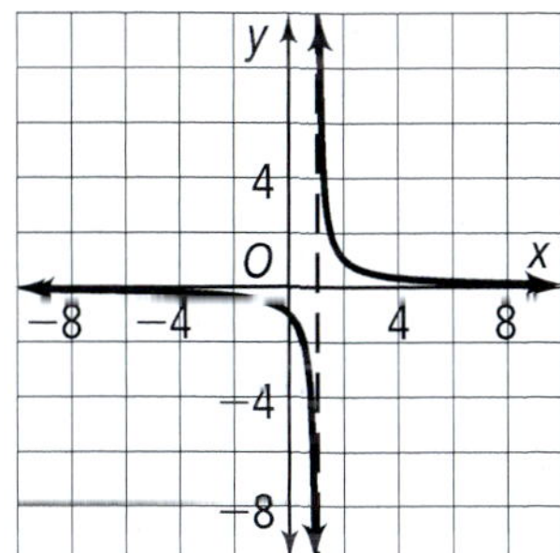

Ⓕ $y = \frac{1}{x - 1}$ Ⓗ $y = \frac{x}{x - 1}$

Ⓖ $y = \frac{1}{x + 1}$ Ⓘ $y = \frac{x}{x + 1}$

13. How many distinct real roots does the equation $x^4 + 3x^3 - 4x = 0$ have?

Ⓐ 1 Ⓑ 2 Ⓒ 3 Ⓓ 4

14. Which function best represents the graph?

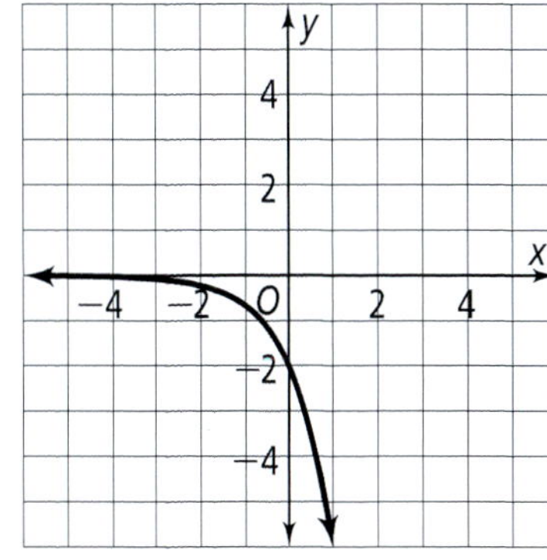

Ⓕ $f(x) = 2 \cdot 3^{-x}$

Ⓖ $f(x) = -2 \cdot 3^{x}$

Ⓗ $f(x) = 2 \cdot 3^{x}$

Ⓘ $f(x) = -2 \cdot 3^{-x}$

15. Consider the piecewise defined function graphed below.

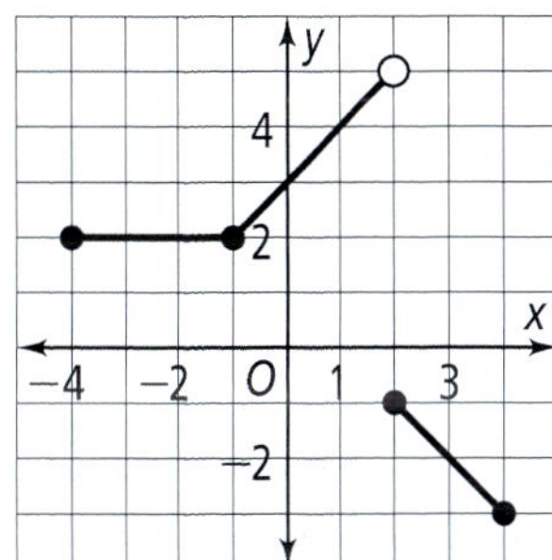

What is the equation for the piecewise defined function?

Ⓐ $f(x) = \begin{cases} 2, \text{ if } -4 \le x < -1 \\ x + 3, \text{ if } -1 \le x < 2 \\ -x + 1, \text{ if } 2 \le x \le 4 \end{cases}$

Ⓑ $f(x) = \begin{cases} 2, \text{ if } -4 \le x \le -1 \\ x + 3, \text{ if } -1 < x \le 2 \\ -x + 1, \text{ if } 2 < x \le 4 \end{cases}$

Ⓒ $f(x) = \begin{cases} 2x, \text{ if } -4 \le x < -1 \\ x + 3, \text{ if } -1 \le x < 2 \\ -x + 1, \text{ if } 2 \le x \le 4 \end{cases}$

Ⓓ $f(x) = \begin{cases} 2x, \text{ if } -4 \le x < -1 \\ x + 3, \text{ if } -1 \le x < 2 \\ -x - 1, \text{ if } 2 \le x \le 4 \end{cases}$

16. The graph of a rational function is shown below.

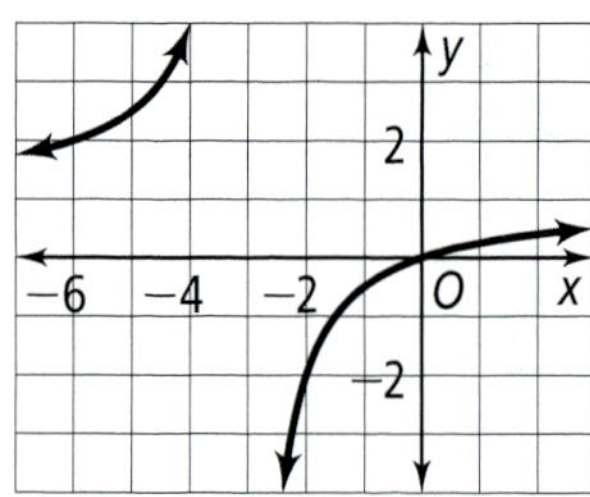

Which function best represents the graph?

F $f(x) = \frac{3}{x-1}$

G $f(x) = \frac{1}{x+3}$

H $f(x) = \frac{3x}{x-1}$

I $f(x) = \frac{x}{x+3}$

17. Consider the polynomial function $f(x) = -2x^4 + 8x^3 + 4x^2 - 3$. What is the end behavior of the graph?

A down and down

B down and up

C up and up

D up and down

18. If $f(x) = (x + 2)^2 - 1$, what is the largest possible domain of f so that its inverse is also a function?

F $x \geq -2$ H $x \geq 0$

G $x \geq -1$ I $x \geq 2$

19. Solve $\frac{3}{2x+10} + \frac{5}{4} = \frac{7}{x+5}$ for x.

A $-\frac{50}{11}$ C $-\frac{9}{5}$

B $-\frac{34}{10}$ D $-\frac{3}{5}$

20. Solve $\sqrt{x-2} - 7 = -4$ for x.

F 5 H 18

G 11 I 25

21. What is the x-coordinate of the vertex of the graph of $f(x) = 2x^2 + 4x - 6$?

A −6 B −1 C 1 D 4

22. The horizontal asymptote of the graph of $y = \frac{4x-4}{2x-6}$ is $y = t$ for a real number t. What is the value of t?

F 1 G 2 H 3 I 4

23. Graph $f(x) = |2x + 6| - 1$.

24. The graph of $y = x^2$ is shown below.

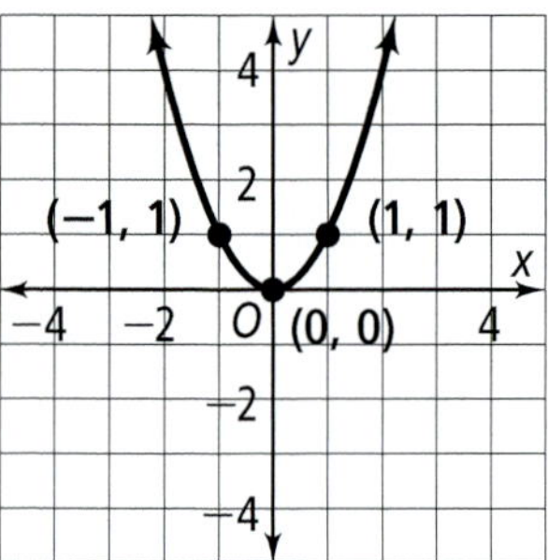

Use transformations to graph $y = -x^2 - 3$.

25. Consider the graph of the function $f(x) = 2(4)^x$. Explain how the graph of the function $g(x) = -2(4)^x + 3$ can be obtained from the graph of $f(x)$.

26. Let $f(x) = \frac{4}{x-1}$.

a. Determine $f^{-1}(x)$. Show or explain your work.

b. Find $f(f^{-1}(x))$ and $f^{-1}(f(x))$. Show your work.

c. How are the domain and range of f and f^{-1} related?

27. Consider the expression $\left(\frac{r^{3m}}{r^{-m}t^{4n}}\right)^{\frac{1}{n}} \cdot \left(\frac{r^{\frac{1}{n}}}{t^{\frac{2}{m}}}\right)^{-m}$.

a. Simplify the expression so that r and t are only written once. Show your work.

b. Using your answer from part (a), evaluate the expression when $m = 1$, $n = 2$, and $t = -3i$. Show your work.

c. For what values of r will the expression you found in part (b) be a real number? Explain your answer.

28. A company needs to ship bags of golf balls that contain 690 golf balls, plus or minus 6 golf balls. If x represents the actual number of golf balls, which inequality can represent this situation?

Ⓐ $|x + 6| \geq 690$ Ⓒ $|x + 690| \geq 6$

Ⓑ $|x - 6| \leq 690$ Ⓓ $|x - 690| \leq 6$

29. A boat took 4 h to make a trip downstream with a current of 6 km/h. The return trip against the same current took 10 h. How far did the boat travel?

Ⓕ 48 km Ⓖ 84 km Ⓗ 160 km Ⓘ 196 km

30. What are all the complex solutions of $x^2 - 4x = -5$?

Ⓐ $x = -1, x = 5$

Ⓑ $x = 1, x = 3$

Ⓒ $x = 2 + i, x = 2 - i$

Ⓓ $x = 2 + 3i, x = 2 - 3i$

31. The function below can be used to find the total amount $C(x)$ an electric company charges a customer who uses x kilowatt-hours (kWh) in a month.

$$C(x) = \begin{cases} 0.07275x + 6.00, \text{ if } 0 \leq x \leq 400 \\ 0.05535x + 35.10, \text{ if } x > 400 \end{cases}$$

If a customer uses 546 kWh in a month, what is the total amount charged?

Ⓕ \$45.72 Ⓖ \$65.32 Ⓗ \$78.28 Ⓘ \$111.04

32. Simplify $\dfrac{r^{\frac{1}{2}}}{r^{-\frac{1}{4}}}$.

Ⓐ $-r^{\frac{1}{4}}$ Ⓑ $-r^2$ Ⓒ $r^{\frac{1}{8}}$ Ⓓ $r^{\frac{3}{4}}$

33. The graph of a quadratic function, $y = ax^2 + bx + c$ passes through the points shown. What is the axis of symmetry of the parabola?

Ⓕ $x = -2$

Ⓖ $x = -1$

Ⓗ $x = 1$

Ⓘ $x = 2$

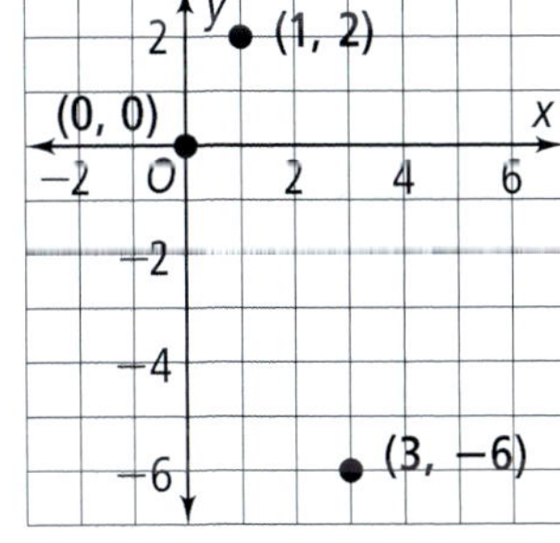

34. What is the end behavior of the graph of the polynomial function $f(x) = -2x^5 + x^4 + 3x^3 - x + 1$?

Ⓐ down and up

Ⓑ up and up

Ⓒ up and down

Ⓓ down and down

35. The principal amount invested in an account with 1.5% interest compounded continuously is \$500. The equation $A(x) = 500e^{0.015x}$ can be used to find the balance in the account after x years. To the nearest year, in how many years will the account have a balance of \$820?

Ⓕ 2 years Ⓗ 72 years

Ⓖ 33 years Ⓘ 109 years

36. The graph of the exponential equation $y = 2^x$ is reflected across the y-axis and moved down 1 unit. What is the equation of the resulting graph?

Ⓐ $y = 2^{-x-1}$ Ⓒ $y = 2^{-x} - 1$

Ⓑ $y = -2^{x-1}$ Ⓓ $y = -2^x - 1$

37. The function $C(x) = \dfrac{10}{2x^2 + 1}$ can be used to find the concentration $C(x)$ in mg/L of a certain drug in the bloodstream of a patient x hours after the injection is given. In approximately how many hours after the injection will the concentration of the drug be 1.3 mg/L?

Ⓕ 0.5 h Ⓗ 1.8 h

Ⓖ 0.7 h Ⓘ 2.3 h

38. The half-life of radium-226 is about 1,600 years. After 4,000 years what percentage of a sample of radium-226 remains?

Ⓐ 2.5% Ⓒ 40.0%

Ⓑ 17.7% Ⓓ 75.8%

39. Solve $8.2(3^{2x-4}) - 11 = 557.1$. Round your answer to the nearest tenth.

Ⓕ 1.8 Ⓖ 2.9 Ⓗ 3.5 Ⓘ 3.9

40. An exponential function is represented in the table below.

x	$f(x)$
−2	12
−1	6
0	3
1	1.5

Which equation best represents the function?

(A) $f(x) = 3(2^{-x})$

(B) $f(x) = 3(2^{x})$

(C) $f(x) = 2^{-x} + 3$

(D) $f(x) = 2^{x} + 3$

41. What is the range of the graph of $f(x) = -ab^x$ if $a > 0$ and $b > 1$?

(F) $f(x) \leq 0$

(G) $f(x) \leq a$

(H) $f(x) \geq a$

(I) All real numbers

42. The characteristics of function $f(x) = ax^n$ are shown below.

Domain: All real numbers

Range: $x \leq 0$

Symmetric with respect to the y-axis

What must be true about the values of a and n?

(A) $a < 0$ and n is even

(B) $a < 0$ and n is odd

(C) $a > 0$ and n is even

(D) $a > 0$ and n is odd

43. A train leaves a city traveling due north. A car leaves the city at the same time traveling due west. The car is traveling 15 mi/h faster than the train. After 2 h they are approximately 150 mi apart. What is the speed of the train?

(F) 30 mi/h

(G) 45 mi/h

(H) 60 mi/h

(I) 75 mi/h

44. A high school sold 800 tickets for a soccer game. Three types of tickets were sold, adult, student and child. There were four times as many adult tickets sold as child tickets. And there were 62 more student tickets sold than adult tickets. How many adult tickets were sold?

(A) 82 (B) 123 (C) 328 (D) 384

45. Let $f(x) = 3x + 5$ and let $g(x) = x^2 + 2x$. What is $f(-3) \cdot g(-3)$?

(F) −32

(G) −12

(H) 9

(I) 60

46. The volume of a square pyramid with a height equal to four less than the length of a side of the base is given by $V(x) = \frac{1}{3}(x^3 + 8x^2 + 16x)$ where x is the height in cm. If the length of a side of the base is 9 cm, what is the volume of the pyramid?

(A) 135 cm^3

(B) 405 cm^3

(C) 507 cm^3

(D) 1,521 cm^3

47. A quadratic function is represented in the table below.

x	$f(x)$
1	−13
2	−3
3	3
4	5
5	3

Which equation best represents the function?

(F) $f(x) = -2(x - 4)^2 + 5$

(G) $f(x) = -2(x - 3)^2 + 3$

(H) $f(x) = 2(x - 4)^2 + 5$

(I) $f(x) = 2(x - 3)^2 + 3$

48. Find the x-value of the solution to the following system of equations.

$$\begin{cases} 3x + y = -3 \\ x + y = 1 \end{cases}$$

(A) −2

(B) −1

(C) $\frac{3}{5}$

(D) 3

49. Solve $4(3^x) = 26$ for x.

(F) 0.3

(G) 1.3

(H) 1.7

(I) 2.2

50. The amount of cesium-137 remaining after x years in an initial sample of 200 milligrams can be found using the equation $C(x) = 200e^{-0.02295x}$. In approximately how many years will the sample contain 120 milligrams of cesium-137?

Ⓐ 13 Ⓑ 22 Ⓒ 26 Ⓓ 39

51. Graph the solution set of the following system of inequalities.

$$\begin{cases} x - 3y \leq 6 \\ 2x + y > 5 \end{cases}$$

52. Simplify the expression. Show your work.

$\sqrt{16x^2y^{12}}$

53. What is the vertex of the graph of $f(x) = a|bx - 1| + c$? Explain your answer.

54. A company produces two types of doghouses, regular and deluxe. A regular doghouse requires 7 hours to build and 3 hours to paint. A deluxe doghouse requires 11 hours to build and 4 hours to paint. The company employs 5 builders and 2 painters. Each employee can work a maximum of 40 hours.

a. Write a system of inequalities that can be used to find the number of each type of doghouse built in a week. Define the variables you use in your system.

b. Graph the solution set of your system of inequalities from part (a). Label each line in your graph.

c. A regular doghouse sells for \$100. A deluxe doghouse sells for \$200. How many of each can be built, painted, and sold in one week to maximize sales?

55. Consider the function $f(x) = \frac{1}{x}$.

a. Graph $f(x)$.

b. Explain how the graph of $g(x) = \frac{4}{x+2}$ compares to the graph of $f(x)$.

c. What is the horizontal asymptote (if any) of the graph of $g(x)$?

d. What is the vertical asymptote of the graph of $g(x)$? Explain how this relates to the domain of $g(x)$.

56. Consider the recursive model shown below.

$$\begin{cases} a_1 = 5 \\ a_{n+1} = a_n - 7 \end{cases}$$

What is an explicit formula for this sequence?

Ⓕ $a_n = -7 + 5n$

Ⓖ $a_n = 5 - 7n$

Ⓗ $a_n = -7 + 5(n - 1)$

Ⓘ $a_n = 5 - 7(n - 1)$

57. An arch in the shape of the upper half of an ellipse supports a bridge that spans a distance of 80 ft. The maximum height of the arch is 30 ft. To the nearest tenth of a foot, what is the height of the arch 28 ft from the center?

Ⓐ 14.4 ft Ⓒ 28.1 ft

Ⓑ 21.4 ft Ⓓ 29.7 ft

58. A scientist wants to study the affects of a new medication on acne. Which type of study would give the most reliable results?

Ⓕ Controlled experiment

Ⓖ Observational study

Ⓗ Survey

Ⓘ Random sample

59. Suppose scores on an entry exam are normally distributed. The exam has a mean score of 140 and a standard deviation of 20. What is the probability that a person who took the test scored between 120 and 160?

Ⓐ 14% Ⓒ 68%

Ⓑ 40% Ⓓ 95%

60. What is $\cos\theta$ when $\sin\theta = \frac{3}{5}$ and θ is in Quadrant II?

Ⓕ $-\frac{4}{5}$ Ⓗ $\frac{2}{5}$

Ⓖ $-\frac{2}{5}$ Ⓘ $\frac{4}{5}$

61. An equation of an ellipse is $9(x + 9)^2 + 4(y + 4)^2 = 36$. What is the y-coordinate of the center of the ellipse?

Ⓐ -9 Ⓑ -4 Ⓒ 4 Ⓓ 9

62. What is the determinant of the matrix?

$$\begin{bmatrix} 1 & 3 & -1 \\ 1 & 2 & 1 \\ -2 & -5 & -4 \end{bmatrix}$$

Ⓕ -8 Ⓗ 0

Ⓖ -4 Ⓘ 4

63. Write the expression as a single logarithm.
$4\log_3 x + \log_3 y - 2\log_3 z$

Ⓐ $\log_3 \frac{x^4 y}{z^2}$ Ⓒ $\log_3 (4x + y - 2z)$

Ⓑ $\frac{\log_3 x^4 y}{\log_3 z^2}$ Ⓓ $\log_3 (x^4 + y - z^2)$

64. A computer manufacturing company sampled two different parts and tested for defects. The results are shown in the table below.

	Part A	Part B
Defective	14	33
Not defective	266	312

What is the probability that if a Part B is randomly chosen, it is defective?

Ⓕ 5.28% Ⓗ 9.57%

Ⓖ 5.71% Ⓘ 10.58%

65. The graph of the hyperbola $\frac{(x-2)^2}{25} - \frac{(y-3)^2}{9} = 1$ is shown at the right.

How does the graph of

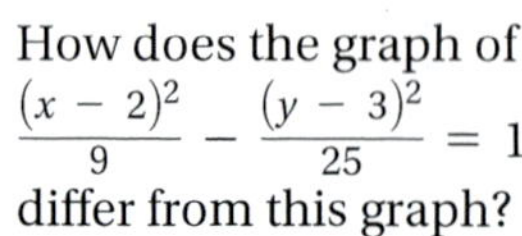

$\frac{(x-2)^2}{9} - \frac{(y-3)^2}{25} = 1$

differ from this graph?

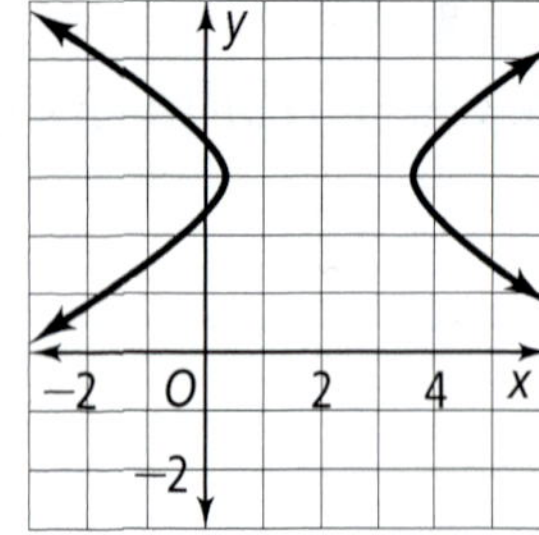

Ⓐ The asymptotes are less steep.

Ⓑ The foci become $(-1, 3)$ and $(5, 3)$

Ⓒ The vertices become $(-1, 3)$ and $(5, 3)$

Ⓓ The transverse axis becomes vertical.

66. What is 64° in radians? Round your answer to the nearest hundredth.

Ⓕ 1.12 Ⓗ 10.19

Ⓖ 5.63 Ⓘ 402.12

67. A set of data has a normal distribution with a mean of 72 and a standard deviation of 6. What percent of data is greater than 84?

Ⓐ 84% Ⓒ 13.5%

Ⓑ 50% Ⓓ 2.35%

68. An employee's initial salary is \$30,000. The person receives a 5% raise at the end of each year. What is the formula for the term s_n which represents the salary at the beginning of the nth year?

Ⓕ $s_n = 30{,}000 + 1.05n$

Ⓖ $s_n = 30{,}000 + 5(n - 1)$

Ⓗ $s_n = 30{,}000(1.05)^{n-1}$

Ⓘ $s_n = 30{,}000(1.05)^n$

69. Use the Change of Base Formula to approximate the value of $\log_2 3.2$ to the nearest tenth.

Ⓐ 0.2 Ⓒ 1.7

Ⓑ 0.8 Ⓓ 9.2

70. If $B = \begin{bmatrix} -2 & 1 \\ 4 & -1 \end{bmatrix}$, what is B^{-1}?

Ⓕ $\begin{bmatrix} -0.5 & 1 \\ 0.25 & -1 \end{bmatrix}$ Ⓗ $\begin{bmatrix} 2 & -1 \\ -4 & 1 \end{bmatrix}$

Ⓖ $\begin{bmatrix} 0.5 & 0.5 \\ 2 & 1 \end{bmatrix}$ Ⓘ $\begin{bmatrix} 4 & -1 \\ -2 & 1 \end{bmatrix}$

71. Which expression is equivalent to $(\sin\theta)(\sec\theta)$?

Ⓐ $\cos\theta$ Ⓒ $\sin\theta$

Ⓑ $\tan\theta$ Ⓓ $\csc\theta$

72. The magnitude M of an earthquake can be found using the equation $M(x) = \log\left(\frac{x}{0.001}\right)$ where x represents the seismograph reading of the earthquake in mm. An earthquake has a magnitude of 6.2. What is the seismograph reading of the earthquake in mm?

Ⓕ 0.0062 Ⓗ 1.014

Ⓖ 0.0008 Ⓘ 1584.9

73. Which function has a period of 4π and an amplitude of 6?

Ⓐ $y = -6 \sin 8\theta$

Ⓑ $y = 6 \sin 2\theta$

Ⓒ $y = 3 \sin 6\theta$

Ⓓ $y = -6 \sin \frac{1}{2}\theta$

74. Which equation represents a circle with center $(-4, -6)$ and radius 6?

Ⓕ $(x - 4)^2 + (y - 6)^2 = 36$

Ⓖ $(x + 4)^2 + (y + 6)^2 = 36$

Ⓗ $(x + 4)^2 + (y + 6)^2 = 6$

Ⓘ $(x - 4)^2 + (y - 6)^2 = 6$

75. What is the exact value of tan 240°?

Ⓐ $\frac{\sqrt{2}}{2}$ Ⓒ 1

Ⓑ $\frac{\sqrt{3}}{3}$ Ⓓ $\sqrt{3}$

76. Multiply $\begin{bmatrix} 4 & -1 \\ 0 & 5 \end{bmatrix} \cdot \begin{bmatrix} 1 & 3 \\ -6 & 1 \end{bmatrix}$.

Ⓕ $\begin{bmatrix} 4 & 14 \\ -24 & 11 \end{bmatrix}$ Ⓗ $\begin{bmatrix} 10 & -30 \\ 11 & 5 \end{bmatrix}$

Ⓖ $\begin{bmatrix} 4 & -3 \\ 0 & 5 \end{bmatrix}$ Ⓘ $\begin{bmatrix} 10 & 11 \\ -30 & 5 \end{bmatrix}$

77. Consider the vectors **u** and **v** below.

a. Show the addition of the two vectors graphically. Label your answer **w**.

b. Using your answer from part (a), find -0.5**w**.

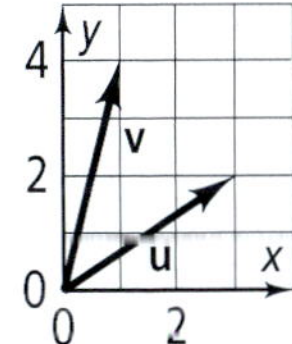

78. Solve for x. Show or explain your work.

$2 \ln 4x + 5 = 8$

79. A pendulum initially swings through an arc that is 20 inches long. On each swing, the length of the arc is 0.85 of the previous swing.

a. Write a recursive model of geometric decay to represent the sequence of lengths of the arc of each swing. Let $p_1 = 20$.

b. Rewrite your model from part (a) using an explicit formula.

c. What is the approximate total distance the pendulum has swung after 11 swings? Show your work.

d. What is the total distance, approximately, that the pendulum has swung when it stops? Show your work.

80. The equation of an ellipse is $4x^2 + 9y^2 + 8x - 54y + 49 = 0$.

a. Write the equation in standard form. Show your work.

b. What are the foci and vertices of the ellipse? Show your work or explain your answer.

c. Graph the ellipse. Label the center of the ellipse on your graph.

81. Consider the following system of equations.

$$\begin{cases} x + 2z = -1 \\ y - 2z = 2 \\ 2x + y + z = 1 \end{cases}$$

a. Represent the system of equations using the matrix equation $AX = B$.

b. Find the determinant of the matrix A.

c. Solve the equation from part (a). If it cannot be solved, use your result from part (b) to explain why.

82. Consider the function $f(x) = 2\cos(4x)$.

a. What are the period and amplitude of the graph of $f(x)$?

b. Graph $f(x)$ over two periods.

c. Solve $f(x) = 0.5$ algebraically. Show your work and give your answer in radians.

Skills Handbook

Skills Handbook

Percents and Percent Applications

Percent means "per hundred." Find fraction, decimal, and percent equivalents by replacing one symbol for *hundredths* with another.

Example 1

Write each number as a percent.

a. $0.082 = 8.2\%$

Move the decimal point two places to the right and write a percent sign.

b. $\frac{3}{5} = \frac{60}{100} = 60\%$

Write the fraction as hundredths. Then replace the hundredths with a percent sign.

c. $1\frac{1}{6} = \frac{7}{6} = 1.166\overline{6} = 116.\overline{6}\%$

First, use $7 \div 6$ to write $1\frac{1}{6}$ as a decimal.

Example 2

Write each percent as a decimal.

a. $50\% = 0.50 = 0.5$

b. $\frac{1}{2}\% = 0.5\% = 00.5\% = 0.005$

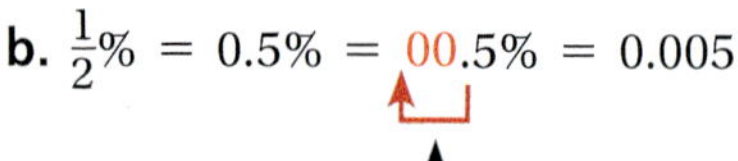

Move the decimal point two places to the left and drop the percent sign.

Example 3

Use an equation to solve each percent problem.

a. What is 30% of 12?

$n = 0.3 \times 12$

$n = 3.6$

b. 18 is 0.3% of what?

$18 = 0.003 \times n$

$\frac{18}{0.003} = \frac{0.003n}{0.003}$

$6000 = n$

c. What percent of 60 is 9?

$n \times 60 = 9$

$60n = 9$

$n = \frac{9}{60} = 0.15 = 15\%$

Exercises

Write each decimal as a percent and each percent as a decimal.

1. 0.46 **2.** 1.506 **3.** 0.007 **4.** 8% **5.** 103.5% **6.** 3.3%

Write each fraction or mixed number as a percent.

7. $\frac{1}{4}$ **8.** $\frac{3}{8}$ **9.** $\frac{2}{3}$ **10.** $\frac{4}{9}$ **11.** $1\frac{3}{20}$ **12.** $\frac{1}{200}$

Use an equation to solve each percent problem. Round your answer to the nearest tenth, if necessary.

13. What is 25% of 50?

14. What percent of 58 is 37?

15. 120% of what is 90?

16. 8 is what percent of 40?

17. 15 is 75% of what?

18. 80% of 58 is what?

Operations With Fractions

To add or subtract fractions, use a common denominator. The common denominator is the least common multiple of the denominators.

Example 1

Simplify $\frac{2}{3} + \frac{3}{5}$.

$\frac{2}{3} + \frac{3}{5} = \frac{2}{3} \cdot \frac{5}{5} + \frac{3}{5} \cdot \frac{3}{3}$ For 3 and 5, the least common multiple is 15.

$= \frac{10}{15} + \frac{9}{15}$ Write $\frac{2}{3}$ and $\frac{3}{5}$ as equivalent fractions with denominators of 15.

$= \frac{19}{15}$ or $1\frac{4}{15}$ Add the numerators.

Example 2

Simplify $5\frac{1}{4} - 3\frac{2}{3}$.

$5\frac{1}{4} - 3\frac{2}{3} = 5\frac{3}{12} - 3\frac{8}{12}$ Write equivalent fractions.

$= 4\frac{15}{12} - 3\frac{8}{12}$ Write $5\frac{3}{12}$ as $4\frac{15}{12}$ so you can subtract the fractions.

$= 1\frac{7}{12}$ Subtract the fractions. Then subtract the whole numbers.

To multiply fractions, multiply the numerators and multiply the denominators. You can simplify by using a greatest common factor.

Example 3

Simplify $\frac{3}{4} \cdot \frac{8}{11}$

Method 1 $\frac{3}{4} \cdot \frac{8}{11} = \frac{24}{44} = \frac{24 \div 4}{44 \div 4} = \frac{6}{11}$

Divide 24 and 44 by 4, their greatest common factor.

Method 2 $\frac{3}{\cancel{4}_1} \cdot \frac{\cancel{8}^2}{11} = \frac{6}{11}$

Divide 4 and 8 by 4, their greatest common factor.

To divide fractions, use a reciprocal to change the problem to multiplication.

Example 4

Simplify $3\frac{1}{5} \div 1\frac{1}{2}$

$3\frac{1}{5} \div 1\frac{1}{2} = \frac{16}{5} \div \frac{3}{2}$ Write mixed numbers as improper fractions.

$= \frac{16}{5} \cdot \frac{2}{3}$ Multiply by the reciprocal of the divisor.

$= \frac{32}{15}$ or $2\frac{2}{15}$ Simplify.

Exercises

Perform the indicated operation.

1. $\frac{3}{5} + \frac{4}{5}$
2. $\frac{1}{2} + \frac{2}{3}$
3. $4\frac{1}{2} + 2\frac{1}{3}$
4. $5\frac{3}{4} + 4\frac{2}{5}$
5. $\frac{2}{3} - \frac{3}{7}$
6. $5\frac{1}{2} - 3\frac{2}{5}$
7. $7\frac{3}{4} - 4\frac{4}{5}$
8. $3\frac{4}{5} \cdot 10$
9. $2\frac{1}{2} \cdot 3\frac{1}{5}$
10. $6\frac{3}{4} \cdot 5\frac{2}{3}$
11. $\frac{1}{2} \div \frac{1}{3}$
12. $\frac{6}{5} \div \frac{3}{5}$
13. $8\frac{1}{2} \div 4\frac{1}{4}$
14. $\frac{8}{9} - \frac{2}{3}$
15. $5\frac{1}{4} \cdot 8$

Ratios and Proportions

A *ratio* is a comparison of two quantities by division. You can write *equal ratios* by multiplying or dividing each quantity by the same nonzero number.

Ways to Write a Ratio

$a : b$ $\quad a$ to b $\quad \frac{a}{b}\ (b \neq 0)$

Example 1

Write $3\frac{1}{3} : \frac{1}{2}$ as a ratio in simplest form.

$3\frac{1}{3} : \frac{1}{2} \rightarrow \dfrac{3\frac{1}{3}}{\frac{1}{2}} = \dfrac{20}{3}$ or $20 : 3$ (× 6 numerator and denominator)

In simplest form, both terms should be integers.
Multiply by the common denominator, 6.

A rate is a ratio that compares different types of quantities. In simplest form for a rate, the second quantity is one unit.

Example 2

Write 247 mi in 5.2 h as a rate in simplest form.

$\dfrac{247 \text{ mi}}{5.2 \text{ h}} = \dfrac{47.5 \text{ mi}}{1 \text{ h}}$ or 47.5 mi/h (÷ 5.2 numerator and denominator)

Divide by 5.2 to make the second quantity one unit.

A proportion is a statement that two ratios are equal. You can find a missing term in a proportion by using the cross products.

Cross Products of a Proportion

$\frac{a}{b} = \frac{c}{d} \rightarrow ad = bc$

Example 3

The Copy Center charges $2.52 for 63 copies. At that rate, how much will the Copy Center charge for 140 copies?

cost → / copies →	$\dfrac{2.52}{63} = \dfrac{c}{140}$	Set up a proportion.
	$2.52 \cdot 140 = 63c$	Use cross products.
	$c = \dfrac{2.52 \cdot 140}{63}$	Solve for c.
	$= 5.6$ or \$5.60	

Exercises

Write each ratio or rate in simplest form.

1. 15 to 20
2. 85 : 34
3. 38 g in 4 oz
4. 375 mi in 4.3 h
5. $\frac{84}{30}$

Solve each proportion. Round your answer to the nearest tenth, if necessary.

6. $\frac{a}{5} = \frac{12}{15}$
7. $\frac{21}{12} = \frac{14}{x}$
8. $8 : 15 = n : 25$
9. $2.4 : c = 4 : 3$
10. $\frac{17}{8} = \frac{n}{20}$
11. $\frac{13}{n} = \frac{20}{3}$
12. $5 : 7 = y : 5$
13. $\frac{0.4}{3.5} = \frac{5.2}{x}$
14. $\frac{4}{x} = \frac{7}{6}$
15. $4 : n = n : 9$
16. A canary's heart beats 130 times in 12 s. Use a proportion to find about how many times its heart beats in 50 s.

Simplifying Expressions With Integers

To add two numbers with the same sign, *add* their absolute values. The sum has the same sign as the numbers. To add two numbers with different signs, find the *difference* between their absolute values. The sum has the same sign as the number with the greater absolute value.

Example 1

Add.

a. $-8 + (-5) = -13$ **b.** $-8 + 5 = -3$ **c.** $8 + (-5) = 3$

To subtract a number, add its opposite.

Example 2

Subtract.

a. $4 - 7 = 4 + (-7)$
$= -3$

b. $-4 - (-7) = -4 + 7$
$= 3$

c. $-4 - 7 = -4 + (-7)$
$= -11$

The product or quotient of two numbers with the same sign is positive. The product or quotient of two numbers with different signs is negative.

Example 3

Multiply or divide.

a. $(-3)(-5) = 15$ **b.** $-35 \div 7 = -5$ **c.** $24 \div (-6) = -4$

Example 4

Simplify $2^2 - 3(4 - 6) - 12$.

$$\begin{aligned} 2^2 - 3(4 - 6) - 12 &= 2^2 - 3(-2) - 12 \\ &= 4 - 3(-2) - 12 \\ &= 4 - (-6) - 12 \\ &= 4 + 6 - 12 = -2 \end{aligned}$$

Order of Operations

1. Perform any operation(s) inside grouping symbols.
2. Simplify any terms with exponents.
3. Multiply and divide in order from left to right.
4. Add and subtract in order from left to right.

Exercises

Simplify each expression.

1. $-4 + 5$ **2.** $12 - 12$ **3.** $-15 + (-23)$ **4.** $4 - 17$ **5.** $-5 - 12$

6. $3 - (-5)$ **7.** $-8 - (-12)$ **8.** $-19 + 5$ **9.** $(-7)(-4)$ **10.** $-120 \div 30$

11. $(-3)(4)$ **12.** $75 \div (-3)$ **13.** $(-6)(15)$ **14.** $(18)(-4)$ **15.** $-84 \div (-7)$

16. $-2(1 + 5) + (-3)(2)$ **17.** $-4(-2 - 5) + 3(1 - 4)$ **18.** $20 - (3)(12) + 4^2$

19. $\frac{-15}{-5} - \frac{36}{-12} + \frac{-12}{-4}$ **20.** $5^2 - 6(5 - 9)$ **21.** $(-3 + 2^3)(4 + \frac{-42}{7})$

Area and Volume

The *area* of a plane figure is the number of square units contained in the figure. The *volume* of a space figure is the number of cubic units contained in the figure. Formulas for area and volume are listed on page 693.

Example 1

Find the area of each figure.

a.

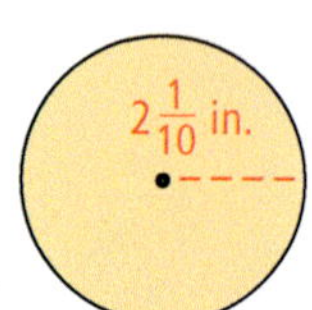

$$A = \pi r^2$$
$$\approx \frac{22}{7} \cdot \left(\frac{21}{10}\right)^2$$
$$= \frac{693}{50} = 13\frac{43}{50} \text{ in.}^2$$

b.

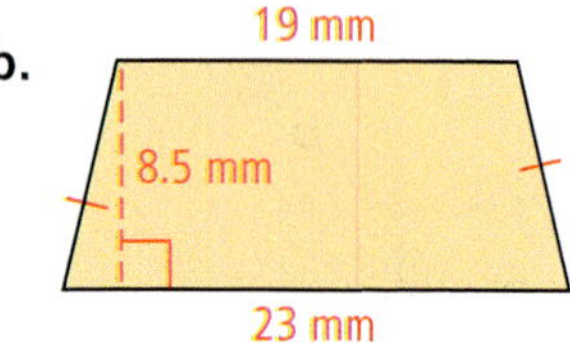

$$A = \frac{1}{2}(b_1 + b_2)h$$
$$= \frac{1}{2}(19 + 23) \cdot 8.5$$
$$= 178.5 \text{ mm}^2$$

Example 2

Find the volume of each figure.

a.

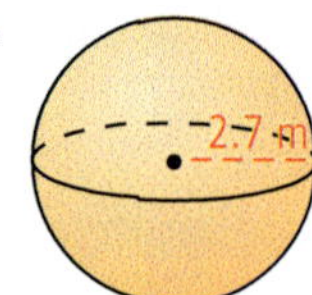

$$V = \frac{4}{3}\pi r^3$$
$$\approx \frac{4}{3} \cdot 3.14 \cdot 2.7^3$$
$$= 82.40616 \approx 82.4m^3$$

b.

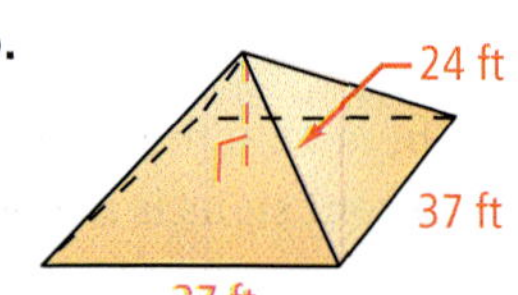

$$V = \frac{1}{3}Bh$$
$$= \frac{1}{3}(37^2) \cdot 24$$
$$= 10{,}952 \text{ ft}^3$$

Exercises

Find the exact area of each figure.

1.

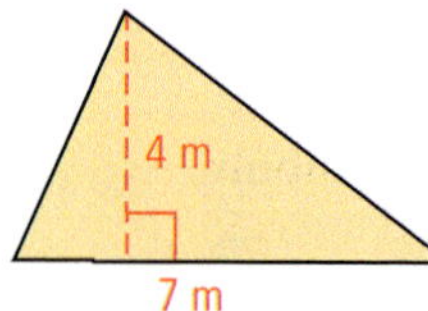

2.

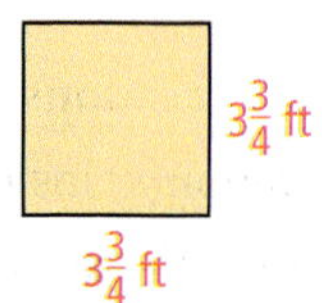

3.

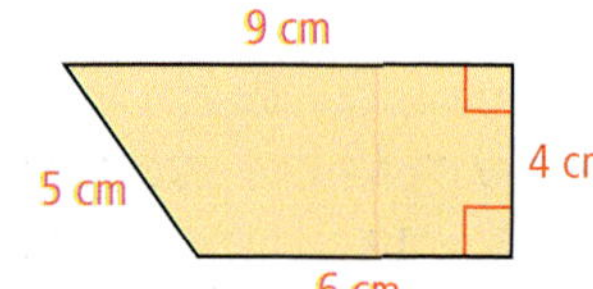

4.

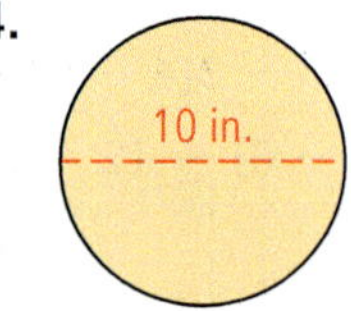

Find the exact volume of each figure.

5.

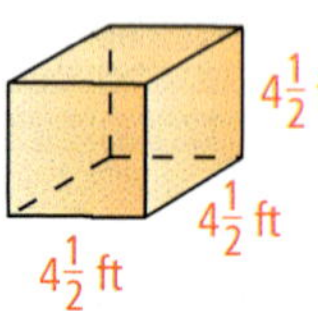

6.

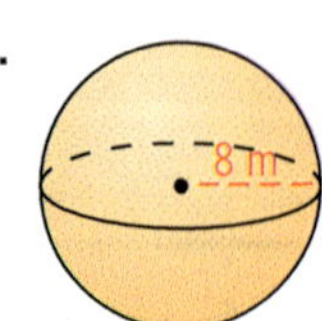

7.

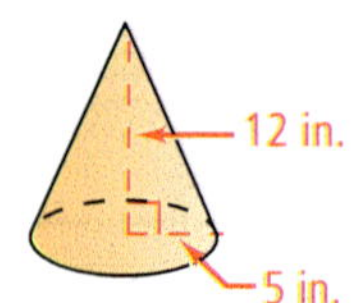

8.

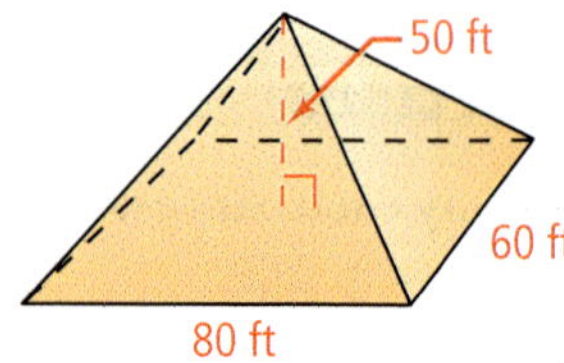

9. Find the area of a triangle with a base of 17 in. and a height of 13 in.

10. Find the volume of a rectangular box 64 cm long, 48 cm wide, and 58 cm high.

11. Find the surface area of the cube in Exercise 5.

The Coordinate Plane, Slope, and Midpoint

The *coordinate plane* is formed when two perpendicular number lines intersect at a point called the origin, forming four quadrants.

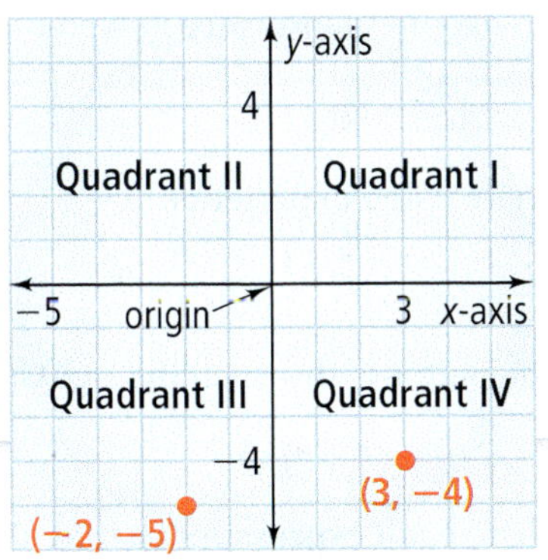

Example 1

In which quadrant would you find each point?

a. (3, −4) Move 3 units right and 4 units down. The point is in Quadrant IV.

b. (−2, −5) Move 2 units left and 5 units down. The point is in Quadrant III.

To find the slope of a line on the coordinate plane, choose two points on the line and use the slope formula.

Example 2

Find the slope of each line.

a.

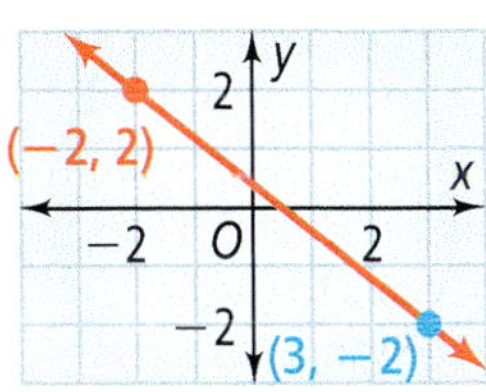

$$m = \frac{y_2 - y_1}{x_2 - x_1} = \frac{2 - (-2)}{-2 - 3} = \frac{4}{-5} \text{ or } -\frac{4}{5}$$

b.

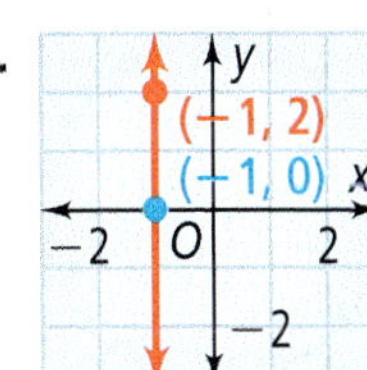

$$m = \frac{y_2 - y_1}{x_2 - x_1} = \frac{2 - 0}{-1 - (-1)} = \frac{2}{0}$$

Since you cannot divide by zero, this line has an undefined slope.

If (x_m, y_m) is the midpoint of the segment joining (x_1, y_1) and (x_2, y_2), then $x_m = \frac{x_1 + x_2}{2}$ and $y_m = \frac{y_1 + y_2}{2}$.

Example 3

Find the coordinates of the midpoint of the segment with endpoints (−2, 5) and (6, −3).

$\frac{-2 + 6}{2} = 2$ and $\frac{5 + (-3)}{2} = 1$ so the midpoint is (2, 1).

Exercises

In which quadrant would you find each point? Graph each point on a coordinate plane.

1. (3, 2) **2.** (−4, 3) **3.** (2, −3) **4.** (4, −2) **5.** (−4, −5) **6.** (−1, −3)

Find the slope of each line.

7.

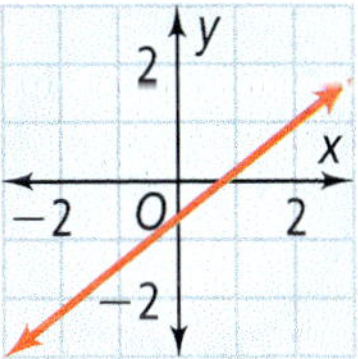

8.

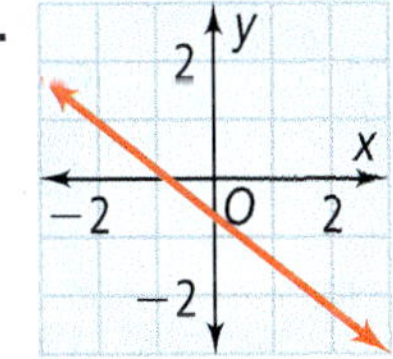

9.

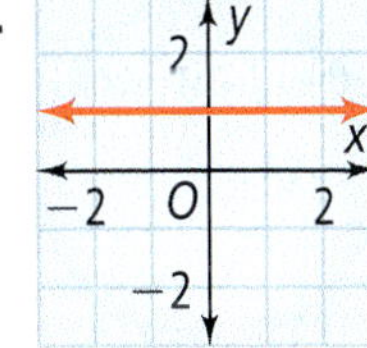

10. the line containing (−3, 4) and (2, −6)

11. the line containing (25, 40) and (100, 55)

Find the midpoint of the segment with the given endpoints.

12. (−4, 4), (2, −5) **13.** (3, 3), (7, −6) **14.** (−1, −8), (0, −3) **15.** (3, 4), (2, −6)

Operations With Exponents

An exponent indicates how many times a number is used as a factor.

$2^n = \blacksquare$	$10^n = \blacksquare$
$2^2 = 4$	$10^2 = 100$
$2^1 = 2$	$10^1 = 10$
$2^0 = 1$	$10^0 = 1$
$2^{-1} = \frac{1}{2}$	$10^{-1} = \frac{1}{10}$
$2^{-2} = \frac{1}{4}$	$10^{-2} = \frac{1}{100}$

Example 1

Write using exponents.

a. $3 \cdot 3 \cdot 3 \cdot 3 \cdot 3 = 3^5$

b. $a \cdot a \cdot b \cdot b \cdot b \cdot b = a^2b^4$

The patterns shown at the right indicate that $a^0 = 1$ and that $a^{-n} = \frac{1}{a^n}$.

Example 2

Write each expression so that all exponents are positive.

a. $a^{-2}b^3 = \frac{1}{a^2} \cdot b^3 = \frac{b^3}{a^2}$

b. $x^3y^0z^{-1} = x^3 \cdot 1 \cdot \frac{1}{z} = \frac{x^3}{z}$

You can simplify expressions that contain powers with the same base.

Example 3

Simplify each expression.

a. $b^5 \cdot b^3 = b^{5+3}$
$= b^8$

Add exponents to multiply powers with the same base.

b. $\frac{x^5}{x^7} = x^{5-7}$
$= x^{-2} = \frac{1}{x^2}$

Subtract exponents to divide powers with the same base.

You can simplify expressions that contain parentheses and exponents.

Example 4

Simplify each expression.

a. $\left(\frac{ab}{n}\right)^3 = \frac{a^3b^3}{n^3}$ Raise each factor in the parentheses to the third power.

b. $(c^2)^4 = c^{2\cdot4} = c^8$ Multiply exponents to raise a power to a power.

Exercises

Write each expression using exponents.

1. $x \cdot x \cdot x$ **2.** $x \cdot x \cdot x \cdot y \cdot y$ **3.** $a \cdot a \cdot a \cdot a \cdot b$ **4.** $\frac{a \cdot a \cdot a \cdot a}{b \cdot b}$

Write each expression so that all exponents are positive.

5. c^{-4} **6.** $m^{-2}n^0$ **7.** $x^5y^{-7}z^{-3}$ **8.** $ab^{-1}c^2$

Simplify each expression. Use positive exponents.

9. d^2d^6 **10.** $\frac{a^5}{a^2}$ **11.** $\frac{c^7}{c}$ **12.** $\frac{n^3}{n^6}$ **13.** $\frac{a^5b^3}{ab^8}$ **14.** $(3x)^2$

15. $\left(\frac{a}{b}\right)^4$ **16.** $\left(\frac{xz}{y}\right)^6$ **17.** $(c^3)^4$ **18.** $\left(\frac{x^2}{y^5}\right)^3$ **19.** $(u^4v^2)^3$ **20.** $(p^5)^{-2}$

21. $\frac{(2a^4)(3a^2)}{6a^3}$ **22.** $(x^{-2})^3$ **23.** $(mg^3)^{-1}$ **24.** $g^{-3}g^{-1}$ **25.** $\frac{(3a^3)^2}{18a}$ **26.** $\frac{c^3d^7}{c^{-3}d^{-1}}$

Skills Handbook

Factoring and Operations With Polynomials

Example 1

Perform each operation.

a. $(3y^2 - 4y + 5) + (y^2 + 9y)$

$= (3y^2 + y^2) + (-4y + 9y) + 5$ To add, group like terms.

$= 4y^2 + 5y + 5$

b. $(n + 4)(n - 3)$

$= n(n) + n(-3) + 4(n) + 4(-3)$ Distribute n and 4.

$= n^2 - 3n + 4n - 12$ Combine like terms.

$= n^2 + n - 12$

To factor a polynomial, first find the greatest common factor (GCF) of the terms. Then use the distributive property to factor out the GCF.

Example 2

Factor $6x^3 - 12x^2 + 18x$.

$6x^3 = 6 \cdot x \cdot x \cdot x$; $-12x^2 = 6 \cdot (-2) \cdot x \cdot x$; $18x = 6 \cdot 3 \cdot x$ List the factors of each term. The GCF is $6x$.

$6x^3 - 12x^2 + 18x = 6x(x^2) + 6x(-2x) + 6x(3)$ Use the distributive property to factor out $6x$.

$= 6x(x^2 - 2x + 3)$

When a polynomial is the product of two binomials, you can work backward to find the factors.

$$x^2 + bx + c = (x + \blacksquare)(x + \blacksquare)$$

The *sum* of these numbers must equal b.

The *product* of these numbers must equal c.

Example 3

Factor $x^2 - 13x + 36$.

Choose numbers that are factors of 36. Look for a pair with the sum -13.

The numbers -4 and -9 have a product of 36 and a sum of -13. The factors are $(x - 4)$ and $(x - 9)$. So, $x^2 - 13x + 36 = (x - 4)(x - 9)$.

Factors	Sum
$-6 \cdot (-6)$	-12
$-4 \cdot (-9)$	-13

Exercises

Perform the indicated operations.

1. $(x^2 + 3x - 1) + (7x - 4)$ **2.** $(5y^2 + 7y) - (3y^2 + 9y - 8)$ **3.** $4x^2(3x^2 - 5x + 9)$

4. $-5d(13d^2 + 7d + 8)$ **5.** $(x - 5)(x + 3)$ **6.** $(n - 7)(n - 2)$

Factor each polynomial.

7. $a^2 - 8a + 12$ **8.** $n^2 - 2n - 8$ **9.** $x^2 + 5x + 4$ **10.** $3m^2 - 9$

11. $y^2 + 5y - 24$ **12.** $s^3 + 6s^2 + 11s$ **13.** $2x^3 + 4x^2 - 8x$ **14.** $y^2 - 10y + 25$

Scientific Notation and Significant Digits

In *scientific notation,* a number has the form $a \times 10^n$, where n is an integer and $1 \leq a < 10$.

Example 1

Write 5.59×10^6 in standard form.

$5.59 \times 10^6 = 5\,590\,000 = 5{,}590{,}000$

A positive exponent indicates a value greater than 1. Move the decimal point six places to the right.

Example 2

Write 0.0000318 in scientific notation.

$0.0000318 = 3.18 \times 10^{-5}$

Move the decimal point to create a number between 1 and 10. Since the original number is less than 1, use a negative exponent.

When a measurement is in scientific notation, all the digits of the number between 1 and 10 are *significant digits*. When you multiply or divide measurements, your answer should have as many significant digits as the least number of significant digits in any of the numbers involved.

Example 3

Multiply (6.71×10^8 mi/h) and (3.8×10^4 h).

$(6.71 \times 10^8\text{ mi/h})(3.8 \times 10^4\text{ h}) = (6.71 \cdot 3.8)(10^8 \cdot 10^4)$	Rearrange factors.
$= 25.498 \times 10^{12}$	Add exponents when multiplying powers of 10.
$= 2.5498 \times 10^{13}$	Write in scientific notation.
$\approx 2.5 \times 10^{13}\text{mi}$	Round to two significant digits.

6.71: three significant digits; 3.8: two significant digits

Exercises

Change each number to scientific notation or to standard form.

1. 1,340,000
2. 6.88×10^{-2}
3. 0.000775
4. 0.0072
5. 1.113×10^5
6. 8.0×10^{-4}
7. 1895
8. 2.3×10^3
9. 123,400
10. 7.985×10^4

Write each product or quotient in scientific notation. Round to the appropriate number of significant digits.

11. $(1.6 \times 10^2)(4.0 \times 10^3)$
12. $(2.5 \times 10^{-3})(1.2 \times 10^4)$
13. $(4.237 \times 10^4)(2.01 \times 10^{-2})$
14. $\dfrac{7.0 \times 10^5}{2.89 \times 10^3}$
15. $\dfrac{1.4 \times 10^4}{8.0 \times 10^2}$
16. $\dfrac{6.48 \times 10^6}{3.2 \times 10^5}$
17. $(1.78 \times 10^{-7})(5.03 \times 10^{-5})$
18. $(7.2 \times 10^{11})(5 \times 10^6)$
19. $(8.90 \times 10^8) \div (2.36 \times 10^{-2})$
20. $(3.95 \times 10^4) \div (6.8 \times 10^8)$
21. $(4.9 \times 10^{-8}) \div (2.7 \times 10^{-2})$
22. $(3.972 \times 10^{-5})(4.7 \times 10^{-4})$

The Pythagorean Theorem and the Distance Formula

In a right triangle, the sum of the squares of the lengths of the legs is equal to the square of the length of the hypotenuse. Use this relationship, known as the Pythagorean Theorem, to find the length of a side of a right triangle.

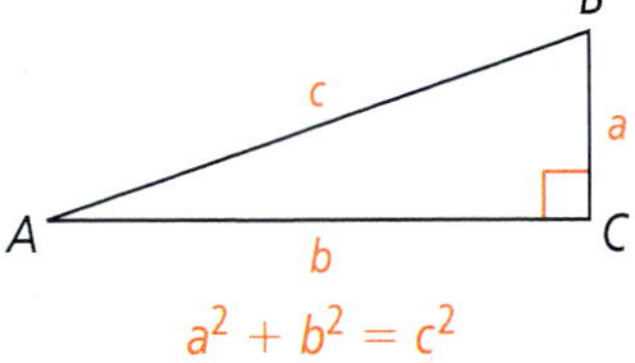

Example 1

Find m in the triangle below, to the nearest tenth.

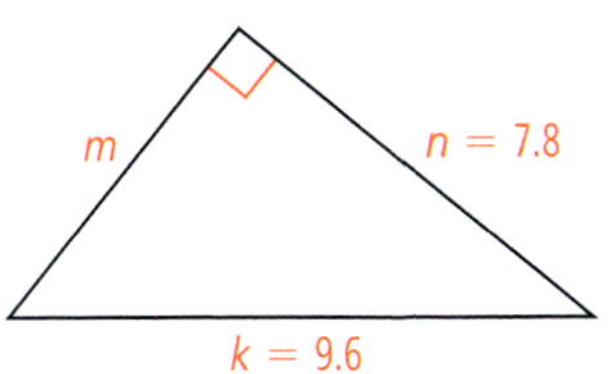

$$m^2 + n^2 = k^2$$
$$m^2 + 7.8^2 = 9.6^2$$
$$m^2 = 9.6^2 - 7.8^2 = 31.32$$
$$m = \sqrt{31.32} \approx 5.6$$

To find the distance between two points on the coordinate plane, use the distance formula.

The distance d between any two points (x_1, y_1) and (x_2, y_2) is

$$d = \sqrt{(x_2 - x_1)^2 + (y_2 - y_1)^2}$$

Example 2

Find the distance between $(-3, 2)$ and $(6, -4)$.

$$d = \sqrt{(6 - (-3))^2 + (-4 - 2)^2}$$
$$= \sqrt{9^2 + (-6)^2}$$
$$= \sqrt{81 + 36}$$
$$= \sqrt{117}$$
$$\approx 10.8$$

Thus, d is about 10.8 units.

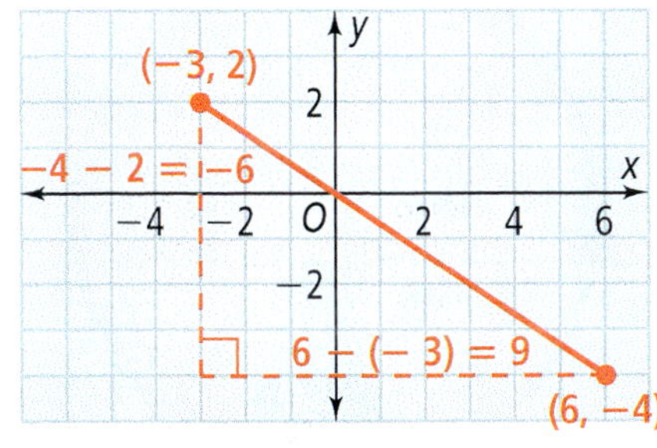

Exercises

In each problem, a and b are the lengths of the legs of a right triangle and c is the length of the hypotenuse. Find each missing length. Round your answer to the nearest tenth.

1. c if $a = 6$ and $b = 8$

2. a if $b = 12$ and $c = 13$

3. b if $a = 8$ and $c = 17$

4. c if $a = 10$ and $b = 3$

5. a if $b = 100$ and $c = 114$

6. b if $a = 12.0$ and $c = 30.1$

Find the distance between each pair of points, to the nearest tenth.

7. $(0, 0), (4, -3)$

8. $(-5, -5), (1, 3)$

9. $(-1, 0), (4, 12)$

10. $(-4, 2), (4, -2)$

11. $(0, 15), (17, 0)$

12. $(-8, 8), (8, 8)$

13. $(-1, 1), (1, -1)$

14. $(-2, 9), (0, 0)$

15. $(-5, 3), (4, 3)$

16. $(2, 1), (3, 4)$

17. $(3, -2), (3, 5)$

18. $(5, 4), (-3, 1)$

Bar and Circle Graphs

Sometimes you can draw different graphs to represent the same data, depending on the information you want to share. A *bar graph* is useful for comparing amounts; a *circle graph* is useful for comparing percents.

Example

Display the 2007 data on immigration to the United States in a bar graph and a circle graph.

Immigration to the United States, 2007

Place of Origin	Immigrants (1000's)
Africa	89.2
Asia	359.4
Europe	120.8
North America	331.7
South America	102.6

Source: Department of Homeland Security

To make a circle graph, first find the *percent* of the data in each category. Then express each percent as a decimal and multiply by 360° to find the size of each *central angle*.

$$\begin{matrix}\text{Africa} \rightarrow \\ \text{Total} \rightarrow\end{matrix} \frac{89.3}{1003.7} \approx 0.09 \text{ or } 9\%$$

$0.09 \times 360^\circ \approx 32^\circ$

Draw a circle and use a protractor to draw each central angle.

Immigration to the United States, 2007

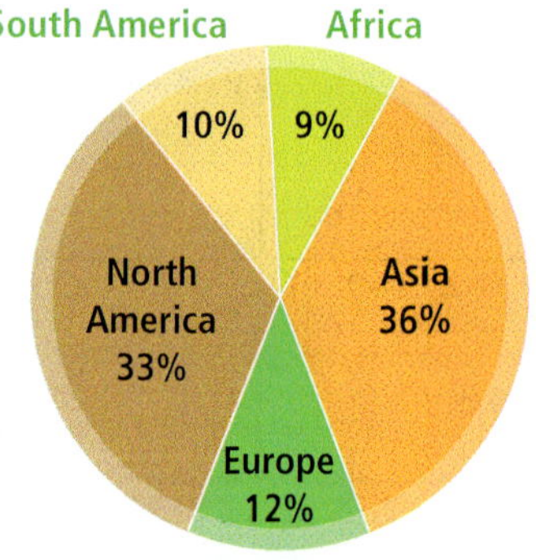

To make a bar graph, place the categories along the bottom axis. Decide on a scale for the side axis. An appropriate scale would be 0–300, marked in intervals of 50. For each data item, draw a bar whose height is equal to the data value.

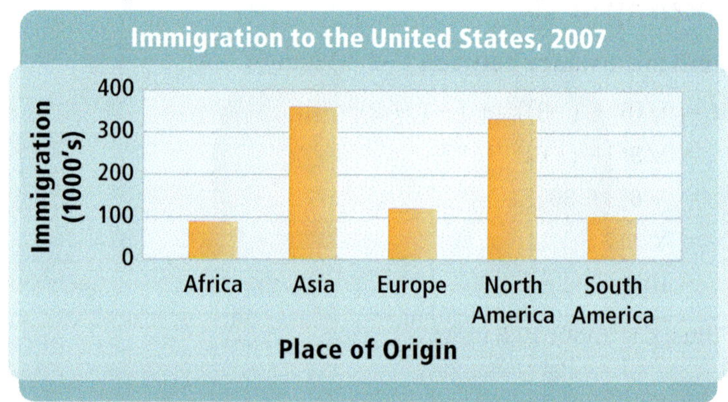

Exercises

Display the data from each table in a bar graph and a circle graph.

1. **NASA Space Shuttle Expenses, 2000**

Operation	Millions of Dollars
Orbiter, integration	698.8
Propulsion	1,053.1
Mission, launch operations	738.8
Flight operations	244.6
Ground operations	510.3

Source: U.S. National Aeronautics and Space Administration

2. **Cable TV Revenue, 2006**

	Millions of Dollars
Airtime	4,566
Basic service	42,918
Pay-per-view, premium services	13,322
Installation	729
Other	27,188

Source: U.S. Census Bureau

Descriptive Statistics and Histograms

For numerical data, you can find the *mean,* the *median,* and the *mode.*

Mean The sum of the data values in a data set divided by the number of data values

Median The middle value of a data set that has been arranged in increasing or decreasing order. If the data set has an even number of values, the median is the mean of the middle two values.

Mode The most frequently occurring value in a data set

Example 1

Find the mean, median, and mode for the following data set. **5 7 6 3 1 7 9 5 10 7**

Mean $\frac{5 + 7 + 6 + 3 + 1 + 7 + 9 + 5 + 10 + 7}{10} = 6$

Median 5, 7, 6, 3, 1, 7, 9, 5, 10, 7 Rearrange the numbers from least to greatest.

1, 3, 5, 5, 6, 7, 7, 7, 9, 10 The median is the mean of the two middle numbers, 6 and 7.

The median is $\frac{6 + 7}{2} = 6.5$.

Mode The most frequently occurring data value is 7.

The frequency of a data value is the number of times it occurs in a data set. A *histogram* is a bar graph that shows the frequency of each data value.

Example 2

Use the survey results to make a histogram for the cost of a movie ticket at various theaters.

Survey of Movie Ticket Prices								
$7	$8	$7	$9	$8	$9	$8	$10	$8

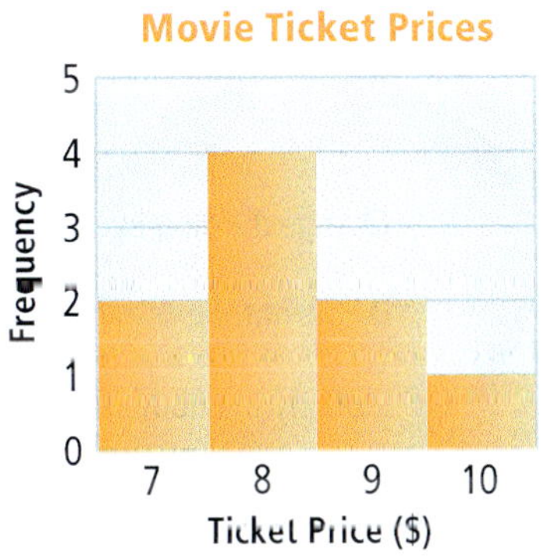

Exercises

Find the mean, the median, and the mode of each data set.

1. −3 4 5 5 −2 7 1 8 9

2. 0 0 1 1 2 3 3 5 3 8 7

3. 2.4 2.4 2.3 2.3 2.4 12.0

4. 1 1 1 1 2 2 2 3 3 4

5. 1.2 1.3 1.4 1.5 1.6 1.7 1.8

6. −4 −3 −2 −1 0 1 2 3 4

Make a histogram for each data set.

7. 7 4 8 6 6 8 7 7 5 7

8. 73 75 76 75 74 75 76 74 76 75

Operations With Rational Expressions

A *rational expression* is an expression that can be written in the form $\frac{\text{polynominal}}{\text{polynominal}}$, where the denominator is not zero. A rational expression is in simplest form if the numerator and denominator have no common factors except 1.

Example 1

Write the expression $\frac{4x + 8}{x + 2}$ in simplest form.

$\frac{4x + 8}{x + 2} = \frac{4(x + 2)}{x + 2}$ Factor the numerator.

$= 4$ Divide out the common factor $x + 2$.

To add or subtract two rational expressions, use a common denominator.

Example 2

Simplify $\frac{x}{2y} + \frac{x}{3y}$.

$\frac{x}{2y} + \frac{x}{3y} = \frac{x}{2y} \cdot \frac{3}{3} + \frac{x}{3y} \cdot \frac{2}{2}$ The common denominator of $3y$ and $2y$ is $6y$.

$= \frac{3x}{6y} + \frac{2x}{6y}$

$= \frac{5x}{6y}$ Add the numerators.

To multiply rational expressions, first find and divide out any common factors in the numerators and the denominators. Then multiply the remaining numerators and denominators. To divide rational expressions, first use a reciprocal to change the problem to multiplication.

Example 3

Simplify $\frac{40x^2}{21} \div \frac{5x}{14}$.

$\frac{40x^2}{21} \div \frac{5x}{14} = \frac{40x^2}{21} \cdot \frac{14}{5x}$ Change dividing by $\frac{5x}{14}$ to multiplying by the reciprocal, $\frac{14}{5x}$.

$= \frac{{}^{8}\cancel{40}\cancel{x^2}^{1}}{{}_{3}\cancel{21}} \times \frac{\cancel{14}^{2}}{\cancel{5x}_{1}}$ Divide out the common factors 5, x, and 7.

$= \frac{16x}{3}$ Multiply the numerators $(8x \cdot 2)$. Multiply the denominators $(3 \cdot 1)$.

Exercises

Write each expression in simplest form.

1. $\frac{4a^2b}{12ab^3}$
2. $\frac{5n + 15}{n + 3}$
3. $\frac{x - 7}{2x - 14}$
4. $\frac{28c^2(d - 3)}{35c(d - 3)}$

Perform the indicated operation.

5. $\frac{3x}{2} + \frac{5x}{2}$
6. $\frac{3x}{8} + \frac{5x}{8}$
7. $\frac{5}{h} - \frac{3}{h}$
8. $\frac{6}{11p} - \frac{9}{11p}$
9. $\frac{3x}{5} - \frac{x}{2}$
10. $\frac{13}{2x} - \frac{13}{3x}$
11. $\frac{7x}{5} + \frac{5x}{7}$
12. $\frac{5a}{b} + \frac{3a}{5b}$
13. $\frac{7x}{8} \cdot \frac{32x}{35}$
14. $\frac{3x^2}{2} \cdot \frac{6}{x}$
15. $\frac{8x^2}{5} \cdot \frac{10}{x^3}$
16. $\frac{7x}{8} \cdot \frac{64}{14x}$
17. $\frac{16}{3x} \div \frac{5}{3x}$
18. $\frac{4x}{5} \div \frac{16}{15x}$
19. $\frac{x^3}{8} \div \frac{x^2}{16}$

Reference

Table 1 Measures

	United States Customary	Metric
Length	12 inches (in.) = 1 foot (ft) 36 in. = 1 yard (yd) 3 ft = 1 yard 5280 ft = 1 mile (mi) 1760 yd = 1 mile	10 millimeters (mm) = 1 centimeter (cm) 100 cm = 1 meter (m) 1000 mm = 1 meter 1000 m = 1 kilometer (km)
Area	144 square inches (in.^{2}) = 1 square foot (ft^2) 9 ft^2 = 1 square yard (yd^2) 43,560 ft^2 = 1 acre (a) 4840 yd^2 = 1 acre	100 square millimeters (mm^2) = 1 square centimeter (cm^2) 10,000 cm^2 = 1 square meter (m^2) 10,000 m^2 = 1 hectare (ha)
Volume	1728 cubic inches (in.^{3}) = 1 cubic foot (ft^3) 27 ft^3 = 1 cubic yard (yd^3)	1000 cubic millimeters (mm^3) = 1 cubic centimeter (cm^3) 1,000,000 cm^3 = 1 cubic meter (m^3)
Liquid Capacity	8 fluid ounces (fl oz) = 1 cup (c) 2 c = 1 pint (pt) 2 pt = 1 quart (qt) 4 qt = 1 gallon (gal)	1000 milliliters (mL) = 1 liter (L) 1000 L = 1 kiloliter (kL)
Weight or Mass	16 ounces (oz) = 1 pound (lb) 2000 pounds = 1 ton (t)	1000 milligrams (mg) = 1 gram (g) 1000 g = 1 kilogram (kg) 1000 kg = 1 metric ton
Temperature	32°F = freezing point of water 98.6°F = normal human body temperature 212°F = boiling point of water	0°C = freezing point of water 37°C = normal human body temperature 100°C = boiling point of water

	Customary Units and Metric Units	
Length	1 in. = 2.54 cm 1 ft ≈ 0.305 m 1mi ≈ 1.61 km	1 cm ≈ 0.39 in. 1 m ≈ 3.28 ft 1 km ≈ 0.62 mi
Area	1 acre = 0.40 ha	1 ha = 2.47 acres
Capacity	1 qt ≈ 0.95 L	1 L ≈ 1.06 qt
Weight and Mass	1 oz ≈ 28.4 g 1 lb ≈ 0.45 kg	1 g ≈ 0.035 oz 1 kg ≈ 2.205 lb

Time		
60 seconds (s) = 1 minute (min)	4 weeks (approx.) = 1 month (mo)	12 months = 1 year
60 minutes = 1 hour (h)	365 days = 1 year (yr)	10 years = 1 decade
24 hours = 1 day (d)	52 weeks (approx.) = 1 year	100 years = 1 century
7 days = 1 week (wk)		

Table 2 **Reading Math Symbols**

Symbols	Words
$\cdot$, $\times$	multiplication sign, times
$\pm$	plus or minus positive or negative
$=$	equals
$\stackrel{?}{=}$	equals?
$\approx$	is approximately equal to
$\neq$	is not equal to
$<$	is less than
$>$	is greater than
$\leq$	is less than or equal to
$\geq$	is greater than or equal to
$\cong$	is congruent to
$\sim$	is similar to
( )	parentheses for grouping
[ ]	brackets for grouping
{ }	braces for a set
%	percent
$\lvert a \rvert$	absolute value of *a*
$-a$	opposite of *a*
$a : b$; $\frac{a}{b}$	ratio of *a* to *b*
$\frac{1}{a}$, a^{-1}, $a \neq 0$	reciprocal of *a*
a^n	*n*th power of *a*
a^{-n}	$\frac{1}{a^n}$, $a \neq 0$
$\sqrt{a}$	nonnegative square root of *a*
$\sqrt[n]{a}$	*n*th root of *a* (nonnegative if *n* even)
$°$ as in $a°$	degree(s)
$\circ$ as in $f \circ g$	composition of functions
π	pi, an irrational number, approximately equal to 3.14
e	an irrational number approximately equal to 2.72
i	the imaginary number $\sqrt{-1}$
$a + bi$, $b \neq 0$	a complex number
∞	infinity
$\sum$	sigma, summation
σ	sigma, standard deviation
σ^2	variance
$\overleftrightarrow{AB}$	line through points *A* and *B*
$\overrightarrow{AB}$	vector *AB*

Symbols	Words
$\overline{AB}$	segment with endpoints *A* and *B*
AB	length of $\overline{AB}$; distance between points *A* and *B*
$\angle A$	angle *A*
$m\angle A$	measure of angle *A*
$\triangle ABC$	triangle *ABC*
(x, y)	ordered pair
$x_1, x_2, \ldots$	specific values of the variable *x*
$y_1, y_2, \ldots$	specific values of the variable *y*
$\bar{x}$	mean of data values x_i
$f(x)$	*f* of *x*; the function value at *x*
f^{-1}	function inverse
log	logarithm
ln	natural logarithm
$\begin{bmatrix} a & b \\ c & d \end{bmatrix}$	matrix
a_{mn}	element in *m*th row, *n*th column of matrix *A*
A^{-1}	inverse of matrix *A*
$\begin{vmatrix} a & b \\ c & d \end{vmatrix}$	determinant of a matrix
det *A*	determinant of matrix *A*
$n!$	*n* factorial
${}_nC_r$	combinations of *n* things chosen *r* at a time
${}_nP_r$	permutations of *n* things arranged *r* at a time
P(event)	probability of the event
$P(A\|B)$	probability of event *A*, given event *B*
sin *A*	sine of $\angle A$
cos *A*	cosine of $\angle A$
tan *A*	tangent of $\angle A$
csc *A*	cosecant of $\angle A$
sec *A*	secant of $\angle A$
cot *A*	cotangent of $\angle A$
^	raised to a power (in a spreadsheet formula)
*	multiply (in a spreadsheet formula)
/	divide (in a spreadsheet formula)
. . .	and so on

Properties and Formulas

Order of Operations
1. Perform any operation(s) inside grouping symbols.
2. Simplify any terms with exponents.
3. Multiply and divide in order from left to right.
4. Add and subtract in order from left to right.

The Pythagorean Theorem
In a right triangle, the sum of the squares of the lengths of the legs is equal to the square of the length of the hypotenuse.

$a^2 + b^2 = c^2$

The Distance Formula
The distance d between any two points (x_1, y_1) and (x_2, y_2) is $d = \sqrt{(x_2 - x_1)^2 + (y_2 - y_1)^2}$.

The Midpoint Formula
The midpoint M of a line segment with endpoints $A(x_1, y_1)$ and $B(x_2, y_2)$ is $\left(\frac{x_1 + x_2}{2}, \frac{y_1 + y_2}{2}\right)$.

Chapter 1 Expressions, Equations, and Inequalities

Closure
For all real numbers a and b, $a + b$ and $a \cdot b$ are real numbers.

The Associative Properties
For all real numbers a, b, and c:

$(a + b) + c = a + (b + c)$
$(a \cdot b) \cdot c = a \cdot (b \cdot c)$

The Commutative Properties
For all real numbers a and b:

$a + b = b + a$ and $a \cdot b = b \cdot a$

The Identity Properties
For every real number a:

$a + 0 = a$ and $0 + a = a$ $a \cdot 1 = a$ and $1 \cdot a = a$
0 is the additive identity   1 is the multiplicative identity.

The Inverse Properties
For every real number a:

$a + (-a) = 0$ and $a \cdot \frac{1}{a} = 1 \quad (a \neq 0)$

The Distributive Properties
For all real numbers a, b, and c:

$a(b + c) = ab + ac$ $(b + c)a = ba + ca$
$a(b - c) = ab - ac$ $(b - c)a = ba - ca$

Multiplication
Let a represent a real number.
Multiplication by 0: $0 \cdot a = 0$.
Multiplication by –1: $-1 \cdot a = -a$

Opposites
Let a and b represent real numbers.
Opposite of a Sum: $-(a + b) = -a + (-b) = -a - b$
Opposite of a Difference: $-(a - b) = -a + b = b - a$
Opposite of a Product: $-(ab) = -a \cdot b = a \cdot (-b)$
Opposite of an Opposite: $-(-a) = a$

Properties of Equality
Assume a, b, and c represent real numbers.

Reflexive: $a = a$
Symmetric: If $a = b$, then $b = a$.
Transitive: If $a = b$ and $b = c$, then $a = c$.
Substitution: If $a = b$, then you can replace a with b and vice versa.
Addition: If $a = b$, then $a + c = b + c$.
Subtraction: If $a = b$, then $a - c = b - c$.
Multiplication: If $a = b$, then $ac = bc$.
Division: If $a = b$ and $c \neq 0$, then $\frac{a}{c} = \frac{b}{c}$.

Properties of Inequality
Let a, b, and c represent real numbers.

Transitive: If $a > b$ and $b > c$, then $a > c$.
Addition: If $a > b$, then $a + c > b + c$.
Subtraction: If $a > b$, then $a - c > b - c$.
Multiplication: If $a > b$ and $c > 0$, then $ac > bc$.
If $a > b$ and $c < 0$, then $ac < bc$.
Division: If $a > b$ and $c > 0$, then $\frac{a}{c} > \frac{b}{c}$.
If $a > b$ and $c < 0$, then $\frac{a}{c} < \frac{b}{c}$.

Chapter 2 Functions, Equations, and Graphs

Direct Variation
$y = kx$ or $\frac{y}{x} = k$, where $k \neq 0$.

Slope of a Line Containing (x_1, y_1) and (x_2, y_2)
$\text{slope} = \frac{\text{vertical change (rise)}}{\text{horizontal change (run)}} = \frac{y_2 - y_1}{x_2 - x_1}$,
where $x_2 - x_1 \neq 0$

Point-Slope Equation of a Line
The equation of the line through point (x_1, y_1) with slope m is $y - y_1 = m(x - x_1)$.

Reference

Reference

Function Families

Assume a, k, and h are positive numbers.

Parent	$y = f(x)$
Reflection across x-axis	$y = -f(x)$
Vertical stretch $(a > 1)$ Vertical shrink $(0 < a < 1)$	$y = af(x)$
Translation	
horizontal to left by h	$y = f(x + h)$
horizontal to right by h	$y = f(x - h)$
vertical up by k	$y = f(x) + k$
vertical down by k	$y = f(x) - k$

Chapter 4 Quadratic Functions and Equations

Quadratic Functions

Parent	$y = x^2$
Reflection across x-axis	$y = -x^2$
Stretch $(a > 1)$ Shrink $(0 < a < 1)$	$y = ax^2$
Translation	
horizontal by h vertical by k	$y = (x - h)^2 + k$
Vertex Form	$y = a(x - h)^2 + k$
Standard Form	$y = f(x) = ax^2 + bx + c$

The graph is a parabola that opens up if $a > 0$ and down if $a < 0$.

The vertex is (h, k) (Vertex Form) and $\left(-\frac{b}{2a}, f\left(-\frac{b}{2a}\right)\right)$ (Standard Form).

The axis of symmetry is $x = h$ (Vertex Form) and $x = -\frac{b}{2a}$ (Standard Form).

Factoring Perfect-Square Trinomials

$a^2 + 2ab + b^2 = (a + b)^2$

$a^2 - 2ab + b^2 = (a - b)^2$

Factoring a Difference of Two Squares

$a^2 - b^2 = (a + b)(a - b)$

Multiplication Property of Square Roots

For any numbers $a \geq 0$ and $b \geq 0$, $\sqrt{ab} = \sqrt{a} \cdot \sqrt{b}$.

Division Property of Square Roots

For any numbers $a \geq 0$ and $b > 0$, $\sqrt{\frac{a}{b}} = \frac{\sqrt{a}}{\sqrt{b}}$.

Zero-Product Property

If $ab = 0$, then $a = 0$ or $b = 0$.

The Quadratic Formula

If $ax^2 + bx + c = 0$, then $x = \frac{-b \pm \sqrt{b^2 - 4ac}}{2a}$

Discriminant

The discriminant of a quadratic equation in the form $ax^2 + bx + c = 0$ is $b^2 - 4ac$.

$b^2 - 4ac > 0 \Rightarrow$ two real solutions

$b^2 - 4ac = 0 \Rightarrow$ one real solution

$b^2 - 4ac < 0 \Rightarrow$ two complex solutions

Square Root of a Negative Real Number

For any positive number a,

$\sqrt{-a} = \sqrt{-1 \cdot a} = \sqrt{-1} \cdot \sqrt{a} = i\sqrt{a}$.

Example: $\sqrt{-5} = i\sqrt{5}$

Note that

$(\sqrt{-5})^2 = (i\sqrt{5})^2 = i^2(\sqrt{5})^2 = -1 \cdot 5 = -5$ (not 5).

Chapter 5 Polynomials and Polynomial Functions

End Behavior of a Polynomial Function

The end behavior of a polynomial function of degree n with leading term ax^n:

a	n	end behavior
positive	even	up and up
positive	odd	down and up
negative	even	down and down
negative	odd	up and down

Factor Theorem

The expression $x - a$ is a linear factor of a polynomial if and only if the value a is a zero of the related polynomial function.

Remainder Theorem

If you divide a polynomial $P(x)$ of degree $n \geq 1$ by $x - a$, then the remainder is $P(a)$.

Factoring a Sum or Difference of Cubes

$a^3 + b^3 = (a + b)(a^2 - ab + b^2)$

$a^3 - b^3 = (a - b)(a^2 + ab + b^2)$

Rational Root Theorem

Let $P(x) = a_n x^n + a_{n-1} x^{n-1} + \cdots + a_1 x + a_0$ be a polynomial with integer coefficients.

Integer roots of $P(x) = 0$ must be factors of a_0.

Rational roots have reduced form $\frac{p}{q}$ where p is an integer factor of a_0 and q is an integer factor of a_n.

Conjugate Root Theorems

Suppose $P(x)$ is a polynomial with *rational* coefficients.

If $a + \sqrt{b}$ is an irrational root with a and b rational, then $a - \sqrt{b}$ is also a root.

Suppose $P(x)$ is a polynomial with *real* coefficients.

If $a + bi$ is a complex root with a and b real, then $a - bi$ is also a root.

Fundamental Theorem of Algebra

If $P(x)$ is a polynomial of degree $n \geq 1$, then $P(x) = 0$ has exactly n roots, including multiple and complex roots.

Binomial Theorem

For every positive integer n, $(a + b)^n =$

$P_0a^n + P_1a^{n-1}b + P_2a^{n-2}b^2 + \cdots + P_{n-1}ab^{n-1} + P_nb^n$

where $P_0, P_1, \ldots, P_n$ are the numbers in the nth row of Pascal's Triangle.

Chapter 6 Radical Functions and Rational Exponents

Properties of Exponents

For any nonzero number a and any integers m and n,

$a^0 = 1$	$(ab)^n = a^nb^n$
$\frac{a^m}{a^n} = a^{m-n}$	$a^m \cdot a^n = a^{m+n}$
$a^{-n} = \frac{1}{a^n}$	$(a^m)^n = a^{mn}$
	$\left(\frac{a}{b}\right)^n = \frac{a^n}{b^n}$

nth Roots of nth Powers

For any real number a,

$$\sqrt[n]{a^n} = \begin{cases} a \text{ if } n \text{ is odd} \\ |a| \text{ if } n \text{ is even} \end{cases}$$

Combining Radical Expressions: Products

If $\sqrt[n]{a}$ and $\sqrt[n]{b}$ are real numbers, then $\sqrt[n]{a} \cdot \sqrt[n]{b} = \sqrt[n]{ab}$.

Combining Radical Expressions: Quotients

If $\sqrt[n]{a}$ and $\sqrt[n]{b}$ are real numbers and $b \neq 0$,

then $\frac{\sqrt[n]{a}}{\sqrt[n]{b}} = \sqrt[n]{\frac{a}{b}}$.

Properties of Rational Exponents

If the nth root of a is a real number and m is an integer, then $a^{\frac{1}{n}} = \sqrt[n]{a}$ and $a^{\frac{m}{n}} = \sqrt[n]{a^m} = \left(\sqrt[n]{a}\right)^m$. If m is negative, $a \neq 0$.

Composition of Inverse Functions

If f and f^{-1} are inverse functions, then $(f^{-1} \circ f)(x) = x$ and $(f \circ f^{-1})(x) = x$ for x in the domains of f and f^{-1}, respectively.

Radical Functions

	Square Root	nth Root
Parent	$y = \sqrt{x}$	$y = \sqrt[n]{x}$
Reflection across x-axis	$y = -\sqrt{x}$	$y = -\sqrt[n]{x}$
Stretch $(a > 1)$ Shrink $(0 < a < 1)$	$y = a\sqrt{x}$	$y = a\sqrt[n]{x}$
Translation horizontal by h vertical by k	$y = \sqrt{x-h} + k$	$y = \sqrt[n]{x-h} + k$

Chapter 7 Exponential and Logarithmic Functions

Exponential Functions

Parent, $b > 0, b \neq 1$	$y = b^x$
Reflection across x-axis	$y = -b^x$
Stretch $(a > 1)$ Shrink $(0 < a < 1)$	$y = ab^x$
Translation horizontal by h vertical by k	$y = b^{x-h} + k$

Continuously Compounded Interest

$A(t) = P \cdot e^{rt}$, where $A(t)$ represents the total, P represents the principal, r represents the interest rate, and t represents time in years.

Logarithmic Functions

	Base b	Base e
Parents, $b > 0, b \neq 1$	$y = \log_b x$	$y = \ln x$
Reflection across x-axis	$y = -\log_b x$	$y = -\ln x$
Stretch $(a > 1)$ Shrink $(0 < a < 1)$	$y = a \log_b x$	$y = a \ln x$
Translation horizontal by h vertical by k	$y = \log_b (x-h) + k$	$y = \ln (x-h) + k$

Properties of Logarithms

For any positive numbers m, n, and b where $b \neq 1$

Product Property:	$\log_b mn = \log_b m + \log_b n$
Quotient Property:	$\log_b \frac{m}{n} = \log_b m - \log_b n$
Power Property:	$\log_b m^n = n \log_b m$

Change of Base Formula

For any positive numbers, m, b, and c, with $b \neq 1$ and $c \neq 1$, $\log_b m = \frac{\log_c m}{\log_c b}$.

Chapter 8 Rational Functions

Inverse Variation

$xy = k$, $y = \frac{k}{x}$, or $x = \frac{k}{y}$, where $k \neq 0$.

Combined Variation

z varies jointly with x and y: $z = kxy$

z varies jointly with x and y and inversely with w: $z = \frac{kxy}{w}$

z varies directly with x and inversely with the product wy: $z = \frac{kx}{wy}$

Reciprocal Functions

Parent	$y = \frac{1}{x}, x \neq 0$
Reflection across x-axis	$y = -\frac{1}{x}, x \neq 0$
Stretch $(a > 1)$ Shrink $(0 < a < 1)$	$y = \frac{a}{x}, x \neq 0$
Translation horizontal by h vertical by k	$y = \frac{a}{x-h} + k, x \neq h$
Asymptotes	$y = k$ (horiz.), $x = h$ (vert.)

Reference

Reference

Chapter 9 Sequences and Series

Arithmetic Mean of Two Numbers

$\frac{x + y}{2}$

Arithmetic Sequence

A recursive definition for an arithmetic sequence with a starting value a and a common difference d has two parts:

$a_1 = a$: initial condition

$a_{n+1} = a_n + d$, for $n \geq 1$: recursive formula

An explicit definition for this sequence is the formula:

$a_n = a + (n - 1)d$ for $n \geq 1$.

Geometric Sequence

A recursive definition for a geometric sequence with a starting value a and a common ratio r has two parts:

$a_1 = a$: initial condition

$a_{n+1} = a_n \cdot r$, for $n \geq 1$: recursive formula

An explicit definition for this sequence is the formula:

$a_n = ar^{n-1}$, for $n \geq 1$.

Sum of a Finite Arithmetic Series

The sum S_n of a finite arithmetic series

$a_1 + a_2 + a_3 + \cdots + a_n$ is $S_n = \frac{n}{2}(a_1 + a_n)$

where a_1 is the first term, a_n is the nth term, and n is the number of terms.

Sum of a Finite Geometric Series

The sum S_n of a finite geometric series

$a_1 + a_1r + a_1r^2 + \cdots + a_1r^{n-1}$ is $S_n = \frac{a_1(1 - r^n)}{1 - r}$

where a_1 is the first term, r is the common ratio, and n is the number of terms.

Sum of an Infinite Geometric Series

An infinite geometric series with $|r| < 1$ converges to the sum S given by the following formula:

$S = \frac{a_1}{1 - r}$.

Chapter 10 Quadratic Relations and Conic Sections

Parabolas

Vertical	Vertex (0, 0)	Vertex (h, k)
Equation	$y = \frac{1}{4c}x^2$	$y = \frac{1}{4c}(x - h)^2 + k$
Focus	$(0, c)$	$(h, c + k)$
Directrix	$y = -c$	$y = -c + k$
Horizontal	**Vertex (0, 0)**	**Vertex (h, k)**
Equation	$x = \frac{1}{4c}y^2$	$x = \frac{1}{4c}(y - k)^2 + h$
Focus	$(c, 0)$	$(c + h, k)$
Directrix	$x = -c$	$x = -c + h$

Circles, radius = r

	Center (0, 0)	Center (h, k)
Equation	$x^2 + y^2 = r^2$	$(x - h)^2 + (y - k)^2 = r^2$

Ellipses

Horizontal, $a > b$	Center (0, 0)	Center (h, k)
Equation	$\frac{x^2}{a^2} + \frac{y^2}{b^2} = 1$	$\frac{(x - h)^2}{a^2} + \frac{(y - k)^2}{b^2} = 1$
Vertices	$(\pm a, 0)$	$(\pm a + h, k)$
Co-Vertices	$(0, \pm b)$	$(h, \pm b + k)$
Foci, $c^2 = a^2 - b^2$	$(\pm c, 0)$	$(\pm c + h, k)$
Major axis	$y = 0$	$y = k$
Minor axis	$x = 0$	$x = h$

Vertical, $a > b$	Center (0, 0)	Center (h, k)
Equation	$\frac{x^2}{b^2} + \frac{y^2}{a^2} = 1$	$\frac{(x - h)^2}{b^2} + \frac{(y - k)^2}{a^2} = 1$
Vertices	$(0, \pm a)$	$(h, \pm a + k)$
Co-Vertices	$(\pm b, 0)$	$(\pm b + h, k)$
Foci, $c^2 = a^2 - b^2$	$(0, \pm c)$	$(h, \pm c + k)$
Major axis	$x = 0$	$x = h$
Minor axis	$y = 0$	$y = k$

Hyperbolas

Horizontal, $a > b$	Center (0, 0)	Center (h, k)
Equation	$\frac{x^2}{a^2} - \frac{y^2}{b^2} = 1$	$\frac{(x - h)^2}{a^2} - \frac{(y - k)^2}{b^2} = 1$
Vertices	$(\pm a, 0)$	$(\pm a + h, k)$
Foci, $c^2 = a^2 + b^2$	$(\pm c, 0)$	$(\pm c + h, k)$
Transverse axis	$y = 0$	$y = k$
Asymptotes	$y = \pm\frac{b}{a}x$	$y = \pm\frac{b}{a}(x - h) + k$

Vertical, $a > b$	Center (0, 0)	Center (h, k)
Equation	$\frac{y^2}{a^2} - \frac{x^2}{b^2} = 1$	$\frac{(y - k)^2}{a^2} - \frac{(x - h)^2}{b^2} = 1$
Vertices	$(0, \pm a)$	$(h, \pm a + k)$
Foci, $c^2 = a^2 + b^2$	$(0, \pm c)$	$(h, \pm c + k)$
Transverse axis	$x = 0$	$x = h$
Asymptotes	$y = \pm\frac{a}{b}x$	$y = \pm\frac{a}{b}(x - h) + k$

Chapter 11 Probability and Statistics

Fundamental Counting Principle

If event M can occur in m ways and is followed by event N that can occur in n ways, then event M followed by event N can occur in $m \cdot n$ ways.

Number of Permutations

The number of permutations of n items of a set arranged r items at a time is

${}_nP_r = \frac{n!}{(n-r)!}$ for $0 \le r \le n$.

Number of Combinations

The number of combinations of n items of a set chosen r items at a time is

${}_nC_r = \frac{n!}{r!(n-r)!}$ for $0 \le r \le n$.

Probability of *A* and *B*

If A and B are independent events, then
$P(A \text{ and } B) = P(A) \cdot P(B)$.

Probability of *A* or *B*

$P(A \text{ or } B) = P(A) + P(B) - P(A \text{ and } B)$
If A and B are mutually exclusive events, then
$P(A \text{ or } B) = P(A) + P(B)$.

Conditional Probability

For any two events A and B with $P(A) \neq 0$, the probability of event B, given event A, is:

$P(B|A) = \frac{P(A \text{ and } B)}{P(A)}$

Mean, Variance, and Standard Deviation

Mean: $\bar{x} = \frac{x_1 + x_2 + x_3 + \cdots + x_n}{n}$

Variance: $\sigma^2 = \frac{\Sigma(x - \bar{x})^2}{n}$

Standard deviation: $\sigma = \sqrt{\frac{\Sigma(x - \bar{x})^2}{n}}$

Binomial Probability

For repeated independent trials, each with a probability of success p and a probability of failure q (with $p + q = 1$), the probability of x successes in n trials is $P(x) = {}_nC_x p^x q^{n-x}$.

Binomial Theorem Using Combinations

For every positive integer n, use the combinations formula ${}_nC_r$ to expand $(a + b)^n$:

$$(a + b)^n = {}_nC_0 a^n + {}_nC_1 a^{n-1}b + {}_nC_2 a^{n-2}b^2 + \cdots + {}_nC_{n-1} ab^{n-1} + {}_nC_n b^n$$

Chapter 12 Matrices

Properties of Matrix Addition

If A, B, and C are $m \times n$ matrices, then

Closure Property:	$A + B$ is an $m \times n$ matrix
Commutative Property:	$A + B = B + A$
Associative Property:	$(A + B) + C = A + (B + C)$
Identity Property:	There is a unique $m \times n$ matrix O such that $O + A = A + O = A$
Inverse Property:	For each A, there is a unique opposite, $-A$, such that $A + (-A) = O$

Properties of Scalar Multiplication

If A and B are $m \times n$ matrices, c and d are scalars, and O is the $m \times n$ zero matrix, then

Closure Property:	cA is an $m \times n$ matrix
Associative Property:	$(cd)A = c(dA)$
Distributive Property:	$c(A + B) = cA + cB$ $(c + d)A = cA + dA$
Identity Property:	$1 \cdot A = A$
Property of Zero:	$0 \cdot A = O$ and $cO = O$

Properties of Matrix Multiplication

If A, B, and C are $n \times n$ matrices and O is the $n \times n$ zero matrix, then

Closure Property:	AB is an $n \times n$ matrix
Associative Property:	$(AB)C = A(BC)$
Distributive Property:	$A(B + C) = AB + AC$ $(B + C)A = BA + CA$
Property of Zero:	$OA = AO = O$

Determinants of 2 × 2 and 3 × 3 Matrices

The determinant of a 2×2 matrix $\begin{bmatrix} a & b \\ c & d \end{bmatrix}$ is $ad - bc$.

The determinant of a 3×3 matrix $\begin{bmatrix} a_1 & b_1 & c_1 \\ a_2 & b_2 & c_2 \\ a_3 & b_3 & c_3 \end{bmatrix}$ is

$a_1b_2c_3 + b_1c_2a_3 + c_1a_2b_3 - (a_3b_2c_1 + b_3c_2a_1 + c_3a_2b_1)$

Inverse of a 2 × 2 Matrix

If $A = \begin{bmatrix} a & b \\ c & d \end{bmatrix}$ and $\det A \neq 0$,

then the inverse of A is

$$A^{-1} = \frac{1}{\det A}\begin{bmatrix} d & -b \\ -c & a \end{bmatrix} = \frac{1}{ad - bc}\begin{bmatrix} d & -b \\ -c & a \end{bmatrix}.$$

Reference

Reference

Chapter 13 Periodic Functions and Trigonometry

Convert Between Radians and Degrees

Use the proportion $\frac{d°}{180°} = \frac{r \text{ radians}}{\pi \text{ radians}}$ to convert between radians and degrees.

To convert degrees to radians, multiply by $\frac{\pi \text{ radians}}{180°}$.

To convert radians to degrees, multiply by $\frac{180°}{\pi \text{ radians}}$.

Length of an Intercepted Arc

For a circle of radius r and a central angle of measure θ (in radians), the length s of the intercepted arc is $s = r\theta$.

Sine and Cosine Functions

	Sine	Cosine
Parents	$y = \sin x$	$y = \cos x$
Reflection across x-axis	$y = -\sin x$	$y = -\cos x$
Amplitude $\lvert a \rvert$	$y = a \sin x$	$y = a \cos x$
Period $\frac{2\pi}{b}$, $b > 0$	$y = \sin bx$	$y = \cos bx$
Translation horizontal by h vertical by k	$y = \sin (x - h) + k$	$y = \cos (x - h) + k$

Tangent Function

Parent	$y = \tan x$
Reflection across x-axis	$y = -\tan x$
Period $\frac{\pi}{b}$	$y = \tan bx$
Translation horizontal by h vertical by k	$y = \tan (x - h) + k$
Asymptotes ($\tan bx$)	$x = n\frac{\pi}{2b}$, n odd

Chapter 14 Trigonometric Identities and Equations

Basic Identities

Reciprocal Identities:

$\csc \theta = \frac{1}{\sin \theta}$ $\quad \sec \theta = \frac{1}{\cos \theta}$ $\quad \tan \theta = \frac{1}{\cot \theta}$

$\sin \theta = \frac{1}{\csc \theta}$ $\quad \cos \theta = \frac{1}{\sin \theta}$ $\quad \cot \theta = \frac{1}{\tan \theta}$

Tangent Identity: $\tan \theta = \frac{\sin \theta}{\cos \theta}$

Cotangent Identity: $\cot \theta = \frac{\cos \theta}{\sin \theta}$

Pythagorean Identities

$\cos^2\theta + \sin^2\theta = 1$ $\quad 1 + \tan^2\theta = \sec^2\theta$ $\quad \cot^2\theta + 1 = \csc^2\theta$

Area of a Triangle

In $\triangle ABC$ with a, b, and c the lengths of the sides opposite $\angle A$, $\angle B$, and $\angle C$, respectively,

Area $\triangle ABC = \frac{1}{2}bc \sin A = \frac{1}{2}ac \sin B = \frac{1}{2}ab \sin C$.

Law of Sines

In $\triangle ABC$ with a, b, and c the lengths of the sides opposite $\angle A$, $\angle B$, and $\angle C$, respectively,

$\frac{\sin A}{a} = \frac{\sin B}{b} = \frac{\sin C}{c}$.

Law of Cosines

In $\triangle ABC$ with a, b, and c the lengths of the sides opposite $\angle A$, $\angle B$, and $\angle C$, respectively,

$a^2 = b^2 + c^2 - 2bc \cdot \cos A$

$b^2 = a^2 + c^2 - 2ac \cdot \cos B$

$c^2 = a^2 + b^2 - 2ab \cdot \cos C$

Negative Angle Identities

$\sin (-\theta) = -\sin \theta$ $\quad \cos (-\theta) = \cos \theta$ $\quad \tan (-\theta) = -\tan \theta$

Cofunction Angle Identities

$\sin\left(\frac{\pi}{2} - \theta\right) = \cos \theta$ $\quad \cos\left(\frac{\pi}{2} - \theta\right) = \sin \theta$ $\quad \tan\left(\frac{\pi}{2} - \theta\right) = \cot \theta$

Angle Difference Identities

$\sin (A - B) = \sin A \cos B - \cos A \sin B$

$\cos (A - B) = \cos A \cos B + \sin A \sin B$

$\tan (A - B) = \frac{\tan A - \tan B}{1 + \tan A \tan B}$

Angle Sum Identities

$\sin (A + B) = \sin A \cos B + \cos A \sin B$

$\cos (A + B) = \cos A \cos B - \sin A \sin B$

$\tan (A + B) = \frac{\tan A + \tan B}{1 - \tan A \tan B}$

Double-Angle Identities

$\cos 2\theta = \cos^2\theta - \sin^2\theta$

$\cos 2\theta = 2\cos^2\theta - 1$

$\cos 2\theta = 1 - 2\sin^2\theta$

$\sin 2\theta = 2\sin\theta \cos\theta$

$\tan 2\theta = \frac{2\tan \theta}{1 - \tan^2 \theta}$

Half-Angle Identities

$\sin \frac{A}{2} = \pm\sqrt{\frac{1 - \cos A}{2}}$

$\cos \frac{A}{2} = \pm\sqrt{\frac{1 + \cos A}{2}}$

$\tan \frac{A}{2} = \pm\sqrt{\frac{1 - \cos A}{1 + \cos A}}$

Formulas of **Geometry**

You will use a number of geometric formulas as you work through your algebra book. Here are some perimeter, area, and volume formulas.

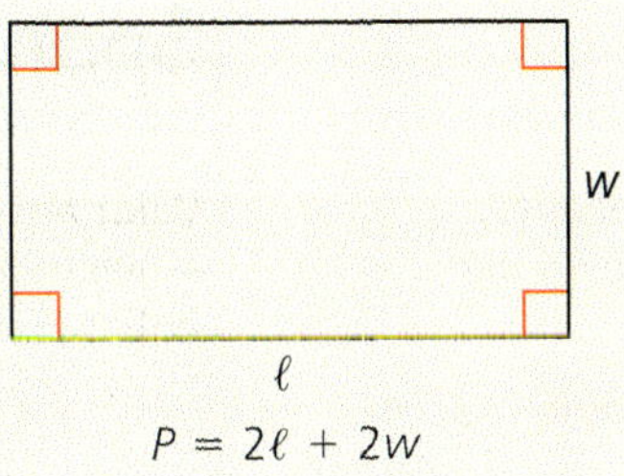

$P = 2\ell + 2w$

$A = \ell w$

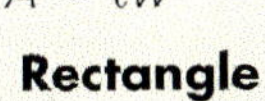

Rectangle

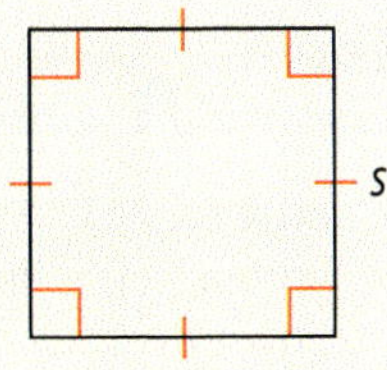

$P = 4s$

$A = s^2$

Square

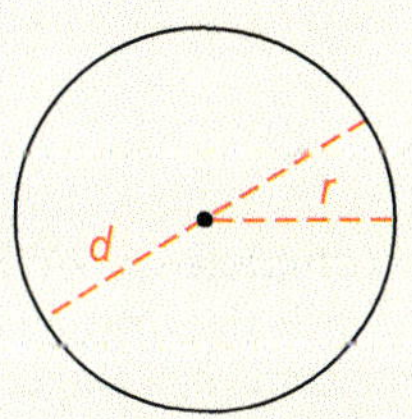

$C = 2\pi r$ or $C = \pi d$

$A = \pi r^2$

Circle

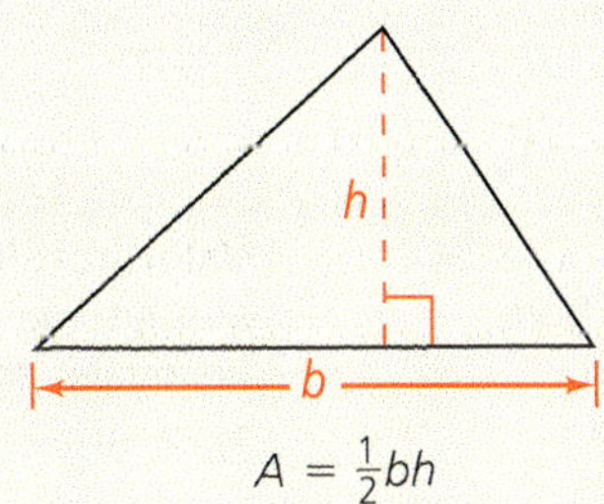

$A = \frac{1}{2}bh$

Triangle

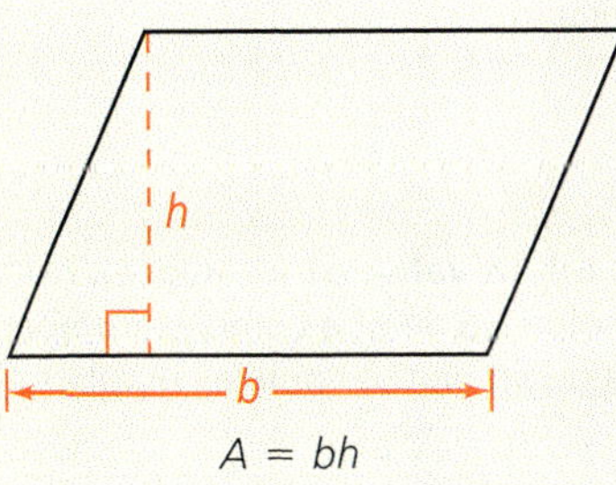

$A = bh$

Parallelogram

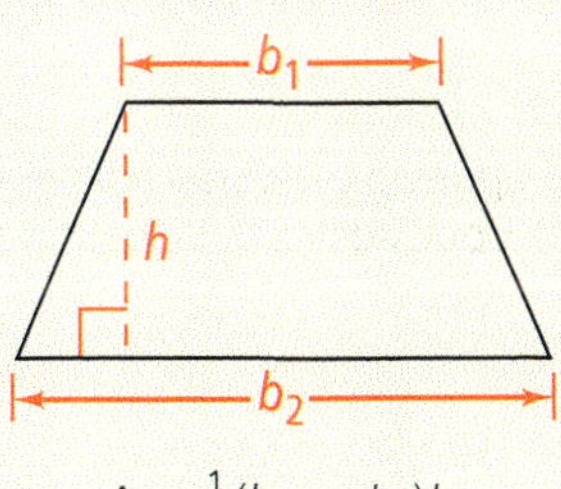

$A = \frac{1}{2}(b_1 + b_2)h$

Trapezoid

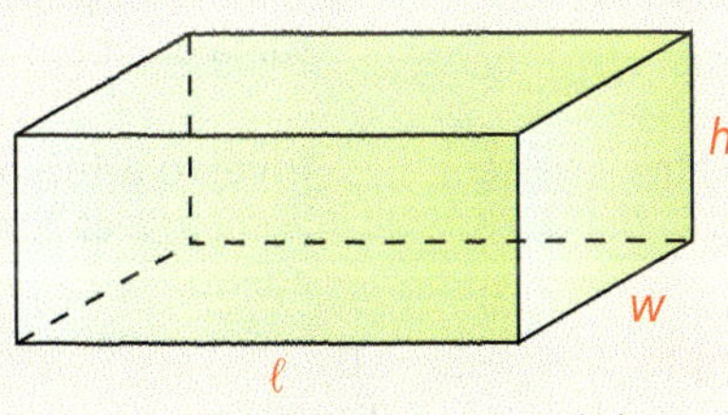

$SA = 2(\ell w + wh + h\ell)$

$V = Bh$

$V = \ell wh$

Right Prism

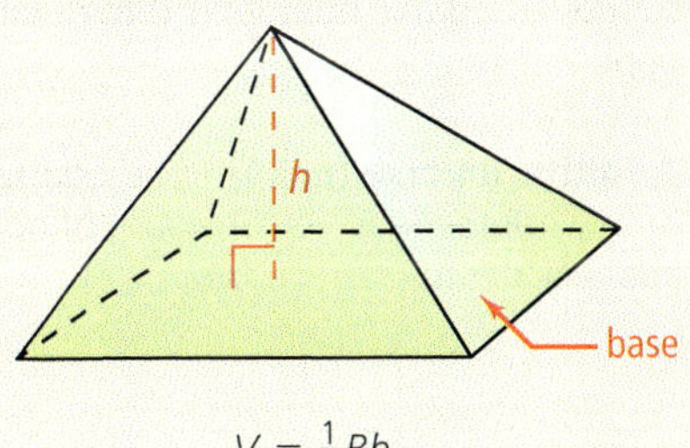

$V = \frac{1}{3}Bh$

Pyramid

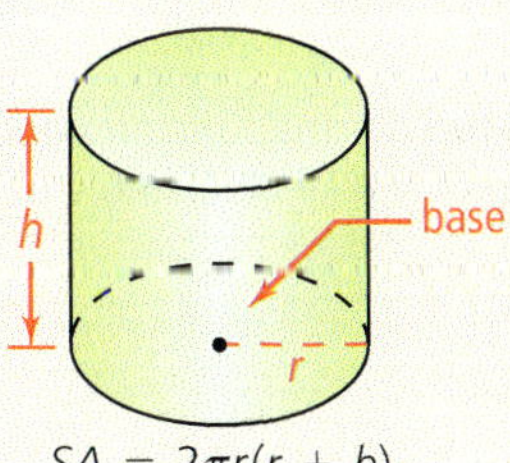

$SA = 2\pi r(r + h)$

$V = Bh$

$V = \pi r^2 h$

Right Cylinder

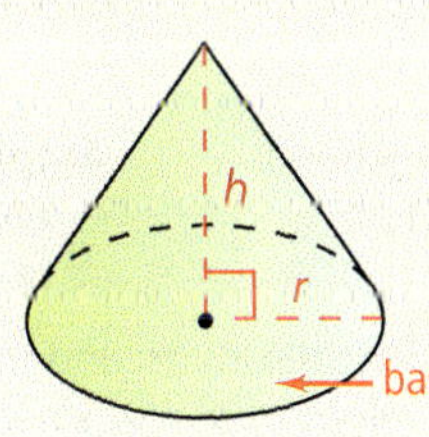

$V = \frac{1}{3}Bh$

$V = \frac{1}{3}\pi r^2 h$

Right Cone

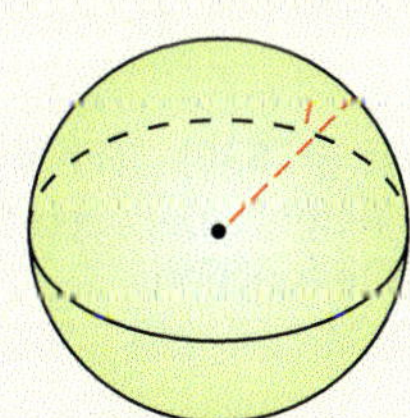

$SA = 4\pi r^2$

$V = \frac{4}{3}\pi r^3$

Sphere

English/Spanish Illustrated Glossary

English | Spanish

Absolute value (p. 41) The absolute value of a real number, x, written $|x|$, is its distance from zero on the number line.

Valor absoluto de un número real (p. 41) El valor absoluto de un número real, x, escrito como $|x|$, es su distancia desde cero en la recta numérica.

Example $|3| = 3$
$|-4| = 4$

Absolute value function (p. 107) A function of the form $f(x) = |mx + b| + c$, where $m \neq 0$, is an absolute value function.

Función de valor absolute (p. 107) Una función de la forma $f(x) = |mx + b| + c$, donde $m \neq 0$, es una función de valor absoluto.

Example $f(x) = |3x - 2| + 3$
$f(x) = |2x|$

Absolute value of a complex number (p. 249) The absolute value of a complex number is its distance from the origin on the complex number plane. In general, $|a + bi| = \sqrt{a^2 + b^2}$.

Valor absoluto de un número complejo (p. 249) El valor absoluto de un número complejo es la distancia a la que está del origen en el plano de números complejo. Generalmente, $|a + bi| = \sqrt{a^2 + b^2}$.

Example $|3 - 4i| = \sqrt{3^2 + (-4)^2} = 5$

Additive identity (p. 14) The additive identity is 0. The sum of 0 and any number is that number. The sum of opposites is 0.

Identidad aditiva (p. 14) La identidad aditiva es 0. La suma de 0 y cualquier número es ese mismo número. La suma de opuestos es 0.

Additive inverse (p. 14) The opposite or additive inverse of any number a is $-a$. The sum of opposites is 0, the additive identity.

Inverso aditivo (p. 14) El opuesto o inverso aditivo de un número a es $-a$. La suma de opuestos es 0, la identidad aditiva.

Example $3 + (-3) = 0$
$5.2 + (-5.2) = 0$

Algebraic expression (p. 5) An algebraic expression is a mathematical phrase that contains one or more variables.

Expresión algebraica (p. 5) Una expresión algebraica es una frase matemática que contiene una o más variables.

Example $2x + 3$
$z - y$

Amplitude (p. 830) The amplitude of a periodic function is half the difference between the maximum and minimum values of the function.

Amplitud (p. 830) La amplitud de una función periódica es la mitad de la diferencia entre los valores máximo y mínimo de la función.

Example The maximum and minimum values of $y = 4 \sin x$ are 4 and -4, respectively.

$$\text{amplitude} = \frac{4 - (-4)}{2} = 4$$

English / Spanish

Arithmetic mean (p. 574) The arithmetic mean, or average, of two numbers is their sum divided by two.

Media aritmética (p. 574) La media aritmética, o promedio, de dos números es su suma dividida por dos.

Example The arithmetic mean of 12 and 15 is $\frac{12 + 15}{2} = 13.5$.

Arithmetic sequence (p. 572) An arithmetic sequence is a sequence with a constant difference between consecutive terms.

Secuencia aritmética (p. 572) Una secuencia aritmética es una secuencia de números en la que la diferencia entre dos números consecutivos es constante.

Example The arithmetic sequence 1, 5, 9, 13, . . . has a common difference of 4.

Arithmetic series (p. 587) An arithmetic series is a series whose terms form an arithmetic sequence.

Serie aritmética (p. 587) Una serie aritmética es una serie cuyos términos forman una progresión aritmética.

Example $1 + 5 + 9 + 13 + 17 + 21$ is an arithmetic series with six terms.

Asymptote (p. 435) An asymptote is a line that a graph approaches as x or y increases in absolute value.

Asíntota (p. 435) Una asíntota es una recta a la cual se acerca una gráfica a medida que x o y aumentan de valor absoluto.

Example The function $y = \frac{x + 2}{x - 2}$ has $x = 2$ as a vertical asymptote and $y = 1$ as a horizontal asymptote.

Axis of symmetry (pp. 107,194) The axis of symmetry is the line that divides a figure into two parts that are mirror images.

Eje de simetría (pp. 107,194) El eje de simetría es la recta que divide una figura en dos partes que son imágenes una de la otra.

Example

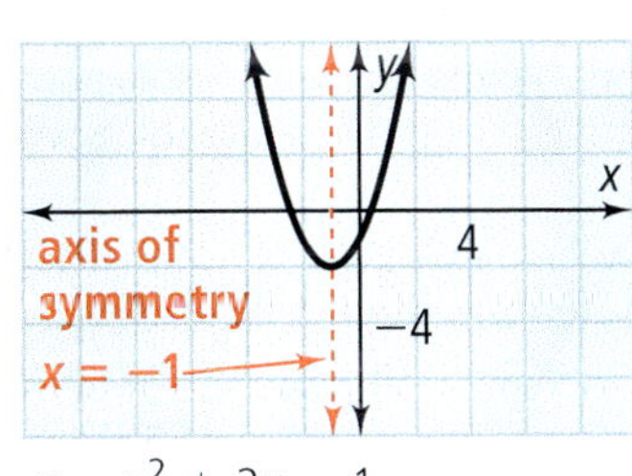

$y = x^2 + 2x - 1$

B

Bias (p. 726) A bias is a systematic error introduced by the sampling method.

Sesgo (p. 726) El sesgo es un error sistemático introducido por medio del método de muestreo.

Bimodal (p. 712) A bimodal data set has two modes.

Bimodal (p. 712) Un conjunto bimodal de datos tiene dos modas.

Example {1, 2, 3, 3, 4, 5, 6, 6}
mode = 3 and 6

English	Spanish
Binomial experiment (p. 731) A binomial experiment is one in which the situation involves repeated trials. Each trial has two possible outcomes (success or failure), and the probability of success is constant throughout the trials.	**Experimento binomial (p. 731)** Un experimento binomial es un experimento que requiere varios ensayos. Cada ensayo tiene dos resultados posibles (éxito o fracaso), y la probabilidad de éxito es constante durante todos los ensayos.
Binomial probability (p. 732) In a binomial experiment with probability of success p and probability of failure q, the probability of x successes in n trials is given by ${}_nC_x p^x q^{n-x}$.	**Probabilidad binomial (p. 732)** En un experimento binomial con una probabilidad de éxito p y una probabilidad de fracaso q, la probabilidad de x éxitos en n ensayos se expresa con ${}_nC_x p^x q^{n-x}$.

Example Suppose you roll a standard number cube and that you call rolling a 1 a success. Then $p = \frac{1}{6}$ and $q = \frac{5}{6}$. The probability of rolling nine 1's in twenty rolls is ${}_{20}C_9\left(\frac{1}{6}\right)^9\left(\frac{5}{6}\right)^{11} \approx 0.0022$.

English	Spanish
Binomial Theorem (pp. 327, 733) For every positive integer n, $(a + b)^n = P_0a^n + P_1a^{n-1}b + P_2a^{n-2}b^2 + \cdots + P_{n-1}ab^{n-1} + P_nb^n$ where $P_0, P_1, \ldots, P_n$ are the numbers in the row of Pascal's Triangle that has n as its second number.	**Teorema binomial (pp. 327, 733)** Para cada número entero positivo n, $(a + b)^n = P_0a^n + P_1a^{n-1}b + P_2a^{n-2}b^2 + \cdots + P_{n-1}ab^{n-1} + P_nb^n$, donde $P_0, P_1, \ldots, P_n$ son los números de la fila del Triángulo de Pascal cuyo segundo número es n.

Example

$$\begin{aligned}(x + 1)^3 &= {}_3C_0(x)^3 + {}_3C_1(x)^2(1)^1 \\ &\quad + {}_3C_2(x)^1(1)^2 + {}_3C_3(1)^3 \\ &= x^3 + 3x^2 + 3x + 1\end{aligned}$$

English	Spanish
Boundary (p. 114) A boundary of the graph of a linear inequality is a line in the coordinate plane. It separates the solutions of the inequality from the nonsolutions. Points of the line itself may or may not be solutions.	**Límite (p. 114)** Un límite de la gráfica de una desigualdad lineal es una línea en el plano de coordenadas. Ésta separa las soluciones de la desigualdad de las no soluciones. Las soluciones pueden ser o no puntos de la línea.
Box-and-whisker plot (p. 714) A box-and-whisker plot is a method of displaying data that uses quartiles to form the center box and the maximum and minimum values to form the whiskers.	**Gráfica de cajas (p. 714)** Una gráfica de cajas es un método para mostrar datos que utiliza cuartiles para formar una casilla central y los valores máximos y mínimos para formar los conectores.

Example

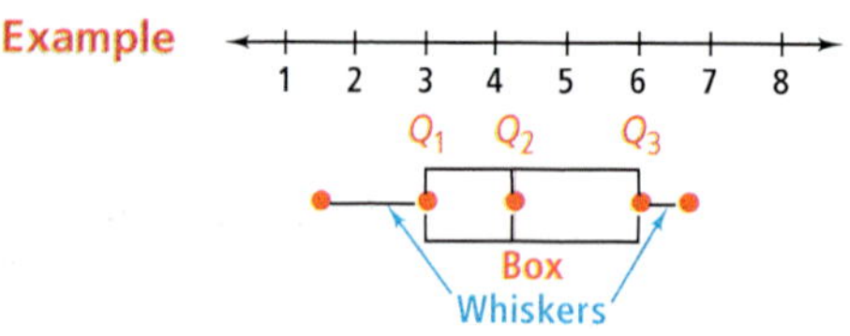

English	Spanish
Branch (p. 508) Each piece of a discontinuous graph is called a branch.	**Rama (p. 508)** Cada segmento de una gráfica discontinua se llama rama.

Example

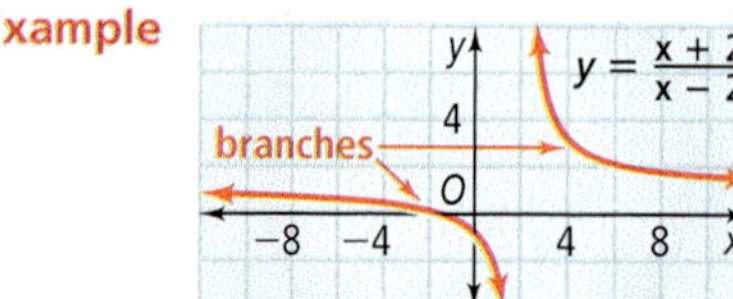

English **C** Spanish

Center of a circle (p. 630) The center of a circle is the point that is the same distance from every point on the circle.

Centro de un círculo (p. 630) El centro de un círculo es el punto que está situado a la misma distancia de cada punto del círculo.

Center of an ellipse (p. 639) The center of an ellipse is the midpoint of the major axis.

Centro de una elipse (p. 639) El centro de una elipse es el punto medio entre los dos ejes mayores.

Center of rotation (p. 804) A center of rotation is the fixed point of a rotation.

Centro de rotación (p. 804) Un centro de rotación es el punto fijo de una rotación.

Example

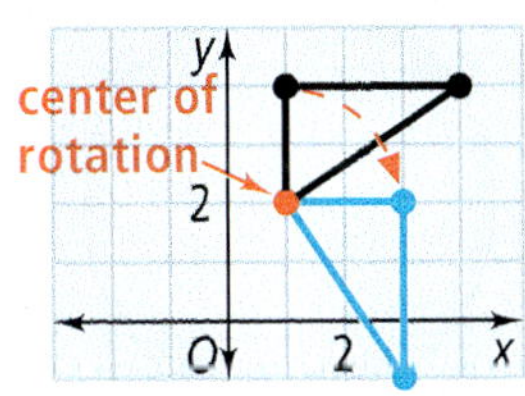

Central angle (p. 844) A central angle of a circle is an angle whose vertex is at the center of a circle.

Ángulo central (p. 844) El ángulo central de un círculo es un ángulo cuyo vértice está situado en el centro del círculo.

Example

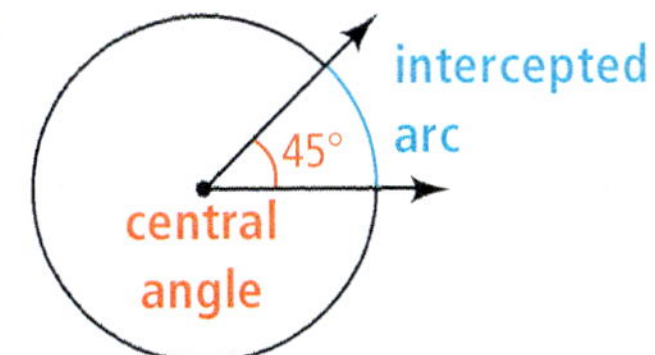

Change of Base Formula (p. 464) $\log_b M = \frac{\log_c M}{\log_c b}$, where M, b, and c are positive numbers, and $b \neq 1$ and $c \neq 1$.

Fórmula de cambio de base (p. 464) $\log_b M = \frac{\log_c M}{\log_c b}$, donde M, b y c son números positivos y $b \neq 1$ y $c \neq 1$.

Example $\log_3 8 = \frac{\log 8}{\log 3} \approx 1.8928$

Circle (p. 630) A circle is the set of all points in a plane at a distance r from a given point. The standard form of the equation of a circle with center (h, k) and radius r is $(x - h)^2 + (y - k)^2 = r^2$.

Círculo (p. 630) Un círculo es el cojunto de todos los puntos situados en un plano a una distancia r de un punto dado. La forma normal de la ecuación cuyo centro es (h, k) y cuyo radio es r es $(x - h)^2 + (y - k)^2 = r^2$.

Example

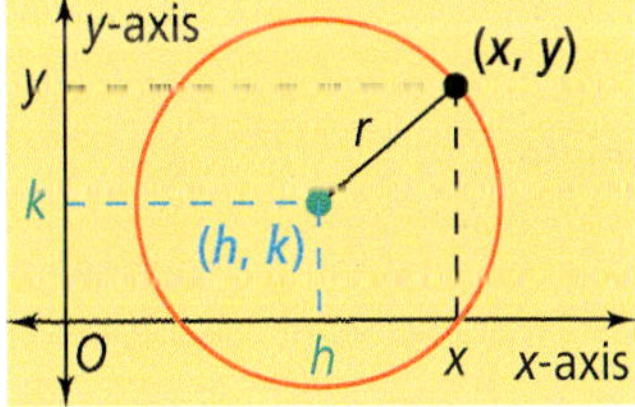

Coefficient (p. 20) The numerical factor in a term.

Coeficiente (p. 20) El factor numérico de un término.

Example The coefficient of $-3k$ is -3.

English

Spanish

Coefficient matrix (p. 793) When representing a system of equations with a matrix equation, the matrix containing the coefficients of the system is the coefficient matrix.

Matriz de coeficientes (p. 793) Al representar un sistema de ecuaciones con una ecuación de matriz, la matriz que contiene los coeficientes del sistema es la matriz de coeficientes.

Example $\begin{cases} x + 2y = 5 \\ 3x + 5y = 14 \end{cases}$

coefficient matrix $\begin{bmatrix} 1 & 2 \\ 3 & 5 \end{bmatrix}$

Combination (p. 676) Any unordered selection of r objects from a set of n objects is a combination. The number of combinations of n objects taken r at a time is

$_nC_r = \frac{n!}{r!(n - r)!}$ for $0 \leq r \leq n$.

Combinación (p. 676) Cualquier selección no ordenada de r objetos tomados de un conjunto de n objetos es una combinación. El número de combinaciones de n objetos, cuando se toman r objetos cada vez, es

$_nC_r = \frac{n!}{r!(n - r)!}$ para $0 \leq r \leq n$.

Example The number of combinations of seven items taken four at a time is

$_7C_4 = \frac{7!}{4!(7 - 4)!} = 35.$

There are 35 ways to choose four items from seven items without regard to order.

Combined variation (p. 501) A combined variation is a relation in which one variable varies with respect to each of two or more variables.

Variación combinada (p. 501) Una variación combinada es una relación en la que una variable varía con respecto a cada una de dos o más variables.

Example $y = kx^2\sqrt{z}$

$z = \frac{kx}{y}$

Common difference (p. 572) A common difference is the difference between consecutive terms of an arithmetic sequence.

Diferencia común (p. 572) La diferencia común es la diferencia entre los términos consecutivos de una progresión aritmética.

Example The arithmetic sequence 1, 5, 9, 13, . . . has a common difference of 4.

Common logarithm (p. 453) A common logarithm is a logarithm that uses base 10. You can write the common logarithm $\log_{10} y$ as $\log y$.

Logaritmo común (p. 453) El logaritmo común es un logaritmo de base 10. El logaritmo común $\log_{10} y$ se expresa como $\log y$.

Example $\log 1 = 0$

$\log 10 = 1$

$\log 50 = 1.698970004 \ldots$

Common ratio (p. 580) A common ratio is the ratio of consecutive terms of a geometric sequence.

Razón común (p. 580) Una razón común es la razón de términos consecutivos en una secuencia geométrica.

Example The geometric sequence 2.5, 5, 10, 20, . . . has a common ratio of 2.

English

Spanish

Completing the square (p. 235) Completing the square is the process of finding a constant c to add to $x^2 + bx$ so that $x^2 + bx + c$ is the square of a binomial.

Completar el cuadrado (p. 235) Completar un cuadrado es el proceso mediante el cual se halla una constante c que se le pueda sumar a $x^2 + bx$, de manera que $x^2 + bx + c$ sea el cuadrado de un binomio.

Example $x^2 - 12x + \blacksquare$

$x^2 - 12x + \left(\frac{-12}{2}\right)^2$

$x^2 - 12x + 36$

Complex conjugates (p. 251) Number pairs of the form $a + bi$ and $a - bi$ are complex conjugates.

Conjugados complejos (p. 251) Los pares de números de la forma $a + bi$ y $a - bi$ son conjugados complejos.

Example The complex numbers $2 - 3i$ and $2 + 3i$ are complex conjugates.

Complex fraction (p. 536) A complex fraction is a rational expression that has a fraction in its numerator or denominator, or in both its numerator and denominator.

Fracción compleja (p. 536) Una fracción compleja es una expresión racional en la que el numerador, el denominador o ambos son una fracción.

Example $\frac{2}{\frac{1}{5}}, \frac{\frac{2}{7}}{\frac{3}{2}}$

Complex number (p. 249) Complex numbers are the real numbers and the imaginary numbers.

Número complejo (p. 249) Los números complejos son los números reales y los números imaginarios.

Example $6 + i$

$7, 2i$

Complex number plane (p. 249) The complex number plane is identical to the coordinate plane except each ordered pair (a, b) represents the complex number $a + bi$. The horizontal axis is the Real axis. The vertical axis is the Imaginary axis.

Plano de números complejos (p. 249) El plano de los números complejos es idéntico al plano de coordenadas, a excepción de que cada par ordenado (a, b) representa el número complejo $a + bi$. El eje horizontal es el eje real. El eje vertical es el eje imaginario.

Example

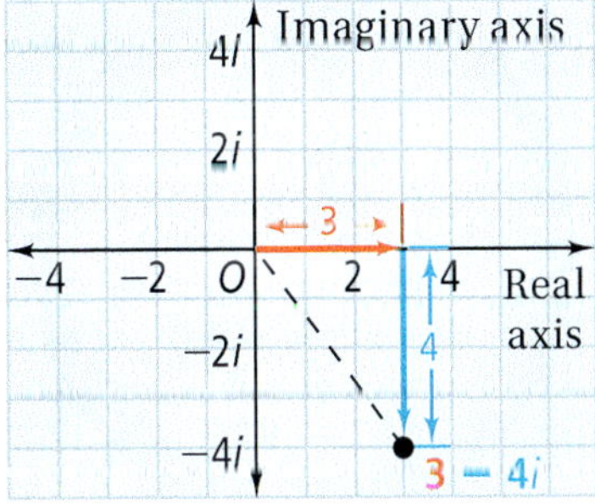

Visual Glossary

English	Spanish

Composite function (p. 399) A composite function is a combination of two functions such that the output from the first function becomes the input for the second function.

Función compuesta (p. 399) Una función compuesta es la combinación de dos funciones. La cantidad de salida de la primera función es la cantidad de entrada de la segunda función.

Example $f(x) = 2x + 1,\ g(x) = x^2 - 1$

$$\begin{aligned}(g \circ f)(5) &= g(f(5)) = g(2(5) + 1)\\ &= g(11)\\ &= 11^2 - 1 = 120\end{aligned}$$

Compound inequality (p. 36) You can join two inequalities with the word *and* or the word *or* to form a compound inequality.

Desigualdad compuesta (p. 36) Puedes unir dos desigualdades por medio de la palabra *y* o la palabra *o* para formar una desigualdad compuesta.

Example $-1 < x$ and $x \leq 3$

$x < -1$ or $x \geq 3$

Conditional probability (p. 696) A conditional probability contains a condition that may limit the sample space for an event. The notation $P(B|A)$ is read "the probability of event B, given event A." For any two events A and B in the sample space, $P(B|A) = \frac{P(A \text{ and } B)}{P(A)}$.

Probabilidad condicional (p. 696) Una probabilidad condicional contiene una condición que puede limitar el espacio muestral de un suceso. La notación $P(B|A)$ se lee "la probabilidad del suceso B, dado el suceso A". Para dos sucesos cualesquiera A y B en el espacio muestral, $P(B|A) = \frac{P(A \text{ y } B)}{P(A)}$.

Example

$$\begin{aligned}&= \frac{P(\text{departs and arrives on time})}{P(\text{departs on time})}\\ &= \frac{0.75}{0.83}\\ &\approx 0.9\end{aligned}$$

Confidence interval (p. 746) Based on the mean of a sample or a sample proportion, the confidence interval indicates the interval in which the population mean or population proportion is likely to lie for a given confidence level.

Intervalo de confianza (p. 746) El intervalo de confianza se basa en la media de una muestra o en la proporción de una muestra, e indica el intervalo en el que probablemente se encuentra dicha media o proporción de la población para un nivel de confianza dado.

Example For an elementary history book, a sample of 30 trials indicates that the mean number of words in a sentence is 12.7. The margin of error at a 95% confidence level is 1.5 words per sentence. The mean number of words μ in all of the sentences in the book at a 95% confidence level is $12.7 - 1.5 \leq \mu \leq 12.7 + 1.5$.

English | Spanish

Conic section (p. 614) A conic section is a curve formed by the intersection of a plane and a double cone.

Sección cónica (p. 614) Una sección cónica es una curva que se forma por la intersección de un plano con un cono doble.

Example

ellipse

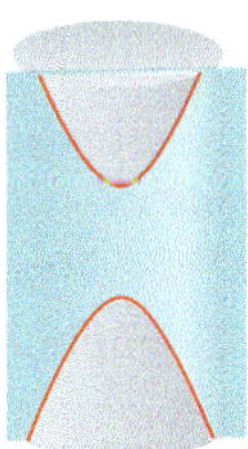
hyperbola

Conjugate axis (p. 646) The conjugate axis for the hyperbola $\frac{x^2}{a^2} - \frac{y^2}{b^2} = 1$, $a > b > 0$, is the segment from $(0, -b)$ to $(0, b)$. For $\frac{y^2}{a^2} - \frac{x^2}{b^2} = 1$, the conjugate axis is the segment from $(-b, 0)$ to $(b, 0)$.

Eje conjugado (p. 646) El eje conjugado de la hipérbola $\frac{x^2}{a^2} - \frac{y^2}{b^2} = 1$, $a > b > 0$, es el segmento desde el punto $(0, -b)$ hasta el punto $(0, b)$. Para $\frac{y^2}{a^2} - \frac{x^2}{b^2} = 1$, el eje conjugado es el segmento desde el punto $(-b, 0)$ hasta el punto $(b, 0)$.

Conjugate Root Theorem (p. 314) If $P(x)$ is a polynomial with rational coefficients, then the irrational roots of $P(x) = 0$ occur in conjugate pairs. That is, if $a + \sqrt{b}$ is an irrational root with a and b rational, then $a - \sqrt{b}$ is also a root. If $P(x)$ is a polynomial with real coefficients, then the complex roots of $P(x) = 0$ occur in conjugate pairs. That is, if $a + bi$ is a complex root with a and b real, then $a - bi$ is also a root.

Teorema de raíces conjugadas (p. 314) Si $P(x)$ es un polinomio con coeficientes racionales, entonces las raíces irracionales de $P(x) = 0$ ocurren en pares conjugados. Es decir, si $a + \sqrt{b}$ es una raíz irracional donde a y b son racionales, entonces $a - \sqrt{b}$ también es una raíz. Si $P(x)$ es un polinomio con coeficientes reales, entonces las raíces complejas de $P(x) = 0$ ocurren en los pares conjugados. Es decir, si $a + bi$ es una raíz compleja donde a y b son reales, entonces $a - bi$ también es una raíz.

Conjugates (p. 314) Number pairs of the form $a + \sqrt{b}$ and $a - \sqrt{b}$ are conjugates.

Conjugados (p. 314) Los pares de números con la forma $a + \sqrt{b}$ y $a - \sqrt{b}$ son conjugados.

Example $5 + \sqrt{3}$ and $5 - \sqrt{3}$ are conjugates.

Consistent system (p. 137) A system of linear equations is consistent if it has at least one solution.

Sistema consistente (p. 137) Un sistema de ecuaciones lineales es consistente si tiene por lo menos una solución.

Constant (p. 5) A constant is a quantity whose value does not change.

Constante (p. 5) Una constante es una cantidad cuyo valor no cambia.

Constant matrix (p. 793) When representing a system of equations with a matrix equation, the matrix containing the constants of the system is the constant matrix.

Matriz de constantes (p. 793) Al representar un sistema de ecuaciones con una ecuación matricial, la matriz que contiene las constantes del sistema es la matriz de constantes.

Example $\begin{cases} x + 2y = 5 \\ 3x + 5y = 14 \end{cases}$

constant matrix $\begin{bmatrix} 5 \\ 14 \end{bmatrix}$

English	Spanish
Constant of proportionality (p. 341) If $y = ax^b$ describes y as a power function of x, then y varies directly with, or is proportional to, the bth power of x. The constant a is the constant of proportionality.	**Constante de proporcionalidad (p. 341)** Si $y = ax^b$ describe a y como una potencia de la función de x, entonces y varía directamente con, o es proporcional a, la b^{ma} potencia de x. La constante a es la constante de proporcionalidad.
Constant of variation (p. 68) The constant of variation is the ratio of the two variables in a direct variation and the product of the two variables in an inverse variation.	**Constante de variación (p. 68)** La constante de variación es la razón de dos variables en una variación directa y el producto de las dos variables en una variación inversa.

Example In $y = 3.5x$, the constant of variation k is 3.5. In $xy = 5$, the constant of variation k is 5.

English	Spanish
Constant term (p. 20) A constant term is a term with no variables.	**Término constante (p. 20)** Un término constante es un término que no tiene variables.
Constraint (p. 157) Constraints are restrictions on the variables of the objective function in a linear programming problem. *See* **Linear programming.**	**Restriccion (p. 157)** Las restricciones son limitaciones a las variables de una función objetiva en un problema de programación lineal. *Ver* **Linear programming.**
Continuous graph (p. 516) A graph is continuous if it has no jumps, breaks, or holes.	**Gráfica continua (p. 516)** Una gráfica es continua si no tiene saltos, interrupciones o huecos.
Continuous probability distribution (p. 739) A continuous probability distribution has as its events any of the infinitely many values in an interval of real numbers.	**Distribución de probabilidad continua (p. 739)** Una distribución de probabilidad continua tiene como sucesos a cualquiera del número infinito de valores en un intervalo de números reales.
Continuously compounded interest (p. 446) When interest is compounded continuously on principal P, the value A of an account is $A = Pe^{rt}$.	**Interés compuesto continuo (p. 446)** En un sistema donde el interés es compuesto continuamente sobre el capital P, el valor de A de una cuenta es $A = Pe^{rt}$.

Example Suppose that $P = \$1200$, $r = 0.05$, and $t = 3$. Then

$$\begin{aligned} A &= 1200e^{0.05 \cdot 3} \\ &= 1200(2.718\ldots)^{0.15} \\ &\approx 1394.20 \end{aligned}$$

English	Spanish
Controlled experiment (p. 726) In a controlled experiment, you divide the sample into two groups. You impose a treatment on one group but not the other "control" group. Then you compare the effect on the treated group to the control group.	**Experimento controlado (p. 726)** En un experimento controlado, se divide la muestra en dos grupos. Uno de los grupos se manipula y el otro grupo "controlado" se mantiene en su estado original. Luego se comparan el estado del grupo manipulado y el estado del grupo controlado.
Convenience sample (p. 725) In a convenience sample you select any members of the population who are conveniently and readily available.	**Muestra de conveniencia (p. 725)** En una muestra de conveniencia se selecciona a cualquier miembro de la población que está convenientemente disponible.

English	Spanish
Converge (p. 598) An infinite series $a_1 + a_2 + \cdots + a_n + \cdots$ converges if the sum $a_1 + a_2 + \cdots + a_n$ get closer and closer to a real number as n increases.	**Convergir (p. 598)** Una serie infinita $a_1 + a_2 + \cdots + a_n + \cdots$ es convergente si la suma $a_1 + a_2 + \cdots + a_n$ se aproxima cada vez más a un número real a medida que el valor de n incrementa.

Example $1 + \frac{1}{2} + \frac{1}{4} + \frac{1}{8} + \cdots$ converges.

English	Spanish
Coordinate space (p. 164) Coordinate space is a three-dimensional space where each point is described uniquely using an ordered triple of numbers.	**Espacio de coordenadas (p. 164)** Un espacio de coordenadas es un espacio tridimensional en el cual cada punto es definido de manera única por una tripleta ordenada de números.

Example

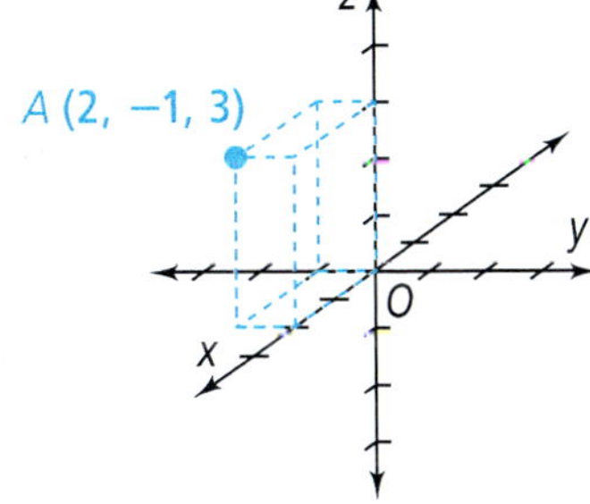

English	Spanish
Correlation (p. 92) A correlation indicates the strength of a relationship between two data sets.	**Correlación (p. 92)** Una correlación indica la fuerza de una relación entre dos conjuntos de datos.
Correlation coefficient (p. 94) The correlation coefficient, r, indicates the strength of the correlation. The closer r is to 1 or -1, the more closely the data resembles a line and the more accurate your model is likely to be.	**Coeficiente de correlación (p. 94)** El coeficiente de correlación, r, indica la fuerza de la correlación. Mientras más cerca está r de 1 ó -1, más se parecen los datos a una línea y será más probable que tu modelo sea preciso.
Corresponding elements (p. 764) Corresponding elements are elements in the same position in each matrix.	**Elementos correspondientes (p. 764)** Los elementos correspondientes son elementos que se encuentran en la misma posición de cada matriz.
Cosecant function (p. 883) The cosecant (csc) function is the reciprocal of the sine function. For all real numbers θ except those that make $\sin \theta = 0$, $\csc \theta = \frac{1}{\sin \theta}$.	**Función cosecante (p. 883)** La función cosecante (csc) se define como el recíproco de la función seno. Para todos los números reales θ, excepto aquéllos para los que $\sin \theta = 0$, $\csc \theta = \frac{1}{\sin \theta}$.

Example If $\sin \theta = \frac{5}{13}$, then $\csc \theta = \frac{13}{5}$.

Visual **Glossary**

English	Spanish

Cosine function, Cosine of θ (pp. 838, 861) The cosine function, $y = \cos \theta$, matches the measure θ of an angle in standard position with the x-coordinate of a point on the unit circle. This point is where the terminal side of the angle intersects the unit circle. The x-coordinate is the cosine of θ.

Función coseno, Coseno de θ (pp. 838, 861) La función coseno, $y = \cos \theta$, empareja la medida θ de un ángulo en posición estándar con la coordenada x de un punto en el círculo unitario. Este es el punto en el que el lado terminal del ángulo interseca al círculo unitario. La coordenada x es el coseno de θ.

Example

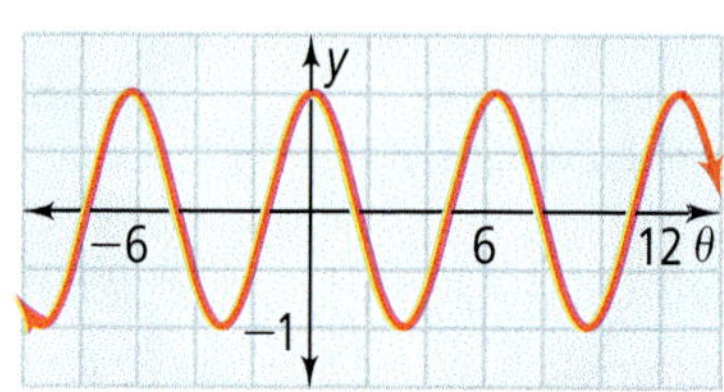

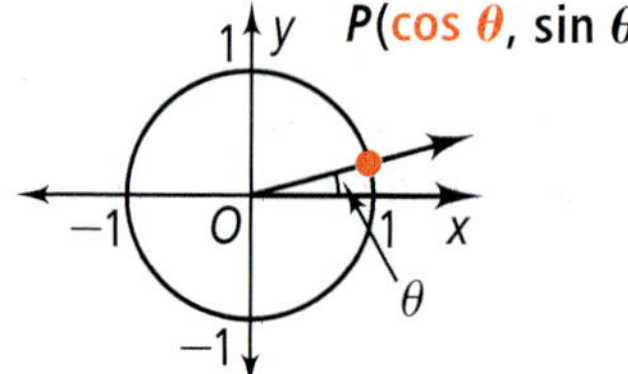

Cotangent function (p. 883) The cotangent (cot) function is the reciprocal of the tangent function. For all real numbers θ except those that make $\tan \theta = 0$, $\cot \theta = \frac{1}{\tan \theta}$.

Función cotangente (p. 883) La función cotangente (cot) es el recíproco de la función tangente. Para todos los números reales θ, excepto aquéllos para los que $\tan \theta = 0$, $\cot \theta = \frac{1}{\tan \theta}$.

Example If $\tan \theta = \frac{5}{12}$, then $\cot \theta = \frac{12}{5}$.

Coterminal angle (p. 837) Two angles in standard position are coterminal if they have the same terminal side.

Ángulo coterminal (p. 827) Dos ángulos que están en posición normal son coterminales si tienen el mismo lado terminal.

Example

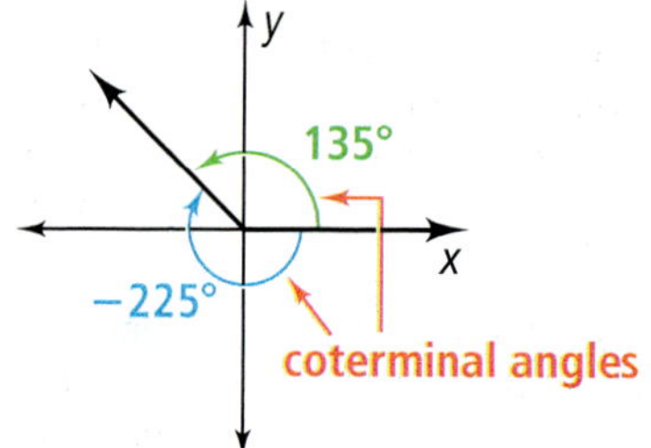

Angles that have measures 135° and −225° are coterminal.

Co-vertices (p. 639) The endpoints of the minor axis of an ellipse are the co-vertices of the ellipse.

Covértices (p. 639) Los puntos de intersección entre una elipse y los ejes menores son los covértices de la elipse.

Example

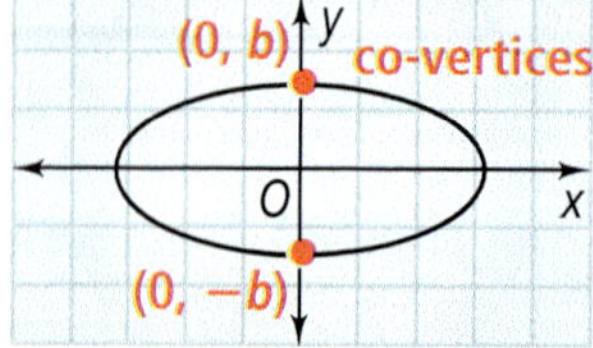

Cumulative probability (p. 695) Probability over a continuous range of events is cumulative probability.

Probabilidad acumulativa (p. 695) La probabilidad que existe a lo largo de una serie continua de sucesos es la probabilidad acumulativa.

English

Spanish

Cycle (p. 828) A cycle of a periodic function is an interval of x-values over which the function provides one complete pattern of y-values.

Ciclo (p. 828) El ciclo de una función periódica es un intervalo de valores de x de los cuales la función produce un patrón completo de valores de y.

Example

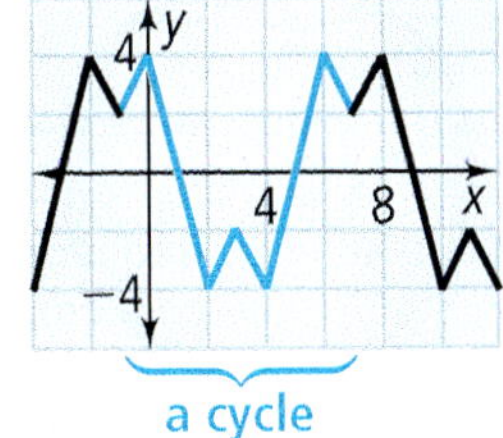

D

Decay factor (p. 436) In an exponential function of the form $y = ab^x$, b is the decay factor if $0 < b < 1$.

Factor de decremento (p. 436) En una función exponencial de la forma $y = ab^x$, b es el factor de decremento si, $0 < b < 1$.

Example In the equation $y = 0.3^x$, 0.7 is the decay factor.

Degree of a monomial (p. 280) The degree of a monomial in one variable is the exponent of the variable.

Grado de un monomio (p. 280) El grado de un monomio en una variable es el exponente de la variable.

Degree of a polynomial (p. 280) The degree of a polynomial is the greatest degree among its monomial terms.

Grado de un polinomio (p. 280) El grado de un polinomio es el grado mayor entre los términos de monomios.

Example $P(x) = x^6 + 2x^3 - 3$ degree 6

Dependent events (p. 688) Two events are dependent if the occurrence of one event affects the probability of the second event.

Sucesos dependientes (p. 688) Cuando el resultado de un suceso influye en la probabilidad de que ocurra el segundo suceso, los dos sucesos son dependientes.

Example You have a bag with red and blue marbles. You draw one marble at random and then another without replacing the first. The colors drawn are dependent events. A red marble on the first draw changes the probability for each color on the second draw.

Dependent system (p. 137) A system of equations that does not have a unique solution is a dependent system.

Sistema dependiente (p. 137) Un sistema de ecuaciones es dependiente cuando no tiene una solución única.

Example $\begin{cases} y = 2x + 3 \\ -4x + 2y = 6 \end{cases}$ represents two equations for the same line, so it has many solutions. It is a dependent system.

English	Spanish

Dependent variable (p. 63) If a function is defined by an equation using the variables x and y, where y represents output values, then y is the dependent variable.

Variable dependiente (p. 63) Si una función es definida por una ecuación que usa las variables x e y, donde y representa valores de salida, entonces y es la variable dependiente.

Example $y = 2x + 1$
y is the dependent variable.

Descartes' Rule of Signs (p. 315) Let $P(x)$ be a polynomial with real coefficients written in standard form.
– The number of positive real roots of $P(x) = 0$ is either equal to the number of sign changes between consecutive coefficients of $P(x)$ or is less than that by an even number;
– The number of negative real roots of $P(x) = 0$ is either equal to the number of sign changes between consecutive coefficients of $P(-x)$ or is less than that by an even number. (Count multiple roots according to their multiplicity.)

Regla de los signos de Descartes (p. 315) Sea $P(x)$ un polinomio con coeficientes reales escritos en forma normal.
– El número de raíces positivas reales de $P(x) = 0$ es igual al número de cambios de signos entre coeficientes consecutivos de $P(-x)$ o es menor que eso en un número par;
– El número de raíces negativas reales de $P(x) = 0$ es igual al número de cambios de signos entre coeficientes consecutivos de $P(-x)$ o es menor que eso en un número par. (Cuenta las raíces múltiples según su multiplicidad).

Determinant (p. 784) The determinant of a square matrix is a real number that can be computed from its elements according to a specific formula.

Determinante (p. 784) El determinante de una matriz cuadrada es un número real que se puede calcular a partir de sus elementos por medio de una fórmula específica.

Example The determinant of $\begin{bmatrix} 3 & -2 \\ 5 & 6 \end{bmatrix}$ is $3(6) - 5(-2) = 28$.

Difference of cubes (p. 297) A difference of cubes is an expression of the form $a^3 - b^3$. It can be factored as $(a - b)(a^2 + ab + b^2)$.

Diferencia de dos cubos (p. 297) La diferencia de dos cubos es una expresión de la forma $a^3 - b^3$. Se puede factorizar como $(a - b)(a^2 + ab + b^2)$.

Example $x^3 - 27 = (x - 3)(x^2 + 3x + 9)$

Difference of two squares (p. 220) A difference of two squares is an expression of the form $a^2 - b^2$. It can be factored as $(a + b)(a - b)$.

Diferencia de dos cuadrados (p. 220) La diferencia de dos cuadrados es una expresión de la forma $a^2 - b^2$. Se puede factorizar como $(a + b)(a - b)$.

Example $25a^2 - 4 = (5a + 2)(5a - 2)$
$m^6 - 1 = (m^3 + 1)(m^3 - 1)$

Dilation (p. 802) A dilation is a transformation that can change the size of a figure. When the center of the dilation is the origin, you can use scalar multiplication to find the coordinates of the vertices of an image.

Dilatación (p. 802) Una dilatación es una transformación que puede cambiar el tamaño de una figura. Cuando el centro de dilatación está en el origen, se hallan las coordenadas de los vértices de la imagen por medio de la multiplicación de escalar.

Example

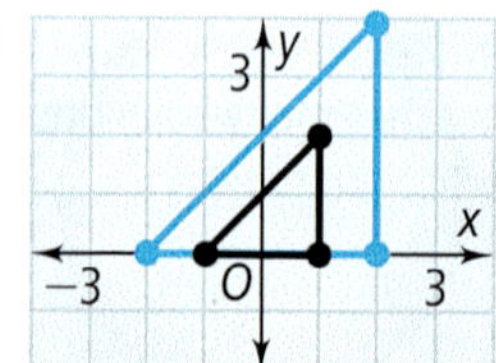

English

Spanish

Direct variation (p. 68) A linear function defined by an equation of the form $y = kx$, where $k \neq 0$, represents direct variation.

Variación directa (p. 68) Una función lineal definida por una ecuación de la forma $y = kx$, donde $k \neq 0$, representa una variación directa.

Example $y = 3.5x$, $y = 7x$, $y = -\frac{1}{2}x$

Directrix (p. 622) The directrix of a parabola is the fixed line used to define a parabola. Each point of the parabola is the same distance from the focus and the directrix.

Directriz (p. 622) La directriz de una parábola es la recta fija con que se define una parábola. Cada punto de la parábola está a la misma distancia del foco y de la directriz.

Example

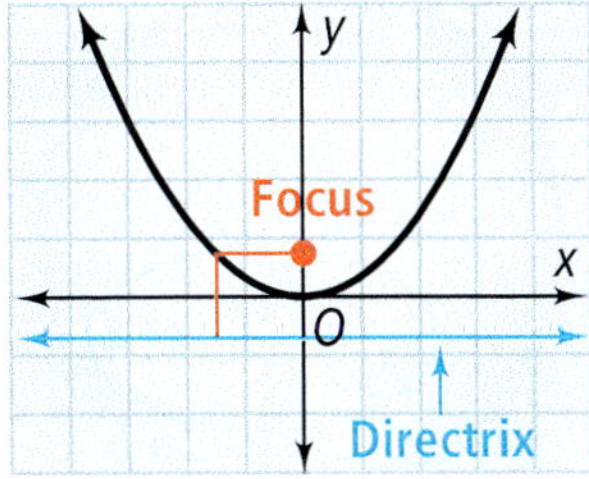

Discontinuous graph (p. 516) A graph is discontinuous if it has a jump, break, or hole.

Gráfica discontinua (p. 516) Una gráfica es discontinua si tiene un salto, interrupción o hueco.

Discrete probability distribution (p. 739) A discrete probability distribution has a finite number of Possible events.

Distribución de probabilidad discreta (p. 739) Una distribución de probabilidad discreta tiene un número finito de sucesos posibles.

Discriminant (p. 242) The discriminant of a quadratic equation in the form $ax^2 + bx + c = 0$ is the value of the expression $b^2 - 4ac$.

Discriminante (p. 242) El discriminante de una ecuación cuadrática en la forma $ax^2 + bx + c = 0$ es el valor de la expresión $b^2 - 4ac$.

Example $3x^2 - 6x + 1$

$$\text{discriminant} = (-6)^2 - 4(3)(1) = 36 - 12 = 24$$

Diverge (p. 598) An infinite series diverges if it does not converge.

Divergir (p. 598) Una serie infinita es divergente si no es convergente.

Example $1 + 2 + 4 + 8 + \cdots$ diverges.

Domain (p. 61) The domain of a relation is the set of all inputs, or x-coordinates, of the ordered pairs.

Dominio (p. 61) El dominio de una relación es el conjunto de todos los valores de entrada, o coordenadas x, de los pares ordenados.

Examples In the relation $\{(0, 1), (0, 2), (0, 3), (0, 4), (1, 3), (1, 4), (2, 1)\}$, the domain is $\{0, 1, 2\}$. In the function $f(x) = x^2 - 10$, the domain is all real numbers.

English	Spanish
Dot product (p. 812) Given vectors $\mathbf{v} = \langle v_1, v_2 \rangle$ and $\mathbf{w} = \langle w_1, w_2 \rangle$, the dot product $\mathbf{v} \cdot \mathbf{w}$ is the quantity $v_1w_1 + v_2w_2$.	**Producto escalar (p. 812)** Dados los vectores $\mathbf{v} = \langle v_1, v_2 \rangle$ y $\mathbf{w} = \langle w_1, w_2 \rangle$, el producto escalar $\mathbf{v} \cdot \mathbf{w}$ es la suma $v_1w_1 + v_2w_2$.

E

Ellipse (p. 638) An ellipse is the set of points P in a plane such that the sum of the distances from P to two fixed points $F1$ and $F2$ is a given constant k. The standard form of the equation of an ellipse with its center at the origin is $\frac{x^2}{a^2} + \frac{y^2}{b^2} = 1$ if the major axis is horizontal and $\frac{x^2}{b^2} + \frac{y^2}{a^2} = 1$ if the major axis is vertical, where $a > b$.

Elipse (p. 638) Una elipse es el conjunto de puntos P situados en un plano tal que la suma de las distancias entre P y dos puntos fijos F_1 y F_2 es una constante dada k. La forma normal de la ecuación de una elipse con su centro en el origen es $\frac{x^2}{a^2} + \frac{y^2}{b^2} = 1$ si el eje mayor es horizontal y $\frac{x^2}{b^2} + \frac{y^2}{a^2} = 1$ si el eje mayor es vertical, donde $a > b$.

Example

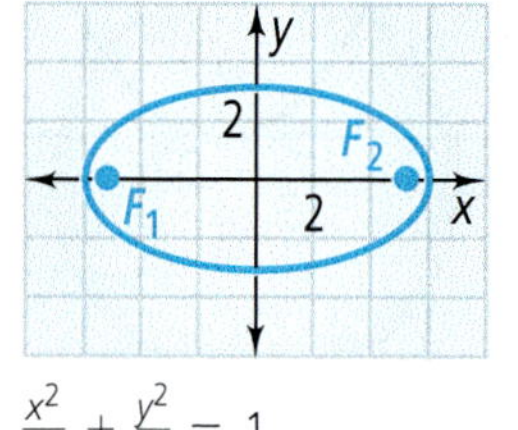

$\frac{x^2}{36} + \frac{y^2}{9} = 1$

$F_1 = (-3\sqrt{3}, 0)$, $F_2 = (3\sqrt{3}, 0)$

End behavior (p. 282) End behavior of the graph of a function describes the directions of the graph as you move to the left and to the right, away from the origin.

Comportamiento extremo (p. 282) El comportamiento extremo de la gráfica de una función describe las direcciones de la gráfica al moverse a la izquierda y a la derecha, apartándose del origen.

Equal matrices (p. 767) Equal matrices are matrices with the same dimensions and equal corresponding elements.

Matrices equivalentes (p. 767) Dos matrices son equivalentes si y sólo si tienen las mismas dimensiones y sus elementos correspondientes son iguales.

Example Matrices A and B are equal.

$$A = \begin{bmatrix} 2 & 6 \\ \frac{9}{3} & 1 \end{bmatrix} \quad B = \begin{bmatrix} \frac{6}{3} & 6 \\ 3 & \frac{-13}{-13} \end{bmatrix}$$

Equally likely outcomes (p. 682) Equally likely outcomes are events in a sample space that have the same chance of occurring.

Resultados igualmente probables (p. 682) Resultados igualmente probables son sucesos en un espacio muestral con la misma probabilidad de ocurrir.

Equation (p. 26) An equation is a statement that two algebraic expressions are equal.

Ecuación (p. 26) Una ecuación es un enunciado que describe dos expresiones algebraicas iguales.

Equivalent systems (p. 144) Equivalent systems are systems that have the same solution(s).

Sistemas equivalentes (p. 144) Sistemas equivalentes son sistemas que tienen la misma solución o las mismas soluciones.

English	Spanish

Evaluate (p. 19) To evaluate an algebraic expression, substitute a number for each variable in the expression. Then simplify using the order of operations.

Evaluar (p. 19) Para evaluar una expresión algebraica, sustituye cada variable de la expresión con un número. Luego, simplifica usando el orden de operaciones.

Example When $x = 2$ and $y = -1$, $2x + 3y$ evaluates to 1.

Expand (p. 326) To expand the power of a binomial, multiply as needed, then write the polynomial in standard form.

Expandir (p. 326) Para expandir la potencia de un binomio, multiplica como sea necesario. Luego, escribe el polinomio en forma normal.

Example

$$\begin{aligned}(x+4)^3 &= (x+4)(x+4)^2\\ &= (x+4)(x^2+8x+16)\\ &= x^3+8x^2+16x+4x^2+32x+64\\ &= x^3+12x^2+48x+64\end{aligned}$$

Experimental probability (p. 681) The experimental probability of an event is the ratio $\frac{\text{number of times the event occurs}}{\text{number of trials}}$.

Probabilidad experimental (p. 681) La probabilidad experimental de un suceso es la razón $\frac{\text{number of times the event occurs}}{\text{number of trials}}$.

Example Suppose a basketball player has scored 19 times in 28 attempts at a basket. The experimental probability of the player's scoring is $P(\text{score}) = \frac{19}{28} \approx 0.68$, or 68%.

Explicit formula (p. 565) An explicit formula expresses the nth term of a sequence in terms of n.

Fórmula explícita (p. 565) Una fórmula explícita expresa el n-ésimo término de una progresión en función de n.

Example Let $a_n = 2n + 5$ for positive integers n. If $n = 7$, then $a_7 = 2(7) + 5 = 19$.

Exponential decay (p. 435) Exponential decay is modeled by a function of the form $y = ab^x$ with $0 < b < 1$.

Decaimiento exponencial (p. 435) El decaimiento exponencial se expresa con una función $y = ab^x$ donde $0 < b < 1$.

Exponential equation (p. 469) An exponential equation contains the form b^{cx}, with the exponent including a variable.

Ecuación exponencial (p. 469) Una ecuación exponencial tiene la forma b^{cx}, y su exponente incluye una variable.

Example

$$\begin{aligned}5^{2x} &= 270\\ \log 5^{2x} &= \log 270\\ 2x \log 5 &= \log 270\\ 2x &= \frac{\log 270}{\log 5}\\ 2x &\approx 3.4785\\ x &\approx 1.7392\end{aligned}$$

English	Spanish

Exponential function (p. 434) The general form of an exponential function is $y = ab^x$, where x is a real number, $a \neq 0$, $b > 0$, and $b \neq 1$. When $b > 1$, the function models exponential growth with growth factor b. When $0 < b < 1$, the function models exponential decay with decay factor b.

Función exponencial (p. 434) La forma general de una función exponencial es $y = ab^x$, donde x es un número real, $a \neq 0$, $b > 0$ y $b \neq 1$. Cuando $b > 1$, la función representa un incremento exponencial con factor de incremento b. Cuando $0 < b < 1$, la función representa el decremento exponencial con factor de decremento b.

Example

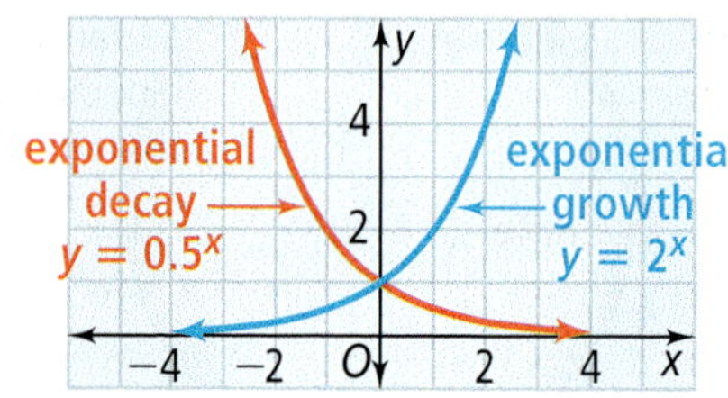

Exponential growth (p. 435) Exponential growth is modeled by a function of the form $y = ab^x$ with $b > 1$.

Crecimiento exponencial (p. 435) El crecimiento exponencial se expresa con una función de la forma $y = ab^x$ donde $b > 1$.

Extraneous solution (p. 42) An extraneous solution is a solution of an equation derived from an original equation but it is not a solution of the original equation.

Solución extraña (p. 42) Una solución extraña es una solución de una ecuación derivada de una ecuación dada, pero que no satisface la ecuación dada.

Example

$$\sqrt{x - 3} = x - 5$$
$$x - 3 = x^2 - 10x + 25$$
$$0 = x^2 - 11x + 28$$
$$0 = (x - 4)(x - 7)$$
$$x = 4 \text{ or } 7$$

The number 7 is a solution, but 4 is not, since $\sqrt{4 - 3} \neq 4 - 5$.

F

Factor Theorem (p. 289) The expression $x - a$ is a linear factor of a polynomial if and only if the value of a is a root of the related polynomial function.

Teorema de factores (p. 289) La expresión $x - a$ es un factor lineal de un polinomio si y sólo si el valor de a es una raíz de la función polinomial con la que se relaciona.

Example The value 2 makes the polynomial $x^2 + 2x - 8$ equal to zero. So, $x - 2$ is a factor of $x^2 + 2x - 8$.

Factoring (p. 216) Factoring is rewriting an expression as the product of its factors.

Descomposición factorial (p. 216) Descomponer en factores es el proceso de escribir de nuevo una expresión como el producto de sus factores.

Example

expanded form	factored form
$x^2 + x - 56$	$(x + 8)(x - 7)$

Feasible region (p. 157) In a linear programming problem, the feasible region contains all the values that satisfy the constraints on the objective function.

Región factible (p. 157) En un problema de programación lineal, la región factible contiene todos los valores que satisfacen las restricciones de la función objetiva.

English	Spanish
Finite Series (p. 587) A finite series is a series with a finite number of terms.	**Serie finite (p. 587)** Una serie finita es una serie con un número finito de términos.
Focal length (p. 622) The focal length of a parabola is the distance between the vertex and the focus.	**Distancia focal (p. 622)** La distancia focal de una parábola es la distancia entre el vértice y el foco.
Focus (plural: foci) of a hyperbola (p. 645) A hyperbola is the set of all points P in a plane such that the difference of the distances from P to two fixed points is constant. Each of the fixed points is a focus of the hyperbola.	**Foco de una hipérbola (p. 645)** Una hipérbola es el conjunto de puntos P en un plano tal que la diferencia de las distancias desde P hasta dos puntos fijos es constante. Cada uno de los puntos fijos es el foco de la hipérbola.
Focus (plural: foci) of a parabola (p. 622) A parabola is the set of all points in a plane that are the same distance from a fixed line and a fixed point not on the line. The fixed point is the focus of the parabola.	**Foco de una parabola (p. 622)** Una parábola es el conjunto de todos los puntos en un plano con la misma distancia desde una línea fija y un punto fijo que no permanece en la línea. El punto fijo es el foco de la parábola.
Focus (plural: foci) of an ellipse (p. 638) An ellipse is the set of all points P in a plane such that the sum of the distances from P to two fixed points is constant. Each of the fixed points is a focus of the ellipse.	**Foco de una elipse (p. 638)** Una elipse es el conjunto de todos los puntos P en un plano en el cual la suma de las distancias desde P hasta dos puntos fijos es constante. Cada uno de estos puntos fijos es un foco de la elipsis.
Frequency table (p. 694) A frequency table is a list of the outcomes in a sample space and the number of times each outcome occurs.	**Tabla de frecuencias (p. 694)** Una tabla de frecuencias es una lista de los resultados de un espacio muestral y el número de veces que cada resultado ocurre.
Function (p. 62) A function is a relation in which each element of the domain corresponds with exactly one element in the range.	**Función (p. 62)** Una función es una relación en la que cada elemento del dominio corresponde exactamente con un elemento del rango.

Example The relation $y = 3x^3 - 2x + 3$ is a function. $f(x) = 3x^3 - 2x + 3$ is the same relation written in function notation.

English	Spanish
Function notation (p. 63) If f is the name of a function, the function notation $f(x)$ shows the function name f and also represents the range value $f(x)$ for the domain value x. You read the function notation $f(x)$ as "f of x" or "a function of x." Note that $f(x)$ does *not* mean "f times x."	**Notacion de una funcion (p. 63)** Si f es el nombre de una función, la notación de la función $f(x)$ indica el nombre de la función y también representa el valor del rango $f(x)$ para el valor del dominio x. La función de la notación $f(x)$ se lee "f de x" o "una función de x." Observa que $f(x)$ *no* significa "f por x".

Example When the value of x is 3, $f(3)$, read "f of 3," represents the value of the function at 3.

English	Spanish
Function rule (p. 63) A function rule represents an output value in terms of an input value.	**Regla de función (p. 63)** Una regla de función representa un valor de salida en función a un valor de entrada.
Fundamental Counting Principle (p. 674) The Fundamental Counting Principle is a tool that you can use to quickly count the number of ways certain things can happen.	**Principio básico de conteo (p. 674)** El principio básico de conteo es una herramienta que se puede utilizar para hacer un conteo rápido del número de formas en que pueden ocurrir ciertas cosas.

English / Spanish

Fundamental Theorem of Algebra (p. 320) If $P(x)$ is a polynomial of degree $n \geq 1$ with complex coefficients, then $P(x) = 0$ has at least one complex root.

Teorema fundamental de álgebra (p. 320) Si $P(x)$ es un polinomio de grado $n \geq 1$ con coeficientes complejos, entonces $P(x) = 0$ tiene por lo menos una raíz compleja.

Example $P(x) = 3x^3 - 2x + 3$ is of degree 3, so $P(x) = 0$ has at least one complex root.

Geometric mean (p. 583) The geometric mean of any two positive numbers is the positive square root of the product of the two numbers.

Media geométrica (p. 583) La media geométrica de dos números positivos es la raíz cuadrada positiva del producto de los dos números.

Example The geometric mean of 12 and 18 is $\sqrt{12 \cdot 18} \approx 14.6969$.

Geometric sequence (p. 580) A geometric sequence is a sequence with a constant ratio between consecutive terms.

Secuencia geométrica (p. 580) Una secuencia geométrica es una secuencia con una razón constante entre términos consecutivos.

Example The geometric sequence 2.5, 5, 10, 20, 40 . . . , has a common ratio of 2.

Geometric series (p. 595) A geometric series is the sum of the terms in a geometric sequence.

Serie geométrica (p. 595) Una serie geométrica es la suma de términos en una progresión geométrica.

Example One geometric series with five terms is $2.5 + 5 + 10 + 20 + 40$.

Greatest common factor (p. 218) The greatest common factor (GCF) of an expression is the common factor of each term of the expression that has the greatest coefficient and the greatest exponent.

Máximo factor común (p. 218) El máximo factor común de una expresión es el factor común de cada término de la expresión que tiene el mayor coeficiente y el mayor exponente.

Example The GCF of $4x^2 + 20x - 12$ is 4.

Greatest integer function (p. 90) The greatest integer function corresponds each input x to the greatest integer less than or equal to x.

Función del entero mayor (p. 90) La función del entero mayor relaciona cada entrada x con el entero mayor que es menor o igual a x.

Growth factor (p. 436) In an exponential function of the form $y = ab^x$, b is the growth factor if $b > 1$.

Factor de incremento (p. 436) En una función exponencial de la forma $y = ab^x$, b es el factor de incremento si $b > 1$.

Example In the exponential equation $y = 2^x$, 2 is the growth factor.

Half-plane (p. 114) A half-plane is the set of points in a coordinate plane that are on one side of the boundary of the graph of a linear inequality.

Semiplano (p. 114) Un semiplano es el conjunto de puntos de un plano de coordenadas que están a un lado del límite de la gráfica de desigualdad lineal.

English	Spanish

Hyperbola (p. 645) A hyperbola is a set of points P in a plane such that the difference between the distances from P to the foci F_1 and F_2 is a given constant k. $|PF_1 - PF_2| = k$ The standard form of an equation of a hyperbola centered at (0, 0) is $\frac{x^2}{a^2} - \frac{y^2}{b^2} = 1$ if the transverse axis is horizontal and $\frac{y^2}{a^2} - \frac{x^2}{b^2} = 1$ if the transverse axis is vertical.

Hipérbola (p. 645) Una hipérbola es un conjunto de puntos P en un plano tal que la diferencia entre las distancias de P a los focos F_1 y F_2 es una constante k dada. $|PF_1 - PF_2| = k$ La forma normal de la ecuación de una hipérbola centrada en (0, 0) es $\frac{x^2}{a^2} - \frac{y^2}{b^2} = 1$, si el eje transversal es horizontal, y $\frac{y^2}{a^2} - \frac{x^2}{b^2} = 1$, si el eje transversal es vertical.

Example

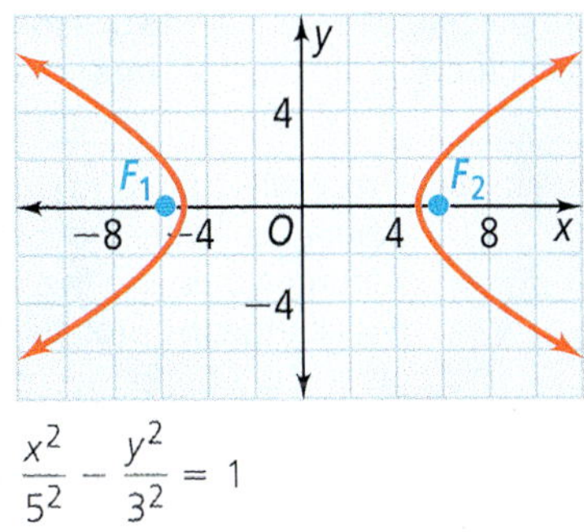

$$\frac{x^2}{5^2} - \frac{y^2}{3^2} = 1$$

I

***i* (p. 248)** The imaginary number i is the principal square root of -1.

***i* (p. 248)** El número imaginario i es la raíz cuadrada principal de -1.

Example $i = \sqrt{-1}$ and $i^2 = -1$.

Identity (p. 28) An equation that is true for every value of the variable is an identity.

Identidad (p. 28) Una ecuación que es verdadera para cada valor de la variable es una identidad.

Image (p. 801) An image is a figure obtained by a transformation of a preimage.

Imagen (p. 801) Una imagen es la figura que resulta después de que la preimagen sufre una transformación.

Example

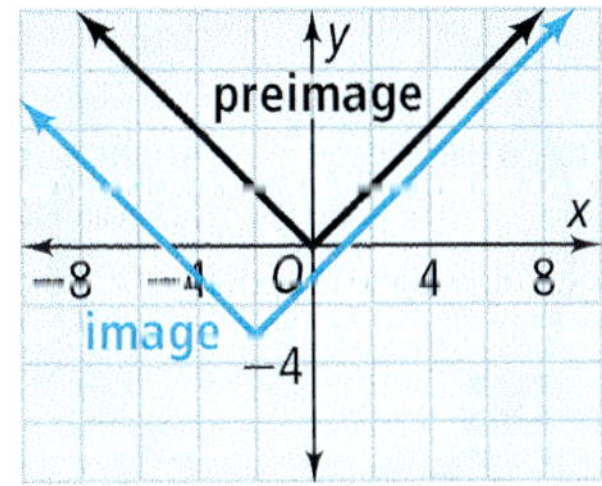

Imaginary number (p. 249) An imaginary number is any number of the form $a + bi$, where a and b are real numbers and $b \neq 0$.

Número imaginario (p. 249) Un número imaginario es cualquier número de la forma $a + bi$, donde a y b son números reales y $b \neq 0$.

Example $2 + 3i$
$7i$
i

Imaginary unit (p. 248) The imaginary unit i is the complex number whose square is -1.

Unidad imaginaria (p. 248) La unidad imaginaria i es el número complejo cuyo cuadrado es -1.

English	Spanish

Inconsistent system (p. 137) A system of equations that has no solution is an inconsistent system.

Sistema incompatible (p. 137) Un sistema incompatible es un sistema de ecuaciones para el cual no hay solución.

Example $\begin{cases} y = 2x + 3 \\ -2x + y = 1 \end{cases}$ is a system of parallel lines, so it has no solution. It is an inconsistent system.

Independent events (p. 688) When the outcome of one event does not affect the probability of a second event, the two events are independent.

Sucesos independientes (p. 688) Cuando el resultado de un suceso no altera la probabilidad de otro, los dos sucesos son independientes.

Example The results of two rolls of a number cube are independent. Getting a 5 on the first roll does not change the probability of getting a 5 on the second roll.

Independent system (p. 137) A system of linear equations that has a unique solution is an independent system.

Sistema independiente (p. 137) Un sistema de ecuaciones lineales que tenga una sola solución es un sistema independiente.

Example $\begin{cases} x + 2y = -7 \\ 2x - 3y = 0 \end{cases}$ has the unique solution $(-3, -2)$. It is an independent system.

Independent variable (p. 63) If a function is defined by an equation using the variables x and y, where x represents input values, then x is the independent variable.

Variable independiente (p. 63) Si una función es definida por una ecuación con las variables x e y, donde x representa los valores de entrada, entonces x es la variable independiente.

Example $y = 2x + 1$
x is the independent variable.

Index (p. 362) With a radical sign, the index indicates the degree of the root.

Índice (p. 362) Con un signo de radical, el índice indica el grado de la raíz.

Example

index 2	index 3	index 4
$\sqrt{16}$	$\sqrt[3]{16}$	$\sqrt[4]{16}$

Infinite series (p. 587) An infinite series is a series with infinitely many terms.

Serie infinita (p. 587) Una serie infinita es una serie con un número infinito de términos.

Initial point (p. 809) The initial point of a vector is the endpoint (not the tip) of a vector arrow.

Punto de inicio (p. 809) El punto de inicio de un vector es el extremo (no la punta) de una flecha vectorial.

English	Spanish

Initial side (p. 836) When an angle is in standard position, the initial side of the angle is given to be on the positive x-axis. The other ray is the terminal side of the angle.

Lado inicial (p. 836) Cuando un ángulo está en posición normal, el lado inicial del ángulo se ubica en el eje positivo de las x. El otro rayo, o semirrecta, forma el lado terminal del ángulo.

Example

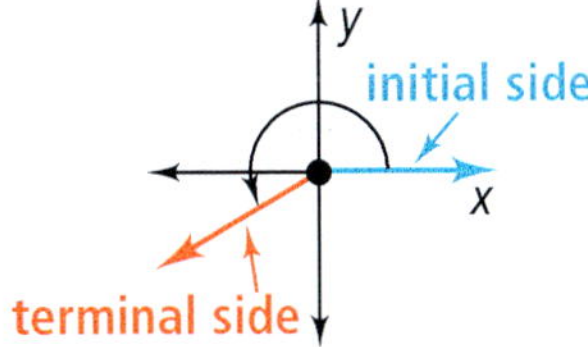

Intercepted arc (p. 844) An intercepted arc is the portion of a circle whose endpoints are on the sides of a central angle of the circle and whose remaining points lie in the interior of the angle.

Arco interceptado (p. 844) Un arco interceptado es la porción de un círculo cuyos extremos quedan sobre los lados de un ángulo central del círculo y cuyos puntos restantes quedan en el interior del ángulo.

Example

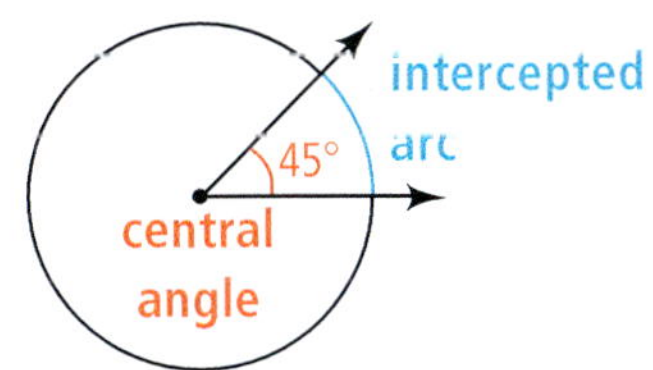

Interquartile range (p. 713) The interquartile range of a set of data is the difference between the third and first quartiles.

Intervalo intercuartil (p. 713) El rango intercuartil de un conjunto de datos es la diferencia entre el tercero y el primer cuartiles.

Example The first and third quartiles of the data set {2, 3, 4, 5, 5, 6, 7, 7} are 3.5 and 6.5. The interquartile range is $6.5 - 3.5 = 3$.

Inverse function (p. 405) If function f pairs a value b with a then its inverse, denoted f^{-1}, pairs the value a with b. If f^{-1} is also a function, then f and f^{-1} are inverse functions.

Funcion inversa (p. 405) Si la función f empareja un valor b con a, entonces su inversa, cuya notación es f^{-1}, empareja el valor a con b. Si f^{-1} también es una función, entonces f y f^{-1} son funciones inversas.

Example If $f(x) = x + 3$, then $f^{-1}(x) = x - 3$.

Inverse operations (p. 27) Inverse operations are operations that undo each other.

Operaciones inversas (p. 27) Operaciones inversas son operaciones que se cancelan mutuamente.

Inverse relation (p. 405) If a relation pairs element a of its domain with element b of its range, the inverse relation "undoes" the relation and pairs b with a. If (a, b) is an ordered pair of a relation, then (b, a) is an ordered pair of its inverse.

Relación inversa (p. 405) Si una relación empareja el elemento a de su dominio con el elemento b de su rango, la relación inversa "deshace" la relación y empareja b con a. Si (a, b) es un par ordenado de una relación, entonces (b, a) es un par ordenado de su inversa.

English	Spanish

Inverse variation (p. 498) An inverse variation is a relation represented by an equation of the form $xy = k$, $y = \frac{k}{x}$, or $x = \frac{k}{y}$, where $k \neq 0$.

Variación inversa (p. 498) Una variación inversa es una relación representada por la ecuación $xy = k$, $y = \frac{k}{x}$, ó $x = \frac{k}{y}$, donde $k \neq 0$.

Example

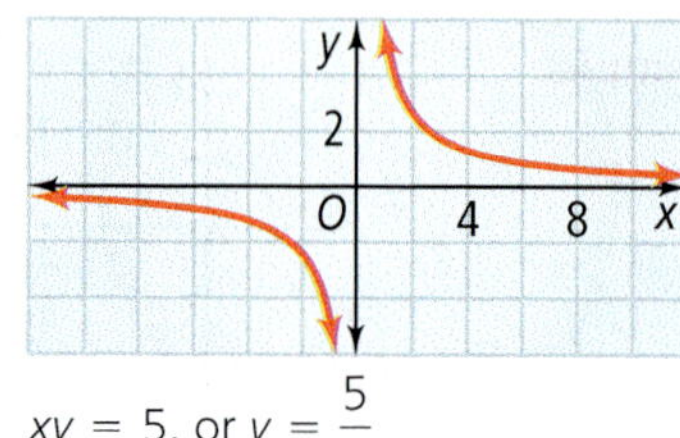

$xy = 5$, or $y = \frac{5}{x}$

J

Joint variation (p. 501) A joint variation is a relation in which one variable varies directly with respect to each of two or more variables.

Variación conjunta (p. 501) Una variación conjunta es una relación en la cual el valor de una variable varía directamente con respecto a cada una de dos o más variables.

Example $z = 8xy$
$T = kPV$

L

Law of Cosines (p. 936) In $\triangle ABC$, let a, b, and c represent the lengths of the sides opposite $\angle A$, $\angle B$, and $\angle C$, respectively. Then
$a^2 = b^2 + c^2 - 2bc \cos A$,
$b^2 = a^2 + c^2 - 2ac \cos B$, and
$c^2 = a^2 + b^2 - 2ab \cos C$

Ley de cosenos (p. 936) En $\triangle ABC$, sean a, b y c las longitudes de los lados opuestos a $\angle A$, $\angle B$ y $\angle C$, respectivamente. Entonces
$a^2 = b^2 + c^2 - 2bc \cos A$,
$b^2 = a^2 + c^2 - 2ac \cos B$ y
$c^2 = a^2 + b^2 - 2ab \cos C$

Example

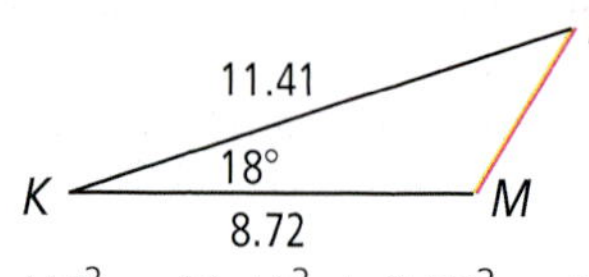

$LM^2 = 11.41^2 + 8.72^2 - 2(11.42)(8.72) \cos 18°$
$LM^2 = 16.9754$
$LM = 4.12$

Law of Sines (p. 929) In $\triangle ABC$, let a, b, and c represent the lengths of the sides opposite $\angle A$, $\angle B$, and $\angle C$, respectively. Then $\frac{\sin A}{a} = \frac{\sin B}{b} = \frac{\sin C}{c}$.

Ley de senos (p. 929) En $\triangle ABC$, sean a, b y c las longitudes de los lados opuestos a $\angle A$, $\angle B$ y $\angle C$, respectivamente. Entonces $\frac{\text{sen } A}{a} = \frac{\text{sen } B}{b} = \frac{\text{sen } C}{c}$.

Example

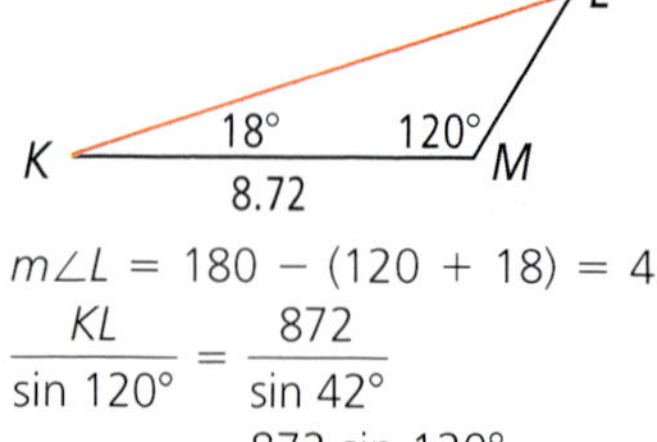

$m\angle L = 180 - (120 + 18) = 42°$
$$\frac{KL}{\sin 120°} = \frac{872}{\sin 42°}$$
$$KL = \frac{872 \sin 120°}{\sin 42°}$$
$$KL = 11.26$$

English | Spanish

Like radicals (p. 374) Like radicals are radical expressions that have the same index and the same radicand.

Radicales semejantes (p. 374) Los radicales semejantes son expresiones radicales que tienen el mismo índice y el mismo radicando.

Example $4\sqrt[3]{7}$ and $\sqrt[3]{7}$ are like radicals.

Like terms (p. 21) Like terms have the same variables raised to the same powers.

Términos semejantes (p. 21) Los términos semejantes tienen las mismas variables elevadas a las mismas potencias.

Limits (p. 589) Limits in summation notation are the least and greatest integer values of the index *n*.

Límites (p. 589) Los límites en notación de sumatoria son el menor y el mayor valor del índice *n* en números enteros.

Example $\sum_{n=1}^{3} (3n + 5)$ (limits: 1 and 3)

Line of best fit (p. 94) The trend line that gives the most accurate model of related data is the line of best fit.

Recta de mayor aproximación (p. 94) La línea de tendencia que representa con mayor precisión los datos relacionado es la recta de mayor aproximación.

Linear equation (p. 75) A linear equation in two variables is an equation that can be written in the form $ax + by = c$. *See also* **Standard form of a linear equation.**

Ecuación lineal (p. 75) Una ecuación lineal de dos variables es una ecuación que se puede escribir de la forma $ax + by = c$. *Ver también* **Standard form of a linear equation.**

Example $y = 2x + 1$ can be written as $-2x + y = 1$.

Linear function (p. 75) A function whose graph is a line is a linear function. You can represent a linear function with a linear equation.

Función lineal (p. 75) Una función cuya gráfica es una recta es una función lineal. La función lineal se representa con una ecuación lineal.

Example

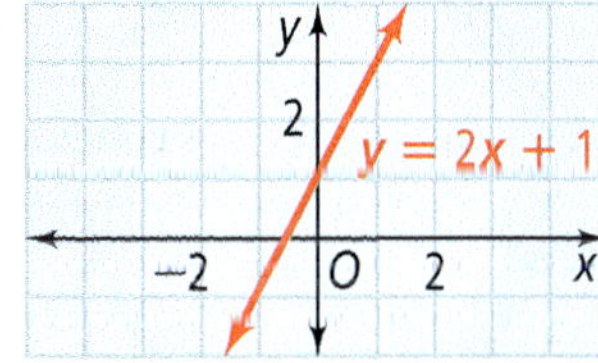

Linear inequality (p. 114) A linear inequality is an inequality in two variables whose graph is a region of the coordinate plane that is bounded by a line.

Desigualdad lineal (p. 114) Una desigualdad lineal es una desigualdad de dos variables cuya gráfica es una región del plano de coordenadas delimitado por una recta.

Example

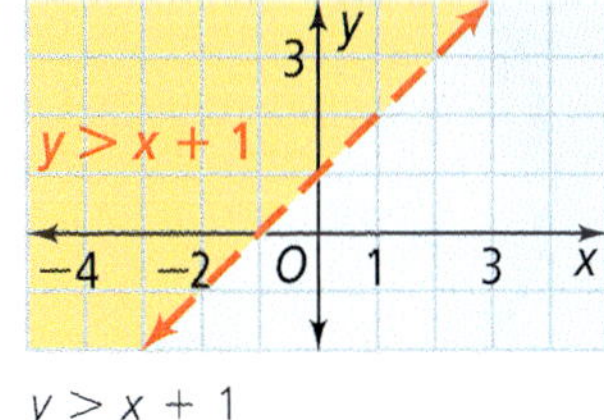

$y > x + 1$

English / Spanish

Linear programming (p. 157) Linear programming is a method for finding a minimum or maximum value of some quantity, given a set of constraints.

Programación lineal (p. 157) Programación lineal es un método para hallar el valor mínimo y máximo de una cantidad que se expresa como un conjunto de limitaciones.

Example Restrictions $x \geq 0$, $y \geq 0$, $x + y \leq 7$, and $y \leq -2x + 8$
Objective function: $B = 2x + 4y$
Evaluate $B = 2x + 4y$ at each vertex.
The minimum value of B occurs when $x = 0$ and $y = 0$. The m aximum value of B occurs when $x = 0$ and $y = 7$.

Linear system (p. 134) A linear system is a set of two or more linear equations that use the same variables.

Sistema lineal (p. 134) Un sistema lineal es un conjunto de dos o más ecuaciones lineales con las mismas variables.

Example

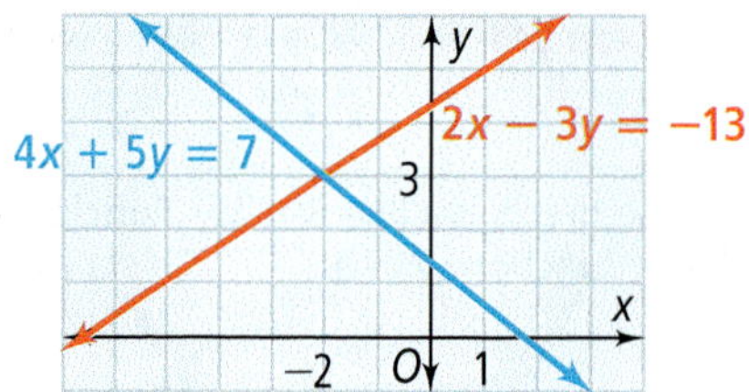

Literal equation (p. 29) A literal equation is an equation that uses more than one letter as a variable.

Ecuación literal (p. 29) Una ecuación literal es una ecuación en la cual más de una letra expresa una variable.

Logarithm (p. 451) The logarithm base b of a positive number x is defined as follows: $\log_b x = y$, if and only if $x = b^y$.

Logaritmo (p. 451) La base del logaritmo b de un número positivo x se define como $\log_b x = y$, si y sólo si $x = b^y$.

Example $\log_2 8 = 3$
$\log_{10} 100 = \log 100 = 2$
$\log_5 5^7 = 7$

English

Logarithmic equation (p. 471) A logarithmic equation is an equation that includes a logarithm involving a variable.

Example $\log_3 x = 4$

Logarithmic function (p. 454) A logarithmic function is the inverse of an exponential function.

Example

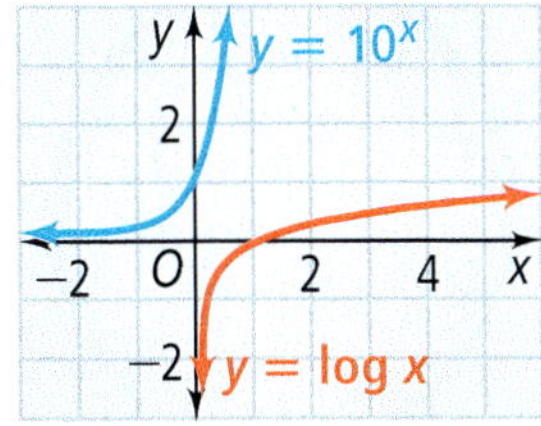

Logarithmic scale (p. 453) A logarithmic scale is a scale that uses the logarithm of a quantity instead of the quantity itself.

M

Magnitude (p. 809) The magnitude of a vector **v** is the length of the arrow.

Major axis (p. 639) The major axis of an ellipse is the segment that contains the foci of the ellipse and has endpoints on the ellipse.

Example

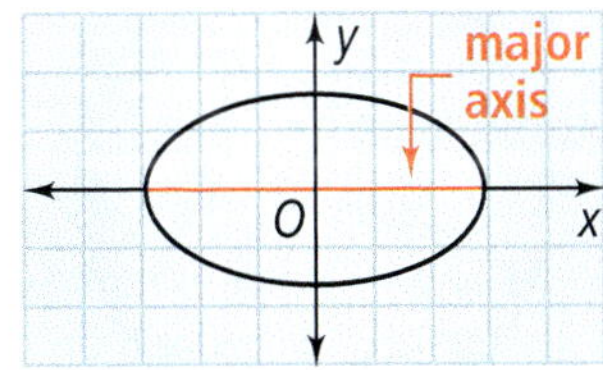

Mapping diagram (p. 60) A mapping diagram describes a relation by linking elements of the domain with elements of the range.

Example

Domain	Range
−5	−10
−1	−2
3	2
	6

Spanish

Ecuación logarítmica (p. 471) Una ecuación logarítmica es una ecuación que incluye un logaritmo con una variable.

Función logarítmica (p. 454) Una función logarítmica es la inversa de una función exponencial.

Escala logarítmica (p. 453) Una escala logarítmica es una escala que usa el logaritmo de una cantidaden vez de la cantidad misma.

Magnitud (p. 809) La magnitud de un vector **v** es la longitud de la flecha.

Eje mayor (p. 639) En una elipsis, el eje mayor es el segmento que contiene los focos de la elipsis y tiene puntos extremos sobre la elipsis.

Mapa (p. 60) Un mapa describe una relación al unir los elementos del dominio con los elementos del rango.

English | Spanish

Margin of error (p. 746) The distance from the sample mean or sample proportion that is used to create a confidence interval for the population mean or the population proportion. For a 95% confidence level, $ME = 1.96 \cdot \frac{s}{\sqrt{n}}$, where ME is the margin of error, s is the standard deviation of the sample data, and n is the number of values in the sample.

Margen de error (p. 746) La distancia desde la media de una muestra o desde la proporción de una muestra que se usa para crear el intervalo de confianza para la media o proporción de una población. Para un nivel de confianza de 95%, $ME = 1.96 \cdot \frac{s}{\sqrt{n}}$, siendo ME el margen de error, s la desviación estándar de los datos de la muestra, y n el número de valores en la muestra.

Example The standard deviation of a sample is 5.0 and the number of trials is 30. The margin of error at a 95% confidence level is $ME = 1.96 \cdot \frac{50}{\sqrt{30}} \approx 1.79$.

Matrix (p. 174) A matrix is a rectangular array of numbers written within brackets.

Matriz (p. 174) Una matriz es un conjunto de números encerrados en corchetes y dispuestos en forma de rectángulo.

Example $A = \begin{bmatrix} 1 & -2 & 0 & 10 \\ 9 & 7 & -3 & 8 \\ 2 & -10 & 1 & -6 \end{bmatrix}$

The number 2 is the element in the third row and first column. A is a 3×4 matrix.

Matrix element (p. 174) Every item listed in a matrix is an element of the matrix. An element is identified by its position in the matrix.

Elemento matricial (p. 174) Cada cifra de una matriz es un elemento de la matriz. El elemento se identifica según la posición que ocupa en la matriz.

Example $A = \begin{bmatrix} 1 & -2 & 0 & 10 \\ 9 & 7 & -3 & 8 \\ 2 & -10 & 1 & -6 \end{bmatrix}$

Element a_{21} is 9, the element in the second row and first column.

Matrix equation (p. 765) A matrix equation is an equation in which the variable is a matrix.

Ecuación matricial (p. 765) Una ecuación matricial es una ecuación en que la variable es una matriz.

Example

Solve $X + \begin{bmatrix} 3 & -2 \\ 5 & 1 \end{bmatrix} = \begin{bmatrix} 4 & 0 \\ 0 & 3 \end{bmatrix}$

$X = \begin{bmatrix} 4 & 0 \\ 0 & 3 \end{bmatrix} - \begin{bmatrix} 3 & -2 \\ 5 & 1 \end{bmatrix} = \begin{bmatrix} 1 & 2 \\ -5 & 2 \end{bmatrix}$

Maximum value (p. 195) The maximum value of a function $y = f(x)$ is the greatest y-value of the function. It is the y-coordinate of the highest point on the graph of f.

Valor máximo (p. 195) El valor máximo de una función $y = f(x)$ es el valor más alto de y de la función. Es la coordenada y del punto más alto de la gráfica de f.

English	Spanish

Mean (p. 711) The sum of the data values divided by the number of data values is the mean. *See also* **Arithmetic mean.**

Media (p. 711) La suma de los valores de datos dividida por el número de valores de datos sumados es la media. *Ver también* **Arithmetic mean.**

Example $\{1, 2, 3, 3, 6, 6\}$

$$\text{mean} = \frac{1 + 2 + 3 + 3 + 6 + 6}{6} = \frac{21}{6} = 3.5$$

Measures of central tendency (p. 711) The mean, the median, and the mode are each central values that help describe a set of data. They are called measures of central tendency.

Medidas de tendencia central (p. 711) La media, la mediana y la moda son los valores centrales que facilitan la descripción de un conjunto de datos. A estos valores se les llama medidas de tendencia central.

Example $\{1, 2, 3, 3, 4, 5, 6, 6\}$
mean = 3.75
median = 3.5
modes = 3 and 6

Measure of variation (p. 719) Measures of variation, such as the range, the interquartile range, and the standard deviation, describe how the data in a data set are spread out.

Medida de dispersión (p. 719) Las medidas de dispersión, tal como el rango, el intervalo intercuartil y la desviación típica, describen cómo se dispersan los datos en un conjunto de datos.

Median (p. 711) The median is the middle value in a data set. If the data set contains an even number of values, the median is the mean of the two middle values.

Mediana (p. 711) La mediana es el valor situado en el medio en un conjunto de datos. Si el conjunto de datos contiene un número par de valores, la mediana es la media de los dos valores del medio.

Example $\{1, 2, 3, 3, 4, 5, 6, 6\}$

$$\text{median} = \frac{3 + 4}{2} = \frac{7}{2} = 3.5$$

Midline (p. 830) The horizontal line through the average of the maximum and minimum values.

Línea media (p. 830) Recta horizontal que pasa a través de la media de los valores máximos y mínimos.

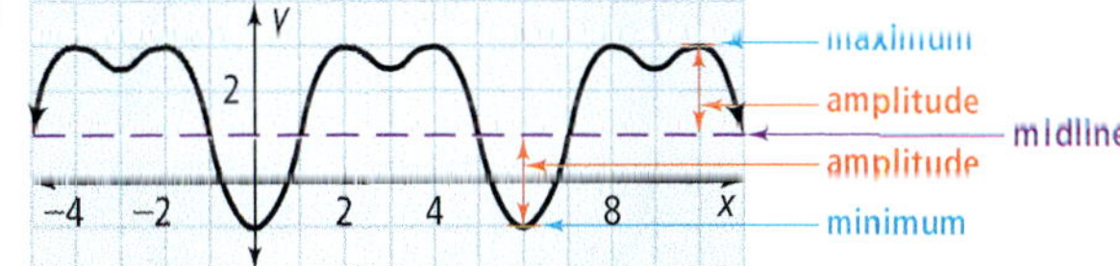

Minor axis (p. 639) The minor axis of an ellipse is the segment that is perpendicular to the major axis at its midpoint and has endpoints on the ellipse.

Eje menor (p. 639) En una elipsis, el eje menor es el segmento perpendicular al eje mayor en su punto medio y que tiene puntos extremos sobre la elipsis.

Example

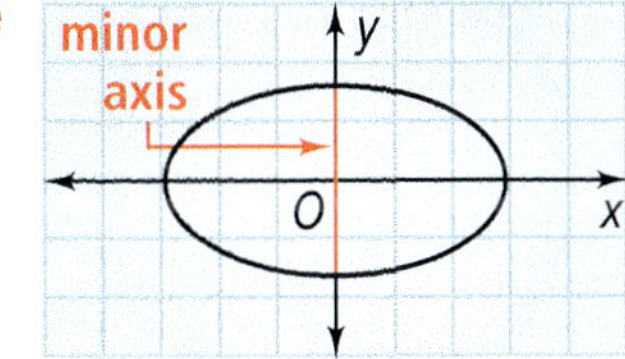

English	Spanish

Minimum value (p. 195) The minimum value of a function $y = f(x)$ is the least y-value of the function. It is the y-coordinate of the lowest point on the graph of f.

Valor mínimo (p. 195) El valor mínimo de una función $y = f(x)$ es el valor más bajo de y de la función. Es la coordenada y del punto más bajo de la gráfica de f.

Mode (p. 711) The mode is the most frequently occurring value (or values) in a set of data.

Moda (p. 711) La moda es el valor o valores que ocurren con mayor frecuencia en un conjunto de datos.

Example {1, 2, 3, 3, 4, 5, 6, 6}
The modes are 3 and 6.

Monomial (p. 280) A monomial is either a real number, a variable, or a product of real numbers and variables with whole number exponents.

Monomio (p. 280) Un monomio es un número real, una variable o un producto de números reales y variables cuyos exponentes son números enteros.

Example $1, x, 2z, 4ab^2$

Multiple zero (p. 291) If a linear factor is repeated in the complete factored form of a polynomial, the zero related to that factor is a multiple zero.

Cero múltiplo (p. 291) Si un factor lineal se repite en la forma factorizada completa de un polinomio, el cero relacionado con ese factor es un cero múltiplo.

Example The zeros of the function $P(x) = 2x(x - 3)^2(x + 1)$ are 0, 3, and -1. Since $(x - 3)$ occurs twice as a factor, 3 is a multiple zero.

Multiplicative identity (p. 14) The multiplicative identity is 1. The product of 1 and any number is that number. The product of reciprocals is 1.

Identidad multiplicativa (p. 14) La identidad multiplicativa es 1. El producto de 1 y cualquier otro número es ese número. El producto del recíproco es 1.

Multiplicative identity matrix (p. 782) For an $n \times n$ square matrix, the multiplicative identity matrix is an $n \times n$ square matrix I, or $I_{n \times n}$, with 1's along the main diagonal and 0's elsewhere.

Matriz de identidad multiplicativa (p. 782) Para una matriz cuadrada $n \times n$, la matriz de identidad multiplicativa es la matriz cuadrada I de $n \times n$, o $I_{n \times n}$, con unos por la diagonal principal y ceros en los demás lugares.

Example $I_{2\times 2} = \begin{vmatrix} 1 & 0 \\ 0 & 1 \end{vmatrix}, I_{3\times 3} = \begin{bmatrix} 1 & 0 & 0 \\ 0 & 1 & 0 \\ 0 & 0 & 1 \end{bmatrix}$

Multiplicative inverse (p. 14) The reciprocal or multiplicative inverse of any nonzero number a is $\frac{1}{a}$. The product of reciprocals is 1, the multiplicative identity.

Inverso multiplicativo (p. 14) El recíproco o inverso multiplicativo de cualquier número a, que no sea cero, es $\frac{1}{a}$. El producto de recíprocos es 1, la identidad multiplicativa.

Example $5 \times \frac{1}{5} = 1$

Multiplicative inverse of a matrix (p. 782) If A and X are $n \times n$ matrices, and $AX = XA = I$, then X is the multiplicative inverse of A, written A^{-1}.

Inverso multiplicativo de una matriz (p. 782) Si A y X son matrices $n \times n$, y $AX = XA = I$, entonces X es el inverso multiplicativo de A, expresado como A^{-1}.

Example $A = \begin{vmatrix} 2 & 1 \\ 4 & 0 \end{vmatrix}, X = \begin{bmatrix} 0 & \frac{1}{4} \\ 1 & \frac{1}{2} \end{bmatrix}$

$AX = \begin{bmatrix} 1 & 0 \\ 0 & 1 \end{bmatrix} = I$, so $X = A^{-1}$

English | Spanish

Multiplicity (p. 291) The multiplicity of a zero of a polynomial function is the number of times the related linear factor is repeated in the factored form of the polynomial.

Multiplicidad (p. 291) La multiplicidad de un cero de una función polinomial es el número de veces que el factor lineal relacionado se repite en la forma factorizada del polinomio.

Example The zeros of the function $P(x) = 2x(x - 3)^2(x + 1)$ are 0, 3, and -1. Since $(x - 3)$ occurs twice as a factor, the zero 3 has multiplicity 2.

Mutually exclusive events (p. 689) When two events cannot happen at the same time, the events are mutually exclusive. If A and B are mutually exclusive events, then $P(A \text{ or } B) = P(A) + P(B)$.

Sucesos mutuamente excluyentes (p. 689) Cuando dos sucesos no pueden ocurrir al mismo tiempo, son mutuamente excluyentes. Si A y B son sucesos mutuamente excluyentes, entonces $P(A \text{ or } B) = P(A) + P(B)$.

Example Rolling an even number E and rolling a multiple of five M on a standard number cube are mutually exclusive events.

$$P(E \text{ or } M) = P(E) + P(M) = \frac{3}{6} + \frac{1}{6} = \frac{4}{6}, \text{ or } \frac{2}{3}$$

N

***n* factorial (*n*!) (p. 675)** For any positive integer n, n factorial is $n(n - 1) \cdot \cdots \cdot 3 \cdot 2 \cdot 1$. Zero factorial $(0!) = 1$.

***n* factorial (*n*!) (p. 675)** Para cualquier entero n, n factorial es $n(n - 1) \cdot \cdots \cdot 3 \cdot 2 \cdot 1$. El cero factorial $(0!) = 1$.

Example $4! = 4 \cdot 3 \cdot 2 \cdot 1 = 24$

***n*th root (p. 361)** For any real numbers a and b, and any positive integer n, if $a^n = b$, then a is an nth root of b.

raíz *n*-ésima (p. 361) Para todos los números reales a y b, y todo número entero positivo n, si $a^n = b$, entonces a es la n-ésima raíz de b.

Example $\sqrt[5]{32} = 2$ because $2^5 = 32$.
$\sqrt[4]{81} = 3$ because $3^4 = 81$.

Natural base exponential function (p. 446) A natural base exponential function is an exponential function with base e.

Función exponencial con base natural (p. 446) Una función exponencial con base natural es una función exponencial con base e.

English | Spanish

Natural logarithmic function (p. 478) A natural logarithmic function is a logarithmic function with base e. The natural logarithmic function, $y = \ln x$, is $y = \log_e x$. It is the inverse of $y = e^x$.

Función logarítmica natural (p. 478) Una función logarítmica natural es una función logarítmica con base e. La función logarítmica natural, $y = \ln x$, es $y = \log_e x$. Ésta es la función inversa de $y = e^x$.

Example

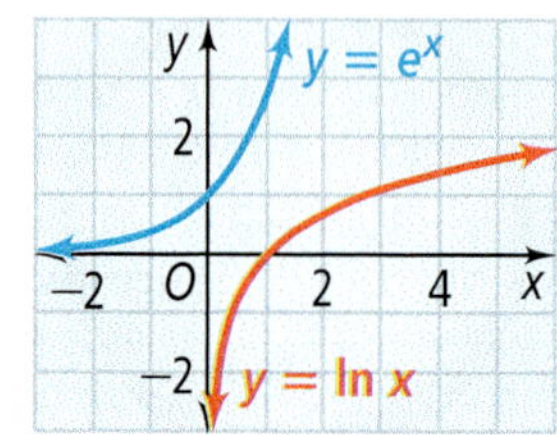

$\ln e^3 = 3$
$\ln 10 \approx 2.3026$
$\ln 36 \approx 3.5835$

Non-removable discontinuity (p. 516) A non-removable discontinuity is a point of discontinuity that is not removable. It represents a break in the graph of f where you cannot redefine f to make the graph continuous.

Discontinuidad irremovible (p. 516) Una discontinuidad irremovible es un punto de discontinuidad que no se puede remover. Representa una interrupción en la gráfica f donde no se puede redefinir f para volverla una gráfica continua.

Normal distribution (p. 739) A normal distribution shows data that vary randomly from the mean in the pattern of a bell-shaped curve.

Distribución normal (p. 739) Una distribución normal muestra, con una curva en forma de campana, datos que varían alcatoriamento respecto de la media.

Example

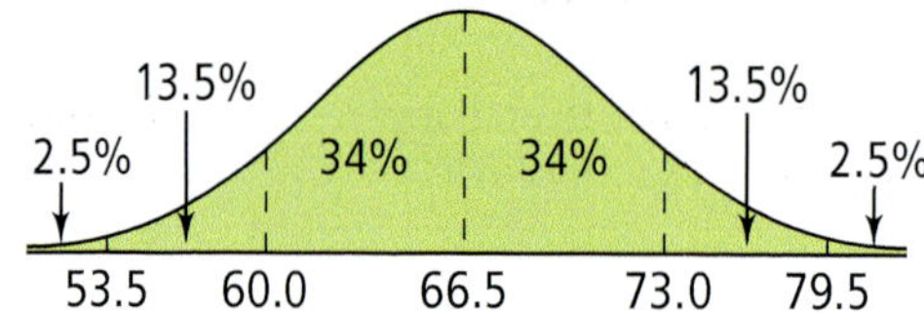

In a class of 200 students, the scores on a test were normally distributed. The mean score was 66.5 and the standard deviation was 6.5. The number of students who scored greater than 73 percent was about 13.5% + 2,5% of those who took the test.
16% of 200 = 32
About 32 students scored 73 or higher on the test.

Normal vectors (p. 812) Normal vectors are perpendicular vectors. Their dot product is 0.

Vectores normales (p. 812) Los vectores normales son vectores perpendiculares. El producto escalar es 0.

Numerical expression (p. 5) A numerical expression is a mathematical phrase that contains numbers and operation symbols.

Expresión numérica (p. 5) Una expresión numérica es una expresión matemática compuesta de números y símbolos de operación.

English

Spanish

Objective function (p. 157) In a linear programming model, the objective function is a model of the quantity that you want to make as large or as small as possible. *See* **Linear programming.**

Función objetiva (p. 157) En un modelo de programación lineal, la función objetiva es un modelo de la cantidad que se quiere aumentar o disminuir cuanto sea posible. *Ver* **Linear programming.**

Observational study (p. 726) In an observational study, you measure or observe members of a sample in such a way that they are not affected by the study.

Estudio de observación (p. 726) En un estudio de observación, se miden u observan a los miembros de una muestra de tal manera que no les afecte el estudio.

One-to-one function (p. 408) A one-to-one function is a function for which each y-value in the range corresponds to exactly one x-value in the domain. A one-to-one function f has an inverse f^{-1} that is also a function.

Función uno a uno (p. 408) Una función uno a uno es una función donde cada valor y que se encuentra en el rango corresponde exactamente a un valor x en el dominio. Una función uno a uno f tiene un inverso f^{-1} que también es una función.

Opposite (p. 14) The opposite or additive inverse of any number a is $-a$. The sum of opposites is zero, the additive identity.

Opuesto (p. 14) El opuesto o inverso aditivo de cualquier número a es $-a$. La suma de opuestos es cero, la identidad aditiva.

Example $3 + (-3) = 0$

$5.2 + (-5.2) = 0$

Ordered triples (p. 164) Ordered triples of the form (x, y, z) represent the location of a point in coordinate space.

Tripletas ordenadas (p. 164) Las tripletas ordenadas de la forma (x, y, z) representan la ubicación de un punto en el espacio de coordenadas.

Example (2, 4, 5)
(0, 1, 2)
(0, 0, 0)

Outlier (p. 712) An outlier is a value substantially different from the rest of the data in a set.

Valor extremo (p. 712) Un valor extremo es un valor considerablemente diferente al resto de los datos de un conjunto.

Example The outlier in the data set {56, 64, 73, 59, 98, 65, 59} is 98

P

Parabola (p. 194) A parabola is the graph of a quadratic function. It is the set of all points P in a plane that are the same distance from a fixed point F, the focus, as they are from a line d, the directrix.

Parábola (p. 194) La parábola es la gráfica de una función cuadrática. Es el conjunto de todos los puntos P situados en un plano a la misma distancia de un punto fijo F, o foco, y de la recta d, o directriz.

Example

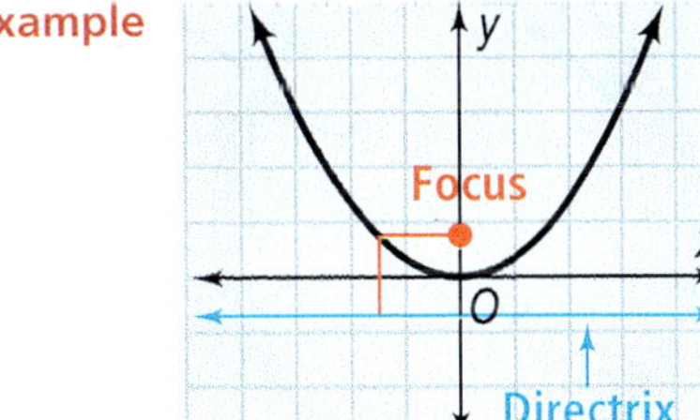

Visual **Glossary**

English	Spanish
Parallel lines (p. 85) Parallel lines are coplanar lines that do not intersect. In the coordinate plane, parallel lines have the same slope.	**Rectas paralelas (p. 85)** Rectas paralelas son líneas coplanares que no se intersecan. En un plano de coordenadas, las rectas paralelas tienen la misma pendiente.
Parent function (p. 99) A parent function is the simplest form of a set of functions that form a family.	**Función elemental (p. 99)** Una función madre es la mínima expresión de un conjunto de funciones que forma una familia.

Example $y = x$ is the parent function for the functions of the form $y = x + k$.

English	Spanish
Pascal's Triangle (p. 327) Pascal's Triangle is a triangular array of numbers in which the first and last number is 1. Each of the other numbers in the row is the sum of the two numbers above it.	**Triángulo de Pascal (p. 327)** El Triángulo de Pascal es una distribución triangular de números en la cual el primer número y el último número son 1. Cada uno de los otros números en la fila es la suma de los dos números de encima.

Example **Pascal's Triangle**

1
1 1
1 2 1
1 3 3 1
1 4 6 4 1
1 5 10 10 5 1

English	Spanish
Percentiles (p. 714) A percentile is a number from 0 to 100 that you can associate with a value x from a data set. It shows the percent of the data that are less than or equal to x.	**Percentiles (p. 714)** Un percentil es un número de 0 a 100 que se puede asociar con un valor x de un conjunto de datos. Éste muestra el porcentaje de los datos que son menores o iguales a x.
Perfect square trinomial (p. 219) A perfect square trinomial is a trinomial that is the square of a binomial.	**Trinomio cuadrado perfecto (p. 219)** Un trinomio cuadrado perfecto es un trinomio que es el cuadrado de un binomio.

Example perfect square trinominal — binominal square

$$16x^2 - 24x + 9 = (4x - 3)^2$$

English	Spanish
Period (p. 828) The period of a periodic function is the horizontal length of one cycle.	**Período (p. 828)** El período de una función periódica es el intervalo horizontal de un ciclo.

Example

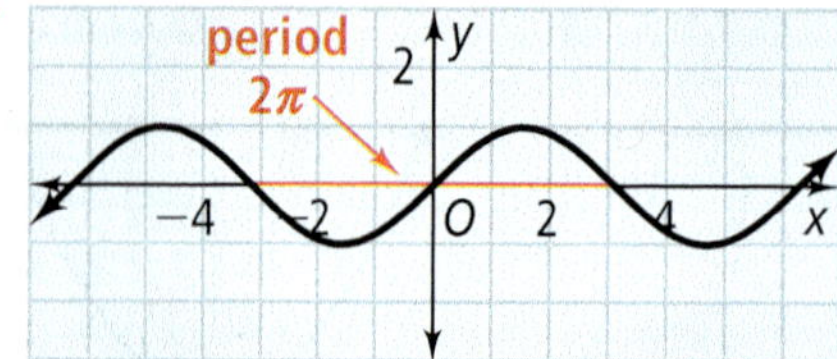

The periodic function $y = \sin x$ has period 2π.

English | Spanish

Periodic function (p. 828) A periodic function repeats a pattern of y-values at regular intervals.

Función periódica (p. 828) Una función periódica repite un patrón de valores y a intervalos regulares.

Example

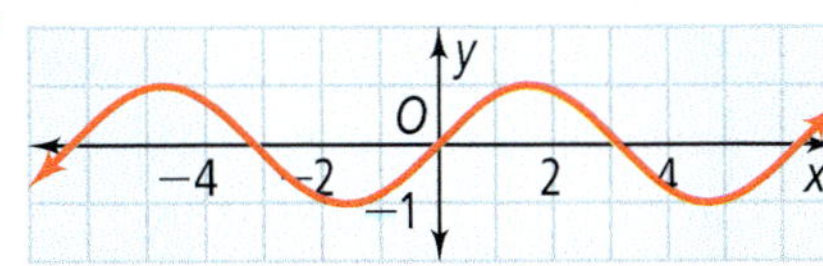

$y = \sin x$

Permutation (p. 675) A permutation is an arrangement of items in a particular order. The number of permutations of n objects taken r at a time is ${}_nP_r = \frac{n!}{(n-r)!}$ for $1 \le r \le n$.

Permutación (p. 675) Una permutación es la disposición de objetos en un orden determinado. El número de permutaciones de n objetos seleccionados r veces es ${}_nP_r = \frac{n!}{(n-r)!}$ para $1 \le r \le n$.

Example

$$\begin{aligned} {}_8P_5 &= \frac{8!}{(8-5)!} \\ &= \frac{8 \cdot 7 \cdot 6 \cdot 5 \cdot 4 \cdot 3!}{3!} \\ &= 8 \cdot 7 \cdot 6 \cdot 5 \cdot 4 \\ &= 6720 \end{aligned}$$

Perpendicular lines (p. 85) Perpendicular lines are lines that intersect to form right angles. In the coordinate plane, perpendicular lines have slopes with product -1.

Rectas perpendiculares (p. 85) Rectas perpendiculares son rectas que se intersecan y forman ángulos rectos. En un plano de coordenadas, las rectas perpendiculares tienen pendientes cuyo producto es -1.

Phase shift (p. 875) A horizontal translation of a periodic function is a phase shift.

Cambio de fase (p. 875) Una traslación horizontal de una función periódica es un cambio de fase.

Example

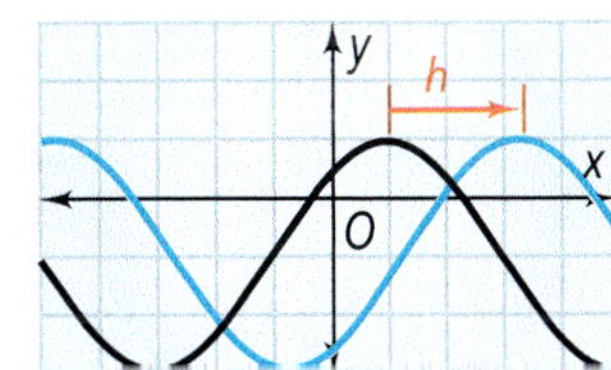

$g(x)$: horizontal translation of $f(x)$
$g(x) = f(x - h)$

Piecewise function (p. 90) A piecewise function has different rules for different parts of its domain.

Función de fragmentos (p. 90) Una función de fragmentos tiene reglas diferentes para diferentes partes de su dominio.

Point of discontinuity (p. 516) A point of discontinuity is the x-coordinate of a point where the graph of $f(x)$ is not continuous.

Punto de discontinuidad (p. 516) Un punto de discontinuidad es la coordenada x de un punto donde la gráfica de $f(x)$ no es continua.

Example $f(x) = \frac{2}{x-2}$ has a point of discontinuity at $x = 2$.

English	Spanish

Point-slope form (p. 81) The point-slope form of an equation of a line is $y - y_1 = m(x - x_1)$, where m is the slope of the line and (x_1, y_1) is a point on the line.

Forma punto-pendiente (p. 81) La forma punto-pendiente de una ecuación lineal es $y - y_1 = m(x - x_1)$, donde m es la pendiente de la recta y (x_1, y_1) es un punto de la recta.

Example $y - 3 = 2(x - 1)$
$y + 4 = 5(x - 2)$
$y - 2 = 3(x + 2)$

Polynomial (p. 280) A polynomial is a monomial or the sum of monomials.

Polinomio (p. 280) Un polinomio es un monomio o la suma de dos o más monomios.

Example $3x^3 + 4x^2 - 2x + 5$
$8x$
$x^2 + 4x + 2$

Polynomial function (p. 280) A polynomial in the variable x defines a polynomial function of x.

Función polinomial (p. 280) Un polinomio en la variable x define una función polinomial de x.

Example $P(x) = a_nx^n + a_n - 1x^{n-1} - 1 + \cdots + a_1x + a_0$ is a polynomial function, where n is a nonnegative integer and the coefficients $a_n, \ldots, a_0$ are real numbers.

Population (p. 725) A population is the members of a set.

Población (p. 725) Una población está compuesta por los miembros de un conjunto.

Power function (p. 341) A power function is a function of the form $y = a \cdot x^b$, where a and b are nonzero real numbers.

Función de potencia (p. 341) Una función de potencia es una función de la forma $y = a \cdot x^b$, donde a y b son números reales diferentes de cero.

Preimage (p. 801) The preimage is the original figure before a transformation.

Preimagen (p. 801) La preimagen es la figura original antes de sufrir una transformación.

Example

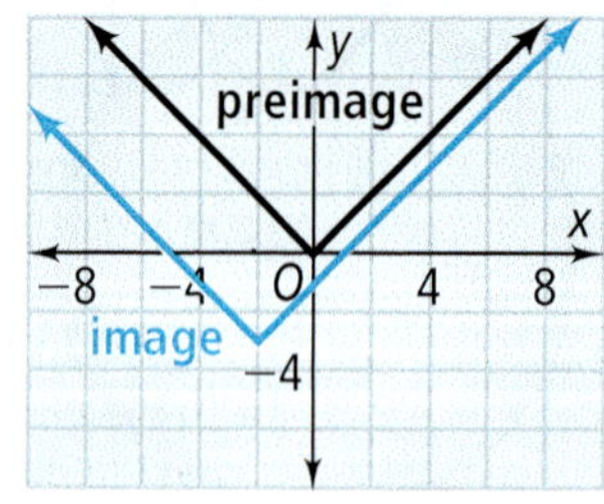

Principal root (p. 361) When a number has two real roots, the positive root is called the principal root. A radical sign indicates the principal root. The principal root of a negative number a is $i\sqrt{|a|}$.

Raíz principal (p. 361) Cuando un número tiene dos raíces reales, la raíz positiva es la raíz principal. El signo del radical indica la raíz principal. La raíz principal de un número negativo a es $i\sqrt{|a|}$.

Example The number 25 has two square roots, 5 and −5. The principal square root, 5, is indicated by $\sqrt{25}$ or $25^{\frac{1}{2}}$.

English	Spanish

Probability distribution (p. 694) A probability distribution is a function that tells the probability of each outcome in a sample space.

Distribución de probabilidades (p. 694) Una distribución de probabilidades es una función que señala la probabilidad de que cada resultado ocurra en un espacio muestral.

Example

Roll	Fr.	Prob.
1	5	0.125
2	9	0.225
3	7	0.175
4	8	0.2
5	8	0.2
6	3	0.075

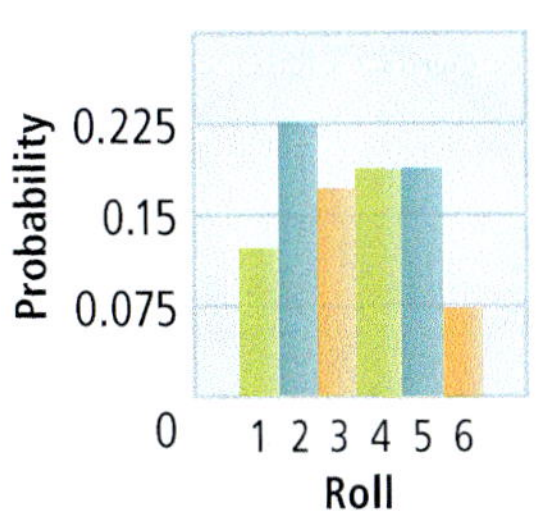

The table and graph both show the experimental probability distribution for the outcomes of 40 rolls of a standard number cube.

Probability model (p. 705) A mathematical representation of a situation in which probabilities are assigned to outcomes.

Modelo de probabilidad (p. 705) Representación matemática de una situación en la que se asignan probabilidades a los resultados.

Example A restaurant gives a randomly selected toy with each child's meal. There are 5 different toys. Getting a certain toy has a probability of $\frac{1}{5}$. A probability model could be a random number table where 0 and 1 represent the first toy, 2 and 3 represent the second toy, and so on. Getting a 0 or a 1 has the probability of $\frac{1}{5}$.

Pure imaginary number (p. 249) If $a = 0$ and $b \neq 0$, the number $a + bi$ is a pure imaginary number.

Número imaginario puro (p. 249) Si $a = 0$ y $b \neq 0$, el número $a + bi$ es un número imaginario puro.

Q

Quadratic equation (p. 226) A quadratic equation is one that can be written in the standard form $ax^2 + bx + c = 0$, where $a \neq 0$.

Ecuación cuadrática (p. 226) Una ecuación cuadrática es una ecuación que se puede expresar en forma normal como $ax^2 + bx + c = 0$, donde $a \neq 0$.

Example $2x^2 + 3x + 1 = 0$

Quadratic Formula (p. 240) The Quadratic Formula is $x = \frac{-b \pm \sqrt{b^2 - 4ac}}{2a}$. It gives the solutions to the quadratic equation $ax^2 + bx + c = 0$.

Fórmula cuadrática (p. 240) La fórmula cuadrática es $x = \frac{-b \pm \sqrt{b^2 - 4ac}}{2a}$. Ésta da las soluciones a la ecuación cuadrática $ax^2 + bx + c = 0$.

Example If $-x^2 + 3x + 2 = 0$, then

$$x = \frac{-3 \pm \sqrt{(3)^2 - 4(-1)(2)}}{2(-1)}$$

$$= \frac{-3 \pm \sqrt{17}}{-2}$$

English	Spanish

Quadratic function (p. 194) A quadratic function is a function that you can write in the form $f(x) = ax^2 + bx + c$ with $a \neq 0$.

Función cuadrática (p. 194) Una función cuadrática es una función que puedes escribir como $f(x) = ax^2 + bx + c$ con $a \neq 0$.

Example

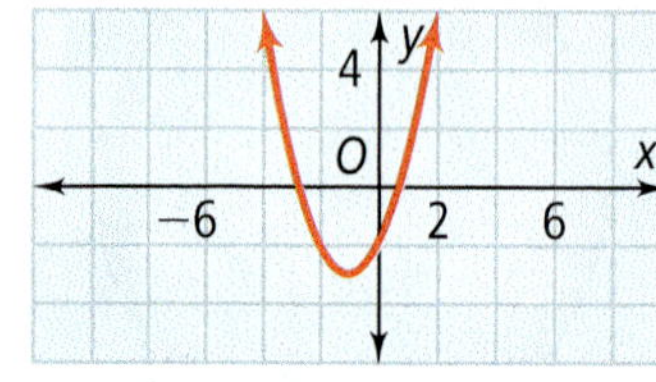

$y = x^2 + 2x - 2$

Quantity (p. 5) A mathematical quantity is anything that can be measured or counted.

Cantidad (p. 5) Una cantidad matemática es cualquier cosa que se puede medir o contar.

Quartile (p. 713) Quartiles are values that separate a finite data set into four equal parts. The second quartile (Q_2) is the median of the data. The first and third quartiles (Q_1 and Q_3) are the medians of the lower half and upper half of the data, respectively.

Cuartil (p. 713) Los cuartiles son valores que separan un conjunto finito de datos en cuatro partes iguales. El segundo cuartil (Q_2) es la mediana de los datos. Los cuartiles primero y tercero (Q_1 y Q_3) son las medianas de la mitad superior e inferior de los datos, respectivamente.

Example $\{2, 3, 4, 5, 5, 6, 7, 7\}$

$Q_1 = 3.5$

Q_2 (median) $= 5$

$Q_3 = 6.5$

Radian (p. 844) $\frac{a°}{180°} = \frac{r \text{ radians}}{\pi \text{ radians}}$

Radián (p. 844) $\frac{a°}{180°} = \frac{r \text{ radianes}}{\pi \text{ radianes}}$

Example $60° \rightarrow \frac{60}{180} = \frac{x}{\pi}$

$x = \frac{60\pi}{180}$

$= \frac{\pi}{3}$

Thus, $60° = \frac{\pi}{3}$ radians.

Radical equation (p. 390) A radical equation is an equation that has a variable in a radicand or has a variable with a rational exponent.

Ecuación radical (p. 390) La ecuación radical es una ecuación que contiene una variable en el radicando o una variable con un exponente racional.

Example $(\sqrt{x})^3 + 1 = 65$

$x^{\frac{3}{2}} + 1 = 65$

Radical function (p. 415) A radical function is a function that can be written in the form $f(x) = a\sqrt[n]{x - h} + k$, where $a \neq 0$. For even values of n, the domain of a radical function is the real numbers $x \geq h$. *See also* **Square root function.**

Función radical (p. 415) Una función radical es una función quepuede expresarse como $f(x) = a\sqrt[n]{x - h} + k$, donde $a \neq 0$. Para n par, el dominio de la función radical son los números reales tales que $x \geq h$. *Ver también* **Square root function.**

Example $f(x) = \sqrt{x - 2}$

English | Spanish

Radicand (p. 362) The number under a radical sign is the radicand.

Radicando (p. 362) La expresión que aparece debajo del signo radical es el radicando.

Example The radicand in $3\sqrt[4]{7}$ is 7.

Radius (p. 630) The radius r of a circle is the distance between the center of the circle and any point on the circumference.

Radio (p. 630) El radio r de un círculo es la distancia entre el centro del círculo y cualquier punto de la circunferencia.

Example

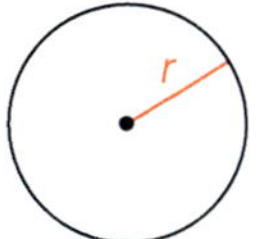

Random sample (p. 725) In a random sample, all members of the population are equally likely to be chosen as every other member.

Muestra aleatoria (p. 725) En una muestra aleatoria, la probabilidad de ser seleccionado es igual para todos los miembros.

Example Let the set of all females between the ages of 19 and 34 be the population. A random selection of 900 females between those ages would be a sample of the population

Range (p. 61) The range of a relation is the set of all outputs or y-coordinates of the ordered pairs.

Rango (p. 61) El rango de una relación es el conjunto de todas las salidas posibles, o coordenadas y, de los pares ordenados.

Example In the relation $\{(0, 1), (0, 2), (0, 3), (0, 4), (1, 3), (1, 4), (2, 1)\}$, the range is $\{1, 2, 3, 4\}$. In the function $f(x) = |x - 3|$, the range is the set of real numbers greater than or equal to 0.

Range of a set of data (p. 713) The range of a set of data is the difference between the greatest and least values.

Rango de un conjunto de datos (p. 713) El rango de un conjunto de datos es la diferencia entre el valor máximo y el valor mínimo de los datos.

Example The range of the set $\{3.2, 4.1, 2.2, 3.4, 3.8, 4.0, 4.2, 2.8\}$ is $4.2 - 2.2 = 2$.

Rational equation (p. 542) A rational equation is an equation that contains a rational expression.

Ecuación racional (p. 542) Una ecuación racional es una ecuación que contiene una expresión racional.

Rational exponent (p. 382) If the nth root of a is a real number and m is an integer, then $a^{\frac{1}{n}} = \sqrt[n]{a}$ and $a^{\frac{m}{n}} = \sqrt[n]{a^m} = (\sqrt[n]{a})^m$. If m is negative, $a \neq 0$.

Exponente racional (p. 382) Si la raíz n-ésima de a es un número real y m es un número entero, entonces $a^{\frac{1}{n}} = \sqrt[n]{a}$ y $a^{\frac{m}{n}} = \sqrt[n]{a^m} = (\sqrt[n]{a})^m$. Si m es negativo, $a \neq 0$.

Example $4^{\frac{1}{3}} = \sqrt[3]{4}$

$5^{\frac{3}{2}} = \sqrt{5^3} = (\sqrt{5})^3$

English	Spanish
Rational expression (p. 527) A rational expression is the quotient of two polynomials.	**Expresión racional (p. 527)** Una expresión racional es el cociente de dos polinomios.
Rational function (p. 515) A rational function $f(x)$ can be written as $f(x) = \frac{P(x)}{Q(x)}$, where $P(x)$ and $Q(x)$ are polynomial functions. The domain of a rational function is all real numbers except those for which $Q(x) = 0$.	**Función racional (p. 515)** Una función racional $f(x)$ se puede expresar como $f(x) = \frac{P(x)}{Q(x)}$, donde $P(x)$ y $Q(x)$ son funciones de polinomios. El dominio de una función racional son todos los números reales excepto aquéllos para los cuales $Q(x) = 0$.

Example

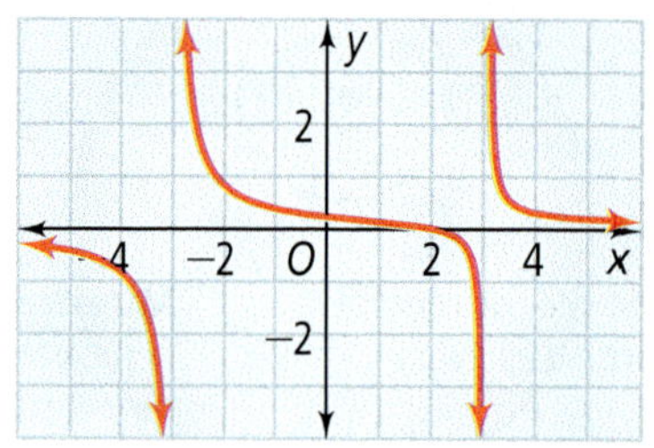

The function $y = \frac{x - 2}{x^2 - 9}$ is a rational function with three branches separated by asymptotes $x = -3$ and $x = 3$.

English	Spanish
Rational Root Theorem (p. 312) Let $P(x) = a_nx^n + a_{n-1}x^{n-1} + \cdots + a_1x + a_0$ be a polynomial with integer coefficients. Then there are a limited number of possible roots of $P(x) = 0$: –Integer roots must be factors of a_0; –Rational roots must have reduced form p/q where p is an integer factor of a_0 and q is an integer factor of a_n.	**Teorema de la Raíz Racional (p. 312)** Sea $P(x) = a_nx^n + a_{n-1}x^{n-1} + \cdots + a_1x + a_0$ un polinomio con enteros como coeficientes. Entonces hay un número limitado de raíces posibles para $P(x) = 0$: –Las raíces enteras deben ser factores de a_0; –Las raíces racionales deben ser de forma simplificada p/q, donde p es un factor entero de a_0 y q es un factor entero de a_n.

Example The polynomial equation $10x^3 + 6x^2 - 11x - 2 = 0$ has leading coefficient 10 (with factors ($\pm 1, \pm 2, \pm 5, \pm 10$) and constant term -2 (with factors ± 1 and ± 2). Its only possible rational roots are $\pm 1, \pm 2, \pm \frac{1}{2}, \pm \frac{1}{5}, \pm \frac{2}{5}, \pm \frac{1}{10}$.

English	Spanish
Rationalize the denominator (p. 369) To rationalize the denominator of an expression, rewrite it so there are no radicals in any denominator and no denominators in any radical.	**Racionalizar el denominador (p. 369)** Para racionalizar el denominador de una expresión, ésta se escribe de modo que no haya radicales en ningún denominador y no haya denominadores en ningún radical.

Example $\frac{1}{\sqrt{2}} = \frac{1}{\sqrt{2}} \times \frac{\sqrt{2}}{\sqrt{2}} = \frac{\sqrt{2}}{2}$

English	Spanish

Reciprocal (p. 14) The reciprocal or multiplicative inverse of any nonzero number a is $\frac{1}{a}$. The product of reciprocals is 1, the multiplicative identity.

Recíproco (p. 14) El recíproco o inverso multiplicativo de un número distinto de cero a es $\frac{1}{a}$. El producto de recíprocos es 1, la identidad multiplicativa.

Example $5 \times \frac{1}{5} = 1$

Reciprocal function (p. 507) A reciprocal function belongs to the family whose parent function is $f(x) = \frac{1}{x}$ where $x \neq 0$. You can write a reciprocal function in the form $f(x) = \left(\frac{a}{x} - h\right) + k$, where $a \neq 0$ and $x \neq h$.

Función recíproca (p. 507) Una función recíproca pertenece a la familia cuya función madre es $f(x) = \frac{1}{x}$ donde $x \neq 0$. Se puede escribir una función recíproca como $f(x) = \left(\frac{a}{x} - h\right) + k$, donde $a \neq 0$ y $x \neq h$.

Example $f(x) = \frac{1}{2x + 5}$

$p(v) = \frac{3}{v} + 5$

Recursive formula (p. 565) A recursive formula defines the terms in a sequence by relating each term to the ones before it.

Fórmula recursiva (p. 565) Una fórmula recursiva define los términos de una secuencia al relacionar cada término con los términos que lo anteceden.

Example Let $a_n = 2.5a_{n-1} + 3a_{n-2}$.

If $a_5 = 3$ and $a_4 = 7.5$, then

$a_6 = 2.5(3) + 3(7.5) = 30$.

Reduced row echelon form (p. 177) A matrix that represents the solution of a system is in reduced row echelon form. The leading 1 in each row has 0's elsewhere in its column.

Forma reducida fila-escalón (p. 177) Una matriz que representa la solución de un sistema está en forma reducida fila-escalón. El 1 principal en cada fila tiene ceros en otras partes de la columna.

Reflection (p. 101) A reflection flips the graph of a function across a line, such as the x- or y-axis. Each point on the graph of the reflected function is the same distance from the line of reflection as is the corresponding point on the graph of the original function.

Reflexión (p. 101) Una reflexión voltea la gráfica de una función sobre una línea, como el eje de las x o el eje de las y. Cada punto de la gráfica de la función reflejada está a la misma distancia del eje de reflexión que el punto correspondiente en la gráfica de la función original.

Example

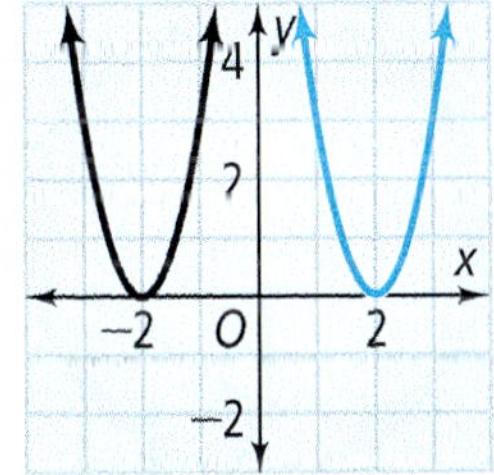

Relation (p. 60) A relation is a set of ordered pairs.

Relación (p. 60) Una relación es un conjunto de pares ordenados.

Example {(0, 1), (0, 2), (0, 3), (0, 4), (1, 3)}

Visual Glossary

English	Spanish

Relative maximum (minimum) (p. 291) A relative maximum (minimum) is the value of the function at an up-to-down (down-to-up) turning point.

Máximo (minimo) relativo (p. 291) El máximo (mínimo) relativo es el valor de la función en un punto de giro de arriba hacia abajo (de abajo hacia arriba).

Example

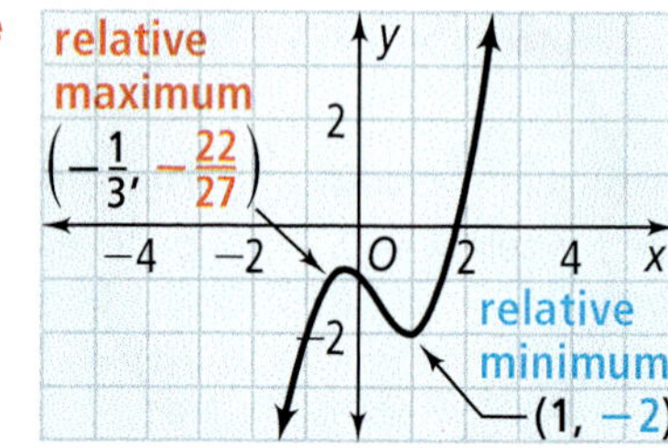

Remainder Theorem (p. 307) If you divide a polynomial $P(x)$ of degree $n > 1$ by $x - a$, then the remainder is $P(a)$.

Teorema del residuo (p. 307) Si divides un polinomio $P(x)$ con un grado $n > 1$ por $x - a$, el residuo es $P(a)$.

Example If $P(x) = x^3 - 4x^2 + x + 6$ is divided by $x - 3$, then the remainder is $P(3) = 3^3 - 4(3)^2 + 3 + 6 = 0$ (which means that $x - 3$ is a factor of $P(x)$).

Removable discontinuity (p. 516) A removable discontinuity is a point of discontinuity, a, of function f that you can remove by redefining f at $x = a$. Doing so fills in a hole in the graph of f with the point $(a, f(a))$.

Discontinuidad removible (p. 516) Una discontinuidad removible es un punto de discontinuidad a en una función f que se puede remover al redefinir f en $x = a$. Al hacer esto, se llena un hueco en la gráfica f con el punto $(a, f(a))$.

Root (p. 232) A root of a function is the input value for which the value of the function is zero. A root of an equation is a value that makes the equation true. *See also* **Zero of a function.**

Raíz (p. 232) La raíz de una función es el valor de entrada para el cual el valor de la función es cero. La raíz de una ecuación es un valor que hace verdadera la ecuación. *Ver también* **Zero of a function.**

Example -2 and 3 are roots of the function $f(x) = (x + 2)(x - 3)$ and the equation $(x + 2)(x - 3) = 0$.

Rotation (p. 804) A rotation is a transformation that turns a figure about a fixed point called the center of rotation.

Rotación (p. 804) Rotación es una transformación que hace girar una figura alrededor de un punto fijo llamado centro de rotación.

Example

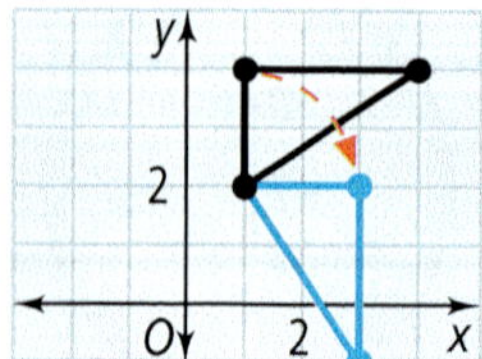

Row operation (p. 176) A row operation on an augmented matrix is any of the following: switch two rows, multiply a row by a constant, add one row to another.

Operación de fila (p. 176) Una operación de fila en una matriz ampliada es cualquiera de las siguientes opciones: el intercambio de dos filas, la multiplicación de una fila por una constante o la suma de dos filas.

English S Spanish

Sample (p. 725) A sample from a population is some of the population.

Muestra (p. 725) Una muestra de una población es una parte de la población.

Example Let the set of all males between the ages of 19 and 34 be the population. A random selection of 900 males between those ages would be a sample of the population.

Sample proportion (p. 747) The ratio $\hat{p}$ compares x to n where x is the number of times an event occurs and n is the sample size. $\hat{p} = \frac{x}{n}$.

Proporción de una muestra (p. 747) La razón $\hat{p}$ compara x a n, siendo x el número de veces que sucede un evento y n el tamaño de la muestra. $\hat{p} = \frac{x}{n}$.

Example In a taste test, 120 persons sampled two types of cola; 40 people preferred cola A. The sample proportion is $\frac{40}{120}$, or $\frac{1}{3}$.

Sample space (p. 682) The set of all possible outcomes of an experiment is called the sample space.

Espacio muestral (p. 682) El espacio muestral es el conjunto de todos los resultados posibles de un suceso.

Example When you roll a number cube, the sample space is $\{1, 2, 3, 4, 5, 6\}$.

Scalar (p. 772) A scalar is a real number factor in a special product, such as the 3 in the vector product 3**v**.

Escalar (p. 772) Un escalar es un factor que es un número real en un producto especial, como el 3 en el producto vectorial 3**v**.

Scalar multiplication (p. 772) Scalar multiplication is an operation that multiplies a matrix A by a scalar c. To find the resulting matrix cA, multiply each element of A by c.

Multiplicación escalar (p. 772) La multiplicación escalar es la que multiplica una matriz A por un número escalar c. Para hallar la matriz cA resultante, multiplica cada elemento de A por c.

Example $$2.5\begin{bmatrix} 1 & 0 \\ -2 & 3 \end{bmatrix} = \begin{bmatrix} 2.5(1) & 2.5(0) \\ 2.5(-2) & 2.5(3) \end{bmatrix} = \begin{bmatrix} 2.5 & 0 \\ -5 & 7.5 \end{bmatrix}$$

Scatter plot (p. 92) A scatter plot is a graph that relates two different sets of data by plotting the data as ordered pairs. You can use a scatter plot to determine a relationship between the data sets.

Diagrama de puntos (p. 92) Un diagrama de puntos es una gráfica que relaciona dos conjuntos de datos presentando los datos como pares ordenados. El diagrama de puntos sirve para definir la relación entre conjuntos de datos.

Example

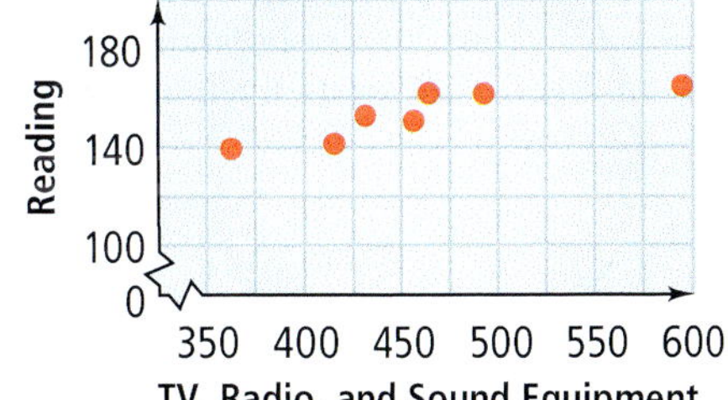

SOURCE: U.S. Bureau of Labor Statistics

English | Spanish

Secant function (p. 883) The secant (sec) function is the reciprocal of the cosine function. For all real numbers θ except those that make $\cos\theta = 0$, $\sec\theta = \frac{1}{\cos\theta}$.

Función secante (p. 883) La función secante (sec) es el recíproco de la función coseno. Para todos los números reales θ, excepto aquéllos para los que $\cos\theta = 0$, $\sec\theta = \frac{1}{\cos\theta}$.

Example If $\cos\theta = \frac{5}{13}$, then $\sec\theta = \frac{13}{5}$.

Self-selected sample (p. 725) In a self-selected sample you select only members of the population who volunteered for the sample.

Muestra de voluntarios (p. 725) En una muestra de voluntarios se seleccionan sólo a los miembros de la población que se ofrecen voluntariamente para ser parte de la muestra.

Sequence (p. 564) A sequence is an ordered list of numbers.

Progresión (p. 564) Una progresión es una sucesión de números.

Example 1, 4, 7, 10, . . .

Series (p. 587) A series is the sum of the terms of a sequence.

Serie (p. 587) Una serie es la suma de los términos de una secuencia.

Example The series $3 + 6 + 9 + 12 + 15$ corresponds to the sequence 3, 6, 9, 12, 15. The sum of the series is 45.

Simplest form of a radical expression (p. 368) A radical expression with index n is in simplest form if there are no radicals in any denominator, no denominators in any radical, and any radicand has no nth power factors.

Mínima expresión de una expresión radical (p. 368) Una expresión radical con índice n está en su mínima expresión si no tiene radicales en ningún denominador ni denominadores en ningún radical y los radicandos no tienen factores de potencia.

Simplest form of a rational expression (p. 527) A rational expression is in simplest form if its numerator and denominator are polynomials that have no common divisor other than 1.

Forma simplificada de una expresión racional (p. 527) Una expresión racional se encuentra en su mínima expresión si su numerador y su denominador son polinomios que no tienen otro divisor aparte de 1.

Example $\frac{x^2 - 7x + 12}{x^2 - 9} = \frac{(x - 4)(x - 3)}{(x + 3)(x - 3)} = \frac{x - 4}{x + 3}$, where $x \neq -3$

Simulation (p. 682) A simulation is a model that imitates one or more events.

Simulación (p. 682) Una simulación es un modelo que imita uno o más sucesos.

Example Suppose a weather forecaster predicts a 50% chance of rain for the next three days. You can use three coins landing heads up to simulate three days in a row of rain.

English | Spanish

Sine curve (p. 852) A sine curve is the graph of a sine function.

Sinusoide (p. 852) Sinusoide es la gráfica de la función seno.

Example

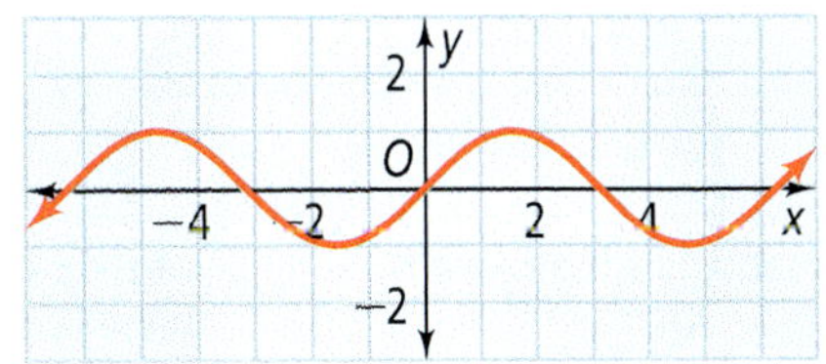

Sine function, Sine of θ (pp. 838, 851) The sine function, $y = \sin\theta$, matches the measure θ of an angle in standard position with the y-coordinate of a point on the unit circle. This point is where the terminal side of the angle intersects the unit circle. The y-coordinate is the sine of θ.

Función seno, Seno de θ (pp. 838, 851) La función seno, $y = \sin\theta$, empareja la medida θ de un ángulo en posición estándar con la coordenada y de un punto en el círculo unitario. Este es el punto en el que el lado terminal del ángulo interseca al círculo unitario. La coordenada y es el seno de θ.

Example

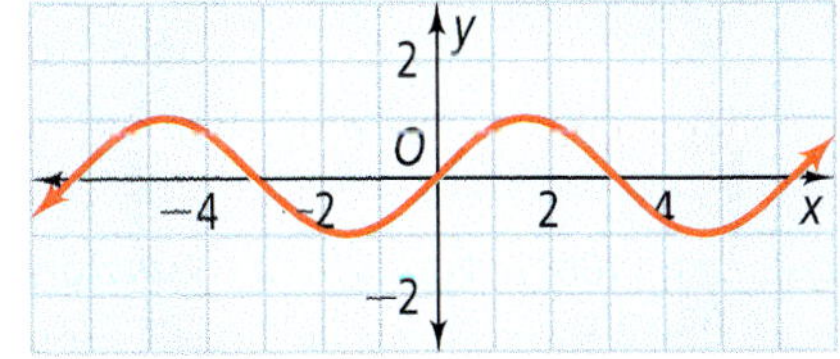

$P(\cos\theta, \sin\theta)$

Singular matrix (p. 785) A singular matrix is a square matrix with no inverse. Its determinant is 0.

Matriz singular (p. 785) Una matriz singular es una matriz al cuadrado que no tiene inverso. El determinante de la matriz es 0.

Slope (p. 74) The slope of a non-vertical line is the ratio of the vertical change to the horizontal change between points. You can calculate slope by finding the ratio of the difference in the y-coordinates to the difference in the x-coordinates for any two points on the line. The slope of a vertical line is undefined.

Pendiente (p. 74) La pendiente de una línea no vertical es la razón del cambio vertical al cambio horizontal entre puntos. Puedes calcular la pendiente al hallar la razón de la diferencia de la coordenada y y la diferencia de la coordenada x para dos puntos cualesquiera de la línea. La pendiente de una línea vertical es indefinida.

Example The slope of the line through points $(-1, -1)$ and $(1, -2)$ is $\frac{-2-(-1)}{1-(-1)} = \frac{-1}{2} = -\frac{1}{2}$.

Slope-intercept form (p. 76) The slope-intercept form of an equation of a line is $y = mx + b$, where m is the slope and b is the y-intercept.

Forma pendiente-intercepto (p. 76) La forma pendiente-intercepto de una ecuación lineal es $y = mx + b$, donde m es la pendiente y b es el intercepto en y.

Example $y = 8x + 2$

$y = -x + 1$

$y = -\frac{1}{2}x - 14$

Solution of a system (p. 134) A solution of a system is a set of values for the variables that makes all the equations true.

Solución de un sistema (p. 134) Una solución de un sistema es un conjunto de valores para las variables que hace que todas las ecuaciones sean verdaderas.

English	Spanish

Solution of an equation (p. 27) A solution of an equation is a number that makes the equation true.

Solución de una ecuación (p. 27) Una solución de una ecuación es cualquier número que haga verdadera la ecuación.

Example The solution of $2x - 7 = -12$ is $x = -2.5$.

Square matrix (p. 782) A square matrix is a matrix with the same number of columns as rows.

Matriz cuadrada (p. 782) Una matriz cuadrada es la que tiene la misma cantidad de columnas y filas.

Example Matrix A is a square matrix.

$$A = \begin{bmatrix} 1 & 2 & 0 \\ -1 & 0 & -2 \\ 1 & 2 & 3 \end{bmatrix}$$

Square root equation (p. 390) A square root equation is a radical equation in which the radical has index 2.

Ecuación de raíz cuadrada (p. 390) Una ecuación de raíz cuadrada es una ecuación radical en la cual el radical tiene índice 2.

Example $\sqrt{x} = 4$

Square root function (p. 415) A square root function is a function that can be written in the form $f(x) = a\sqrt{x - h} + k$, where $a \neq 0$. The domain of a square root function is all real numbers $x \geq h$.

Función de raíz cuadrada (p. 415) Una función de raíz cuadrada es una función que puede ser expresada como $f(x) = a\sqrt{x - h} + k$, donde $a \neq 0$. El dominio de una función de raíz cuadrada son todos los números reales tales que $x \geq h$.

Example $f(x) = 2\sqrt{x - 3} + 4$

Standard deviation (p. 719) Standard deviation is a measure of how much the values in a data set vary, or deviate, from the mean, $\bar{x}$. To find the standard deviation, follow five steps:

- Find the mean of the data set.
- Find the difference between each data value and the mean.
- Square each difference.
- Find the mean of the squares.
- Take the square root of the mean of the squares. This is the standard deviation.

Desviación típica (p. 719) La desviación típica denota cuánto los valores de un conjunto de datos varían, o se desvían, de la media, $\bar{x}$. Para hallar la desviación típica, se siguen cinco pasos:

- Se halla la media del conjunto de datos.
- Se calcula la diferencia entre cada valor de datos y la media.
- Se eleva al cuadrado cada diferencia.
- Se halla la media de los cuadrados.
- Se calcula la raíz cuadrada de la media de los cuadrados. Ésa es la desviación típica.

Example {0, 2, 3, 4, 6, 7, 8, 9, 10, 11}

$\bar{x} = 6$

standard deviation $= \sqrt{12} \approx 3.46$

Standard form of a circle (p. 630) *See* **Circle**.

Forma normal de un círculo (p. 630) *Ver* **Circle**.

Example $(x - 3)^2 + (y - 4)^2 = 4$

English	Spanish

Standard form of a linear equation (p. 82) The standard form of a linear equation is $Ax + By = C$, where A, B, and C are real numbers, and A and B are not *both* zero.

Forma normal de una ecuación lineal (p. 82) La forma normal de una ecuación lineal es $Ax + By = C$, donde A, B y C son números reales, y A y B no son cero *ambos*.

Example In standard form, the equation $y = \frac{4}{3}x - 1$ is $4x + (-3)y = 3$.

Standard form of a polynomial function (p. 281) The standard form of a polynomial function arranges the terms by degree in descending numerical order. A polynomial function, $P(x)$, in standard form is $P(x) = a_nx^n + a_{n-1}x^{n-1} + \cdots + a_1x + a_0$, where n is a nonnegative integer and $a_n, \ldots, a_0$ are real numbers.

Forma normal de una función polinomial (p. 281) La forma normal de una función polinomial organiza los términos por grado en orden numérico descendiente. Una función polinomial, $P(x)$, en forma normal es $P(x) = a_nx^n + a_{n-1}x^{n-1} + \cdots + a_1x + a_0$, donde n es un número entero no negativo y $a_n, \ldots, a_0$ son números reales.

Example $2x^3 - 5x^2 - 2x + 5$

Standard form of a quadratic function (p. 202) The standard form of a quadratic function is $f(x) = ax^2 + bx + c$ with $a \neq 0$.

Forma normal de una función cuadrática (p. 202) La forma normal de una función cuadrática es $f(x) = ax^2 + bx + c$ con $a \neq 0$.

Example $f(x) = 2x^2 + 5x + 2$

Standard position (p. 836) An angle in the coordinate plane is in **standard position** when the vertex is at the origin and one ray is on the positive x-axis.

Posición estándar (p. 836) Un ángulo en el plano de coordenadas se encuentra en **posición estándar** si el vértice se encuentra en el origen y una semirrecta se encuentra en el eje x positivo.

Example

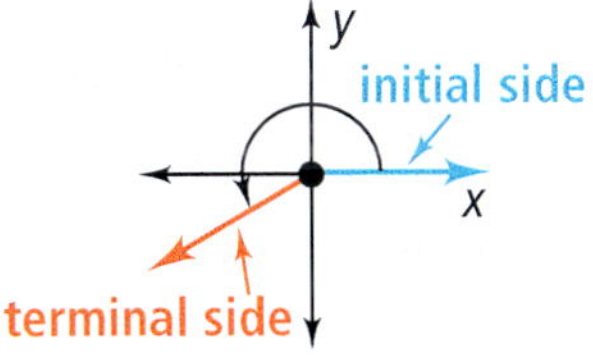

Step function (p. 90) A step function pairs every number in an interval with a single value. The graph of a step function can look like the steps of a staircase.

Función escalón (p. 90) Una función escalón empareja cada número de un intervalo con un solo valor. La gráfica de una función escalón se puede parecer a los peldaños de una escalera.

Sum of cubes (p. 297) The sum of cubes is an expression of the form $a^3 + b^3$. It can be factored as $(a + b)(a^2 - ab + b^2)$.

Suma de dos cubos (p. 297) La suma de dos cubos es una expresión de la forma $a^3 + b^3$. Se puede factorizar como $(a + b)(a^2 - ab + b^2)$

Example $x^3 + 27 = (x + 3)(x^2 - 3x + 9)$

Survey (p. 726) In a survey, you ask every member of a sample the same set of questions.

Encuesta (p. 726) En una encuesta, se le hace a cada miembro de una muestra la misma serie de preguntas.

English / Spanish

Synthetic division (p. 306) Synthetic division is a process for dividing a polynomial by a linear expression $x - a$. You list the standard-form coefficients (including zeros) of the polynomial, omitting all variables and exponents. You use a for the "divisor" and add instead of subtract throughout the process.

División sintética (p. 306) La división sintética es un proceso para dividir un polinomio por una expresión lineal $x - a$. En este proceso, escribes los coeficientes de forma normal (incluyendo los ceros) del polinomio, omitiendo todas las variables y todos los exponentes. Usas a como "divisor" y sumas, en vez de restar, a lo largo del proceso.

Example

$$\begin{array}{r|rrrrr} -3 & 2 & 5 & 0 & -2 & -8 \\ & & -6 & 3 & -9 & 33 \\ \hline & 2 & -1 & 3 & -11 & 25 \end{array}$$

Divide $2x^4 + 5x^3 - 2x - 8$ by $x + 3$. $2x^4 + 5x^3 - 2x - 8$ divided by $x + 3$ gives $2x^3 - x^2 + 3x - 11$ as quotient and 25 as remainder.

System of equations (p. 134) A system of equations is a set of two or more equations using the same variables.

Sistema de ecuaciones (p. 134) Un sistema de ecuaciones es un conjunto de dos o más ecuaciones que contienen las mismas variables.

Example $\begin{cases} 2x - 3y = -13 \\ 4x + 5y = 7 \end{cases}$

Systematic sample (p. 725) In a systematic sample you order the population in some way, and then select from it at regular intervals.

Muestra sistemática (p. 725) En una muestra sistemática se ordena la población de cierta manera y luego se selecciona una muestra de esa población a intervalos regulares.

Tangent function, Tangent of θ (pp. 868, 869) The tangent function, $y = \tan\theta$, matches the measure θ, of an angle in standard position with the y/x ratio of the (x, y) coordinates of a point on the unit circle. This point is where the terminal side of the angle intersects the unit circle. y/x is the tangent of θ.

Función tangente, Tangente de θ (pp. 868, 869) La función tangente, $y = \tan\theta$, empareja la medida θ, de un ángulo en posición estándar con la razón y/x de las coordenadas (x, y) de un punto en el círculo unitario. Este es el punto en el que el lado terminal del ángulo interseca al círculo unitario. y/x es la tangente de θ.

Example

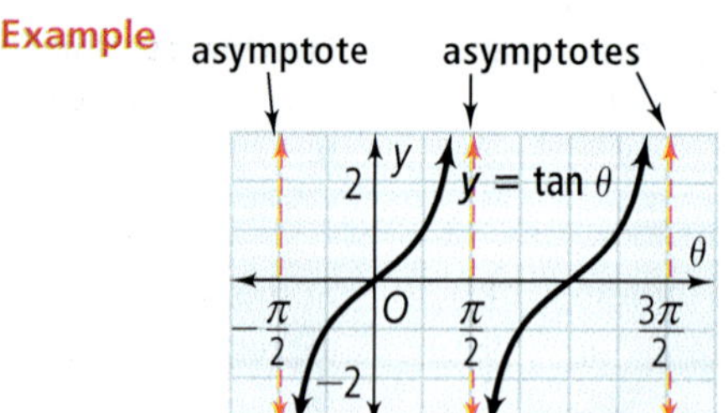

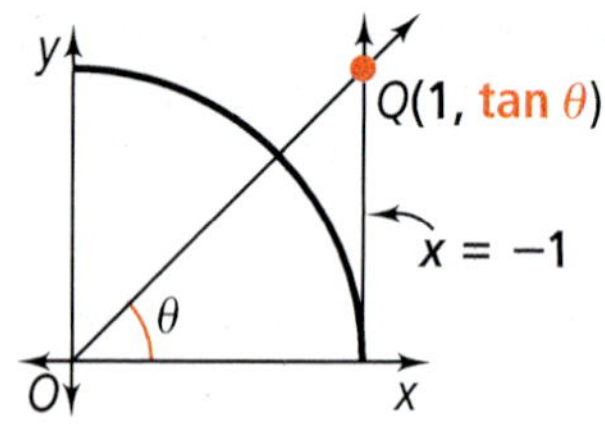

Term of a sequence (p. 564) Each number in a sequence is a term.

Término de una progresión (p. 564) Cada número de una progresión es un término.

Example 1, 4, 7, 10, . . .
The second term is 4.

English | Spanish

Term of an expression (p. 20) A term is a number, a variable, or the product of a number and one or more variables.

Término de una expresión (p. 20) Un término es un número, una variable o el producto de un número y una o más variables.

Example The expression $4x^2 - 3y + 7.3$ has 3 terms.

Terminal point (p. 809) The terminal point of a vector is the tip (not the endpoint) of a vector arrow.

Punto terminal (p. 809) El punto terminal de un vector es la punta (no el extremo) de una flecha vectorial.

Terminal side (p. 836) *See* **Initial side.**

Lado terminal (p. 836) *Ver* **Initial side.**

Test point (p. 115) A test point is a point that you pick on one side of the boundary of the graph of a linear inequality. If the test point makes the inequality true, then all points on that side of the boundary are solutions of the inequality. If the test point makes the inequality false, then all points on the other side are solutions.

Punto de prueba (p. 115) Un punto de prueba es un punto que escoges a un lado del límite de la gráfica de una desigualdad lineal. Si el punto de prueba hace que la desigualdad sea verdadera, entonces todos los puntos en ese límite son soluciones de la desigualdad. Si el punto de prueba hace que la desigualdad sea falsa, entonces todos los puntos del otro lado del límite son soluciones.

Theoretical probability (p. 683) If a sample space has n equally likely outcomes, and an event A occurs in m of these outcomes, then the theoretical probability of event A is $P(A) = \frac{m}{n}$.

Probabilidad teórica (p. 683) Si un espacio muestral tiene n resultados con la misma probabilidad de ocurrir, y ocurre un suceso A en m de estos resultados, entonces la probabilidad teórica del suceso A es $P(A) = \frac{m}{n}$.

Example Use the set {1, 4, 9, 16, 25, 36, 49, 64, 81, 100}. The probability that a number selected at random is greater than 50 is $P(A) = \frac{3}{10} = 0.3$.

Transformation (p. 99) A transformation of a function $y = af(x - h) + k$ is a change made to at least one of the values a, h, and k. The four types of transformations are dilations, reflections, rotations, and translations.

Transformación (p. 99) Una transformación de una función $y = af(x - h) + k$ es un cambio que se le hace a por lo menos uno de los valores a, h y k. Hay cuatro tipos de transformaciones: dilataciones, reflexiones, rotaciones y traslaciones.

Example $g(x) = 2(x - 3)$ is a transformation of $f(x) = x^2$.

Translation (p. 99) A translation shifts the graph of the parent function horizontally, vertically, or both without changing its shape or orientation.

Traslación (p. 99) Una traslación desplaza la gráfica de la función madre horizontalmente, verticalmente o en ambas direcciones, sin cambiar su forma u orientación.

Example

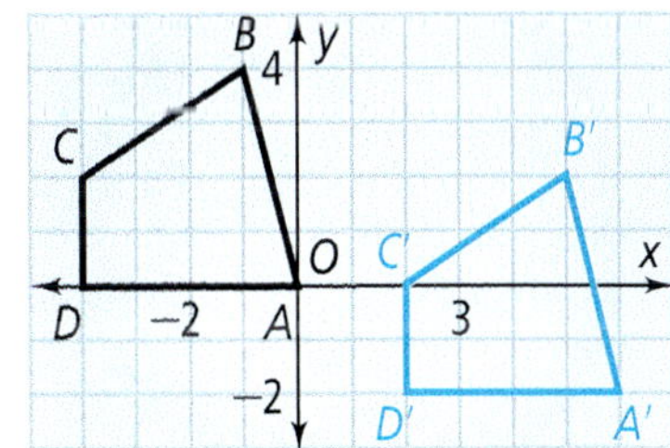

Visual **Glossary**

English / Spanish

Transverse axis (p. 646) The transverse axis of a hyperbola is the segment that is on the line containing the foci and has endpoints on the hyperbola.

Eje transversal (p. 646) El eje transversal de una hipérbola es el segmento que se encuentra sobre la línea que contiene los focos y tiene sus puntos extremos sobre la hipérbola.

Example

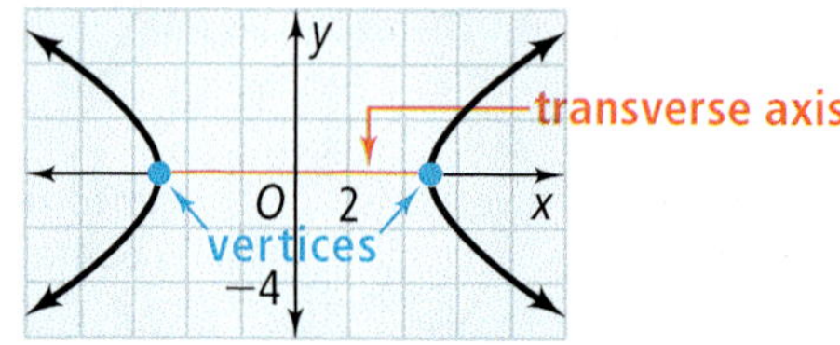

Trend line (p. 93) A trend line is a line that approximates the relationship between two variables, or data sets, of a scatter plot.

Línea de tendencia (p. 93) Una línea de tendencia es una línea que aproxima la relación entre dos variables o conjuntos de datos de un diagrama de dispersión.

Example

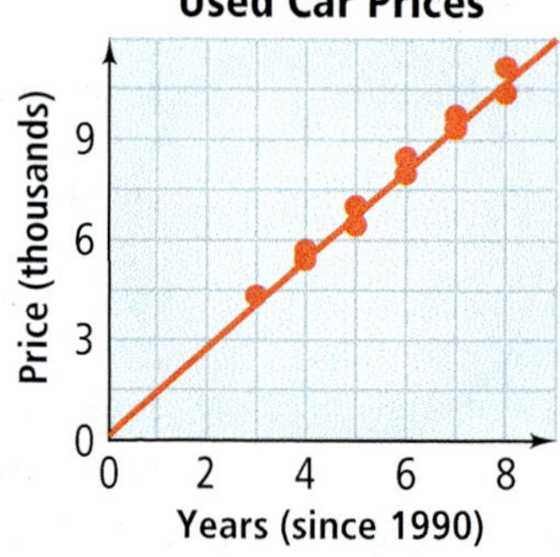

Trigonometric identity (p. 904) A trigonometric identity in one variable is a trigonometric equation that is true for all values of the variable for which both sides of the equation are defined.

Identidad trigonométrica (p. 904) Una identidad trigonométrica en una variable es una ecuación trigonométrica que es verdadera para todos los valores de la variable para los cuales se definen los dos lados de la ecuación.

Example $\tan\theta = \frac{\sin\theta}{\cos\theta}$

Trigonometric ratios (p. 920) If θ is an acute angle of a right triangle, x is the length of the adjacent leg (ADJ), y is the length of the opposite leg (OPP), and r is the length of the hypotenuse (HYP), then the trigonometric ratios of θ are

$\sin\theta = \frac{y}{r} = \frac{\text{OPP}}{\text{HYP}}$ $\quad \csc\theta = \frac{r}{y} = \frac{\text{HYP}}{\text{OPP}}$

$\cos\theta = \frac{x}{r} = \frac{\text{ADJ}}{\text{HYP}}$ $\quad \sec\theta = \frac{r}{x} = \frac{\text{HYP}}{\text{ADJ}}$

$\tan\theta = \frac{y}{x} = \frac{\text{OPP}}{\text{ADJ}}$ $\quad \cot\theta = \frac{x}{y} = \frac{\text{ADJ}}{\text{HYP}}$

Razones trigonométricas (p. 920) Si θ es un ángulo agudo de un triángulo, x es la longitud del cateto adyacente (ADJ), y es la longitud del cateto opuesto (OPP) y r es la longitud de la hipotenusa (HYP), entonces las razones trigonométricas son:

$\sin\theta = \frac{y}{r} = \frac{\text{OPP}}{\text{HYP}}$ $\quad \csc\theta = \frac{r}{y} = \frac{\text{HYP}}{\text{OPP}}$

$\cos\theta = \frac{x}{r} = \frac{\text{ADJ}}{\text{HYP}}$ $\quad \sec\theta = \frac{r}{x} = \frac{\text{HYP}}{\text{ADJ}}$

$\tan\theta = \frac{y}{x} = \frac{\text{OPP}}{\text{ADJ}}$ $\quad \cot\theta = \frac{x}{y} = \frac{\text{ADJ}}{\text{HYP}}$

Turning point (p. 282) A turning point of the graph of a function is a point where the graph changes direction from upwards to downwards or from downwards to upwards.

Punto de giro (p. 282) Un punto de giro de la gráfica de una función es un punto donde la gráfica cambia de dirección de arriba hacia abajo o vice versa.

English

Spanish

Uniform Distribution (p. 694) A uniform distribution is a probability distribution that is equal for each event in the sample space.

Distribución uniforme (p. 694) Una distribución uniforme es una distribución de probabilidad que es igual para cada suceso en el espacio muestral.

Unit circle (p. 838) The unit circle has a radius of 1 unit and its center is at the origin of the coordinate plane.

Círculo unitario (p. 838) El círculo unitario tiene un radio de 1 unidad y el centro está situado en el origen del plano de coordenadas.

Example

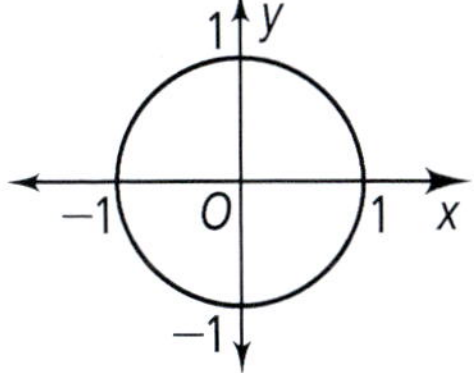

Value (p. 5) The value of a quantity is its measure or the number of items that you count.

Valor (p. 5) El valor de una cantidad es su medida o el número de datos que cuentas.

Variable (p. 5) A variable is a symbol, usually a letter, that represents one or more numbers.

Variable (p. 5) Una variable es un símbolo, generalmente una letra, que representa uno o más valores.

Variable matrix (p. 793) When representing a system of equations with a matrix equation, the matrix containing the variables of the system is the variable matrix.

Matriz variable (p. 793) Al representar un sistema de ecuaciones con una ecuación de matricial, la matriz que contenga las variables del sistema es la matriz variable.

Example $\begin{cases} x + 2y = 5 \\ 3x + 5y = 14 \end{cases}$

variable matrix $\begin{bmatrix} X \\ Y \end{bmatrix}$

Variable quantity (p. 5) A variable quantity can have values that vary.

Cantidad variable (p. 5) Una cantidad variable puede tener valores que varían.

Variance (p. 719) Variance is the square of the standard deviation. $\sigma^2 = \frac{(\Sigma(x - \bar{x})^2)}{n}$.

Varianza (p. 719) La varianza es el cuadrado de la desviación estándar. $\sigma^2 = \frac{(\Sigma(x - \bar{x})^2)}{n}$.

Vector (p. 809) A vector is a mathematical object that has both magnitude and direction.

Vector (p. 809) Un vector es un objeto matemático que tiene tanto magnitud como dirección.

Vertex (p. 107) A vertex of a function is a point where the function reaches a maximum or a minimum value.

Vértice (p. 107) El vértice de una función es el punto donde la funcion alcanza un valor máximo o mínimo.

English	Spanish

Vertex form of a quadratic function (p. 194) The vertex form of a quadratic function is $f(x) = a(x - h)^2 + k$, where $a \neq 0$ and (h, k) is the coordinate of the vertex of the function.

Forma del vértice de una función cuadrática (p. 194) La forma vértice de una función cuadrática es $f(x) = a(x - h)^2 + k$, donde $a \neq 0$ y (h, k) es la coordenada del vértice de la función.

Example $f(x) = x^2 + 2x - 1 = (x + 1)^2 - 2$
The vertex is $(-1, -2)$.

Vertex of a parabola (p. 194) The vertex of a parabola is the point where the function for the parabola reaches a maximum or a minimum value. The parabola intersects its axis of symmetry at the vertex.

Vértice de una parábola (p. 194) El vértice de una parábola es el punto donde la función de la parábola alcanza un valor máximo o mínimo. La parábola y su eje de simetría se intersecan en el vértice.

Example

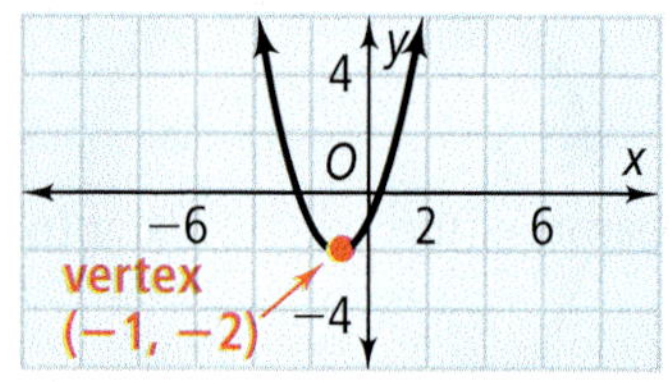

The vertex of the quadratic function $y = x^2 + 2x - 1$ is $(-1, -2)$.

Vertical compression (p. 102) A vertical compression reduces all y-values of a function by the same factor between 0 and 1.

Compresión vertical (p. 102) Una compresión vertical reduce todos los valores de y de una función por el mismo factor entre 0 y 1.

Vertical stretch (p. 102) A vertical stretch multiplies all y-values of a function by the same factor greater than 1.

Estiramiento vertical (p. 102) Un estiramiento vertical multiplica todos los valores de y por el mismo factor mayor que 1.

Vertical-line test (p. 62) You can use the vertical-line test on the graph of a relation to tell whether the relation is a function. If a vertical line passes through more than one point on the graph of a relation, then the relation is *not* a function.

Prueba de la recta vertical (p. 62) Puedes usar la prueba de la línea vertical en la gráfica de una relación para saber si la relación es una función. Si una línea vertical pasa por más de un punto de la gráfica de la relación, entonces la relación *no* es una función.

Example

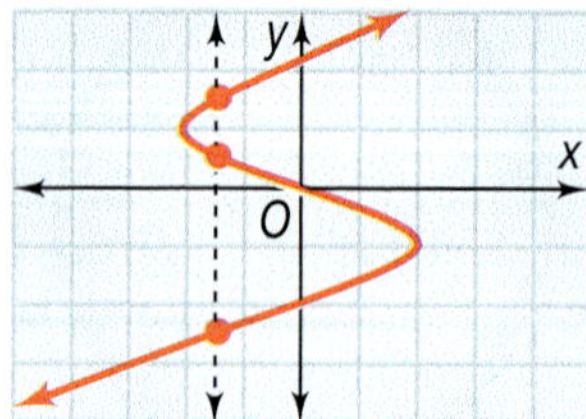

This relation is not a function.

English | Spanish

Vertices of a hyperbola (p. 646) The endpoints of the transverse axis of a hyperbola are the vertices of the hyperbola.

Vértices de una hipérbola (p. 646) Los dos puntos de intersección de la hipérbola y su eje mayor son los vértices de la hipérbola.

Example

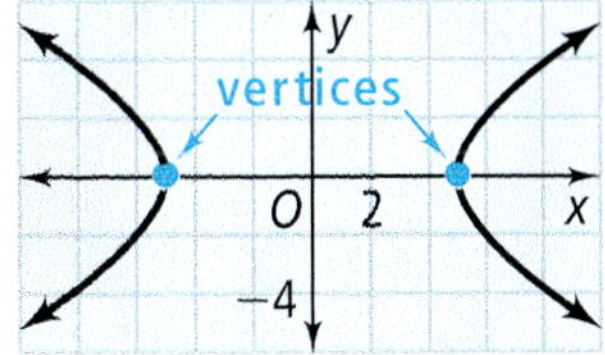

Vertices of an ellipse (p. 639) The endpoints of the major axis of an ellipse are the vertices of the ellipse.

Vértices de una elipse (p. 639) Los dos puntos de intersección de la elipse y su eje mayor son los vértices de la elipse.

Example

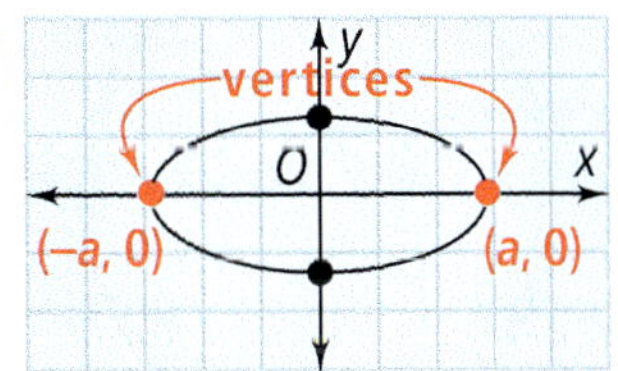

X

x-intercept, y-intercept (p. 76) The point at which a line crosses the x-axis (or the x-coordinate of that point) is an x-intercept. The point at which a line crosses the y-axis (or the y-coordinate of that point) is a y-intercept.

Intercepto en x, intercepto en y (p. 76) El punto donde una recta corta el eje x (o la coordenada x de ese punto) es el intercepto en x. El punto donde una recta cruza el eje y (o la coordenada y de ese punto) es el intercepto en y.

Example

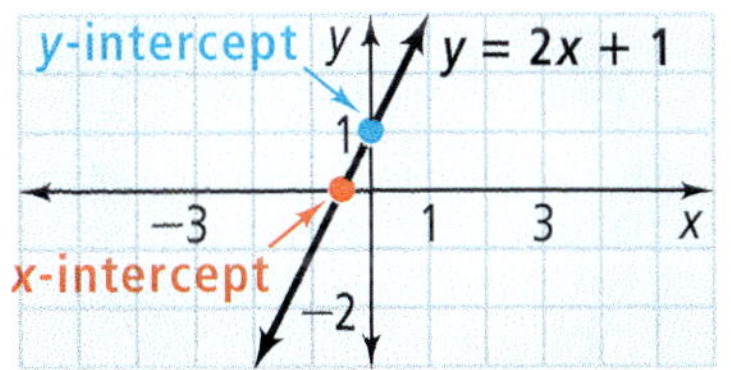

The x-intercept of $y = 2x + 1$ is $\left(-\frac{1}{2}, 0\right)$ or $-\frac{1}{2}$.
The y-intercept of $y = 2x + 1$ is $(0, 1)$ or 1.

Z

Zero matrix (p. 766) The zero matrix O, or $O_{m \times n}$, is the $m \times n$ matrix whose elements are all zeros. It is the additive identity matrix for the set of all $m \times n$ matrices.

Matriz cero (p. 766) La matriz cero, O, o $O_{m \times n}$, es la matriz $m \times n$ cuyos elementos son todos ceros. Es la matriz de identidad aditiva para el conjunto de todas las matrices $m \times n$.

Example $\begin{bmatrix} 1 & 4 \\ 2 & -3 \end{bmatrix} + O = \begin{bmatrix} 1 & 4 \\ 2 & -3 \end{bmatrix}$

English	Spanish

Zero of a function (p. 226) A zero of a function $f(x)$ is any value of x for which $f(x) = 0$.

Cero de una función (p. 226) Un cero de una función $f(x)$ es cualquier valor de x para el cual $f(x) = 0$.

Example

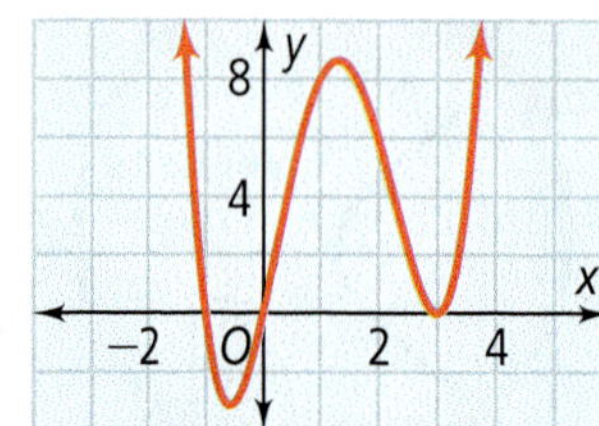

Zero-Product Property (p. 226) If the product of two or more factors is zero, then one of the factors must be zero.

Propiedad del cero del producto (p. 226) Si el producto de dos o más factores es cero, entonces uno de los factores debe ser cero.

Example $(x - 3)(2x - 5) = 0$
$x - 3 = 0$ or $2x - 5 = 0$

***z*-score (p. 748)** The z-score of a value is the number of standard deviations that the value is from the mean.

Puntaje *z* (p. 748) El puntaje z de un valor es el número de desviaciones normales que tiene ese valor de la media.

Example $\{0, 2, 3, 4, 6, 7, 8, 9, 10, 11\}$
$\bar{x} = 6$
standard deviation $= \sqrt{12} \approx 3.46$
For 8, z-score $= \frac{8 - 6}{\sqrt{12}} \approx 0.58$.

Selected Answers

Chapter 1

Get Ready! pp. 1 **1.** 0 **2.** −2 **3.** −2.09 **4.** 8.05 **5.** $-\frac{3}{4}$ **6.** $\frac{11}{12}$ **7.** $10\frac{7}{10}$ **8.** $3\frac{1}{2}$ **9.** −42 **10.** 72 **11.** 9 **12.** −9.8 **13.** $-3\frac{1}{3}$ **14.** $-5\frac{1}{2}$ **15.** $-4\frac{2}{3}$ **16.** $-\frac{3}{4}$ **17.** −21 **18.** 7.35 **19.** $-\frac{1}{6}$ **20.** $-\frac{3}{5}$ **21.** −20 **22.** 8 **23.** 0.97 **24.** −5 **25.** 55 **26.** 3 **27.** because the placement of the parentheses changes the order of operations **28.** 3 **29.** 3 terms **30.** Calculate the answer numerically. **31.** $\frac{3}{n-3}$

Lesson 1-1 pp. 4–10

Got It? 1. The pattern shows a center square and a yellow square added to each side with the number of squares per side increasing by one.

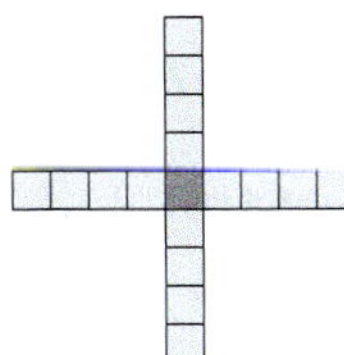

2. 52 tiles **3. a.** $12 **b.** $20 **c.** The number of platys must be a whole number whereas the length of fish can be a fraction or a decimal.

Lesson Check 1. add 35 **2.** rotate 90° clockwise

3.

Input	Process Column	Output
1	2(1)	2
2	2(2)	4
3	2(3)	6
4	2(4)	8
⋮	⋮	⋮
n	$2(n)$	$2n$

4.

Input	Process Column	Output
1	3(1)	3
2	3(2)	6
3	3(3)	9
4	3(4)	12
⋮	⋮	⋮
n	$3(n)$	$3n$

5. Answers may vary. Sample: Look for the same type of change between consecutive figures.
6. Answers may vary. Sample: Tables of values and pictorial representations are both convenient ways to organize data and discover patterns. Tables give more detail. Pictorial representations are visual. **7.** No; the output is $\frac{1}{2}$ the input for all values except the first (Input: 3; Output: 2).

Exercises 9. Base of 3 squares with the number of squares increasing vertically by one on each of the outer squares of the base.

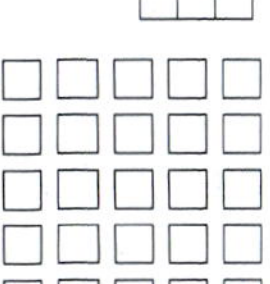

11. One square, then 2^2 or 4 squares, then 3^2 or 9 squares, then 4^2 or 16 squares. In general, the number of squares is $n \times n$ or n^2.

13. $2n$

Input	Process Column	Output
1	2(1)	2
2	2(2)	4
3	2(3)	6
4	2(4)	8
⋮	⋮	⋮
n	$2(n)$	$2n$

15. $4n - 1$

Input	Process Column	Output
1	4(1) − 1	3
2	4(2) − 1	7
3	4(3) − 1	11
4	4(4) − 1	15
⋮	⋮	⋮
n	$4(n) - 1$	$4n - 1$

17. 7; 8; $n + 2$
19. Output = Input + 1

Input	Process Column	Output
1	(1) + 1	2
2	(2) + 1	3
3	(3) + 1	4
4	(4) + 1	5
5	(5) + 1	6
⋮	⋮	⋮
n	$(n) + 1$	$n + 1$

21. Output = Input − 1

Input	Process Column	Output
1	(1) − 1	0
2	(2) − 1	1
3	(3) − 1	2
4	(4) − 1	3
5	(5) − 1	4
⋮	⋮	⋮
n	$(n) - 1$	$n - 1$

23. 40 **25.** add 6 or $6n$; 30, 36, 42 **27.** add 3, then add 4, then add 5, and so on; 21, 28, 36 **29.** multiply by 3; 243, 729, 2187 **31.** The black square and dot each move clockwise one block **33.** 9216 in.^{3} **35.** $n + 10$, where n is the number of months **37.** 21; $4n + 1$ **39.** −13; $7 - 4n$; or $-4n + 7$ **41.** Answers may vary. Sample: Jesse will not grow at the same rate between the ages of 15 and 20 as he has during the 4 years prior to age 15. **43.** D **45. a.** Each number is a result of the division of the previous number by 2. **b.** $36 \div 2 = 18$, $18 \div 2 = 9$, $9 \div 2 = 4.5$, so 4.5 is the first noninteger number. **46.** 1.9 **47.** −3.8 **48.** 27 **49.** 0 **50.** −0.4 **51.** 7 **52.** 50% **53.** 25% **54.** 33.33% **55.** 140% **56.** 172% **57.** 123%

Lesson 1-2 pp. 11–17

Got It? 1. rational numbers
2. $\frac{1}{3}$ 1.4 $\sqrt{3}$
−2 −1 0 1 2

3. a. $\sqrt{26} < 6.25$ or $6.25 > \sqrt{26}$ **b.** $a < c$; a will be to the left of c on the number line.
4. a. Distr. Prop.
b. $a + [3 + (-a)]$
$= a + [(-a) + 3]$ Comm.
$= [a + (-a)] + 3$ Assoc.
$= 0 + 3$ Inverse
$= 3$ Identity

Lesson Check 1. Answers may vary. Sample: the number of times a cricket chirps **2.** Answers may vary. Sample: the change in number of people on a bus after a stop **3.** Answers may vary. Sample: the outdoor temperature in tenths of a degree **4.** Inv. Prop. of Add. **5.** Assoc. Prop. of Mult. **6.** multiplicative inverse **7.** Both properties result in the original term; 0 is the additive identity, whereas 1 is the multiplicative identity. **8.** The equation illustrates the Comm. Prop. of Add. **9.** Answers may vary. Sample: $\sqrt{2}$ is not a rational number because it cannot be written as a quotient of integers.
Exercises 11. y, natural numbers; p, rational numbers
13. −3 −2 −1 0 1 2
15. −4 −3 −2 −1 0 1 2
17. −6 −5 −4 −3 −2 −1 0 1
19. −3 −2 −1 0 1
21. −4 −3 −2 −1 0 1
23. > **25.** < **27.** > **29.** > **31.** > **33.** < **35.** Distr. Prop. **37.** Assoc. Prop. of Mult. **39.** Ident. Prop. of Add. **41–48.** Answers may vary. Samples are given. **41.** −5 **43.** $-1\frac{1}{4}$ **45.** $1\frac{2}{3}$ **47.** 4 **49.** $\sqrt{50}$ in. × $\sqrt{50}$ *in.* × $\sqrt{50}$ *in.* **51.** natural numbers **53.** irrational numbers **55.** irrational numbers **57.** 8, 1, $\frac{1}{3}$, $-\sqrt{2}$, −3 **59.** 5.73, $\frac{1}{4}$, −0.06, $-3\sqrt{3}$, −17 **61.** Answers may vary. Sample: 7 **63.** Answers may vary. Sample: $\sqrt{2}$ and $\sqrt{2}$ **65.** Answers may vary. **67.** Answers may vary. **69.** $5(x + 2y - 7)$ **71.** No; $\frac{1}{0}$ is undefined. **73.** H **75.** add 4; 20, 24, 28 **76.** add 1; 12, 13, 14 **77.** add 1; 0, 1, 2 **78.** $2\frac{1}{4}$ **79.** $11\frac{2}{3}$ **80.** $1\frac{1}{2}$ **81.** 5 **82.** 38 **83.** 15

Lesson 1-3 pp. 18–24

Got It? 1. H **2.** $150 - 2d$, with d = the number of days **3. a.** 18 **b.** Yes; the numerator will become $2x^2 - y^2$, not $2x^2 - 2y^2$. **4.** Let x = the number of two-point shots, y = the number of three-point shots, z = the number of one-point free throws, $2x + 3y + 1z$; 42 points **5. a.** $-3j^2 - 7k + 5j$ **b.** $12a - 53b$
Lesson Check 1. $\frac{2 + b}{3}$ **2.** $4k + m$ **3.** 12 **4.** 13 **5.** −5 **6.** −5 **7.** The student did not distribute the −1. $3p^2q + 2p - (5q + p - 2p^2q) = 3p^2q + 2p - 5q - p + 2p^2q = 5p^2q + p - 5q$ **8.** A constant is a term with no variables, whereas a coefficient is the numerical factor in a term. **9.** Answers may vary. Sample: Both algebraic expressions and numerical expressions represent a quantity using numbers, operations and grouping symbols. An algebraic expression includes variables when representing a quantity. Examples: numerical expression: $3 + 6(5 - 2)$; algebraic expression: $2z + 3z(6 + 5z)$.
Exercises 11. $8(x + 3)$ **13.** $130 - 10w$, with w = number of weeks **15.** $250 - 60w$, with w = number of weeks **17.** −16 **19.** −12 **21.** 4 ft **23.** 1600 ft **25.** \$1331 **27.** \$1610.51 **29.** Let x = the number of 3-run home runs and y = the number of 2-run hits; $3x + 2y$; 14 **31.** $2s + 5$ **33.** $6a + 3b$ **35.** $-0.5x$ **37.** $4g - 2$ **39.** 3 **41.** 37 **43.** 10 **45.** $\frac{\$84}{m}$ **47.** $\frac{5x^2}{2}$ **49.** y **51.** $-2x^2 + 2y^2$ **53.** $8.5x - 15$ **55.** No; John did not use the opposite of a sum correctly; $-(x + y) + 3(x - 4y)$; $-x - y + 3x - 12y$; $2x - 13y$ **57.** Distr. Prop. **59.** Opposite of a Difference

61. Answers may vary. Sample:

$$\begin{aligned} & 2(b-a)+5(b-a) \\ &= (2+5)(b-a) && \text{Distr. Prop.} \\ &= 7(b-a) && \text{Add.} \\ &= 7b-7a && \text{Distr. Prop.} \end{aligned}$$

63. A **65.** C **67.** $-1.5, -\sqrt{2}, -1.4, -0.5$
68. $-\frac{5}{6}, -\frac{3}{4}, -\frac{3}{8}, \frac{1}{2}$ **69.** $-20, 0.2, \frac{1}{2}, \sqrt{2}$
70. $-3, -0.5, -\frac{1}{4}, \frac{3}{4}$ **71.** $7x - 4$ **72.** $-p - \frac{2q}{3}$
73. $2b - 28$ **74.** $2k - 2m$

Lesson 1-4 pp. 26–32

Got It? 1. $\frac{3}{2}$ **2.** -1 **3.** 40 m $\times$ 120 m **4. a.** never **b.** always **5. a.** $C = K - 273$ **b.** always
Lesson Check 1. 23.2 **2.** -90 **3.** 12
4. $k = \frac{1}{2}(r - 15)$ **5.** $k = \frac{1}{3}(z + 6)$
6. $k = -\left(\frac{1}{6}\right)(h + 14)$ **7.** To find a solution of an equation means to find the value of the variable that makes the equation true. **8.** Four buses are not enough. The number of buses must be a whole number, so round the number of buses to 5. **9.** The 2nd line is incorrect; subtract 10 from both sides: $12x = -12$ $x = -1$
Exercises 11. -81 **13.** 14 **15.** 8 **17.** -5
19. $-\frac{1}{9}$ **21.** $\frac{3}{2}$ **23.** -6 **25.** 0 **27.** 300 mi/h; 600 mi/h
29. sometimes **31.** sometimes **33.** $h = \frac{2A}{b}$ **35.** $w = \frac{V}{\ell h}$
37. $x = \frac{c}{a+b}, a \neq -b$ **39.** $x = 2(m + n) + 2$ **41.** 1.5
43. $\frac{23}{3}$, or $7\frac{2}{3}$ **45.** 34° and 56° **47.** $b_2 = \frac{2A}{h} - b_1$
49. $v = \frac{h + 5t^2}{t}$ **51.** $r_2 = \frac{Rr_1}{r_1 - R}$ **53.** 40°, 140°
55. $x = \frac{3b + 2c - 5}{b - c}, \frac{b}{c}$° **57.** $x = \frac{4a - 3bc}{aq - 5bp}, 5bp \neq aq$
59. $x = \frac{10c}{a}, a \neq 0$ **61.** Let c = number of swim days; $3c = 82 + c$; 41 days **63.** No; $n = \frac{s}{1-s}$ ***not*** $\frac{s}{s-1}$
65. 264 ft **67.** 3200 **69.** 7 ft **70.** -7 **71.** $-\frac{16}{3}$
72. -20 **73.** $-\frac{11}{2}$ **74.** $x + 5$ **75.** $16x$
76. $3(12 - x)$ **77.** true **78.** false **79.** true

Lesson 1-5 pp. 33–40

Got It? 1. $\frac{x}{3} \leq 15$
2. $x \leq -8$ 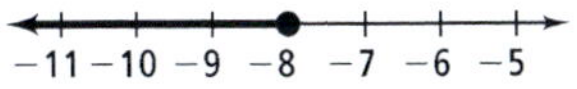
3. more than 32 songs **4.** always
5. a. $x \geq 2$ and $x < 6$
b. sometimes; The compound inequality is true when $x = 5$ and not true when $x = 7$.

6. a. $w < -3$ or $w > \frac{8}{7}$

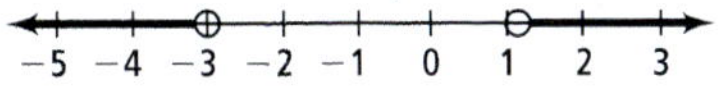

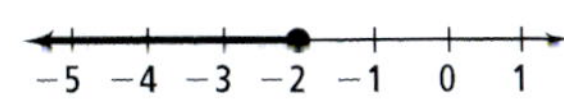

b. $x < -1$ or $x > 3$

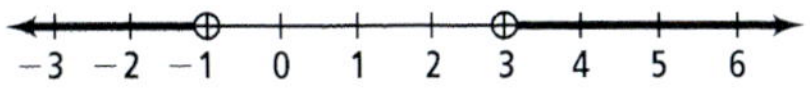

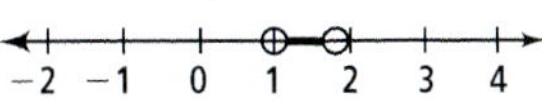

Lesson Check 1. $R \geq J$ **2.** $w \geq 40$ and $w < 74$
3. $x \leq -2$
4. $1 < x < \frac{9}{5}$
5. $x \leq 0$ or $x > 3$ 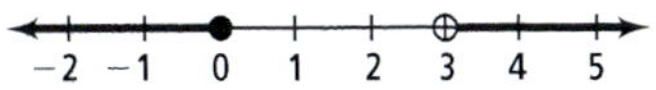
6. Answers may vary. Sample: $5 < 6$, but $-5 > -6$.
7. The transitive, addition and subtraction properties of inequality are similar to the properties of equality. The multiplication and division properties differ. Multiplying or dividing each side of an inequality by a negative number reverses the direction of the inequality symbol.
8. Answers may vary. Sample: $3x + 5 < 3(x + 5)$
9. No; Answers may vary. Sample: $2x < x + 1$ and $x + 1 > 3$
Exercises 11. $8x \geq 25$ **13.** $\frac{x}{12} \leq 6$
15. $k > -9$ 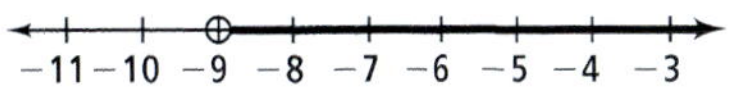
17. $t \leq 11$ 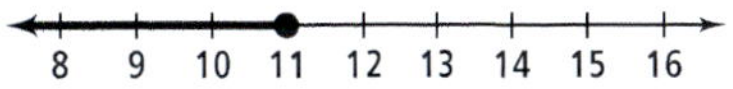
19. $y \leq -6$ 
21. $m < 10$
23. $w > 6$ 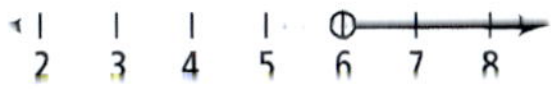
25. The longest side is less than 21 cm. **27.** at most 40 students **29.** always **31.** never **33.** sometimes
35. sometimes
37. $-4 \leq x \leq 2$

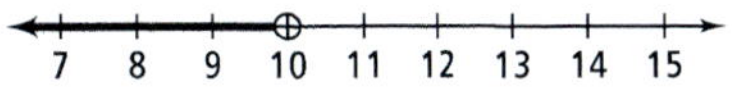

39. $-5 < x \leq 6$

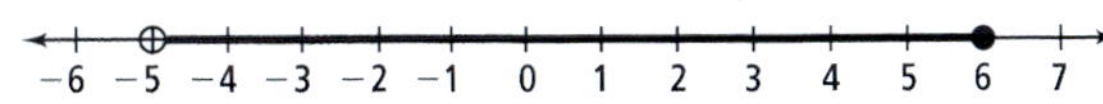

41. all real numbers

43. $x \leq -3$ or $x \geq 9$

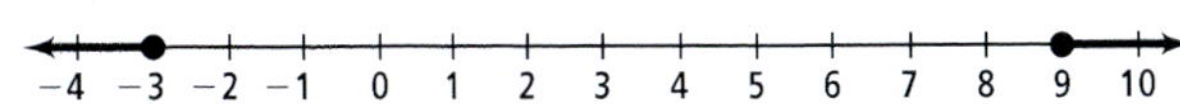

45. $z \geq 6$

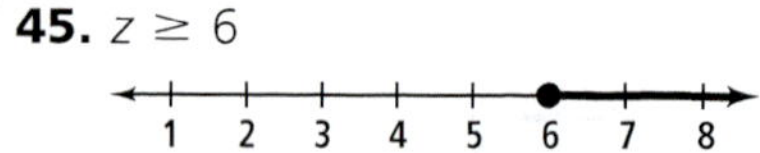

47. $x \geq -48$

49. no solution **51.** 98 **53.** $2 < AB < 6$ **55.** The classmate reversed the direction of the $\geq$ symbol to $\leq$ incorrectly. The correct answer is $y \leq -20$. **57.** Distr. Prop.; arithmetic; Subtr. Prop. of Inequality; Mult. Prop. of Inequality

59. $-1 < x < 8$

61. $x < -2$ or $x > 2$

63. Answers may vary. Sample: $-3x + 1 > 4$

65. Answers may vary. Sample: $2x + 4 \leq 0$ or $-3x - 3 \leq 0$

67. D **69.** D **71.** $7a + 5$ **72.** $-2x + 14y$

73. $\frac{b}{12} + 1$ **74.** $1.61 - 0.1k$ **75.** 4

76. no solution **77.** $\frac{9}{10}$ **78.** -20

Lesson 1-6 pp. 41–48

Got It?

1. $\frac{2}{3}, -2$

2. $-7, -11$

3. -1 **4.** $-\frac{4}{3} \leq x \leq 4$

5. a. $x < -5$ or $x > 1$

b. The graph will have two closed circles with an arrow extending to the left of one and to the right of the other. **6.** $|h - 52.5| \leq 0.5$

Lesson Check 1. $-4, 4$ **2.** $-12, 4$ **3.** $-\frac{6}{5}$

4. $-11 < x < 9$

5. $x \leq -1$ or $x \geq 4$

6. A solution of an eq. is extraneous if it is a solution to a derived eq., but is not a solution to the original eq.

7. when the number is positive or 0 **8.** Answers may vary. Sample: $d < -5$ and $5d > 25$ **9.** Answers may vary. Sample: An absolute value equation or inequality represents two equations or inequalities; each equation or inequality is solved in the same manner as a linear equation or inequality.

Exercises 11. $-8, 8$ **13.** $-\frac{5}{3}, 3$ **15.** no solution

17. $-7, 17$ **19.** $-\frac{3}{2}$ **21.** $\frac{3}{2}$ **23.** $-1, \frac{3}{2}$

25. $0 < y < 18$

27. $-2 < x < 6$

29. $-3\frac{1}{2} \leq w \leq \frac{1}{2}$

31. $x < -12$ or $x > 6$

33. $y \leq -9$ or $y \geq 15$

35. $x \leq -3$ or $x \geq 4$

37. $|h - 1.4| \leq 0.1$ **39.** $|C - 27.5| \leq 0.25$

41. $|m - 1250| \leq 50$ **43.** no solution **45.** $-\frac{14}{3}, \frac{16}{3}$

47. no solution **49.** $\frac{11}{8}$ **51.** $-\frac{71}{36}$ **53.** $|c - 28.75| \leq 0.25$; 0.25; $28.50 \leq c \leq 29.00$ **55.** $|x| < 4$

57. $-6 \leq x \leq 8\frac{2}{3}$

59. all real numbers

61. all real numbers

63. $x \leq -8.4$ or $x \geq 9.6$

65. $-5 < x < 11$

67. The graph of $|x| < a$ is the set of all points on the number line that lie between a and $-a$. The graph of $|x| > a$ has two parts; the left part consists of the points to the left of $-a$, and the right part consists of the points to the right of a. **69.** $|t - 350| \leq 5$ **71.** $|t - 15| \leq 30$ **73.** $|x - 9.55| \leq 0.02$; $9.53 \leq x \leq 9.57$

75. never; absolute value is nonnegative **77.** sometimes; $|5| = 5$ but $|-5| \neq -5$ **79.** sometimes; $|-4 + 2| \neq -4 + 2$ **81.** The "3" in the second set of equations should be "-3."

$-4x + 1 < -3$

$-4x < -4$

$x > 1$ *not* $x > -\frac{1}{2}$

83. $\frac{ab+d}{c}$, $\frac{-ab+d}{c}$, $c \neq 0$, $ab \geq 0$

85. $(-6 \leq x \leq -5)$ or $(5 \leq x \leq 6)$

87. $x \geq \frac{5}{2}$

89. Use *and* if the absolute value is less than a value and use *or* if the absolute value is greater than a value.

91. 0 **93.** 0.04

94. $y < 6$

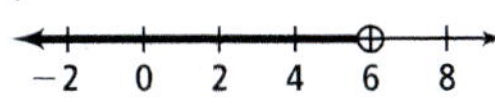

95. $s < \frac{2}{15}$

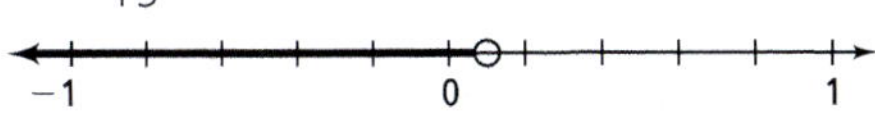

96. $a > 4$

97. Each figure has 4 more squares than the previous figure.

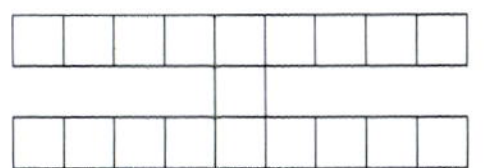

98. Each figure n has n more circles than the previous figure.

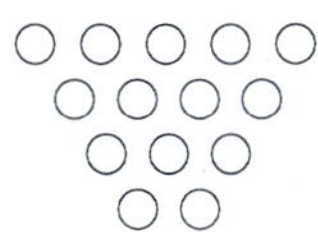

99.

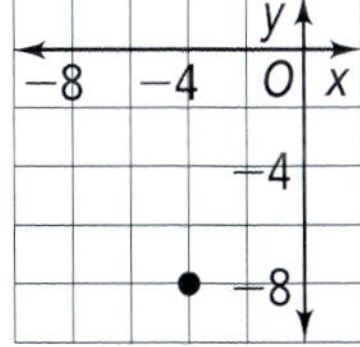

100.

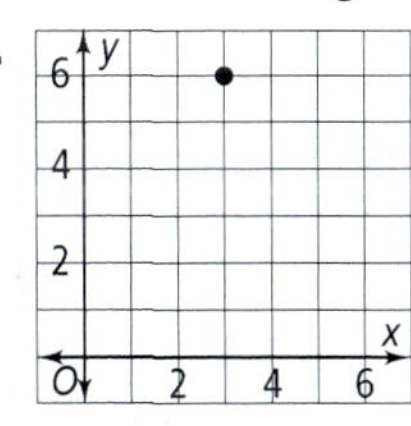

101.

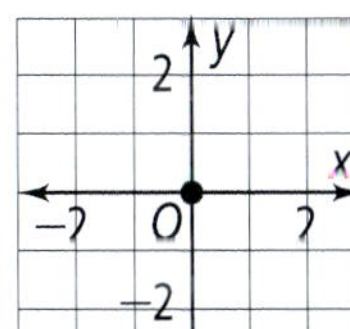

102.

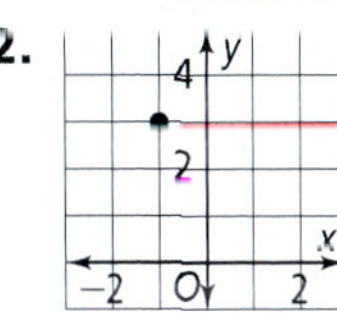

Chapter Review pp. 50–52

1. solution of the equation **2.** absolute value **3.** reciprocal **4.** compound inequality **5.** add 5; *25, 30, 35* **6.** add 1; 7, 8, 9 **7.** 12; $n + 8$ **8.** 76; $19n$ **9.** $\$20n$ **10.** irrational numbers **11.** rational numbers, integers **12.** rational numbers, integers, whole numbers, natural number **13.** real numbers, rational numbers **14.** $-\sqrt{60} > -8$ or $-8 < -\sqrt{60}$ **15.** $5 < \sqrt{32}$ or $\sqrt{32} > 5$ **16.** Inv. Prop. of Mult. **17.** Assoc. Prop. of Mult. **18.** 114 **19.** $5b$ **20.** 11 **21.** 6

22. $z \leq \frac{2}{5}$

23. $x > 2$

24. no solution

25. $x \leq \frac{3}{2}$ or $x > 6$

26. 10 cm, 6 cm **27.** 1 **28.** no solution **29.** $x = -8$ or $x = -12$ **30.** no solution

31. $-\frac{1}{3} \leq x \leq \frac{5}{3}$

32. $y < 0$ or $y > 18$

33. $-\frac{18}{7} \leq x \leq \frac{18}{7}$

34. $x < -14$ or $x > 10$

35. $|x - 43.6| \leq 0.1$

Chapter 2

Get Ready! p. 57 **1.** $6s$ **2.** $4a + b$ **3.** $xy - y + x$ **4.** $1.5g$ **5.** 0 **6.** $3b - 2c - 2$ **7.** $6f - 5d$ **8.** $3h + 3g$ **9.** $-2z + 5$ **10.** $2g - 4dg - 12d$ **11.** $8v - 6$ **12.** $7t - 3st - 5s$ **13.** −56 **14.** 80 **15.** −10 **16.** −24 **17.** 1075 **18.** 5 **19.** −1.75 **20.** 1.5 **21.** 20 **22.** 2 **23.** 5 **24.** 4 **25.** $-2 < x < 8$

26. $a \leq 0$

27. $x > -1$

28. $x < -4$ or $x > \frac{10}{3}$

29. $-\frac{1}{2} \leq d \leq \frac{25}{4}$

30. $-24 \leq f \leq 18$

31. Answers may vary. Sample: the Civil War, the Great Depression, the Louisiana Purchase **32.** Answers may vary. Sample: From 1 to 2 years of age; a person has usually stopped growing by age 30, but a baby is still growing at age 1. **33.** Answers may vary. Sample: The image is a reflection, left to right, of what other people see; the size is the same. **34.** Answers may vary. Sample: An inequality determines the limit of a value, or a boundary, for the solution on the number line.

Lesson 2-1 pp. 60–67

Got It?

1. Let Jan = 1, Feb = 2, Mar = 3, and Apr = 4.

Input	Output
1 →	69
2 →	70
3 →	75
4 →	78

{(1, 69), (2, 70), (3, 75), (4, 78)}

x Month	y Temperature (°F)
1	69
2	70
3	75
4	78

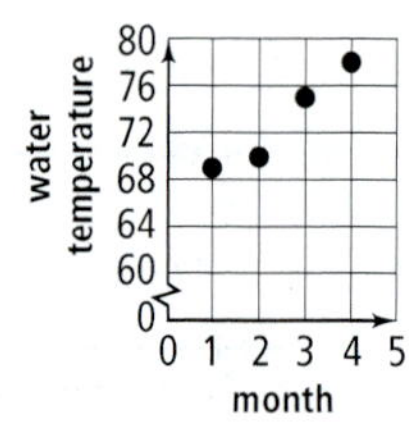

2. domain: {−3, 0, 2, 9, 23}, range: {−99, −18, 0, 7, 14} **3. a.** no **b.** yes **c.** In a mapping diagram for a relation that is not a function, there is at least one element in the domain that has more than one arrow originating from it. In a mapping diagram for a function, each element in the domain has at most one arrow originating from it. **4.** b and c **5. a.** 9 **b.** 1 **c.** −19 **6.** Let x = number of bottles purchased and C = total cost; $C(x) = 1.19x$; \$17.85

Lesson Check 1. domain: {0, 3, 4}, range: {−2, 1, 2, 4} **2.** domain: {−4, −3, 0, 4}, range: {−4, −3, 0, 4} **3.** no **4.** yes **5.** Yes; a relation is any set of pairs of input and output values. No; a function is a relation in which each element of the domain is paired with exactly one element of the range. **6.** *Every* vertical line does not need to intersect a function. Rewrite as: "In a function, every vertical line must intersect the graph in *at most* one point." **7.** A horizontal-line test checks the pairing of one element of the range with one or more elements of the domain. A function can have a pairing of one element of the range with *one or more* elements of the domain. A horizontal-line test cannot determine whether a relation is a function.

Exercises 9. domain: {1, 2, 3, 4, 5, 6}, range: {6, 7, 8, 9, 11} **11.** yes **13.** yes **15.** yes **17.** 71; (4, 71) **19.** −15; (9, −15) **21.** −2; (3, −2) **23.** −132; (−11, −132) **25.** $C(m) = 4.52 + 0.12m$; \$34.52 **27.** 13.5 cm^2 **29.** domain: all real numbers, range: $y \geq 0$; yes **31.** ≈4849 cm^3

33. a. 109.4 **b.** 10.4 **c.** −11.1 **d.** −7.2

35. $f(x) - g(x) = (3x - 21) - (3x + 21)$
$= 3x - 21 - 3x - 21 = -42$

37. No; each $x > \frac{7}{3}$ is paired with two y values. **39. a.** yes **b.** No; 6 would be paired with both 2.5 and 3.

41. H **43.** 1.9, $\sqrt{3}$, $\frac{5}{4}$, −1.2 **44.** $\frac{2}{3}$, $-\frac{20}{3}$ **45.** −13, 15 **46.** $x > -3$ **47.** $x \leq \frac{3}{2}$ **48.** $-\frac{3}{2} < x < \frac{3}{2}$ **49.** $x \geq -3$ **50.** $\frac{1}{4}x$ **51.** $-\frac{1}{2}x$ **52.** $20x$

Lesson 2-2 pp. 68–73

Got It? 1. a. yes; −7, $y = -7x$ **b.** no **2. a.** yes; $-\frac{5}{3}$ **b.** yes; $\frac{1}{9}$ **3.** 60 **4. a.** 280 **b.** No; if $y^2 = kx^2$, then $y = \pm\sqrt{kx}$. So, $\frac{y}{x}$ could be $+\sqrt{k}$ for one pair of values and $+\sqrt{k}$ for another pair. Then y would not vary directly with x.

5. a. **b.**

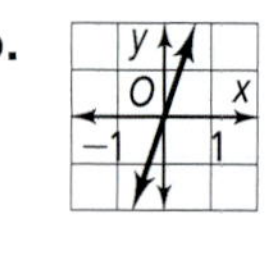

Lesson Check 1. $y = -\frac{1}{2}x$ **2.** $\frac{3}{2}$ **3.** $\frac{5}{4}$ **4.** Answers may vary. Sample: Two variables are directly related when the ratio of the output to the input is a constant value. **5.** For a direct variation, $y = kx$ where k is the constant of variation. If $x = 0$, then $y = 0$ and the graph of $y = kx$ passes through the origin. **6.** Answers may vary. Sample: $y = -8x$.

Exercises 7. yes; 7, $y = 7x$ **9.** no **11.** yes; 12 **13.** yes; −2 **15.** no **17.** yes; 6 **19.** −3 **21.** $\frac{6}{7}$ **23.** 21 **25.** 4 min **27.**

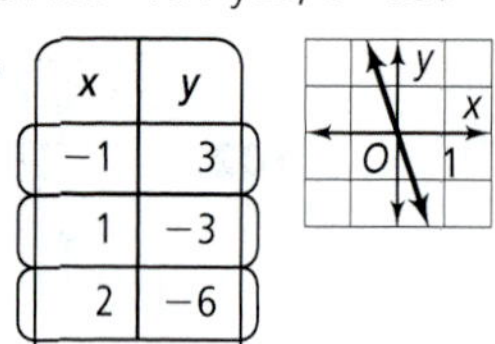

x	y
−1	3
1	−3
2	−6

29.

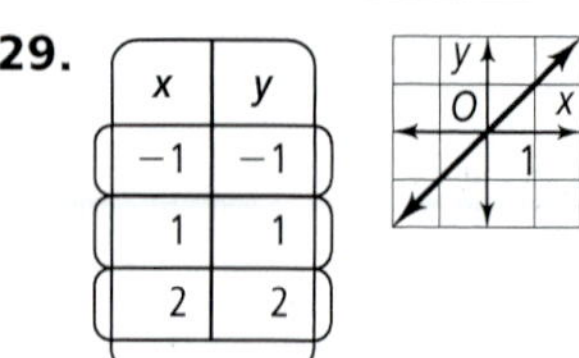

x	y
−1	−1
1	1
2	2

31. yes; $k = \frac{2}{3}$, $y = \frac{2}{3}x$ **33.** no

35. $y = 2x$

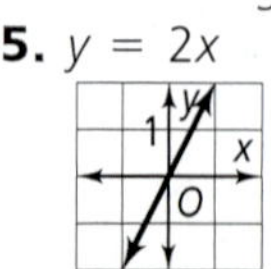

37. $y = -4.5x$

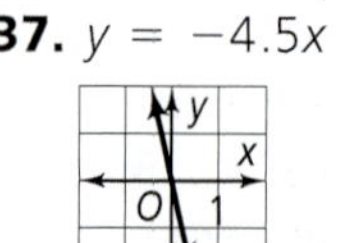

39. $y = \frac{3}{5}x$

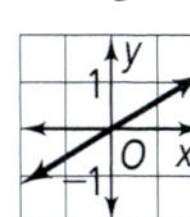

41. $y = \frac{2}{7}x$

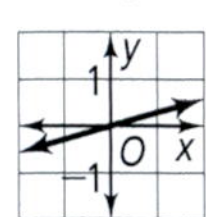

43. 0.625 **45.** 0.225 **47.** First, it does not say that y varies directly with x. Second, every direct variation includes the point (0, 0), so x cannot be determined because k could be any value.

49. Answers may vary. Sample: $y = 3.2x$

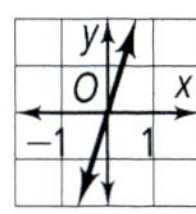

51. Answers may vary. Sample: If y varies directly with x^2, and $y = 2$ when $x = 4$, then $y = \frac{81}{8}$ when $x = 9$.

53. $c = 0, a \neq 0, b \neq 0$

55. y is divided by 7; $y = kx$, so if x is divided by 7, $y = k\left(\frac{x}{7}\right)$ or $\frac{1}{7}$ the original value of y.

57. 1091 **59.** 0 **61.** −5

62.

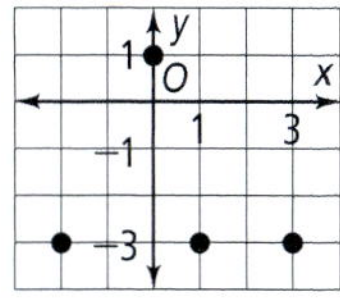

domain: {−2, 0, 1, 3}; range: {−3, 1}

63.

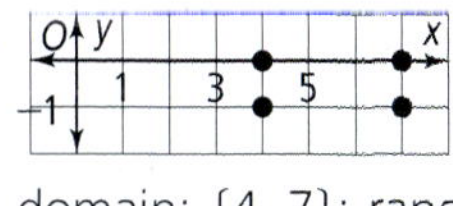

domain: {4, 7}; range: {−1, 0}

64.

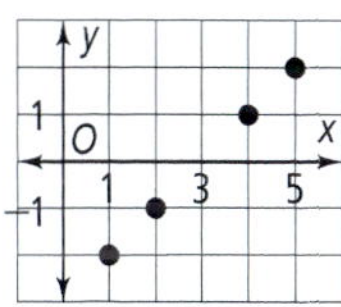

domain: {1, 2, 4, 5}; range: {−2, −1, 1, 2}

65.

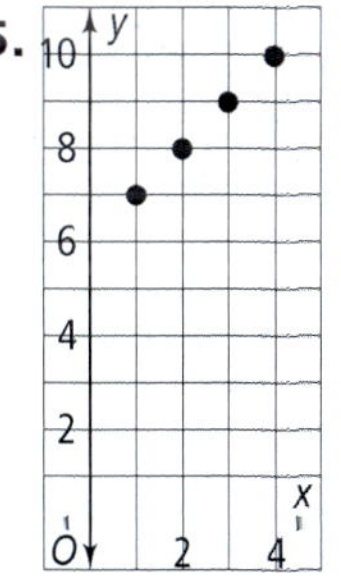

domain: {1, 2, 3, 4}; range: {7, 8, 9, 10}

66. $8n$; 40, 48, 56 **67.** $7 - 2n$; −3, −5, −7 **68.** $12(13 - n)$; 96, 84, 72 **69.** $15(n + 1)$; 90, 105,120 **70.** $\frac{17}{3}$; 7; $\frac{23}{3}$; $\frac{29}{3}$ **71.** −3.2; −2; −1.4; 0.4 **72.** −5; 1; 4; 13 **73.** −9; −8; −7.5; −6

Lesson 2-3 pp. 74–80

Got It? 1. a. −1 **b.** 1 **c.** undefined **d.** $\frac{1-4}{8-5} = \frac{-3}{3} = -1 = \frac{3}{-3} = \frac{4-1}{5-8}$ **2. a.** $y = 6x + 5$ **b.** $y = -\frac{1}{2}x - 3$ **c.** No; any two points on a line can be used to calculate the slope. **3. a.** $y = -\frac{3}{2}x + 9$; $-\frac{3}{2}$; (0, 9) **b.** $y = -\frac{7}{5}x - 7$; $-\frac{7}{5}$; (0, −7) **4.** $y = \frac{4}{7}x - 2$

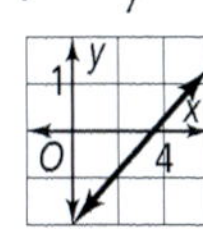

Lesson Check 1. $y = \frac{1}{2}x + 1$ **2.** $y = \frac{4}{3}x + \frac{1}{3}$ **3.** −1 **4.** 1 **5.** The y-intercept of a line is the point at which the line crosses the y-axis. The x-intercept is the point at which the line crosses the x-axis. **6.** Since division by zero is undefined, the slope of a vertical line that passes through (a, b) and (a, c), $\frac{c-b}{0}$, is undefined. **7.** She subtracted the x-coordinates in the wrong order. The x- and y-coordinates of each point must be subtracted consistently.

Exercises 9. −2 **11.** $\frac{4}{11}$ **13.** 1 **15.** 0 **17.** $y = 3x + 2$ **19.** $y = \frac{5}{6}x + 12$ **21.** $y = -5x - 7$ **23.** $y = \frac{3}{2}x + \frac{7}{2}$; $\frac{3}{2}$, $\left(0, \frac{7}{2}\right)$ **25.** $y = -\frac{4}{3}x + \frac{5}{6}$; $-\frac{4}{3}$, $\left(0, \frac{5}{6}\right)$ **27.** $y = 7$; 0, (0, 7)

29.

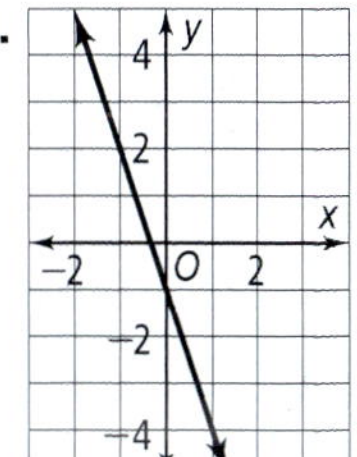

31.

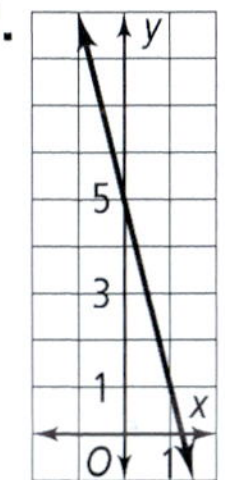

33.

35.

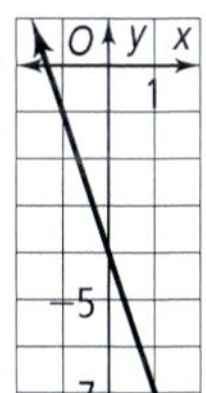

37.

39.

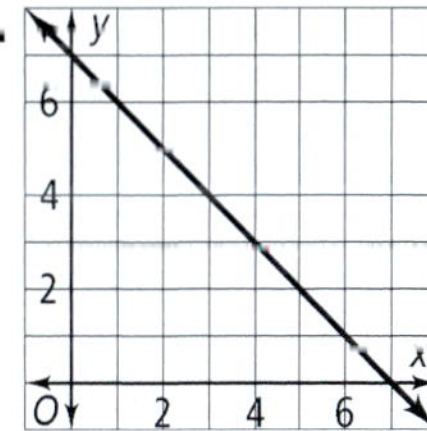

41.

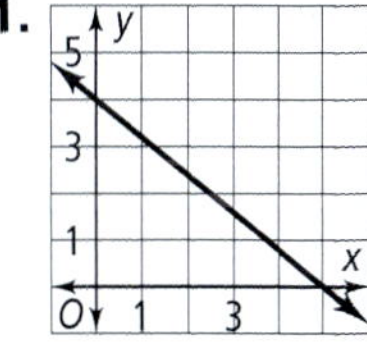

43.

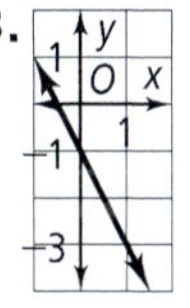

45.

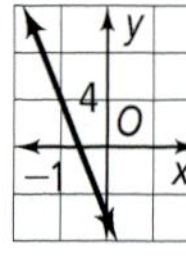

47. 0; (0, 3) **49.** $-\frac{1}{4}$; (0, 3) **51.** undefined slope; no y-intercept **53.** $-\frac{1}{2}$; $\left(0, -\frac{5}{2}\right)$ **55.** $-\frac{A}{B}$; $\left(0, \frac{C}{B}\right)$

57. a. 1 **b.** 1 **c.** 1 **d.** 1 **e.** Any two points on a line can be used to find the slope of the line.
59. $-\frac{5}{13}$ **61.** $\frac{15}{2}$ **63.** B **65.** C
67. domain: {−2, 1, 2, 3, 4}, range: {−2, −1, 2, 3}; no
68. domain: all real numbers, range: $y \geq -2$; no
69. domain: {−5, 0, 2, 9}, range: {−3, −1, 5, 15}; no
70. 2 **71.** 8 **72.** 13

Lesson 2-4 pp. 81–88

Got It? 1. $y + 1 = -3(x - 7)$ **2. a.** $y - 7 = \frac{7}{5}x$
b. $y = \frac{7}{5}(x + 5)$; Either point can be used to put the equation of the line in point-slope form.
3. $-91x + 10y = 36$
4. (0, −2), (4, 0);

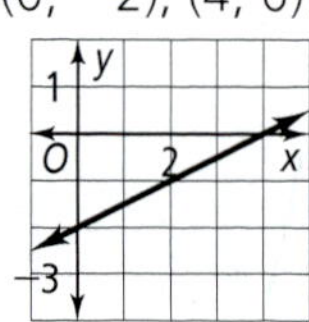

5. a.

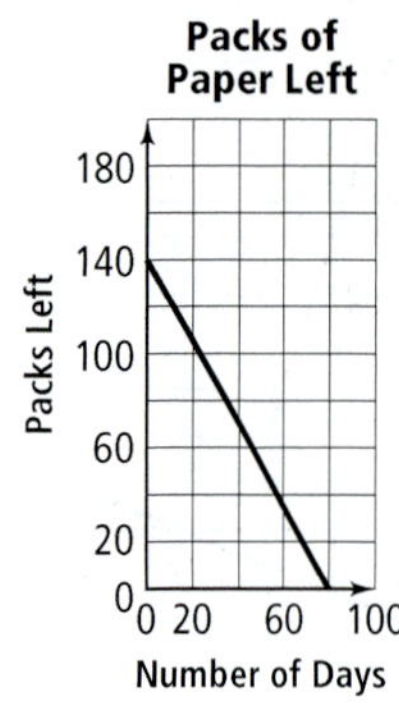

b. $7x + 4y = 560$ **c.** 87.5 **6. a.** $y = -2x + 6$
b. $y = -\frac{3}{2}x + 6$

Lesson Check 1. $y = -3x - 1$ **2.** $y = \frac{1}{2}x + 2$
3. (0, 6), (2, 0)

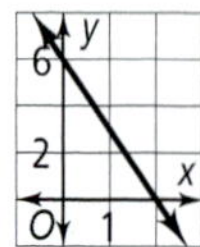

4. $3x + y = -1$ **5.** $3x + 2y = -9$ **6. a.** point-slope
b. slope-intercept **c.** standard **d.** point-slope
7. Point-slope form; since the x-intercept is the point where y is zero, you know the point on the line, $(x, 0)$, and you know the slope. **8.** $-\frac{b}{a}$ **9.** No; one line has a slope of −2 and the other line has a slope of $-\frac{1}{2}$. $-\frac{1}{2}$ is the reciprocal of −2, not the *negative* reciprocal.

Exercises 11. $y - 12 = \frac{5}{6}(x - 22)$ **13.** $y + 2 = 0$

15. $y - 2 = 5x$ **17.** Answers may vary. Sample: $y = \frac{5}{4}(x - 1)$ or $y - 5 = \frac{5}{4}(x - 5)$ **19.** Answers may vary. Sample: $y + 1 = -\frac{4}{3}x$ or $y + 5 = -\frac{4}{3}(x - 3)$
21. Answers may vary. Sample: $y - 9 = -\frac{7}{5}(x - 1)$ or $y - 2 = -\frac{7}{5}(x - 6)$ **23.** $7x + y = -9$
25. $-42x + 10y = 79$
27. (0, −2), (−5, 0) **29.** (0, 2), $\left(\frac{14}{5}, 0\right)$

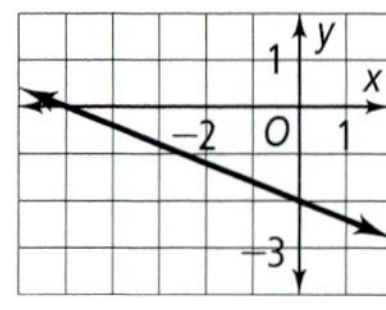

31. $y = -2.5x + 20$

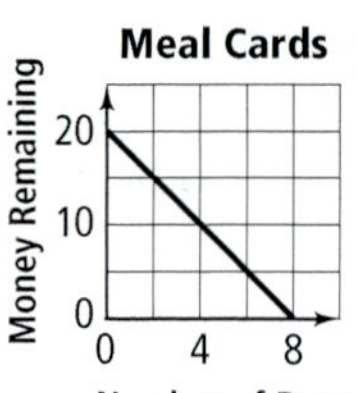

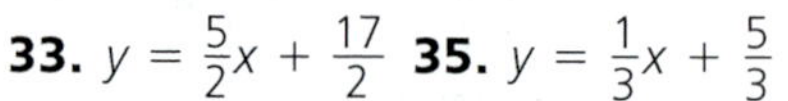

33. $y = \frac{5}{2}x + \frac{17}{2}$ **35.** $y = \frac{1}{3}x + \frac{5}{3}$

37.

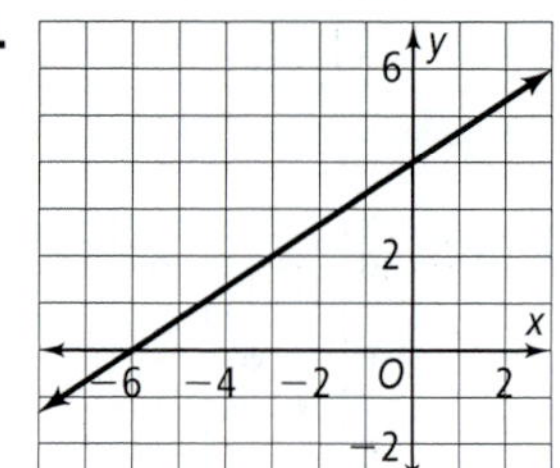

39.

41.

43. $y + 4 = \frac{7}{5}\left(x + 3\right)$ or $y + \frac{1}{2} = \frac{7}{5}\left(x + \frac{1}{2}\right)$
45. $y = \frac{3}{4}x + 3$ **47.** $y = -\frac{3}{2}x - \frac{1}{2}$
49. a. $y - 12 = -\frac{4}{3}(x + 3)$
b. $y + 4 = -\frac{4}{3}(x - 9)$ **c.** They have the same standard form: $4x + 3y = 24$.
51. $y = -\frac{3}{2}x - 1$

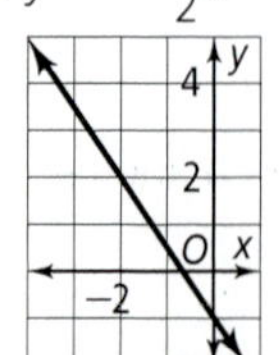

53. yes

55. The eq. of the line connecting (1, 3) to (−2, 6) is $y = -x + 4$. The eq. of the line connecting (1, 3) to (3, 5) is $y = x + 2$. The slopes are neg. reciprocals so the lines are perpendicular. Therefore, by def. of a rt. triangle, it is a rt. triangle. **57.** B

59. $|x - 3| \geq 5$

$x - 3 \geq 5$ or $x - 3 \leq -5$

$x \geq 8$ or $x \leq -2$

60. domain: {−3, −1, 0, 1, 2}, range: {−2, 0, 2, 4}; yes **61.** domain: all real numbers, range: {−2}; yes **62.** domain: {−18, −2, 0, 3, 39}, range: {−1, 3, 17, 28, 32}; yes **63.** Multiplicative Inv. **64.** Distr. Prop. **65.** Add. Inv., Add. Ident. **66.** $y = 3x - 5$ **67.** $y = \frac{1}{2}x$ **68.** $y = \frac{4}{5}x + 7$ **69.** $y = -\frac{3}{8}x + 12$

Lesson 2-5 pp. 92–98

Got It?

1. a. strong negative correlation

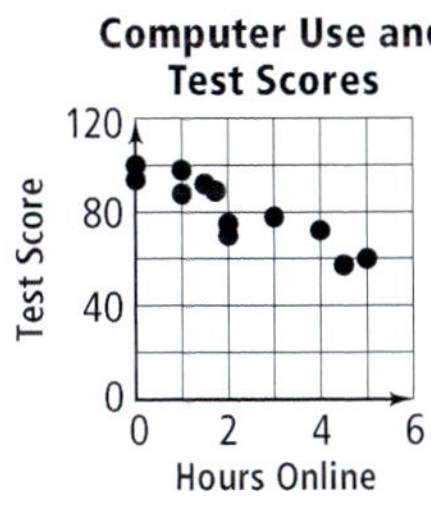

b. about \$170 **2.** Answers may vary. Sample: $y = 3575x + 19354$

3. $y = 0.09x + 2.44$, where 1997 is year 0; \$4.96

Lesson Check

1. strong positive correlation

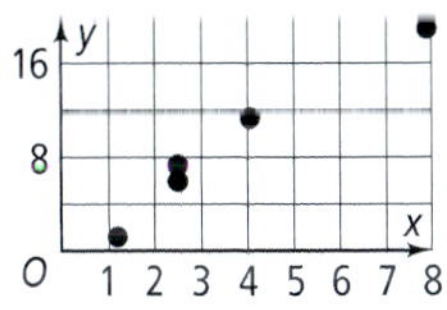

2. no correlation

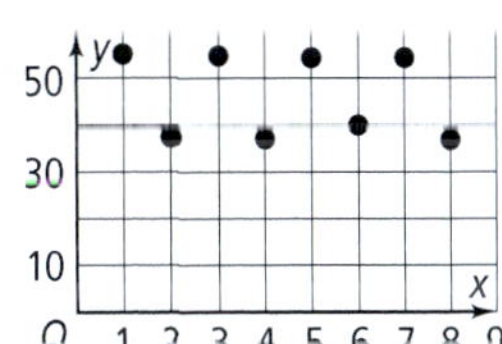

3. strong positive correlation

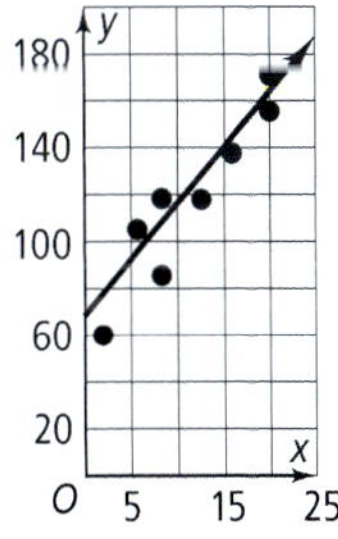

4. Plot the data points in a scatter plot to determine the correlation. The closer the data points fall along a line with a positive or negative slope, the stronger the correlation. **5.** No; answers may vary. Sample: A trend line is determined by using two pts. close to the line drawn through the data sets of the scatter plot. The line of best fit is the most accurate of the trend lines because it uses all the data pts. **6.** The slope of the trend line or line of best fit is positive for data pts. with positive correlation and negative for data pts. with negative correlation. The constant of variation for a direct variation is positive for data pts. with positive correlation.

Exercises

7. strong negative correlation

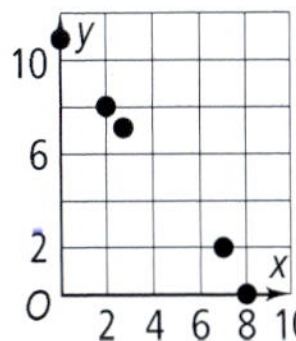

9. Answers may vary. Sample: $y = -0.7x - 4$

11. Answers may vary. Sample: $y = 4.47x + 33.31$

13. 6,055,359 tonnes **15.** no; Answers may vary.

17. yes

19. a.

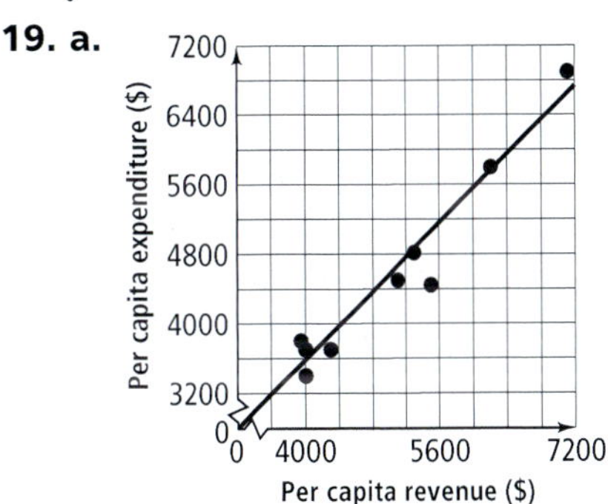

b. about \$2506 **c.** about \$5610 **d.** Answers may vary. Sample: Yes; expenditure would be predicted to be between \$5300 and \$5400.

21. G

23.
$$(y - 1) = \frac{2}{3}(x + 1)$$
$$(-3 - 1) = \frac{2}{3}(a + 1)$$
$$-4 = \frac{2}{3}a + \frac{2}{3}$$
$$-\frac{14}{3} = \frac{2}{3}a$$
$$-7 = a$$

24.

25.

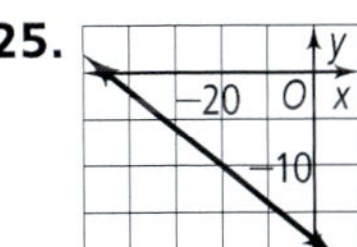

26.

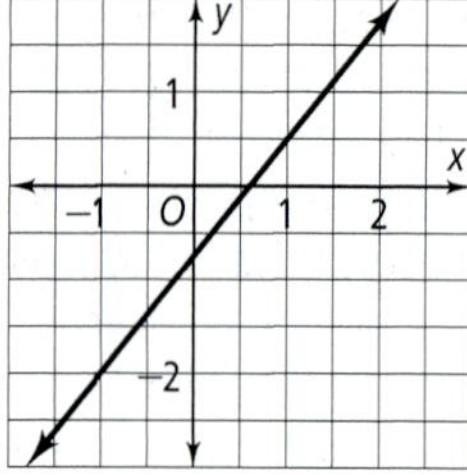

27. $-2x + y = 2$ **28.** $x + y = 0$ **29.** $y = 2$

30.

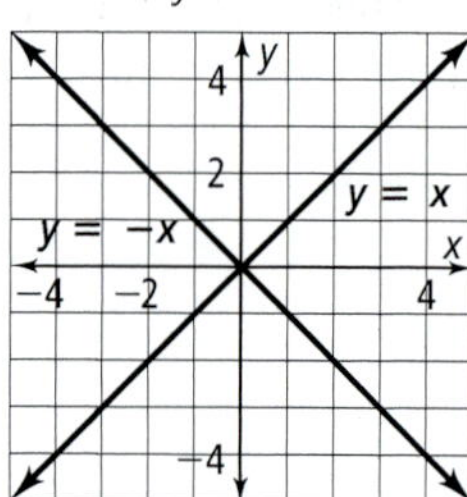

31.

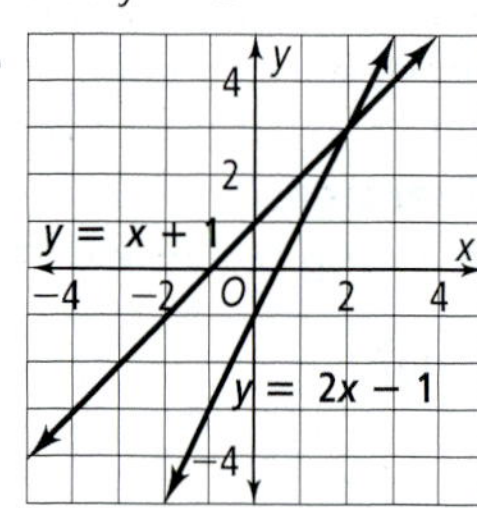

32.

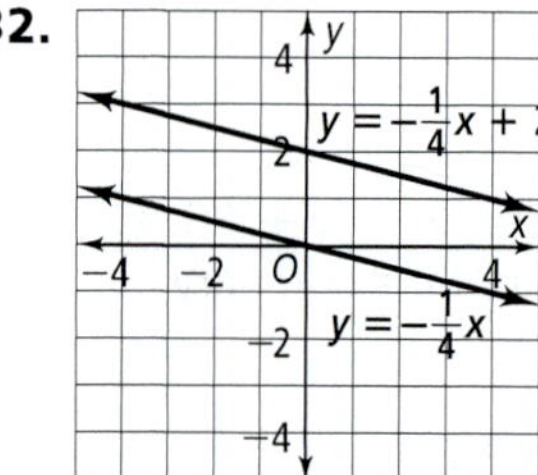

Lesson 2-6 pp. 99–106

Got It? 1. a. Each output for $y = 2x - 3$ is three less than the corresponding output for $y = 2x$. The graph of $y = 2x - 3$ is the graph of $y = 2x$ translated down three units.

b.

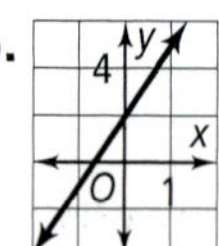

2. $f\left(x + \frac{1}{2}\right)$ **3.** $h(x) = -3x - 3$

4. a.

x	y
-5	$\frac{2}{3}$
-2	$\frac{2}{3}$
0	-1
3	$\frac{1}{3}$
5	$-\frac{2}{3}$

b. Sometimes; Answers may vary. Sample: switching the order of a horizontal translation and a reflection in the y-axis will change the resulting graph, but switching the order of a horizontal and vert. translation will not.

5. a. $g(x) = 2x - 3$ **b.** $g(x) = f(x + 4) - 2$; translated left 4 units and translated down 2 units

Lesson Check 1. translated 6 units up **2.** compressed vertically by a factor of 0.25 **3.** translated 4 units to the rt. **4.** reflected over y-axis **5.** translated 1 unit to the left and 2 units down

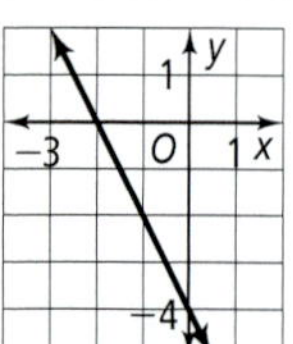

6. stretched vertically by a factor of 2 and translated 1 unit up

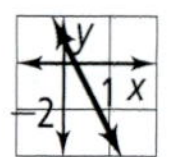

7. $g(x)$ is the graph of $h(x)$ translated 2 units up

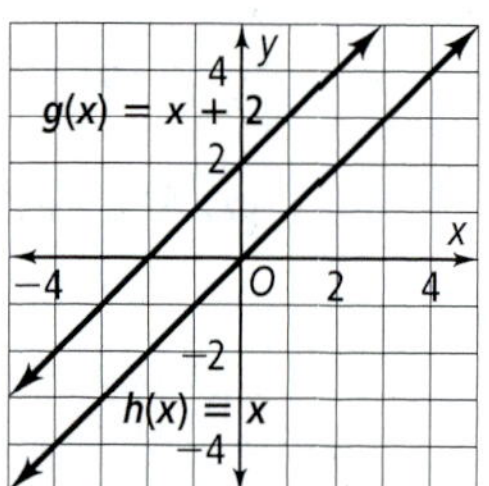

8. Answers may vary. Sample: $f(x) = x$, $f(x - 2) = f(x) - 2$ **9.** $f(x) = -x - 2$; $g(x) = f(-x) = x - 2$

Exercises

11. The function is $y = x$ translated 4.5 units up.

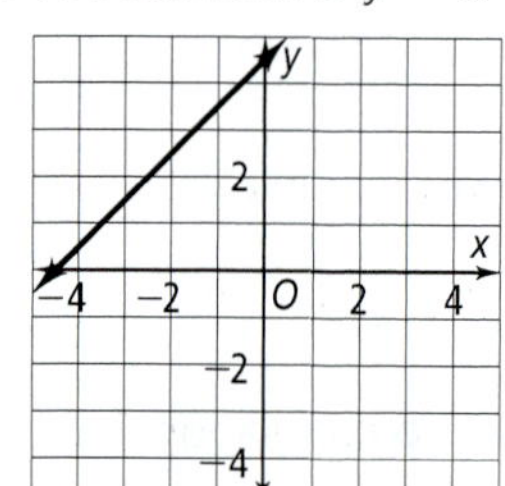

13.

x	$f(x) + 3$
-2	6
0	4
1	1
3	2

15.

x	$f(x) + 4$
-3	5
-1	2
1	4
4	7

17. $y = f(x) + 4$

19. translated rt. 4 units

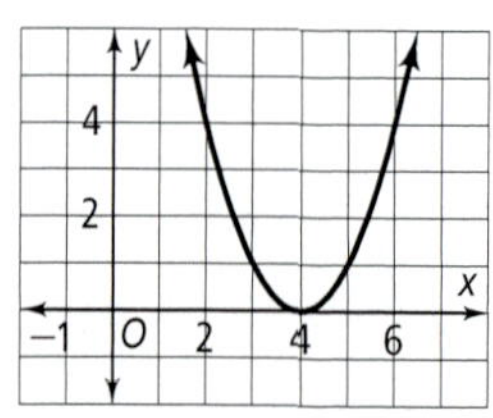

21. translated left 3 units

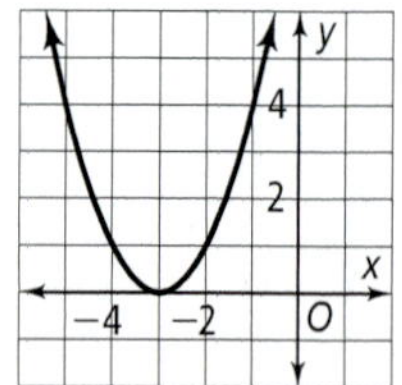

23. $g(x) = -x - 1$ **25.** $g(x) = -2x + 4$

27. $y = 2x$ **29.** $y = \frac{1}{4}x$ **31.** $g(x) = -0.5x$

33. vertically compressed by a factor of $\frac{1}{4}$ and translated down 2 units

35. translate to the right 10 s

37. $f(x) = -\frac{1}{3}x - 1$; $g(x) = \frac{1}{3}x + 1$; $g(x) = -f(x)$

41. translated 6 units down **43.** translated 4 units up

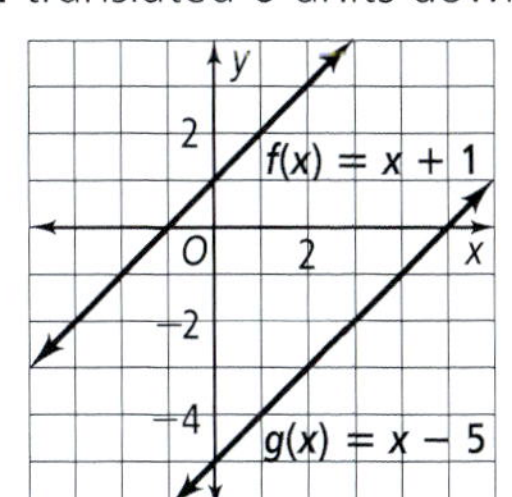

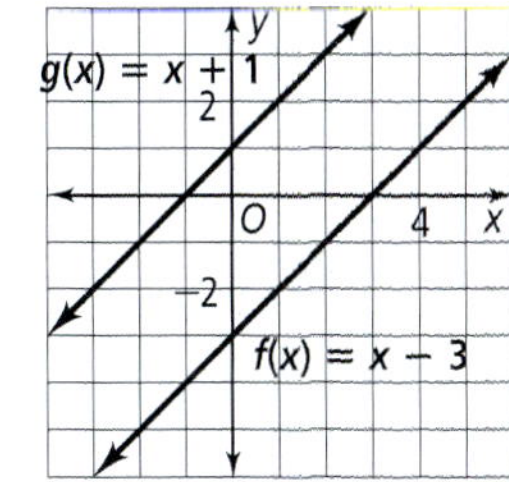

45. The first two steps are incorrect; the transformations should be: shift 1 unit left, vertically stretch by a factor of 2, and shift 3 units down.

47.

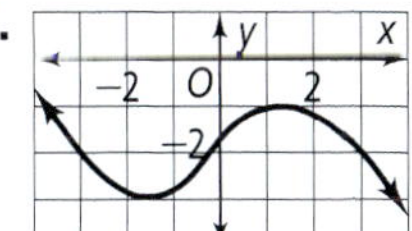

49.

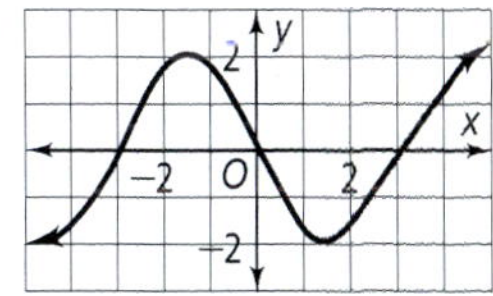

51. D **53.** A

55. $y = -15.82x + 914.59$

56. $-2, 8$ **57.** $-12, 11$ **58.** $-3, \frac{21}{5}$

Lesson 2-7 pp. 107–113

Got It?

1. a.

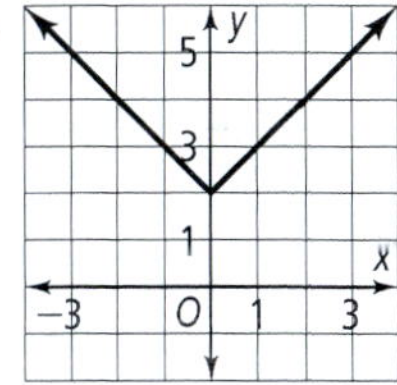

vertex at (0, 2); translated up 2 units from the parent function

b. No; transformations of this form move the vertex up or down along the axis of symmetry, so the axis stays the same.

2.

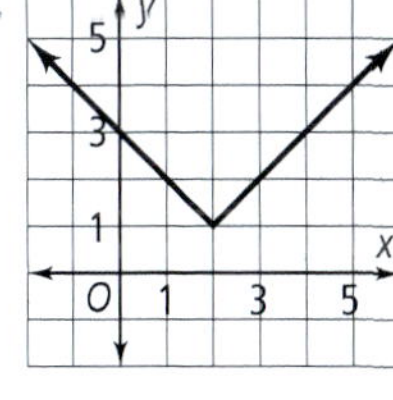

3. a.

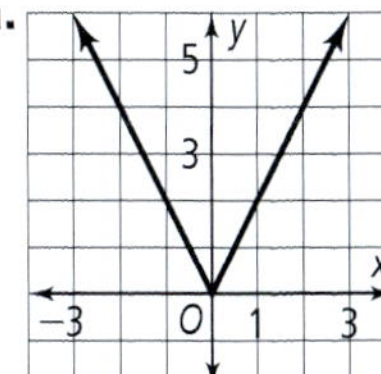

b.

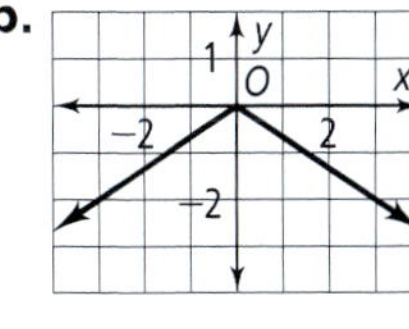

4. $(1, -3)$; $x = 1$; translated 1 unit to the rt., vertically stretched by a factor of 2, then reflected over the x-axis and translated down 3 units **5.** $y = \frac{1}{4}|x - 2| - 1$

Lesson Check **1.** $(-4, -3)$; $x = -4$ **2.** $(-3, 9)$; $x = -3$ **3.** vert. stretch **4.** vert. stretch **5.** Yes; you can determine the position of a graph of an absolute value function by identifying the vertex, axis of symmetry and the transformation of the absolute value parent function. Answers may vary. Sample: $y = -\frac{1}{2}|x + 3| - 5$; vertex $(-3, -5)$; axis of symmetry, $x = -3$; translated 3 units to the left, vertically compressed by a factor of $\frac{1}{2}$, then reflected over the x-axis and translated down 5 units. **6.** Answers may vary. Sample: $y = |x + 1| - 2$ and $y = -|x + 1| - 2$ **7.** $y = |x|$ is the same as $y = x$ when $x \geq 0$ and is the reflection of $y = x$ across the x-axis when $x < 0$.

Exercises

9.

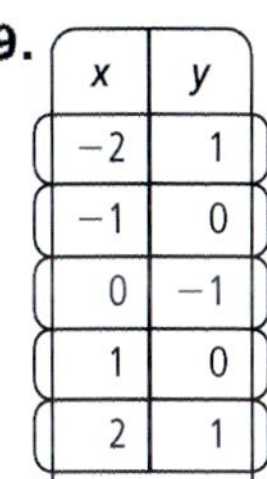

x	y
−2	1
−1	0
0	−1
1	0
2	1

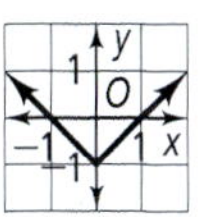

11.

x	y
−4	2
−3	1
−2	0
−1	1
0	2

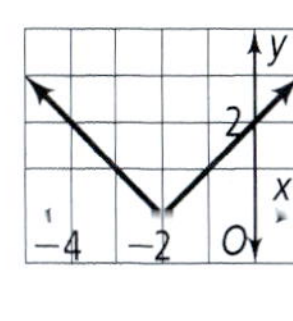

13.

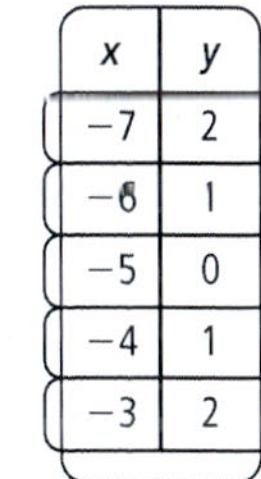

x	y
−7	2
−6	1
−5	0
−4	1
−3	2

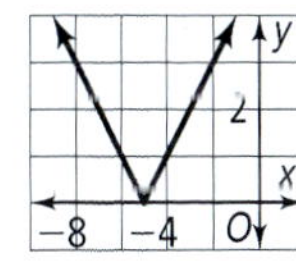

15.

x	y
−8	1
−7	0
−6	−1
−5	0
−4	1

17. 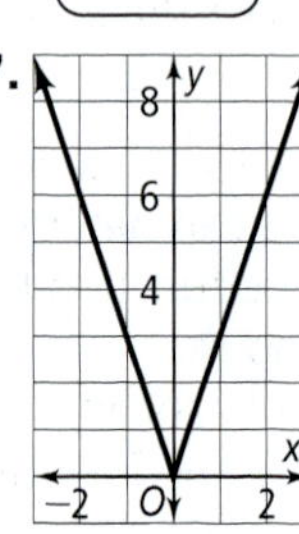

vertically stretched by a factor of 3

19. 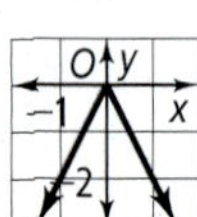

vertically stretched by a factor of 2 and reflected across the *x*-axis

21. 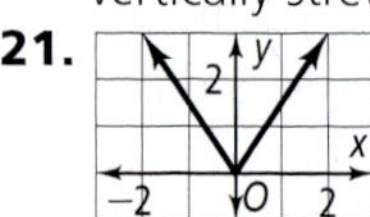

vertically stretched by a factor of $\frac{3}{2}$

23. $(-2, -4)$; $x = -2$; translate 2 units to the left and 4 units down **25.** $(-6, 0)$; $x = -6$; translate 6 units to the left and vertically stretch by a factor of 3 **27.** $(5, 0)$; $x = 5$; translate 5 units to the rt. and reflect across the *x*-axis **29.** $y = -2|x - 5| + 1$

31. 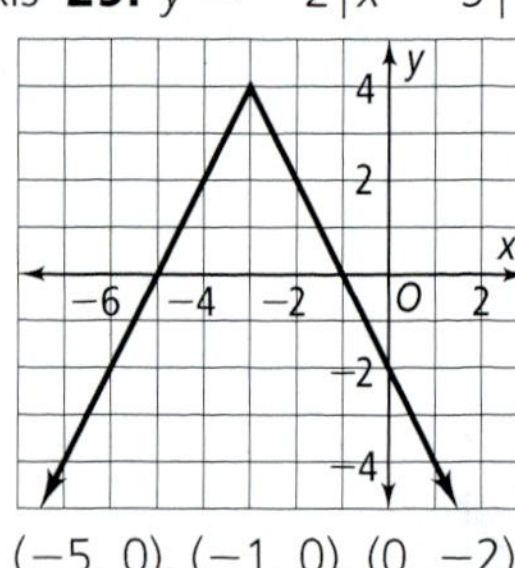

$(-5, 0)$, $(-1, 0)$, $(0, -2)$

33. 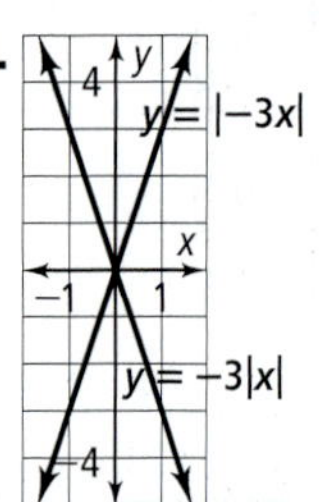

The graphs are not identical; one is the reflection across the *x*-axis of the other.

35. a. Answers may vary. Sample: reflection across the *x*-axis, vert. compression by a factor of $\frac{1}{2}$, translation down $\frac{1}{2}$ unit, translation rt. 6 units
b. No; changing the order of the transformations can change the graph.

37. 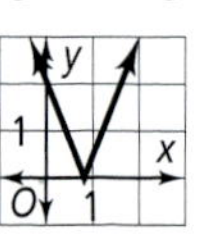

39. 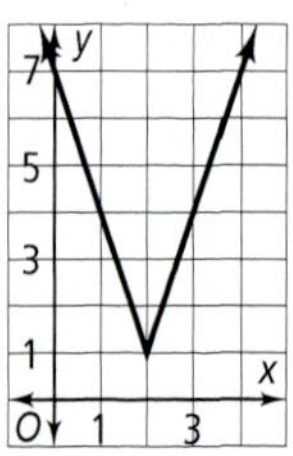

41. 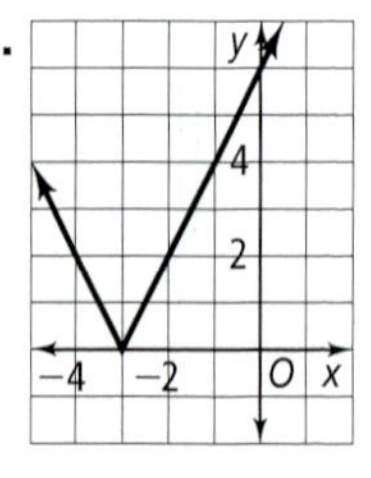

43. 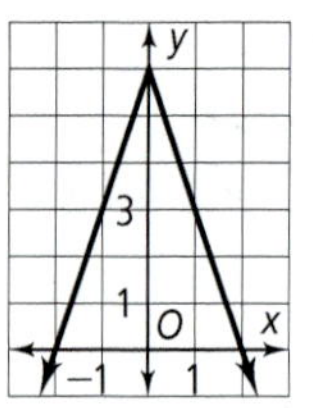

45. a. 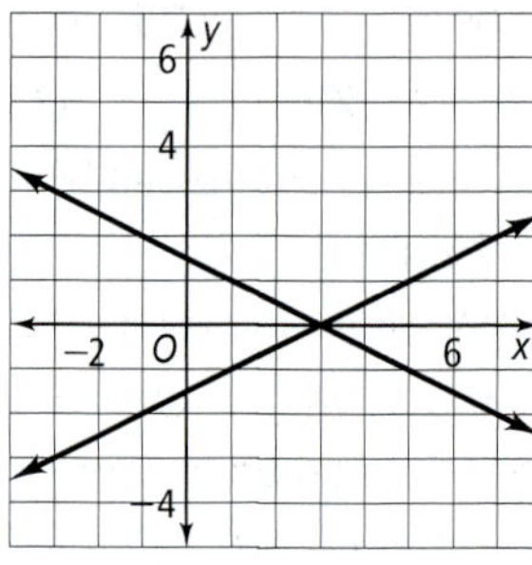

b. No; $f(x) = g(x)$ only for $x = 3$.

47. 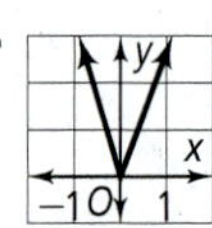

49. 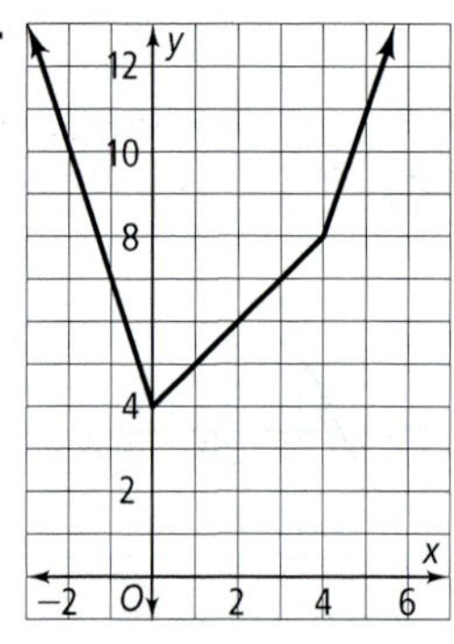

51. C **53.** A **55.** $y = x + 1$ **56.** $y = -\frac{1}{2}x + 2$
57. $g(x) = -x - 7$ **58.** $g(x) = -2x - 6$
59. $g(x) = -4 - x$ **60.** Answers may vary. Sample: $y = \frac{4}{5}x + 1$ 61. Answers may vary. Sample: $y = -\frac{4}{5}x + 8$

62. 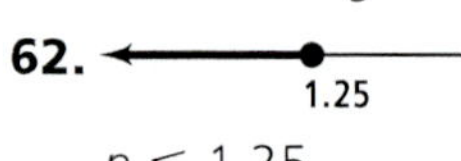

$p \leq 1.25$

63. 13

$t > 13$

64. 

$t \leq -3$

Lesson 2-8 pp. 114–120

Got It?

1. a. 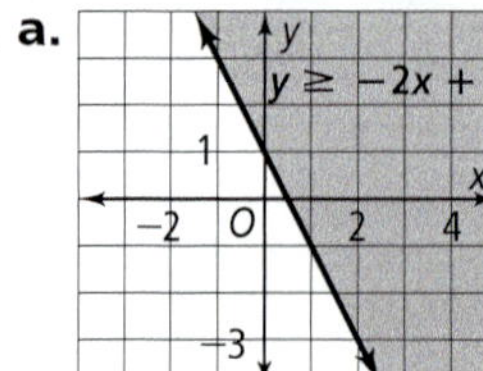

b. 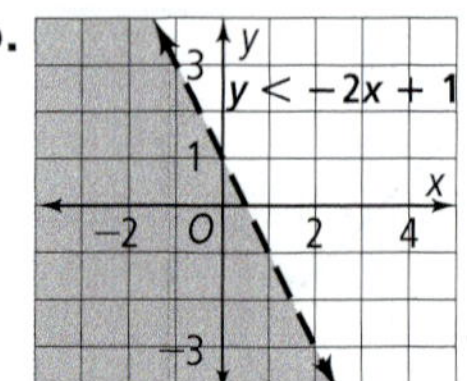

2. a.

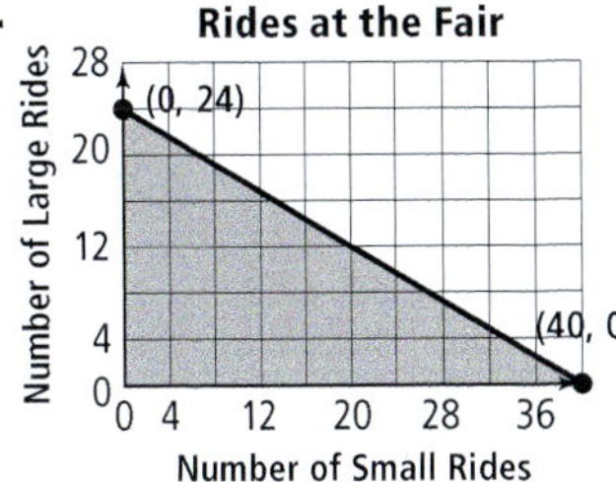

b. The number of rides cannot be neg.

3.

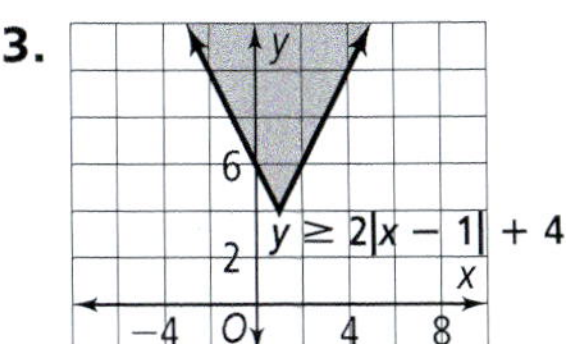

4. a. $y > -|x + 4| + 3$ **b.** No; when you solve the ineq. for y, you multiply both sides by -1. This changes the ineq. sign from $>$ to $<$.

Lesson Check

1.

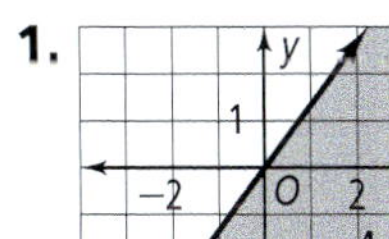

2.

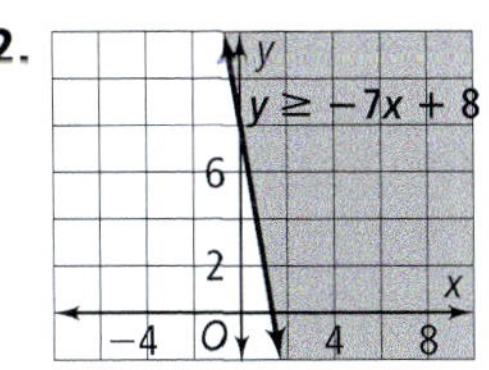

3.

4.

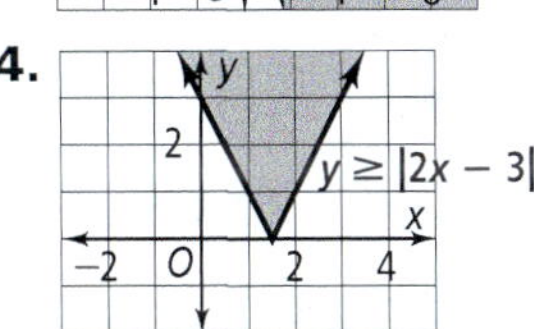

5. No; the boundary consists of points where *equality* holds, while the shaded region consists of points where the *inequality* holds. **6.** Graphing a linear inequality in two variables includes first graphing the boundary line, which is a linear eq. in two variables, and then shading the half-plane. **7.** No; $\left(\frac{3}{4}, 0\right)$ does not satisfy the inequality because 2.25 > 3 is false.

Exercises

9.

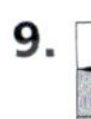

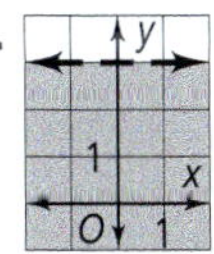

11.

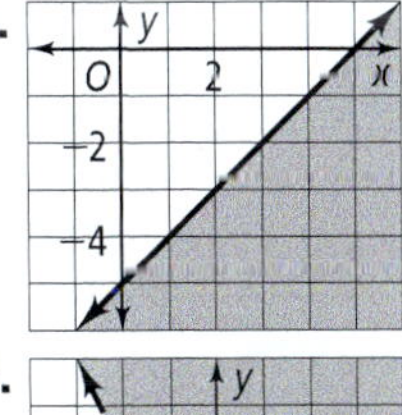

13.

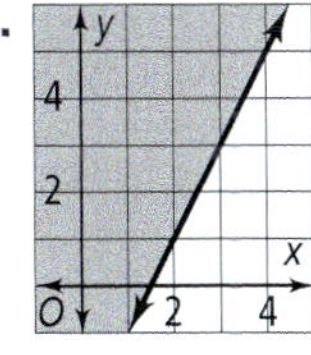

15.

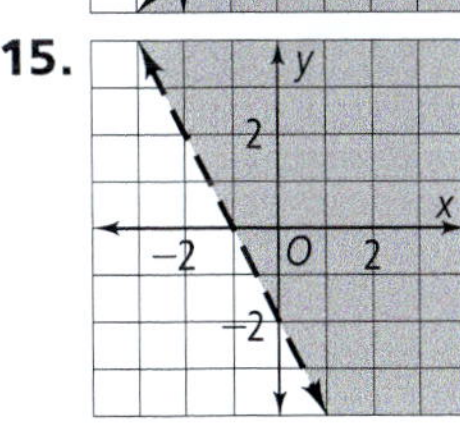

17. a. $y \geq 20x$ if $x \leq 6$; $y \geq 15x$ if $x > 6$

b.

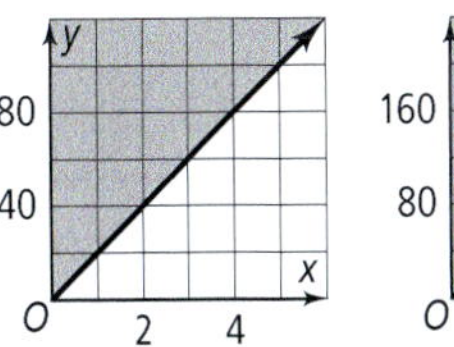

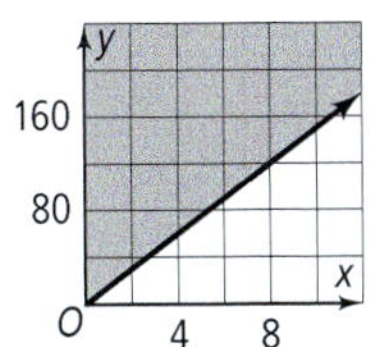

19.

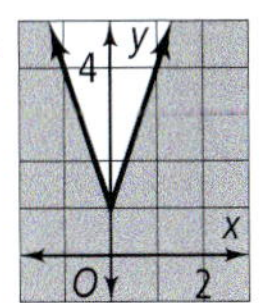

21.

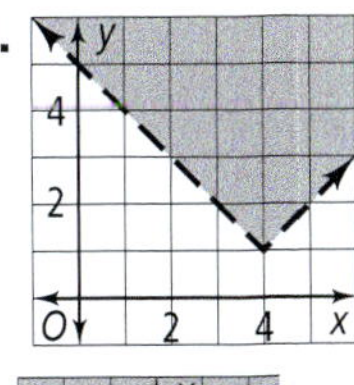

23.

25.

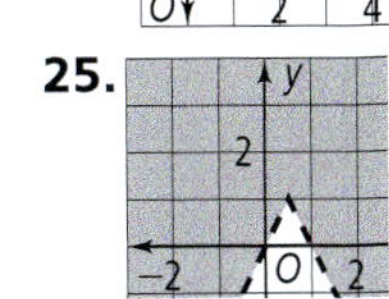

27. $y < -x - 2$ **29.** $2y \geq |2x + 6|$

31.

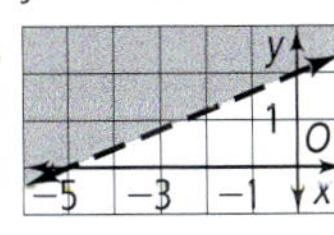

33.

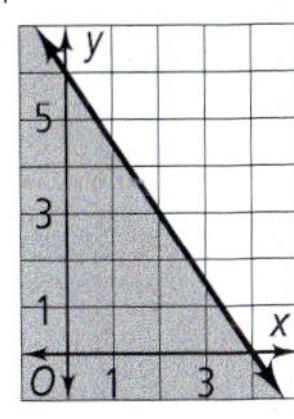

35.

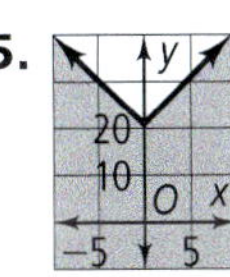

37.

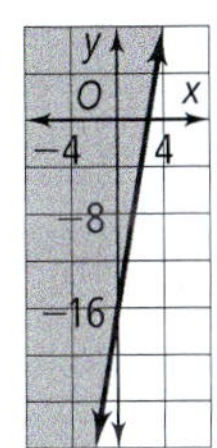

39. $x > -3$ **41.** $y \geq -2x + 4$

43. $y < -|x - 4|$ **45.** C

47. when the origin lies on the boundary line

49.

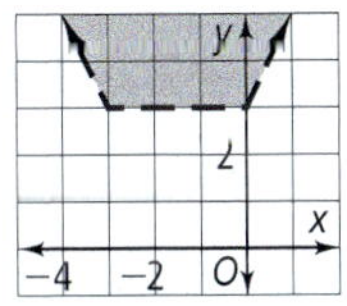

51.

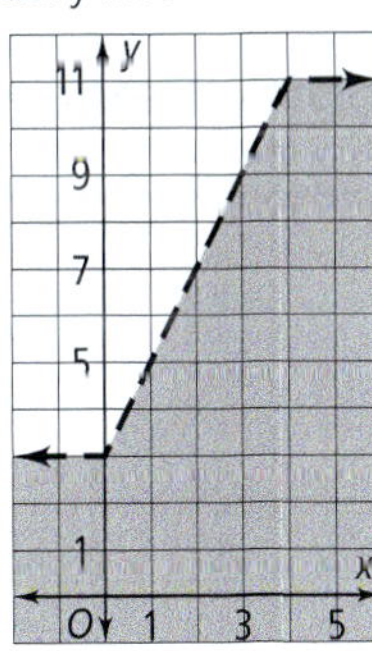

53. I

55.
let c = amnt. of a commission,
let s = amnt. of a sale

$$c = ks$$
$$(\$48{,}000) = k(\$800{,}000)$$
$$0.06 = k$$
$$c = 0.06\,(\$650{,}000) = \$39{,}000$$

(OR sltn. by another appropriate method)

Selected Answers

56.

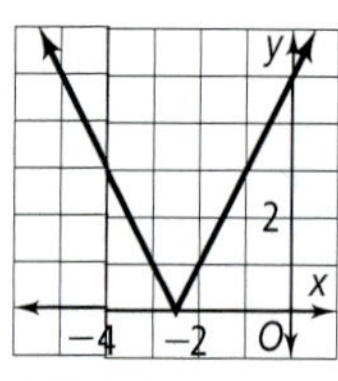

57.

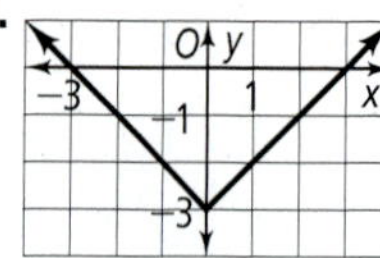

58.

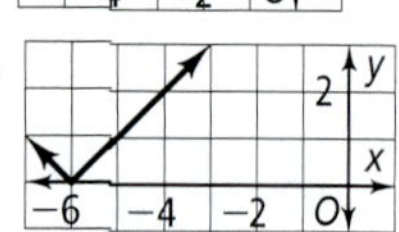

59.

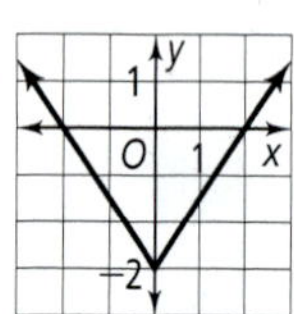

60.

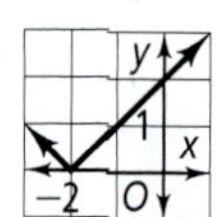

61.

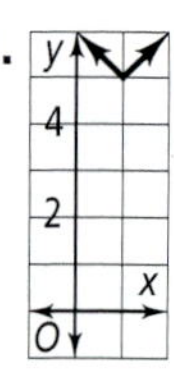

62. no **63.** yes; 100 **64.** yes; -5 **65.** no **66.** yes; 3 **67.** no **68.** yes; -10 **69.** no **70.** strong neg. correlation

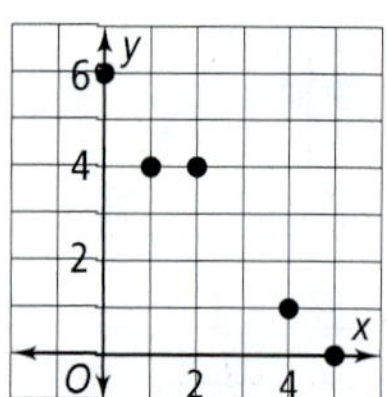

71. no correlation

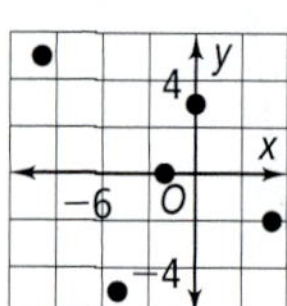

72–74.

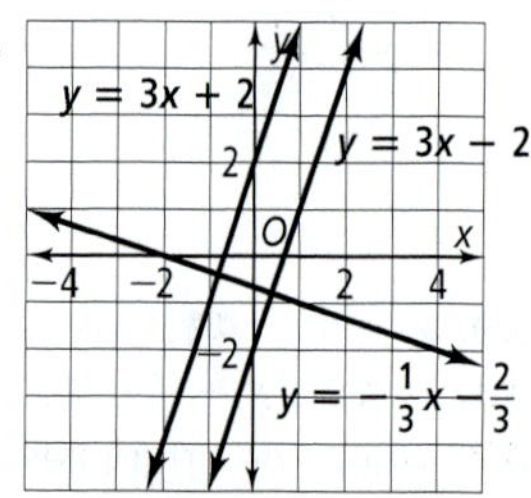

Chapter Review pp. 122–126

1. sometimes **2.** point-slope **3.** yes; domain: $\{-10, -6, 5, 6, 10\}$, range: $\{2, 3, 4, 7\}$ **4.** no; domain: $\{1, 3, 4, 10\}$, range: $\{5, 6, 8, 12\}$ **5.** no; domain: $\left\{-2, -\frac{3}{2}, -1, \frac{1}{2}, 1, 2, 3\right\}$, range: $\left\{-\frac{7}{2}, -\frac{1}{2}, 0, \frac{1}{2}, \frac{3}{2}, 2, \frac{5}{2}\right\}$ **6.** yes; domain: $\{-2, -1, \frac{1}{2}, 3\}$, range: $\{2\}$ **7.** 6, 4.5, 1 **8.** $-3\frac{3}{4}$, $-3\frac{3}{16}$, $-1\frac{7}{8}$ **9.** no **10.** no **11.** yes; 1; $y = x$ **12.** -4; 1.2 **13.** $\frac{10}{3}$; -1 **14.** $\frac{7}{2}$, $-1\frac{1}{20}$ **15.** $-\frac{4}{3}$; 0.4 **16.** $-\frac{2}{5}$ **17.** $\frac{7}{6}$ **18.** $\frac{2}{3}$ **19.** $-\frac{4}{9}$ **20.** $y = -3x + 4$ **21.** $y = \frac{1}{2}x + 6$

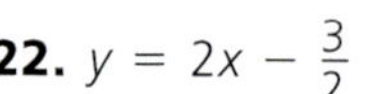

22. $y = 2x - \frac{3}{2}$

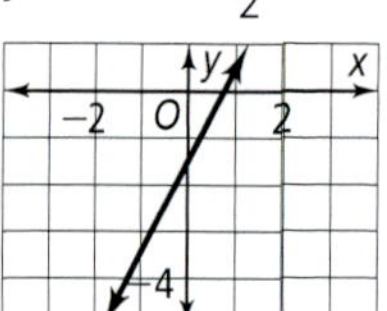

23. $y = \frac{2}{3}x + 3$

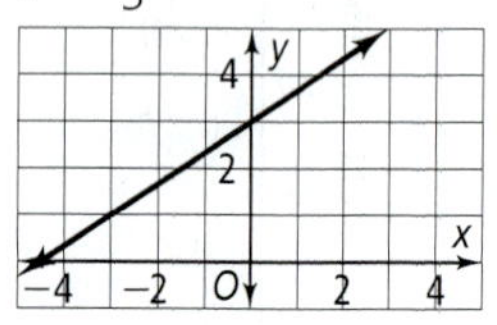

24. $y = -x + 5$

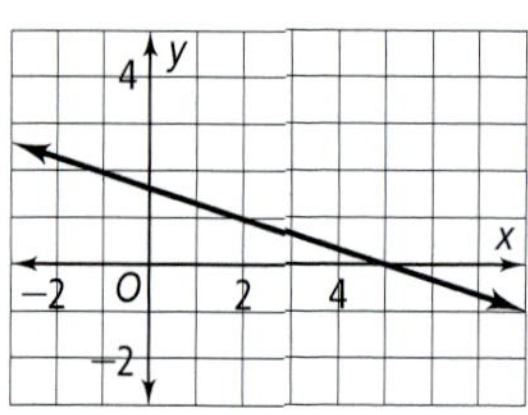

25. $y = -\frac{1}{3}x + \frac{5}{3}$

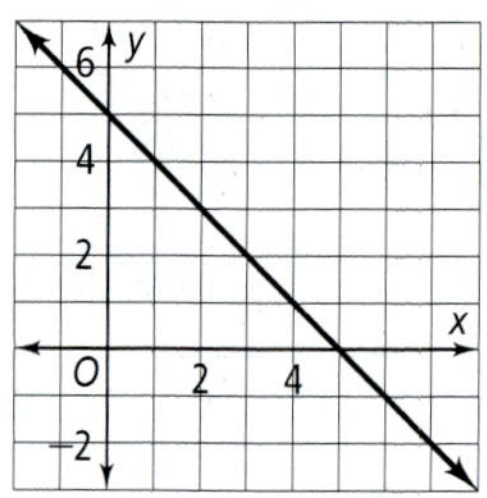

26. $y = -3(x - 4)$; $3x + y = 12$
27. $y + 1 = 5(x - 1)$; $5x - y = 6$
28. $y + 7 = -\frac{7}{3}(x - 3)$; $7x + 3y = 0$
29. $y - 3 = 2(x - 2)$; $2x - y = 1$
30. a. $y = -\frac{1}{2}x + 7$ **b.** $y = 2x - 13$
c.

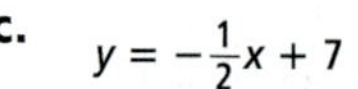

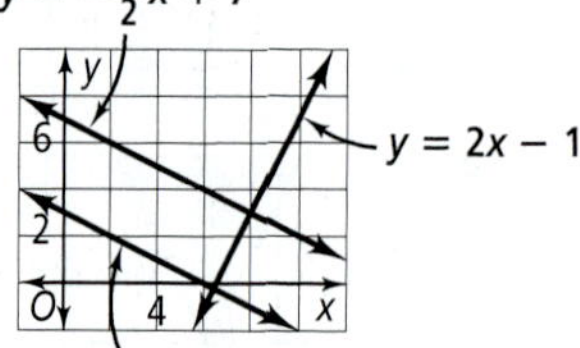

31.

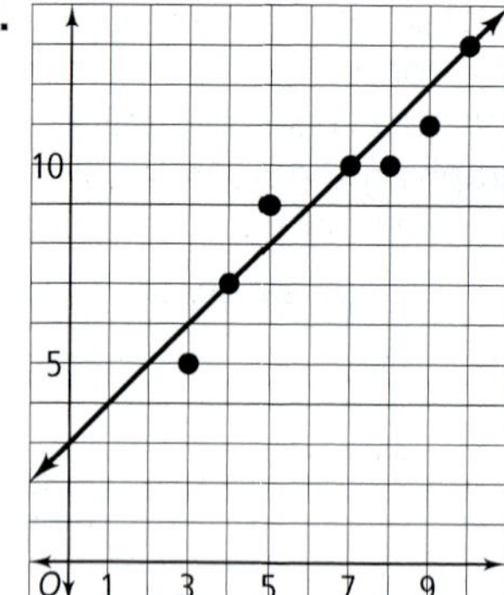

strong pos. correlation; Answers may vary. Sample: $y = x + 3$; 18

32.

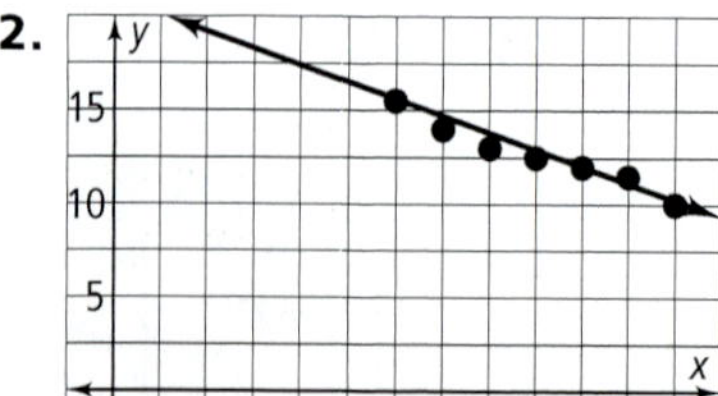

strong neg. correlation; Answers may vary. Sample: $y = -0.9x + 21$; 7.5

33. 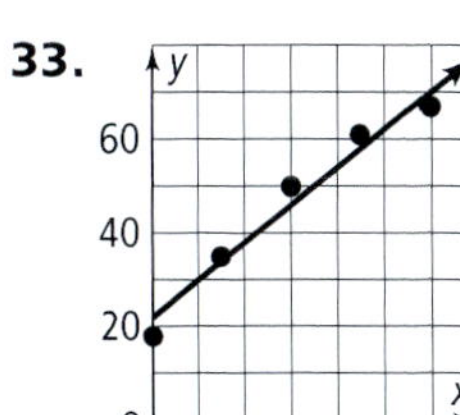

strong pos. correlation; Answers may vary. Sample: $y = 4x + 22$; 82

34. $y = f(x + 2) - 7$ **35.** $y = -f(x - 5)$ **36.** $y = f(-x) + 3$ **37.** translated 4 units down **38.** vertically stretched by a factor of 12, translated 2 units up **39.** vertically stretched by a factor of 2, reflected across the y-axis, reflected across the x-axis

40. $y = |x - 2| + 4$ **41.** $y = |x + 3|$

42. $y = |x - 5| + 2$ **43.** $y = |x - 4| + 1$

44. 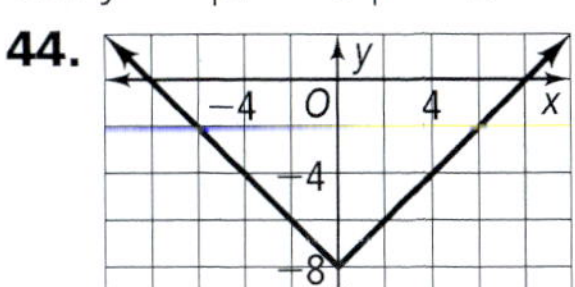 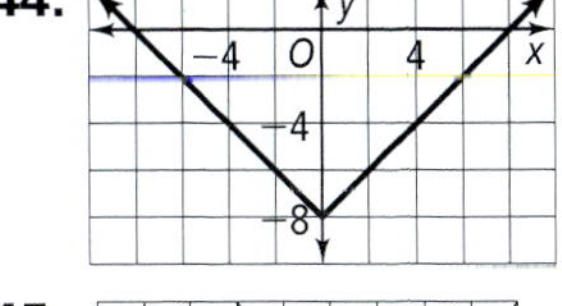

45. 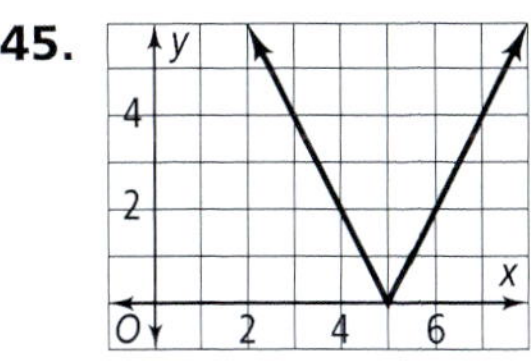

46. 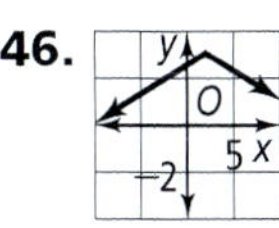

47.

48. (4, 0); $x = 4$

49. (0, 2); $x = 0$

50. 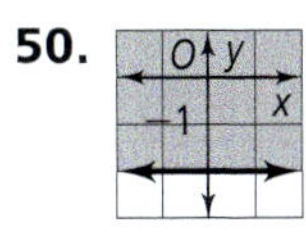

51. 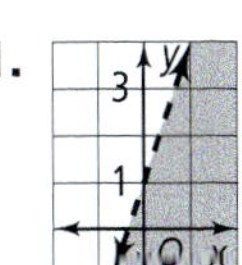

52. 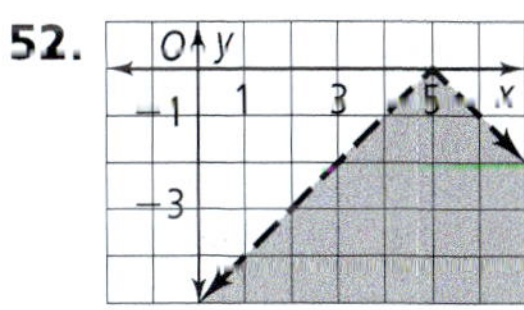

53. 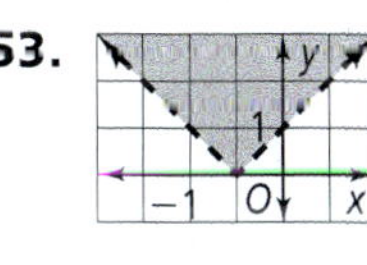

54. a. Answers may vary. Sample. $x + 3y \leq 15$

b. Answers may vary. Sample: domain: {0, 1, 2, 3, 4, 5, 6, 7, 8, 9, 10, 11, 12, 13, 14, 15}, range: {0, 1, 2, 3, 4, 5}

c. 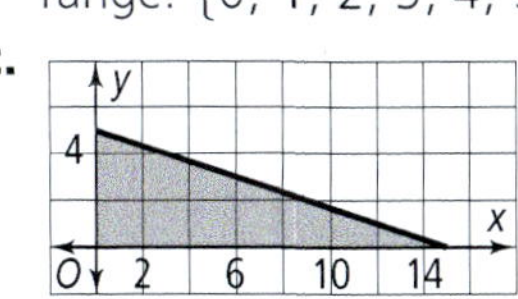

55. Answers may vary. Sample: $y \leq -|x| - 1$

Chapter 3

Get Ready! p. 131 **1.** 28 **2.** 33 **3.** $-\frac{15}{2}$ **4.** 15 **5.** $y = \frac{1}{2}x - \frac{5}{2}$ **6.** $y = -2x - 3$ **7.** $y = 5x + 16$ **8.** $y = 3x - 7$ **9.** $y = \frac{2}{5}x - \frac{3}{2}$ **10.** $y = 4x + 11$ **11.** $y = -6x - 8$ **12.** $y = -3x + 18$

13. 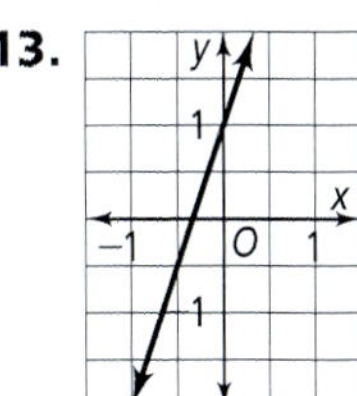

14. 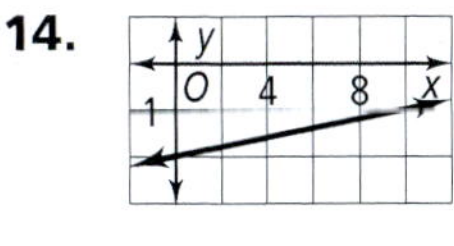

15. 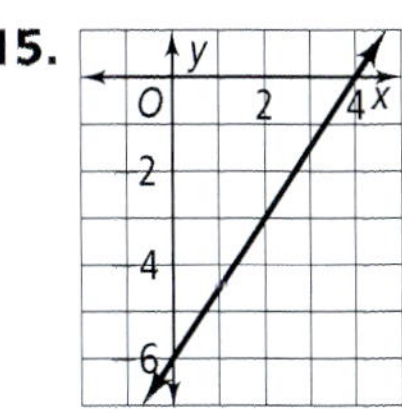

16. 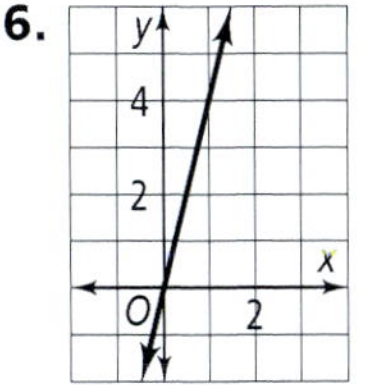

17. 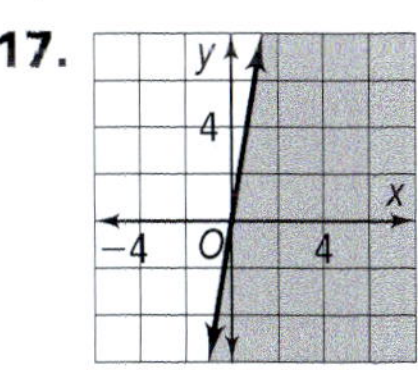

18. 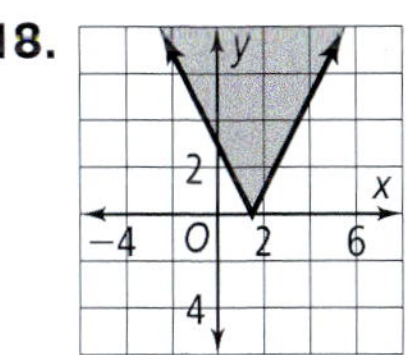

19. 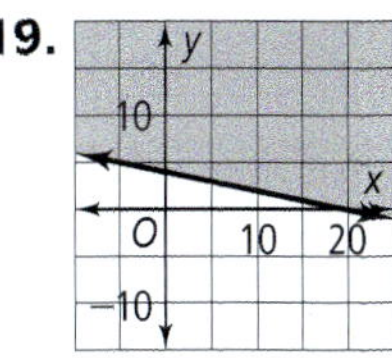

20. 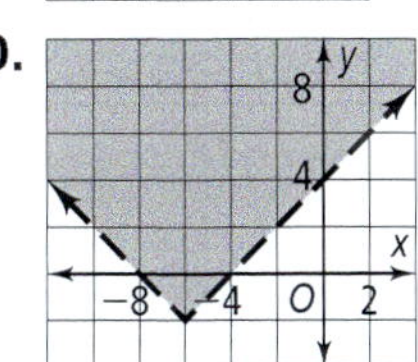

21. Answers may vary. Samples: Rocky Mountains, Appalachian Mountains **22.** Answers may vary. Sample: Your actions are consistent with your words when your actions show what you are saying. For example, if you say you are happy and you are laughing or smiling. **23.** 15 books or more; 15 books or more but less than 19, i.e., 15, 16, 17, or 18 books

Lesson 3-1 pp. 134–141

Got It? **1.** (2, −1) **2. a.** Spiny Dogfish: 59.5 cm; Greenland: 55.75 cm **b.** Each species of shark has a maximum total length; growth rates decrease with increase in age. **3.** in the yr 1990; about 1,100,000 **4. a.** inconsistent **b.** independent **c.** dependent

Lesson Check

1. (2, 1)

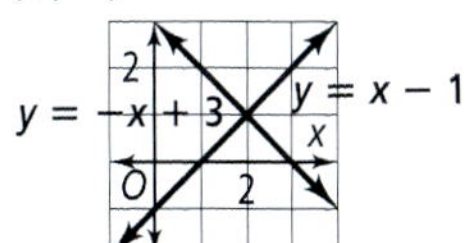

2. (2, 0)

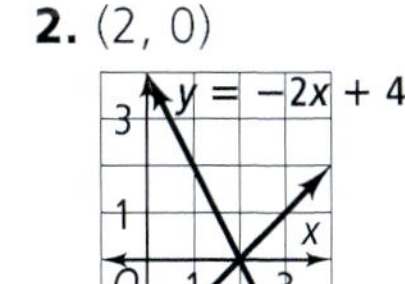

3. 2 pens; 4 pencils **4.** No; an independent system has a unique solution whereas an inconsistent system has no solution.
5. Answers may vary. Sample: $\begin{cases} y = 2x + 1 \\ y = 2x - 3 \end{cases}$
6. Independent; if the slope of one equation is the negative reciprocal of the slope of the other equation, the lines are perpendicular and intersect at a unique point.
Exercises 7–11. How solutions are determined may vary (graphing or using a table).
7. (3, 1)

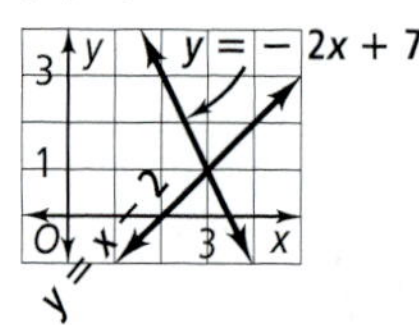

9. (−2, 4)

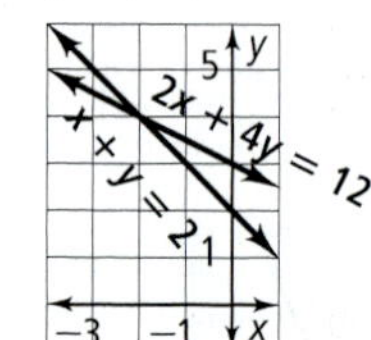

11. no solution

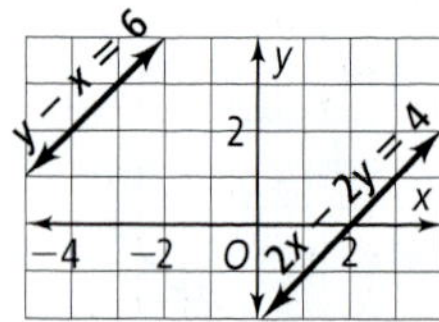

13. 2 small; 4 large
15. Models may vary. Sample: Use 0 for 1970.
$\begin{cases} y = 0.22x + 67.5 \\ y = 0.15x + 75.507 \end{cases}$
Around 2085, the quantities will be equal.
17. dependent **19.** inconsistent **21.** independent
23. dependent **25.** independent **27.** dependent
29. infinitely many solutions **31.** no solution

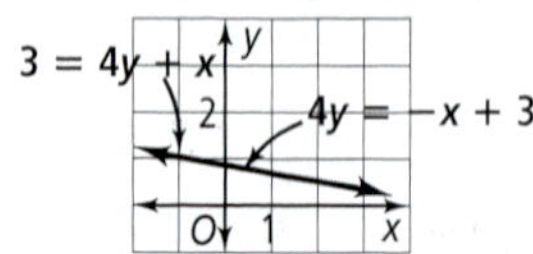

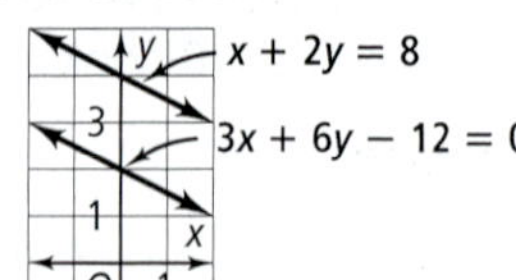

33. $\left(\frac{28}{5}, \frac{26}{5}\right)$

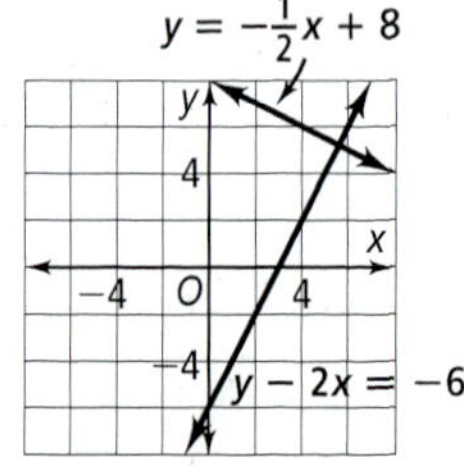

35. inconsistent **37.** inconsistent

39.

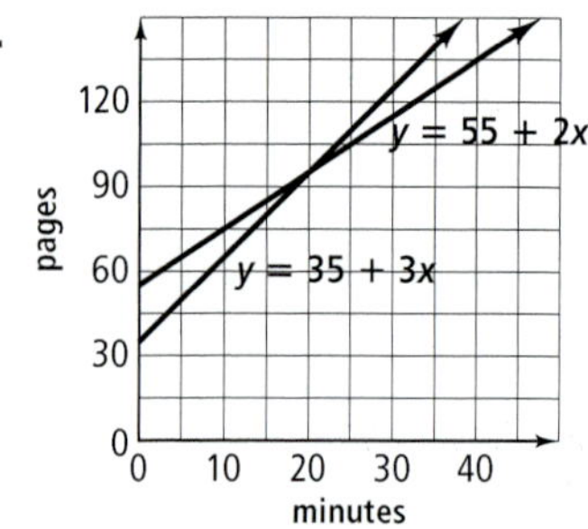

After 20 min you and your friend will have read the same number of pages.
41. My friend did not extend the table of values far enough. Scrolling down will show that when $x = 4$, then Y1 = Y2 = −2. So, the system has a solution, (4, −2).
43. An independent system has one solution. The slopes are different, but the *y*-intercepts could be the same. An inconsistent system has no solution. The slopes are the same and the *y*-intercepts are different. A dependent system has an infinite number of solutions. The slopes and *y*-intercepts are the same. **45.** sometimes **47.** never
49. Answers may vary. Sample: $5x + 2y = 5$
51. They are equivalent eqs. **53.** A **55.** C

57.

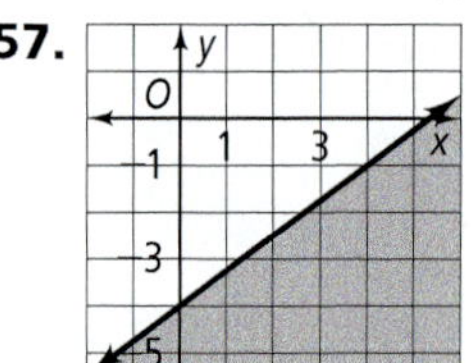

58.

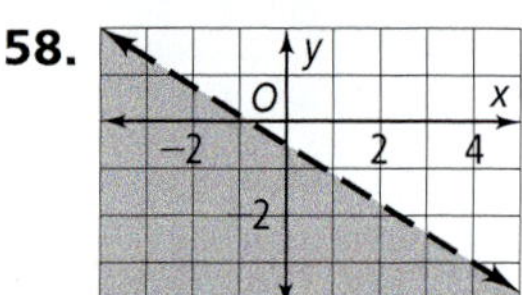

59.

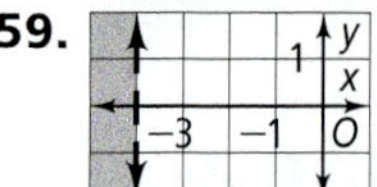

60. $n < -\frac{8}{7}$ **61.** $x \geq -\frac{29}{2}$ or −14.5 **62.** $x > \frac{5}{8}$
63. 2 **64.** $-\frac{3}{5}$ **65.** 2 **66.** 10 **67. a.** −3 **b.** 8 **c.** −10

Lesson 3-2 pp. 142–148

Got It? 1. (−2.5, 2.5) **2.** $.95 per download; $5.50 one-time registration fee **3.** (4, 0) **4. a.** (−2, 3) **b.** Yes; the solution (−5, 2) is a solution to both eqs. in the system, so substituting $y = 2$ into either equation will result in $x = -5$. **5. a.** no solution; The eq. is always false. **b.** infinite number of solutions; The eq. is always true.
Lesson Check 1. (1, 2) **2.** (−6, −6) **3.** (2, 1)
4. (5, −3) **5.** $\left(-\frac{1}{5}, \frac{19}{5}\right)$ **6.** (2, 1)

7. Answers may vary. Sample:
$\begin{cases} 4x - 3y = -2 \\ 3x - 2y = -1 \end{cases}$
$\begin{cases} -8x + 6y = 4 \\ 9x - 6y = -3 \end{cases}$
8. In the substitution method of solving a system of equations, you first solve one equation for one of the variables. Then substitute for this variable in the other equation and solve for the other variable. In the elimination method, you create an equivalent system of equations that contain a pair of additive inverses so that you can eliminate one variable and solve for the remaining variable.
9. Let r = number of regular cups of coffee and c = number of large cups of coffee. First, $r + c = 5$: because a total of 5 cups of coffee were purchased. Second, $r + 1.5c = 6$: because each regular cup of coffee is \$1, each large cup is \$1.50, and the total spent is \$6. Then, solve the system of equations using elimination by subtracting the first equation from the second to eliminate r and solve for c. $c = 2$; 2 large cups
Exercises **11.** $(-2, 4)$ **13.** $(0.75, 2.5)$ **15.** $(8, -1)$ **17.** $(-2, -5)$ **19.** seven \$1-bills; eight \$5-bills **21.** 3 vans and 2 sedans **23.** $(2, 4)$ **25.** $(2, -2)$ **27.** $(4, 1)$ **29.** $(1, 1)$ **31.** infinite number of solutions; $\{(x, y) | -2x + 3y = 13\}$ **33.** $(3, 2)$ **35.** $(5, 4)$ **37.** $\left(\frac{20}{17}, \frac{19}{17}\right)$ **39.** $(4, 1)$ **41.** no solution **43.** 10 deliveries **45.** $(4, -3)$ **47.** $(-3, 4)$ **49.** $(300, 150)$ **51.** $(0.5, 0.25)$
53. Error in 5th line: $-4(-7 - x) = 28 + 4x$ not $-28 - 4x$; Lines 5–9 should be: $3x + 28 + 4x = 14$; $7x = -14$; $x = -2$; $y = -7 - (-2)$; $y = -5$
55. Answers may vary. Sample:
$\begin{cases} -3x + 4y = 12 \\ 5x - 3y = 13 \end{cases}$; $(8, 9)$
57. In determining whether to use substitution or elimination to solve an equation, look at the equations to determine if one is solved or can be easily solved for a particular variable. If that is the case, substitution can easily be used. Otherwise, elimination might be easier. **59.** Substitution; the second equation is solved for y; $(-7, -26)$ **61.** Elimination; substitution would be difficult since no coefficient is 1 in the original system. Dividing the first equation by 3 and dividing the second equation by 5 results in an equivalent system where y would be eliminated from the system if the equations were subtracted; $(-1, -3)$
63. yes; -40 degrees **65.** 0 **67.** 2 **69.** 6 **71.** 4
72. no solution

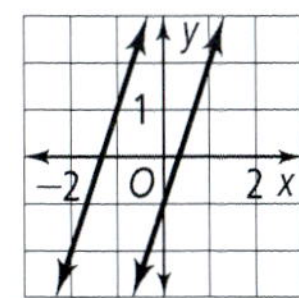

73. infinite number of solutions, $\{(x, y) | -9x - 3y = 1\}$

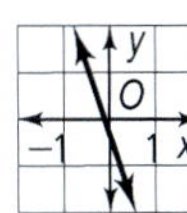

74. no solution

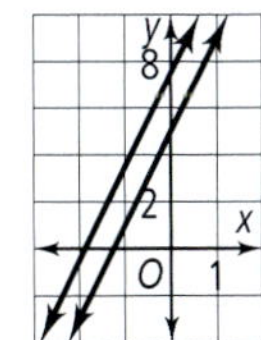

75. function **76.** function **77.** not a function
78. $x < -1$

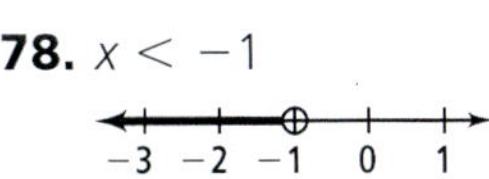

79. $x > -\frac{1}{2}$

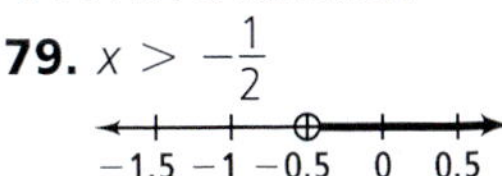

80. $y < 1$

Lesson 3-3 pp. 149–155

Got It? **1.** $(4, 1)$, $(5, 0)$, $(6, 0)$, $(7, 0)$
2.

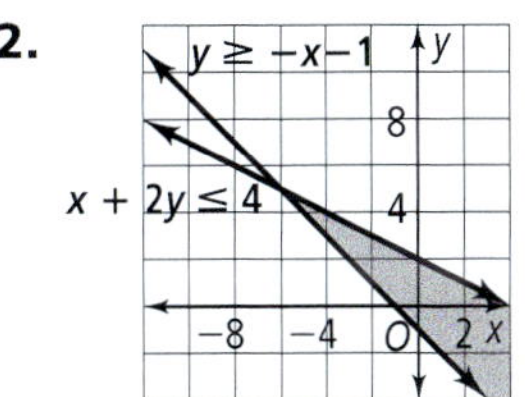

3. 5 meats and no vegetables; 4 meats and 1 or 2 vegetables; 3 meats and 2, 3, or 4 vegetables; 2 meats and 3, 4, 5, or 6 vegetables; 1 meat and 4 − 8 vegetables; no meat 5 − 10 vegetables
4. $y > 2|x - 1|$

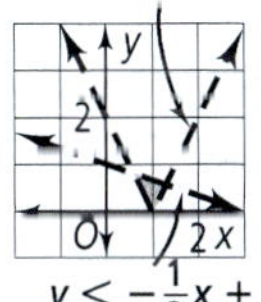

Lesson Check

1.

$2x + y \leq 5$
$x + y \geq 2$

2. $y < x + 1$ $y > x$

3.

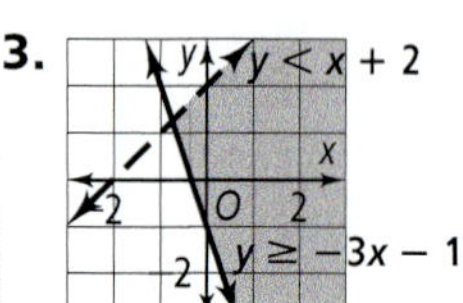

4. 0 h TV and 1, 2, or 3 h football; 1 h TV and 1 or 2 h football; or 2 h TV and 1 h football **5.** Intersection; the solution of two inequalities is the overlap or the intersection of the graphs of the individual inequalities. **6.** The graphical solution of a system of inequalities consists of the overlap or intersection of the individual half-planes and corresponding boundary lines (either dotted or solid). The graphical solution of a system of equations includes only the intersection of the lines, not the half-planes. **7.** For each inequality, the wrong half-plane has been shaded. The half-plane below $y = \frac{1}{2}x - 1$ should be shaded and the half-plane above $y = -3x + 3$ should be shaded. Also, both boundary lines should be dashed because the inequalities are $<$ and $>$.

Exercises 9. (0, 0), (0, 1), (0, 2), (0, 3), (0, 4), (0, 5), (0, 6), (0, 7), (1, 0), (1, 1), (1, 2), (1, 3), (1, 4), (1,5), (1, 6), (2, 0), (2, 1), (2, 2), (2, 3), (2, 4), (2, 5), (3, 3), (3, 4)

11. $y < -x + 1$ $y \le 2x + 2$

13.

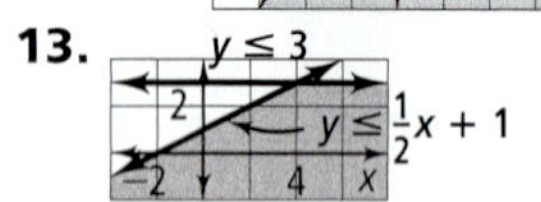

15.

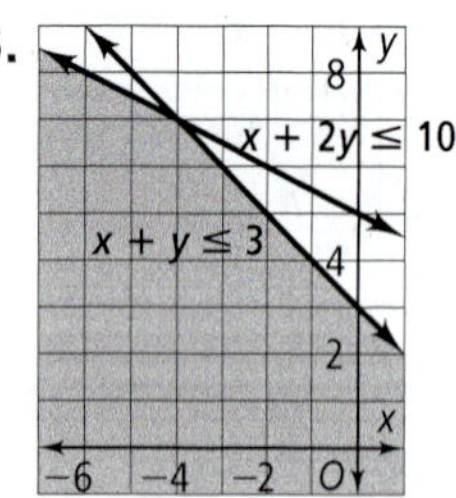

17. $y > -2x$ $-2x + y \le -2$

19. no solution

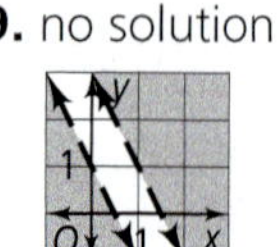

21. Let r = number of rose plants and t = number of tulip plants.
$t + r \ge 50$
$r \le 80$
Because the number of plants must be a whole number, only the points in the overlap that represent whole numbers are solutions of the problem.

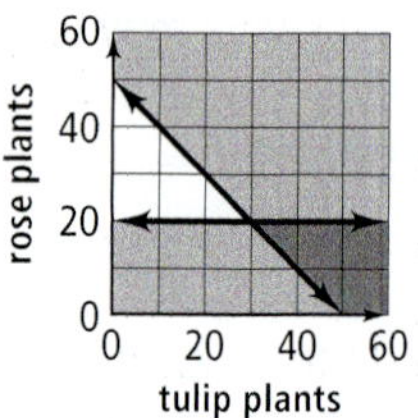

23. $y < -\frac{1}{3}x + 1$ $y < |2x - 1|$

25.

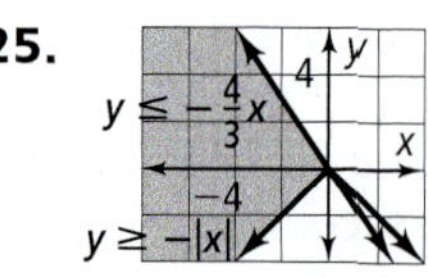

27.

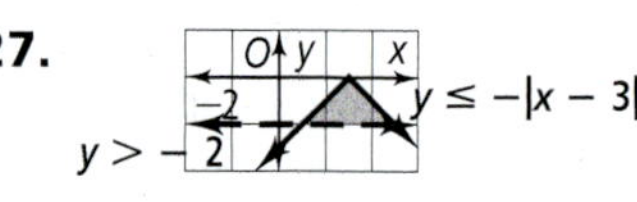

29.

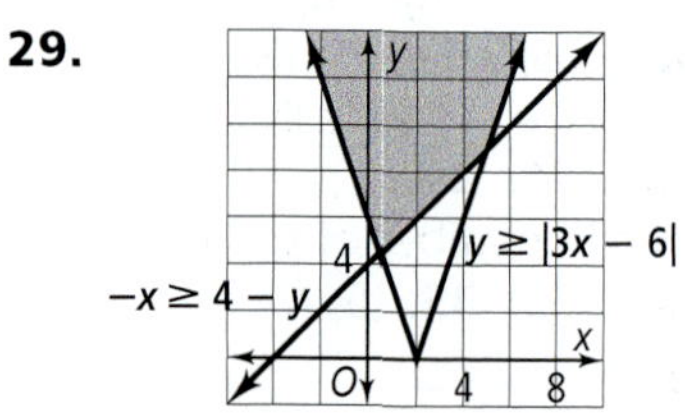

31. (0, 4), (0, 5), (0, 6), (1, 3), (1, 4), (1, 5), (2, 3), (2, 4); the sum of the servings must be greater than or equal to 4 and less than or equal to 6.

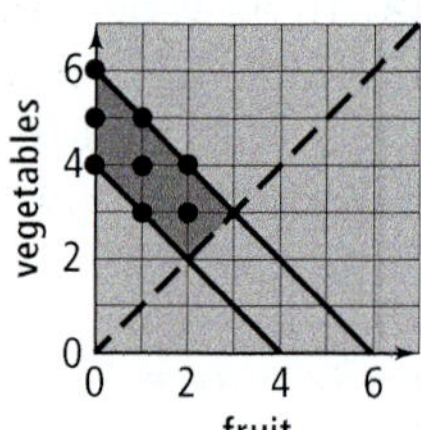

33. Answers may vary. Sample: $\begin{cases} x < 5 \\ y \ge 1 \end{cases}$

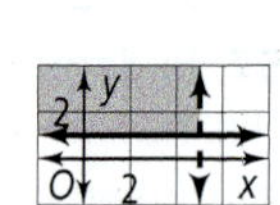

35. Use test pts. that are not on either of the boundary lines and that make the calculations as easy as possible (e.g., the origin). **37.** A, B **39.** A, B **41.** B, C **43.** A **45.** A

47.

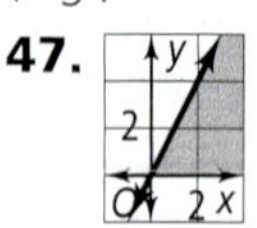

49.

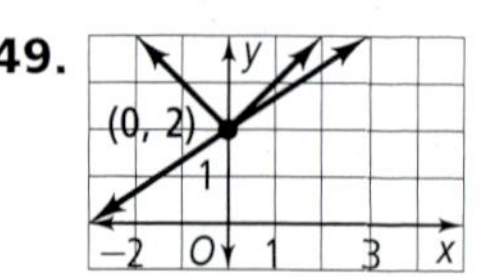

51. 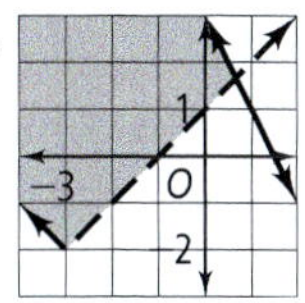

53. $y > |x - 1| + 1$

$y \leq -|x - 3| + 4$

55. $\begin{cases} y \geq |x| - 2 \\ y \leq -|x| + 2 \end{cases}$

57. $\begin{cases} y \leq 4 \\ y \geq 0 \\ y \leq 2x \\ y \geq 2x - 8 \end{cases}$

59. C **61.** B **63.** $(-9, -26)$ **64.** $\left(\frac{23}{14}, -\frac{13}{14}\right)$
65. no solution **66.** $(-2, -1)$ **67.** $(-1, 2)$
68. $\left(-\frac{4}{7}, \frac{1}{14}\right)$ **69–72.** Answers may vary. Samples are given for each exercise. **69.** $(0, 3)$ **70.** $(0, 3)$
71. $(2, -1)$ **72.** $(1, -1)$

Lesson 3-4 pp. 157–162

Got It? 1. a. *P* has a maximum value of 7.5 at (0, 2.5). **b.** Yes; Answers may vary. Sample: $P = 5$ has the same (maximum) value at all four vertex points. $P = x + 2y$ has maximum value 5 at *R* and *S*. **2.** 100 T-shirts and 10 sweatshirts.

Lesson Check

1. 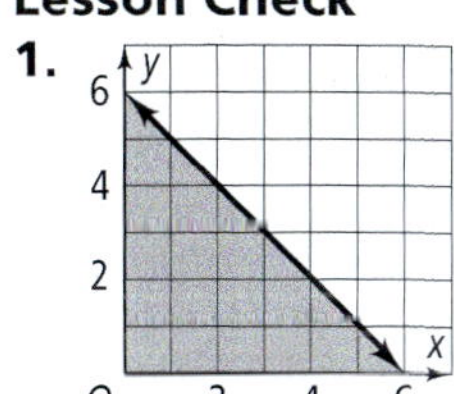

2. 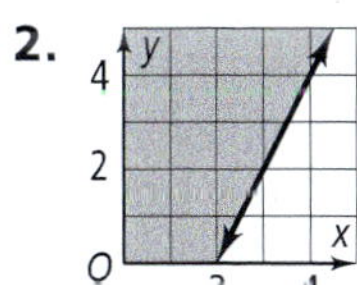

3. 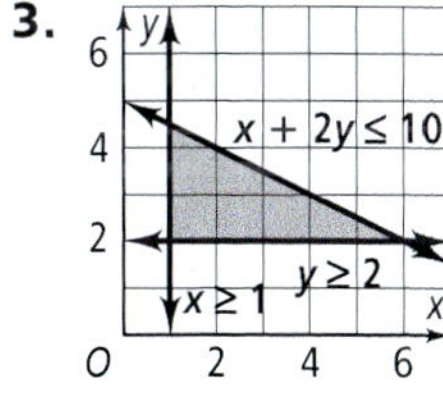

4. 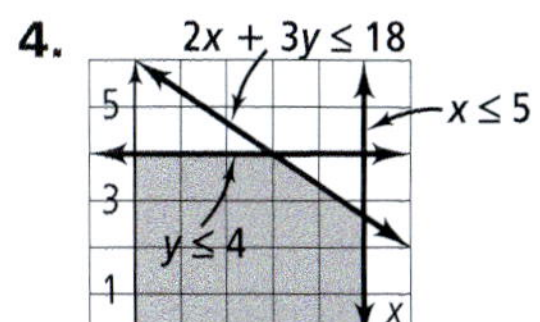

5. (0, 0), (0, 4), (5, 0), (5, 4)

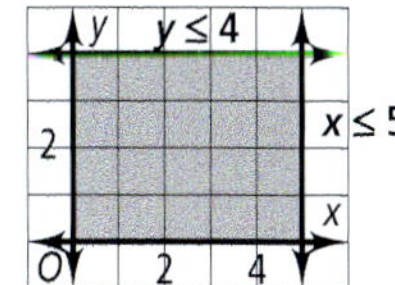

6. (0, 8), (3, 5), (0, 5)

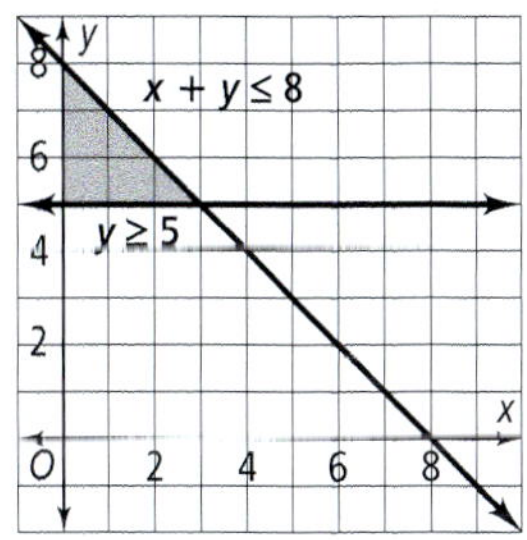

7. Constraints are limits or restrictions on the variables in the objective function in a linear programming problem. These constraints are written as linear inequalities.
8. Linear programming is an extension of solving linear inequalities. For each, you are given constraints represented by linear inequalities that are graphed. All the points in the overlapping region are solutions, but linear programming problems are usually looking for maximum or minimum values of some quantity modeled with an objective function.

9. Answers may vary. Sample: $\begin{cases} y \leq x \\ y \leq -x + 4 \\ 0 \leq y \leq 1 \end{cases}$

$P = 2x + 3y$, $P(0, 0) = 0$, $P(1, 1) = 5$, $P(3, 1) = 9$, $P(4, 0) = 8$; maximum value of *P* is 9 at (3, 1)

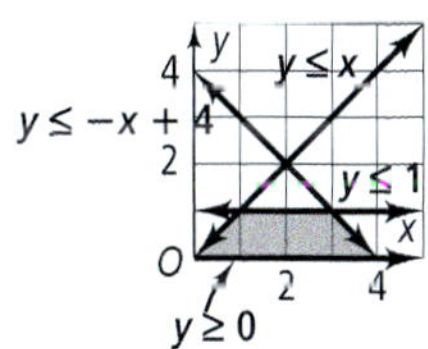

Exercises

11. 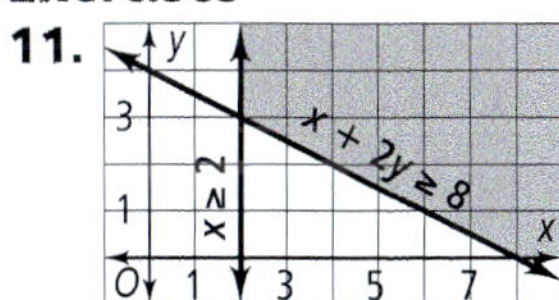

vertices: (8, 0), (2, 3); minimized at (8, 0)

13. Let s = number of Spruce trees and m = number of Maple trees.

a. $\begin{cases} 30s + 40m \leq 2100 \\ 600s + 900m \leq 45{,}000 \\ s \geq 0, m \geq 0 \end{cases}$

b. $P = 650s + 300m$

c.

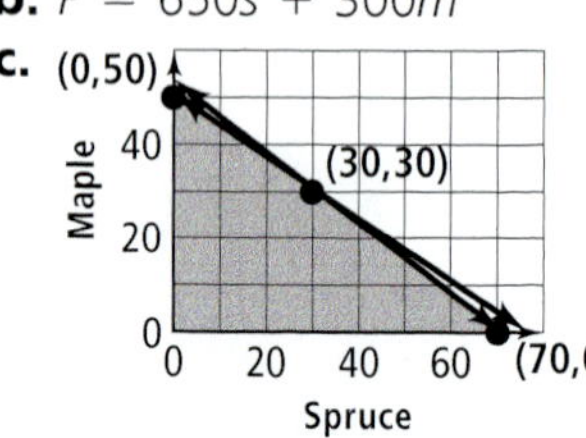

vertices: (0, 0), (0, 50), (30, 30), (70, 0)

d. 70 spruce trees and 0 maple trees

15. He is not considering the constraint $y \leq x + 3$; maximize when $P = 11$ at (1, 4)

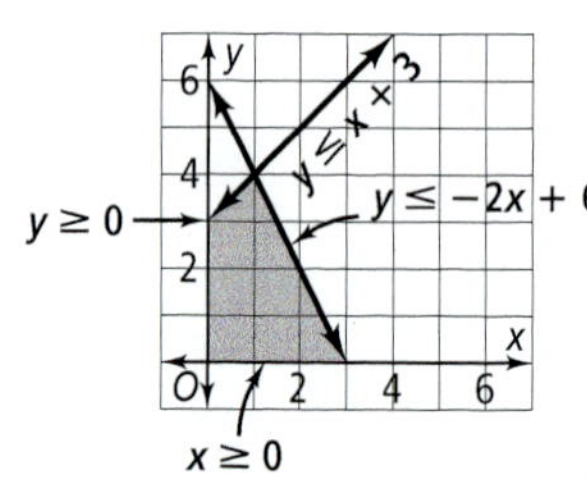

17. vertices: (0, 0), (1, 4), (0, 4.5), $\left(\frac{7}{3}, 0\right)$; maximized when $P = 6$ at (1, 4)

19.

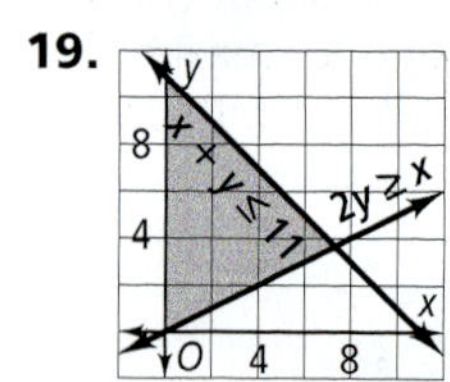

vertices: (0, 0), $\left(7\frac{1}{3}, 3\frac{2}{3}\right)$, (0, 11); maximized when $P = 29\frac{1}{3}$ at $\left(7\frac{1}{3}, 3\frac{2}{3}\right)$.

21.

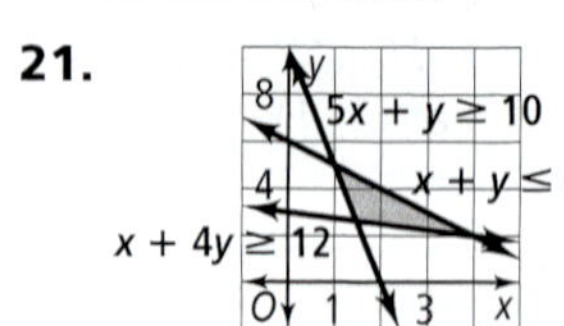

vertices: $\left(\frac{28}{19}, \frac{50}{19}\right)$, (4, 2) (1,5); minimized when $C = 67{,}370$ at $\left(\frac{28}{19}, \frac{50}{19}\right)$

23.

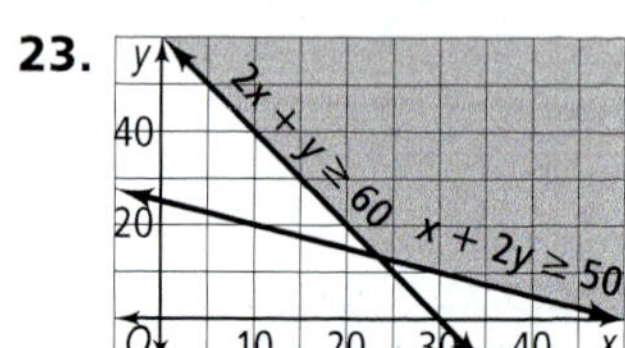

vertices: (0, 60), $\left(23\frac{1}{3}, 13\frac{1}{3}\right)$, (50, 0); minimized when $x = 23\frac{1}{3}$ and $y = 13\frac{1}{3}$; Round to (23, 14) and (24, 13); (24, 13) gives a minimum cost of $261

25. C

27.

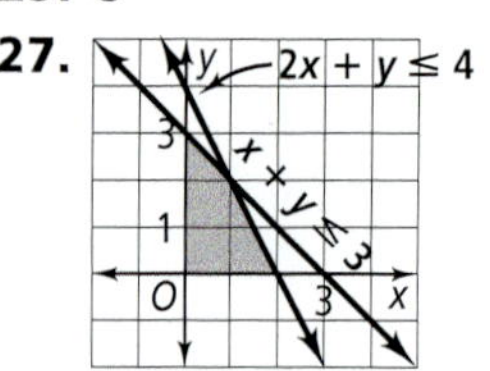

vertices: (0, 0), (2, 0), (0, 3), (1, 2)

28.

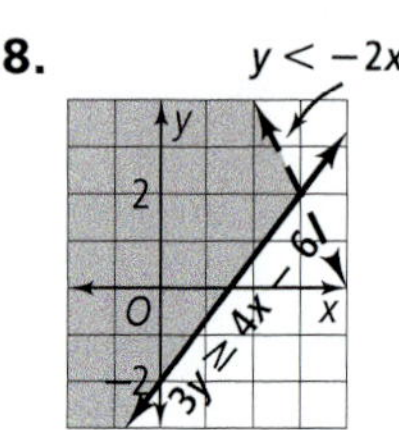

29.

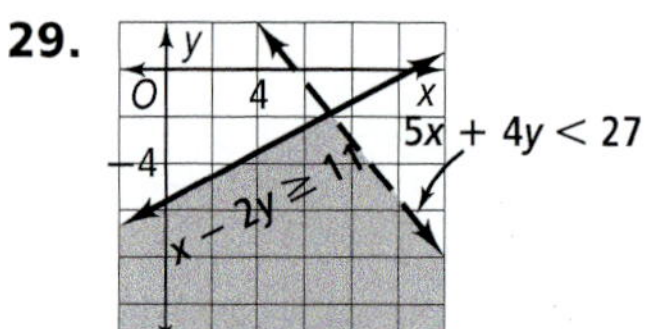

30.

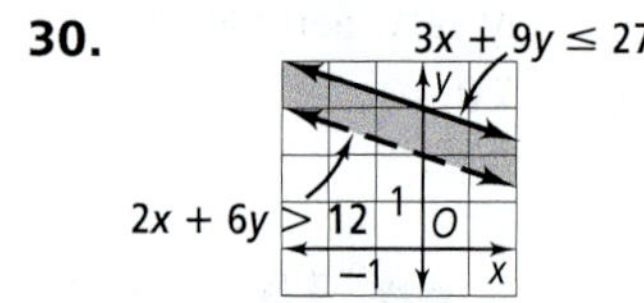

31. 1 **32.** −34 **33.** 24 **34.** 65 **35.** (0, 6), (−3, 0) **36.** (0, 4), (18, 0) **37.** (0, −1), (1, 0)

Lesson 3-5 pp. 166–173

Got It? 1. (4, 2, −3) **2. a.** (4, −1, 2) **b.** Answers may vary. Sample: Yes; you can choose to eliminate either x, y, or z resulting in a system of equations in 2 variables. **3. a.** (2, 1, −4) **b.** No; in Step 1 we solved for x in terms of y only. Therefore, once we found the value of y we could have substituted that value into the equation we wrote in Step 1 and solved for x without ever finding the z-value. **4.** 50 T-shirts, 50 polo shirts, and 100 rugby shirts

Lesson Check 1. (5, −3, −2) **2.** (6, 0, −2) **3.** (0, 3, 4) **4.** (2, 1, −5) **5.** Answers may vary. Sample: Substitution is the best method to use when one of the equations can be solved easily for one variable. **6.** Answers may vary. Sample: (0, 0, 0) is a unique solution to a system of three variables. The planes intersect at one common point. When a system has no solution, no point lies in all three planes. **7.** No solution since no point lies in all three planes. The three planes are parallel. **8.** infinitely many solutions

Exercises 9. (4, 2, −3) **11.** (2, 1, −5) **13.** $\left(\frac{1}{2}, -3, 1\right)$ **15.** (1, −4, 3) **17.** (4, −1, 2) **19.** $\left(-\frac{10}{13}, -\frac{2}{13}, \frac{4}{13}\right)$ **21.** (8, −4, 2) **23.** (−2, −1, −3) **25.** (0, 1, 7)

27. (5, −2, 0) **29.** (1, 3, 2) **31.** $m\angle P = 32°$; $m\angle Q = 96°$; $m\angle R = 52°$ **33.** (8, 1, 3)
35. $\left(\frac{1}{2}, 2, -3\right)$ **37.** no solution **39.** (2, 4, 6)
41. (0, 2, −3)
43. Answers may vary. Sample: Solution is (1, 2, 3).
$$\begin{cases} x + y + z = 6 \\ 2x - y + 2z = 6 \\ 3x + 3y + z = 12 \end{cases}$$
45. Let *E*, *F*, and *V* represent the number of edges, faces, and vertices, respectively. From the first statement, $E = \frac{5}{2}F$. From the second statement, $V = \frac{5}{3}F$. From the third statement, $V + F = E + 2$. Solving this system of 3 equations yields $E = 30$, $F = 12$, and $V = 20$.
47. $\frac{3}{2}$
49. 144 mezzanine seats
50. $P = 12$ is maximized at (0, 4).
51. $x \geq -\frac{3}{2}$; 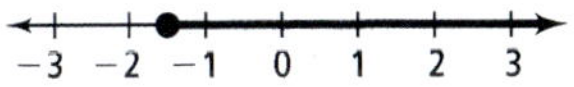

52. $x \leq -18$; 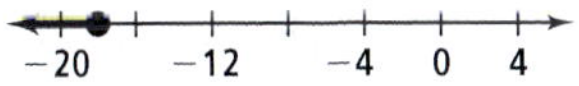

53. $x < -1$; 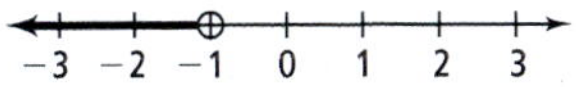

54. $\left(7, \frac{5}{4}\right)$ **55.** dependent system; infinite number of solutions, $\left\{(x, y) \mid y = -\frac{1}{2}x - \frac{3}{4}\right\}$
56. inconsistent system; no solution

Lesson 3-6 pp. 174–181

Got It? 1. 17

2. a. $\left[\begin{array}{cc|c} -4 & -2 & 7 \\ 3 & 1 & -5 \end{array}\right]$ **b.** $\left[\begin{array}{ccc|c} 4 & -1 & 2 & 1 \\ 0 & 1 & 5 & 20 \\ 2 & 1 & 0 & 7 \end{array}\right]$

3. $\begin{cases} 2x = 6 \\ 5x - 2y = 1 \end{cases}$

4. a. (1, 2) **b.** elimination; you use the same steps to solve **5.** $\left(1, \frac{1}{2}, 3\right)$

Lesson Check 1. 2 × 1 **2.** 2 × 4

3. $\left[\begin{array}{cc|c} 3 & 5 & 0 \\ 1 & 1 & 2 \end{array}\right]$ **4.** $\left[\begin{array}{ccc|c} 1 & 3 & -1 & 2 \\ 1 & 0 & 2 & 8 \\ 0 & 2 & -1 & 1 \end{array}\right]$

5. 16 **6.** a_{21} is 0, the element in row 2, column 1. a_{12} is −9, the element in row 1 and column 2. **7.** Answers may vary. Sample: The entry fee to a school play is $2 for adults. Jamie paid a total of $8 for 4 student entry fees and 2 adult entry fees. What is the student entry fee?

Exercises 9. 1 **11.** 8

13. $\left[\begin{array}{cc|c} 3 & 2 & 16 \\ 0 & 1 & 5 \end{array}\right]$ **15.** $\left[\begin{array}{ccc|c} 1 & -1 & 1 & 150 \\ 2 & 0 & 1 & 425 \\ 0 & 1 & 3 & 0 \end{array}\right]$

17. $\left[\begin{array}{ccc|c} 1 & -1 & 1 & 0 \\ 1 & -2 & -1 & 5 \\ 2 & -1 & 2 & 8 \end{array}\right]$ **19.** $\begin{cases} 5x + y = -3 \\ -2x + 2y = 4 \end{cases}$

21. $\begin{cases} 2x + y + z = 1 \\ x + y + z = 2 \\ x - y + z = -2 \end{cases}$ **23.** $\begin{cases} 5x + 2y + z = 5 \\ 4x + y + 2z = 8 \\ x + 3y - 6z = 2 \end{cases}$

25. (−1, 0) **27.** (4, 6) **29.** (2, 3)
31. $10,000 at 4% and $15,000 at 6%, Let x = amount invested at 4% and y = amount invested at 6%.
$$\begin{cases} x + y = 25{,}000 \\ 0.04x + 0.06y = 1300 \end{cases}$$
$$\left[\begin{array}{cc|c} 1 & 1 & 25000 \\ 0.04 & 0.06 & 1300 \end{array}\right] = \left[\begin{array}{cc|c} 1 & 0 & 10000 \\ 0 & 1 & 15000 \end{array}\right]$$
33. (3, 1,1) **35.** (35, −22, −16) **37.** (1, 1, 1, 1)
39. (2, 3) **41.** 1 qt. of red paint: $7.75; 1 qt. of yellow paint: $5.75 **43.** Answers may vary. Sample: 0; 0
45. (8, 2) **47.** $\left(\frac{1}{8}, -\frac{1}{17}\right)$ **49.** G
51. $x \leq -\frac{3}{2}$; 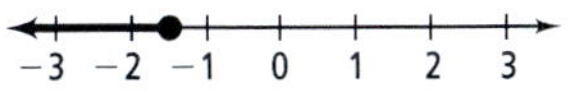

52. $x \geq -35$; 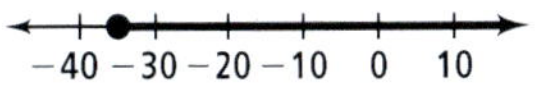

53. $x \geq 4$; 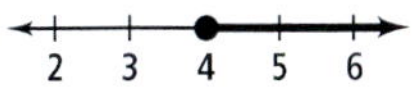

54. $\frac{15}{2}, -\frac{9}{2}$ **55.** 10, −10 **56.** 10, −6 **57.** $y = 2x$
58. $y = \frac{1}{3}x$

Chapter Review pp. 183–186

1. independent system **2.** Linear programming; constraints
3. 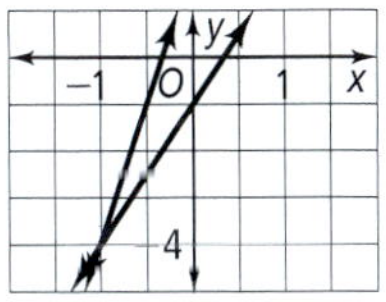

independent; (−1, −4)
4. dependent **5.** inconsistent **6.** dependent
7. independent; (−4, 6) **8.** independent; (1, 0) v
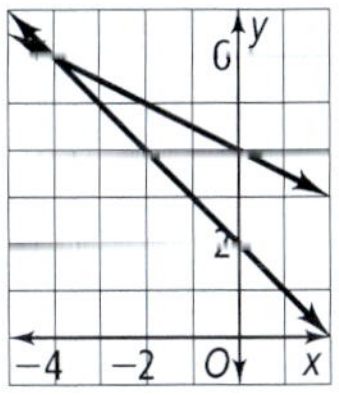

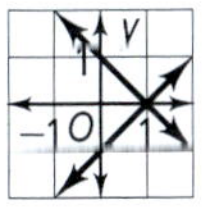

9. 3 pens **10.** (−1, −2) **11.** (0, −5) **12.** (−2, 3)
13. inconsistent; no solution **14.** 1 serving of roast beef and 2 servings of mashed potatoes

15. 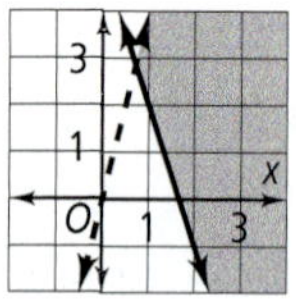

16. 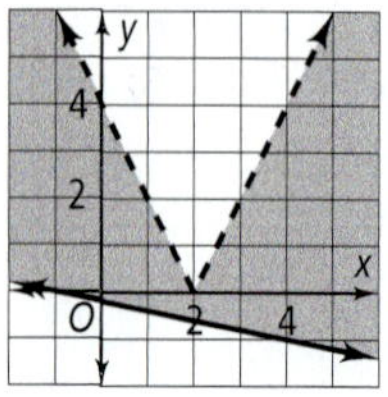

17. 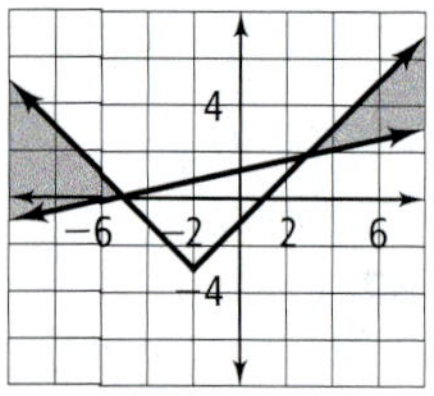

18. 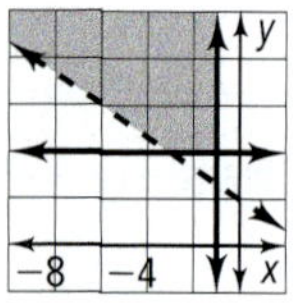

19. 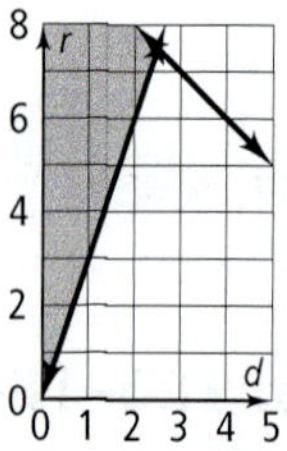

Let r = amount of regular coffee and d = amount of decaffeinated coffee

$$\begin{cases} r + d \leq 10 \\ r \geq 3d,\ r \geq 0,\ d \geq 0 \end{cases}$$

20. 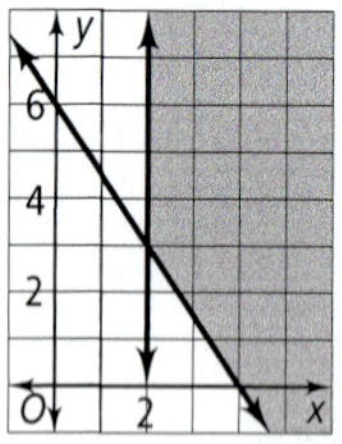

vertices: (4, 0) and (2, 3); $C = 4$ is minimized at (4, 0).

21. 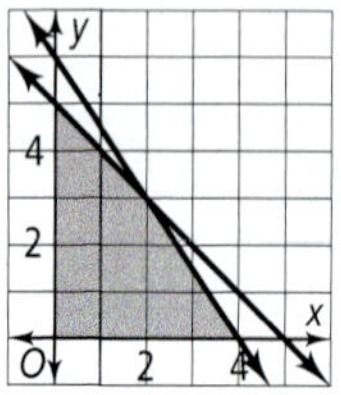

vertices: (0, 0), (4, 0), (2, 3), (0, 5); $P = 25$ is maximized at (0, 5).

22. 50 chef's salads and 50 Caesar salads
23. (1, 3, −2) **24.** (−4, 1, −5) **25.** (6, 0, −2)
26. no solution **27.** $\left(\frac{1}{2}, \frac{1}{4}\right)$ **28.** (1, −1)
29. (2, −4, 6) **30.** (5, 2, −3)

Chapter 4

Get Ready! p. 191

1. 6 **2.** 4
3. $-2 < x < 6$ 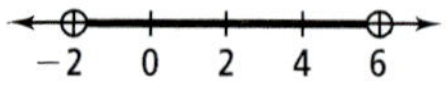

4. $y \leq -\frac{5}{4}$ or $y \geq \frac{9}{4}$ 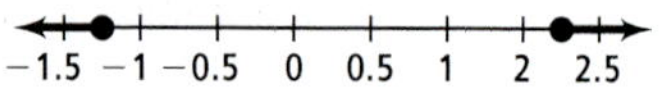

5. $y = 9x - 10$

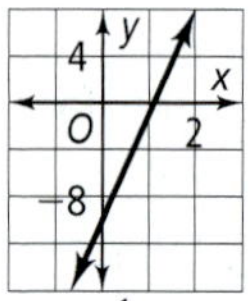

6. $y = \frac{1}{2}x + 8$

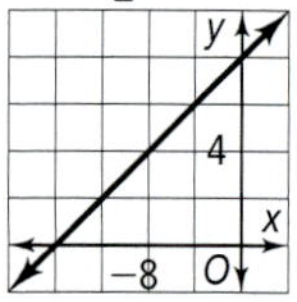

7. translated 4 units to the right and 2 units up
8. translated 10 units to the left and 3 units down
9. (10, −1) **10.** (−6, −6) **11.** Answers may vary. Sample: application forms, registration forms, tests
12. Answers may vary. Sample: monsters, ghosts, tooth fairy **13.** writing

Lesson 4-1 pp. 194–201

Got It?

1. a. 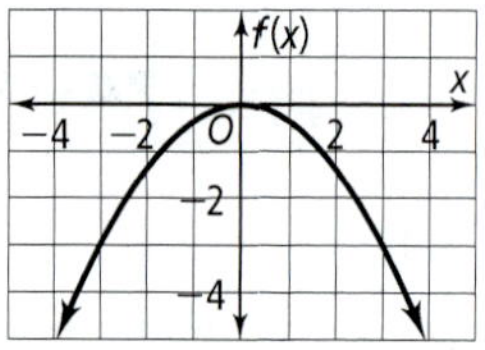

b. If a is a negative number, the parabola will open downward. There will be a maximum value for y at the vertex of the parabola.

2. a. translated 3 units up

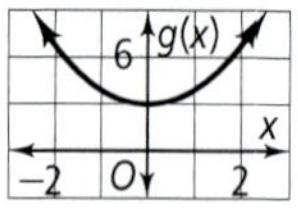

b. translated 1 unit to the left

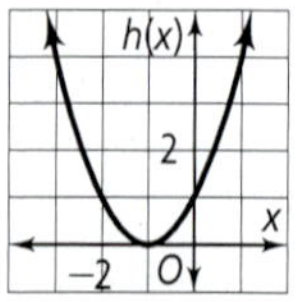

3. vertex: (−1, 4); axis of symmetry: $x = -1$; maximum: 4; domain: all real numbers, range: $y \leq 4$ **4.** stretch by the factor 2, translate 2 units to the left and 5 units down.
5. $f(x) = -\frac{2}{9}(x - 2)^2 + 7$

Lesson Check

1. 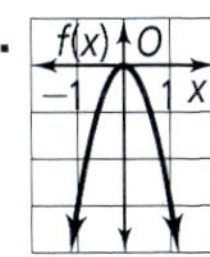

2. minimum **3.** $y = -2(x - 0)^2 + 35$ **4.** when $a > 0$
5. No; a must be > 0 or < 0. **6.** $y = (x + 6)^2$ is the graph of $y = x^2$ translated 6 units to the left and has a minimum at $(-6, 0)$; $y = (x - 6)^2 + 7$ is the graph of $y = x^2$ translated 6 units to the right and 7 units up, with a minimum at $(6, 7)$.

Exercises

7.  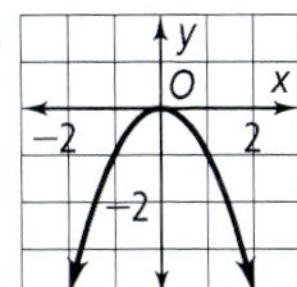

9. 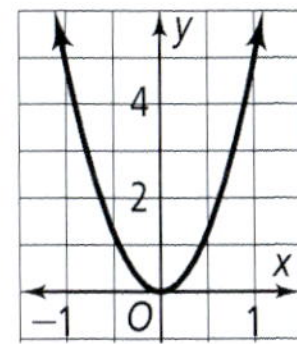

11. 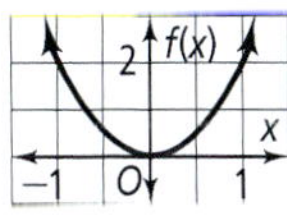

13. 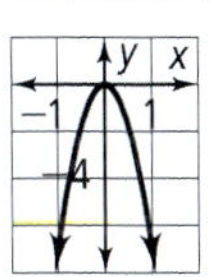

15. translated 3 units up

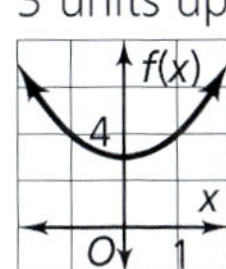

17. translated 6 units down

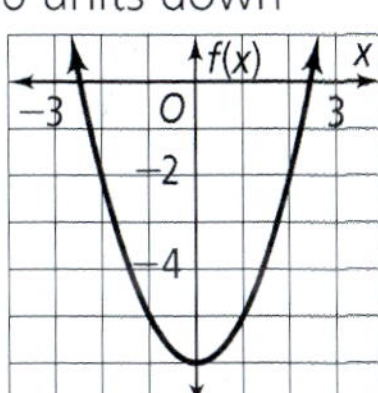

19. translated 9 units down

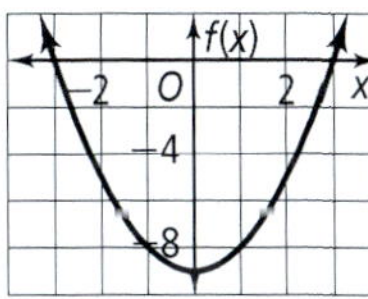

21. translated 1.5 units up

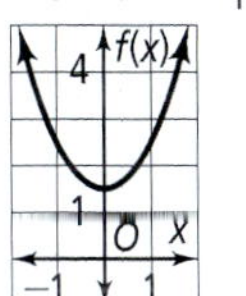

23. vertex: $(-20, 0)$; axis of symmetry: $x = -20$, maximum: 0; domain: all real numbers, range: $y \leq 0$
25. vertex: $(-5.5, 0)$; axis of symmetry: $x = -5.5$; minimum: 0; domain: all real numbers, range: $y \geq 0$
27. vertex: $(4, -25)$; axis of symmetry: $x = 4$; maximum: -25; domain: all real numbers, range: $y \leq -25$

29. $x = 1$

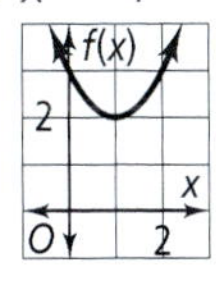

31. $x = 2$

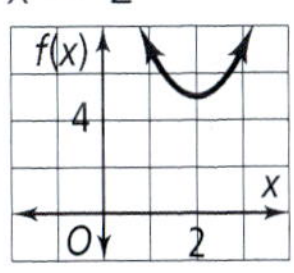

33. $x = 1$

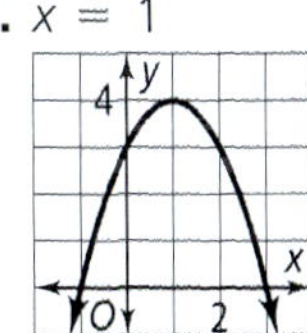

35. $y = (x - 2)^2 + 5$ **37.** $y = -\frac{1}{2}(x + 4)^2 + 2$
39. 1000 chips; $20
41. similar: same vertex $(-1, -2)$ and open upward, same domain (all real numbers), same range $(y \geq -2)$, same y-intercept, $(0, 1)$; different: $y = 3|x + 1| - 2$ is an absolute value function with x-intercepts $\left(-\frac{1}{3}, 0\right)$ and $\left(-\frac{5}{3}, 0\right)$, and $y = 3(x + 1)^2 - 2$ is a quadratic function with x-intercepts $(-1 \pm \sqrt{\frac{2}{3}}, 0)$

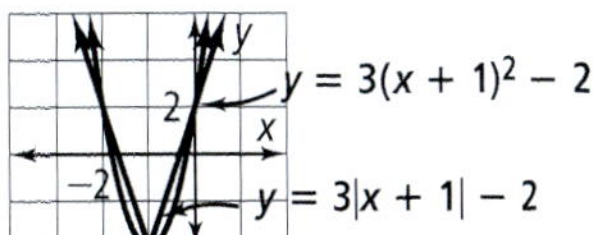

43. similar: same vertex $(-3, 0)$ and open upward, same domain (all real numbers), same range $(y \geq 0)$, same x-intercept, $(-3, 0)$; different: $y = |x + 3|$ is an absolute value function with y-intercept $(0, 3)$, and $y = (x + 3)^2$ is a quadratic function with y-intercept $(0, 9)$

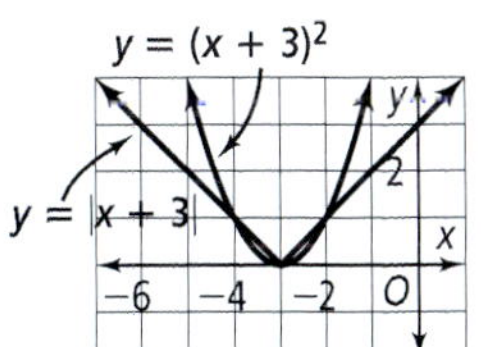

45. stretch vertically by a factor of 2, reflect across the x-axis, translate 1 unit to the left and 1 unit up

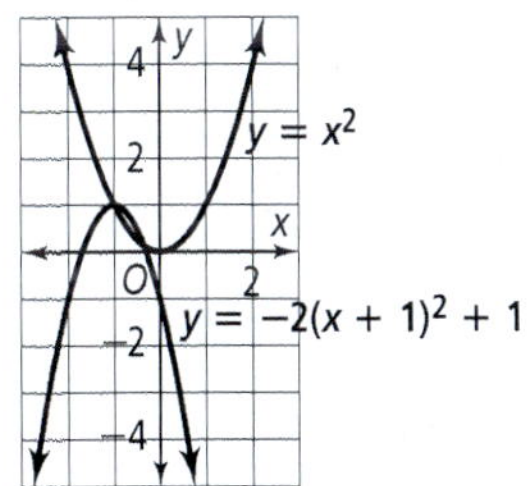

47. a. 10; 9; −1 **b.** 9; 6; −3 **c.** 6; 1; −5
d. The rate of change decreases at a greater rate.
e. Yes; the rates of change increase as you move away from the origin on the negative x-axis.
49. $y = -7(x - 1)^2 + 2$
51. $y = -7x^2 + 5$
53. Answers may vary. Sample: $y = (x + 10)^2 - 4$
55. Answers may vary. Sample: $k = -2, a = 3; y = 3(x - 1)^2 - 2$
57. Answers may vary. Sample: $a = -6, k = 35; y = -6(x + 1)^2 + 35$
59. a. $ah^2 + k$ **b.** when $h = 0$
61. $y = \frac{1}{2}(x + 3)^2$ **63.** $y = -\frac{1}{4}(x - 4)^2$
65. $y = -4(x + 3)^2$ **67.** G

69. $2x + 2y = 225$

$$2y = 225 - 2x$$

$$y = \frac{225}{2} - x$$

$$A = xy = x\left(\frac{225}{2} - x\right)$$

$$A = -x^2 + 112.5x$$

Graph the function. There is a max. at about (56.25, 3164.06), so the max. area is about 3164.06 ft^2 and the length of each side is 56.25 ft.

70. (3, 2) **71.** (−10, 6) **72.** (1, 0, 3) **73.** (0, 0) **74.** (−1, 0) **75.** (5, 0)

Lesson 4-2 pp. 202–208

Got It? 1. vertex: $\left(-\frac{2}{3}, 7\frac{1}{3}\right)$; axis of symmetry: $x = -\frac{2}{3}$; maximum: $7\frac{1}{3}$; range: $y \leq 7\frac{1}{3}$

2.

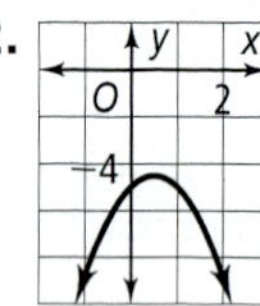

3. $y = -(x - 2)^2 - 1$ **4. a.** 4 ft **b.** because the y-intercept is (0, 0)

Lesson Check 1. vertex: (0, −4); axis of symmetry: $x = 0$; minimum: −4

2.

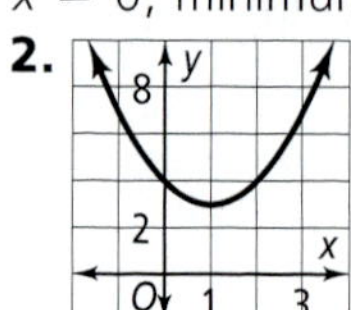

3.

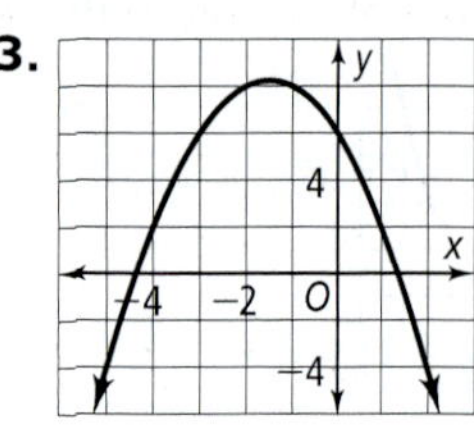

4. $y = (x - 1)^2 + 8$ **5.** $y = -\left(x - \frac{3}{2}\right)^2 + \frac{5}{4}$

6. Error in calculation of x. The correct calculation is:

$$x = \frac{-(-4)}{2(2)} = 1$$

$$y = 2(1) - 4(1) - 3$$
$$= 2 - 4 - 3$$
$$= -5$$

Vertex: (1, −5)

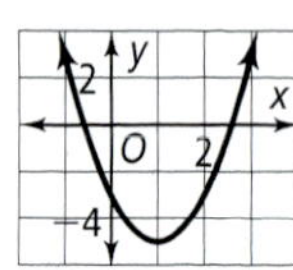

7. The vertex of a function written in vertex form can easily be determined. It is (h, k) where $f(x) = a(x - h)^2 + k$. The vertex of a function in standard form is $\left(\frac{-b}{2a}, f\left(\frac{-b}{2a}\right)\right)$ where $f(x) = ax^2 + bx + c$.

Exercises 9. vertex: (1, 2); axis of symmetry: $x = 1$; maximum: 2; range: $y \leq 2$ **11.** vertex: (1, 6); axis of symmetry: $x = 1$; maximum: 6; range: $y \leq 6$

13. vertex: $\left(-\frac{3}{4}, 5\frac{1}{8}\right)$; axis of symmetry: $x = -\frac{3}{4}$; maximum: $5\frac{1}{8}$; range: $y \leq 5\frac{1}{8}$ **15.** vertex: $\left(-\frac{1}{2}, \frac{1}{4}\right)$; axis of symmetry: $x = -\frac{1}{2}$; maximum: $\frac{1}{4}$; range: $y \leq \frac{1}{4}$

17.

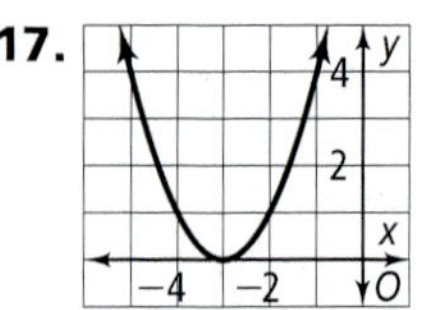

19.

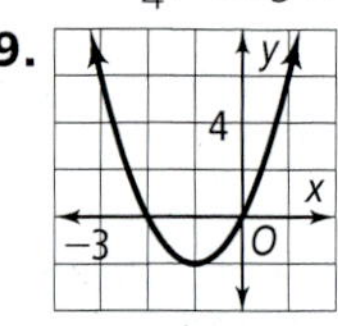

21.

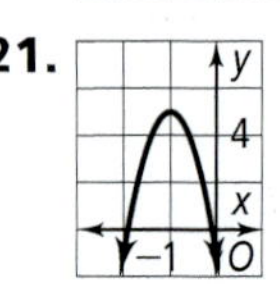

23.

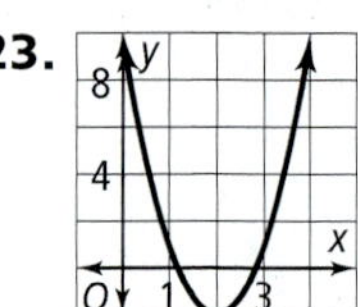

25.

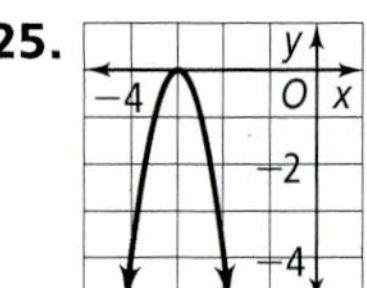

27. $y = (x + 1)^2 + 4$ **29.** $y = 2\left(x - \frac{5}{4}\right)^2 + \frac{71}{8}$

31. $y = \frac{9}{4}\left(x + \frac{2}{3}\right)^2 - 2$

33.

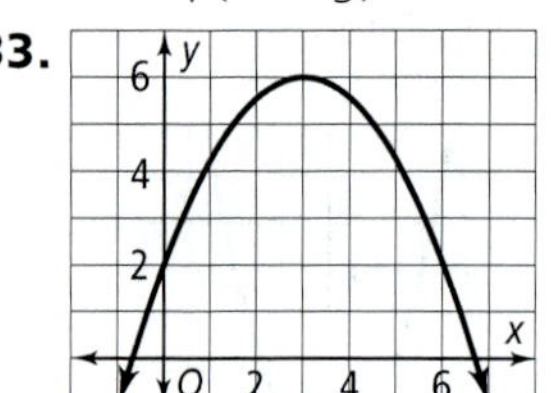

35.

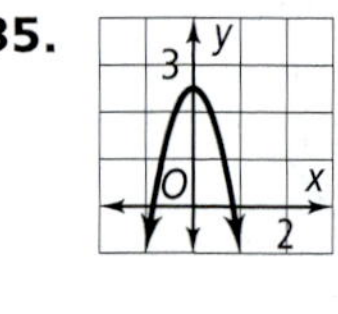

37. 4 cm × 9 cm × 9 cm **39.** $b = -6$; $c = 5$

41. $a = 1$; $c = -2$ **43. a.** The projectile represented by the equation goes higher by 28 feet. **b.** equation: $t = 0.27$ s and $t = 3.73$ s; table: $t = 0.38$ s and $t = 2.62$ s

45. (0, 3) **47.** (0, −54) **49.** $a = -6$, $b = 24$

51. $a = 3$, $b = -12$

53.

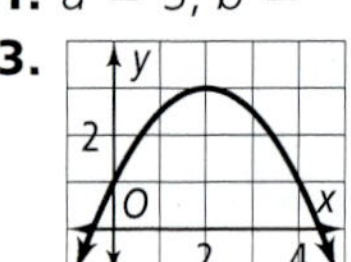

55. 0 **57.** $\frac{1}{4}$ **58.** −2.5

59. Explanations may vary. Sample: Elimination because the x's have opposite signs; (7, −1).

60. Explanations may vary. Sample: Substitution because the first equation is already solved for y; (27, 15).

61. Explanations may vary. Sample: Elimination because you can multiply the second equation by −3 to eliminate n; (3, 2).

62. vertex: (−2, −1); axis of symmetry: $x = -2$; minimum at (−2, −1); domain: all real numbers, range: $y \geq -1$

63. vertex: (1, 3); axis of symmetry: $x = 1$; maximum at (1, 3); domain: all real numbers, range: $y \leq 3$
64. vertex: (3, −2); axis of symmetry: $x = 3$; minimum at (3, −2); domain: all real numbers, range: $y \geq -2$
65. vertex: (−3, 5); axis of symmetry: $x = -3$; maximum at (−3, 5); domain: all real numbers, range: $y \leq 5$
66. vertex: (−4, 0); axis of symmetry: $x = -4$; minimum at (−4, 0); domain: all real numbers, range: $y \geq 0$
67. vertex: (4, 6); axis of symmetry: $x = 4$; maximum at (4, 6); domain: all real numbers, range: $y \leq 6$

Lesson 4-3 pp. 209–214

Got It? 1. $y = -3x^2 + x$ **2a.** Rocket 2 **b.** Rocket 1: D: $0 \leq t \leq 9.4$, R: $0 \leq h \leq 352.6$; Rocket 2: D: $0 \leq t \leq 12$, R: $0 \leq h \leq 580$ **c.** The domains tell you how many seconds the rockets were in the air. The domains are different because the rockets were in the air for different amounts of time. **3.** $y = -0.329x^2 + 9.798x + 15.571$; 88.5°F at 2:53 P.M. (although the meteorologist's prediction is 89° at 3 P.M.)
Lesson Check 1. $y = -2x^2 + 3x - 1$ **2.** $y = 2x^2 + 6x + 7.5$ **3.** $y = -2x^2 + 10x - 13.5$ **4.** Answers may vary. Sample: A rough plot of the data will indicate whether the data are collinear (linear regression) or non-collinear where the data follows a curve (quadratic regression). **5.** A parabola that opens up always attains greater values than one that opens down. **6.** y is not a function of x since for one value of x, "3," there are 2 values of y, "4" and "0."
Exercises 7. $y = -x^2 + 3x - 4$ **9.** $y = 2x^2 - x + 3$ **11.** $y = x^2 - 6x + 3$ **13.** $y = x^2 + 2x$ **15.** $y = -x^2 + x - 2$ **17. a.** $y = -16x^2 + 33x + 46$, where x is the number of seconds after release and y is the height in ft **b.** 28.5 ft **c.** about 63 ft **19.** yes; $y = -2x^2 + 3x + 5$ **21.** yes; $y = 0.625x^2 - 1.75x + 1$ **23.** $y = 0.005x^2 - 1.95x + 120$; 66 mm **25. a.** $y = -0.004x^2 + 0.859x + 27.53$ **b.** Answers may vary. Sample: domain: integers from −10 to 17; range: whole numbers from 18 to 42 **c.** the year 2003 **d.** The year 2022; the year is outside the domain of the data pts. **27. a.** (3, 5) **b.** 3; substitute x- and y-values in the general form of a quadratic, then solve the resulting linear system for the coefficients. **29.** A **31.** D

33.
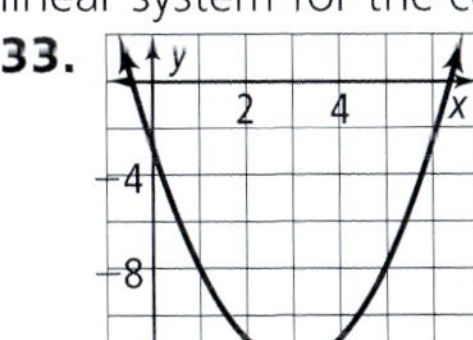

34.
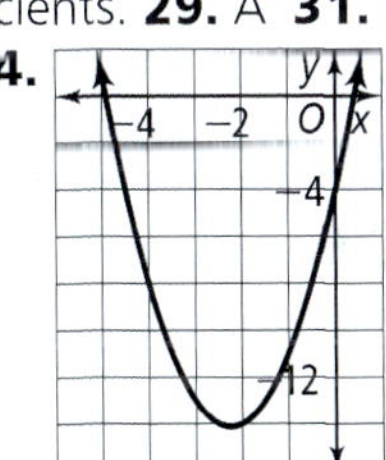

35.
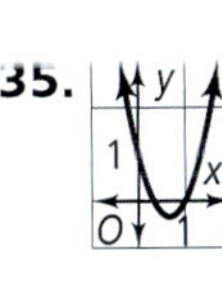

36. (2, 5) **37.** (5, 8) **38.** (−1, −1) **39.** $\frac{4}{5}$ **40.** $-\frac{7}{2}$ **41.** $x^2 + 5x - 1$ **42.** $6x^2 - 10x - 3$ **43.** $4x^2 - x - 10$

Lesson 4-4 pp. 216–223

Got It? 1. a. $(x + 10)(x + 4)$ **b.** $(x - 5)(x - 6)$ **c.** $-(x + 2)(x - 16)$ **2. a.** $7(n^2 - 3)$ **b.** $9(x + 2)(x - 1)$ **c.** $4(x^2 + 2x + 3)$ **3. a.** $(x + 1)(4x + 3)$ **b.** $(x - 2)(2x - 3)$ **c.** No; $2x^2 + 2x + 2 = 2(x^2 + x + 1)$, there are no real factors of a and c whose product is 1 and whose sum is 1. **4.** $(8x - 1)^2$ **5.** $(4x - 9)(4x + 9)$
Lesson Check 1. $(x + 4)(x + 2)$ **2.** $(x - 12)(x - 1)$ **3.** $(x - 9)(x + 9)$ **4.** $(5y - 6)(5y + 6)$ **5.** $(y - 3)^2$ **6.** $(2x - 1)^2$ **7.** $5x$ **8.** $4a^2$ **9.** 6 **10.** $7h$ **11.** No; the middle term is not twice the product of the square root of the end terms. **12.** For $a \neq 1$, look for two factors whose sum is b and whose product is ac. For $a = 1$, look for two factors whose sum is b and whose product is c.
13. $a^2 - 2ab + b^2 - 25$ Group the first 3 terms.
$= (a^2 - 2ab + b^2) - 25$
$= (a - b)^2 - 5^2$
$= (a - b - 5)(a - b + 5)$
Exercises 15. $(x + 2)(x + 3)$ **17.** $(x + 2)(x + 8)$ **19.** $(x + 2)(x + 20)$ **21.** $-(x - 1)(x - 12)$ **23.** $(x - 4)(x - 6)$ **25.** $(x - 4)(x - 9)$ **27.** $-(x - 4)(x + 5)$ **29.** $(c - 7)(c + 9)$ **31.** $-(t - 11)(t + 4)$ **33.** $5b$; $5b(5b - 4)$ **35.** 5; $5(t + 1)(t - 2)$ **37.** 9; $9(3p^2 - p + 2)$ **39.** $(x - 8)(2x - 3)$ **41.** $(m - 3)(2m - 5)$ **43.** $(x - 12)(2x - 3)$ **45.** $(y + 4)(5y - 8)$ **47.** $(x + 1)^2$ **49.** $(k - 9)^2$
51. cannot be factored; 8 is not a perfect square and there are no positive factors of 32 that have a sum of 16.
53. $(x - 2)(x + 2)$ **55.** cannot be factored; This is a sum of squares, not a difference of squares.
57. $(5x - 1)$ cm by $(5x - 1)$ cm **59.** $2(3z + 2)(3z - 2)$
61. $16(2t + 1)(2t - 1)$ **63.** $3(y + 5)(y + 3)$
65. $3(x + 1)(x - 9)$ **67.** $2(x - 5)(2x - 1)$
69. $-\left(\frac{1}{4}s - 1\right)\left(\frac{1}{4}s + 1\right)$
71. The third line should be $x(2x - 5) - (2x - 5)$, and the final line should be $(x - 1)(2x - 5)$. **73.** y; $y(y - 1)$ **75.** 10; $10(x - 3)(x + 3)$ **77.** 2; $2(x^2 - 37x + 6)$ **79.** D
81. Answers may vary. **83.** $(0.5t + 0.4)(0.5t - 0.4)$
85. $(x + 12)(x - 3)$ **87.** $(2x + 9)(3x + 14)$
89. Factor the GCF, $4x^2$, from the terms to get $4x^2(x^2 + 6x + 8)$. Look for numbers whose product is 8 and whose sum is 6. The numbers 4 and 2 work. The complete factorization is $4x^2(x + 4)(x + 2)$.
91. $(2x - 5y)(2x + 5y)(4x^2 + 25y^2)$
93. $(x - 8)(x + 3)$ **95.** B **97. a.** By entering the given lists into a graphing calculator and then calculating the quadratic regression, you get $h = -16t^2 + 22t + 3$ as the quadratic model for the ball's height as a function of time.

b. $h = -16t^2 + 22t + 3$
$h = -[8t(2t - 3) + 1(2t - 3)]$
$h = -(2t - 3)(8t + 1)$

98. $y = -0.149x^2 + 5.171x + 16.971$ **99.** penny: 2.5g, nickel: 5g, dime: 2.3g

100.

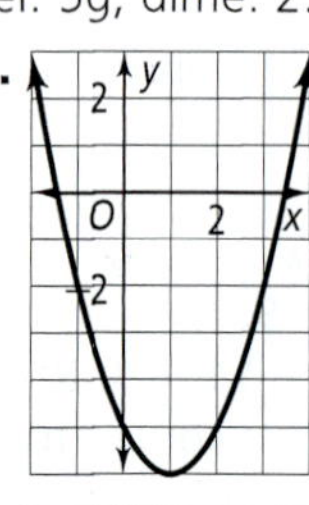

101.

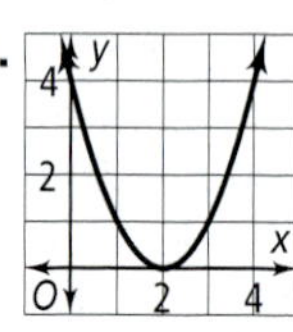

102.

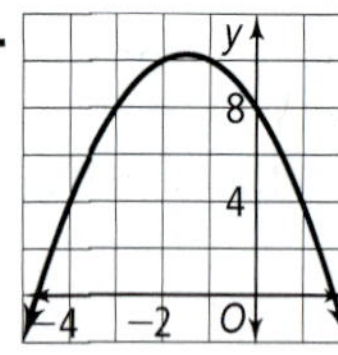

Algebra Review p. 225 1. $3\sqrt{2}$ **3.** $-4\sqrt{2}$ **5.** $\frac{-\sqrt{91}}{13}$ **7.** $-10\sqrt{2}$ **9.** 108 **11.** $|xy|$ **13.** $-\frac{|x|\sqrt{35}}{5}$ **15.** $\frac{5\sqrt{14}}{7}$

Lesson 4-5 pp. 226–231

Got It? 1. 3, 4 **2.** 3, $\frac{1}{4}$ **3.** −6, 4 **4. a.** $53\frac{1}{3}$ m; $21\frac{1}{3}$ m; answers may vary. Sample: domain: $0 \le x \le 60$, range: $0 \le y \le 30$ **b.** No; domains and ranges are constrained by real-world limits.

Lesson Check 1. 3, −3 **2.** −4, −9 **3.** $-\frac{2}{3}$, 1 **4.** 4.372, −1.372 **5.** 2.608, −2.108 **6.** −1; since $y = 0$ when $x = 5$, substitute these values into the equation to find b. **7.** when the coefficients are integers and a recognizable pattern of factoring is evident **8.** One solution: when the table's range consists of zero and all positive numbers or zero and all negative numbers. No solution: when the table does not include zero and the y-values are either all positive or all negative numbers

Exercises 9. −4, −2 **11.** −1, $\frac{3}{2}$ **13.** −2, −1 **15.** 0, 4 **17.** 0, 4 **19.** 3, 8 **21.** −0.78, 1.28 **23.** 4, −10 **25.** 0.8, −1 **27.** −1.67, −1.5 **29.** −0.94, 2.34 **31.** −1, 0.25 **33.** −1.46, 5.46 **35.** −1.16, 2.16 **37.** 3 in. **39.** about 3.6 ft **41.** Answers may vary. Samples are given: **a.** $x^2 - 8x + 15 = 0$ **b.** $x^2 + x - 6 = 0$ **c.** $x^2 + 7x + 6 = 0$ **43.** $-\frac{3}{2}, -\frac{2}{3}$ **45.** $-4, \frac{5}{2}$ **47.** −1, 4 **49.** −4, 0 **51.** 4.37, −1.37 **53.** $-1, \frac{10}{3}$ **55.** (0, −2), (2, 2) **57.** Solve $(x - 4)(x - 6) = 0$ to find that the zeros of $y = x^2 - 10x + 24$ are 4 and 6. Average 4 and 6 to get 5. This is the x-coordinate of the vertex. Substitute 5 for x in $x^2 - 10x + 24$ to find that −1 is the y-coordinate of the vertex. The vertex is (5, −1).

59. a. 2.45m **b.** ≈1.3s **61.** H **63.** reflection across the x-axis, followed by a vertical translation 2 units up, and stretched by a factor of 3

64. $(4x - 1)(4x + 1)$ **65.** $(5x - 1)(x - 5)$ **66.** $(2x - 1)(x + 7)$ **67.** (2, 0, −2) **68.** (−2, 1, 5) **69.** (7, 1, −1) **70.** vertex: (−9, 4); axis of symmetry: $x = -9$; translation 9 units to the left and 4 units up **71.** vertex: $\left(\frac{7}{2}, 0\right)$; axis of symmetry: $x = \frac{7}{2}$; stretch by a factor of 2, translation $3\frac{1}{2}$ units to the right **72.** vertex: (0, −1); axis of symmetry: $x = 0$; vert. compression by a factor of $\frac{3}{4}$, translation 1 unit down **73.** $x^2 + 8x + 13$ **74.** $4x^2 - 4x + 1$ **75.** $x^2 - 6x + 9$

Lesson 4-6 pp. 233–239

Got It? 1. a. $\sqrt{5}, -\sqrt{5}$ **b.** $\sqrt{2}, -\sqrt{2}$ **2.** 42 in. × 67.2 in. **3.** 2, 12 **4. a.** 9 **b.** No; $\left(\frac{b}{2}\right)^2 = \frac{b^2}{4}$, which is a function of b. **5.** $\frac{1}{2} \pm \frac{\sqrt{13}}{2}$ **6.** $y = \left(x + \frac{3}{2}\right)^2 - \frac{33}{4}$; vertex: $\left(-\frac{3}{2}, -\frac{33}{4}\right)$; y-intercept: (0, −6)

Lesson Check 1. 6, −6 **2.** 3, −3 **3.** 1 **4.** 25 **5.** 4 **6.** 36 **7.** 2500 **8.** 256 **9.** First, you rewrite the eq. to get all terms with x on one side. Then, you find $\left(\frac{b}{2}\right)^2$ and add it to both sides of the eq. Then, you factor the resulting trinomial.

10. $x^2 + 12x + 5 = 3$

$x^2 + 12x = -2$	Rewrite to get all terms with x on one side of the eq.
$\left(\frac{12}{2}\right)^2 = 6^2 = 36$	Find $\left(\frac{b}{2}\right)^2 = 36$.
$x^2 + 12x + 36 = -2 + 36$	Add 36 to each side.
$(x + 6)^2 = 34$	Factor the trinomial.

11. Your friend should also have subtracted 49; $(x^2 - 14x + 49) + 36 - 49 = (x - 7)^2 - 13$

Exercises 13. 2, −2 **15.** $\frac{5}{3}, -\frac{5}{3}$ **17.** $2\sqrt{2}, -2\sqrt{2}$ **19.** −4, −2 **21.** −1, 3 **23.** −4, 3 **25.** $-\frac{4}{5}, \frac{2}{5}$ **27.** $-\frac{10}{3}, \frac{2}{3}$ **29.** $\frac{1}{4}$ **31.** 100 **33.** 4 **35.** $6 \pm \sqrt{29}$ **37.** $1 \pm \sqrt{6}$ **39.** $5 \pm \sqrt{13}$ **41.** $3 \pm \sqrt{11}$ **43.** $-\frac{5}{4} \pm \frac{1}{4}\sqrt{37}$ **45.** $-\frac{3}{5} \pm \frac{\sqrt{21}}{5}$ **47.** $y = 2(x - 2)^2 - 7$ **49.** $y = (x + 2)^2 - 11$ **51.** $y = -(x - 2)^2 + 3$ **53.** 10, −10 **55.** 22, −22 **57.** 18, −18 **59.** 1, −1 **61.** 84, −84 **63.** $\frac{-5 \pm \sqrt{37}}{2}$ **65.** $\frac{1 \pm \sqrt{21}}{2}$ **67.** $\frac{2 \pm \sqrt{10}}{3}$

69. $\frac{-3 \pm \sqrt{41}}{8}$ **71.** $\frac{1}{3}, -\frac{2}{3}$ **73.** $-3 \pm \sqrt{7}$
75. a. $-.01(x - 59)^2 + 36.81$; 36.81 ft **b.** 7.65 ft
c. about 120 ft **77.** $\frac{-a \pm a\sqrt{13}}{6}$
79. $-\frac{3}{2a}, -\frac{1}{2a}, a \neq 0$ **81.** $-\frac{2}{3a}, \frac{5}{2a}, a \neq 0$
83. $y = -4\left(x + \frac{5}{8}\right)^2 + \frac{73}{16}; \left(-\frac{5}{8}, \frac{73}{16}\right)$
85. $y = -\frac{1}{5}(x - 2)^2 + 3$; $(2, 3)$ **87.** G
89. $x^2 - 9 = -8x$

$x^2 + 8x = 9$	Move the variables to the left side and the constant to the right by using the Add. Prop. of Eq.
$\left(\frac{8}{2}\right)^2 = 4^2 = 16$	Find $\left(\frac{b}{2}\right)^2 = 16$.
$x^2 + 8x + 16 = 9 + 16$	Add 16 to each side.
$(x + 4)^2 = 25$	Factor left side. Simplify right side.
$x + 4 = \pm 5$	Take the square root of each side.
$x = -4 \pm 5$	Add -4 to each side.
$x = -9, 1$	Simplify.

90. $\frac{1}{2}, 1$ **91.** $-4, 1$ **92.** $8, -\frac{2}{3}$
93. yes; $y = \frac{1}{2}x^2 + \frac{7}{2}x + 9$ **94.** yes;
$y = -\frac{1}{2}x^2 + x + 2$ **95.** yes; $y = 3x^2 - 5x + 2$
96. (2, 0) **97.** (3, 1) **98.** (3, 1) **99.** 24 **100.** 84

Lesson 4-7 pp. 240–247

Got It? 1. a. -2 **b.** $-2 \pm \sqrt{7}$ **2. a.** \$10.74 **b.** Yes; a neg. profit means more money was spent than earned.
3. a. no real solutions **b.** two real solutions **4.** Yes;
$b^2 - 4ac = (85)^2 - 4(-16)\left(-109\frac{11}{12}\right) = 190\frac{1}{3}$.
The discriminant is pos. So the eq. has two real solutions.
Lesson Check 1. $\frac{5 \pm \sqrt{53}}{2}$ **2.** $\frac{-3 \pm \sqrt{61}}{2}$ **3.** $3, -\frac{1}{2}$
4. no real solutions **5.** -32; no real solutions **6.** 273; two real solutions **7.** 0; one real solution **8.** $k = \pm 6$ for one real solution; $k > 6$ or $k < -6$ for two real solutions
9. Answers may vary. Sample: The discriminants of eqs with one real solution are all zero and thus equal, but the solutions may or may not be equal. An example is $x^2 - 8x + 16$ and $x^2 - 4x + 4$. Each has a discriminant of zero, but the solutions are 4 and 2. **10.** Yes; the eqs. can share common factors such as for $x^2 + 2x - 8$ where the discriminant is 36 and the solutions are 2 and -4, and $x^2 - 4x + 4$ where the discriminant is zero and the solution is 2.
Exercises 11. 1, 3 **13.** $-\frac{7}{2}, 1$ **15.** -5 **17.** $\frac{3 \pm \sqrt{5}}{2}$
19. $\frac{2 \pm \sqrt{10}}{3}$ **21.** 1, 4 **23.** \$16.34 **25.** -4; no real solutions **27.** 0; one **29.** 169; two **31.** 1; two
33. 0; one **35.** -23; no real solutions **37.** no
39. 2.29 in. × 15.71 in. **41.** $-\frac{1}{6}, 1$ **43.** $-2.49, 0.89$
45. $-0.19, 2.69$ **47.** 1, 10 **49.** $-\frac{3}{2}, \frac{1}{2}$ **51.** $-1.70, 4.70$
53. $-8.47, 0.47$ **55.** $1.47, -7.47$ **57.** about 1.89 s
59. one **61.** two **63.** two **65.** two **67. a.** k such that $|k| < 12$ **b.** 12 or -12 **c.** k such that $|k| > 12$
69. a. $x^2 = 100\pi$ **b.** 17.72 cm
71. Answers may vary. Sample: $x^2 + 5x + 3 = 0$
73. $0, \pm\sqrt{5}$ **75.** $-5, 1$ **77.** The absolute value of $\frac{\sqrt{b^2 - 4ac}}{2a}$ is the distance from the axis of symmetry to the x-intercepts, if the discriminant is nonnegative.
79. 3 **81.** 15 **82.** $-2, 10$ **83.** $\frac{2 \pm \sqrt{2}}{2}$ **84.** $\frac{3 \pm \sqrt{41}}{2}$
85. $9z^2 + 3z$ **86.** $4x + k$ **87.** $2y - 8x$
88. $2\sqrt{17}$ **89.** 5 **90.** 13

Lesson 4-8 pp. 248–255

Got It? 1. a. $2i\sqrt{3}$ **b.** $5i$ **c.** $i\sqrt{7}$ **d.** $8i \neq -8$
2. a. 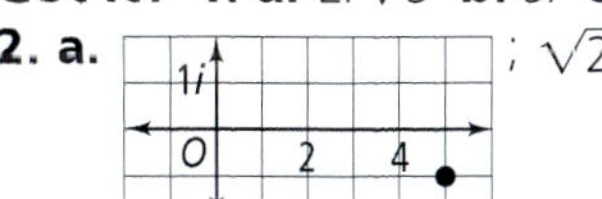

; $\sqrt{26}$

b. 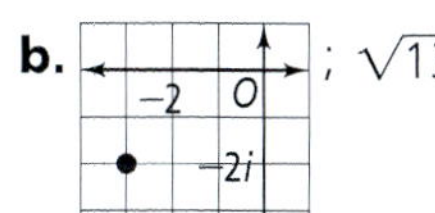

; $\sqrt{13}$ **c.** 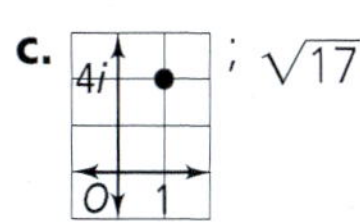

; $\sqrt{17}$

3. a. $4 - i$ **b.** $-2 + 7i$ **c.** $12i$ **d.** $18i$ **4. a.** -21
b. $23 - 2i$ **c.** 41 **5. a.** $\frac{7}{25} - \frac{26}{25}i$ **b.** $-\frac{1}{6} - \frac{2}{3}i$
c. $\frac{15}{113} - \frac{112}{113}i$ **6. a.** $5(x + 2i)(x - 2i)$
b. $(x + 9i)(x - 9i)$ **7. a.** $\frac{1 \pm i\sqrt{23}}{6}$
b. $2 \pm i$
Lesson Check 1. $5i\sqrt{3}$ **2.** 5 **3.** $(x + 4i)(x - 4i)$
4. $7 - 3i$ **5.** $13 - 6i$ **6.** The add. inv. of a complex number, $a + bi$, is the opposite of the complex number, or $-a - bi$. The complex conjugate of a complex number, $a + bi$, is the real part plus the opposite of the imaginary part of the complex number, or $a - bi$. **7.** error in the sign of the last term of the first line, which carries through to the end of the calculation; the line should be: ". . .
$= 16 + 28i - 28i - 49i^2$
$= 16 + 49$
$= 65$."
Exercises 9. $i\sqrt{7}$ **11.** $9i$
13. 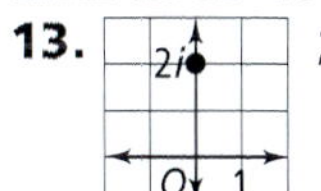

; 2 **15.** 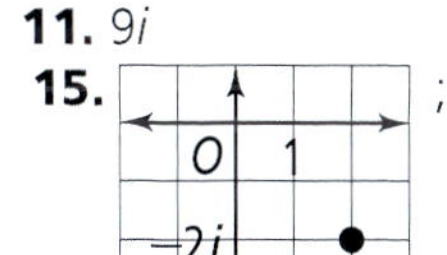

; $2\sqrt{2}$

17. ; $3\sqrt{5}$ **19.** $1 - 7i$ **21.** $10 + 6i$

23. $9 + 58i$ **25.** -36 **27.** $-\frac{2}{5} - \frac{3}{5}i$ **29.** $\frac{8}{17} + \frac{19}{17}i$
31. $\frac{8}{13} + \frac{12}{13}i$ **33.** $(x + 5i)(x - 5i)$

35. $3(s + 5i)(s - 5i)$ **37.** $(2b + i)(2b - i)$

39. $-1 \pm i\sqrt{2}$ **41.** $1 \pm i\frac{\sqrt{10}}{2}$ **43.** $2 \pm i$

45. a. A: -5; B: $3 + 2i$; C: $2 - i$; D: $3i$; E: $-6 - 4i$; F: $-1 + 5i$ **b.** A: 5; B: $-3 - 2i$; C: $-2 + i$; D: $-3i$; E: $6 + 4i$; F: $1 - 5i$ **c.** A: -5; B: $3 - 2i$; C: $2 + i$; D: $-3i$; E: $-6 + 4i$; F: $-1 - 5i$ **d.** A: 5; B: $\sqrt{13}$; C: $\sqrt{5}$; D: 3; E: $2\sqrt{13}$; F: $\sqrt{26}$ **47.** $-5, 5$ **49.** $-1 + 5i$

51. $8 - 2i$ **53.** $6 + 10i$ **55.** $10 + 11i$ **57.** trapezoid
59. $\frac{1}{26} + \frac{3}{52}i$ **61.** sum: 2, product: 3

63. sum: $\frac{3}{2}$, product: $\frac{3}{2}$ **65.** Answers may vary. Sample:

$x^2 - 4x + 29 = 0$ **67.** $x = -7, y = 3$

69. $x = -7, y = -3$ **71.** all nonzero numbers x and y such that $|x| = |y|$ **73.** B **75.** A

77. $\frac{-3 \pm \sqrt{41}}{4}$ **78.** $\frac{-1 \pm \sqrt{17}}{8}$ **79.** $\frac{-7 \pm \sqrt{17}}{2}$

80. 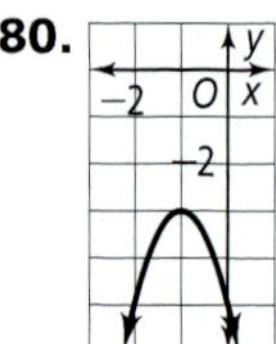 ; axis of symmetry: $x = -1$

81. 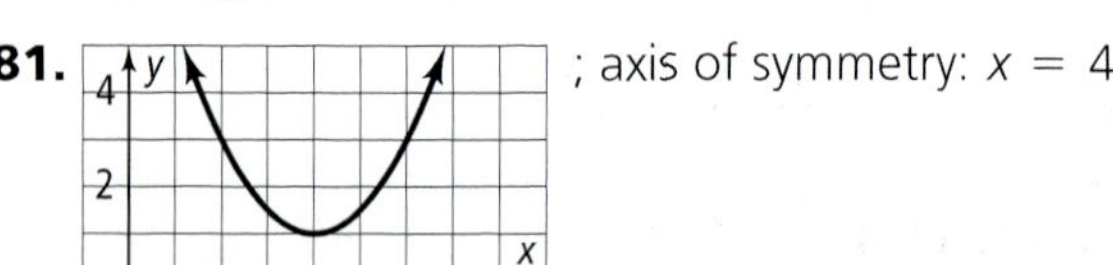 ; axis of symmetry: $x = 4$

82. 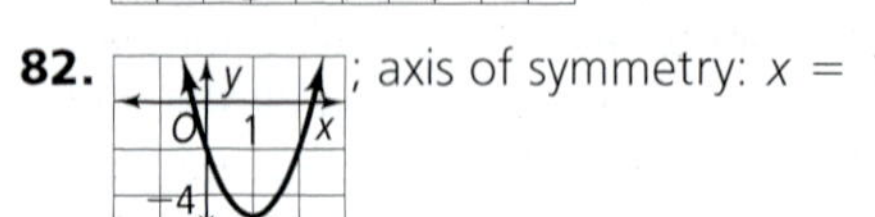 ; axis of symmetry: $x = 1$

83. $y = 3x - 4$ **84.** $y = -0.5x - 2$
85. $y = -7x + 10$ **86.** $y = 2x + 8$

87. 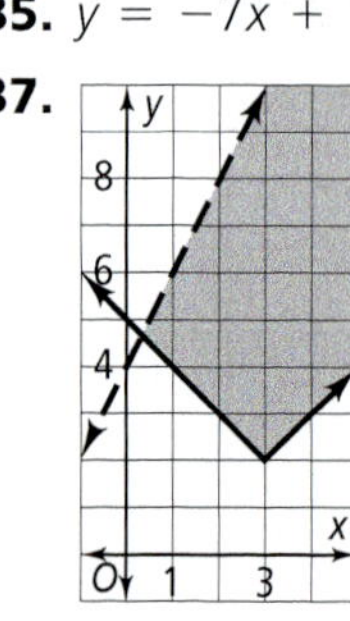

88. 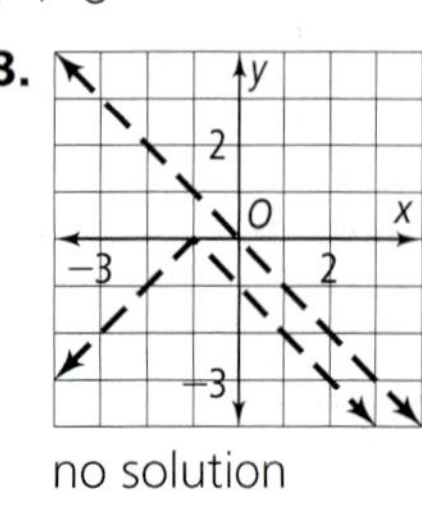

no solution

89. 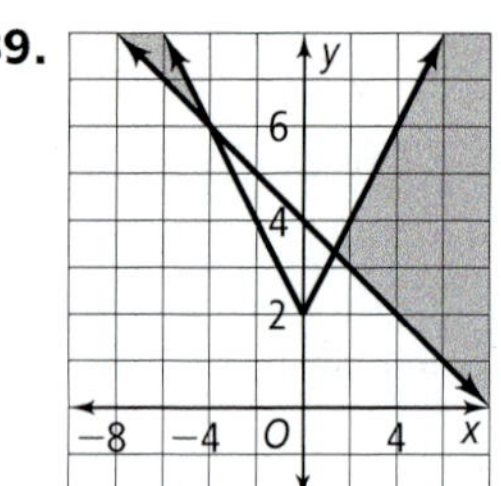

Lesson 4-9 pp. 258–264

Got It? 1. $(-3, 0), (-2, 1)$ **2.** $(-5, 0), (1, 6)$
3. a. $(0, 5), (2, 1)$ **b.** no solution
4. a. 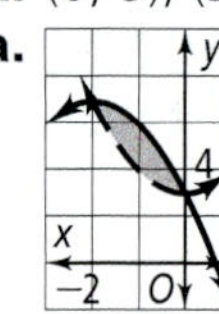 **b.** infinite, one, or none

Lesson Check 1. $(1, 2), (2, 3)$ **2.** $(1, -1), (2, 0)$
3. $(2, -5), \left(-\frac{4}{3}, \frac{25}{9}\right)$

4. 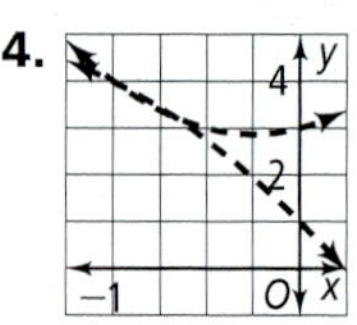

5. 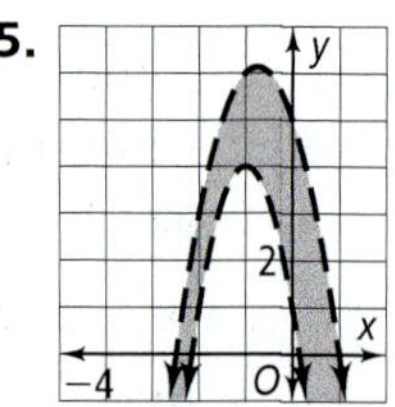

6. For each system of eqs., linear or quadratic, to solve the system you need to find the pt. (or pts.) of intersection or, in the case of inequalities, the regions of intersection. A linear system of eqs. can have one, infinite, or no solutions, whereas a quadratic systems of eqs. can have one, two, infinite, or no solutions.
7. a. two, one, or zero

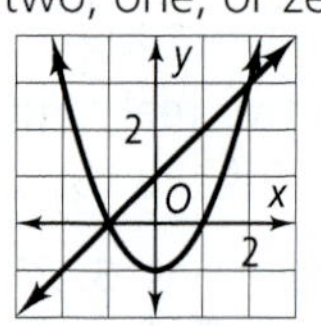

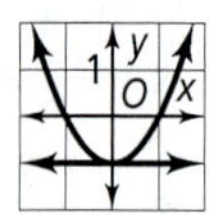

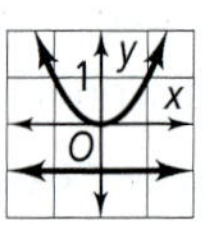

b. two, one, or zero

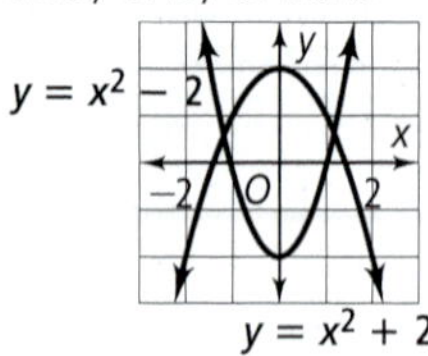

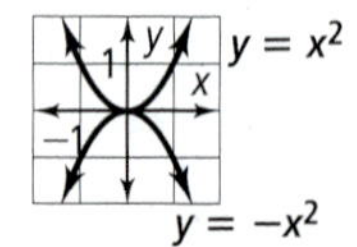

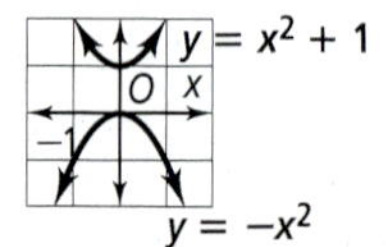

c. four, three, two, one, or zero

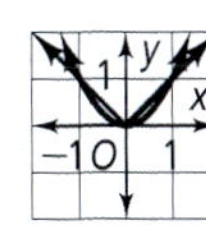

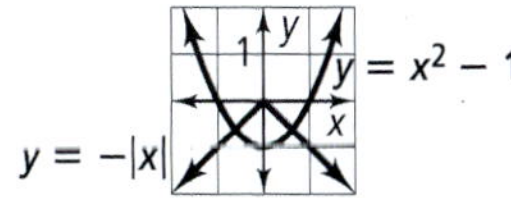

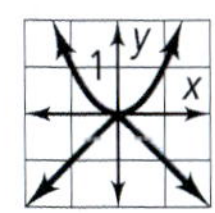

Exercises

9. (0, 1), (4, 9);

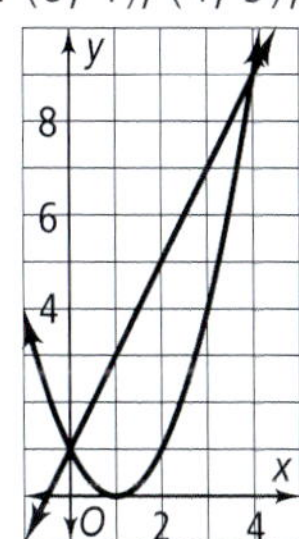

11. (0, 1), $\left(-\frac{5}{2}, 6\right)$;

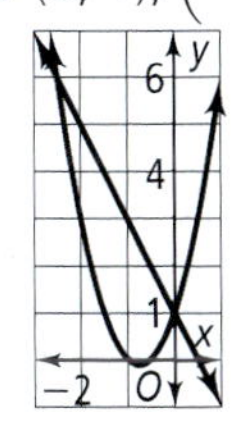

13. $\left(\frac{-3 + \sqrt{17}}{2}, \frac{-11 + \sqrt{17}}{2}\right), \left(\frac{-3 - \sqrt{17}}{2}, \frac{-11 - \sqrt{17}}{2}\right)$ $\approx$ (0.56, −3.44), (−3.56, −7.56)

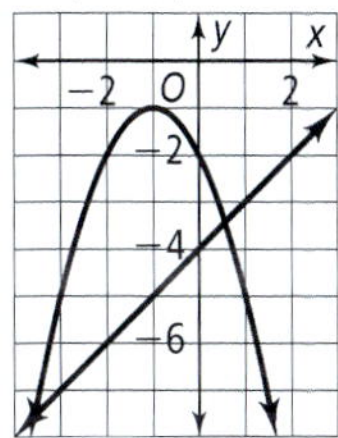

15. (3, 7), (−2, 2) **17.** (1, −2) **19.** $(-3 + \sqrt{6}, -1 + \sqrt{6}), (-3 - \sqrt{6}, -1 - \sqrt{6})$ **21.** (0, −1)
23. (0, −3), $\left(\frac{1}{3}, -\frac{31}{9}\right)$ **25.** no solution
27.

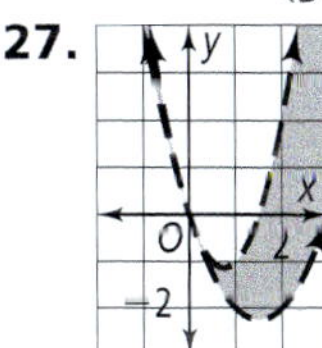

29. width = 20.45 in. and length = 21.45 in.; or width = 6.55 in. and length = 7.55 in.
31. (8.42, −5.42), (−1.42, 4.42) **33.** $\left(-1, -\frac{3}{2}\right)$
35. (0, −4), (2, −2)
37.

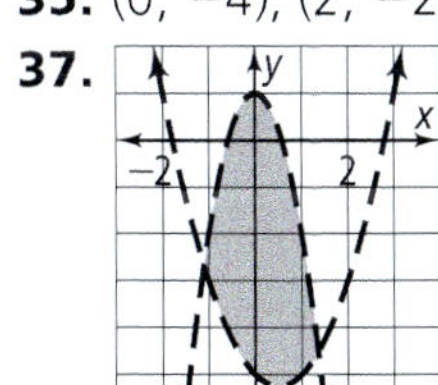

39.

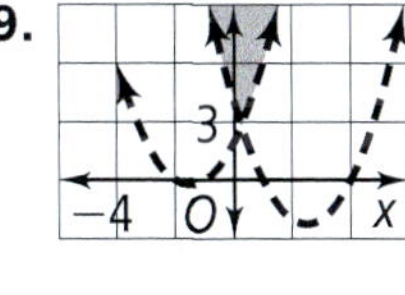

41. (0, −1), (1, 0) **43.** (10, 68), (−3, 16)

45. (−8, −16), $\left(-\frac{4}{3}, 4\right)$ **47. a.** $0 < p < 25$ **b.** 13 **49.** no solution **51.** $\left(\frac{1}{2}, \frac{7}{2}\right)$

53.

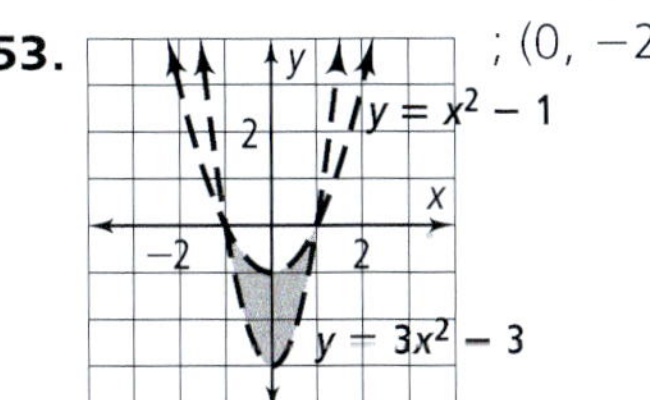

; (0, −2)

55.

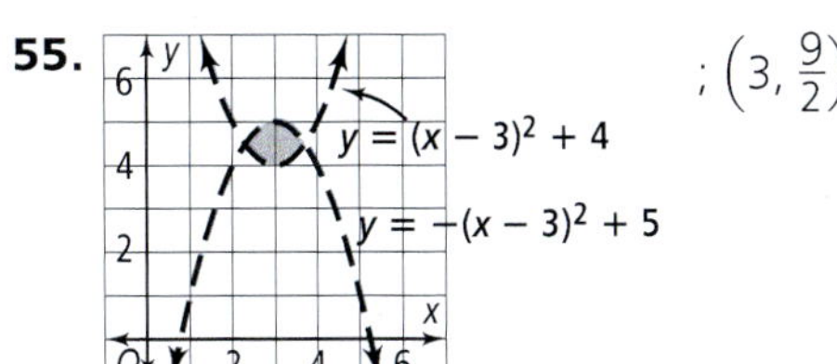

; $\left(3, \frac{9}{2}\right)$

57. never; $x^2 + c \neq x^2 + d$
59. always; $\left(\frac{-a - b}{2}, \frac{a^2 - 2ab + b^2}{2}\right)$
61. $\sqrt{5} - 1$ units **63.** F
65. $x = \frac{-b \pm \sqrt{b^2 - 4ac}}{2a}$

$x = \frac{-(5) \pm \sqrt{(5)^2 - 4(-3)(4)}}{2(-3)}$

$x = \frac{5 \pm \sqrt{73}}{6}$

66. $-4 + 3i$ **67.** $7 + 7i$ **68.** $3 + 3i$ **69.** $-1, -\frac{3}{2}$
70. 1, 3 **71.** $\frac{3}{5}$ **72.** $y = -(k - 2)^2 + 10$
73. $y = (x + 3)^2 - 8$ **74.** $y = 2(n - 2)^2 - 11$
75. $11q$ **76.** $2a^2b + ab^2$ **77.** $-y^2 + 2y$

Chapter Review pp. 267–272

1. standard **2.** discriminant **3.** complex **4.** vertex: (−2, −6); axis of symmetry: $x = -2$; minimum: −6; domain: all real numbers; range: $y \geq -6$ **5.** vertex: (3, 2); axis of symmetry: $x = 3$; maximum: 2; domain: all real numbers; range: $y \leq 2$ **6.** vertex: (1, 5), axis of symmetry: $x = 1$; minimum: 5; domain: all real numbers, range: $y \geq 5$ **7.** vertex: (−9, −4); axis of symmetry: $x = -9$; minimum: −4 domain: all real numbers; range: $y \geq -4$

8.

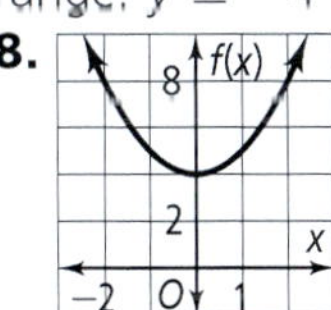

translation 4 units up

9. translation 9 units to the right and 2 units up

10. 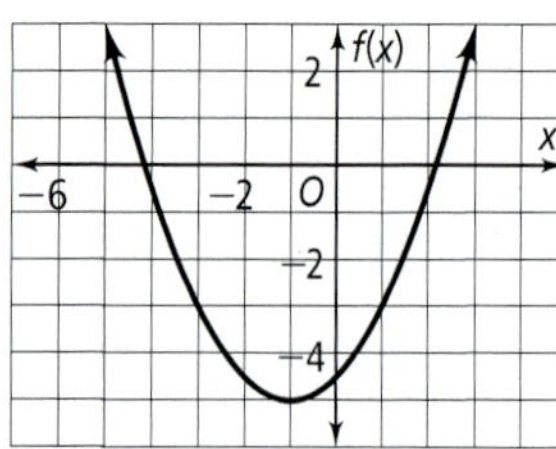
vert. compression by a factor of $\frac{1}{2}$, translation 1 unit to the left and 5 units down

11. $y = 2(x - 2)^2 + 1$ **12.** $y = \left(-\frac{1}{3}\right)(x + 5)^2 + 4$

13.

14. 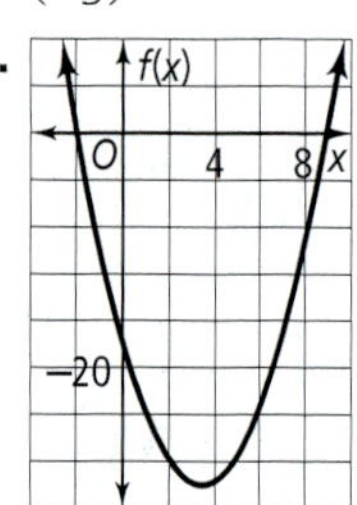

15.

16. 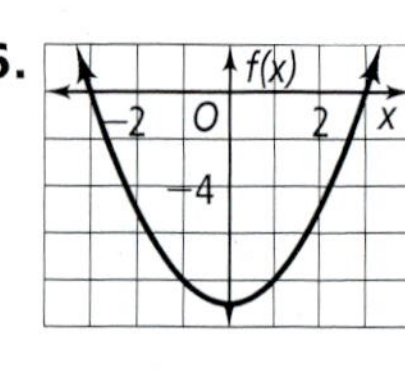

17. $f(x) = 4(x - 1)^2 - 2$ **18.** $f(x) = (x - 4)^2 - 4$
19. $f(x) = 8\left(x + \frac{1}{2}\right)^2 - 14$ **20.** $f(x) = -2\left(x + \frac{3}{2}\right)^2 + \frac{29}{2}$
21. 1s; 25 ft **22.** $y = x^2 - 6x + 5$
23. $y = -2x^2 + 8x - 8$ **24.** $y = x^2 + 3x - 18$
25. $y = -0.5x^2 + 2.5x - 7$
26. $y = -0.0043x^2 + 0.3521x + 0.3691$
27. $(x - 6)(x - 2)$ **28.** $(3x - 4)(x + 5)$
29. $-2(2x - 1)(x - 3)$ **30.** $(x + 10)(x + 4)$
31. $(x - 7)^2$ **32.** $(3x + 5)^2$ **33.** $4(3x - 2)(3x + 2)$
34. $(5x - 2)(5x + 2)$ **35.** $6x$; $6x(x - 4)$
36. 7; $-7(2x^2 + 7)$ **37.** $-2, 6$ **38.** $-2, \frac{7}{2}$
39. $-4, 2$ **40.** $-9, 2$
41. 1, -2.6;

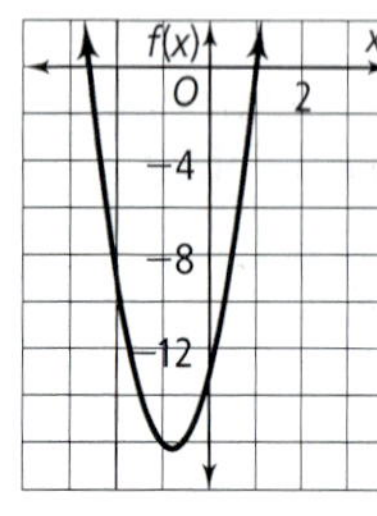

42. 1.345, -3.345;

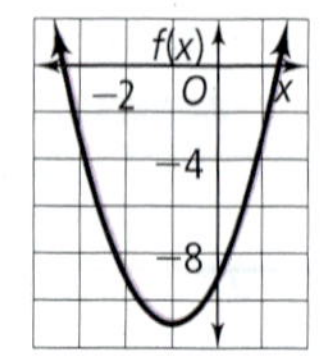

43. 1.618, -0.618;

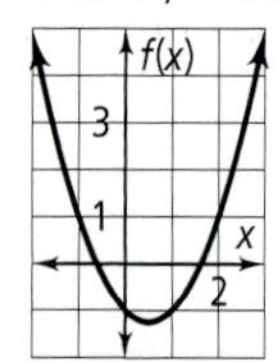

44. 3.236, -1.236;

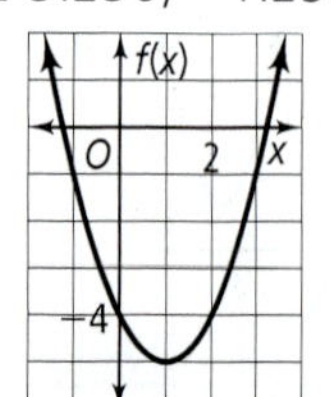

45. 2, 4 **46.** no real solutions **47.** 4.56, 0.44 **48.** 2, 4
49. ± 2 **50.** $\pm\sqrt{5}$ **51.** ± 3 **52.** $\pm 2\sqrt{3}$ **53.** 9 **54.** $\frac{9}{4}$
55. $-4 \pm \sqrt{10}$ **56.** $5 \pm \sqrt{38}$ **57.** $-1, \frac{1}{3}$ **58.** $1 \pm i\sqrt{3}$
59. $\frac{-3 \pm i\sqrt{91}}{2}$ **60.** $1, -\frac{3}{4}$ **61.** $1, -\frac{8}{3}$ **62.** 3 **63.** 4, -1
64. 1.744, -0.344 **65.** 164; two **66.** -3; none
67. 0; one **68.** 233; two **69.** 10.2736 ft × 16.5472 ft
70. $2i\sqrt{6}$ **71.** $-3 + i\sqrt{2}$ **72.** $-50 + 40i$
73. $6 + 4i\sqrt{6}$ **74.** $3 + 9i$ **75.** $13 + 20i$ **76.** $21 - 25i$
77. $-12 - 15i$ **78.** $-3 - 2i$ **79.** $-\frac{1}{2} - \frac{1}{2}i$ **80.** $\pm 3i$
81. $\frac{1}{5} \pm \frac{2}{5}i$ **82.** $2 \pm i\sqrt{6}$ **83.** $\frac{-4 \pm i\sqrt{26}}{7}$
84. $(7, -6), (-2, 12)$ **85.** $(5, -27), (-2, 8)$
86. $(-5, 37), (3, -27)$;

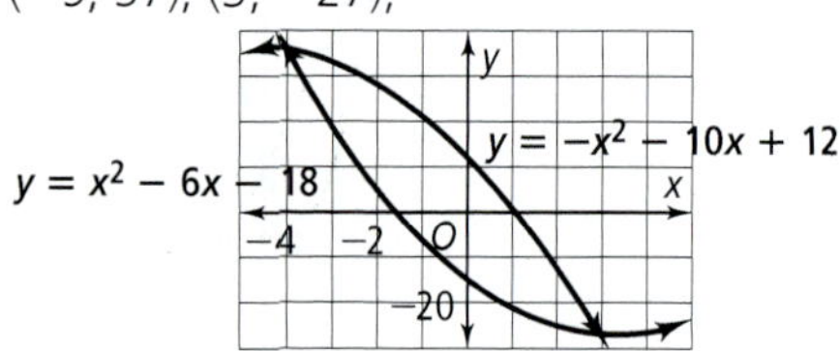

87. $(6.32, 15.64), (-3.32, -3.64)$;

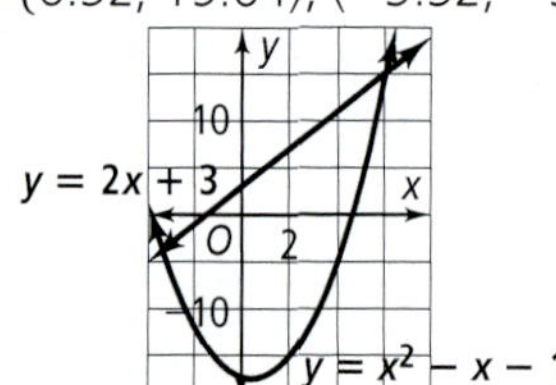

88. 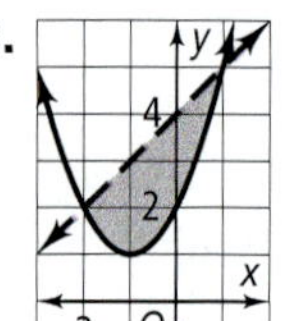

89. 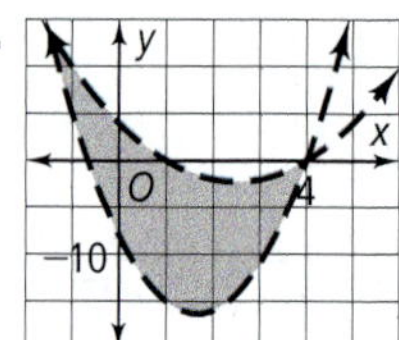

Chapter 5

Get Ready! p. 277

1. 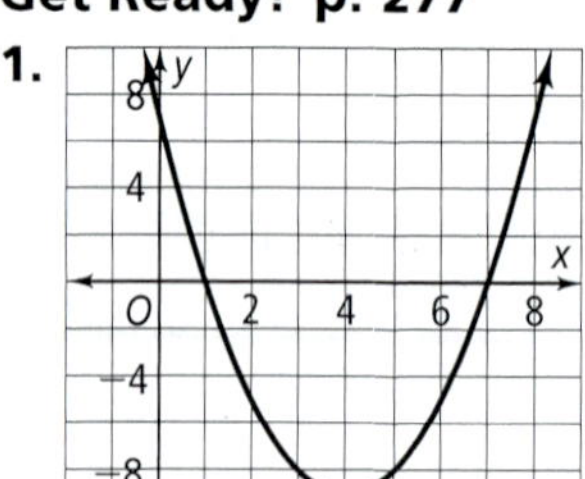

2. 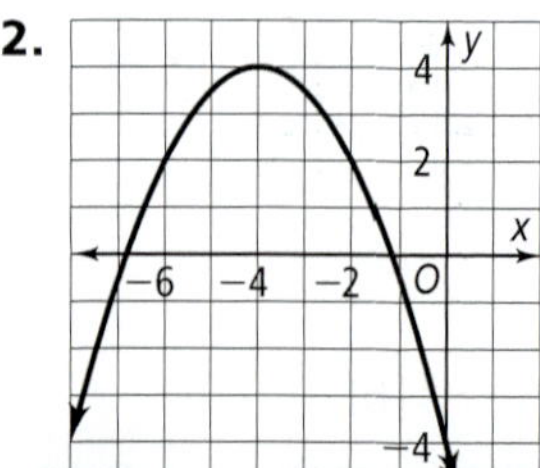

3. 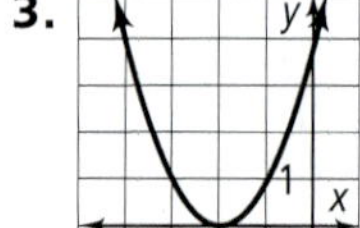

4. $y = x^2 + 3x - 4$ **5.** $y = x^2 - 2x + 1$
6. $y = -x^2 - x + 12$ **7.** 0.25, −1 **8.** −6.39, 4.39
9. −6.38, 0.38 **10.** −4, 5 **11.** −9, 3 **12.** 1, 2
13. 24; two real solutions **14.** 0; one real solution
15. 0; one real solution **16.** south **17.** The highest pt. in Maine may be lower than the highest pt. in the United States. The relative maximum of a graph for a given region is the maximum for that region only, whereas the maximum of the graph may be greater than or equal to the relative maximum for the region.
18. $4x^2 + 4x + 1$

Lesson 5-1 pp. 280–287

Got It? 1. a. $5x^4 + 3x^3 - x$; quartic trinomial
b. $-4x^5 + 2x^2 + 13$; quintic trinomial **2.** up and down
3. a. end behavior: up and down; two turning points (0.33, −2.15) and (1, −2); decreases from $-\infty$ to $\frac{1}{3}$, increases from $\frac{1}{3}$ to 1, and decreases from 1 to ∞

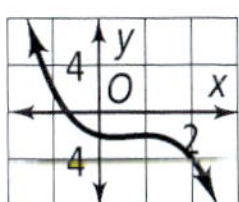

b. end behavior: down and up; no turning points; increases from $-\infty$ to ∞

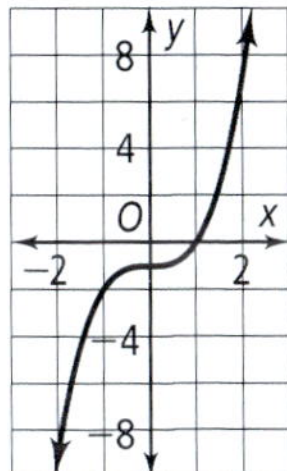

4. a. degree: 4 **b.** Answers may vary. Sample: $y = x^5$
Lesson Check 1. cubic monomial **2.** quadratic trinomial **3.** $5x^2 + 7x + 3$ **4.** $9x - 3$ **5.** up and down
6. Yes; the graph of a linear binomial or linear monomial is a straight line; $f(x) = 2x$ **7.** The graph of $y = 4x^3 + 4$ has no turning points, *not* one turning point.
Exercises 9. $-3x + 5$; linear binomial
11. $x^4 - x^3 + x$; quartic trinomial **13.** $3a^3 + 5a^2 + 1$; cubic trinomial **15.** $12x^4 + 3$; quartic binomial
17. $-2x^3$; cubic monomial **19.** $-x^4 + 3x^2$; quartic binomial **21.** up and up **23.** down and up **25.** down and down **27.** up and down **29.** up and down **31.** up and up
33. end behavior: up and down; two turning pts. (−0.51, 1.12) and (0.36, 4.12); decreases from $-\infty$ to −0.51, increases from −0.51 to 0.36, and decreases from 0.36 to ∞ **35.** end behavior: down and up; no turning pts.; increases from $-\infty$ to ∞ **37.** end behavior: down and up; no turning pts.; increases from $-\infty$ to ∞ **39.** 3
41. $-4a^4 + a^3 + a^2$; quartic trinomial
43. $6x^2$; quadratic monomial **45.** $-9d^3 - 13$; cubic binomial **47.** negative; 3 **49.** positive; 4

51. For $f(x) = x^3 - 3x^2 - 2x - 6$,

x	f(x)	1st diff	2nd diff	3rd diff
0	−6	−4		
1	−10	−4	0	6
2	−14	2	6	6
3	−12	14	12	6
4	2	32	18	
5	34			

For $f(x) = ax^3 + bx^2 + cx + d$,

x	f(x)	1st diff	2nd diff	3rd diff
0	d	$a + b + c$		
1	$a + b + c + d$	$7a + 3b + c$	$6a + 2b$	$6a$
2	$8a + 4b + 2c + d$	$19a + 5b + c$	$12a + 2b$	$6a$
3	$27a + 9b + 3c + d$	$37a + 7b + c$	$18a + 2b$	$6a$
4	$64a + 16b + 4c + d$	$61a + 9b + c$	$24a + 2b$	
5	$125a + 25b + 5c + d$			

53. a.

x	y	1st diff	2nd diff
−2	8	−6	
−1	2	−2	4
0	0	2	4
1	2	6	4
2	8		

b.

x	y	1st diff	2nd diff
−2	20	−15	
−1	5	−5	10
0	0	5	10
1	5	15	10
2	20		

c.

x	y	1st diff	2nd diff
−2	18	−15	
−1	3	5	10
0	−2	5	10
1	3	15	10
2	18		

d.

x	y	1st diff	2nd diff
−2	28	−21	
−1	7	−7	14
0	0	7	14
1	7	21	14
2	28		

e.

x	y	1st diff	2nd diff
−2	29		
		−21	
−1	8		14
		−7	
0	1		14
		7	
1	8		14
		21	
2	29		

f.

x	y	1st diff	2nd diff
−2	23		
		−18	
−1	5		14
		−4	
0	1		14
		10	
1	11		14
		24	
2	35		

Second differences of quadratic functions are constant.
55. missing values: y: 0, −6, −4; first differences: −6, 2
57. −35; Let $y = 2x^3 + bx^2 + cx + 7$. Evaluate $f(1) = 7$ and $f(2) = 9$ to find a system of equations for b and c. $b = -5$ and $c = 3$, so $f(-2) = -35$. **59.** F
61. $x^2(3x^2 - 2x) - 3x^4 = 3x^4 - 2x^3 - 3x^4 = -2x^3$; The degree is 3 and there is 1 term, so this is a cubic monomial. **62.** (3.7, −4.4), (−2.7, 8.4) **63.** (−20, 360)
64. (−3, 0), (−1, −8) **65.** $35x - 5y = -2$
66. $6x + 2y = -5$ **67.** $2x + 7y = 28$
68. $12x - 10y = 5$ **69.** $(x + 4)(x + 3)$
70. $(x + 10)(x - 2)$ **71.** $(x - 12)(x - 2)$

Lesson 5-2 pp. 288–295

Got It? 1. $x(x - 4)(x + 3)$
2. 0, 3, −5;

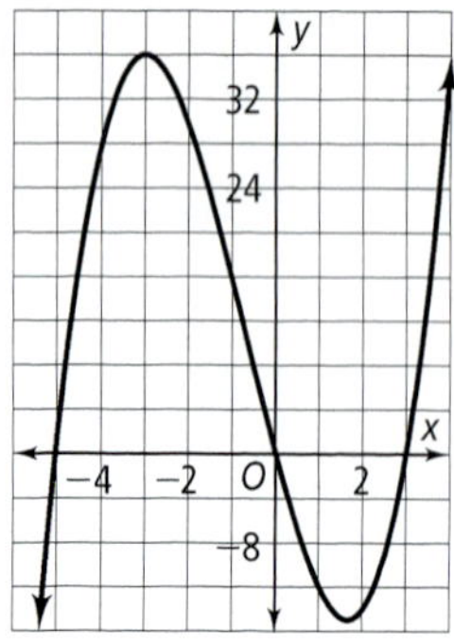

3. a. $f(x) = x^2 - 9$ **b.** $P(x) = x^3 - 3x^2 - 9x + 27$

c.

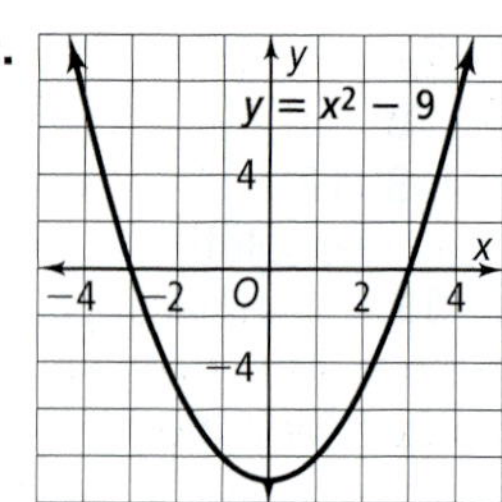

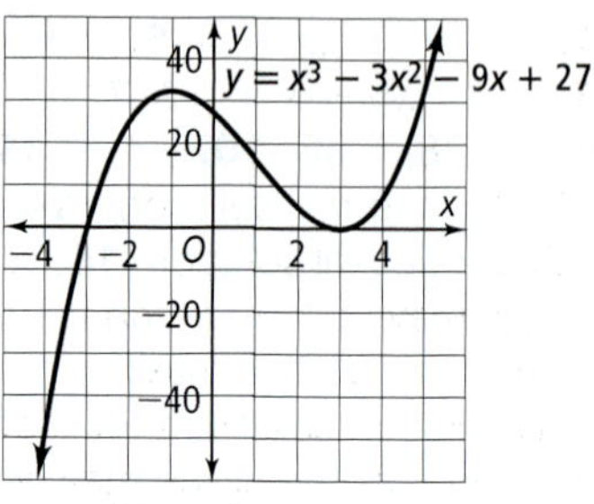

Both graphs have x-intercepts of 3 and −3. The quadratic has up and up end behavior and one turning pt., and the cubic has down and up end behavior and two turning pts.
4. 0 is a zero of multiplicity 1, the graph looks close to linear at $x = 0$; 2 is a zero of multiplicity 2, the graph looks close to quadratic at $x = 2$. **5.** relative maximum: (−0.86, 3.13); relative minimum: (0.64, −2) **6.** 2.28 in.^{3}
Lesson Check 1. 0, 6 **2.** −4, 5 **3.** −12, 7, 9
4. $f(x) = x^3 - x$ **5.** $h(x) = x^4 + 4x^3 - 26x^2 - 60x + 225$
6. Error in writing the factors: a function that has zeros at 3 and −1 has factors of $x - 3$ and $x + 1$ *not* $x + 3$ and $x - 1$, so $f(x) = x^2 - 2x - 3$ *not* $x^2 + 2x - 3$.
Exercises 7. $x(x + 5)(x + 2)$ **9.** $x(x - 7)(x + 3)$
11. $x(x + 4)^2$
13. 1, −2;

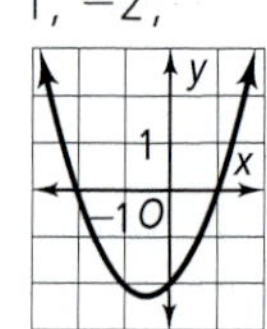

15. 0, −5, 8;

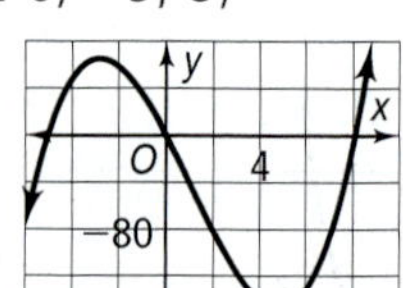

17. −1, 1, 2;

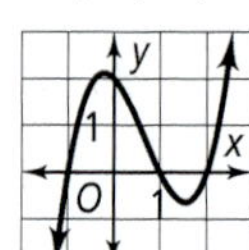

19. $y = x^3 - 18x^2 + 107x - 210$ **21.** $y = x^3 + 9x^2 + 15x - 25$ **23.** $y = x^3 + 2x^2 - x - 2$
25. $y = x^4 - 5x^3 + 6x^2$ **27.** −3 (multiplicity 3)
29. $-1, 0, \frac{1}{2}$ **31.** 4 (multiplicity 2) **33.** $-\frac{3}{2}$, 1 (multiplicity 2) **35.** relative maximum: (−3.19, 24.19); relative minimum: (0.52, −1.38) **37.** relative maximum: (2.15, 12.32); relative minimum: (−0.15, −12.32)
39. a. $\ell = 16 - 2x$; $w = 12 - 2x$; $h = x$
b. $V = x(16 - 2x)(12 - 2x)$
c.

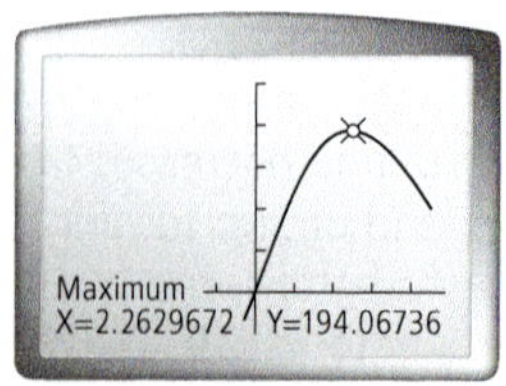

194 in.^{3}, 2.26 in.

41. $y = -2x(x + 5)(x - 4)$ **43.** 1 ft increase in each dimension **45.** $V = 12x^3 - 27x$ **47.** relative maximum: (2.53, 10.51); relative minimum: (5.14, −7.14); $\frac{3}{2}$, 4, 6 **49.** no relative maximum; relative minimum: (−1, −1); −2, 0 **51.** Answers may vary. Sample: The linear factors can be determined by examining the x-intercepts of the graph. **53.** −1, 4, $\frac{3}{2}$ **55.** $y = x^4 - 1$ **57.** B **59.** B **61.** $-3x^5 + 3x^2 - 1$; quintic trinomial **62.** $-7x^4 - x^3$; quartic binomial **63.** $x^3 - 2$; cubic binomial **64.** $(x + 4)(x + 1)$ **65.** $(x - 5)(x + 3)$ **66.** $(x - 6)^2$ **67.** −3, 2 **68.** $\frac{1}{2}$, 3 **69.** $\pm\frac{5}{2}$

Lesson 5-3 pp. 296–302

Got It? 1. a. $\pm 1, \pm 2i$ **b.** 0, 2, 3 **2. a.** $\pm 2, \pm 2i$ **b.** 0, −4, 2 **c.** 2, $-1 \pm i\sqrt{3}$ **3. a.** −1.84 **b.** The second method seems to be a more reliable way to find the solutions because you do not risk missing a pt. of intersection. **4.** 7, 8, 9

Lesson Check 1. $(x - 6)(x + 3)$ **2.** $(x - 3)(x^2 + 3x + 9)$ **3.** $(x^2 + 4)(x + 3)$ **4.** $(x - 2)(x + 2)(x^2 + 2)$ **5.** −4, $\frac{1}{2}$ **6.** −2, 0, 1 **7. a.** difference of squares **b.** sum of cubes **c.** difference of cubes **d.** difference of squares **8.** Graphing; imaginary numbers don't exist on the x-axis. **9.** Method 1: Graph $y = x^6 - x^2$. Find the zeros for the real solutions.

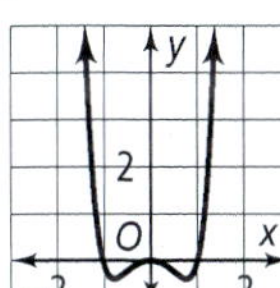

Method 2: Factor and solve x for $x^6 - x^2 = 0$.
$x^2(x^4 - 1) = x^2(x^2 - 1)(x^2 + 1) =$
$x^2(x - 1)(x + 1)(x^2 + 1) = 0$
$x = 0, \pm 1$

Exercises 11. 10, $-5 \pm 5i\sqrt{3}$ **13.** $\frac{1}{4}$, $\frac{-1 \pm i\sqrt{3}}{8}$ **15.** $\frac{1}{3}$, $-\frac{5}{2}$ **17.** 4, $-2 \pm 2i\sqrt{3}$ **19.** $-\frac{1}{2}$, $\frac{1 \pm i\sqrt{3}}{4}$ **21.** ± 2 **23.** $\pm\sqrt{2}$, $\pm 3i$ **25.** −2, 1, 5 **27.** 0, 1 **29.** 0, −1, −2 **31.** 0, −0.5, 1.5 **33.** 1, 7 **35.** −2, 5 **37.** 16 yrs old; $x^2(x + 2) = 3x + 4560$;

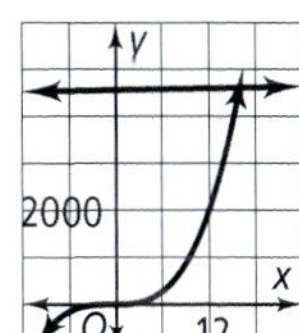

39. 0, $5 \pm 2\sqrt{3}$ **41.** 0, $\frac{5 \pm 5\sqrt{2}}{2}$ **43.** $\frac{4}{3}$, $\frac{-2 \pm 2i\sqrt{3}}{3}$ **45.** $1.9(1 \pm i)$, $-1.9(1 \pm i)$ **47.** 0, ±1, ±2 **49.** −1, $\pm i$ **51.** 2 ft × 3 ft × 6 ft **53.** 5 m

55. ±3, ±1; $y = (x - 1)(x + 1)(x - 3)(x + 3)$;

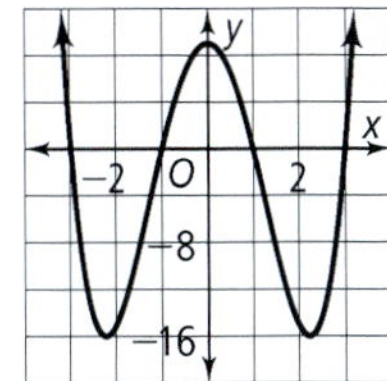

57. Answers may vary.
59. Answers may vary. Sample:
$f(x) = 4x(x + 12)\left(x - \frac{1}{4}\right)\left(x - \frac{1}{6}\right)$
$p(x) = x(x + 12)(4x - 1)(6x - 1)$
61. C **63.** B **65.** $3(x - 4)(x - 2)$ **66.** $2x(x + 3)^2(x - 3)$ **67.** $x^2(x - 5)(x + 1)$ **68.** −2, 6 **69.** ±6 **70.** $-\frac{1}{2}$, 3 **71.** 12 **72.** $-\frac{1}{5}$

Lesson 5-4 pp. 303–310

Got It? 1. $3x - 8$, R 0 **2. a.** yes; $P(x) = (x + 5)(x^4 - 1)$ **b.** $(x + 2)(3x + 1)$ **3.** $x^2 + 7x - 8$, R 0 **4.** width: $(x + 1)$ in.; height: $(x + 2)$ in.; length: $(x + 3)$ in. **5.** 0

Lesson Check 1. $2x + 3$, R 5 **2.** $x^2 + 2x + 5$ **3.** $x^2 + x - 2$ **4.** $4x^2 + x - 6$, R 6 **5.** $9x^2 + 12x + 40$, R 120 **6.** $x - a$ is a factor of $P(x)$. **7.** The polynomials need to be written in standard form since the leading coefficient of both polynomials determines the leading term of the quotient. **8.** Answers may vary. Check student's work.

Exercises 9. $x - 8$ **11.** $x^2 + 4x + 3$, R 5 **13.** $3x^2 + 3x + 2$ **15.** $x - 10$, R 40 **17.** no **19.** yes **21.** $x^2 + 4x + 3$ **23.** $x^2 - 11x + 37$, R −128 **25.** $x + 1$, R 4 **27.** $x^2 - 3x + 9$ **29.** $y = (x + 1)(x + 3)(x - 2)$ **31.** width = x; length = $x + 3$; height = $x - 2$ **33.** 0 **35.** 12 **37.** 10 **39.** 0 **41.** The constant term of the dividend is missing and the divisor is −1 *not* 1:
$x^3 - x^2 - 2x = (x + 1)(x^2 - 2x) = x(x + 1)(x - 2)$
43. $x + 2$ **45.** $x^3 - 3x^2 + 12x - 35$, R 109 **47.** $x + 4$ **49.** no **51.** yes **53.** yes **55.** no **57.** $x^3 - x^2 + 1$ **59.** $x^3 + 7x + 5$ **61.** $x^3 - 2x^2 - x + 6$ **63. a.** $x + 1$ **b.** $x^2 + x + 1$ **c.** $x^3 + x^2 + x + 1$ **d.** $(x - 1)(x^4 + x^3 + x^2 + x + 1)$ **65.** $x + 2i$ **67.** D **69.** A **71.** 0, −1 **72.** 0, 1 **73.** −5, 0, 5 **74.** $\frac{-3 \pm \sqrt{17}}{2}$ **75.** $-1 \pm \sqrt{3}$ **76.** 1, $-\frac{5}{7}$ **77.** $\frac{5 \pm \sqrt{5}}{2}$ **78.** $3 \pm \sqrt{2}$ **79.** $\frac{-7 \pm \sqrt{5}}{2}$

80.

81.

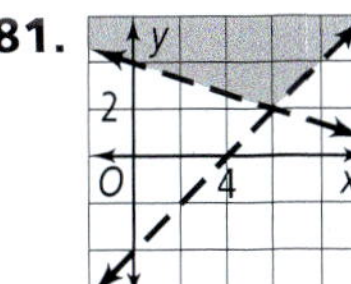

82.

83. 24 **84.** 5 **85.** $23 - 11i$

Lesson 5-5 pp. 312–317

Got It? 1. $\frac{2}{3}$ **2.** $2, -1, -\frac{3}{2}$ **3.** $3 + 2i$
4. $P(x) = x^4 - 14x^3 + 69x^2 - 194x + 208$ **5. a.** There are three or one positive real roots and one negative real root. The graph confirms one negative and one positive real root. **b.** Real roots can be confirmed graphically because they are x-intercepts. Complex roots cannot be confirmed graphically because they have an imaginary component.

Lesson Check 1. $\pm 1, \pm 2$ **2.** $\pm 1, \pm 2, \pm 3, \pm 6, \pm\frac{1}{2}, \pm\frac{3}{2}$ **3.** $\pm 1, \pm 2, \pm 3, \pm 4, \pm 6, \pm 12, \pm\frac{1}{3}, \pm\frac{2}{3}, \pm\frac{4}{3}$
4. $P(x) = x^2 - 14x + 45$ **5.** $P(x) = x^3 + 4x^2 + 4x + 16$
6. Answers may vary. Samples: $1 + 2i, 1 - 2i$; $1 + \sqrt{2}, 1 - \sqrt{2}$ **7. a.** never; 5 does not divide 8 evenly **b.** always; -2 divides 8 evenly **8.** Complex number roots come in pairs; if $-4i$ is a root, so is $4i$.

Exercises 9. ± 1; no rational roots **11.** $\pm 1, \pm 2, \pm 2, \pm 4, \pm\frac{1}{2}$; no rational roots **13.** $\pm 1, \pm 2, \pm 3, \pm 4, \pm 6, \pm 12, \pm\frac{1}{4}, \pm\frac{1}{2}, \pm\frac{3}{4}, \pm\frac{3}{2}$; no rational roots **15.** $\pm 1, \pm 2, \pm 5, \pm 10, \pm\frac{1}{7}, \pm\frac{2}{7}, \pm\frac{5}{7}, \pm\frac{10}{7}$; no rational roots **17.** $\pm 1, \pm 2, \pm 5, \pm 10, \pm\frac{1}{2}, \pm\frac{5}{2}, \pm\frac{1}{5}, \pm\frac{2}{5}, \pm\frac{1}{10}$; no rational roots
19. $14 + \sqrt{2}, 6i$ **21.** $\sqrt{3}, 5 + \sqrt{11}$
23. $P(x) = x^2 + 24x + 135$ **25.** $P(x) = x^2 - 18x + 90$
27. $P(x) = x^4 - 10x^3 + 294x^2 - 1690x + 21{,}125$
29. $P(x) = x^4 - 58x^3 + 1290x^2 - 13{,}066x + 51{,}545$
31. two or no positive real roots; one negative real root
33. no rational roots **35.** no rational roots **37.** no rational roots **39.** $P(x) = x^4 + 3x^3 + 207x^2 + 675x - 4050$
41. $P(x) = x^4 + 6x^3 + 27x^2 - 366x - 518$ **43.** Error in second line, sign of second term; the line should be: $P(-1) = -x^3 + x^2 - x + 1$. Since there are three sign changes in $P(-x)$, there are three or one negative real roots.
45. height: 5 ft; bases: 10 ft, 14 ft
47. Answers may vary. Samples: **a.** $x - 1 - \sqrt{2} = 0$
b. $x^2 - 2(1 + \sqrt{2})x + (1 + \sqrt{2})^2 = 0$ **c.** -1
49. Answers may vary. Sample: You cannot use the Conjugate Root Theorem for irrational roots unless the equation has rational coefficients.
51. -4 **53.** $\frac{1}{6}$ **55.** $x^2 + 6x + 6$, R 3
56. $8x^2 - 36x + 216$, R -1289 **57.** $7x - 3$, R 2
58. $\pm 3i$ **59.** $\pm 9i$ **60.** $\pm 12i$ **61.** $-5x^4 + 6x^2 + 9x + 11$; quartic polynomial of four terms
62. $-4x^5 + 7x^3 + 13x + 2$; quintic polynomial of four terms

Lesson 5-6 pp. 319–324

Got It? 1. $0, 1, -5, 2$ **2. a.** $-1, 2, \frac{1 \pm i\sqrt{23}}{4}$
b. i. A 5th degree polynomial function has four, two, or no turning pts. Three turning pts. are visible, so there must be a fourth one. This will turn the graph back across the x-axis. **ii.** The Fundamental Thm. of Algebra states there will be five roots, and the Conjugate Root Thm. requires pairs of irrational or complex roots. Only two zeros appear in the graph, so there are three zeros remaining. Of the remaining roots, either there are three real roots, or one real and two complex roots. Either way, there is at least one real root that does not appear in the graph.

Lesson Check 1. four roots **2.** fourteen roots
3. $5, \pm 4i$ **4.** $0, 2, \pm i$ **5.** By the Fundamental Thm. of Algebra, polynomial equation of degree n has exactly n roots. **6.** Answers may vary. Sample: $y = x^4 + 8x^2 + 16$ **7.** Use synthetic division to test for and factor out linear factors until a quadratic factor is obtained. Then use the Quadratic Formula if the quadratic factor cannot be factored further.

Exercises 9. $-1, \pm 2i$ **11.** $-1, 2, 4$ **13.** $2, \pm\sqrt{3}$
15. $0, \pm 3, \pm i$ **17.** $3, \pm i$ **19.** $2, \pm\sqrt{3}$ **21.** $\pm 2, \pm i$
23. $-6, \pm i$ **25.** $\pm 4, \frac{-1 \pm i\sqrt{3}}{2}$ **27.** five complex roots; one, three, or five real roots; possible rational roots: $\pm 1, \pm 2, \pm 3, \pm 6, \pm 9, \pm 18$ **29.** six complex roots; zero, two, four, or six real roots; possible rational roots: $\pm\frac{1}{4}, \pm\frac{1}{2}, \pm\frac{3}{4}, \pm 1, \pm\frac{3}{2}, \pm 2, \pm 3, \pm 4, \pm 6, \pm 8, \pm 12, \pm 24$ **31.** $-2, \pm\sqrt{5}$ **33.** $-2, \frac{4}{3}, 3$
35. $3, -1 \pm i\sqrt{2}$ **37.** $-\frac{1}{2}, -1, \pm 3$ **39.** 3 bridges
41. sometimes **43.** always **45.** Answers may vary. Sample: $y = x^4 + 3x^2 + 2$
47. Given any polynomial eq. of odd degree $n \geq 1$, the eq. has exactly n roots. Since imaginary roots occur in pairs, a polynomial of odd degree n will have an even number of imaginary roots and thus an odd number of real roots. So, any odd degree polynomial eq. with real coefficients has at least one real root.
49. 5th degree; rational zero:
$$-\frac{5}{6}(x - \sqrt{2})(x + \sqrt{2})(x - \sqrt{3})(x + \sqrt{3})(x + \frac{5}{6})$$
$$= x^5 + \frac{5}{6}x^4 - 5x^3 - \frac{25}{6}x^2 + 6x + 5.$$
51. G **53.** Substitute $(2, -2)$ into both inequalities to see if the pt. satisfies both inequalities. If it does, then $(2, -2)$ is a solution of the system. If one or both inequalities are not satisfied by the pt. $(2, -2)$, then $(2, -2)$ is not a solution of the system.

54. $x^4 + 6x^3 + 14x^2 + 24x + 40 = 0$
55. $3 \pm 2\sqrt{2}$ **56.** $\frac{-5 \pm i\sqrt{47}}{4}$
57. $\frac{3 \pm i\sqrt{23}}{4}$ **58.** $f(x) = -x^2 + 2x + 3$
59. $f(x) = 2x^2 + 24x + 75$ **60.** $x^3 + 3x^2 + 3x + 1$
61. $x^3 - 9x^2 + 27x - 27$ **62.** $x^4 - 8x^3 + 24x^2 - 32x + 16$ **63.** $x^2 - 2x + 1$ **64.** $x^3 + 15x^2 + 75x + 125$
65. $-x^3 + 12x^2 + 48x + 64$

Lesson 5-7 pp. 326–330

Got It? 1. $a^8 + 8a^7b + 28a^6b^2 + 56a^5b^3 + 70a^4b^4 + 56a^3b^5 + 28a^2b^6 + 8ab^7 + b^8$ **2. a.** $16x^4 - 96x^3 + 216x^2 - 216x + 81$ **b.** If you express 11 as $(10 + 1)$ and calculate the powers using Pascal's triangle, it will be the coefficients.
Lesson Check 1. $x^3 + 3x^2a + 3xa^2 + a^3$
2. $x^5 - 10x^4 + 40x^3 - 80x^2 + 80x - 32$
3. $4x^2 + 16x + 16$ **4.** $27a^3 - 54a^2 + 36a - 8$
5. a. yes **b.** yes **c.** no **6.** The coefficients for the expansion of $(a + b)^n$ are equal to the numbers in the nth row of Pascal's Triangle, respectively. **7.** 13; $n + 1$
Exercises 9. $a^4 + 8a^3 + 24a^2 + 32a + 16$
11. $x^3 - 15x^2 + 75x - 125$ **13.** $x^{10} + 20x^9 + 180x^8 + 960x^7 + 3360x^6 + 8064x^5 + 13{,}440x^4 + 15{,}360x^3 + 11{,}520x^2 + 5120x + 1024$ **15.** $b^9 + 27b^8 + 324b^7 + 2268b^6 + 10{,}206b^5 + 30{,}618b^4 + 61{,}236b^3 + 78{,}732b^2 + 59{,}049b + 19{,}683$
17. $a^4 + 12a^3b + 54a^2b^2 + 108ab^3 + 81b^4$
19. $65{,}536 - 131{,}072x + 114{,}688x^2 - 57{,}344x^3 + 17{,}920x^4 - 3584x^5 + 448x^6 - 32x^7 + x^8$
21. $27a^3 - 189a^2 + 441a - 343$
23. $81y^4 - 1188y^3 + 6534y^2 - 15{,}972y + 14{,}641$
25. a. 6 **b.** 489,888 **27.** $135x^4$ **29.** $625b^8$ **31.** The challenge of the Binomial Theorem occurs when there is a coefficient with the x. However, it is much more efficient to use the Binomial Theorem than FOIL when expanding a binomial that is raised to a high power. **33.** $x^{20} + 40x^{18} + 720x^{16} + 7680x^{14} + 53{,}760x^{12} + 258{,}048x^{10} + 860{,}160x^8 + 1{,}966{,}080x^6 + 2{,}949{,}120x^4 + 2{,}621{,}440x^2 + 1{,}048{,}576$ **35.** $a^5 - 5a^4b^2 + 10a^3b^4 - 10a^2b^6 + 5ab^8 - b^{10}$ **37.** $256x^4 - 1792x^3y + 4704x^2y^2 - 5488xy^3 + 2401y^4$
39. $4096x^{18} + 12{,}288x^{15}y^2 + 15{,}360x^{12}y^4 + 10{,}240x^9y^6 + 3840x^6y^8 + 768x^3y^{10} + 64y^{12}$
41. $125a^3 + 150a^2b + 60ab^2 + 8b^3$ **43.** $-32y^{10} + 80y^8x - 80y^6x^2 + 40y^4x^3 - 10y^2x^4 + x^5$ **45.** Answers may vary. Sample: Since one of the terms is negative $(-y)$ and it is alternately raised to odd and even powers; the term is negative when raised to an odd power and positive when raised to an even power.

47. $-29{,}113 + 17{,}684i$
49. $(x^{14} - 21x^{10} + 35x^6 - 7x^2) + i(-7x^{12} + 35x^8 - 21x^4 + 1)$ **51.** $\frac{1}{64}$
53. $(-1 + i\sqrt{3})^3 = -1 + 3i\sqrt{3} + 9 - 3i\sqrt{3} = 8$
55. G **57.** Let c_a = the first company's cost per month and let c_b = the second company's cost per month: $c_a = 2.25t + 7.95$ and $c_b = 2.75t$

$$c_a = c_b$$
$$2.25t + 7.95 = 2.75t$$
$$7.95 = 0.5t$$
$$15.9 = t$$

The cost will be equal after 15.9 hours of use.

58. $-3, -1, \frac{-3 \pm i\sqrt{11}}{2}$ **59.** $1, \pm i, \pm 3i$
60. $-4, \frac{-3 \pm i\sqrt{7}}{4}$ **61.** $-1, 1, 2, 7$ **62.** $-18 + 43i$
63. $600i$ **64.** -2 **65.** $\frac{7}{5} + \frac{31}{5}i$ **66.** $2x^3 + 5x^2 - x + 9$; cubic polynomial of 4 terms **67.** $-7x^2 + 4x + 1$; quadratic trinomial **68.** $12x^4 - 3x^3 - 9x^2 + x - 8$; quartic polynomial of 5 terms

Lesson 5-8 pp. 331–338

Got It? 1. $y = 1.667x^3 + 1.3 \times 10^{-12}x^2 - 4.667x + 5$ **2.** 22.52 billion lbs **3.** Answers may vary. Sample: The cubic model would fit the data better than the linear model because of the $(n + 1)$ Pt. Principle. Both models have down and up end behavior and increasing growth. The cubic shows slowing growth followed by rapidly increasing growth.
4. a. $y = 0.269867411x - 3.919692952$

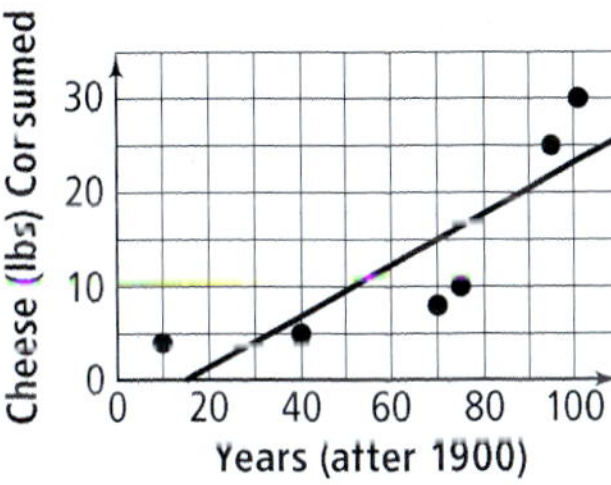

b. 1980: 17.7 lbs; 2000: 23.07 lbs; 2012: 26.31 lbs; most confident for the yrs 1980 and 2000, since they are within the domain of the data set; least confident for the yr 2012, since it is outside the domain of the data set
Lesson Check 1. linear **2.** quadratic **3.** cubic
4. quartic **5.** interpolation since the data point is within the domain of the data set **6.** yes, since the four pts. pass the vertical-line test, a cubic function will fit the pts.; $y = -x^3 - 3x^2 - 3x$ **7.** cubic model; the closer R^2 is to 1, the better the fit

Exercises 9. $y = \frac{1}{2}x - 3$ **11.** $y = -0.929x^2 + 7.786x + 4$ **13.** $y = x^2 - 6x + 1$ **15.** $y = 2x^3 + x^2 - 4x + 6$ **17.** (where x = yrs after 1900) quadratic: $y = -1.25 \times 10^{-4}x^2 - 0.003x + 2.804$, cubic: $y = -8.33 \times 10^{-6}x^3 + 0.002x^2 - 0.190x + 8.142$; cubic; cubic **19.** linear: $y = -0.057x + 19.93$, quadratic: $y = -0.025x^2 + 0.14x + 19.595$; quadratic; quadratic
21. 1950: 2.60%; 1988: 1.23%; 2010: 0.35%
23. January: 19.714 millions of barrels/day; March: 19.8 millions of barrels/day; October: 18.535 millions of barrels/day **25.** cubic: $y = 10.25x^3 + 5x^2 - 2.25x - 8.2$; quartic: $y = 2.042x^4 + 10.25x^3 - 4.042x^2 - 2.25x - 4$; quartic, ($R^2 = 1$) **27.** $y = -0.275x^4 + 0.85x^3 - 4.025x^2 - 8.15x + 7$ **29.** $y = 0.0611511911x^3 - 0.9276466231x^2 + 6.184642324x + 1.750778723$; $R^2 = 0.9994739763$; good fit **31.** $y = 0.111x^4 - 45.618x^3 + 6997.73x^2 - 476{,}931.355x + 12{,}185{,}696.59$ **33.** A quadratic model would be more appropriate, given the real world context. According to the cubic model, there would be a negative number of students enrolled in the course in the year 2024.
35. a. –0.48; –0.75
b.

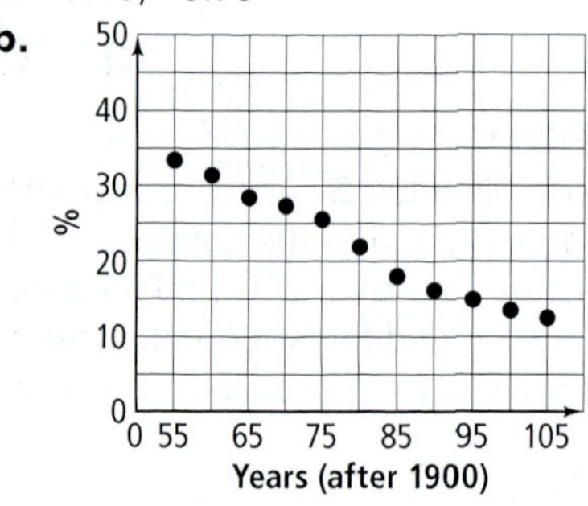

A linear model seems to be most appropriate.
c. Answers may vary. Sample: $y = -0.4464x + 57.77$
d. 4.2% **e.** No; although R^2 is close to 1, the model is not realistic since it predicts that the percentage will eventually become 0, and then negative.

37. $y = \frac{1}{8}(x^2 - 4)^2$

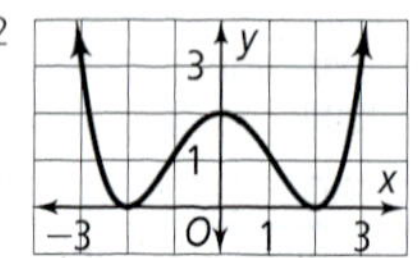

39. 1 **41.** $-\frac{7}{8}$
43. $32x^5 + 240x^4 + 720x^3 + 1080x^2 + 810x + 243$ **44.** $1331x^3 - 363x^2 + 33x - 1$
45. $4096 - 6144x + 3456x^2 - 864x^3 + 81x^4$
46. $64 + 432x + 972x^2 + 729x^3$
47. $|x - 8| < 1$ **48.** $|x - \frac{3}{8}| \le \frac{1}{8}$
49. $|y - 2.8| < 1.1$ **50.** $|t - 750| < 250$
51. $s = \sqrt{A}$ **52.** $\ell = \frac{P}{2} - w$ **53.** $r = \frac{C}{2\pi}$ **54.** $b = \frac{A}{h}$

55.

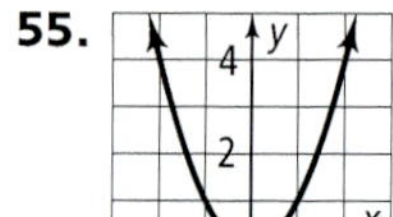

56.

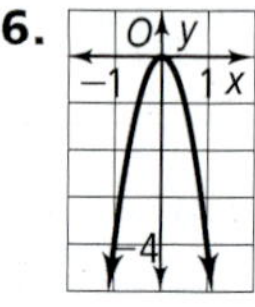

57.

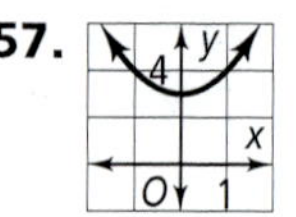

58.

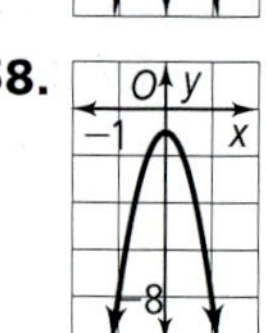

Lesson 5-9 pp. 339–345

Got It? 1. $y = 2(x + 3)^3 - 4$ **2.** $1 - \sqrt[3]{2}$
3. a. Answers may vary. Sample: $y = x^4 - 6x^3 + x^2 - 6x$ **b.** Yes; $-f(x)$ is the function reflected across the x-axis, so the zeros will stay the same. **4.** 972 kW
Lesson Check 1. –2 **2.** 3 **3.** $\frac{1}{3}$ **4.** No; a power function is of the form $y = ax^b$, where y varies directly with the bth power of x. **5.** $y = x^3$ has end behavior of down and up with no turning pt. Thus, at most, there is one real root. **6.** Both $y = x^3$ and $y = 4x^3$ pass through the origin, have the same end behavior of down and up, and no turning pts. $y = 4x^3$ is $y = x^3$ stretched vertically by a factor of 4.
Exercises 7. $y = -3(x - 1)^3 + 2$
9. $y = -(x + 5)^3 - 1$ **11.** $y = -3\left(x + \frac{1}{2}\right)^3 + \frac{3}{4}$
13. $\frac{8}{3}$ **15.** $-\frac{2}{15}$ **17.** $1 - \frac{1}{2}\sqrt[3]{20}$ **For Exercises 19–24, answers may vary. Samples: 19.** $x^4 - x^3 - x^2 - x - 2$ **21.** $x^4 - 2x^3 - 2x^2 - 2x - 3$
23. $x^4 + 5x^3 + 5x^2 + 5x + 4$ **25.** 38.4 ≈ 38 slices
27. 90 lb·ft²/s² **29.** Yes; using parent function $y = x^2$, stretch vertically by a factor of 2, then translate 5 units up and 3 units to the right. **31.** Yes; using parent function $y = x^2$, translate 9 units down and 4 units to the right.
33. reflection across the x-axis, vert. stretch by a factor of 2, translation 1 unit up and 1 unit to the right
35. vert. stretch by a factor of 3, translation 2 units down and 1 unit to the right **37.** reflection across the x-axis, translation 2 units up and 4 units to the right
39. 40 lb · ft²/s² **41.** Some quartic polynomials have four x-intercepts and cannot be written in that form; $y = x^4 - 20x^2 + 64$; $y = x^4 - 5x^2 + 4$.
43. 150 watts **45.** B
47.
$$\begin{aligned} y &= (x - 2)^2 - 1 \\ &= x^2 - 4x + 4 - 1 \\ &= x^2 - 4x + 3 \end{aligned}$$
48. $y = -2x^3 + 3x^2 - x - 2$ **49.** $y = 3x^3 - 5x - 3$
50. $y = -\frac{4}{5}x + \frac{16}{5}$ **51.** $y = -3x + 5$ **52.** yes **53.** no

54. $x^2(x^8 + 1)$ **55.** $(x - y)(x + y)(x^2 + y^2)$
56. $13x^3y^6(13x^3y^6 - 1)$

Chapter Review pp. 347–352

1. D **2.** B **3.** C **4.** A **5.** $y = -x^4 + 12$; quartic binomial; down and down **6.** $y = x^2 - x + 7$; quadratic trinomial; up and up **7.** $y = -x^4 + 2x^3 + 3x^2 - 6x + 12$; quartic polynomial of five terms; down and down **8.** $y = x^3 + 2x^2 - 4x + 8$; cubic polynomial of four terms; down and up **9.** $y = x^4 - 3x^3 + 3x^2 + 10$; quartic polynomial of four terms; up and up **10.** 3
11. If n is even there are an odd number of turning points; if n is odd there are an even number of turning points.
12. $f(x) = x^3 - 4x^2 - 11x - 6$ **13.** $f(x) = x^3 - x^2 - 2x$ **14.** $f(x) = x^3 - 6x^2 + 11x - 6$
15. $f(x) = x^3 - 3x^2 - 6x + 8$ **16.** 0, −2 (multiplicity 3)
17. 2 (multiplicity 2), −2 (multiplicity 2) **18.** 0, $-\frac{1}{2}$, 1
19. 5, −2 (multiplicity 2) **20.** relative maximum: (0.8672, −1.9351); relative minimums: (0, −3), (2.8828, −12.1704); zeros: $x = 0.5992$, $x = 3.7115$
21. relative maximum: (−0.8441, 9.3023); relative minimum: (0.7108, −0.0964); zeros: $x = -1.6180$, $x = 0.6180$, $x = 0.8$ **22.** relative minimum: (1, −4); zeros: $x = -0.2491$, $x = 1.6633$
23. relative maximum: (−0.4142, −3.3432); relative minimum: (2.4142, −14.6569); zero: $x = 4$ **24.** 3, 8
25. $-\frac{1}{2}$ **26.** 0, $\frac{-1 \pm \sqrt{37}}{2}$ **27.** $\frac{2 \pm i\sqrt{2}}{2}$
28. no real roots;

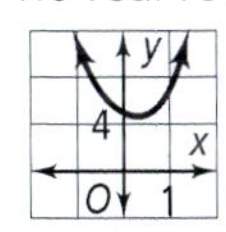

29. 2.3949;

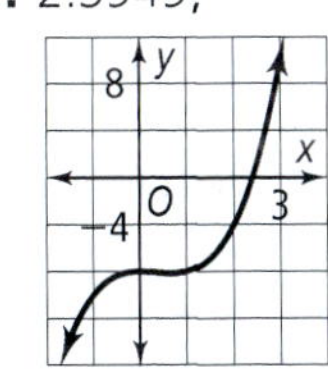

30. 4.87 in. × 2.87 in. × 2.87 in. **31.** $x^2 + 6x + 9$
32. $2x^2 + x - 3$, R 1 **33.** yes **34.** no **35.** $x^2 - 1$
36. $2x^2 - 6x + 2$, R −20 **37.** $5x^2 + 18x + 36$, R 12
38. −14 **39.** 2 **40.** ±1, ±2, ±3, ±6
41. ±1, ±2, $\pm\frac{1}{3}$, $\pm\frac{2}{3}$ **42.** ±1, ±2, ±3, ±4, ±6, ±12, $\pm\frac{1}{4}$, $\pm\frac{1}{7}$, $\pm\frac{3}{4}$, $\pm\frac{3}{2}$ **43.** ±1, ±7, $\pm\frac{1}{3}$, $\pm\frac{7}{3}$ **44.** −3
45. −5 **46.** 1, −4, $-\frac{1}{2}$ **47.** 1, −2, $-\frac{2}{3}$ **48.** $1 + i$
49. $5 - \sqrt{3}$, $\sqrt{2}$ **50.** $3i$, $-7i$ **51.** $-2 - \sqrt{11}$, $-4 + 6i$
52. $y = x^2 - 17x + 70$ **53.** $y = x^3 + 3x^2 + 25x + 75$
54. $y = x^2 - 12x + 37$ **55.** $y = x^4 - 4x^3 - 10x^2 + 68x - 80$ **56.** one positive real zero; two or no negative real zeros **57.** two or no positive real zeros; one negative real zero **58.** four, two, or no positive real zeros; no negative real zeros **59.** two or no positive real zeros; two or no negative real zeros **60.** 3 **61.** 4 **62.** 5 **63.** 6

64. 1, $-3 \pm \sqrt{7}$ **65.** 2, $\pm\sqrt{5}$ **66.** 1, $\frac{1 \pm i\sqrt{15}}{4}$
67. −3, 6, $\frac{1 \pm \sqrt{5}}{2}$ **68.** 9 **69.** 1, 8, 28, 56, 70, 56, 28, 8, 1
70. 16 **71.** 105 **72.** $x^3 + 27x^2 + 243x + 729$
73. $b^4 + 8b^3 + 24b^2 + 32b + 16$
74. $27a^3 + 27a^2 + 9a + 1$ **75.** $x^3 - 15x^2 + 75x - 125$ **76.** $x^3 - 6x^2y + 12xy^2 - 8y^3$
77. $243a^5 + 1620a^4b + 4320a^3b^2 + 5760a^2b^3 + 3840ab^4 + 1024b^5$ **78.** $x^6 + 6x^5 + 15x^4 + 20x^3 + 15x^2 + 6x + 1$ **79.** $64x^6 - 192x^5 + 240x^4 - 160x^3 + 60x^2 - 12x + 1$ **80.** 108 **81.** $6a^2c^2$
82. $y = 3.5x^2 - 4.5x + 5$
83. $y = 2.082999x - 2.475234$; $y = 0.086232x^2 + 0.008929x + 5.963724$; $y = -0.002554x^3 + 0.178307x^2 - 0.913645x + 8.205128$; cubic is best fit since $R^2 = 1$.
84. $y = 5.8667x^3 + 120.5333x^2 - 629.2667x + 1421$; $1097.2 \approx 1097$ **85.** $y = -(x - 2)^3 + 1$
86. $y = 6(x + 3)^3$ **87.** Answers may vary. Sample: $y = x^4 - 10x^3 + 25x^2 - 10x + 24$ **88.** $y = 0.3x^5 + 3$

Chapter 6

Get Ready! p. 357 **1.** domain: {1, 2, 3, 4}, range: {2, 3, 4, 5} **2.** domain: {1, 2, 3, 4}, range: {2}
3. domain: all real numbers, range: $y \geq -8$ **4.** domain: all real numbers, range: $y \geq 3$

5.

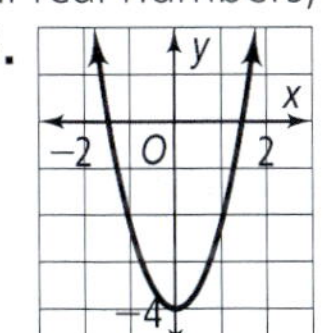

6.

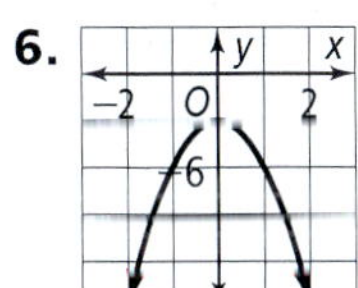

7.

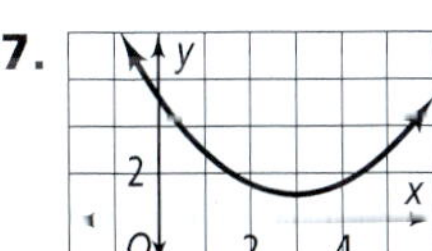

8. $3y^2 - 14y + 8$ **9.** $49a^2 - 100$
10. $x^3 + 4x^2 - 15x - 18$ **11.** −2, 7 **12.** $\frac{5}{2}$, 3
13. −4, $\frac{2}{3}$ **14.** $\frac{1}{2}$ **15.** $\pm\frac{7}{2}$ **16.** $\pm\sqrt{7}$ **17.** Yes; it is a better deal to first take 50% off the shirt and then use the \$10 coupon. **18.** A "one-to-one function" is a function where there is exact correspondence of every element of the domain with exactly one element of the range.
19. the non-negative root

Lesson 6-1 pp. 361–367

Got It? 1. a. 0; −1; 2 **b.** ±0.1; no real square root; $\pm\frac{6}{11}$ **c.** Any negative number multiplied by itself an even number of times will always be positive. Therefore, there can be no real nth roots (where n is even) for a negative number b. **2. a.** −3 **b.** no real root **c.** 7 **d.** no real root **3. a.** $9x^2$ **b.** a^4b^5 **c.** $|x^3|y^4$ **4.** 0; 100
Lesson Check 1. ±5 **2.** ±0.4 **3.** no real root **4.** $3|b|$ **5.** $a^4|b^9|$ **6.** $-5a$ **7.** 16 has two real fourth roots, ±2. **8.** The real roots of a number are the positive and negative (but not imaginary) roots of the number; the principal root of a number is the nonnegative root of the number. **9.** n is odd.
Exercises 11. ±0.07 **13.** $\pm\frac{8}{13}$ **15.** 0.5 **17.** 0.07 **19.** none **21.** $\pm\frac{10}{3}$ **23.** 0.5 **25.** −3 **27.** $3y^2$ **29.** $2y^2$ **31.** ±10 **33.** ±0.5 **35.** about 0.8 in. **37. a.** about 79.01 ft **b.** about 44.44 ft **39.** $\frac{1}{3}$ **41.** $\frac{1}{4}$ **43.** Answers may vary. Sample: $\sqrt[3]{-8x^6}$, $-\sqrt[4]{16x^8}$, $\sqrt[5]{-32x^{10}}$ **45.** always; x^2 is always nonnegative **47.** some; they are equal for $x = -1, 0, 1$ **49.** even: $|m|$; odd: m **51.** even: $|m^3|$; odd: m^3 **53.** 48 **55.** a diagonal of a square with side 5 **57.** G **59.** System: $-35 = (-3)^3a + (-3)^2b + (-3)c + d$, $1 = (0)^3a + (0)^2b + (0)c + d$, $3 = (2)^3a + (2)^2b + (2)c + d$, $7 = (4)^3a + (4)^2b + (4)c + d$ (OR equivalent system); solution: $a = 0.35$, $b = -1.85$, $c = 3.3$, $d = 1$, cubic polynomial: $y = 0.35x^3 - 1.85x^2 + 3.3x + 1$. **60.** $y = (x + 2)^3 + 3$ **61.** $y = \frac{1}{2}x^3 - 2$ **62.** $1, \frac{3}{4}$ **63.** $\frac{5 \pm i\sqrt{11}}{6}$ **64.** $\frac{11}{6}$ **65.** $2x^3y^3$ **66.** $\frac{ac}{3}$ **67.** $\frac{4}{x^2}$

Lesson 6-2 pp. 367–373

Got It? 1. a. No; the indexes are different. **b.** Yes; $\sqrt[5]{10}$ **2.** $4x^2\sqrt[3]{2x}$ **3.** $15x^3y^3\sqrt{7y}$ **4. a.** $5|x|$ **b.** yes; $\frac{3x^2\sqrt{2x}}{x\sqrt{2x}} = 3x$ **5. a.** $\frac{\sqrt[3]{175xy}}{5y}$ **b.** D; there is no y in the expression.
Lesson Check 1. $\sqrt{10}$ **2.** $-3\sqrt[3]{4}$ **3.** Not possible, the indexes are different. **4.** not possible; $\sqrt{-4}$ is not a real number. **5.** $\sqrt[3]{3x}$ **6.** $x^2\sqrt{3x}$ **7.** $2x\sqrt[3]{4x}$ **8.** $x \leq 0$; for $x \leq 0$, $-4x^3 \geq 0$ and $\sqrt{-4x^3}$ is real. **9.** error in line 1: $\frac{\sqrt[7]{x^5}}{\sqrt[4]{x^2}} \neq \sqrt[7-4]{\frac{x^5}{x^2}}$
Exercises 11. 4 **13.** not possible **15.** 5 **17.** 6 **19.** $2x\sqrt{5x}$ **21.** $5x^2\sqrt{2x}$ **23.** $3y^3\sqrt[3]{2y}$ **25.** $-5x^2y\sqrt[3]{2y^2}$ **27.** $-2xy\sqrt[5]{xy^2}$ **29.** $8y^3\sqrt{5y}$ **31.** $40x|y|\sqrt{3}$ **33.** $-2x^2y\sqrt[3]{30x}$ **35.** $4xy^2\sqrt[3]{y}$ **37.** 10 **39.** $2x^2y^2\sqrt{2}$ **41.** $\frac{2\sqrt[3]{x^2y}}{x}$ **43.** $\frac{\sqrt{2x}}{2}$ **45.** $\frac{\sqrt[3]{4x}}{2}$ **47.** $\frac{\sqrt[4]{250}}{5}$ **49.** $\frac{\sqrt{15y}}{5y}$ **51.** $\frac{\sqrt[3]{150ab^2c}}{5a}$ **53.** 6 cm^2 **55.** about 212 mi/h **57.** $5\sqrt{10}$ **59.** $3x^6y^5\sqrt{2y}$ **61.** $10 + 7\sqrt{2}$ **63.** $\frac{|x|\sqrt{10y}}{2y^2}$ **65.** $\frac{\sqrt[3]{3x^2}}{3x}$ **67.** $\frac{\sqrt[3]{2xy^2}}{xy}$ **69.** 4 g/cm^3 **71.** Check students' work. **73.** always **75.** sometimes **77.** $2\sqrt{5}$ **79.** $a = -2c$, $b = -6d$ **81.** H **83.** F **85.** $11|a^{45}|$ **86.** $9c^{24}d^{32}$ **87.** $4a^{27}$ **88.** $2y^5$ **89.** $y^2 - 4y + 16$, $R -128$ **90.** $6a^2 - 5a + 4$ **91.** 25 **92.** 25 **93.** $\frac{121}{4}$ **94.** $\frac{121}{4}$ **95.** $\frac{3}{5} + \frac{1}{5}i$ **96.** $\frac{10}{13} - \frac{15}{13}i$ **97.** $\frac{16}{17} - \frac{4}{17}i$ **98.** $-\frac{7}{74} - \frac{5}{74}i$

Lesson 6-3 pp. 374–380

Got It? 1. a. The indexes are different. You cannot combine the expressions. **b.** $7x\sqrt{xy}$ **c.** $2\sqrt[5]{3x^2}$ **2. a.** about 84.9 in. **b.** The length of the diagonal of a square of side 6 can be found using the Pythagorean Thm. to be $\sqrt{6^2 + 6^2} = \sqrt{72}$. Using this information you can calculate the perimeter of the window and simplify the expression at the end. **3.** $6\sqrt[3]{2}$ **4.** $46 + 16\sqrt{5}$ **5. a.** 24 **b.** 1 **6. a.** $-\sqrt{21} - \sqrt{35}$ **b.** $\frac{1}{3}(12x + 4x\sqrt{6})$ **c.** after rationalizing; When the numerator is multiplied by the conjugate of the denominator it is more convenient if $\sqrt{8}$ is not yet simplified.
Lesson Check 1. $12\sqrt{6}$ **2.** cannot combine **3.** $3\sqrt{3x}$ **4.** $7\sqrt{3}$ **5.** 13 **6.** $75 + 34\sqrt{5}$ **7.** $-16 - 3\sqrt{2}$ **8. a.** not like radicals **b.** like radicals; $9\sqrt{3xy}$ **c.** not like radicals **9.** They are alike in that you can also use the FOIL method and Distr. Prop. to multiply binomial radical expressions; they are different in that you cannot multiply like radicands together if they do not have the same index.
Exercises 11. $4\sqrt[3]{3}$ **13.** $-2\sqrt{x}$ **15.** $5\sqrt[3]{x^2}$ **17.** $33\sqrt{2}$ **19.** $7\sqrt{2}$ **21.** $9\sqrt[3]{3} - 6\sqrt[3]{2}$ **23.** $8 + 4\sqrt{5}$ **25.** $63 - 38\sqrt{2}$ **27.** $49 + 12\sqrt{13}$ **29.** 14 **31.** −40 **33.** $-2 + 2\sqrt{3}$ **35.** $13 + 7\sqrt{3}$ **37.** 140.3 in.^{2} **39.** $8\sqrt{3}$ **41.** $5\sqrt{3} - 4\sqrt{2}$ **43.** $-2\sqrt[3]{2}$ **45.** $-11 + \sqrt{21}$ **47.** $84 + 24\sqrt{6}$ **49.** 2 **51.** $4x\sqrt{3}$ **53.** Answers may vary. Sample: Without simplifying first, you must estimate three square roots and then add the estimates. If they are first simplified, then they can be combined as $13\sqrt{2}$. Then only one square root need be estimated. **55.** Answers may vary. Sample: $(\sqrt{7} + 2)(\sqrt{7} - 2)$, $(2\sqrt{2} + \sqrt{5})(2\sqrt{2} - \sqrt{5})$ **57.** $2\sqrt{3} - \sqrt{2}$ **59.** $11|x| - 3|x|\sqrt{11}$ **61.** $\frac{3\sqrt{5} + 2\sqrt{3}}{3}$ **63.** $\frac{x + 5\sqrt[4]{x^3}}{x}$ **65.** $-\frac{1}{2}$ **67.** $a = 0$ and $b \geq 0$, or $b = 0$ and $a \geq 0$ **69.** 13 **71.** $\frac{15}{7}$ **73.** 9 **74.** $3\sqrt[3]{2}$ **75.** $\frac{2\sqrt[3]{x^2}}{x}$ **76.** 4 **77.** 6 **78.** $2x$

79. $7x^2\sqrt{2}$ **80.** $x\sqrt{15}$ **81.** $15x^2$ **82.** $2, -1 \pm i\sqrt{3}$ **83.** $-10, 5 \pm 5i\sqrt{3}$ **84.** $\frac{1}{5}, \frac{-1 \pm i\sqrt{3}}{10}$ **85.** $\sqrt{7}$ (multiplicity 2), $-\sqrt{7}$ (multiplicity 2) **86.** $\frac{2\sqrt{5}}{5}$ (multiplicity 2), $-\frac{2\sqrt{5}}{5}$ (multiplicity 2) **87.** $\pm\frac{1}{3}, \pm\frac{1}{3}i$ **88.** x^6 **89.** p^5q^5 **90.** 2^9 or 512 **91.** 3^3 or 27

Lesson 6-4 pp. 381–388

Got It? 1. a. 8 **b.** 11 **c.** 6 **2. a.** $\frac{\sqrt[8]{w^3}}{w}, \sqrt[5]{w}$ **b.** $x^{\frac{3}{4}}, y^{\frac{4}{5}}$ **c.** If m is negative, a is in the denominator and $\frac{1}{a}$ is undefined when $a = 0$. **3. a.** The length of a Venusian year is about 0.61 Earth years. **b.** The length of a Jovian year is about 12.76 Earth years. **4. a.** $\sqrt[4]{27}$ **b.** $\sqrt[6]{x^5}$ **c.** $\sqrt[6]{16{,}807}$ **5. a.** $\frac{1}{8}$ **b.** 8 **c.** $\frac{1}{2187}$ **6. a.** $\frac{1}{2x^5}$ **b.** $27x\sqrt[8]{x^4y^3}$

Lesson Check 1. 5 **2.** 5 **3.** $\frac{1}{125}$ **4.** $\frac{1}{128}$ **5.** $\sqrt[4]{11^3}$ **6.** $\frac{\sqrt{x}}{x}$ **7.** $(1 + \sqrt{2})$ or any nonzero number times $(1 + \sqrt{2})$ **8.** error in third line, second term; $5(5^{\frac{1}{2}}) = 5^{\frac{3}{2}}$. The third and fourth lines should be: $20 - 5^{\frac{3}{2}}$ / $20 - 5\sqrt{5}$ **9.** $(-64)^{\frac{1}{3}} = \sqrt[3]{-64} = -4$ and $-64^{\frac{1}{3}} = -\sqrt[3]{64} = -4$; $(-64)^{\frac{1}{2}} = \sqrt{-64}$, is not a real number, but $-64^{\frac{1}{2}} = -\sqrt{64} = -8$ is a real number.

Exercises 11. 3 **13.** 10 **15.** $7\sqrt{3}$ **17.** 3 **19.** $\sqrt[6]{x}$ **21.** $\sqrt[7]{x^2}$ or $(\sqrt[7]{x})^2$ **23.** $\frac{1}{\sqrt[8]{y^9}}$ or $\frac{1}{(\sqrt[8]{y})^9}$ **25.** $\sqrt{x^3}$ or $(\sqrt{x})^3$ **27.** $(-10)^{\frac{1}{2}}$ **29.** $(7x)^{\frac{3}{2}}$ **31.** $a^{\frac{2}{3}}$ **33.** $c^{\frac{1}{2}}$ **35.** ≈72.8 m **37.** ≈7.9 m **39.** $\sqrt[12]{6^7}$ **41.** $\sqrt[10]{5^7}$ **43.** $\frac{\sqrt[3]{4}}{2}$ **45.** $\frac{\sqrt[6]{7776}}{6}$ **47.** 4 **49.** 4 **51.** $\frac{1}{16}$ **53.** 64 **55.** $\frac{1}{x^2}$ **57.** $\frac{x^{\frac{1}{3}}}{3x}$ **59.** $-\frac{3}{x^3}$ **61.** $\frac{y^4}{x^3}$ **63.** $\frac{1}{x}$ **65.** x^3y^9 **67.** about 78%; 61%; 37% **69.** −7 **71.** 64 **73.** 2,097,152 **75.** $-\frac{1}{81}$ **77.** 125 **79.** $x^{\frac{1}{2}}$ **81.** $x^{\frac{3}{10}}$ **83.** $x^{\frac{1}{6}}y^{\frac{1}{4}}$ **85.** $\frac{4x^7}{9y^9}$ **87.** $\frac{2x^2}{3y^3}$ **89. a.** $(\sqrt{x})^4 = \sqrt{x} \cdot \sqrt{x} \cdot \sqrt{x} \cdot \sqrt{x} = x \cdot x = x^2$, so $\sqrt[4]{x^2} = \sqrt{x}$. **b.** $\sqrt[4]{x^2} = (x^2)^{\frac{1}{4}} = x^{\frac{2}{4}} = x^{\frac{1}{2}} = \sqrt{x}$ **91.** 49 **93.** $x^{2\pi}$ **95.** $3^{\sqrt{2}}$ **97.** 33.13 mi/h **99.** 12 **101.** 3 **102.** $4\sqrt[3]{3}$ **103.** $21\sqrt{2}$ **104.** $1 + 3\sqrt{5}$ **105.** −7 **106.** $-8\sqrt{3}$ **107.** $9\sqrt[4]{2}$ **108.** $4x(x^2 - 2x + 4)$ **109.** $(x + 2)^2$ **110.** $(x - 9)^2$ **111.** $(4a - 3b)(4a + 3b)$ **112.** $(5x - 4y)^2$ **113.** $(3x + 8)^2$ **114.** −3, 2 **115.** 7, −2 **116.** $-\frac{3}{2}, 1$ **117.** $-\frac{1}{3}, 2$ **118.** $-\frac{5}{2}, \frac{1}{2}$ **119.** $-\frac{2}{3}, \frac{3}{2}$

Lesson 6-5 pp. 390–397

Got It? 1. 6 **2.** 5, −11 **3.** 37,500,000 m^3 **4. a.** 10 **b.** when you raise each side of an equation to a power **5.** 9

Lesson Check 1. 12 **2.** 27 **3.** $\frac{1}{25}$ **4.** 4 **5.** 1 **6.** 512 **7.** 3; The solution of 3 yields a negative value for $x - 6$, but the right side of the equation $(\sqrt{3(3)})$ cannot be negative. **8.** Solving square root equations is different from solving absolute value equations in that you use a different technique to isolate the variable. In square root equations, you square each side. In absolute value equations, you write two new equations and solve both. Solving square root equations is similar to solving absolute value equations in that both can introduce extraneous solutions.

Exercises 9. 16 **11.** 22 **13.** 5 **15.** 4 **17.** $\frac{2}{3}$ **19.** −29, 25 **21.** 78 **23.** 0 **25.** about 4 in. **27.** 1 **29.** 3 **31.** −3, −4 **33.** 1 **35.** 3 **37.** 1 **39.** −2 **41.** 1 **43.** 5 **45.** $10\sqrt[4]{3}$ cm or about 13.16 cm **47.** 5 **49.** 8 **51.** 5 **53.** 1 **55.** 9, −7 **57.** 9 **59.** $x = 4$ is a solution, but $x = 1$ is an extraneous solution. **61.** Answers may vary. Sample: $\sqrt{x - 3} = \sqrt{3x + 5}$ **63.** C **65.** 0, 2 **67.** 0 **69.** $\sqrt{6}$ **71.** $\sqrt{10}$ **73.** B **75.** D **77.** 3 **78.** 2 **79.** 625 **80.** 512 **81.** $\frac{1}{1000}$ **82.** 16 **83.** 125 **84.** $6\sqrt{2}$ **85.** 3, 4 **86.** 3, 5 **87.** −5, −4 **88.** $-2, -\frac{2}{3}$ **89.** $-\frac{1}{3}, -\frac{4}{3}$ **90.** $-2, -\frac{3}{4}$
91. domain: {0, 2, 4}, range: {−5, −3, −1}; yes
92. domain: {−1, 0, 1}, range: {2, 0, 1}; yes
93. domain: {−2, 0, 1}, range: {−2, 0, 1}; yes
94. domain: {3, 4, 5}, range: {−1}; yes
95. domain: {0, 1, 2}, range: {0, 1, 2}; no
96. domain: {0}, range: {−2, 0, 2}; no

Lesson 6-6 pp. 398–404

Got It? 1. $(f + g)(x) = 2x^2 + x + 5$, domain: all real numbers; $(f - g)(x) = 2x^2 - x + 11$, domain: all real numbers **2.** $(f \cdot g)(x) = 9x^3 - 30x^2 - 23x - 4$, domain: all real numbers; $\left(\frac{f}{g}\right)(x) = x - 4$, domain: all real numbers except $x = -\frac{1}{3}$ **3.** 4 **4.** Let $D(x)$ = cost after applying the 15% store discount, $E(x)$ = cost after applying the 20% employee discount, and x = cost of items. Then $D(x) = 0.85x$ and $E(x) = 0.80x$. **a.** $(E \circ D)(x) = 0.68x$ **b.** $(D \circ E)(x) = 0.68x$ **c.** The total discounts are the same.

Lesson Check 1. $3x^3 - 2x^2 + 3x - 2$ **2.** $-x^2 + 3x - 3$ **3.** $3x^2 + 1$ **4.** $x^2 + 3x - 1$

Selected Answers

5. $x^2 - 3x + 3$ **6.** $-x^2 + 3x - 3$ **7.** Answers may vary. Sample: $f(x) = 3x^2 + 1$, $g(x) = 2x + 1$; $(f \circ g)(x) = 12x^2 + 12x + 4$; $(g \circ f)(x) = 6x^2 + 3$ **8.** Answers may vary. Sample: $f(x) = 2x$, $g(x) = 0.5x$; $f(g(x)) = x$

Exercises 9. $x^2 + 7x + 5$; domain: all real numbers
11. $x^2 - 7x - 5$; domain: all real numbers
13. $\frac{7x + 5}{x^2}$; domain: all real numbers except $x = 0$
15. $2 - x + \frac{1}{x}$; domain: all real numbers except $x = 0$
17. $\frac{1}{x} + x - 2$; domain: all real numbers except $x = 0$
19. $2x - x^2$; domain: all real numbers except $x = 0$
21. $2x^2 + 2x - 4$; domain: all real numbers
23. $-2x^2 + 2$; domain: all real numbers
25. $2x + 3$; domain: all real numbers except $x = 1$
27. 8 **29.** 20 **31.** 8 **33.** $4a$ **35.** $4a^2 + 4$ **37.** 25
39. 9 **41.** 0.25 **43.** $a^2 - 3$ **45. a.** $f(x) = 0.95x$ **b.** $g(x) = x - 200$ **c.** \$1225 **d.** \$1235
47. $x^2 - x + 7$; domain: all real numbers
49. $x^2 - 5x - 3$; domain: all real numbers
51. $-x^2 + 5x + 13$; domain: all real numbers
53. $4x^2 - 14x + 3$; domain: all real numbers
55. $2x^3 - x^2 - 11x + 10$; domain: all real numbers
57. $\frac{2x + 5}{x^2 - 3x + 2}$; domain: all real numbers except $x = 1$ and 2 **59.** Substitute $5995x$ for y; \$79,850 **61. a.** $g(x)$ is the bonus earned when x is the amount of sales over \$5000. $h(x)$ is the excess sales over \$5000. **b.** $(g \circ h)(x)$; you first need to find the excess sales over \$5000 to calculate the bonus. **63.** 1 **65.** 0 **67.** 8 **69.** -2
71. a. ≈ 1963; The area after 2 seconds is about 1963 in.^2 **b.** $\approx 7854\ \text{in.}^2$ **73.** $x - 2$; $x - 2$ **75.** $x - 3$; $x - 6$
77. $\frac{x^2 + 5}{2}$; $\frac{x^2 + 10x + 25}{4}$
79. $x^7 - x^6 - 16x^5 + 10x^4 + 85x^3 - 25x^2 - 150x$; domain: all real numbers **81.** $\frac{x - 3}{x^2 + 2x}$; domain: all real numbers except $x = 0$, -2, $\sqrt{5}$, and $-\sqrt{5}$ **83.** 2
85. $8a + 4h$ **87.** H **89.** Look at the 5th number in Row 7 of Pascal's triangle to find the coefficient of the x^3y^4 term in the expansion of $(x + y)^7$. $35(3x)^3(-y)^4 = 945x^3y^4$, so 945 is the coefficient.
90. 1 **91.** -3 **92.** 4 **93.** 3 **94.** 2 **95.** 3
96. $x^8 + 32x^7 + 448x^6 + 3584x^5 + 17{,}920x^4 + 57{,}344x^3 + 114{,}688x^2 + 131{,}072x + 65{,}536$ **97.** $x^6 + 6x^5y + 15x^4y^2 + 20x^3y^3 + 15x^2y^4 + 6xy^5 + y^6$ **98.** $16x^4 - 32x^3y + 24x^2y^2 - 8xy^3 + y^4$
99. $128x^7 - 1344x^6y + 6048x^5y^2 - 15{,}120x^4y^3 + 22{,}680x^3y^4 - 20{,}412x^2y^5 + 10{,}206xy^6 - 2187y^7$
100. $59{,}049 - 65{,}610x + 29{,}160x^2 - 6480x^3 + 720x^4 - 32x^5$ **101.** $1024x^5 - 1280x^4y + 640x^3y^2 - 160x^2y^3 + 20xy^4 - y^5$ **102.** $x^8 + 4x^7 + 6x^6 + 4x^5 + x^4$ **103.** $x^{12} + 12x^{10}y^3 + 60x^8y^6 + 160x^6y^9 + 240x^4y^{12} + 192x^2y^{15} + 64y^{18}$
104. no solution

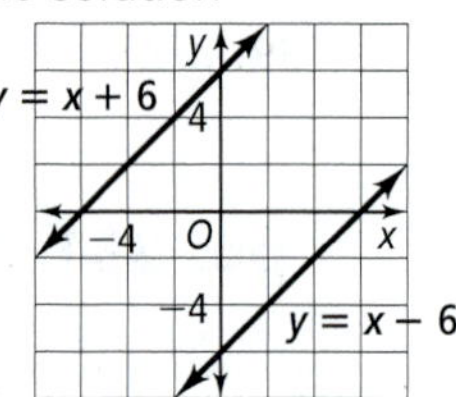

105. (2, 2)

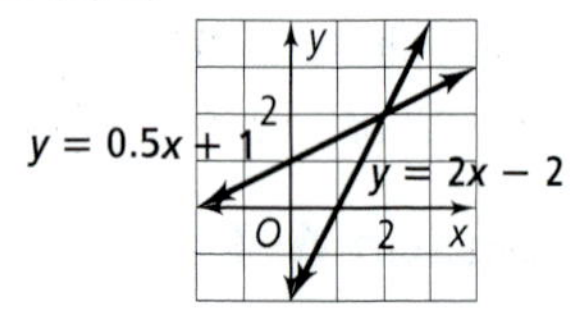

106. (1, 1)

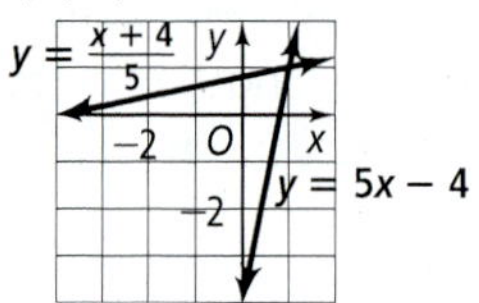

Lesson 6-7 pp. 405–412

Got It?

1. a.

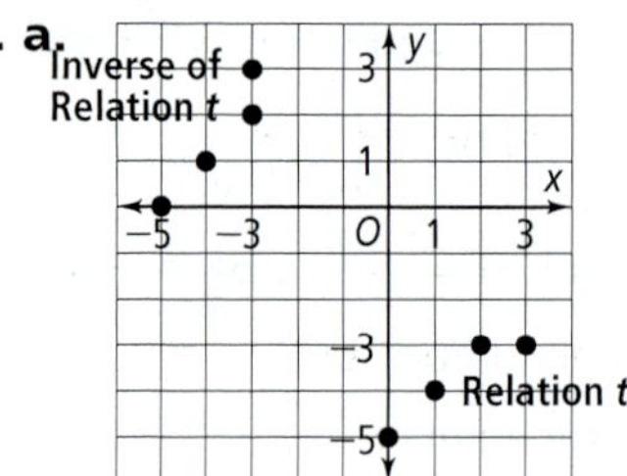

b. t is a function; the inverse of t is not a function; there are 2 y-values for one x-value **2.** $y = \frac{x}{2} - 4$
3.

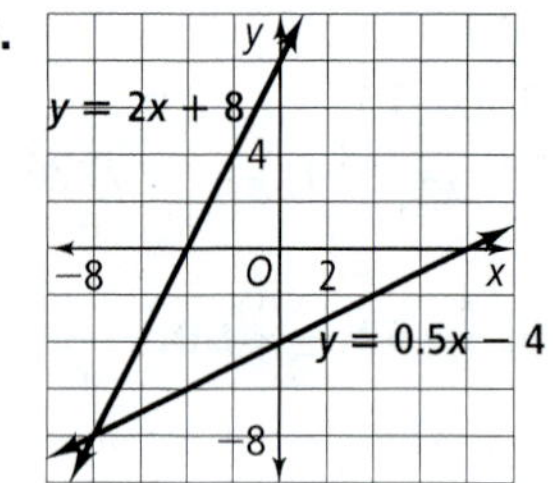

4. a. domain: all real numbers; range: all real numbers **b.** $g^{-1}(x) = -\frac{1}{4}x + \frac{3}{2}$ **c.** domain: all real numbers; range: all real numbers **d.** Yes; for each x in the domain of g^{-1}, there is only one value of y in the range. **5.** $v = \sqrt{19.6d}$; 21.7 m/s **6. a.** $g^{-1}(x) = \frac{4 - 2x}{x}$ **b.** 0 is not in the domain of g^{-1} so $(g \circ g^{-1})(0)$ does not exist. **c.** 0

Lesson Check 1. $f^{-1}(x) = \frac{x - 3}{4}$; yes

2. $f^{-1}(x) = \pm\sqrt{x + 1}$; no **3.** $f^{-1}(x) = -1 \pm \sqrt{x}$; no
4. a. $h^{-1}(x) = -\frac{1}{x} - 2$ **b.** -2.25 **c.** 0 **5.** no; yes
6. 2, 5 **7.** Answers may vary. Samples: $f(x) = 2x + 1$ and $g(x) = x - 2$; $f(x) = x^2$ and $g(x) = x + 1$

Exercises

9.

x	0	1	2	3
y	1	2	3	4

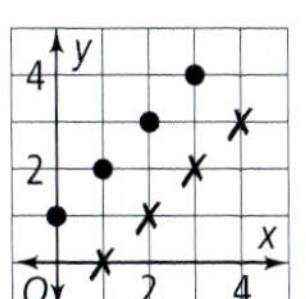

11.

x	2	2	2	2
y	−3	−2	−1	0

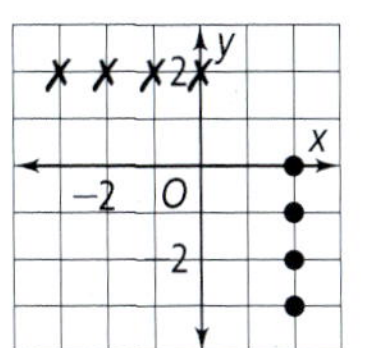

13. $y = \frac{1}{2}x + \frac{1}{2}$; yes **15.** $y = \pm\sqrt{\frac{5 - x}{2}}$; no
17. $y = \pm\sqrt{\frac{x + 5}{3}}$; no **19.** $y = \frac{4 \pm \sqrt{x}}{3}$; no

21.

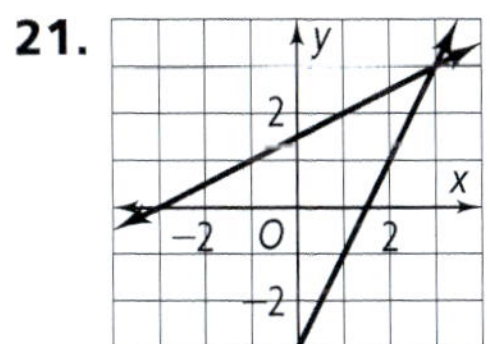

23.

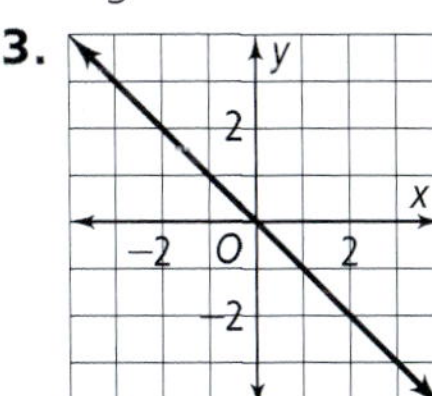

25.

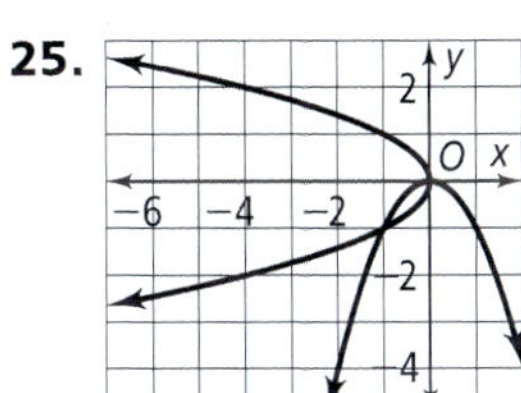

27.

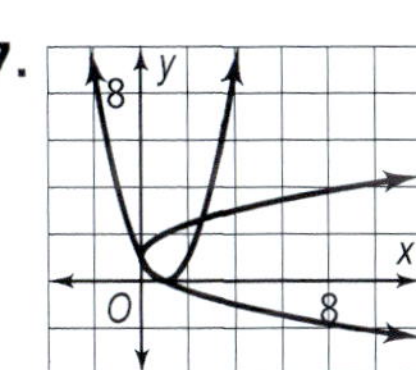

29.

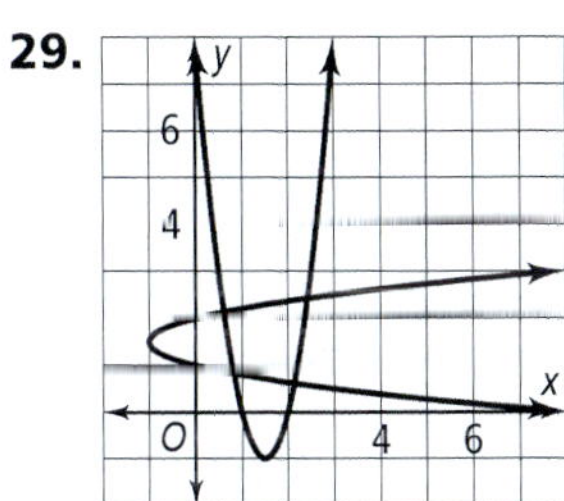

31. $f^{-1}(x) = x^2 + 5$, $x \geq 0$, domain of f: $x \geq 5$, range of f: $y \geq 0$, domain of f^{-1}: $x \geq 0$, range of f^{-1}: $y \geq 5$; f^{-1} is a function **33.** $f^{-1}(x) = \frac{3 - x^2}{2}$, $x \geq 0$, domain of f: $x \leq \frac{3}{2}$, range of f: $y \geq 0$, domain of f^{-1}: $x \geq 0$, range of f^{-1}: $y \leq \frac{3}{2}$; f^{-1} is a function **35.** $f^{-1}(x) = \pm\sqrt{1 - x}$, domain of f: all real numbers, range of f: $y \leq 1$, domain of f^{-1}: $x \leq 1$, range of f^{-1}: all real numbers; f^{-1} is not a function **37. a.** $r = \sqrt[3]{\frac{3V}{4\pi}}$; yes **b.** 20.29 ft **39.** -10
41. *d* **43.** $f^{-1}(x) = \pm\sqrt[4]{x}$; no **45.** $f^{-1}(x) = \pm\sqrt{\frac{2x + 8}{3}}$; no

47. $f^{-1}(x) = \frac{x^2 - 6x + 10}{2}$, $x \geq 3$; yes **49.** -1
51. $f^{-1}(x) = x^2$, $x \leq 0$, domain of f: $x \geq 0$, range of f: $y \leq 0$, domain of f^{-1}: $x \leq 0$, range of f^{-1}: $y \geq 0$; f^{-1} is a function **53.** $f^{-1}(x) = 3 - x^2$, $x \geq 0$, domain of f: $x \leq 3$, range of f: $y \geq 0$, domain of f^{-1}: $x \geq 0$, range of f^{-1}: $y \leq 3$; f^{-1} is a function **55.** $f^{-1}(x) = \pm\sqrt{2x}$, domain of f: all real numbers, range of f: $y \geq 0$, domain of f^{-1}: $x \geq 0$, range of f^{-1}: all real numbers; f^{-1} is not a function **57.** $f^{-1}(x) = \pm\sqrt{x} + 4$, domain of f: all real numbers, range of f: $y \geq 0$, domain of f^{-1}: $x \geq 0$, range of f^{-1}: all real numbers; f^{-1} is not a function
59. $f^{-1}(x) = \pm\frac{1}{\sqrt{x}} - 1$, domain of f: $x \neq -1$, range of f: $y > 0$, domain of f^{-1}: $x > 0$, range of f^{-1}: $y \neq -1$; f^{-1} is not a function **61.** $f^{-1}(x) = \left(\frac{3}{x}\right)^2$, $x \geq 0$, domain of f: $x > 0$, range of f: $y > 0$, domain of f^{-1}: $x > 0$, range of f^{-1}: $y > 0$; f^{-1} is a function
63. a–b. Answers may vary. Samples are given.
a.

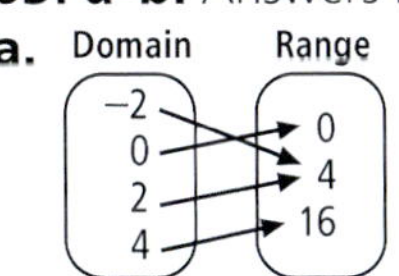

b.

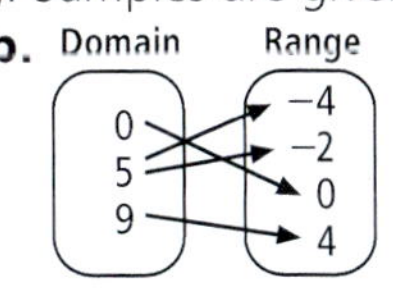

65. $h = s\sqrt{2}$; $s = \frac{h\sqrt{2}}{2} = 3\sqrt{2} \approx 4.2$ in. **67. a.** The horizontal line test tells you if there is more than one x-value for every y-value. Since the graph of f^{-1} interchanges the x and y values of f, if f passes the horizontal line test, f^{-1} will pass the vertical line test and it will be a function. **b.** no **69.** $f^{-1}(x) = x^3 + 5$; yes
71. $f^{-1}(x) = 2 + \sqrt[3]{x}$; yes
73. $f^{-1}(x) = \pm\sqrt[4]{\frac{5x}{6}}$; no **75.** C **76.** F **77.** B
79. $2x + 7$ **80.** $-x - 10$ **81.** $-\frac{3}{2}x + 11$
82. $2x^2 + 28x$ **83.** 32 **84.** $2x + 28$ **85.** -2
86. no real root **87.** 3 **88.** -3 **89.** -3
90. 0.4 **91.** 30 **92.** 0.05
93.

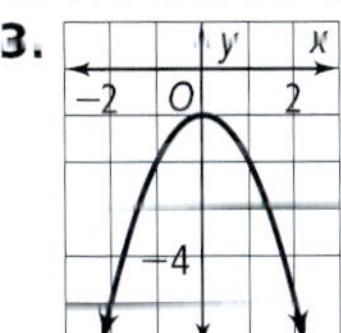

94.

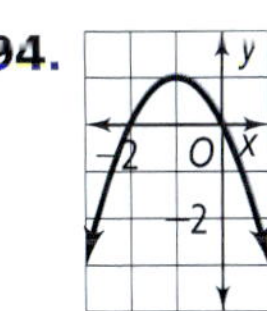

95.

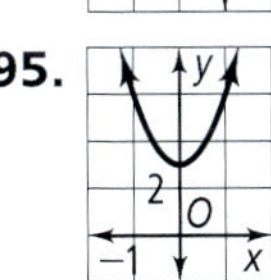

Lesson 6-8 pp. 414–420

Got It?

1.

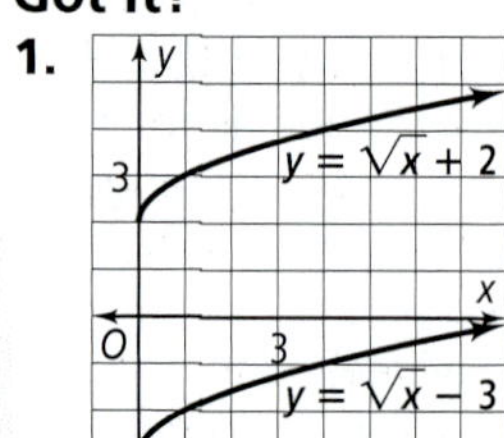

2.

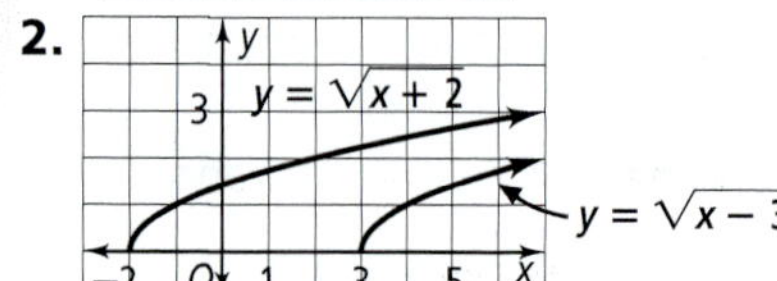

3.

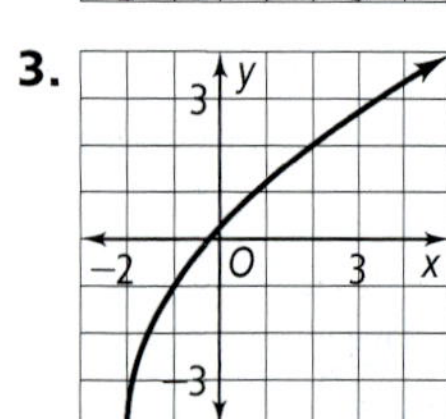

4. 1999 **5.**

6. a. $y = \sqrt[3]{8x + 32} - 2$ is the graph of $y = 2\sqrt[3]{x}$ translated 4 units to the left and 2 units down.
b. $y = 9|x + 2|$; the graph of $y = 9|x + 2|$ is the graph of $y = 9|x|$ translated 2 units to the left; You are rewriting the function so that x has a coefficient of 1.

Lesson Check

1.

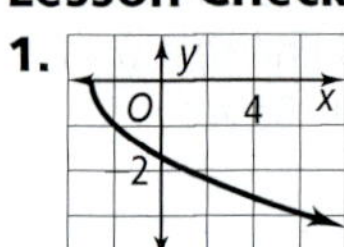

2.

3. $y = 2\sqrt{x - 1}$; the graph of $y = 2\sqrt{x}$ translated 1 unit to the right **4.** $y = 2\sqrt[3]{x + 2}$; the graph of $y = 2\sqrt[3]{x}$ translated 2 units to the left **5.** When $|a| < 1$, a will vertically compress $y = a\sqrt{x}$ and when $|a| > 1$, a will vertically stretch $y = a\sqrt{x}$; this is similar to its effect on other functions. **6.** $g(x)$ is the reflection of $f(x)$ across the x-axis and again across $x = -1$.

Exercises

7.

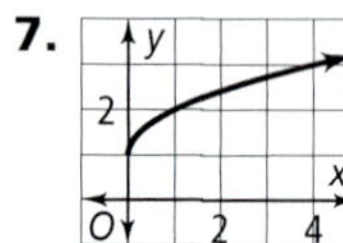

9.

11.

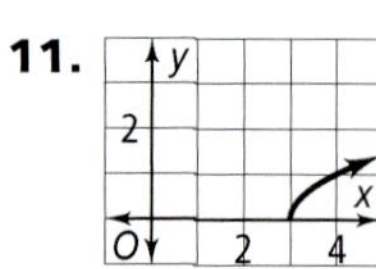

13.

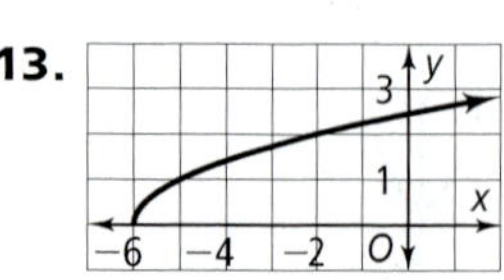

15.

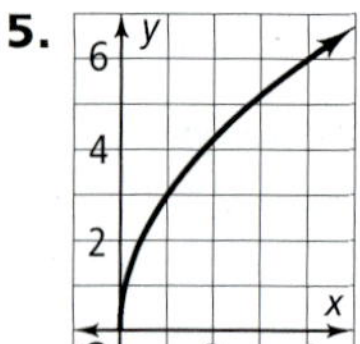

17.

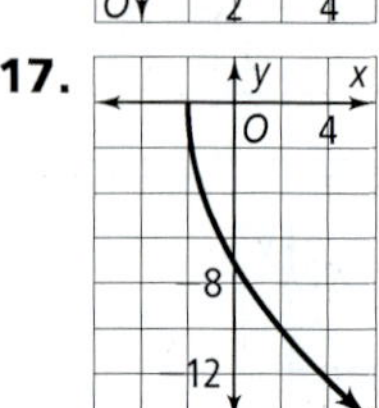

19.

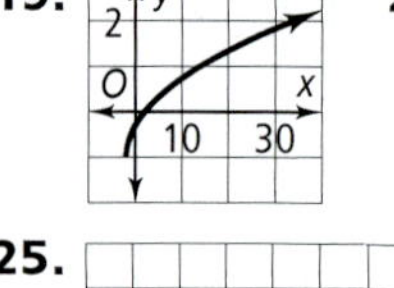

21. 147 **23.** −1

25.

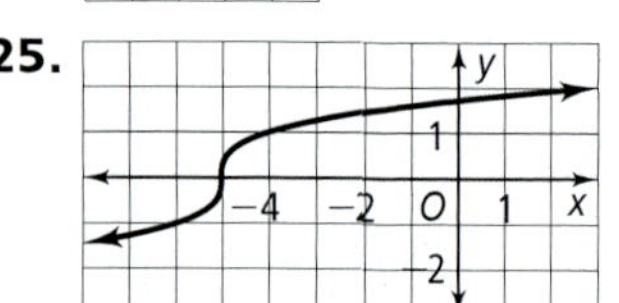

27.

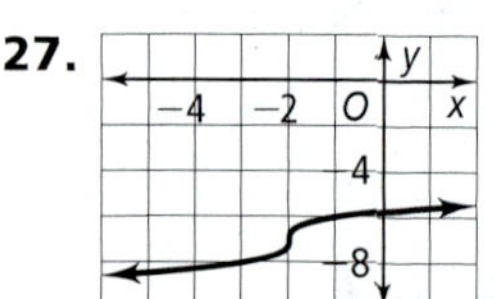

29.

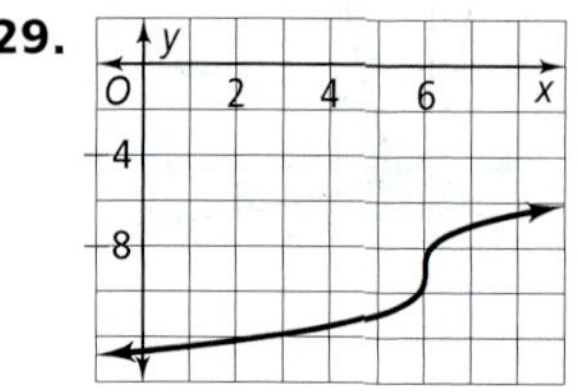

31. $y = 3\sqrt{x - 1}$; the graph of $y = 3\sqrt{x}$ translated 1 unit to the right **33.** $y = -4\sqrt{x + 4}$; the graph of $y = -4\sqrt{x}$ translated 4 units to the left
35. $y = 5\sqrt{x + 5} - 3$; the graph of $y = 5\sqrt{x}$ translated 5 units to the left and 3 units down
37. ≈16.44 ft; ≈29.22 ft
39.

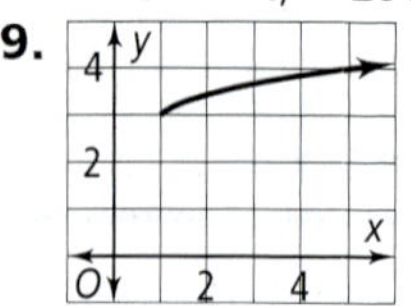

domain: $x \geq 1$, range: $y \geq 3$

41. a. $y = \sqrt{x - 2} - 2$ **b.** domain: $x \geq 2$, range: $y \geq -2$ **c.** No; the function pairs the number 3 with the number −1, which is not a non-negative real number.
43. a. $y = 5\sqrt{x - 4} - 1$; the graph of $y = 5\sqrt{x}$, translated 4 units to the right and 1 unit down
45. $y = -2\sqrt[3]{x - \frac{1}{4}}$; the graph of $y = -2\sqrt[3]{x}$, translated $\frac{1}{4}$ unit to the right **47.** $y = 10 - \frac{1}{3}\sqrt[3]{x + 3}$; the graph of $y = -\frac{1}{3}\sqrt[3]{x}$, translated 3 units to the left and 10 units up

49. $\frac{1}{3}$ **51.** 0, 1, 9

53. a. **b.** $15\sqrt{2}$ in. ≈ 21.2 in.

55. $y = -\sqrt{8}\sqrt{x - \frac{3}{4}}$; the graph of $y = -\sqrt{8x}$, translated $\frac{3}{4}$ unit to the right; domain: $x \geq \frac{3}{4}$, range: $y \leq 0$

57. $y = -\sqrt{12}\sqrt{x + \frac{3}{2}} - 3$; the graph of $y = -\sqrt{12x}$, translated $\frac{3}{2}$ units to the left and 3 units down; domain: $x \geq -\frac{3}{2}$, range: $y \leq -3$

59. for all odd positive integers **61.** F

63. For $f(x) = \sqrt{x - 1}$ the domain is $x \geq 1$ and range is $y \geq 0$. For $g(x) = \sqrt{x} - 1$ the domain is $x \geq 0$ and range is $y \geq -1$.

64. $f^{-1}(x) = \frac{3(x + 3)}{2}$; yes

65. $f^{-1}(x) = (x + 4)^2 - 3, x \geq -4$; yes

66. $f^{-1}(x) = \frac{-1 \pm \sqrt{x}}{2}$; no **67.** $\frac{x\sqrt{3xy}}{y}$ **68.** $\frac{\sqrt{6xy}}{2y}$

69. $\frac{\sqrt[3]{9xy^2}}{3y}$ **70.** $\frac{\sqrt[5]{48x^3y^4}}{2y}$ **71.** $\frac{9 \pm \sqrt{21}}{2}$

72. $\frac{-3 \pm 3\sqrt{5}}{2}$ **73.** $\frac{-1 \pm \sqrt{61}}{10}$ **74.** 8 **75.** 16 **76.** 2

Chapter Review pp. 422–426

1. radicand **2.** radical functions **3.** rational exponent **4.** composite function **5.** 5 **6.** 0.7 **7.** −2 **8.** −2 **9.** $9|x|$ **10.** $4x^2$ **11.** $2|x^3|$ **12.** $0.2x$ **13.** $\frac{x^2}{2}$ **14.** $5x^2y^3$ **15.** 3 **16.** −7 **17.** 4 **18.** $4x^2$ **19.** $30y$ **20.** 4 **21.** $3xy$ **22.** $\frac{3|x|}{y^2}$ **23.** $\frac{2\sqrt{3}}{3}$ **24.** $\frac{\sqrt{3x}}{8}$ **25.** $\frac{y\sqrt[3]{150x}}{10x^2}$ **26.** $22\sqrt{3}$ **27.** $26\sqrt{5x}$ **28.** $x\sqrt[3]{2}$ **29.** $14 + 7\sqrt{2}$ **30.** −6 **31.** $100 + 10\sqrt{6} - 10\sqrt{3} - 3\sqrt{2}$ **32.** $\frac{5 + 2\sqrt{5}}{5}$ **33.** $\frac{9 + 3\sqrt{2}}{7}$ **34.** 5 **35.** 3 **36.** 4 **37.** 25 **38.** x **39.** $-2y^3$ **40.** $81x^2y^4$ **41.** $\frac{1}{x^3y^6}$ **42.** $\frac{1}{x}$ **43.** x^3y^6 **44.** −1 **45.** 15 **46.** 5 **47.** 10, −8 **48.** 2, −1 **49.** −2 **50.** 0, 16 **51.** 0, 36 **52.** 9.05 W **53.** $x^2 + x - 20$; domain: all real numbers **54.** $x^2 - x - 12$; domain: all real numbers **55.** $x^3 - 4x^2 - 16x + 64$; domain: all real numbers **56.** $x + 4$; domain: all real numbers except $x = 4$ **57.** 50 **58.** 5 **59.** 23 **60.** $5a^2 + 3$ **61.** $D(C(x)) = 0.5x - 0.5$, $C(D(x)) = 0.5x - 1$; use the coupon after the store discount.

62. $f^{-1}(x) = \pm\sqrt{\frac{x + 8}{2}}$; no **63.** $f^{-1}(x) = 5 - \frac{1}{3}x$; yes

64. $f^{-1}(x) = x^2 - 6, x \geq 0$; yes **65.** $f^{-1}(x) = \frac{3 \pm \sqrt{x}}{2}$; no

66. domain of f: all real numbers, range of f: all real numbers, domain of f^{-1}: all real numbers, range of f^{-1}: all real numbers

67. domain of f: all real numbers, range of f: $y \geq 0$; domain of f^{-1}: $x \geq 0$, range of f^{-1}: all real numbers

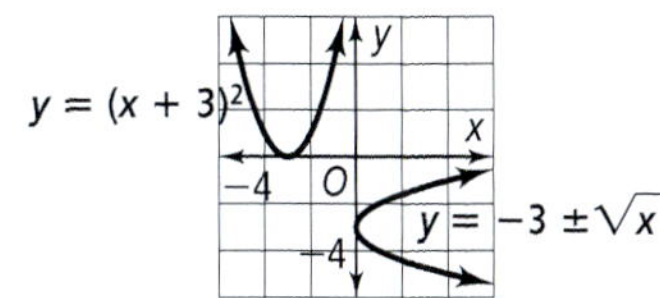

68. domain of f: $x \geq 3$, range of f: $y \geq 0$, domain of f^{-1}: $x \geq 0$, range of f^{-1}: $y \geq 3$

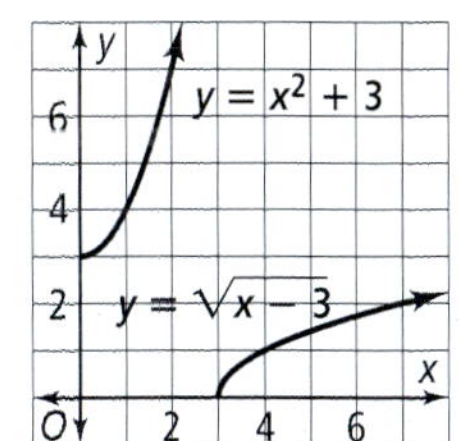

69. domain of f: all real numbers, range of f: $y \leq 6$, domain of f^{-1}: $x \leq 6$, range of f^{-1}: all real numbers

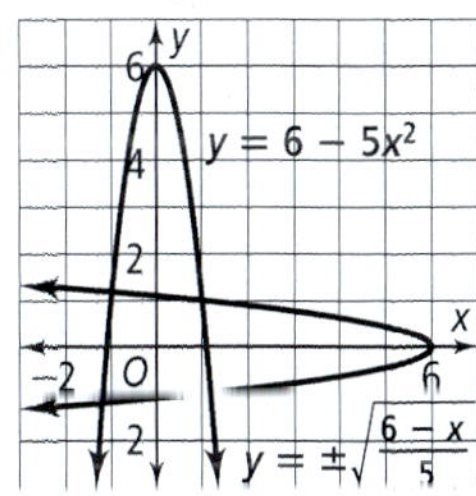

70. $s = \sqrt[3]{V}$; 4 ft

71. domain: $x \geq 0$, range: $y \geq -5$

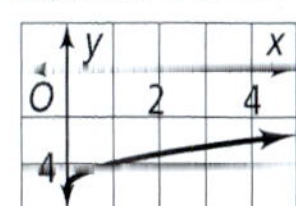

72. domain: $x \geq -8$, range: $y \geq 0$

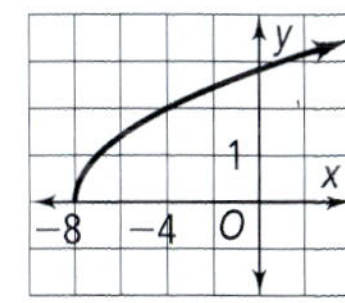

73. domain: $x \geq 0$, range: $y \geq 9$

74. domain: $x \geq 4$, range: $y \leq 0$

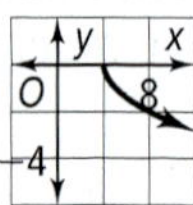

75. domain: all real numbers, range: all real numbers

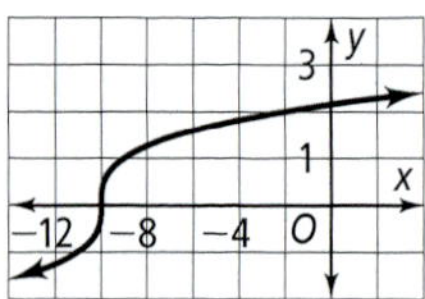

76. domain: all real numbers, range: all real numbers

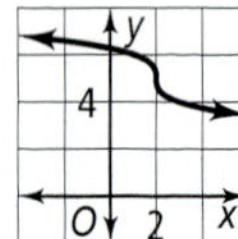

77. $y = 3\sqrt{x-3} + 4$; the graph of $y = 3\sqrt{x}$ translated 3 units to the right and 4 units up **78.** $y = -6\sqrt{x-4}$; the graph of $y = -6\sqrt{x}$ translated 4 units to the right **79.** $y = 2\sqrt[3]{x+3}$; the graph of $y = 2\sqrt[3]{x}$ translated 3 units to the left **80.** $y = \frac{1}{2}\sqrt{x-4} + 6$; the graph of $y = \frac{1}{2}\sqrt{x}$ translated 4 units to the right and 6 units up
81. no solution **82.** 6

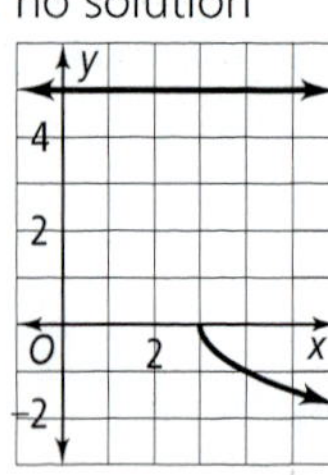

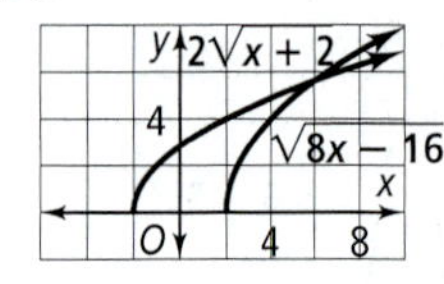

Chapter 7

Get Ready! p. 431 **1.** 0.1; 10; 1000 **2.** $\frac{4}{9}$; 1; $\frac{9}{4}$
3. $-\frac{1}{625}$; $-\frac{1}{25}$; -1 **4.** $-\frac{1}{3}$; -1; -3

5. ; $y = 2x + 2$

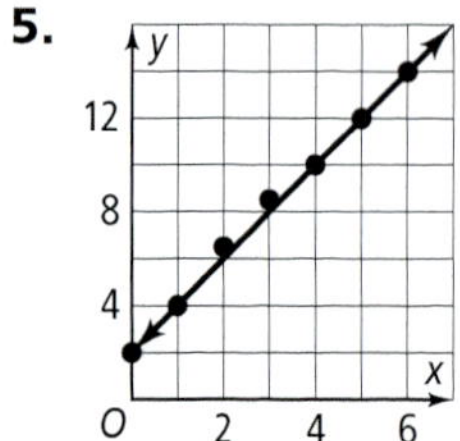

6. ; $y = 25x + 25$

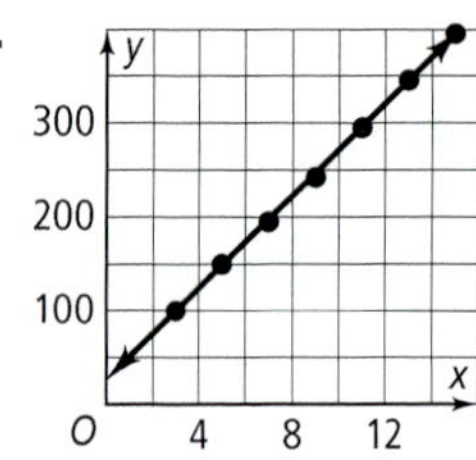

7. $y = x^2$

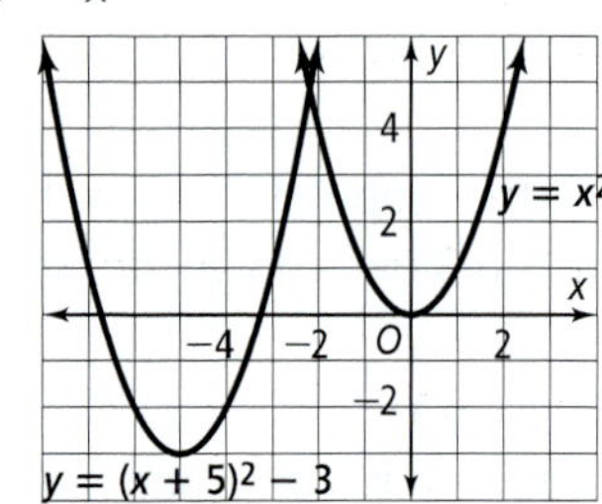

8. $y = x^3$

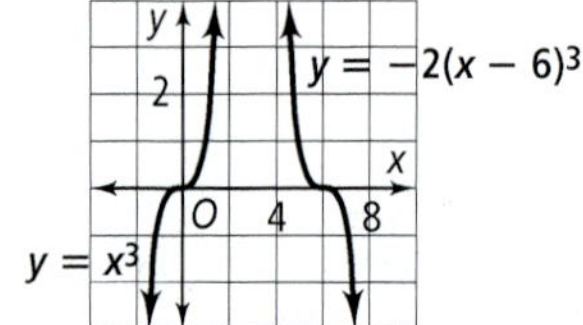

9. x^2 **10.** $16x^4$ **11.** $y = \pm\sqrt{\frac{10-x}{2}}$; no
12. $y = -4 + \sqrt[3]{x+1}$; yes **13.** decrease
14. exponential **15.** no

Lesson 7-1 pp. 434–441

Got It?

1. a.

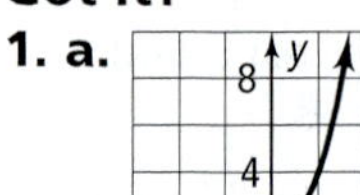

b.

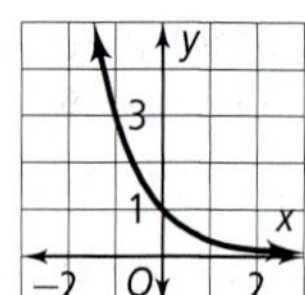

c.

d. domain: all real numbers, range: $y > 0$; y-intercept: $(0, a)$ where $y = ab^x$

2. a. exponential growth; 3 **b.** exponential decay; 11
c. exponential growth; 2000 **3.** $593.84 **4. a.** after 8 yrs
b. after 11 yrs; then the account contains $1710.34.
5. a. ≈3 **b.** No; the function is asymptotic to the x-axis.
Lesson Check **1.** decay; 10 **2.** growth; 0.75
3. growth; 1 **4.** decay; 1

5.

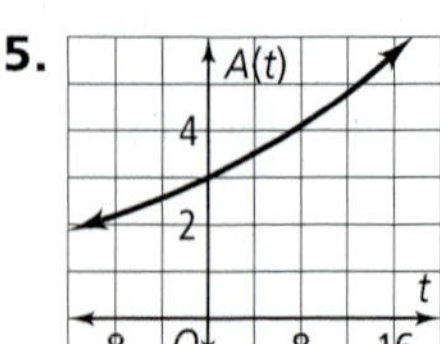

6.

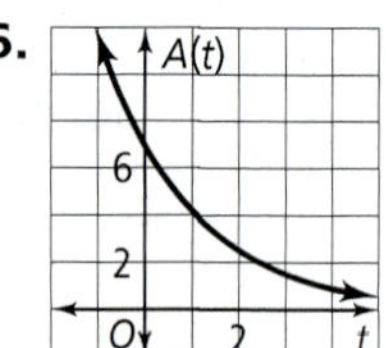

7. If $a > 0$ and $b > 1$, then the function represents exponential growth; if $a > 0$ and $0 < b < 1$, then the function represents exponential decay. **8. a.** quadratic; degree 2 with $3x^2$ as the leading term **b.** exponential; the equation is of the form $y = ab^x$ **c.** linear; degree 1 with x as the leading term **d.** exponential; the equation is of the form $y = ab^x$ **9.** $0.3 < 1$, so 0.3 is the decay factor

Exercises

11. 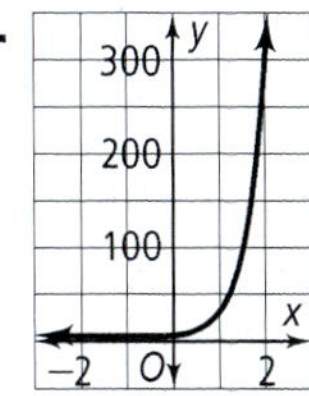

13. 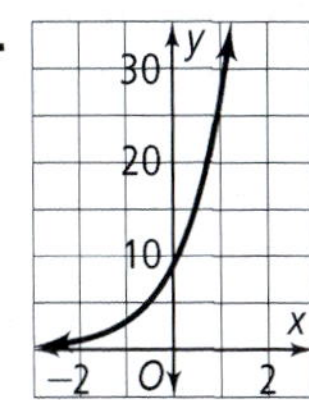

15. 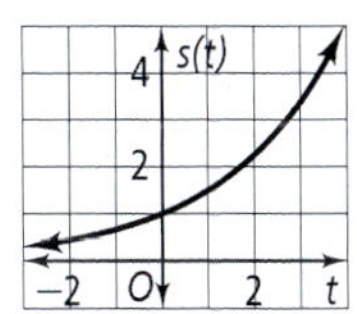

17. 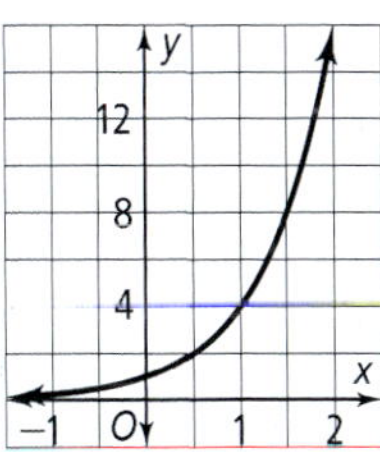

19. exponential decay; 2 **21.** exponential decay; 0.8 **23.** exponential growth; 0.45 **25.** exponential decay; 1 **27.** $y = 120{,}000(1.012)^x$; 143,512 **29. a.** $y = 72(\frac{1}{2})^x$ **b.** 2.25 in. **31. a.** about 5.6% **b.** about 0.0017% **33.** Answers may vary. Sample: $y = 59.5(0.6)^x$ **35.** 6 **37.** 0.45 **39.** 0.999 **41.** 2 **43.** C **45.** B; The graph shows a decreasing function, which eliminates A. The y-values are all positive, which eliminates C. **47.** G **49.** G

51. 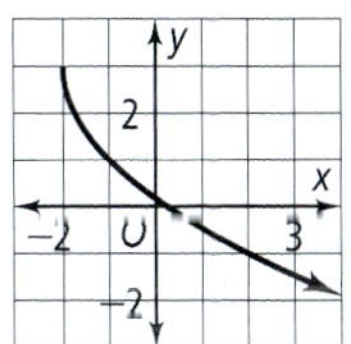

52. 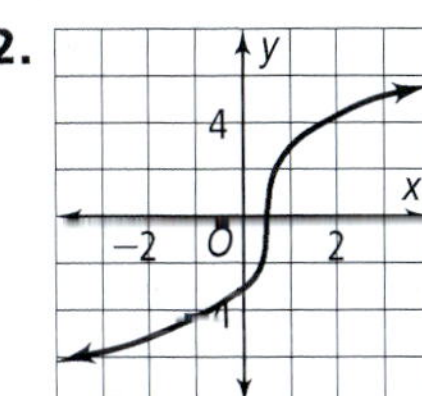

53. 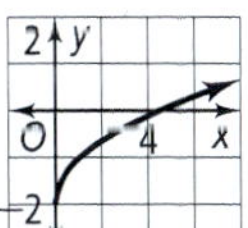

54. $(2 + 3x)(4 - 6x + 9x^2)$ **55.** $(3x - 1)(x + 4)$ **56.** $(4x - 5)(4x - 5)$ **57.** $(1, -1)$ **58.** $(0, 0)$ **59.** $(3, -3, 9)$

60. 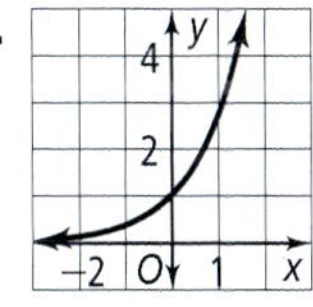

61. 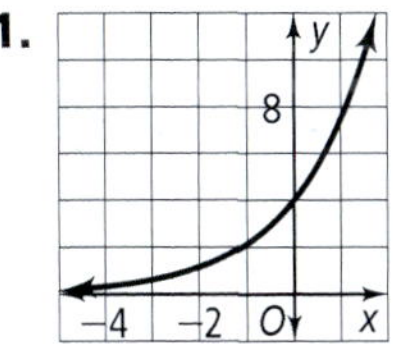

62. 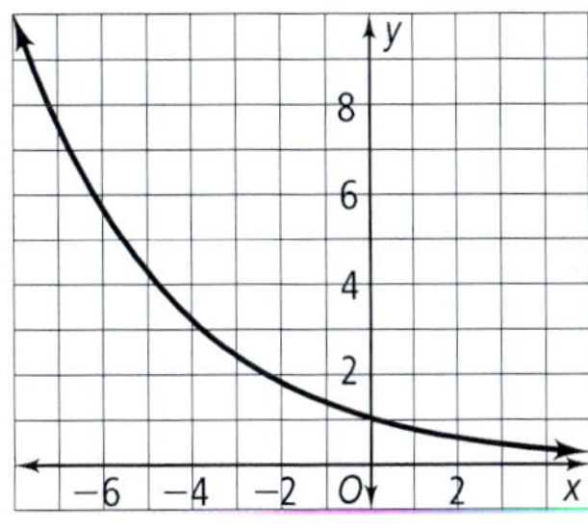

63. 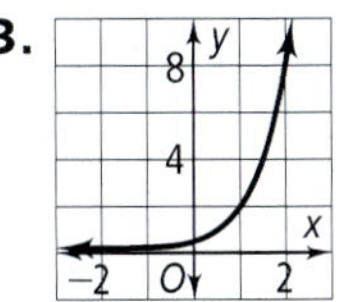

Lesson 7-2 pp. 442–450

Got It?

1. reflects across the x-axis; compresses by a factor of 0.5

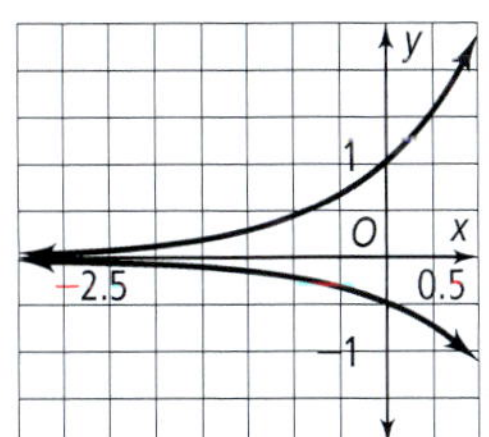

2. a. translate 2 units to the left; the y-intercept becomes 16 **b.** Stretch the graph of $y = (0.25)^x$ by a factor of 5 and translate the graph of $y = 5 \cdot 0.25^x$ 5 units up

3. a. about 31.9 min **b.** No; a hot coffee cannot cool below room temperature. So, to use exponential data, it is important to translate the data by 68 units.

4. $e^8 \approx 2980.957987$; three methods: use the e^x key, $x = 8$; graph $y = e^x$ and find y for $x = 8$; or use the table of values for $y = e^x$ and find y for $x = 8$

5. about $4475

Lesson Check 1. stretch by a factor of 2 and reflection across the x-axis **2.** compress by a factor of $\frac{1}{2}$ **3.** translate 5 units to the right **4.** translate 3 units up **5.** yes **6.** no; $2000e^{0.05t} \neq 1000(e^{0.04t} + e^{0.06t})$

Exercises

7. 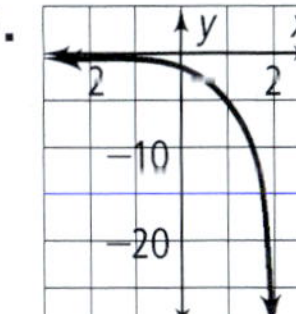

9. 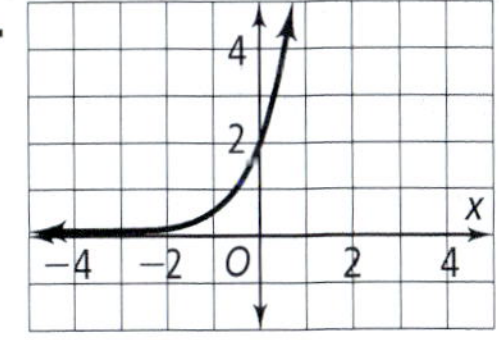

11. 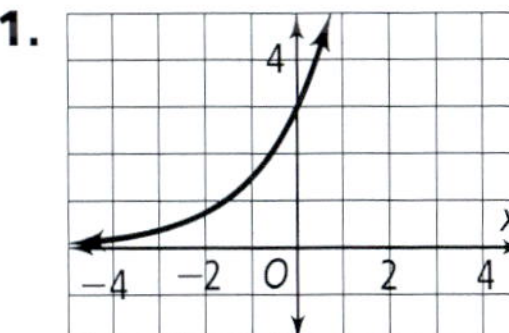

13. 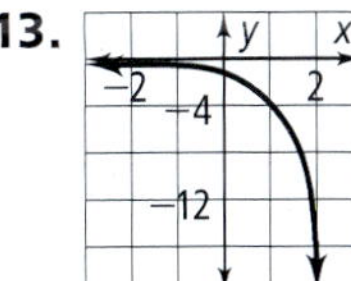

15.

17. 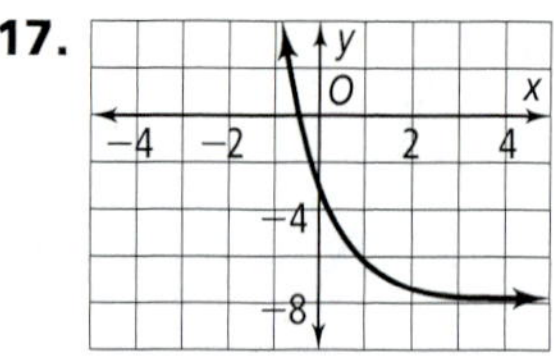

19. 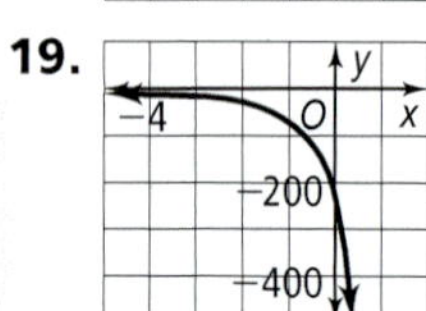

21. 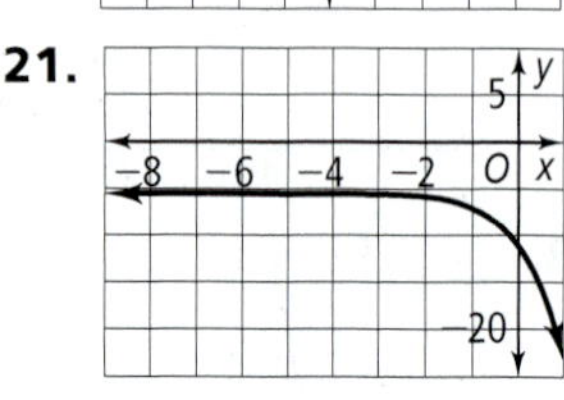

23. 403.4288 **25.** 1 **27.** 15.1543 **29.** \$448.30
31. \$6168.41 **33.** graph is a shift of the parent function 2 units to the left and 1 unit up

35. 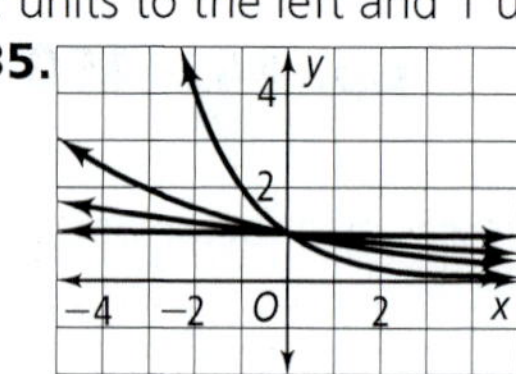

As the value of b approaches 1, the graph comes closer to being a straight line.

37. $y = 24\left(\frac{1}{2}\right)^{\frac{1}{5730}x}$; 0.64 mg **39.** $y = -3^x$; $y = -3^{x-8} + 2$ **41.** $y = -3\left(\frac{1}{3}\right)^x$; $y = -3\left(\frac{1}{3}\right)^{x+15} - 1$ **43. a.** about 8 names; about 20 names **b.** Graphically, it will never happen; the graph has $y = 25$ as an asymptote. (In reality, you would be close to knowing all the names in about 21 days.) **c.** Answers may vary. Sample: My learning rate might be higher since I can learn names quickly. **45.** G **47.** H

49.
$$A(t) = Pe^{rt}$$
$$\$8000 = Pe^{(0.06)(4)}$$
$$P = \frac{\$8000}{e^{(0.06)(4)}}$$
$$P = \$6293.02$$

50. exponential growth; 23 **51.** exponential growth; 3
52. exponential decay; 2 **53.** exponential growth; 5
54. $6\sqrt{5}$ **55.** $-\sqrt[3]{4}$ **56.** $5(\sqrt{3} + \sqrt{5})$
57. $2(\sqrt[4]{2} + \sqrt[4]{8})$ **58.** $\sqrt{3}$ **59.** $11\sqrt{7}$
60. $f^{-1}(x) = \frac{x+1}{4}$; yes **61.** $f^{-1}(x) = x^{\frac{1}{7}}$; yes
62. $f^{-1}(x) = \left(\frac{x-1}{5}\right)^{\frac{1}{3}}$; yes

Lesson 7-3 pp. 451–458

Got It? 1. a. $\log_6 36 = 2$ **b.** $\log_{\frac{2}{3}} \frac{8}{27} = 3$
c. $\log_3 1 = 0$ **2. a.** 3 **b.** $\frac{5}{2}$ **c.** $-\frac{5}{6}$ **3.** $\approx$16 times
4. a. domain: $x > 0$; range: all real numbers; no y-intercept; vertical asymptote: $x = 0$

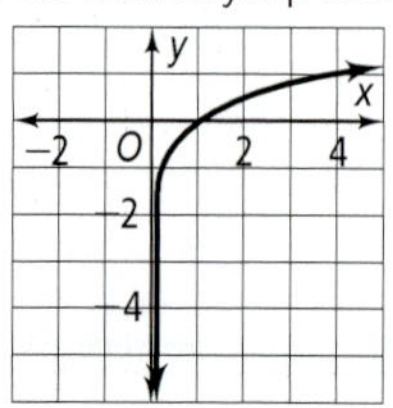

b.

x	$2^y = x$	y
−1	$2^y = -1$	undefined
0	$2^y = 0$	undefined
1	$2^y = 1$	0
2	$2^y = 2$	1

5. a. translates the graph of the parent function 3 units to the right and 4 units up; The asymptote changes from $x = 0$ to $x = 3$. The domain changes from $x > 0$ to $x > 3$. The range remains all real numbers. **b.** stretch the graph of the parent function by a factor of 5; The asymptote, domain, and the range remain the same.
Lesson Check 1. $\log_5 25 = 2$ **2.** $\log_4 64 = 3$ **3.** $\log_3 243 = 5$ **4.** $\log_5 25 = 2$ **5.** 3 **6.** 1 **7.** 2 **8.** −2 **9. a.** no **b.** yes **c.** yes **d.** no **10.** Choose a few points on the graph of $y = 6^x$, reverse their coordinates, and plot them. **11.** $y = \log_2(x + 4)$ translates the graph of $y = \log_2 x$ 4 units to the left. Asymptote changes from $x = 0$ to $x = -4$. Domain changes from $x > 0$ to $x > -4$. Range remains the same.
Exercises 13. $\log 1000 = 3$ **15.** $\log \frac{1}{10} = -1$ **17.** $\log_{\frac{1}{2}} 4 = -2$ **19.** $\log 0.01 = -2$ **21.** $\frac{1}{2}$ **23.** $\frac{3}{2}$ **25.** $\frac{1}{2}$ **27.** 2 **29.** 1 **31.** 3 **33.** The earthquake in Chile was about 39.81 times more intense. **35.** The earthquake in Missouri was about 10 times more intense.

36–39. 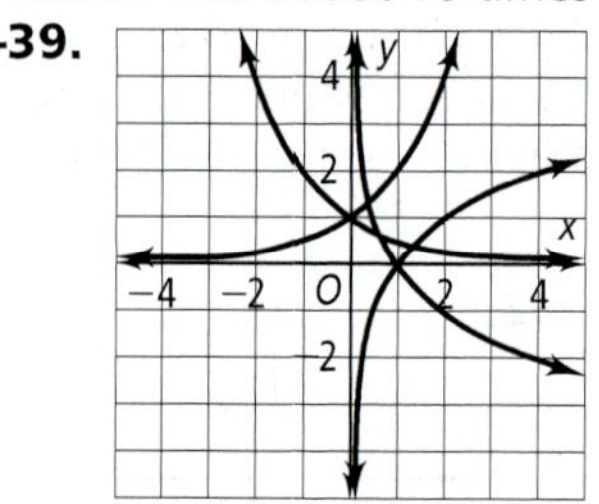

41. translate the graph 2 units to the right **43.** translate the graph 2 units to the left and 1 unit down
45. $\approx 3.16 \times 10^{-9}$ **47.** $10^{-4} = 0.0001$ **49.** $4^0 = 1$ **51.** $2^{-1} = \frac{1}{2}$ **53.** $10^1 = 10$ **55.** −2 **57.** 7
59. $(2x + 1)^5 = (a + b)$ **61.** $y = 4^x$ **63.** $y = 10^x$
65. $y = 10^x - 1$ **67.** $y = 2^{x-2}$

69. 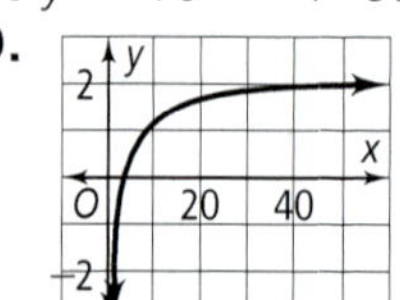

71. 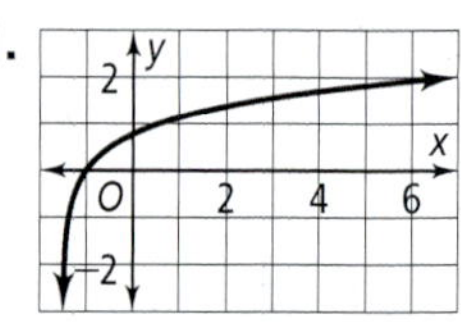

73. domain $x > 0$, range: all real numbers **75.** domain $x > 3$, range: all real numbers **77.** $4 = \log_3(81)$ **79.** $8 = \log_6(a + 1)$ **81.** 3 **83.** −2 **85.** D **87.** C

89.

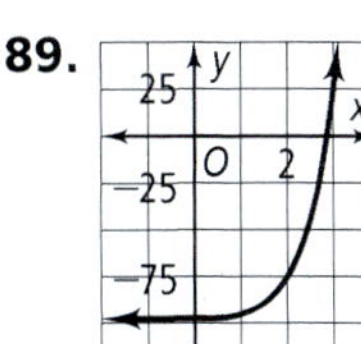

90.

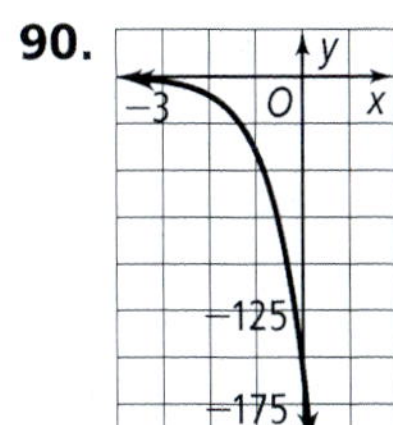

91.

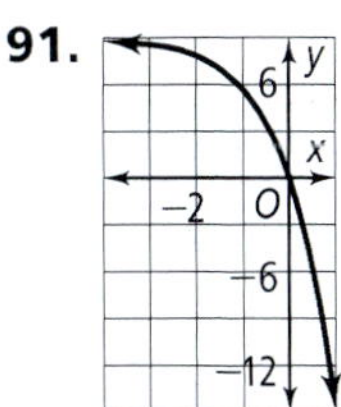

92. $(2x - 3)(2x - 1)$ **93.** $4(b - 5)(b + 5)$ **94.** $(5x - 2)(x + 3)$ **95.** 2 **96.** 256 **97.** $\frac{1}{4}$ **98.** 12

Lesson 7-4 pp. 462–468

Got It? 1. a. $\log_4 15x^2$ **b.** 1 **2. a.** $\log_3 2 + 3\log_3 5 - \log_3 37$ **b.** $2 + 5\log_3 x$ **3. a.** $\frac{5}{3}$ **b.** ≈ 2.085 **4.** Substance B; log 2; $-\log[H^+_B] + \log[H^+_B] = \log 2$

Lesson Check 1. $\log_4 16$ **2.** $\log_6 6$ **3.** $\log_3 x - \log_3 y$ **4.** $2\log m + 5\log n$ **5.** $\frac{1}{2}\log_2 x$ **6. a.** Product Prop. and Power Prop. **b.** Quotient Prop. **7.** 0.00001 **8.** Answers may vary. Samples: $\log 150 = \log 25 + \log 6$

Exercises 9. log 14 **11.** log 972 **13.** $\log \frac{m^4}{n}$ **15.** $\log_6 5x$ **17.** $\log_3 32xy$ **19.** $2 + \log_7 x + \log_7 y + \log_7 z$ **21.** $2\log a$ **23.** $2\log_3 2 + 2\log_3 x$ **25.** $2\log a + 3\log b - 4\log c$ **27.** $1 + \frac{1}{2}\log_8 3 + \frac{5}{2}\log_8 a$ **29.** $1 + 4\log m - 2\log n$ **31.** ≈ 1.2 **33.** ≈ 1.43 **35.** ≈ 3.631 **37.** ≈ 3.183 **39.** −2 **41.** 1 **43.** 2 **45.** Yes, because the loudness of the sound is 102 dB. **47.** The coefficient $\frac{1}{2}$ is missing in $\log_4 s$;

$$\log_4 \sqrt{\frac{t}{s}} = \frac{1}{2}\log_4 \frac{t}{s}$$
$$= \frac{1}{2}(\log_4 t - \log_4 s)$$
$$= \frac{1}{2}\log_4 t - \frac{1}{2}\log_4 s$$

49. The log of a product is equal to the sum of the logs. $\log(MN) = \log M + \log N$. **51.** false; $\frac{1}{2}\log_3 3 = \log_3 3^{\frac{1}{2}}$, not $\log_3 \frac{3}{2}$ **53.** false; $\log_b \frac{x}{y} = \log_b x - \log_b y$ **55.** false; $\log_4 7 - \log_4 3 = \log_4 \frac{7}{3}$ not $\log_4 4$. **57.** $\log_x \frac{2\sqrt{y}}{z^3}$ **59.** $\log_b \frac{\sqrt[3]{x^2}\sqrt[4]{y^3}}{z^5}$ **61.** $\log s + \frac{1}{2}\log 7 - 2\log t$ **63.** $3\log m - 4\log n + 2\log p$ **65.** $\frac{1}{2}\log_b x + \frac{2}{3}\log_b y - \frac{2}{5}\log_b z$ **67.** $\frac{1}{2}\log(x + 2) + \frac{1}{2}\log(x - 2) - 2\log(x + 3)$ **69.** $\frac{\log 8}{\log 3}$ **71.** $\frac{\log 3.3}{\log 9}$ **73.** A 1.0 magnitude star is about 2.5 times brighter than a 2.0 magnitude star.

75. $\frac{1}{2}\log x + \frac{1}{4}\log 2 - \log y$

77. $\frac{1}{2}\log_7(r + 9) - 2\log_7 s - \frac{1}{3}\log_7 t$ **79.** 0 **81.** 1

83. $\log 18 = \log \frac{36}{2} = \log 36 - \log 2$;
Quotient Prop.
$= \log 2 \cdot 9 = \log 2 + \log 9$;
Product Prop.
$= \log 324^{\frac{1}{2}} = \frac{1}{2}\log 324$;
Power Prop.
$= \log 2 \cdot 3^2 = \log 2 + 2\log 3$;
Product and Power Prop.

84. $\log_7 49 = 2$ **85.** $\log_8 \frac{1}{4} = -\frac{2}{3}$ **86.** $-3 = \log_5 \frac{1}{125}$ **87.** ± 8 **88.** $\frac{64}{7}$ **89.** 2 **90.** $x^3 + 5x^2 - 3x - 15$ **91.** $x^4 + 17x^2 + 16$ **92.** $x^4 - 2x^3 - 2x^2 + 14x - 35$ **93.** 2 **94.** 3 **95.** $\frac{1}{3}$

Lesson 7-5 pp. 469–476

Got It? 1. $\frac{4}{9}$ **2. a.** ≈ 1.5122 **b.** because the terms cannot be written with a common base **3. a.** ≈ 0.8588 **b.** ≈ 1.2114 **4.** ≈ 13.51 yrs **5.** 1.45 **6.** 200

Lesson Check 1. 2 **2.** ≈ 3.6439 **3.** 25 **4.** 2000 **5.** The log bases are not equal.

$\log_2 x = 2\log_3 9$
$\log_2 x = \log_3 9^2$
$\log_2 x = 4$
$x = 2^4$
$x = 16$

6. Yes; $5^x = 0$ has no solution.

Exercises 7. 3 **9.** 1 **11.** $\frac{4}{5}$ **13.** 2 **15.** 1.5850 **17.** 3 **19.** 0.9534 **21.** 0.2720 **23.** 0.5690 **25.** 4.7027 **27.** 6 **29.** 0.64 **31.** about the yr 2012 **33.** $\frac{\sqrt{10}}{10}$ or about 0.3162 **35.** 10,000 **37.** $\sqrt{10}$ or ≈ 3.1623 **39.** 2 **41.** $100{,}000\sqrt{5}$ or $\approx 223{,}606.8$ **43.** $\frac{1}{4}$ **45.** 7 **47. a.** 18.9658 **b.** 18.9658 **c.** Answers may vary. Sample: You don't have to use the Change of Base Formula with the base 10 method, but there are fewer steps with the base-2 method. **49.** ≈ 7.6 yrs **51.** 3 **53.** 3 **55.** −2 **57.** $-\frac{1}{2}$ **59.** Answers may vary. Sample: $\log x = 1.6$; $x \approx 39.81$ **61.** 143.6 **63. a.** top up: 10^{-5} W/m^2 top down: $10^{-2.5}$ W/m^2 **b.** 99.68% **65.** 625 **67.** 10 **69.** 1.5 **71.** 2.7944 **73.** 500 **75.** $114.\overline{3}$ **77.** $x = y = 2$ **79.** −2, 5 **81.** 1 **83 a.** bassoon, guitar, harp, violin, viola, cello **b.** bassoon, guitar, harp, cello, bass **c.** harp, violin **d.** harp **85.** 333 **87.** 4 **89.** $\log 2 + 3\log x - 2\log y$ **90.** $\log_3 x - \log_3 y$ **91.** $1 + \frac{1}{2}\log_3 x$ **92.** $x^2 - 3x - 1$ **93.** $3x^2 - 3$

94. $9x^2 - 1$ **95.** $1, \pm i$ **96.** $\pm 2, \pm 2i$
97. $\pm\sqrt{3}, \pm\sqrt{2}$ **98.** $\log_2 3$ **99.** $\log 3x^4$ **100.** $\log_7 \frac{32}{y^2}$

Lesson 7-6 pp. 478–483

Got It? 1. a. $\ln 175$ **b.** $\ln \frac{x}{4}$ **c.** $\ln 5x^3y^2$ **2. a.** e^2, or about 7.39 **b.** $\frac{-5 \pm e^2}{2}$, or about 0.8 or −4.13 **c.** $\frac{e^2}{6}$, or about 1.23 **3. a.** $\ln 2 + 2$, or about 4.48 **b.** $-\ln 10$, or about −2.3 **c.** $\frac{\ln 10}{3}$, or about 0.77 **4. a.** No; the maximum velocity of 5.4 km/s is less than the 7.7 km/s needed for a stable orbit. **b.** Yes; if, R could be changed so that $V > 7.7$.
Lesson Check 1. ln 81 **2.** ln 1.8 **3.** ln 12 **4.** −ln 4 **5.** ≈10.9 **6.** ≈14.4 **7.** ≈7.39 **8.** ≈−0.718
9. error in 3rd line: $4x = 5$
should be: $4x = e^5$
$x = \frac{e^5}{4}; x \approx 37.1$
10. No; ln 5 has base e and $\log_2 10$ has base 2.
Exercises 11. ln 125 **13.** ln 4 **15.** $\ln \frac{\sqrt[3]{xy}}{z^4}$ **17.** ln 40,960 **19.** ln 1 **21.** 0.135 **23.** ±11.588 **25.** ±2.241 **27.** 1488.979 **29.** ≈2.890 **31.** ≈1.242 **33.** ≈2.401 **35.** 0 **37.** ≈2.2 **39.** at least 25 s **41.** ≈11,552 yrs **43.** $\frac{1}{4}$ **45.** 83 **47.** 2 **49.** 10 **51.** $\frac{1}{2}$ **53.** ≈301 days **55.** never **57.** 10.8 **59.** ≈19.8 h **61.** 78.342
63. Because the function is simplified in the beginning and the sq. root of the exponential function is not calculated.
65. a. ≈43 min
b. $t = -\frac{1}{0.041}\ln\left(\frac{T - 72}{164}\right)$

time (min)
Temperature (°F)

Temperature (°F)	225	200	175	150	125	100	75
Minutes Later	1.7	6.0	11.3	18.1	27.6	43.1	97.6

67. 4 **69.** 0.2975 **71.** 3 **72.** 4 **73.** 2.846 **74.** 0.272 **75.** $3333.\overline{3}$ **76.** 1.002 **77.** 9.0×10^{-5}
78. $y = \frac{x-7}{5}$; yes **79.** $y = \sqrt[3]{\frac{x-10}{2}}$; yes **80.** $y = \pm\sqrt{5-x}$; no **81.** $y = \frac{x-2}{3}$; yes **82.** 10 **83.** 15 **84.** $\frac{6}{5}$

Chapter Review pp. 487–490

1. exponential decay; exponential growth **2.** asymptote **3.** logarithm; natural logarithm function **4.** continuously compounded interest **5.** natural logarithmic function **6.** exponential growth; (0, 1) **7.** exponential growth; (0, 2) **8.** exponential growth; (0, 0.2) **9.** exponential decay; (0, 3) **10.** exponential growth; $\left(0, \frac{25}{7}\right)$ **11.** exponential growth; (0, 0.0015) **12.** exponential decay; (0, 2.25) **13.** exponential decay; (0, 0.5) **14.** $y = 12{,}500(0.91)^x$; \$7800 **15.** $y = 50(1.03)^x$; \$58 **16.** The parent graph $y = 2^x$ is stretched by a factor of 5, translated 1 unit to the left, and 3 units up. **17.** The parent graph $y = \left(\frac{1}{3}\right)^x$ is reflected across the x-axis, stretched by a factor of 2, and translated 2 units to the right. **18.** \$1100.76 **19.** \$291.91 **20.** 0.0498 **21.** 0.3679 **22.** 148.4132 **23.** 0.6065 **24.** $2 = \log_6 36$ **25.** $-3 = \log_2 0.125$ **26.** $3 = \log_3 27$ **27.** $-3 = \log 0.001$ **28.** 6 **29.** −2 **30.** −5 **31.** 0

32.
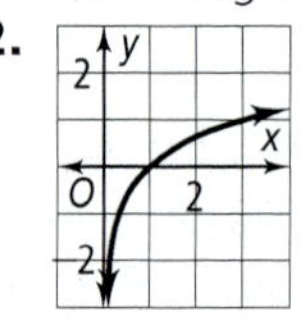

33.
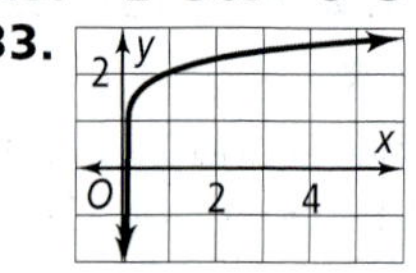

34.
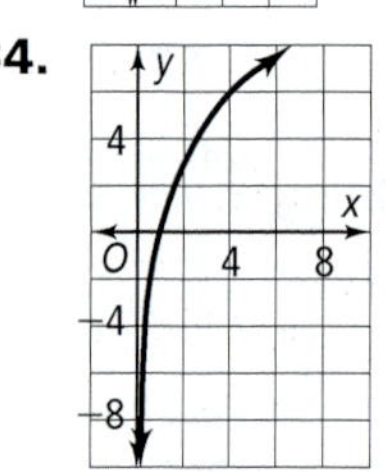

35.
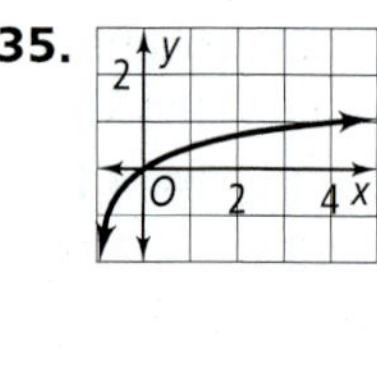

36. The parent graph $y = \log_4 x$ is stretched by a factor of 3 and translated 1 unit to the left.
37. The parent graph $y = \ln x$ is reflected across the x-axis and translated 2 units up. **38.** log 24; Product Prop. **39.** $\log_2 \frac{5}{3}$; Quotient Prop. **40.** $\log_3 7x^4$; Power and Product Prop. **41.** $\log \frac{x}{y}$; Quotient Prop. **42.** $\log \frac{5}{x^2}$; Power and Quotient Prop. **43.** $\log_4 x^5$; Power and Product Prop. **44.** $2\log_4 x + 3\log_4 y$; Product and Power Prop. **45.** $\log 4 + 4\log s + \log t$; Product and Power Prop. **46.** $\log_3 2 - \log_3 x$; Quotient Prop.
47. $2\log(x + 3)$; Power Prop.
48. $3\log_2 2 + 3\log_2 (y - 2)$; Power and Product Prop. **49.** $2\log z - \log 5$; Power and Quotient Prop.
50. ≈2.8 **51.** ≈2.1 **52.** 0.75 **53.** 3.2619 **54.** 4.6542 **55.** 1.3652 **56.** 3.3333 **57.** 8 **58.** 50 **59.** 7.6256×10^{12} **60.** 0.9307 **61.** 0.6599 **62.** 0.6658 **63.** 3.0589 **64.** ≈18.2 h **65.** ≈0.83 **66.** ≈2.26 **67.** ≈4.31 **68.** ≈0.54 **69.** ≈3.77 **70.** ≈6.03 **71.** ≈3.4%

Chapter 8

Get Ready! p. 495 **1.** $\frac{4}{3}$; -4 **2.** $-\frac{2}{3}$; 2 **3.** $-\frac{10}{3}$; 10 **4.** $-\frac{16}{7}$; $\frac{48}{7}$ **5.** $(x + 3)(x - 2)$ **6.** $(4x + 5)(x + 3)$ **7.** $(3x - 5)(3x + 5)$ **8.** $(x - 6)^2$ **9.** $(3x + 4)(x + 2)$ **10.** $(x - 3)(x - 2)$ **11.** 1, -8 **12.** -6, -8 **13.** 4, 2 **14.** 0, $-\frac{2}{3}$ **15.** 8, $\frac{1}{2}$ **16.** 15, -2 **17.** Answers may vary. Sample: Inverse is used when one quantity increases as the other quantity decreases.
18. Answers may vary. Sample:

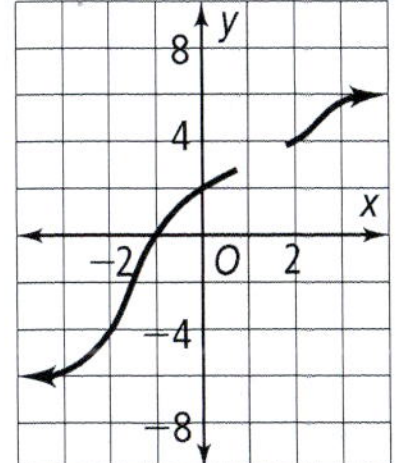

Lesson 8-1 pp. 498–505

Got It? **1. a.** direct; $y = 40x$ **b.** inverse; $y = \frac{8}{x}$ **c.** neither **2. a.** $y = -\frac{56}{x}$
b.

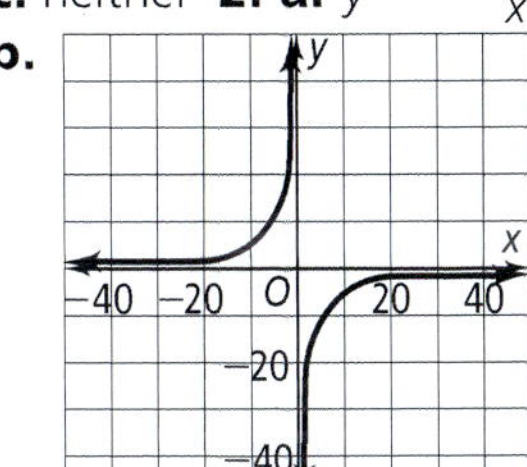

c. -28

3. a. $t = \frac{225}{n}$ **b.** 9 students **4.** 23 bags **5. a.** 4018 joules **b.** 12 m; No, you need not calculate PE to find the height. Substitute the mass and height of the first diver, and the mass of the second diver in $PE = mgh$ and set the two expressions equal. Solve the equation for h to calculate the height of the second diver.
Lesson Check **1.** inverse; $y = \frac{6}{x}$ **2.** direct; $y = 5x$ **3.** In direct variation, two positive quantities either increase together or decrease together. In an inverse variation, as one quantity increases, the other quantity decreases and vice versa. **4.** p varies directly with q, r, and t and inversely with s. **5.** d varies directly with the cube root of r and inversely with the square of t.
Exercises **7.** neither **9.** inverse; $y = \frac{0.3}{x}$

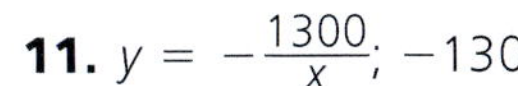
11. $y = -\frac{1300}{x}$; -130

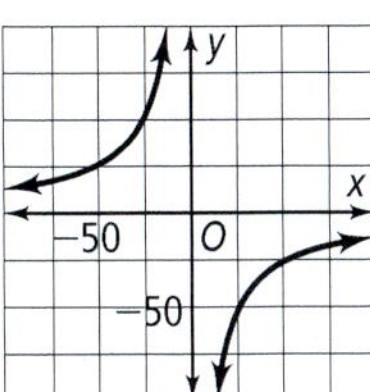

13. $y = \frac{5}{x}$; $\frac{1}{2}$

15. $y = \frac{250}{x}$; 25

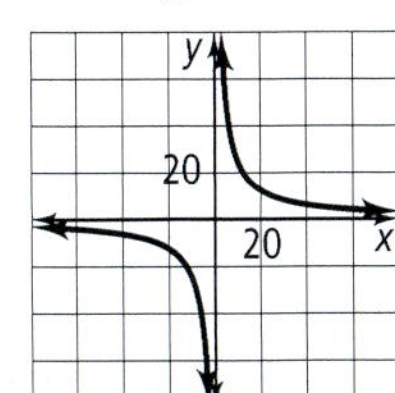

17. $y = -\frac{5}{3x}$; $-\frac{1}{6}$

19. a. $s = 2.5m$ **b.** 100 muffins **21. a.** $PE = 2gh$ **b.** 2 m **23.** $F = \frac{km}{d^2}$; $m = \frac{Fd^2}{k}$ **25. a.** $V = \frac{nRT}{P}$ **b.** ≈76.58 L **c.** ≈20 moles **27.** $z = 10xy$; 360 **29.** 10 **31.** $18\frac{2}{3}$ **33.** 18 **35.** $\frac{1}{4}$ **37.** 2.5 **39.** 2.625 **41.** Div. by zero is undefined. **43.** Answers may vary. Sample: Quadruple the volume and leave the radius constant, halve the radius and leave the volume constant, multiply the volume by 16 and double the radius, and multiply the volume and radius by $\frac{1}{4}$. **45.** G **47.** H **49.** $\frac{e^5}{4} \approx 37.1$ **50.** $3e^4 \approx 163.79$ **51.** $\frac{e^2}{8} \approx 0.92$ **52.** $-90x^2$ **53.** $84x^2$ **54.** $10x^2y^3\sqrt{2y}$
55. $y = |x|$ translated 2 units up

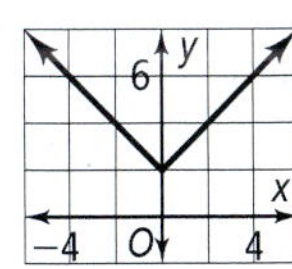

56. $y = |x|$ translated 2 units to the left

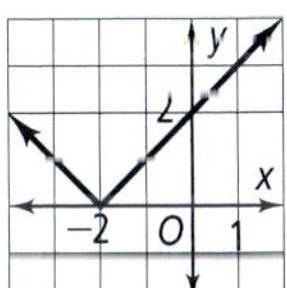

57. $y = |x|$ translated 3 units down

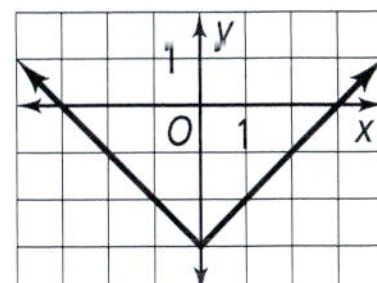

58. $y = |x|$ translated 3 units to the rt.

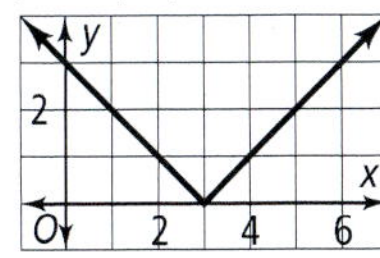

Selected Answers

59. $y = |x|$ translated 4 units to the left and 5 units down

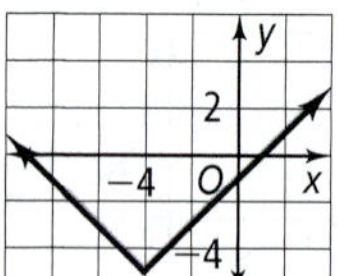

60. $y = |x|$ translated 10 units to the rt. and 7 units up

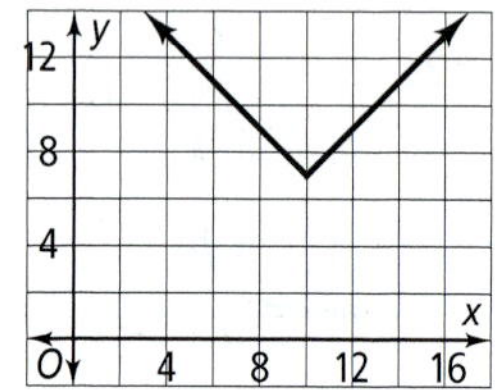

Lesson 8-2 pp. 507–514

Got It?

1. a.

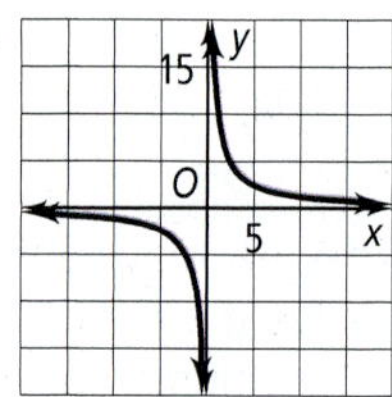

no x- or y-intercept; horizontal asymptote: $y = 0$; vertical asymptote: $x = 0$; domain: all real numbers except $x = 0$, range: all real numbers except $y = 0$

b. Yes; because all the functions have similar graphs. **2. a.** $y = \frac{1}{2x}$ is a shrink of the graph of $y = \frac{1}{x}$ by a factor of $\frac{1}{2}$. **b.** $y = \frac{2}{x}$ is a stretch of the graph of $y = \frac{1}{x}$ by a factor of 2. **c.** $y = -\frac{1}{2x}$ is a reflection across the x-axis and a shrink of the graph of $y = \frac{1}{x}$ by a factor of $\frac{1}{2}$.

3. domain: all real numbers except $x = 4$, range: all real numbers except $y = 6$

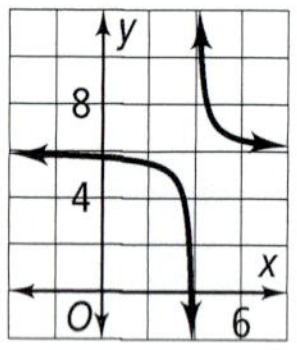

4. $y = \frac{2}{x - 1} - 4$ **5. a.** $C = \frac{1200}{n}$; domain: whole numbers from 1 to 312; 160 students **b.** $C = \frac{1200}{n - 30}$; Domain: whole numbers from 1 to 282; 190 students

Lesson Check

1.

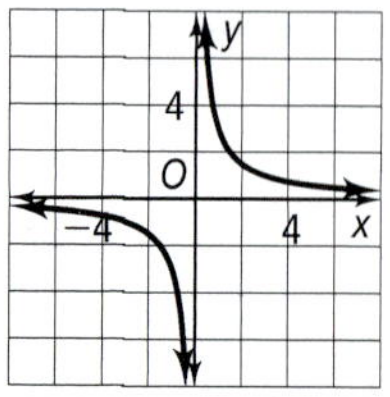

2. $y = \frac{1}{x}$ translated 5 units up **3.** $y = \frac{1}{x}$ reflected across the x-axis and stretched by a factor of 4 **4.** horizontal asymptote: $y = -7$, vertical asymptote: $x = -2$ **5.** shrink of the graph of $y = \frac{1}{x}$ by a factor of $\frac{1}{2}$

6. Answers may vary. Sample: $y = -\frac{2}{x}$ **7.** For $y = \frac{2}{x}$: stretch if $|a| > 1$ and shrink if $0 < |a| < 1$

Exercises

9. no x- or y-intercept; horizontal asymptote: $y = 0$, vertical asymptote: $x = 0$; domain: all real numbers except $x = 0$, range: all real numbers except $y = 0$

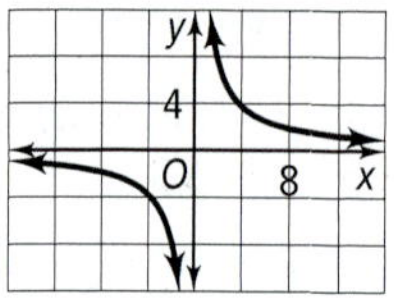

11.

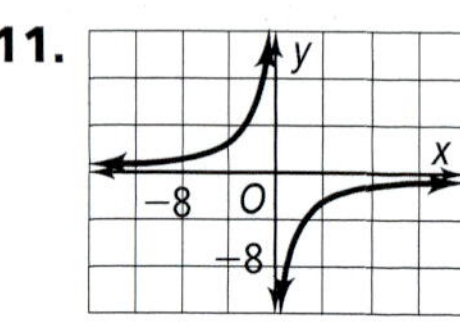

no x- or y-intercept; horizontal asymptote: $y = 0$, vertical asymptote: $x = 0$; domain: all real numbers except $x = 0$, range: all real numbers except $y = 0$

13.

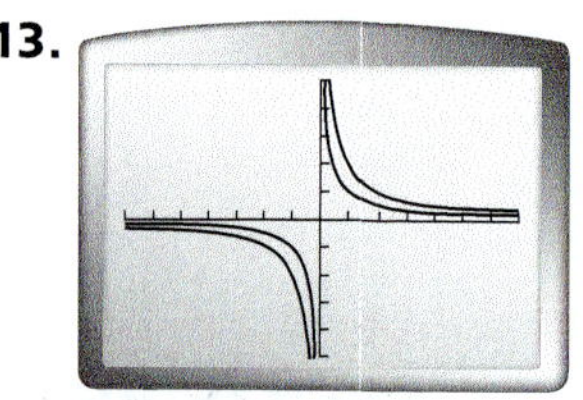

stretch by a factor of 2

15.

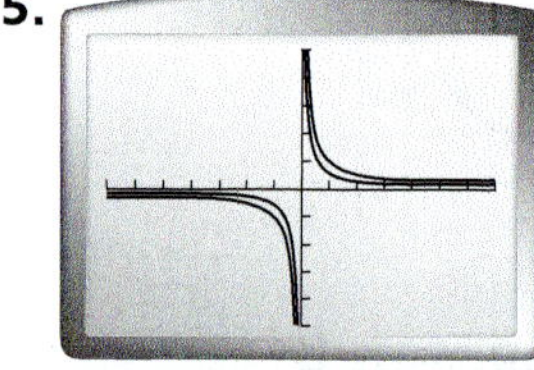

compression by a factor of 0.5

17.

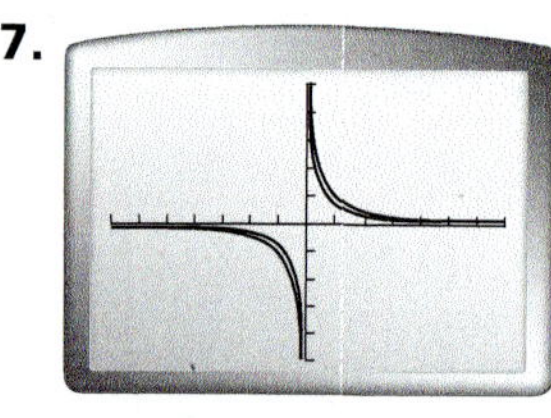

compression by a factor of 0.75

19. domain: all real numbers except $x = 0$, range: all real numbers except $y = -3$

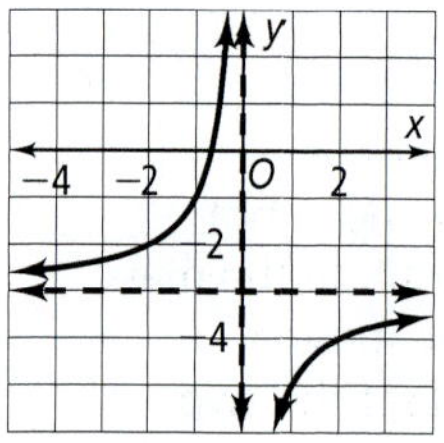

21.

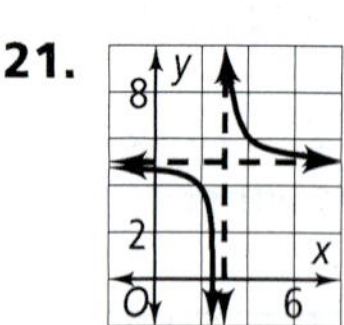

domain: all real numbers except $x = 3$, range: all real numbers except $y = 4$

23.

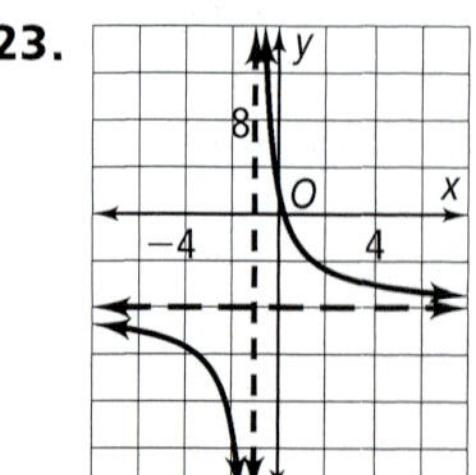

domain: all real numbers except $x = -1$, range: all real numbers except $y = -8$

25. 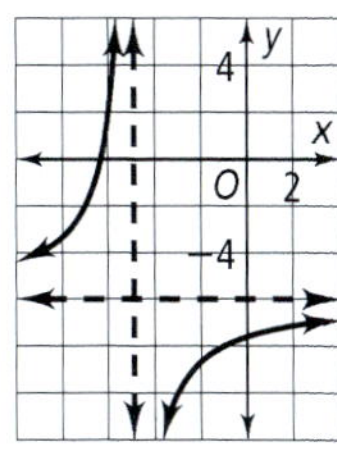
domain: all real numbers except $x = -5$, range: all real numbers except $y = -6$

27. $y = \frac{2}{x + 2} + 3$ **29.** 7.67 ft

33. 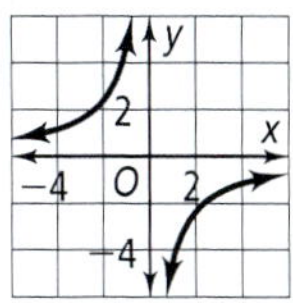

35. 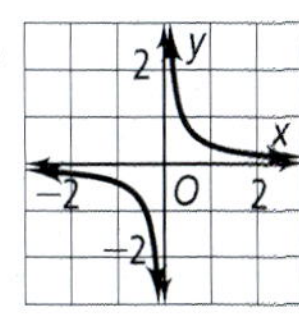

37. Answers may vary. Sample: The graph of the translation looks similar to the graph of $y = \frac{1}{x}$, so knowing the asymptotes helps to position the translation; check students' work.

39. 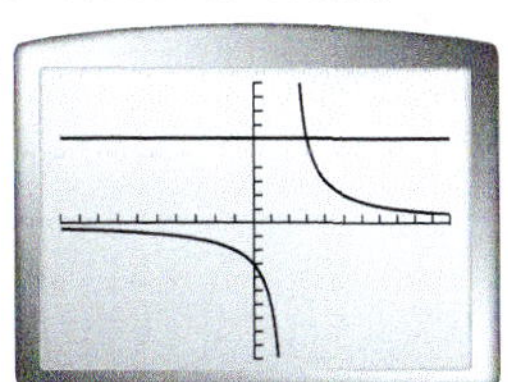 ; (3, 6)

41. ; (−1.75, −4)

43. a. $m = \frac{10{,}000}{g}$

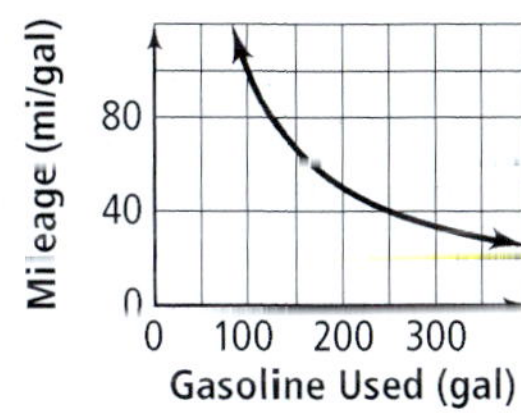

b. $m = \frac{10{,}000}{g - 50}$ **c.** 25 mi/gal; 28.57 mi/gal

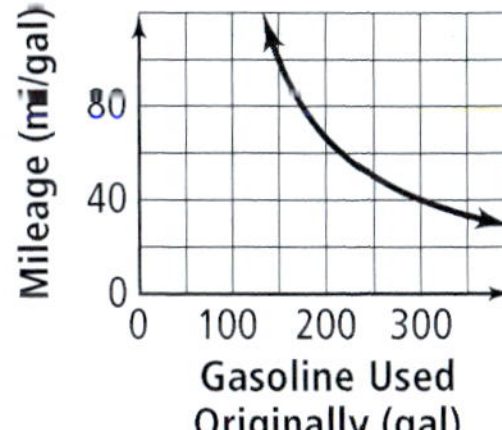

45. The branches of $y = \frac{1}{x^2}$ are in Quadrants I and II. The branches of $y = \frac{1}{x}$ are in Quadrants I and III. The graphs intersect at (1, 1). The graph of $y = \frac{1}{x^2}$ is closer to the x-axis for $x > 1$, and the graph of $y = \frac{1}{x}$ is closer to the y-axis for $0 < x < 1$. **47.** $y = \frac{16}{x}$, $y = -\frac{16}{x}$
49. A **51.** A
53. $81 = 27b^{-1}$

$81 = \frac{27}{b}$

$b = \frac{27}{81}$

$b = \frac{1}{3}$

54. $y = \frac{24}{x}$; $-\frac{24}{5}$ **55.** $y = \frac{50}{x}$; -10 **56.** $y = \frac{48}{x}$; $-\frac{48}{5}$
57. exponential growth; 3 **58.** exponential growth; 0.1
59. exponential decay; 5 **60.** exponential decay; 3
61. $79 - 20\sqrt{3}$ **62.** 6 **63.** −2 **64.** $(x - 4)(x - 2)$
65. $(x + 9)(x - 3)$ **66.** $(2x - 7)(x + 4)$
67. $(2x - 3)(x - 8)$

Lesson 8-3 pp. 515–523

Got It? 1. a. domain: all real numbers except $x = 4$ and $x = -4$; points of discontinuity: non-removable at $x = 4$ and $x = -4$; no x-intercept, y-intercept: $\left(0, -\frac{1}{16}\right)$
b. domain: all real numbers; no points of discontinuity; x-intercepts: (1, 0) and (−1, 0), y-intercept: $\left(0, -\frac{1}{3}\right)$
c. domain: all real numbers except $x = -2$ and $x = -1$; points of discontinuity: non-removable at $x = -2$, removable at $x = -1$; no x-intercept, y-intercept: $\left(0, \frac{1}{2}\right)$ **2. a.** $x = 1$ and $x = -3$ **b.** $x = -3$ **c.** no vertical asymptotes **3. a.** $y = -2$ **b.** $y = 0$ **c.** no horizontal asymptote

4. 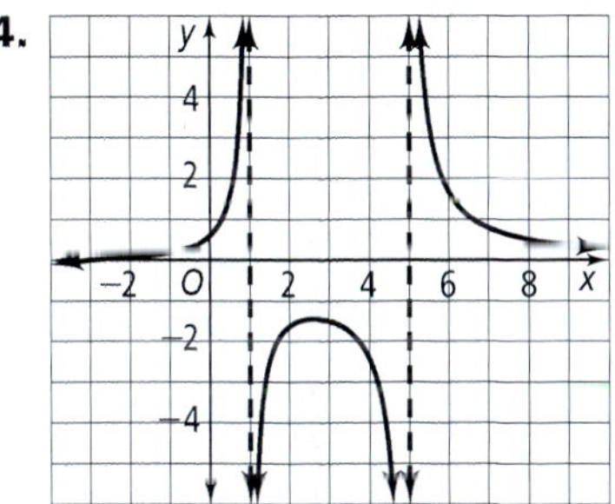

5. a. 4 gal **b.** No, because the graph changes when $y_1 = 0.8$ and intersects the graph of $y_2 = \frac{2 + (0.1)x}{2 + x}$ at $x \approx 0.6$. So, to have 80% orange juice, about 0.6 gal should be added.
Lesson Check 1. $x = -5$ and $x = -4$ **2.** $x = 9$ and $x = -2$ **3.** $x = -1$ **4.** $x = \frac{1}{3}$ and $x = 2$ **5.** $x = -5$

6. $x = -2$ and $x = -3$ **7.** $x = 1$
8. $x = 1$ and $x = -3$
9. 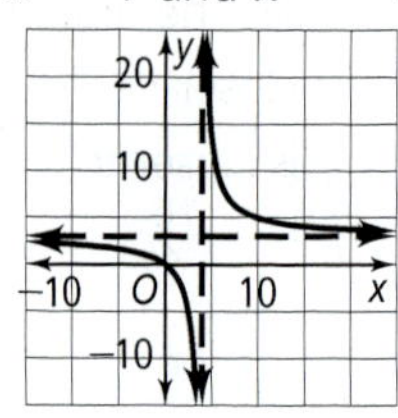 **10.** 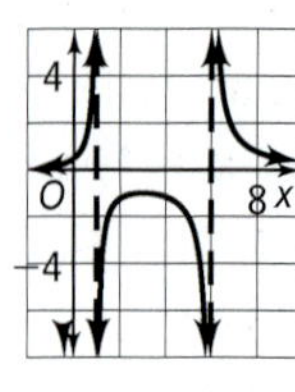

11. the function is undefined at $x = 1$ and $x = -3$
12. degree 2; function is discontinuous at 2 values of x
Exercises 13. domain: all real numbers except $x = 0$ and $x = 2$; points of discontinuity: non-removable at $x = 0$ and $x = 2$; no x- or y-intercept **15.** domain: all real numbers except $x = \pm 1$; pts. of discontinuity: non-removable at $x = -1$, removable at $x = 1$; no x-intercept; y-intercept: (0, 3) **17.** vertical asymptote at $x = -2$
19. vertical asymptotes at $x = -\frac{3}{2}$ and $x = 1$
21. hole at $x = -2$ **23.** $y = 0$ **25.** $y = 1$ **27.** $y = 0$
29. 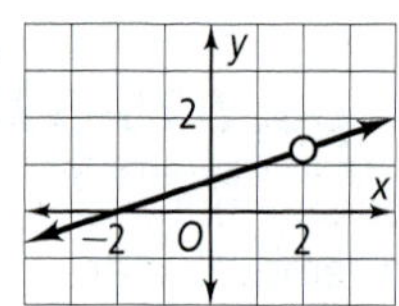 **31.** 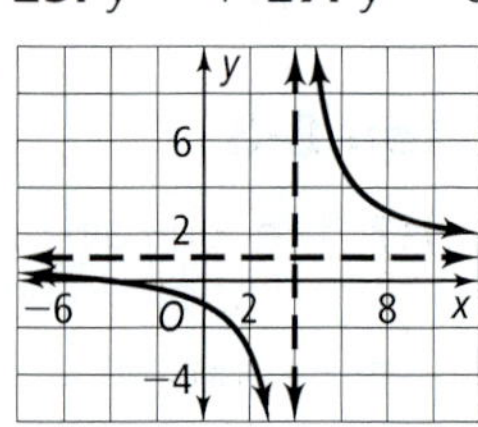

33. 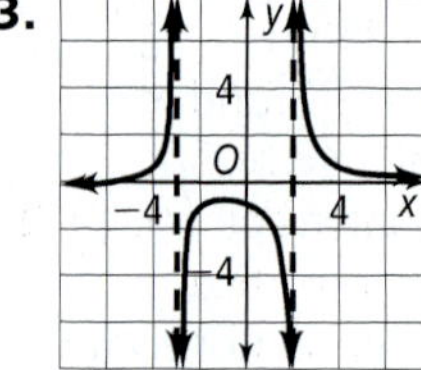

35. 900 ml **37.** vertical asymptote at $x = -2$
39. 6 free throws
41. correct answer: vertical asymptotes: $x = -5$ and $x = -1$, horizontal asymptote: $y = 1$
43. 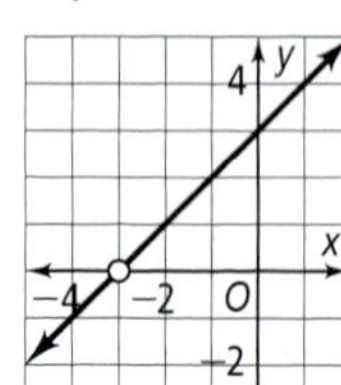 **45.** 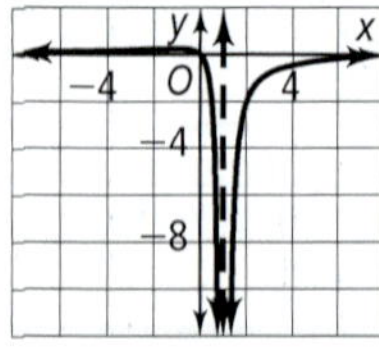

47. Answers may vary. Sample: There is no value of x for which the denominator equals 0.
49. Answers may vary. Samples: **a.** $y = \frac{x^2 - 7x + 12}{x^2 + 2x - 3}$
b. $y = \frac{x + 4}{x^2 - 3x}$ **c.** $y = \frac{3(x + 1)^2}{x^2 - 4}$ **51.** 8 **53.** $\frac{2}{3}$

55. 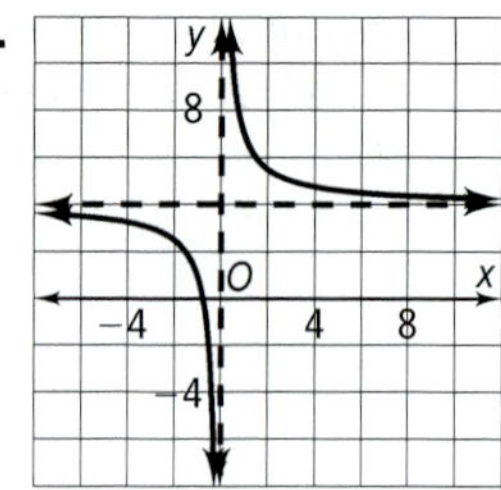
domain: all real numbers except $x = 0$, range: all real numbers except $y = 4$

56. 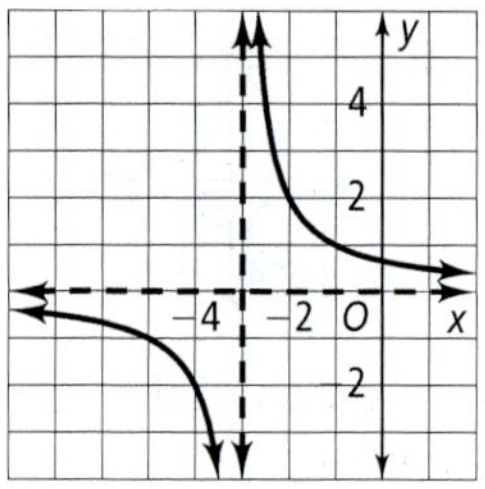 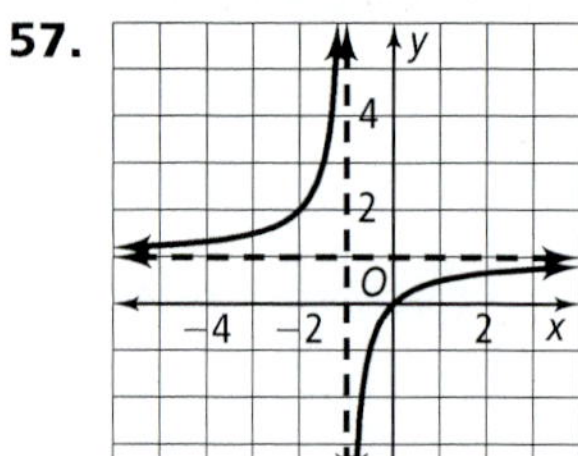
domain: all real numbers except $x = -3$, range: all real numbers except $y = 0$

57. 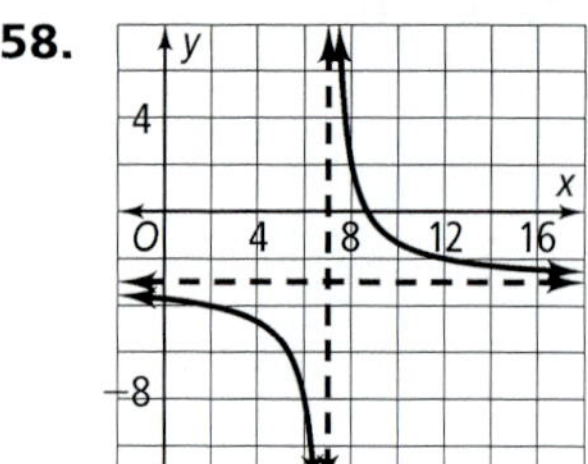
domain: all real numbers except $x = -1$, range: all real numbers except $y = 1$

58. 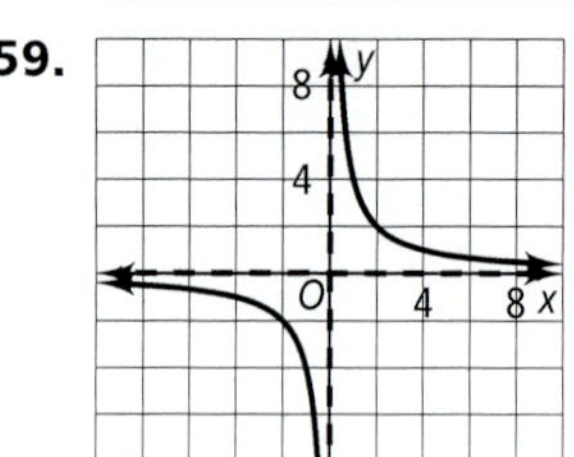
domain: all real numbers except $x = 7$, range: all real numbers except $y = -3$

59. 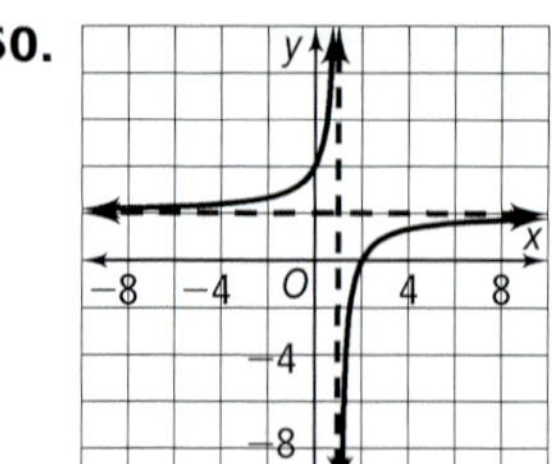
domain: all real numbers except $x = 0$, range: all real numbers except $y = 0$

60.
domain: all real numbers except $x = 1$, range: all real numbers except $y = 2$

61. $y = \frac{x+3}{2}$; yes **62.** $y = 6 - x$; yes **63.** $y = \pm\sqrt{\frac{x}{2}}$; no **64.** $y = \pm\sqrt{5x}$; no **65.** $y = \frac{1}{x} - 2$; yes
66. $y = (x-1)^2 + 2$; yes
67. $a < 10\frac{2}{3}$

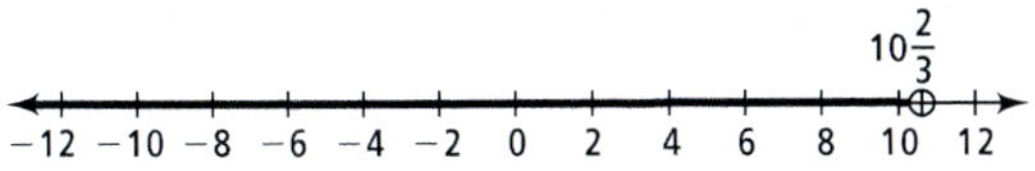

68. $x \geq 36$

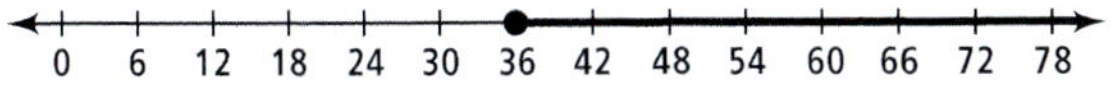

69. $x \geq 17\frac{4}{5}$

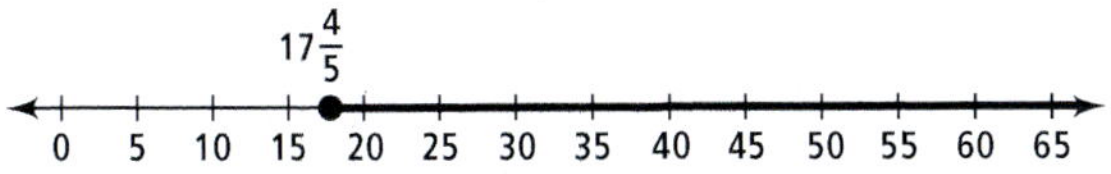

70. $y > 4$

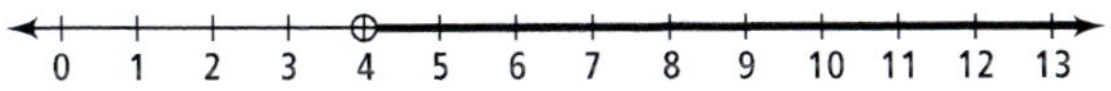

71. $x < 3$

72. $b < 5$

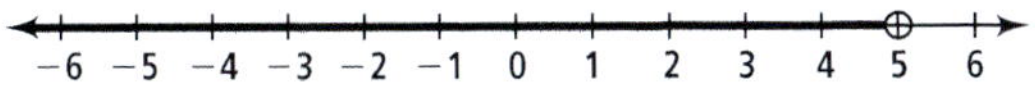

73. $(2x-1)(x-1)$ **74.** $(2x-3)(2x+3)$
75. $(5x+1)(x+1)$ **76.** $10(x-1)(x+1)$

Lesson 8-4 pp. 527–533

Got It? 1. a. $-\frac{4x}{y}$; $x \neq 0, y \neq 0$ **b.** $\frac{x+4}{x-3}$; $x \neq 2$ or 3 **c.** $-\frac{4}{x+3}$; $x \neq \pm 3$ **2.** $\frac{2(x+1)}{(x+4)^2}$; $x \neq \pm 4$ **3. a.** $\frac{2x}{x-1}$; $x \neq 1, -1, -4, 0$, or 3 **b.** 6 restrictions; 2 in each of the original denominators, and 2 in the denominator of the reciprocal of the second rational expression. **4.** a square

Lesson Check 1. $\frac{z-3}{2(z+3)}$; $z \neq -3$ **2.** $\frac{3}{x}$; $x \neq 0$ or 1 **3.** $\frac{3(x+5)}{x+3}$; $x \neq -3, -6$, or 2 **4.** $-\frac{x+6}{x+2}$; $x \neq -6, -2, 2, 3$, or 5 **5.** Yes; the numerator and denominator are polynomials with no common factor. **6.** No; $x = 2$ will make the denominator of $\frac{x}{x-2}$ 0, so $x = 2$ is not a solution. There is no solution to the eq.

7. Length $= \frac{2(a+8)}{a+5}$, $-10 < a < -5$, $a \neq -8$

Exercises 9. $\frac{1}{2x-1}$; $x \neq 0$ or $\frac{1}{2}$ **11.** $7 - z$; $z \neq -7$ **13.** $-\frac{x+4}{x-5}$; $x \neq 5$ or 3 **15.** $\frac{xy^5}{4}$; $x \neq 0, y \neq 0$ **17.** $-\frac{4(x+6)}{3(3x+8)}$; $x \neq 3$ or $-\frac{8}{3}$ **19.** 1; $x \neq -2, -1, 2$, or 3

21. $\frac{y}{2x^2}$; $x \neq 0, y \neq 0$ **23.** 1; $y \neq -2$ or 4 **25.** $\frac{4(y-3)}{y(y+5)}$; $y \neq 2, -5$, or 0 **27.** $\frac{x-8}{x-10}$; $x \neq -3$ or 10
29. $\frac{y(y+3)}{12(y+4)}$; $x \neq 0, y \neq -4$ or 3
31. $R_{cylinder} = \frac{V_{cylinder}}{SA_{cylinder}} = \frac{rh}{2(r+h)}$; $R_{cube} = \frac{V_{cube}}{SA_{cube}} = \frac{s}{6}$; if $r = h = s$, then $R_{cylinder} = \frac{r}{4}$ and $R_{cube} = \frac{r}{6}$. $R_{cylinder} > R_{cube}$. The cylindrical shaped box is more efficient. If $s = h = 2r$ (diameter), then $R_{cylindrical} = R_{cube}$. The boxes are equally efficient.
33. $\frac{18x}{(x+9)(x+3)}$; $x \neq -9, -3$, or 3 **35.** $\frac{x+1}{x-1}$; $x \neq -\frac{1}{2}, \frac{1}{2}, 1$, or -2 **37.** They are equally efficient.
39. never **41.** never **43.** 2; $x \neq -3$ or 1
45. a. $2x^n + 1$ **b.** 2 is a factor of $2x^n$, so $2x^n$ is even and $2x^n + 1$ is odd. **47.** $\frac{-3a^2b^2}{4}$; $a \neq 0, a \neq b, b \neq 0$
49. $\frac{(x+1)(x+5)}{(x-3)(x+4)}$; $x \neq 3, -3, -4, 1$, or -5 **51.** H
53.
$$-x \log 3 = \log \frac{1}{243}$$
$$-x = \frac{\log \frac{1}{243}}{\log 3}$$
$$-x = -5$$
$$x = 5$$
54. hole at $x = 3$ **55.** vertical asymptotes at $x = -\frac{2}{3}$ and $x = -1$ **56.** hole at $x = 4$, vertical asymptote at $x = -3$ **57.** 3 **58.** -5 **59.** $\frac{3}{2}$ **60.** $\frac{3}{4}$ **61.** 49 **62.** 168 **63.** 2 **64.** $\frac{17}{38}$ **65.** $\frac{19}{75}$ **66.** $\frac{11}{72}$ **67.** $\frac{137}{180}$

Lesson 8-5 pp. 534–541

Got It? 1. a. $2(x+2)(x-3)$ **b.** $(x-1)(x-2)^2(x+4)$ **2. a.** $\frac{x+2}{x}$, $x \neq 1$ or 0 **b.** $\frac{2(x-1)}{x^2-4}$; $x \neq \pm 2$ **c.** Yes, however the denominator could have to be factored more and there would be additional, incorrect limitations on x. **3. a.** $\frac{x-2}{x-1}$, $x \neq 1$ or 2 **b.** $\frac{x^2-x-4}{x^2+6x+5}$; $x \neq -5$ or -1 **4. a.** $\frac{x^2y}{x+y}$ **b.** $\frac{(x-1)^2}{2x}$; $x \neq 0, -2$, or ± 1 **5.** Option 1 still gives the better combined mpg since Option 3 gives 18.46 mpg.

Lesson Check 1. $\frac{2a-10}{3a-5}$; $a \neq \frac{5}{3}$ **2.** $\frac{6x-11}{x^2-4}$; $x \neq \pm 2$ **3.** $\frac{-11m}{3m+6}$; $m \neq -2$ **4.** $\frac{-4(2b-5)}{(b-4)(b+4)(b-2)}$; $b \neq 2$ or ± 4

5. error in finding a common denominator:

$$\frac{1+\frac{1}{x}}{\frac{3}{x}} = \frac{\frac{x+1}{x}}{\frac{3}{x}}$$

$$= \frac{x+1}{x} \cdot \frac{x}{3}$$

$$= \frac{x+1}{3}$$

6. Answers may vary. Sample: $\frac{x^2-1}{x^2-6x+5}, \frac{x^2+6x+5}{x^2-25}$

Exercises **7.** $9(x+2)(2x-1)$ **9.** $5(y+4)(y-4)$ **11.** $\frac{1}{x}$; $x \neq 0$ **13.** $\frac{-3}{x}$; $x \neq 0$ **15.** $\frac{xy+8y+4}{2xy^2}$; $x \neq 0, y \neq 0$ **17.** $\frac{y-6}{2(y+2)}$; $y \neq -2$ **19.** $\frac{-x+6}{(x-3)(x+3)}$; $x \neq \pm 3$ **21.** $\frac{-2x(x+3)}{(x-2)(x-1)(x+1)}$; $x \neq \pm 1$ or 2 **23.** $\frac{15}{28}$ **25.** $\frac{b}{9}$ **27.** $\frac{3x}{2+xy}$ **29.** $\frac{3}{x-6}$ **31.** $\frac{3x-8}{4x^2}$; $x \neq 0$ **33.** $\frac{7x-17}{(x-3)(x+3)}$; $x \neq \pm 3$ **35.** $\frac{x(3x^2+x-1)}{x^2-2}$; $x \neq \pm\sqrt{2}$ **37.** 3.84 in. **39.** Yes; when you add, subtract, multiply, or divide rational expressions you get another rational expression. The restriction is that you must divide by a nonzero rational expression. **41.** $\frac{3x+2y}{7x-5y}$ **43.** x **45. a.** $\frac{2}{3}$ **b.** $\frac{3}{5}$ **c.** $\frac{2}{3}$ **d.** $\frac{1}{3}$ **47. a.** $\approx$1.18 ohms **b.** 6 ohms, 6 ohms, 3 ohms **49.** G **51.** F **53.** $\frac{12x}{x+3}$; $x \neq 2$ or ± 3 **54.** $\frac{3(x+2)}{4(x-3)}$; $x \neq \pm 2$ or 3 **55.** $\frac{3(x+1)}{2(x+3)}$; $x \neq \pm 1$ or -3 **56.** $\log_3 yt^4$ **57.** $\log p^7q^2$ **58.** $\log_5 \frac{x}{\sqrt[5]{y}}$ **59.** 30 **60.** 82 **61.** $\frac{15}{4}$ **62.** 101 **63.** $-\frac{4}{5}$ **64.** 21 **65.** 18

Lesson 8-6 pp. 542–548

Got It? **1. a.** 1 **b.** 0 **2. a.** $\approx$ 4.47 m/h **b.** The direction of wind affects the speed (rate) of the bike. Since the speed is inversely related to time, change in speed will lead to change in time. Since there is no wind, the speed of the bike will remain same to and from the store, hence the time to and from the store will remain the same. **3.** $0.\overline{27}$

Lesson Check **1.** 5 **2.** -1 **3.** -2 **4.** 310 mi/h

5. LCD was not found. The correct answer is

$$\frac{35+9x}{7x} = \frac{28(7)}{7x}, x \neq 0$$

$$9x = 161$$

$$x = \frac{161}{9} = 17.\overline{8}$$

6. Answers may vary. Sample: $\frac{2}{x-3} + \frac{1}{x+3} = \frac{5x}{x^2-9}$

7. Answers may vary. Sample: (1) Substitute the solution into the original equation. (2) Check to see if the solution is in the domain of the graph of the original equation.

Exercises **9.** 10 **11.** 2 **13.** -1, 12 **15.** ≈ -1.45, ≈ 1.65 **17.** -3, -2 **19.** 1 **21.** 0.6 **23.** 1.5 **25.** 1.75 **27.** ± 2 **29.** ≈ 1.69, ≈ -0.44 **31.** $E = mc^2$ **33.** $c = \pm\sqrt{a^2 - b^2}$ **35.** $B = \pm\sqrt{\frac{2Vm}{r^2q}}$ **37.** $1\frac{5}{7}$ h **39.** 4 test scores **41. a.** \$2250 **b.** $\frac{15{,}000}{24+x}(3.60)$ **c.** $2250 - \frac{15{,}000}{24+x}(3.60)$ **d.** $\approx 32.\dot{7}$ mpg **43.** 3 **45.** no solution **47.** no solution **49.** no solution **51.** 1, $-\frac{2}{3}$ **53.** Answers may vary. **55.** Answers may vary. **57.** D **59.** B **61.** $\frac{-y-13}{4(y+1)}$ **62.** $\frac{5xy-12}{2y(y+2)}$ **63.** $\frac{x^2+3}{2(x-1)(x+3)}$ **64.** $x = -3$ **65.** $x = -1$ **66.** $x = -0.875$ **67.** $y = \frac{5-x}{2}$; yes **68.** $y = \pm\sqrt{x-1}$; no **69.** $y = \sqrt[3]{x+4}$; yes **70.** add 2; 9, 11, 13 **71.** subtract 2; -10, -12, -14 **72.** multiply by 5; 625, 3125, 15625 **73.** subtract 5; 30, 25, 20 **74.** multiply by 2; 128, 256, 512 **75.** subtract 4; -19, -23, -27

Chapter Review pp. 553–556

1. simplest form **2.** combined variation **3.** complex fraction **4.** point of discontinuity **5.** branch **6.** 12 **7.** $y = \frac{72}{x}$ **8.** $y = 6x$ **9.** $z = \frac{7}{4}xy$; 56 **10.** $z = \frac{4x}{y}$; 2

11. no x- or y-intercept; vert. asymptote: $x = 0$, horizontal asymptote: $y = 0$

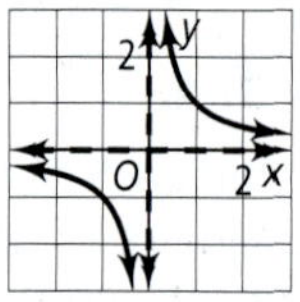

12.

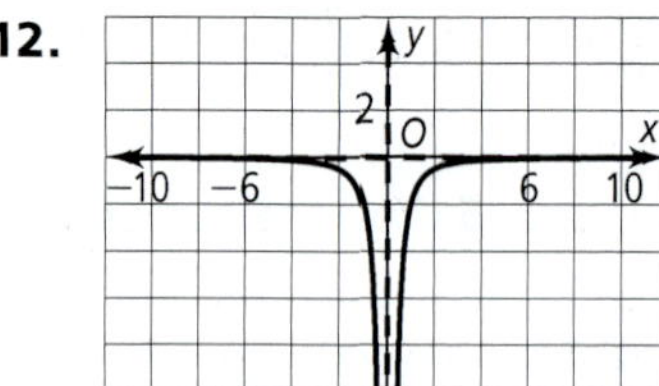

no x- or y-intercept; vert. asymptote: $x = 0$, horizontal asymptote: $y = 0$

13.

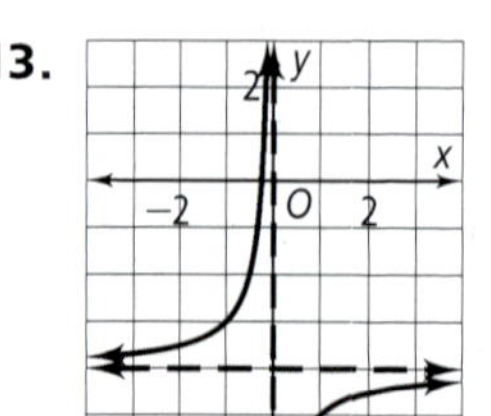

x-intercept: $(-0.25, 0)$, no y-intercept; vert. asymptote: $x = 0$, horizontal asymptote: $y = -4$

14. 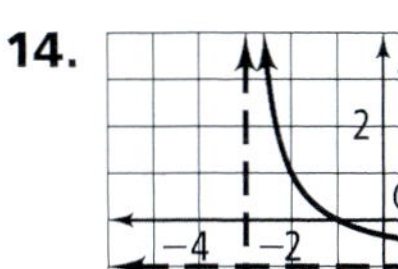

x-intercept: $(-1, 0)$, y-intercept: $(0, -\frac{1}{3})$; vert. asymptote: $x = -3$, horizontal asymptote: $y = -1$

15. $y = \frac{4}{x} + 3$ **16.** $y = \frac{4}{x - 2} + 2$ **17.** $y = \frac{4}{x + 3} - 4$
18. $y = \frac{4}{x - 4} - 3$
19. pts. of discontinuity: $x = -2, 1$;

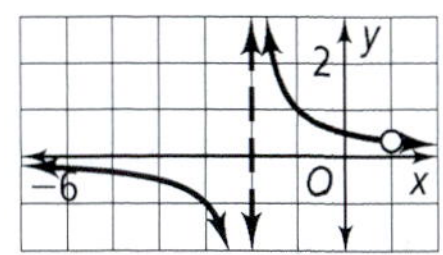

vert. asymptote: $x = -2$, horizontal asymptote: $y = 0$; hole at $x = 1$
20. 1, −1

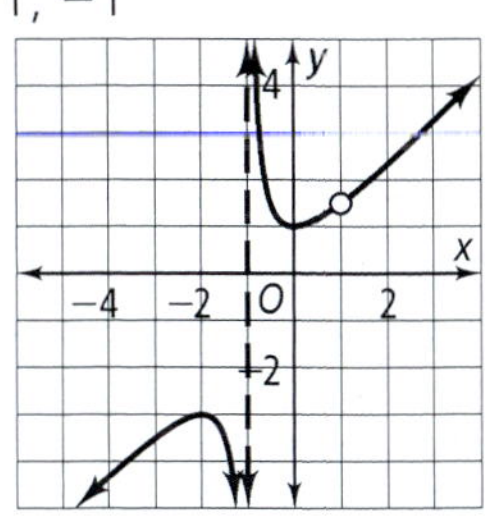

vert. asymptote: $x = -1$; hole at $x = 1$

21. no pts. of discontinuity

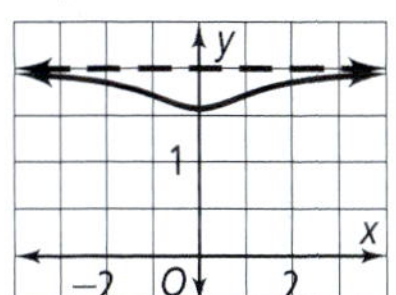

horizontal asymptote: $y = 2$

22. 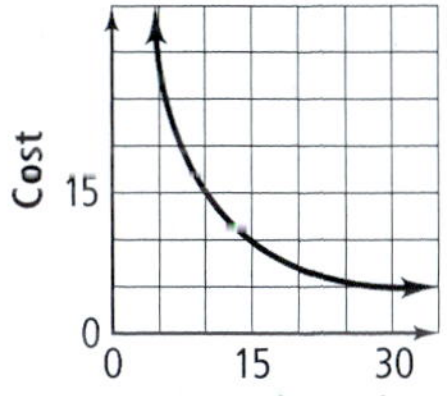

≈31,056 headsets

23. $\frac{x + 5}{x + 4}$; $x \neq -4$ or -5 **24.** $\frac{(x - 1)(x + 1)}{x + 3}$; $x \neq -4, -3$, or 6 **25.** $\frac{(2x - 1)(x + 1)}{x + 4}$; $x \neq -4, -1$, or 0
26. $\frac{r}{3}$, where r is the radius **27.** $\frac{3(3x - 4)}{(x - 2)(x + 2)}$; $x \neq \pm 2$
28. $\frac{-x^2 + 3x + 2}{x(x + 1)(x - 1)(x + 3)}$; $x \neq \pm 1$, 0, or −3

29. $\frac{2(x - 1)}{3x - 1}$ **30.** $\frac{1}{4(x + y)}$ **31.** −1 **32.** no solution
33. −12, 9 **34.** you: 10 mi/h friend: 8 mi/h

Chapter 9

Get Ready! p. 561 **1.** 9, 11, 13, 15 **2.** 1, 6, 11, 16 **3.** 0.9, 1.1, 1.3, 1.5 **4.** −2, −7, −12, −17 **5.** $3\frac{1}{3}, 7\frac{1}{3}, 11\frac{1}{3}, 15\frac{1}{3}$ **6.** −12, −15, −18, −21 **7.** subtract 5; −11, −16, −21 **8.** mult. by 2; 16, 32, 64 **9.** alternate subtract 9 and add 1; −7, −6, −15 **10.** add 3; 19, 22, 25 **11.** $\frac{4}{3}$ **12.** $\frac{3}{4}$ **13.** $\frac{5}{3}$ **14.** $\frac{5}{18}$ **15.** Answers may vary. Sample: $f(x) = 2x - 1$; 1, 3, 5, 7, 9 **16.** Answers may vary. Sample: $g(x) = 1 - 2x$; −1, −3, −5, −7, −9; yes; common difference: −2 **17.** Answers may vary. Sample: $h(x) = 5(2)^x$; 10, 20, 40, 80, 160; yes; common ratio: 2

Lesson 9-1 pp. 564–571

Got It? **1.** 147 **2. a.** $a_1 = 1$ and $a_n = na_{n-1}$ **b.** $a_1 = 1$ and $a_n = a_{n-1} + n^2$ **3. a.** $a_n = n^2 - 1$; 399 **b.** To find the nth term using an explicit formula, you simply substitute for n in the formula. To find the nth term using a recursive definition may require many iterations. **4.** 18 months
Lesson Check **1.** 2, 7, 12, 17, 22 **2.** −1, 0, 3, 8, 15 **3.** $a_1 = 3$ and $a_n = 2a_{n-1}$ **4.** $a_n = 2 + 3n$ **5.** A recursive formula defines the terms in a sequence by relating each term after the first term to the one before it and requires that the previous term be known to find a given term. An example of a recursive formula for the sequence 8, 4, 2, 1, . . . is $a_1 = 8$ and $a_n = \frac{1}{2}a_{n-1}$. An explicit formula describes the nth term of a sequence using the variable n and only requires the number of the term to be known. An example of an explicit formula for the sequence 1, 3, 5, 7, . . . is $a_n = 2n - 1$. **6.** The "+ 1" in $a_n = 3n + 1$ is incorrect for the sequence 1, 4, 7, 10, ... The correct explicit formula is $a_n = -2 + 3n$.
Exercises **7.** 5, 8, 11, 14, 17, 20 **9.** $\frac{1}{2}, 1, \frac{3}{2}, 2, \frac{5}{2}, 3$ **11.** 2, 10, 24, 44, 70, 102 **13.** $-\frac{1}{2}, 3, \frac{25}{2}, 31, \frac{123}{2}, 107$ **15.** $a_1 = 80$ and $a_n = a_{n-1} - 3$ **17.** $a_1 = 0$ and $a_n = a_{n-1} + (n + 1)$ **19.** $a_1 = 100$ and $a_n = \frac{1}{10}a_{n-1}$ **21.** $a_1 = 4$ and $a_n = -2a_{n-1}$ **23.** $a_1 = 1$ and $a_n = a_{n-1} + n^2$ **25.** $a_n = 3n + 1$; 31 **27.** $a_n = \frac{n - 6}{2}$; 2 **29.** $a_n = n^2 + 1$; 101 **31.** $a_n = 3^{n-1}$; 19,683 **33.** 5 **35.** $\frac{5}{16}$ **37.** $\frac{9}{1024}$ **39.** −47 **41.** $-\frac{9}{8}$ **43.** recursive; 3, 9, 21, 45, 93 **45.** explicit; −24, −21, −16, −9, 0 **47.** explicit; −6, −18, −38, −66, −102 **49.** 25, 36, 49, 64

Selected Answers

51. $\frac{16}{5}, \frac{25}{6}, \frac{36}{7}, \frac{49}{8}$ **53.** \$140 **55.** 20, 23; $a_n = 3n + 2$, explicit OR $a_n = a_{n-1} + 3$; $a_1 = 5$, recursive
57. 216, 343; $a_n = n^3$, explicit
59. 144, 169; $a_n = (n + 6)^2$, explicit OR $a_n = a_{n-1} + 2n + 11$, $a_1 = 49$, recursive **61.** $-1, -\frac{1}{2}$; $a_n = \frac{-32}{2^n}$, explicit OR $a_n = \frac{a_{n-1}}{2}$, $a_1 = -16$, recursive
63. $-11, -19$; $a_n = 29 - 8n$, explicit OR $a_n = a_{n-1} - 8$, $a_1 = 21$, recursive **65. a.** 25 boxes **b.** 110 boxes **c.** 9 levels **67.** $a_n = 10 \cdot 2^{n-1}$
69. $a_n = 1 + 4(n - 1)$ **71.** 34.9 **73.** 2.19 **75.** 36.5
76. 2 **77.** 4 **78.** −5 **79.** −1 **80.** 1 **81.** 2
82. subtract 2; −2, −4, −6
83. add 17; 185, 202, 219 **84.** add $\frac{3}{7}$; $\frac{17}{7}, \frac{20}{7}, \frac{23}{7}$

Lesson 9-2 pp. 572–577

Got It? 1. a. not arithmetic **b.** arithmetic **2. a.** 93 **b.** 95, 110 **3. a.** 115 **b.** yes, use the formula for arithmetic mean and solve for a_7, $a_7 = 2a_6 - a_5$ **4.** 65 seats
Lesson Check 1. 56 **2.** 87 **3.** 13 **4.** 39 **5.** In an arithmetic sequence, the diff. between any two consecutive terms is always the same number. **6.** Answers may vary. Sample: 2, 4, 8, 16, 32, . . .
Exercises 7. yes; 10 **9.** yes; 3 **11.** yes; 4 **13.** 127
15. 240 **17.** 12.5 **19.** −7 **21.** 13 **23.** 7.5 **25.** \$135
27. 18 **29.** 36 **31.** 2 **33.** The student multiplied the third term by 2 instead of adding 2. The correct answer is 6. **35.** 120 **37.** 1.1 **39.** 0
41. $a_n = 2 + 2(n - 1)$; $a_n = a_{n-1} + 2$, $a_1 = 2$
43. $a_n = -5 + 1(n - 1)$; $a_n = a_{n-1} + 1$, $a_1 = -5$
45. $a_n = -5 + 1.5(n - 1)$; $a_n = a_{n-1} + 1.5$, $a_1 = -5$
47. $a_n = 1 + \frac{1}{3}(n - 1)$; $a_n = a_{n-1} + \frac{1}{3}$, $a_1 = 1$
49. $a_n = 27 - 12(n - 1)$; $a_n = a_{n-1} - 12$, $a_1 = 27$
51. Answers may vary. Sample: An advantage of a recursive formula is that only the preceding term must be known to find the next term; a disadvantage is that many calculations may be required to find a term. An advantage of an explicit formula is that it is easy to find any term. Use the recursive formula when the previous term and common diff. are known. Use the explicit formula when the term number and common diff. are known. **53.** −4, −10, −16
55. −8, −17, −26 **57.** 17, 17, 17 **59.** −12.5, −8, −3.5
61. \$5055 **63.** 21st term **65.** 54 **67.** $a_1 = -1, d = 3$
69. $a_1 = 52, d = -10$ **71.** $a_1 = -100.5, d = 22$
73. $9k + 32$ **75.** D

77.

$$\frac{3}{(x-1)(x+1)} + \frac{4x(x-1)}{(x-1)(x+1)} = \frac{1.5(x+1)}{(x-1)(x+1)}, x \neq \pm 1$$
$$3 + 4x^2 - 4x = 1.5x + 1.5$$
$$4x^2 - 5.5x + 1.5 = 0$$
$$x^2 - \frac{11}{8}x + \frac{3}{8} = 0$$
$$(x-1)\left(x - \frac{3}{8}\right) = 0$$
$$x = \frac{3}{8}$$

(Reject $x = 1$ because 1 is not in the domain.)
78. recursive; −2, −7, −12, −17, −22
79. explicit; 6, 18, 36, 60, 90 **80.** explicit; 0, 3, 8, 15, 24
81. recursive; −121, −108, −95, −82, −69
82. $y - 3 = \frac{8}{3}x$ or $y - 11 = \frac{8}{3}(x - 3)$
83. $y - 6 = 4(x - 4)$ or $y - 30 = 4(x - 10)$
84. $y - 10 = 8(x - 1)$ or $y - 42 = 8(x - 5)$
85. $r = \frac{\sqrt[3]{6\pi^2 V}}{2\pi}$ **86.** 32 **87.** 625 **88.** −81

Lesson 9-3 pp. 580–586

Got It? 1. a. yes; $a_1 = 2, r = 2$ **b.** no **c.** yes; $a_1 = 2^3$, $r = 2^4$ **2.** 6 or −6 **3. a.** explicit; it is easier to use because only one calculation is needed. **b.** about 16.8 cm, about 4 cm **4.** ±60
Lesson Check 1. no **2.** yes; 2 **3.** 729 **4.** 0.0064
5. The third term would be the geometric mean of 5 and 80 which is 20. Since a is pos. and r^2 is always pos., the third term, ar^2, cannot be neg.
6. For both the arithmetic mean and the geometric mean, the middle term of any three consecutive terms can be determined using the first and last of the three terms. The arithmetic mean is the sum of the first and last terms divided by 2, whereas the geometric mean is the square root (or its opposite) of the product of the first and the last terms.
Exercises 7. yes; 2 **9.** yes; −2 **11.** yes; 0.4
13. yes; $-\frac{1}{3}$ **15.** yes; 1.5 **17.** yes; 6 **19.** 6561
21. 0.078125 **23.** $\frac{-3}{2048}$ **25.** about 656.1 g; about 182.5 g; about 96.2 g
27. ±1530 **29.** ±1.5 **31.** ±6 **33.** $a_n = 100(-20)^{n-1}$; (100, −2000, 40,000, −800,000, 16,000,000
35. $a_n = 1024(0.5)^{n-1}$; 1024, 512, 256, 128, 64
37. $a_n = 10(-1)^{n-1}$; 10, −10, 10, −10, 10 **39.** arithmetic; 125, 150 **41.** geometric; −80, 160 **43.** neither; 25, 36
45. 7.5, 22.5, 67.5 or −7.5, 22.5, −67.5
47. −6.64, −11.02, −18.30 or 6.64, −11.02, 18.30
49. about 74.3 mi **51.** 768 **53.** 3×4^{19} or 824,633,720,832 **55.** 4 **57.** 10 **59.** Both the common diff. and the common ratio are used to find the next term in a sequence, but a common diff. is added and a common ratio is multiplied. **61.** 7 **63.** B **65.** C **67.** $x \neq -1, -5$; there's a hole in the graph at $x = -1$. There's a vert. asymptote at $x = -5$.
68. $a_n = -3 + 3(n - 1)$; $a_n = a_{n-1} + 3$, $a_1 = -3$
69. $a_n = 17 - 9(n - 1)$; $a_n = a_{n-1} - 9$, $a_1 = 17$
70. $a_n = -2 - 11(n - 1)$; $a_n = a_{n-1} - 11$, $a_1 = -2$

71. $7\sqrt{14}$ **72.** $\frac{3\sqrt{3x}}{7x}$ **73.** $\frac{x}{y}$ **74.** $5\sqrt[3]{6}$ **75.** vert. asymptote: $x = -3$ **76.** vert. asymptote: $x = -1$ **77.** vert. asymptotes: $x = 0, 1$ **78.** vert. asymptote: $x = 3$; hole at $x = -3$ **79.** $a_n = a_{n-1} + n,\ a_1 = 1$ **80.** $a_n = a_{n-1} + (2n - 1),\ a_1 = 1$ **81.** $a_n = a_{n-1} + n^2,\ a_1 = 1$

Lesson 9-4 pp. 587–593

Got It? 1. a. 1030 **b.** Yes; no; the sum of any number of even numbers is always even. The sum of an odd number of odd numbers is odd, but the sum of an even number of odd numbers is even. **2.** 59 sales; 1725 sales

3. a. $\sum_{n=1}^{40}(-12 + 7n)$ **b.** $\sum_{n=1}^{50}(510 - 10n)$ **4. a.** 2140 **b.** 100 **c.** 1 **5.** 41,650

Lesson Check 1. 91 **2.** 780 **3.** $\sum_{n=1}^{7} 3n$ **4.** $\sum_{n=1}^{12}(-3 + 4n)$ **5.** An arithmetic sequence is a list of numbers for which successive numbers have a common difference. **6.** The lower limit should not be 3, it should be one and the expression $5n - 2$ should be in parentheses. The correct summation notation is $\sum_{n=0}^{8}(3 + 5n)$. **7.** Yes; $44 = 2(a_1 + a_4)$, so any combination of a_1 and a_4 with a sum of 22 is a possible series.

Exercises 9. 92 **11.** 176 **13.** −165 **15.** $\sum_{n=1}^{5} 4n$ **17.** $\sum_{n=1}^{12}(2 + 3n)$ **19.** $\sum_{n=1}^{10}(-3n)$ **21.** 25 **23.** 20 **25.** −2 **27.** 2400 **29.** 682 **31.** −8556 **33.** 432 seats **35.** sequence; finite **37.** series; infinite **39.** series; finite **41.** −48 **43.** 35 **45.** −146 **47. a.** $a_n = n + 1$ **b.** $\sum_{n=1}^{9}(n + 1)$ **c.** 18 cans **d.** No; no; 13 rows have 104 cans, 14 rows have 119 cans, 15 rows have 135 cans, and 16 rows have 152 cans. The number of rows would not be an integer for 110 cans or 140 cans. **49.** −765 **51.** 300 **53.** 34 **55.** $10x + 45y$ **57.** C **59.** B

61.
$$\left(\frac{b}{2}\right)^2 = \left(\frac{10}{2}\right)^2 = 25$$
$$x^2 + 10x + 25 = -35 + 25$$
$$(x + 5)^2 = -10$$
$$x + 5 = \pm i\sqrt{10}$$
$$x = -5 \pm i\sqrt{10}$$

62. $a_n = 2^{n-1}$; 1, 2, 4 **63.** $a_n = -1(-1)^{n-1}$; −1, 1, −1 **64.** $a_n = 3\left(\frac{3}{2}\right)^{n-1}$; 3, $\frac{9}{2}$, $\frac{27}{4}$ **65.** $\frac{x + 3}{x - 4}$; $x \neq 4$, $x \neq -1$ **66.** $\frac{c - 2}{c - 5}$; $c \neq 5$, $c \neq 6$ **67.** $\frac{z^2 + 12z + 20}{z - 1}$; $z \neq 1$, $z \neq 0$ **68.** $-\frac{1}{3}$ **69.** $\frac{3}{4}$ **70.** $-\frac{1}{2}$

Lesson 9-5 pp. 595–601

Got It? 1. a. 315 **b.** −1705 **2.** about \$2138.43 **3. a.** diverges **b.** converges; $\frac{1}{4}$ **c.** converges; 2 **d.** Yes; if $|r| < 1$, the series converges. If $|r| \geq 1$, the series diverges.

Lesson Check 1. $\frac{31}{80}$ **2.** $\frac{55}{9}$ **3.** converges **4.** diverges **5.** Since $r = 1.1 > 1$, the series diverges and does not have a sum. **6.** An infinite geometric series has a sum only when the series converges, which is when $|r| < 1$. **7.** The sum of a finite arithmetic series is $S_n = \frac{n}{2}(a_1 + a_n)$. The sum of finite geometric series is $S_n = \frac{a_1(1 - r^n)}{1 - r}$. The formulas are similar in that each sum requires the first term and the number of terms in the series. The formulas are different in that the sum of a finite arithmetic series needs the last term, while the sum of a finite geometric series needs the common ratio.

Exercises 9. 1456 **11.** −5115 **13.** $\frac{15}{32}$ **15.** $\frac{121}{81}$ **17.** converges; $\frac{4}{3}$ **19.** converges; 8 **21.** diverges; no sum **23.** diverges; no sum **25.** diverges; no sum **27.** 1 **29.** $\frac{9}{2}$ **31.** $\frac{9}{5}$ **33.** arithmetic; 420 **35.** geometric; about 96.47 **37.** geometric; about 121.5 **39. a.**

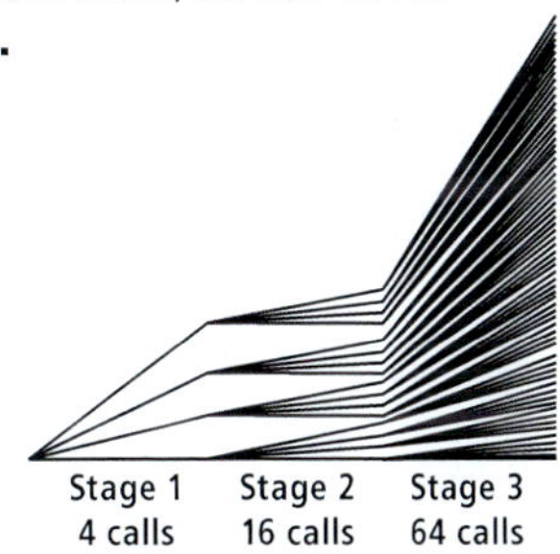

b. 4 + 16 + 64 + 256 + 1024 + 4096 **c.** 5460 employees **41.** $\frac{5}{4}$ **43.** $\frac{3}{4}$ **45.** $0.8\overline{3}$ **47. a.** $\frac{7}{8}$ **b.** 10 **49. a.** Answers may vary. Sample: The student used $r - 1$ instead of $1 - r$ in the formula for the sum of an infinite geometric series. **b.** $\frac{1}{2}$

51. a. $rS_n = r(a_1 + a_1r + \cdots + a_1r^{n-1}) = a_1r + a_1r^2 + \cdots + a_1r^n$

b.
$$S_n - rS_n = a_1 + a_1r + a_1r^2 + \cdots + a_1r^{n-1} - a_1r - a_1r^2 - \cdots - a_1r^{n-1} - a_1r^n$$
$$= a_1 - a_1r^n$$

c.
$$S_n - rS_n = a_1 - a_1r^n$$
$$S_n(1 - r) = a_1 - a_1r^n$$
$$S_n = \frac{a_1 - a_1r^n}{1 - r} = \frac{a_1(1 - r^n)}{1 - r}$$

53. $a_1 = \frac{9}{10}$, $r = \frac{1}{10}$; $S = \frac{\frac{9}{10}}{1 - \frac{1}{10}} = 1$ **55.** $\frac{2}{3}$ **57.** 1.11 **59.** 140 **60.** −825 **61.** $\frac{7c - 4}{2c^2}$ **62.** $\frac{10(2y + 3)}{(y + 3)(y - 3)}$

63. $\frac{x^2 + 6x + 4}{(x + 6)(x - 6)}$ **64.** 0 **65.** 2 **66.** 1

67.

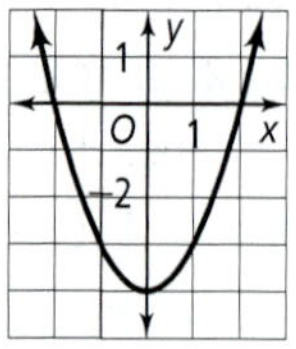

68.

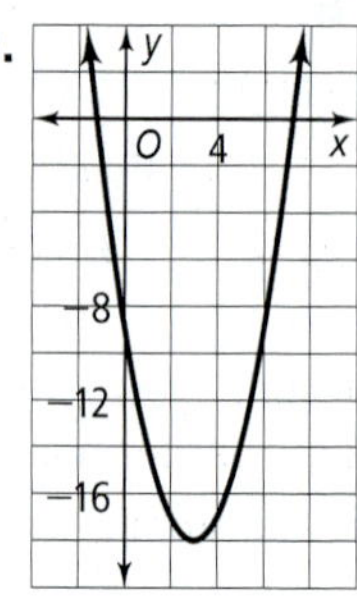

69.

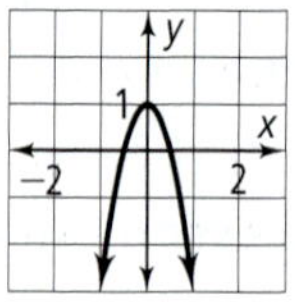

Chapter Review pp. 603–606

1. limits **2.** sequence **3.** converges **4.** common ratio **5.** explicit formula **6.** 1, −1, −3, −5, −7 **7.** 1, 0, −3, −8, −15 **8.** 2, 3, 5, 9, 17 **9.** 20, 10, 5, 2.5, 1.25 **10.** $a_n = a_{n-1} + 17$, $a_1 = 5$ **11.** $a_n = a_{n-1} + 9$, $a_1 = -2$ **12.** $a_n = 3n - 2$ **13.** $a_n = 6.5 - 2.5n$ **14.** no **15.** yes; $d = 15$, $a_{32} = 468$ **16.** yes; $d = 3$, $a_{32} = 100$ **17.** no **18.** 5 **19.** 101.5 **20.** 5 **21.** −4.9 **22.** −10.5, −8, −5.5 **23.** 1.4, 0.8, 0.2 **24.** $a_n = -2 + 9(n - 1)$ **25.** $a_n = 62 - 3(n - 1)$ **26.** yes; $r = \frac{1}{2}$; $\frac{1}{16}$, $\frac{1}{32}$ **27.** no **28.** yes; $r = 1.2$; 6.2208, 7.46496 **29.** ±6 **30.** ±0.04 **31.** ±10, −5, ±2.5 **32.** $a_n = 2^{n-1}$ **33.** $a_n = 25\left(\frac{1}{5}\right)^{n-1}$ **34.** 2560 **35.** 1536 **36.** $\sum_{n=1}^{5}(13 - 3n)$; 20 **37.** $\sum_{n=1}^{7}(45 + 5n)$; 455 **38.** $\sum_{n=1}^{11}(4.6 + 1.4n)$; 143 **39.** $\sum_{n=1}^{8}(23 - 2n)$; 112 **40.** 3; −8, 26; 27 **41.** 9; 4, 8; 54 **42.** 31 **43.** $53\frac{1}{8}$ **44.** $14\frac{7}{18}$ **45.** converges; $S = 187.5$ **46.** diverges **47.** diverges **48.** converges; $S = 2$

Chapter 10

Get Ready! p. 611

1.

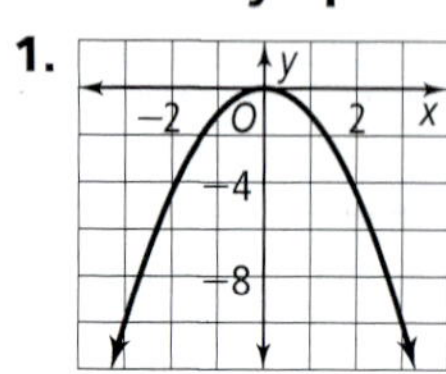

2.

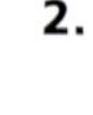

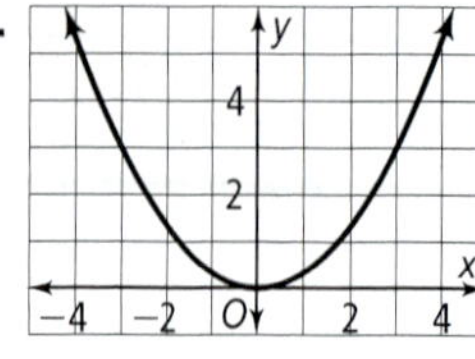

3.

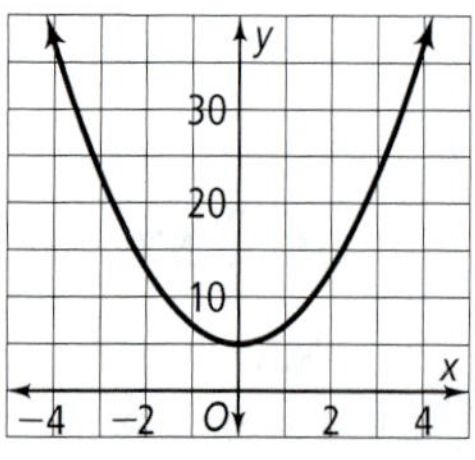

4.

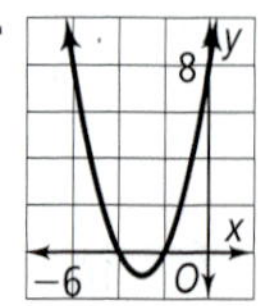

5. quadratic; $-x^2$, $6x$, 1 **6.** linear; none, $-12x$, −18 **7.** linear; none, x, $-\frac{13}{2}$ **8.** quadratic; $-8x^2$, $28x$, none **9.** quadratic; $-2x^2$, $-3x$, 6 **10.** linear; none, $-x$, −10 **11.** 16 **12.** $\frac{25}{4}$ **13.** 49

14. $y = (x + 3)^2 - 2$

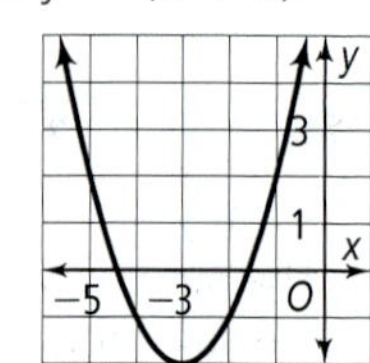

15. $y = 2(x - 1)^2 + 8$

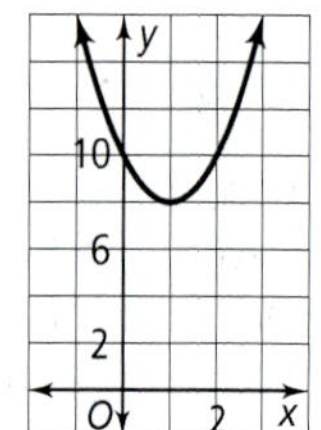

16. $y = -3\left(x - \frac{1}{6}\right)^2 + \frac{1}{12}$

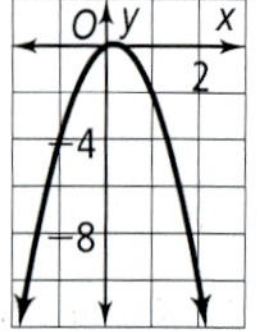

17.

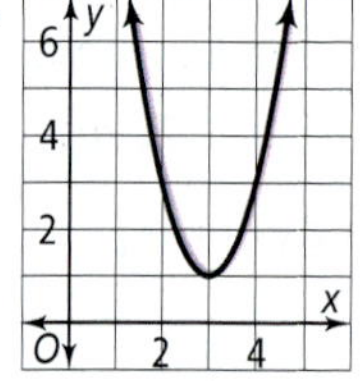

18.

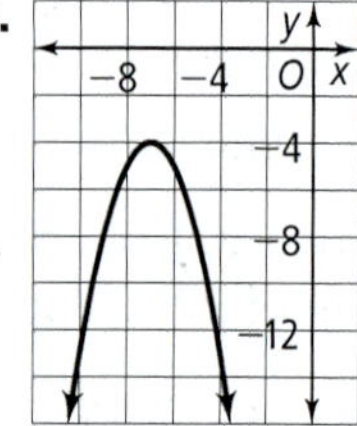

19.

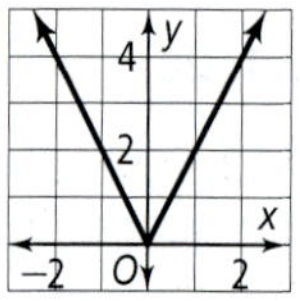

20.

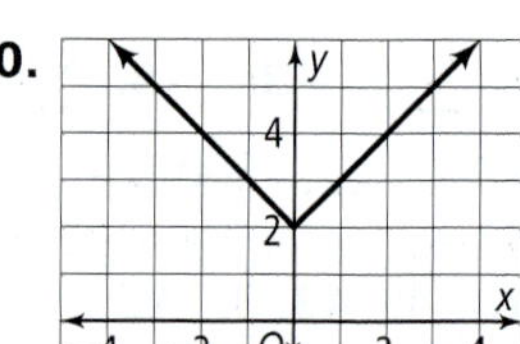

21. The radius of a circle is the distance from the center of the circle to any pt. on the circle. The radius extends in every direction from the center and ends on the circle. All radii of the same circle are equal. **22.** The vertex of a parabola is the lowest or highest pt. of a parabola; it is the pt. where the parabola changes direction.

Lesson 10-1 pp. 614–621

Got It?

1. a.

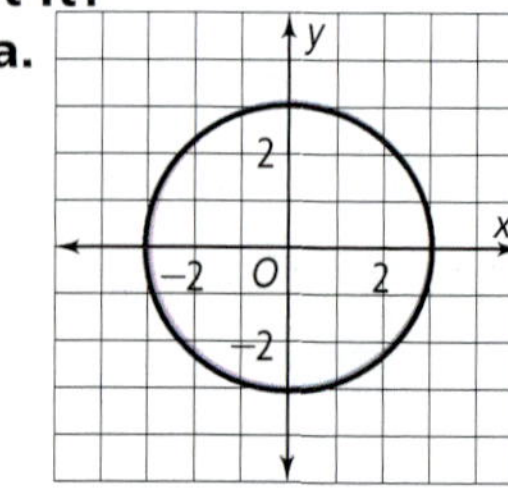

circle: center (0,0); radius 3; lines of sym.: every line through the origin; domain: $-3 \leq x \leq 3$, range: $-3 \leq y \leq 3$

b. 6 is outside the domain of x.

2. 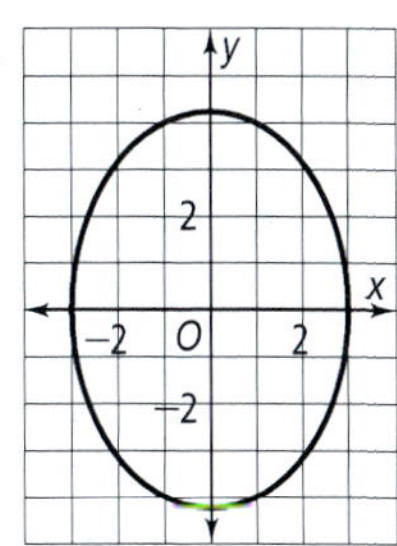

ellipse: center (0,0); lines of sym.: x-axis and y-axis; domain: $-3 \leq x \leq 3$, range: $-3\sqrt{2} \leq y \leq 3\sqrt{2}$

3. 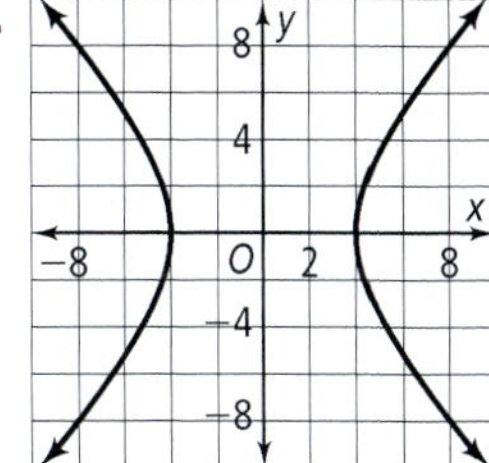

hyperbola: center (0,0); lines of sym.: x-axis and y-axis; domain: $x \leq -4$ or $x \geq 4$, range: all real numbers

4. center of hyperbola: (0, 0); no x-intercepts, y-intercepts: (0, 1), (0, −1); domain: all real numbers, range: $y \leq -1$ or $y \geq 1$ **5. a.** Ellipse; the eq. $9x^2 + 25y^2 = 225$ represents a conic section with two sets of intercepts, (±3, 0) and (0, ±5). Since the intercepts are not equidistant from the center, the eq. models an ellipse.
b. Hyperbola; the eq. $x^2 - y^2 = 1$ represents a conic section with one set of intercepts, (±1, 0), so the eq. must model a hyperbola.

Lesson Check

1. 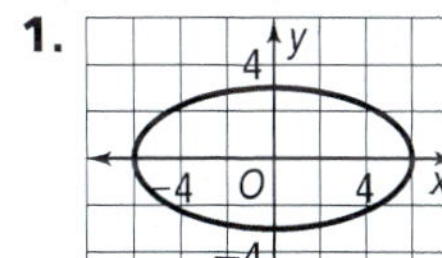

lines of sym.: x-axis and y-axis; domain: $-6 \leq x \leq 6$, range: $-3 \leq y \leq 3$

2. 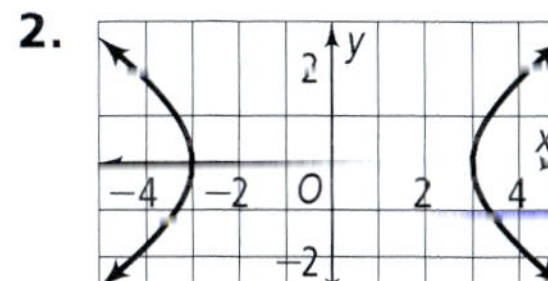

lines of sym.: x-axis and y-axis; domain: $x \leq -3$ or $x \geq 3$, range: all real numbers

3. domain: $x \leq -2.5$ or $x \geq 2.5$, range: all real numbers
4. domain: $-6 \leq x \leq 6$, range: $-1.5 \leq y \leq 1.5$
5. a. hyperbola **b.** circle **6.** Answers may vary. Sample answer: The domain of an ellipse is an interval between two real numbers, such as $-a \leq x \leq a$. The domain of a hyperbola is two intervals, such as $x \leq -a$ or $x \geq a$, if there are x-intercepts, or all real numbers if there are no x-intercepts.

Exercises

7. 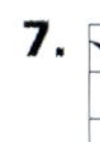 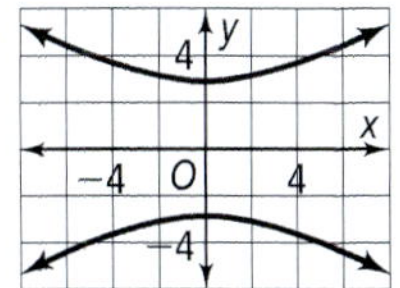

hyperbola; center: (0, 0); no x-intercepts, y-intercepts: $\left(0, \pm\frac{5\sqrt{3}}{3}\right)$; lines of sym.: x-axis and y-axis; domain: all real numbers, range: $y \leq -\frac{5\sqrt{3}}{3}$ or $y \geq \frac{5\sqrt{3}}{3}$

9. 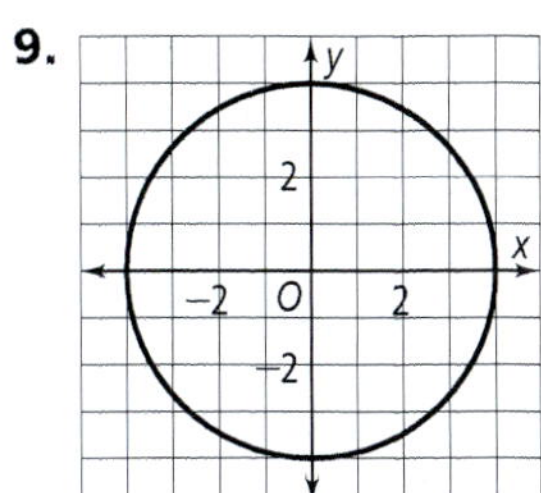

circle; center: (0, 0); radius: 4; x-intercepts: (±4, 0), y-intercepts: (0, ±4); infinitely many lines of sym.; domain: $-4 \leq x \leq 4$, range: $-4 \leq y \leq 4$

11. 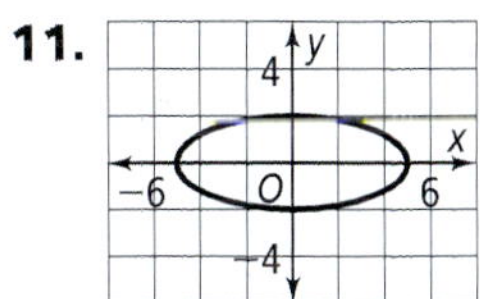

ellipse; center: (0, 0); x-intercepts: (±5, 0), y-intercepts: (0, ±2); lines of sym.: x-axis and y-axis; domain: $-5 \leq x \leq 5$, range: $-2 \leq y \leq 2$

13. 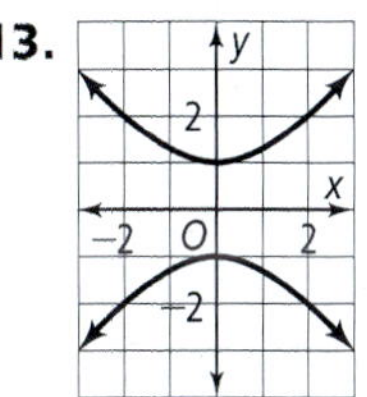

hyperbola; center: (0, 0); no x-intercepts, y-intercepts: (0, ±1); lines of sym.: x-axis and y-axis; domain: all real numbers, range: $y \leq -1$ or $y \geq 1$

15. 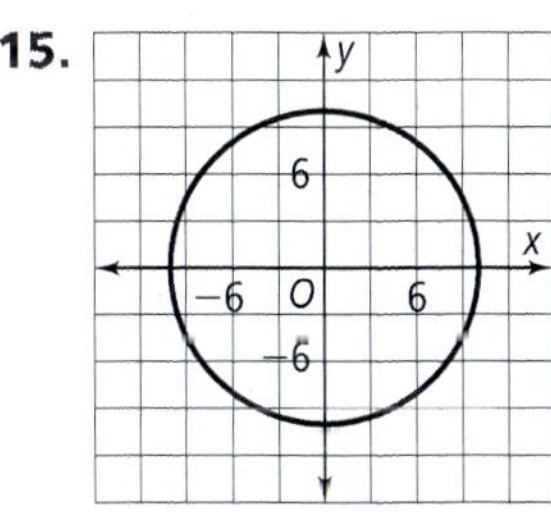

circle; center: (0, 0); radius: 10; x-intercepts: (±10, 0), y-intercepts: (0, ±10); infinitely many lines of sym.; domain: $-10 \leq x \leq 10$, range: $-10 \leq y \leq 10$

17. 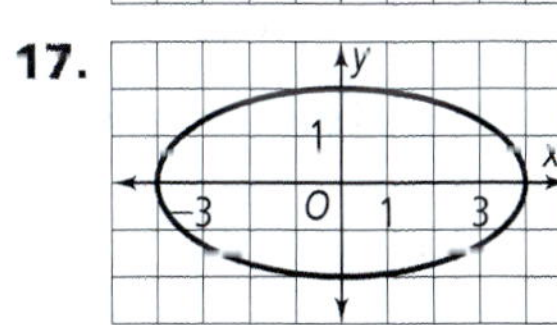

ellipse; center: (0, 0); x-intercepts: (±4, 0), y-intercepts: (0, ±2); lines of sym.: x axis and y-axis; domain: $-4 \leq x \leq 4$, range: $-2 \leq y \leq 2$

19. 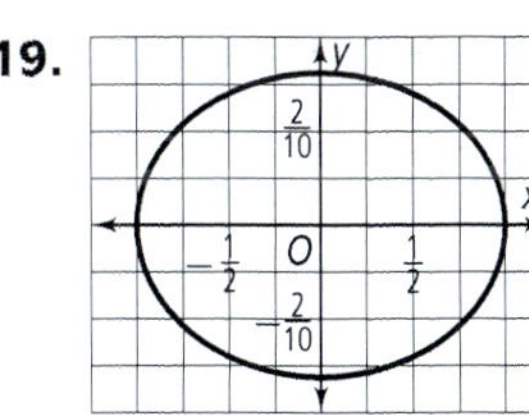

ellipse; center: (0, 0); x-intercepts: (±1, 0), y-intercepts: $\left(0, \pm\frac{1}{3}\right)$; lines of sym.: x-axis and y-axis; domain: $-1 \leq x \leq 1$, range: $-\frac{1}{3} \leq y \leq \frac{1}{3}$

21. hyperbola; center: (0, 0); no x-intercepts, y-intercepts: $\left(0, \pm\frac{1}{2}\right)$; lines of sym.: x-axis and y-axis; domain: all real numbers, range: $y \le -\frac{1}{2}$ or $y \ge \frac{1}{2}$

23. hyperbola; center: (0, 0); no x-intercepts, y-intercepts: $(0, \pm 2)$; domain: all real numbers, range: $y \le -2$ or $y \ge 2$ **25.** hyperbola; center: (0, 0); x-intercepts: $(\pm 3, 0)$, no y-intercepts; domain: $x \le -3$ or $x \ge 3$, range: all real numbers **27.** hyperbola; center: (0, 0); no x-intercepts, y-intercepts: $(0, \pm 3)$; domain: all real numbers, range: $y \le -3$ or $y \ge 3$ **29.** 22 **31.** 24 **33.** 27

35.

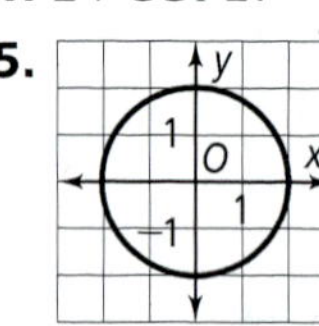

circle; center: (0, 0); radius: 2; x-intercepts: $(\pm 2, 0)$, y-intercepts: $(0, \pm 2)$; infinitely many lines of sym.; domain: $-2 \le x \le 2$, range: $-2 \le y \le 2$

37.

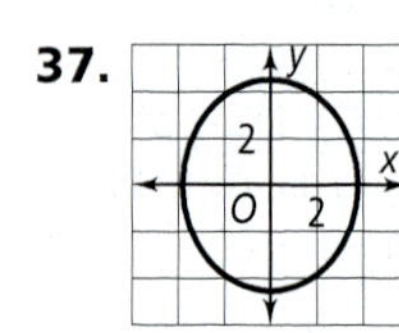

ellipse; center: (0, 0); x-intercepts: $\left(\pm\frac{8\sqrt{5}}{5}, 0\right)$, y-intercepts: $(0, \pm 2\sqrt{5})$; lines of sym.: x-axis and y-axis; domain: $-\frac{8\sqrt{5}}{5} \le x \le \frac{8\sqrt{5}}{5}$, range: $-2\sqrt{5} \le y \le 2\sqrt{5}$

39. a. All lines in the plane that pass through the center of a circle are axes of sym. of the circle. **b.** The axes of sym. of an ellipse intersect at the center of the ellipse. The same is true for a hyperbola. This can be confirmed using, for example, $4x^2 + 9y^2 = 36$ and $4x^2 - 9y^2 = 36$.

41.

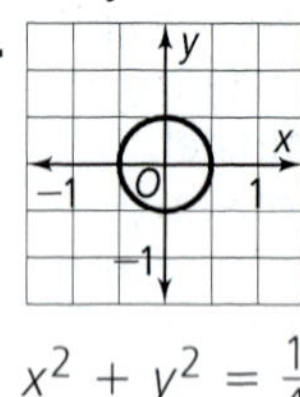

$x^2 + y^2 = \frac{1}{4}$

43.

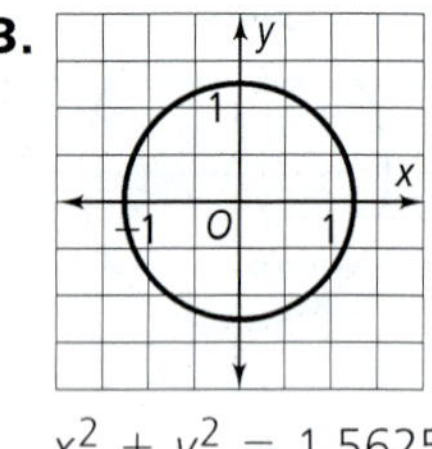

$x^2 + y^2 = 1.5625$

45. Sample: $(\sqrt{2}, 1)$ **47.** Sample: (2, 0) **49.** $(0, -\sqrt{7})$ **51.** Answers may vary.

53. a.

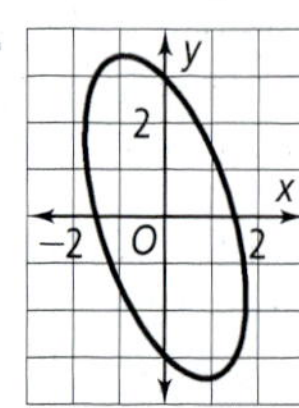

b.

55. I **57.** H **59.** diverges **60.** diverges **61.** converges **62.** $x^3 - 3x^2y + 3xy^2 - y^3$

63. $p^6 + 6p^5q + 15p^4q^2 + 20p^3q^3 + 15p^2q^4 + 6pq^5 + q^6$ **64.** $x^4 - 8x^3 + 24x^2 - 32x + 16$ **65.** $243 - 405x + 270x^2 - 90x^3 + 15x^4 - x^5$

66.

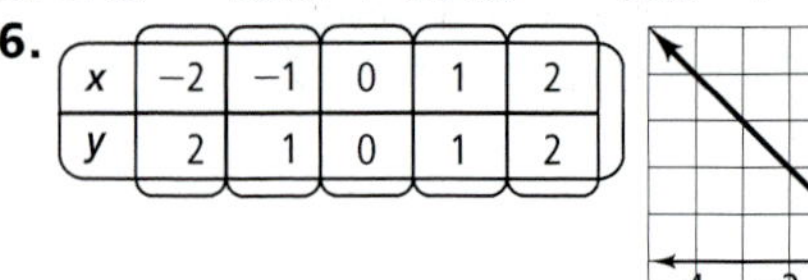

x	−2	−1	0	1	2
y	2	1	0	1	2

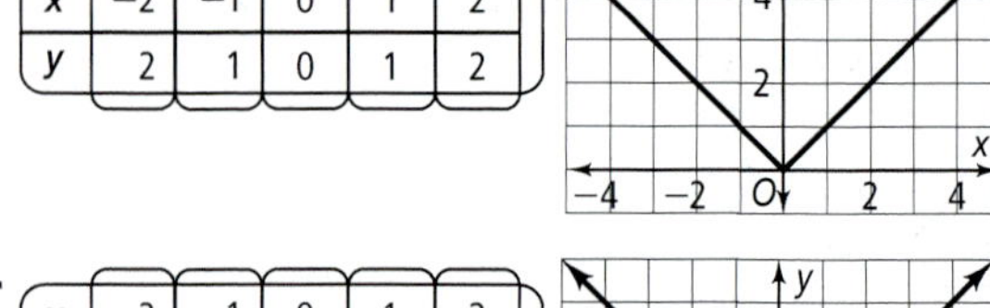

67.

x	−2	−1	0	1	2
y	5	4	3	4	5

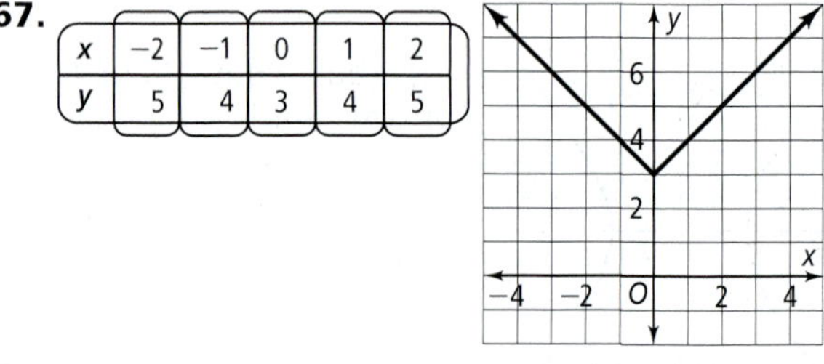

68.

x	0	1	2	3	4
y	2	1	0	1	2

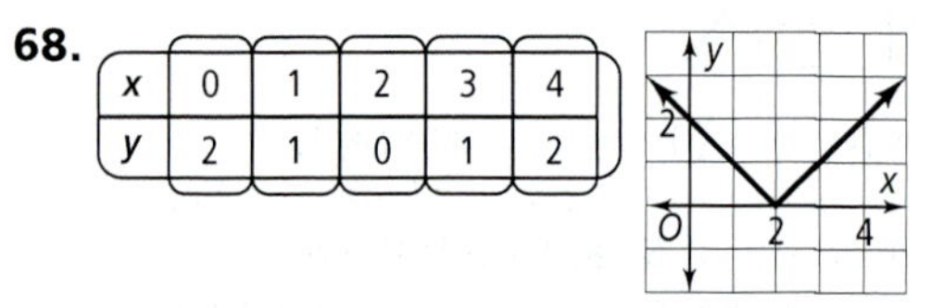

69.

x	−3	−2	−1	0	1
y	−2	−3	−4	−3	−2

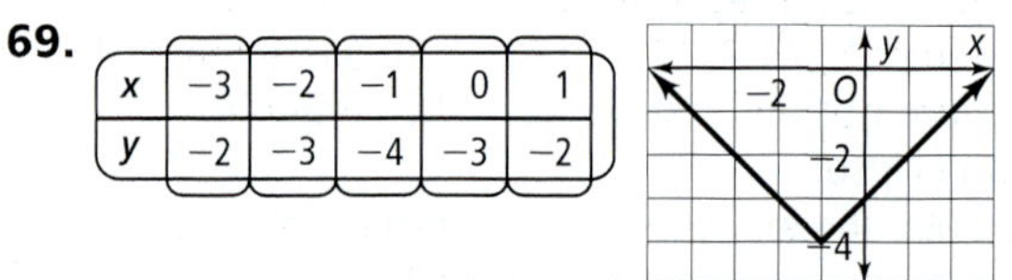

Lesson 10-2 pp. 622–629

Got It? 1. a. $y = -\frac{1}{6}x^2$ **b.** vertex: (0, 0); focus: (0, 1); directrix: $y = -1$ **c.** As the distance between the vertex and focus increases, the width of the parabola increases. **2. a.** $x = \frac{1}{10}y^2$ **b.** vertex: (0, 0); focus: $\left(-\frac{1}{16}, 0\right)$; directrix: $x = \frac{1}{16}$ **3.** 1 cm **4.** vertex: $(-4, 2)$; focus: $\left(-4, 2\frac{1}{4}\right)$; directrix: $y = 1\frac{3}{4}$ **5.** $y = \frac{1}{8}(x - 1)^2 + 4$

Lesson Check 1. $y = \frac{1}{2}x^2$ **2.** $x = \frac{1}{4}(y - 2)^2 + 3$ **3.** vertex: (0, 0); focus: (4, 0); directrix: $x = -4$ **4.** vertex: $(-3, -4)$; focus: $(-3, -3.75)$; directrix: $y = -4.25$ **5.** 6 units **6.** With the focus one unit away from the vertex of a parabola at the origin, $c = \pm 1$. Given this information, the student cannot tell whether the parabola opens in the vert. direction, with one of the eqs. $y = \frac{1}{4}x^2$ or $y = -\frac{1}{4}x^2$, or whether the parabola opens in the horizontal direction, with one of the eqs. of $x = \frac{1}{4}y^2$ or $x = -\frac{1}{4}y^2$.

Exercises 7. $x = \frac{1}{24}y^2$ **9.** $y = \frac{1}{28}x^2$ **11.** $x = \frac{1}{8}y^2$

13. vertex: (0, 0);
focus: $\left(0, \frac{1}{16}\right)$;
directrix: $y = -\frac{1}{16}$

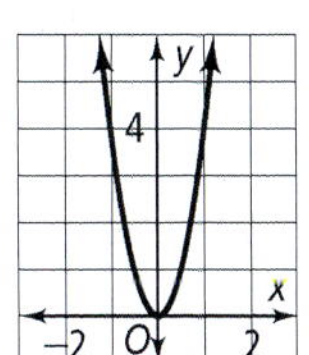

15. vertex: (0, 0);
focus: $\left(\frac{1}{4}, 0\right)$;
directrix: $x = -\frac{1}{4}$

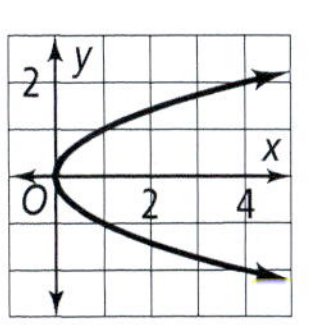

17. $x = \frac{1}{12}y^2$ **19.** $y = \frac{3}{4}x^2$ **21.** $y = -\frac{5}{56}x^2$ **23.** Answers may vary. Sample: $y = x^2$. The light produced by the bulb will reflect off the parabolic mirror in parallel rays.

25. vertex: (3, 2);
focus: $\left(3, \frac{9}{4}\right)$;
directrix: $y = \frac{7}{4}$

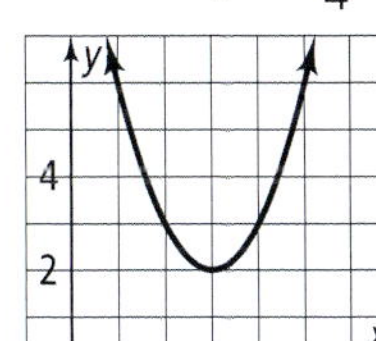

27. vertex: (1, −5);
focus: $\left(1, -4\frac{3}{4}\right)$;
directrix: $y = -5\frac{1}{4}$

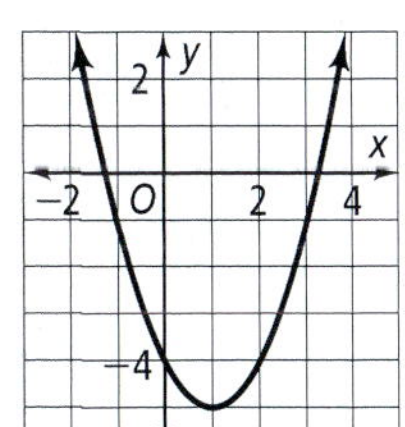

29. vertex: (−1, −4);
focus: $\left(-1, -3\frac{7}{8}\right)$;
directrix: $y = -4\frac{1}{8}$

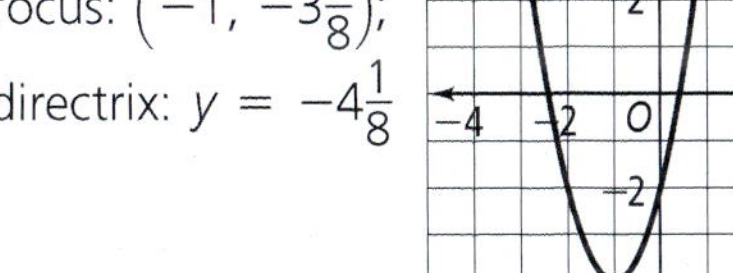

31. $x = -\frac{1}{32}(y - 3)^2$ **33.** $y = -\frac{1}{16}(x - 7)^2 + 2$

35. vertex: (0, 0); focus: (0, −1); directrix: $y = 1$

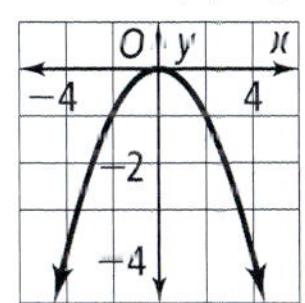

37. vertex: (0, 0);
focus: (−2, 0);
directrix: $x = 2$

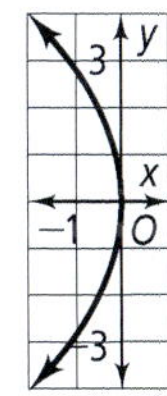

39. vertex: (4, 0);
focus: (4, −6);
directrix: $y = 6$

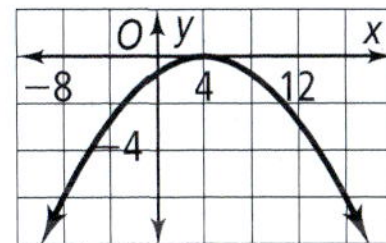

41.

Tsunami Speed (m/s)
Ocean Depth (m)

43. $y = \frac{1}{4}x^2$ **45.** 3.5 in.

47.

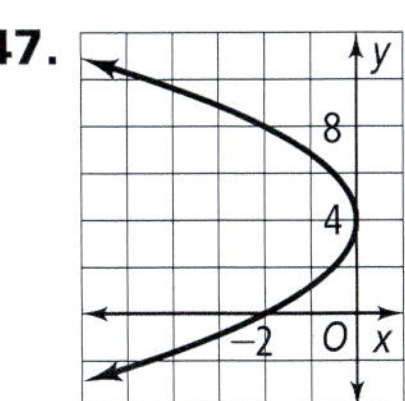

49.

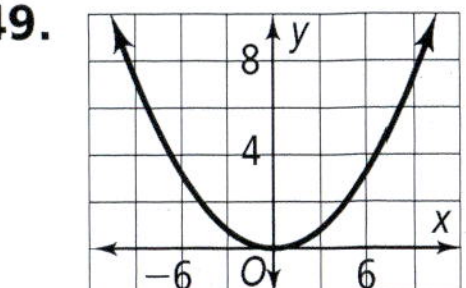

51.

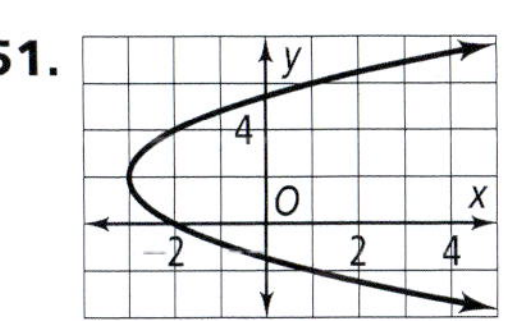

53. $x = -\frac{1}{2}(y - 1)^2 + 1$

55. Answers may vary. Sample: Write the eq. in the form $x = \frac{1}{4\left(\frac{1}{8}\right)}y^2$. The distance from the focus to the directrix is $2\left(\frac{1}{8}\right)$, or $\frac{1}{4}$. **57. a.** the top half of the parabola, $y^2 = x$ **b.** domain: $x \geq 0$, range: $y \geq 0$ **59.** D **61.** B

63.

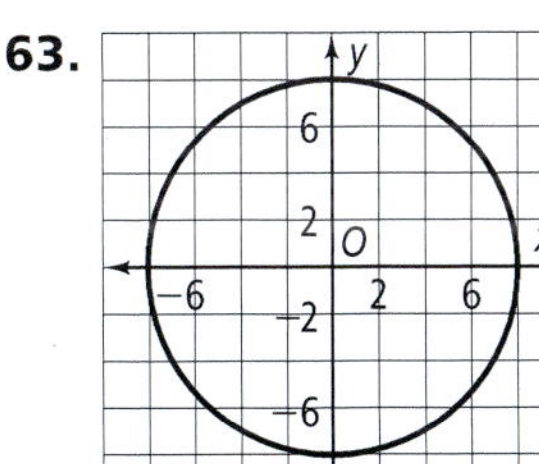

circle: center (0, 0), radius 8; x-intercepts: (±8, 0), y-intercepts: (0, ±8); infinitely many lines of sym.; domain: $-8 \leq x \leq 8$, range: $-8 \leq y \leq 8$

64.

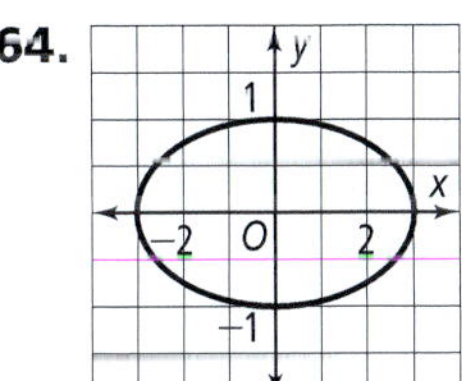

ellipse: center (0, 0); x-intercepts: (±3, 0), y-intercepts: (0, ±1); lines of sym.: x-axis and y-axis; domain: $-3 \leq x \leq 3$, range: $-1 \leq y \leq 1$

65.

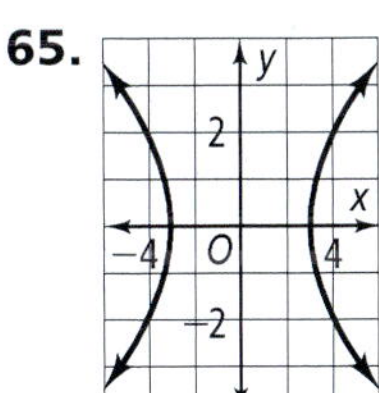

hyperbola: center (0, 0); x-intercepts (±3, 0), no y-intercept; lines of sym.: x-axis and y-axis; domain: $x \leq -3$ or $x \geq 3$, range: all real numbers

66. 1 **67.** 4 **68.** 25 **69.** 9

Lesson 10-3 pp. 630–636

Got It? 1. $(x-5)^2+(y+2)^2=64$ **2. a.** $(x+5)^2+(y+3)^2=1$ **b.** $(x-2)^2+(y-3)^2=9$
3. a. $(x-7)^2+(y+10)^2=144$ **b.** Yes; the values h and k determine the position of the circle and r determines the size. **4. a.** center $(-8,-3)$, radius 11
b. center $(3,-7)$, radius $\sqrt{66}$
5.

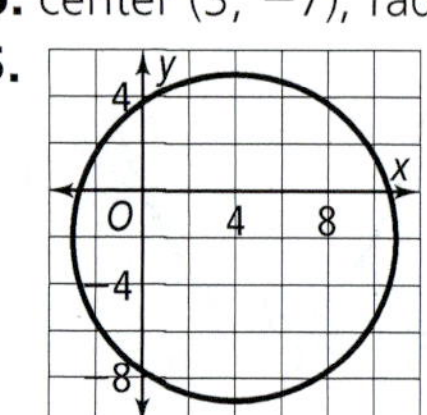

Lesson Check 1. $(x+1)^2+(y+5)^2=4$
2. $x^2+y^2=36$ **3.** $x^2+(y-3)^2=121$
4. $(x+5)^2+(y+3)^2=16$ **5.** The circle with equation $(x+7)^2+(y-7)^2=8$ is a translation of the circle with equation $x^2+y^2=8$ as 7 units left and 7 units up, not right and down. **6.** If $P(x, y)$ is one of the pts. $(r, 0)$, $(-r, 0)$, $(0, r)$, or $(0, -r)$, subst. shows that $x^2+y^2=r^2$. If $P(x, y)$ is any other pt. on the circle, drop a perpendicular $\overline{PK}$ from P to K on the x-axis. $\triangle OPK$ is a rt. triangle with legs of lengths $|x|$ and $|y|$ and with hypotenuse of length r. By the Pythagorean Thm., $|x|^2+|y|^2=r^2$, so $x^2+y^2=r^2$.

Exercises 7. $x^2+y^2=100$ **9.** $(x-2)^2+(y-3)^2=20.25$ **11.** $(x-1)^2+(y+3)^2=100$
13. $x^2+(y+1)^2=9$ **15.** $(x-2)^2+(y+4)^2=25$
17. $x^2+(y+5)^2=100$
19.

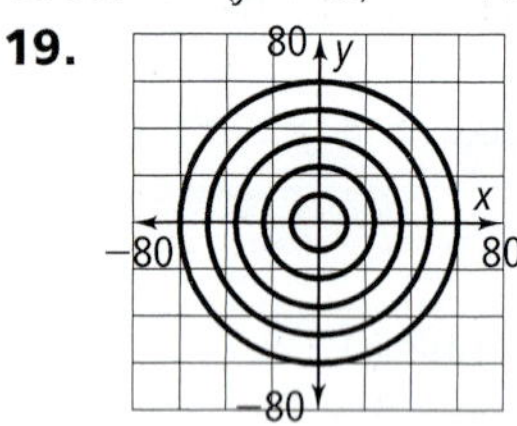

; $x^2+y^2=144$, $x^2+y^2=576$, $x^2+y^2=1296$, $x^2+y^2=2304$, $x^2+y^2=3600$

21. $(x-2)^2+(y+6)^2=16$ **23.** center $(-2, 10)$, radius 2 **25.** center $(-3, 5)$, radius 9 **27.** center $(-6, 0)$, radius 11

29.

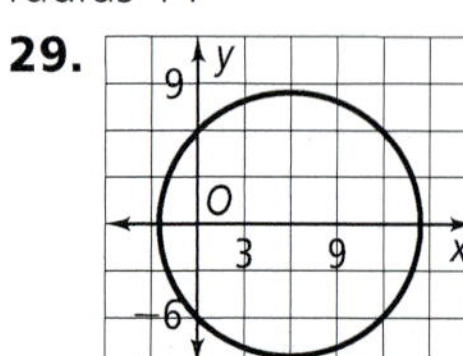

31.

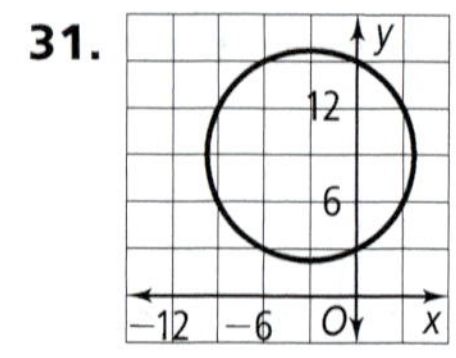

33.

35. $x^2+y^2=9$ **37.** $x^2+y^2=3$ **39.** $x^2+y^2=169$
41. $x^2+y^2=26$ **43.** $x^2+y^2=36$, $(x-8)^2+(y-6)^2=16$, $(x-8)^2+(y)^2=4$ **45.** $(x+6)^2+(y-13)^2=49$ **47.** $(x+2)^2+(y-7.5)^2=2.25$
49. $(x-2)^2+(y-1)^2=25$ **51.** $(x+1)^2+(y+7)^2=36$ **53.** center $(0, 0)$, radius $\sqrt{2}$ **55.** center $(0, 0)$, radius $\sqrt{14}$ **57.** center $(-5, 0)$, radius $3\sqrt{2}$
59. center $(-3, 5)$, radius $\sqrt{38}$ **61.** center $(3, 1)$, radius $\sqrt{6}$

63.

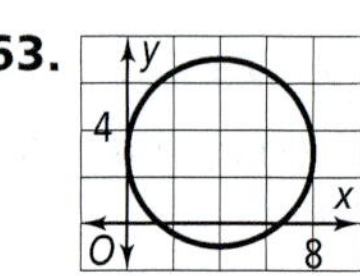

circle; $(x-4)^2+(y-3)^2=16$

65. a.

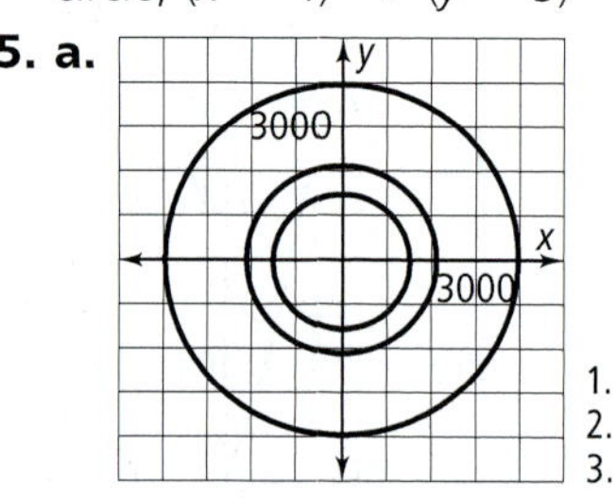

b. Earth: $x^2+y^2=15{,}705{,}369$
Mars: $x^2+y^2=4{,}456{,}321$
Mercury: $x^2+y^2=2{,}296{,}740$

67. 12 **69.** 5 **71.** 0.42 **72.** $x=-\frac{1}{12}y^2$
73. at $x=-1$ **74.** at $x=2$, and $x=3$ **75.** no points of discontinuity **76.** 4 **77.** 2 **78.** −3 **79.** 4 **80.** $\frac{1}{2}$
81. 1 **82.** −2, −9 **83.** 0, ±4, $\pm 4i$ **84.** ±2, $\pm 2\sqrt{2}$

Lesson 10-4 pp. 638–644

Got It? 1. $\frac{x^2}{4}+\frac{y^2}{25}=1$
2. a. (±8, 0).

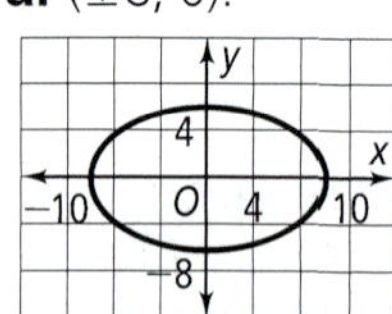

b. The vertex and co-vertex approach the same distance from the center of the ellipse; a circle
3. 24 ft **4.** $\frac{x^2}{36}+\frac{y^2}{53}=1$

Lesson Check 1. $\frac{x^2}{64}+\frac{y^2}{36}=1$ **2.** $(\pm\sqrt{21}, 0)$
3. $\frac{x^2}{169}+\frac{y^2}{25}=1$ **4.** $6\sqrt{23}$ ft $\approx$ 28.77 ft **5.** The student used a and b instead of a^2 and b^2; $\frac{x^2}{1681}+\frac{y^2}{841}=1$

6. The eq. of an ellipse with center at the origin is $\frac{x^2}{a^2} + \frac{y^2}{b^2} = 1$. For a circle, the major axis and the minor axis are of equal length such that $a = b = r$. Thus, by subst., $\frac{x^2}{r^2} + \frac{y^2}{r^2} = 1$ or $x^2 + y^2 = r^2$.

Exercises **7.** $\frac{x^2}{16} + \frac{y^2}{9} = 1$ **9.** $\frac{x^2}{9} + y^2 = 1$

11. $\frac{x^2}{16} + \frac{y^2}{49} = 1$ **13.** $\frac{x^2}{81} + \frac{y^2}{4} = 1$

15. $(0, \pm\sqrt{5})$

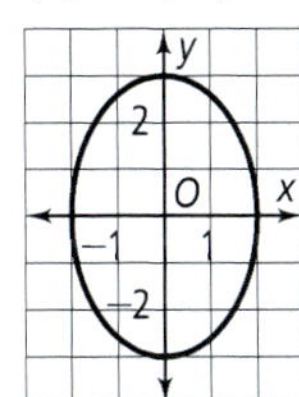

17. $(\pm 4\sqrt{2}, 0)$

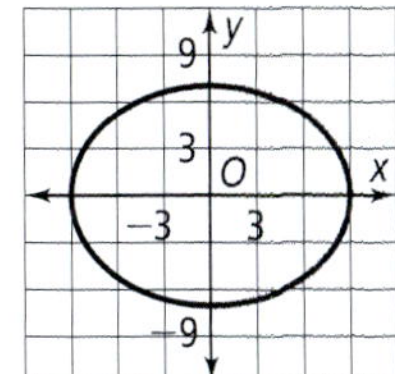

19. $(0, \pm 6)$

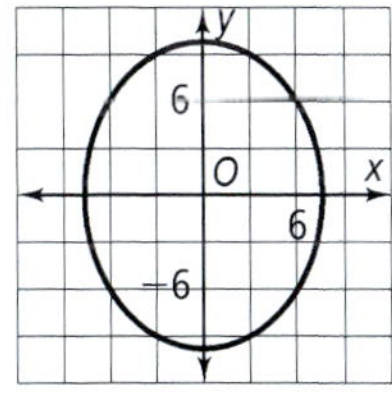

21. $(\pm 2\sqrt{3}, 0)$

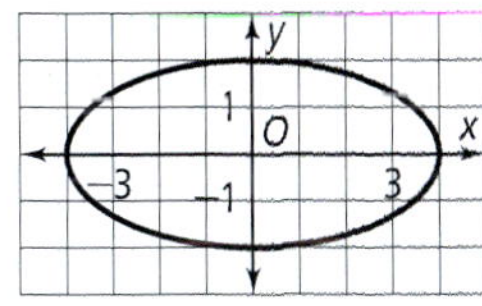

23. 32 **25.** 6 **27.** 12 **29.** $24\sqrt{2}$ **31.** $\frac{x^2}{100} + \frac{y^2}{64} = 1$
33. $\frac{x^2}{89} + \frac{y^2}{64} = 1$ **35. a.** about 22.25 ft **b.** Due to the reflective prop. of an ellipse, you can aim your putt at any part of the border. The ball will reflect off the border and go directly into the hole. **37.** $(0, \pm 2\sqrt{3})$ **39.** $(0, \pm\sqrt{21})$ **41.** $(0, \pm 1)$ **43. a.** 0.9 **b.** 0.1 **c.** The shape is close to a circle. **d.** The shape is close to a line segment.

45. $\frac{x^2}{16} + y^2 = 1$ **49.** $\frac{x^2}{25} + \frac{y^2}{4} = 1$

51. $\frac{x^2}{702.25} + \frac{y^2}{210.25} = 1$ **53.** $\frac{x^2}{256} + \frac{y^2}{324} = 1$

55. $\frac{x^2}{16} + \frac{y^2}{12} = 1$ **57.** $\frac{x^2}{36} + \frac{y^2}{27} = 1$ **59.** $\frac{x^2}{20} + \frac{y^2}{18} = 1$
61. a. 3×10^6 min **b.** about 0.016
c. $\frac{x^2}{8.649 \times 10^{15}} + \frac{y^2}{8.64675 \times 10^{15}} = 1$ **63.** B **65.** A
67. $(x - 1)^2 + (y + 5)^2 = 9$ **68.** $(x + 2)^2 + (y - 4)^2 = 81$ **69.** $\frac{1}{2x - 3x^4}$; $x \neq 0, x \neq \sqrt[3]{\frac{2}{3}}$ **70.** $\frac{x - 6}{x - 1}$; $x \neq 1, x \neq -6$ **71.** $\frac{x - 5}{x^2 - 2x + 4}$; $x \neq -2$ **72.** log 15
73. $\log_3 6$ **74.** log 2 **75.** $y = 2x + 4$ **76.** $y = \frac{1}{3}x$

Lesson 10-5 pp. 645–652

Got It? 1. a. $\frac{y^2}{16} - \frac{x^2}{9} = 1$

b.

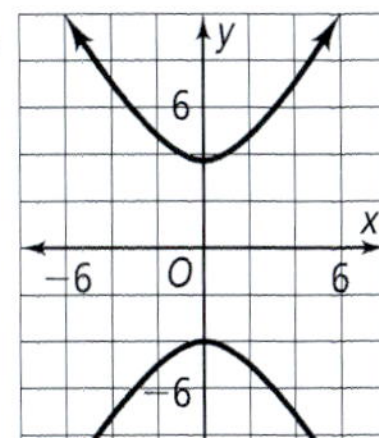

c. when $a = b$

2. vertices: $(\pm 2, 0)$; foci: $(\pm\sqrt{13}, 0)$; asymptotes: $y = \pm\frac{3}{2}x$

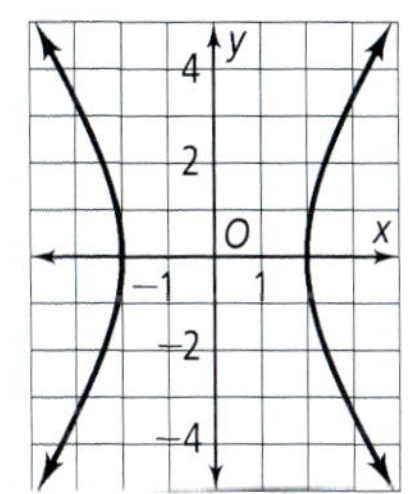

3. $\frac{x^2}{49} - \frac{y^2}{72} = 1$

Lesson Check

1. vertices: $(\pm 6, 0)$; foci: $(\pm\sqrt{61}, 0)$; slopes of asymptotes: $\pm\frac{5}{6}$

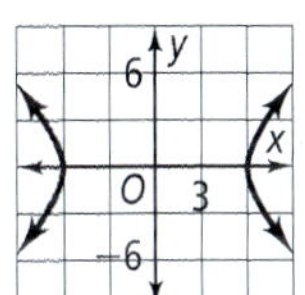

2. vertices: $(0, \pm 4)$; foci: $(0, \pm\sqrt{41})$; slopes of asymptotes: $\pm\frac{4}{5}$

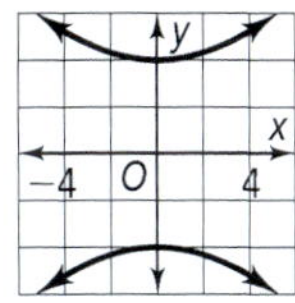

3. vertices: $(0, \pm 2)$; foci: $(0, \pm 2\sqrt{5})$; slopes of asymptotes: $\pm\frac{1}{2}$

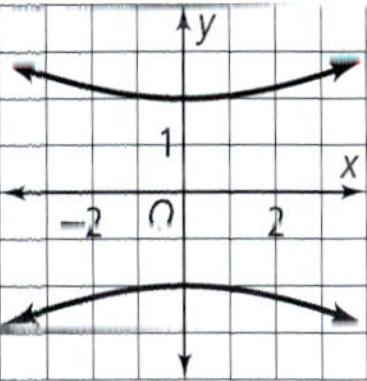

4. vertices: $(\pm 5, 0)$; foci: $(\pm\sqrt{41}, 0)$; slopes of asymptotes: $\pm\frac{4}{5}$

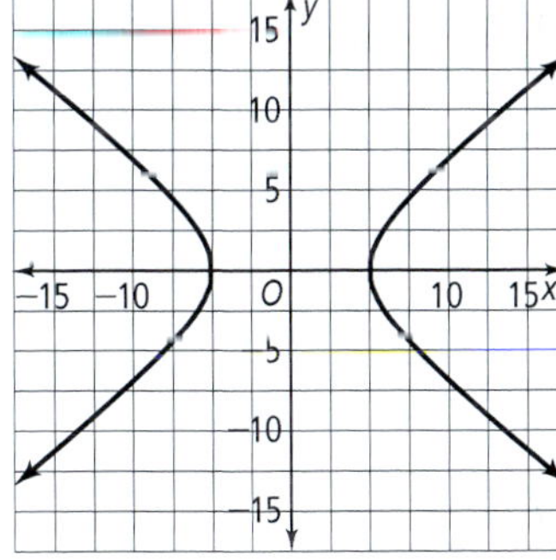

5. $\frac{x^2}{25} - \frac{y^2}{24} = 1$ **6.** Answers may vary. Sample: Similarities—Both have two axes of sym. that intersect at the center of the figure and two foci that lie on the

Selected **Answers**

same line as the two "principal" vertices. Differences—An ellipse consists of pts. whose distances from the foci have a constant sum, whereas a hyperbola consists of pts. whose distances from the foci have a constant diff. **7.** Answers may vary. Sample: A hyperbola is vert. or horizontal depending on whether it has a positive coefficient not because the larger denominator is under the y^2 term.

Exercises 9. $\frac{x^2}{144} - \frac{y^2}{25} = 1$ **11.** $\frac{x^2}{49} - \frac{y^2}{121} = 1$

13. $\frac{x^2}{4} - \frac{y^2}{5} = 1$

15. vertices: $(0, \pm 7)$; foci: $(0, \pm\sqrt{113})$; asymptotes: $y = \pm\frac{7}{8}x$

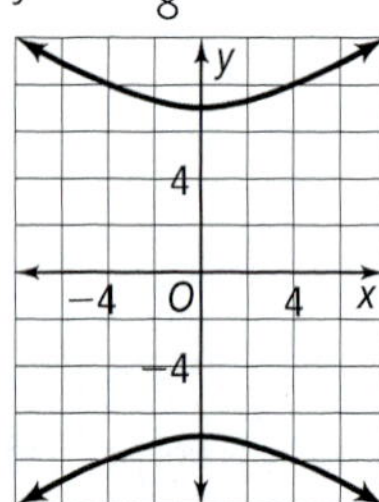

17. vertices: $(\pm 8, 0)$; foci: $(\pm 10, 0)$; asymptotes: $y = \pm\frac{3}{4}x$

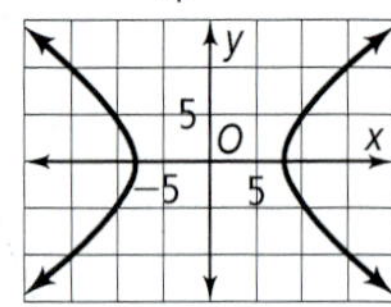

19. vertices: $(0, \pm 3)$; foci: $(0, \pm 3\sqrt{10})$; asymptotes: $y = \pm\frac{1}{3}x$

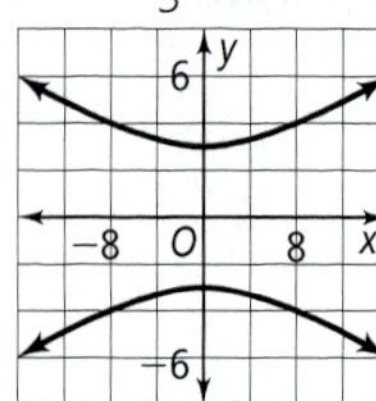

21. vertices: $(\pm 2\sqrt{2}, 0)$; foci: $(\pm 2\sqrt{11}, 0)$; asymptotes: $y = \pm\frac{3\sqrt{2}}{2}x$

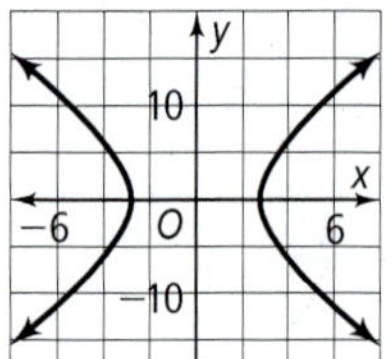

23. $x^2 - \frac{y^2}{6} = 1$ **25.** $\frac{x^2}{9} - \frac{y^2}{16} = 1$

27. $y^2 - \frac{x^2}{3} = 1$ **29.** $\frac{y^2}{20.25} - \frac{x^2}{4} = 1$

31. $\frac{x^2}{32} - \frac{y^2}{64} = 1$

33. $y = \pm\sqrt{x^2 - 1}$; $(\pm 1, 0)$

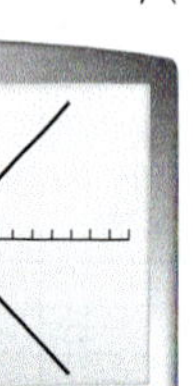

35.

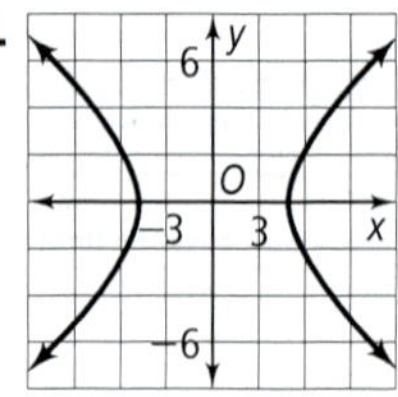

37.

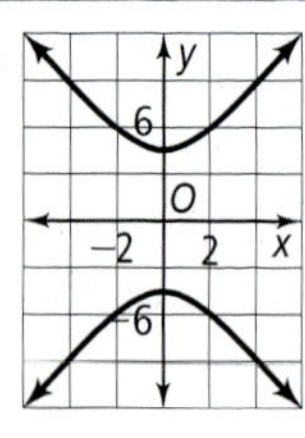

39. Answers may vary.

41. a. For the x-values in those rows, the value of $x^2 - 9$ is neg., so $\sqrt{x^2 - 9}$ is not a real number. **b.** As x increases, y increases, but the diff. between x and y gets closer to zero. **c.** No; for positive values of x greater than 3, $x = \sqrt{x^2}$ and $\sqrt{x^2} \neq \sqrt{x^2 - 9}$.

d. $y = x$, $y = -x$

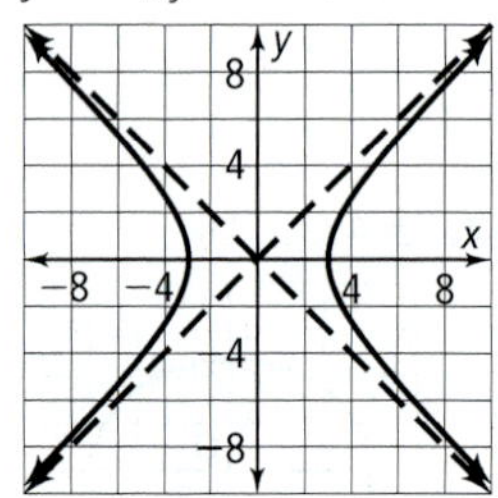

43. A **45.** A

47. foci: $(\pm 3, 0)$

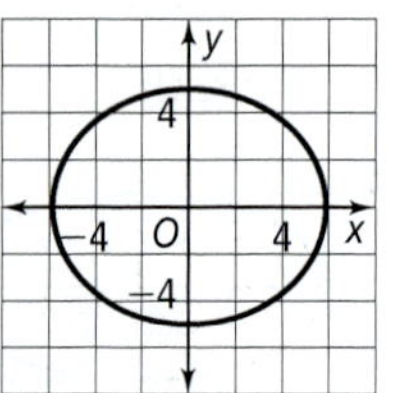

48. foci: $(0, \pm 1)$

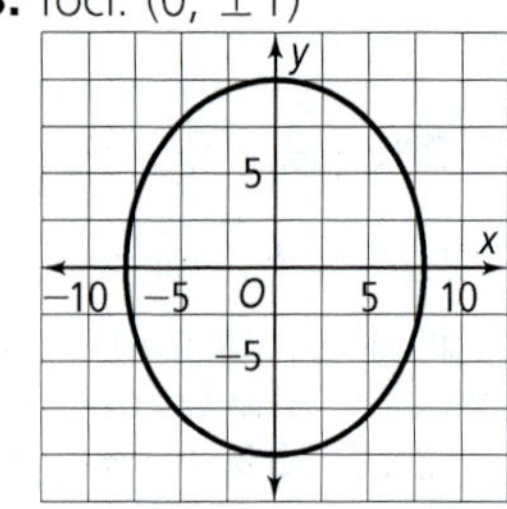

49. foci: $(0, \pm 6)$

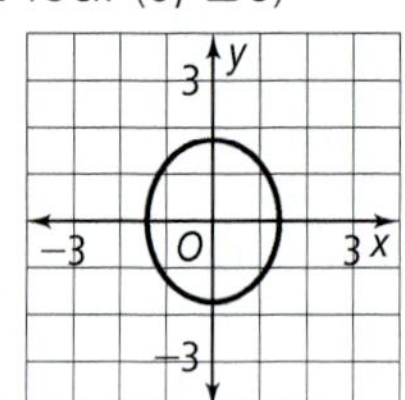

50. $\frac{1}{3}$ **51.** 125 **52.** 6 **53.** $y = (x - 3)^2 - 8$
54. $y = 2(x + 3)^2 - 18$ **55.** $y = 3(x + 4)^2 - 50$

Lesson 10-6 pp. 653–660

Got It?

1. $\frac{(x - 12)^2}{100} + \frac{(y - 3)^2}{64} = 1$;

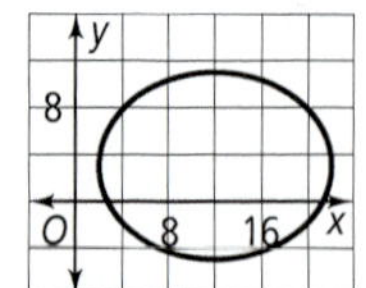

2. center (2, −2); vertices: (−4, −2), (8, −2); foci: (−8, −2), (12, −2); asymptotes: $y + 2 = \pm\frac{4}{3}(x - 2)$

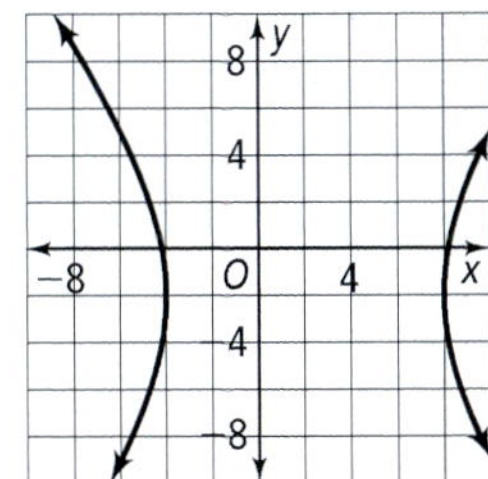

3. a. circle with center (6, −2) and radius $4\sqrt{3}$

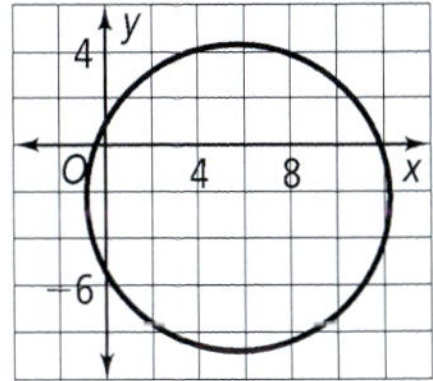

b. Replace $+y^2$ with $-y^2$ to get a new eq., $4x^2 - y^2 - 24x + 6y + 9 = 0$. The standard eq. of the hyperbola is $\frac{(x - 3)^2}{4.5} - \frac{(y - 3)^2}{18} = 1$.

4. $\frac{(x + 1)^2}{9} + \frac{y^2}{8} = 1$

Lesson Check 1. center (−7, −1); vertices: (−22, −1), (8, −1); foci: (−16, −1), (2, −1) **2.** center (1, 3); vertices: (4, 3), (−2, 3); foci: $(1 \pm \sqrt{13}, 3)$ **3.** $\frac{x^2}{48} + \frac{(y - 4)^2}{64} = 1$ **4.** $\frac{(y + 1)^2}{49} - \frac{(x + 3)^2}{51} = 1$ **5.** ellipse and hyperbola **6.** Your friend used the center (−2, 1) instead of (2, −1). The vertices are (−10, −1) and (14, −1). **7.** The student didn't write the standard-form equation correctly. The standard-form equation is $(x + 4)^2 + (y - 5)^2 = 0$. This is an equation of a circle with center at (−4, 5) and radius 0. Since the radius is 0, the graph is a single point, (−4, 5).

Exercises

9. $\frac{(x - 3)^2}{21} + \frac{(y + 6)^2}{25} = 1$

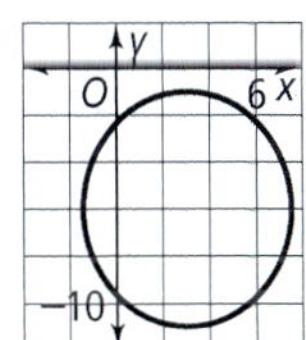

11. $\frac{(x - 1.5)^2}{42.25} + \frac{(y - 4)^2}{12} = 1$

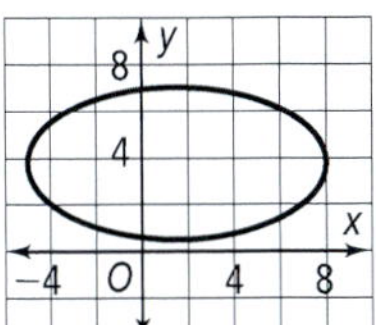

13. center: (3, 4); vertices: (3, 1), (3, 7); foci: $(3, 4 \pm \sqrt{13})$

15. $y = (x - 4)^2 + 3$; parabola; vertex: (4, 3)

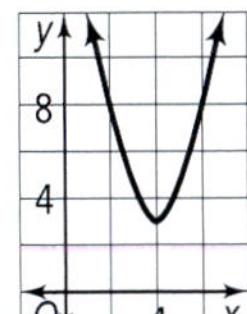

17. $(y - 2)^2 - (x - 3)^2 = 1$; hyperbola; center: (3, 2), foci: $(3, 2 \pm \sqrt{2})$;

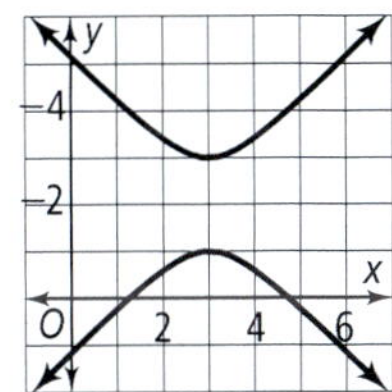

19. $x = \frac{1}{2}(y - 2)^2 + 3$; parabola; vertex: (3, 2)

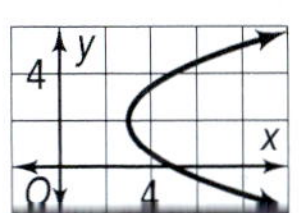

21. a. hyperbola **b.** one focus; the other focus **c.** with the center at the origin, $x^2 - \frac{y^2}{3} = 1$

23. a. hyperbola **b.** horizontal line

25.

$\frac{(y + 3)^2}{4} - \frac{(x - 6)^2}{5} = 1$

27. $\frac{(x - 1)^2}{9} + \frac{(y + 1)^2}{16} = 1$ **29.** $x^2 + y^2 = 16$

31. $y = 2(x + 2)^2 + 4$

33.

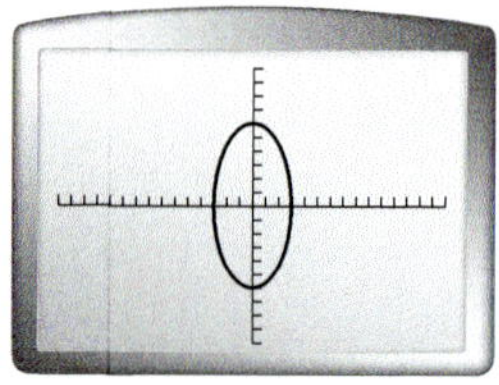

ellipse; $\frac{x^2}{9} + \frac{y^2}{36} = 1$

35.

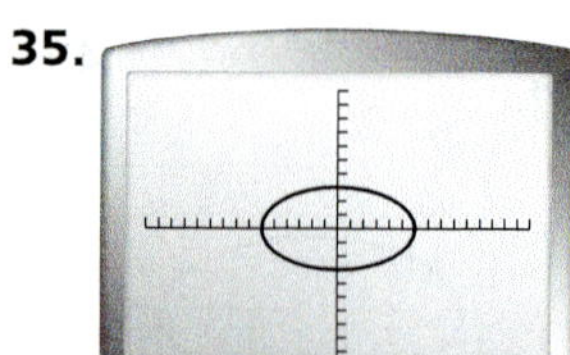

ellipse; $\frac{x^2}{36} + \frac{y^2}{9} = 1$

37. a. Earth: $\frac{x^2}{(149.60)^2} + \frac{y^2}{(149.58)^2} = 1$;

Mars: $\frac{x^2}{(227.9)^2} + \frac{y^2}{(226.9)^2} = 1$;

Mercury: $\frac{x^2}{(57.9)^2} + \frac{y^2}{(56.6)^2} = 1$

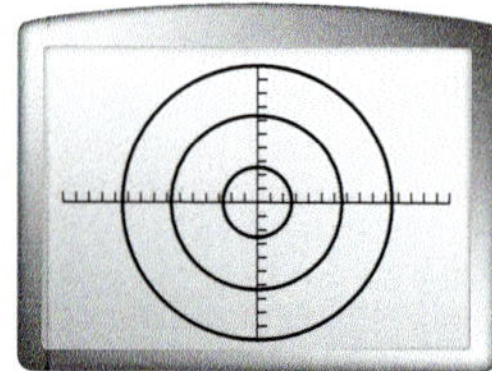

WINDOW FORMAT
Xmin = −379.0322...
Xmax = 379.03225...
Xscl = 25
Ymin = −250
Xmax = 250
Yscl = 25

b. Earth: $\frac{a}{b}$ is closest to 1 **39.** I **41.** Let a_1, a_2, and a_3 represent the missing terms in the arithmetic sequence: 15, a_1, a_2, a_3, 47. a_2 is the arithmetic mean of 15 and 47, so $a_2 = \frac{15 + 47}{2} = 31$. Likewise, $a_1 = \frac{15 + a_2}{2} = \frac{15 + 31}{2} = 23$, and $a_3 = \frac{a_2 + 47}{2} = \frac{31 + 47}{2} = 39$. The missing terms are 23, 31, and 39.

42. foci: $(\pm\sqrt{85}, 0)$

43. foci: $(0, \pm\sqrt{21})$

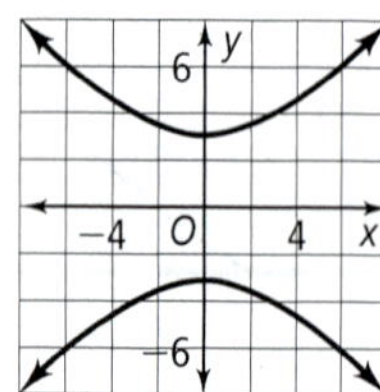

44. foci: $(0, \pm 2\sqrt{26})$

45. $\frac{3 \pm \sqrt{17}}{2}$ **46.** 5 **47.** $\frac{9 \pm \sqrt{201}}{12}$ **48.** 1 **49.** 2 **50.** 3 **51.** 8 **52.** −19 **53.** 6

Chapter Review pp. 663–666

1. directrix **2.** major axis **3.** standard form of an eq. of a circle **4.** radius **5.** transverse axis

6.

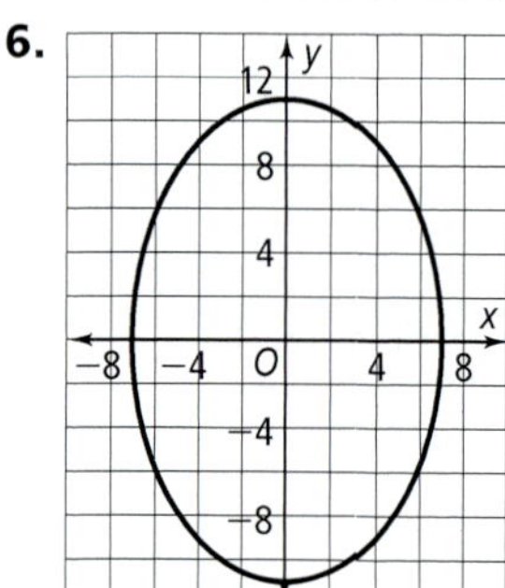

ellipse; lines of sym.: x-axis and y-axis; domain: $-7 \leq x \leq 7$, range: $-11 \leq y \leq 11$

7.

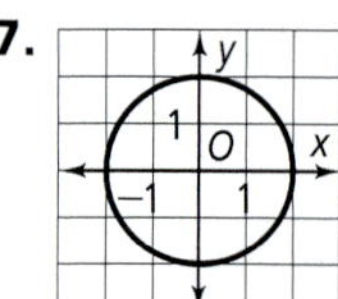

circle; lines of sym.: every line through the center; domain: $-2 \leq x \leq 2$, range: $-2 \leq y \leq 2$

8.

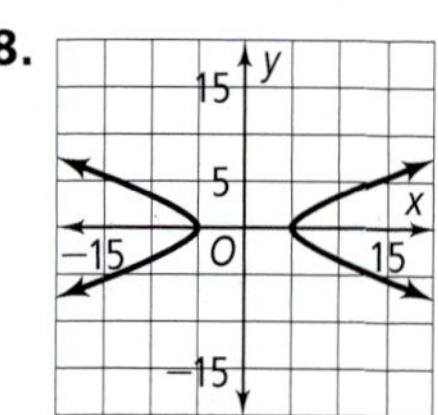

hyperbola; lines of sym.: x-axis and y-axis; domain: $x \leq -5$ or $x \geq 5$, range: all real numbers

9.

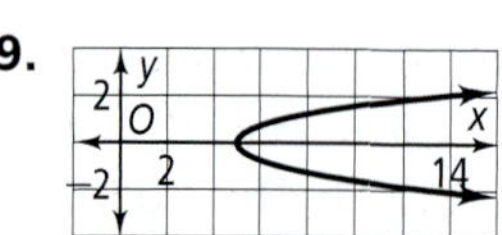

parabola; line of sym.: x-axis; domain: $x \geq 5$, range: all real numbers

10. center (0, 0); domain: $x \leq -4$ or $x \geq 4$, range: all real numbers **11.** center (0, 0); domain: $-3 \leq x \leq 3$, range: $-2 \leq y \leq 2$ **12.** $x = \frac{1}{20}y^2$ **13.** $y = -\frac{1}{20}x^2$ **14.** $y = \frac{1}{24}x^2$ **15.** $y = \frac{1}{10}x^2$ **16.** $y = 3x^2$ **17.** $y = \frac{1}{8}x^2 + 1$ **18.** $x = -\frac{1}{12}y^2 + 1$

19. focus: $\left(0, \frac{1}{20}\right)$, directrix: $y = -\frac{1}{20}$

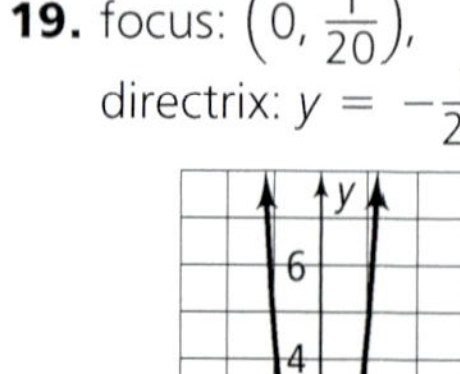

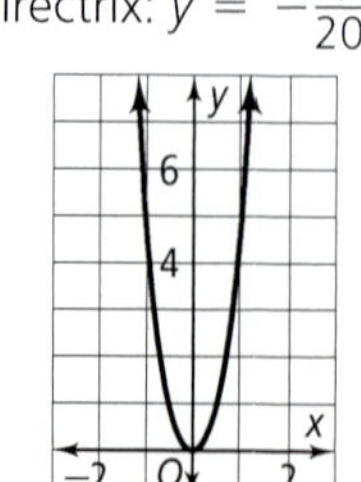

20. focus: $\left(\frac{1}{8}, 0\right)$, directrix: $x = -\frac{1}{8}$

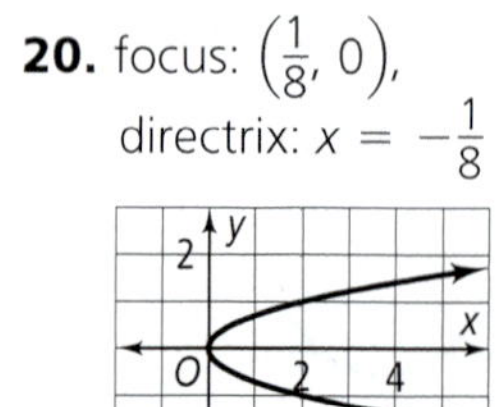

21. focus: $(-2, 0)$, directrix: $x = 2$

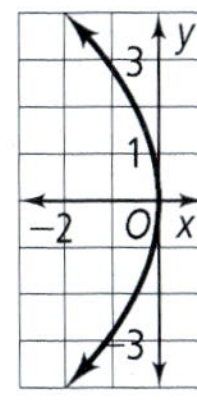

22. $x^2 + y^2 = 16$ 23. $(x - 8)^2 + (y - 1)^2 = 25$
24. $(x + 3)^2 + (y - 2)^2 = 100$
25. $(x - 5)^2 + (y + 3)^2 = 64$
26. center $(1, 0)$, radius 8

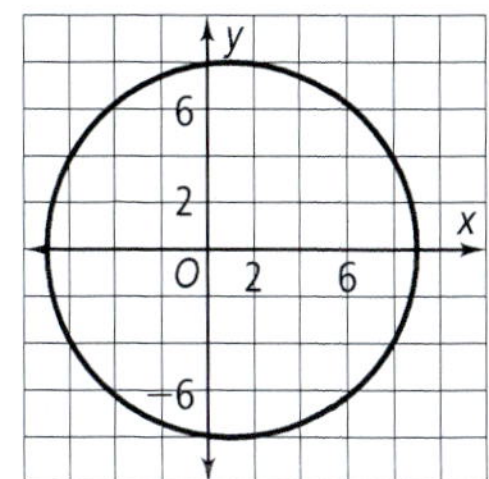

circle with radius 8 translated 1 unit to the rt.

27. center $(-7, -3)$, radius 7

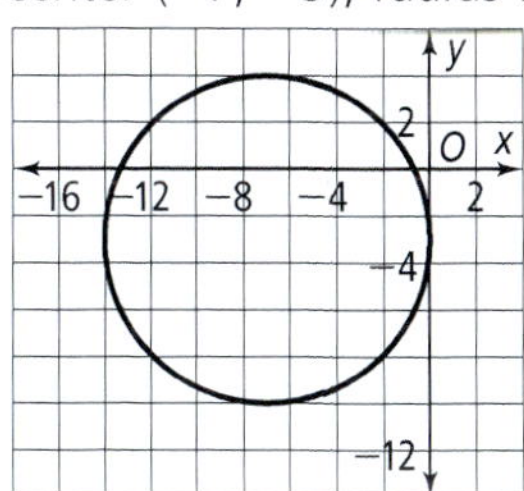

circle with radius 7 translated 7 units to the left and 3 units down

28. $\frac{x^2}{17} + \frac{y^2}{16} = 1$ 29. $\frac{x^2}{25} + \frac{y^2}{29} = 1$ 30. $\frac{x^2}{9} + \frac{y^2}{10} = 1$

31. $\frac{x^2}{40} + \frac{y^2}{36} = 1$ 32. $\frac{x^2}{64} + \frac{y^2}{16} = 1$

33. foci: $(0, \pm\sqrt{5})$

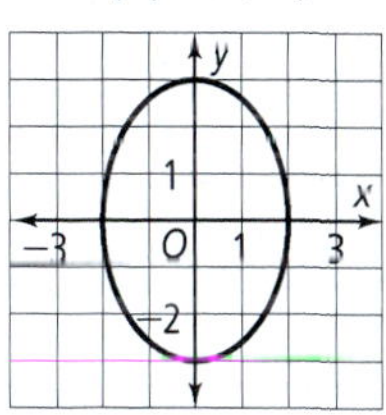

34. foci: $(\pm 3\sqrt{29}, 0)$

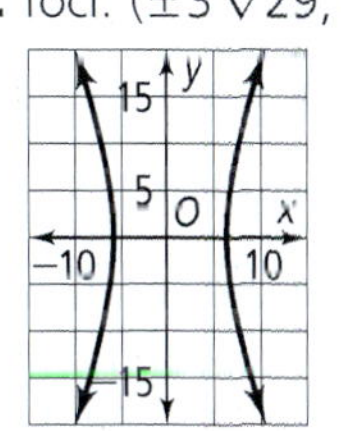

35. foci: $(0, \pm\sqrt{569})$

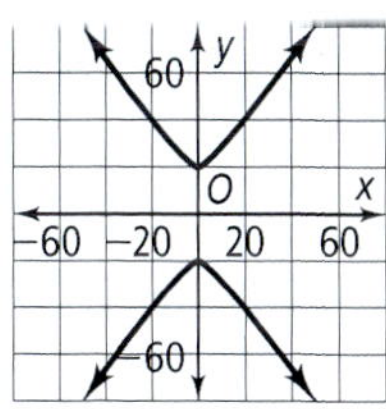

36. foci: $(\pm\sqrt{202}, 0)$

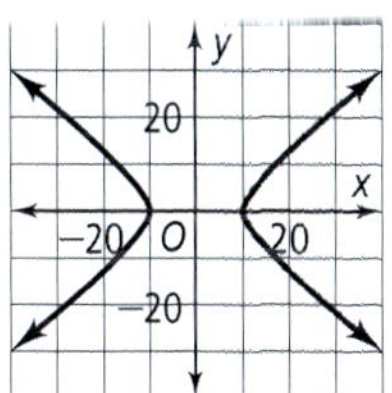

37. $\frac{x^2}{64} - \frac{y^2}{225} = 1$ 38. $\frac{y^2}{49} - \frac{x^2}{576} = 1$

39. $\frac{x^2}{1.148 \times 10^{10}} - \frac{y^2}{3.395 \times 10^{10}} = 1$

40. $(x - 1)^2 + (y - 1)^2 = 25$

41. $\frac{(x - 3)^2}{4} + \frac{(y + 2)^2}{9} = 1$

42. $\frac{(x - 6)^2}{9} - \frac{(y - 3)^2}{16} = 1$

43. hyperbola; center $(0, -2)$, foci: $(0, -2 \pm 2\sqrt{10})$

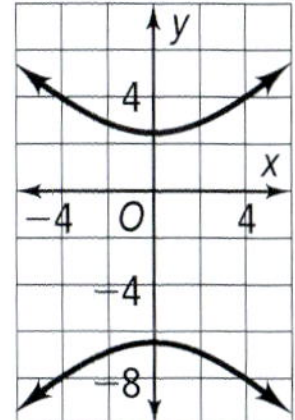

44. circle; center $\left(-\frac{3}{2}, 2\right)$, radius $\frac{\sqrt{61}}{2}$

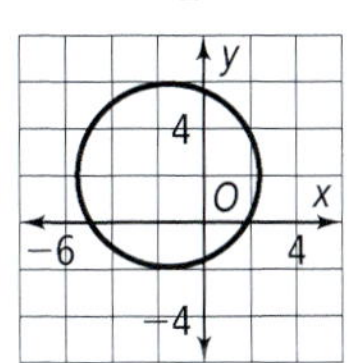

45. parabola; vertex: $\left(-\frac{1}{2}, \frac{169}{4}\right)$

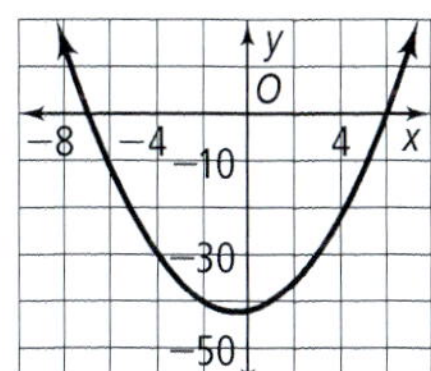

Chapter 11

Get Ready! p. 671

1. $83.\overline{3}\%$ 2. $19.\overline{4}\%$ 3. $\approx 92.308\%$ 4. 30.56%
5. 6720 6. 22,100 7. 10 8. $a^5 + 5a^4b + 10a^3b^2 + 10a^2b^3 + 5ab^4 + b^5$ 9. $j^3 + 9j^2k + 27jk^2 + 27k^3$
10. $m^2 + 1.4m + 0.49$ 11. $16 + 32t + 24t^2 + 8t^3 + t^4$ 12. $m^2 + 2mn + n^2$ 13. $x^4 + 12x^3y + 54x^2y^2 + 108xy^3 + 81y^4$ 14. $\pm\frac{1}{10}$ 15. $\pm\frac{1}{20}$
16. $\pm\frac{1}{14}$ 17. $\pm\frac{1}{2}$ 18. $\pm\frac{1}{3}$ 19. $\pm\frac{1}{24}$ 20. In a math class, when actual trials are difficult to conduct, you can find experimental probability by using a simulation which is a model of one or more events. 21. It rained today.
22. The mean, 12.6; the data are fairly evenly distributed around the mean which makes the mean the best representation of the data given.

Lesson 11-1 pp. 674–680

Got It? 1. 6,760,000 2. 40,320 3a. 2730 b. Yes; because $n = 10$ and $r = 3$ in the formula ${}_nP_r$ for both cases. 4a. 56 b. 36 c. 3003 5. 1680

Lesson Check 1. 120 **2.** 3024 **3.** 10 **4.** 21 **5.** 4,151,347,200 **6.** A permutation is an arrangement of items in a particular order; order is important. An arrangement in which order does not matter is a combination. **7.** ${}_nP_r = \frac{n!}{(n-r)!}$; substituting $n = r$, we get ${}_nP_n = \frac{n!}{(n-n)!} = \frac{n!}{0!}$; but ${}_nP_n = n! = \frac{n!}{1}$; substituting values again, we get $\frac{n!}{1} = \frac{n!}{(n-n)!}$, and so $0! = 1$.

Exercises 9. 20 **11.** 12 **13.** 3,628,800 **15.** 720 **17.** 120 **19.** 3003 **21.** 8 **23.** 336 **25.** 6 **27.** 60,480 **29.** 10,897,286,400 **31.** 56 **33.** 4 **35.** 15 **37.** $\frac{5}{18}$ **39.** combination; 4368 **41.** combination; 70 **43.** True because of the Assoc. Prop. of Mult. **45.** False; answers may vary. Sample: $(3 \cdot 2)! = 6! = 720$ and $3! \cdot 2! = 6 \cdot 2 = 12$ **47.** False; answers may vary. Sample: $(3!)^2 = 6^2 = 36$ and $3^{(2!)} = 3^2 = 9$ **49.** C **51.** Two ways, because order matters. **53a.** 2 **b.** 6 **c.** $(n-1)!$ **55. a.** 35 **b.** 6 **c.** ${}_7C_3 = \frac{7!}{3!4!}$, so ${}_7C_3 \cdot 3! = \frac{7!}{4!}$, which is the permutation formula for ${}_7P_3$. **57.** H **59.** G **61.** center: (2, 1); vertices: (2, 6), (2, −4); co-vertices: (5, 1), (−1, 1); foci: (2, 5), (2, −3) **62.** $(x-1)^2 + (y-1)^2 = 36$ is a circle (not an ellipse) with center (1, 1) and radius 6. **63.** $4(x-1)^2$ **64.** $-(x+3)^2$ **65.** $3(x-5)(x+5)$ **66.** 30,240 **67.** $\frac{4}{5}$ **68.** 210

Lesson 11-2 pp. 681–687

Got It? 1. 0.40 or 40% **2.** 0.20 or 20% **3a.** $\frac{1}{2}$ **b.** The likelihood of getting an even or odd is the same, i.e. $\frac{1}{2}$. **4.** $\frac{48}{2{,}598{,}960}$ or 0.0000184689 or ≈0.00185% **5.** 0.05 or 5%

Lesson Check 1. 0.75 or 75% **2.** 0.80 or 80% **3.** $\frac{1}{6}$ **4.** $\frac{1}{3}$ **5.** Experimental probabilities are calculated on the basis of data from an experiment, actual or simulated. Given equally likely outcomes, the basis for calculating theoretical probability is being able to determine the no. of ways that an event can occur within these outcomes. Comparisons of measures such as length and area are the basis of geometric probability. **6.** Answers may vary. Samples: Flip a coin; generate random numbers on a calculator; roll a die with odd numbers as true and even numbers as false. **7.** Because you are averaging over more samples, you are getting a more accurate average.

Exercises 9. the number 1: $\frac{21}{134} \approx 15.7\%$; the number 2: $\frac{11}{67} \approx 16.4\%$; the number 3: $\frac{45}{268} \approx 16.8\%$; the number 4: $\frac{11}{67} \approx 16.4\%$; the number 5: $\frac{47}{268} \approx 17.5\%$; the number 6: $\frac{23}{134} \approx 17.2\%$ **11.** Answers may vary. Sample: Toss 5 coins. Keep a tally of the times three or more heads are tossed. (A head represents a correct answer.) Do this 100 times. The total number of tally marks, as a percent, gives the experimental probability. The simulated probability should be about 50%. **13.** $\frac{3}{10}$, or 30% **15.** $\frac{4}{5}$, or 80% **17.** $\frac{48}{125}$, or 38.4% **19.** $\frac{103}{125}$, or 82.4% **21.** $\frac{77}{125}$, or 61.6% **23.** $\frac{{}_{30}C_3 \cdot {}_{120}C_6}{{}_{150}C_9} \approx 0.17879 \approx 17.9\%$ **25.** $\frac{5}{8}$, or 62.5% **27.** $\frac{3}{4}$, or 75% **29.** $\frac{116}{147} \approx 78.9\%$ **31.** $\frac{43}{147} \approx 29.3\%$ **33.** 1 chance in 2,869,685 or ≈ 0.00003485% **35.** if there are any restrictions on the last digit of a ZIP code **37.** B **39.** B

41.

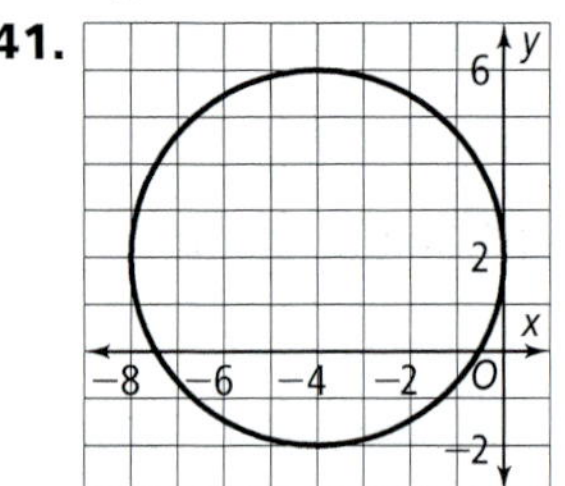

42. 20 **43.** 840 **44.** 10 **45.** 45 **46.** $\frac{25b - 7a^3}{5a^2b^2}$ **47.** $\frac{3q + 7p}{pq}$ **48.** 0 **49.** $\frac{7}{36} = 19.\overline{4}\%$ **50.** $\frac{25}{36} = 69.\overline{4}\%$ **51.** $\frac{1}{2}$, or 50%

Lesson 11-3 pp. 688–693

Got It? 1. Independent; the number of coins is the same after the coin is replaced. **2.** 0.20, or 20% **3a.** Not mutually exclusive; 2 is a prime number and an even number. **b.** Mutually exclusive; there is no even number less than 2 in the roll of a number cube. **4a.** 0.61, or 61% **b.** Yes; the percentage of students tells which language is chosen by more students. **5a.** $\frac{5}{9}$ **b.** $\frac{5}{9}$

Lesson Check 1. $\frac{1}{15}$, or $6.\overline{6}\%$ **2.** $\frac{27}{80}$, or 33.75% **3.** 1, or 100% **4.** $\frac{7}{8}$, or 87.5% **5.** $\frac{5}{8}$, or 62.5% **6.** Events A and B are independent if the outcomes of A do not affect the outcomes of B. The events are mutually exclusive if A and B cannot occur at the same time. For independent events, $P(A \text{ and } B) = P(A) \cdot P(B)$. For mutually exclusive events, $P(A \text{ and } B) = 0$. For any events, $P(A \text{ or } B) = P(A) + P(B) - P(A \text{ and } B)$. **7.** Since these are not mutually exclusive events, $P(A \text{ and } B) \neq 0$. The student should have multiplied to get the correct answer, which is 0.21 or 21%.

Exercises 9. independent **11.** dependent **13.** $\frac{1}{6}$ **15.** 0.54 **17.** $\frac{9}{25}$ **19.** mutually exclusive; if the numbers are equal, then the sum is even **21.** $\frac{3}{4}$ **23.** 39% **25.** $\frac{1}{2}$

27. $\frac{5}{6}$ **29.** $\frac{2}{3}$ **31.** $\frac{2}{5}$ **33.** 14.5% **35.** 87.4% **37.** $\frac{4}{15}$ **39.** $\frac{8}{15}$ **41.** not mutually exclusive **43.** $\frac{4}{9}$ **45.** $\frac{8}{11}$ **47.** 8 **49.** 5 **51.** $\frac{1}{6}$ **52.** $\frac{1}{2}$ **53.** $\frac{1}{2}$ **54.** $\frac{3}{7}$ **55.** $-\frac{3}{2}$, 2 **56.** $\frac{1}{6}$ **57.** $\frac{1}{2}e^3 \approx 10.04$ **58.** $\frac{1}{2}e^6 \approx 201.71$ **59.** $\pm e^2 \approx \pm 7.39$ **60.** $\frac{1}{16}$ **61.** $\frac{1}{16}$ **62.** $\frac{3}{16}$

Lesson 11-4 pp. 696–702

Got It? 1a. $\approx$0.57355 or $\approx$57.355% **b.** Female; there are more females enrolled. **2a.** $\approx$0.026448 or $\approx$2.64% **b.** $\approx$0.040302 or $\approx$4.03% **3.** 0.2 **4.** 9%

Lesson Check 1. $\frac{1}{2}$ **2.** $\frac{1}{13}$, or 7.7% **3.** 0% **4.** 50%

5. The sum of the probability of an event happening and the probability of an event not happening is 1. Each branch represents either the event happening or the event not happening. **7.** Answers may vary. Sample: Tree diagrams apply to cases in which more than one event occurs in a sequence. The Fundamental Counting Principle applies to situations in which there are multiple outcomes of a single event. With a tree diagram, but not with the Fundamental Counting Principle, you can determine probabilities of dependent events, or conditional probabilities.

Exercises 9. 0.6 **11.** $\approx$0.085 **13.** $\approx$0.682 **15.** $\approx$0.709 **17.** $\approx$23%

19.

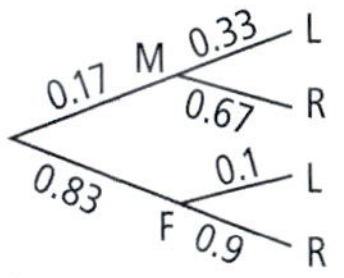

M = male
F = female
R = right-handed
L = left-handed

$P(L|F) = 10\%$, $P(M \text{ and } R) \approx 11.4\%$

21. 75% **23.** P(S and W) **25.** $\frac{2}{3}$, or 66.67% **27.** 0.08, or 8% **29.** 0.84 **31.** 0.16

33.

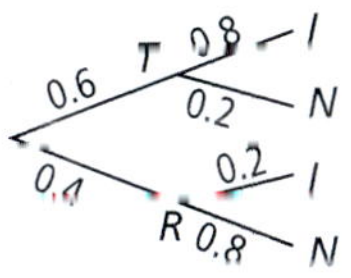

T = representative that completed training seminars
R = representative that didn't complete a training seminar
I = representative with increased sales
N = representative without increased sales

$P(I|N) = 0.2$

35. H **37.** (1, 1), (1, 2), (1, 3), (2, 1), (2, 2), (2, 3), (3, 1), (3, 2), (3, 3); yes **38.** $\frac{1}{3} = 33.\overline{3}\%$ **39.** $\frac{17}{76} \approx 0.22368 \approx 22.37\%$ **40.** $x = \frac{1}{4}(y-2)^2 + 5$ **41.** $y = \frac{1}{12}(x+2)^2 + 3$ **42.** 2 **43.** 0.830 **44.** 1.404 **45.** 3.465 **46.** $\frac{1}{7}$ **47.** $\frac{3}{7}$ **48.** $\frac{2}{7}$ 49. $\frac{1}{7}$

Lesson 11-5 pp. 703–709

Got It? 1. Answers may vary. Sample answer: No, it is not likely that both siblings have an equal chance of winning the race. **2. a.** 1, 6, 8, 9, 3 **b.** Yes, each student has an equal chance of being selected for either team. **3.** Answers may vary. Sample answer: Roll the cube until you get a 6. Keep track of the results. Repeat several times and take the average number of rolls needed. **4.** Answers may vary. Sample answer: No, almost as many volunteers who received the placebo reported improvement as received the drug. Fewer than half of those who received the drug reported improvement.

Lesson Check 1. about 0.89 **2.** about 0.86 **3.** Answers may vary. Sample answer: Flipping a coin to decide who has to wash the dishes is a fair decision. Arm wrestling to see who has to wash the dishes might be unfair if one brother is stronger than the other. **4.** Answers may vary. Sample answer: He only conducted 1 trial of the simulation, which is not enough to arrive at an accurate prediction. He should conduct the simulation at least 25 times and find the average number of boxes needed. **5.** Answers may vary. Sample answer: A simulation is an imitation or way of acting something out. In a mathematical simulation, a probability model is used to act out a situation that would be difficult or impractical to actually perform.

Exercises 7. Answers may vary. Sample answer: This will not result in a fair decision because the first person chooses the second person and might favor someone over someone else. **9.** 01, 05, 16, 03, 08 **11. a.** about 0.82 **b.** about 0.35 **c.** Sample answer: Yes, a high percentage of students who took the class passed the board exams on their first attempt so the class appears to be beneficial. **13.** 28 trials; about 1.14 correct answers per trial **15.** Answers may vary. Sample answer: Yes, the defensive driving course appears to be very effective and should be offered again. None of the drivers who took the course were involved in a major accident in the previous year. **17.** Answers may vary. Sample answer: Use a graphing calculator to generate random integers from 1 to 5. Let the integers 1, 2, 3, and 4 represent a made field goal, and let 5 represent a missed field goal. Generate random integers in groups of 3 to simulate the attempts in the next game. Perform the simulation at least 20 or 25 times and find the average number of field goals made. **19.** A **21.** C **23.** 0.35 **24.** 0.52 **25.** 0.7 **26.** 18

Lesson 11-6 pp. 711–718

Got It? 1. mean: 5.25, median: 5, mode: 5 **2a.** Yes; it is unlikely that the water temperature of a lake would change by 25 degrees. **b.** No; 98 would represent the busiest night of the week and it may relate to a weekly event. **3.** Dauphin Island: mean: $69.08\overline{3}$, mode: 84, range: 33, $Q_1 = 58$, median: 71, $Q_3 = 81$, interquartile range: 23; Grand Isle: mean: $73.41\overline{6}$, modes: 61, 70, 77, 83, 85, range: 24, $Q_1 = 64.5$, median: 73.5, $Q_3 = 83$, interquartile range: 18.5; The range and the interquartile range show the temperatures varying less at Grand Isle than at Dauphin Island. Also, the temperatures at Grand Isle are generally higher. **4a.** Use STAT PLOT, select a box-and-whisker plot. Enter data for the three remaining Gulf Coast sites. Enter the window values. Draw the box-and-whisker plots. Use TRACE on the plot to find quartiles Q_1, Q_2 and Q_3.

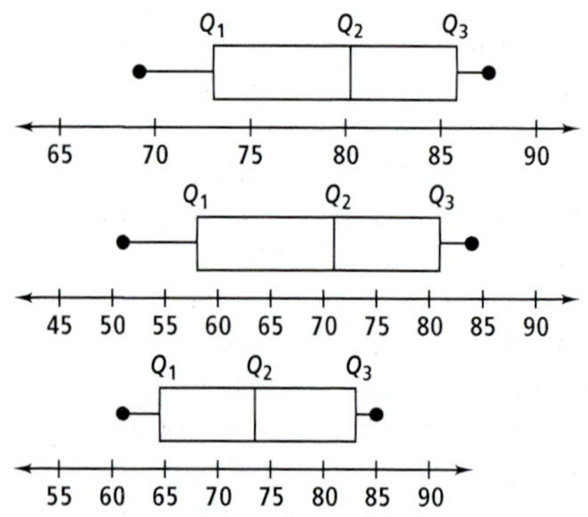

b. Yes, a box-and-whisker plot uses minimum and maximum values, the median, and the first and third quartiles to display the variability in a data set. **5a.** 79 **b.** 98

Lesson Check 1. outlier: 54; outlier included: mean: 22.8, median: 19.5, mode: 18; outlier not included: mean: $19.\overline{3}$, median: 19, mode: 18 **2.** outlier: 40; outlier included: mean: $92.\overline{6}$, median: 98, mode: 90; outlier not included: mean: 99.25, median: 99, mode: 90 **3.** the mean because the sum of the data values is affected and the mean depends on the sum **4.** 40%: 49 and below; 80%: 58 and below **5.** the mean; when data are somewhat sym, the best representation is the mean **6.** The error is in how to calculate the median. The median is the middle value or the 11th value which is 90.

Exercises 7. mean: $112.\overline{3}$, median: 95, mode: none **9.** 9.8 **11.** Jacksonville: mean: $67.991\overline{6}$, mode: none, range: 29.2, $Q_1 = 58.15$, median: 68.4, $Q_3 = 78.6$, interquartile range: 20.45; Austin: mean: $68.58\overline{3}$, mode: none, range: 36, $Q_1 = 56.85$, median: 70.5, $Q_3 = 80.75$, interquartile range: 23.9; the range and the interquartile range show the temperatures varying less at Jacksonville than at Austin.

13.

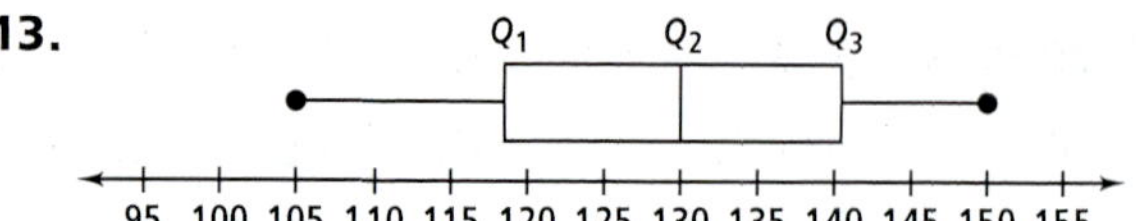

15. 5; 17 **17.** outlier: 381; outlier included: mean: ≈161.214, median: 158, mode: none; outlier not included: mean: ≈144.308, median: 142, mode: none

19.

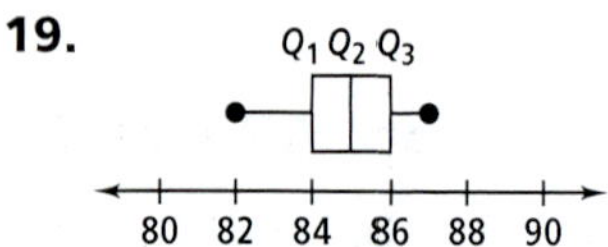

21. 30th **23.** 89 is at the 100th percentile, since 100% of the values are less than or equal to 89. **25.** The median; a few outliers can heavily influence the mean without drastically affecting the median. **27.** 83.9 **29.** Answers may vary. Sample: The range for women's shot put is greater than that for men's. The men are more consistent, as indicated by the shorter box and whiskers. Overall the men tend to throw farther. **31.** G

33. P(H | I) = 0.40, P(H and I) = 0.20

$$P(H|I) = \frac{P(H \text{ and } I)}{P(I)}$$
$$0.40 = \frac{0.20}{P(I)}$$
$$P(I) = 0.50$$

34. 0.20 **35.** 0.56 **36.** yes; −9 **37.** yes; 17 **38.** no **39.** yes; 0 **40.** ±16 **41.** ±0.09 **42.** $\pm\frac{11}{4}$ **43.** $\pm\frac{19}{5}$

Lesson 11-7 pp. 719–724

Got It? 1. $\bar{x} = 69.8\overline{3}$, $\sigma^2 = 115.1389$, $\sigma = 10.7303$ **2.** $\bar{x} = 7.2\overline{6}$, $\sigma = 3.316$ **3a.** within 3 standard deviations of the mean **b.** FEMA can expect that the no. of hurricanes for a 15-year period will fall within 3 standard deviations of the mean.

Lesson Check 1. $\bar{x} = 10$, $\sigma^2 = 19.8$, $\sigma = 4.45$ **2.** within 2 standard deviations of the mean **3.** Measures of central tendency are specific data pts. which give a summary of the middle of the data set, whereas the measures of variation give a summary of the variation of the data set within the range of distribution. **4.** Standard deviation measures how widely spread the data values are. If the data pts. are close to the mean, the standard deviation is small; if the data pts. are far from the mean, the standard deviation is large. The data pts. of Set B are closer to the mean of 70 than the data pts. of Sets A and C; likewise, the data pts. of Set A are closer to 70 than the data pts. of Set C. **5.** The effect of an outlier on the standard deviation is to increase the standard deviation.

Exercises 7. $\bar{x} \approx 15.1$, $\sigma^2 \approx 12.4$, $\sigma \approx 3.5$ **9.** $\bar{x} = 43.8$, $\sigma^2 = 75.76$, $\sigma \approx 8.7$ **11.** $\bar{x} \approx 12320.00$, $\sigma \approx 273.71$ **13.** 3 standard deviations **15.** $\bar{x} = 53.8$, $\sigma \approx 3.4$; 1σ: 7; 2σ: 9; 3σ: 10 **17.** Overall farm income increased slightly, but there was less variability among the states in 2002. The income in 2001 clustered more tightly around the mean. (2001: $\sigma_x \approx 2679$, 2002: $\sigma_x \approx 2758$) **21.** Your first friend; one standard deviation encompasses all values within one standard deviation above and below the mean. The graph shows that all values are within 3 standard deviations of the mean. **23. a.** no change to σ **b.** σ increases by a factor of 10 **25.** 13 **27.** $\frac{5}{6}$

28. Q_1 Q_2 Q_3

25 30 35 40 45 50 55 60

29. Q_1 Q_2 Q_3

20 25 30 35 40 45 50 55

30. center (2, −1); radius 6 **31.** center (1, 1); radius 2 **32.** $\frac{1}{2}$ **33.** $-\frac{1}{3}$ **34.** $\frac{1}{6}$ **35.** $-\frac{1}{11}$ **36.** $-\frac{1}{9}$ **37.** $\frac{1}{7}$

Lesson 11-8 pp. 725–730

Got It? 1a. convenience sample; yes; since the location is at the food court in the mall, the sample may over-represent food court or fast food supporters. **b.** Answers may vary. Sample: population data for the US census **2.** Controlled study; If other factors of the volunteers are random, like age, gender, and overall health, are known, the results can be used to make a general conclusion. **3.** Answers may vary. Sample: Use a systematic sample. Go to every fifth house in your neighborhood. State the first and last names of the governor and ask a household member to identify the named person. A possible unbiased survey question is, "Who is this person?". **Lesson Check 1a.** convenience sample **b.** Yes; since the location is near the exit of a history museum, the sample may over represent people who enjoy learning history and the results will have a bias. **2.** Yes; the question is leading and loaded. It suggests the person wants a particular answer. **3.** All members of the set are the population. A sample is a subset of the population. Answers may vary. Sample: population: students in a high school; sample: students who like to snowboard **4.** It is important to have as little error as poss. in a sample, thus giving an unbiased sample. An unbiased sample is more representative of an entire population. **5.** A large sample size would give a better estimate. The size of the sample is important to the reliability of the sample. **Exercises 7.** systematic sampling; no bias **9.** Survey; the statistics can be used to make a general conclusion about the population because the sample is randomly generated, and the survey question does not introduce a bias into the study. **11.** Controlled experiment; the statistics from this study can be used to make a general conclusion about the effectiveness of the plant food for this particular plant type as compared with giving no plant food at all. **13.** Answers may vary. Sample: Convenience sampling; interview students at a local high school. **15.** Answers may vary. Sample: Self-selected sampling; a newspaper article invites females over the age of 21 to call the paper and express their opinions. **17.** self-selected sampling; biased because only those who spend time online will respond. **19. a.** all students at the school **b.** every tenth student who enters the school building the day of the survey **c.** Answers may vary. Sample: A little over half of students favor the new dress code. **21.** Answers will vary. Sample: No, because you would have to assume that all registered voters will actually vote on Election Day. **23. a.** convenience sample **b.** observational study **c.** Answers may vary. Sample: The statistics do not necessarily represent the school population because a random sample was not used to conduct the study. **25.** Yes, the question is leading the respondent to a particular desired answer, and it gives statistics that may elicit a strong reaction. Also, it requires the respondent to answer a question about whether a person *should* wear a safety belt, which may not necessarily influence whether they support the law. **27.** G **29.** $\bar{x} \approx 2.83$, $\sigma \approx 2.54$ **30.** $\bar{x} \approx 5.62$, $\sigma \approx 3.67$ **31.** $y = \frac{1}{2}(x - 5)$; yes **32.** $y = \pm\sqrt{x}$; no **33.** $y = \pm\sqrt{\frac{9x}{5}}$; no **34.** $y = \frac{x^2}{9}$, $x \geq 0$; yes **35.** 6 **36.** 1 **37.** 10 **38.** 792

Lesson 11-9 pp. 731–738

Got It?
1. $P(0) = 0.07776$; $P(1) = 0.2592$; $P(2) = 0.3456$; $P(3) = 0.2304$; $P(5) = 0.01024$
2. $81x^4 + 108x^3y + 54x^2y^2 + 12xy^3 + y^4$
3. ≈0.1035, or about 10.4%
Lesson Check 1. ≈0.3110, or ≈31.10% **2.** ≈0.1641, or ≈16.41% **3.** $20c^3d^3$ **4.** $-10x^4y$ **5.** 0.2646, or 26.46% **6.** Answers may vary. Sample: A binomial experiment has three important features: **a.** The situation involves repeated trials; flipping a coin 10 times has 10 trials. **b.** Each trial has two possible outcomes; in this case, heads or tails. **c.** The probability of success is constant throughout the trials; the trials of flipping a coin, are independent. **7.** The student wrote "5" instead of "4". It should be: ${}_nC_{(5-1)}a^{n-4}b^4 = {}_7C_4 j^3(-k)^4 = 35j^3k^4$
Exercises 9. ≈0.1361, or ≈13.61%

11. ≈0.0015, or ≈0.15%
13. $a^4 + 4a^3b + 6a^2b^2 + 4ab^3 + b^4$
15. $243x^5 + 810x^4y + 1080x^3y^2 + 720x^2y^3 + 240xy^4 + 32y^5$ **17.** $896g^6h$ **19.** e^6
21. $P(0) \approx 0.1176$, $P(1) \approx 0.3025$, $P(2) \approx 0.3241$, $P(3) \approx 0.1852$, $P(4) \approx 0.0595$, $P(5) \approx 0.0102$, $P(6) \approx 0.0007$
23. $P(0) \approx 0.000001$, $P(1) \approx 0.000054$, $P(2) \approx 0.0012$, $P(3) \approx 0.0146$, $P(4) \approx 0.0984$, $P(5) \approx 0.3543$, $P(6) \approx 0.5314$
25. 0.99328 **27.** ≈0.2824 **29.** ≈0.1109
31. ≈0.2461 **33.** ≈0.6230 **35a.** 0.0914 **b.** The probability that three boxes would be underweight is 0.0001. You can conclude that there might be a malfunction in the machinery or that the company's claim may be false. **37.** The probability of a group of 30 students having 4 or fewer left-handed students is about 77.05%. This means that more than three quarters of the classes will have enough left-handed desks; 4 is an adequate no.
39a. $P(0) = 0.001$, $P(1) = 0.027$, $P(2) = 0.243$, $P(3) = 0.729$

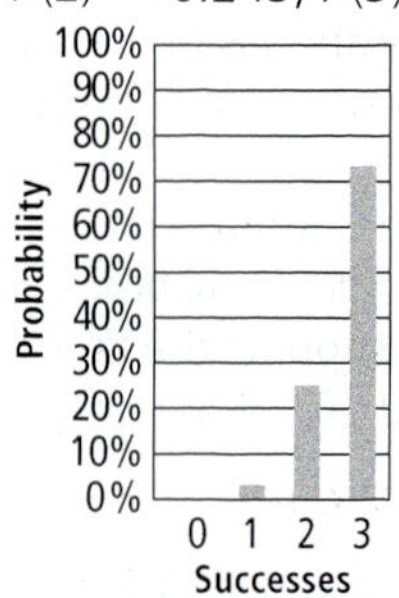

b. $P(0) = 0.166375$, $P(1) = 0.408375$, $P(2) = 0.334125$, $P(3) = 0.091125$

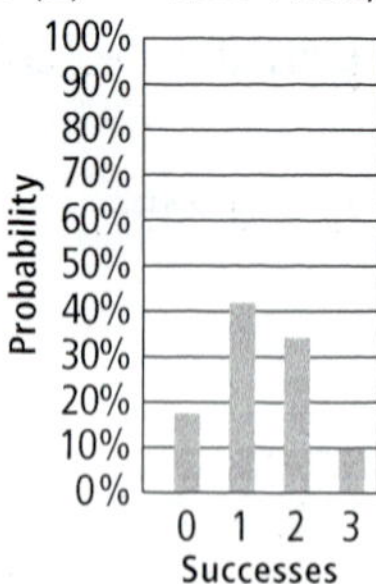

c. The probabilities of each graph sum to 1; $P(0) + P(1) + P(2) + P(3) = 1$. The probabilities of part (a) increase with increasing success numbers; the maximum probability occurring at $P(3)$. The probabilities of part (b) peak with a maximum at $P(1)$ and then decrease with increasing success numbers. **41.** Answers may vary.
43. a. The graph is sym. about the line $x = 3.5$.

b.

x	y
0	0.0078
1	0.0547
2	0.1641
3	0.2734
4	0.2734
5	0.1641
6	0.0547
7	0.0078

c. No; the bulge in the graph has shifted rt. **45.** G **47.** H **49.** loaded and leading question by the use of the word "beautiful" and the phrase "Do you agree" **50.** does not provide enough information about the amendments to make a decision **51.** vertices: $(0, \pm 7)$; foci: $(0, \pm\sqrt{74})$; asymptotes: $y = \pm\frac{7}{5}x$ **52.** vertices: $(0, \pm 3)$; foci: $(0, \pm\sqrt{13})$; asymptotes: $y = \pm\frac{3}{2}x$ **53.** vertices: $(0, \pm 3)$; foci: $(0, \pm 5)$; asymptotes: $y = \pm\frac{3}{4}x$ **54.** $\frac{2}{3}$ **55.** $\frac{1}{2}$
56. $\frac{2}{3}$ **57.** $\bar{x} = 24.4$, $\sigma \approx 5.04$ **58.** $\bar{x} = 81.8$, $\sigma \approx 4.77$
59. $\bar{x} = 8.6$, $\sigma \approx 0.47$ **60.** $\bar{x} \approx 24.74$, $\sigma \approx 2.046$

Lesson 11-10 pp. 739–745

Got It? 1a. 71% **b.** 88%
2.

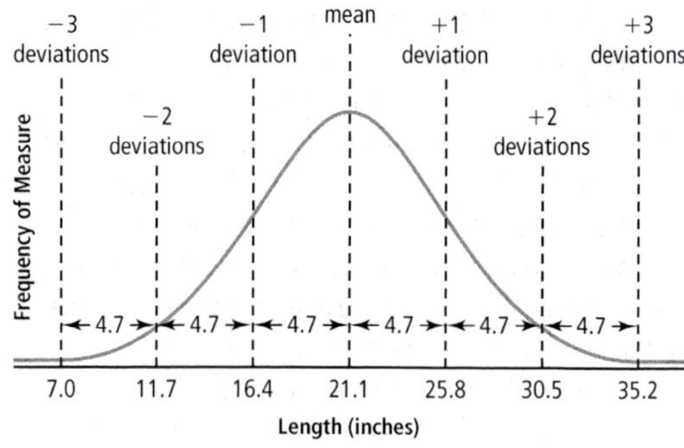

3a. 2.5% **b.** 210 students **c.** The students that received a B had scores between 165 and 180.
Lesson Check 1. 94%
2.

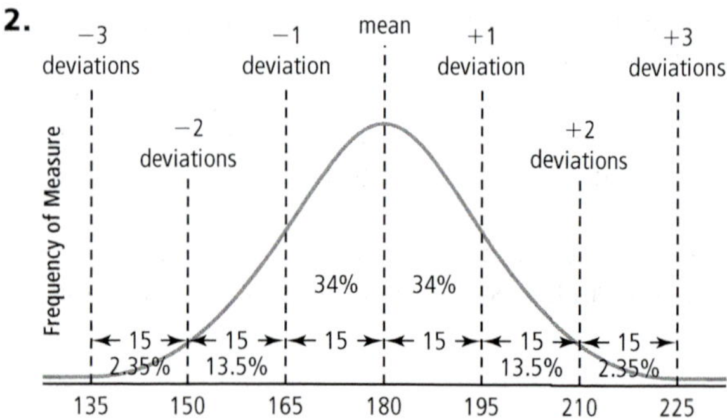

3. 47.5% **4.** Normal distribution means that most of the examples in a data set are close to the mean; the distribution of the data is within 1, 2, or 3 standard deviations of the mean. **5.** The mean and median are equivalent in a normal distribution. **6.** mean increases by 10: the bell curve is translated 10 units to the rt.; standard deviation increases by 10: the bell curve is stretched out by a factor of 10; the bell curve will be less steep due to the larger standard deviation.

Exercises **7.** ≈43% **9.** ≈43 men

11.

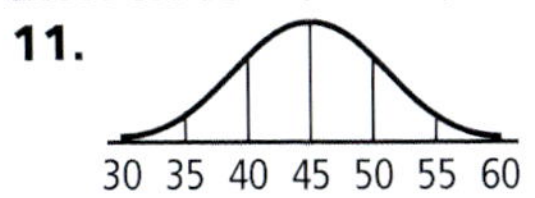

13.

39 41 43 45 47 49 51

15. 68% **17.** 50% **19a.** Set 2

b. and c.

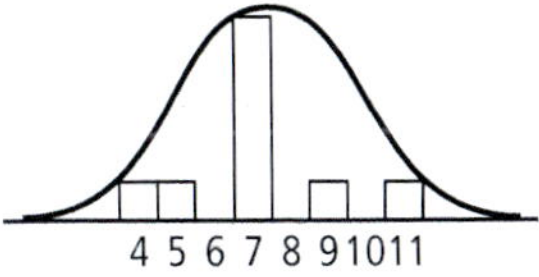

21. 59 min **23.** 47.5% **25.** 81.5% **27.** 84%

29a.

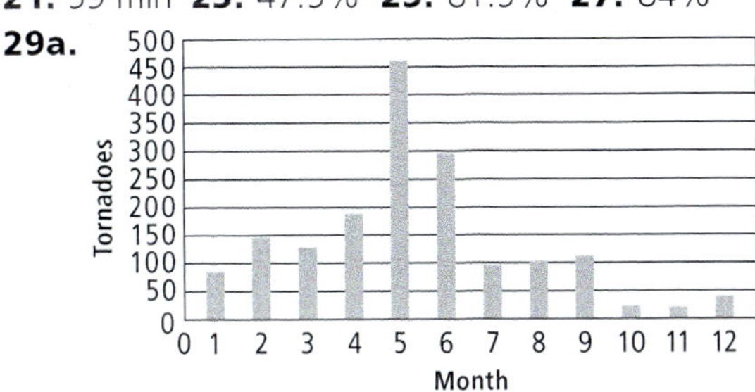

b. No; the curve is skewed to the left. **c.** No, mean and standard deviation are appropriate only for measuring normally distributed data. **31.** Yes; Elena scored within the top 10% of her group. Her score is 2.75 standard deviations above the mean, which places her in the top 1%. Jake did not score in the top 10%. His score is 1.16 standard deviations above the mean, or at the 88th percentile. **33.** A binomial distribution has a finite no. of probabilities, which sum to 1 and are a subset of a larger normal distribution. For example, using $n = 6$, $p = 0.5$, the binomial distribution probabilities are $P(0) \approx 0.0156$, $P(1) \approx 0.0938$, $P(2) \approx 0.2344$, $P(3) \approx 0.3125$, $P(4) \approx 0.2344$, $P(5) \approx 0.0938$, $P(6) \approx 0.0156$.

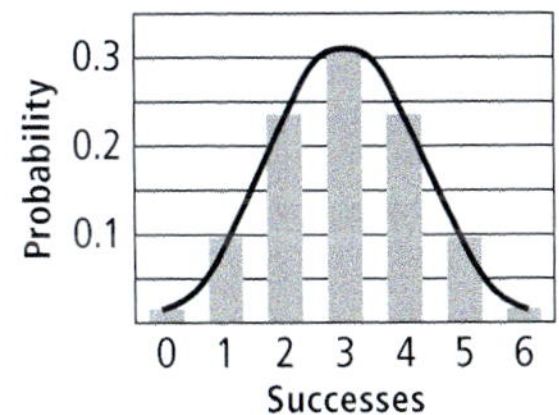

35. l **37.** For Distribution A with 50 data values, 25 values are at or below 40, which is the mean. For Distribution B with 30 data values, 15 values are at or below the mean 40. So Distribution A has more values at or below 40. **38.** 0.02867 **39.** 0.1612 **40.** 0.03676

41.

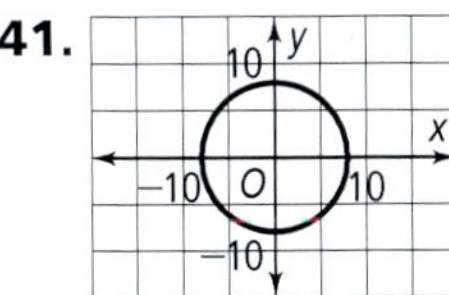

circle; center: (0, 0), radius: 8; lines of sym.: all lines through the center; domain: $-8 \le x \le 8$, range: $-8 \le y \le 8$

42.

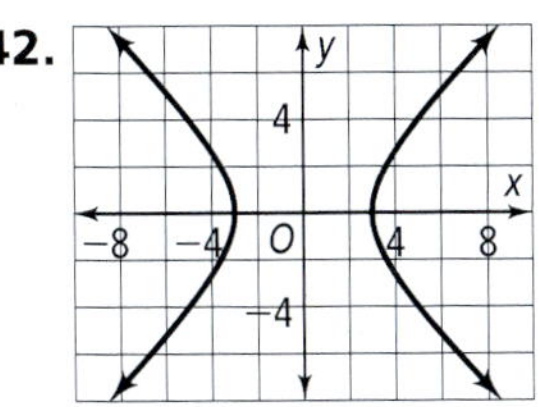

hyperbola; center: (0, 0), foci: $(\pm 3\sqrt{2}, 0)$; lines of sym.: $x = 0$, $y = 0$; domain: $x \le -3$ or $x \ge 3$; range: all real numbers

43.

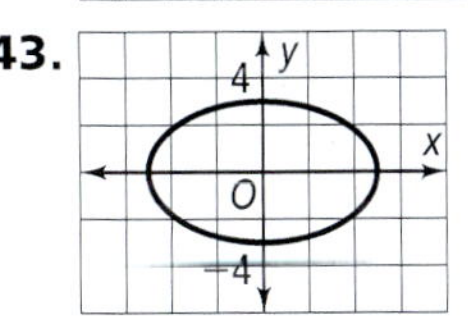

ellipse; center: (0, 0), foci: $(\pm 4, 0)$; lines of sym.: $x = 0$, $y = 0$; domain: $-5 \le x \le 5$, range: $-3 \le y \le 3$

44. $y = x - 3$;

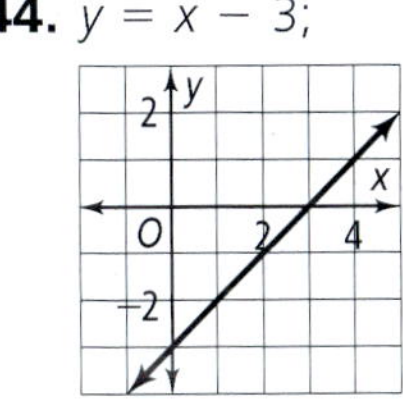

45. $y = x$;

2 y x −2 O 2 −2

46. $y = x - \frac{5}{4}$;

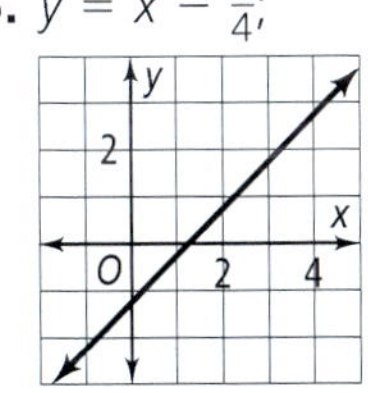

Chapter Review pp. 751–756

1. sample **2.** outlier **3.** probability distribution **4.** range of a set of data **5.** 6 **6.** 362,880 **7.** 12 **8.** 30 **9.** 21 **10.** 10 **11.** 30 **12.** 744 **13.** 220; 84; 20; 1 **14.** 3.315312×10^9 **15.** 216 **16.** $\frac{47}{70}$ **17.** 0 **18.** $\frac{2}{5}$ **19.** Not necessarily; you may pick a 5 zero times, one time, or more than once. Each time you pick, the prob. that it will be a 5 is $\frac{1}{20}$. **20.** dependent **21.** independent **22.** 0.21 **23.** 0.79 **24.** 0.3 **25.** 0.7 **26.** $\frac{1}{4}$ **27.** $\frac{1}{5}$ **28.** $\frac{1}{8}$ **29.** This will not necessarily result in a fair decision, because the principal may aim at a particular name, which means that not all students have an equally likely chance of being chosen. **30.** Yes, this will result in a fair decision, because the probabilities of each goalie being chosen are the same. **31.** Answers may vary. **32.** 9 **33.** mean: 6,

median: 6, mode: 9 **34.** mean: $10.\overline{6}$, median: 7, modes: 3 and 7 **35.** mean: 15, median: 15, mode: 18 **36.** mean: 9.5, median: 9.5, mode: none **37.** range: 35; $Q_1 = 30$; $Q_3 = 55$ **38.** range: 35; $Q_1 = 25$; $Q_3 = 50$ **39.** range: 65; $Q_1 = 42$; $Q_3 = 87$ **40.** heights of 3 people **41.** ages of thirty college students **42.** gas mileage of 18 automobiles of various types
43. $\bar{x} \approx 6.64$, $\sigma \approx 5.12$ **44.** $\bar{x} \approx 17.14$, $\sigma \approx 3.52$
45. $\bar{x} = 7.5$, $\sigma \approx 2.67$ **46.** not a random sample; they will all begin with the letter "a" **47.** not a random sample; the lawyers will choose jurors that are likely to support their side **48.** random sample; all students have an equal chance to be chosen **49.** not a random sample; the five with the largest (or smallest) circulation size will be picked **50.** People at the bus station may be less likely to own a car and therefore less likely to be in favor of a new garage. **51.** $\frac{1}{2}$ **52.** $\frac{1}{3}$
53. ≈ 0.14 **54.** ≈ 0.0710 **55.** ≈ 0.2066 **56.** ≈ 0.1766
57. $21a^5b^2$ **58.** $56a^3b^5$ **59.** continuous **60.** discrete
61. discrete **62.** continuous **63.** 16%; 2.5%

Chapter 12

Get Ready! p. 761

1. 6 **2.** $\frac{1}{3}$ **3.** $-\frac{3}{8}$ **4.** $\frac{7}{72}$ **5.** −9 **6.** 11 **7.** 3 **8.** $\left(\frac{1}{2}, -4\right)$
9. $\left(-\frac{2}{3}, -\frac{2}{3}\right)$ **10.** $\left(\frac{3}{4}, \frac{11}{4}\right)$ **11.** (7, 9, −6) **12.** (0, 0, 8)
13. (7, 5, 0)

Lesson 12-1 pp. 764–770

Got It?

1a. $\begin{bmatrix} -15 & 25 \\ -1 & 1 \\ -2 & 15 \end{bmatrix}$ **b.** $\begin{bmatrix} -9 & 23 \\ -5 & 9 \\ 0 & 5 \end{bmatrix}$

c. Yes; it does not matter in which order you add matrices.

2. $A = \begin{bmatrix} 3 & 6 & -1 \\ 1 & 3 & 7 \\ 1 & 1 & 3 \end{bmatrix}$ **3a.** $\begin{bmatrix} 0 & 0 \\ 0 & 0 \end{bmatrix}$ **b.** $\begin{bmatrix} -1 & 10 & -5 \\ 0 & 2 & -3 \end{bmatrix}$

4a. $x = 6$, $y = -6$ **b.** $x = 4$, $y = -3$, $z = 2$

Lesson Check

1. $\begin{bmatrix} 1 & 1 \\ -2 & 8 \end{bmatrix}$ **2.** $\begin{bmatrix} 1 & -9 & 8 \\ -3 & -1 & 8 \end{bmatrix}$ **3.** $\begin{bmatrix} -3 & 4 \\ -5 & 11 \end{bmatrix}$ **4.** $\begin{bmatrix} 6 & 10 \\ 13 & -4 \end{bmatrix}$

5. Yes; the elements in each of the corresponding positions are equal.

6. The elements were not subtracted. The correct answer is $\begin{bmatrix} 6 \\ 5 \end{bmatrix} - \begin{bmatrix} 3 \\ 7 \end{bmatrix} = \begin{bmatrix} 3 \\ -2 \end{bmatrix}$

Exercises

7. $\begin{bmatrix} 6 & 5 & 4 \\ 2 & -1 & 7 \end{bmatrix}$ **9.** $\begin{bmatrix} 3.9 & -2.3 \\ -0.6 & 9.1 \end{bmatrix}$ **11.** $\begin{bmatrix} 4 & -8 \\ -1 & -1 \\ 11 & 1 \end{bmatrix}$

13. $\begin{bmatrix} 6 & 2 \\ -1 & 3 \end{bmatrix}$ **15.** $\begin{bmatrix} 2 & -3 & 4 \\ 5 & 6 & -7 \end{bmatrix}$

17. $x = -2$, $y = 3$, $z = 1$ **19.** $\begin{bmatrix} 0 & 5 \\ 8 & -6 \\ 0 & 5 \end{bmatrix}$ **21.** $\begin{bmatrix} 6 & 3 \\ -3 & 3 \end{bmatrix}$

23. $\begin{bmatrix} -4 & 1 \\ -3 & -1 \end{bmatrix}$ **25a.** $\begin{bmatrix} 952 \\ 720 \\ 1108 \\ 1172 \\ 1044 \end{bmatrix}$; $\begin{bmatrix} 760 \\ 832 \\ 1252 \\ 1144 \\ 1064 \end{bmatrix}$

b. Allen: 4996; Iagorashvili: 5052

27. Matrix B would have the same dimensions as A. Its elements would be the opposites of the corresponding elements in A.

29. $c = \frac{5}{2}$, $d = \frac{2}{5}$, $f = 7$, $g = 5$, $h = -1$

31. Consider any two 2 × 2 matrices, $A = \begin{bmatrix} a & b \\ c & d \end{bmatrix}$ and $B = \begin{bmatrix} w & x \\ y & z \end{bmatrix}$. By the definition of matrix addition and the Comm. Prop. of Add.

$$A + B = \begin{bmatrix} a & b \\ c & d \end{bmatrix} + \begin{bmatrix} w & x \\ y & z \end{bmatrix} = \begin{bmatrix} a + w & b + x \\ c + y & d + z \end{bmatrix}$$
$$= \begin{bmatrix} w + a & x + b \\ y + c & z + d \end{bmatrix} = \begin{bmatrix} w & x \\ y & z \end{bmatrix} + \begin{bmatrix} a & b \\ c & d \end{bmatrix}$$
$$= B + A$$

33. B **35.** B **37.** 68% **38.** 97.5% **39.** 47.5%
40. 2, −6 **41.** $\frac{2}{3}$, 2 **42.** $-\frac{1}{2}$, −6 **43.** 5, 0
44. $\begin{bmatrix} 9 & 15 \\ 6 & 24 \end{bmatrix}$ **45.** $\begin{bmatrix} -20 \\ 35 \end{bmatrix}$

Lesson 12-2 pp. 772–779

Got It?

1. $\begin{bmatrix} 8 & 24 & -19 \\ -3 & 9 & 10 \end{bmatrix}$ **2.** $\begin{bmatrix} 5 & -1 \\ \frac{7}{3} & 0 \end{bmatrix}$ **3a.** $\begin{bmatrix} -6 & 0 \\ -9 & 11 \end{bmatrix}$

b. $\begin{bmatrix} -3 & 7 \\ 6 & 8 \end{bmatrix}$ **c.** No; explanations may vary. Sample: For the matrices in parts (a) and (b), $AB = \begin{bmatrix} -6 & 0 \\ -9 & 11 \end{bmatrix}$ and $BA = \begin{bmatrix} -3 & 7 \\ 6 & 8 \end{bmatrix}$, so $AB \neq BA$. **4.** player from 1994: 100 pts., player from 2006: 81 pts. **5a.** no **b.** yes **c.** yes **d.** no **e.** yes

Lesson Check

1. $\begin{bmatrix} 6 & -2 \\ 4 & 0 \end{bmatrix}$ **2.** $\begin{bmatrix} -3 & 11 \\ -10 & 6 \end{bmatrix}$ **3.** $\begin{bmatrix} 5 & 7 \\ 2 & 6 \end{bmatrix}$ **4.** $\begin{bmatrix} 9 & -1 \\ -2 & 2 \end{bmatrix}$

5. Scalar; repeated matrix addition is repeated addition of each element of the matrix, which is the same as scalar multiplication of the matrix. **6.** The product of two matrices A and B exists only if the number of columns of A is equal to the number of rows of B. Since A is a 2×4 matrix with 4 columns and B is a 3×6 matrix with 3 rows and $4 \neq 3$, the product AB does not exist. Likewise, since $6 \neq 2$, the product BA does not exist.

Exercises

7. $\begin{bmatrix} 9 & 12 \\ 18 & -6 \\ 3 & 0 \end{bmatrix}$ **9.** $\begin{bmatrix} -3 & -6 \\ 9 & -3 \end{bmatrix}$ **11.** $\begin{bmatrix} 9 & 2 \\ 2 & 6 \\ 3 & -10 \end{bmatrix}$

13. $\begin{bmatrix} 19 & 11 \\ -12 & 10 \end{bmatrix}$ **15.** $\begin{bmatrix} 8 & -2.5 \\ -1.5 & -1 \end{bmatrix}$ **17.** $\begin{bmatrix} -4 & 8 \\ -22 & 2 \end{bmatrix}$

19. $\begin{bmatrix} 5 & -12 \\ 9 & -6 \end{bmatrix}$ **21.** $\begin{bmatrix} -8 & 0 \\ 0 & -8 \end{bmatrix}$ **23.** $[34 \quad 0]$

25. $\begin{bmatrix} -15 & 0 \\ 25 & 0 \end{bmatrix}$ **27.** $\begin{bmatrix} -2 \\ 5 \end{bmatrix}$ **29.** yes **31.** yes **33.** yes

35a. River's Edge: 99 pts.; West River: 97 pts. **b.** West River

37. $\begin{bmatrix} 9 & -6 \\ 15 & -3 \\ -6 & -12 \end{bmatrix}$ **39.** $\begin{bmatrix} 17 & -24 \\ -33 & -7 \\ 69 & -18 \end{bmatrix}$ **41.** $\begin{bmatrix} 34 & -1 \\ 6 & -13 \\ -7 & 16 \end{bmatrix}$

43. $\begin{bmatrix} -90 & 0 \\ -78 & 42 \\ -30 & -30 \end{bmatrix}$ **45.** yes **47.** yes **49.** B **51.** C

53. Since the center is at the origin, the vertices are $\left(\pm\frac{50}{2}, 0\right)$ and the co-vertices are $\left(0, \pm\frac{40}{2}\right)$. Using $\frac{x^2}{a^2} + \frac{y^2}{b^2} = 1$, $a = \pm 25$ and $b = \pm 20$, so $\frac{x^2}{625} + \frac{y^2}{400} = 1$.

54. $\begin{bmatrix} -33 & -12 \\ 6 & 27 \end{bmatrix}$ **55.** $\begin{bmatrix} 9 & -6 & 12 \\ 2 & 20 & 12 \end{bmatrix}$

56a. 12 **b.** 12 **c.** 0 **57a.** −12 **b.** −12 **c.** 0

Lesson 12-3 pp. 782–790

Got It? 1a. yes **b.** yes **c.** No; no matrix that is multiplied by the zero matrix will give an identity matrix. **2a.** 3 **b.** 0 **c.** −48 **3a.** 12 units^2 **b.** 28 units^2

4a. yes; $\begin{bmatrix} 1 & -1 \\ -\frac{3}{2} & 2 \end{bmatrix}$ **b.** no **c.** yes; $\begin{bmatrix} 3 & -4 \\ -5 & 7 \end{bmatrix}$

5a. 88, 68, 84, 60, 12, 32, 52, 72, 28, 30, 14, 18, 2, 8, 14, 20 **b.** Multiply the coded information by the inverse of the coding matrix: $\begin{bmatrix} 4 & 1 & 7 & 3 & 1 & 2 & 3 & 4 \\ 9 & 8 & 7 & 6 & 1 & 3 & 5 & 7 \end{bmatrix}$

Lesson Check 1. 16 **2.** 7 **3.** does not exist

4. yes; $\begin{bmatrix} 3 & -2 \\ -7 & 5 \end{bmatrix}$

5. The student did not subtract correctly.

$$\det\begin{bmatrix} 2 & 5 \\ -3 & 1 \end{bmatrix} = (2)(1) - (-3)(5) = 2 - (-15) = 2 + 15 = 17$$

6. A 2×3 matrix does not have a multiplicative inverse because it is not a square matrix. The number of rows must equal the number of columns for a multiplicative inverse to be possible.

Exercises 7. yes **9.** yes **11.** no **13.** 0 **15.** $-\frac{11}{40}$ **17.** 11 **19.** −6 **21.** −5 **23.** 106 **25.** 6 **27.** 466,250 mi^2

29. yes; $\begin{bmatrix} -1 & 3 \\ 1 & -2 \end{bmatrix}$ **31.** yes; $\begin{bmatrix} 0 & \frac{1}{2} \\ \frac{1}{3} & -\frac{1}{6} \end{bmatrix}$

33. yes; $\begin{bmatrix} -\frac{1}{8} & -\frac{1}{2} \\ \frac{3}{16} & \frac{1}{4} \end{bmatrix}$ **35.** yes; $\begin{bmatrix} 0 & \frac{1}{3} \\ -\frac{1}{2} & \frac{1}{6} \end{bmatrix}$

37. 2, 10, 10, 6, 9, 55, 15, 15, 9, 20 **39.** −120 **41.** 9 **43.** −3 **45.** 1 **47.** Answers may vary. Sample: Form a new matrix by switching the element in row 1, column 1 with the element in row 2, column 2. Then replace the other two elements with their opposites. Finally, divide each element by the determinant of the original matrix. **49.** 38 units^2

51. yes; $\begin{bmatrix} -5 & 7 \\ 3 & -4 \end{bmatrix}$ **53.** yes; $\begin{bmatrix} 0.5 & 0 \\ 0 & 0.5 \end{bmatrix}$

55. yes; $\begin{bmatrix} 0.4 & 0.4 & 0.2 \\ -0.6 & -0.6 & 0.2 \\ -0.2 & 0.8 & 0.4 \end{bmatrix}$

57. no inverse because the determinant equals zero

59. 6 **61.** $MN = \begin{bmatrix} ae + bg & af + bh \\ ce + dg & cf + dh \end{bmatrix}$

$$\begin{aligned} \det MN &= (ae + bg)(cf + dh) - (af + bh)(ce + dg) \\ &= acef + adeh + bcfg + bdgh \\ &\quad - acef - adfg - bceh - bdgh \\ &= adeh + bcfg - adfg - bceh \end{aligned}$$

$$\begin{aligned} \text{Also, } \det M \cdot \det N &= (ad - bc)(eh - fg) \\ &= adeh - adfg - bceh + bcfg. \end{aligned}$$

So, $\det M \cdot \det N = \det MN$

63. $\frac{1}{2}$ **65.** 15 **67.** $\begin{bmatrix} 2 & 5 \\ 1 & 1 \end{bmatrix}$ **68.** $\begin{bmatrix} -10 & 19 \\ -20 & 7 \end{bmatrix}$ **69.** 720

70. 362,880 **71.** $1.08972864 \times 10^{10}$ **72.** 110,880 **73.** no solution **74.** (6, 0, −3) **75.** (3, −3, 9) **76.** (−2, −1, −3)

Lesson 12-4 pp. 792–800

Got It?

1a. $\begin{bmatrix} -8 \\ 9 \end{bmatrix}$ **b.** $\begin{bmatrix} -14 & -20 \\ 19 & 28 \end{bmatrix}$

c. Since matrix A has no inverse, the eq. has no solution.

2a. $\begin{bmatrix} 3 & -7 \\ 5 & 1 \end{bmatrix}\begin{bmatrix} x \\ y \end{bmatrix} = \begin{bmatrix} 8 \\ -2 \end{bmatrix}$

b. $\begin{bmatrix}1 & 3 & 5\\ -2 & 1 & -4\\ 7 & -2 & 0\end{bmatrix}\begin{bmatrix}x\\ y\\ z\end{bmatrix}=\begin{bmatrix}12\\ -2\\ 7\end{bmatrix}$

c. $\begin{bmatrix}2 & -8\\ -1 & 1\end{bmatrix}\begin{bmatrix}x\\ y\end{bmatrix}=\begin{bmatrix}-3\\ -4\end{bmatrix}$

3a. (5, −21) **b.** no solution **4.** run: 32 min; jog: 8 min

Lesson Check

1. $\begin{bmatrix}-6 & 3\\ 4 & -2\end{bmatrix}\begin{bmatrix}x\\ y\end{bmatrix}=\begin{bmatrix}8\\ 10\end{bmatrix}$ **2.** $\begin{bmatrix}2 & 3 & 0\\ 1 & -2 & 1\\ 0 & 6 & -4\end{bmatrix}\begin{bmatrix}x\\ y\\ z\end{bmatrix}=\begin{bmatrix}12\\ 9\\ 8\end{bmatrix}$

3. (5, 3) **4.** (−6, −6) **5.** The student did not separate the coefficient matrix and the variable matrix. The matrix eq. should be written as $\begin{bmatrix}2 & 3\\ -4 & 5\end{bmatrix}\begin{bmatrix}x\\ y\end{bmatrix}=\begin{bmatrix}5\\ 1\end{bmatrix}$.

6. Use matrix multiplication to combine the coefficient matrix and the variable matrix into a product matrix. Then set the first element in the product matrix equal to the first element in the constant matrix and set the second element in the product matrix equal to the second element in the constant matrix. The result will be a system of equations.;
$-2p + 3q = 2$
$4p + q = -5$

Exercises

7. $\begin{bmatrix}-15 & -17\\ 26 & 29\end{bmatrix}$ **9.** $\begin{bmatrix}\frac{29}{31}\\ -\frac{66}{217}\\ \frac{34}{217}\end{bmatrix}$ **11.** $\begin{bmatrix}1 & 1\\ 1 & -2\end{bmatrix}\begin{bmatrix}x\\ y\end{bmatrix}=\begin{bmatrix}5\\ -4\end{bmatrix}$; coefficient matrix: $\begin{bmatrix}1 & 1\\ 1 & -2\end{bmatrix}$, variable matrix: $\begin{bmatrix}x\\ y\end{bmatrix}$, constant matrix: $\begin{bmatrix}5\\ -4\end{bmatrix}$ **13.** $\begin{bmatrix}3 & 5\\ 1 & 1\end{bmatrix}\begin{bmatrix}a\\ b\end{bmatrix}=\begin{bmatrix}0\\ 2\end{bmatrix}$; coefficient matrix: $\begin{bmatrix}3 & 5\\ 1 & 1\end{bmatrix}$, variable matrix: $\begin{bmatrix}a\\ b\end{bmatrix}$, constant matrix: $\begin{bmatrix}0\\ 2\end{bmatrix}$

15. $\begin{bmatrix}1 & -1 & 1\\ 2 & 0 & 1\\ 0 & 1 & 3\end{bmatrix}\begin{bmatrix}r\\ s\\ t\end{bmatrix}=\begin{bmatrix}150\\ 425\\ 0\end{bmatrix}$; coefficient matrix: $\begin{bmatrix}1 & -1 & 1\\ 2 & 0 & 1\\ 0 & 1 & 3\end{bmatrix}$, variable matrix: $\begin{bmatrix}r\\ s\\ t\end{bmatrix}$, constant matrix: $\begin{bmatrix}150\\ 425\\ 0\end{bmatrix}$

17. (2, 1) **19.** $\left(\frac{1}{2}, 20\right)$ **21.** (3, 2) **23.** (2, −1, 3) **25.** (1, 2, −2) **27.** 2.5 lb of almonds, 3.5 lb of peanuts, and 3 lb of raisins **29.** (−2, −1) **31.** (−1, 0) **33.** (5, 0, 1) **35.** (1, 0, 3) **37.** (1, 1, 1, 1) **39.** (6, 2) **41.** (16, −22)

43. (5.4, 7.4) **45.** (6, 1) **47.** (2, −1, 4) **49.** length = 280 ft, width = 140 ft **51.** $\begin{bmatrix}-3 & 2\\ -5 & 8\end{bmatrix}$ **53.** $\begin{bmatrix}10\\ 3\\ -2\end{bmatrix}$ **55.** 14

57. Answers may vary. Sample: $y + z = 0$; $y + z = 1$

59. B **61.** A **63.** −44 **64.** 4913 **65.** −218

66. $34.\overline{4}$; 30.9; 5.56 **67.** 4.17; 1.32; 1.15 **68.** $19.\overline{6}$ m; $22.\overline{2}$ m; 4.7 m **69.** 57.4 mi; 345.44 mi^2; 18.6 mi

70.

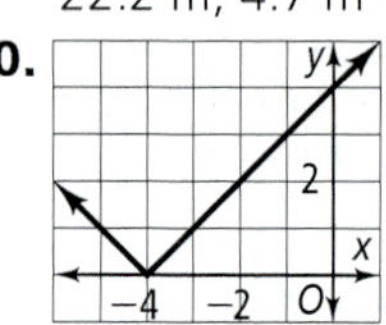
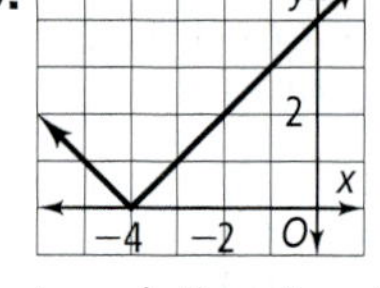

translation 4 units to the left

71.

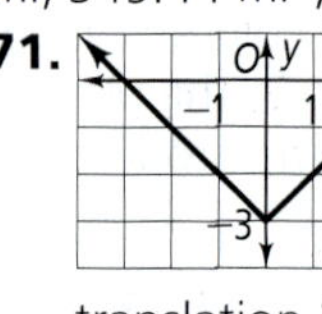

translation 3 units down

72.

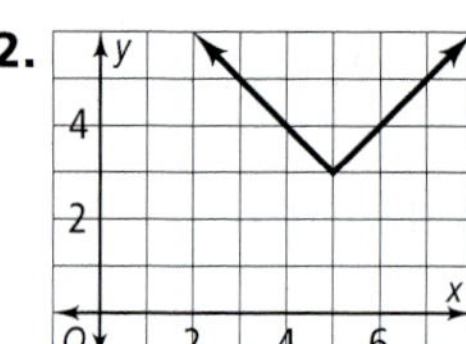

translation 5 units to the right and 3 units up

Lesson 12-5 pp. 801–808

Got It? 1a. Subtract 8 from each x-coordinate and add 5 to each y-coordinate.

b. $\begin{bmatrix}0 & -1 & -5 & 1 & 4\\ -5 & -1 & 0 & 3 & 0\end{bmatrix}+\begin{bmatrix}-3 & -3 & -3 & -3 & -3\\ 2 & 2 & 2 & 2 & 2\end{bmatrix}$
$=\begin{bmatrix}-3 & -4 & -8 & -2 & 1\\ -3 & 1 & 2 & 5 & 2\end{bmatrix}$; (−3, −3), (−4, 1), (−8, 2), (−2, 5), (1, 2)

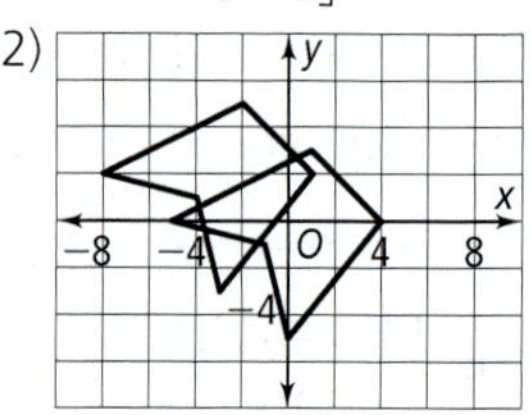

2. Answers may vary. Samples:

a. $\begin{bmatrix}0 & 5 & 5 & 0\\ 0 & 0 & 3 & 3\end{bmatrix}$

b. $2\begin{bmatrix}0 & 5 & 5 & 0\\ 0 & 0 & 3 & 3\end{bmatrix}=\begin{bmatrix}0 & 10 & 10 & 0\\ 0 & 0 & 6 & 6\end{bmatrix}$; (0, 0), (10, 0), (10, 6), (0, 6)

c. 4

3a. (0, 3), (4, 4), (1, −1);

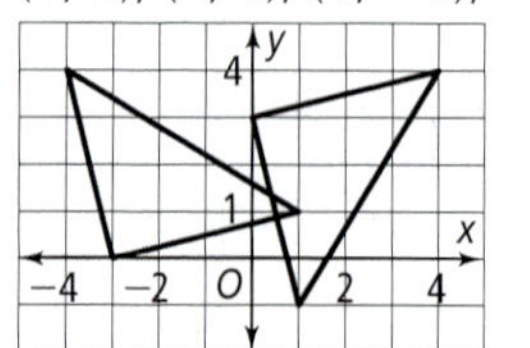

b. (−3, 0), (−4, 4), (1, 1);

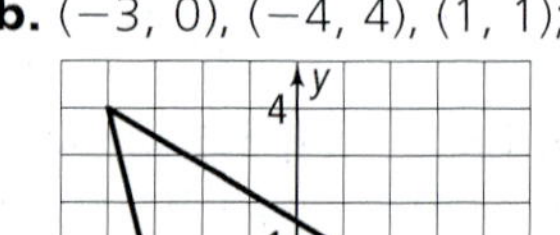
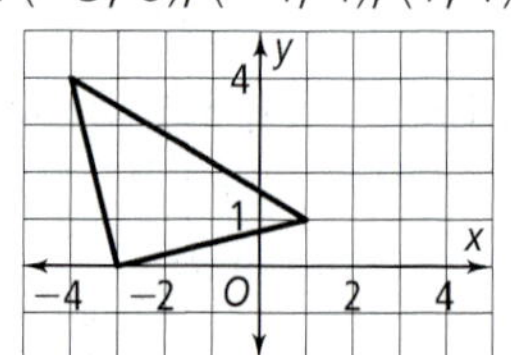

4a. (1, −1), (3, −1), (6, −4), (1, −3);

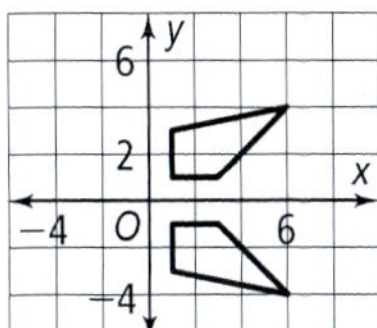

b. (1, 1), (1, 3), (4, 6), (3, 1);

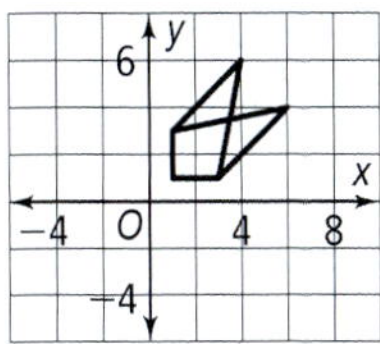

Lesson Check 1. $A'\left(\frac{1}{2}, \frac{1}{2}\right)$, $B'(1, 2)$, $C'\left(2, -\frac{1}{2}\right)$ **2.** $A'(1, -1)$, $B'(4, -2)$, $C'(-1, -4)$ **3.** $A'(1, 1)$, $B'(4, 2)$, $C'(-1, 4)$ **4.** Translations, rotations and reflections leave the size of the figure unchanged. Translation moves the figure to a new location. Rotation turns the figure about a fixed pt. Reflection maps a figure in the coordinate plane to its mirror image using a specific line as its mirror. **5.** Answers may vary. Sample: reflection across $y = x$; translation 4 units to the right and 4 units down

6. They are equal.

$$\begin{bmatrix} -3 & 0 \\ 0 & -3 \end{bmatrix}\begin{bmatrix} 1 & -2 & 4 \\ 1 & -1 & 2 \end{bmatrix} = -3\begin{bmatrix} 1 & 0 \\ 0 & 1 \end{bmatrix}\begin{bmatrix} 1 & -2 & 4 \\ 1 & -1 & 2 \end{bmatrix} = -3\begin{bmatrix} 1 & -2 & 4 \\ 1 & -1 & 2 \end{bmatrix}$$

Exercises 7. (−2, 2), (−2, 6), (2, 6), (2, 2);

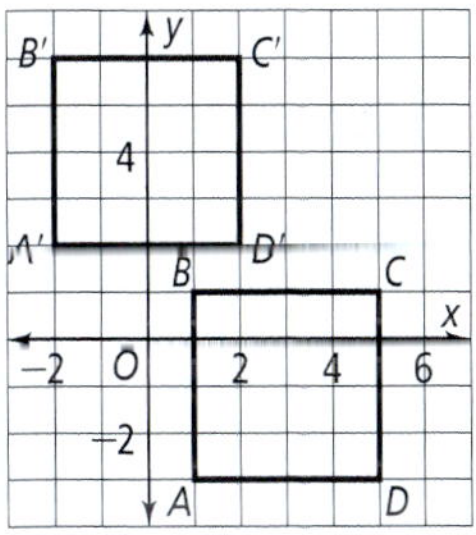

9. (−13, 7), (−19, 6), (9, 0);

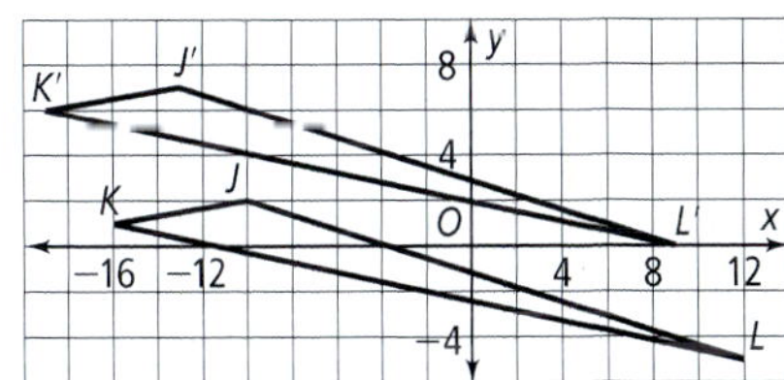

11. (0, 0), (4, 8), (10, 10), (16, 2)
13. (−12, 9), (3, 6), (4.5, 0), (1.5, −6), (−3, 0)

15.

17. (−3, 3), (3, −6), (3, −3), (−3, −6)
19. (−1, −3), (−2, −2), (−3, −2), (−4, −3), (−2.5, −5)

21.

23. (3, −3), (−3, −6), (−3, −3), (3, −6)
25. (3, 1), (2, 2), (2, 3), (3, 4), (5, 2.5)

27. $\begin{bmatrix} -8 & -5 & -2 & -5 \\ -8 & -5 & -8 & -11 \end{bmatrix}$ **29.** $\begin{bmatrix} -5 & -1 & -3 \\ -2 & -1 & 1 \end{bmatrix}$

31. $\begin{bmatrix} 3 & -1 & 1 \\ -1 & -2 & -4 \end{bmatrix}$ **33.** Apply a 90° rotation matrix $\begin{bmatrix} 0 & -1 \\ 1 & 0 \end{bmatrix}$ three times to determine the image matrix for each frame. The fourth application of the 90° rotation matrix would show the gymnast at the starting position of Frame 1.

37. f: $\begin{bmatrix} -5 & -2 & 1 \\ 3 & 0 & 3 \end{bmatrix}$, g: $\begin{bmatrix} -1 & 2 & 5 \\ 1 & -2 & 1 \end{bmatrix}$ translation

39. $\begin{bmatrix} -1.5 & 0.25 & -2.5 \\ 0 & 1.5 & 1.5 \end{bmatrix}$

41. Answers may vary. Sample: The reflection of a matrix of pts. from a function table across the line $y = x$ interchanges the values of y and x in the function table. Finding the inverse of the matrix of pts. of a function from a function table also results in the interchanging of the values of y and x.

43. H **45.** G **47.** (3, −2) **48.** (1, −1, 2) **49.** (0, 1, −2)

50. $\begin{bmatrix} 16 \\ 12 \end{bmatrix}$ **51.** [12] **52.** [22]

Lesson 12-6 pp. 809–815

Got It? 1. **u** = ⟨3, 4⟩; **v** = ⟨−1, −6⟩ **2a.** ⟨5, 3⟩
b. translation, reflection and dilation **3.** $\sqrt{74} \approx 8.60$

4a. 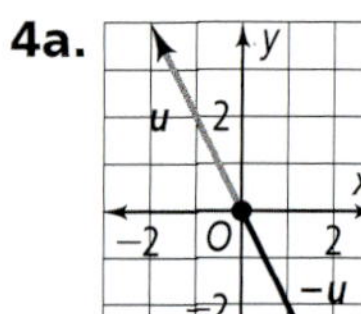 **b.** 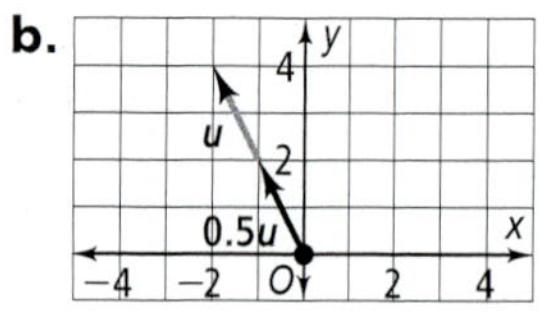

5a. not normal **b.** normal

Lesson Check **1.** ⟨5, 2⟩ **2.** ⟨9, −11⟩ **3.** ⟨−1, −12⟩ **4.** ⟨0, −15⟩ **5.** The magnitudes of vectors **a**, **b**, and **c** are the same: 5. **6.** Although the *x*-component of ⟨8, 3⟩ is 4 times the *x*-component of ⟨2, 1⟩, the *y*-component and the magnitude of ⟨8, 3⟩ are not 4 times those of ⟨2, 1⟩, $3 \neq 4 \times 1$ and $\sqrt{73} \neq 4 \times \sqrt{5}$.

Exercises **7.** ⟨4, 1⟩ **9.** ⟨4, −2⟩ **11.** ⟨0, 2⟩ **13.** ⟨−1, 5⟩ **15.** ⟨2, 0⟩ **17.** ⟨3, 0⟩ **19.** ⟨3, −4⟩ **21.** ⟨4, −1⟩ **23.** ⟨−3, 8⟩ **25.** ⟨−8, 20⟩ **27.** ⟨−6, −12⟩ **29.** not normal **31.** normal **33.** ⟨0, −14⟩ **35.** ⟨18, −13⟩ **37.** about 304 mi/h **39.** $\overrightarrow{AB} = \langle 3, 1\rangle$, $\overrightarrow{BC} = \langle -2, 3\rangle$, $\overrightarrow{CA} = \langle -1, -4\rangle$ **41.** ⟨6, −3⟩ **43.** ⟨−4, −2⟩ **45.** **v** − **v** = ⟨0, 0⟩; ⟨0, 0⟩, the zero vector, is the additive identity for the set of all vectors and −**v** is the additive inverse of any given vector **v**; so **v** + ⟨0, 0⟩ = **v** and **v** + (−**v**) = ⟨0, 0⟩.

47. yes; Distributive Prop.

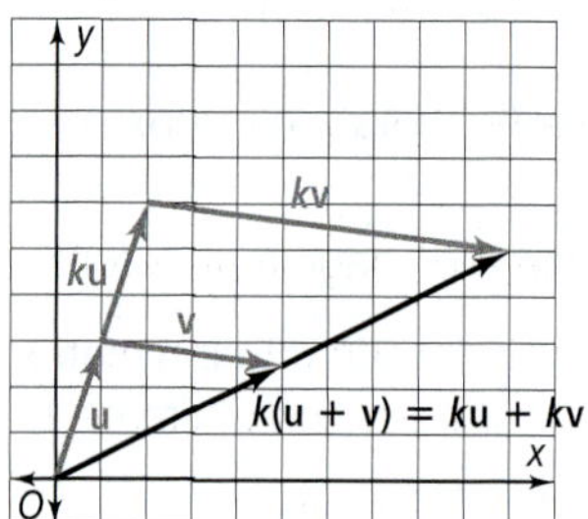

49. yes; Assoc. Prop. of Add.

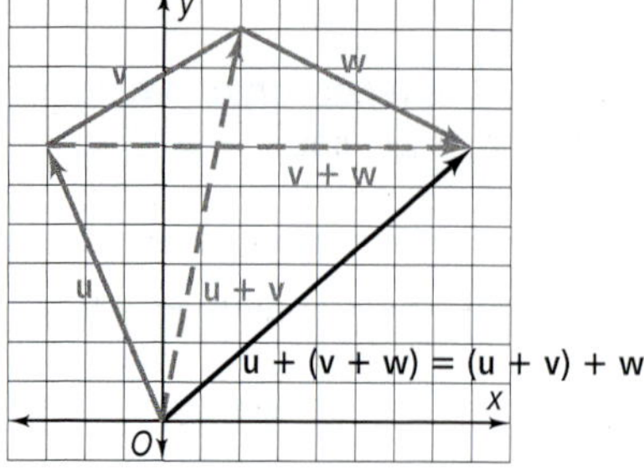

51. **a** and **d** are parallel; **a** and **b** are perpendicular; **b** and **d** are perpendicular **55.** −3 **57.** $\frac{1}{2}$ **59.** 5

60. $\begin{bmatrix} 1 & -2 & 3 \\ 1 & 5 & 3 \end{bmatrix}$; 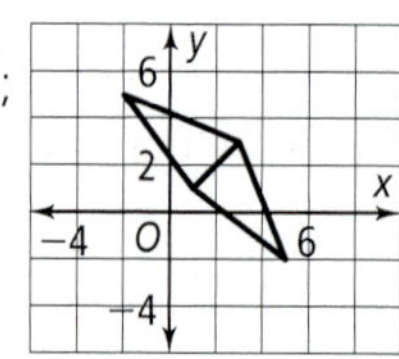

61. $\begin{bmatrix} -3 & 0 & 2 & -1 \\ -2 & -3 & 0 & 2 \end{bmatrix}$; 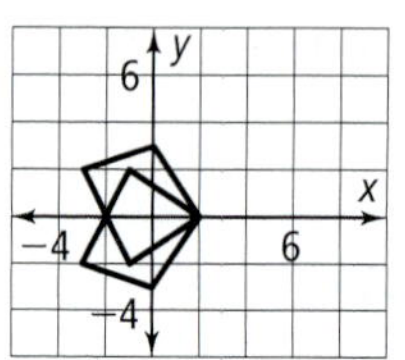

62.

63. 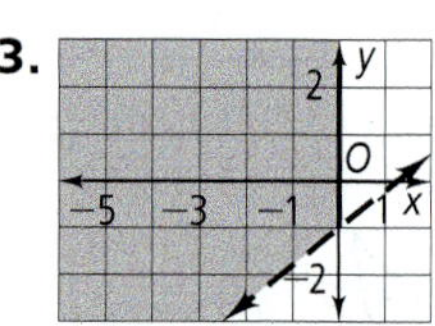

64.

65. $y - 1 = -3x$ OR $y + 5 = -3(x - 2)$

66. $y + 4 = -\frac{7}{5}(x + 4)$ OR $y - 3 = -\frac{7}{5}(x + 9)$

67. $y - 2 = -(x - 7)$ OR $y - 8 = -(x - 1)$ **68.** yes **69.** no **70.** yes **71.** no

Chapter Review pp. 817–820

1. equal matrices **2.** zero matrix **3.** matrix equation **4.** square matrix

5. $\begin{bmatrix} -1 & 9 & -8 \\ 4 & 0 & 6 \end{bmatrix}$ **6.** $\begin{bmatrix} 5 & -4 \\ 5 & 0 \end{bmatrix}$ **7.** $[1 \quad -8 \quad 12]$

8. $\begin{bmatrix} -3 & 10 \\ -3 & 3 \end{bmatrix}$ **9.** $x = -2, w = 8, r = 4, t = -1$

10. $t = -4, y = \frac{11}{2}, r = 4, w = 4$

11. $\begin{bmatrix} 18 & 3 & 0 & 24 \\ -12 & 9 & 21 & 33 \end{bmatrix}$ **12.** undefined **13.** undefined

14. $\begin{bmatrix} -6 & 10 & 21 & 41 \\ -28 & 10 & 28 & 28 \end{bmatrix}$ **15.** $\begin{bmatrix} -14 & -2 \\ 43 & -7 \end{bmatrix}$

16. $\begin{bmatrix} -11 & 18 \\ -17 & -2 \end{bmatrix}$ **17.** 24; $\begin{bmatrix} \frac{1}{6} & -\frac{1}{24} \\ 0 & \frac{1}{4} \end{bmatrix}$

18. 0; does not exist

19. 42; $\begin{bmatrix} \frac{5}{42} & -\frac{1}{42} \\ -\frac{4}{21} & \frac{5}{21} \end{bmatrix}$ **20.** 6; $\begin{bmatrix} \frac{1}{3} & -\frac{2}{3} & 0 \\ -\frac{1}{6} & \frac{1}{3} & -\frac{1}{2} \\ \frac{1}{3} & \frac{1}{3} & 0 \end{bmatrix}$

21. $\begin{bmatrix} 1 & 2 \\ -1 & 0 \end{bmatrix}$ **22.** (−4, −7) **23.** $\begin{bmatrix} 2 \\ 2 \end{bmatrix}$ **24.** $\begin{bmatrix} 2 & 1 \\ 3 & 2 \end{bmatrix}$

25. no unique solution **26.** no unique solution

27. $\begin{bmatrix} 0 & -5 & -2 \\ 5 & 4 & 9 \end{bmatrix}$ **28.** $\begin{bmatrix} -3 & 2 & -1 \\ 1 & 0 & 5 \end{bmatrix}$ **29.** $\begin{bmatrix} 1 & 0 & 5 \\ 3 & -2 & 1 \end{bmatrix}$
30. $\begin{bmatrix} 1.5 & -1 & 0.5 \\ 0.5 & 0 & 2.5 \end{bmatrix}$ **31.** $\begin{bmatrix} 6 & -4 & 2 \\ 2 & 0 & 10 \end{bmatrix}$ **32.** $\begin{bmatrix} 1 & 0 & 5 \\ -3 & 2 & -1 \end{bmatrix}$
33. $\langle -1, 8 \rangle$; about 8.1 **34.** $\langle 7, -5 \rangle$; about 8.6
35. $\langle -9, 12 \rangle$; 15 **36.** $\langle -2, 14 \rangle$; about 14.1
37. $\langle -8, 14 \rangle$; about 16.1 **38.** $\langle -4, 21 \rangle$; about 21.4
39. 0; normal **40.** 0; normal

Chapter 13

Get Ready! p. 825

1. vert. asymptote: $x = 3$ **2.** vert. asymptotes: $x = -\frac{1}{2}$ and $x = 4$ **3.** $\frac{2b}{a}$ **4.** $\frac{55}{18}$ **5.** $\frac{3}{2(c + d)}$ **6.** $\frac{c}{16}$
7. $\frac{c + 4}{9}$ **8.** $\frac{16}{x}$ **9.** $\frac{15}{7}$ **10.** 6 **11.** 4, 1; $a_n = 19 - 3n$, explicit or $a_1 = 16$, $a_n = a_{n-1} - 3$, recursive
12. -216, -343; $a_n = -n^3$, explicit
13. $(x - 1)^2 + (y + 4)^2 = 16$;

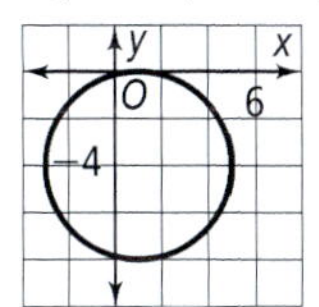

14. $\frac{(x - 2)^2}{9} + \frac{(y - 5)^2}{4} = 1$;

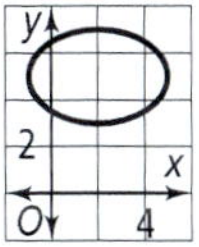

15. $y = \frac{1}{32}x^2 - 3$;

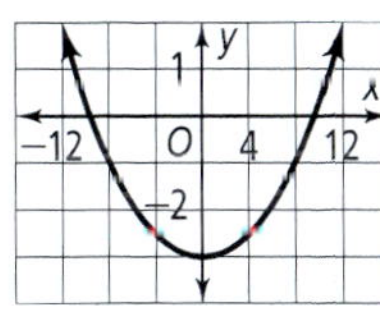

16. $\frac{(y - 1)^2}{9} - \frac{(x - 6)^2}{16} = 1$;

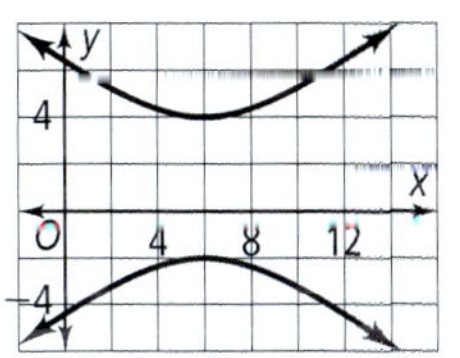

17. Answers may vary. Sample: Similar data tends to recur after a certain period has lapsed. In this case, 12 months.

Lesson 13-1 pp. 828–834

Got It? 1a. from $x = -3$ to $x = 1$ or from $x = 0$ to $x = 4$; 4 **b.** from $x = -4$ to $x = -1$ or from $x = 0$ to $x = 3$; 3 **2a.** no **b.** yes; 4 **c.** 15 cycles; $\frac{1}{3}$ s; $\frac{1}{440}$ s **3a.** 1.5; $y = -0.5$ **b.** 1.5; $y = 0.5$
4. period: 0.006; amplitude: 0.25; $y = -0.75$
Lesson Check 1. periodic; 5 **2.** no **3.** Answers may vary. Sample: hands of a clock, phases of the moon
4. The amplitude is not 2, but $\frac{2}{2} = 1$. **5.** $f(6) = f(11) = 2$; for any x, $f(x + 5)$ will always equal $f(x)$ because the period is 5. **6.** -4
Exercises 7. $x = -2$ to $x = 3$, $x = 2$ to $x = 7$; 5
9. $x = 0$ to $x = 4$, $x = 2$ to $x = 6$; 4 **11.** periodic; 12
13. not periodic **15.** periodic; 7 **17.** 3; $y = -1$
19.

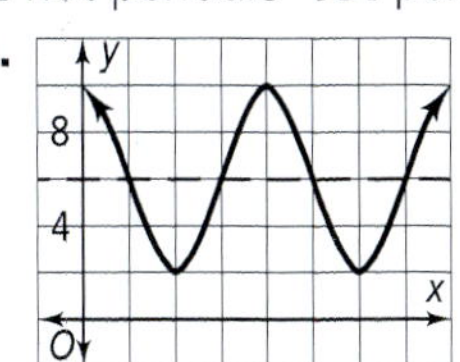

1 unit on the x-axis is 0.005 s.
21. a. y **b.** x **23.** repeating of a pattern at regular intervals **25. a.** 1 s **b.** 1.5 mV
27. 3, -3, 4;

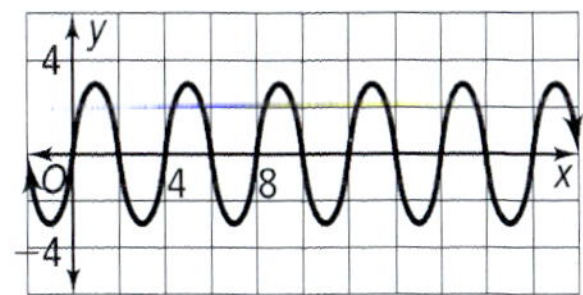

29. 4, -4, 8;

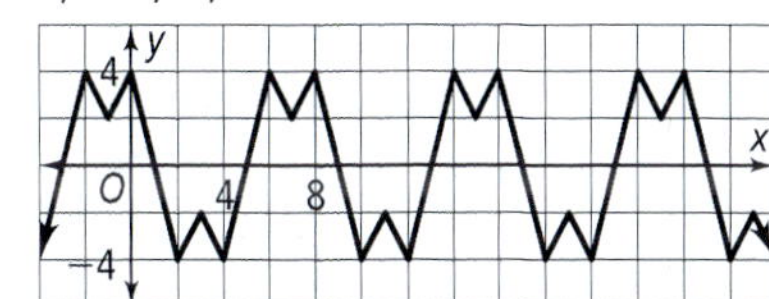

31. 2 weeks **33.** 1 hr **35. a.** 67 **b.** 70 **c.** 70 **d.** 67
37. C **39.** B **41.** 64 s; The first two functions are at the beginning of their cycles together every $6 \cdot 7 = 42$ seconds: 42, 84, 126, . . . The third function is at the beginning of its cycle every 8 seconds, starting at $(42 + 20)$ seconds: 62, 70, 78, 86, 94, 102, 110, 118, 126, . . . The three functions are all at the beginning of their cycles at 126 seconds, which is 64 seconds after the third function achieves its first maximum. **42.** about 302 mph
43.

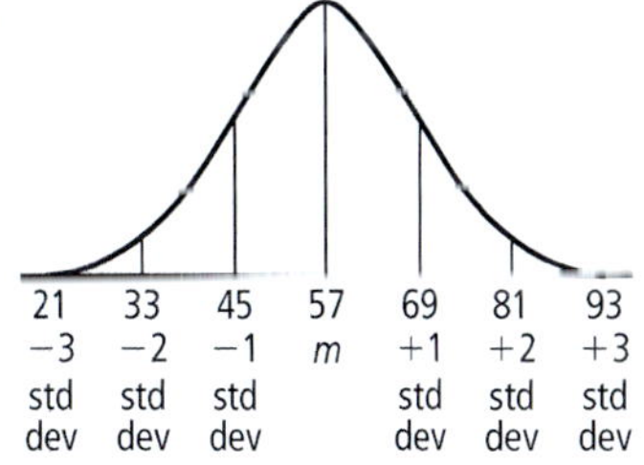

44. $x^2 + y^2 = 1$ **45.** $x^2 + y^2 = 13$ **46.** $x^2 + y^2 = 25$
47. $x^2 + y^2 = 2$ **48.** $x^2 + y^2 = 1$ **49.** $x^2 + y^2 = 1$

Geometry Review p. 835

1. $\frac{\sqrt{2}}{2}$ in. **3.** $\frac{\sqrt{6}}{2}$ ft **5.** hypotenuse: 6 in., longer leg: $3\sqrt{3}$ in. **7.** shorter leg: $\frac{1}{2}$ ft, longer leg: $\frac{\sqrt{3}}{2}$ ft

Lesson 13-2 pp. 836–842

Got It? 1. 225°
2a.

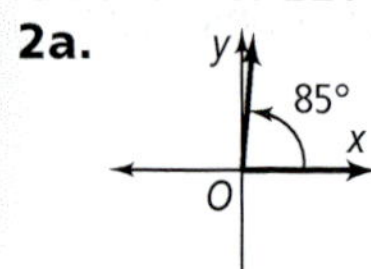

b.

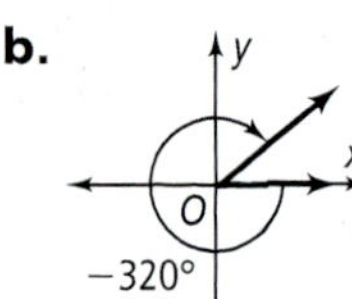

c.

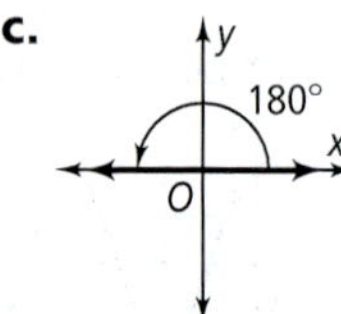

3. −315°, 45°, 405°
4a. cos (−90°) = 0, sin (−90°) = −1; cos (360°) = 1, sin (360°) = 0; cos (540°) = −1, sin (540°) = 0
b. $\cos -90° = \frac{0}{1} = 0$, $\sin -90° = \frac{-1}{1} = -1$
$\cos 360° = \frac{1}{1} = 1$, $\sin 360° = \frac{0}{1} = 0$
$\cos 540° = \frac{-1}{1} = -1$, $\sin 540° = \frac{0}{1} = 0$

5a. $\frac{\sqrt{2}}{2}, -\frac{\sqrt{2}}{2}$ **b.** $-\frac{\sqrt{3}}{2}, \frac{1}{2}$ **c.** Yes; for example, when $\theta = 45°$, $\sin\theta = \cos\theta$.

Lesson Check 1. 135° **2.** 240°
3.

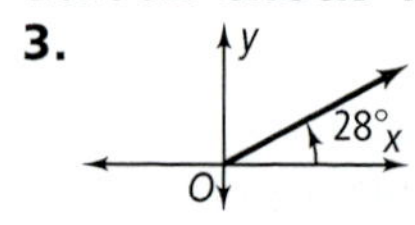

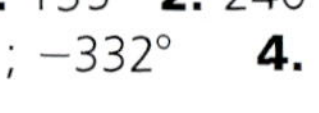

; −332° **4.**

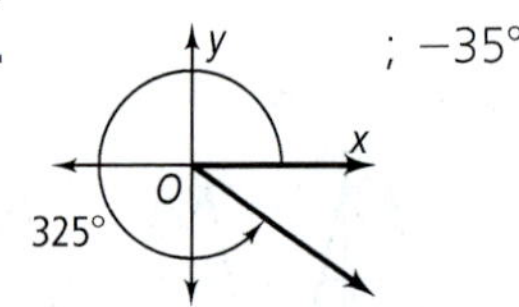

; −35°

5. Answers may vary. Sample: 45° and −315° **6.** The measure of the coterminal angle is not 310°; the measure of the coterminal angle is 50° − 360° = −310°.
Exercises 7. −315° **9.** 240° **11.** −30°
13.

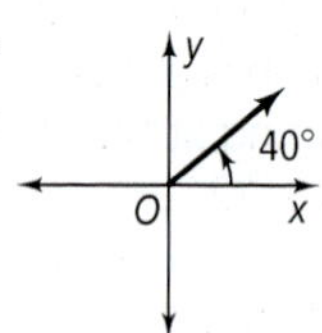

15.

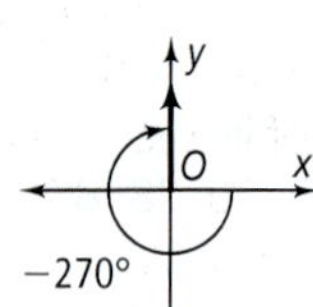

17.

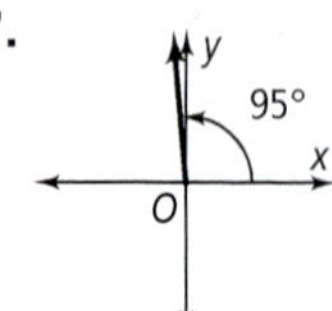

19. 215° **21.** 4° **23.** 150°

25. 180° **27.** $-\frac{\sqrt{2}}{2}, \frac{\sqrt{2}}{2}$; −0.71, 0.71

29. $-\frac{1}{2}, \frac{\sqrt{3}}{2}$; −0.50, 0.87 **31.** $\frac{\sqrt{2}}{2}, -\frac{\sqrt{2}}{2}$; 0.71, −0.71 **33.** $-\frac{\sqrt{2}}{2}, \frac{\sqrt{2}}{2}$; −0.71, 0.71 **35.** 0.98, −0.17 **37.** 0.00, 1.00 **39.** 5

41–43. Answers may vary. Samples:
41. 370°, −350° **43.** 40°, −320° **45.** II
47. negative *x*-axis **49.** positive *x*-axis
51a.

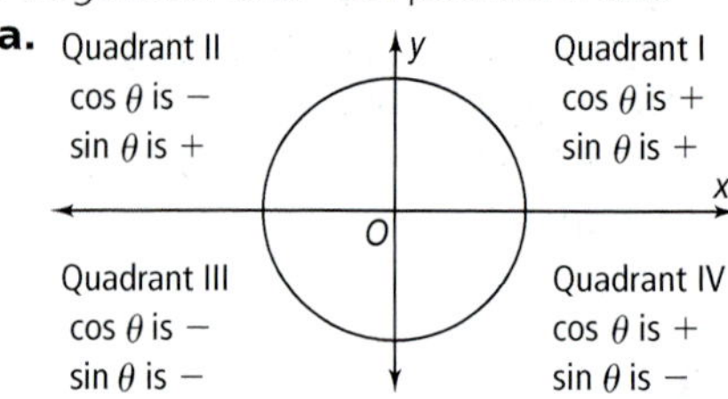

b. II **c.** If the terminal side of an angle is in Quadrants I or II, then the sine of the angle is positive. If the terminal side of an angle is in Quadrants I or IV, then the cosine of the angle is positive.
53.

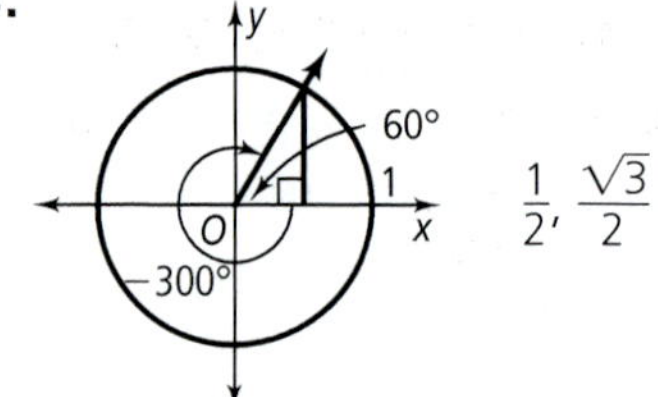

$\frac{1}{2}, \frac{\sqrt{3}}{2}$

55.

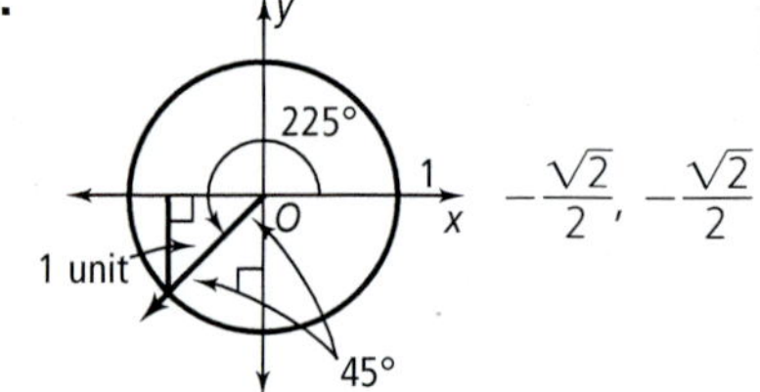

$-\frac{\sqrt{2}}{2}, -\frac{\sqrt{2}}{2}$

57.

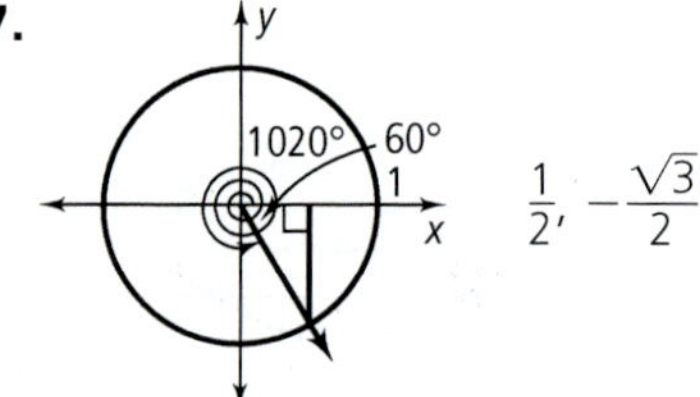

$\frac{1}{2}, -\frac{\sqrt{3}}{2}$

59. No; yes; if the sine and cosine are both negative, the angle is in Quadrant III. 60° is in Quadrant I and −120° is in Quadrant III;

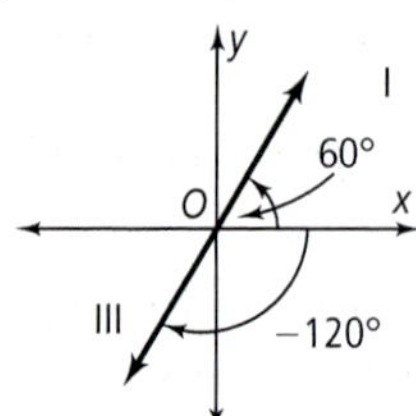

61. H

63.

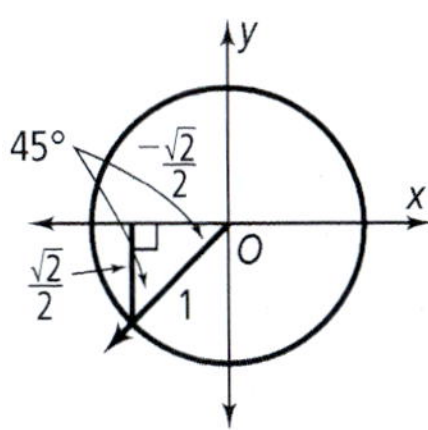

The terminal side forms an angle of 45° with the negative x-axis, so: $\sin(-135°) = -\frac{\sqrt{2}}{2}$ and $\cos(-135°) = -\frac{\sqrt{2}}{2}$. Then $[\sin(-135°)]^2 + [\cos(-135°)]^2$
$= \left(-\frac{\sqrt{2}}{2}\right)^2 + \left(-\frac{\sqrt{2}}{2}\right)^2 =$
$\frac{2}{4} + \frac{2}{4} = \frac{4}{4} = 1.$

64. periodic; 3 **65.** not periodic **66.** periodic; 6

67. $(0, 2\sqrt{5}), (0, -2\sqrt{5})$;

68. $(0, 5\sqrt{5}), (0, -5\sqrt{5})$;

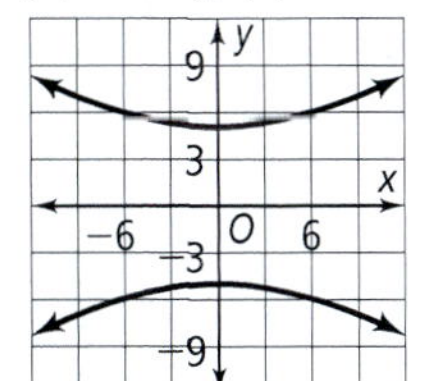

69. $(\sqrt{85}, 0), (-\sqrt{85}, 0)$;

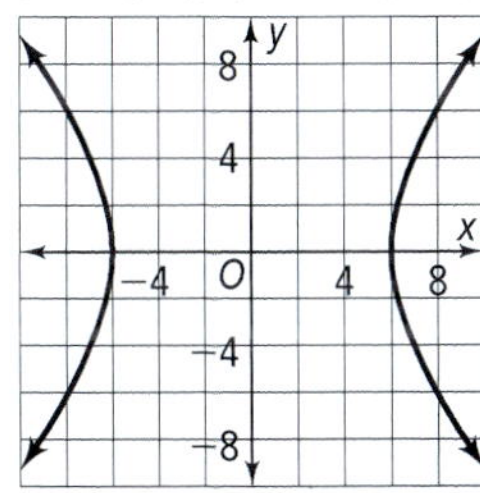

70. $(\sqrt{145}, 0), (-\sqrt{145}, 0)$;

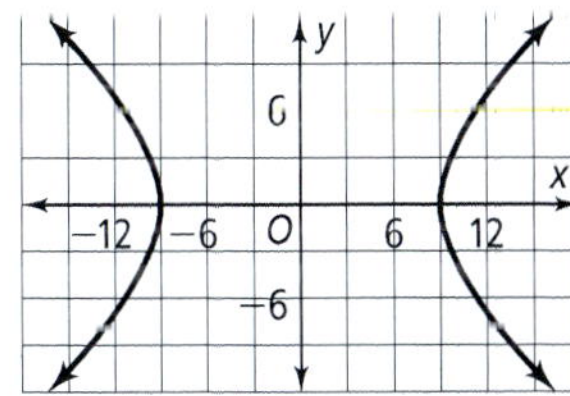

71. 50.24 in.^{2} **72.** 3846.5 m^2 **73.** 200.96 mi^2
74. 0.0746 ft^2

Lesson 13-3 pp. 844–850

Got It? 1a. 90° **b.** $\frac{5\pi}{4}$ radians **c.** $\frac{360°}{\pi} \approx 114.59°$ **d.** $\frac{5\pi}{6}$ radians **2.** $-\frac{\sqrt{3}}{2}, -\frac{1}{2}$ **3a.** 6.3 in. **b.** Arc length would also double. **4.** ≈15,708 km

Lesson Check 1. $\frac{5\pi}{3}$ radians ≈ 5.24 radians **2.** 135° **3.** $\frac{20\pi}{3} \approx 20.94$ in. **4.** 1 radian **5.** 6 "perfect" slices

Exercises 7. $\frac{5\pi}{6}$, 2.62 **9.** $-\frac{\pi}{3}$, −1.05 **11.** $\frac{\pi}{9}$, 0.35 **13.** 198° **15.** −172° **17.** 270° **19.** $\frac{1}{2}, \frac{\sqrt{3}}{2}$ **21.** $\frac{\sqrt{2}}{2}, -\frac{\sqrt{2}}{2}$ **23.** 0, −1 **25.** $-\frac{\sqrt{3}}{2}, -\frac{1}{2}$ **27.** 10.5 m **29.** 25.1 in. **31.** 43.2 cm **33.** ≈31.9 ft **35.** ≈42.2 in. **37.** III **39.** negative x-axis

41. 0.71, −0.71

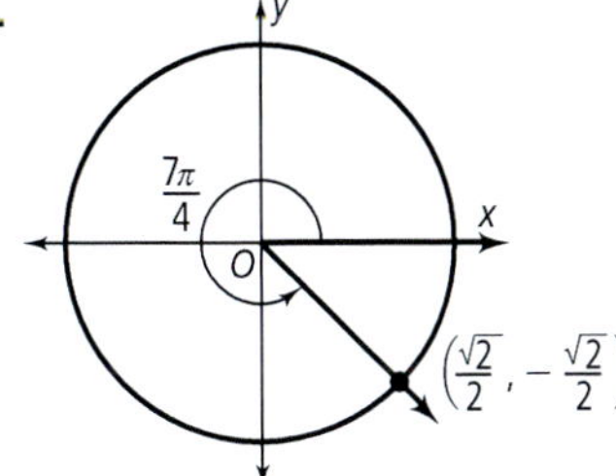

43. 0.00, 1.00

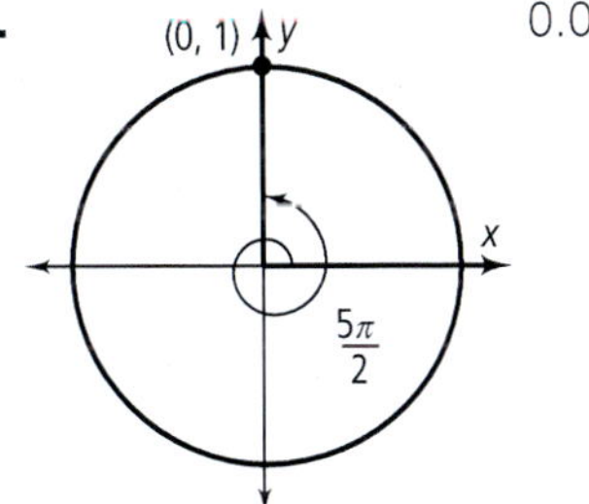

45. If two angles measured in radians are coterminal, the difference of their measures will be evenly divisible by 2π.

47. ≈11 radians **49.** ≈6.3 cm **51.** $-\frac{3\pi}{2}$ radians

53. $\frac{\theta}{s} = \frac{2\pi}{2\pi r}$
$\frac{\theta}{s} = \frac{1}{r}$
$\theta r = s$
$s = \theta r$

55. G **57.** For a central angle of 1 radian, the length of the intercepted arc is the length of the radius.

58.

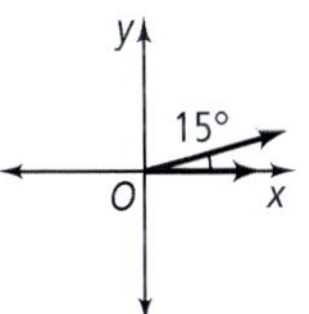

59.

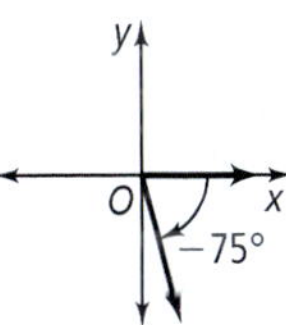

60.

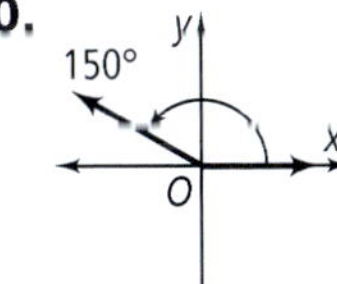

61.

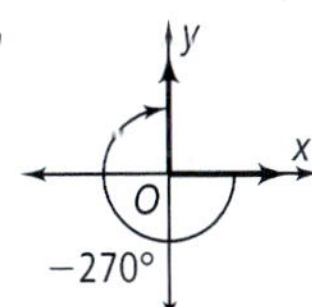

62.

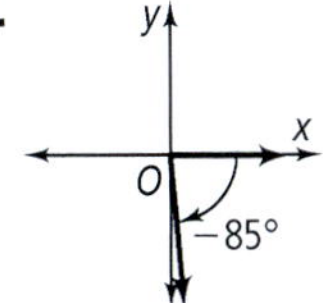

63. mean ≈ 12.9, s.d. ≈ 3.53 **64.** mean = 30, s.d. ≈ 8.09 **65.** 2 **66.** all real numbers **67.** 1 **68.** $y = 0$

Lesson 13-4 pp. 851–858

Got It? 1a. ≈ 0.1411; estimates may vary. **b.** −1
2a. 2; 2π **b.** 3; $\frac{4\pi}{3}$ **3a.** 3; −3 **b.** 0.6; 0.6

4.

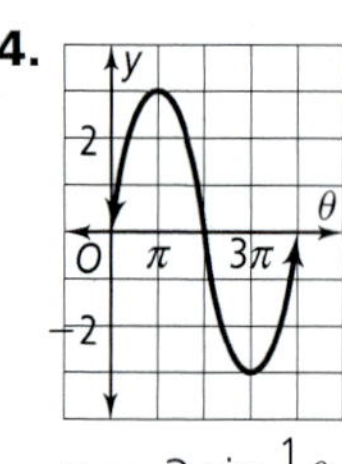

$y = 3 \sin \frac{1}{2}\theta$

5a.

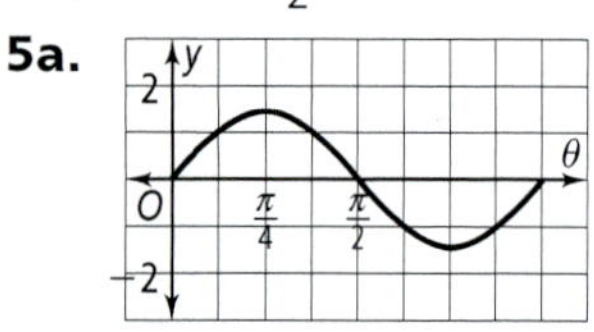

b.

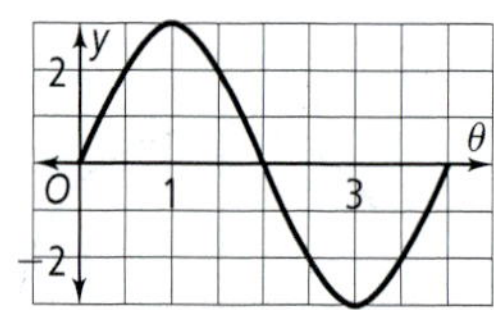

6. $y = \sin \frac{\pi}{320}\theta$

Lesson Check 1a. 2 **b.** 3; π **c.** $y = 3 \sin 2\theta$

2.

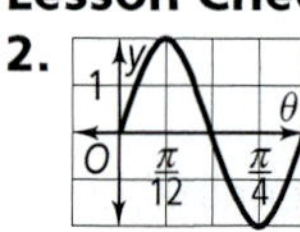

3. One cycle of a sine function is an interval on the x-axis with length equal to the period. The period is the length of one cycle. **4.** Answers may vary. Sample: $y = 5 \sin \frac{\theta}{3}$

5. The amplitude is 3, but since $a < 0$, the graph is reflected across the x-axis. Also, the period is 2, not π.

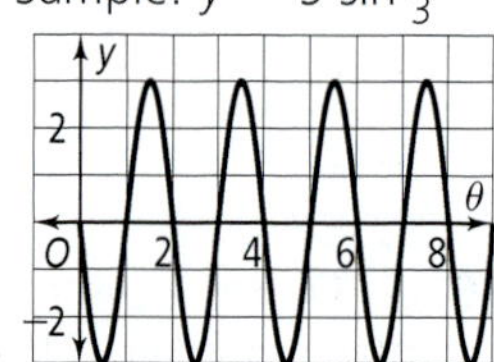

Exercises 7. ≈0.1 **9.** ≈−1 **11.** ≈−0.7 **13.** $\frac{1}{2}$; 1, 4π

15.

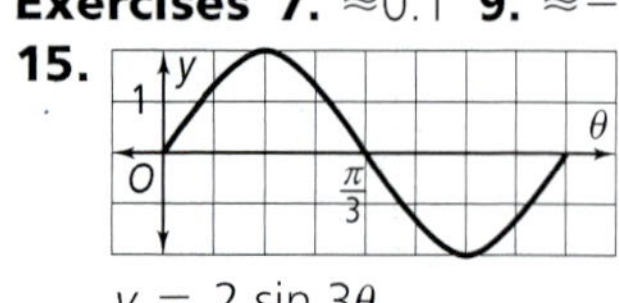

$y = 2 \sin 3\theta$

17.

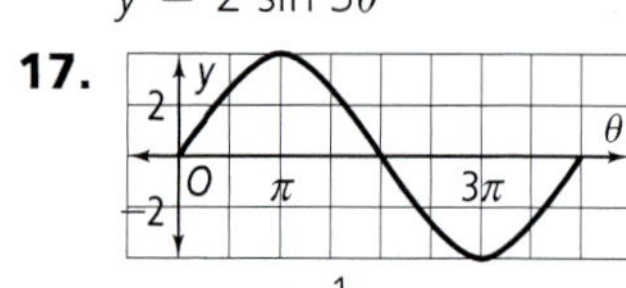

$y = 4 \sin \frac{1}{2}\theta$

19.

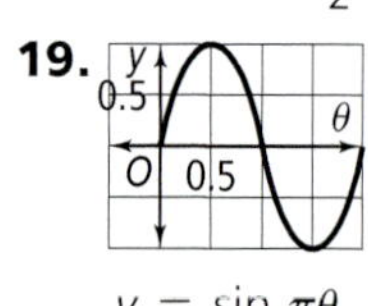

$y = \sin \pi\theta$

21.

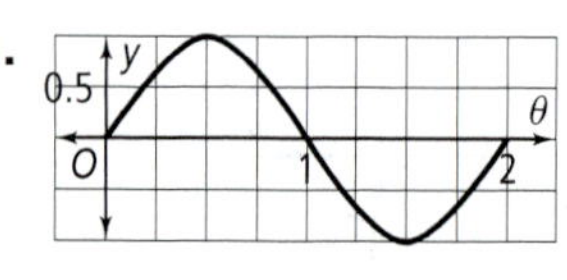

23.

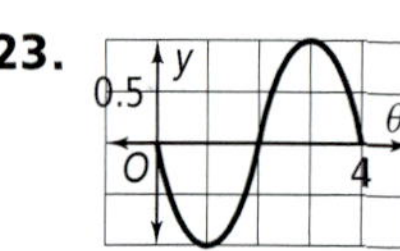

25.

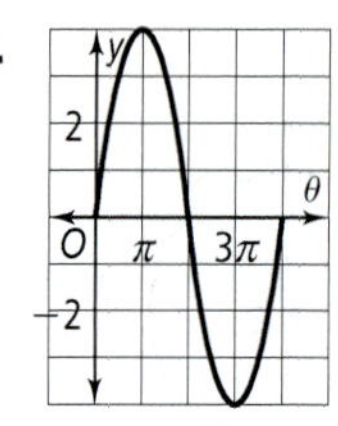

27. 2π; $y = 2 \sin \theta$ **29.** π; $y = \frac{5}{2} \sin 2\theta$ **31.** 1; 1, 2π
33. π; 1, 2 **35.** 1; 5, 2π

37.

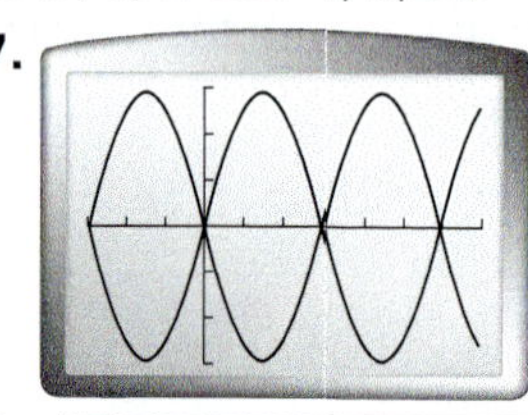

They are reflections of each other across the x-axis. When a is replaced by its opposite, the graph is a reflection of the original graph across the x-axis.

39.

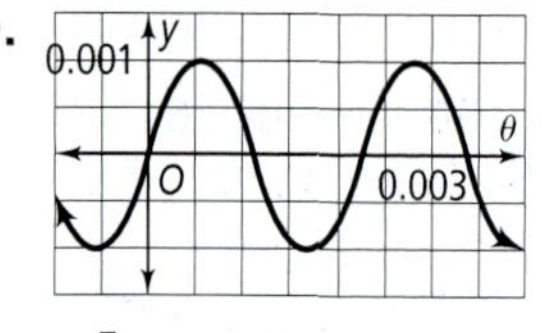

41. π, $\frac{5}{2}$;

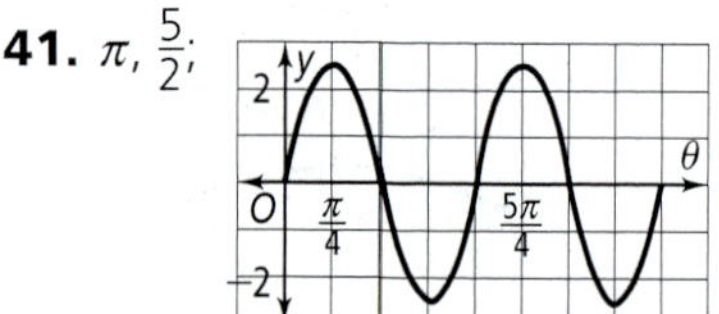

43. $\frac{2\pi}{3}$, 0.4;

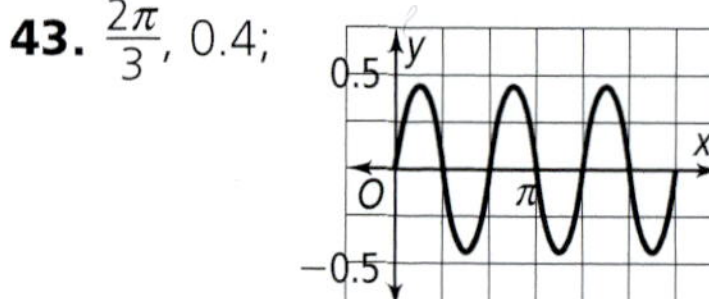

45. $\frac{12}{5}$, 1.2;

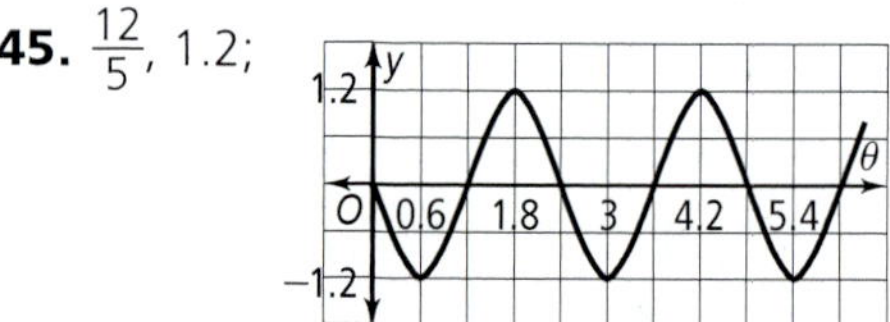

47.

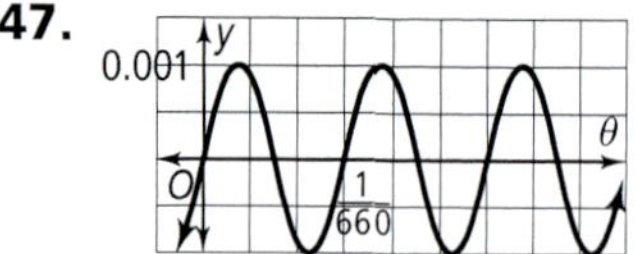

49. $y = \sin 60\pi\theta$ **51.** $y = \sin 240{,}000\pi\theta$

53. 2π, 1;

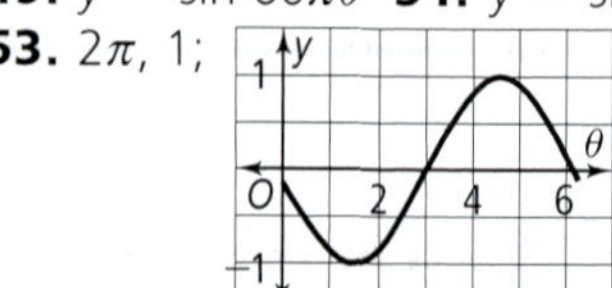

55. C
57. C
59. 120°; consider the point where a 60° angle intersects the unit circle. Reflect this point across the y-axis. The image is the intersection of a 120° angle and the unit

circle. These two points have the same y-coordinate. Therefore $\sin 120° = \sin 60°$.

60. $-\frac{4\pi}{9}$ radians, -1.40 radians **61.** $\frac{5\pi}{6}$ radians, 2.62 radians **62.** $-\frac{4\pi}{3}$ radians, -4.19 radians **63.** $\frac{16\pi}{9}$ radians, 5.59 radians **64.** $-\frac{5\pi}{2}$ radians, -7.85 radians **65.** ≈49% **66.** 1 **67.** 0 **68.** −1 **69.** 0

Lesson 13-5 pp. 861–867

Got It? 1. domain: all real numbers; period: 2π; range: $-1 \leq y \leq 1$; amplitude: 1 sine: max at $\frac{\pi}{2}$; min at $\frac{3\pi}{2}$; zeros at 0, π, 2π

2.

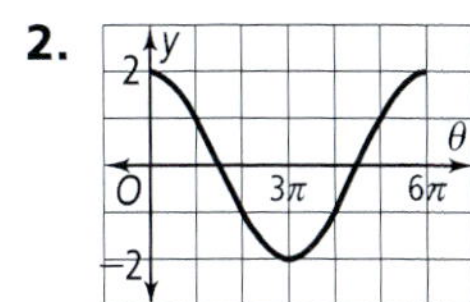

3a. $f(t) = -35 \cos\left(\frac{4\pi}{25}t\right)$ **b.** The function would cross the midline at 3 hours, 7 minutes, 30 seconds. The midline represents average water level. **4a.** 1.15, 1.99, 4.29, 5.13 **b.** 2.21, 4.06 **c.** $0 \leq \theta < 2.21$ and $4.06 < \theta \leq 2\pi$; $2.21 < \theta < 4.06$

Lesson Check

1.

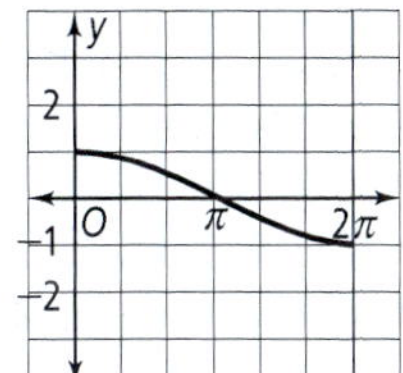

2.

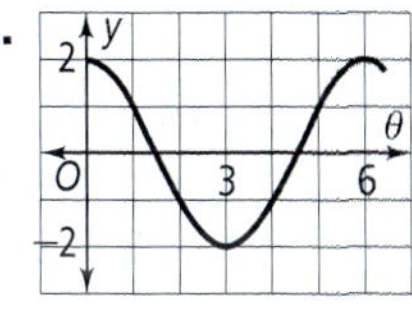

3. $y = 3 \cos \theta$ **4.** $y = 1.5 \cos 2\theta$ **5.** Answers may vary. Sample: $y = 5 \cos\left(\frac{5}{2}\theta\right)$ **6a.** $0 \leq \theta < \frac{\pi}{2}$, $\frac{3\pi}{2} < \theta \leq 2\pi$ **b.** $\pi < \theta < 2\pi$ **c.** $y = 3 \sin\left(\frac{2\pi}{3}x - \frac{\pi}{2}\right)$

Exercises 7. 2π, 3; max: 0, 2π; min: π; zeros: $\frac{\pi}{2}$, $\frac{3\pi}{2}$ **9.** π, 1; max: 0, π, 2π; min: $\frac{\pi}{2}$, $\frac{3\pi}{2}$; zeros: $\frac{\pi}{4}$, $\frac{3\pi}{4}$, $\frac{5\pi}{4}$, $\frac{7\pi}{4}$

11.

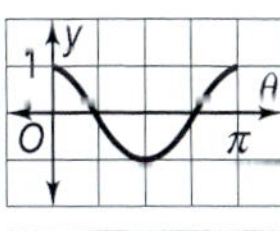

13.

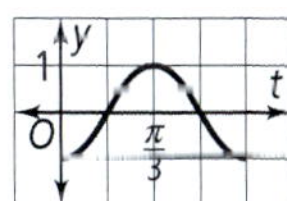

15.

17. $y = \frac{\pi}{2} \cos \frac{2\pi}{3}\theta$ **19.** $y = -3 \cos 2\theta$
21. 0.52, 2.62, 3.67, 5.76
23. 0.55, 1.45, 2.55, 3.45, 4.55, 5.45 **25.** 0.00
27. 2π, $-3 \leq y \leq 3$, 3 **29.** 4π, $-2 \leq y \leq 2$, 2
31. 6π, $-3 \leq y \leq 3$, 3 **33.** $\frac{4}{3}$, $-16 \leq y \leq 16$, 16
35. $y = 70 + 13 \cos \frac{\pi}{6}(x - 1)$ where x represents the months of the year with January as 1, February as 2, March as 3, etc. **37.** 0.64, 2.50

39. 0.50, 2.50, 4.50

41a.

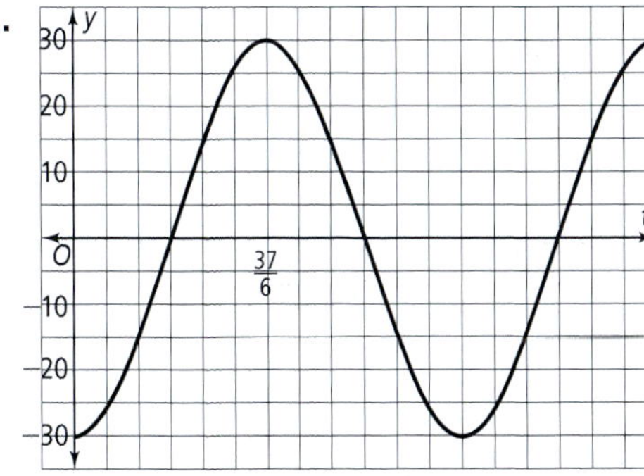

b. 4:40 P.M.; 5:00 A.M. (next day); 5:20 P.M. (next day); 5:40 A.M. (2 days after time 0) **c.** 6 h 10 min; 6 h 10 min

43. On the unit circle, the x-values of $-\theta$ are equal to the x-values of θ, so $\cos(-\theta) = \cos \theta$. $-\cos \theta$ is the opposite of $\cos \theta$, so these graphs are reflections of each other across the x-axis.

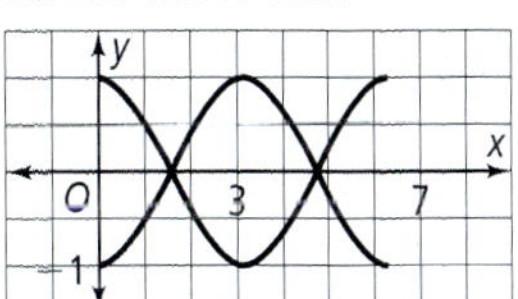

45. a.

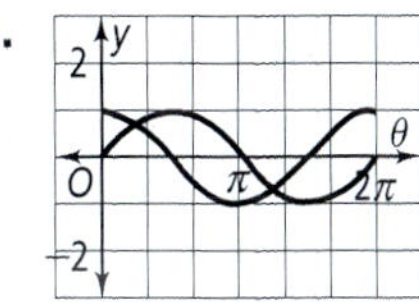

shift of $\frac{\pi}{2}$ units to the right

b.

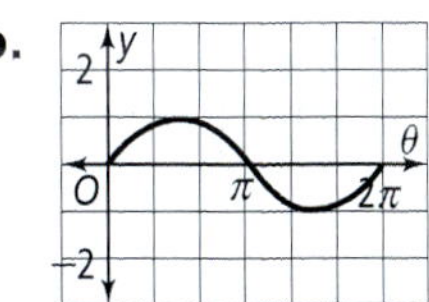

They are the same.

c. To write a sine function as a cosine function, replace sin with cos and replace θ with $\theta - \frac{\pi}{2}$.

47. F **49.** G

51.

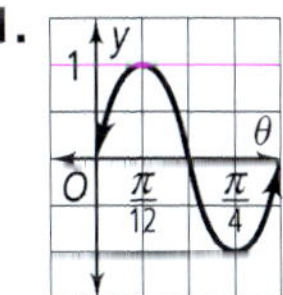

$y = \sin 6\theta$

52.

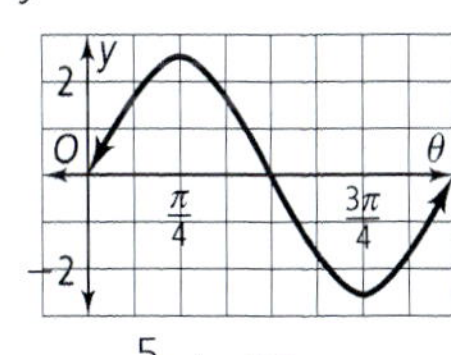

$y = \frac{5}{2} \sin 2\theta$

53. 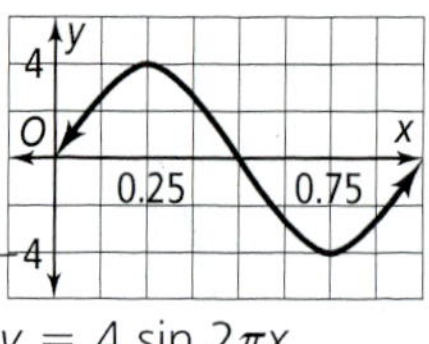

$y = 4 \sin 2\pi x$

54. about 1111 **55.** about 204 **56.** about 83
57. $a_n = 10 \cdot 3^{n-1}$; 10, 30, 90, 270, 810
58. $a_n = 12(-0.3)^{n-1}$; 12, −3.6, 1.08, −0.324, 0.0972 **59.** $a_n = 900\left(-\frac{1}{3}\right)^{n-1}$; 900, −300, 100, $-\frac{100}{3}$, $\frac{100}{9}$ **60.** 0.866, 0.5, 1.732
61. 0.5, 0.866, 0.577
62. 1, 0, undefined **63.** 0.5, −0.866, −0.577
64. 1, 0, undefined **65.** 0, 1, 0

Lesson 13-6 pp. 868–874

Got It? 1a. not defined **b.** $\sqrt{3}$ **c.** −1
2a. 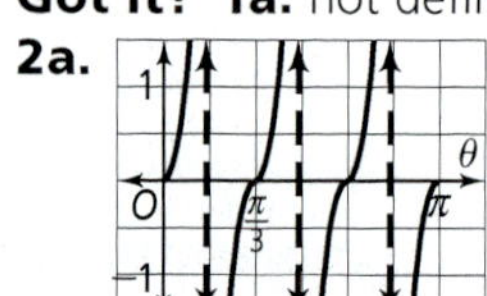
b.

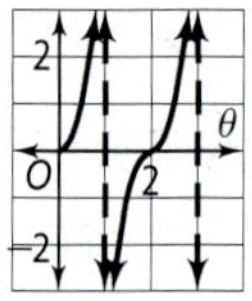

3a. ≈46.6 ft **b.** ≈8°
Lesson Check 1. 1 **2.** $\frac{\sqrt{3}}{3}$ **3.** −1 **4.** 0 **5.** $\frac{2\pi}{3}$ **6.** The student found 3θ instead of θ. Divide each side of the equation $3\theta = -1.33$ by 3 to get $\theta = -0.44$.
7. $y = \tan 2\theta$; the period of the function $y = \sin 4\theta$ is $\frac{\pi}{2}$. For a tangent function to have the same period, $\frac{\pi}{2}$, b must equal 2.
Exercises 9. 0 **11.** undefined **13.** 0 **15.** undefined
17. $\frac{\pi}{2}$ **19.** $\frac{2\pi}{3}$; $\theta = -\frac{\pi}{3}$ and $\theta = \frac{\pi}{3}$ **21.** $\frac{3\pi^2}{2}$; $\theta = -\frac{3\pi^2}{4}$ and $\theta = \frac{3\pi^2}{4}$

23. 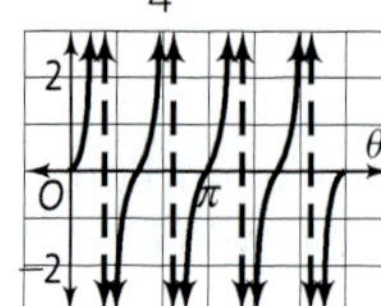
25. 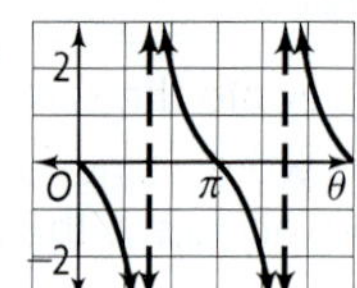

27. 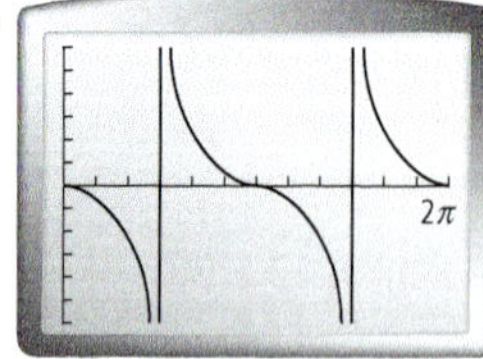

−100, undefined, 100

29a. 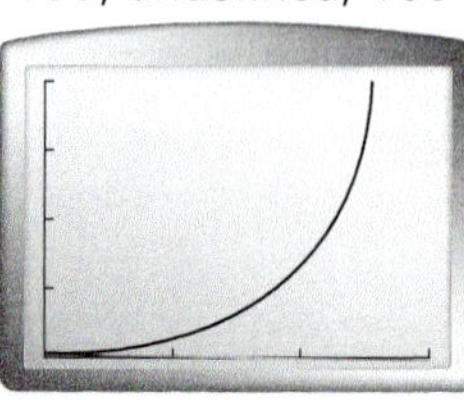
b. ≈14.3 ft
c. ≈20.2 ft

31. $\frac{2\pi}{5}$; 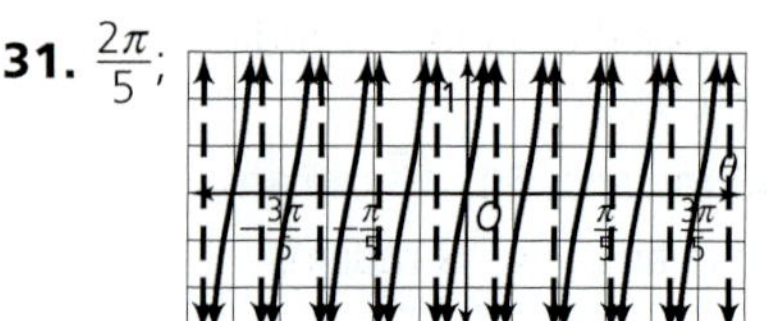

33. 1.11, 4.25 **35.** 0.08, 1.65, 3.22, 4.79
37. 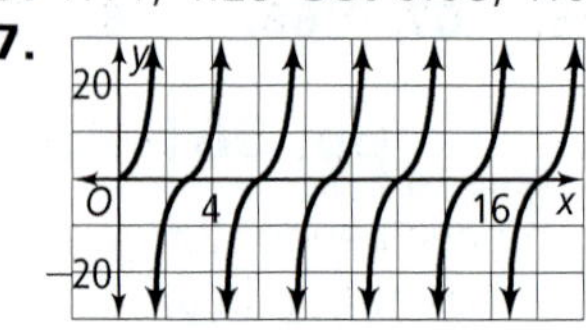

$150\sqrt{3}$ in.^{2} ≈ 260 in.^{2}

39. 200 **41.** 135 **43.** 70 **45.** $y = -\tan\left(\frac{1}{2}x\right)$
47a.

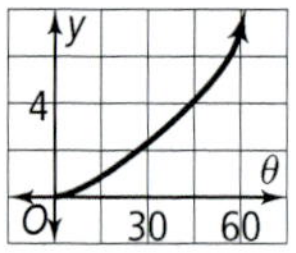

≈6.9 ft
b. ≈27.7 ft^2
c. ≈166.3 ft^2

49. Answers may vary. Sample: Triangles *OAP* and *OBQ* both share the angle θ and each triangle has a right angle, so they are similar by AA. $\frac{\sin\theta}{\cos\theta} = \frac{AP}{OA} = \frac{BQ}{OB} = \frac{\tan\theta}{1}$. Thus $\frac{\sin\theta}{\cos\theta} = \tan\theta$. **51.** 2; for $0 \le x < 2\pi$, x is nonnegative and there are only 2 sections of the graph of the tangent function on or above the x-axis. **53.** H **55.** I **57.** 1.32, 4.97 **58.** 1.77, 4.51 **59.** 6.15 **60.** 0.44, 1.56, 2.44, 3.56, 4.44, 5.56 **61.** mean ≈ 5.9, median = 6, modes = 4 and 6 **62.** 83 **63.** −227 **64.** 145 **65.** −332 **66.** 2 units to the right and up 5 units **67.** 5 units to the left and down 4 units **68.** 2 units to the left and up 1 unit

Lesson 13-7 pp. 875–882

Got It? 1a. 5; 5 units to the right **b.** −3; 3 units to the left
2a. 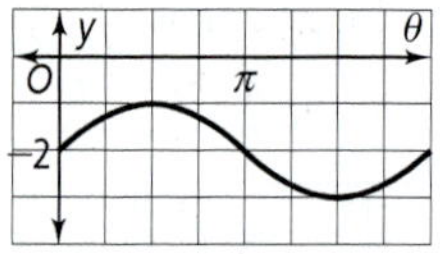
b. 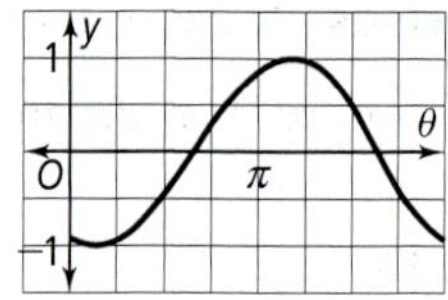

c. $y = \sin(x - 2)$ **d.** $y = \sin x - 2$
3a. 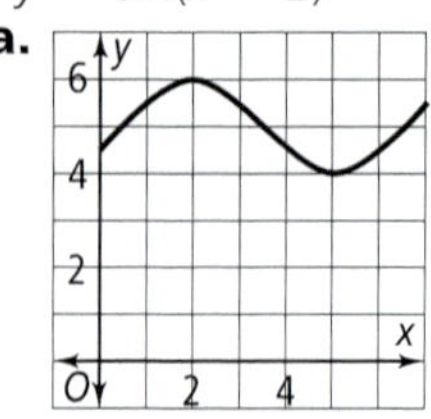
b. 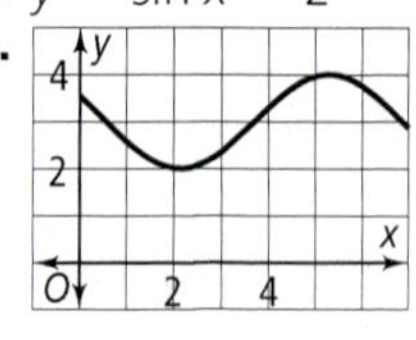

4a. 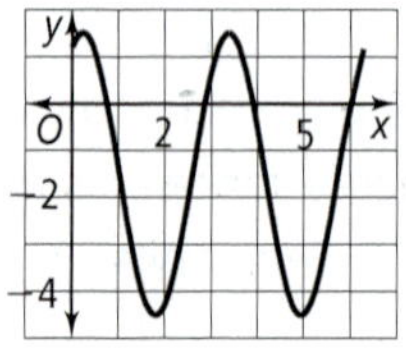
b. 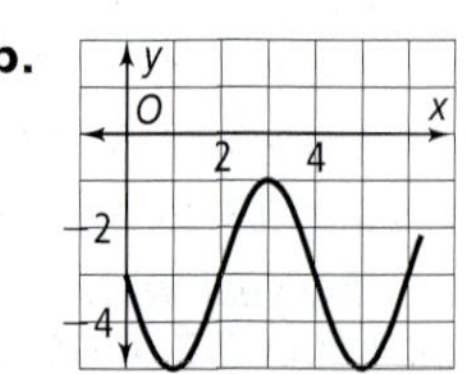

5a. $y = \cos x + \frac{\pi}{2}$ **b.** $y = 2\sin\left(x - \frac{\pi}{4}\right)$ **6a.** 69.9°F **b.** the average of the highest and lowest temperatures **c.** Yes; data can be extrapolated to calculate for next year.

Lesson Check

1.

2. phase shift: 2 units to the right; vertical shift: 9 units up **3.** $y = \cos\left(x - \frac{2\pi}{3}\right) + 3$ **4.** Answers may vary. Sample: $y = 4\sin\frac{2}{3}(x - \pi) - 5$ **5.** Scott is correct; $y = a\cos b(x - h) + k = \cos 3\left(x + \frac{\pi}{6}\right)$ where $h = -\frac{\pi}{6}$. The phase shift is $\frac{\pi}{6}$ units to the left of $y = \cos 3x$.

Exercises 7. −2; 2 units to the left **9.** 3; 3 units to the right **11.** $\frac{5\pi}{7}$; $\frac{5\pi}{7}$ units to the right

13.

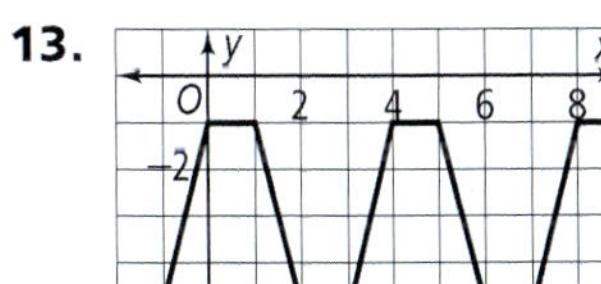

15.

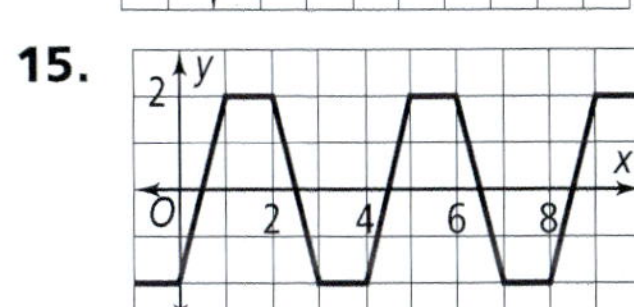

17.

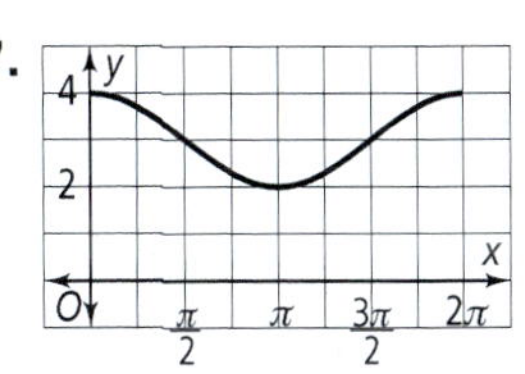

19.

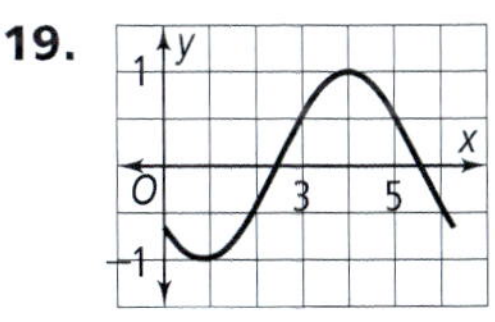

21.

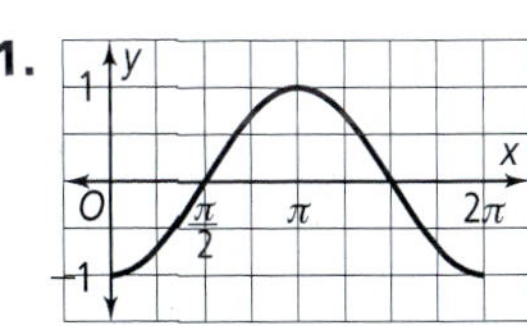

23. 1 unit to the left and 2 units down **25.** 3 units to the right and 2 units up

27.

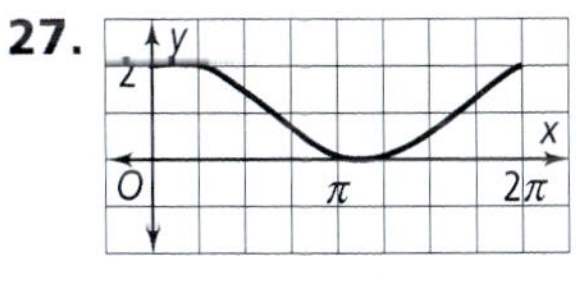

29.

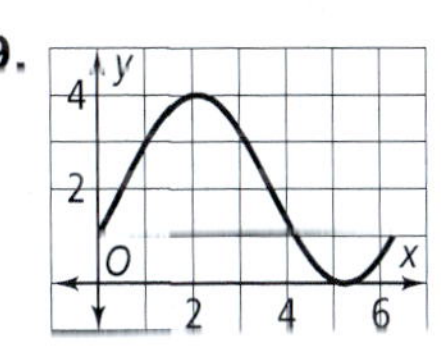

31.

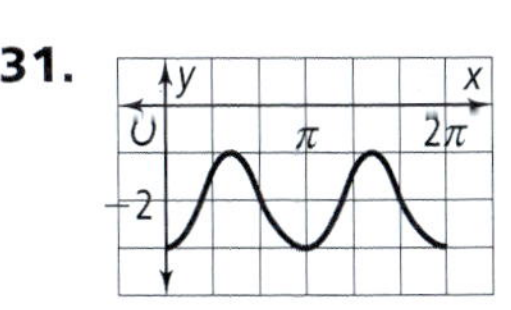

33.

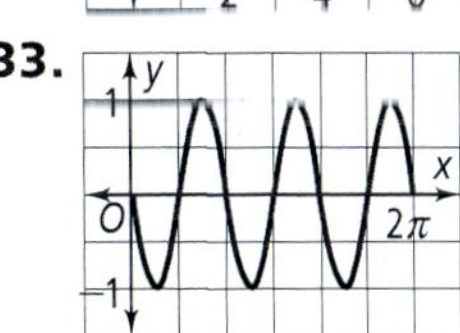

35.

37. $y = \cos x - \frac{\pi}{2}$ **39.** $y = \cos(x - 1.5)$ **41.** $y = \cos(x + 3) + \pi$ **43.** $y = 1.5\cos\left[\frac{\pi}{6}(x - 6) - \frac{\pi}{2}\right] + 2$

45. $y = -10\cos\frac{\pi}{10}x$; $y = 10\sin\left(\frac{\pi}{10}x - \frac{\pi}{2}\right)$

49.

51.

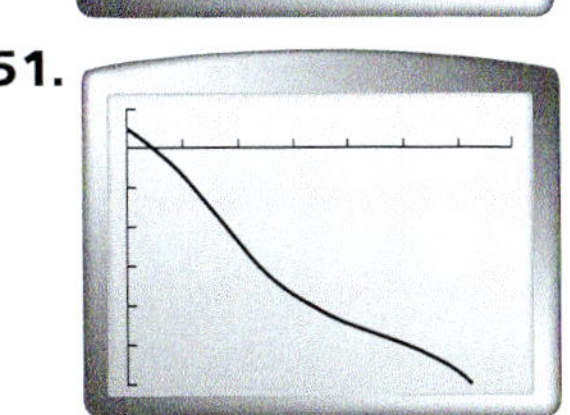

53.

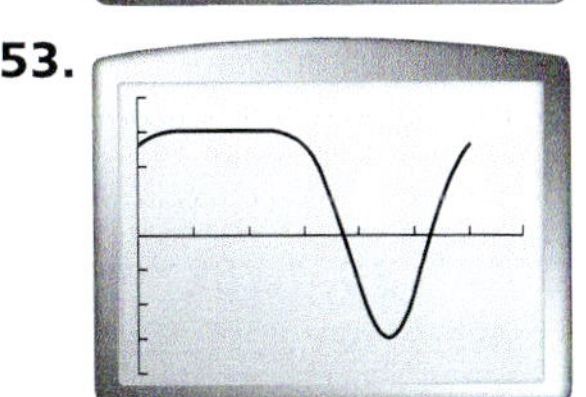

55. C **57.** B **59.** $\frac{\pi}{6}$; $\theta = -\frac{\pi}{12}, \frac{\pi}{12}$ **60.** 4π; $\theta = -2\pi, 2\pi$ **61.** $\frac{2\pi}{3}$; $\theta = -\frac{\pi}{3}, \frac{\pi}{3}$ **62.** 6π; $\theta = -3\pi, 3\pi$ **63.** 0.0064 **64.** 0.3456 **65.** ≈0.136 **66.** ≈0.198 **67.** $\frac{13}{9}$ **68.** $-\frac{8}{5}$ **69.** 2π **70.** $\frac{15}{4m}$ **71.** $-\frac{t}{14}$

Lesson 13-8 pp. 883–890

Got It? 1a. $\frac{2\sqrt{3}}{3}$ **b.** −1 **c.** −1 **d.** $\frac{3}{4}, \frac{5}{4}, \frac{5}{3}$ by the definition of a unit circle, the length of the hypotenuse of the right triangle is 1; by the Pythag. Thm., the length of the unlabeled leg is $\frac{4}{5}$. So the triangle is similar to a 3-4-5 right triangle. **2a.** ≈2.16 **b.** ≈4.649 **c.** ≈1.035 **d.** undefined **e.** use $\frac{\cos x}{\sin x}$

3.

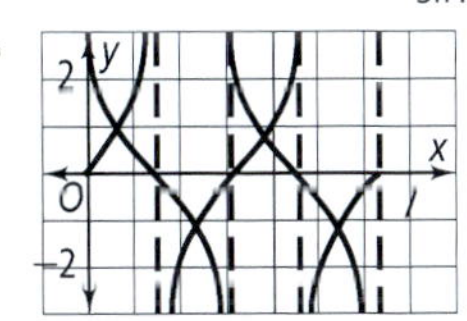

4. ≈1.4142 **5.** about 860 ft away and 3185 ft away

Lesson Check 1. 1 **2.** $\frac{2\sqrt{3}}{3}$ **3.** ≈1.003 **4.** ≈1.346 **5.** 29.4 ft **6.** $y = 5\sec\theta = \frac{5}{\cos\theta}$; there is no value of θ that will make $\frac{5}{\cos\theta}$ equal to zero. **7.** The student found the reciprocal of $(1 + \cos 20°) = \frac{1}{1 + \cos 20°} \approx 0.5155$. The answer should be: $\sec 20° + 1 = \frac{1}{\cos 20°} + 1 \approx$

$1.0642 + 1 \approx 2.0642$. **8.** The graphs have the same period and range. The domain of $y = \sec x$ is all real numbers except $n\pi + \frac{\pi}{4}$ (where n is an integer), which are its asymptotes. The domain of $y = \csc x$ is all real numbers except $\frac{n\pi}{2}$ (where n is an integer), which are its asymptotes. The graph of $y = \csc x$ can be obtained as a translation of $y = \sec\left(x - \frac{\pi}{4}\right)$ of the parent function $y = \sec x$.

Exercises **9.** -1 **11.** $\frac{-\sqrt{3}}{\sqrt{3}}$ **13.** 0 **15.** $-\sqrt{2}$
17. ≈ -1.248 **19.** ≈ 0.675 **21.** ≈ -1.6 **23.** undefined

25.

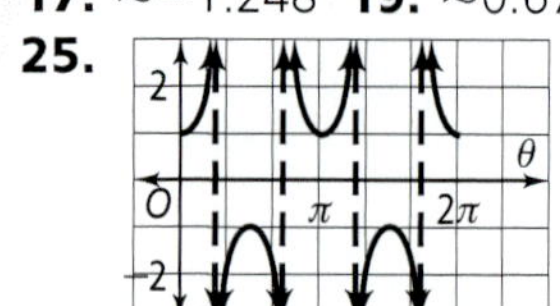

27.

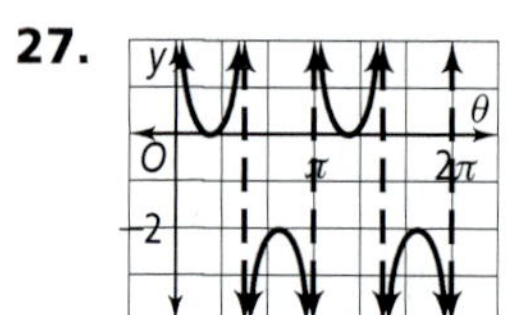

29. 1.1547 **31.** -2.9238 **33.** 1.0642 **35.** 1.7321
37. ≈ 104 ft and ≈ 164 ft **39.** Answers may vary. Sample: $y = \csc\left(\theta + \frac{\pi}{2}\right)$ **41.** C

43.

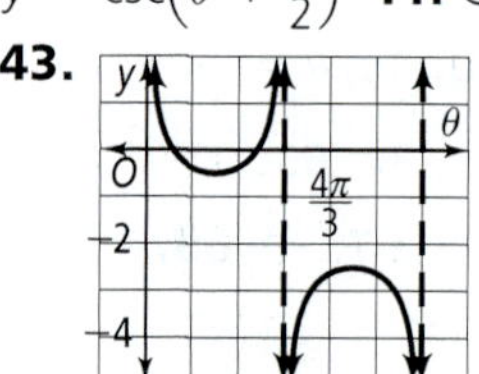

45.

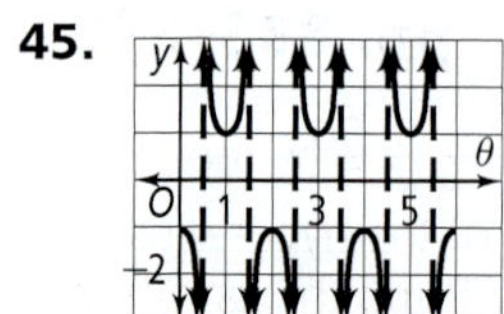

47a. domain: all real numbers except multiples of π, range: $y \geq 1$ or $y \leq -1$; period: 2π **b.** 1 **c.** -1
49. csc 180° is undefined because sin 180° = 0 and $\csc\theta = \frac{1}{\sin\theta}$. **51.** cot 0° is undefined because sin 0° = 0 and $\cot\theta = \frac{\cos\theta}{\sin\theta}$.

53a.

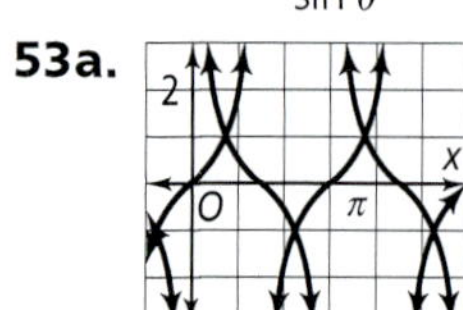

b. The domain of $y = \tan x$ is all real numbers except odd multiples of $\frac{\pi}{2}$, where its asymptotes occur. The domain of $y = \cot x$ is all real numbers except multiples of π, where its asymptotes occur. The range of both functions is all real numbers. **c.** The graphs have the same period and range. Their asymptotes are shifted $\frac{\pi}{2}$ units.
d. Answers may vary. Sample: $x = \frac{\pi}{4}, x = \frac{3\pi}{4}$

55.

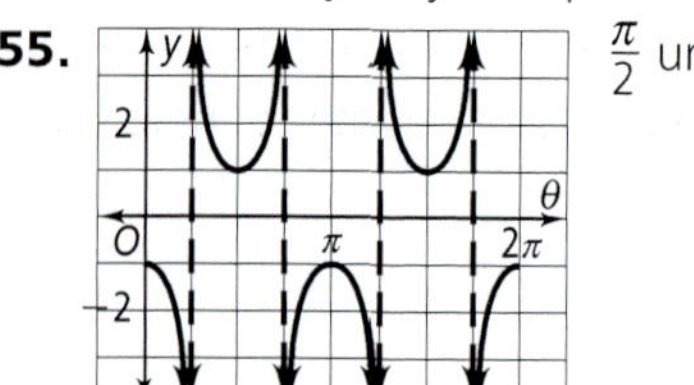

$\frac{\pi}{2}$ units to the left

57. 2 units to the left and 1 unit down

59.

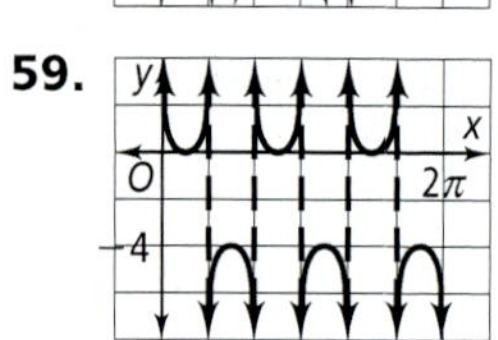

$\frac{\pi}{6}$ units to the right and 2 units down

61a. II **b.** I
63. $y = \cos 3x$ cycles 3 times for each cycle of $y = \cos x$. Thus, for each cycle of $y = \sec x$, $y = \sec 3x$ cycles 3 times, and each cycle of $y = \sec 3x$ is $\frac{1}{3}$ as wide as one cycle of $y = \sec x$.
65. $\frac{4}{3}$ **69.** 1.4 **71.** 2, 2π; 5 units down
72. 1, 2π; 4 units left, 7 units down
73. 3, 2π; $\frac{\pi}{6}$ units to the left, 4 units up
74. 5, 2; 1.5 units to the right, 8 units down

75.

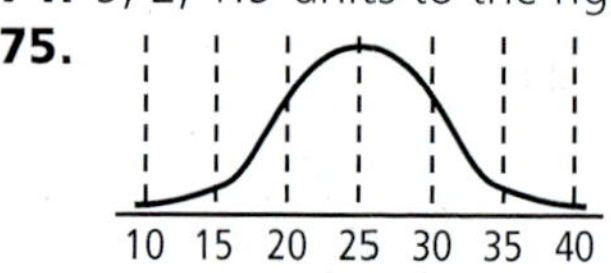

76.

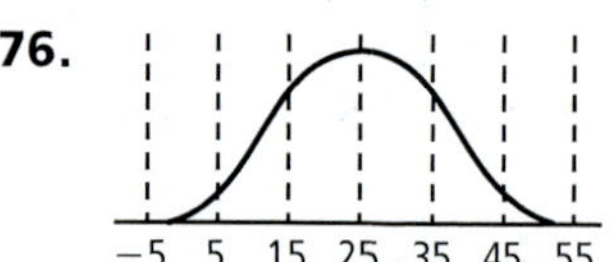

77. true; Distr. Prop. **78.** true; Distr. Prop. and Comm. Prop. of Add. **79.** not true

Chapter Review pp. 892–896

1. period **2.** unit circle **3.** tangent function
4. phase shift **5.** secant function
6. periodic; from 0 to 4 or from 4 to 6; 4; 2
7. Answers may vary. Sample:

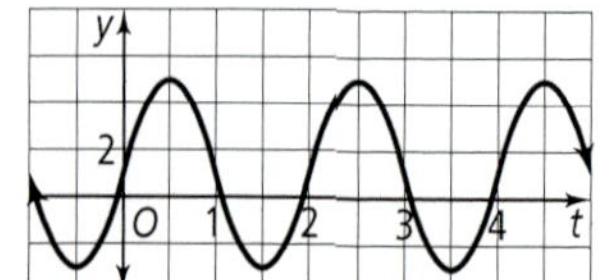

8.

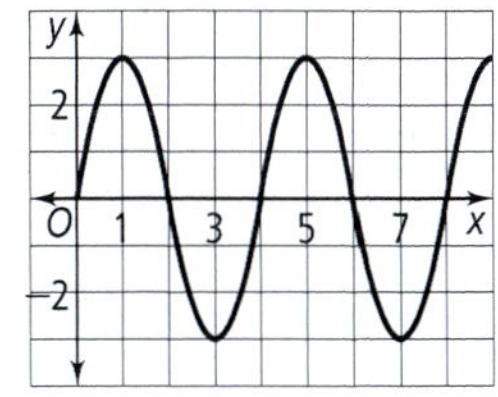

9. $-225°$

10.

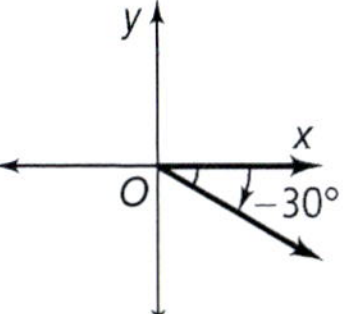

11. 240° **12.** $\sin(315°) = -\frac{\sqrt{2}}{2} \approx -0.71$, $\cos(315°) = \frac{\sqrt{2}}{2} \approx 0.71$; $\sin(-315°) = \frac{\sqrt{2}}{2} \approx 0.71$, $\cos(-315°) = \frac{\sqrt{2}}{2} \approx 0.71$ **13a.** $\frac{\pi}{3}$ **b.** $\frac{1}{2}, \frac{\sqrt{3}}{2}$

14a. $-\frac{\pi}{4}$ **b.** $\frac{\sqrt{2}}{2}, -\frac{\sqrt{2}}{2}$ **15a.** π **b.** $-1, 0$

16a. 360° **b.** 1, 0 **17a.** 150° **b.** $-\frac{\sqrt{3}}{2}, \frac{1}{2}$

18a. $-135°$ **b.** $-\frac{\sqrt{2}}{2}, -\frac{\sqrt{2}}{2}$

19. 26.2 ft

20.

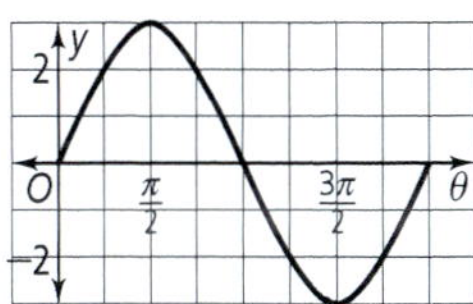

21.

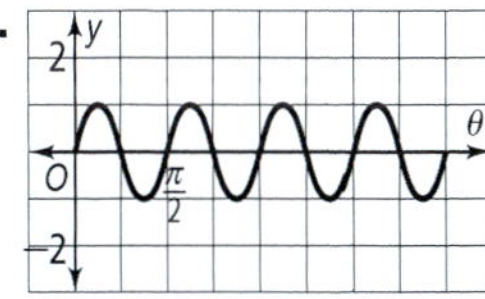

22. $y = 4 \sin 4\theta$

23.

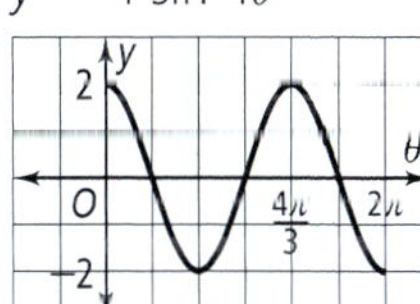

24.

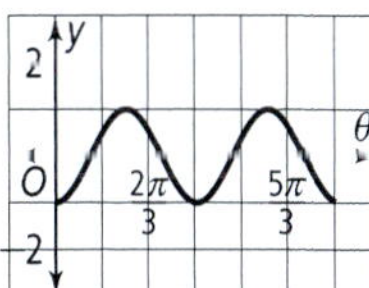

25. $y = 3 \cos 2\theta$ **26.** 0.58, 1.00, 2.15, 2.57, 3.72, 4.14, 5.29, 5.71 **27.** 0.70, 1.30, 2.70, 3.30, 4.70, 5.30

28.

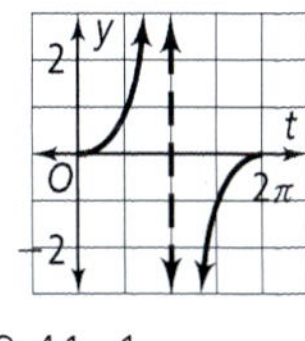

0.41, 1

29.

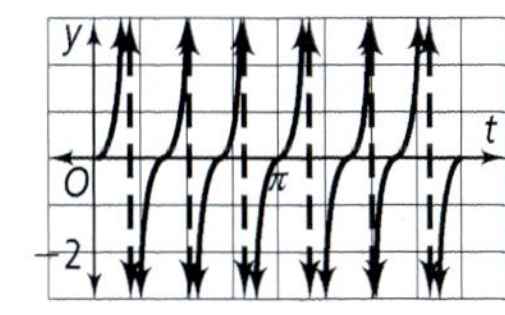

-1, undefined

30.

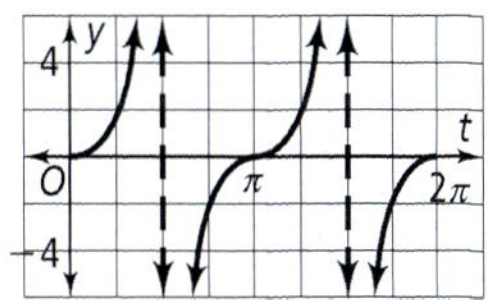

2, undefined

31.

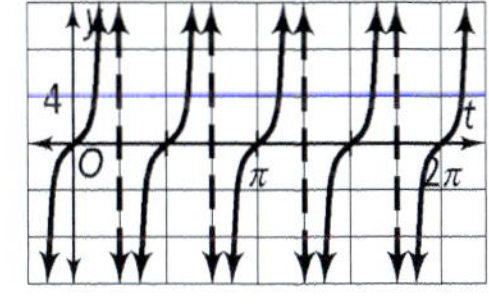

undefined, 0

32.

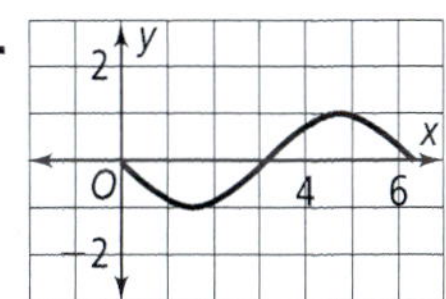

33.

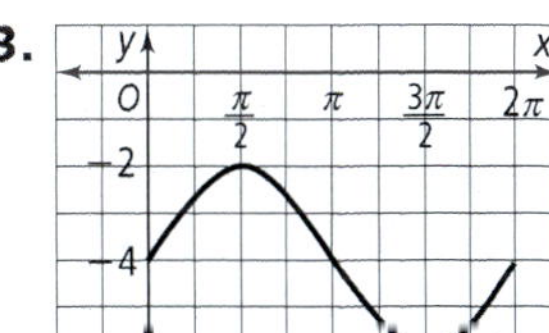

34.

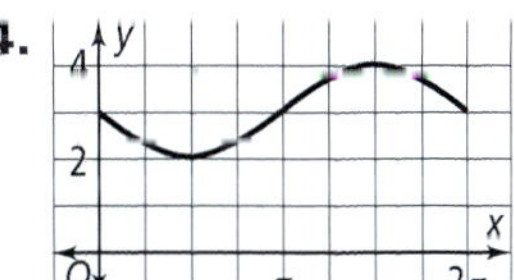

35.

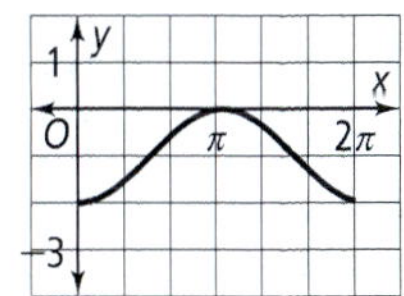

36. $y = \sin\left(x - \frac{\pi}{4}\right)$ **37.** $y = \cos x - 2$ **38.** $\sqrt{2}$

39. $-\frac{\sqrt{3}}{3}$ **40.** 2 **41.** $\sqrt{3}$

42.

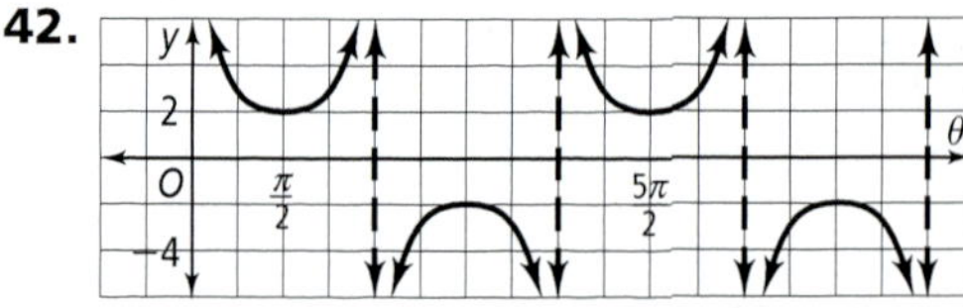

43.

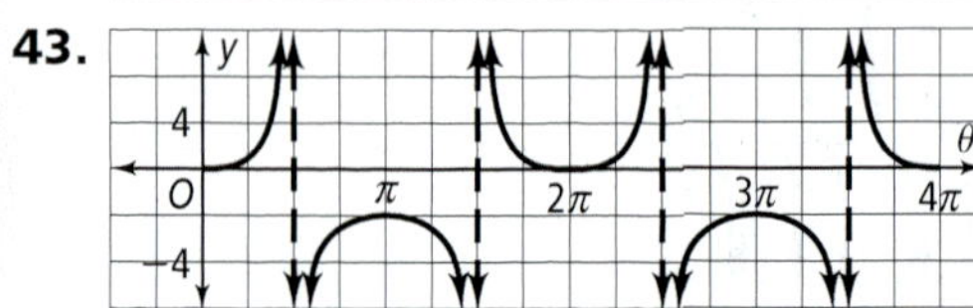

44.

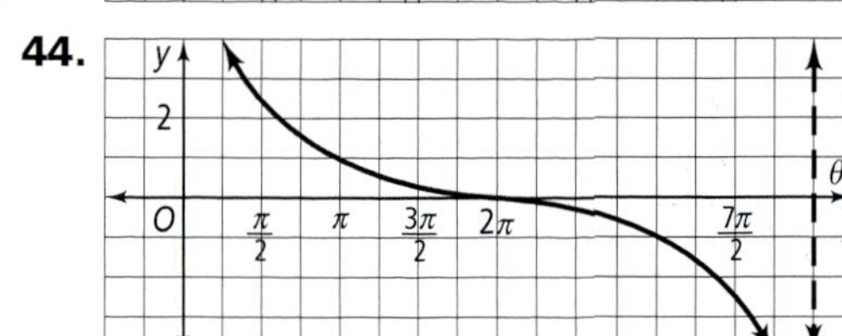

45.

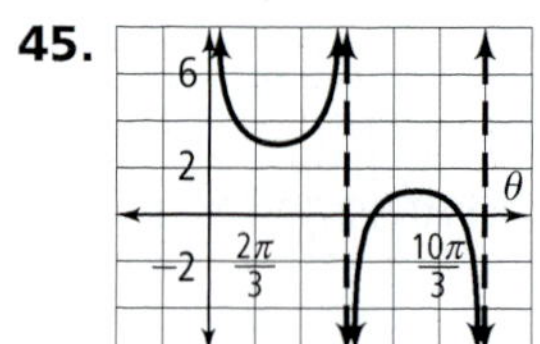

Chapter 14

Get Ready! p. 901

1. $x = \pm\frac{5}{2}$ **2.** $x = \pm\sqrt{23}$ **3.** $x = \pm 4\sqrt{\frac{5}{3}}$ **4.** $x = \pm\sqrt{\frac{11}{2}}$ **5.** $x = \pm\sqrt{30}$ **6.** $x = \pm 2$ **7.** $f^{-1}(x) = \frac{x-2}{5}$; domain of f and range of f^{-1}: all real numbers, range of f and domain of f^{-1}: all real numbers; yes **8.** $f^{-1}(x) = x^2 - 3$; domain of f and range of f^{-1}: all real numbers ≥ -3, domain of f^{-1}: all real numbers; range of f: all real numbers ≥ 0; yes **9.** $f^{-1}(x) = \frac{x^2+4}{3}$; domain of f and range of f^{-1}: all real numbers $\geq \frac{4}{3}$, domain of f^{-1}: all real numbers; range of f: all real numbers ≥ 0; yes **10.** $f^{-1}(x) = \frac{5}{x}$; domain of f and range off^{-1}: all real numbers except 0, domain of f^{-1} and range of f: all real numbers except 0; yes **11.** $f^{-1}(x) = \frac{10}{x} + 1$; domain of f and range of f^{-1}: all real numbers except 1, domain of f^{-1} and range of f: all real numbers except 0; yes **12.** $f^{-1}(x) = \frac{10}{x+1}$; domain of f and range of f^{-1}: all real numbers except 0, domain of f^{-1} and range of f: all real numbers except -1; yes **13.** $x = -\frac{3}{2}$ **14.** $x = 0.002$ **15.** $x = 1.0646$ **16.** $x = 18257.4$ **17.** $x = 0.00003$ **18.** $x = 5$ **19.** 0.67; 0.74; 1.11 **20.** −0.26; −0.97; 3.73 **21.** 0.96; 0.28; 0.29 **22.** −0.87; 0.50; −0.58 **23.** Answers may vary. Sample: The eq. is true for all values of θ for which $\tan^2\theta$ and $\sec^2\theta$ are defined. **24.** Answers may vary. Sample: the lengths of the sides of rt. triangles

Lesson 14-1 pp. 904–910

Got It? 1. all real numbers except multiples of π

2. $\frac{\csc\theta}{\sec\theta} = \frac{\left(\frac{1}{\sin\theta}\right)}{\left(\frac{1}{\cos\theta}\right)} = \frac{\cos\theta}{\sin\theta} = \cot\theta$; all real numbers except multiples of $\frac{\pi}{2}$

3a.
$$\begin{aligned} 1 + \cot^2\theta &= 1 + \left(\frac{\cos\theta}{\sin\theta}\right)^2 \\ &= 1 + \frac{\cos^2\theta}{\sin^2\theta} \\ &= 1 + \frac{1-\sin^2\theta}{\sin^2\theta} \\ &= 1 + \frac{1}{\sin^2\theta} - \frac{\sin^2\theta}{\sin^2\theta} \\ &= 1 + \csc^2\theta - 1 \\ &= \csc^2\theta \end{aligned}$$

b. No; the domains of $\sin\theta$ and $\cos\theta$ are all real numbers, but the domains of $\tan\theta$, $\cot\theta$, $\sec\theta$, and $\csc\theta$ have restrictions.

4. $\sec^2\theta - \sec^2\theta\cos^2\theta$
$$\begin{aligned} &= \left(\frac{1}{\cos\theta}\right)^2 - \left(\frac{1}{\cos\theta}\right)^2\cos^2\theta \\ &= \frac{1}{\cos^2\theta} - \frac{1}{\cos^2\theta}\cdot\cos^2\theta \\ &= \frac{1}{\cos^2\theta} - \frac{\cos^2\theta}{\cos^2\theta} \\ &= \frac{1-\cos^2\theta}{\cos^2\theta} \\ &= \frac{\sin^2\theta}{\cos^2\theta} \\ &= \tan^2\theta \end{aligned}$$

5. $\csc\theta$

Lesson Check

1. $\tan\theta\csc\theta$
$$\begin{aligned} &= \frac{\sin\theta}{\cos\theta}\cdot\frac{1}{\sin\theta} \\ &= \frac{1}{\cos\theta} \\ &= \sec\theta \end{aligned}$$

2. $\csc^2\theta - \cot^2\theta$
$$\begin{aligned} &= \left(\frac{1}{\sin\theta}\right)^2 - \left(\frac{\cos\theta}{\sin\theta}\right)^2 \\ &= \frac{1}{\sin^2\theta} - \frac{\cos^2\theta}{\sin^2\theta} \\ &= \frac{1-\cos^2\theta}{\sin^2\theta} \\ &= \frac{\sin^2\theta}{\sin^2\theta} \\ &= 1 \end{aligned}$$

3. $\sin\theta\tan\theta$

$= \sin\theta \cdot \dfrac{\sin\theta}{\cos\theta}$

$= \dfrac{\sin^2\theta}{\cos\theta}$

$= \dfrac{1-\cos^2\theta}{\cos\theta}$

$= \dfrac{1}{\cos\theta} - \dfrac{\cos^2\theta}{\cos\theta}$

$= \sec\theta - \cos\theta$

4. $\tan\theta\cot\theta - \sin^2\theta$

$= \tan\theta\dfrac{1}{\tan\theta} - \sin^2\theta$

$= \dfrac{\tan\theta}{\tan\theta} - \sin^2\theta$

$= 1 - \sin^2\theta$

$= \cos^2\theta$

5. Answers may vary. Sample: Letting a and b be the legs, and c the hypotenuse of a right triangle, the Pythagorean Theorem states that $a^2 + b^2 = c^2$. Dividing both sides by c^2, then $\frac{a^2}{c^2} + \frac{b^2}{c^2} = \left(\frac{a}{c}\right)^2 + \left(\frac{b}{c}\right)^2 = 1$. Calling the angle between a and c θ, then $\sin\theta = \frac{b}{c}$ and $\cos\theta = \frac{a}{c}$. By substitution, $\cos^2\theta + \sin^2\theta = 1$. **6.** wrong calculation: $2 - \cos^2\theta = 2 - (1 - \sin^2\theta) = 2 - 1 + \sin^2\theta = 1 + \sin^2\theta$

Exercises

7. $\cos\theta\cot\theta$

$= \cos\theta\left(\dfrac{\cos\theta}{\sin\theta}\right)$

$= \dfrac{1-\sin^2\theta}{\sin\theta}$

$= \dfrac{1}{\sin\theta} - \sin\theta$; all real numbers except multiples of π

9. $\cos\theta\tan\theta$

$= \cos\theta\left(\dfrac{\sin\theta}{\cos\theta}\right) = \sin\theta$; all real numbers except odd multiples of $\frac{\pi}{2}$

11. $\cos\theta\sec\theta$

$= \cos\theta\left(\dfrac{1}{\cos\theta}\right) = 1$; all real numbers except odd multiples of $\frac{\pi}{2}$

13. $\sin\theta\csc\theta$

$= \sin\theta\left(\dfrac{1}{\sin\theta}\right) = \dfrac{\sin\theta}{\sin\theta} = 1$; all real numbers except multiples of π

15. $\csc\theta - \sin\theta$

$= \dfrac{1}{\sin\theta} - \sin\theta$

$= \dfrac{1-\sin^2\theta}{\sin\theta}$

$= \dfrac{\cos^2\theta}{\sin\theta}$

$= \cot\theta\cos\theta$; all real numbers except odd multiples of $\frac{\pi}{2}$

17. $\sin^2\theta$ **19.** $-\cot^2\theta$ **21.** $\sin\theta$ **23.** 1 **25.** 1 **27.** 1 **29.** $\sec\theta$ **31.** $\sec^2\theta$ **33.** $\csc\theta$ **35.** $\sin^2\theta$ **37.** 1 **39.** 1

41. $\pm\sqrt{1-\cos^2\theta}$ **43.** $\pm\dfrac{\sqrt{1-\sin^2\theta}}{\sin\theta}$

45. $\pm\sqrt{\csc^2\theta - 1}$

47. $\sin^2\theta\tan^2\theta = \sin^2\theta\left(\dfrac{\sin^2\theta}{\cos^2\theta}\right)$

$= (1-\cos^2\theta)\left(\dfrac{\sin^2\theta}{\cos^2\theta}\right)$

$= \dfrac{\sin^2\theta - \sin^2\theta\cos^2\theta}{\cos^2\theta}$

$= \dfrac{\sin^2\theta}{\cos^2\theta} - \dfrac{\sin^2\theta\cos^2\theta}{\cos^2\theta}$

$= \tan^2\theta - \sin^2\theta$

49. $\sin\theta\cos\theta(\tan\theta + \cot\theta)$

$= \sin\theta\cos\theta\left(\dfrac{\sin\theta}{\cos\theta} + \dfrac{\cos\theta}{\sin\theta}\right)$

$= \dfrac{\sin^2\theta\cos\theta}{\cos\theta} + \dfrac{\cos^2\theta\sin\theta}{\sin\theta}$

$= \sin^2\theta + \cos^2\theta = 1$

51. $\dfrac{\sec\theta}{\cot\theta + \tan\theta}$

$= \dfrac{\frac{1}{\cos\theta}}{\frac{\cos\theta}{\sin\theta} + \frac{\sin\theta}{\cos\theta}} \cdot \dfrac{\sin\theta\cos\theta}{\sin\theta\cos\theta}$

$= \dfrac{\sin\theta}{\cos^2\theta + \sin^2\theta} = \dfrac{\sin\theta}{1} = \sin\theta$

53. $\dfrac{1-\sin^2\theta}{\sin^2\theta}$

55. $\sin^2\theta + \cos^2\theta = 1$

$(0.5)^2 + \cos^2\theta = 1$

$0.25 + \cos^2\theta = 1$

$\cos^2\theta = 1 - 0.25$

$\cos^2\theta = 0.75$

$\cos\theta = \pm\sqrt{0.75}$

Since θ is in the first quadrant, $\cos\theta$ is positive; $\cos\theta = 0.866025404$

$\tan\theta = \dfrac{\sin\theta}{\cos\theta}$

$= \dfrac{0.5}{0.866025404}$

$= 0.577350269$

57. $\sin^2\theta + \cos^2\theta = 1$

$\sin^2\theta + (-0.6)^2 = 1$

$\sin^2\theta + 0.36 = 1$

$\sin^2\theta = 1 - 0.36$

$\sin^2\theta = 0.64$

$\sin\theta = \pm\sqrt{0.64}$

Since θ is in the third quadrant, $\sin\theta$ is negative; $\sin\theta = -0.8$

$$\tan\theta = \frac{\sin\theta}{\cos\theta}$$
$$= \frac{-0.8}{-0.6}$$
$$= 1.333333333$$

59. $\tan\theta = \frac{\sin\theta}{\cos\theta}$, $\sin^2\theta + \cos^2\theta = 1$, which can be rewritten as $\sin^2\theta = 1 - \cos^2\theta$

$$1.2 = \frac{\sin\theta}{\cos\theta}$$
$$1.2^2 = \frac{\sin^2\theta}{\cos^2\theta}$$
$$1.44(\cos^2\theta) = \sin^2\theta$$
$$1.44(\cos^2\theta) = 1 - \cos^2\theta$$
$$1.44(\cos^2\theta) + \cos^2\theta = 1$$
$$2.44(\cos^2\theta) = 1$$
$$\cos^2\theta = \frac{1}{2.44}$$
$$\cos^2\theta = 0.409836066$$
$$\cos\theta = \pm\sqrt{0.409836066}$$
Since θ is in the first quadrant, $\cos\theta$ is positive;
$\cos\theta = 0.640184400$
$$\tan\theta = \frac{\sin\theta}{\cos\theta}$$
$$1.2 = \frac{\sin\theta}{0.640184400}$$
$$\sin\theta = 0.76822128$$

61. $\sin^2\theta + \cos^2\theta = 1$
$$(0.2)^2 + \cos^2\theta = 1$$
$$0.04 + \cos^2\theta = 1$$
$$\cos^2\theta = 1 - 0.04$$
$$\cos^2\theta = 0.96$$
$$\cos\theta = \pm\sqrt{0.96}$$
Since $\sin\theta$ is positive and $\tan\theta$ is negative, θ is in the fourth quadrant, so $\cos\theta$ is positive;
$\cos\theta = 0.97979590$

63. $\cos(\theta + \pi) = |\cos\theta|$, but is also in Quadrant III and is negative, so $\cos(\theta + \pi) = -\cos\theta$ **65.** 1

67. If $n_2 > n_1$, then $\theta_1 > \theta_2$; if $n_2 < n_1$, then $\theta_1 < \theta_2$; if $n_2 = n_1$, then $\theta_2 = \theta_1$. **69.** H **71.** F

73. By the Difference of Squares Property and the second Pythagorean Identity: $(\sec\theta + 1)(\sec\theta - 1) = \sec^2\theta - 1$
$$= \tan^2\theta$$

74.

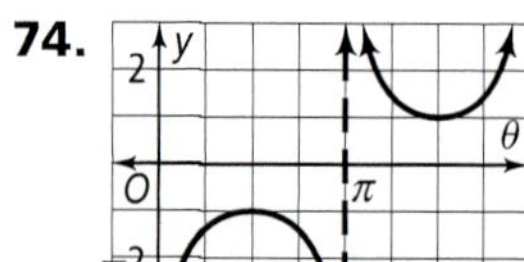

75.

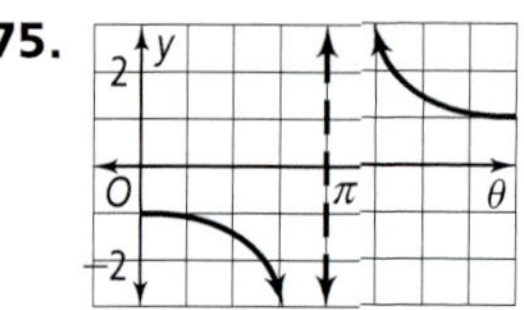

76.

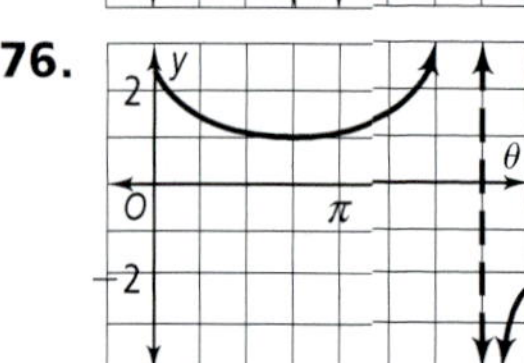

77.

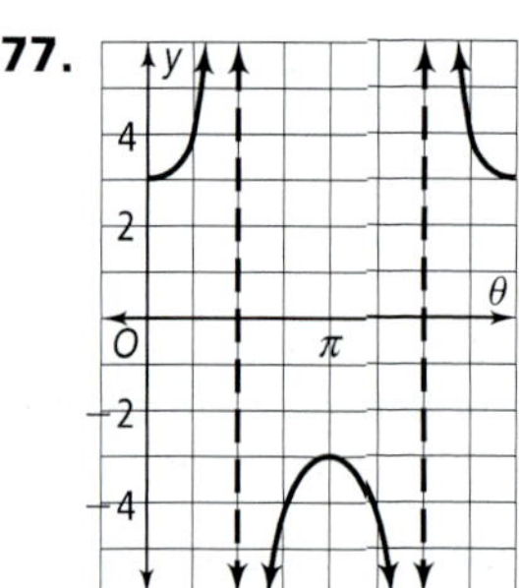

78. 35° **79.** 45° **80.** 135° **81.** 211°
82. $f^{-1}(x) = x - 1$ **83.** $f^{-1}(x) = \frac{x + 3}{2}$
84. $f^{-1}(x) = \pm\sqrt{x - 4}$

Lesson 14-2 pp. 911–918

Got It? 1a. 120° **b.** 30° **c.** 45° **2a.** $0.46 + 2\pi n$ and $2.69 + 2\pi n$ **b.** $-0.82 + 2\pi n$ and $3.96 + 2\pi n$
3a. $0.41 + 2\pi n$ and $3.56 + 2\pi n$, or just $0.41 + \pi n$
b. $-0.63 + 2\pi n$ and $2.51 + 2\pi n$, or just $-0.63 + \pi n$
c. $\tan(\theta + \pi) = \tan\theta$ **4.** $\frac{11\pi}{6}$ and $\frac{7\pi}{6}$ **5.** $\frac{\pi}{2}$ and $\frac{3\pi}{2}$
6. The air conditioner comes on about 7 hours after midnight, 7 A.M., and goes off about 7 hours before midnight, 5 P.M.
Lesson Check 1. $-30° + 360° \cdot n$ and $210° + 360° \cdot n$
2. $60° + 360° \cdot n$ and $120° + 360° \cdot n$ **3.** 2.30, 3.98
4. 0 **5.** Answers may vary. Sample: To find the inverse of $y = 3x - 4$, you interchange x and y and solve for y: $x = 3y - 4$, $y = \frac{x + 4}{3}$. Replace y with $f^{-1}(x)$ to find $f^{-1}(x) = \frac{x + 4}{3}$. To find the inverse of $y = 3\sin\theta - 4$, you interchange θ and y and solve for $\sin y$ and then solve for y: $\theta = 3\sin y - 4$, $\sin y = \frac{\theta + 4}{3}$ and $y = \sin^{-1}\left(\frac{\theta + 4}{3}\right)$. Replace y with $f^{-1}(\theta)$ to find $f^{-1}(\theta) = \sin^{-1}\left(\frac{\theta + 4}{3}\right)$. The procedure for finding the inverse is the same; for $y = 3\sin\theta - 4$ you will also use the inverse sine function.
6. The student divided both sides of the equation by $\sin\theta$, which in the given interval can be equal to zero. There is an error in that the student failed to take into account the fact that division by zero is not possible.

Exercises **7.** 90° + $n \cdot$ 360° **9.** 240° + $n \cdot$ 360° and 300° + $n \cdot$ 360° **11.** 90° + $n \cdot$ 360° and 270° + $n \cdot$ 360°, or just 90° + $n \cdot$ 180° **13.** 0.79 + πn, or just $\frac{\pi}{4} + \pi n$ **15.** −0.89 + $2\pi n$ and 4.04 + $2\pi n$ **17.** 2.67 + $2\pi n$ and 3.62 + $2\pi n$ **19.** $\frac{\pi}{6}, \frac{5\pi}{6}$ **21.** $\frac{\pi}{4}, \frac{5\pi}{4}$ **23.** 0.46, 3.61 **25.** no solution **27.** $\frac{\pi}{2}, \pi, \frac{3\pi}{2}$ **29.** $\frac{\pi}{4}, \frac{3\pi}{4}, \frac{5\pi}{4}, \frac{7\pi}{4}$ **31.** 0, π **33.** $\frac{7\pi}{6}, \frac{11\pi}{6}$ **35.** 30° + $n \cdot$ 360° and 150° + $n \cdot$ 360° **37.** 210° + $n \cdot$ 360° and 330° + $n \cdot$ 360° **39.** $\frac{3\pi}{2}$ **41.** 3.04, 6.18 **43.** 0.0028 s; 0.019 s **45.** $0 + 2\pi n, \frac{2}{3}\pi + 2\pi n, \frac{4}{3}\pi + 2\pi n$ **47.** $\frac{\pi}{6} + 2\pi n, \frac{5\pi}{6} + 2\pi n, \frac{3\pi}{2} + 2\pi n$ **49.** $\frac{\pi}{6} + 2\pi n, \frac{5\pi}{6} + 2\pi n, \frac{\pi}{2} + \pi n$ **51.** $\frac{\pi}{2} + 2\pi n, \frac{7\pi}{6} + 2\pi n, \frac{11\pi}{6} + 2\pi n$ **53.** $\frac{\pi}{6} + 2\pi n, \frac{5\pi}{6} + 2\pi n$ **55.** $\frac{\pi}{4} + \frac{\pi n}{2}$ **57.** $\frac{\pi}{4} + \frac{\pi n}{2}$ **59.** $\frac{2\pi}{3} + 2\pi n, \frac{4\pi}{3} + 2\pi n$ **61a.** Answers may vary. Sample: $\cos\theta = -1$, $2\cos\theta = -2$, $3\cos\theta = -3$ **b.** Start with $\cos\theta = -1$, and then multiply both sides of the eq. by any nonzero number. **63.** $\theta = \sin^{-1}\left(\frac{y}{3}\right) - 2$ **65.** $\theta = \cos^{-1}\left(\frac{y-1}{2}\right)$ **67.** D **69.** C **71.** D **73.** $\cot\theta$ **74.** $\tan^2\theta$ **75.** 1 **76.** 1 **77.** $\sin\theta$ **78.** $\tan\theta$ **79.** $y = 4\cos\frac{\pi}{4}\theta$ **80.** $y = 3\cos\theta$ **81.** $y = \frac{\pi}{4}\cos\frac{2}{3}\theta$ **82.** 4 **83.** 21 **84.** $52\frac{1}{2}$

Lesson 14-3 pp. 919–926

Got It? **1.** $\sin\theta = \frac{12}{13}$, $\cos\theta = -\frac{5}{13}$, $\tan\theta = -\frac{12}{5}$, $\csc\theta = \frac{13}{12}$, $\sec\theta = -\frac{13}{5}$, $\cot\theta = -\frac{5}{12}$ **2a.** 27.1 m **b.** 32.3 m **3.** $\sin E = \frac{3}{5}$, $\sec F = \frac{5}{3}$ **4.** ≈76,430 ft ≈ 14.5 mi **5a.** 23.58° **b.** 56.25° **6a.** 72 ft **b.** Answers may vary. Sample: Build the ramp in 3 sections, each of which is 24 ft, and with landings between sections.
Lesson Check **1.** $\sin 57° = \frac{b}{c}$, $\cos 57° = \frac{a}{c}$, $\tan 57° = \frac{b}{a}$ **2.** 15.4 **3.** $\sin 33° = \frac{a}{c} = 0.5$, $\cos 33° = \frac{b}{c} = 0.8$, $\tan 33° = \frac{a}{b} = 0.6$ **4.** 36.9° **5.** Answers may vary. Sample: Using the inverse of cosine, you can find the acute angle between the shortest side and the hypotenuse, $\theta = \cos^{-1}\left(\frac{8.4}{12.9}\right) \approx 49.4°$. Because the triangle is a right triangle, the remaining acute angle is ≈90° − 49.4° = 40.6°. **6.** The student confuses $\sin^{-1}\theta$ with $\frac{1}{\sin\theta}$. He or she should have divided by sin 0.45. $x = \frac{4}{\sin 0.45} \approx 9.20$ cm

Exercises
7. $\sin\theta = \frac{3}{5}$, $\cos\theta = -\frac{4}{5}$, $\tan\theta = -\frac{3}{4}$, $\csc\theta = \frac{5}{3}$, $\sec\theta = -\frac{5}{4}$, $\cot\theta = -\frac{4}{3}$
9. $\sin\theta = -\frac{5\sqrt{26}}{26}$, $\cos\theta = \frac{\sqrt{26}}{26}$, $\tan\theta = -5$, $\csc\theta = -\frac{\sqrt{26}}{5}$, $\sec\theta = \sqrt{26}$, $\cot\theta = -\frac{1}{5}$
11. $\sin\theta = \frac{\sqrt{7}}{4}$, $\cos\theta = -\frac{3}{4}$, $\tan\theta = -\frac{\sqrt{7}}{3}$, $\csc\theta = \frac{4\sqrt{7}}{7}$, $\sec\theta = -\frac{4}{3}$, $\cot\theta = -\frac{3\sqrt{7}}{7}$
13a. $\frac{15}{17} \approx 0.88$ **b.** $\frac{17}{8} \approx 2.13$ **c.** $\frac{8}{15} \approx 0.53$ **d.** $\frac{17}{8} \approx 2.13$ **e.** $\frac{17}{15} \approx 1.13$ **f.** $\frac{8}{15} \approx 0.53$ **15.** 41.8 **17.** 25.2 **19.** $a \approx 8.7$, $m\angle A = 60.0°$, $m\angle B = 30.0°$ **21.** $a = 9.0$, $m\angle A \approx 36.9°$, $m\angle B \approx 53.1°$ **23.** $a \approx 8.0$, $m\angle A \approx 61.8°$, $m\angle B \approx 28.2°$ **25a.** $m\angle A = \cos^{-1}\left(\frac{1200}{d}\right)$ **b.** 37° **c.** 53°
27.

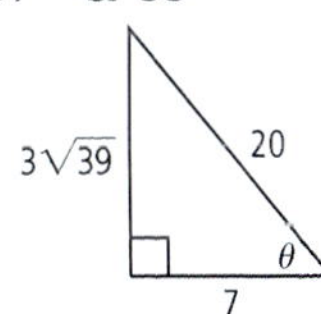

$\sin\theta = \frac{3\sqrt{39}}{20}$, $\tan\theta = \frac{3\sqrt{39}}{7}$, $\csc\theta = \frac{20\sqrt{39}}{117}$, $\sec\theta = \frac{20}{7}$, $\cot\theta = \frac{7\sqrt{39}}{117}$

29.

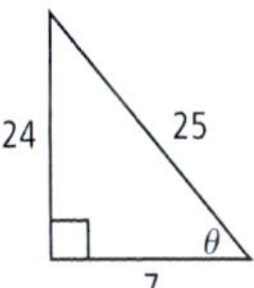

$\sin\theta = \frac{24}{25}$, $\cos\theta = \frac{7}{25}$, $\csc\theta = \frac{25}{24}$, $\sec\theta = \frac{25}{7}$, $\cot\theta = \frac{7}{24}$

31.

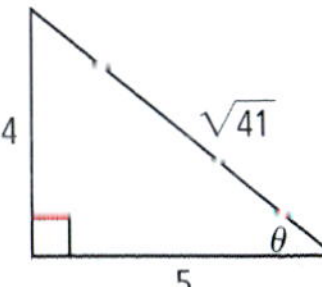

$\sin\theta = \frac{4\sqrt{41}}{41}$, $\cos\theta = \frac{5\sqrt{41}}{41}$, $\tan\theta = \frac{4}{5}$, $\csc\theta = \frac{\sqrt{41}}{4}$, $\sec\theta = \frac{\sqrt{41}}{5}$

33.

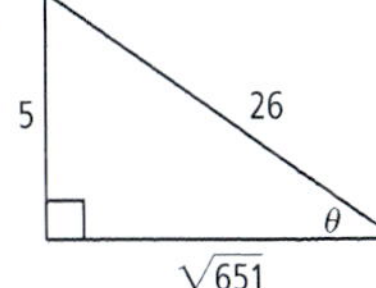

$\sin\theta = \frac{5}{26}$, $\cos\theta = \frac{\sqrt{651}}{26}$, $\tan\theta = \frac{5\sqrt{651}}{651}$, $\sec\theta = \frac{26\sqrt{651}}{651}$, $\cot\theta = \frac{\sqrt{651}}{5}$

35. 33.4 ft **37.** 20.3 m^2 **39.** $c \approx 12.2$, $m\angle A \approx 35.0°$, $m\angle B \approx 55.0°$ **41.** $a \approx 3.9$, $c \approx 6.9$, $m\angle B = 55.8°$ **43.** $a \approx 19.8$, $b \approx 2.9$, $m\angle A = 81.7°$ **45.** Using inverse sine, you can find that $\theta = 30°$. Since sine is positive in the first and second quadrants, another solution is 150°. All the solutions would be $30° + n \cdot 360°$ and $150° + n \cdot 360°$.

47. $\sec A = \frac{c}{b} = \frac{1}{\left(\frac{b}{c}\right)} = \frac{1}{\cos A}$

49. $\cos^2 A + \sin^2 A = \left(\frac{b}{c}\right)^2 + \left(\frac{a}{c}\right)^2$
$= \frac{b^2 + a^2}{c^2} = 1$

51. $y \approx 61.7$ m **53.** G **54.** C **55.** I

57. $180° + n \cdot 360°$ **58.** $45° + n \cdot 180°$ and $90° + n \cdot 180°$ **59.** $0° + n \cdot 180°$

60.

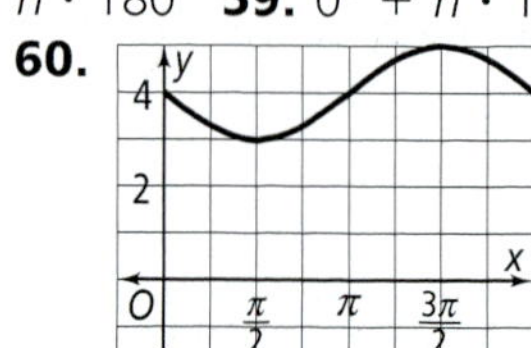

61.

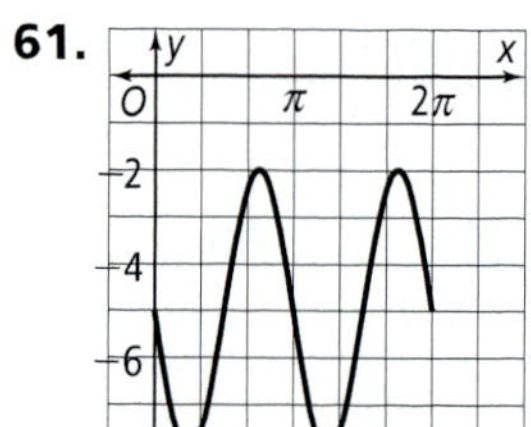

62.

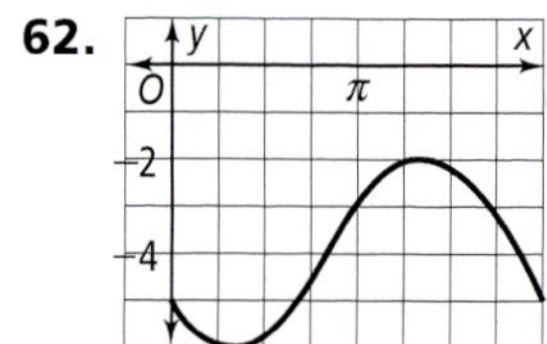

63. 6 cm^2 **64.** 45 $in.^2$ **65.** 32.76 mm^2

Lesson 14-4 pp. 928–934

Got It? 1. 36.6 $in.^2$ **2.** 31.0 yd **3a.** 68.4° **b.** yes; $\frac{\sin T}{\text{height}} = \frac{\sin 90°}{9}$; height $\approx 9 \sin 47° \approx 6.6$

4. 104.7 ft

Lesson Check 1. $\approx$10.7 square units **2.** $\approx$19.8 **3.** $\approx$26.3° or $\approx$153.7° **4.** AAS, ASA **5.** No; For $\frac{\sin 22°}{\sin 45°}$ you find the sine of each numerator and denominator, sin 22° and sin 45°; for $\sin\left(\frac{22°}{45°}\right)$ you find the sine of the quotient of $\left(\frac{22°}{45°}\right)$.

Exercises 7. 9.1 $in.^2$ **9.** 10.9 **11.** 7.4 **13.** 33.5° **15.** 31.7° **17.** 32 cm **19.** 66° **21.** $m\angle E \approx 40.3°$, $m\angle F \approx 85.7°$, $f \approx 12.3$ m **25.** 44.4 **27.** 49.4

29. 28.0 ft **31.** 4.0 cm **33a.** 56.4°, 93.6°, 26.4° **b.** No; $\triangle EFG$ could be congruent to $\triangle ABC$ instead of $\triangle ABD$.

35. No; you need at least one side in order to set up a proportion you can then solve.

37. I

39. $A = \frac{1}{2} ab \sin C$

$\sin C = \frac{2(A)}{ab} = \frac{2(31.5)}{9(14)} = 0.5$

$m\angle C = \sin^{-1} 0.5 = 30°$

So, the measure of the included angle for the given sides is 30° or 150°.

40.

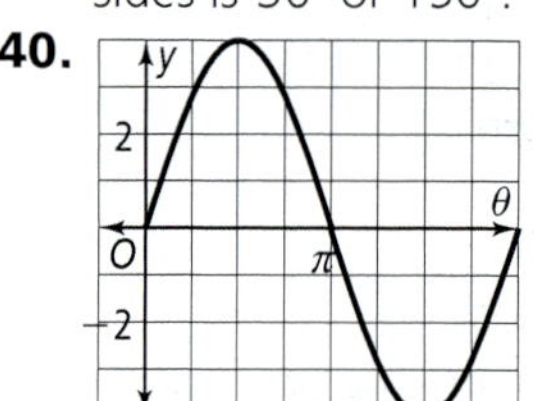

41.

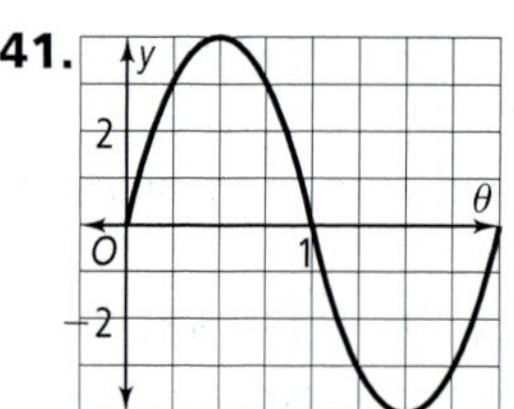

42.

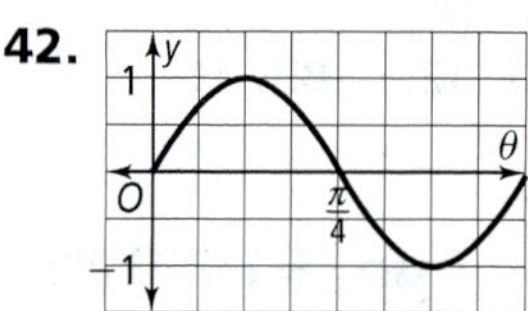

43. $\langle -1, 7\rangle$ **44.** $\langle 5, 3\rangle$ **45.** $\langle -3, -1\rangle$ **46.** $\langle -6, 4\rangle$ **47.** 53.1° **48.** 24.6° **49.** 38.7° **50.** 54.7°

Lesson 14-5 pp. 936–942

Got It? 1a. 6.4
b. 1 mi to 6 mi;
$a^2 = 2.5^2 + 3.5^2 - 2(2.5)(3.5) \cos A$
$a^2 = 18.5 - 17.5 \cos A$;
if $A = 0°$, then $\cos A = 1$ and $a = 1$; if $A = 180°$, then $\cos A = -1$ and $a = 6$. **2.** 75.3° **3.** 30.7°

Lesson Check 1. $\approx$11.85 in. **2.** $\approx$52.4° **3.** $\approx$34.1° **4.** $\approx$60.3° **5.** Use the Law of Sines when you have two sides and a non-included angle or two angles and a side; use the Law of Cosines when you have two sides and an included angle or three sides. **6.** The denominator should be negative. The answer should be:

$\cos C = \frac{15^2 - 11^2 - 17^2}{-2(11)(17)} \approx 0.495$

$C = \cos^{-1}(0.495) \approx 60.3°$

Exercises 7. 37.1 **9.** 13.7 **11.** 27.0 **13.** 33.7° **15.** 47.2° **17.** 50.8 **19.** 27.0° **21.** $b^2 = a^2 + c^2 - 2ac \cos B$

23. $\frac{\sin B}{b} = \frac{\sin C}{c}$ **25.** $\frac{\sin C}{c} = \frac{\sin A}{a}$ **27.** ≈ 59.1 nautical miles **29.** $b \approx 34.7$, $m\angle A \approx 26.7°$, $m\angle C \approx 33.3°$
31. $m\angle A \approx 56.1°$, $m\angle B \approx 70.0°$, $m\angle C \approx 53.9°$
33. For any two side lengths a and b, the ratio $\frac{a}{b}$ is equal to the ratio $\frac{\sin A}{\sin B}$, which can be found since A and B are given.
35a. ≈ 45.4 mi **b.** 14.4° left; 4.4° west of north
37. 11.0 cm **39.** 27.0° **41.** 13.0 cm **43.** 21.5°
45. 8.3 ft **47.** 79.6° **49.** 18 cm
51. a. 2.1 m
b. 9.8 m^2
53. a. $\cos A > 0$ if $b^2 + c^2 > a^2$;
$\cos A = 0$ if $b^2 + c^2 = a^2$;
$\cos A < 0$ if $b^2 + c^2 < a^2$
b. acute △ if $\cos A > 0$; right △ if $\cos A = 0$; obtuse △ if $\cos A < 0$
55. 85.4 **57.** 27.1° **59.** 24.1 units^2 **60.** 17.1 in. **61.** 8.9 m **62.** 26.3 in. **63.** 54.0° **64.** 2π, $x = \pm\pi$
65. $\frac{2}{3}$, $x = \pm\frac{1}{3}$ **66.** $\frac{\pi}{3}$, $x = \pm\frac{\pi}{6}$ **67.** 1, $x = \pm\frac{1}{2}$
68. $\sin\theta$ **69.** $\cos\theta$ **70.** 1 **71.** $\csc^2\theta$

Lesson 14-6 pp. 943–950

Got It?

1. $\cos\left(\theta - \frac{\pi}{2}\right) = \cos\left(-\left(\frac{\pi}{2} - \theta\right)\right)$
$= \cos\left(\frac{\pi}{2} - \theta\right)$
$= \sin\theta$

2. $\sec(90° - A) = \frac{1}{\cos(90° - A)} = \frac{1}{\sin\theta} = \csc\theta$
$\sec(90° - A) = \csc\theta$

3a. $0, \pi$ **b.** yes; πn

4. $\frac{\sqrt{6} - \sqrt{2}}{4}$

5. $\sin(A + B) = \sin(A - (-B))$
$= \sin A\cos(-B) - \cos A\sin(-B)$
$= \sin A\cos B - \cos A(-\sin B)$
$= \sin A\cos B + \cos A\sin B$

6. $-2 - \sqrt{3}$

Lesson Check

1. $\sin\left(\frac{\pi}{2} + \theta\right) + \sin\left(\frac{\pi}{2} - \theta\right)$
$= \sin\left(\frac{\pi}{2} - (-\theta)\right) + \sin\left(\frac{\pi}{2} - \theta\right)$
$= \cos(-\theta) + \cos\theta$
$= \cos\theta + \cos\theta$
$= 2\cos\theta$

2. $\frac{\pi}{4}, \frac{5\pi}{4}$ **3.** $\frac{\sqrt{2}}{2}$ **4.** $-\frac{\sqrt{2} + \sqrt{6}}{4}$

5. There are 2 solutions, $\frac{\pi}{2}$ and $\frac{3\pi}{2}$, between 0 and 2π because: $-\cos\theta = \cos\theta$
$2\cos\theta = 0$
$\cos\theta = 0$; $\theta = \frac{\pi}{2}, \frac{3\pi}{2}$

6. $\sin\left(\frac{\pi}{2} - \theta\right) = \sin\frac{\pi}{2}\cos\theta - \cos\frac{\pi}{2}\sin\theta$
$= (1)\cos\theta - (0)\sin\theta$
$= \cos\theta$

Exercises

7. $\csc\left(\theta - \frac{\pi}{2}\right) = \frac{1}{\sin\left(\theta - \frac{\pi}{2}\right)}$
$= \frac{1}{\sin\left(-\left(\frac{\pi}{2} - \theta\right)\right)}$
$= \frac{1}{-\sin\left(\frac{\pi}{2} - \theta\right)}$
$= \frac{1}{-\cos\theta}$
$= -\sec\theta$

9. $\cot\left(\frac{\pi}{2} - \theta\right) = \frac{\cos\left(\frac{\pi}{2} - \theta\right)}{\sin\left(\frac{\pi}{2} - \theta\right)}$
$= \frac{\sin\theta}{\cos\theta}$
$= \tan\theta$

11. $\tan\left(\theta - \frac{\pi}{2}\right) = \tan\left(-\left(\frac{\pi}{2} - \theta\right)\right)$
$= -\tan\left(\frac{\pi}{2} - \theta\right)$
$= -\cot\theta$

13. $\tan(90° - A) = \cot A$ **15.** $\cot(90° - A) = \tan A$
17. $\frac{\pi}{2}, \frac{3\pi}{2}$ **19.** π **21.** $\frac{\pi}{2}, \frac{3\pi}{2}$ **23.** $\frac{\sqrt{2}}{2}$ **25.** 0
27. $-\sqrt{3} - 2$ **29.** $\frac{\sqrt{2} + \sqrt{6}}{4}$
31. $-2 + \sqrt{3}$ **33.** $-\frac{1}{2}$ **35.** $\frac{1}{2}$

37. $\sin(A - B) = \cos\left[\frac{\pi}{2} - (A - B)\right]$
$= \cos\left[\left(\frac{\pi}{2} - A\right) + B\right]$
$= \cos\left(\frac{\pi}{2} - A\right)\cos B - \sin\left(\frac{\pi}{2} - A\right)\sin B$
$= \sin A\cos B - \cos A\sin B$

39. $\tan(A + B) = \frac{\sin(A + B)}{\cos(A + B)}$
$= \frac{\sin A\cos B + \cos A\sin B}{\cos A\cos B - \sin A\sin B}$
$= \frac{\frac{\sin A\cos B + \cos A\sin B}{\cos A\cos B}}{\frac{\cos A\cos B - \sin A\sin B}{\cos A\cos B}}$
$= \frac{\frac{\sin A\cos B}{\cos A\cos B} + \frac{\cos A\sin B}{\cos A\cos B}}{\frac{\cos A\cos B}{\cos A\cos B} - \frac{\sin A\sin B}{\cos A\cos B}}$
$= \frac{\tan A + \tan B}{1 - \tan A\tan B}$

41. $(5\cos\theta - 5\sqrt{3}\sin\theta, 5\sin\theta + 5\sqrt{3}\cos\theta)$
43. $\sin 5\theta$ **45.** $\cos 5\theta$ **47.** $\tan 2\theta$

49. a. even: cosine, secant; odd: sine, cosecant, tangent, cotangent

b. No; answers may vary. Sample:
$y = \sin x - \cos x$. For $x = \frac{\pi}{4}$, $f(x) = \sin\frac{\pi}{4} - \cos\frac{\pi}{4} = 0$, and $f(-x) = \sin\left(-\frac{\pi}{4}\right) - \cos\left(-\frac{\pi}{4}\right) = -\sqrt{2}$. Because $f(x) \neq f(-x)$, and $-f(x) \neq f(-x)$, the function is neither even nor odd.

51. $\sin(\pi - \theta) = \sin\pi\cos\theta - \cos\pi\sin\theta$
$= 0 - (-1)\sin\theta = \sin\theta$

53. $\cos(\pi + \theta) = \cos\pi\cos\theta - \sin\pi\sin\theta$
$= (-1)\cos\theta - 0 = -\cos\theta$

55. $\cos\left(\theta + \frac{3\pi}{2}\right) = \cos\theta\cos\frac{3\pi}{2} - \sin\theta\sin\frac{3\pi}{2}$
$= \cos\theta(0) - \sin\theta(-1) = \sin\theta$

57. I

59. $\sin(165°) = \sin(15°)$
$= \sin(45° - 30°)$
$= \sin 45°\cos 30° - \cos 45°\sin 30°$
$= \frac{\sqrt{2}}{2}\cdot\frac{\sqrt{3}}{2} - \frac{\sqrt{2}}{2}\cdot\frac{1}{2}$
$= \frac{\sqrt{6}}{4} - \frac{\sqrt{2}}{4} = \frac{\sqrt{6} - \sqrt{2}}{4}$

60. $\approx$16.34 ft **61.** $\approx$10.0 cm **62.** $\frac{4\pi}{9}$ and 1.40
63. $-\frac{5\pi}{18}$ and -0.87 **64.** $-\frac{\pi}{12}$ and -0.26
65. $\frac{7\pi}{18}$ and 1.22 **66.** $\frac{19\pi}{18}$ and 3.32
67. $\cos A\cos B - \sin A\sin B$
68. $\sin A\cos B + \cos A\sin B$ **69.** $\frac{\tan A + \tan B}{1 - \tan A\tan B}$

Lesson 14-7 pp. 951–957

Got It?

1. $\cos 2\theta = \cos^2\theta - \sin^2\theta$
$= \cos^2\theta - (1 - \cos^2\theta)$
$= \cos^2\theta - 1 + \cos^2\theta$
$= 2\cos^2\theta - 1$

2. $\frac{\sqrt{3}}{2}$ **3.** $2\cos 2\theta = 2(2\cos^2\theta - 1) = 4\cos^2\theta - 2$

4a. $\frac{1}{2}$ **b.** $-\frac{\sqrt{3}}{3}$

5a. $-\frac{3}{5}$ **b.** $-\frac{4}{3}$ **c.** If $270° < \theta < 360°$, $135° < \frac{\theta}{2} < 180°$ and $\frac{\theta}{2}$ is also in Quadrant II. The answers will remain the same.

Lesson Check 1. $\frac{\sqrt{3}}{2}$ **2.** 1 **3a.** $-\frac{5}{13}$ **b.** $\sqrt{\frac{13 + 2\sqrt{13}}{26}}$
c. $-\sqrt{\frac{13 - 2\sqrt{13}}{26}}$

4. The student did not correctly determine in which quadrant $\frac{\theta}{2}$ will be. If $180° < \theta < 270°, 90° < \frac{\theta}{2} < 135°$, then $\frac{\theta}{2}$ is in Quadrant II and the tangent will be negative. **5.** $\sin 4A$

6. $\sin\frac{5A}{2} = -\sqrt{\frac{1 - \cos 5A}{2}}$ if $360° < 5A < 450°$, $180° < \frac{5A}{2} < 225°$ and the sine is negative in Quadrant III.

Exercises

7. $\sin 2\theta = \sin(\theta + \theta)$
$= \sin\theta\cos\theta + \cos\theta\sin\theta$
$= 2\sin\theta\cos\theta$

9. $-\frac{\sqrt{3}}{2}$ **11.** $-\sqrt{3}$ **13.** $-\frac{1}{2}$ **15.** $-\frac{1}{2}$ **17.** $\frac{\sqrt{2 + \sqrt{3}}}{2}$
19. $\frac{\sqrt{2 - \sqrt{3}}}{2}$ **21.** $\frac{\sqrt{2 + \sqrt{2}}}{2}$ **23.** 0 **25.** $\frac{3\sqrt{10}}{10}$ **27.** 3
29. $\frac{4\sqrt{17}}{17}$ **31.** -4

33. $\cos B = 2\cos^2\frac{B}{2} - 1$

$2\cos^2\frac{B}{2} = \cos B + 1$

$\cos^2\frac{B}{2} = \frac{\cos B + 1}{2}$

Since $\cos B = \frac{a}{c}$,

$\cos^2\frac{B}{2} = \frac{\frac{a}{c} + 1}{2} = \frac{a + c}{2c}$.

35. $\cos 2R = \cos^2 R - \sin^2 R$
$= \left(\frac{s}{t}\right)^2 - \left(\frac{r}{t}\right)^2$
$= \frac{s^2}{t^2} - \frac{r^2}{t^2}$
$= \frac{s^2 - r^2}{t^2}$

37. $\sin^2\frac{S}{2} = \left(\sin\frac{S}{2}\right)^2$
$= \left(\pm\sqrt{\frac{1 - \cos S}{2}}\right)^2$
$= \frac{1 - \cos S}{2} = \frac{1 - \frac{r}{t}}{2}$
$= \frac{1}{2} - \frac{r}{2t} = \frac{t - r}{2t}$

39. $\tan^2\frac{S}{2} = \left(\tan\frac{S}{2}\right)^2$
$= \left(\pm\sqrt{\frac{1 - \cos S}{1 + \cos S}}\right)^2 = \frac{1 - \cos S}{1 + \cos S} = \frac{1 - \frac{r}{t}}{1 + \frac{r}{t}}$
$= \frac{t - r}{t + r}$

41. $-\frac{24}{25}$ **43.** $\frac{24}{7}$ **45.** $\frac{\sqrt{5}}{5}$ **47.** $-\frac{1}{2}$
49. $\cos\theta(8\sin\theta - 3) = 0$; $\frac{\pi}{2}, \frac{3\pi}{2}$, 0.384, 2.757
51. $\cos\theta(2\sin^2\theta - 1) = 0$; $\frac{\pi}{2}, \frac{3\pi}{2}, \frac{\pi}{4}, \frac{3\pi}{4}, \frac{5\pi}{4}, \frac{7\pi}{4}$
53. 1 **55.** $\cos\theta - \sin\theta$ **57.** Answers may vary. Sample:
a. $\sin 60° = \frac{\sqrt{3}}{2}$, $\cos 60° = \frac{1}{2}$ **b.** $\sin 120° = \frac{\sqrt{3}}{2}$
c. $\cos 30° = \frac{\sqrt{3}}{2}$

59. $4\sin\theta\cos\theta(\cos^2\theta - \sin^2\theta)$

61. $\dfrac{4\tan\theta(1-\tan^2\theta)}{\tan^4\theta - 6\tan^2\theta + 1}$

63. $\pm\sqrt{\frac{1}{2} \pm \frac{1}{2}\sqrt{\frac{1}{2} + \frac{1}{2}\cos\theta}}$

65. a.
$$\tan\frac{A}{2} = \pm\sqrt{\frac{1-\cos A}{1+\cos A}}$$
$$= \pm\sqrt{\frac{1-\cos A}{1+\cos A}} \cdot \sqrt{\frac{1+\cos A}{1+\cos A}}$$
$$= \pm\sqrt{\frac{1-\cos^2 A}{(1+\cos A)^2}}$$
$$= \pm\sqrt{\frac{\sin^2 A}{(1+\cos A)^2}}$$
$$= \frac{\sin A}{1+\cos A}$$

Since $\tan\frac{A}{2}$ and $\sin A$ have the same sign wherever $\tan\frac{A}{2}$ is defined, only the positive sign occurs.

67. G **69.** I **71.** $\frac{\sqrt{2}}{2}$ **72.** $\frac{\sqrt{3}}{2}$ **73.** $\sqrt{3}$ **74.** about 4.1; 12 **75.** 2; 4 **76.** 45.2 **77.** 26.6 **78.** 57.6

Chapter Review pp. 959–962

1. Law of Sines **2.** trig. ratios **3.** Law of Cosines **4.** trig. ident. **5.** Law of Sines **6.** $\sin\theta\tan\theta = \sin\theta\frac{\sin\theta}{\cos\theta} = \frac{\sin^2\theta}{\cos\theta} = \frac{1-\cos^2\theta}{\cos\theta} = \frac{1}{\cos\theta} - \cos\theta$; domain of validity: all real numbers except odd multiples of $\frac{\pi}{2}$

7. $\cos^2\theta\cot^2\theta = \cos^2\theta\frac{\cos^2\theta}{\sin^2\theta} = \frac{\cos^2\theta(1-\sin^2\theta)}{\sin^2\theta} = \frac{\cos^2\theta - \cos^2\theta\sin^2\theta}{\sin^2\theta} = \cot^2\theta - \cos^2\theta$; domain of validity: all real numbers except 0 and multiples of π

8. $\cos^2\theta$ **9.** 1 **10.** 1 **11.** $-\sin^2\theta$ **12.** $\cos^2\theta$ **13.** 1 **14.** 60° **15.** 60° **16.** −30° **17.** 30° **18.** 0.34 **19.** −1.11 **20.** 2.27 **21.** 0.20 **22.** $\frac{\pi}{3}, \frac{5\pi}{3}$ **23.** $\frac{\pi}{6}, \frac{7\pi}{6}$ **24.** $0, \frac{\pi}{2}, \pi$ **25.** $\frac{\pi}{3}, \frac{5\pi}{3}$ **26.** $\sin\theta = \frac{4}{5}$, $\cos\theta = -\frac{3}{5}$, $\tan\theta = -\frac{4}{3}$, $\csc\theta = \frac{5}{4}$, $\sec\theta = -\frac{5}{3}$, $\cot\theta = -\frac{3}{4}$ **27.** $\sin\theta = -\frac{15}{17}$, $\cos\theta = -\frac{8}{17}$, $\tan\theta = \frac{15}{8}$, $\csc\theta = -\frac{17}{15}$, $\sec\theta = -\frac{17}{8}$, $\cot\theta = \frac{8}{15}$ **28.** $\frac{3}{5}$, 0.6 **29.** $\frac{4}{5}$, 0.8 **30.** $\frac{3}{4}$, 0.75 **31.** $\frac{5}{3}$, $1.\overline{6}$ **32.** $g \approx 9.5$, $\angle F \approx 18.4$, $\angle H \approx 71.6$ **33.** $h = 16$, $\angle F \approx 36.9$, $\angle H \approx 53.1$ **34.** $f \approx 37.7$, $\angle F \approx 43.3$, $\angle H \approx 46.7$ **35.** $g \approx 6.4$, $\angle F \approx 51.3$, $\angle H \approx 38.7$ **36.** 13.7 **37.** 29.4 **38.** 13.14 m^2 **39.** 92.12 ft^2 **40.** 7.1 in.

41. 43.9° **42.** 52.2°

43.
$$\cos\left(\theta + \frac{\pi}{2}\right) = \cos\theta\cos\frac{\pi}{2} - \sin\theta\sin\frac{\pi}{2}$$
$$= \cos\theta \times 0 - \sin\theta \times 1 = -\sin\theta$$

44.
$$\sin^2\left(\theta - \frac{\pi}{2}\right) = \left[\sin\left(\theta - \frac{\pi}{2}\right)\right]^2$$
$$= \left[\sin\theta\cos\frac{\pi}{2} - \cos\theta\sin\frac{\pi}{2}\right]^2$$
$$= [\sin\theta \times 0 - \cos\theta \times 1]^2$$
$$= (-\cos\theta)^2 = \cos^2\theta$$

45. $2 - \sqrt{3}$ **46.** $-\frac{\sqrt{3}}{2}$ **47.** $\frac{\sqrt{2}-\sqrt{6}}{4}$ **48.** $-2 - \sqrt{3}$ **49.** $\frac{\sqrt{3}}{2}$ **50.** $\frac{\sqrt{3}}{2}$ **51.** $-\sqrt{3}$ **52.** $-\frac{\sqrt{3}}{2}$

Skills Handbook

p. 972 **1.** 46% **3.** 0.7% **5.** 1.035 **7.** 25% **9.** $66.\overline{6}$% **11.** 115% **13.** 12.5 **15.** 75 **17.** 20%

p. 973 **1.** $1\frac{2}{5}$ **3.** $6\frac{5}{6}$ **5.** $\frac{5}{21}$ **7.** $2\frac{19}{20}$ **9.** 8 **11.** $1\frac{1}{2}$ **13.** 2 **15.** 42

p. 974 **1.** 3 to 4 **3.** 19 g in 2 oz **5.** $\frac{14}{5}$ **7.** 8 **9.** 1.8 **11.** 1.95 **13.** 45.5 **15.** ±6

p. 975 **1.** 1 **3.** −38 **5.** −17 **7.** 4 **9.** 28 **11.** −12 **13.** −90 **15.** 12 **17.** 19 **19.** 9 **21.** −10

p. 976 **1.** 14 m^2 **3.** 30 cm^2 **5.** $91\frac{1}{8}$ ft^3 **7.** 100π $in.^3$ **9.** 110.5 $in.^2$ **11.** $121\frac{1}{2}$ ft^2

p. 977 **1.** I **3.** IV **5.** III **7.** $\frac{4}{5}$ **9.** 0 **11.** $\frac{1}{5}$ **13.** $(5, -\frac{3}{2})$ **15.** $(\frac{5}{2}, -1)$

p. 978 **1.** x^3 **3.** a^4b **5.** $\frac{1}{c^4}$ **7.** $\frac{x^5}{y^7z^3}$ **9.** d^8 **11.** c^6 **13.** $\frac{a^4}{b^5}$ **15.** $\frac{a^4}{b^4}$ **17.** c^{12} **19.** $u^{12}v^6$ **21.** a^3 **23.** $\frac{1}{mg^3}$ **25.** $\frac{a^5}{2}$

p. 979 **1.** $x^2 + 10x - 5$ **3.** $12x^4 - 20x^3 + 36x^2$ **5.** $x^2 - 2x - 15$ **7.** $(a-6)(a-2)$ **9.** $(x+4)(x+1)$ **11.** $(y+8)(y-3)$ **13.** $2x(x^2 + 2x - 4)$

p. 980 **1.** 1.34×10^6 **3.** 7.75×10^{-4} **5.** 111,300 **7.** 1.895×10^3 **9.** 1.234×10^5 **11.** 6.4×10^5 **13.** 8.52×10^2 **15.** 17.5 **17.** 8.95×10^{-12} **19.** 3.77×10^{10} **21.** 1.8×10^{-6}

p. 981 **1.** 10 **3.** 15 **5.** 54.7 **7.** 5 **9.** 13 **11.** 22.7 **13.** 2.8 **15.** 9 **17.** 7

p. 982 **1.**

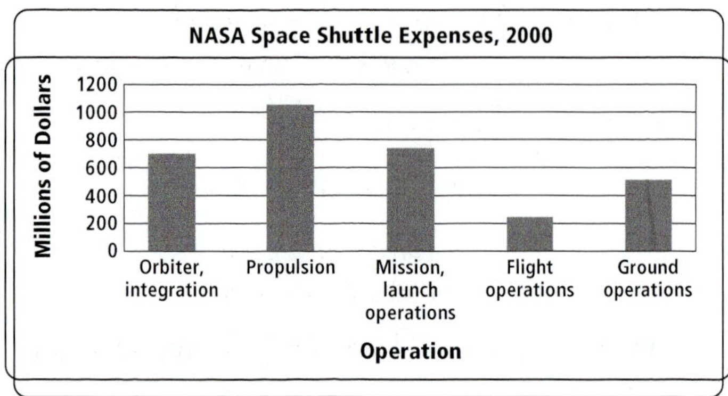

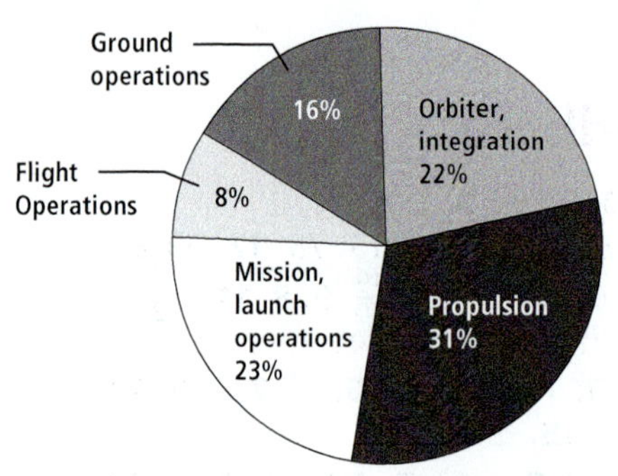

p. 983 **1.** $3.\overline{7}$; 5; 5 **3.** $3.9\overline{6}$; 2.4; 2.4 **5.** 1.5; 1.5; no mode

7.

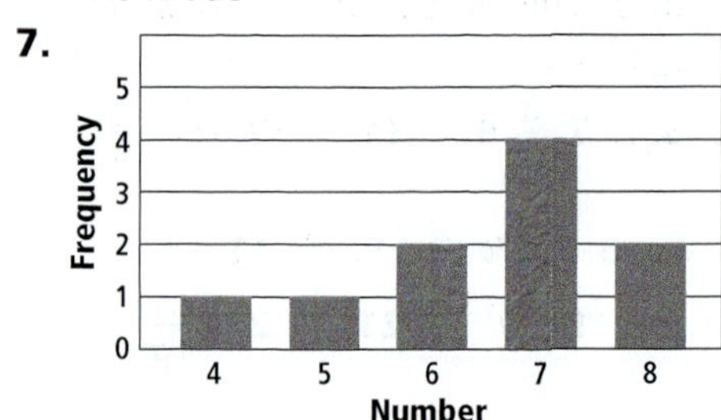

p. 984 **1.** $\frac{a}{3b^2}$ **3.** $\frac{1}{2}$ **5.** $4x$ **7.** $\frac{2}{h}$ **9.** $\frac{x}{10}$ **11.** $\frac{74x}{35}$ **13.** $\frac{4x^2}{5}$ **15.** $\frac{16}{x}$ **17.** $\frac{16}{5}$ **19.** 2x

Index

E

F

Index

Index

H

I

J

K

L

M

N

O

P

Index

R

Index

S

Index

T

U

Acknowledgments

Staff Credits

The people who made up the High School Mathematics team—representing composition services, core design digital and multimedia production services, digital product development, editorial, editorial services, manufacturing, marketing, and production management—are listed below.

Dan Anderson, Scott Andrews, Christopher Anton, Carolyn Artin, Michael Avidon, Margaret Banker, Charlie Bink, Niki Birbilis, Suzanne Biron, Beth Blumberg, Kyla Brown, Rebekah Brown, Judith Buice, Sylvia Bullock, Stacie Cartwright, Carolyn Chappo, Christia Clarke, Mary Ellen Cole, Tom Columbus, Andrew Coppola, AnnMarie Coyne, Bob Craton, Nicholas Cronin, Patrick Culleton, Damaris Curran, Steven Cushing, Sheila DeFazio, Cathie Dillender, Emily Dumas, Patty Fagan, Frederick Fellows, Jorgensen Fernandez, Mandy Figueroa, Suzanne Finn, Sara Freund, Matt Frueh, Jon Fuhrer, Andy Gaus, Mark Geyer, Mircea Goia, Andrew Gorlin, Shelby Gragg, Ellen Granter, Jay Grasso, Lisa Gustafson, Toni Haluga, Greg Ham, Marc Hamilton, Chris Handorf, Angie Hanks, Scott Harris, Cynthia Harvey, Phil Hazur, Thane Heninger, Aun Holland, Amanda House, Chuck Jann, Linda Johnson, Blair Jones, Marian Jones, Tim Jones, Gillian Kahn, Matthew Keefer, Brian Keegan, Jonathan Kier, Jennifer King, Tamara King, Elizabeth Krieble, Meytal Kotik, Brian Kubota, Roshni Kutty, Mary Landry, Christopher Langley, Christine Lee, Sara Levendusky, Lisa Lin, Wendy Marberry, Dominique Mariano, Clay Martin, Rich McMahon, Eve Melnechuk, Cynthia Metallides, Hope Morley, Christine Nevola, Michael O'Donnell, Michael Oster, Ameer Padshah, Stephen Patrias, Jeffrey Paulhus, Jonathan Penyack, Valerie Perkins, Brian Reardon, Wendy Rock, Marcy Rose, Carol Roy, Irene Rubin, Hugh Rutledge, Vicky Shen, Jewel Simmons, Ted Smykal, Emily Soltanoff, William Speiser, Jayne Stevenson, Richard Sullivan, Dan Tanguay, Dennis Tarwood, Susan Tauer, Tiffany Taylor-Sullivan, Catherine Terwilliger, Mark Tricca, Maria Torti, Leonid Tunik, Ilana Van Veen, Lauren Van Wart, John Vaughan, Laura Vivenzio, Samuel Voigt, Kathy Warfel, Don Weide, Laura Wheel, Eric Whitfield, Sequoia Wild, Joseph Will, Kristin Winters, Allison Wyss, Dina Zolotusky

Additional Credits: Michele Cardin, Robert Carlson, Kate Dalton-Hoffman, Dana Guterman, Narae Maybeth, Carolyn McGuire, Manjula Nair, Rachel Terino, Steve Thomas

Illustration

Stephen Durke: 574; **Phil Guzy:** 596, 597; **Rob Schuster:** 4, 5, 11, 18, 26, 33, 39, 41, 48, 60, 68, 74, 81, 84, 99, 107, 114, 116, 134, 142, 143, 149, 157, 165, 166, 168, 171, 174, 194, 202, 207, 216, 226, 233, 240, 258, 280, 288, 294, 296, 303, 308, 312, 326, 331, 367, 374, 375, 381, 390, 395, 398, 405, 414, 429, 434, 449, 451, 469, 498, 515, 522, 527, 534, 542, 547, 564, 566, 567, 570, 571, 572, 580, 587, 595, 609, 614, 617, 619, 630, 638, 641, 653, 865; **Ted Smykel:** 209; **Pearson Education:** 596, 597; **Judi Pinkham:** 230; **Pronk&Associates:** 12, 362, 399, 462, 589; **XNR Productions:** 788

Technical Illustration

GGS Book Services

Photography

All photographs not listed are the property of Pearson Education

Back Cover: Klein J.-L & Hube/Biosphoto

Page 3, ©Franck Seguin/Corbis; **28,** ©BL Images Ltd/Alamy; **39,** ©Richard Wahlstrom/JupiterImages; **45,** ©UPI Photo/Roger Williams/Newscom; **59,** ©Joe McBride/Getty Images; **61,** ©JUPITERIMAGES/Brand X/Alamy; **61,** ©JUPITERIMAGES/Brand X/Alamy; **61,** ©JUPITERIMAGES/Brand X/Alamy; **61,** ©JUPITERIMAGES/Brand X/Alamy; 61, ©Roberto Mettifogo/Getty Images; **84,** ©www.indepthexposure.com; **84,** ©Zen Shui/SuperStock; **133,** ©Hisham Ibrahim/Getty Images; **135,** ©Doug Perrine/Peter Arnold Inc.; **135,** ©PAUL NICKLEN/National Geographic Stock; **159,** ©Andy Crawford/Dorling Kindersley; **159,** ©Thomas Northcut/Photodisc/Getty Images; **159,** ©Steve Gorton/Dorling Kindersley; **164,** ©Image Source/Getty Images; **193,** ©Atlantide Phototravel/Corbis; **198,** ©Stuart Westmorland/Getty Images; **205,** Jeff Greenberg/PhotoEdit Inc.; **210,** ©Thomas Barwick/Getty Images; **228,** ©Andy Harmer/Photo Researchers, Inc.; **228,** ©Harvey Lloyd/Getty Images; **234,** ©Rolf Hicker Photography/Alamy; **279,** ©Claudius/zefa/Corbis; **306,** ©James Baigrie/Botanica/Jupiterimages; **333,** ©D. Hurst/Alamy; **333,** ©Peter Cade/Getty Images; **333,** ©D. Hurst/Alamy; **333,** ©FOOD DRINK AND DIET/MARK SYKES/Alamy; **333,** ©Andre Gallant/Getty Images; **342,** ©Ed Darack/Getty Images; **359,** ©Jake Norton/Getty Images; **375,** ©Panoramic Images/Getty Images; **383,** ©Detlev van Ravenswaay/Photo Researchers, Inc.; **392,** ©Bob Llewellyn/Jupiterimages; **408,** ©Bob Krist/CORBIS; **433,** ©Peter Mason/Getty Images; **438,** ©Jose B. Ruiz/npl/Minden Pictures; **453,** ©Earth Imaging/Getty Images; **467,** ©Jerry Lodriguss/Photo Researchers, Inc.; **476,** ©Dave King/Dorling Kindersley; **480,** ©NASA; **497,** ©Stephen Dalton/Minden Pictures; **502,** ©Corbis Super RF/Alamy; **502,** ©Chase Jarvis/Photolibrary; **513,** ©NASA-HQ-GRIN; **544,** ©NASA - JPL; **563,** ©Yu Xiangquan/Xinhua Press/Corbis; **578,** ©John Glover/Alamy; **578,** ©Bob Gibbons/Alamy; **578,** ©Courtesy of Unwins/Dorling Kindersley; **578,** ©Peter Anderson/Dorling Kindersley; **582,** ©Thomas J. Peterson/Alamy; **613,** ©Iain Masterton/Alamy; **618,** ©B.A.E. Inc./Alamy; **625,** ©Roger Ressmeyer/CORBIS; **632,** ©George Steinmetz/Corbis; **641,** ©Museum of Science and Industry; **673,** Photolibrary.com; **675,** Ron Chapple Stock/Corbis; **684,** Koji Aoki/Getty Images; **763,** Jim Sanborn; **775,** Jason Lugo/iStockphoto; **803,** Wolfgang Spunbarg/PhotoEdit; **827,** Ron Watts/Getty Images; **847 t,** Steve Gorton/Dorling Kindersley; **847 b,** European Space/Photo Researchers Inc.; **889,** Demetrio Carrasco/Dorling Kindersley; **903,** AFP Photo/Fabrice/Newscom; **921,** age footstock/Superstock; **931,** mediacolor's/Alamy Images